Ryan

THE DEVELOPMENT OF
THE VERTEBRATE SKULL

THE DEVELOPMENT OF THE VERTEBRATE SKULL

Gavin R. de Beer

With a new Foreword by
Brian K. Hall and James Hanken

The University of Chicago Press
Chicago & London

The University of Chicago Press, Chicago, IL 60637
The University of Chicago Press, Ltd., London

First published 1937 by Oxford University Press
and reprinted 1971 with additional bibliography

94 93 92 91 90 89 88 87 86 85   5 4 3 2 1

Library of Congress Cataloging in Publication Data

De Beer, Gavin, Sir, 1899–1972.
The development of the vertebrate skull.

Reprint. Originally published: Oxford [Oxfordshire] :
Clarendon Press, 1971.
Bibliography: p.
1. Skull. 2. Vertebrates. I. Title.
QL822.D35 1985 596'.0471 85–1081
ISBN 0–226–13958–1 (cloth)
ISBN 0–336–13960–3 (paper)

TO

THE MEMORY OF

THOMAS HENRY HUXLEY

WILLIAM KITCHEN PARKER

ERNST GAUPP

JAN WILLEM VAN WIJHE

AS A MONUMENT TO

THEIR LABOURS IN THE STUDY OF

THE DEVELOPMENT

OF THE

VERTEBRATE SKULL

FOREWORD

THIS volume commemorates the approaching fiftieth anniversary of Sir Gavin Rylands de Beer's *The Development of the Vertebrate Skull,* originally published in 1937. In an age when many scientific works are virtually outdated as soon as they appear in print, its republication is indeed a testament to the tremendous impact de Beer's book had on the field of evolutionary vertebrate morphology, and to the high regard in which it continues to be held.

It is a truly monumental work: 515 text pages supplemented with 143 plates, 205 entries in the genera index to the Systematic Section alone, and 1,076 literature references dating from as early as 1717. By de Beer's own admission, it is "the outcome of some fifteen years' work devoted to the study of the development of the skull in all the vertebrate groups" (p. xxix).

The years since 1937 have seen a tremendous increase in the number of published studies that bear on aspects of vertebrate skull development of interest to de Beer. They range from clinical to evolutionary in orientation. Yet, *The Development of the Vertebrate Skull* remains a valuable reference. It still contains the most up-to-date and complete descriptions of skull ontogeny for numerous taxa. In other fields, especially developmental mechanics, its accounts have been superseded; nevertheless, it continues to provide a vital historical perspective on the evolution of ideas.

In the following pages, we highlight developments in several areas that were of especial concern to de Beer, and which remain of primary importance to questions of vertebrate structure and evolution. These center on the fundamental organization of the head; the origin and differentiation of the cells which form the skull; the nature of the primary skeletal tissues, cartilage and bone; and the evolution of skull diversity. Where possible, we have arranged these topics in the same sequence used in *The Development of the Vertebrate Skull.* By limits of space alone, we have been obliged to omit other topics equally appropriate. Readers interested in details concerning skull development in particular vertebrate groups are referred to recent reviews (e.g., Moore 1981, on mammals; Bellairs and Kamal 1981, on reptiles; other references below).

It is unlikely that anyone will ever again be able to prepare such a comprehensive and up-to-date summary of vertebrate skull development within a single volume. We can only agree with the anonymous reviewer for *Nature* (2 July 1938) who concluded: "The writing of such a book must have entailed an immense amount of careful work, and students of the morphology of vertebrates owe a debt of gratitude to the author for having carried it out so successfully."

Head Segmentation

There is perhaps no topic pertaining to the fundamental organization and structure of the vertebrate head that is more "classical" a problem than segmentation. The idea that the head is organized around a fundamental plan of serial segments dates back at least two centuries. It figured prominently in debates over vertebrate origins and relationships in the latter half of the nineteenth century, and this controversy carried on into the early part of the twentieth century.

De Beer considered segmentation sufficiently important to devote approximately half of his Introductory Section to it. He explicitly accepted two claims: first, following from the work of Balfour and others, that "in early stages of development the head is segmented in a manner precisely similar to the trunk" (p. 15); second, based on the observations of Marshall, that segmentation of somites arises independently of that of ventral mesoderm (i.e., visceral arches). The primary, somitic segmental pattern comprised three anterior, or prootic, somites, and a variable number of posterior, or metotic, somites (the first two metotic somites typically were transitory and contributed nothing to the adult skull). In fact, de Beer apparently considered these aspects sufficiently resolved that to him the primary remaining issues concerning segmentation were the exact number of segments incorporated into the skull in different vertebrate groups, and the causes for this variation; he reviewed the evidence accordingly. While accepting T. H. Huxley's rejection of any vertebral contribution to the skull proper (with the possible exception of the extreme posterior region), he devoted little attention to the possible segmental arrangement of cranial bones in the *adult* skull. Instead, citing the work of Severtsov, he concluded that a primitive, simple segmental pattern of dermal bones "has been obscured" (p. 33). Yet even this topic was discussed only in the context of the number of segments that contribute to the adult skull.

The certainty with which de Beer presented cranial segmentation belied both the limited state of knowledge at the time and the lack of consensus among anatomists concerning just how segmented the head really is. The picture portrayed by de Beer may have been representative of a majority view, but divergent opinions were strongly maintained in 1937 and have continued to the present day. Kingsbury (1926) and later Romer (1972) rejected the classical scheme of cranial segmentation, stressing instead the degree to which the organization of cranial tissues is fundamentally different than those behind the head. Concurrently, the Swedish school of lower vertebrate paleontology and comparative morphology (Stensiö, Holmgren, Jarvik, Bjerring) developed an extreme view in which virtually *all* cranial components may be identified with a single, unifying scheme of cranial segmentation that literally is continuous with that of the trunk (Bjerring 1977; Jarvik 1980).

These views represent radically different opinions as to the nature of cranial segmentation, yet they do share a fundamentally similar methodology: conclusions are drawn from static descriptions of relatively late-stage embryos. And it is here that research conducted in only the last few years promises to finally resolve this age-old dispute over the basic organization of the head. These studies use a vastly different methodology: detailed, electron-microscopic examination of early embryos (e.g., primitive streak, gastrula), and experimental analysis of tissue origins using permanent cell markers in chimeric grafts between related species. Two discoveries are particularly important.

The first is the unequivocal demonstration that paraxial mesoderm in the cranial region of early embryos *is* segmented. The segments, termed somitomeres, are not visible by light microscopy but are revealed clearly by the scanning electron microscope (Meier 1984). They have been observed in salamanders (Jacobson and Meier 1984), snapping turtles (Meier and Packard 1984), chicks (Meier 1981), and mice (Meier and Tam 1982) and thus would seem to be a basic feature of vertebrate head organization. In the posterior region of the head, as well as the trunk, somitomeres are the precursors of somites (Jacobson and Meier 1984; Tam et al. 1982). Anterior cranial somitomeres fail to give rise to somites, but they do contribute to a number of adult tissues, namely, virtually all of the myogenic component of voluntary muscles, including the branchiomeric musculature which traditionally has been considered a derivative of unsegmented lateral plate mesoderm, or hypomere (Noden 1983*a, b*). In addition, there may be a fundamental difference between amniotes and anamniotes in terms of both the number of segments that contribute to the head and the number of cranial somitomeres that fail to differentiate into somites (Jacobson and Meier 1984).

The second is the demonstration that, at least in the chick, most cartilage and bone of the skull and lower jaw, and nearly all of the connective tissue of voluntary muscles, are derived not from mesoderm, whether segmented or unsegmented, but instead from cranial neural crest (Noden 1982, 1983*a,b*). The neural crest is initially unsegmented, but soon conforms to the segmental pattern of the somitomeres and pharyngeal pouches during migration. (For more on the neural crest, see below.)

These discoveries support parts of both the segmentalist and nonsegmentalist views described above. Yet they present an overall picture of the organization and development of the head that is radically different from any proposed earlier (Jacobson and Meier 1984; Meier 1981; Noden 1984). Resolution of several additional questions is still to come. For example, what is the relationship between segmentation of paraxial mesoderm and that of the neural tube? Neural tube segmentation with respect to the pattern of motor axon outgrowth and the distribution of sensory nerves is not intrinsic to the tube, but instead is imposed by the adjacent somites (Detwiler 1934; Keynes and Stern 1984). In the head, division of the brain anlage into segmented neuromeres comforms initially to the segmental pattern of adjacent somitomeres, but this correspondence is only transitory and is obscured during subsequent development (Meier 1982). And what is the relationship, topographically and causally, between segmentation of paraxial mesoderm and that of the visceral arches? In any event, it is clear that the revised view of head segmentation now being assembled will present an entirely new picture of the organization of the vertebrate head, including the skull.

Skull Diversity

The Systematic Section represents de Beer's attempt to amass an integrated review of all descriptive data pertaining to skull development in vertebrates. This entailed employing a standard terminology throughout, and drawing nearly all of the 143 plates for illustrated comparisons. Subsequent discussions of skull structure, homology, and evolution then were built on this base.

In the Preface, de Beer confessed the fear that his lengthy tome would have a "sterilizing" (p. xxxi) effect on the field, whereas his ultimate objective in writing the book was to provide a summary of knowledge that would facilitate or even stimulate subsequent work on the skull. While it is not at all clear that publication of *The Development of the Vertebrate Skull* is even partly responsible, there is no denying that de Beer's fear was realized in part: skull development and structure as a discipline did not maintain the central position in considerations of vertebrate biology after 1937 that it held for preceding decades. As cited by Barrington (1973), the Additional Bibliography of the 1971 edition, which de Beer himself compiled, contains fewer than 30 papers on skull development that appeared after the book's original publication. We, however, have identified at least 125 *additional* papers published between 1937 and 1971, and 65 papers since 1971, that describe new and significant aspects of skull development in vertebrates other than standard laboratory species. Thus, whereas de Beer's Additional Bibliography is indicative of a general decline in interest in skull development after 1937, it is not an adequate representation of the studies that have been performed, including many significant contributions.

Studies of skull development published after 1937 can be considered broadly representative of three topics: vertebrate phylogeny, developmental mechanics, and evolutionary mechanisms. Each topic has been emphasized to varying degrees among the major vertebrate groups, reflecting the appropriateness of particular taxa for examining certain problems and, in some cases, their suitability for experimental manipulation.

Vertebrate Phylogeny. At the time of publication of *The Development of the Vertebrate Skull* many fundamental aspects of vertebrate relationships were unresolved. The relationships of cyclostomes to fossil jawless, as well as jawed, fishes; the closest tetrapod ancestor among lobe-finned fishes; the relationships of the three orders of modern amphibians, both among themselves and with respect to archaic taxa; and relationships among ratite birds and other flightless species are examples. Skull development is a potential source of data for establishing phylogenetic affinity and defining phylogenetic trends, but de Beer was extremely reluctant to draw phylogenetic inferences from the data he amassed; the few he offered (pp. 456–69) were intended "at least as a basis for further work" (p. xxx).

"Further work," however, has been slow in coming. The number of developmental studies applied to phylogenetic questions is slight in comparison with the number based on adult features, so that the potential use of skull development in phylogenetic analysis remains largely untapped. Nevertheless, their implications can be large. Extensive descriptions of head development in the bowfin, *Amia calva,* by the Swedish school have been used to support a radical hypothesis of polyphyletic origin of tetrapods (reviewed in Jarvik 1980). Features of cranial skeletal and circulatory development in a wide variety of Recent mammals have challenged several long-held views concerning both the reptilian-mammalian transition and the subsequent radiation of modern mammals (Presley 1979, 1981; Presley and Steel 1976, 1978). Contrasting patterns of ossification of the orbitosphenoid have been used to emphasize the phylogenetic distinction of amphisbaenid reptiles from remaining squamates, the lizards and snakes (Bellairs and Gans 1983). Yet, nearly fifty years after the publication of *The Development of the Vertebrate Skull,* many of the phylogenetic questions prominent in 1937 remain unanswered.

Two phenomena, however, may qualify the use of development in phylogenetic analysis. One is variation. Use of features of development as characters when defining phylogenetic relationships among major taxa assumes that variation at lower taxonomic levels is slight in comparison with that at higher levels. Substantial variation among closely related taxa, and certainly within species, restricts the taxonomic utility of any character. For example, development and adult morphology of the auditory bulla have been used as characters for establishing phylogenetic relationships among certain mammalian taxa. Yet, recent studies of the ontogeny of the auditory bulla reveal a diversity of developmental patterns even among genera within the same order (Novacek 1977; Presley 1978; reviewed in Moore 1981). This variation at lower taxonomic levels may limit the future use of characters of the auditory bulla in defining relationships at higher taxonomic levels (Moore 1981).

The second potential qualification is adaptation. Development itself can evolve, especially in response to selection for adaptive features. And while altered patterns of development need not affect adult features (de Beer was well aware of such embryonic or larval adaptations), they may severely restrict the use of development in phylogenetic analysis. For example, in some caecilians—limbless, viviparous amphibians which make up the order Gymnophiona—the pattern of cranial ossification is an integral component of a unique specialization for fetal maintenance via maternal oviducal secretions; it provides little information of phylogenetic value at the level of relationships among Recent or archaic amphibian orders (Wake and Hanken 1981).

Developmental Mechanics. De Beer's personal research interests and expertise lay primarily in descriptive studies (Barrington 1973). He had, however, a sincere interest and training in experimental approaches to development, as attested by two books, *An Introduction to Experimental Embryology* and *The Elements of Experimental Embryology* (the latter with J. S. Huxley). It's therefore not surprising that whereas the Systematic Section of *The Development of the Vertebrate Skull* comprises static descriptions almost exclusively, de Beer elsewhere cited the promise offered by the experimental method for revealing "causal relationships in the development of the skull," information about which was "meagre" (p. xxx).

The subsequent period has witnessed a burgeoning application of experimental approaches to elucidating the mechanics of skull development (Hoyte 1966; Moffett 1972; Dixon and Sarnat 1982). For the most part, however, these studies have been confined to only a handful of species. Thus, whereas processes of skull differentiation probably are best known for the domestic fowl (i.e., "the chick"), there have been virtually no experimental studies that have addressed skull differentiation in other birds. Extrapolation of conclusions based on only a few species to vertebrates generally is justified for certain fundamental features; developmental processes are largely conservative. Focus on a limited array of species, however, has inevitable drawbacks.

For example, hormonal mediation of epithelial-mesenchymal interactions plays a key role in the development of several nonskeletal organ systems (Cunha et al. 1983). In amphibians, initial formation of most, and in some cases all, cranial ossification centers accompanies other, widespread anatomical changes that constitute metamorphosis from an aquatic to a terrestrial form—changes which are largely under hormonal control (White and Nicoll 1981). Yet, by focusing studies of skull development on amniotes (e.g., chick, mouse), which lack metamorphosis, researchers may have unwittingly avoided revealing a possible role of hormones in mediating epithelial-mesenchymal interactions that precede cranial osteogenesis in at least some vertebrates.

Broadening the systematic spectrum of vertebrates for laboratory experiments may also have practical benefits, as in the recent development of procedures for rearing alligator cranial explants in vitro as a model system for studying neural crest differentiation (Ferguson et al. 1983).

Evolutionary Mechanisms. Several authors recently have addressed the question of why morphology (including descriptive embryology) was, at least in the West, "excluded" from the evolutionary synthesis of the 1930s and 1940s (e.g., Churchill 1980; Coleman 1980; Ghiselin 1980; Hamburger 1980; Lauder 1982; Wake 1982). De Beer is a case in point. From E. S. Goodrich, his mentor at Oxford, he inherited a primary interest in documenting *patterns* of morphological evolution. Indeed, the Systematic Section of *The Development of the Vertebrate Skull* reflects de Beer's preoccupation, derived from nineteenth century embryology, with two themes, homology and a refutation of the biogenetic law (Churchill 1980). He, like most descriptive embryologists and morphologists at the time, was not in a position to provide the "theory of development" (Waddington 1941) that might have included morphology in the synthesis—with its emphasis on *mechanisms* of evolutionary change at the level of populations—from its earliest stages. (Lost in the lament is the fact that several outstanding descriptive embryologists and morphologists during and soon after the synthesis did continue their studies with an eye toward evolutionary mechanisms, including work on the skull. Most notable here are the Russian school of lower vertebrate morphology, particularly I. I. Schmalhausen's extensive studies of head development in salamanders of the primitive family Hynobiidae [reviewed in Schmalhausen 1968; see also Adams 1980]; and, in the United States, D. D. Davis [e.g., Davis 1964].)

One phenomenon that did receive de Beer's attention was heterochrony: variation in the timing of a developmental event. In 1930 de Beer had published *Embryology and Evolution* (later revised and issued as *Embryos and Ancestors,* 1940) in which he set forth several processes which could either advance or retard the development of a character in an individual in comparison with its ancestors and so bring about morphological change during evolution. In *Development of the Vertebrate Skull,* however, heterochrony was discussed only briefly in two contexts: the allometric relationship between facial and skull growth (pp. 471–72), and identification of the proper ontogenetic stage for making phylogenetic comparisons (pp. 447–48). In the former case, de Beer pointed to the existence of two distinct growth rates in mandibular growth—an initial rapid rate and a subsequent slower rate—the changeover often coinciding with onset of ossification. In the latter, de Beer gave some suggestion as to the mechanism of heterochronic change ("the manifestation of different rates of histogenetic activity in different regions," p. 448) and commented that evolution may make use of the effects of heterochrony ("embryonic variation may be the starting-point of variations also affecting the fully formed structure," p. 448). In this regard he described one form of altered timing of development, namely, caenogenesis (embryonic adaptation), but had little else to say. Despite de Beer's obvious interest in heterochrony as an important mechanism of evolutionary change generally, he refrained from presenting it as more than an interesting but minor theme in skull development.

The last few years have seen a surge in the number of studies that incorporate development into analyses of mechanisms of morphological evolution. The importance of heterochronic processes was re-emphasized by Gould (1977) as acceleration or retardation of development which result in recapitulation and paedomorphosis, respectively—the earlier or later appearance of a character in a descendant than in its ancestors. Alberch et al. (1979) subsequently proposed a scheme of ontogenetic trajectories to quantify heterochrony. With this theoretical foundation, various authors have renewed the search for evidence of heterochrony and are applying such analysis to skull growth in order to relate skull ontogeny and phylogeny. Data are now available from recent studies of amphibians (Alberch and Alberch 1981; Alberch 1983; Hanken 1984; Hanken and Hall 1984; Travis 1980; Trueb 1985; Wake 1980), snakes (Haluska and Alberch 1983), birds (Grant 1981), rodents (Atchley 1983; Brylski 1985), living and fossil horses (Radinsky 1983), fossil rhinoceroses (Prothero and Sereno 1982), primates (Shea 1983), and man (Buschang et al. 1982; Gould 1977). In addition, the degree to which developmental processes may in some instances restrict variation available for natural selection, and thus constrain evolutionary change, yet in other instances facilitate the appearance of novel morphological arrangements, has been emphasized (Alberch 1982, 1983; Hanken 1983). Finally, documentation of extensive skull polymorphisms within natural populations of groups as different as cichlid fishes (Liem and Kaufman 1984), ambystomatid salamanders (Collins and Cheek 1983), and rodents (Berry and Searle 1963) further demonstrates the enormous potential for rapid and extensive morphological and ecological, and thus evolutionary, divergence conferred by heterochrony and by epigenetic developmental processes.

Despite this accumulating knowledge documenting the importance of heterochronic and allometric changes, we have only recently gained insight into the developmental processes that produce such effects. Katz (1980) related the constants of the allometric formula, $Y = bX^k$, to the relative number of cell division centers (b) and the difference between the rates of cell division (k) of two parts, Y and X; such an approach "explains" differences intrinsic in eye growth rates between the salamanders *Ambystoma tigrinum* and *A. punctatum* (= *maculatum*), documented by Twitty and Schwind (1931) more than fifty years ago. Genes are now being identified which inhibit growth of some bones while accelerating the growth of others *in the same organism* (Forsthoefel et al. 1983; Hall 1984*a*). Variation in the timing of epigenetic tissue interactions and in the initial size of skeletal blastemata have been identified as potentially important mechanisms for effecting heterochronic change (Hall 1984*b*), adding flesh to the bones of de Beer's notion of heterochrony as a "manifestation of different rates of histogenetic activity in different regions" (p. 448).

To date, however, these approaches have involved few taxa. Indeed, we still know little of the nature and extent of natural variation of skull development parameters in any vertebrate. Experimental approaches for analyzing both developmental mechanics and developmental genetics remain underutilized in the context of skull evolution. Thus, the potential contribution of development to an understanding of the mechanisms of skull evolution remains largely untapped. The time is now ripe for new insights into heterochrony and allometry as processes guiding development and evolution of the skull.

Role of the Neural Crest

The germ layer theory was firmly entrenched at the time de Beer was writing *The Development of the Vertebrate Skull.* It held that cranial skeletal tissues (cartilage and bone) and dentine were mesodermal derivatives. Any statements to the contrary were treated as heresy. "Heretical" statements, however, were accumulating, particularly in relation to the embryological origin of the visceral cartilages. De Beer reviewed the evidence (pp. 472–76), mostly from anuran and urodele amphibians, for a neural crest (i.e., ectodermal) origin of these cartilages. Neural crest is a population of cells which break away from the neural epithelium during neurulation and assume a mesenchymal morphology as they migrate throughout the early embryo. Descriptive histology, extirpation, and grafting of neural crest cells pointed to their exclusive contribution to the visceral cartilages, but it was with evident reluctance that de Beer accepted the evidence: "It will be seen, therefore, that from the existing state of knowledge it is difficult to resist the conclusion that the cartilage of the trabeculae and visceral arches is derived from the neural crest, strange as it may seem, and great as are the difficulties presented by an attempt to frame a general theory of the origin of cartilaginous material in terms of the germ-layers" (p. 476)—not an enthusiastic embrace of a new paradigm but a grudging acceptance that a favorite theory could not accommodate new facts! (Romer [1972, p. 129] provided a similar story concerning evidence that neural crest cells contribute to the jaw cartilages in mammals, evidence "discovered in 1911, by a student working under Professor J. P. Hill of University College, London, but the fact was not published for nearly half a century [Hill and Watson, 1958]—a delay due, it seems, to reluctance of the professor supervising the student's work to commit treason to the germ layer theory!")

These and other observations soon led to rejection of a strict application of the germ layer theory (Oppenheimer 1940). In 1947 de Beer himself published a seminal paper showing that visceral arch cartilages, odontoblasts, and probably osteoblasts of dermal bone arose from neural-crest-derived (ecto)mesenchyme in *Ambystoma mexicanum*. The last sentence of the abstract reads: "The germ layer theory in its classical form must therefore be abandoned." Three years later Hörstadius (1950) published his important monograph on the neural crest.

Since then, interest in the neural crest origin of craniofacial tissues has mushroomed, both in the study of normal development and in craniofacial anomalies and neurocristopathies (tumors of neural crest derivatives). Weston's (1970) critical review stimulated much interest in migration of neural crest cells. The subsequent discovery, by Le Douarin, of a permanent marker in the form of differing patterns of heterochromatin in cells of the Japanese quail and domestic fowl opened up experimental work on higher vertebrates (reviewed in Le Douarin 1982) and the prospect of accurately following cells in interspecific chimeras (Lemus et al. 1983; Noden 1982). It is now firmly established from such experimental evidence that the branchial basket of the ammocoete larva, the trabecular, visceral, hyoid, and mandibular cartilages of urodele amphibians and of larval anuran amphibians, and all of the facial and much of the cranial bone and cartilage in birds are of neural crest origin (reviewed in Hall 1980; Le Douarin 1982; Noden 1984). Control of initiation of skeletogenesis within neural-crest-derived mesenchyme, however, resides in extracellular matrices associated with adjacent epithelia (Hall 1983). Much less experimental evidence is available for the mammalian craniofacial skeleton (Hall 1980; Le Douarin 1982; Morriss and Thorogood 1978) but the evidence coming out is congruent (Erickson and Weston 1983; Flint 1983; Verwoerd and van Oostrom 1979).

It has been persuasively argued that the evolutionary appearance of the neural crest was a major event in the evolution of the vertebrate head (Gans and Northcutt 1983; Northcutt and Gans 1983). We thus today find neural crest cells finally translated from heretical outsiders knocking on the door of respectability to key members of the congregation of cells that form the vertebrate skull.

Membrane, Dermal, and Cartilage Bones

De Beer was concerned with distinguishing cartilage from bone as skeletal tissues, and membrane bones from cartilage bones as skeletal units (pp. 1–7, 495–502), the former requiring knowledge of histogenesis and ontogeny and the latter knowledge of ontogeny and phylogeny.

Russell (1916) provided a useful historical review of membrane versus cartilage bones; more recently, Delaporte (1983) surveyed the history of periosteal and endochondral ossification. We now have biochemical markers for cartilage and bone (type I and type II collagen, osteocalcin, cartilage-specific proteoglycan), so that distinguishing between the two tissues is no longer the problem it was for de Beer. Reconciliation of ontogenetic and phylogenetic data with respect to cartilage bone and membrane bone, however, remains a problem. As echoed by de Beer, joining a debate that went back to the early 1800s, "histogenesis cannot be regarded as an infallible guide to phylogeny" (p. 6). He cites T. H. Huxley's summary of both the accepted view and the resulting conundrum, namely: "It is highly probable that throughout the vertebrate series, certain bones are always, in origin, cartilage-bones, while certain others are always, in origin, membrane bones." But what "if a membrane-bone is found in the position ordinarily occupied by a cartilage-bone, is it to be regarded merely as the analogue and not as the homologue of the latter?" (p. 4). Smith (1947) dealt with the problem simply by categorizing it, establishing separate ontogenetic and phylogenetic classifications of bone. Patterson (1977) and Reif (1982) faced it head on by following von Baer (1826) in making a fundamental distinction between the inner (endo-) and outer skeletons—the endoskeleton consisting of membrane, perichondral, and endochondral bone and cartilage, the outer skeleton of dermal bone and, in birds and mammals, secondary cartilage.

Patterson, like de Beer and Huxley before him, was particularly concerned with the question of inconvertibility of membrane and cartilage bones. All three had made detailed studies of the head skeleton in fishes, where vertebrate skeletogenesis is at its most complex and skeletal tissues are the most difficult to classify. De Beer believed that "the membrane bones in the skin should be distinguished as dermal bones" (p. 7), and Patterson (1977) argued cogently for an absolute distinction between dermal bones and bone and cartilage of the endoskeleton. For Patterson, fusion or loss explained supposed examples of dermal bones invading the endoskeleton—an interpretation shared by Huysseune et al. (1981) in their studies of the cichlid head skeleton; by Jollie (1975) who studied *Esox;* and by Bellairs and Gans (1983) in a study of the reptilian orbitosphenoid, a membrane bone with associated cartilage nodules in the amphisbaenian, *Leposternon microcephalum,* but a cartilage bone in other reptiles. Thus, while no distinction can be made in the process of osteogenesis (ontogeny) of membrane and cartilage bones, the two ought clearly be regarded as phylogenetically separate skeletal systems. Recently, Ruibal and Shoemaker (1984) described extensive dermal bones (osteoderms) in tropical anurans, ossifications which are quite independent of the underlying endoskeleton and which appear late in postmetamorphic life. Such structures are ideal objects of further experimental study of the relationship between bone in the inner and outer skeletons.

The Relationship of Dermal Bones to Lateral Line Canals

De Beer devoted a few pages (pp. 6, 489–90, 508) and four Agenda items (**ii.**6, p. 513; **iii.**10–12, p. 514) to the problem of the relationship between dermal bones and lateral line canals. Some dermal bones (he cited nasals in flatfish; nasal, frontal, intertemporal, and postparietal in *Amia*) lie in close association with sense organs (neuromasts) of the lateral line canals, both ontogenetically and phylogentically. Is this association a causal one, with sense organs of the lateral line inducing osteogenesis, or is it only the topographical consequence of dermal bones secondarily associating with lateral line canals? De Beer saw it as secondary. He argued that many dermal bones in fishes, including several in the skull, have no association with lateral line canals, and that the homologues of canal-associated bones in fishes arise in higher vertebrates in the absence of lateral line canals. His presumption was that a causal relationship must be constant throughout all vertebrates if it is to exist in a single vertebrate group. Subsequent studies, however, show this not to be true.

For example, Meckel's cartilage is induced by embryonic epithelia in all vertebrates, but the specific epithelium which performs the induction varies from group to group: pharyngeal endoderm in amphibians, cranial ectoderm in the domestic fowl, and mandibular epithelium in the laboratory mouse (Hall 1983). Thus, the requirement for induction is constant but the specific epithelial inductor varies, and so it may be also for dermal bones in fishes. Not all dermal bones need be induced by elements of lateral line canals for some to be so induced. As noted by de Beer, "The matter could, however, be settled by the extirpation of the lateral line placode in a young embryo" (p. 489, and see Agenda, items **iii.**10–12, pp. 514–15). Only two such experimental studies are known to us.

Moy-Thomas (1941) removed the primordium of the lateral line from one side of the head of a single rainbow trout embryo prior to ossification and found that a frontal bone developed just as on the operated side. He concluded that no induction normally occurred. (He also showed that development of the frontal proceeded in embryos whose brain had been extirpated, thus ruling out any induction between brain and overlying osteogenic mesenchyme.) However, Westoll (1941) objected, arguing that dermal bones in teleosts were known to be independent of lateral line canals (e.g., regeneration of the frontal occurs independently of neuromasts [Pinganaud-Perrin 1973]), and that the experiments should be repeated using a species such as *Amia*. This challenge has been accepted only recently by Meinke (1982), who has begun studies of the role of epithelial-mesenchymal interactions in development of the dermal skeleton in fishes.

The second study is that of Devillers (1947, 1965) using *Salmo*. He found that neuromasts did act as centers of aggregation for osteogenic mesenchymal cells, the primordium of dermal canal bone. Later the cells separate from the epithelium, as does the neuromast, following which further ossification occurs around the neuromast. (A very similar series of events occurs in the development of scleral bones under epithelial scleral papillae in birds [Fyfe and Hall 1983; Murray 1943].) Devillers emphasized, as did de Beer, the dual composition of dermal canal bones: a laterosensory (tubular) and a basal (membrane) component. According to Devillers (1965), however, both components are induced by neuromasts in the salmon but only the tubular component is so induced in cyprinids. Devillers also cited unpublished work by Guinnebault showing that the parietal in the salmon, although not directly associated with a lateral line canal, "depends to some extent on the induction of the lateral line system" (1965, p. 265). The contradiction between these conclusions and those of Moy-Thomas (1941) and Westoll (1941) remains unresolved.

A different type of experimental study, by Merrilees (1975), dealt with continued deposition after skeletogenesis had been initiated. Using transplantation techniques, he showed that a specialized cord of epithelial cells in the lateral line canals of the goldfish, *Carassius auratus*, controls cavitation of the tubular component by *inhibiting* deposition of bone, cartilage, and scale. There are other skeletal systems where an epithelium initially stimulates but later inhibits skeletogenesis, as in chondrogenesis in the embryonic limb bud (Hall 1983; Solursh et al. 1984). This raises the possibility, yet uninvestigated, of initial stimulation by neuromasts and later inhibition by epithelia of the lateral line canals.

Recently, Jollie (1984*a,b,c*) provided detailed descriptions of cranial osteogenesis with respect to lateral line development in *Salmo, Polypterus,* and *Lepisosteus*. We know of no additional, experimental studies on origin of canal bone even though several authors have raised the neuromast-dermal bone relationship as an important problem to probe experimentally (Graham-Smith 1978; Northcutt and Gans 1983; Patterson 1977; Schaeffer 1977). Patterson (1977) also has addressed the related problem of sensory canal cartilage in elasmobranchs and its possible induction by elements of the sensory canals. What we now need are further experimental studies on elasmobranch, chondrostean, holostean, and teleost fishes, the fifty years since de Beer's *Development of the Vertebrate Skull* having produced only three reports for teleosts and none for remaining groups.

Secondary Cartilage

De Beer's concern with the relationship of membrane bone to cartilage bone led him to consider the significance and consequences of the presence of secondary cartilage on membrane bones (pp. 2, 38, 502, 514). Membrane and dermal bones ought not to have any association with cartilage, but scattered reports in the literature since the 1850s had pointed to nodules of cartilage on membrane bones *after* ossification had commenced (contrast with primary cartilage which *precedes* endochondral ossification), and developing quite independently of the primary cartilaginous skeleton from which it also differed histologically. Earlier, Schaffer (1930) had reviewed the known distribution of secondary cartilages which had been seen in the mandibular symphysis, on mandibular processes (condylar, coronoid, and angular), and in sutures between dermal skull bones in mammals, and on membrane bones of the craniofacial skeleton in birds. Secondary cartilage received little subsequent attention, however, until dental anatomists began to investigate the growth of mandibular processes, especially the condylar. It was then firmly established that condylar cartilage was indeed secondary and derived from periosteal cells; that it was subsequently replaced by a modified process of endochondral ossification; and that it was adaptive in origin, forming only in response to the biomechanical forces imposed by muscle action (reviewed in Beresford 1981; Durkin et al. 1973; Hall 1978; Koski 1975; Vinkka 1982). De Beer asked in his Agenda of problems for study in the future whether secondary cartilage cells could be demonstrated to turn into osteoblasts in vitro. Melcher (1971) and Silbermann et al. (1983) have demonstrated that condylar chondrocytes can.

Reinvestigation of avian secondary cartilages did not begin until the 1950s when P. D. F. Murray began a series of studies both to describe the nature and distribution of avian secondary (adventitious) cartilages (Murray 1957, 1963) and to show that mechanical forces were required to evoke them from otherwise osteogenic periosteal cells (Murray and Smiles 1965). One of us (B. K. H.) was the last of Murray's graduate students, and took up the study of avian secondary cartilage stimulated both by Murray's studies and by the Agenda set out by de Beer (reviews in Hall 1970, 1978, 1979).

Secondary cartilage is not an oddity, anomaly, or occasional aberration. It is a regular and predictable feature of both avian and mammalian membrane bones, and not just those of the craniofacial skeleton for both the avian and mammalian clavicle possess secondary cartilage (Beresford 1981; Gardner 1968). Moreover, it reflects the ability of periosteal cells to form either membrane bone or cartilage, the latter representing an adaptation to resist local mechanical stresses—at sutures and articulations, under muscle and ligament insertions (Hall 1968), or during repair of fractures (Hall and Jacobson 1975). Secondary cartilage is now beginning to take a more prominent place in basic texts (Ham and Cormack 1979) and has been the subject of a recent, authoritative, and comprehensive monograph (Beresford 1981).

But what of other vertebrates? Membrane bones (dermal in the sense of Patterson 1977) are found in all vertebrates, but secondary cartilage is restricted to birds and mammals. Cartilage in fish that superficially resembles secondary cartilage has been shown to represent nodules of primary cartilage secondarily fused to membrane bones; it also differs both histologically and histochemically from secondary cartilage (Huysseune et al. 1981; Ismail et al. 1982; Patterson 1977). Nor does secondary cartilage form during the repair of fractured membrane bones in fishes (Moss 1962). Secondary cartilage has never been described during normal development of the cranial skeleton of amphibians or reptiles (Bellairs and Kamal 1981; Hall 1984*c*), nor does it form when their membrane bones are fractured (Ferguson, personal communication; Hall and Hanken 1985), although fracture provides an environment conducive to secondary chondrogenesis in birds and mammals (Jolly 1961; Hall and Jacobson 1975). Development of the ability of periosteal cells of membrane bones to form secondary cartilage thus appears to have been a late event in vertebrate evolution (Hall 1984*c;* Hall and Hanken 1985). More experimental studies on fish, amphibian, and reptilian membrane bones would greatly aid in identifying how the skeletogenic potential of their periostea evolved.

Cartilage in the Cyclostomes

Skeletal tissues of agnathans always have been central to debates concerning the origin of vertebrates from non-vertebrate chordates, and concerning the relationships among jawless and jawed fishes. This has never been more true than the last few years, during which time several authors have evaluated the dramatic possibility of a close phylogenetic relationship between lampreys and gnathostomes, distinct from hagfish (Jensen 1963; Schaeffer and Thomson 1980; Hanken and Hall 1983; Løvtrup 1984; Mallatt 1984). Major differences in the published descriptions of the skeletal tissues in the two modern-day agnathan lineages are consistent with this idea of their phylogenetic independence.

De Beer briefly discussed a peculiar skeletal (connective?) tissue found in the head of the ammocoetes (lamprey) larva and known as mucocartilage (pp. 38–39, 41–46). It has a basophilic extracellular matrix like cartilage, and is rich in elastic fibers like elastic cartilage. Yet, it is restricted to the larval stage and was variously regarded as being either transformed into or replaced by the "true" cartilage of the adult. For de Beer, mucocartilage "does not deserve the name of cartilage at all . . . [and is] nothing but a particular kind of connective tissue" (p. 45).

Wright and Youson (1982) recently reexamined mucocartilage in *Petromyzon marinus* and described a surrounding, vascularized "perichondrium" rich in collagen and elastic-like microfibrils. The enclosed mucocartilage is avascular, lacks collagen, is sparsely cellular, and, like the investing membrane, contains elastic-like microfibrils. They side with de Beer and with Hardisty (1979) in regarding mucocartilage as a specialized larval connective tissue. Its developmental fate, however, remains uncertain and requires further detailed study, for adult cartilage apparently may develop where no mucocartilage existed, within degenerating mucocartilage, or even within degenerating larval muscle.

Wright and her colleagues also have investigated the structure and biochemistry of adult cartilage, in both the lamprey, *Petromyzon marinus* (Wright and Youson 1983; Wright et al. 1983), and the hagfish, *Myxine glutinosa* (Wright et al. 1984). Lamprey cartilage is highly cellular with a central zone of hypertrophic chondrocytes. It is bounded by a vascular perichondrium which contains collagen fibrils, and has an extracellular matrix which consists of a dense network of branched, non-collagenous fibrils composed of the protein, lamprin. Lamprin, which comprises approximately one-half of the dry weight of the annular cartilage (glycosaminoglycans, the normal major constituent of cartilage, make up less than 5 percent of the dry weight of lamprey cartilage), has only traces of hydroxyproline, no hydroxylysine, and much tyrosine and histidine. The hagfish has two different types of cartilage, one with branched fibrils of a protein similar to lamprin, termed myxinin, the other with hypertrophic cells filled with cytoplasmic filaments and very similar to some of the invertebrate cartilages (Wright et al. 1984).

We see encapsulated here the difficulty and inconsistency in the classification of these skeletal tissues. Mucocartilage is bounded by a perichondrium but is not considered cartilage because it lacks collagen. In the adult tissue the matrix also lacks collagen and has minimal glycosaminoglycans, but it is regarded as cartilage! (A similar difficulty arises with the invertebrate cartilages, many of which lack type II—"cartilage-type"—collagen; see Person [1983] for a very useful discussion on how to define cartilage.)

Notwithstanding these questions concerning appropriate terminology, published descriptions of the skeletal tissues of lampreys and hagfish are consistent with the idea of an early separation and long, independent evolution of the two groups. We urge caution, however, as many differences may just reflect the stages studied. Very recently, Robert Langille, working in the laboratory of B. K. Hall, has shown considerable similarity in the ultrastructural organization of cartilage of spawning adult lamprey and the type I hagfish cartilage published by Wright et al. (1984). Such similarity, seen at only certain phases of the life cycle, may further complicate the use of developmental data in the interpretation of the exact relationships among cyclostomes and gnathostomes.

This brings us full circle to the use of developmental studies to understand phylogeny, a major aim of de Beer's fifty years ago, and a coupling that once again has brought the skull into center stage.

Brian K. Hall
Dalhousie University
Halifax, Nova Scotia

James Hanken
University of Colorado
Boulder, Colorado

LITERATURE CITED

Adams, M. B. 1980. Severtsov and Schmalhausen: Russian morphology and the evolutionary synthesis. In *The Evolutionary Synthesis* (E. Mayr and W. B. Provine, eds.), pp. 193–225. Harvard University Press, Cambridge.

Alberch, P. 1982. Developmental constraints in evolutionary processes. In *Evolution and Development* (J. T. Bonner, ed.), pp. 313–32. Springer-Verlag, Berlin.

———. 1983. Morphological variation in the neotropical salamander genus *Bolitoglossa*. *Evolution* 37:906–19.

Alberch, P., and J. Alberch. 1981. Heterochronic mechanisms of morphological diversification and evolutionary change in the neotropical salamander, *Bolitoglossa occidentalis* (Amphibia: Plethodontidae). *J. Morphol.* 167:249–64.

Alberch, P., S. J. Gould, G. F. Oster, and D. B. Wake. 1979. Size and shape in ontogeny and phylogeny. *Paleobiology* 5:296–317.

Atchley, W. R. 1983. A genetic analysis of the mandible and maxilla in the rat. *J. Craniofac. Gen. Devel. Biol.* 3:409–22.

Baer, K. E. von. 1826. Ueber das äussere und innere Skelet. *Meckel's Arch. Anat. Physiol. Leipzig.* 9:327–76.

Barrington, E. J. W. 1973. Gavin Rylands de Beer. *Biograph. Mem. Fellows Roy. Soc.* 19:65–93.

Bellairs, A. d'A., and C. Gans. 1983. A reinterpretation of the amphisbaenian orbitosphenoid. *Nature* 302:243–44.

Bellairs, A. d'A., and A. M. Kamal. 1981. The chondrocranium and the development of the skull in recent reptiles. In *Biology of the Reptilia* (C. Gans, ed.) 11:1–263. Academic Press, New York.

Beresford, W. A. 1981. *Chondroid Bone, Secondary Cartilage and Metaplasia*. Urban and Schwarzenberg, Munich and Baltimore.

Berry, R. J., and A. G. Searle. 1963. Epigenetic polymorphism of the rodent skeleton. *Proc. Zool. Soc. Lond.* 140:577–615.

Bjerring, H. C. 1977. A contribution to structural analysis of the head of craniate animals. *Zool. Scripta* 6:127–83.

Brylski, P. 1985. Developmental allometry and heterochrony in geomyid and heteromyid rodents. *J. Zool., Lond.* Forthcoming.

Buschang, P. H., G. G. Nass, and G. F. Walker. 1982. Principal components of craniofacial growth for white Philadelphia males and females between 6 and 22 years of age. *Am. J. Orthod.* 82:508–12.

Churchill, F. B. 1980. The modern evolutionary synthesis and the biogenetic law. In *The Evolutionary Synthesis* (E. Mayr and W. B. Provine, eds.), pp. 112–22. Harvard University Press, Cambridge.

Coleman, W. 1980. Morphology in the evolutionary synthesis. In *The Evolutionary Synthesis* (E. Mayr and W. B. Provine, eds.), pp. 174–180. Harvard University Press, Cambridge.

Collins, J. P., and J. E. Cheek. 1983. Effect of food and density on development of typical and cannibalistic salamander larvae in *Ambystoma tigrinum nebulosum*. *Am. Zool.* 23:77–84.

Cunha, G. R., J. M. Shannon., O. Taguchi, H. Fujii, and B. A. Meloy. 1983. Epithelial-mesenchymal interactions in hormone-induced development. In *Epithelial-mesenchymal Interactions in Development* (R. H. Sawyer and J. F. Fallon, eds.), pp. 51–74. Praeger Pubs., New York.

Davis, D. D. 1964. The giant panda—a morphological study of evolutionary mechanisms. *Fieldiana Zool. Mem.* 3:1–339.

de Beer, G. R. 1947. The differentiation of neural crest cells into visceral cartilages and odontoblasts in *Amblystoma,* and a re-examination of the germ-layer theory. *Proc. Roy. Soc. Lond.* 134B:377–98.

Delaporte, F. 1983. Theories of osteogenesis in the eighteenth century. *J. Hist. Biol.* 16:343–60.

Detwiler, S. R. 1934. An experimental study of spinal nerve segmentation in *Amblystoma* with reference to the plurisegmental contribution to the brachial plexus. *J. Exp. Zool.* 67:395–441.

Devillers, C. 1947. Recherches sur la crâne dermique des téléostéens. *Ann. Paleont.* 33:1–94.

Devillers, C. 1965. The role of morphogenesis in the origin of higher levels of organization. *Syst. Zool.* 14:259–71.

Dixon, A. D., and B. G. Sarnat (eds.) 1982. *Factors and Mechanisms Influencing Bone Growth.* A. R. Liss, Inc., New York.

Durkin, J. F., J. Heeley, and J. T. Irvine. 1973. The cartilage of the mandibular condyle. *Oral Sci. Rev.* 2:29–99.

Erickson, C. A., and J. A. Weston. 1983. An SEM analysis of neural crest migration in the mouse. *J. Embryol. Exp. Morphol.* 74:97–118.

Ferguson, M. W. J., L. S. Honig, P. Bringas, Jr., and H. C. Slavkin. 1983. Alligator mandibular development during long-term organ culture. *In Vitro* 19:385–93.

Flint, O. P. 1983. A micromass culture method for rat embryonic neural cells. *J. Cell Sci.* 61:247–62.

Forsthoefel, P. F., P. B. Kanjananggulpan, and S. Harmon. 1983. Developmental interactions of cells mutant for strong luxoid gene with normal cells in chimeric mice. *J. Hered.* 74:153–62.

Fyfe, D. M., and B. K. Hall. 1983. The origin of the ectomesenchymal condensations which precede the development of the bony scleral ossicles in the eyes of embryonic chicks. *J. Embryol. Exp. Morphol.* 73:69–86.

Gans, C., and R. G. Northcutt. 1983. Neural crest and the origin of vertebrates: A new head. *Science* 220:268–74.

Gardner, E. 1968. The embryology of the clavicle. *Clin. Orthop. Rel. Res.* 58:9–16.

Ghiselin, M. T. 1980. The failure of morphology to assimilate Darwinism. In *The Evolutionary Synthesis* (E. Mayr and W. B. Provine, eds.), pp. 180–93. Harvard University Press, Cambridge.

Gould, S. J. 1977. *Ontogeny and Phylogeny.* Harvard University Press, Cambridge.

Graham-Smith, W. 1978. On the lateral lines and dermal bones in the parietal region of some Crossopterygian and Dipnoan fishes. *Phil. Trans. R. Soc. Lond.* 282B:41–105.

Grant, P. R. 1981. Patterns of growth in Darwin's finches. *Proc. R. Soc. Lond.* 212B:403–32.

Hall, B. K. 1968. The fate of adventitious and embryonic articular cartilage in the skull of the common fowl, *Gallus domesticus* (Aves: Phasianidae). *Aust. J. Zool.* 16:795–806.

———. 1970. Cellular differentiation of skeletal tissues. *Biol. Rev.* 45:455–84.

———. 1978. *Developmental and Cellular Skeletal Biology.* Academic Press, New York.

———. 1979. Selective proliferation and accumulation of chondroprogenitor cells as the mode of action of biomechanical factors during secondary chondrogenesis. *Teratology* 20:81–92.

———. 1980. Chondrogenesis and osteogenesis in cranial neural crest cells. In *Current Research Trends in Prenatal Craniofacial Development* (R. M. Pratt and R. L. Christiansen, eds.), pp. 47–63. Elsevier/North Holland, New York.

———. 1983. Tissue interactions and chondrogenesis. In *Cartilage,* vol. 2 (B. K. Hall, ed.), pp. 187–222. Academic Press, New York.

———. 1984*a*. Developmental mechanisms underlying the formation of atavisms. *Biol. Rev.* 59:89–124.

———. 1984*b*. Developmental processes underlying heterochrony as an evolutionary mechanism. *Can. J. Zool.* 62:1–7.

———. 1984*c*. Developmental processes underlying the evolution of cartilage and bone. In *The Structure, Development and Evolution of Reptiles* (M. W. J. Ferguson, ed.), pp. 155–76. Academic Press, London.

Hall, B. K., and J. Hanken. 1985. Repair of fractured lower jaws in the spotted salamander: Do amphibians form secondary cartilage? *J. Exp. Zool.* 233:359–68.

Hall, B. K., and H. N. Jacobson. 1975. The repair of fractured membrane bones in the newly hatched chick. *Anat. Rec.* 181:55–70.

Haluska, F., and P. Alberch. 1983. The cranial development of *Elaphe obsoleta* (Ophidia, Colubridae). *J. Morphol.* 178:37–56.

Ham, A. W., and D. H. Cormack. 1979. *Histophysiology of Cartilage, Bone, and Joints.* J. B. Lippincott Co., Philadelphia.

Hamburger, V. 1980. Embryology and the modern synthesis in evolutionary theory. In *The Evolutionary Synthesis* (E. Mayr and W. B. Provine, eds.), pp. 97–112. Harvard University Press, Cambridge.

Hanken, J. 1983. Miniaturization and its effects on cranial morphology in plethodontid salamanders, genus *Thorius* (Amphibia: Plethodontidae). II. The fate of the brain and sense organs and their role in skull morphogenesis and evolution. *J. Morphol.* 177:255–68.

———. 1984. Miniaturization and its effects on cranial morphology in plethodontid salamanders, genus *Thorius* (Amphibia: Plethodontidae). I. Osteological variation. *Biol. J. Linn. Soc.* 23:55–75.

Hanken, J., and B. K. Hall. 1983. Evolution of the skeleton. *Nat. Hist.* 92(4):28–39.

———. 1984. Variation and timing of the cranial ossification sequence of the Oriental fire-bellied toad, *Bombina orientalis* (Amphibia, Discoglossidae). *J. Morphol.* 182:245–55.

Hardisty, M. W. 1979. *Biology of the Cyclostomes.* Chapman and Hall, London.

Hill, J. P., and K. M. Watson. 1958. The early development of the brain in marsupials. *J. Anat.* 92:493–97.

Hörstadius, S. 1950. *The Neural Crest: Its Properties and Derivatives in the Light of Experimental Research.* Oxford University Press, London.

Hoyte, D. A. N. 1966. Experimental investigations of skull morphology and growth. *Int. Rev. Gen. Exp. Zool.* 2:345–407.

Huysseune, A., M. H. Ismail, and W. Verraes. 1981. Some histological and ultrastructural aspects of the development of the articulation between neurocranial base and upper pharyngeal jaw in *Haplochromis elegans* (Teleostei: Cichlidae). *Verh. Anat. Ges.* 75S:499–500.

Ismail, M. H., W. Verraes, and A. Huysseune. 1982. Developmental aspects of the pharyngeal jaws in *Astatotilapia elegans* (Trewavas, 1933) (Teleostei: Cichlidae). *Neth. J. Zool.* 32:513–43.

Jacobson, A. G., and S. Meier. 1984. Morphogenesis of the head of a newt: Mesodermal segments, neuromeres, and distribution of neural crest. *Devel. Biol.* 106:181–93.

Jarvik, E. 1980. *Basic Structure and Evolution of Vertebrates,* vols. 1–2. Academic Press, London.

Jensen, D. D. 1963. Hoplonemertines, myxinoids and vertebrate origins. In *The Lower Metazoa: Comparative Biology and Phylogeny* (E. C. Dougherty, ed.), pp. 113–26. University of California Press, Los Angeles.

Jollie, M. 1975. Development of the head skeleton and pectoral girdle in *Esox*. *J. Morphol.* 147:61–88.

———. 1984*a*. Development of the head skeleton and pectoral girdle of salmons, with a note on the scales. *Can. J. Zool.* 62:1757–78.

———. 1984*b*. Development of the head and pectoral skeleton of *Polypterus* with a note on scales (Pisces: Actinopterygii). *J. Zool., Lond.* 204:469–507.

———. 1984*c*. Development of cranial and pectoral girdle bones of *Lepisosteus* with a note on scales. *Copeia* 1984:476–502.

Jolly, M. T. 1961. Condylectomy in the rat. An investigation of the ensuing repair processes in the region of the temperomandibular articulation. *Aust. Dental. J.* 6:243–56.

Katz, M. J. 1980. Allometry formula: A cellular model. *Growth* 44:89–96.

Keynes, R. J., and C. D. Stern. 1984. Segmentation in the vertebrate nervous system. *Nature* 310:786–89.

Kingsbury, B. F. 1926. Branchiomerism and the theory of head segmentation. *J. Morphol.* 42:83–109.

Koski, K. 1975. Cartilage in the face. *Birth Defects Orig. Article Ser.* 11(7):231–54.

Lauder, G. V. 1982. Introduction to E. S. Russell, *Form and Function,* pp. xi–xiv. University of Chicago Press, Chicago.

Le Douarin, N. 1982. *The Neural Crest.* Cambridge University Press, London.

Lemus, D., M. Fuenzalida, J. Illanes, and Y. Paz de la Vega. 1983. Ultrastructural aspects of dental tissues and their behavior in xenoplastic associations (Lizard:quail). *J. Morphol.* 176:341–50.

Liem, K. F., and L. S. Kaufman. 1984. Intraspecific macroevolution: functional biology of the polymorphic cichlid species *Cichlasoma minckleyi*. In *Evolution of Fish Species Flocks* (A. A. Echelle and I. Kornfield, eds.), pp. 203–15. University of Maine at Orono Press.

Løvtrup, S. 1984. Ontogeny and phylogeny. In *Beyond Neo-Darwinism* (M. W. Ho and P. T. Saunders, eds.), pp. 159–90. Academic Press, London.

Mallatt, J. 1984. Early vertebrate evolution: pharyngeal structure and the origin of gnathostomes. *J. Zool., Lond.* 204:169–83.

Meier, S. 1981. Development of the chick embryo mesoblast: Morphogenesis of the prechordal plate and cranial segments. *Dev. Biol.* 83:49–61.

Meier, S. P. 1982. The development of segmentation in the cranial region of vertebrate embryos. *Scanning Electron Microsc.* 3:1269–82.

Meier, S. 1984. Somite formation and its relationship to metameric patterning of the mesoderm. *Cell Diff.* 14:235–43.

Meier, S., and D. S. Packard, Jr. 1984. Morphogenesis of the cranial segments and distribution of neural crest in the embryos of the snapping turtle, *Chelydra serpentina. Dev. Biol.* 102:309–23.

Meier, S., and P. P. L. Tam. 1982. Metameric pattern development in the embryonic axis of the mouse. I. Differentiation of the cranial segments. *Differentiation* 21:95–108.

Meinke, D. 1982. A light and scanning electron microscope study of microstructure, growth and development of the dermal skeleton of *Polypterus* (Pisces: Actinopterygii). *J. Zool., Lond.* 197:355–82.

Melcher, A. H. 1971. Behaviour of cells of condylar cartilage of foetal mouse mandible maintained *in vitro. Archs. Oral Biol.* 16:1379–91.

Merrilees, M. J. 1975. Tissue interactions: morphogenesis of the lateral line system and labyrinth of vertebrates. *J. Exp. Zool.* 192:113–18.

Moffett, B. C. 1972. *Mechanisms and Regulation of Craniofacial Morphogenesis.* Swets & Zeitlinger B. V., Amsterdam.

Moore, W. J. 1981. *The Mammalian Skull.* Cambridge University Press, Cambridge.

Morriss, G., and P. V. Thorogood. 1978. An approach to cranial neural crest cell migration and differentiation in mammalian embryos. In *Development in Mammals,* vol. 3 (M. H. Johnston, ed.), pp. 363–412. Elsevier North Holland, Amsterdam.

Moss, M. L. 1962. Studies of the acellular bone of teleost fish. II. Response to fracture under normal and acalcemic conditions. *Acta Anat.* 48:46–60.

Moy-Thomas, J. A. 1941. Development of the frontal bones of the rainbow trout. *Nature* 147:681–82.

Murray, P. D. F. 1943. The development of the conjunctival papillae and of the scleral bones in the chick embryo. *J. Anat.* 77:225–40.

———. 1957. Cartilage and bone—a problem in tissue differentiation. *Aust. J. Sci.* 14:65–73.

———. 1963. Adventitious (secondary) cartilage in the chick embryo and the development of certain bones and articulations in the chick skull. *Aust. J. Zool.* 11:368–430.

Murray, P. D. F., and M. Smiles. 1965. Factors in the evocation of adventitious (secondary) cartilage in the chick embryo. *Aust. J. Zool.* 13:351–81.

Noden, D. M. 1982. Patterns and organization of craniofacial skeletogenic and myogenic mesenchyme: A perspective. In *Factors and Mechanisms Influencing Bone Growth* (A. Dixon and B. Sarnat, eds.), pp. 167–203. A. R. Liss, Inc., New York.

———. 1983*a*. The role of the neural crest in patterning of avian cranial skeletal, connective, and muscle tissues. *Dev. Biol.* 96:144–65.

———. 1983*b*. The embryonic origins of avian cephalic and cervical muscles and associated connective tissues. *Am. J. Anat.* 168:257–76.
———. 1984. Craniofacial development: New views on old problems. *Anat. Rec.* 208:1–13.
Northcutt, R. G., and C. Gans. 1983. The genesis of neural crest and epidermal placodes: A reinterpretation of vertebrate origins. *Quart. Rev. Biol.* 58:1–28.
Novacek, M. 1977. Aspects of the problem of variation, origin and evolution of the eutherian auditory bulla. *Mammal Rev.* 7:131–49.
Oppenheimer, J. M. 1940. The non-specificity of the germ layers. *Quart. Rev. Biol.* 15:1–27.
Patterson, C. 1977. Cartilage bones, dermal bones and membrane bones, or the exoskeleton versus the endoskeleton. In *Problems in Vertebrate Evolution* (S. M. Andrews, R. S. Miles, and A. D. Walker, eds.), pp. 77–122. Academic Press, New York.
Person, P. 1983. Invertebrate cartilages. In *Cartilage,* vol. 1, *Structure, Function, and Biochemistry* (B. K. Hall, ed.), pp. 31–57. Academic Press, Inc., New York.
Pinganaud-Perrin, G. 1973. Conséquences de l'ablation de l'os frontal sur la forme des os du toit crânien de la truite (*Salmo irideus* Gib, Pisces-Teleostei). *C. R. Acad. Sci.* 276:2809–11.
Presley, R. 1978. Ontogeny of some elements of the auditory bulla in mammals. *J. Anat.* 126:428.
———. 1979. The primitive course of the internal carotid artery in mammals. *Acta Anat.* 103:238–44.
———. 1981. Alisphenoid equivalents in placentals, marsupials, monotremes and fossils. *Nature* 294:668–70.
Presley, R., and F. L. D. Steel. 1976. On the homology of the alisphenoid. *J. Anat.* 121:441–59.
———. 1978. The pterygoid and ectopterygoid in mammals. *Anat. Embryol.* 154:95–110.
Prothero, D. R., and P. C. Sereno. 1982. Allometry and paleoecology of medial Miocene dwarf rhinoceroses from the Texas Gulf Coastal Plain. *Paleobiology* 8:16–30.
Radinsky, L. 1983. Allometry and reorganization in horse skull proportions. *Science* 221:1189–91.
Reif, W.-E. 1982. Evolution of dermal skeleton and dentition in vertebrates: The odontode regulation theory. *Evol. Biol.* 15:287–368.
Romer, A. S. 1972. The vertebrate animal as a dual animal—somatic and visceral. In *Evolutionary Biology,* vol. 6 (T. Dobzhansky, M. K. Hecht, and W. C. Steere, eds.), pp. 121–56. Appleton-Century-Crofts, New York.
Ruibal, R., and V. Shoemaker. 1984. Osteoderms in anurans. *J. Herpetol.* 18:313–28.
Russell, E. S. 1916. *Form and Function: A Contribution to the History of Animal Morphology.* John Murray, London. (Paperback reprint: University of Chicago Press, Chicago, 1982.)

Schaeffer, B. 1977. The dermal skeleton in fishes. In *Problems in Vertebrate Evolution* (S. M. Andrews, R. S. Miles, and A. D. Walker, eds.), pp. 25–52. Academic Press, New York.

Schaeffer, B., and K. S. Thomson. 1980. Reflections on agnathan-gnathostome relationships. In *Aspects of Vertebrate History* (L. L. Jacobs, ed.), pp. 19–33. Museum of Northern Arizona Press, Flagstaff.

Schaffer, J. 1930. Die Stutzgewebe. In *Handbuch der Mikroskopischen Anatomie des Menschen* (W. von Mollendorff, ed.), Julius Springer, Berlin 2(2):338–50.

Schmalhausen, I. I. 1968. *The Origin of Terrestrial Vertebrates*. Academic Press, New York.

Shea, B. T. 1983. Allometry and heterochrony in the African apes. *Am. J. Phys. Anthropol.* 62:275–89.

Silbermann, M., D. Lewinson, H. Gonen, M. A. Lizarbe, and K. von den Mark. 1983. *In vitro* transformation of chondroprogenitor cells into osteoblasts and the formation of new membrane bone. *Anat. Rec.* 206:373–83.

Smith, H. M. 1947. Classification of bone. *Turtox News* 25:234–36.

Solursh, M., K. L. Jensen, N. C. Zanetti, T. F. Linsenmayer, and R. S. Reiter. 1984. Extracellular matrix mediates epithelial effects on chondrogenesis *in vitro*. *Devel. Biol.* 105:451–57.

Tam, P. P. L., S. Meier, and A. G. Jacobson. 1982. Differentiation of the metameric pattern in the embryonic axis of the mouse. II. Somitomeric organization of the presomitic mesoderm. *Differentation* 21:109–22.

Travis, J. 1980. Genetic variation for larval specific growth rate in the frog *Hyla gratiosa*. *Growth* 44:167–81.

Trueb, L. 1985. A summary of osteocranial development in anurans with notes on the sequence of cranial ossification in *Rhinophrynus dorsalis* (Anura: Pipoidea: Rhinophrynidae). *Proc. Fifth Symp. African Anurans*. Forthcoming.

Twitty, V. C., and J. L. Schwind. 1931. The growth of eyes and limbs transplanted heteroplastically between two species of *Amblystoma*. *J. Exp. Zool.* 59:61–86.

Verwoerd, C. D. A., and C. G. van Oostrom. 1979. Cephalic neural crest and placodes. *Adv. Anat. Embryol. Cell Biol.* 58:1–80.

Vinkka, H. 1982. Secondary cartilages in the facial skeleton of the rat. *Proc. Finn. Dental Soc.* 78(suppl. 7):1–137.

Waddington, C. H. 1941. Evolution of developmental systems. *Nature* 147:108–10.

Wake, D. B. 1980. Evidence of heterochronic evolution: a nasal bone in the Olympic salamander *Rhyacotriton olympicus*. *J. Herpetol.* 14:292–95.

———. 1982. Functional and evolutionary morphology. *Persp. Biol. Med.* 25:603–20.

Wake, M. H., and J. Hanken. 1981. Development of the skull of *Dermophis mexicanus* (Amphibia: Gymnophiona), with comments on skull kinesis and amphibian relationships. *J. Morphol.* 173:203–23.

Westoll, T. S. 1941. Latero-sensory canals and dermal bones. *Nature* 148:168.

Weston, J. A. 1970. The migration and differentiation of neural crest cells. *Adv. Morphog.* 8:41–114.

White, B. A., and C. S. Nicoll. 1981. Hormonal control of amphibian metamorphosis. In *Metamorphosis, A Problem of Developmental Biology,* 2d ed. (L. I. Gilbert and E. Frieden, eds.), pp. 363–96. Plenum Press, New York.

Wright, G. M., and J. H. Youson. 1982. Ultrastructure of mucocartilage in the larval anadromous sea lamprey, *Petromyzon marinus*. *Am. J. Anat.* 165:39–51.

———. 1983. Ultrastructure of cartilage from young adult sea lamprey, *Petromyzon marinus* L.: A new type of vertebrate cartilage. *Am. J. Anat.* 167:59–70.

Wright, G. M., F. W. Keeley, and J. H. Youson. 1983. Lamprin: A new vertebrate protein comprising the major structural protein of adult lamprey cartilage. *Experientia* 39:495–96.

Wright, G. M., F. W. Keeley, J. H. Youson, and D. L. Babineau. 1984. Cartilage in the Atlantic hagfish, *Myxine glutinosa*. *Am. J. Anat.* 169:407–24.

PREFACE

THIS book is the outcome of some fifteen years' work devoted to the study of the development of the skull in all the vertebrate groups. To pretend that I have personally controlled all the statements made herein would be a gratuitous and extravagant exaggeration. But I have had the privilege of studying and researching on embryonic material of all the groups of Vertebrates, and where I have had to make use of them I have interpreted the findings and descriptions of other workers in the light of my own experience, not hesitating to modify or even to reverse their conclusions if I considered it necessary.

My reasons for preparing this book have been numerous.

In the first place, I must claim (or confess) that I have always been attracted, aesthetically I suppose, by the actual shapes and transformations presented by the embryonic skulls of different forms at different stages, and the collection of information concerning them (which has formed the major part of my own practical work during the past fifteen years) has been a work of delight.

Next, from the didactic point of view, I have been dismayed at the lack of co-ordination displayed by the numerous workers in this field, leading as it has to a most confused and redundant nomenclature, and to neglect of a number of morphological principles and features of interest which escaped recognition largely because the objects of comparison bore different names. This difficulty is felt in acute form when comparing descriptions of the embryonic human skull with that of other Vertebrates.

To remedy this, I have devoted myself to a systematic description of all the vertebrate types using a uniform nomenclature, and stressing the points of comparison as far as possible. This programme has also entailed the redrawing of nearly all the figures so as to present them under a uniform method of treatment.

I hope, therefore, that this work may be consulted with profit by vertebrate morphologists, palaeontologists, and human anatomists for purposes of both study and research: consulted, I say advisedly, for nobody will choose to read such a work from cover to cover. This fact has had an important bearing on the method of treatment of the subject, for it was obviously essential that each section should be reasonably complete and completely intelligible in itself, without the necessity for constant reference to some other place in the text. At the same time, mere repetition was to be avoided.

It may be objected that I have fallen a victim to the tendency to incorporate too much detail in my work. My answer is that in morphology general principles are founded on matters of quite intricate detail and require detail for their illustration, just as the geologist has found that in order to interpret the structure of a range of mountains, or of the whole earth, he has to look through a microscope. In morphology it is true that we have as yet but few general principles of wide application: it is precisely in the hope that readers and future workers may be able to formulate more that I have sifted and included such matters of detail as I considered to be least unlikely to further this end.

The modern tendency is, quite rightly in my opinion, to appeal to the experimental method to advance knowledge in all departments. A glance at my chapter on 'Causal relationships in the development of the skull' will show how meagre is the information in this field. And this has been another incentive to the preparation of this work. Logical and fruitful experimentation must start from a factual basis of observation. It was the observation of the singleness of the lens in cyclopic eyes that suggested to Herbst that the correlation between eyecup and lens might be a causal one. As is well known, this hypothesis has formed the basis from which the magnificent experimental analyses of Spemann and Harrison have started.

The material presented by the development of the skull is so rich and varied that I venture to hope that it will be possible to suspect and subsequently to establish further correlations of this type. There are, for instance, the questions of the relations between the lateral line sense-organs and the bones which surround them, or the relations between a cartilage-bone and the cartilage whose place it takes. These hypotheses and doubtless others are susceptible of experimental verification or refutation, and to demonstrate our present ignorance and encourage such work is another reason why I have written this book.

A close study of the development of such a structure as the skull in a group as compact as the Vertebrates might be expected to yield at any rate some conclusions concerning the phylogeny of Vertebrates, and, in general, the problem of the relation of embryology to evolution. Such as they are, my conclusions on this subject and on other general morphological considerations will, I hope, be found acceptable at least as a basis for further work; for the reader will find himself confronted with new problems of unexpectedly profound nature. He will see, for instance, that having followed the development of some bones in nearly all vertebrate groups, we are in the absurd position of still lacking a satisfactory definition of what is meant by the expression 'a bone'.

I may perhaps be permitted to stress the fact that I have found it necessary to make a distinction between the *developmental history* of structures and the *morphology* of embryonic structures. The latter is remarkably uniform throughout the Vertebrates, whereas the former is very inconstant, owing to variations in time at which the different parts of a structure may appear.

At the same time I must make it clear that this work in no way pretends to be a study of the complete morphology of the skull; if the comparative anatomy of living and fossil adult skulls had to be included the size of the work would have had to be five times what it is.

A few words may, perhaps, not be out of place as regards the construction of this book. The Introductory, Comparative, and General Sections are intended for 'general consumption' by biologists; the Systematic Section is a work of reference designed to illustrate in greater detail the principles and points raised in the other sections, and to provide information concerning the extent of knowledge (stages studied, times at which the various structures arise, progress of architectural craniogenesis, morphology) concerning the development of the skull in the various groups.

In each group, the first form is described fairly fully, so as to serve as a type for that group. Thus, the trout and the rabbit serve as general types for Teleostei and Placentalia respectively, and obviate the necessity for special sections on the 'Teleost skull' or the 'Placental skull'. The diagnostic features of the skulls of the various groups are considered together in the General Section.

Lastly, I wish to express the hope that this work, for all its size, will escape the lamentable fate of being regarded as exhaustive as regards the subject and exhausting to those who study it. In some cases, no doubt, the appearance of large works on particular subjects has had the effect of sterilizing them. Nothing could be more distant from my mind than to convey the impression that no further profitable work remains to be done on the development of one of the most important and interesting structures in the animal kingdom. For this reason, following the example of H.-B. de Saussure, I end my book with a list of a few of the more important problems requiring solution.

G. R. DE B.

ACKNOWLEDGEMENTS

MY first duty and pleasure is to acknowledge the debt which I owe to Professor E. S. Goodrich, in whose Department the work for this book was done, and from whom I have unfailingly received encouragement, criticism, and advice. He it was who introduced me to the interests of morphology, and I hope that he will be good enough to regard the following pages as, in some sense, an attempt to acquit myself partially of the obligations under which I stand to him, by applying his teaching and methods to the problems which have occupied me for some years.

I have indeed been fortunate in the friends and colleagues with whom I have discussed certain aspects of this work, either in conversation, or by correspondence, and from whom I have received material, information (in many cases unpublished), figures to reproduce in this book, reprints of their works or assistance in connexion with publications inaccessible to me. Among all these persons, I would especially like to record my indebtedness to Dr. M. Augier, Mr. F. H. Aumonier, Miss G. T. Brock, Dr. R. Broom, Dr. A. Chabanaud, Prof. W. E. Le Gros Clark, Prof. E. Fawcett, Miss H. B. Fell, Prof. R. A. Fisher, Mr. R. W. Haines, Prof. R. G. Harrison, Prof. J. P. Hill, Prof. N. Holmgren, Prof. C. J. van der Klaauw, Prof. E. Matthes, Prof. B. Matveiev, Mr. J. A. Moy-Thomas, Prof. A. Meek, Mr. J. R. Norman, Mr. H. W. Parker, Mr. H. K. Pusey, Prof. A. N. Sewertzoff, Mr. E. L. Seyd, Miss M. Tribe, Prof. D. M. S. Watson, Dr. T. S. Westoll, Prof. P. Wintrebert, and the late Prof. J. W. van Wijhe.

In the preparation of the Plates for this book, I have enjoyed the invaluable assistance of my wife, without whose help the task of redrawing and relabelling all the figures would have been formidable indeed. I wish also to acknowledge the help of the Drawing Office of the Clarendon Press, which has constantly afforded me every facility and courtesy, as has every other Department and Officer of that great institution.

For permission to reproduce figures, I am indebted to Messrs MacMillan & Co., Ltd. for a figure (Plate 69), relabelled, from E. S. Goodrich's *Studies on the Structure and Development of Vertebrates*, and to MM. Masson et Cie. for some figures (Plate 135), relabelled, from M. Augier's section on 'Le Squelette Céphalique' in Poirier and Charpey's *Traité d'Anatomie Humaine*. Use has been made of figures by Dr. de Burlet in M. Weber's *Die Säugetiere* for figs. 5 and 6, Plate 113; and of figures in Gaupp's article in O.Hertwig's *Handbuch der Entwicklungslehre* for fig. 4, Plate 78; Plate 86; figs. 1, 2, 3, Plate 99; and figs. 4, 5, 6, Plate 134. Both these works are published by Herr Gustav Fischer in Jena. The authorship of figures redrawn and relabelled from current journals is acknowledged in the underline in each case. The journals, which are, of course, referred to in the Bibliography, are the following:

Acta Zoologica.
American Journal of Anatomy.
Anatomical Record.
Anatomische Hefte.
Anatomischer Anzeiger.
Archives de Biologie.

Archiv für Mikroskopische Anatomie.
Archives d'Anatomie, d'Histologie et d'Embryologie.
Archives Russes d'Anatomie, d'Histologie et d'Embryologie.
Bijdragen tot de Dierkunde.
Bulletin Biologique de la France et de la Belgique.
Bulletin of the Harvard Museum of Comparative Zoology.
Comptes Rendus de la Société des Anatomistes.
Contributions to Embryology of the Carnegie Institution of Washington.
Denkschriften der Medizinischen Naturwissenschaftlichen Gesellschaft, Jena.
Jenaische Zeitschrift für Naturwissenschaften.
Journal of Anatomy.
Journal of Morphology.
Journal of the Linnean Society of London.
Morphologische Arbeiten
Morphologisches Jahrbuch.
Petrus Camper.
Philosophical Transactions of the Royal Society.
Proceedings of the Linnean Society of New South Wales.
Quarterly Journal of Microscopical Science.
Société des Naturalistes de Moscou.
Transactions and Proceedings of the Zoological Society of London.
Transactions of the Royal Society of Edinburgh.
Travaux de la Société Impériale des Naturalistes de Moscou.
University of Illinois Biological Monographs.
University of Washington Studies.
Voeltzkow's Reise in Ostafrika.
Zeitschrift für Morphologie und Anthropologie.
Zeitschrift für Anatomie und Entwicklungsgeschichte.
Zeitschrift für Angewandte Anatomie.
Zoologica.
Zoologische Jahrbücher.

To the Council of the Royal Society of London, and to the Editor of the *Quarterly Journal of Microscopical Science*, I am indebted for permission to reproduce figures from my own papers in their publications. To the latter, to the Council of the Zoological Society of London, and to Professor Goodrich, I am indebted for the blocks for Plates 5 and 7.

Lastly, I wish to record and acknowledge gratefully the help received from Merton College, by means of which it has been possible to provide a number of Plates adequate for the illustration of the text of this work.

CONTENTS

I. INTRODUCTORY SECTION

II. SYSTEMATIC SECTION

III. COMPARATIVE SECTION

IV. GENERAL SECTION

LIST OF PLATES

[At the end of the volume]

I. INTRODUCTORY SECTION

1. HISTORICAL

THE history of the study of the development of the vertebrate skull is bound up with the recognition of the two materials, cartilage and bone, of which the skull may be composed, and with the theories concerning the relations in which these tissues stand in regard to one another, both anatomically and developmentally. It will be convenient to deal first with this aspect of the problem before proceeding to consider the historical development of the theories of the skull, and of the improvements in technique which rendered available the data on which such theories are based. (For earlier historical reviews of these matters, see Gaupp, 1897, 1901, 1906*a*; Matthes, 1921*c*, 1922.)

i. Cartilage and Bone

The distinction between cartilage and bone as skeletal materials was, of course, known to Aristotle, who contrasted the Chondrichthyes with the Osteichthyes; but it was a long time before closer study of developmental stages revealed the fact that some bones were preceded by cartilaginous structures of similar shape which they replaced, while other bones arose directly from connective tissue membranes, without cartilaginous precursors. The former are customarily referred to as cartilage-bones, and the latter as membrane-bones.

The relationship between cartilage and bone may be considered under two aspects: (*a*) histological, and (*b*) morphological.

1. Histological relationship between cartilage and bone.

Having regard to the methods of investigation then available, it is not surprising that the early observers considered the transition from cartilage into bone as quite normal. Theodore Kerckring, in 1670, expresses his views: 'in toto corpore nobis observatum est, plurimas partes, quae passim cartilagineae dicuntur, primum membranosas esse, ac paulatim abire in cartilaginis consistentiam, antequam perveniant ad ossium soliditatem' (p. 213 of 1717 edition).

Some few years later, in 1736, Robert Nesbitt, of Basinghall Street, London, complained of Kerckring that he had followed his predecessors Riolanus, Eyssonius, and Coiter into error. Nesbitt then proceeded to relate (in his own words)

> 'what is visible of the manner of ossification, and from that shew the ancient and common notion of all bones being originally cartilaginous, to be a vulgar error;' (p. 4).

He continues (p. 10):

> 'The bony particles in *foetuses* begin to be deposited or to shoot either between membranes or within cartilages.
>
> 'Those which shoot between membranes are what form most of the hardest and

most solid of foetal bones, and appear much sooner than the others, which compose all *epiphyses*, and such bones only, whose places are supplied for some time by cartilages, which have nearly the same shape those parts are naturally of, when they become bony.'

After going on to describe the process of ossification in membrane and in cartilage (in which latter case he recognized the part played by the invasion of blood-vessels into the cartilage) Nesbitt laid stress on the independence of the cartilage-bone from the cartilage in the midst of which it is formed.

'How little *foetal* bones are dependent on the cartilages, in which they are generated, may be made apparent, if any of them, while the cartilages entirely or almost surround the bone, be kept a sufficient time in water; for then as you see, . . . on only slitting the cartilage, the bone will, as soon as the large vessels that enter its substance are divided, slip as easily, if not easier, from it, than an acorn does out of its cup' (p. 21).

Nesbitt's view on the histological relationship of cartilage and bone have stood the test of time and were placed on a firm footing by Sharpey (1848), who developed the theory of substitution as regards cartilage and cartilage-bones. He further showed that in cartilage-bones the outer layers arose beneath the perichondrium or periosteum (perichondral or subperiosteal ossifications) while the inner layers may arise in the body of the cartilage (endochondral ossifications), in neither case by conversion of existing cartilage, but by fresh development of bone, absorption of cartilage, and the substitution of the former for the latter. The absorption of the cartilage accompanies its invasion by blood-vessels, bringing in chondroclasts which further break down the cartilage, and osteoblasts which secrete the bone.

These conclusions, supported and amplified by Gegenbaur (1864), may be described as the orthodox view, which is now generally accepted. Cases have, however, been cited and claimed to prove the possibility of actual conversion of cartilage into bone (Gegenbaur, 1867; Schmidt-Monnard, 1883). The difficulty of establishing the truth of this possibility is great, especially in view of the fact, pointed out by Stephan (1900), that the perichondrium actually becomes the periosteum, and that histological preparations may therefore show a continuous transition from cartilage into bone. This is, however, no universally accepted evidence of the conversion of cartilage into bone by the assumption of bone-secreting properties on the part of cells which have already secreted cartilage. This latter contingency is, however, claimed as a possibility by Weidenreich (1930) in his monograph on bone and ossification, and by Fell (1933) as a result of her tissue-culture investigations, in which explants of apparently pure bone cells, devoid of the possibility of invasion by blood-vessels bringing in other cell-types, can develop into cartilage of so-called 'hypertrophic' type (characteristic of cartilage in which ossification is occurring) which then may undergo direct transformation into bone. The problem is, however, complicated by the fact that the hypertrophic cartilage thus obtained is very similar to the tissue known as 'secondary cartilage' (characterized, like hypertrophic cartilage, by large spaces surrounding the cells and very little intercellular matrix) which is commonly found associated with undoubted membrane-bones, and differs considerably from the 'primary cartilage' of the chondrocranium with its rich intercellular matrix closely surrounding the small contained cells.

The occurrence of secondary cartilage in the normal development of membrane-bones is fairly widespread (F. Strasser, 1905; Fuchs, 1909*a*; G. M. Levi, 1930; Schaffer, 1930), and there is no doubt that such cartilage has nothing to do with the true primary cartilage of the embryonic skeleton. It remains to be proved, however, even in the case of this secondary cartilage that it becomes directly transformed into bone, though the possibility of this also is claimed by Fell (personal communication). A second problem requiring solution is whether this secondary cartilage of the membrane-bones is comparable in any way with those degenerating regions of primary cartilage (forming so-called 'hypertrophic' cartilage) in which endochondral ossification is proceeding. Lastly, there is the question whether the conversion of secondary cartilage into bone in tissue-culture, if it does take place, may be taken as evidence of a similar conversion in the normal developing organism. These and kindred matters are reserved for later discussion (p. 482).

2. Morphological relationships of cartilage and bone: cartilage-bone and membrane-bone.

Although certain cartilaginous constituents of the skull were already known, among them 'Meckel's cartilage, described by Meckel in 1820, the cartilaginous skull, or chondrocranium (as Huxley ultimately called it), as a whole was specifically recognized for the first time by E. Arendt in 1822 in the pike, and the fact noted that some of the bones were embedded in the cartilage (cartilage-bones) while others lay superficially to it (membrane-bones).

Shortly afterwards (1826) K. E. von Baer put forward the view of 'inner' and 'outer skeleton', based on his studies of fishes. The 'inner skeleton' was composed of the chondrocranium together with its embedded bones and covering bones, and was contrasted with the 'outer skeleton', represented by the bony dermal scales such as are possessed by the sturgeons. In the light of subsequent work it will be noticed that the separation of the covering bones from the 'outer skeleton' and their inclusion with the 'inner skeleton' was not a particularly felicitous step on von Baer's part.

The next advance was made by A. Dugès in 1834 who, as a result of his study of the development of the skull in Anura, recognized the cartilage-bones ('primordial ossifications') and the covering bones (which develop 'not in cartilage'), in addition to the chondrocranium, which latter he compared with the permanently cartilaginous skull of the Selachian. He further showed that the chondrocranium persists to some extent in the skulls of bony fish.

The existence of a chondrocranium in early stages of all vertebrates was recognized by C. Reichert (1838), who followed the development of the visceral arch skeleton, and further stressed the distinction between cartilage-bones (or replacing bones), and membrane-bones (which latter he was the first to call 'Belegknochen', covering bones, or investing bones). He observed that the mammalian jaw articulation between squamosal and dentary involves two membrane-bones, and is not comparable with the jaw articulation of non-mammalian vertebrates, which involves two cartilage-bones, the quadrate and articular. Following from this, a brilliant insight led him to recognize the articular, quadrate, and columella auris in the mammalian malleus, incus, and stapes respectively.

In the following year (1839) H. Rathke published a detailed account of the development of the skull of the grass-snake. He recognized and named the trabeculae cranii in the chondrocranium, and in the skull as a whole he thought he could distinguish one portion comparable (in type of origin, at all events) with the vertebral column, and another portion not so comparable, and to which the covering bones belong.

To L. Jacobson (1842) comes the merit of spreading the conception of 'primordial cranium' for the chondrocranium. But while this term soon acquired the meaning of 'primitive skull', Jacobson used it in the sense of 'provisional skull', as contrasted with the bony skull or 'permanent skull'. He described it in cow and man, and recognized, as did Spöndli (1846) from his work on pig and man, that while some parts of it might disappear altogether, others might persist modified, and others again be replaced by cartilage-bone.

Many observers about this time (Johannes Müller, 1843; H. Stannius, 1846; R. Owen, 1846; C. Bergmann, 1846) were agreed that the concept of the primordial cranium (chondrocranium) was of universal application in the vertebrates, and that as contrasted with the 'secondary cranium' (bony skull) it provided a sound morphological criterion for the establishment or rejection of homologies between bones of different forms.

At about the same time (1844) L. Agassiz classified the bones of fish skulls into ossifications of the chondrocranium and protective plates, which latter include not only the superficial dermal ossifications, but also more deep-seated bones surrounding the chondrocranium, and therefore represent the entire category of membrane-bones.

These views were more widely and generally formulated by A. v. Kölliker (1849). The questions of morphology or homology of any given element were to be decided by its development. Kölliker called the cartilage-bones primary and the membrane-bones secondary, and regarded the distinction between them as fundamental.

The justification of drawing so hard a distinction between bones merely because they happen to develop inside or outside the perichondrium was considered by T. H. Huxley (1864), who concluded: 'It is highly probable that, throughout the vertebrate series, certain bones are always, in origin, cartilage-bones, while certain others are always, in origin, membrane-bones.' He was, however, troubled by the question 'if a membrane-bone is found in the position ordinarily occupied by a cartilage-bone, is it to be regarded merely as the analogue and not as the homologue of the latter?'

However, in 1858, H. Müller had shown that if by origin is meant histological differentiation, there is no difference between cartilage-bones and membrane-bones, the osteoblasts being identical cells in each case. The developmental difference between cartilage-bones and membrane-bones is thus solely a topographical one.

Gegenbaur's work constitutes one of the most important contributions made to these matters. On the one hand, he subjected the chondrocrania of Selachians to a detailed and systematic study, resulting in a classical monograph (1872) which is with advantage appealed to for information to this day.

But no less important were his conclusions (1864, 1867, 1870) relating to cartilage-bones (or 'autostoses') and membrane-bones (or 'allostoses': the 'parostoses' of W. K. Parker, 1870). In this connexion Gegenbaur observed that cartilage-bones make their first appearance (both in ontogeny and in phylogeny) in the form of perichondral lamellae which, though subperichondral in position, nevertheless lie over more or less intact cartilage. (These are the 'ectostoses' of W. K. Parker 1870). Only later may intracartilaginous endochondral ossification take place (Parker's 'endostoses'). These two types of ossification are, however, not distinct in respect of any fundamental difference, but may best be regarded as different stages of ossification. It is worth noting that if the appellations 'primary' and 'secondary' were to be applied here, primary would have to denote the perichondral and secondary the endochondral type. Thus Kölliker's designation of 'primary ossifications' as applied to cartilage-bones on account of their connexion with the primordial or chondrocranium is seen to be incapable of withstanding more detailed analysis. Further, as will shortly appear, there is another objection to the use of the word 'primary'.

With regard to the membrane-bones, Gegenbaur, in company with Leydig (1854) and Huxley (1864), went directly counter to von Baer's views and regarded them as of originally dermal origin, or ossifications in the skin, directly comparable with the bony scales of fishes. As to the relation between cartilage-bones and membrane-bones, Gegenbaur (1878) concluded that the primitive position of bone was superficial to the cartilage, and the invasion of the latter, secondary. Thus, in phylogeny, membrane-bones may have become cartilage-bones; and the distinction between membrane-bones and the perichondral ossifications of cartilage-bones, viz. whether the bone is or is not separated from the cartilage by a layer of connective tissue (the perichondrium), loses much of its absolute value and significance. Indeed, Walther (1882) regarded some perichondral ossifications as membrane-bones and others as cartilage-bones, while Göldi (1884) tried to establish a third category of perichondral bones, intermediate between membrane-bones and endochondral cartilage-bones. Something similar had been attempted by Vrolik (1871). Reverting again to the classification of bones as applied respectively to cartilage-bones and membrane-bones, it will be seen that the 'secondary' bones are 'primary' in phylogeny on Gegenbaur's view.

In place of the embryological criterion Gegenbaur therefore substituted the morphological criterion, based on phylogeny, to establish the homology of bones in differing forms; and he recognized that while some bones could be demonstrated by comparative anatomy of a series of adult living or fossil forms to have originated in the skin or in membranes as dermal ossifications or membrane-bones, other bones could not, and these are the cartilage-bones. The line between them is not hard and fast, but to be decided in each case on its own merits. Sometimes (e.g. supra-ethmoid of certain Teleosts) it seems that the transition takes place during ontogeny, and a membrane-bone invades the cartilage. Conversely, it may happen that a cartilage-bone (e.g. vertebra in Teleosts; metapterygoid of Syngnathus) finds itself developing as an intramembranous ossification, presumably because the cartilage in which it 'ought' to be preformed simply does not arise.

On this matter, therefore, Gegenbaur's conclusions form the accepted general opinion to-day, and, as von Ebner (1911), Weidenreich, (1923), and Edinger (1929) have recognized, histogenesis cannot be regarded as an infallible guide to phylogeny.

There is a matter connected with the origin of the membrane-bone that demands further consideration, and this is the relation between membrane-bones and the denticles or teeth that become attached to many of them.

Williamson (1850 and 1851) described the teeth in fishes as arising independently and subsequently becoming fused on to underlying plates of membrane-bone. The conjoint mass then formed a scale which, with certain modifications, might present the appearance of one or other of the cosmoid, palaeoniscoid, lepidosteoid, or teleost types (see Goodrich, 1909). The scales and the dermal bones of the fishes show the same structure.

O. Hertwig (1876, 1879) attempted to develop this theory by proposing the view that the plate of membrane-bone is itself the result of lateral extension and fusion of the bases of the denticles. In other words, membrane-bones would be phylogenetically and ontogenetically nothing but fused teeth, and of fundamentally different origin from cartilage-bones, which have always been deep and never borne teeth.

Simple and seductive as this view may appear on the surface, it is open to objections so grave that it is surprising to find it still entertained. These objections have been aptly summarized by Goodrich (1907, 1909): (i) in every case that has been properly studied the denticles arise independently from the bony plate; if only later stages are studied when the denticles and bony plate have fused, an erroneous impression is obtained; (ii) denticles are composed not of bone but of dentine, so that even if their bases were to fuse, they would not form plates of bone; (iii) membrane-bones and scales sink beneath the epidermis and grow in thickness by addition of material over the outer surface; this is unintelligible if the bony plate is fused denticles, since the latter can only grow from beneath.

The conclusion, as evidenced also in the most recent work of Sewertzoff (1925*a*) and Moy-Thomas (1934), is that membrane-bones arise in topographical relation to and underlying but not out of denticles.

Another attempt has been made to explain the phylogenetic origin of dermal bones by assuming that they arose in connexion with sense-organs of the lateral line canal, as they do often arise in ontogeny (Vrolik, 1871; Pehrson, 1922). But comparative anatomy shows that the tube-like 'canal bones' are a secondary condition, and their early origin in ontogeny is clearly an embryonic adaptation. Besides, little would be gained on this view even if it were well founded, for there are countless dermal bones in the skull, to say nothing of scales on the body, which never have any relations with lateral line canals at all.

The phylogenetic picture of the history of ossification in vertebrates is therefore one of superficial origin of membrane-bones, bearing teeth, some of them entering into relations with lateral line canals; followed by an invasion of ossification beneath the perichondrium of the underlying cartilage and in connective tissue. Whether this invasion was the result of immigration of osteoblasts into the cartilage, or of an extension or 'infection' of the zone or site of origin and differentiation of osteoblasts, it is difficult to decide. Recent experimental work on the location of the power of phosphatase synthesis in cartilage by Fell and Robison (1929, 1930) favours the latter alternative, for the phosphatase-producing power of cartilage which will subsequently ossify does not owe this property (as contrasted with its absence in non-ossifying cartilage) to any migration of cell elements, but to regional embryonic differentiation in situ.

Lastly, as van Wijhe (1882*b*) pointed out, the classification of bones into cartilage-bones and actually or originally tooth-bearing bones in the skin does not exhaust the possibilities, for some bones arise in tendons and membranes of connective tissue, under the skin, but outside the cartilage. These are also membrane-bones, and the membrane-bones in the skin should be distinguished as dermal bones.

ii. Theories of the skull

In 1790 Goethe's servant picked up a dried and dislocated sheep's skull in the Jewish cemetery in Venice, and held it out for his master to look at. At the sight of it the idea promptly flashed through Goethe's mind that the skull was made up of a number of bony rings which he decided must represent vertebrae. Thus originated the famous vertebral theory of the skull, which for Goethe was but an application to vertebrates of the principle of repetition of form within the body which he had applied with conspicuous success to plants, by showing that in the flower the sepals, petals, and stamens are but repetitions and modifications of a common standard form, the leaf. Goethe's vertebral theory of the skull was not published until 1820, and the same idea had meanwhile occurred to other men, among them Oken (1807), Spix (1815), Bojanus (1819), Etienne Geoffroy St. Hilaire (1818), Meckel (1820), and Owen (1846). Among these protagonists of *anatomie philosophique* or *Naturphilosophie*, the vertebral theory of the skull is almost the classical example of the fantastic lengths to which speculation, unchecked by scientific evidence, can go in its frantic effort to find, at all costs, the reflection in different parts of the body of some idealized form. The skull was vertebrae, the jaws were limbs, the teeth were claws, the palate was a diaphragm, the sternum a second ventral vertebral column. The climax of absurdity was reached by Carus (1828), for whom the whole body was nothing but variations on the one theme, the vertebra!

As might be expected from the nature of the case, unanimity was hard to achieve concerning the number of vertebrae that the skull was supposed to contain. Geoffroy St. Hilaire claimed 7, Goethe 6, Bojanus, Oken, and Owen 4, and Spix 3; 4 was the favourite number, the 'centra' of these vertebrae being represented by the vomer, presphenoid, basisphenoid, and basioccipital: the 'rings' being completed by nasals and lateral ethmoids, frontals and orbitosphenoids, parietals and alisphenoids, supraoccipital and exoccipitals respectively. Von Baer, Reichert, and Rathke gave the theory qualified support, and it was still adhered to by Kölliker and Johannes Müller. The opposition to the vertebral theory of the skull was slow in crystallizing. Cuvier (1836) would admit nothing beyond the similarity of the posterior part of the skull (the occipital region) with a vertebra, which similarity he attributed to similar function. The first real attack was launched by C. Vogt (1842), who in his works on the development of Corregonus and Alytes, pointed out that only that part of the skull which was developed round the notochord could lay any claim to comparison with a vertebra. He therefore concluded that except for an occipital vertebra, 'there are no cranial vertebrae at all'.

Vogt's friend and companion on the celebrated glacier sojourns of the 'Hôtel des Neuchâtelois', Louis Agassiz, joined Vogt in his rejection of the vertebral theory, but it was T. H. Huxley who in his Croonian Lecture of 1858 gave the theory its death-blow. Huxley demonstrated for the bony skull of all groups of vertebrates a common plan of structure which had nothing in common with that of a vertebra at all; he further reminded his listeners that the bony skull was preceded by a solid and unsegmented chondrocranium in which no one would attempt to look for bony vertebrae, and which was the only skull in Cyclostomes, Selachii, and Holocephali; and by carefully comparing the development of the skull with that of the vertebral column, he showed that each had become specialized along its own line, and that there was no more justification in claiming the skull to be modified vertebrae than there would be in stating that the vertebral column was a modified skull. The repetitive nature of the vertebrae is a primary feature of those structures, whereas the multiplicity of bones in the bony skull is secondarily superimposed on the undivided chondrocranium.

It was now clearly ridiculous to regard the bones of the skull as arranged according to any vertebral pattern. But the old vertebral theory contained in it the germ of another concept, that of repetition of parts, or segmentation, which was cloaked beneath the stultifying specification of bony vertebrae as the repeated parts. This concept, which was destined to culminate in the segmental theory of the skull, developed along two lines. On the one hand, it was becoming realized, owing to the works of Balfour (1876) and van Wijhe (1882*a*), that the mesoderm of the head right up to the anterior end is segmented into somites, perfectly continuous with those of the trunk. Previously it had been thought by Remak (1850) that the series of somites did not extend farther forward than the hind end of the skull. That was why he rejected the vertebral theory: for the wrong reason, as it turned out. These somites produce sclerotomes, so that the chondrocranium to which they give rise is at least made up of material which was segmental in origin. This aspect of the problem will be referred to again below. The other aspect of the problem concerns the possibility that although the major part of the skull has nothing to do with vertebrae, may not the hindermost part of the chondrocranium which in many forms is pierced by a number of hypoglossal ventral nerve-roots contain the cartilaginous rudiments of a number of vertebrae which were once independent in previous phylogeny, but which have become attached to the hind end of the true skull?

To this question Gegenbaur (1872, p. 294), who had made a close study of the skulls of all Selachians, replied in the affirmative. He distinguished in the chondrocranium a posterior 'vertebral' portion, traversed by the notochord (the sheath of which may be invaded) and showing evidence of segmentation in the various dorsal and ventral nerve-roots (and in the visceral arches which he erroneously considered to be ribs), from an anterior 'prevertebral' portion corresponding to the region of the optic and olfactory nerves. The vertebral portion for Gegenbaur was coextensive with the notochord, parachordals, and basal plate, and represented phylogenetically free vertebrae (at least nine in number) which had become incorporated in the skull from behind.

The next contribution was that of Stöhr (1879, 1881, 1882), who had studied the chondrocrania of Teleostei, Urodela, and Anura, and seen that the occipital arch in Urodela is a neural arch which becomes incorporated in the skull, and had come to the conclusion that the posterior end of the skull, the occipital arch, does not occupy the same numerical segmental position in different vertebrate groups, since additional (originally free) vertebrae have been and are still being incorporated. The correctness of this view was soon demonstrated by Rosenberg (1886), who found that the skull of Carcharias contained one segmental element (vertebra) more than that of Mustelus, and that no constant position of the occipital arch could be claimed in Selachians.

So far the segmental theory of the skull was on solid ground, but it soon began to go adrift as a result of unfortunate assumptions, to which, however, careful attention must be given because of the credence accorded to the errors into which the theory was led.

The matter was next taken up by Froriep (1882). Like Gegenbaur, he distinguished in the skull an anterior non-segmental from a posterior segmented portion, but instead of placing the limit between them at the front of the basal plate, Froriep placed it farther back in the region of the vagus nerve. Behind the vagus there was evidence of segmentation in the form of the hypoglossal roots, for some of which Froriep discovered rudimentary dorsal root ganglia, thus assimilating the hypoglossal roots to the type of the spinal nerves behind them. To this portion of the skull Froriep applied the term 'spinal', in contrast to the more anterior 'prespinal' or 'cerebral' region, in front of the vagus, and in which he could find no evidence of segmentation.

Sagemehl (1884, 1891) put forward the view that the skulls of Selachii and Amphibia were coextensive and primitive, lacking the supra- and basioccipital bones, and to be contrasted with those of Teleostomi and Amniota which contained these bones as a consequence of three additional vertebral elements having been plastered on behind. For the former group Sagemehl introduced the expression 'protometameric', while 'auximetameric' was used to describe the latter. But while the auximetameric cranium owed its posterior elongation to the assimilation of originally free vertebrae (brought about by the posterior extension of the parasphenoid and consequent loss of mobility), the 'vertebral' portion of the protometameric cranium was composed of fused segmental elements derived from the somites, but which had never been true vertebrae. For Sagemehl, therefore, the skull was composed of two fundamentally distinct regions: (1) anteriorly, the segmented but non-vertebral region, forming the protometameric cranium, and (2) posteriorly, the segmented vertebral region forming together with (1) the auximetameric cranium.

Gegenbaur, returning to the problem in 1887 and 1888, recognized the distinction of two regions meeting at the level of the vagus. The anterior region he called palingenetic and he now regarded it as of segmental origin (unlike Froriep's view of his 'cerebral' region; like Sagemehl's view of the posterior part of his protometameric cranium), though not of vertebral origin, all the way to the front of the basal plate. Gegenbaur was confirmed in his opinion as regards this by Balfour's and van Wijhe's discoveries of somites in the head right up to the anterior end. Gegenbaur called the post-vagal portion of the skull caenogenetic, of obvious segmental origin and recent phylogenetic incorporation in the skull.

The problem now presented was to explain the variation in the number of hypoglossal roots which different skulls might show. Instead of simply accepting the number of segments in the skull as variable and the position of the occipital arch as requiring to be determined in each case, recourse was had to an assumption which, as will be seen, proved fatal. This was to postulate the loss and disappearance of whole segments at the limit between the prespinal and spinal regions of the skull, brought about by destruction and degeneration according to Froriep, or to amalgamation and fusion of segments according to Dohrn (1901).

This was the state of knowledge, or rather of thought, concerning the theory of the skull when M. Fürbringer published his monumental work in 1897, embodying some of the conclusions of Stöhr, Sagemehl, and Froriep.

Fürbringer agreed with Stöhr that the posterior limit of the skull differed in different forms, and he explained this (*a*) by accepting Sagemehl's auximetameric cranium as containing three vertebrae added to the protometameric cranium, and (*b*) by following Froriep's suggestion that in the region of the vagus a varying number of segments had vanished.

Fürbringer called Froriep's prespinal or cerebral region, in front of the vagus, the 'palaeocranium', represented to-day by the Cyclostomes. Behind the palaeocranium in other vertebrates there is added the 'neocranium' (Froriep's spinal region), composed of originally free segmental elements or vertebrae. But, following Sagemehl, Fürbringer distinguished an anterior, older addition, the 'protometameric neocranium' (exemplified in Selachii and Amphibia) from a posterior, more recent addition, the 'auximetameric neocranium' (higher fish, and Amniota). For Fürbringer's view the distinction between the protometameric addition (perhaps of elements which though segmental, never were actual vertebrae) and the auximetameric addition (of true, originally free vertebrae) is fundamental, and he thought he could demonstrate the limit between these two regions with the help of the nerves. The ventral roots of the protometameric region he called 'occipitals', and regarded them as lacking corresponding dorsal roots; the ventral roots of the auximetameric region, however, which he called 'occipitospinals', in principle possessed corresponding dorsal roots, which they had not had time to loose since the comparatively recent incorporation of their corresponding vertebrae in the skull. The occipital roots were enumerated from *behind forwards* with the last letters of the alphabet: *z*, *y*, *x*, *w*, *v*, for as many hypoglossal roots as the Selachii could show, the spinal roots in such protometameric crania continuing backwards 1, 2, 3, 4, &c. But in auximetameric crania, spinal nerves 1, 2, and 3 have become occipitospinals *a*, *b* and *c*; the first free spinal nerve being now 4.

Lastly, by accepting Sagemehl's conclusion that the hinder limits of the skulls in Selachii and Amphibia were coextensive (a conclusion supported by Gegenbaur (1872) and Gaupp (1893) based on alleged resemblances of the cranio-vertebral joint and of the occipital arch in the chondrocrania of these two groups), Fürbringer postulated a fixed position for occipital nerve *z*, and simply assumed where necessary the disappearance of the required number of segments anterior to it in order to bring nerve *z* into the correct position. Thus, since by hypothesis the hind limit of the skull in Selachii and in Amphibia is in the same place, and since the Amphibia have no cranial hypoglossal nerve-roots while the Selachii may have as many as five, therefore *at least* five segments have vanished in the region of the vagus in Amphibia! Indeed, this caused no misgivings, for taking Amphioxus and its numerous gill-slits as the starting-point, Fürbringer was prepared for a loss of some thirty segments! Further details concerning Fürbringer's theory must be sought in his own work, and in Gaupp's (1897) historical sketch of the problem of the segmentation of the head.

This elaborate theory can be attacked and exposed along several lines. It may be asked what evidence is there (1) that a large number of segments vanish in the region of the vagus; (2) that the occipital arches of Selachii and Amphibia are in the same position, and (3) that any difference exists between the occipital roots of the protometameric and the occipitospinal roots of the auximetameric region? The answer in each case is *none at all*; the matter must be treated in detail, however, owing to the general reluctance shown in abandoning Fürbringer's untenable theory, and each of the above questions will be dealt with in order.

(1) The attack on this side of the problem is due to Goodrich (1911, 1918), who showed as a result of a careful, stage-by-stage study of the development of the segments and skull in dogfish and axolotl that there is no evidence for this holocaust of segments at all. In Cyclostomes every segment of the body is present and represented in the adult by a somite or myomere, and no segment is lost. In Gnathostomes at most two and sometimes only one segment is lost in the region underlying the auditory capsule.

(2) Fürbringer's efforts to make the hind ends of the skull in Selachii and Amphibia coincide can be shown to be wrong in fact and unnecessary in theory. Here again it is Goodrich who has pointed out the error, for the number of segments in the skull can be ascertained simply by counting, and while the number in the dogfish is seven, in the axolotl it is six. In other Selachii the number may even be eight or nine.

As to the desire to make the occipital arch in all Selachii and Amphibia occupy numerically the same segment, it is necessary to point out, as Goodrich (1913) has done, that metameric segmentation and homology have nothing to do with one another. Forelimbs and pectoral fins are none the less homologous for arising from numerically different segments of the body: the structure can be transposed up or down the scale of segments without interfering with its phylogenetic successional history; the numerical values of the segments from which it arises reflect the exigencies of embryonic development. The same is true of the occipital arch, which may certainly be homologous in dogfish and axolotl in spite of arising from material of a different segment, for the only practical definition of homology is whether or not the structures in question can be traced to one and the same representative in a common ancestor. It is impossible to deny this in the case of Scyllium with its seven segments and the closely related Squalus with its nine, any more than it is possible to deny the homology of pectoral fins and forelimbs of different animals, whatever be the segments from which they arise.

Lastly, the supposed similarity between Selachii and Amphibia in lacking a basioccipital bone is destroyed by the recognition of this bone in fossil Amphibia, and even in living forms, e.g. Triton (Stadtmüller, (1929*b*), Hypogeophis (Marcus, Stimmelmayr and Porch, 1935), Bufo marinus (Debierre, 1895).

(3) The distinction between occipital (protometameric) and occipitospinal (auximetameric) nerves is supposed by Fürbringer to rest on the fact that the former are the ventral nerve-roots which have lost their corresponding dorsal roots, whereas the latter, which were originally postvagal spinal nerves, should each possess a corresponding dorsal root. The arbitrary and unreal nature of this distinction may be made apparent both from Fürbringer's own work and from that of others. For instance, Fürbringer is forced to admit that in Hexanchus, his nerve *z*, the last of the occipitals, may have a dorsal root corresponding to it. The distinction between occipital and occipitospinal roots breaks down at once, and is seen to be sheer nonsense from the fact that, as Froriep, van Wijhe, and Goodrich showed, rudimentary dorsal root ganglia may be seen corresponding to all of the last three segments of the head. The vagus has been formed, not by total polymerization of the four segments corresponding to its branchial branches, but by partial polymerization involving certain components such as the special cutaneous and leaving behind certain other elements (e.g. general cutaneous) which form the dorsal ganglia in question (see Goodrich, 1930).

No reliance may be placed on the intracranial dorsal root ganglia to distinguish between occipital and occipitospinal nerves, and consequently between Fürbringer's protometameric and auximetameric regions of the skull. This emerges plainly from another of Fürbringer's own descriptions. Hexanchus has five hypoglossal roots, which are regarded by him as occipitals and numbered from *v* to *z*. Chimaera likewise has five roots, but these Fürbringer considers to be nerves *y*, *z*, *a*, *b*, *c*, regardless of the fact that he could find no dorsal roots corresponding to the last three. It will further be remembered that the last nerve in Hexanchus *has* a corresponding dorsal root. And so Chimaera is considered by Fürbringer to have an auximetameric neocranium with three segments more plastered on to the hind end as compared with Hexanchus.

The conclusion is obvious: there is no distinction to be drawn between occipital and occipitospinal roots or between protometameric and auximetameric neocrania; there is no reason to suppose that the hind limits of the skull coincide in Selachii and Amphibia, or for that matter, within the Selachii; and there is no necessity or evidence for assuming a destruction of segments in order to make the occipital arches fall into line.

The theory of the skull must revert to the views of earlier workers in order to avoid the impasse of Fürbringer's speculations, and foremost among these workers are Balfour (1876) and van Wijhe (1882*a*). They showed that the head region was segmented into somites right up to the anterior end, and that the chondrocranium, or most of it, arises from the segmented sclerotomes. This, the major part of the chondrocranium, is coextensive with the basal plate, and may be called the 'chordal' region, as contrasted with the 'prechordal' region, in front of the notochord, and which as Sewertzoff (1916), Matveiev (1925), and de Beer (1931*b*) believe, contains the trabeculae cranii. The chordal and prechordal regions correspond more or less with Gegenbaur's vertebral and prevertebral regions, but Gegenbaur's terms prejudge the question as to whether the *whole* of the chordal region ever was composed of vertebrae, or only the hinder part. Here, Froriep's (1882) observations regarding the absence of evidence for segmentation *in the cartilage* (for Balfour and van Wijhe proved that there was primary segmentation of the mesoderm) in front of the vagus, may be accepted. Froriep's 'cerebral' or 'prespinal' region (Fürbringer's palaeocranium), is now known to occupy the first four segments of the body, and there is reason to believe that the cartilaginous skull in this region never was (phylogenetically) composed of vertebrae. The reasons for this view, put forward afresh by de Beer and Barrington (1934), are that (1) this region together with the prechordal structures represents the entire skull in Cyclostomes, where the glossopharyngeal and vagus emerge behind the skull, and there are no properly formed vertebrae at all. In other words, the palaeocranium is older than the vertebral column, as van Wijhe (1889) pointed out. (2) The evidence in favour of the formerly vertebral nature of the hinder part of the gnathostome chondrocranium begins in the 5th segment, and never farther forward. This evidence includes the close apposition of the parachordals on each side of the notochord (since farther forward they diverge away from it); the invasion of the notochordal sheath; the preoccipital arches separating the hypoglossal nerve-roots; and true cranial ribs.

Froriep's 'spinal' region (Fürbringer's 'neocranium') of the skull therefore extends from the 5th segment backwards, but the degree to which it extends varies in different forms, and there is no basis whatever for any distinction between 'protometameric' and 'auximetameric' constituents. The number of segments concerned in the neocranial addition to the palaeocranium is:

Cyclostomata	0	Squalus	5
Anura	1	Mammalia	5
Urodela	2	Aves	5½
Scyllium	3 or 4	Spinax	6
Salmo	4	Acipenser	14

The details of the determination of these numbers may be found below, in the section specially concerned with the segmentation of the head. For the moment it must suffice to notice how at long last, and in spite of being side-tracked on more than one occasion, the only spark of truth which the old vertebral theory of the skull contained, viz. the repetitive or segmental nature of a part of the cartilaginous skull, has emerged. And, curiously enough, it is probable that this segmental aspect was unknown to the propounders of the original vertebral theory, for they were concerned with a formal explanation of the skull on transcendental lines using the repetition of a known form—the vertebra—and knew nothing of the mesodermal somites of the head.

iii. Technique of study of the development of the skull

For the earlier investigators the methods available for the study of developmental stages of the skull were those of maceration and dissection. Compared with modern results the wonder is, not that these early workers made mistakes, but that they made so few. Particularly is this the case with the man who more than any other devoted his life and work to the study of the development of the skull in practically all groups of vertebrates—William Kitchen Parker. His series of monographs published between 1862 and 1891 represent a landmark in the comparative study of the development of homologous structures in a large number of different forms, and some of his figures are used to this day. But the study of hand-cut thick sections, studied with a lens, could not give really detailed and accurate information concerning the features which are now regarded as of the greatest importance in the study of the developing skull, viz. the relations, both morphological and topographical, of the cartilages and bones to neighbouring membranes, nerves, arteries, and veins.

Parker's work, written under the influence of Huxley's destruction of the vertebral theory of the skull, suffers further in that positive guiding principles are hard to find in the enormous mass of described facts that it presents; and, from a practical point of view, it suffers most from the absence of tables of explanation of lettering for his numerous and beautiful figures.

The progress of histological technique, involving the differential staining of tissues, and the cutting of thin serial sections capable of study under the microscope, was soon to remedy these drawbacks in Parker's work. But for the study of a three-dimensional structure altering in shape through time, like the skull, series of sections are insufficient to convey the adequate stereoscopic impression of the object studied to the observer, and to meet this difficulty G. Born (1876, 1883*a*, 1888) introduced the method of wax model reconstructions, made by drawing accurate enlargements of the several single sections on plates made up to the correct thickness, cutting them out, and building them together again. A device ensuring accurate register has been invented by de Burlet (1913*b*). A modification of the method, using plaster of Paris has been elaborated by Lewis (1915); while de Beer's models have been constructed of blotting paper soaked in wax. Peter (1922) has devoted a manual to the methods of reconstruction.

Born applied his method for the first time to the study of the nasal capsule in Amphibia, published in 1876, and it was then applied to the investigation of the development of the whole skull (Urodela) by Stöhr in 1879.

The method once established, and with the help of the invention of the microtome cutting a ribbon of serial sections (Caldwell and Threlfall in 1882), a series of researches was initiated, devoted to all groups of vertebrates, and chief among this army of workers was Ernst Gaupp. It was he who exploited the wax model reconstruction method to the full, and by providing a body of accurate and detailed information concerning developing cartilages and bones, including their relations to neighbouring structures, paved the way for the application of sound morphological criteria to the developing skull.

Nearly all groups of vertebrates have now been studied in the light of Born's method and of Gaupp's classical descriptions of the developing skulls of Rana (1893), Lacerta (1900), and Echidna (1908).

The method of wax reconstruction, however, is laborious and time-absorbing, and its application to any particular form means in practice that at most two or three models are reconstructed of carefully selected stages: the intermediate stages are either mentally or graphically reconstructed from serial sections or omitted. It should, perhaps, be mentioned that the method of graphic reconstruction (either as described by Kastschenko (1886, 1887) using squared paper, or by Graham Kerr (1902) using glass plates immersed in cedarwood oil, or by Pusey (in press) using greaseproof paper immersed in xylol), is very useful for circumscribed and definite purposes, but it is not to be preferred to the wax model method for general application to the study of skull development. It is particularly suitable for the study of the intermediate stages, between two stages modelled in wax. The glass plate method is almost as laborious as wax reconstruction, but it has the advantage of transparency which is helpful in tracing the course of nerves and blood-vessels through the cartilage.

These requirements are, however, admirably met by an elaboration of the process of selective staining of whole mounts, devised for cartilage by van Wijhe (1902, 1922). After suitable fixation, staining in Victoria blue and clearing in xylol, transparent and permanent preparations are obtained in which nothing but the cartilage is stained, and visible under the binocular microscope from all sides. By this means so many stages can easily be prepared that a continuous series may be studied.

An analogous method for the selective staining of bone by means of alizarin has been perfected by P. Gray (1929); Lundvall (1905) has devised a method of staining both cartilage and bone in the same preparation.

It is obvious that such preparations, selectively stained, do not give the required information concerning the relations of the skeletal elements to their neighbouring structures. Therefore, the modern practice in the study of the development of the skull in any given form is to combine the wax model method with that of total selectively stained preparations, assisted where necessary by mental or graphic reconstruction of serial sections.

2. THE SEGMENTAL COMPOSITION OF THE SKULL

Since Balfour's discovery of the fact that in early stages of development the head is segmented in a manner precisely similar to the trunk, the question soon arose as to how many segments of the body are involved in the formation of the skull. In order to give a satisfactory answer to this question it will be necessary to give a brief account of the development of head segments, so far as it is known in different vertebrates, and then to consider the method of determining the segmental position of the posterior limit of the skull in the various groups.

i. The segmentation of the head

Since the Selachii have supplied the material on which the classical researches on this subject were performed and most of the more recent work has been devoted to them, it will be convenient to consider the Selachii first, although they do not represent the simplest conditions found in vertebrates. Subsequent sections will then be devoted to Cyclostomes, Teleostomes, Dipnoi, Urodela, Anura, and Amniota.

1. Selachii.

Balfour (1876) showed in early embryos of Scyllium and other Selachians that the mesoderm in the head is segmented into somites similar to and continuous with those of the trunk. Indeed, it is not possible when studying such stages to make out the position of the future posterior limit of the head. He further observed that the visceral clefts pierced the mesoderm beneath the somites (behind the first) intersegmentally so that the visceral arches corresponded to the segments; and related to each of these visceral arches he described the series of dorsal cranial nerve-roots—trigeminal, acustico-facial, glossopharyngeal, and vagus—as comparable with the dorsal nerve-roots of the trunk. Each of the cranial dorsal nerve-roots runs over the hinder part of the mesodermal somite corresponding to the visceral arch to which that nerve is distributed. He suggested that the first or premandibular somite gave rise to some of the extrinsic eye-muscles, and recognized that the mesoderm of the visceral arches produced visceral muscles.

Altogether, Balfour's results are so epoch-making and possessed of true morphological insight that it will be appropriate to quote his conclusions in full:

'The morphological importance of the sections of the body-cavity [*i.e.*, somites] in the head cannot be overestimated, and the fact that the walls become developed into the muscular system of the head renders it almost certain *that we must regard them as equivalent to the muscle-plates of the body, which originally contain, equally with those of the head, sections of the body-cavity*. If this determination is correct, there can be no doubt that they ought to serve as valuable guides to the number of segments which have coalesced to form the head' (1878 edition, p. 208).

The next contribution was that of Milnes Marshall (1881), who traced the origin of the inferior oblique and superior, inferior and internal recti muscles from the first or premandibular somite. The oculomotor nerve which supplies these muscles is thus proved to be the ventral nerve-root of the 1st segment. Similarly, he traced the external rectus muscle from the 3rd somite and recognized its nerve, the abducens, as the ventral root of the 3rd segment. He also showed that although the segmentation of the somites and of the visceral arches correspond, their segmentations arise independently since the visceral clefts pierce the mesoderm ventrally to the line of segmented somites, in the region of the unsegmented and continuous lateral plate.

In the following year (1882*a*) there appeared the work of van Wijhe which has contributed as much, if not more than any other, to clarify the problem presented by the segmentation of the vertebrate head. He established the distinctions between the visceral muscles of the visceral arches, innervated by the dorsal cranial nerve-roots, and the myotomic muscles of the segmented somites innervated by ventral nerve-roots (including the eye-muscles). He traced the formation of the superior oblique muscle from the 2nd somite, its nerve, therefore, being the trochlear, and thus the segmentation of the first three segments and somites (which because of their position in front of the auditory vesicle may be called the prootic somites) is unravelled; their dorsal roots being the profundus (V1), trigeminal (V2 and V3), and facial respectively, and corresponding to the oculomotor, troclear, and abducens. For the hinder or metotic somites, van Wijhe recognized the various hypoglossal roots as the ventral roots, to which the vagus represents the corresponding and partially polymerized dorsal roots, leaving rudiments of the unpolymerized segmentally situated dorsal root ganglia. The 4th segment lacks a ventral root. As is well known, the corresponding dorsal and ventral nerve-roots of a segment in the head do not join to form a mixed nerve. In this the cranial nerves resemble the spinal nerves at early stages, and the nerves of Amphioxus throughout life.

The scheme of head segmentation worked out by van Wijhe is, however, probably erroneous in one respect (see below p. 18), and the first scheme which may be regarded as correct was produced by Goodrich (1918).

The foregoing points may now be illustrated with reference to their development in Squalus following the account given by de Beer (1922), and figured on Plates 1, 2, and 3.

In a 4·5 mm. embryo all the somites are visible: the 1st somites communicate with one another across the middle line by a transverse commissure immediately in front of the tip of the notochord. The first two visceral clefts (spiracle and 1st gill-slit) pierce the unsegmented lateral plate.

At the 5 mm. stage it may be seen that muscle-fibres have appeared in the 5th and following somites; it is also to be noticed that the trigeminal nerve is related to the mandibular arch and 2nd somite while the facial nerve is similarly related to the hyoid arch and 3rd somite: in each case the nerve overlies the hind part of its corresponding somite.

At the 6 mm. stage the 2nd and 3rd gill-slits have been pierced, and it may be seen that the glossopharyngeal nerve is related to the 1st branchial arch and 4th somite (1st metotic somite), and the 1st and 2nd branches of the vagus to the 2nd and 3rd branchial arches and the 5th and 6th somites (2nd and 3rd metotic somites) respectively. Again, the nerves overlie the hind parts of their respective corresponding somites. At a later stage the 3rd and 4th branches of the vagus will be found to be related to the 4th and 5th gill-slits and the 7th and 8th somites respectively.

The orderly sequence of somites is seen to be complete and undisturbed. The 1st somite produces a pouch, the so-called anterior head-cavity (found only in Squalus and Galeus), which is of no morphological significance and ultimately disappears. It is important to notice that the 5th somite is the most anterior of those in which muscle-fibres appear, and that the 6th develops a dorsal extension which projects upwards immediately behind the vagus, which nerve lies on the outer side of the 5th and of the anterior part of the 6th somite. The 6th and following somites are each innervated by a ventral nerve-root: those corresponding to the 4th and 5th somites seem to have disappeared. Rudimentary dorsal root ganglia are present corresponding to the 7th, 8th, and 9th somites: the 10th somite is innervated by a ventral root forming part of the most anterior mixed nerve, since this ventral root joins the dorsal root of this, the most anterior, spinal ganglion.

These relations may conveniently be set out in tabular form.

Segments	*Somites*	*Derivatives*	*Ventral nerve-root*	*Dorsal nerve-root*	*Visceral arch*
1	1 or premandibular	Inferior oblique, superior, inferior, and internal recti	Oculomotor	Profundus, V1	Premandibular
2	2 or mandibular	Superior oblique	Trochlear	Trigeminal, V2 & 3	Mandibular
3	3 or hyoid	External rectus	Abducens	Facial	Hyoid
4	4 or 1st metotic	Disappears	..	Glosso-pharyngeal	1st branchial
5	5 or 2nd metotic	Disappears	..	Vagus 1st branchial branch	2nd branchial
6	6 or 3rd metotic	1st permanent myomere	Occipital	Vagus 2nd branch	3rd branchial
7	7 or 4th metotic	2nd myomere	Occipital (hypoglossal)	Vagus 3rd branch and rudimentary ganglion	4th branchial
8	8 or 5th metotic	3rd myomere	Occipital (hypoglossal)	Vagus 4th branch and rudimentary ganglion	5th branchial
9	9 or 6th metotic	4th myomere	Occipital (hypoglossal)	Rudimentary ganglion	
10	10 or 7th metotic	5th myomere, first trunk myomere	First mixed spinal		

In view of subsequent events it is important to notice that there are two somites and two only, the 3rd and 4th, between the 2nd which is related to the trigeminal and the 5th which is related to the 1st branch of the vagus. The importance of this lies in the fact that in Selachii the 4th and subsequently the 5th somites disappear (crushed beneath the large auditory capsule), leaving the 6th somite, the 3rd metotic somite, to produce the foremost permanent myomere, which laps round the vagus in the manner described above. Had earlier stages not been followed there would be little certainty regarding the segmental number of this first permanent myomere.

The results just described, which are based on studies of Squalus acanthias, are practically in accord with those obtained by Ziegler (1908) in Torpedo ocellata, by Brohmer (1909) in Spinax (both of whom, however, reckon only three segments to the vagus), and by Goodrich (1918) in Scyllium canicula. Unfortunately, however, they differ in one small respect from the original scheme put forward by van Wijhe (1882*a*), working on Scyllium catulus and Scyllium canicula. The difference consists in the fact that van Wijhe intercalated one somite between the somites here described as 3rd and 4th. The result is that Balfour's (Ziegler's, Goodrich's and de Beer's) 4th somite (1st metotic somite) becomes van Wijhe's 5th, and so on, all down the line. Van Wijhe's 1st metotic somite is his 5th of the whole series, while it is Balfour's 4th. Van Wijhe regarded his 3rd and 4th somites as related to the hyoid arch and assumed that in this region one complete segment, and the corresponding nerves, visceral arch, and visceral cleft had disappeared. Further, this necessitated his attributing the glossopharyngeal nerve to the 5th segment and the vagus to segments 6 to 9, and these nerves would then overlie the anterior part of their corresponding somites. Van Wijhe's 4th, 5th, and 6th somites disappear.

Van Wijhe's scheme has been followed by Platt (1891), Hoffmann (1897), Neal (1897, 1898, 1918), Sewertzoff (1898), and Braus (1899). It can be brought into line with Balfour's scheme only by assuming that one or the other sets of authors has committed an error of observation, of the number of somites present on the one hand, and of the number of somites broken down and lost on the other. As regards the first point, Ziegler emitted the opinion that what van Wijhe had taken for his 4th somite was really only a part of the 3rd. The discrepancy on the latter point is still greater now that in a reaffirmation of his scheme, van Wijhe (1922) believes that the 7th somite also disappears and that the 8th contributes the first permanent myomere. This conflict of evidence can only be resolved by further work, but, for the present, the Balfour scheme, as developed by Goodrich (1918), will be adhered to, not only because there is quite insufficient evidence of a whole segment (corresponding to van Wijhe's 4th somite) having been lost, but also because it accords entirely with the work of Koltzoff on Cyclostomes which will be discussed below. Meanwhile, the complete agreement between all the above-mentioned authors, based on positive evidence of the strongest kind, renders unnecessary a consideration of the opinions of Kastschenko (1888), Rabl (1892, 1897), Froriep (1902), and de Lange (1936), who denied the existence of true somites anterior to the ear, or of Edgeworth (1899, 1903, 1907, 1911), who included visceral muscles in his 'myotomes'.

On the other hand, another group of authors, among whom may be mentioned Platt (1891), Killian (1891), Sewertzoff (1898), Dohrn (1901), and Gast (1909), claimed the existence of a far larger number of somites in the head than allowed for by Balfour's or van Wijhe's schemes. Their results, even on the same material, are very discrepant. In Torpedo Sewertzoff claimed thirteen, Dohrn over fifteen, and Killian at least eighteen somites. Ziegler (1908) pointed out that Torpedo embryos may show an inconstant and variable number of small cavities in the mesoderm, but that there is no good evidence for regarding them as somites. The same may be said of the eight little cavities which Boeke (1904) found in the prootic region of Muraena. These subsequently become reorganized into three, which then lose their individuality, so that the eye-muscles arise from mesenchyme.

Matveiev (1915, 1925), working on Pristiurus, has come to the conclusion that the diplospondyly characteristic of the sclerotomes of the trunk, is traceable in the head. By counting two sclerotomites to each segment (except the 1st) he has arrived at a scheme which resembles that of van Wijhe except that van Wijhe's segments 2 and 3 are each regarded as double, and called II_1, II_2, and III_1, III_2. Matveiev consequently regards van Wijhe's 4th segment as in reality the 6th (which he, however, calls number IV in order to avoid confusing the enumeration). With the help of these two extra segments, and the subdivision of all his segments (except the 1st) into two sclerotomites, Matveiev succeeds in correlating the apparent discrepancies of other authors by assuming that the structures which some of them took to be somites were really sclerotomites or half-segments.

The existence of sclerotomites in the hinder region of the head is quite possible; but while this view provides a formal explanation of the state of affairs presented by the segmented mesoderm of the head, by intercalating two extra segments it fails to account for, and actually upsets, the regular segmental arrangement of dorsal and ventral nerve-roots, intersegmental blood-vessels, and the parallel arrangement of the visceral clefts. If van Wijhe's (and Balfour's) segments 2 and 3 really represent four segments between them, then two more dorsal and ventral nerve-roots must be found for them, in addition to two visceral arches, clefts and blood-vessels, if the regular correspondence is to be maintained. And of the existence of such structures there is no evidence at all. As to the subdivision of the anterior somites, it is to be observed that their position and their fate, in producing six eye-muscles between them may be responsible for much modification. It is, therefore, more profitable to adhere to Balfour's original and simple scheme, especially, as will shortly be seen, it is applicable to other vertebrates, and accounts more fully for the facts than any other.

Meanwhile, there is one point of great importance to be noted. It has been shown in Torpedo and in the duck by Goodrich (1917) and de Beer (1926*c*) that the premandibular somites enter into relations with the hypophysis in a manner exactly comparable to that in which the 1st coelomic pouch in Amphioxus enters into relations with the preoral pit. This fact establishes the exact correspondence between the premandibular somite of Craniates and the foremost coelomic pouch in Amphioxus. There is, therefore, definite evidence that the premandibular somite represents the original 1st segment of the body, and there is no justification for the assumption that there may originally have been segments in front of it. Further, while there are indications of the former existence of a mandibular visceral cleft (between the mandibular and premandibular arches), the correspondence between the premandibular arch and the premandibular segment necessitates the conclusion that the only opening in front of the premandibular arch was the mouth, and not another visceral cleft separate from the mouth. This means that the labial cartilages found in some forms cannot represent visceral arches. (van Wijhe, 1906; see also p. 409).

2. Cyclostomata.

The segmentation of the head of Petromyzon planeri has been the subject of an extremely thorough investigation by Koltzoff (1902), from whose work the figures on Plate 4 are taken.

At the 8-day stage the series of somites are plainly visible; only somite 1 has not yet appeared. The cavities of somites 2 and 3 arise as enterocoels in a manner similar to the foremost somites of Amphioxus. In embryos eleven days old the 1st somite is visible, and the 4th and following somites produce muscle-fibres. A reconstruction of an embryo at the time of hatching shows that the 4th somite is related to the glossopharyngeal nerve, the 5th to the 1st branch of the vagus. It is thus obvious that these somites conform exactly to Balfour's scheme. The 1st and 2nd permanent myomeres (4th and 5th of the whole series) become split into dorsal and ventral positions, between which the vagus passes outwards and backwards. Thus, the 6th somite forms the myomere immediately behind the vagus, as in Squalus according to de Beer.

The fate of the three prootic somites in Petromyzon is interesting. The 1st somite gives rise to the inferior (anterior) oblique, superior and internal recti, innervated by the oculomotor: the 2nd somite gives rise to the superior (posterior) oblique innervated by the trochlear: the 3rd somite gives rise to the external rectus innervated by the abducens; but this nerve also innervates the inferior rectus, the development of which Koltzoff was unable to determine with certainty. But Addens (1928) has found that the so-called abducens in Petromyzon contains, in addition to the true abducens fibres, some oculomotor fibres for the inferior rectus muscle, which have taken this line of exit from the brain instead of accompanying the other oculomotor fibes. There is therefore no need in comparing the innervation of the eye-muscles of Cyclostomes with those of Gnathostomes to assume any change of peripheral innervation, nor a different method of origin for the inferior rectus muscle.

The ventral nerve-roots supplying the 1st and 2nd myomeres join and run backwards close to the spinal cord before emerging from its sheath, which they do behind the vagus (de Beer, 1924*c*).

3. Urodela.

The remaining vertebrates present a difficulty in the enumeration of their segments from the fact that with increasing specialization the series of somites does not appear in complete and unbroken sequence. The 1st somite is commonly recognizable as a head-cavity, but the 2nd and 3rd often do not present a simple appearance; the eye-muscles to which they give rise may condense straightaway out of mesenchyme, which stretches as a sheet from the 1st somite to the better developed somites immediately behind the auditory vesicle. In the circumstances it is extremely difficult to assign a segmental number to the first visible metotic somite except in favourable material of which a very close series of stages has been studied. This has been done in the axolotl, to which attention will now be turned.

In axolotl embryos 3 mm. long (Plate 5) Goodrich (1911) found that the row of somites in the trunk was continuous with those of the head as far forward as a point immediately beneath the auditory vesicle and above the 1st branchial arch. The somite formed here is medial to the glossopharyngeal nerve, and is clearly the 4th of the whole series, the 1st metotic. The prootic somites are represented in mesenchyme, but the eye-muscles to which they give rise and the ventral nerve-roots which supply them show, on comparison with Cyclostomes and Selachians, that their segmental value is not impaired by the fact that they have lost their simple arrangement as a row of somites.

The 2nd metotic somite, 5th of the whole series, is related to the 1st branch of the vagus and the 2nd branchial arch, and so on.

The 1st metotic somite disappears completely; the 2nd divides into dorsal and ventral portions, of which the former forms a myomere and persists, becoming fused on to the myomere behind it (3rd metotic, 6th of whole series). In early stages this myomere of the 3rd metotic somite possesses a ventral nerve-root, which, however, disappears when the dorsal portion of this myomere (together with the attached portion of that of the 2nd metotic somite) fuses with the next posterior myomere (4th metotic to form the temporal longitudinal muscle, innervated by the ventral nerve-root of the 4th metotic somite. The dorsal root corresponding to this segment becomes absorbed in the vagus, and the 1st complete mixed nerve is found in the 5th metotic segment, the 8th of the whole series. From the foregoing description it will be clear that the 1st permanent myomere arises from the 2nd metotic somite, the 5th of the whole series, and that only one somite, the 1st metotic, 4th of the whole series, disappears.

These results are supported by the investigation of Froriep (1917) on Salamandra, where the foremost myomere arises from the 2nd postotic somite.

It should be mentioned that in the axolotl, Matveiev (1922) has described the subdivision of the segments (except the 1st) into two sclerotomites each, and, as in the Selachian, he further regards segments 2 and 3 as double and representing four segments. The objection to this, as already mentioned (see p. 20), is that it fails to account for the segmentation of the nerves and other structures. The subdivision into sclerotomites in the hinder region of the skull is, however, quite possible.

4. Aves.

The segmentation of the head of the duck has been studied by de Beer and Barrington (1934), who have arrived at the following conclusions (Plate 6). The premandibular somite is a well-formed head-cavity, with a transverse commissure to its fellow of the opposite side, and relations with the hypophysis which establish the fact, as shown by Goodrich (1917), that this somite is homologous with the similarly named somite of Selachii. The muscles to which it gives rise and their innervation are typical. The 2nd and 3rd somites are represented only in mesenchyme, giving rise to their respective eye-muscles, but it is possible that a trace of the cavity of the 3rd somite may for a time be present.

Behind the prootic segments, the foremost somite at the 7-somite stage is a transient structure, situated beneath the auditory placode. It contains a small myocoelic cavity, and its anterior border is somewhat indistinct. It seems, however, to represent the 1st metotic somite, the 4th of the whole series. At the 14-somite stage it has disappeared, and the foremost metotic somite is now the 2nd situated immediately behind the auditory placode and the glosso-pharyngeal nerve. This 2nd metotic somite develops some muscle-fibres, but also disappears. The 3rd metotic somite lies beneath and behind the vagus, and gives rise to the 1st permanent myomere and possesses a transient ventral nerve-root. This corresponds to the 6th segment of the whole series.

5. Other vertebrates.

The literature on this subject contains a number of descriptions of occipital somites, some of which disappear and others persist. Unless, however, a complete series of stages have been studied, and unless the prootic as well as the metotic somites have been investigated, an element of uncertainty must exist as to whether the foremost metotic somite observed is really the 1st, a doubt which naturally affects the segmental number of the somite alleged to form the 1st permanent myomere.

Pending further research and in the interests of caution, these results are considered below as affording provisional evidence.

(a) Acipenser. Sewertzoff (1895) reported in the sturgeon the existence of a long row of occipital somites of which the foremost seven disappear, leaving the 8th to form the foremost permanent myomere. This is presumably the 8th metotic and 11th of the whole series. Rudimentary dorsal ganglia are present in the 8th, 9th, 10th, and 11th segments of the whole series. The reason for the large number of disappearing somites is doubtless due to the number of vertebral elements rendered immobile by their attachment to the hind end of the skull. In the prootic region Neumayer (1932) has described a large head-cavity.

(b) Amia. Schreiner (1902) and de Beer (1924*a*) found that the two foremost somites of the metotic series disappear, leaving the 3rd to produce the 1st permanent myomere. This may be the 6th of the whole series, if the foremost somite to disappear was the 1st metotic, which, however, is unlikely since it seems during the time of its existence to be better developed than the 1st metotic of any other Gnathostome. The 1st permanent myomere possesses a ventral but no dorsal root; the next (7th?) has a ventral root and an isolated dorsal ganglionic rudiment, while the 8th (?) possesses a complete mixed nerve.

Of the prootic series, the 1st somite forms a well-developed head-cavity.

(c) Lepidosteus. Schreiner (1902) found that of the series of metotic somites the foremost three disappear, leaving the 4th to form the 1st permanent myomere; presumably the 7th of the whole series. This segment has a ventral nerve-root and a transient dorsal ganglionic rudiment. The next segment (8th) has a dorsal and ventral roots.

(d) Salmo. In the trout, the problem has been studied by Harrison (1895), Willcox (1899), and Beccari (1922). According to Willcox, the two foremost somites of the metotic series disappear, leaving the 3rd to give rise to the 1st permanent myomere, possibly representing the 6th segment of the whole series. The ventral root of this 1st and of the 2nd permanent myomere combine, but pass out in the region of the 3rd permanent myomere (i.e. behind the septum separating myomeres 2 and 3).

(e) Ceratodus. All three prootic somites form well-developed head-cavities according to E. H. Gregory (1904). (Edgeworth (1926), however, believes that *all* the extrinsic eye-muscles are derived from the premandibular somite.) Sewertzoff (1902) found that of the series of metotic somites the foremost two disappeared; the 3rd (6th of the whole series?) seems to persist but lacks a ventral root: the 7th and 8th have ventral roots, and the 9th and following segments have dorsal roots as well.

Krawetz (1911) and Greil (1913) likewise noted the disappearance of the two foremost somites. Greil (l.c., p. 755) shows the 1st metotic somite immediately behind the otic vesicle, which makes it probable that it is the 4th of the whole series.

(f) Pelobates. Pelobates was studied by Sewertzoff (1895), who found three metotic somites, containing myocoelic cavities, the middle one of which is opposite the hinder part of the ganglionic rudiment of the vagus, and the front one opposite the middle of the ganglion. But at the next stage another small somite appears opposite the glossopharyngeal ganglion, and this is presumably the 1st metotic (4th segment) which, however, soon disappears, as does the following one, leaving the 3rd metotic somite (6th segment) to contribute the 1st permanent myomere.

(g) Alytes. In Alytes van Seters (1922) found a series of metotic somites the foremost of which he regards as corresponding to the vagus ganglion: this would represent the 2nd metotic or 5th of the whole series. It forms a few muscle-fibres but disappears, as does the 3rd metotic somite leaving the 4th (7th of the whole series) to form a permanent myomere, with a ventral nerve-root. The 6th and 7th segments have transient dorsal ganglionic rudiments, and segment 8 has a complete spinal nerve.

(h) Microhyla. Of the row of metotic somites found in this form by van der Steen (1930), the 1st, which lies medially to the auditory vesicle, produces no muscle-fibres. The 2nd, 3rd, and 4th of these somites do and give rise to myomeres which are transient, leaving the 5th of van der Steen's somites to produce the foremost permanent myomere. This segment is also the foremost one to possess a complete spinal nerve.

However, as van der Steen specifies that the 1st of his somites corresponds to the ganglion of the facial nerve, his 2nd to the glossopharyngeal, &c., it would seem that his 2nd somite is really the 1st metotic or 4th of the whole series; the 1st permanent myomere would then be formed by the 4th metotic somite, the 7th of the whole series. Van der Steen's 1st somite may therefore be the 3rd *prootic*, and this view is supported by his recognition (albeit tentative) of two somites in front of it.

(i) Rana. From the brief description given by Elliot (1907), all that can be deduced for certain is that two metotic somites disappear at an early stage, and that a 3rd somite corresponding to the 1st spinal nerve becomes reduced in post-larval life. Presumably this 3rd metotic somite is the 6th of the whole series, leaving the 7th to provide the foremost permanent myomere. This is also the foremost segment to possess a complete spinal nerve.

These conditions would seem to be not dissimilar from those described in Bufo by Chiarugi (1890), but the data are insufficient to allow of certainty. Similarly, it is not possible to equate the early results of Goette (1875) on Bombinator satisfactorily with those obtained in other Anura.

(j) Gymnophiona. The segmentation of the head of Hypogeophis has been studied by Marcus (1910) whose results are in many ways difficult to interpret. He describes premandibular, mandibular, and hyoid somites, but the order and manner in which he regards the eye-muscles as derived from them and innervated is quite unlike anything else in the whole of the vertebrates. A reinvestigation might help to clear up some of the discrepancies.

Behind the hyoid or 3rd somite, Marcus describes a whole row of somites of which the 5th would produce the first permanent myomere (his somites *v* to *z*). He describes ventral roots corresponding to all of these, with the proviso that those supplying *v* and *w* are very dubious. If the doubt may be extended to these somites as well, the 1st permanent myomere would be formed from the 3rd metotic somite, or 6th segment of the whole series.

(k) Lacertilia. Van Bemmelen (1889), Hoffmann (1889), Chiarugi (1890), Sewertzoff (1897), and Corning (1900) have all recognized a number of metotic somites in the forms studied by them, but it is not clear from their accounts how many transient somites may have been present in front of the foremost one described by them. For this reason it is not possible to determine for certain the segmental number of the 1st permanent myomere. The foremost myomere of Chiarugi possesses a ventral nerve-root, and therefore can hardly represent any segment more anterior than the 2nd metotic, since the 1st does not possess a ventral root in any Gnathostome.

(l) Chelonia. The segmentation of the head of Emys lutaria has been studied by Filatoff (1908). The three prootic somites are present; they contain head-cavities (though that of the 3rd is small) and give rise to the eye-muscles in the typical way. This is confirmed in Chelydra by Johnson (1913).

Between the 3rd somite and the 1st postotic myomere is a zone of mesoderm in which the traces of segmentation have vanished. Having regard to the fact that the 1st myomere in other forms is commonly formed from the 3rd metotic somite (6th of the whole series), it seems probable that two somites, the 1st and 2nd metotic, are represented in the unsegmented zone, which is situated immediately beneath and behind the ear vesicle. All the postotic myomeres, except the 1st, possess ventral nerve-roots.

(m) Mammalia. The three prootic somites were found to be developed in Marsupials by Fraser (1915). In the rat, Butcher (1929) made out that the 1st myomere arose from the 2nd postotic somite found by him, the 1st being transient and apparently missed by Dawes (1930) in the mouse and Muggia (1931*a*) in man; for the metotic somites which they describe (five in the mouse, four in man) all develop myomeres, and there is no evidence of the 1st metotic somite (4th somite of the whole series) doing this, even temporarily, in any form above the Cyclostomes. But the 5th somite (2nd metotic) is not known to do it except temporarily in Selachii, and in Urodela. Therefore, there must be some doubt whether Butcher's 1st myomere is really the 2nd metotic somite. Dawes's 1st myomere (probably the 2nd metotic somite) breaks down, and allowing that he did not see the 1st metotic somite, his 2nd to 5th (permanent) myomeres correspond to metotic somites 3 to 6, or segments 6 to 9. In the rabbit, Hunter (1935) described three occipital myomeres, but it is not clear which metotic somites they represent.

6. Conclusion.

It will be seen, therefore, that with the exception of the rather special case presented by the Acipenseroidea, Lepidosteus, and Anura (and van Wijhe's latest interpretation of the Scyllioidei), there is little evidence for the disappearance of more than two somites: the 1st and 2nd metotic, or 4th and 5th of the whole series.

ii. The determination of the position of the hind limit of the skull

On considering the question of the number of segments contained in the skull, a distinction must first be drawn between the neurocranium and the visceral arch skeleton or splanchnocranium, for the latter, varying with the number of visceral clefts present, will represent a number quite independent of that contained in the former.

1. The visceral arch skeleton.

The visceral arch skeleton is commonly held to begin with the mandibular arch, which corresponds in position with the 2nd segment. But if, as is possible, the trabeculae cranii represent the skeleton of a premandibular visceral arch, then the splanchnocranium may be said to start from segment 1.

The hyoid arch corresponds to the 3rd segment and the 1st branchial arch to the 4th. Thereafter the number of segments composing the splanchnocranium varies with the number of branchial arches, being that number plus three, and may be tabulated as follows (see also Gaupp, 1905*c*):

Petromyzon	10 or 11	Anura, Urodela	7
Bdellostoma	17	Lacertilia	5
Heptanchus	10	Birds	4
Hexanchus, Pliotrema	9	Mammals	6
Scyllium, Salmo	8		

It must be remembered, however, when stating that the visceral arch skeleton contains so many segments that the segmentation of the visceral arches is, strictly, independent of that of the somites, but corresponds with it (see above, p. 16).

2. The neurocranium.

The methods of determining the extent of the neurocranium and the number of segments which compose it are three in number.

(i) Taking advantage of the fact that the somites give rise to sclerotomes, skeletogenous tissue which migrates medianwards round the notochord and gives rise to vertebral centra, basidorsals and basiventrals and to the parachordals, it is sometimes possible to observe which are the somites whose sclerotomes produce the parachordals. It is sometimes necessary to count by half-sclerotomes, or even by quarters, so that the number of segments in the neurocranium may not be a whole number, but may end in a fraction.

(ii) Since the hind end of the parachordals is marked by the occipital arches (in all forms above Cyclostomes), which, like the neural arches that they are, arise close to the septum separating two myomeres, it is possible by determining the position of this septum to observe which is the myomere that lies immediately behind it. For this reason it is essential to know the segmental number of the first permanent myomere. The same technique may be applied to cranial ribs, where present.

In some forms preoccipital arches may likewise be assigned to a septum whose number is known. It often happens in fish that one or more vertebral neural arches are attached behind the proper occipital arch: these should be called occipitospinal arches. In other cases the centra of vertebrae but not their neural arches may be attached to the hind end of the parachordal. This may be ascertained by method (i) above.

(iii) The ventral nerve-roots which emerge from the skull immediately in front of the occipital arch supply the occipital myomeres, and by knowing the segmental number of the hindmost myomere innervated by the occipital nerves, the number of segments in the neurocranium may be ascertained. It is common to apply the term 'hypoglossal' to the foramina through which the occipital ventral nerve-roots emerge.

An attempt will now be made to apply these methods to the neurocrania of the different groups.

(a) **Petromyzon.** Koltzoff (1902) and Filatoff (in Sewertzoff 1916) have shown that the anterior parachordals develop from the sclerotomes of the first three (prootic) segments. Sewertzoff (1916) found that the posterior parachordals develop in the region of the 4th and 5th segments, though they have not been traced to their sclerotomes. There is no occipital arch, and the glossopharyngeal and vagus nerves emerge behind the auditory capsule, which, dorsally, represents the hind limit of the skull.

The neurocranium is thus composed of four (or perhaps five) segments. Behind the skull are rudiments of neural arches, and the homologues of these are to be found in the occipital region of the skulls of higher forms.

(b) **Scyllium** (Plate 7). Goodrich (1918) showed that the sclerotomes of segments 3 and 4, and probably also 1 and 2, contribute to the formation of the parachordals at their first appearance, and that the products of segments 5, 6, and 7 are added on behind, while the condyles are formed from the sclerotome of segment 8. The occipital arch is related to segment 7, and in front of it are some preoccipital arches corresponding to segments 6 and 5. Between the preoccipital arches and the occipital arch the ventral nerve-roots (hypoglossal) emerge, so that if these arches are comparable with neural arches the hypoglossal foramina are comparable with the intervertebral spaces through which the spinal nerves emerge.

If the occipital arch be regarded as the posterior limit of the skull, then the skull of Scyllium occupies seven segments. The 3rd permanent myomere is innervated by the 1st postoccipital ventral root, corresponding to the 8th segment, the dorsal nerve-root of which is vestigial.

But some of the 8th segment also contributes to the skull, viz. the condyles, which may represent part of a centrum of a vertebra, and therefore the total number of segments which contribute to the nemocranium should perhaps be expressed as 7+, or 8 if it be understood that the 8th segment contributes only a centrum.

(c) **Squalus** (Plate 3). According to de Beer (1922) the 1st postoccipital ventral nerve-root supplies the 5th permanent myomere, which corresponds to the 10th of the whole series. This gives nine as the number of segments included in the skull of Squalus on the Balfour scheme. This is in agreement in fact with the results of Hoffmann (1897–8) and Sewertzoff (1899) for, while they claim ten segments, they are counting on the van Wijhe scheme.

Since the postoccipital nerve emerges through a notch on the hind face of the occipital arch, it is probable that a condyle has been added from the 10th segment. This would make Squalus' skull composed of 9+, or 10, remembering that the contribution of the 10th segment is only a centrum. Its corresponding neural arch seems to have disappeared.

(d) Pristiurus. Matveiev (1925) considers that the three prootic somites (his I, II_1, II_2, and III_1) produce sclerotomes which give rise to the polar cartilages, pilae antoticae and orbital cartilages, while the metotic sclerotomes (III_2 to VII according to him; 5 to 9 according to van Wijhe; 4 to 8 according to Balfour) give rise to the parachordals. Further, according to Matveiev, the sclerotomes are divisible into anterior and posterior sclerotomites, producing interdorsals and basidorsals, interventrals and basiventrals respectively, so that the occipital arch and the preoccipital arches present a diplospondylous appearance.

Converting Matveiev's results to the Balfour scheme of somites, then, Pristiurus has a skull composed of eight segments. This is in agreement in fact with the results of Sewertzoff (1899), who found that the skull of Pristiurus contained one segment less than that of Squalus.

(e) Acipenser. Concerning the fates of sclerotomes Sewertzoff (1916) has stated that the sclerotome of the 1st or premandibular segment plays no part in the formation of the trabeculae, but contributes to the polar cartilages and parachordals. In the occipital region Sewertzoff (1895) described three fused pairs of occipital and preoccipital arches, corresponding to the septa between the 5th, 6th, 7th, and 8th metotic somites. More recently (1928) he has described a further preoccipital arch, in front of the others, and presumably corresponding to the septum between the 4th and 5th metotic somites.

These preoccipital and occipital arches form a compact mass (though pierced by hypoglossal foramina) which carries the posterior limit of the skull back to the 7th metotic (10th of the whole series) segment inclusive.

But farther posteriorly the parachordals are extended backwards by the equivalents of five vertebrae fused together, bearing each of them a pair of neural arches (now to be known as occipitospinal arches) and a pair of ribs of which the first pair later becomes aborted.

The skull of Acipenser, therefore, shows evidence of secondary prolongation by the addition of vertebrae, and the number of segments of which it is composed is 10+5, or 15.

(f) Amia. The occipital arch in Amia may be seen (Schreiner, 1902; de Beer, 1924*a*) to lie immediately behind a hypoglossal foramen for a ventral nerve-root which supplies the 2nd permanent myomere, which represents the 4th metotic somite and (presumably) the 7th segment. A preoccipital arch separates the foramen for this nerve from that for the nerve which supplies the 1st permanent myomere. But behind the occipital arch the parachordals are prolonged backwards and bear two pairs of occipitospinal arches (neural arches of fused vertebrae). The result is that the number of segments contributing to the formation of the skull in Amia is 7+2, or 9.

(g) Lepidosteus. According to Schreiner (1902) the foremost occipital ventral nerve-root innervates the 1st permanent myomere, which is (presumably) the 4th metotic and the 7th segment. This nerve, and the one immediately following it, emerge through foramina situated between the two preoccipital arches and the occipital arch. The occipital arch, therefore, marks the posterior limit of the 8th segment. Later on, however, vertebral elements with occipitospinal arches are fused on to the hind end of the skull, to the number of three, thus enclosing three more ventral roots in foramina. Indeed, the parachordals are in cartilaginous connexion with the basidorsals and basiventrals of vertebrae still farther back, but Veit (1911) has shown that the limit between skull and vertebral column coincides with the position of the 6th ventral root, which is thus the 1st spinal (although it may be caught up in a groove on the hind face of the last occipitospinal arch).

The skull of Lepidosteus is thus composed of 8+3, or 11 segments.

(h) Salmo. The 1st permanent myomere is apparently formed from the 3rd metotic somite or 6th segment. The septum between the 5th and 6th metotic somites coincides with an arch which Willcox (1899) took to be the occipital arch, but which Beccari (1922) has shown to be a detached occipitospinal arch, immediately behind the occipital arch and in front of the 1st vertebra. The first three permanent myomeres are supplied by nerves which emerge between the occipital arch and the occipitospinal arch; the nerve for the 4th myomere emerges between the occipitospinal arch and the neural arch of the 1st vertebra. The occipital arch, therefore, must correspond to the septum between the 4th and 5th metotic somites, marking the posterior limit of the 7th segment. But it is probable that the centrum corresponding to the detached occipitospinal arch is fused on to the hind end of the parachordal, forming a sort of condyle. On the analogy of the method followed in the case of Scyllium, the number of segments in the head of Salmo may be designated $7+c$, or 8 segments, if it be remembered that the neural arch of the 8th segment is still free.

(i) Ceratodus. Sewertzoff (1902) noted that the ventral nerve-root immediately in front of the occipital arch innervates the 3rd permanent myomere, presumably derived from the 5th metotic somite or 8th segment. The occipital arch thus lies in the septum between segments 8 and 9. Behind this occipital arch there are two occipitospinal arches included in the skull, corresponding to the septa between the 9th and 10th and the 10th and 11th segments. Corresponding to the 2nd occipitospinal arch is a cranial rib. Between the occipital and the two occipitospinal arches are two foramina for the hindmost hypoglossal ventral nerve-roots (corresponding to which dorsal roots are also present). In front of the occipital arch there emerge the two hypoglossal roots corresponding to the 2nd and 3rd permanent myomeres, metotic somites 4 and 5: segments 7 and 8.

These results are in conformity with those of Krawetz (1911), who found, however, that occasionally the occipital arch may arise in the septum between segments 9 and 10, instead of 8 and 9 as described by Sewertzoff. In such cases the two occipitospinal arches lie one segment farther back, the whole skull is one segment longer, and there may be three ventral roots in front of the occipital arch, making five in all instead of four.

In Protopterus and Lepidosiren the position of the occipital arch, according to Agar's (1906) work, seems to be in the same position as in Ceratodus, but Protopterus has one and Lepidosiren no occipitospinal arch fused on to it behind.

The number of segments composing the skull of Ceratodus is therefore ten or eleven, but it is possible that this number may be increased by the subsequent addition of vertebral segments, which may be detected in old specimens by the presence of additional pairs of cranial ribs. The number of segments in the skulls of Protopterus and Lepidosiren is one and two less, respectively, than in Ceratodus.

(j) Urodela. In Amblystoma (axolotl) Goodrich (1911) showed that the ventral nerve-root immediately behind the occipital arch supplies the 3rd permanent myomere which, since the 1st permanent myomere arises from the 2nd metotic somite, represents the 7th segment. The occipital arch therefore lies close to the septum separating the 6th and 7th segments. In front of the occipital arch is a preoccipital arch corresponding to the septum between the 5th and 6th segments. A transient ventral nerve-root emerges between the preoccipital and occipital arch, so that, at early stages at least, this amphibian possesses a hypoglossal foramen. This may persist into late stages in Cryptobranchus.

The position of the occipital arch shows that there are here six segments in the skull, but attention must be paid to the condyles. In Megalobatrachus Fortman (1918) showed that the anterior sclerotomite of the segment immediately behind the occipital arch produces a cartilage (intervertebral cartilage) which itself becomes split in such a way that the anterior half of it becomes attached to the occipital arch in front while the posterior half becomes attached to the atlas vertebra behind. The results have been confirmed by Mookerjee (1930) in Triton vulgaris, and probably apply to Amblystoma, where the number of segments concerned in the skull would be $6\frac{1}{4}$.

(k) Anura. In Alytes, according to van Seters (1922), the 1st permanent myomere corresponds to the 4th metotic somite (7th segment of the whole series) and is innervated by the 2nd spinal nerve. This places the atlas vertebra's neural arch in the septum between the 6th and 7th segments and the occipital arch in the septum between the 5th and 6th. The same follows from Elliot's (1907) account of Rana, where there is said also to be a cartilaginous preoccipital arch corresponding to the septum between metotic somites 1 and 2 or segments 4 and 5. But in Urodela the preoccipital arch lies between segments 5 and 6, and Elliot's results require confirmation.

In Microhyla, however, van der Steen (1930) places the foremost permanent myomere between the 1st and 2nd vertebrae, which would make the occipital arch coincide with the septum between the 6th and 7th segments if the 1st permanent myomere arises from the 8th segment, or between the 5th and 6th if it arises from the 7th (see p. 24) which is more probable.

The skull in Anura would thus contain five (or possibly six) segments. This number is without allowing for the condyles. The latter, as in Urodela, seem according to Mookerjee (1931) to arise in Rana, Bufo, and Xenopus from the anterior half of an intervertebral cartilage, itself the product of the anterior sclerotomite of the 1st trunk segment. If this is so, then $5\frac{1}{4}$ (or alternatively $6\frac{1}{4}$) segments are concerned in the Anuran neurocranium.

(l) Gymnophiona. In Hypogeophis Marcus (1910) showed that the ventral root innervating the 1st permanent myomere emerges immediately in front of the occipital arch. If the myomere is formed from the 6th segment, there are six segments included in the skull of Hypogeophis. It seems that the condyles are formed as in other Amphibia, so that counting them, $6\frac{1}{4}$ is the complete number.

(m) Lacertilia. Chiarugi (1890) in Lacerta places the hind limit of the skull between his 5th and 6th myomeres: so far as can be determined without further (much-needed) investigation, this corresponds to the septum between the 6th and 7th metotic somites, so that the skull would contain nine segments.

Chiarugi's number of occipital segments is larger by one than Corning's (1900), but in such cases, it is the positive evidence that counts, and Corning may have missed Chiarugi's 1st myomere, just as the latter probably missed the 1st metotic somite.

(n) Chelonia. Filatoff (1908) counted four metotic myomeres in the occipital region, and noted that the anterior sclerotomite of the 5th also contributed to the skull. If the 1st myomere is formed from the 3rd metotic segment, or 6th of the whole series, there would be $9\frac{1}{2}$ segments in the skull of Emys.

(o) Aves. The conditions in the duck have been investigated by de Beer and Barrington (1934), who find that the hindmost cranial ventral nerve-root corresponds to the 4th permanent myomere, or 6th metotic somite, or 9th segment. The occipital arch is formed close to the septum separating the 9th and 10th segments, and anterior to it are five preoccipital arches (enclosing five hypoglossal foramina) and four cranial ribs. The number of hypoglossal foramina eventually becomes reduced to two.

The condyle of the skull is formed from the anterior sclerotomite of the 10th segment, with the result that $9\frac{1}{2}$ segments are included in the skull of the duck. Of these $9\frac{1}{2}$ segments, however, two are visibly added on to the hind end of the parachordals during development, by the attachment of two vertebrae.

(p) Mammalia. In the mouse Dawes (1930) has shown that the occipital region of the skull is derived from five metotic somites, probably numbers 2 to 6 corresponding to segments 5 to 9. The occipital arch, between the last occipital myomere and the 1st cervical myomere (which latter is innervated by the 1st spinal or suboccipital nerve), therefore falls between segments 9 and 10: i.e. there are nine segments included in the skull. In Felis there are indications of four pairs of preoccipital and occipital arches, and four hypocentral elements, presumably corresponding to segments 6 to 9 (Kernan, 1915).

In the mammals it seems that the so-called condylar foramen may represent more than one hypoglossal foramen (see p. 384). Thus, in the mouse the four occipital myomeres are innervated by three ventral nerve-roots (the root to the foremost myomere having disappeared) which all emerge through the same single condylar foramen. In the rabbit, on the other hand, there are two condylar foramina.

iii. Conclusion

The variation in position of the occipital arch in different groups of vertebrates is shown in Plate 8, from which it will be obvious how erroneous was the attempt to fix hard-and-fast positions for the hinder limit of the neurocranium. The truth is, as Goodrich (1913) showed for the paired and median fins of fish as well as for the occipital arch, that structures may vary their position in respect of the segments which give rise to them, without thereby impairing their homology: the structure can be transposed up and down the keys in phylogeny without ceasing to be the same 'tune'.

As to the reasons for the variation in position of the occipital arch, much may be speculated, but little can be said of any value. The progress of the process of cephalization and perfection of the brain during evolution would be expected to occupy more and more space, so that it is not surprising to find the occipital arch farther posterior in mammals than it is in Cyclostomes. Yet, on the other hand, it is difficult to see why there should be more cranial segments in Selachii than in Anura or Urodela, which, however degenerate their modern representatives may be, nevertheless represent an undoubtedly higher grade of evolution than the Selachii. The tendency to extreme reduction of the number of segments of the trunk (i.e. vertebrae) exhibited by Anura (in particular, Xenopus and Hymenocheirus) can hardly be responsible for the shortness of the skull, for the Urodela retain a large number of segments in their trunk.

Attempts have been made to explain the addition of segments to the hind region of the skull of Amniota on Lamarckian principles, by MacBride (1932). This author sees a cause of this addition in the cervical flexure of the embryo: this flexure in turn being supposed to be the embryonic reflection of the assumption by the adult Amniotes of an altered habitual posture of the head and method of eating. Quite apart from the many and serious difficulties which make it hard to accept the Lamarckian principle, this explanation fails to account for a number of facts. First, the embryonic cervical flexure is not confined to Amniota, for it is exhibited in a marked degree by Selachii, and it is inconceivable that any similarity of posture of the head or method of feeding could account for the flexure of these two groups. Second, in many fish, such as Acipenser, Amia, Lepidosteus, Ceratodus, &c., the number of neural and occipitospinal arches tacked on to the hind end of the skull is very large, although they show little or no cervical flexure in the embryo. Thirdly, the idea of extra segments having been *added* to the head in Amniota is derived only from a comparison between Amniota and *living* Amphibia, where the number of segments of the head is very small, and must be the result of a secondary reduction, seeing that the number is larger in all fish, and was probably larger in fossil Amphibia where there is a hypoglossal nerve foramen. Indeed, there is the possibility that the Amniote line of vertebrate evolution was characterized by the same number of segments in the head (9 or 9½ according as to whether the body of the proatlas centrum fuses with the odontoid process of the 1st vertebra or with the parachordal of the skull) ever since the fish stage. It is noticeable that the number of segments of the head is very variable in the fish, even between quite closely related forms (cf. Scyllium and Squalus), while it has become fixed in the progressive Amniote line of evolution. This may be compared with the fixity in number of the cervical segments in the highest vertebrates—the mammals—or in a general way with the fixity in total number of segments in the highest groups of Arthropoda.

Lastly, no evidence has been adduced to show that the incorporation of extra segments in the back of the skull has any connexion, mechanical or otherwise, with the phenomenon of embryonic cranial flexure.

It must be admitted, therefore, that there is ignorance concerning the causes of the variation in position of the hinder limit of the skull. In the bony fish it may be pointed out that the plastering on of vertebral elements is usually correlated with a posterior extension of the parasphenoid bone, imparting rigidity to the whole structure. It would, however, be more than rash to assert that either of these features was the cause of the other.

There is one other possibility of unravelling the segmentation of the head, although its application is by no means easy. It is based on the facts that (*a*) the dermal bones of the head are similar in structure and origin to the scales on the trunk of bony fish, and (*b*) the scales on the trunk are primitively in segmental arrangement, although they can secondarily become increased in number (Koumans, 1936).

It must be concluded that originally the dermal bones on the head were also arranged in some simple segmental way, but the arrangement has been obscured (cf. Sewertzoff, 1925*c*). For one thing, many of the head bones are much larger than scales. They may represent enlarged single scales, or several scales fused. To obtain information on this point is not easy, but it is worth while turning to the lateral line system for a possible clue.

Along the main trunk lateral line in bony fish the sense-organs correspond to the scales, and must, therefore, be regarded as segmental in position. The question therefore arises, can a segmental arrangement of sense-organs be made out in the head? The matter is complicated by the fact that whereas some of the cranial lateral line canals appear to be longitudinal, others seem to represent vestiges of transverse canals. To determine the number of segments revealed by the sense-organs, it is obviously necessary to consider only the longitudinal lines and to neglect the transverse ones. There is further difficulty in deciding which are the transverse lines.

Following Allis (1934), the main infraorbital line may be regarded as composed of three sections: a posterior longitudinal (supratemporal) portion; a transverse (postorbital) portion, and an anterior (suborbital and antorbital) portion. The supraorbital line is a longitudinal one.

If now the sense-organs in the longitudinal portions of the canals of Amia (Allis, 1889) are counted, they lead to an interesting result. The suborbital portion of the infraorbital lateral line shows seven sense-organs (organs numbers 3, 4, 6, 7, 8, 9, and 10) which may be regarded as representing a segmental series. (Organs numbers 1 and 2 may be regarded as belonging to the transverse rostral canal, organ 5 is out of alinement and seems to belong to a portion of a transverse preorbital canal; organs 11, 12, 13, and 14 belong to the transverse postorbital canal.)

The supraorbital canal also shows seven organs. Coming now to the supratemporal canal, it possesses three organs between the point where it is joined by the postorbital canal and that where it gives off the occipital transverse canal.

So far as can be made out by counting the organs situated on longitudinal stretches of canals, there are ten as far back as the occipital canal which lies in the postparietal or extrascapular bone and coincides more or less in position with the hind limit of the skull. It is, therefore, of great interest to find that nine is the number of segments attributed to the skull in Amia from study of the somites and nerves.

Now if the head of Amia occupies nine or ten segments and the dermal bones were originally similar to segmental trunk scales, it is clear that either some of the scales on the head have become lost and others enlarged, or that fusion of scales has occurred to give rise to the dermal bones such as exist now. Both possibilities may have occurred, but there seems to be a certain amount of evidence for the latter since Pehrson (1922) has found that in some cases at least, the rudiments of the bones in Amia correspond to the lateral line sense-organs. Thus, along the supraorbital line, each of the seven organs becomes surrounded by a bony rudiment, of which the first three fuse to form the nasal and the remaining four give rise to the frontal. The nasal would then correspond to three original scales (which remain separate in Polypterus, p. 85), while the frontal would correspond to four. In the suborbital row each of the two suborbital bones would correspond to a scale, the lachrymal would correspond to two scales, and the antorbital to three. The inter-temporal-supratemporal bone would correspond to three scales.

While the general principle here invoked (originally segmental arrangement of head-scales and lateral line organs) may be sound, it is, of course, possible that its application to this or similar cases is erroneous. The hope remains that palaeontology will one day provide a form in which the segmental nature of the head-scales is clearly revealed. Meanwhile, attention must be drawn to some points of difficulty.

In the first place it would be erroneous to conclude that a separate ossification must always represent an original segmental scale in a linear row: if this were so, the number of segments in the head of Lepidosteus as estimated by the number of rostral ossicles, would be ridiculous.

Similarly, it would be erroneous to conclude that a bone which represents the fusion of several scales *must* always have several centres of ossification. The frontal of Amia is presumably homologous with that of Salmo, which, so far as is known, appears to ossify from one centre.

Lastly, it should be noted that the innervation of the sense-organs of the lateral line on the head does not correspond at all with the Balfour scheme of head segmentation. The seven organs of the supraorbital line, the seven organs (segmentally counted) of the suborbital line, and two organs of the supratemporal line are all innervated via the facial nerve, the dorsal root of the 3rd segment. The glossopharyngeal, the dorsal root of the 4th segment, innervates the sense-organ in the supratemporal line, which, on the scheme here put forward, represents the 10th segment.

This fact would indeed be serious were it not for the utter disregard of segmental correspondence exhibited in the innervation of the main lateral line of the trunk, for all its segmental organs, as far as the tail, are innervated by the vagus, the dorsal roots of the 5th to 8th segments. It is clear, therefore, that the lateral line sense-organs are not innervated by nerves corresponding to their segments. This, however, while increasing the difficulty of study, need not invalidate the use of the lateral-line sense-organs in an attempt to provide additional evidence concerning the segmentation of the head.

3. THE CRANIOGENIC MATERIALS

The materials which go to the formation of the skull are two varieties of connective tissue—cartilage and bone—the former occurring alone in the Cyclostomes and Elasmobranchii, and both together being concerned in the remaining vertebrates to relative extents which vary in the different groups.

It is not the purpose of this work to give a detailed account of the histological characteristics of the various kinds of cartilage and of bone, but a few remarks on the origin of these tissues in the head should properly precede a consideration of the details of skull development in the various groups of vertebrates.

From the fact that the skull is a compound structure consisting of the skeletal elements of the brain-case and sense-capsules (neurocranium) on the one hand and of the visceral arches (splanchnocranium) on the other, it will be convenient to subdivide the subject-matter to be considered in the same sense.

i. The craniogenic materials of the neurocranium

The skeleton of the brain-case (and of the sense-capsules) is a specialization of the connective-tissue membranes surrounding it, and which include the meningeal membranes. In Amphioxus these membranes remain in the form of connective tissue, but in Craniates the outermost of these membranes acquires chondrogenic and osteogenic properties, becoming, in fact, the perichondrium and periosteum of the cartilages and bones of the neurocranium. It is, therefore, of importance to review the conditions of the meningeal membranes in the various groups of Craniates, following the recent accounts given by van Gelderen (1924, 1925), and Ariëns Kappers (1926, 1929).

In all Craniates at early stages the neural tube is surrounded by a relatively undifferentiated sheath of mesenchyme, the meninx primitiva. This subsequently becomes split into an inner endomeninx and an outer ectomeninx. The former is thin and vascular and closely surrounds the neural tube; the latter is stout and fibrous, and its outermost portion gives rise to cartilage: i.e. may be regarded as a perichondrium.

The ectomeninx becomes separated from the endomeninx by a varying amount of vacuolated and sometimes fatty tissue, as in Cyclostomes, where this zone is so extensive that some authorities (Sagemehl, Kappers) deny that the perichondrium outside it is a meningeal membrane at all.

The Selachii also have a fairly wide zone between the ectomeninx and endomeninx. In Teleostei the ectomeninx is in part perichondral and in part periosteal, but a specialization has taken place in the endomeninx which becomes split into an inner layer which may be called the pia mater, and an outer layer which corresponds to an arachnoid membrane. This specialization is not found in the lower Tetrapods. In Urodeles the endomeninx remains simple, but the ectomeninx begins to split into an outer perichondral layer and an inner layer which may be called the dura mater. The epidural space between them is restricted, and is occupied by venous sinuses. In Anura the split is much more extensive and the epidural space is wide, and occupied by diverticula of the endolymphatic sac.

In the Amniotes the same split occurs in the ectomeninx between the perichondrium and the dura mater, but in the region of the skull, owing to the great expansion of the brain, the two layers become apposed to one another again, forming a secondary dura mater. This apposition does not occur in the region of the trunk, and for this reason the epidural space which is obliterated in the skull is present in the vertebral column.

In addition, in birds and in mammals, the endomeninx splits into an inner pia mater and an outer arachnoid membrane.

There may, therefore, be two, three, or as many as four connective-tissue layers surrounding the neural tube, and the outermost of these is concerned with the formation of the skull.

The meninx primitiva is commonly regarded as derived from the sclerotomes of the mesodermal somites. In Amphibia it has been claimed by Harvey and Burr (1924), as a result of experimental work, that the endomeninx is derived from the cells of the neural crest. Be this as it may, the matter is irrelevant to the present subject, since the endomeninx plays no part in the formation of the skull.

The outermost layer of the ectomeninx, on the other hand, is the site of formation of cartilage, or of bone, forming the neurocranium.

The trabeculae cranii, although contributing to the neurocranium, seem at least in the lower vertebrates to arise externally to the ectomeninx.

It must be added that in Urodela, Anura, and perhaps in Lepidosteus the cells of the notochord (within its unbroken sheath) seem capable of producing a tissue resembling cartilage, which may contribute to the basal plate.

ii. The craniogenic materials of the visceral arch skeleton

The origin of the tissue which ultimately gives rise to the skeletal elements of the visceral arches (and, it may be added, in all probability, of the trabeculae cranii) is uncertain. The classical view is that the condensations of mesenchyme from which these skeletal structures arise are derived from the splanchnic layer of mesoderm surrounding the alimentary canal, and which, in the region of the head, are localized in the various visceral arches and in the upper jaw.

The view has, however, repeatedly been expressed that the cartilage of the visceral arch skeleton and trabeculae are formed from cells which have been derived from the neural crest, and which have migrated into their respective positions. As the evidence in support of this contention is mainly experimental, it will be reserved for that section of this work (p. 472) in which the experimental work on the development of the skull is reviewed.

But whatever the ultimate source of the cells which ultimately compose the visceral arch skeleton, it is clearly not the same as that of the neurocranium.

iii. Chondrification

For the proper appreciation and comparison of descriptions of chondrocrania it is necessary to make a distinction between dense mesenchyme (procartilage) destined to become cartilage, and true cartilage itself. Further, it is essential to distinguish between three types of cartilaginous tissue, viz. (i) normal hyaline 'primary' cartilage, of which the chondrocranium of Gnathostomes is composed; (ii) the so-called mucocartilage of Cyclostomes; and (iii) the 'secondary' cartilage which is found in the rudiments of certain membrane-bones, especially in Mammals.

1. 'Primary' cartilage.

The first origin of a future cartilage is usually to be found in a condensation of mesenchyme cells, not sharply marked off from their surroundings, and which may be so dense that the nuclei almost appear to be in contact with one another, as seen in sectional preparations. According to Schaffer (1930), the rudiment may actually be a syncytium, in which, subsequently, boundaries appear dividing up the cytoplasm into as many territories as there are nuclei, and thus bringing about the first appearance of the intercellular substance in the form of thin capsules. The intercellular substance is at first oxyphil, but soon becomes basophil as more material is added by secretion from the cells.

In other cases the condensation of mesenchyme cells may be less dense, and the intercellular substance is laid down simply between the cells. The intercellular substance is, of course, responsible for giving the tissue the consistency which enables it to function as a skeleton. It is, however, not rigid, and, as in the case of the quadrate in Anura (p. 206), can become buckled and bent in consequence of the changes which the jaws undergo at metamorphosis.

A characteristic feature of cartilage is its power of expansive or intussusceptive or interstitial growth, by means of a limited amount of cell-division, enlargement of the cell-capsules, and increase of intercellular substance. The result of this is to stretch the immediately surrounding layer of mesenchyme into a membrane which becomes the perichondrium. The characteristic of fully formed cartilage is therefore not only the presence of chondrin but also a well-marked perichondrium. Procartilage, on the other hand, grades off more or less insensibly into the surrounding mesenchyme.

In addition to the intussusceptive method, cartilage may grow by apposition, by the acquisition of chondrin-producing properties on the part of neighbouring cells, either mesenchyme cells *in situ*, or new cells formed from the innermost layer of the perichondrium, which thus become added on at the surface of the cartilage. Cartilage may thus grade insensibly into procartilage. Further, cartilage may become eroded, as Streeter (1917, 1918) showed in the development of the auditory capsule in Man, and the cartilage-cells then dedifferentiate into procartilage and then into mesenchyme.

While as a rule cartilage results from a differentiation of mesenchyme *in situ*, there are cases in which chondrification follows a migration of cells; e.g. after invasion of the notochordal sheath in Selachii, Holocephali, Acipenseroidei, and Dipnoi.

Although in most cases each procartilaginous condensation corresponds to a subsequent single chondrification, it may happen that two or more separate centres of chondrification may arise in a single procartilaginous rudiment, and subsequently either fuse (e.g. trabecula and polar cartilage) or remain separate (e.g. pharyngo-, epi-, cerato-, and hypobranchials). The fusion of different cartilaginous rudiments is very common in the formation of the neurocranium.

In Selachii, where, of course, the skeleton is entirely cartilaginous, calcification of the cartilage normally takes place. In higher vertebrates calcification of the cartilage also occurs, but principally in connexion with the changes in the cartilage which precede ossification. These changes involve enlargement of the cavities containing the cells and their arrangement in columns or rows, and the reduction and ultimate destruction of the intercellular material by humoral absorption and by chondroclasts. Such cartilage is known as *hypertrophic* cartilage. The space previously occupied by the cartilage may then become occupied by newly formed bone (see also p. 481).

According to Ziba (1911), Retterer (1917), Levi (1930), and Fell (1933), the possibility exists that some of the cartilage cells may actually become osteoblasts. It appears, further, according to Zawisch-Ossenitz (1927), that the periosteum and osteoblasts are capable of giving rise in addition to bone to nodules of 'secondary' cartilage; a type of tissue which will be referred to below.

The Cyclostomes, and especially Myxinoids, present numerous histological variations of cartilage, which have been exhaustively studied by Cole (1905).

2. 'Secondary' cartilage.

Nodules of a peculiar kind of cartilage-like tissue, resembling hypertrophic (primary) cartilage, are commonly found in the rudiments of developing membrane-bones, especially in mammals. This tissue, termed by Schaffer (1930) 'vesicular chondroid bone', is characterized by large capsules, with calcified walls, and is said by Weidenreich (1930) to play the same part in ossification as the strands of calcified connective-tissue fibres in the formation of the perichondral bone of a diaphysis, or as the cartilage of an ossifying epiphysis, and the calcifications form a framework which may become incorporated in the bone.

According to F. Strasser (1905) secondary cartilage may arise in those regions of the rudiments of membrane-bones which are subjected to precocious strains and stresses. Among mammals Gaupp (1907*a*) has described in Galeopithecus the existence of secondary cartilage in the angular, coronoid, and condylar regions of the dentary as well as in its main shaft, in the main portion and in the palatine process of the maxilla, in the palatine, pterygoid, squamosal, jugal, vomer, and frontal. Some of these nodules are found more or less regularly in all mammals (except, apparently, in Monotremes), especially the nodule in the pterygoid.

Among birds, F. Strasser (1905) has described a nodule of secondary cartilage in the pterygoid. In general, secondary cartilage appears to be associated with, and of the same nature as, callus-formation and sesamoids. The relation of secondary cartilage to hypertrophic (primary) cartilage is by no means clear, but here again, according to Fell (personal communication), it appears to be not impossible that some of the cells of the secondary cartilage may become osteoblasts.

3. Mucocartilage.

The term mucocartilage (*Schleimknorpel*) was applied by Schneider (1879) to certain structures in the head-region of the Ammocoete larva of Petromyzon, characterized by the possession of a homogeneous ground-substance which is basiphil like true cartilage, but differs from the latter in having affinity for such stains as mucicarmine and mucihaematin, as shown by Schaffer (1930). Collagen connective-tissue fibres may be seen wandering through the ground-substance. Damas (1935), after demonstrating the existence of elastic fibres permeating the ground-substance, calls the tissue simply elastic connective tissue. Contrary to previous belief, it is now clear that mucocartilage is *not* a precursor of true cartilage (which Petromyzon also possesses), and it appears to represent a variety of connective tissue *sui generis*.

iv. Ossification

From the point of view of the origin of their constituent materials all bones, whether preformed in cartilage (cartilage-bones) or not (membrane-bones), are alike in owing their formation to the activities of a particular kind of connective-tissue cells, the osteoblasts, which are of mesodermal origin. These become aggregated in regions occupied by either ordinary connective tissue (where they give rise to membrane-bones), or in the perichondrium immediately overlying the cartilage (where they give rise to perichondral ossifications of cartilage-bones). Further, the cartilage may become invaded by connective tissue and blood-vessels, and osteoblasts brought in in this way may then give rise to endochondral centres of ossification of cartilage-bones. The thickness of the bone may thus come to be occupied by sponge-like bone, or spongiosa, which may later be partly absorbed by the action of osteoclasts. In the higher vertebrates the membrane-bones may similarly come to consist of compact and smooth external and internal lamellae (the outer and inner tables) enclosing between them the spongy diploe.

In the case of cartilage-bones generally, the centres of ossification often coincide in position with previous centres of chondrification, especially where the cartilages remain separate. But in the neurocranium, where the rule is for the independently formed cartilages eventually all to fuse together, the centres of ossification need not bear any relation to the positions of the original centres of chondrification.

It is probable that bone is incapable of intussusceptive growth (although Thoma (1923) and Lexer (1924) claim that it can grow in such a manner). The increase in size of a bone must therefore be due to appositional growth at the surface and edges. At the same time, its shape may be altered (e.g. degree of curvature of the parietal) by differential apposition in different places. According to Mair (1926) little or no absorption takes place on the external and internal surfaces, but only in the diploe.

It is characteristic of bones to remain separated from one another by intervening non-ossified regions (which thus play the same part as the cartilaginous region between the diaphysis and epiphyses of a long bone) during development until at such time as they meet, the suture between them becomes obliterated, and power of growth ceases. This is to be contrasted with the habit of cartilage to fuse early into a single structure which then grows by intussusception and apposition. A comparison between these two methods reveals the greater and more rapid ultimate expansive power of bone, and it is doubtless for this reason that the brain-vault in higher vertebrates has to depend less and less on cartilage and more and more on bone for its protection (see p. 454).

It should be added that in some primitive bony fish the dermal bones of the skull possess the same structure as the scales of the body, and among living forms, these conform to the palaeoniscoid type in Polypterus and to the lepidosteoid type in Lepidosteus (see Goodrich, 1909). In so far as the cosmoid layer of the palaeoniscoid scale represents originally free denticles, these may be said to contribute to the formation of the skull in this fish.

II. SYSTEMATIC SECTION

TO speak of a skull in the Acrania is, of course, paradoxical. Nevertheless, mention must be made of Amphioxus for two reasons. First, it (together with Balanoglossus and the Ascidians) possesses a visceral arch skeleton, formed of coagulated elastic fibres, entirely separate and independent from the axial skeleton (here, the notochord). Secondly, Amphioxus possesses true cartilage in its buccal cirrhi (van Wijhe, 1904*b*), which shows that 'visceral' cartilage may have evolved before 'axial' cartilage. Attempts have been made (Pollard, 1894, 1895) to see in the buccal cirrhi of Amphioxus representatives of the labial cartilages of Craniates (see p. 409).

CRANIATA

4. CYCLOSTOMATA. PETROMYZONTIA

PETROMYZON

The first studies on the development of the skull in lampreys are due to W. K. Parker (1883*a*). Subsequently, the early stages were studied by von Kupffer (1894, 1895), Sewertzoff (1914, 1916, 1917), and Schalk (1913). Gaskell (1908) reconstructed the skull of the Ammocoete, while the metamorphosis has been investigated by Bujor (1881), Kaensche (1890), Tretjakoff (1929), and Damas (1935). Certain features in the development have been followed by Nestler (1890). The development of the ear and auditory capsule has been described by Studnička (1912). The contribution of the segmental sclerotomes to the axial portion of the skull was followed by Koltzoff (1902) and Filatoff (in Sewertzoff 1916); the histological nature and provenance of the mucocartilage, of which a part of the skull is composed, have been studied by Schneider (1879), Schaffer (1896), Dohrn (1884*b*, 1902) and Goette (1901), while the metamorphosis of the skull of the Ammocoete into that of the lamprey and the fate of the mucocartilage have been described by Tretjakoff (1929), Balabai (1935), and Damas (1935). The problem of the trabeculae in Cyclostomes has been studied by de Beer (1931*b*). The morphology of the skull in the adult lamprey has been the subject of researches by T. H. Huxley (1876*a*), Hatschek (1892), Neumayer (1898), Allis (1903*b*, 1924, 1926*a*), and Tretjakoff (1926); while Stensiö (1927) has had occasion to compare it with that of Cephalaspidae.

The development of the skull of Myxinoids is unfortunately unknown: its morphology in the adult has been described by Cole (1905, 1909), and the development of the visceral arches of Bdellostoma briefly treated by Neumayer (1910)

In the following account the early stages are based on the work of Sewertzoff, and the later on that of Damas, verified and amplified here and there by Seyd (unpub.) and by the present writer.

i. Early development of the skull of Petromyzon planeri

(i) The first part of the skull to appear is the visceral arch skeleton, in the form of slender vertical struts of procartilage situated between the branchial sacs, and corresponding in position to the branchial branches of the glossopharyngeal and vagus nerves. The foremost of these struts, lying between the 1st and 2nd branchial sacs and corresponding to the glossopharyngeal nerve, represents the 1st branchial arch. At this stage, six branchial arches are laid down and each is free above and below. The neurocranium has not yet appeared. (Plate 9, fig. 1.)

(ii) The first elements of the neurocranium to chondrify are the paired anterior parachordals (trabeculae of the earlier authors). These lie one on either side of the notochord, their anterior end slightly *behind* the tip of the notochord, their posterior end extending to a point between the anterior parts of the auditory vesicles. The hind part of the anterior parachordals is in contact with the lateral edge of the notochord; the front part diverges slightly from it. (Plate 9, fig. 2.) From Koltzoff's and Filatoff's studies it appears that the anterior parachordals are formed from the sclerotomes of the first three segments of the body.

In the visceral arch skeleton the first six branchial arches are chondrified, the 7th is still mesenchymatous. The dorsal and ventral ends of all the branchial arches from the 2nd to the 6th are beginning to turn forwards. On the other hand, the dorsal and ventral ends of the 1st branchial arch are turned backwards. On the anterior border of the 1st and 2nd branchial arches are seen the rudiments of the epitrematic and hypotrematic commissures, which enclose the outer opening of the branchial sacs and eventually become fused on to the visceral arch in front, thus forming the lateral framework of the branchial basket. On each side, the ventral ends of all the branchial arches are interconnected by a longitudinal strand of mesenchyme which forms the rudiment of the hypobranchial commissure. Dorsally, the ends of the branchial arches are separated from the notochord by the axial (epibranchial) muscles.

(iii) At a slightly later stage the anterior parachordals extend backwards between the auditory sacs and forwards round the forebrain. Behind them and separate from them there have appeared the elements which Sewertzoff calls the posterior parachordals; on each side of the notochord and close up against it between the hinder parts of the auditory sacs. The posterior parachordals appear from the start in mesenchymatous continuity with the dorsal ends of the 1st branchial arches, but their nature as parts of the axial skeleton is proved by their apparent origin from the sclerotomes of the 4th and 5th segments. Their cartilage differs histologically from that of the branchial arch. (Plate 9, fig. 3.)

The posterior parachordals are nevertheless rather puzzling structures, and their connexion with the 1st branchial arch led Hatschek (1892) to the view that they, like the subchordal rods (developed later, from the dorsal ends of the 2nd–7th branchial arches) were really part of the visceral skeleton. However, the posterior parachordals differ from the subchordal rods in that they lie against the lateral surfaces of the notochord (like the anterior parachordals) instead of the ventrolateral surfaces, and that they lie dorsally (instead of ventrally) to the axial musculature. Further, the origin of the posterior parachordals from the sclerotomes is, if true, proof positive of their axial nature, and their fusion with the 1st branchial arch must be ascribed to the very general occurrence of fusion between elements in the skull of the lamprey.

(iv) At the 26-day stage (Plate 9, fig. 4) the anterior and posterior parachordals are still separate. Each anterior parachordal now possesses a laterally directed process between the trigeminal and facial nerves. This is Huxley's 'palatoquadrate', Parker's 'pedicle and pterygoid portion of the mandibular arch', Sewertzoff's dorsal portion of the mandibular arch, and Damas's 'primitive velar bar'. At first sight Parker's term 'pedicle' seems to be the most appropriate, for he thought that by 'pedicle' he was describing what is now known as the basitrabecular process; he was, however, in fact referring to the ascending process of the frog. The process in Petromyzon will here be called the mandibular process.

The dorsal ends of the branchial arches (from the 2nd to the 6th) are now T-shaped, forming for each arch the anterior and posterior subchordal processes, which in the case of the 2nd and 3rd arches are becoming apposed to the ventrolateral surfaces of the notochord. Epitrematic and hypotrematic processes are now present on all the branchial arches (except the last), while on the 1st arch these processes have fused with one another in front of the opening of the 1st branchial sac. The cartilaginous hypobranchial rods now connect the ventral ends of the branchial arches on each side, lying ventrolaterally to the ventral aorta.

(v) At the 41-day stage the anterior and posterior parachordal cartilages on each side have fused, forming a continuous base to the skull between the auditory capsules, and, of course, attached posteriorly to the dorsal end of the 1st branchial arch. Anteriorly, the divergent ends of the anterior parachordals are beginning to curl in towards one another beneath the forebrain. There is still no chondrification in either nasal or auditory capsules.

The anterior and posterior subchordal processes of the 2nd–6th branchial arches are now interconnected by a band of mesenchyme which forms the rudiment of the subchordal rod, on each side.

(vi) At the 45-day stage (Plate 9, fig. 5) rudiments have appeared of three rods of so-called mucocartilage ('Schleimknorpel', Schneider 1879), on each side. The hindmost of these lies as far in front of the 1st branchial arch as the latter is from the 2nd, in the wall of the pharynx immediately behind the disappearing spiracular pouch and in front of the hyomandibular branch of the facial nerve. It is clear, therefore, that their structure lies in the hyoid arch, but it will be termed the mucohyoid arch since it does *not* become the definitive hyoid arch skeleton of the adult. (It is Sewertzoff's 'hyoid arch', and Damas's 'velar process'.)

The next mucocartilaginous rod lies in the base of the velum, in front of the spiracular pouch and close to the mandibular branch of the trigeminal nerve. It will be called the mucomandibular arch (Damas's 'vestibular process.)' The foremost of the three rods lies in the side wall of the buccal cavity (Damas's 'vestibular process'). It will be called the mucopremandibular arch. At this stage all these mucocartilaginous rods end freely above and below.

(vii) In the fully formed Ammocoete (6–8 cm. long) (Plate 9, fig. 6) there has been a further development of both true cartilaginous and of mucocartilaginous structures.

As regards the elements of true cartilage, the anterior ends of the parachordals have joined one another in a transverse bar in front of the hypophysial sac. The parachordals thus enclose a fenestra into which the hypophysis grows *downwards* and backwards. This disposition in the fully formed Ammocoete is the result of the displacement of the hypophysial rudiment, itself the consequence of the enormous expansion of the upper lip which carries the combined hypophysial and nasal aperture to the dorsal surface of the head.

The sense-capsules are now chondrified. The nasal capsule is closely moulded round the conjoint nasal sacs and glands, and lies quite free from the anterior parachordals or any other cartilage or mucocartilage. Internally, the nasal capsule shows a feeble subdivision into right and left halves (de Beer, 1924*d*).

The auditory capsules are ovoid with an aperture on the median side for the entry of the auditory nerve, and attached by a very slender cartilaginous (anterior basicapsular) commissure to the lateral edge of the parachordal of their own side. It is probable, therefore, that the auditory capsules start by independent chondrification. Unlike those of Gnathostomes the auditory capsules of Petromyzon do not fit the contained auditory vesicle at all closely, and it is not possible to make out the shape of the vesicle and the configuration of its (two) semicircular canals from the outside. Posteriorly, a cartilaginous subchordal road on each side joins the dorsal ends of the branchial arches from the 2nd to the 7th. The hypotrematic process of each branchial arch (except the 1st) has become attached to the branchial arch next in front.

Among the mucocartilaginous elements, the chief novelty is the formation of a large ventral shield underlying the anterior part of the head, close to the skin. The ventral ends of the mucopremandibular, mucomandibular, and mucohyoid arches are attached to the inner surface of the ventral shield, which, posteriorly, comes into contact with the ventral portion of the 1st branchial arch. Anteriorly, the ventral shield is prolonged forwards in the midline into a labial lobe, dorsally to which is found a transverse bar.

The lining of the buccal cavity possesses on its floor a median longitudinal ridge, which is the posterior prolongation of the large central papilla on the ventral border of the mouth. This ridge contains a strip of mucocartilage, termed by Damas the 'ventral crest'. Anteriorly, this is attached on each side to the ventral shield and transverse bar. The central papilla also contains a core of mucocartilage.

The upper lip is protected by a small mucocartilaginous shield, the prenasal plate, the lateral edges of which are connected with the dorsal ends of the mucopremandibular arches. In the velum an additional bar of mucocartilage has appeared, situated medially to the dorsal end of the mucomandibular arch; this is Damas's internal velar bar, which appears to be attached to the end of the mandibular process of the parachordal. The dorsal end of the mucohyoid arch has forked, the anterior prong becoming attached to the mandibular process of the parachordal, and the posterior prong to the dorsal end of the 1st branchial arch. The facial nerve runs down ventrally and medially to this posterior prong which, in Sewertzoff's view, represents the subchordal rod in the region between the hyoid and 1st branchial arches.

ii. Metamorphosis

To the earlier workers it did not appear to be in doubt that the mucocartilaginous structures gave rise directly to the definitive cartilages found in the corresponding positions in the adult. It is important to realize, however, that following the work of Bujor (1881), Kaensche (1890), Tretjakoff (1929), Damas (1935), and unpublished observations of Seyd, this is not the case. The mucocartilage is a special provisional tissue, the histological characteristics of which have been described on page 38. It does not deserve the name of cartilage at all, disappears before or during metamorphosis, and the structures which it forms do not persist into the adult. The adult structures arise partly from the products of break-down of the mucocartilage in their place, partly alongside, and partly in completely new positions. Mucocartilage appears to be nothing but a particular kind of connective tissue, composed of elastic fibres and other anastomosing fibres in a homogeneous ground-substance, filling in spaces between other structures, and not bounded by definite borders or limiting membranes. It is presumably developed as a provisional stiffening for those parts of the structure of the Ammocoete which will undergo change or reduction in the course of metamorphosis. This fact is of great importance in view of the weight which has been attached to comparisons between certain mucocartilaginous structures of Petromyzon, such as the prenasal plate, and the cephalic shield of Cephalaspis.

The fate of the various mucocartilaginous structures, so far as it is known, appears to be as follows.

The ventral shield disappears as such, and its place is occupied by ordinary connective tissue. The disappearance seems to be brought about by invasion by blood sinuses, the activity of leucocytes, and the reversion of the elastoblasts and fibroblasts to ordinary connective-tissue cells, which multiply greatly. The prenasal plate is displaced forwards and gradually breaks down into ordinary connective tissue, penetrated by blood sinuses: it does *not* give rise to any cartilage in this region (posterodorsal or anterodorsal).

The mucopremandibular, mucomandibular, and mucohyoid arches are replaced by the definitive cartilaginous premandibular, mandibular, and hyoid arches which arise *medially* to them. This replacement is betrayed by the different relations which the hyomandibular branch of the facial nerve bears to the mucohyoid and hyoid arches.

The mucocartilaginous ventral crest and the core of the central papilla break down into connective tissue in which the cartilages of the rasping tongue then proceed to chondrify. These are the lingual cartilage which forms the shaft and the apical cartilage which forms the tip. Paired processes on the apical cartilage become detached and give rise to the styliform cartilages. Similarly, the mucocartilaginous transverse bar (situated dorsally to the anterior edge of the original ventral crest and ventrally to the front end of the ventral crest) breaks down and is replaced by the cartilaginous median ventral copula.

In addition to this complete disappearance of some parts of the mucocartilaginous skeleton and the replacement by true cartilage of other parts, the metamorphosis of the lamprey is characterized by the chondrification (*de novo* in connective tissue) of a number of pieces of cartilage which protect the roof and sides of the now enormous sucking mouth. These are the posterodorsal, anterodorsal, posterolateral, and anterolateral paired cartilages, and the annular cartilage. Cartilages also appear in the sides of the sucker (three pairs of 'dental' cartilages), and in the side wall of the brain-case, rising up from the anterior parachordal (see Plate 10).

iii. The morphology of the chondrocranium of Petromyzon

The base of the brain-case is formed of the parachordals, which lie on each side of the persistent and well-developed notochord, extend forward beyond the tip of the notochord, join one another, and enclose a fenestra through which the internal carotid arteries enter the cranial cavity, and the hypophysial duct passes downwards and backwards. The side wall of the brain-case is formed posteriorly by the auditory capsule which is fused along its whole length to the parachordal, and anteriorly by the orbital cartilage. This is attached posteriorly to the auditory capsule, and is connected with the anterior parachordal by two pillars. The posterior of these possibly represents the pila antotica, for it forms the anterior boundary to the foramen through which the trigeminal and facial nerves emerge from the skull. The anterior pillar possibly represents the preoptic root of the orbital cartilage, for it forms the anterior border to the optic foramen and the lateral border to the combined olfactory foramina.

The olfactory capsule lies above the anterior transverse connexion between the anterior parachordals, but is only attached to the brain-case by connective tissue. Externally, the olfactory capsule appears to be a median and unpaired spherical structure, but internally it possesses a sagittal cartilaginous partition which forms a partial division of the cavity into right and left portions.

Posteriorly, the skull ends with the auditory capsules, behind which the glossopharyngeal and vagus nerves emerge. The facial nerve seems to run through the foremost part of the auditory capsule and emerges through a foramen in its floor, but this appears to be because the front wall of the capsule contains the prefacial commissure. There are no cartilaginous septa separating the semicircular canals from the main capsular cavity. The roof of the skull is non-existent except for the tectum synoticum.

The brain-case and the visceral arch skeleton are fused together in many places. The dorsal end of the premandibular arch is attached to the anterior end of the (anterior) parachordal, and the dorsal end of the mandibular arch is attached to the mandibular process. A horizontal cartilaginous bar stretches between the premandibular and mandibular arches. This is the subocular bar, which the earlier authors homologized with the pterygoid process of the pterygoquadrate, attached to the brain-case as in Amphibia (see p. 188). But this subocular bar lies dorsally instead of ventrally to the maxillary and mandibular branches of the trigeminal nerve, and therefore cannot represent the pterygoquadrate. Sewertzoff regards it as the epitrematic process of the mandibular arch which has become attached to the premandibular arch, just as the epitrematic processes of the branchial arches become attached to the arch in front of them.

The mandibular arch extends down beneath the subocular bar as a vertical rod (Parker's 'epihyal') which ends ventrally in a hook-shaped process directed forwards (the 'cornual cartilage', Parker's 'ceratohyal') which forms a guide medially to which the lingual cartilage slides backwards and forwards. Sewertzoff considers this hook-shaped process to represent the hypotrematic process of the mandibular arch.

The hyoid arch (Parker's 'extrahyal') is attached to the dorsal end of the mandibular arch by a connexion which Sewertzoff considers to represent the epitrematic process of the hyoid arch. Ventrally, the hyoid arch is attached to the anterior end of the hypobranchial rod of its side, while half-way up, it effects a connexion with the hypotrematic process of the 1st branchial arch, immediately behind it.

The branchial arches preserve the same general relations as at previous stages, but become complicated by the formation of additional processes which fill up the gap between them and complete the formation of the branchial basket. Ventrally, the two hypobranchial rods are more or less fused in the midline.

It is to be noted that the branchial arches lie in a very superficial position in the side wall of the body, being situated externally to the branchial sacs, muscles, and dorsal and ventral aortae. For this reason it has been held that the branchial arches of Petromyzon are homologous with the extra-branchial cartilages and not with the branchial arches of Selachii (see p. 62). But as Goodrich (1909) has pointed out, the relations of the Cyclostome branchial basket to the branchial nerves are similar to those found in Selachii, and the superficial position of the branchial arches in Petromyzon is probably associated with the peculiar development of the branchial sacs in this form. It is, therefore, probable that the branchial arches of Cyclostomes and those of Gnathostomes are derived from some common ancestral form, though each type has become modified along its own line. In particular, the attachment of the arches in Petromyzon to the axial skeleton (to the notochord by the sub-chordal rod in the case of the branchial arches, and to the neurocranium in the case of the hyoid, mandibular, and premandibular arches) must be regarded as a specialization in view of the original separation which must have existed between the brain-case and the visceral arch skeleton.

The skeleton of the tongue (lingual and apical cartilages) was regarded by Huxley (1876*a*) and Parker (1883*a*) as the hyoid copula, or 'basihyal'. It is doubtful whether an exact homologization of this peculiar Cyclostome feature with structures in Gnathostomes is feasible. It may, however, be suggested that the tongue skeleton represents the hypobranchial rods corresponding to the premandibular, mandibular, and hyoid arches, fused together in the mid-line, and separated off from the wide posterior pair of hypobranchial rods in connexion with the acquisition of mobility. For the same reason the tongue is free from the premandibular, mandibular, and hyoid arches, which are free ventrally, though the mandibular arch comes in moveable contact with the lingual cartilage. The tongue is activated by muscles innervated by the mandibular branch of the trigeminal nerve, which suggests that the lingual cartilage forms part of the mandibular arch skeleton, as indeed it may if it represents the mandibular section of the hypobranchial rod. It can, however, be said to correspond only in a very general way with any part of the mandibular arch skeleton of Gnathostomes. The same applies to the median ventral copula of Petromyzon (Huxley's and Parker's 'basimandibular').

There remain in the roof and sides of the sucker of Petromyzon a number of cartilages which appear to be quite peculiar to Cyclostomes, and developed in relation to their very specialized method of feeding. These are: the posterodorsal cartilage, which is attached by its hind end to the anterior end of the anterior parachordal; the anterodorsal cartilage which is overlapped by the posterodorsal cartilage; the annular cartilage which surrounds the sucker; the posterolateral cartilage which is attached by its ventral end to the premandibular arch, and the anterolateral cartilage.

Sewertzoff regards the posterolateral and anterolateral cartilages as each representing a visceral arch. This view is, however, in conflict with all the evidence derived from the study of the segmentation of the head (see p. 20) which points clearly to the premandibular arch as the foremost visceral arch, since it is related to the premandibular segment which is clearly the foremost segment of the body. Further, it is to be noticed that although its skeleton is mucocartilaginous, the premandibular arch is the foremost arch in the Ammocoete, which in many respects retains more primitive features (e.g. ciliary method of feeding) than the somewhat degenerate adult.

It is, therefore, highly probable that the posterolateral and anterolateral cartilages (like the posterodorsal, anterodorsal and annular cartilages) are specialized structures, developed, as Balabai (1935) has pointed out, in connexion with the formation of the sucking mouth and rasping tongue, and devoid of phylogenetic significance. Holmgren (personal communication) has expressed the view that the anterolateral cartilage is formed from the premandibular arch: this is at variance with the observations of Seyd and of the present writer who find (as stated above) that the anterolateral cartilage arises *de novo*.

Two features of the chondrocranium of Petromyzon are of particular morphological significance. In the first place, it is to be noticed that it is the shortest of all skulls, occupying only four or five segments of the body. It has no occipital arch. The occipital region of the skull of Gnathostomes is thus represented in Petromyzon by some of the separate neural arches which flank the notochord behind the skull.

The other feature refers to the apparent absence from the skull of Petromyzon of the elements found in all Gnathostomes under the name of trabeculae. The base of the skull in Petromyzon appears to be solely parachordal in origin, although the anterior parachordals eventually project some way in front of the tip of the notochord. The chief evidence for this is the fact that at early stages the so-called anterior parachordals lie entirely behind the tip of the notochord, they flank the notochord closely, and extend back between the auditory vesicles. Such a position is unknown for a trabecula. The question therefore arises whether trabeculae are represented in the Cyclostome skull, and, if so, where.

Sewertzoff (1917), by following the displacement which is undergone (owing to the formation of the huge upper lip) by that region of the head which first lies just above the mouth, flanking the hypophysis (where trabeculae arise in Gnathostomes), concludes that the posterodorsal cartilage represents the trabeculae, and in this Sewertzoff is followed by Balabai (1935). The posterodorsal cartilage is attached to the anterior end of the anterior parachordals. But if the displacement of this region of the head is followed carefully, it will be seen that the *posterior* part of the posterodorsal cartilage represents the *anterior* end of the original region where trabeculae are found in Gnathostomes. The relations are thus reversed, and the attachment of the posterodorsal cartilage to the parachordals cannot be held to support the view that it represents the trabeculae. In addition, there is the objection that the posterodorsal cartilage is median and unpaired, whereas the trabeculae are always of paired origin, even if they fuse later.

The attempt has been made by de Beer (1931*b*) to show that the most likely claimants in Petromyzon to represent the trabeculae are the premandibular arches. This view, which is an outcome of Huxley's conclusion that the trabeculae are visceral structures, is reserved for discussion below (p. 375) after consideration of the relations of the trabeculae in Gnathostomes.

5. CHONDRICHTHYES. SELACHII

Among the workers who have investigated the development of the Selachian skull, priority goes to W. K. Parker (1879*b*). Next come Sewertzoff (1897, 1899), van Wijhe (1904*a*, 1922), Mori (1924), Matveiev (1925), and de Beer, (1924*b*, 1931*a*). The development of the visceral arches has been studied by Dohrn (1884*a*), Gibian (1913), and Sewertzoff (1927). While many of these authors have investigated more than one form, it will be convenient to take the types separately.

I. Squalus acanthias

The early development of the skull of Squalus has been studied by means of graphic reconstructions by Sewertzoff (1897, 1899), and with the help of whole preparations and models by van Wijhe (1904*a*, 1922): later stages have been investigated by Mori. The following account is based on the descriptions of these workers.

The degree of development in embryos of similar length varies according to their provenance. Those from Heligoland are farther advanced than those from Helder, while material from Boston, Mass., is still more retarded in chondrification.

i. The development of the chondrocranium of Squalus acanthias.

Stage 1. (23 mm. Heligoland, van Wijhe.) PLATE 11, FIGS. 1, 2.

The first chondrifications are the paired parachordals which reach forward to a point level with the root of the abducens nerve, but a long way behind the tip of the notochord, which touches the commissural strand between the two premandibular somites. In the region between the auditory vesicles, the lateral edges of the parachordals extend farther to the side than elsewhere, forming the laminae basioticae, which begin to push underneath the auditory sacs.

Posteriorly, each parachordal extends back to a point level with the ventral root supplying the myomere of the 9th segment (de Beer's enumeration), and is in contact with the notochord all the way. It is prolonged backwards as a thin sheet of cartilage level with the dorsal surface of the notochord to a point beneath the exit of the 1st spinal (mixed) nerve.

The dorsal surface of the cartilage in the hinder part of the parachordal is as flat as elsewhere. Sewertzoff reported indications of segmentation in the form of rudimentary preoccipital arches; these must, however, be only mesenchymatous and transient for they were not found by van Wijhe.

STAGE 2. (28 mm. Heligoland, van Wijhe.) PLATE 11, FIGS. 3, 4.

The auditory capsule has two independent centres of chondrification—the anterolateral and the posterior cartilages. At this stage the former has fused on to the lamina basiotica, thus forming an anterior basicapsular commissure.

Posteriorly, each parachordal sends back along the notochord two delicate bands of cartilage in which segmental swellings indicate the rudimentary basidorsals and basiventrals. Interdorsals are already present, corresponding to the first four vertebrae.

The occipital arch is of great interest: it is made up of an anterior portion which is in continuity with the parachordal, and a posterior portion which is free from the parachordal. The two elements appear to correspond respectively to a neural arch (basidorsal) and intercalary (interdorsal). At this stage the last cranial hypoglossal root is still free in front of the occipital arch.

Anteriorly, the orbital cartilage and pila antotica appear as separate cartilages, which soon fuse (forming van Wijhe's pleurosphenoid cartilage) to form a plate which is, however, still free from the anterior end of the parachordal. The pila antotica and orbital cartilages arise in mesenchyme derived from lateral proliferations of the commissures between the premandibular somites.

Ventrally to the anterior end of the parachordal there appears on each side a polar cartilage and a trabecula: all separate from one another.

The external rectus muscle is attached to the parachordal and runs forwards over the polar cartilage, without, however, being attached to it.

The pterygoquadrate is now chondrified as in Meckel's cartilage, but the latter has two centres of origin on each side, one anterior and the other posterior. The anterior end of the pterygoquadrate at this stage is the basal process which is attached by mesenchymatous tissue to the lateral edge of the trabecula of its side. (The basal process has usually been termed orbital process by previous authors.) In the hyoid arch the hyomandibula chondrifies progressively from ventral to dorsal end, and thus grows upwards towards the anterolateral cartilage of the auditory capsule. The ceratohyals are present, and the basihyal.

STAGE 3. (39·5 mm. Heligoland, van Wijhe.) PLATE 11, FIGS. 5, 6, 7, 8.

The separate cartilages of previous stages have now joined up: the posterior cartilage of the auditory capsule on to the anterolateral and on to the parachordal; the trabeculae with one another anteriorly, with the polar cartilages posteriorly, and these in turn with the ventral surface of the parachordals at their anterior end; the pilae antoticae are fused on to the dorsal surface of these same ends of the parachordals; Meckel's cartilage is now in one piece.

The trabeculae and polar cartilages now form cartilaginous bars which meet the parachordals almost at right angles.

The parachordals are separated from one another by the notochord, but in the hinder part of their length the parachordals are continuous with cartilage cells which have invaded the fibrous sheath of the notochord, thus forming a cylindrical cartilaginous commissure between the parachordals. This cylinder is continued backwards at least as far as the 8th vertebra. Outside the notochordal sheath the two bands of cartilage of the previous stage fuse into one band on each side, bearing a number of transverse ridges each of which represents a basidorsal fused to a basiventral. Later on, with the swelling and stretching of the notochord, these elements become separate. At this stage, however, there is no cranio-vertebral joint at all. The invasion of the notochordal sheath, which is similar to that which takes place in the formation of centra of the vertebral column, reaches no farther forward than the auditory region. The elastica externa of the notochord in the region of the parachordals persists for some time but eventually disappears.

The occipital arch is a unified structure, pierced at its base by a foramen for the last hypoglossal root. The remaining hypoglossal roots anterior to this, and which may be as many as four in number, emerge through the fissura metotica which is bordered medially by the parachordal, laterally by the auditory capsule, and posteriorly by the occipital arch. Soon, the penultimate hypoglossal root becomes enclosed in a foramen as the cartilage of the occipital arch grows forward and envelops it. There are now, therefore, two hypoglossal foramina on each side. Through the fissura metotica also emerge the vagus nerve, and the glossopharyngeal which later on runs beneath the ventral edge of the auditory capsule.

The posterior surface of the occipital arch bears a notch which lodges the 1st spinal ventral root, thus suggesting that an additional vertebral centrum has become fused on to the hind end of the parachordals. There is apparently no (basidorsal) neural arch formed corresponding to this vertebra. At all events, the next arch immediately behind the occipital arch (which contains an intercalary) is an intercalary.

The medial wall of the auditory capsule is foreshadowed by an upgrowing dorsal process on the lamina basiotica. Situated behind the entry of the auditory nerve into the cavity of the capsule, and in front of the exit of the glossopharyngeal, the capsule now has septa for the semicircular canals.

The original transverse commissure between the premandibular somites has given rise to a transverse bolster of connective tissue, the acrochordal, which, however, does not chondrify. At first at the tip of the notochord this acrochordal slides back along it until it touches the anterior end of the parachordals, or basal plate which the parachordals constitute.

The attachment of the pila antotica to the parachordal forms the anterior margin to the incisura prootica, through which the trigeminal, abducens, and facial nerves emerge. The trochlear nerve pierces a foramen in the orbital cartilage.

The attachment of the trabeculae and polar cartilages to the parachordals encloses the hypophysial fenestra, through which the vestiges of the hypophysial stalk pass down to the epithelium of the stomodaeum, and, more posteriorly, the internal carotid arteries enter the cranial cavity. The portion of the hypophysial fenestra through which the internal carotids enter becomes separated off as a special foramen by a transverse bar of cartilage forming a precarotid commissure. Posterodorsally, the ventral surface of the parachordals is excluded from bordering the carotid foramen by the formation of another transverse bar of cartilage, derived by ingrowth from each polar cartilage towards the middle line, which gives rise to the dorsum sellae, or crista sellaris.

Anterolaterally, each trabecula gives off a process, the lamina orbitonasalis (ethmoid process of Sewertzoff), which is still free from the anterior end of the orbital cartilage. The side wall of the skull in this region therefore presents an incisure, bordered dorsally by the orbital cartilage, posteriorly by the pila antotica, ventrally by the polar cartilage and trabecula, and anteriorly (incompletely) by the lamina orbitonasalis. Through this incisure the oculomotor and optic nerves leave the skull, and the pituitary vein and efferent pseudobranchial artery enter it.

Anteriorly, the trabecular plate gives off the median rostrum, between which and the lamina orbitonasalis on each side is a wide incisure through which the olfactory nerve reaches the nasal sac.

In the visceral arch skeleton the anterior end of the pterygoquadrate has grown forward beyond the basal process, which latter is movably articulated on the lateral surface of the trabecula. According to Sewertzoff and Disler (1924), a separate nodule of cartilage, which they regard as a pharyngomandibular element, becomes fused on to the medial side of the basal process.

Epibranchials and ceratobranchials are present in each branchial arch (those of the first four arches perforated by holes formed in connexion with the differentiation of the adductor muscles). As Sewertzoff (1923, 1927) has shown, the muscle first lies in a notch at the ventral end of the epibranchial and at the dorsal end of the ceratobranchial. These notches subsequently become enclosed by cartilage, and thus give rise to the holes in question. Pharyngobranchials are present in the first four arches. It is important to notice that these arches are of the sigma or Σ-type, for the angle between the pharyngobranchial and the epibranchial points *forwards*, as does that between the ceratobranchial and the hypobranchial when the latter has appeared. On the other hand, the angle between the epibranchial and the ceratobranchial points backwards.

All the cartilages of the branchial arches (with the exception of the epi- and ceratobranchial of the 5th) chondrify from independent centres as Braus (1905) and van Wijhe (1904*a*) showed, and do not result from the fragmentation of continuous cartilaginous bars as Dohrn (1884*a*) suggested.

Stage 4. (44 mm. Mori.) PLATE 12, FIG. 1.

The chief progress made at this stage concerns the enclosure of the orbital incisure by the attachment of the anterior end of the orbital cartilage to the trabecula of its side by means of the preoptic root of the orbital cartilage. The optic and oculomotor nerves, the efferent pseudobranchial artery and the pituitary vein now each have separate foramina, the latter leading into the cartilaginous interorbital canal, which runs just beneath the dorsum sellae, behind the internal carotid foramen.

The lateral wall of the skull in the orbital region bears the eyestalk, which is inserted near the oculomotor foramen and abuts against the eyeball, round which a sclerotic cartilage is now formed. The sclerotic cartilage may be regarded as a movable sense-capsule, comparable with the nasal and auditory capsules. The eyestalk projects out between the profundus and maxillary branches of the trigeminal nerves. Between the preoptic root of the orbital cartilage and the lamina orbitonasalis is a gap which marks the distinction between the true side wall of the cranial cavity (preoptic root of the orbital cartilage) and the hind wall of the nasal capsule (lamina orbitonasalis) which is not strictly a part of the wall of the brain-case. The dorsal portion of this gap has become narrowed down to a foramen orbitonasale through which the profundus nerve leaves the orbit. The ventral portion of the gap remains for a time as a vacuity in the cartilage. It seems that a small part of it must persist as the orbitonasal canal conveying a branch of the facial vein out of the orbit. It is, however, not figured by Mori or Sewertzoff at later stages, although it is present in the adult.

The foramen prooticum is in the process of subdivision by the prefacial commissure.

In the visceral arch skeleton the spiracular cartilage is present. It arises in connexion with the posterior part of the pterygoquadrate, of which it represents the otic process found in other forms. The basibranchials (copulae) arise according to Gibian (1913) from tissue which appears to be derived from the (mesenchymatous) ventral ends of the hypobranchials.

The labial cartilages, consisting of two dorsal and one ventral element on each side, are now in process of chondrification.

Stage 5. (50 mm. Sewertzoff.) PLATE 12, FIG. 2.

The skull is now taking on the appearance of that of the adult, and the angle between the parachordals and the trabeculae is becoming smoothed out by the upward curvature of the latter at the front end.

In the auditory region the glossopharyngeal nerve is separated from the vagus by becoming enclosed in the hinder part of the auditory capsule. The prefacial commissure separates the facial and trigeminal nerves.

Anteriorly, a roof and a lateral wall are present round the nasal capsules, for the base of the rostrum forms an internasal septum or plate the dorsal edge of which is in cartilaginous continuity with the lamina orbitonasalis on each side. The result is the formation of a pair of short cylinders, the posterior apertures of which are the large foramina olfactoria through which the olfactory nerves reach the nasal sacs. The anterior or external narial apertures are widely open between the internasal plate and the ventral edges of the lateral wall and the front end of the roof of each capsule.

Stage 6. (90 mm. Mori.) PLATE 12, FIG. 3.

The chief interest of this stage concerns the anterior region of the skull. A supraorbital cartilage is attached to the dorsal edge of the orbital cartilage, the line of attachment being indicated by a row of foramina (supraorbital foramina) through which twigs from the superficial ophthalmic nerve supply the sense-organs of the supraorbital lateral line canal. Anterolaterally, the supraorbital cartilage runs on to the dorsal edge of the lamina orbitonasalis, leaving, however, a gap—the preorbital canal—through which the superficial ophthalmic nerve itself leaves the orbit and emerges on to the dorsal surface of the roof of the nasal capsule. From the preorbital canal the ridge of the supraorbital cartilage is prolonged downwards over the lateral surface of the lamina orbitonasalis to the anteroventral corner of the orbit forming the preorbital ridge. The lateral wall of the nasal capsule is that part of the lamina orbitonasalis that lies in front of this ridge: the remaining (hinder) part of the lamina orbitonasalis forms the hind wall of the nasal capsule. Ventral and anterior to the preorbital canal is the orbitonasal foramen through which the profundus nerve leaves the orbit, and finds itself beneath the cartilage of the roof of the nasal capsule, but outside the membrane (dura mater) covering the olfactory bulb. The profundus nerve does not, therefore, enter the cranial cavity and it soon emerges again on to the dorsal surface of the roof of the nasal capsule by passing through the epiphanial foramen.

The orbital region of the skull is now roofed over, leaving anteriorly a fenestra precerebralis, from the hinder part of which a small epiphysial fenestra is separated off. The rostrum has elongated into a keel-like structure, and its base—the internasal plate—no longer shares in the formation of the medial wall of the nasal capsule, for a new medial wall to each capsule has appeared on each side of the septum. This new medial wall is a curved plate of cartilage which arises from the internasal plate and runs outwards and downwards, beneath the olfactory bulb to join the lateral wall of the nasal capsule forming a rudimentary floor to the hinder part of the nasal capsule.

At its anteroventral corner, the medial wall gives off a curved process which helps to form the ventral border of the external narial aperture, and appears to correspond to the cornu trabeculae of other forms. In this way, the original wide foramen olfactorium is divided (by the new medial wall of the capsule) into two: a laterally situated definitive foramen olfactorium leading from the cranial cavity into that of the nasal capsule, and a medially situated aperture between the internasal plate and the medial wall of the nasal capsule, corresponding to Gegenbaur's basal canal.

The external or narial aperture (fenestra narina) of the nasal capsule is irregular in shape and has on its ventral border a deep notch, situated between the cornu trabeculae on the medial side, and the lateral wall of the capsule; this notch, or subnasal incisure, is of interest since it probably corresponds to the fenestra choanalis of Tetrapoda, as Allis (1913) has argued.

In addition, however, to the free anterior edges of the medial and lateral walls and roof of the nasal capsule, the fenestra narina is bordered by two nasal cartilages. Of these, one is shaped very roughly like the figure 3, the anterior half-hoop of the '3' coming almost to surround the incurrent aperture into the nasal sac, while the other half-hoop almost surrounds the excurrent aperture. The other nasal cartilage is a small strut which lies along the lateral edge of the medial wall of the capsule.

The lateral edge of the trabeculae is produced to form a so-called subocular shelf, which extends posteriorly (as the projecting lateral edge of the parachordal) as far as the auditory capsule. A foramen in this subocular shelf serves for the passage of the orbital artery, and suggests that the subocular shelf is really a separate element attached to the edge of trabecula and parachordal.

ii. The morphology of the chondrocranium of Squalus acanthias

Contrary to the general principle of this work, which is to avoid general descriptions of the adult skull, that of Squalus must be dealt with since the fully formed chondrocranium, to be compared with that of other forms, is that of the adult. To various aspects of its morphology, in addition to the classical work of Gegenbaur (1872), certain works of Allis have been devoted (1913, 1923*a*), while a general description has been given by Wells (1917). Variations in the visceral arch skeleton have been studied by van Deinse (1916).

The relations of the glossopharyngeal nerve to the auditory capsule and of the endolymphatic duct to the parietal fossa will be described in connexion with Scyllium (p. 63). Here, attention will be paid to certain morphological considerations connected with the nasal capsule, the so-called acustico-trigemino-facialis recess, the subpituitary space, and the cranio-vertebral joint.

The nasal capsule has been treated by Allis (1913, 1923*a*), who showed that the large fontanelle in the anterior part of the dorsal surface of the skull (called by Gegenbaur the prefrontal fontanelle) is not a part of the roof but of the anterior wall of the cranial cavity, homologous with the fenestra precerebralis of Urodela, for which reason it is here given the same name.

The fenestra precerebralis is occluded by the membranous dura mater, which thus forms the morphological anterior wall of the cranial cavity. In front of this the cavum precerebrale is the cavity lodged on the dorsal side of the spout-shaped rostral cartilage of Squalus, or enclosed between the rostral processes of Scyllium.

But Allis's chief contributions to the morphology of the Selachian skull have been the recognition of the existence within the cartilaginous skull-walls of certain spaces which lie externally to the dura mater or limiting membrane of the cranial cavity, and which are therefore strictly not intracranial but rather intramural, since these spaces would be obliterated if chondrification spread to the dura mater from the existing cartilaginous wall in these regions.

1. The ectethmoid chamber. This term has been given to a space situated beneath the membranous lining to the nasal sac and above the cartilaginous floor of the hinder part of the capsule. Posteriorly, the ectethmoid chamber is continuous with a canal—the orbitonasal canal—which, passing between the external surface of the preoptic root of the orbital cartilage and the medial surface of the lamina orbitonasalis, communicates with the orbit. Through the orbitonasal canal a branch of the facial vein from the orbital sinus enters the ectethmoid chamber. (Gegenbaur describes the profundus nerve as passing through the same canal, but here there has been confusion; the foramen for the profundus nerve, though part of the same original orbitonasal fissure between the lamina orbitonasalis and the preoptic root of the orbital cartilage, is more dorsal, and is here given the name of orbitonasal foramen.)

From the ectethmoid chamber the vein emerges through the subnasal incisure. In Chlamydoselachus this subnasal incisure is converted into a closed nasal fontanelle by cartilaginous fusion between the cornu trabeculae to the anteroventral corner of the lateral wall of the nasal capsule, forming a subnasal plate or lamina transversalis anterior. The nasal fontanelle corresponds in position to the fenestra choanalis of Tetrapods.

The existence of the ectethmoid chamber and of the orbitonasal foramen is evidence of the fact that the lamina orbitonasalis is not an original part of the side wall of the cranial cavity, but is the original posterolateral wall of the nasal capsule, which has secondarily come into relation with the wall of the cranial cavity. Thus, the passage of the profundus nerve and of a branch of the facial vein through the cartilage and apparently into the cavity of the nasal capsule is not to be mistaken for an entry of these structures into the cranial cavity.

For the sake of clearness it should be added that the nasal fontanelle of Chlamydoselachus, or the subnasal incisure of Squalus, is quite different from the basal canal, which is the medial portion of the original olfactory foramen, cut off from the definitive foramen by the formation of the medial wall of the nasal capsule.

2. The acustico-trigemino-facialis recess. The recognition of the morphological significance of this recess is due to Allis (1914*a*), who showed that the foramina in the cartilaginous wall through which the trigeminal, facial, and auditory nerves pass (the first two into the orbit: the last into the auditory capsule) do not lead directly out of the cranial cavity but out of a chamber which is separated from the true cranial cavity medial to it by the membranous wall formed of the dura mater, pierced by the above-mentioned nerves. There is thus a recess in the side wall of the skull, the acustico-trigemino-facialis recess, lodging the proximal portions of the trigeminal, facial, and auditory nerve ganglia in a space which may be called intramural.

The trigeminal foramen is close to that of the pituitary vein, the two often being sunk in a common depression (Allis's trigemino-pituitary fossa) in the hinder part of the orbit, and in which the external rectus muscle commonly takes its origin.

In Squalus (especially in old specimens) the facial foramen leading out of the recess is prolonged into a canal which forks and has divided external openings: the anterior for the palatine branch and the posterior for the hyomandibular branch of the nerve.

The head vein (vena capitis lateralis) runs back in this region from the orbital sinus to the anterior jugular, and passes ventromedially to the trigeminal nerve and dorsolaterally to the facial nerve. In Squalus, however, in the region immediately opposite the exits of the acustico-trigemino-facialis recess, the head vein is for a short distance enclosed in a cartilaginous tunnel or jugular canal, the medial wall of which is simply the prefacial commissure, while the lateral wall (which corresponds to the lateral commissure of bony fish) is formed by the chondrification of a membrane which extends from the edge of the subocular shelf to the lateral wall of the auditory capsule. It is through a portion of this shelf that the orbital artery pierces a foramen on its course from the internal carotid to the orbit.

The importance of the relations of the acustico-trigemino-facialis recess and jugular canal in Selachii resides in the fact that these structures, together with the subpituitary space, play a leading part in the formation of the posterior myodome and trigemino-facialis chamber of bony fishes, where the morphological relations are, however, not so clear. It will be seen that the trigemino-facialis recess corresponds to the pars ganglionaris of the trigemino-facialis chamber, while the jugular canal corresponds to its pars jugularis. (See p. 429).

3. The subpituitary space. The floor of the skull in the hinder part of the trabecular region forms a depression, the pituitary fossa, in which the pituitary body is lodged. The pituitary body is surrounded by the dura mater which also forms the perichondral membrane on the upper surface of the cartilaginous floor. However, in the hindmost part of the pituitary fossa, beneath the dorsum sellae and between the regions originally formed from the polar cartilages, the dura mater is not closely adpressed to the cartilage, leaving a subdural space which Allis (1914*a*, 1919*a*, 1928) has described as the subpituitary space. It is through this space that the pituitary vein passes, from one orbit to the other (via the interorbital canals); and it is likewise into this space that the internal carotid arteries enter (passing up between the trabeculae) and join the efferent pseudobranchial arteries (which have entered by passing inwards dorsally to the trabeculae), before penetrating through the membranous lining of the pituitary fossa and thus entering the true cranial cavity.

In connexion with Allis's view that the trabeculae are of visceral, extracranial origin, he regards the subpituitary space as extracranial. (See p. 428).

4. The cranio-vertebral limit. From the foregoing description it will be clear that (according to van Wijhe), at early stages, there is cartilaginous fusion between the hind ends of the parachordals and the basal elements of the anterior vertebrae, by means of continuous bands of cartilage stretching back laterally to the notochordal sheath, and there is also a continuous tubular cartilage, the result of invasion of the fibrous sheath, reaching back to the 8th vertebra. Subsequently, the different centra become separated from one another, and the centrum of the foremost free vertebra becomes separated from that which is absorbed into the hindmost part of the skull. This latter centrum develops no neural arch, with the result that the first arch behind the occipital arch is an intercalary (interdorsal). The conditions here are therefore paralleled by those found in certain Amniotes where the 1st arch behind the occipital is the proatlas interdorsal arch. But whereas in the Amniote the centrum corresponding to this proatlas vertebra is a pleurocentrum (intercalary) which may (Aves) or may not (Lacertilia, Mammalia) fuse on to the hind end of the skull, in Selachii the proatlas centrum is a hypocentrum (basal) which fuses on to the hind end of the parachordal.

Whatever the composition of the Selachian centrum may be (i.e. whether it arises solely from the posterior sclerotomite or half-sclerotome of a segment or not), the limit between the skull and the 1st vertebra is an intervertebral one. The skull has well-marked paired condyles, formed from the extended basiventrals (haemal arches) of the absorbed centrum, which articulate with the corresponding structures of the 1st free vertebra. The persistence of the intervertebral notochord, between the skull (absorbed centrum) and the 1st vertebra, maintains the attachment of the skull to the vertebral column and reduces the flexibility of the cranio-vertebral joint to a small amount, which is no greater than that found between adjacent vertebrae of the vertebral column. There is thus here no specialized occipito-atlantal articulation, such as is found in rays, Holocephali and Tetrapoda.

5. The notochord. The fate of the notochord in the region of the skull varies in different Selachii. The cartilage formed in the invaded part of the sheath becomes indistinguishable from the cartilage of the neighbouring parachordal after disappearance of the elastica externa. The notochord itself persists as a slender strand embedded in the basal plate in Squalus; in Mustelus, on the other hand, the notochord is said to have disappeared in the adult. The parachordals are interconnected across the midline not only by the cartilage in the invaded notochordal sheath but also by chondrification above and below the sheath.

II. Scyllium (Scyliorhinus) canicula

The development of the skull of the common dogfish was described by Parker (1879*b*). More recently, certain aspects have been investigated by van Wijhe (1904*a*) and Goodrich (1918), and a complete study was made by de Beer (1931*a*), whose work is the basis of the following account.

It was found that embryos of Scyllium from the English Channel showed a different degree of development from embryos of the same length from the Mediterranean, in consequence of which the provenance of the material is added to the length of the embryos at the various stages.

i. The development of the chondrocranium of Scyllium canicula

Stage 1. (24 mm. Plymouth.) PLATE 13, FIG. 1.

The first appearance of chondrification results in the formation of the parachordals, which extend in contact with the notochord from a point just behind the front wall of the auditory vesicle to a point overlying the 1st gill-slit. Goodrich (1918) has shown that the sheets of mesenchyme from which the parachordals arise are derived from the sclerotomes of segments 3 and 4, and possibly also 1 and 2. Later on, the parachordals are extended backwards by further sclerotomal additions.

Stage 2. (29 mm. Naples.) PLATE 13, FIG. 2.

The lateral edge of the anterior half of the parachordal extends sideways to form a lamina basiotica which, anteriorly, is continuous (by means of the anterior basicapsular commissure) with the cartilage of the auditory capsule which starts beneath the anterior ampulla. Here, then, the auditory capsule appears to chondrify in continuity with the parachordal.

The hind corner of the lamina basiotica is marked by the rudiment of the dorsal process which will form part of the medial wall of the auditory capsule.

In front of the anterior basicapsular commissure the cartilage of the parachordals extends a short way farther forward but is here no longer in contact with the notochord. The space between the front ends of the parachordals and the notochord will ultimately become the basicranial fenestra.

Beneath the anterior tips of the parachordals and in continuity with them a pair of cartilaginous nodules are present, which represent the polar cartilages.

In the visceral arches the pterygoquadrate, Meckel's cartilage, the hyomandibula, and ceratohyal have appeared as independent chondrifications.

The vertebral column is of interest, for it shows segmental and independent basidorsal cartilages. There is still, however, some doubt about them, for van Wijhe (1922) claims to have found continuous cartilaginous bands involving the basidorsals in Scyllium as well as in Squalus. Further work is needed on this point.

STAGE 3. (29·5 mm. Naples.) PLATE 13, FIG. 3.

The trabeculae have now appeared. They arise, as de Beer (1931*b*) showed, as condensations of mesenchyme in the maxillary process, on each side of the hypophysis, and their rudiments are associated with those of the pterygoid process of the pterygoquadrates. At this stage the trabeculae are still free from the polar cartilages, and make an angle with the parachordals less acute than that seen in Squalus.

The orbital cartilages are also present and independent, being still unattached by a pila antotica to the parachordal. The parachordal now extends back to a point behind the level of the 2nd gill-slit, and this posterior extension is due to sclerotomal contribution which Goodrich has shown to emanate from the 5th to 8th segments inclusive. The rudiment of the occipital arch is visible.

Two neural arches have appeared, separate and independent of their respective basidorsals.

Epibranchial and ceratobranchial cartilages have chondrified independently in the 1st and 2nd branchial arches. They are U-shaped, the troughs of the U's being occupied by the ends of the adductor muscles.

STAGE 4. (30 mm. Naples.) PLATE 13, FIG. 4.

The orbital cartilages are attached to the foremost point of the parachordals of their own side by a pila antotica, and the trabeculae are attached to the polar cartilages. The lateral wall of the auditory capsule is now chondrified, but an independent origin of the posterior portion has not been observed. Between the auditory capsule and the lateral edge of the parachordal is a slit, the basicapsular fenestra, which is continuous posteriorly with the fissura metotica, bounded behind by the occipital arch and through which the glossopharyngeal and vagus nerves emerge. The dorsal process of the parachordal marks the approximate line of demarcation between the morphological constituents of the gap.

The base of the occipital arch is pierced by a hypoglossal foramen. Van Wijhe (1904*a*) considered that this condition indicated that the occipital arch represented a single neural arch, pierced by a ventral nerve-root foramen as are the vertebral neural arches of Squalus, but not of Scyllium. Goodrich (1918), on the other hand, has shown that it is much more probable that the occipital arch at this stage represents two arches, the occipital and the preoccipital, related to the 7th and 6th sclerotomes respectively, and enclosing a ventral nerve-root between them. Later on, another preoccipital arch is indicated, farther forward (see Plate 16, fig. 2).

In the visceral arch skeleton, pharyngobranchials are present in the first three branchial arches, while in the vertebral column, basiventral cartilages have appeared, as independent chondrifications.

STAGE 5. (30 mm. Plymouth.) PLATE 13, FIGS. 5, 6. PLATE 15, FIGS. 1, 2.

Anteriorly, the trabeculae converge towards, but do not touch, an isolated rostral cartilage which runs forwards between the nasal sacs. In front of the nasal sacs a pair of cartilages appear representing the isolated cornua trabecularum, for these cartilages, the rostrum and the trabeculae are but separate chondrifications in what is really (and will soon become) a continuous structure.

Behind the cornua trabecularum are the nasal cartilages, each shaped like the figure 6, with the ring wrapped round the inhalant nasal aperture.

Anterolaterally, each trabecula gives off a cartilaginous process which is the rudiment of the lamina orbitonasalis.

Although the polar cartilage, attached as it is to trabecula and parachordal, has lost its independence, its position is clearly marked by notches on the dorsal surface of the now continuous trabeculo-polar bar. The anterior notch lodges the efferent pseudobranchial artery, while the posterior notch lodges the pituitary vein. Further, each polar cartilage is connected with its fellow of the opposite side by a transverse cartilaginous postpituitary commissure, which is situated beneath the level of the parachordals, and behind the internal carotid arteries.

The orbital cartilage has extended and lapped round the trochlear nerve, which is thus enclosed in a foramen. From the dorsoposterior corner of the orbital cartilage a process is directed backwards towards the auditory capsule, and will ultimately convert the incisura prootica into the foramen prooticum.

The auditory capsule is now fairly well developed and possesses three septa for the semicircular canals. In its hindmost region, however, the cavity of the auditory capsule has only a membranous floor, beneath which is a cartilaginous lateral extension from the parachordal, the lamina hypotica. Between this lamina hypotica and the membranous floor of the capsule, the glossopharyngeal nerve runs out, and as the capsule subsequently becomes attached to the dorsal surface of the lamina hypotica, the glossopharyngeal nerve appears to run through the capsule. It is, however, excluded from the true cavity of the capsule by the membranous floor mentioned above. One result of this development is that the glossopharyngeal is thus separated from the vagus, which alone emerges through the fissura metotica.

The notochordal sheath has now begun to be invaded by the cartilage cells of the parachordals in its hindmost part. This is the origin of the chordal commissure which interconnects the hitherto separate parachordals. This invasion of the fibrous sheath appears to give rise to a continuous cylinder of chordal cartilage (see Plate 16, fig. 2), which will ultimately become segmented. The parachordals now constitute the basal plate.

In the vertebral column interdorsal (intercalary) cartilages have appeared, the first being behind the 1st basidorsal neural arch. Here then, unlike Squalus, the 1st arch behind the occipital arch is a basal arch, and not an intercalary.

In the visceral arch skeleton the pterygoquadrate shows a thickening on its dorsal surface near the anterior end, which foreshadows the basal process. The hyomandibula has extended dorsally and abuts against the lateral wall of the auditory capsule. Ceratobranchial cartilages are present in the 4th and 5th branchial arches.

STAGE 6. (36 mm. Plymouth.) PLATE 13, FIGS. 7, 8. PLATE 14, FIG. 4.

The median rostral cartilage has fused with the trabeculae on each side behind, forming the trabecular plate and thus enclosing the hypophysial fenestra. The cornua trabecularum have become attached to the rostral cartilage on each side, and are continuous with a thin film of cartilage which passes over the nasal sac and joins the lamina orbitonasalis, thus forming the rudiment of the front wall and roof of the nasal capsule.

Each capsule is now a very short cylinder, the hind aperture being the large olfactory foramen through which the olfactory nerve reaches the nasal sac. From the ventro-posterior corner of each olfactory foramen a small foramen is marked off by a bar of cartilage which runs from the rostrum (just behind the attachment of the cornu trabeculae) to the anteroventral edge of the lamina orbitonasalis. The foramen so formed is the basal canal, while the bar of cartilage forms a rudimentary floor to the hind part of the nasal capsule.

Behind the lamina orbitonasalis (though largely in continuity with it) the preoptic root of the orbital cartilage has risen up from the trabecula. The position of a foramen (orbitonasal canal, through which a vein leaves the orbit) marks the line of distinction between the lamina orbitonasalis (side and hind wall of the nasal capsule) and the preoptic root of the orbital cartilage (side wall of the cranial cavity). Dorsally, the orbital cartilage has almost but not quite met the lamina orbitonasalis.

The foramen prooticum is divided up. The superficial ophthalmic branches of the trigeminal and facial nerves have separate foramina. The oculomotor is also enclosed in a foramen, in front of the pila antotica.

A number of important structures have appeared in the hinder part of the hypophysial fenestra. One of these is the precarotid commissure, which encloses the internal carotid foramen between itself and the postpituitary commissure. Then, from the region of each polar cartilage, a cartilaginous process grows inwards towards the tip of the notochord. When they meet, these processes will form the dorsum sellae, which is the posterior edge of the pituitary fossa and the anterior boundary of the basicranial fenestra. The relation of these structures is seen in Plate 16, figs. 3 and 4.

The supraorbital cartilage appears first in the posterior region, apposed to the dorsolateral wall of the auditory capsule, where it forms a half-ring enclosing the infraorbital lateral line canal. The subocular shelf, or lateral extension of the trabeculae and anterior basal plate, arises from two centres, one posterior and the other anterior. The former encloses the orbital artery in a foramen, while the latter, developed in relation to the lamina orbitonasalis, is in mesenchymatous connexion with the basal process of the pterygoquadrate. This connexion may for a time be to some extent cartilaginous, as Edgeworth (1925) showed. Eventually the two portions of the subocular shelf join and become attached to the lateral edge of the trabecula and anterior basal plate.

The transparency of whole mounts prepared by the van Wijhe technique enables the relations of the septa and spaces for the semicircular canals in the auditory capsule to be clearly seen (Plate 14, fig. 4). It will also be noticed that the first two sets of basidorsals are free from their corresponding neural arches.

STAGE 7. (45 mm. Naples.) PLATE 14, FIGS. 1, 2, 3, 5.

The supraorbital cartilage is attached to the dorsal edge of the orbital cartilage, leaving gaps (supraorbital foramina) for twigs of the superficial ophthalmic nerve supplying the sense-organs of the supraorbital lateral line canal. Anteriorly, the superficial ophthalmic nerve itself leaves the orbit through the preorbital canal, enclosed between the supraorbital cartilage and the lamina orbitonasalis. The profundus nerve in Scyllium is reduced and sometimes absent (Young, 1933). The orbitonasal foramen is therefore either absent or so small as to be unrecognizable in whole mounts. Where present, however, it corresponds exactly to its homologue in Squalus.

The aperture in the side wall of the skull through which the optic nerve emerges is large. Later, the foremost portion of it is separated off as the preoptic foramen, which allows a vein from the dorsal part of the brain to join the orbital sinus.

In the nasal capsule a further contribution has been made to the medial wall by a cartilage which leaves the median rostral cartilage and slopes forwards and upwards (dorsally to the cornua trabeculae) to meet the front wall of the capsule, and then continues forwards as the lateral rostral process. The median rostral process is the anterior prolongation of the rostral cartilage. The nasal cartilage has become dissociated into two pieces owing to the discontinuity of the outer prong from its base.

The hypophysial fenestra is obstructed by islets of cartilage which will contribute to the formation of the trabecular plate. The basicranial fenestra is reduced by further extension of the parachordals (basal plate).

A roof has now appeared in the form of a square plate of cartilage, the posterior corners of which are in continuity with the tips of the occipital arches forming a tectum posterius, while the anterior corners are almost continuous with processes which extend towards them from the roofs of the auditory capsules. On each side of this square plate are the fontanelles of the parietal fossae, through which pass the endolymphatic ducts on their emergence from the auditory capsules.

The medial wall of the auditory capsule is still incomplete, and represented by the dorsal process of the lamina basiotica of the parachordal, which separates the auditory and glossopharyngeal nerves (Plate 14, fig. 5; Plate 16, fig. 1).

The spiracular cartilage is present, dorsally to the hinder part of the pterygoquadrate. The basal process is a well-marked cone, connected by a ligament to the base of the lamina orbitonasalis. As in the cases of Squalus and Mustelus, Sewertzoff and Disler (1924) claim that in Scyllium an independent pharyngomandibular cartilage becomes fused on to the median side of the basal process. In the hyoid arch the basihyal cartilage has a foramen through which passes the duct of the thyroid gland. This foramen is evidence for the originally paired nature of the basihyal. In addition, it seems possible that the hypohyals are also fused to the basihyal, contributing to the posterolateral corners of the cartilage.

Hypobranchials, hyal and branchial rays, dorsal and ventral extrabranchial cartilages are present. The hypobranchials of the 4th and 5th arches fuse to form a median 'cardiobranchial'. The hyal rays are of interest, for they are grouped together at their base forming two branching twigs of cartilage, one attached to the hyomandibula and the other to the ceratohyal. It will be seen that these structures go far to explain the structure of the skeleton of the hyoid arch in rays and skates.

ii. The morphology of the chondrocranium of Scyllium canicula

In the main, the morphology of the adult chondrocranium of Scyllium is sufficiently similar to that of Squalus for the descriptions given of that form to apply (see p. 55). In a few respects, however, knowledge of the Selachian chondrocranium can be extended from investigations on Scyllium, relating to the wall of the auditory capsule.

1. The course of the glossopharyngeal nerve. Gegenbaur (1872) described the glossopharyngeal nerve as passing through the cavity of the auditory capsule, as indeed it does if the dry skull only is considered, and the membranes neglected. Since primitively the nerve must have passed out behind the capsule (as it does in Cyclostomes), the condition in Selachians needs explanation and presents a problem similar to the passage of the facial nerve through the capsule in Urodeles.

The key to the problem is the formation in Selachii of the lamina hypotica, a lateral extension of the hinder part of the parachordal, beneath the auditory capsule. Originally the glossopharyngeal must have passed in between, in a slit which is morphologically part of the fissura metotica. Then the floor of the hinder part of the capsule became reduced and fenestrated, and here the lamina hypotica serves as a cartilaginous floor, which becomes attached to the walls of the overlying capsule except for two places, which serve for the entry and exit of the glossopharyngeal nerve into and out of its canal. The true floor of the capsule is membranous, and discernible in sections, where it may be seen to separate the glossopharyngeal nerve from the posterior semicircular canal, the saccule, and the auditory nerve. Strictly, therefore, the glossopharyngeal nerve does not traverse the cavity of the auditory capsule, but only appears to do so in consequence of the above-mentioned modification.

2. The parietal fossa. The parietal fossa was described by Gegenbaur (1872, p. 49) as a depression in the roof of the skull above the auditory capsules, and at the bottom of which were two apertures which seemed to him to lead into the auditory capsules on each side. These apertures are Parker's (1879*b*, p. 208) 'aqueducts of the vestibule'.

To appreciate the morphology of this region it is necessary to consider the medial wall of the auditory capsule and the structure of the membranous labyrinth. The latter in Elasmobranchs is peculiar in that the posterior semicircular canal forms practically a complete circle or ring, apposed to the postero-medial surface of the utricle and saccule. The lateral portion of the posterior 'semicircular' canal is lodged in the cartilaginous capsule, but the medial portion protrudes through an extensive vacuity in the medial wall of the capsule, seen in Plate 14, fig. 6.

This posterior canal vacuity places the cavity of the auditory capsule in communication with surrounding regions through the aperture in the parietal fossa, but not with the cranial cavity from which it is marked off by the dura mater, in this region unchondrified. Morphologically, therefore, the parietal fossa apertures lead into the auditory capsules, and it is only in the dry skull that a seeker may be passed through these apertures into the cranial cavity.

Owing to the existence of the posterior canal vacuity the cartilaginous medial wall of the capsule is incomplete for a short distance, and here there is no contribution from the wall of the capsule to the formation of the skull-roof. Behind the vacuity the roof is formed by the tectum posterius (joining the tips of the occipital arches), in front of the vacuity it is formed by the tectum synoticum. The apertures in the parietal fossa thus owe their existence to the posterior canal vacuity.

The foramen endolymphaticum is in Scyllium quite distinct from and situated anteriorly to the posterior canal vacuity, but like it, it opens into the aperture of the parietal fossa. In Heterodontus, Norris (1929) has shown that they are confluent, but the mistake is often made of confusing the foramen endolymphaticum, either for its neighbour the posterior canal vacuity, or for the aperture in the parietal fossa which is, of course, not a foramen in the wall of the capsule at all, but is traversed by the ductus endolymphaticus on its way to the parietal sac, which in turn opens to the exterior on the dorsal surface of the head.

3. The rostral processes. A comparison between Squalus and Scyllium reveals the fact that in place of the three rostral processes of the latter, the former possesses a spout-shaped snout, presenting the form of a V as seen in transverse section, and flanked by the basal communicating canals. Gegenbaur (1872) was of opinion that an extension of the latter canals would lead to a fenestration of Squalus's rostrum and give rise to the three-pronged rostral basket as found in Scyllium and other forms. Allis (1923*a*), however, has shown that this explanation will not hold, for the basal communicating canals are apertures situated behind the anterior limiting membrane of the cranial cavity, while the spaces between the prongs or rostral processes are anterior to that membrane. Further, as has been seen in the description of the development of Scyllium, basal communicating canals are present in that form, a long way behind the rostral processes.

It is necessary to conclude, therefore, that the spaces between the rostral processes have nothing to do with basal communicating canals, but are perforations of the rostral cartilage developed in connexion with the presence of large numbers of sensory ampullae, as Sewertzoff and Allis suggest.

4. The hypophysial and basicranial fenestrae. The distinction between these fenestrae is a matter of importance owing to the form which they may take in higher forms. The hypophysial fenestra is the space enclosed between the trabeculae and polar cartilages; anteriorly it is bounded by the trabecular plate, but posteriorly its limits are harder to define, and change with progressive stages of development. In general, however, it may be said that the hind border of the hypophysial fenestra is formed by the anteroventral edge of the basal or parachordal plate.

The basicranial fenestra is the unchondrified portion of the basal plate, behind the acrochordal cartilage and in front of the parachordals. It lies in a different plane to the hypophysial fenestra, although the hind borders of the two fenestrae may coincide.

III. Pristiurus melanostomus (plate 17, figs. 1, 2)

Early stages in the development of Pristiurus have been studied with the help of graphic reconstructions by Sewertzoff (1899), while Matveiev (1925) has devoted attention to the segmental nature of the occipital region.

Sewertzoff finds in Pristiurus one segment less included in the occipital region as compared with Squalus, i.e. nine instead of ten, but as he counts according to van Wijhe's scheme, the number of segments in the skull of Pristiurus according to Balfour's plan would be eight. These results agree in fact with those of Matveiev, who likewise finds nine (van Wijhe) segments.

But the chief interest of Matveiev's investigation relate to the evidence, not only of segmentation, but of diplospondyly in the occipital region of the skull of Pristiurus. It is well known since von Ebner's work that in the trunk regions the sclerotomes are subdivided into anterior sclerotomites which give rise to intercalaria and posterior sclerotomites which produce the basalia of the vertebral column. Matveiev claims that these conditions prevail in the head region also, and that basidorsals and interdorsals, basiventrals and interventrals may be recognized in the parachordals, occipital and preoccipital arches, corresponding to his segments 6–9 (i.e. 5 to 8 on Balfour's scheme). These results are of sufficient interest and importance to be worthy of confirmation.

In the anterior region of the skull Matveiev has described the trabeculae as arising from prechordal visceral mesenchyme, while the sclerotomes of his first four segments give rise to the orbital cartilage, pila antotica, and polar cartilage. The sclerotomes of the next posterior segments contribute to the formation of the parachordals.

As regards the later development, what little that is known indicates that it follows very much the same lines as Scyllium.

IV. Heterodontus (Cestracion) philippi (plate 17, fig. 3; plate 18)

The development of the skull of the Port Jackson shark has been studied by de Beer (1924*b*). In general, it follows the same lines as Squalus, and therefore needs no redescription, except for one or two details. The first concerns the very broad articulation of the pterygoquadrate with the side of the braincase. The spiracular cartilage arises, as Edgeworth (1925) has shown, in procartilaginous continuity with the hinder part of the pterygoquadrate before becoming distinct. This is, in addition to its morphological relationship to other structures, further evidence in favour of the view that the spiracular cartilage is the homologue of the otic process.

The number of segments included in the skull is not known for certain, since it has not been found out which segment contributes the 1st permanent myomere. The hindmost of the three hypoglossal ventral roots innervates the 3rd myomere, so that if the 1st is formed from the 6th segment, the occipital arch would correspond to the septum between the 8th and 9th segments and there would be eight segments included in the skull. In addition, it would seem that the centrum corresponding to the 9th segment is attached to the hind end of the skull, for the ventral nerve-root innervating the 4th myomere emerges through a notch on the hinder side of the occipital arch. As in Squalus, the 1st arch following after the occipital arch is an interdorsal arch.

Daniel (1915) has shown that rudiments are present of 6th epibranchials and ceratobranchials, which in the adult become attached to the 5th.

But the chief interest of Heterodontus resides in the fact that in this form the relations of the cartilages to the nerves and blood-vessels have been fully worked out, and form a plan which is applicable to all Selachii.

i. Relations of the skull to the arteries

The internal carotids, or anterior prolongations of the dorsal aorta, run towards one another, and each gives off the orbital artery (external carotid of some authors) which pierces the cartilage of the subocular shelf to run into the orbit. The internal carotids then meet in the middle line and pass through the internal carotid foramen (posterior portion of the hypophysial fenestra) to enter the subpituitary space. In this space they diverge again, and each receives the efferent pseudobranchial artery (anterior carotid of T. J. Parker). This artery leaves the pseudobranch, passing on the medial side of the spiracular cartilage (otic process) and of the orbital artery, and on the lateral side of the palatine nerve; it gives off the ophthalmica magna artery which runs laterally (dorsally to the orbital artery), and then pierces the skull-wall by a foramen *dorsally* to the trabecula, to reach the internal carotid. The internal carotid then pierces the membranous floor of the pituitary fossa, enters the true cranial cavity, and runs forwards. It gives off the arteria centralis retinae which accompanies the optic nerve out through the optic foramen, and then divides to form the cerebral arteries and the basilar artery.

Following the internal carotid backwards from the point where it gives off the orbital artery, it passes on the medial side of the hyomandibula (forming the posterior carotid of T. J. Parker) and receives the efferent hyoidean artery, behind which point it is the lateral dorsal aorta, which continues backwards ventrally to the pharyngobranchials. The efferent hyoidean artery lies on the posterior side of the ceratohyal and hyomandibula, and gives off the afferent pseudobranchial artery which runs laterally to the ceratohyal and hyomandibula to the pseudobranch. The afferent pseudobranchial artery is a commissural vessel comparable to those in the branchial arches which interconnect the efferent branchial arteries of the arch, passing laterally to the cartilaginous arch skeleton. The afferent pseudobranchial artery is medial to the facial nerve and afferent hyoidean artery, just as the cross commissural vessel in the 1st branchial arch is medial to the glossopharyngeal nerve and the afferent branchial artery.

The true mandibular arch vessel of early embryos is represented by the efferent pseudobranchial artery and by some small vessels connected with the base of the efferent hyoidean artery.

ii. Relations of the skull to the veins

The head vein, or vena capitis lateralis, runs forwards from the anterior cardinal sinus, passing dorsally to the roots of the vagus, glossopharyngeal, and facial nerves, and to the hyomandibula. The vein laps round the roots of the trigeminal nerve and runs into the orbital sinus. The portion that lies medially to the trigeminal gives off the pituitary vein which pierces the cartilage by the interorbital canal, dorsally to the trabecula and in front of the pila antotica. The pituitary vein is then in the subpituitary space, where it joins its fellow from the opposite side, beneath the notochord. That part of the head vein which gives off the pituitary vein is a relic of the vena capitis medialis which at earlier stages ran all the way ventro-medially to the dorsal nerve-roots and to the auditory sac, became interrupted by the development of the auditory capsule, and has disappeared, being replaced by the vena capitis lateralis.

From the orbital sinus two veins are given off which pierce the cartilage of the skull. One of these passes through a preoptic foramen just in front of the optic foramen, and runs between the internal surface of the cartilage and the dura mater to the basal communicating canal, through which it emerges into the space beneath the olfactory foramen. The other vein penetrates through the same (orbitonasal) foramen as the profundus nerve, but instead of running out again on to the external surface of the cartilage like the nerve, the vein runs upwards to the dorsal part of the forebrain.

The relations of the veins in this region appear to differ in the various Selachii. Thus, in Scyllium the vein draining the region beneath the olfactory foramen passes back to the orbit through the orbitonasal canal between the side wall of the cranial cavity and the lamina orbitonasalis. On the other hand, the vein from the dorsal side of the brain penetrates the skull-wall by the preoptic foramen, on its course back to the orbital sinus.

Thus, the basal communicating canal in Scyllium transmits no vein, nor is the orbitonasal foramen (always minute if not absent) traversed by a vein; Heterodontus, on the other hand, has no orbitonasal canal. The reason for this is presumably to be found in the fact that the place in the anterior wall of the orbit where the orbitonasal canal would be expected to be is in contact with the large articular facet of the pterygoquadrate, so that the venous drainage from the nasal capsule has been forced to take a different course.

iii. Relations of the skull to the nerves

The olfactory nerve passes from the cranial cavity into the cavity of the nasal capsule by piercing the membrane which spans the olfactory foramen.

The optic nerve runs outwards dorsally to the inferior branch of the oculomotor and ventrally to the profundus. The oculomotor forks soon after emerging from its foramen (which is anterior to the pila antotica), the two branches of the fork enclosing the profundus nerve between them.

The trochlear nerve runs ventrally to the superficial ophthalmic nerve and dorsally to the profundus. Of the trigeminal group, the profundus runs forwards in the orbit from the trigeminal foramen (which is posterior to the pila antotica), between the two branches of the oculomotor, dorsally to the optic nerve and ventrally to the trochlear, until it leaves the orbit by the orbitonasal foramen. It then lies between the internal surface of the cartilage of the lamina orbitonasalis and the membrane lining the cavity of the nasal capsule, and emerges again on to the dorsal surface of the cartilage through an epiphanial foramen.

The maxillary and mandibular branches of the trigeminal run outwards, behind the orbital artery but anterodorsally to the efferent pseudobranchial artery.

The abducens runs forwards within the cranial cavity and emerges through the foremost portion of the trigeminal foramen. In some forms it has a foramen of its own, in the base of the pila antotica.

Of the facial nerve-roots the superficial ophthalmic and buccal branches emerge through the trigeminal foramen (i.e. anteriorly to the prefacial commissure), the former running forward in the orbit dorsally to the oculomotor and trochlear nerves. In many Selachians (but not in Heterodontus) it pierces the preorbital canal to get to the dorsal surface of the snout. The buccal branch accompanies the maxillary branch of the trigeminal.

The palatine and hyomandibular branches of the facial nerve emerge through the facial foramen (i.e. behind the prefacial commissure). The palatine runs ventrally to the side of the subocular shelf and runs forwards beneath the the articulation of the basal process of the pterygoquadrate with the side wall of the brain-case. On its course the palatine nerve lies laterally to the internal carotid and medially to the efferent pseudobranchial artery.

The hyomandibular branch of the facial runs outwards beneath the head vein, and turns backwards dorsally to the hyomandibula. It then runs ventrally, passing laterally to the afferent pseudobranchial artery, and divides into three. One branch is the ramus mandibularis externus, which runs forwards on the outer side of Meckel's cartilage; another is the ramus mandibularis internus, which slips in behind Meckel's cartilage on to its medial surface where it runs forwards; the last is the ramus hyoideus, which runs down on the outer side of the ceratohyal.

The auditory nerve is distributed to the various sense-organs of the membranous labyrinth within the auditory capsule in a manner which is of little moment from the present point of view.

The glossopharyngeal nerve passes through a canal which, as described above (see p. 63), appears to traverse the cavity of the auditory capsule. After its emergence the glossopharyngeal runs underneath the head vein and divides into the pre-and post-trematic branches. The pretrematic branch runs down laterally to the lateral dorsal aorta (posterior carotid) and then medially to the efferent hyoidean artery, on the posterior side of the hyoid arch. The post-trematic branch itself divides into anterior and posterior branches: the former running downwards medially to the lateral dorsal aorta, while the latter runs on the outer side of that artery and of the epibranchial cartilage which it pierces to supply the adductor muscle. The nerve then continues running downwards in the 1st branchial arch, passing laterally to the cross commissural vessel between the two efferent branchial arteries.

The vagus nerve, issuing through the jugular foramen, behaves like the glossopharyngeal in respect of its pre- and post-trematic branches.

V. Torpedo ocellata

With the exception of Parker's (1879*b*) early work, little has been done on the development of the skull of the Batoid Selachii beyond a description of a few stages by de Beer (1926*b*), and investigations into the structure of the hyoid arch by Dohrn (1886), Krivetski (1917), Edgeworth (1931) and de Beer (1932).

i. Morphology of the chondrocranium of Torpedo ocellata (24 mm.)

(PLATES 19, 20.)

At this stage (as studied by van Wijhe preparation) all the main cartilages have appeared and joined up, and the skull presents the characteristic Selachian appearance, though it is much flattened. The position of the polar cartilage is still plainly visible. The pituitary vein and efferent pseudobranchial artery are enclosed in foramina, and just dorsally to them, the eyestalk is attached to the anterior side of the pila antotica. The orbitonasal canal which gives passage to the nasal branch of the facial vein is present, but the preoptic and optic foramina are still confluent. The hypophysial fenestra has been reduced to the passage for the hypophysial duct and the internal carotid foramen. A small basicranial fenestra is present, bounded anteriorly by the dorsum sellae. The foramen prooticum is remarkably large, but the palatine branch of the facial nerve is enclosed in a little foramen of its own.

The parachordals, on either side of the notochord, are interconnected by an extensive chordal commissure, which is discontinuous posteriorly with the large plate formed by the fused vertebrae of the neck region. The cranio-vertebral joint is therefore present.

The auditory capsules are well formed, and posteriorly, the glossopharyngeal nerve has just become separated from the vagus by the obliteration of a part of the fissura metotica as the lamina hypotica comes into contact with the auditory capsule.

The skull is markedly hyostylic, and the pterygoquadrate is separated by a wide space from the trabecula.

The hyomandibula is large, and behind it is a cartilaginous bar bearing hyal rays which has usually been called the ceratohyal but, as will be seen below, without justification. It will here be termed the pseudohyoid.

Pharyngobranchials are present in the first four branchial arches; epibranchials and ceratobranchials in all five.

Nothing is known of the origin of M. Fürbringer's (1903) extraseptal cartilages.

ii. Relations of the skull to nerves and blood-vessels

The relations of the skull to the nerves and blood-vessels have been studied with the help of a wax-model reconstruction in Torpedo marmorata (24 mm.) by de Beer (1926*b*).

The cartilages in this embryo are similar to the one described above, except that the olfactory, preorbital, and trochlear foramina are enclosed, a prefacial commissure is present, the spiracular cartilage has appeared, and there is an antorbital cartilage which appears to be a detached portion of the lamina orbitonasalis.

It may be said at once that the general relations of the arteries, veins, and nerves are the same as in pleurotrematous Selachii, as described in Heterodontus (see above, p. 66). The vein from the dorsal side of the forebrain pierces the preoptic foramen to reach the orbital sinus, into which a vein from the nasal capsule runs by piercing the orbitonasal canal. There is no orbitonasal foramen, nor a preorbital canal, with the result that neither the superficial ophthalmic nor the profundus have to pass through any foramina on their way from the orbit to the snout. This is presumably due to the absence of a supraorbital cartilage, and the lack of a subocular shelf involves the absence of a foramen for the orbital artery.

The eyestalk is situated ventrally to the profundus, and dorsally to the inferior branch of the oculomotor nerve and the head vein.

The lateral dorsal aorta (posterior carotid of T. J. Parker) is interrupted between the efferent hyoidean and first efferent branchial arteries.

But an important difference is seen in the relations of the afferent pseudobranchial artery, which here runs medially to the pseudohyoid, instead of laterally to the ceratohyal, which indicates that the pseudohyoid is not the ceratohyal.

iii. The skeleton of the hyoid arch

Krivetski (1917) drew attention to the above-mentioned difference between the relations of the afferent pseudobranchial artery to the ceratohyal and to the pseudohyoid, and suggested that the pseudohyoid is the product of the fusion of the hyal rays, the ceratohyal having disappeared.

The development of the pseudohyoid has been studied by de Beer (1932) in Torpedo ocellata and marmorata, Raja blanda, Rhynchobatus sp., and Pristis sp.

The afferent pseudobranchial artery is the same vessel in Heterodontus (typical of sharks) and Torpedo (typical of rays), for in both cases it runs forwards from the efferent hyoidean artery passing medially to the afferent hyoidean artery and to the hyomandibular branch of the facial nerve, and laterally to the hyomandibula. The afferent pseudobranchial artery is therefore a safe guide in establishing the difference between the ceratohyal and the pseudohyoid.

In Torpedo embryos 21 mm. long the skeleton of the hyoid arch is in the form of a bar of dense mesenchyme, medial to the afferent pseudobranchial artery. The dorsal half of this bar chondrifies to form the hyomandibula; the ventral half remains unchondrified save at the extreme ventromedian end, where two small centres of chondrification represent the ceratohyal and hypohyal elements. Meanwhile, behind this bar, a second bar of dense mesenchyme is formed, parallel with the first, but situated laterally to the afferent pseudobranchial artery, and it is continuous with the bases of the hyal rays. It is from this second bar that the pseudohyoid will chondrify, and there is no doubt that Krivetski was correct in his estimation of its nature.

A comparison with sharks will show that the pseudohyoid is foreshadowed in them. In Scyllium (Plate 14, fig. 1) the dorsal rays of the hyoid arch are attached to one another by their base, and the branching bar so formed is attached to the hyomandibula. Similarly, the ventral hyal rays are attached to one another and to the ceratohyal. These are really dorsal and ventral pseudohyoid bars, and they bear the correct relations to the afferent pseudobranchial artery, viz. they lie laterally to it. Such a state of affairs also occurs in Carcharodon; Luther (1909) has described it in Stegostoma, and K. Fürbringer (1903) in Chlamydoselachus, Odontaspis, Laemargus, Heterodontus, Spinax, Echinorhinus, and Scymnus.

What has happened in the rays, therefore, is that the ceratohyal has become much reduced, and its function in providing the skeletal support for the anterior wall of the 1st gill-slit has been taken over by the pseudohyoid, which becomes jointed. In Rhynchobatus the reduction of the ceratohyal has not proceeded so very far, and the ceratohyal still bears four hyal rays; in Torpedo, Raja, and Pristis the reduction has gone further, the ceratohyal bears no hyal rays at all, and is attached to the lateral end of the transverse hyoid copula.

The reasons for this modification are probably to be sought in the fact that in the rays, with their dorsal spiracle and ventral gill-slits, the distance between the spiracle and the 1st gill-slit (i.e. the width of the hyoid arch) is too great for the hyomandibula to serve as the support of the hind wall of the spiracle and of the front wall of the 1st gill-slit. The hyal rays have coalesced in the performance of the latter function, and the ceratohyal has become progressively reduced. These relations are shown in Plate 20, figs. 2, 3.

The elucidation of the relations of the skeleton in the hyoid arch of rays is of some theoretical importance, for the supposed double nature of this skeleton (hyomandibula plus pseudohyoid) was appealed to by Dohrn (1886) in support of the view that one segment and visceral cleft had been lost in this region, which would, of course, be in accordance with van Wijhe's (1882*a*) scheme (see above, p. 19) of counting the segments of the head. It is at any rate clear that no such support is forthcoming from the skeleton of the hyoid arch in rays and skates.

Allis (1915) has attempted to show, on grounds which, however, do not seem to be adequate, that the hyomandibula of Batoid Selachii is a pharyngohyal, while that of non-Batoids is an epihyal. There is no good reason to doubt the homology of the hyomandibula in these closely related forms.

iv. The cranio-vertebral joint

The vertebrae of the anterior end of the vertebral column in the rays are commonly fused together into a rigid mass, and it is presumably in association with these conditions that a true joint is developed between the skull and the anterior end of the column. This joint is made up of three articular facets: one median, representing the articulation of the posterior surface of the centrum of the hindmost vertebral element absorbed in the skull with the anterior surface of the centrum of the 1st free vertebra, and two lateral formed by the articulation of the hinder surface of each occipital arch with the anterior surface of the neural arches of the 1st vertebra.

In the formation of this joint, it seems that some absorption of cartilage takes place, for Braus (1899) found in Torpedo narce that the 1st spinal nerve-root is released from its foramen in the 1st vertebra by the disappearance of that vertebra, and the nerve subsequently becomes enclosed in the developing articular process of the occipital arch. This matter might well be reinvestigated.

6. HOLOCEPHALI

Of all groups of vetebrates the Holocephali are those concerning which the least is known as regards the development of the skull. Their bibliography in this respect is therefore short and simple. Two late stages of Callorhynchus have been described by Schauinsland (1903), one still later stage has been investigated by de Beer and Moy-Thomas (1935), while of Chimaera, one advanced stage has been studied by Dean (1906).

Since in the Holocephali the skull remains cartilaginous throughout life, the adult condition must here be taken into account for comparison with other forms. Various features in the morphology of the adult Holocephalian skull have been treated by Hubrecht (1877), Fürbringer (1904), Luther (1909) and Allis (1915, 1917, 1919*a*, and 1926*b*).

The following account applies to Callorhynchus antarcticus, and is based on the works of Schauinsland, and of de Beer and Moy-Thomas; reference to Dean's work on Chimaera is made where what is known of that form assists in the interpretation of the Holocephalian skull.

i. The chondrocranium of Callorhynchus antarcticus

1. STAGE 1. (60 mm.) PLATE 21, FIGS. 5, 6.

The skull of an embryo of this stage has been studied and figured by Schauinsland (1903), but with scant identification and labelling of the various parts. Some of the regions shown are procartilaginous, but the fact that they have not yet chondrified is of little importance from the morphological point of view.

At this comparatively late stage all the cartilages have fused up, in some cases to the extent of almost obscuring their true relations. The floor is continuous, being composed of equally broad trabecular and basal plates, disposed at right angles to one another, with the anterior edge of the basal or parachordal plate projecting forwards and upwards as an accentuated dorsum sellae, or acrochordal cartilage, into which the notochord extends. The parachordals are continuous with one another ventrally to the notochord, and, in addition, their cartilage cells invade the notochordal sheath so that a tubular chordal commissure is also formed. The notochord seems to persist in the basal plate throughout life. The trabecular plate is perforated by a small median hypophysial foramen, but there are no internal carotid foramina, for these arteries are degenerate and fail to enter the cranial cavity.

The auditory capsules and occipital arches are attached to the lateral edges of the basal plate. The medial wall of the auditory capsule remains unchondrified, but its floor is complete, with the result that the glossopharyngeal nerve does not pass (or appear to pass) through the cavity of the auditory capsule. The fissura metotica between the auditory capsule and the occipital arch is subdivided into two foramina, through which the glossopharyngeal and vagus nerves emerge.

Projecting forwards from each auditory capsule is a post-orbital process which is continuous with the hind end of the orbital cartilage of its side. This is connected with the basal plate by the pila antotica, which encloses between itself and the auditory capsule the foramen prooticum and a special foramen for the superficial ophthalmic nerve. The orbital cartilage is attached anteriorly by means of its preorbital root to the lateral edge of the trabecular plate. The ventral part of the preorbital root of the orbital cartilage is continuous with the lamina orbitonasalis which lies in front of it and forms the side wall of the nasal capsule. The distinction between the lamina orbitonasalis and the preoptic root of the orbital cartilage is marked by the orbitonasal canal (through which a branch of the facial vein leaves the orbit to reach the space beneath the nasal capsule), and farther dorsally by a V-shaped notch in the apex of which the profundus ophthalmic nerve passes forwards out of the orbit, medially to a dorsal process of the lamina orbitonasalis, and so reaches the dorsal surface of the snout. The profundus nerve is accompanied forwards by the superficial ophthalmic nerve, which runs in a notch between the orbital cartilage and the supraorbital cartilage. The latter is attached to the dorsal edge of the orbital cartilage anteriorly. At later stages it curves outwards to form the preorbital process.

Anteriorly in the middle line, the trabecular plate is continued into a nasal septum which forms a vertical plate, the dorsal edge of which is continuous with the roof of the nasal capsule on each side, which in turn fuses with the anterodorsal edge of the lamina orbitonasalis of its side.

It is important to note that there is at this stage no cartilaginous roof whatever to the cranial cavity, and that the brain-case is low and very broad, with the eyes wide apart and the brain between them.

In the visceral arch skeleton the pterygoquadrate cartilage already shows the extensive fusion with the neurocranium which is expressed in Gregory's (1904) useful term holostylic. The fusion involves the region from the ethmoid process to the basal process, and then, after a gap (the cranio-quadrate passage), the otic process.

It looks from Schauinsland's reconstruction as though the contact between the otic process and the auditory capsule had only recently taken place, and that study of an earlier stage would show them separate. This is borne out by a consideration of Dean's reconstruction of a 45-mm. embryo of Chimaera. This shows a deep notch immediately beneath the point of exit of the trigeminal nerve, in the place where the otic process should be.

Meckel's cartilages are joined anteriorly in a symphysis involving a very broad and complete cartilaginous fusion. Ventrally to the symphysis is a median basimandibular cartilage. The remainder of the visceral arch skeleton may be reserved for the description of the next stage.

2. Stage 2. (90 mm.) PLATE 22.

Considerable changes have taken place between the last stage and this. The skull is now roofed over. Posteriorly, a gap is left through which the ductus endolymphaticus from each auditory vesicle pierces the skull-roof and communicates with the outside. Anteriorly, the skull-roof is continued forwards over a canal, the ethmoidal canal, which opens on the dorsal surface of the nasal capsule, near its front end. The anterior region of the nasal capsule is surrounded by a number of rostro-labial cartilages of doubtful significance, and by the median and the paired lateral rostral processes, which may be homologous with the prongs of the rostral basket found in some Selachii. The eye is surrounded by a well-developed sclerotic cartilage. In addition, the male possesses a median frontal clasper, articulated against the roof of the ethmoidal canal.

In the visceral arch skeleton the feature of chief importance is the apparent presence of pharyngohyal, epihyal, and ceratohyal cartilages in the hyoid arch. It may be noted that the arches are of the sigma type, with the pharyngal element pointing backwards from the epal element.

ii. The morphology of the skull of Callorhynchus

1. The ethmoidal canal. The ethmoidal canal is a longitudinal and horizontal tunnel entirely enclosed by cartilage, situated immediately above the roof of the nasal capsules, and opening anteriorly on to the dorsal surface of the snout and posteriorly into a space (the interorbital space) which is continuous with the orbits on each side. The interorbital space is outside (dorsal to) the dura mater, and therefore it is really extracranial in spite of the fact that it is covered over by the cartilaginous cranial roof. In the adult the skull has a large and high interorbital septum, situated dorsally to the brain-case. At the stage here studied, however, the interorbital septum has not yet been formed, and the eyes are still widely separated, with the brain in between. The interorbital septum will eventually be formed at the expense of the interorbital space, which will become higher and narrower as the roof of the skull gets lifted higher and higher up.

The interorbital space, then, leads into the hinder part of the ethmoidal canal (which, like it, is strictly extracranial) and gives passage to the superficial ophthalmic nerve which runs forwards in the ethmoidal canal. The hind opening of this canal thus represents the preorbital canals of Selachii, confluent with one another in the middle line. A little further forward the ophthalmicus profundus nerve leaves the orbit and enters the ethmoidal canal by piercing the cartilage of its side wall. This foramen represents the orbitonasal foramen of Selachii. Both superficial and profundus nerves then run forwards within the ethmoidal canal to its anterior opening, through which they emerge on to the snout.

Throughout its length, the ethmoidal canal is dorsal to the anterior part of the cranial cavity containing the forebrain, and to the cavity of the nasal capsules, and separated from them by a cartilaginous floor which at this stage is complete. These conditions can be explained along the following lines.

The Selachii are the nearest relatives of the Holocephali, and the interpretation of the conditions in the latter can be found in a consideration of the relations of the ophthalmic nerves in the former.

The profundus nerve traverses the orbit (even in Scyllium where, as Young (1933) has shown, it is often present although reduced) and pierces the cartilaginous anterior wall of the orbit by the orbitonasal foramen. Morphologically (de Beer, 1931*a*) the point of entry represents the original gap (orbitonasal fissure) between the lamina orbitonasalis (hind wall of the nasal capsule) and the true side wall of the skull (preoptic root of the orbital cartilage). As in other forms, however, the profundus nerve never really enters the cranial cavity since it always lies outside the dura mater. For a distance which varies in different forms, the profundus runs forwards in this position and emerges again through the cartilage by an epiphanial foramen on to the dorsal surface of the nasal capsule.

The superficial ophthalmic nerve leaves the orbit by a foramen (the pre-orbital canal) distinct from that of the profundus, and which represents the gap between the supraorbital cartilage and the true side wall of the skull. The superficial ophthalmic nerve then finds itself directly on the dorsal surface of the nasal capsule.

Another point to notice in Selachii is the universal existence of an unchondrified portion of the skull in the anterior region, forming a precerebral or epiphysial fontanelle, immediately beneath which the epiphysis is to be found. In the Holocephalian the ophthalmic nerves leave the orbit by two foramina (on each side) which have every appearance of being homologous with those in Selachii. Only, instead of emerging freely on to the dorsal surface of the nasal capsules, these nerves here find themselves enclosed in the ethmoidal canal. The latter, then, must represent an originally external space, which has been secondarily roofed over.

This modification must have been associated with the development of the interorbital septum. The formation of this structure in Holocephali *above* the brain results in a lifting up of the roof of the skull to an extent that is quite remarkable. One effect of this has been to produce an upward distortion of the lamina orbitonasalis: this structure in Selachii forms the hind wall of the nasal capsule; in Holocephali the nasal capsules have no cartilaginous hind wall at all, for they are directly continuous by wide apertures with the anterior part of the cranial cavity. The lamina orbitonasalis here, therefore, is displaced dorsally and forms the side wall of the ethmoidal canal as well as the side wall of the nasal capsule.

Another effect of the formation of the interorbital septum is seen by a consideration of what would happen if the skull still possessed the large, open, and unchondrified epiphysial fontanelle of Selachii. The fontanelle would now point forwards, and by an increasingly wide gap would give vulnerable access to the brain-cavity behind. It is presumably as an increased mechanical security for the skull with its lifted roof that the ethmoidal canal becomes roofed over. But the distinction between this secondary roofing and the true roof of the skull is shown by the persistence up to a certain stage in Callorhynchus (Plate 23, fig. 3) of what is really a pineal foramen overlying the epiphysis, and which represents a part of the original epiphysial fontanelle. It may also be noticed that while the true roof of the skull is closely associated with the dura mater, the roof of the ethmoidal canal is not.

At later stages in Callorhynchus the pineal foramen is occluded and the roof of the ethmoidal canal is complete as far forward as its anterior paired opening, where the ophthalmic nerves run out on to the dorsal surface of the nasal capsule, passing laterally to the rostral processes. The original epiphysial fontanelle now follows a curved line, along the anterior part of which the floor of the ethmoidal canal is formed. This floor represents an extension backwards of the cartilage which forms the roof of the nasal capsules. An attempt is made to show this in a diagram (Plate 23, fig. 2) of a hypothetical intermediate stage, to which, however, some fossils (Helodus) approximate closely (Moy-Thomas, 1936).

2. The attachment of the jaws to the brain-case. It is well known that the pterygoquadrate of Holocephali is firmly fused on to the cartilage of the brain-case by its own processes, in a manner which Huxley (1876*b*) imagined as approximating to his 'autostylic' type. It is now, however, clear that this attachment differs from the autostyly found in Dipnoi and Amphibia (Goodrich, 1909), and the expression 'holostylic' may be used to denote the Holocephalian condition (Gregory, 1904).

The essential feature of the attachment in these animals is its completeness. The front part of the pterygoquadrate is fused with the nasal capsule and ethmoidal region of the skull; the hind part is fused with the otic capsule, and the intermediate part of the pterygoquadrate is fused with the lateral edge of the trabecular plates of the skull, leaving only three gaps or foramina, through which certain nerve-roots and blood-vessels pass.

The anterior region of this extensive attachment represents a fusion between the ethmoid process of the pterygoquadrate and the ethmoid region of the skull.

The hindmost region of the attachment of the upper jaw to the skull bears all the typical relations of an otic process to the vena capitis lateralis and to the trigeminal and facial nerves, and it is in perfect cartilaginous continuity with the remainder of the pterygoquadrate cartilage.

The possibility has been envisaged (Goodrich, 1909) that the otic process is accompanied by and fused with some part of the hyomandibula. It is all the more necessary to decide this point, since in such a form as Acipenser, the hyomandibula bears the same relations to surrounding structures as does the otic process, *except* that it is posterior instead of anterior to the spiracle. The question must therefore be decided by the position of the spiracle.

Unfortunately, in the Holocephali, the spiracle becomes closed and disappears at an early stage: there is nothing left of it at the stages here studied, and the figures of complete embryos of earlier stages shown by Schauinsland and Dean are insufficient to give any clue as to the position which the spiracle would have occupied in later stages, had it persisted. Pending the study of the much-wanted intermediate stages, there is, however, one clue which indicates the probable former position of the spiracle, and this is the position of the spiracular sense-organ. This is found in the form of a clump of a few ampullae, innervated by the facial nerve, and it lies ventro-posteriorly to the pterygoquadrate and otic process: if this is so, the latter cannot contain any admixture of hyomandibular cartilage.

The fusion of the pterygoquadrate with the trabecular plate in the intermediate region, between the ethmoid fusion in front and the otic fusion behind; is interrupted by three foramina, representing the space originally existing between the pterygoquadrate and the neurocranium, and through which pass, in order from front to back, the efferent pseudobranchial artery, the palatine nerve and orbital artery together, and the hyomandibular nerve and vena capitis lateralis together.

The fusion of the pterygoquadrate with the brain-case immediately in front of the foramen through which the palatine nerve passes answers exactly to the definition of a basal connexion, and the fact that the orbital artery runs up through the same foramen as that which allows the palatine nerve to run down, is exactly paralleled in Lepidosteus and Salmo. As to the relative participation of basal process (of the pterygoquadrate) and basitrabecular process (of the neurocranium) little can be said. Judging from the condition in Selachii where the basitrabecular process is only feebly developed, and is represented by a part of the subocular shelf, it is probable that the major part of the basal connexion in Callorhynchus is formed by the basal process.

In most forms which possess both otic and basal connexions there is an open passage—the cranio-quadrate passage (Goodrich, 1930)—between the neurocranium and the pterygoquadrate, and through this passage the palatine and hyomandibular branches of the facial nerve run downwards, the vena capitis lateralis runs backwards, and the orbital artery runs forwards.

In Callorhynchus the two foramina which transmit, the one the palatine nerve and the orbital artery, the other the hyomandibular nerve and the vena capitis lateralis, are separated by a short region of fusion between the pterygoquadrate and the neurocranium. As a result, the cranio-quadrate passage is divided into two, and is therefore not easily recognizable at first sight. Its true morphological relations are, however, typical; and the cartilage which thus divides it in two bears all the relations of a postpalatine commissure, which is of wide occurrence in other forms.

Attention may now be turned to the relation of the efferent pseudobranchial artery. This artery, which represents the dorsal portion of the original mandibular vessel (de Beer 1924*c*), is of interest because of the difference in relations which it shows in Selachii and in Teleostomi.

In the latter group the efferent pseudobranchial artery runs dorsally and inwards beneath the pterygoquadrate, and then runs *underneath* (ventrally to) the trabeculae to join the internal carotid artery just as it enters the cranial cavity through the hypophysial fenestra. In the Selachii, on the other hand, the efferent pseudobranchial artery again passes ventro-medially to the pterygoquadrate, but it then runs above (dorsally to) the trabecula and thus enters the cranial cavity independently and from the side, before joining the internal carotid (Allis, 1923*b*; de Beer, 1924*c*).

It is therefore of great importance that in Callorhynchus the relations of the efferent pseudobranchial artery are similar to those shown by Selachians. The only differences are matters of detail, and concerned with the facts that in Callorhynchus (i) there is no pseudobranch (doubtless connected with the early closure of the spiracle); (ii) the internal carotid arteries are completely aborted and do not even enter the skull (presumably consequent on (i), since there is now no interruption to the flow of blood in the efferent 'pseudobranchial' artery, which in Holocephali thus furnishes the sole supply of blood to the brain); and (iii) the arteria ophthalmica magna is absent.

In the Holocephali, therefore, owing to the extensive fusion of the pterygoquadrate with the brain-case, the efferent pseudobranchial artery as it runs upwards finds itself caught in between the ptergyoquadrate and the trabecular plate, and passes through a canal. The upper opening of this canal leads into the orbit, but the artery immediately penetrates the dura mater and enters the cranial cavity; its passage through the orbit, which is quite long in Selachii, is extremely short in Holocephali.

From what has already been said, it should be clear that the morphological relations of the pterygoquadrate in Holocephali are quite typical and similar in principle to those found in Selachii.

3. The skeleton of the hyoid arch. In the skeleton of the hyoid arch Hubrecht (1877) found three cartilages on each side, which Schauinsland (1903) identified as pharyngohyal, epihyal, and ceratohyal elements; these elements corresponding serially to the pharyngobranchials, epibranchials, and ceratobranchials behind them. This view is supported by de Beer and Moy-Thomas (1935). The interest of this state of affairs is clear, for if this is so, then the Holocephali present a case of a hyoid arch skeleton which is scarcely if at all modified from the condition of the branchial arches, and in respect of the hyoid arch they would be the most primitive living forms known. The only alternative is to regard the subdivision of the hyoid arch as secondary, with the hyomandibula either absent or represented by part or all of the 'otic process' (see above, p. 76). This matter can only be settled definitely when the relations of the spiracle are known.

4. The occipital region. The occipital region of the skull in Holocephali is of interest for two reasons. One is the large number of foramina for roots of the hypoglossal nerve, namely, five. Unfortunately, as the segmentation has not been worked out in early stages, it is impossible to say which is the segment to which the last root corresponds, or to determine the number of segments included in the skull (see p. 12).

The other feature of importance is the existence of a special cranio-vertebral joint. The posterior edge of the basal plate forms a saddle-shaped surface, concave from side to side and convex dorsoventrally, and articulating with the centrum of the 1st vertebra. At the same time, the posterior surface of each occipital arch bears a facet which articulates with the anterior surface of the neural arch of the 1st vertebra. There is thus a point of special flexibility between the skull and the vertebral column, which is probably all the more necessary in view of the fact (as in Batoid Selachii) that the vertebrae of the anterior end of the column are immovably fused together.

7. TELEOSTOMI. POLYPTERINI

Polypterus

The development of the chondrocranium of Polypterus is only imperfectly known. A few stages have been described by Winslow (1898), and by Budgett (1902) and Graham Kerr (1907), and these latter stages have been reinvestigated in the light of more modern knowledge by Moy-Thomas (1933). The chondrocranium of older and adult stages has been studied by Pollard (1892), Lehn (1918), and Allis (1922), while some points in its morphology have been investigated by de Beer (1926*a*) and Sewertzoff (1926*b*).

Equally little is known concerning the development of the bony skull; the dermal bones related to the lateral line canals have been figured by Pehrson (1922); the cartilage-bones of the neurocranium were described by Lehn (1918), while Traquair (1871) and Allis (1922) have subjected most of the bones of the adult to morphological analysis. The development of the bones of the lower jaw has been the subject of a special study by Schmäh (1934).

i. The development of the chondrocranium of Polypterus senegalus.

The following account is based mainly on the works of Moy-Thomas and Budgett.

1. STAGE 1. (6·75 mm.) PLATE 24, FIGS. 1, 2.

The parachordals are chondrified on each side of the notochord and in contact with it for the hind three-quarters of their length; anteriorly the parachordals diverge away from the notochord, and end freely slightly in front of the level of the notochord's tip. Laterally each parachordal projects in the direction of the auditory capsule.

At some distance behind the hinder ends of the parachordals and quite separate from them, the occipital arches are chondrified in the form of tall thin neural arches, rising up from the dorsolateral sides of the notochord, and slanting forwards and upwards.

Except for the ceratohyals which are chondrified, the remaining elements are in a procartilaginous state. Among these are the trabeculae, in the form of straight rods, wide apart, and situated in the anterior line of prolongation of the ventral surface of the parachordals.

The rudiments of the auditory capsule and hyosymplectic are confluent, as are those of the pterygoquadrate and Meckel's cartilage. Other rudiments represent the ceratobranchials.

2. STAGE 2. (8 mm.) PLATE 24, FIGS. 3, 4, 5.

The trabeculae are connected in front by a broad trabecular or ethmoid plate, and posteriorly each trabecula is attached to the anterior end of the parachordal of its side, so that a large hypophysial fenestra is enclosed, the hinder part of which is partly confluent with the basicranial fenestra (allowing, of course, for the fact that these fenestrae have the same posterior border—the anterior edge of the parachordals—but are not in the same plane). No polar cartilage is discernible.

On the dorsal surface of the ethmoid plate the nasal septum rises up as a crest in the middle line, while on each side, the lamina orbitonasalis forms as it were the upturned lateral edge of the ethmoid plate.

The auditory capsule is now fairly well chondrified and attached to the parachordal by both anterior and posterior basicapsular commissures, enclosing the basicapsular fenestra between them. The posterior basicapsular commissure lies behind the glossopharyngeal nerve, with the result that that nerve is separated from the vagus, and passes through a foramen of its own which has been marked off from the basicapsular fenestra.

The facial nerve emerges immediately in front of the anterior basicapsular commissure, but in front of the facial nerve there is another bar of cartilage—the prefacial commissure—connecting the auditory capsule with the parachordal, and thus enclosing the facial foramen. It is to be noted that the prefacial commissure is a part of the true side wall of the skull, and is not to be mistaken for the lateral commissure, the rudiments of which may also be seen at this stage.

Laterally to the prefacial commissure the prootic process projects downwards from the side wall of the auditory capsule, and towards the postpalatine process which projects sideways from the parachordal just beneath the facial foramen. When the prootic and postpalatine processes join, they form the lateral commissure, or lateral wall to the jugular canal. Already at this stage it may be noted that the head vein runs dorsally to the postpalatine process while the internal carotid artery runs ventrally to it.

In the occipital region the occipital arches, tall and slender, and still free from the parachordals, slope upwards and forwards from the notochord, but do not yet touch the auditory capsule. The fissura metotica through which the vagus nerve emerges, is therefore still bounded medially by the notochord, and is not yet enclosed to form a vagus foramen.

The pterygoquadrate has a well-developed pterygoid process articulating with the side of the ethmoid plate, and, posteriorly a large blunt process pointing towards the auditory capsule; this appears to be a vestige of the otic process, or metapterygoid process as figured by Sewertzoff (1926*b*) in Polypterus delhesi.

Meckel's cartilage is also well developed, meeting but not fused to its fellow of the opposite side. The hyosymplectic forms a single cartilage, articulated with but not fused to the auditory capsule. An interhyal is now present between the hyosymplectic and the ceratohyal, at the ventral end of which the paired hyohyal has appeared.

Ceratobranchial cartilages are present in the first three branchial arches and hypobranchials in the first two: a broad median cartilage represents the basibranchial of the first two arches: there is no basihyal.

3. Stage 3. (9·3 mm.) PLATE 25, FIGS. 1, 2, 3.

The chief advances shown by this stage are the completion of the lateral commissure: the appearance of the orbital cartilages, as independent bars midway between the laminae orbitonasales and the auditory capsules; the appearance of the rudiments of the tectum synoticum; the attachment of the dorsal end of the occipital arches to the auditory capsule; and the fusion of certain cartilages previously separate. Thus, the two Meckel's cartilages fuse at the median symphysis; the hyosymplectic fuses with the pterygoquadrate, and the hypobranchials fuse with the ceratobranchials, of which there are now four, which is the total number of branchial arches in Polypterus. The basibranchial has elongated posteriorly and articulates with all four cerato-hypobranchials.

4. Stage 4. (30 mm.) PLATES 26, 27, 29.

The chondrocranium has now practically achieved the form of the adult. The orbital cartilage has become attached anteriorly to the trabecula by means of its preoptic root, to the dorsal edge of the nasal septum (by the spheno-septal commissure), and to the lamina orbitonasalis. Posteriorly, the orbital cartilage is attached to the postorbital process of the auditory capsule and to the supraorbital cartilage. The orbital cartilages are also interconnected by a slender transverse epiphysial bar.

The gap between the preoptic root of the orbital cartilage and the posterior edge of the nasal septum is the large foramen olfactorium evehens, through which the olfactory nerve passes from the cranial cavity into that of the nasal capsule. The line of fusion between the preoptic root of the orbital cartilage and the lamina orbitonasalis is marked by the position of the orbitonasal foramen, through which the profundus nerve leaves the orbit, runs for a short distance beneath the cartilaginous roof of the nasal capsule, and reaches the dorsal surface of the snout. The superior oblique muscle is attached to the orbital cartilage in the mouth of the orbitonasal foramen, which thus comes to function as a rudimentary anterior myodome.

The nasal capsules are well developed. Each one has a floor (the ethmoid plate), a median wall (the nasal septum), a hind wall (the lamina orbitonasalis), a roof and front wall (formed by extension of cartilage between the nasal septum and lamina orbitonasalis), and a large lateral nasal opening.

The supraorbital cartilage appears to be developed as an anterior prolongation of the horizontal crest or crista parotica which projects on the lateral wall of the auditory capsule. But instead of remaining in contact with the dorsolateral edge of the orbital cartilage all the way along (except, of course, for the foramina for the twigs of the superficial ophthalmic nerve innervating the sense-organs of the supraorbital lateral line canal), the supraorbital cartilage becomes separated from the orbital cartilages by an elongated space, the supraorbital fontanelle, which lodges the temporal muscle.

Cartilage has now appeared in the side wall of the skull in the hinder part of the orbital region, and there are now foramina for the oculomotor and profundus, for the trigeminal and abducens and for the superficial ophthalmic (facial) nerves, and also for the pituitary vein. However, the cartilage which forms the anterior boundary to the trigeminal foramen does not answer the requirements of a pila antotica, for the profundus which should emerge behind the pila antotica, emerges through the same foramen as the oculomotor (which always emerges in front of the pila antotica in other forms). The conclusion which must be drawn is that there is no pila antotica in Polypterus, and that the cartilage which forms the side wall of the skull in this region is perhaps a secondary development.

The auditory capsules are now interconnected by a tectum synoticum, while in the region of the basal plate the parachordals are connected with one another by cartilage passing under the ventral surface of the notochord. As in all Teleostomes, the medial wall of the auditory capsule remains unchondrified.

The occipital region presents some features of importance. Each parachordal sends back two strips of cartilage, one along the dorsolateral and another along the ventrolateral sides of the notochord; the former joins on to the base of the occipital arch, so that the vagus foramen is completely encircled by cartilage. It is also to be noted that the occipital arch is but the enlarged anterior member of the series of vertebral neural arches. The arch immediately behind the occipital arch will become the proatlas arch of the adult, leaving the following one to be the neural arch of the 1st vertebra.

The ventrolateral cartilaginous strips are interconnected by a transverse arch of cartilage, ventral to the notochord, resembling a haemal arch and forming the posterior opening to the aortic canal. Anteriorly, the aortic canal forks into two because the centre of the haemal arch is connected with the under surface of the parachordal plate by a piece of cartilage running forwards and upwards in the middle line. There are, therefore, paired anterior openings to the aortic canal.

A pair of labial cartilages is now present, lying dorsally to Meckel's cartilages.

The pterygoquadrate has again become free from the hyosymplectic, and the otic process has become greatly reduced. The two ends of the hyosymplectic have become swollen, giving it the characteristic appearance of the adult. To the middle of the hyosymplectic on the posterior side is attached the cartilage which supports the external gill.

The hypobranchials and ceratobranchials of the first three branchial arches are now separate again.

ii. Morphology of the chondrocranium of Polypterus (76 mm.), and relations of blood-vessels and nerves

(PLATES 28, 29.)

Anteriorly, the nasal septum is produced into a short rostrum with two processes, one above the other, which almost surround the ethmoid transverse commissure between the anterior ends of the infraorbital lateral line canals.

The composition of the nasal capsules has been described in connexion with the 30-mm. stage. It is, however, necessary to insist on them because, with the exception of Acipenser, a complete nasal capsule is not formed in the higher bony fish. This fact appears to be correlated with the dorsal position of the nasal apertures in those forms, resulting in a total reduction of the roof of the capsule. In Polypterus (as in Acipenser) the nasal apertures are lateral, in which position they do not interfere with the roof or front wall of the capsule.

The orbitonasal foramen is worthy of consideration, for as already explained, it marks the line of attachment of the lamina orbitonasalis to the preoptic root of the orbital cartilage, and serves for the exit of the superficial ophthalmic and profundus nerves and of the ophthalmic artery from the orbit, at the same time as forming the anterior myodome for the superior oblique muscle. The cartilage overlying the orbitonasal foramen corresponds to the sphenethmoid commissure of other forms.

As soon as they have passed through the mouth of the orbitonasal foramen, the nerves find themselves in a short cartilaginous canal. Proceeding forward, the cartilaginous floor ceases and the nerves seem to lie inside the cavity of the nasal capsule, from which they emerge partly through an epiphanial foramen in the roof and partly through the lateral narial opening (fenestra narina).

The pterygoid processes of the pterygoquadrates are articulated with the posterolateral corners of the ethmoid plate.

The eye is surrounded by a well-developed sclerotic cartilage.

The foramina for the optic nerve, oculomotor and profundus nerves, superficial ophthalmic nerve, trigeminal and abducens nerves have already been mentioned in connexion with the absence of a proper pila antotica.

The facial foramen (between the prefacial commissure and the anterior basicapsular commissure) allows the facial nerve (palatine and hyomandibular branches) to leave the cranial cavity and enter the jugular canal. There, the palatine branch runs forwards emerging from the anterior opening, while the hyomandibular branch turns backwards and emerges through the posterior opening of the jugular canal. The canal is also traversed by the head vein (which runs dorsally to the nerves), with the result that the conditions in Polypterus are very similar to those found in Squalus, and the jugular canal corresponds to the pars jugularis of the trigeminofacialis chamber of higher bony fish.

A short distance in front of the jugular canal the pituitary vein is given off by the head vein and pierces a foramen in the side wall of the skull. Having passed through this foramen, the pituitary vein finds itself in the subpituitary space or cavum sacci vasculosi, which now has a cartilaginous roof in the form of the prootic bridge, or dorsum sellae. Two small foramina indicate remnants of the basicranial fenestra between the prootic bridge and the parachordals. The relation of the subpituitary space are thus very similar to those found in Selachii and to those found in higher fish possessing a posterior myodome. In Polypterus the hypophysial fenestra remains widely open. In the basal plate the notochord becomes suppressed except at the hind end. The parachordals fuse with one another in the midline both above and beneath the notochord.

Attention must now be turned to the remarkable course of the dorsal aorta and internal carotid (Allis, 1908*a*). The dorsal aorta enters the hind opening of the aortic canal and forks, the lateral dorsal aortae emerging from the paired anterior apertures of the aortic canal, immediately beneath the basal plate. Each lateral dorsal artery then continues forwards beneath the chondrocranium and dorsally to the parasphenoid bone, in a canal which may be called the parabasal canal. There it receives the 1st efferent branchial and the efferent hyoidean arteries, and continues forwards as the common carotid. Opposite the anterior opening of the jugular canal the common carotid gives off the orbital artery (ophthalmic) which runs upwards on the medial side of the palatine nerve which has emerged from the jugular canal and runs down to accompany the carotid forwards, lying laterally to it. The carotid then gives off a few arteries (ophthalmica magna, maxillo-mandibular branches of the orbital) and runs inwards to enter the cranial cavity passing *dorsally* to the trabeculae (through a foramen at first confluent with the optic foramen, and subsequently isolated but close to it).

This exceptional arrangement is already present at the 30-mm. stage, but unfortunately no information is available as to the condition in earlier stages. It is, therefore, not possible to say definitely whether this is the true internal carotid artery bearing atypical relations to the trabeculae, or whether (as in the other cases where this relation is found) the so-called carotid in Polypterus is a substituted vessel. The latter alternative is the more probable in view of the fact that the efferent pseudobranchial artery seems to have been lost, and that the lateral wall of the skull in this region shows traces of secondary modifications (absence of a pila antotica) probably connected with the loss of a functional myodome.

Palaeontological investigations (e.g. those of Watson, 1925*b*) have rendered it almost certain that all the living, bony fish which lack a functional posterior myodome (Polypterus, Acipenser, Lepidosteus, Amiurus, Gadus, &c.) have lost it, in the sense that the external rectus muscles no longer penetrate through the cartilaginous cranial wall into the subpituitary space. If, as is supposed, the muscle penetrated through the pituitary vein foramen, the result must have been to disturb the pila antotica. Hence, a reformation of the side wall of the skull in this region might be expected to show an atypical condition.

Polypterus presents a further point of interest in that the pituitary body retains connexion with the roof of the mouth, by means of a persistent hollow hypophysis or Rathke's pocket (de Beer, 1926*c*), which perforates the parasphenoid bone. It will be remembered that the hypophysial fenestra is not occluded by cartilage. Such an arrangement with the pituitary body lodged in the pituitary fossa could not exist in a fish with a functional myodome and the eye-muscles occupying the subpituitary space. If, therefore, as it seems necessary to believe, Polypterus has lost its myodome, the retention throughout life of a hollow hypophysis, which is a transient embryonic condition in other forms, is in Polypterus a secondary feature: a failure in this respect to develop beyond the embryonic condition: a condition which is the reverse of phylogenetically primitive.

The crista parotica projects as a well-marked horizontal crest on the lateral wall of the auditory capsule, and posteriorly it envelopes the supratemporal branch of the glossopharyngeal nerve in a foramen.

In the occipital region one vertebral centrum has become absorbed into the hind end of the skull, leaving its neural arch free to form a proatlas basidorsal arch, between the occipital arch and the 1st or atlas vertebra. At the 76-mm. stage the occipital arch is pierced by two foramina for hypoglossal ventral nerve-roots: one near its anterior and the other near its posterior border. The anterior nerve at previous stages passed through the vagus foramen in front of the occipital arch; the posterior nerve has corresponding to it a dorsal nerve-root which passes through a notch in the posterior surface of the occipital arch. The proatlas arch, like the neural arch of the 1st vertebra, is pierced by both a ventral and a dorsal nerve-root foramen.

The roof of the chondrocranium is represented only by the epiphysial bar and the tectum synoticum.

In the mandibular arch the pterygoquadrate at this stage is somewhat featureless; the otic process has disappeared, and the only contact with the neurocranium is by means of the pterygoid process which articulates with the posterolateral corner of the ventral surface of the ethmoid plate.

In the lower jaw the labial cartilage resembles in some respects the large coronoid process which characterizes Meckel's cartilage in Amia and Lepidosteus.

The hyoid arch is of importance because of the relations of the hyomandibula to the hyomandibular branch of the facial nerve. After emerging from the posterior opening of the jugular canal, this nerve splits into its mandibular and hyoidean rami which run in such a way that the ramus mandibularis passes outwards *in front* of the hyomandibula, while the ramus hyoideus passes outwards *behind* that cartilage. This relation is quite unique.

The first three branchial arches present the interesting feature of a subdivision of their dorsal ends into so-called infrapharyngobranchials and suprapharyngobranchials (van Wijhe, 1882*b*). Of these, the infrapharyngobranchials represent the true pharyngobranchials, while the suprapharyngobranchials are processes of the epibranchials, directed dorsally as Allis (1918) showed. The glossopharyngeal nerve runs outwards posteriorly to both pharyngo- and epibranchials of the 1st arch.

There appear to be no dorsal elements in the 4th branchial arch.

The branchial arches conform to the V-type, characteristic of all Teleostomi, i.e. the arch is bent into a V with the apex between the epi- and ceratobranchials. The pharyngo- and hypobranchials project forwards instead of backwards as in the sigma type (see p. 52).

Branchial rays are present and they become fused together at their base producing comb-like structures.

iii. The development of the osteocranium of Polypterus

(PLATE 29, FIGS. 5, 6, 7.)

1. The membrane-bones. The development of those bones which are related to the lateral line canals has been studied in 35-mm. larvae of Polypterus bichir by Pehrson (1922). The bones, however, are already well formed, and nothing is known as to their earliest development.

These bones include:

The *post-temporal*; the *extrascapulars* (all three (on each side) apparently corresponding to the so-called postparietal of Amia and traversed by the transverse occipital lateral line canal); the *parietal*, through the lateral edge of which the lateral line canal passes (for which reason certain authorities prefer to regard it as the *intertemporal*, see p. 508); the *frontal*; the *nasal*, *adnasal*, and *terminale* (all three apparently corresponding to the so-called nasal of Amia); the *supra-ethmoid* (traversed by the transverse canal joining the anterior ends of the infraorbital lateral line canals); the *postfrontal* (which is fused to the sphenotic cartilage-bone); the *postorbital*; the *maxilla* or *suborbital* (tooth-bearing); the *lachrymal*; the *premaxilla* (tooth-bearing); the *preopercular*; the *angular*; and the *dentary* (tooth-bearing).

It is possible that part of the dentary which lodges the lateral line canal may represent a separate independent ossification, the *splenial*, in which case the whole bone would be a dentalosplenial.

A curious feature of these bones of Polypterus, as compared with those of other fish (e.g. Amia), is that in some cases they are separate where in other fish they are fused together (terminale, adnasal, and nasal; postparietals and extrascapulars); while in other cases Polypterus has only one bone where other fish have two or more. This has been held to mean that in Polypterus there has been fusion between bones which in other fish are separate. This, for instance, may be the case in the maxilla, premaxilla, and parietal, which in other fish are not as a rule traversed by lateral line canals. The maxilla is therefore by some authors regarded as having fused with two *suborbitals* (one of them probably a jugal). Because it does not lie in the lateral labial fold, Sewertzoff (1925*b*) does not think that it is a maxilla at all, but a suborbital to which teeth are attached. This is discussed on page 508. The premaxilla may similarly have fused with the *antorbital* and the lateral portion of the supra-ethmoid of Amia; the *parietal* with the *intertemporal*. (See Allis, 1900, 1919c, 1922). The *preopercular* has been regarded as including the *squamosal*, which is unlikely, since the squamosal seems to be lacking in Palaeoniscoids.

Further investigation on this point, to see whether bones which topographical comparison suggests as compound really do have a compound developmental origin, is badly needed and would assuredly give interesting results. The possibility must be borne in mind that instead of two bones fusing, one may have disappeared and become replaced by extension of the other, and, on the whole, this is the most probable explanation for the suborbital and intertemporal.

In addition to the above mentioned, the superficial membrane-bones include a row of *spiracular* ossicles, a variable number of cheek-bones, the *opercular*, *subopercular*, and *lateral gulars*. Bones usually known as 'Q_1' and 'Q_2', probably represent the maxilla, if, as just explained, the bone usually called the maxilla is the suborbital.

Lastly, the membrane-bones include the following:

The *parasphenoid*, a very large flat bone underlying the neurocranium almost throughout its length, it is perforated by the hypophysial stalk, and gives off extensive lateral wings which are closely applied to the cartilage of the auditory capsule, even invading it in the adult according to Allis (its anterior portion which bears teeth has been held to represent the fused prevomer).

The *subrostral*, a small median toothed bone, was regarded by Allis (1898) as the fused (pre)vomers, but probably represents a basirostral of Acipenser.

The *prevomer*, represented by the paired toothed bone which Allis has called 'mesial dermopalatine'.

Among the bones of the visceral arch skeleton, the premaxilla and maxilla have already been dealt with.

The *dermopalatine*, tooth-bearing ('lateral dermopalatine' of Allis), is fused on to the anterior end of the ectopterygoid, the line of fusion being indicated by a foramen through which a branch of the palatine nerve passes. The dermopalatine is attached also to a cartilage-bone, the autopalatine.

The *ectopterygoid* (tooth-bearing) is developed as a sheath on the median ventral side of the pterygoquadrate cartilage. As already mentioned, its anterior end is held to represent the dermopalatine.

The *endopterygoid* (tooth-bearing) lies dorsally to the ectopterygoid, covering the dorso-median surface of the pterygoquadrate cartilage.

The *metapterygoid*, curiously enough, seems here to be a membrane-bone, developed outside the perichondrium of the posterodorsal region of the pterygoquadrate cartilage, and bears teeth.

An alternative interpretation of these bones is that the whole of the so-called ectopterygoid is the dermopalatine, and that the so-called metapterygoid is the ectopterygoid, the metapterygoid being absent.

The bones of the lower jaw have been studied by Schmäh (1934). The dentary and angular have already been dealt with.

The *prearticular* is a large tooth-bearing bone flanking the inner side and part of the upper surface of Meckel's cartilage. It is produced into a marked uprising coronoid process which Allis regards as representing the *coronoid* bone of reptiles. At all events, this process of the prearticular is a purely membranous ossification and has nothing to do with the lateral cartilage, or cartilaginous coronoid process of Amia.

The *coronoids* (splenials of some authors), anterior and posterior, are small tooth-bearing bones on the dorsal side of the anterior part of Meckel's cartilage.

A *pharyngeum inferius* is developed over the 4th ceratobranchial, and bears teeth.

The *parahyoid* (Fuchs 1929) is a Y-shaped bone lying between the coraco-hyoid muscles ventrally to the copula, the paired prongs directed forwards. This bone, probably of paired origin, may be homologous with the 'urohyal' of Teleostei (p. 130) and the parahyoid of Anura (p. 212). It is a tendon ossification and does not properly belong to the hyobranchial skeleton.

2. The cartilage-bones (PLATE 28). Information concerning the origin of cartilage-bones of the neurocranium is restricted to Lehn's (1918) studies of 55-mm., 76-mm., and 90-mm. larvae, and Allis's (1922) of a 75-mm. specimen, and to inferences drawn from the morphology of the adult, based on the works of van Wijhe (1882*b*), Pollard (1892), and Allis (1922). Schmäh (1934) has followed the development of the lower jaw.

The *occipital* is a large ossification in the hinder part of the basal plate, (enclosing the aortic canal) and in the occipital arches meeting in the middle line above the foramen magnum. It represents a fusion between the *basioccipital* and *exoccipitals* of other forms; even at 56 mm. the bone is a single one.

The so-called *opisthotic* is a perichondral bone at the posterolateral corner of the auditory capsule, on both external and internal surfaces. Fairly small at 75 mm., it becomes very large in the adult, and has been held to incorporate the *epiotic* of other fish, together with the *pterotic* and even the prootic. The latter is very unlikely. It is, however, clear that the posterior part of the opisthotic contains regions of intramembranous origin which may correspond to the *intercalary* of other fish. It is not, therefore, possible to claim the so-called opisthotic of Polypterus as a proof of the perichondral origin of the (restricted) opisthotic of other bony fish, since its perichondral portion may represent the epiotic or pterotic or both.

The *prootic* is absent, but its place has been taken by the lateral ascending wing of the parasphenoid.

The *sphenotic* ossifies in the supraorbital cartilage and postorbital process, and forms the lateral border to the supraorbital fontanelle. It is fused with the post-frontal, but it is unknown whether this fusion takes place after a separate origin, or whether the two elements are connected at the outset of ossification.

The *sphenoid* is a large bone in the lateral wall of the orbital region of the skull. Mostly ossifying in cartilage, a certain portion of it ossifies in the membrane spanning the optic foramen. In the adult the sphenoid occupies the entire side wall of the orbital region, the region of the paired trabeculae, the prootic bridge, and the hind part of the nasal septum. It seems, therefore, to represent a fusion between the orbitosphenoid, pleurosphenoid, basi-sphenoid, and perhaps part of the prootic, which latter bone always ossifies in the prootic bridge in higher bony fish.

The *lateral ethmoid* develops late (90-mm. larvae) as a perichondral ossifica-tion in the hind surface of the lamina orbitonasalis.

In the pterygoquadrate cartilage two perichondral ossifications are found:

The *autopalatine*, which arises in the anterior part of the pterygoid process and eventually fuses with that anterior portion of the ectopterygoid which is held to represent the dermopalatine.

The *quadrate*.

In the lower jaw, there are:

the *mentomeckelian*, near but not quite at the tip of Meckel's cartilage;

the *mediomeckelian* (mediomandibular of Schmäh), which ossifies about half-way along Meckel's cartilage;

the *articular*, which ossifies at the articular facet near the hind end of Meckel's cartilage, and in the retroarticular process according to Haines (in press). A separate *retroarticular* capping the retroarticular process was described by van Wijhe, but not found by Allis or Schmäh in their specimens. These four Meckelian ossifications convey the impression of representing fragments of an originally continuous Meckelian bone, as found in Ospia (Stensiö, 1932). The same impression is gained from a study of Amia.

The visceral arches are extensively ossified.

In the hyoid arch there are: *hyomandibula*, *interhyal*, *ceratohyal*, and *hypohyal* (the so-called accessory hyomandibula appears to be a spiracular ossicle—therefore dermal) attached to the posterior side of the dorsal end of the hyomandibula; though it may come into relation with the cartilage which in the larva supports the external gill.

The ossifications in the branchial arches are: 4 *ceratobranchials*, 3 *hypobranchials*, 2 *pharyngobranchials* (infrapharyngobranchials), 1 *epibranchial* (suprapharyngobranchial), counting in each case from the front; and the basibranchial. The ceratobranchial of the 4th arch is covered by a pharyngeum inferius bearing teeth.

8. CHONDROSTEI. *ACIPENSEROIDEI*

Acipenser ruthenus

The development of the skull of the sterlet has been studied by Salensky (1881) in a monograph which has fortunately been reprinted. W. K. Parker with his usual energy and skill, but with the drawbacks attendant on his methods, described the development (1882*a*), in which same year van Wijhe (1882*b*) worked out the morphology of the visceral arches. Since that date a few stages of development of the chondrocranium have been studied with the help of wax-model reconstructions by de Beer (1925), while a more thorough investigation based on graphic reconstructions has been contributed by Sewertzoff (1928), who has also described the relations of the nerves (1911), and the development of the dermal bones (1926*a*). The morphology of certain structures in the head has been the subject of studies by Luther (1913), and Kurz (1924).

i. The development of the chondrocranium of Acipenser ruthenus

The following account is largely based on Sewertzoff's and de Beer's work. Unfortunately, Sewertzoff does not indicate the precise ages or lengths of his specimens: the various stages can therefore only be described as 'early', 'later', &c.

1. EARLIEST STAGE. PLATE 30, FIG. 1.

The first part of the skull to chondrify is the parachordals. These extend by the side of the notochord (and in contact with it all along) from a point just behind its tip to the level of the 1st myomere. The foremost ends of the parachordals are interconnected by a bar of cartilage passing ventrally to the notochord.

The trabeculae and polar cartilages are not yet chondrified, but are represented by paired bars of dense tissue extending forwards and continuing forward in the same line as the ventral edge of the parachordals. According to Sewertzoff (1916) the sclerotome of the 1st or premandibular segment contributes not to the trabecula, but to the polar cartilage and parachordal. It is important to note that the internal carotid arteries enter the cranial cavity by passing, not medially but *laterally* to the trabeculae. This most exceptional relation is due to the fact that even at very early stages, as described by Neumeyer (1932), the internal carotids are remarkably wide apart, as indeed they have to be in the adult owing to the great width of the parasphenoid, over the edge of which they pass.

Each auditory capsule is represented by two centres of chondrification: the anterior one is connected with the lateral edge of the parachordal by procartilage (rudiment of the anterior basicapsular commissure), while the posterior one is isolated.

The rudiments of the mandibular and hyoid arch skeletal elements are still mesenchymatous.

2. STAGE 2. PLATE 30, FIGS. 2, 3.

At this slightly older stage the polar cartilages are chondrified in connexion with the ventral edge of the anterior end of the parachordals. The anterior end of the parachordals is remarkably thick, and its dorsal edge forms a transverse ridge, now in front of the tip of the notochord, and projecting forwards over the subpituitary space.

The trabeculae are also chondrified but are connected with the polar cartilages only by mesenchyme. The gap between them is indicated by a notch in the lateral edge of the trabeculopolar bar, through which the internal carotid artery runs upwards and inwards into the cranial cavity.

Sewertzoff lays great stress on the fact that although the trabeculopolar bar is in the direct line of prolongation of the parachordals, as in Holostei and Teleostei, it is attached to the ventral surface of the anterior edge of the parachordals as in Selachii. It will, however, be seen that in such forms as Amia and Salmo, the trabeculopolar bar runs forwards at a level which is more ventral than that of the main parachordal plate and prootic bridge. This depression of the region of the polar cartilage is associated with the formation of the posterior myodome, for the external rectus muscle passes in over the polar cartilage, and presses it down. It must be confessed that in this connexion, although Acipenser has no myodome, the cartilages present much more resemblance to the conditions in Holostei and Teleostei than to the Selachii.

The two rudiments of each auditory capsule have now joined, and each capsule is attached to the parachordal by an anterior and a posterior basicapsular commissure. In this way a basicapsular fenestra is enclosed, together with a portion of the fissura metotica, for the glossopharyngeal nerve is trapped on the anterior side of the posterior basicapsular commissure.

The parachordals now extend much farther back than at the previous stage. In the hinder part of the parachordals, the rudiments are seen of four occipital arches, corresponding in position to the septa separating the 4th and 5th, 5th and 6th, 6th and 7th, 7th and 8th metotic somites. The hindmost of these arches therefore corresponds to the hinder limit of the 10th segment.

3. Stage 3. PLATE 30, FIGS. 4, 5.

The trabeculae and polar cartilages are now in cartilaginous connexion, and a plate of cartilage has spread across the middle line forming a complete floor to the subpituitary space. As the lateral edges of the trabecular and polar cartilages extend still farther to the side, so the notch for the internal carotid becomes deeper.

Each trabecula gives off laterally a lamina orbitonasalis, which projects behind the nasal sac. The orbital cartilages are now present as independent chondrifications at the sides of the forebrain, dorsally to the optic nerves.

The auditory capsules now show a side wall with well-formed anterior and posterior cupolae. From the anterior cupola a large postorbital process projects forward towards, but does not reach, the orbital cartilage. In the floor of the auditory capsule the foramen for the glossopharyngeal nerve is reduced to small dimensions.

The four occipital arches of the previous stage have joined dorsally, above the three thereby enclosed hypoglossal foramina, to form a broad occipital arch which rises up behind the auditory capsule, from which it is separated by the fissura metotica through which the vagus nerve emerges. Behind the occipital arches some rudiments of neural (occipitospinal) arches may be seen, which will ultimately become attached to the skull.

In the visceral arch skeleton several cartilages have appeared. The pterygoquadrates meet one another in a symphysis in the middle line, ventrally to the trabeculae. Meckel's cartilages call for no special comment, except that their distal ends are in contact with a procartilaginous median basimandibular element.

At early stages the mesenchymatous rudiment of the hyoid arch skeleton does not reach the auditory capsule, and the hyomandibular branch of the facial nerve runs over its dorsal end. Later, however, the mesenchymatous rudiment establishes contact with the auditory capsule in such a way that the hyomandibular nerve passes outwards behind it. Five separate elements become chondrified in the hyoid arch: the hyomandibula, symplectic, interhyal, ceratohyal, and hyohyal. There is no basihyal.

4. Stage 4. PLATE 30, FIGS. 6, 7, 8.

A great advance has now been made in the ethmoid region of the skull. The trabeculae have become joined anteriorly, forming a trabecular plate, and enclosing the hypophysial fenestra. On each side of the trabecular plate a process is given off laterally: the cornua trabecularum. The orbital cartilage on each side has become attached to the postorbital process, to the parachordal by means of the pila antotica, and to the trabecula by means of the preoptic root of the orbital cartilage. The latter is an extensive plate, forming the side wall to the anterior region of the cranial cavity and the medial wall to the hinder part of the nasal capsule. The anterior margin of the preoptic root of the orbital cartilage forms the posterior border of the foramen (or at this stage still the incisura) olfactorium evehens.

The hind wall of the nasal capsule is formed by the lamina orbitonasalis, which rises up with its medial edge in contact with the preoptic root of the orbital cartilage: the floor of the nasal capsule is formed by a spread of cartilage between the ventral edge of the lamina orbitonasalis and the cornu trabeculae.

There is as yet no median nasal septum nor a rostrum. The anterior borders of the foramina olfactoria evehentia are formed by paired cartilaginous pillars (pilae ethmoidales) between which is a fenestra precerebralis, looking straight back into the front end of the cranial cavity. These pillars, which correspond to the medial walls of the nasal capsules of Selachii and of Urodela, are ultimately merged in the large median nasal septum and rostrum.

The internal carotid arteries are now enclosed in foramina, owing to the lateral development of the cartilage of the trabeculae: these foramina are, however, almost in the side wall of the skull instead of being in its floor, and reflect the atypical course taken by the internal carotid arteries at early stages.

Anteroventrally to the carotid foramen, the edge of the trabecula is produced into the basitrabecular process, which is connected by a ligament with the pterygoquadrate. (In the figures of chondrocrania given by Edgeworth (1935) the basitrabecular process reaches into close proximity with the pterygoquadrate: the latter has not yet dropped away from the chondrocranium.)

The parachordals are interconnected by cartilage beneath the notochord which appears to persist in the basal plate throughout life. It would be interesting to know whether the notochordal sheath is invaded by the cartilage cells of the parachordals.

Immediately in front of the articular facet for the hyomandibula, a short tunnel or canal is formed by the junction between a down-growing prootic process from the wall of the auditory capsule and an out-growing postpalatine process from the edge of the basal (parachordal) plate. Through this jugular canal there run the hyomandibular branch of the facial nerve, the head vein, and the orbitonasal artery.

The auditory capsule is complete, and has the septa enclosing the semicircular canals. The medial wall remains membranous. Externally, the posterolateral corner of the capsule is produced backwards into a process (corresponding to the epiotic region of other fish) to which the post-temporal ligament is attached.

The auditory capsule and occipital arch are connected over the fissura metotica which is thus converted into the vagus foramen, and the occipital arches are also connected with their fellows of the opposite side by a cartilaginous bridge over the cranial cavity, representing the tectum posterius. A tectum synoticum is formed later, farther forward.

In the visceral arch skeleton some important features may be noticed. Immediately behind the symphysis of the pterygoquadrates, in the roof of the mouth, there appear a number of cartilages to which Parker applied the term metapterygoid, while Sewertzoff calls them postpalatine cartilages. They appear to be peculiar to these fish, and are presumably correlated with the fact that the roof of the mouth is a long way ventral to the trabecular plate. The cartilages in question may be called simply the mouth-roof cartilages. In certain species of Acipenser, Bugajew (1929) has described a varying number of cartilaginous nodules in the ligament connecting the basitrabecular process with the pterygoquadrate. These he regards as representatives of the pharyngomandibular element.

The skeleton of the branchial arches is represented by pharyngal, epal, and hypal elements in the first four arches, and ceratal elements in all five. The 4th hypo- and ceratobranchials subsequently fuse, as also do the 4th pharyngo- and epibranchials. A modification arises in connexion with the first two branchial arches. Here, the pharyngobranchials touch the auditory capsule, but so also do the epibranchials by means of a process directed backwards and upwards. The dorsal end of these branchial arches are thus forked into the so-called infra- and suprapharyngobranchials. The latter are the processes of the epibranchials while the former are the true pharyngobranchials. The glossopharyngeal and vagus nerves pass in front of both pharyngo- and epibranchial of their respective arches.

In the mid-ventral line, four median copulae are present, representing the basibranchials. There is no basihyal, but a basimandibular is present.

ii. Morphology of the chondrocranium and relations of the blood-vessels and nerves of Acipenser stellatus (10 days)

(PLATE 31.)

Anteriorly the skull is produced into a median rostrum, the hinder part of which forms a septum between the anterior portions of the nasal capsules. The rostrum is a massive structure of cartilage, with a ridge or carina in the middle line of its ventral surface, and lateral ridges which are continuous with the cornua trabecularum farther back. This rostrum is thus quite different from that of Selachii. The fenestra precerebralis of earlier stages is now completely obliterated. The nasal capsules may be likened to square boxes with the sides left open, forming the wide fenestrae narinae lodging both incurrent and excurrent nasal apertures. The floor is formed by the ethmoid plate of cartilage stretched between the cornu trabeculae and the lamina orbitonasalis; the hind wall is formed by the lamina orbitonasalis; the medial wall is formed by the preoptic root of the orbital cartilage behind, and the nasal septum in front, the foramen olfactorium evehens being situated between them; the roof and front wall are formed by cartilaginous extensions from the orbital cartilage and the nasal septum.

The line of demarcation between the lamina orbitonasalis and the preoptic root of the orbital cartilage (the latter representing the true side wall of the cranial cavity) is indicated by the position of two foramina or passages: the orbitonasal foramen and the orbitonasal canal. The former lies in the antero-medio-dorsal corner of the orbit and transmits the ophthalmic branch of the trigeminal nerve: the posterior mouth of the orbitonasal foramen is occupied by the attachment of the superior oblique muscle, forming an incipient anterior myodome. The orbitonasal canal conveys a vein from the orbital sinus to the region beneath the nasal capsule.

The eye is surrounded by a well-developed sclerotic cartilage.

The basitrabecular process bears typical relations, being situated over and in front of the palatine nerve and beneath the head vein.

The side wall of the orbital region is pierced by a number of foramina, for the trochlear, optic, and oculomotor nerves, and for the pituitary vein: all these lie in front of the pila antotica, which separates them in typical manner from the prootic foramen. There is also the foramen for the internal carotid artery, which is situated in the lowest part of the side wall. Through the prootic foramen emerge all the branches of trigeminal and facial nerves, and the abducens. There appears to be no prefacial commissure. The ophthalmic branch of the facial runs forwards and upwards in the orbit and reaches the dorsal surface of the snout. The ophthalmic branch of the trigeminal is not a profundus, for it passes dorsally to all branches of the oculomotor and dorsally to the trochlear. It leaves the orbit by the orbitonasal foramen. The maxillary and mandibular branches of the trigeminal and the buccal branch of the facial call for no comment. The otic branch of the facial nerve runs in a dorsolateral direction, and is enclosed (together with the dorsal spiracular recess, or Wright's organ) in a foramen between the base of the postorbital process and the cartilage of the crista parotica, which forms a horizontal ridge on the lateral wall of the auditory capsule, and is continued as a sort of supraorbital cartilage along the dorsolateral edge of the orbital cartilage. The palatine branch of the facial emerges from the prootic foramen and runs down laterally to the edge of the parachordal and forwards beneath the basitrabecular process. On its way, the nerve runs laterally to the internal carotid and medially to the efferent pseudobranchial artery.

The hyomandibular branch of the facial nerve runs backwards from the prootic foramen and threads its way through the jugular canal, in company with the head vein and orbital artery. The jugular canal is bounded laterally by a lateral commissure, formed by the prootic process of the auditory capsule growing downwards and inwards to meet the postpalatine process jutting out from the parachordal. Schmahlhausen (1923*a*) has attempted to regard the wall of the jugular canal as a pharyngohyal cartilage, but this cannot be accepted.

The jugular canal, like that of Polypterus (and differing from that form only in that here the orbital artery also traverses the canal), corresponds to the pars jugularis of the trigeminofacialis chamber of higher fish.

The articular facet for the head of the hyomandibula is just dorsal to the jugular canal. When, therefore, the hyomandibular nerve emerges from the hind opening of the jugular canal, it finds itself behind the hyomandibula, and runs outwards and downwards, both its mandibular and hyoidean branches curving round the posterior surface of the hyomandibula. The ramus hyoideus runs downwards to the region of the ceratohyal, while the ramus mandibularis passes on to the lateral surface of the hyomandibula and then runs on the medial side of the symplectic and Meckel's cartilage, thereby showing itself to be the ramus mandibularis internus facialis. There is apparently no ramus externus. It is to be noticed that the relations of the facial nerve to the hyomandibula are *not* the same as those of the glossopharyngeal to the 1st branchial arch.

Internally, an interesting development is shown at the base of each pila antotica, for here the rudiments of the dorsum sellae project from the side inwards towards the middle line: when they meet there will be formed a cartilaginous roof to the cavum sacci vasculosi or subpituitary space. It is into this space that the foramina for the internal carotid arteries and pituitary veins open. The previously membranous roof of the subpituitary space has therefore occupied the morphological position of a basicranial fenestra.

The cavum sacci vasculosi thus corresponds to that of Polypterus, and to the dorsal compartment of the posterior myodome of Teleostei.

In the auditory region the capsules are interconnected by a tectum synoticum. The glossopharyngeal and vagus foramina are separate, these nerves having been separated by the posterior basicapsular commissure.

The later stages of larval and adult life are characterized by a progressive development and thickening of cartilage. This shows itself in the formation of a complete cartilaginous roof to the brain-case, and in the plastering on to the hind end of the skull of some five vertebrae, the neural arches of which thus become occipitospinal arches, and the ribs cranial ribs. This addition brings the number of segments which contribute to the skull up to fifteen. The spinal nerves enclosed between these vertebrae preserve their dorsal roots.

It may be noted that these added vertebrae are composed of basal (posterior sclerotomite) and intercalary (anterior sclerotomite) elements. The occipital arches preceding them, however, do not exhibit this diplospondyly but are formed solely from basalia, the reason being, according to Sewertzoff, that they arise much earlier than the time of appearance of the intercalaries.

Another sign of the excessive development of cartilage may be seen on the ventral surface of the rostrum and trabecular plate. Two membrane-bones develop close to this surface—the prevomer and the parasphenoid. The hind end of the former and the front end of the latter become completely covered by cartilage, developed from each side and fusing ventrally to the bones to form the blunt processus basalis medialis of Sewertzoff's descriptions.

It may also be noticed that the ethmoidal transverse commissure between the anterior ends of the infraorbital lateral line canals is surrounded by the cartilage of the tip of the rostrum, much as in Polypterus. In Acipenser Mulder (1928) has shown that the forward extension of the rostrum takes place by means of the chondrification in situ of a curious kind of gelatinous connective tissue which is permeated by blood-vessels. These vessels thus come to be enclosed in the cartilage.

iii. The development of the osteocranium of Acipenser

The development of the bones has not been studied with the systematic thoroughness that could be desired. The early origin of the membrane-bones has been described by Sewertzoff (1926*a*); while the only account of the cartilage-bones remains that of Parker (1882*a*). (Plate 32).

1. The membrane-bones of Acipenser ruthenus. The first bones to develop are the *premaxillae*, *maxillae*, and *dentaries*. They arise some little distance in front of the pterygoquadrates and Meckel's cartilages respectively, being separated from them by the mandibular muscles.

Teeth develop quite independently of the bones and subsequently become attached to them, those in the upper jaw being restricted to the premaxillae. Very soon, the premaxillae fuse with the maxillae, and also with another ossification, the *dermopalatine*, which arises close to the ventrolateral edge of the pterygoquadrate cartilage. The so-called maxilla of the adult is thus a triple compound bone.

Ventrally to the pterygoquadrate cartilage, the *ectopterygoid* appears as a plate of bone, quite independently and at some distance from the teeth which will become attached to it.

Teeth also develop in the 1st branchial arch; under the pharyngobranchial and over the hypobranchial elements of that arch. These teeth do not seem to be attached to an underlying plate, but become fused together by their bases.

As is well known, the teeth are functional only in the larval stages, and disappear in the adult. In the case of the teeth on the jaws Sewertzoff observed that the points of the teeth fall off, while their bases fuse with each other and are overgrown by the bony structure of the growing bone. This being so, such bone should resemble somewhat the structure of the palaeoniscoid scale.

In addition to the bones above described in Acipenser ruthenus, Parker in A. sturio has figured a small 'mesopterygoid' (of uncertain significance (endopterygoid?), and apparently detached from the ectopterygoid) and a *coronoid*.

A small bone in the shape of a vertical strut is found close to the lateral edge of the pterygoquadrate cartilage, immediately dorsal to the articulation between the quadrate and Meckel's cartilage. Parker called this bone the preoperculum; Sewertzoff prefers the less committal but somewhat unfortunate term dermoquadrate, and it has been regarded as a jugal. Allis (1904) has described a row of little tubular bones enclosing the preopercular lateral line canal, and until the relation of this bone to the preopercular lateral line can be shown, it is simpler to call it a *supramaxillary*, like that of many Teleostei.

The remaining dermal bones develop at later stages.

The *parasphenoid* is large: its anterior end becomes attached to the prevomer underneath the cartilage of the processus basalis medialis. On each side the parasphenoid gives off an ascending wing which rises up behind the prootic foramen and in front of the anterior opening of the jugular canal as far as the crista parotica. Posteriorly, the parasphenoid forks into two strips which extend back as far as the last incorporated vertebra. The parasphenoid bears a few teeth. Curiously enough, the relations of the parasphenoid to the dorsal aorta and internal carotid have never been described.

The *prevomer* is unpaired and toothless: it becomes attached to the parasphenoid behind and is largely overgrown by cartilage. (Sewertzoff is dubious of its being a prevomer and calls it a subrostral.)

Of the bones related to the lateral line canals, a number develop in the form of gutters which subsequently become tubular canal bones. These are the *lateral basirostrals* (which, although situated on the under surface of the snout, appear to correspond to the suborbitals, antorbitals, and lachrymals of other forms, and especially to the rostrals of Lepidosteus, like them lodging the infraorbital canal), the bones in the internarial bar (separating incurrent and excurrent nasal apertures), and the *postorbital* bones.

The remainder of the bones related to the lateral line canals develop like trunk scales, with a median crest, between which and the upturned lateral edge of the bone the lateral line canal is enclosed (just as in those trunk scales through which the lateral line passes). Among these are:

nasal, *frontal* (each of which Sewertzoff considers to be made up of separate medial and lateral components, because they have two crests);
postfrontal;
intertemporal, *supratemporal*, *postparietal* (paired and subsequently fused together in the middle line to form the dermosupraoccipital), *post-temporal*.

The remaining dermal bones, unrelated to lateral line canals, develop simply like trunk scales, with a median crest. Among these are:

rostrals, upwards of a dozen in number, on the dorsal surface of the rostrum;
basirostrals, about half a dozen in the middle line on the ventral surface of the rostrum (the hindmost covering the prevomer);
prefrontal, overlying the dorsal end of the lamina orbitonasalis;
suborbital, situated laterally to the hindmost lateral basirostral, from which the infraorbital lateral line canal is passed back to the postorbitals;
supraorbital, lying between the frontals and the dorsal border of the orbit;
parietal and *postoccipital* which form the direct anterior continuation of the line of median dorsal scutes of the trunk.

The so-called *opercular*, *subopercular*, and *interopercular* are also present, but they appear to lack the median crest. According to Tatarko (1936) the opercular is absent and the bone so called is the subopercular, the so-called subopercular is the interopercular, and the so-called interopercular is the uppermost of the branchiostegal rays.

In the adult, the so-called *postfrontal* is composed of the true postfrontal fused to the two dorsalmost postorbitals: the so-called *suborbital* is made up of the true suborbital fused to the three ventralmost postorbitals.

It is further to be noted that in the adult the median crests on the bones can no longer be distinguished when the bones develop their characteristic sculptured external surface.

The *sclerotic* bones are two in number: one dorsal and the other ventral. According to Edinger (1929) they are separated from the sclerotic cartilage by connective tissue.

2. The cartilage-bones in Acipenser sturio. The ossifications in the chondrocranium of sturgeons are few. Their development can hardly be said to have been studied in any detail; but Parker (1882*a*) has figured them and they comprise:

the *lateral ethmoid*, developed in the ventral part of the posterior surface of the lamina orbitonasalis, quite separate from the prefrontal;
the *orbitosphenoid*, developed in the side wall of the skull, above the optic foramen;

the *pleurosphenoid*, developed in the posterodorsal corner of the orbit; and the *prootic*, ossified in the anterior wall of the auditory capsule and surrounding the foramen prooticum.

In the visceral arch skeleton, Parker shows ossifications in the *hyomandibula*, *ceratohyal*, *epibranchials* 1 and 2; *ceratobranchials* 1–4, and the *posterior copula*.

The ossification of the skull in sturgeons is clearly in a degenerate state, and this degeneracy appears to have been brought about by a retardation of development, for it seems that bones may appear in adult and old specimens when they reach a certain age and size. Correlated with the deficiency of bone is the high degree of development of cartilage already mentioned.

9. HOLOSTEI. *AMIOIDEA*

Amia calva

The development of the chondrocranium of Amia has been very thoroughly studied by Pehrson (1922), two stages have been described by Hague (1924), while the morphological relations of the cartilages have been worked out by de Beer (1924*c*, 1926*a*). The segmentation of the occipital region has been studied by Schreiner (1902). As regards the bony skull, in addition to Bridge (1877), van Wijhe (1882 *b*), and Sagemehl (1884), Allis (1897, 1898, 1899) has devoted invaluable monographs to the elucidation of its morphology, but little is known concerning the origins of the various bones, with the exception of certain dermal bones studied by Pehrson (1922) and Sewertzoff (1925*a*).

i. Development of the chondrocranium of Amia

The following account is based chiefly on the work of Pehrson, supplemented here and there by de Beer.

1. Stage 1. (8 mm. length.) PLATE 33, FIGS. 1, 2.

Trabeculae, polar cartilages, and parachordals are present, attached to one another, more or less in line. The limits of the polar cartilage are easily discernible, for immediately in front of it, on the median side of the trabecula, there is a notch lodging the internal carotid artery, while behind the polar cartilage a shallow notch on the lateral surface of the parachordal lodges the palatine branch of the facial nerve. Immediately in front of this notch, the lateral edge of the polar cartilage projects slightly to the side; this is the rudiment of the basitrabecular process. The external rectus muscle for a time has its origin on the dorsal surface of the polar cartilage, and this is the first part of the neurocranium to chondrify, the process of chondrification then spreading forwards and backwards to the trabeculae and parachordals respectively.

The anterior ends of the trabeculae are procartilaginous, and connected with the likewise procartilaginous anterior ends of the pterygoquadrate cartilages.

The parachordals are of considerable interest, for they do not yet anywhere touch the notochord, being separated from it by a wide space, which is, however, narrower posteriorly.

The auditory capsule is represented by a small plate of cartilage which is in procartilaginous connexion with the lateral edge of the parachordal. There is as yet no occipital region to the skull at all.

The pterygoquadrate cartilage lies close to a line of tooth-rudiments on the roof of the mouth, and has a well-developed pterygoid process, connected by procartilage with the anterior end of the trabeculae.

Meckel's cartilage lies almost in the transverse plane, the mouth occupying a ventral position beneath the snout, as in Selachii.

In the hyoid arch, the hyomandibula is chondrified and perforated by the hyomandibular branch of the facial nerve. It may be noted that the hyomandibula and the pterygoquadrate are completely separate and distinct.

2. STAGE 2. (9·5 mm. long.) PLATE 33, FIGS. 3, 4.

Anteriorly, the trabeculae have fused with one another forming a remarkably broad trabecular plate, the lateral corners of which are slightly turned outward forming cornua trabecularum. With the ventrolateral surface of the latter, the now chondrified anterior end of the pterygoid process of the pterygoquadrate is connected.

The basitrabecular process (lateral extension of the polar cartilage) has grown sideways and fused with the inwardly-growing prootic process of the auditory capsule, thus forming the lateral commissure, or lateral wall of the trigeminofacialis chamber. The auditory capsule is now firmly attached to the parachordal by the anterior basicapsular commissure, between which and the lateral commissure is the posterior opening of the trigeminofacialis chamber. The relations of the cartilages in this region to the neighbouring nerves and blood-vessels are shown in Plate 35, figs. 1–6; Plate 143, fig. 2. It will be noticed that the palatine nerve is typical in its relations, passing down behind and forwards beneath the basitrabecular process, and then running laterally to the internal carotid but medially to the efferent pseudobranchial artery, being caught in the fork where these two vessels join. The head vein and hyomandibular nerve emerge behind the lateral commissure, while the orbital artery enters the trigeminofacialis chamber by a small foramen of its own, near the anterior edge of the anterior basicapsular commissure, and runs forwards laterally to the palatine nerve.

The hinder end of the external rectus muscle has grown inwards, over the dorsal surface of the polar cartilage, towards the tip of the notochord, and thus already occupies the position of the posterior myodome.

Behind the anterior basicapsular commissure there is a gap between the auditory capsule and the parachordal—the basicapsular fenestra, which is open posteriorly and confluent with the fissura metotica. Through this confluent slit the glossopharyngeal and vagus nerves emerge, but they will become separated by the posterior basicapsular commissure when it is formed.

The hind ends of the parachordals are now in contact with the notochord, being apposed to its ventrolateral surfaces. The hypophysial fenestra is very large.

Behind the parachordals there are now a pair of cartilaginous plates lying on the dorsolateral surface of the notochord, and bearing the occipital arches. The occipital region thus betrays its origin from originally separate vertebral neural arches, which become attached to the hind ends of the parachordals.

The occipital arch lies in the septum between the 2nd and 3rd permanent myomeres which seem according to Schreiner (1902) to arise from the 4th and 5th metotic somites. The ventral nerve-roots supplying these myomeres emerge in front of the occipital arch, which latter therefore presumably marks the hinder limit of the 7th segment of the whole series. Later, two more vertebral bodies, bearing neural (occipitospinal) arches become attached behind the occipital arches, so that the skull then occupies nine segments. This is shown in Plate 34, fig. 7.

As already mentioned, the extreme anterior end of the pterygoid process of the pterygoquadrate chondrifies from a centre separate from the main cartilage, though connected with it by procartilage.

3. STAGE 3. (10–10·6 mm.) PLATE 33, FIGS. 5, 6.

The ethmoid plate has become a broad plate of cartilage, on the lateral edges of which the laminae orbitonasales have arisen in the form of cartilaginous pillars, which point vertically upwards (dorsally) and end freely. The pterygoid process of the pterygoquadrate is in cartilaginous connexion with the ventrolateral surface of the ethmoid plate by an elongated surface of contact, thus conforming to Swinnerton's 'panartete' type.

In the wall of the trigeminofacialis chamber the exits of the palatine and hyomandibular branches of the facial nerve become separated by a small bar of cartilage—the postpalatine commissure. This has occurred on the left side of the embryo figured in Plate 33, fig. 6, but not yet on the right. The relations of the nerves to the cartilage are seen in Plate 35, fig. 1.

In the later embryo of this stage the orbital cartilages have appeared as independent short bars, half-way between the laminae orbitonasales and the auditory capsules.

The auditory capsules are now attached to the parachordal by posterior as well as anterior basicapsular commissures enclosing the basicapsular fenestra between them. This has the effect of separating the glossopharyngeal from the vagus nerve, since the former passes out in front of the posterior basicapsular commissure. That the glossopharyngeal foramen is, however, not strictly a part of the basicapsular fenestra is shown by the existence of a strut of cartilage (seen on the right side in Plate 33, fig. 6) which separates them, and corresponds in position with the basivestibular commissure of some Teleostei (e.g. Gasterosteus) and of Tetrapoda.

The auditory capsules themselves are now moulded round all sides of the auditory vesicle except the medial side, where no cartilage develops at all.

Posteriorly, the parachordals have extended towards but not yet reached the occipital arches.

4. STAGE 4. (14 mm. length.) PLATE 34, FIGS. 1, 2.

Important developments have taken place in the ethmoid region. Medially to the laminae orbitonasales, the preoptic roots of the orbital cartilages have arisen, attached to the dorsal surface of the ethmoid plate. Dorsally these preoptic roots are attached to the anterior ends of the orbital cartilages, which in turn are connected with the dorsal end of the lamina orbitonasalis of their own side by the sphenethmoid commissure. In this manner the orbitonasal foramen is enclosed, in the form of a short tunnel (bounded ventrally by the ethmoid plate, laterally by the lamina orbitonasalis, dorsally by the spheneth-moid commissure, and medially by the preoptic root of the orbital cartilage) through which the profundus nerve passes into the cavity of the nasal capsule out of the orbit, and from the medial side of which the oblique eye-muscles originate. The orbitonasal foramen thus forms the rudiment of the anterior myodome.

The nasal septum rises up on the dorsal surface of the ethmoid plate in the middle line, but ends freely, being as yet unconnected with the orbital cartilages by sphenoseptal commissures. The foramina olfactoria evehentia, through which the olfactory nerves pass out of the cranial cavity into the nasal capsule, have therefore as yet no dorsal border.

The eye is surrounded by a sclerotic cartilage.

The orbital cartilages are interconnected by two transverse bars, the paraphysial and the epiphysial cartilages. Each orbital cartilage is perforated by a foramen through which a vein from the dorsal regions of the cranial cavity joins the superior orbital sinus, and posteriorly is attached to the postorbital process of the auditory capsule. The base of the postorbital process, when it springs from the anterodorsal wall of the auditory capsule, shows two perforations. One of these is a true foramen which leads into the trigeminofacialis chamber and serves for the exit of the otic branch of the facial nerve; the other is not a foramen at all but a tunnel which surrounds the dorsal spiracular recess (Wright's organ). This spiracular canal is enclosed laterally by a bar of cartilage projecting forwards from the crista parotica, the horizontal ridge on the external surface of the lateral wall of the auditory capsule, and beneath which is the articular facet for the hyomandibula.

Posteriorly, the extensions of the parachordals have become connected with the bases of the occipital arches; but the dorsal ends of the latter are still free, so that the vagus or jugular foramen is not yet completely enclosed.

In the visceral arch skeleton, the pterygoquadrate is now a substantial cartilage, bearing rudiments of basal and otic processes. Anteriorly, the cartilaginous connexion between its pterygoid process and the ethmoid plate has been severed by the formation of distinct limiting membranes. Meckel's cartilage shows its characteristic huge coronoid process which, according to van Wijhe (1882*b*), may have had a separate origin, and may thus represent a labial. The symplectic process of the hyosymplectic is elongated and slender, and reaches to the point of junction between quadrate and Meckel's cartilage.

The interhyal is present between the hyosymplectic and the ceratohyal, and large hypohyals have also appeared.

5. Stage 5. (19·5 mm. long.) PLATE 34, FIGS. 3, 4.

The chief interest of this stage lies in the development of the cranial roof. The space between the paraphysial and epiphysial bars of the previous stage is now completely covered by cartilage (except for a minute epiphysial foramen) forming the tectum cranii. The front edge of this tectum is, however, still unconnected with the nasal septum.

The dorsal ends of the occipital arches are fused with the posterior walls of the auditory capsules, thus enclosing the vagus foramen on each side. Extensions of cartilage towards the middle line from the roofs of the auditory capsules and from the dorsal ends of the occipital arches form the rudiments of the tectum synoticum and tectum posterius respectively.

The crista parotica now extends outwards behind the articulation of the hyomandibula, and encloses the supratemporal branch of the glossopharyngeal nerve in a foramen. This foramen (which of course has nothing to do with the exit of the nerve from the cranial cavity) is visible from both dorsal and ventral aspects of the chondrocranium.

6. STAGE 6. (31·5 mm.) PLATE 34, FIG. 5.

The chondrification of the ethmoid region has now more or less achieved the adult condition. The fossae nasales on each side of the nasal septum have a complete median and posterior wall, now that the nasal septum is connected with the orbital cartilages by the sphenoseptal commissures on each side, leaving only a small foramen olfactorium evehens in the posteromedian corner of each nasal fossa. Each foramen olfactorium evehens, through which the olfactory nerve leaves the cranial cavity for the nasal capsule, is therefore bounded, medially by the nasal septum, ventrally by the ethmoid plate, laterally by the preoptic root of the orbital cartilage, and dorsally by the sphenoseptal commissure, and is a perforation of the true cranial wall.

In the posterolateral corner of the fossa nasalis is the orbitonasal foramen, leading into the orbit. This is bounded medially by the preoptic root of the orbital cartilage, ventrally by the ethmoid plate, laterally by the lamina orbitonasalis, and dorsally by the sphenethmoid commissure, and is not a perforation of the true cranial wall at all. The oblique eye-muscles push forwards from the orbit into the hinder opening of the orbitonasal foramen on each side, and are attached to the lateral edge of the nasal septum, beneath the foramen olfactorium evehens. The profundus nerve runs forwards out of the orbit through the orbitonasal foramen, but it acquires a small foramen of exit into the nasal capsule of its own.

The configuration of the cartilages in this region is such that the orbitonasal foramen looks like a short tube into the side of which, near its anterior end, the foramen olfactorium evehens opens. In other words, the anterior openings of the foramen olfactorium evehens and of the orbitonasal foramen are practically confluent, merging into what may be called a single aperture in the hind wall of the fossa nasalis—the foramen olfactorium advehens. As the morphological relations of this latter foramen play an important part in the interpretation of the conditions in those Teleostei which have a well-developed anterior myodome and whose olfactory nerves pass through the orbits, it will be useful to define the limits of the foramen olfactorium advehens in Amia. It is bounded medially by the nasal septum, ventrally by the ethmoid plate, laterally by the lamina orbitonasalis, and dorsally by the sphenoseptal and sphenethmoid commissures. The plane of the foramen olfactorium advehens is not the same as that of the orbitonasal foramen or foramen olfactorium evehens; the plane of the foramen olfactorium advehens is more or less transverse, that of the foramen olfactorium evehens faces anterolaterally, that of the orbitonasal foramen faces anteromedially. Between the planes of these three apertures, therefore, there is a triangular space, the cavum orbitonasale, which in Amia forms part of the fossa nasalis or cavity of the nasal capsule; in Teleostei such as Salmo (see p. 123), however, it is incorporated in the orbit.

Behind the foramen olfactorium evehens the preoptic root of the orbital cartilage extends backwards, thus reducing the size of the sphenoidal fissure, through which all nerves from the optic to the abducens inclusive, in addition to the ophthalmic and buccal branches of the facial nerve, leave the cranial cavity. At this stage, the size of the sphenoidal fissure is in process of reduction from behind also, for the rudiment of the pila lateralis has appeared, which will enclose the superficial ophthalmic, maxillary, and mandibular branches of the trigeminal nerve and the ophthalmic and buccal branches of the facial nerve, and thus separate them from the optic, oculomotor, trochlear, abducens, and profundus nerves.

The cartilaginous roof over the skull is now complete. The sclerotic cartilage surrounding the eye is well formed, but on the medial side it retains a large membranous and unchondrified circular zone.

7. Stage 7. (41 mm.) — plate 35, figs. 5, 6.

The chief interest of this stage centres about the trigeminofacialis chamber. The pila lateralis is now complete, situated in front of the lateral commissure and connecting the cartilage in the region of the basitrabecular process (anterolateral corner of the floor of the trigeminofacialis chamber) with the under side of the postorbital process.

At previous stages the internal carotid artery received the efferent pseudobranchial artery of its own side, and then ran upwards in the notch between the trabecula and polar cartilage and gave off the ophthalmica magna artery. Now, however, the internal carotid and the efferent pseudobranchial arteries run upwards separately, each in its own foramen, and the efferent pseudobranchial is continuous with the ophthalmica magna. Between the internal carotid and efferent pseudobranchial foramina, a short cartilaginous process projects dorsally and laterally, and eventually becomes ossified by the basisphenoid.

The prootic bridge is now present as a square plate of cartilage in front of the tip of the notochord, and above the level of the hypophysial fenestra. On each side, the parachordals slope medially and ventrally, so that the actual medial edges (bordering the hypophysial fenestra) are at a lower level than the cartilage further to the side, and it is here that the posterolateral corners of the prootic bridge are attached to the parachordals. The anterolateral corners are free, there being a notch on each side between the anterior part of the proötic bridge and the parachordal, through which the abducens nerve runs downwards to the external rectus muscle. The latter has now commenced to occupy the space of the posterior myodome; i.e. the space comprised beneath the prootic bridge and the level of the hypophysial fenestra.

Behind the prootic bridge is a space—the basicranial fenestra. This, of course, lies at a more dorsal level than the hypophysial fenestra, although the hind border of both fenestrae coincide. Ultimately, the basicranial fenestra becomes obliterated as the chondrification of the prootic bridge extends.

It is also worthy of note that the basal and otic processes of the pterygoquadrate are well developed, almost but not quite touching the basitrabecular process (lateral commissure) and auditory capsule respectively. The otic process lies dorsally to the efferent pseudobranchial artery as does the otic process of Selachians.

ii. Morphology of the fully formed chondrocranium of Amia

PLATES 36, 37, FIG. 1.

The persistence of cartilage in the adult skull of Amia makes it possible to study the morphology of the chondrocranium in the adult, as has been done by Allis in a number of works (1897, 1914*a*, 1919*a*).

The points of chief interest concern the posterior myodome and the trigeminofacialis chamber. The latter is a space lodging the ganglia of the trigeminal and facial nerves, bounded medially by the membranous dura mater and laterally by the cartilaginous lateral commissure and pila lateralis, which subsequently become ossified (in whole or part) by the prootic and pleurosphenoid bones respectively.

To enter the trigeminofacialis chamber, the roots of the trigeminal and facial nerves pierce its membranous medial wall. Thence, the hyomandibular branch of the facial emerges through the posterior aperture (between the anterior basicapsular commissure and the lateral commissure); the palatine branch pierces its own foramen in the cartilaginous floor; the otic branch of the facial nerves pierces a foramen in the roof of the chamber, near the base of the postorbital process; the buccal branch of the facial and the maxillary and mandibular branches of the trigeminal emerge through the trigeminal foramen (between the lateral commissure and the pila lateralis). The superficial ophthalmic nerves originally emerge by this same foramen, but subsequently become enclosed in a separate foramen. The profundus nerve runs forwards into the orbit medially to the pila lateralis.

The trigeminofacialis chamber of Amia is not divided into ganglionic and jugular portions; the ganglia are therefore in contact with the two blood-vessels which pass through the chamber. One of these is the head vein which runs back from the orbital sinus passing medially to the pila lateralis, ventrally to the trigeminal and buccal nerves, dorsally to the hyomandibular nerve, and emerges through the posterior aperture of the chamber. The other is the orbital artery, which leaves the internal carotid and runs upwards to pierce its own foramen in the floor of the chamber, turns forwards passing laterally to the palatine nerve, and emerges through the trigeminal foramen.

The posterior opening of the chamber and the orbital artery foramen become surrounded by the prootic bone. The foramina for the otic and palatine branches of the facial remain cartilaginous; the foramen for the superficial ophthalmic nerve becomes surrounded by the pleurosphenoid bone, which also forms the anterior border of the trigeminal foramen. In the ventro-medial corner, the trigeminofacialis chamber is continuous with the myodome, for there is here a gap between the cartilaginous floor of the chamber and the membranous medial wall, and the lateral parts of the prootic bridge (laterally to the abducens nerves) remain unchondrified and unossified. (Plate 143, fig. 2.)

The myodome is short and lodges only the external rectus muscles, which pass backwards and inwards, dorsally to the trabeculae (strictly, polar cartilages), and the membranous floor of the cranial cavity (dura mater), ventrally to the pituitary vein and to the prootic bridge. The hinder part of the myodome thus occupies the hind part of the hypophysial fenestra.

The whole myodome of Amia corresponds to the so-called dorsal myodomic compartment of Teleostei and may be derived, as Allis has shown, from some Selachian-like form as a result of enlargement of the foramina for the pituitary veins, throwing the cavity of the subpituitary space open to the orbit, dorsally to the trabeculae. The absence of a prefacial commissure and of a pila antotica and deficient chondrification of the lateral regions of the dorsum sellae (prootic bridge) would place this myodome in communication with the trigeminofacialis recess, converted into a trigeminofacialis chamber by the formation of the lateral commissure.

The notochord is reduced to a slender rod in the hind part of the basal plate, but here it still separates the parachordals completely from one another. Eventually the notochord is enclosed in the basioccipital bone.

iii. The development of the osteocranium of Amia

The development of the bones of the skull of Amia has not been systematically studied. Pehrson (1922) has investigated the dermal bones related to the lateral line canals; Allis (1897, 1899) has considered the morphology of certain cartilage-bones, but further work is needed on their development. The enumeration of bones in the following section does not therefore deal with all the bones present in Amia (See Plate 34, fig. 6; Plate 37, figs. 2, 3, 4).

1. The membrane-bones, i.e. bones developing outside the perichondrium. Beginning with the bones related to the lateral line canals:

The *parasphenoid* arises from accumulations of osteoblasts which may be observed at the 9·5 mm. stage, stretching from the level of the optic nerve to that of the pituitary. It eventually extends forwards and backwards beneath the skull and laterally it gives off a pair of wings which underlie the floor of the trigeminofacialis chamber. Between each of these wings and the overlying cartilaginous floor is a small longitudinal tunnel, the parabasal canal. The internal carotid artery enters this canal at the hind end and traverses it from back to front on its way to its cartilaginous foramen, and half-way along its course it is joined in the canal by the palatine nerve, which issues straight into it from its (cartilaginous) foramen. (The space of the parabasal canals corresponds to that of the ventral myodomic compartment of Teleostei, lodging the internal recti muscles.)

The *prevomers* arise from paired accumulations of cells beneath the ethmoid plate, at the 10-mm. stage. The prevomers bear teeth.

The *antorbital* ('adnasal' of some authors) and *lachrymal* arise at the 12-mm. stage from groups of osteoblasts underlying the preorbital section of the infraorbital lateral line canal. The bones soon form gutters and then tubes. From the lachrymal at the 14-mm. stage lamellae extend away from the tube beneath the skin.

The *suborbitals* and *postorbital* 1 arise in the same way slightly later, each bone being related to one sense-organ of the infraorbital lateral line canal; they are in the form of gutters at the 14-mm. stage. *Postorbital* 2 and the *postfrontal* arise slightly later again (19·5 mm.). The aliform extensions of the postorbitals develop into the large plates characteristic of these bones. In some specimens the area of the two postorbitals may be occupied by a single bone, but it is unknown whether this condition is due to fusion of two centres or to the presence of only a single centre of ossification.

The postfrontal is always free from the sphenotic.

The *supra-ethmoid* ('rostral' autt.) arises from four groups of osteoblasts underlying the four sense-organs of the ethmoid commissure between the infraorbital canals at the 12-mm. stage, and soon presents the form of a transverse tubular bone. Pehrson is of opinion that it represents the numerous separate rostral bones of primitive forms.

The *nasal* arises at 14 mm. from three pairs of rudiments related to the first three sense-organs of the supraorbital lateral line canal. They soon give rise to a single tube with lateral lamellae. The tripartite nature of the rudiment leads Pehrson to conclude that the nasal of Amia corresponds to the separate terminale, adnasal and nasal of Polypterus.

The *frontal* arises like the nasal, but from four paired groups of osteoblasts underlying the 4th–7th sense-organs of the supraorbital lateral line canal. They soon form a single tube with a wide medial lamelliform extension which gives rise to the main body of the frontal. The groups of osteoblasts in each pair are situated medially and laterally to their respective sense-organs. The medial groups are, according to Pehrson, the rudiments of the frontal proper, while the lateral groups he regards as representing the separate supraorbital bones of other forms.

It is difficult to see why the paired nature of the rudiments should be significant as indicative of the presence of medial and lateral bones in the case of the frontal but not the nasal; and further why the tripartite grouping of the rudiments in the case of the nasal should indicate the existence of three bones, while the quadripartite grouping of rudiments in the case of the frontal represents no known (though postulated) transverse subdivision in the frontal bone.

The *intertemporal* and *supratemporal* ('squamosal') arises at 14 mm. from two paired groups of osteoblasts related to the temporal section of the infraorbital lateral line canal. The anterior group seems to represent the intertemporal and the posterior group the supratemporal.

The *postparietal* (*extrascapular*) arises from three groups of osteoblasts underlying the three sense-organs of the transverse occipital lateral line canal. Pehrson considers that they represent the three separate extrascapulars of Polypterus and Lepidosteus.

The *post-temporal* (supracleithrum) connects the postparietal with the pectoral girdle and lodges the post-temporal portion of the lateral line canal.

The *preopercular* arises from two paired groups of osteoblasts underlying the sense-organs of the opercular lateral line canal. It might perhaps be suggested that the more dorsal rudiment represents the squamosal and the more ventral the simple preopercular of other forms. This is, however, certainly incorrect as the Amioidei, and indeed all Actinopterygii, have no squamosal. Pehrson regards these rudiments as representing the preopercular and quadratojugal, which is still less likely, for the latter is never known to harbour a lateral line canal.

The *dentary* arises at 11 mm., ensheaths Meckel's cartilage, and becomes attached to teeth. Subsequently, independent bony rudiments, apparently corresponding to the *splenial*, underlying the mandibular lateral line canal become attached to the dentary. The combined bone is probably therefore a dentalosplenial.

The *angular* (dermarticular, autt.), like the dentary arises, independently of the sense-organs of the mandibular canal, and only later comes into relation with them.

Sewertzoff (1925*a*) has also examined the development of certain of the dermal bones of Amia. He agrees that the postorbitals, suborbitals, lachrymal, antorbital, and supra-ethmoid arise like gutters and subsequently acquire a lamelliform extension. The latter three bones he homologizes with the lateral basirostrals of Acipenser. But he regards the nasal, frontal, intertemporal, and post-temporal as arising like bony trunk scales, the lateral line canal being enclosed between the median crest and the upturned lateral edge of the bone.

The so-called *opisthotic* (*intercalary* of Allis) ossifies as a lamella *outside* the perichondrium of the posterior projection of the crista parotica. Posteriorly, the bone ends in a point to which the supracleithrum is attached by ligament. The opisthotic is pierced by the supratemporal branch of the glossopharyngeal nerve.

The *premaxilla* arises at the 12-mm. stage, and bears teeth. Its posterior processes lie on the dorsal surface of the cartilage (in contact with the perichondrium) of the ethmoid plate, and surround the foramen olfactorium advehens.

The *maxilla* arises in the maxillary fold of the upper jaw at the 9·5-mm. stage. It bears teeth.

The *coronoid* (presplenial of some authors) arises over the dorsal surface of the anterior end of Meckel's cartilage at the 10·6 mm. stage. It also bears teeth. The development of the *prearticular* (splenial of some authors) and of the supra-angular has not been followed.

The *opercular* arises at the 19·5-mm. stage.

The *dermopalatine* is ossified at the 14-mm. stage in the form of a plate of bone bearing teeth lying under the ventral surface of the anterior part of the pterygoid process of the pterygoquadrate cartilage. It becomes contiguous with the autopalatine, a cartilage bone.

The *ectopterygoid* is ossified at the 14-mm. stage as a small plate of bone underlying the ventral surface of the posterior part of the pterygoid process of the pterygoquadrate cartilage. Teeth develop later and become attached to the anterior part of the ectopterygoid.

The *endopterygoid* arises at the 18-mm. stage in the form of a plate of bone situated on the medial side of the posterior part of the pterygoid process of the pterygoquadrate cartilage. It therefore lies between the ectopterygoid and the parasphenoid and forms the dorsolateral skeleton to the roof of the mouth.

The *parietal* is ossified at the 25-mm. stage as a plate of bone overlying the tectum synoticum, and medial to the intertemporal.

The *pharyngeum superius* and *inferius* are present, bearing teeth, underlying the 3rd pharyngobranchial and overlying the 5th ceratobranchial respectively, but apparently free from the cartilage-bones developed in those cartilages.

There are no sclerotic bones.

2. The cartilage-bones, i.e. bones developing inside the perichondrium. The development of the cartilage bones of Amia has not been properly studied. According to Pehrson at the 31–4-mm. stage, ossification has begun in the *exoccipital*, *basioccipital*, *hyomandibula*, *symplectic*, *metapterygoid*, *quadrate*, and *retroarticular*. Little need be said about them except to note that the exoccipitals meet in the middle line and form the dorsal border to the foramen magnum. There is no supraoccipital. The metapterygoid ossifies in both otic and basal processes, the latter being connected by a ligament with the lateral commissure (basitrabecular process).

The *pre-ethmoid* (Gaupp, Swinnerton; septomaxillary Bridge, Allis) is a small bone developed late in the cartilage of the posterior surface of the cornu trabeculae. It is still absent at the 50-mm. stage. Allis (1898) considers the possibility that it may represent the ethmoid (a cartilage-bone).

The *lateral ethmoid* arises late in the cartilage of the lamina orbitonasalis, and according to Allis, has no dermal addition to its surface; nor, according to him, is there a (dermal) prefrontal bone in Amia.

The *autopalatine* arises late in the anterior region of the pterygoid process of the pterygoquadrate cartilage, and becomes contiguous with the dermo-palatine.

The *sphenotic* is beginning to ossify at the 50-mm. stage round the lateral prominence of the postorbital process. It remains separate from the post-frontal.

The *prootic* is ossified at the 50-mm. stage and takes the form of lamellae on the internal and external surface of the auditory capsule in the neighbour-hood of the posterior aperture of the trigeminofacialis chamber, in the lateral commissure, and at the lateral side of the prootic bridge.

The *epiotic* is ossified at the 50-mm. stage on the dorsal surface of the audi-tory capsule, immediately in front of the exoccipital.

The *pleurosphenoid* ('alisphenoid' of Allis) develops later in the pila lateralis and surrounds the exits of the ophthalmic nerves.

The *basisphenoid* (propituital, Chabanaud 1936) is paired, and ossifies late: it has not yet appeared at the 50-mm. stage. It develops in the region separat-ing the foramina for the internal carotid and the efferent pseudobranchial arteries. Allis regards it as being for the most part a membrane-bone, but it seems to occupy exactly the position of the cartilaginous process present in this region at certain stages.

A *mentomeckelian* is present at the tip of each Meckel's cartilage, but it develops late. Other ossifications of cartilage-bones in Meckel's cartilage are the *articular* (Bridge's ossicles *b* and *c*) the *coronomeckelian* (Bridge's ossicle *d*), and the already mentioned *retroarticular* (Bridge's ossicle *a*). All except the latter develop late, and appear to represent fragments of an originally completely ossified Meckelian bone, occupying the whole of Meckel's cartilage, as found in such fossil forms as Ospia (Stensiö, 1932).

The *retroarticular* (angular, Ridewood, 1904*a*, *et al.*) may be referred to again here, for it ossifies in the cartilage of the tip of the retroarticular process of Meckel's cartilage, and extends backwards into the mandibulo-hyoid ligament which is attached to it.

10. HOLOSTEI. *LEPIDOSTEOIDEI*

Lepidosteus osseus

The development of the chondrocranium of Lepidosteus was the object of study of one of Parker's (1882*b*) series of papers. It has been studied by means of wax models by Veit (1907, 1911), and by means of graphic reconstructions by Hammarberg (1937). Several points in the morphology have been treated by Allis (1909, 1920) and some by de Beer (1926*a*). The segmentation of the occipital region has been worked out by Schreiner (1902), while the morphology of the bony skull has been studied by Regan (1923) and Mayhew (1924), and the development of the dermal bones has been followed by Hammarberg (1937) and Aumonier (in press).

i. The development of the chondrocranium of Lepidosteus

The following account is based chiefly on the work of Veit and of the present writer.

1. Stage 1. (10–11 mm. length.) PLATE 38, FIGS. 1, 2.

The skull at this stage shows three independent pairs of cartilages: the parachordals, polar cartilages, and trabeculae.

The parachordals lie on either side of the notochord with which the hinder two-thirds of their length are in contact, the anterior third diverging somewhat to the side. The parachordals extend from the level of the trigeminus ganglion to that of the glossopharyngeal. At the level where they part company with the notochord, the anterior basicapsular commissure connects the lateral edge of each parachordal with the rudiment of the auditory capsule, at this stage a small plate underlying the anterior part of the auditory vesicle. Parachordal and auditory capsule appear to chondrify together.

In the direct line of prolongation of each parachordal, but separated from it by a narrow space, is the polar cartilage, on to which the external rectus muscle is attached.

In front of the polar cartilages, and likewise in the direct line of the prolongation, though separate from them, are the trabeculae, wide apart and converging slightly anteriorly.

2. Stage 2. (11–12 mm.) PLATE 38, FIGS. 3, 4, 5.

The trabeculae have become interconnected by their anterior ends, while their posterior ends are now attached to the polar cartilages, which in turn are attached to the anterior ends of the parachordals. The hypophysial and basicranial fenestrae are thus enclosed, but they are widely confluent. Posteriorly, each parachordal sends back two extensions along the side of the notochord, one dorsal and the other ventral. These extensions represent a pair of lines of basidorsals and basiventrals respectively, fused longitudinally. The dorsal edge of the dorsal extensions exhibits a number of prominences representing neural arches, to which intermyotomal septa are attached. These neural or occipital arches are three in number, and behind each one lies a ventral nerve-root. The hindmost of these roots supplies the 3rd permanent myomere, or 6th metotic somite according to Schreiner's observations, from which it follows that the three occipital arches mark the hinder limits of the 6th, 7th, and 8th segments of the whole series, respectively.

The auditory capsules are further developed, each one extending back beneath the auditory vesicle to its hinder end, and up on its side, especially anteriorly. Between the lateral edge of the parachordal and the auditory capsule, behind the anterior basicapsular commissure, is the basicapsular fenestra which is open posteriorly and thus confluent with the fissure metotica. The glossopharyngeal and vagus nerves are therefore not yet separated by any cartilage.

Immediately in front of the anterior basicapsular commissure, a medially directed prootic process grows from the anterior wall of the auditory capsule towards a laterally directed postpalatine process which extends from the lateral edge of the parachordal. Where these join (as on the right in Plate 38, fig. 4), a lateral commissure is formed, separated from the anterior basicapsular commissure by a foramen through which the hyomandibular branch of the facial nerve and the head vein emerge from the trigeminofacialis chamber, of which the lateral commissure is the lateral wall. The palatine nerve at this stage runs down by the side of the parachordal, in front of the postpalatine process. It soon becomes lodged in a notch, for the lateral edge of what originally was the polar cartilage projects to the side, forming the rudiment of the basitrabecular process, and the notch is formed between this and the postpalatine process.

In the visceral arch skeleton, the pterygoquadrate has chondrified as a longitudinal bar of cartilage, parallel with and ventrolateral to the trabecula, and independent of any other cartilage. Meckel's cartilage lies ventrally to the pterygoquadrate, with which it does not yet articulate. The hyomandibula is attached by its dorsal end to the lateral wall of the auditory capsule by procartilage, and is pierced by the hyomandibular nerve. It forms a more or less rectangular plate, and projects backwards from its attachment to the auditory capsule; there is no indication of a symplectic process as yet, unless by chance it be represented by a small nodule of cartilage which Veit has interpreted as the interhyal, which is situated between the hyomandibular and the ceratohyal.

Ceratobranchial cartilages are present in the first 3 branchial arches, and a hypobranchial in the 1st.

There is as yet little in the chondrocranium at this stage to suggest the future elongation of the snout, so characteristic of the adult Lepidosteus.

3. Stage 3. (14 mm. long.) PLATE 39, FIGS. 1, 2.

The anterior fusion between the trabeculae has developed into an extensive ethmoid plate, the anterolateral corners of which project slightly to the side, representing the cornua trabecularum, while on the dorsal surface of the posterolateral corners of the ethmoid plate, the laminae orbitonasales rise up as short pillars.

The orbital cartilages are present, each one attached posteriorly to the postorbital process of the auditory capsule. The development of the postorbital process and of the basitrabecular process have resulted in an extension in an anterior direction of the lateral commissure. The amount of the addition thus made to the floor and side wall of the trigeminofacialis chamber is indicated by the position of the foramina for the otic and palatine branches of the facial nerve: the former pierces the base of the postorbital process, while the latter represents the notch of the previous stage behind the basitrabecular process and in front of the postpalatine process, now converted into a foramen by the fusion of the basitrabecular process with the lateral commissure. The palatine nerve thus runs down behind and forwards beneath the basitrabecular process in typical manner. The basitrabecular process is connected by procartilage with the basal process of the pterygoquadrate.

The basicapsular fenestra is obliterated, and a posterior basicapsular commissure attaches the hind part of the auditory capsule to the parachordal, behind the glossopharyngeal nerve. The latter is therefore enclosed in a foramen in the floor of the capsule, and is separated from the vagus which emerges through the fissura metotica.

The wall of the auditory capsule is now complete except dorsally, where the roof is still somewhat undeveloped, and medially, where it is never formed except in the extreme anterior and posterior regions.

In the occipital region, the three rudiments of occipital arches borne upon the dorsal posterior extensions of the parachordals have risen up and enclosed two hypoglossal foramina between them, for the nerve-roots supplying the 1st and 2nd myomeres (7th and 8th segments). The foremost of these three occipital arches (corresponding to the hind limit of the 6th segment) is the best developed and rises up immediately behind the auditory capsule, thus forming the hind border to the fissura metotica.

The ventral posterior extension of the parachordal is separate from the dorsal except anteriorly, so that the notochord is visible from the side through a V-shaped notch. Dorsally, the notochord is visible between the parachordals throughout their length, but ventrally, at the level of the glossopharyngeal nerve, the parachordals have become united by a cartilaginous hypochordal commissure.

In the visceral arch skeleton, the pterygoquadrate has a low basal process which is united (fused) by procartilage to the basitrabecular process of the neurocranium. Posteriorly, a short blunt otic process lies dorsally to the symplectic process of the hyosymplectic; anteriorly, the pterygoid process extends forwards ventrolaterally to the trabecula, but does not reach the ethmoid plate. More than half the length of the pterygoquadrate lies behind the articulation with Meckel's cartilage, where temporary fusion occurs.

The hyosymplectic is now well developed, the symplectic process extending forwards beneath the posterior end of the pterygoquadrate. In addition to the interhyal and ceratohyal, there are now hypohyals, ceratobranchials in all five branchial arches, and a median basibranchial corresponding to the first three.

4. STAGE 4. (20 mm. length.) PLATE 39, FIGS. 3, 4.

The chondrocranium now begins to show elongation, especially in the anterior region where an extensive snout has been formed at the expense of the ethmoid plate. (For a mathematical consideration of the elongation of the snout, see p. 471.) On its dorsal surface the ethmoid plate bears the nasal septum which is continuous posteriorly on each side with the anterior end of the orbital cartilage, by means of the sphenoseptal commissure. The orbital cartilage in this region is connected with the trabecula by a broad preoptic root, and also with the dorsal end of the lamina orbitonasalis by means of the sphenethmoid commissure. The various foramina in this region are closely comparable to those of Amia. The foramen olfactorium evehens (between nasal septum, sphenoseptal commissure, preoptic root of the orbital cartilage and ethmoid plate) serves for the exit of the olfactory nerve from the cranial cavity to the fossa nasalis; the orbitonasal foramen (between lamina orbitonasalis, sphenethmoid commissure, preoptic root of the orbital cartilage, and ethmoid plate) serves for the passage of the orbitonasal artery, out of the orbit into the nasal fossa, while the oblique eye-muscles are attached to the preoptic root of the orbital cartilage close to the posterior opening of the orbitonasal foramen. The configuration of the cartilages in this region is such that the orbitonasal foramen forms a tunnel into which the foramen olfactorium evehens opens; the space contained in the tunnel is the cavum orbitonasale, and its anterior opening into the nasal fossa is the foramen olfactorium advehens (combining the foramen olfactorium evehens and the orbitonasal foramen).

The anterior end of the pterygoid process of the pterygoquadrate is in contact with the lateral edge of the ethmoid plate.

The orbital cartilages are interconnected by two transverse bars of cartilage: an anterior paraphysial and a posterior epiphysial bar. The paraphysial bar is further connected with the nasal septum by a longitudinal strut of cartilage. The cartilaginous roofing over the anterior region of the cranial cavity is therefore fairly well developed.

The lateral commissure shows an important development of the basitrabecular process, which as compared with the previous stage has extended anteriorly and laterally, and is now articulated with the basal process of the pterygoquadrate. The lateral wall of the auditory capsule bears a horizontal crest, the crista parotica, which is prolonged anteriorly as a bar of cartilage, attached to the base of the postorbital process in such a way as to surround the dorsal spiracular recess (or Wright's organ), and posteriorly the supratemporal branch of the glossopharyngeal nerve becomes surrounded by cartilage in a similar manner. Beneath the crista parotica is the articular facet for the head of the hyomandibula. Posteriorly, the auditory capsules are connected with one another by means of the tectum synoticum, and with the occipital arch of their own side, with the result that the fissura metotica is reduced to the small vagus foramen.

The dorsal and ventral posterior extensions of the parachordals on each side of the notochord have developed still farther back, the dorsal extensions bearing rudiments of neural arches of which the foremost three are occipitospinal arches which will become included in the skull, behind which point the neural arches belong to the eventually free anterior vertebrae. The three occipitospinal arches lie respectively behind the ventral nerve-roots which supply the 3rd, 4th, and 5th myomeres (9th, 10th, and 11th segments of the whole series). The position of the hindmost occipitospinal arch therefore indicates that eleven segments are included in the skull of Lepidosteus.

Meckel's cartilage now possesses a fairly well-developed coronoid process, not, however, so large as that of Amia. Van Wijhe (1882*b*) found this process to be an independent cartilage in the adult, where it corresponds in position to the labial cartilage in Polypterus, which it may represent. On the other hand, there is no indication of the independence of this cartilage in early stages.

ii. Morphology of the chondrocranium of Lepidosteus osseus (15 cm. long). PLATE 40.

The cartilaginous roof of the chondrocranium is complete except for a pair of lateral vacuities left on either side of the taenia tecti medialis which connects the epiphysial bar with the tectum synoticum.

The side wall of the chondrocranium is also more complete; the oculomotor, trochlear, and abducens nerves and the pituitary vein are enclosed in separate foramina, leaving the optic nerve and trigeminal nerve to emerge alone through their foramina. (The trigeminal foramen also transmits the ophthalmic and buccal branches of the facial nerve, the head vein and the orbital and orbitonasal arteries, and is also known as the orbital opening of the trigeminofacialis chamber.)

The cartilage which forms the posterior border of the pituitary vein foramen and the anterior border of the trigeminal foramen is partly in the position of a pila antotica. It is, however, not typical, for it lies a considerable distance in front of the prootic bridge (cf. the pila antotica and dorsum sellae or acrochordal or crista sellaris of other forms). If, as seems likely, Lepidosteus has lost the posterior myodome, the cartilaginous side wall in this region must be a new formation, for which reason the cartilage in question may be called a pila antotica secundaria.

The internal carotid arteries which at previous stages entered the cranial cavity by passing simply through the hypophysial fenestra between the trabeculae, now each have their own foramen as the cartilage extends and the hypophysial fenestra becomes reduced. The internal carotids receive the efferent pseudobranchial arteries before piercing their foramina, there are consequently no foramina for the efferent pseudobranchial arteries.

The region of the trigeminofacialis chamber is of great interest. The anterolateral part of its floor, the basitrabecular process, forms a massive knob which articulates with the basal process of the pterygoquadrate. With the possible exception of Gymnarchus, this contact between basal and basitrabecular process has been lost in all other Teleostomi.

Through the posterior aperture of the trigeminofacialis chamber the hyomandibular branch of the facial nerve emerges and the head vein and the orbital artery enter. The otic branch of the facial nerve emerges through a foramen in the roof of the chamber, while the palatine branch of the facial nerve emerges through the same foramen in the floor as is traversed by the orbitonasal artery in entering the chamber: the artery passing on the medial side of the nerve. The orbital and orbitonasal arteries of Lepidosteus correspond to branches of the undivided orbital artery of Amia.

The medial wall of the trigeminofacialis chamber is formed at early stages by the membranous dura mater; at later stages this becomes ossified by the prootic so that the trigeminofacialis chamber is entirely cut off from the cranial cavity, except for the passages out of the cavity traversed by the trigeminal and facial nerves.

The trigeminofacialis chamber in Lepidosteus is not subdivided into pars jugularis and pars ganglionaris; the ganglia of the nerves and the head vein are in contact. This is in contrast with the condition in Teleostei (see p. 128).

Between the parachordals the prootic bridge appears. It is present in 11-cm. larvae, as a rectangular plate in front of the notochord, and still free on each side from the parachordals, to which, however, it soon becomes attached. Behind the prootic bridge at the 11-cm. stage is the basicranial fenestra, which soon becomes obliterated by growth of the cartilage. The prootic bridge then forms the roof of a space, the cavum sacci vasculosi, which corresponds to the subpituitary space of Selachians[1] and to the posterior part of the myodome of Amia. In Lepidosteus, however, this space lodges no eye-muscles, for these do not pass into the cranial cavity at all, and it may be said, therefore, as with Polypterus and Acipenser, that Lepidosteus possesses a non-functional myodomic space. The conditions as found in Amia would result if in Lepidosteus the trigeminofacialis chamber were to acquire a communication with the myodomic space, and if the latter were occupied by the external recti muscles. The notochord becomes reduced to a slender thread in the hinder part of the basal plate. In this region, however, the cells of the notochord give rise within the notochordal sheath to a tissue which much resembles cartilage. The parachordals meet one another in the midline ventrally to the notochord.

In the occipital region the foramina for the ventral roots of the 7th and 8th segments eventually become surrounded by the exoccipital; those of the 9th and 10th segments are lodged in notches in the basioccipital which also ossifies in the occipitospinal arches: that of the 11th is eventually enclosed in a foramen on the posterior side of the last occipitospinal arch. The bases of the occipitospinal arches meet one another in the midline above the notochord and the parachordals. In this region, therefore, the skull seems to have two cartilaginous floors, one above the other.

The sclerotic cartilage is particularly well formed, like a kettle drum.

A pharyngomandibular cartilage is described by Hammarberg (1937).

The dorsal ends of the first two branchial arches are forked into 'infra-' and 'suprapharyngobranchials'. The former are the true pharyngobranchials: the latter are the anterior ends of the epibranchials separated off. The glossopharyngeal and vagus nerves pass behind the infra-, but in front of the suprapharyngobranchials of their respective arches.

ii. The development of the osteocranium of Lepidosteus

A complete study of the development of the bones of Lepidosteus has not been made. The development of the dermal bones has been studied by Hammarberg (1937) and Aumonier (in press), but apart from this, knowledge is restricted to descriptions of young specimens and adults by van Wijhe (1882*b*), Veit (1907), Tate Regan (1923), and Mayhew (1924). (See Plate 41.)

The cartilage-bones of the neurocranium are few in number: *orbitosphenoid*,

[1] To correspond exactly with the Selachian subpituitary space, the cavum sacci vasculosi ought to communicate with the orbits through the pituitary vein foramina. This in Lepidosteus is not the case, for these foramina are here situated farther forwards, in which position they afford additional evidence for the supposition that the cartilaginous side wall of the skull in Lepidosteus has been secondarily developed.

prootic, *sphenotic*, *epiotic*, *exoccipital*, and *basioccipital*, the latter also ossifying in the occipitospinal arches. Later, the *pleurosphenoid* ossifies between the orbitosphenoid and the prootic. There is no *pterotic*, and no separate *opisthotic*, but it may have fused with or its place may have been taken by the exoccipital.

Of the membrane-bones, the *parasphenoid* is worthy of attention. Between it and the cartilaginous base of the skull the two parabasal canals are enclosed; the internal carotid arteries entering their hind ends, the efferent pseudobranchial arteries entering them half-way along, and the palatine nerves (which enter the canals through their foramina in the floor of the trigeminofacialis chambers) emerging from their anterior openings. Laterally to the parabasal canals, the parasphenoid gives off a pair of wings which underlie and partly enclose the anterior surface of the basitrabecular processes. Here there is very close connexion between the cartilage and the bone, spicules of which may be seen to enclose broken-down cartilage cells and capsules. It seems on careful investigation, however, that these cartilage cells are not primary cartilage, but partake of the nature of secondary cartilage, such as is found in higher vertebrates—e.g. Mammals. It is with these lateral wings of the parasphenoid that the metapterygoid bones are in contact.

The *intertemporal*, developed in close relation to the sense-organs of the temporal section of the infraorbital lateral line, is very large, and unconnected with any cartilage-bone. The *postfrontal*, likewise, is quite independent of the sphenotic.

The *maxilla* is a small bone in the maxillary fold; it is the hindmost of the tooth-bearing bones of the margin of the upper jaw, but its teeth are lost in the adult. In front of it are a number of *rostrals* (suborbitals) which bear teeth and develop each bone in relation to one sense-organ of the infraorbital lateral line canal. They correspond to the lateral basirostrals of Acipenser. The hinder members of the suborbital series, and the *postorbitals* and *supraorbitals* develop later and surround the orbit.

The *frontal* arises in fairly close relation with the sense-organs of the supraorbital lateral line canal, but separate centres corresponding to each organ have not been demonstrated.

The *premaxilla* ('intermaxillary') at its first origin is deep-seated, and its relations with the supraorbital lateral line canal are acquired later. According to Hammarberg it is fused with the nasal which is absent according to Aumonier.

The *adnasal* arises in relation to four organs of the supraorbital lateral line.

The *supra-ethmoid* arises paired, in relation to the ethmoid cross commissure.

The *antorbital* (lateral prenasal) arises in relation to the foremost two organs of the infraorbital lateral line, and is the foremost rostral or suborbital. The three foregoing bones constitute the 'os mobiles du nez'.

The *parietal* arises late, the temporal lateral line running along its lateral border, and the occipital commissure runs between the parietal and the *extrascapular*.

The *opercular* and *subopercular* arise in the hinder part of the opercular fold without relation to lateral line organs.

The *preopercular* arises in close relation with the opercular lateral line canal. At later and adult stages the dorsal portion of the preopercular becomes overlapped by the hindmost of the irregular and numerous *opercular ossicles*. The pores of the lateral line canal emerge between the opercular ossicles which thus do not enter into relations with them.

The *interopercular* is probably represented by one of the three branchiostegal rays which are developed.

The bones of the palate consist of the (paired) *prevomers, palatines, ectopterygoids*, and the *endopterygoids* ('mesopterygoid'), the last only a thin splint.

The bones of the lower jaw consist of the *dentary* (which lodges the lateral line canal and therefore may be a dentalo-splenial), anterior and posterior *coronoids, angular, supra-angular*, and *prearticular*.

No sclerotic bones are present.

11. TELEOSTEI. *CLUPEIFORMES*

I. Salmo

The development of the cartilaginous skull of the salmon and trout was the subject of an important paper by W. K. Parker (1873*a*); by the method of wax-model reconstructions it has been studied by Stöhr (1882), Winslow (1896), Gaupp (1906*a*), Böker (1913), and de Beer (1926*a*). The early stages have been investigated with the van Wijhe technique by de Beer (1927) and by Saunderson (1935), whose results form the basis of the following account. Certain regions have been studied by Willcox (1899), Harrison (1895), Zanichelli (1909), Rutherford (1909), and Makushok (1925, 1928).

On the development of the bony skull, information is obtainable from papers by W. K. Parker (1873*a*), Schleip (1904), Gaupp (1903, 1906*a*), and Böker (1913). Tchernavin (1937) has studied the changes in the skull during breeding.

i. The development of the chondrocranium of Salmo fario

Cartilage appears first in embryos about 9·3 mm. long, some six days before hatching (at room temperature), in the hindmost region of the parachordals. From these, chondrification spreads rapidly forwards on each side of the notochord, and the parachordal cartilages are then almost complete (Plate 42 fig. 1). Their anterior portions are slender and diverge forwards away from the notochord and from one another; their middle (or mesotic) portions are thick both dorsoventrally and mediolaterally, and lie medially to the central portions of the otic sacs; the posterior (or occipital) portions of the parachordals are thin, flat plates, bearing the rudiments of the occipital arches at their posterolateral corners.

According to Stöhr (1882), the posterior and anterior portions of the parachordals in Salmo salar and S. trutta have separate centres of chondrification, which spread towards one another. This was not observed in S. fario by van Wijhe (personal communication), or de Beer (1927), nor indeed did Tichomiroff (1885) observe it in his work on S. salar. Separate centres of chondrification have, however, been found in other Teleostei (e.g. Gasterosteus). This method of origin of the parachordals appears to be a variation of little theoretical importance. It is tempting to consider the possibility that the anterior parachordals of Stöhr might represent polar cartilages, fused on to the anterior ends of the true parachordals (as in Scyllium). The morphology of later embryos of Salmo, however, tells against this view, for the anterior region of the parachordals enters into relation with the auditory capsule (by means of the anterior basicapsular commissure), which true polar cartilages do not, being situated further forward and giving off the basitrabecular processes on each side. However, the difficulty of establishing limits and points of reference in developing stretches of cartilage is great, and it is just possible that Stöhr's anterior parachordals represent (in whole or in part) the polar cartilages of other forms.

The parachordals, as chondrified, present no evidence of metameric segmentation, and the segmentation of this region of the Teleostean skull has to be determined with reference to the relations of the septa or myocommata of the occipital myotomes. According to Willcox (1899), so far as can be made out from her paper, the structure which she identifies as the occipital arch (in S. salar and S. fario) is related to the septum between the 5th and 6th postotic segments. If the 1st permanent postotic myomere is formed from the 6th segment, as it seems to be, for the 1st and 2nd postotic somites (4th and 5th segments) disappear, then the skull of Salmo occupies eight segments. A curious feature of this (and other Teleostean skulls), however, is the absence of cranial hypoglossal nerve-roots, in front of the occipital arch. The question therefore arises, What happens to the nerve-roots of the alleged four occipital myomeres? According to Willcox (1899) the first two of these myomeres disappear, together with their ventral nerve-roots and the nerve-root of the 3rd; the nerve-roots of the 4th (which then also innervates the 3rd) and 5th myomeres join and emerge in front of the septum separating the 5th and 6th myomeres. These nerves form the so-called hypoglossal nerve-root in the fully formed skull, and emerge from the spinal cord *behind* the occipital arch (Makushok, 1928), enclosed in a foramen by the exoccipital bone. (Harrison, 1895, made out only three occipital somites in the trout, but he must have missed the foremost one described by Willcox for he only noticed the disappearance of one, and not of two as Willcox did.)

There is, therefore, some discrepancy between the conditions in the adult and those in the embryo as described by Willcox, for in the latter, if she is right, these hypoglossal roots emerge *in front* of what she calls the occipital arch. The tangle seems to have been unravelled by Beccari (1922), who showed that the cartilaginous arch developed in the septum between the 5th and 6th postotic somites is a sort of 'proatlas', behind the occipital arch (which is in the septum between the 4th and 5th somites) and the neural arch of the 1st free vertebra (in the septum between the 6th and 7th somites). This proatlas basidorsal arch, the existence of which de Beer has confirmed, is separate from its centrum, which seems to fuse on to the hind end of the parachordals. It is by the extension of the exoccipital into the proatlas arch that that bone comes to enclose the hypoglossal roots.

There are, therefore, eight segments in the skull of Salmo, but the neural arch of the 8th segment is still free, and may be called a proatlas basidorsal arch.

Soon after the appearance of the parachordals, chondrification sets in in some of the elements of the visceral arch skeleton, i.e. Meckel's cartilages, the ceratohyal, and hyomandibula (Plate 43, fig. 3).

Slightly later (9·6 mm., Plate 42, figs. 2, 3) the trabeculae appear, as independent rod-like cartilages fairly wide apart, and some distance in front of the parachordals, more or less in their direct line of prolongation. The floor of the skull is, therefore, not subjected to a cranial flexure. The trabeculae lie beneath and to the side of the forebrain and close to the skin of the underside of the snout; no evidence has, however, been found to support Lundborg's (1899) view that the ectoderm contributes to their formation.

In embryos about 10·5 mm. long (five days before hatching) the hind end of each trabecula is joined on to the front end of the parachordal of its side, forming a rod, kinked outwards in two places, one behind the other: the junction occurs between the kinks. The rudiments of the auditory capsules have appeared, flanking the anterolateral corners of the otic sacs: each of these cartilaginous rudiments is connected with the anterior portion of the parachordal of its side by dense procartilage, representing the future anterior basicapsular commissure (Plate 43, fig. 1).

In lateral view (Plate 43, fig. 2) an embryo of this stage presents a number of interesting features in connexion with the visceral arch skeleton. The hyomandibular and symplectic cartilages are present and distinct, though they soon become connected by procartilage. The dorsal edge of the hyomandibula is indented by a deep notch in which the hyomandibular branch of the facial nerve runs outwards. In front of, and quite separate from, the symplectic is the quadrate. Meckel's cartilage and the ceratohyal are present, as are also the 1st and 2nd ceratobranchials.

Slightly later (Plate 43, fig. 4), the visceral arch skeleton makes further progress in chondrification by the appearance of the hypohyals and of the 3rd and 4th ceratobranchials.

In embryos 11·4 mm. long (four days before hatching) the auditory capsule extends back under the otic sac and up its lateral surface, and is united with the parachordal at two points: in front by the anterior basicapsular commissure, situated just behind the facial nerve; and behind by the posterior basicapsular commissure which is situated between the exits of the glossopharyngeal and vagus nerves (Plate 44, fig. 1). A space is thus left between the basicapsular commissures, the ventral edge of the capsule and the lateral edge of the parachordal, the basicapsular fenestra. In Salmo (as in most Teleostomi) the passage of the glossopharyngeal nerve through the hindmost part of this basicapsular fenestra shows that it includes a small portion of space representing morphologically the anterior part of the fissura metotica. The posterior part of the fissura metotica (or jugular foramen), enclosed between the posterior basicapsular commissure and the occipital arch (but still open dorsally), is traversed by the vagus nerve.

The trabeculae are still distinct, and their anterior ends are slightly expanded and turned outwards (the cornua trabecularum).

At the time of hatching (embryos 12·3 mm. long) the trabeculae are interconnected, and an anterior limit is thus formed to the hypophysial fenestra (Plate 44, fig. 2). The rudiments of the lateral commissures are now present, in the form of two processes on each side. One of these, the prootic process, arises from the middle of the anterior edge of the auditory capsule and grows medially, on the ventral side of the head vein and in front of the hyomandibular nerve. The other, or postpalatine process, arises from the edge of the parachordal, in the corner formed between the latter and the front edge of the anterior basicapsular commissure, and grows towards the tip of the prootic process. As its name implies, the postpalatine process grows out behind the palatine branch of the facial nerve. The glossopharyngeal nerve now passes through a small foramen of its own, cut off by cartilage from the remainder of the basicapsular fenestra.

A lateral view of an embryo at this stage is instructive (Plate 44, fig. 3). The prootic and postpalatine processes have practically met, forming the lateral commissure. In the visceral arch skeleton the quadrate now shows a short pterygoid process; the symplectic and hyomandibula have joined to form a hyosymplectic which lies immediately behind and in contact with the quadrate. The head of the hyomandibula surrounds the hyomandibular branch of the facial nerve in a foramen. The interhyal is present, between the hind edge of the hyomandibula and the hind end of the ceratohyal.

In embryos 14·2 mm. long (five days after hatching), the anterior portions of the trabeculae have fused to form an extensive trabecula communis, expanded in front into the ethmoid plate (Plate 45, fig. 1). The orbital cartilages have chondrified independently in the form of thin rods, extending from a point above the lateral corners of the ethmoid plate almost as far as the front walls of the auditory capsules.

The lateral commissure is now fully developed, and forms the side wall of the trigeminofacialis chamber; it lies laterally and ventrally to the head vein which enters the chamber through its posterior aperture, i.e. between the lateral commissure and the anterior basicapsular commissure. The hyomandibular branch of the facial nerve passes out through the same aperture. The palatine branch of the facial nerve runs down in front of the lateral commissure, to the side of the trabecula. The basicapsular fenestra is divided into two by cartilaginous development of the floor of the auditory capsule. The anterior remnant of the basicapsular fenestra, entirely surrounded by capsular cartilage (i.e. no longer bounded on the medial side by the parachordal), may be regarded as the analogue of the fenestra vestibuli (ovalis) of Tetrapoda.

The side wall of the auditory capsule has grown up and arched over to form the roof, and the septum of the posterior semicircular canal has appeared, separating the cavity of that canal from the main cavity of the auditory capsule. The latter has no medial wall whatever. The hind wall of the auditory capsule has fused with the dorsal end of the occipital arch, with the result that the jugular foramen (for the exit of the vagus nerve) is completely enclosed.

Further development, as seen in embryos 15·5 mm. long (seven days after hatching), largely concerns the ethmoid and orbitotemporal regions, and the visceral arches (Plate 45, fig. 2). From the posterolateral corners of the ethmoid plate, the lamina orbitonasalis arises on each side as a short pillar pointing towards the front end of the orbital cartilage. The hind end of the latter, which may also be called the taenia marginalis, has fused with the front of the auditory capsule, which is produced into a postorbital process. The pterygoid process of the quadrate has extended forwards, and its anterior end underlies the lateral edge of the ethmoid plate. Epibranchials have chondrified in the first three branchial arches, and a ceratobranchial in the 5th arch.

A little later (embryos twelve days after hatching) the orbital cartilage forks at its anterior end (Plate 45, fig. 3); the lateral prong (future sphenethmoid commissure) points towards the lamina orbitonasalis; the median prong (future sphenoseptal commissure) points towards the nasal septum. The latter has now appeared as a median process sticking up dorsally from the centre of the ethmoid plate.

In embryos fifteen days after hatching (Plate 45, fig. 4) fusion has taken place between the sphenethmoid commissure and the lamina orbitonasalis on the one hand, and between the sphenoseptal commissure and the nasal septum on the other. This latter fusion represents the roof of the nasal capsule: indeed, all the cartilaginous roof the nasal capsule will ever have. In this way a large foramen is enclosed on each side of the nasal septum. This is the foramen olfactorium advehens, through which pass the olfactory nerve, the orbitonasal artery, and the ophthalmic nerve: the latter soon acquiring a small foramen of its own. Since there is as yet no preoptic root to the orbital cartilage it is impossible to define the limits of the foramen olfactorium evehens, or of the orbitonasal fissure.

The chondrocranium has by now (embryos 16·5 mm. long) almost reached its full development (Plate 46, fig. 1). The anterior end of the pterygoid process of the pterygoquadrate shows two facets for articulating with the under surface of the ethmoid plate: an anterior rostropalatine and a posterior ethmopalatine articular facet. The two taeniae marginales are interconnected about the middle of their length by a slender cartilaginous rod, the epiphysial bar, which is the rudiment of the future extensive roof over the anterior part of the cranial cavity. In the auditory capsule, septa for all these semicircular canals are present, and in whole preparations of the chondrocranium they can be distinguished by their opacity from the cavities of the canals themselves. The visceral arch skeleton is almost complete (Plate 46, fig. 2). In addition to ceratobranchials in all five branchial arches, hypobranchials in the first four, epibranchials in the first three, and pharyngobranchials in the first two, the median ventral elements have appeared in the form of the basihyal and, behind it, the unpaired median basibranchials which soon become fused to give rise to the longitudinal copula, as Tichomiroff (1885) showed.

ii. Morphology of the chondrocranium of Salmo fario (14 mm. stage) and of S. salar (25 mm. and 33 mm. stages)

(PLATES 47, 48.)

1. Basal plate and notochord. For the hinder half of their length the parachordals lie with their medial edges close to the notochord, which separates them. Seen in transverse section, this region of the basal plate shows a median hogs-back (convex dorsally) flanked by a pair of depressions (concave dorsally). The latter lodge the saccules which now have a practically complete cartilaginous floor beneath them (except for the glossopharyngeal nerve foramen) owing to the closing of the basicapsular fenestrae. The median hogs-back forms a groove on the ventral surface of the basal plate, which will become the posterior portion of the myodome.

A short distance in front of the place where the parachordals diverge away from the notochord, their medial edges become depressed to a level more ventral than that of the basal plate as a whole. In this region the two parachordals are interconnected by a flat horizontal and transverse shelf of cartilage: the prootic bridge. The lateral edges of this bridge are, however, attached not to the medial (ventrally depressed) edges of the parachordals, but bridge across at the more dorsal general level of the basal plate. The medial edges of the parachordals then extend downwards and inwards, beneath the prootic bridge. This region will give rise to the middle portion of the myodome. Behind the prootic bridge a small basicranial fenestra is enclosed, bordered laterally by the parachordals, and traversed by the notochord. The foremost point reached by the notochord is *relatively* farther back than at previous stages, which is due to the more rapid growth of the parachordals. Eventually the notochord becomes enclosed in the basioccipital bone.

It is hard to speak of an occipito-atlantal joint, for there is no contact whatever at this stage between the skull and the vertebral column. The latter is represented by the basidorsal elements. The hind end of the basal plate seems to be represented by the basidorsal and basiventral elements of the occipital vertebra. Seen in lateral view, these give rise to two bosses, one above the other, projecting backwards along the notochord on each side.

2. The occipital region. One pair of occipital arches is present, and there are no hypoglossal foramina, or any other evidence of segmentation in the cartilage. The dorsal ends of the occipital arch fuse with the hind wall of the auditory capsule, enclosing the jugular foramen. The occipital arches do not appear to contribute much to the tectum synoticum: there is thus little or no tectum posterius.

3. The auditory region. The absence of any medial wall to the auditory capsule makes it difficult to define the limit between the lateral edge of the parachordal and the medial edge of the cartilaginous capsular floor. The lateral edge of the parachordal may be taken as the medial border of the basicapsular fenestra. However, the inclusion of the glossopharyngeal nerve in the basicapsular fenestra, and the passage of this nerve through the cavity of the auditory capsule show that this cavity has encroached on space which is essentially extracapsular, in particular, the anterior portion of the fissura metotica.

By its grooves and swellings, the external surface of the capsule reflects the position of the anterior, posterior, and lateral semicircular canals, and of the utricle. The head of the hyosymplectic fits in a groove on the ventral side of the prominence of the lateral semicircular canal. The only internal partitions to be found are the septa of the three semicircular canals. The roof of the capsule ends medially with a free edge, except posteriorly where it is connected with its fellow of the opposite side by the tectum synoticum. Anteriorly, the capsule gives off a postorbital process which is continuous with the posterior end of the taenia marginalis. The root of the postorbital process is perforated by a foramen for the otic branch of the facial nerve.

4. The orbitotemporal region. The hinder ends of the trabeculae are attached to the anterior ends of the parachordals at the level of the depressed medial edges of the latter; the trabeculae are, therefore, at a more ventral level than the basal plate and prootic bridge. On the other hand, the trabecula communis, formed by the fusion of the anterior halves of the trabeculae, is raised dorsally to the same level as the basal plate. The (paired) trabeculae, therefore, present a depressed region, and it is here, over the trabeculae, that the external and internal recti muscles enter the cranial cavity from the orbit. This region is thus the anterior portion of the myodome. The trabeculae enclose a large hypophysial fenestra, through the anterior part of which the internal carotid arteries enter the skull. In later stages the trabecula communis becomes discontinuous by the absorption of the cartilage about half-way along its length.

The lateral commissure is a plate of cartilage connecting the lateral edge of the basal plate at its foremost end (immediately behind the palatine nerve) with the anterior wall of the auditory capsule (near the root of the postorbital process). It is formed from the postpalatine and prootic processes. In addition, a process grows out from the lateral edge of the trabeculae in their hindmost region, in front of the palatine nerve, and fuses with the lateral commissure. The palatine nerve is thus enclosed in a foramen, and the process last mentioned is the reduced basitrabecular process. The palatine foramen is also pierced by the orbitonasal artery, which runs upwards on the medial side of the palatine nerve, forwards dorsally to the basitrabecular process, and through the orbit to the nasal capsule.

The complete lateral commissure, as composed of the postpalatine, prootic, and basitrabecular processes, forms the lateral wall of the trigeminofacialis chamber. Through the hinder opening of this chamber (between the lateral commissure and the anterior basicapsular commissure) the head vein and the orbital artery enter and the hyomandibular branch of the facial nerve emerges. (In Salmo salar and S. fontinalis, the nerve and the vein are separated by an extra strut of cartilage. It may also be noted that the orbitonasal and orbital arteries of Salmo correspond to the undivided orbital artery in Amia.)

The anterior aperture of the trigeminofacialis chamber opens widely into the orbit, and through it pass the head vein, the orbital artery, and the branches of the trigeminal nerve. (For the ultimate subdivision of the chamber, see p. 128.)

The palatine nerve runs forwards dorsally to the postpalatine portion of the lateral commissure, pierces its foramen and continues forwards ventrally to the basitrabecular process. The position of the palatine foramen is an important landmark, indicating the limit of the lateral edge of the basal plate. Eventually, as the anterior edge of the prootic bridge extends forwards, it entraps and surrounds the palatine nerve in a foramen internum before it gets to its true foramen (externum). The abducens nerve likewise becomes enclosed in a foramen by the advancing edge of the prootic bridge.

There is no true prefacial commissure, nor a pila antotica, nor a pila metoptica, nor (until later) a preoptic root to the orbital cartilage. The result is that the entire side wall of the orbitotemporal region is occupied by a large fenestra, the sphenoid fissure, through which all the cranial nerves from the Ist to the Vth, and the ophthalmic artery leave the skull, and the pituitary vein and external rectus (later also the internal rectus) enter it.

Later on, the anterior part of the sphenoid fissure becomes reduced by the chondrification of the preoptic root of the orbital cartilage, and of the interorbital septum. The posterior edge of the interorbital septum then forms an anterior border of (and separation between) the foramina through which the optic and oculomotor nerves pass.

The abducens nerve does not strictly leave the skull at all, since the external rectus comes into the myodome where it is innervated. The efferent pseudobranchial artery runs inwards ventrally to the trabecula, before joining the internal carotid of its own side.

While the orbital region of the skull has for some time no side wall at all, it has an extensive roof, the tectum cranii, which stretches between the orbital cartilages. The anterior edge of the tectum grows forwards, gradually restricting the size of the epiphysial fenestra, until it fuses with the sphenoseptal commissures (roof of nasal capsule). The hind edge of the tectum grows back, eventually reaching the level of the auditory capsules, and from it in the middle line a long process—the taenia tecti medialis, extends backwards almost as far as the occipital region.

The various regions of the posterior myodome have already been described. It may be recalled that its anterior portion is represented by the region of the trabeculae, which form part of its floor, over which the external and internal recti muscles push inwards, lying ventrally to the dura mater, and lifting the floor of the brain and the pituitary up from the hypophysial fenestra. The middle portion of the myodome lies ventrally to the prootic bridge, actually in the hinder part of the hypophysial fenestra; it has no cartilaginous floor but is bounded ventrally by the parasphenoid bone. The posterior portion of the myodome lies in the groove on the ventral side of the basal plate. The internal recti in the so-called ventral compartment of the myodome, are separated by a horizontal membrane from the external recti in the dorsal compartment. This dorsal compartment is traversed by the pituitary vein, and corresponds to the entire myodome of Amia. The ventral compartment is traversed by the internal carotid arteries and palatine nerves, and is regarded by Allis as derived from the space occupied by the parabasal canals in Amia.

The eye is surrounded by a sclerotic cartilage which takes the form of a ring, for the medial side remains unchondrified.

5. The ethmoid region. The ethmoid region is somewhat difficult to interpret owing to the facts that the nasal capsule is very incomplete, and the preoptic root of the orbital cartilage very late in chondrifying.

Instead of being a six-sided cartilaginous box, each nasal capsule in Salmo is nothing but a pit, the fossa nasalis, with a floor (the ethmoid plate), a median wall (the nasal septum), and a hind wall (the lamina orbitonasalis). The roof of the nasal capsule is non-existent except for its extreme hind edge, represented by the sphenoseptal commissure and forming the dorsal border to the foramen olfactorium advehens, through which the olfactory nerve, the orbitonasal artery, and the ophthalmic nerve enter. (The latter nerve becomes enclosed in a small foramen distinct from the olfactorium advehens.) The sphenoseptal commissure at the same time forms the anterior edge of the roof of the cranial cavity when the tectum cranii fuses with it. These relations are shown in Plate 46, fig. 3.

The wall of the anterior part of the cranial cavity remains membranous for a long time, but the preoptic root of the orbital cartilage (its ventral portion fused with its fellow of the opposite side, to form the interorbital septum, rising up from the dorsal surface of the anterior part of the trabecula communis) eventually chondrifies (Plate 46, fig. 4), and a foramen olfactorium evehens is formed (Salmo fario, 30 mm. long). The lamina orbitonasalis is situated considerably farther forward than the foramen olfactorium evehens, with the result that the olfactory nerve has a long way to run before it reaches the foramen olfactorium advehens. During this part of its course the olfactory nerve presents the appearance of running through the orbit; in fact, it is running in the cavum orbitonasale, which is abnormally large owing to the forward position of the lamina orbitonasalis. The large gap between the hinder surface of the lamina orbitonasalis and the preoptic root of the orbital cartilage, limited above by the sphenethmoid commissure, is the fissura orbitonasalis, leading from the orbit to the cavum orbitonasale. This cavum is traversed not only by the olfactory nerve issuing from the foramen olfactorium evehens, and the ophthalmic nerve and orbitonasal artery coming from the orbit, but also by the superior and inferior oblique muscles, the points of origin of which push forwards and inwards from the orbit, through the cavum orbitonasale to a canal hollowed out in the thickness of the nasal septum, thus forming the anterior myodome. At their extreme front end the two anterior myodomic canals are confluent.

6. The visceral arch skeleton. The pterygoquadrate is supported anteriorly by the rostropalatine and ethmopalatine articulations with the ethmoid plate, and posteriorly by the hyomandibula. The basal process is represented by a stump pointing from the root of the pterygoid process inwards towards the basitrabecular process, which it does not, however, reach. The otic (metapterygoid) process is also small; it lies along the anterior edge of the hyomandibula, and the afferent pseudobranchial artery runs ventrally to it.

The pterygoid process articulates with the ethmoid plate by two processes, the rostropalatine and the ethmopalatine, thus conforming to the type which Swinnerton (1902) has called 'disartete'.

Meckel's cartilage presents no special features. It may be noticed, however, that the processus retroarticularis extends back behind the articulation with the quadrate, and that the processus prearticularis rises in front of it. The hyosymplectic is articulated with the auditory capsule, laterally to the head vein and adductor muscle. It is pierced by a foramen for the hyomandibular branch of the facial nerve. In late embryos this foramen becomes emarginated from behind, and the nerve then runs in a notch on the posterior edge of the hyosymplectic.

The symplectic forms the direct anterior prolongation of the hyomandibula, and underlies the posteroventral edge of the pterygoquadrate with which it is in contact. The interhyal connects the ventral edge of the hyomandibula (at the point where the latter passes into the symplectic) with the posterodorsal corner of the large ceratohyal. The anterior end of the latter is in contact with the hypohyal, which in turn abuts against the side of the median basihyal.

The 5th branchial arch is represented by a pair of ceratobranchials; all the other branchial arches possess the full complement of pharyngo-, epi-, cerato-, and hypobranchial elements. The basibranchials are fused into an unpaired rod, attached to the hind end of the basihyal, forming the so-called copula.

In late embryos paired premaxillary cartilages arise close to the hind ends of the premaxillae. They probably correspond to the anterior upper labial cartilages of Selachians.

iii. Ossification of the skull of Salmo fario and S. salar

(PLATE 49)

The development of the bones of the skull in the trout and the salmon (which appear to follow identical lines) has been studied by W. K. Parker (1873*a*), Schleip (1904), Gaupp (1906*a*), and Böker (1913).

The distinction between cartilage- and membrane-bones in the ossification of the skull of Salmo is in some cases hard to draw. Several of the cartilage-bones have intra-membranous extensions; one (the basisphenoid), though probably phylogenetically a cartilage-bone, develops solely in membrane; a membrane-bone (the prevomer) becomes perichondral, as does also the angular. In addition, many cartilage-bones and membrane-bones fuse to form composite bones. The stages given are those at which the bones arise.

1. The membrane-bones, i.e. bones arising without relation to or contact with cartilage. Among the membrane-bones two types of origin may be discerned, and these are represented by: (*a*) the bones related to the lateral line canals; (*b*) the bones underlying the skin, not related to lateral line canals; and the bones related to the mouth, palate, and alimentary canal.

(*a*) The membrane-bones related to the lateral line canals are: frontal, nasal, lachrymal, suborbitals, postorbitals, intertemporal, postparietal, subtemporal, preopercular, dentary, angular. They originate as small gutter-shaped plates of bone medially to (but quite separate from) the sense-organs of the lateral line system. As the latter becomes converted from a system of grooves to a system of tubes, the bone-rudiments grow into hollow cylinders, surrounding the tubes, and perforated at intervals by the tubules connecting the lateral line canals with the exterior. This condition is that of the so-called canal-bones (Goodrich, 1909), exemplified in Salmo at early stages by the suborbitals, postorbital, intertemporal, postparietal, and subtemporal. In the case of the frontal, nasal, lachrymal, preopercular, dentary, and prearticular, however, development continues beyond the tubular canal-bone stage, and plates of bone extend away from the tube, beneath the skin. A complete series can be made out showing all grades of development of the plate, from the intertemporal where it is practically non-existent to the frontal where it is very large.

The *frontal* (S. fario, 16 mm.) lodges part of the supraorbital lateral line canal, and ossifies over the tectum cranii, the taenia marginalis, and the anterior part of the auditory capsules. Each bone extends medially until it meets its fellow from the opposite side in the middle line. The lateral edge of the frontals is the dorsal border of the orbit.

The *nasal* (S. fario, 18 mm.) lodges the anterior part of the supraorbital lateral line canal, and ossifies over the fossa nasalis.

The *lachrymal* (S. fario, 18 mm. = 2nd orbital of Gaupp) lodges the anterior part of the infraorbital lateral line canal, and lies in the anteroventral corner of the orbit.

The *suborbitals* (S. salar, 26 mm.) and *postorbital* (S. salar, 33 mm.) lodge the infraorbital lateral line canal between the lachrymal and the intertemporal, and form the ventral and posterior borders of the orbit. They are little more than tubular canal-bones. One of the suborbitals probably represents the *jugal* and the upper part of the postorbital may represent the *postfrontal.*

The *intertemporal* (S. fario, 17 mm. = dermosquamosal of Gaupp) lodges the lateral line canal from the junction between supra- and infraorbital canals to the point where the occipital canal is given off. The intertemporal is little more than a tubular canal-bone, and it fuses with the underlying pterotic.

The *postparietal* (= extrascapula) lodges the transverse occipital lateral line canal, and lies dorsally to the tectum synoticum, running inwards and forwards towards its fellow from the opposite side. It is little more than a tubular canal-bone.

The *subtemporal* (= supratemporal of Parker) is a very small tubular canal-bone between the intertemporal and the preopercular and lodging the dorsal portion of the mandibular lateral line canal. It fuses with the preopercular.

The *preopercular* lodges part of the mandibular lateral line canal, and lies posterolaterally to the hyomandibula.

The *angular* (dermarticular of Gaupp) receives the mandibular lateral line canal from the preopercular, and passes it on to the dentary. It takes the place of the articular and comes into relations with Meckel's cartilage.

The *dentary* lodges the anterior portion of the mandibular lateral line canal, and lies laterally to Meckel's cartilage. Anteriorly, the dentary is continuous with a cartilage-bone, the mentomeckelian, so that the so-called dentary is really a mixed ossification. The dentary is also a tooth-bearing bone. The teeth appear independently of the bone, and only subsequently become attached to it. Similarly, it is some time after the appearance of the bone that it acquires its relations to the lateral line canal. In the absence of evidence of separate ossification of the bone surrounding the canal, the whole bone cannot be called a dentalo-splenial, but must be regarded as a dentary which has acquired relations with the lateral line canal, or a splenial which has taken the dentary's place.

(*b*) The membrane-bones not related to lateral line canals are: suprascapular, parietal, supraorbital, orbital, supra-ethmoid, supramaxilla, prevomer, parasphenoid, premaxilla, maxilla, dermopalatine, ectopterygoid, endopterygoid, operculum, suboperculum, interoperculum, branchiostegal rays, dermentoglossum, supracopulare, pharyngeum superius and inferius.

These bones arise simply as flat lamellae, often very close to the cartilage of the chondrocranium. Indeed, the internal periosteum of the bone may actually be in contact with the external perichondrium of the cartilage.

Many of these bones enter into relation with teeth: such in Salmo are: premaxilla, maxilla, prevomer, dermopalatine, pharyngeum superius and inferius, dermentoglossum, and the dentary which is also a lateral line bone. In all cases the teeth and the bone arise independently and the former may subsequently become attached to the latter.

The *suprascapular* (supracleithrum), which overlies the hinder part of the epiotic, connects the hind part of the skull with the pectoral girdle.

The *parietal* (S. fario, 19 mm.) overlies the hinder part of the auditory capsule; medially it comes into contact with the lateral edge of the supraoccipital and the posterolateral edge of the frontal. The periosteum of the parietal touches the perichondrium of the underlying cartilage.

The *supraorbital* is a small bone in the anterodorsal corner of the orbit close to the anterolateral corner of the frontal.

The *orbital* forms the anterior border of the orbit and lies laterally to the lateral ethmoid.

The *supra-ethmoid* (S. salar, 31 mm.) ossifies over the dorsal edge of the nasal septum; its inner periosteum is indistinguishable from the perichondrium of the cartilage. It may, therefore, contain an element of the (perichondral) ethmoid ossification.

The *premaxilla* ossifies laterally to the ethmoid plate. The teeth appear before the bone, and subsequently become attached to its horizontal portion, while its ascending facial portion covers the edge of the ethmoid plate. Later still, the labial cartilage chondrifies between the premaxilla and the ethmoid plate but is separated from both by connective tissue.

The *maxilla* ossifies behind the premaxilla, with its anterior portion close to the anterior end of the pterygoid process of the pterygoquadrate, with which it articulates by a cartilaginous facet. Teeth are borne only on the anterior part of the maxilla.

The *supramaxilla* (jugal autt.) arises laterally to the dorsal edge of the maxilla. This bone can hardly be a jugal since it does not house part of the infraorbital lateral line canal.

The *prevomer* (paired rudiment, afterwards fused) enters into relations with the cartilage of the ethmoid plate, for which reason it is dealt with below.

The *parasphenoid* (S. fario, 13 mm.) arises on the ventral side of the skull, and soon extends from the prevomer almost to the hind end. Anteriorly, the parasphenoid underlies the interorbital septum; posteriorly, it forms the floor of the (posterior) myodome. The lateral edges of the parasphenoid are produced into a pair of processes (the ascending processes) directed towards the pleurosphenoids. Behind these, the edges of the parasphenoid meet those of the prootic and basioccipital. The hind end of the parasphenoid is forked, the dorsal aorta passing between the prongs and forwards ventrally to the parasphenoid.

The ascending process of the parasphenoid underlies the lateral commissure, leaving a space, the parabasal or Vidian canal, on each side. The palatine nerve enters the canal from above and runs forwards, issuing from its anterior opening. The internal carotid artery enters the parabasal canal by its posterior opening, traverses it, and runs dorsally through the hypophysial fenestra.

The *dermopalatine* arises ventrally to the pterygoid process of the pterygoquadrate, separated from the cartilage by connective tissue. Teeth, which had appeared before the bone, subsequently become attached to it, and ultimately the bone fuses with the autopalatine dorsal to it, to form a mixed ossification.

The *ectopterygoid* ossifies as a plate ventrally to the cartilage of the middle portion of the pterygoquadrate, and dorsally to the mouth epithelium.

The *endopterygoid* resembles the ectopterygoid, lying medially to it.

The *opercular*, *subopercular*, *interopercular*, and *branchiostegal rays* ossify as plates behind the preopercular, beneath the skin of the opercular fold.

The *dermentoglossum* arises as a plate lying dorsally to the basihyal, and ventrally to a set of teeth to which it soon becomes attached. The dermentoglossum and the basihyal appear to remain separate.

The *supracopulare* arises as a plate lying dorsally to the basibranchials, with the 2nd of which (symbranchial III) it fuses to form a mixed ossification.

The *pharyngeum superius* arises as a plate lying ventrally to the 4th pharyngobranchial and epibranchial, and bears teeth.

The *pharyngeum inferius* arises as a number of platelets of bone lying dorsally to the 5th ceratobranchial, and ventrally to a set of teeth which arise in this region of the pharynx. At very early stages the teeth fuse with the bony plates, which in turn fuse with one another and with the underlying 5th ceratobranchial to form a mixed bone.

2. Cartilage-bones, i.e. bones arising as perichondral ossifications, with subsequent endochondral ossification. The *basioccipital* (S. salar, 24 mm.) makes its appearance in Salmo salar embryos 24 mm. long, in the form of four perichondral lamellae; one each on the dorsal and ventral surfaces of the hinder part of the basal plate, on each side of the notochord. Eventually the notochordal sheath itself becomes surrounded by bone, and the paired rudiments of the basioccipital are then joined to form one bone, which, with the help of endochondral centres of ossification, replaces the cartilage of the basal plate. Anteriorly, the basioccipital reaches as far as the basicranial fenestra; posteriorly, its dorsal lamella is excluded from the ventral border of the foramen magnum by the exoccipitals. The basioccipital forms the roof of the posterior portion of the myodome.

The *exoccipital* ossifies as a pair of perichondral lamellae on the outer and inner surfaces of the occipital arch. Ossification spreads round the ventral border of the foramen magnum as far as the notochord, and anteriorly in the ventral wall of the auditory capsule: a region normally ossified by the opisthotic which is absent (see p. 130). The exoccipital thus surrounds the jugular foramen (vagus) and the glossopharyngeal foramen, underlies the saccule and forms part of the walls of the posterior semicircular canal. A posterior extension of ossification behind the occipital arch and involving the proatlas arch, entraps the 1st spinal ventral nerve-root, which thus emerges through a foramen as the so-called hypoglossal.

The *supraoccipital* (S. fario, 20 mm.) arises as unpaired dorsal and ventral perichondral lamellae on the tectum synoticum, and on the taenia tecti medialis. The median occipital spine is the result of ossification in the connective tissue septum between the anterior trunk-myotomes. This spine becomes attached to the supraoccipital.

The *epiotic* appears first as a perichondral lamella on the outer surface of the lateral wall of the posterior semicircular canal. A lamella on the inner surface ossifies later, as does the base of the tendon by which the epiotic is attached to the supracleithrum.

The *pterotic* (S. salar, 26 mm.) ossifies as perichondral lamellae on the outer and inner surfaces of the lateral wall of the lateral semicircular canal. Eventually it fuses with an overlying dermal ossification (the intertemporal, dermosquamosal of Gaupp) to form a composite bone.

The *sphenotic* (S. salar, 30 mm.) ossifies as perichondral lamellae on the outer and inner surfaces of the lateral wall of the anterior semicircular canal and on the postorbital process. On its external surface the sphenotic bears a facet for the articulation of the head of the hyomandibula.

The *prootic* (S. salar, 24 mm.) appears in the form of outer and inner perichondral lamellae on the anterior wall of the auditory capsule, and on the lateral commissure. Ossification spreads inwards in the anterior region of the basal plate and prootic bridge, towards the anterior ends of the parachordals, and also anteriorly to the lateral commissure in the membrane surrounding the exit of the trigeminal nerve.

The cartilage of the lateral commissure is absorbed, and in the space thus provided, the head vein runs. This space thus forms the pars jugularis of the trigeminofacialis chamber, enclosed between the outer and inner lamellae of the prootic, as distinct from the pars ganglionaris, which lodges the trigeminal and facial ganglia and lies medially to the inner lamella of the prootic and laterally to the dura mater (which marks the position of the true side wall of the skull, here unossified).

The ossification of the prootic in the prootic bridge and in the anterior portions of the parachordals provides the bony roof and side walls of the middle portion of the myodome. The anterior edge of the prootic bridge entraps the palatine and abducens nerves which thus penetrate foramina through it.

The *pleurosphenoid* (S. fario, 40 mm. 'alisphenoid' autorum) arises late in the form of outer and inner perichondral lamellae in the hinder part of the taenia marginalis (now enveloped in the tectum cranii). Ossification spreads ventrally in the membrane, and the pleurosphenoid thus comes to enclose the trochlear nerve and the ophthalmic branch of the trigeminal nerve in separate nerve foramina.

The *orbitosphenoid* (S. salar, 33 mm.) ossifies as perichondral lamellae on the interorbital septum and in the cartilage (preoptic root of the orbital cartilage) surrounding the foramen olfactorium evehens. The paired orbitosphenoids eventually fuse.

The *lateral ethmoid* (S. salar, 33 mm.) arises as a perichondral lamella on the lateral surface of the lamina orbitonasalis.

The *quadrate* (S. fario, 17 mm.) ossifies in the pterygoquadrate cartilage as perichondral lamellae in the region of the articulation with the lower jaw; the actual articular facet is still, however, cartilaginous. The hind edges of the quadrate forms a groove in which the anterior end of the symplectic bone is firmly attached. An ossified ligament connects the quadrate with the opercular.

The *metapterygoid* ossifies as perichondral lamellae in the posterodorsal portion of the pterygoquadrate cartilage, comprising the diminutive otic process. The fibres of the adductor mandibulae and levator arci palatine muscles take their origin on the metapterygoid, which bone is increased by the ossification of the base of the tendons of these muscles.

The *autopalatine* is a perichondral ossification on the ventral surface of the pterygoid process of the pterygoquadrate cartilage. It undergoes fusion with a membrane-bone, the dermopalatine, to form a mixed bone.

The *articular* is usually regarded as fusing with the angular, a membrane-bone. But Haines (in press) has shown that the articular is absent and that the *angular* takes its place and comes into relation with Meckel's cartilage.

The *mentomeckelian* ossifies as a perichondral lamella round the anterior end of Meckel's cartilage. It is nowhere continuous posteriorly with the dentary.

The *retroarticular* (angular autt.) ossifies where the fibres of the mandibulohyoid ligament are attached to the processus retroarticularis of Meckel's cartilage, and the bone spreads as a perichondral lamella over the cartilage.

The *coronomeckelian* arises where the fibres of the adductor mandibulae muscle are inserted on the processus prearticularis of Meckel's cartilage, as an ossification of these fibres and of a perichondral lamella which spreads over the cartilage.

The *hyomandibula* ossifies in the proximal portion of the hyosymplectic cartilage from lateral and medial perichondral lamellae. The hyomandibular branch of the facial nerve runs out beneath the ventrolateral edge of the hyomandibular bone.

The *symplectic* ossifies in the distal portion of the hyosymplectic cartilage.

The *interhyal* ossifies in the interhyal cartilage.

The *epihyal* ossifies in the posterodorsal portion of the ceratohyal cartilage.

The *ceratohyal* ossifies in the anteroventral portion of the ceratohyal cartilage.

The *hypohyals* ossify in the similarly named cartilages, from dorsal and ventral perichondral lamellae.

The *basihyal* (copula I, glossohyal, or symbranchial I) possesses a perichondral lamella in its hinder region.

The *pharyngo-*, *epi- cerato-*, and *hypobranchials* ossify in the first three branchial arches; in addition, the ceratobranchials of the 4th and 5th branchial arches ossify, and the epibranchial of the 4th. All arise as perichondral lamellae surrounding the cartilaginous shafts. The 5th ceratobranchial subsequently fuses with a membrane-bone, the pharyngeum inferius, to form a mixed ossification.

The *basibranchials* (copula II, symbranchials II–IV) arise in the continuous copula cartilage as three perichondral centres of ossification. The 2nd basibranchial 'symbranchial III) subsequently fuses with a membrane-bone, the supracopulare, to form a mixed ossification.

In all these cases (except when specifically stated) the cartilage enclosed between the perichondral lamellae becomes absorbed, and endochondral ossification completes the formation of the bones.

For the *basisphenoid* and *opisthotic*, see below.

3. Bones of peculiar formation. The *basisphenoid* (propituital, Chabanaud, 1936) arises from three centres of ossification. One lies medially in the membrane forming the interorbital septum and the actual floor of the anterior part of the cranial cavity, dorsal to the anterior part of the hypophysial fenestra. This is the region in which the internal and external recti enter the myodome, and the basisphenoid extends downwards in the vertical longitudinal membrane in the middle line which separates the eye-muscles of the two sides. This unpaired rudiment of the basisphenoid, Chabanaud's 'belophragm', is thus T-shaped, the cross-piece of the T being in the cranial floor, the vertical piece forming a median septum to the anterior portion of the (posterior) myodome. The two other centres of ossification are paired plates of bone lying in the dura mater on each side of the cross-piece of the median T-shaped basisphenoid rudiment, and in contact with the anterior edge of the lateral commissure (prootic bone). These lateral rudiments, Chabanaud's 'meningost', fuse on to the ends of the cross-piece of the median rudiment, forming the roof of the opening into the orbit of the myodome on each side. By stretching from the prootic to the hind edge of the interorbital septum beneath the optic chiasma, the basisphenoid thus separates the opening of the myodome from the sphenoid fissure (through which the optic and oculomotor nerves emerge).

The basisphenoid of Salmo thus presents the case of a bone which phylogenetically was presumably a cartilage-bone, but which ontogenetically ossifies in the membranous wall of the cranial cavity, without cartilaginous preformation. The hinder parts of the trabeculae do not ossify, but disappear.

The *angular* has been dealt with above.

The so-called *opisthotic* (S. salar, 33 mm. *intercalary*, Vrolik, 1871) ossifies in the ligament which runs from the ventral surface of the exoccipital, passing to the supracleithrum round the hind wall of the auditory capsule, laterally to the jugular foramen. According to Sagemehl (1891) the opisthotic was phylogenetically a cartilage-bone with extension of ossification into the above-mentioned ligament. In Salmo, this ligamentous ossification is the only representative of the opisthotic, the place of which in the cartilaginous wall of the auditory capsule has been usurped by extensions of the pterotic and exoccipital.

The *prevomer* (S. fario, 17 mm.) rudiment is paired anteriorly, unpaired posteriorly. It ossifies in membrane, ventrally to the ethmoid plate. Subsequently, teeth become attached to the ventral surface of the prevomer, the ossification of which then extends dorsally so that it comes into actual contact with the cartilage, i.e. assumes perichondral relations.

The median sagittal membrane-bone situated beneath the copula and between the sternohyoid muscles (the 'episternal' of E. Geoffroy St. Hilaire, the 'interclavicle' of Cole and Johnstone, the 'urohyal' of Owen and Günther) appears to be a tendon-ossification, without affinity to either pectoral girdle of hyobranchial skeleton, but homologous perhaps with the *parahyoid* of Polypterus and of Amphibia (see pp. 87, 212).

The *sclerotic* bones are two in number in each eye, and consist of small perichondral lamellae on the anterior and posterior sides of the rim of the sclerotic cartilage.

II. Exocoetus sp.

The development of the skull of the flying fish has been studied by Lasdin (1913) and presents a number of features of interest.

i. Development of the chondrocranium of Exocoetus

1. Stage 1. (2·25 mm. stage.) PLATE 50, FIGS. 1, 2.

The skull at this stage consists of the trabeculae, separate from one another and from the parachordals, which latter appear from the first to be fused with the auditory capsules, and the occipital arches. The trabeculae are at a more ventral level than the parachordals, as seen in lateral view.

2. Stage 2. (2·4 mm. stage.) PLATE 50, FIGS. 3, 4.

At this stage the hind ends of the trabeculae are attached to the front end of the parachordals, which appear to be interconnected by a prootic bridge.

In the visceral arch skeleton the quadrate and Meckel's cartilages (the latter fused together at the symphysis) and the hyosymplectic appear all separately and independently.

3. STAGE 3. (5 mm. stage.) PLATE 50, FIGS. 5, 6.

At this stage the definitive form of the Teleostean chondrocranium has been reached. Anteriorly, the trabeculae fuse to form a trabecula communis and an ethmoid plate, in front of which, anterior extensions of the trabeculae project forwards and inwards and meet, thus enclosing a fontanelle (as in Solea, p. 162). On each side of the ethmoid plate a lamina orbitonasalis rises up and is attached to the anterior end of the orbital cartilage. The two orbital cartilages are interconnected anteriorly by the sphenoseptal commissures, and a little farther back by the epiphysial bar. Posteriorly, each orbital cartilage is continuous with the postorbital process of the auditory capsule of its own side. The pterygoid process of the quadrate is attached to the posterolateral corner of the ethmoid plate. The separate trabeculae behind the trabecula communis are now discontinuous with the parachordals.

A lateral commissure is present, forming the side wall of the trigeminofacialis chamber, the posterior opening of which is divided into two by a strip of cartilage, for the exit of the hyomandibular branch of the facial nerve and the head vein.

The auditory capsules are well formed and almost interconnected by a tectum synoticum.

4. STAGE 4. (23 mm. stage) PLATE 50. FIGS. 7, 8.

The chief difference at this stage (at which ossification is now well advanced) concerns the completion of the cartilaginous skeleton of the ethmoid region. There is now a well-formed nasal septum, separating the two foramina olfactoria advehentia. A taenia tecti medialis extends back in the middle line from the epiphysial bar, and the tectum synoticum is complete.

The trabeculae are now fused to form a trabecula communis which projects backwards from the ethmoid plate and ends freely. There is also a cartilaginous interorbital septum.

Altogether, the chondrocranium of Exocoetus bears a strong resemblance to that of Salmo.

ii. Development of the osteocranium of Exocoetus

According to Lasdin the only membrane-bones which have no contact at all with cartilage are the *nasal* and the *supra-ethmoid*; the only cartilage-bone which has no intramembranous extensions is the *exoccipital*. There is, however, no reason to apply too strict a classification to the cartilage-bones and membrane-bones; and in fact the bones of Exocoetus appear to develop much as in Salmo, where some undoubted membrane-bones invade the cartilage and nearly all the cartilage-bones have extramembranous extensions.

The *prevomer* is toothless and invades the cartilage of the ethmoid plate.

The *parasphenoid* appears to be somewhat short, and to fill in the gap between the trabecula communis and the parachordals.

The *basisphenoid* contains a core of cartilage, which is of interest, seeing that in most Teleostei this original cartilage-bone ossifies more and more as a membrane-bone.

The *frontal*, lodging part of the supraorbital lateral line canal, fuses with the *pleurosphenoid*, a cartilage-bone ossifying in the postorbital process.

The *pterotic* and *sphenotic* ossify in the cartilaginous lateral wall of the auditory capsule and also have intramembranous extensions. These bones (at all events at this stage) have no connexions with the lateral line canal which is lodged in tubular superficial bones which correspond to the *supratemporal* and/or *intertemporal*.

The *epiotic*, ossifying in the posterolateral corner of the cartilaginous auditory capsule, has extensions into the ligamentous attachment of the pectoral girdle.

The *basioccipital*, ossifying in the posterior part of the parachordals, has laminar extensions beneath the notochord.

The *prootic* ossifies in the cartilage of the anterior wall of the auditory capsule and of the lateral commissure. It also ossifies in the membranous medial wall of the trigeminofacialis chamber.

III. Clupea

PLATE 51, FIGS. 1, 2.

The development of the cartilaginous skull of the herring (Clupea harengus) is the subject of a short paper by F. R. Wells (1923); certain aspects of it are treated by de Beer (1926*a*) and Norman (1926).

Apart from certain matters of detail, the development of the chondrocranium of Clupea is not sufficiently different from that of Salmo to warrant a separate complete description. In a few respects, however, there are differences of importance, which will be considered in the following section.

i. The fissura metotica (jugular foramen)

The fissura metotica is the aperture typically bordered in front by the auditory capsule and behind by the occipital arch. In Clupea, however, the auditory capsule extends backwards laterally to the occipital arch, with the result that the fissura metotica is converted into a tunnel enclosed on the medial side by the occipital arch, and laterally by the medial wall of the posterior extension of the auditory capsule. This medial wall is here only membranous, with the result that the cavity of the tubular fissura metotica appears to be confluent with that of the auditory capsule: the two cavities are, however, quite distinctly separated by membrane. A somewhat similar state of affairs is found in Selachians. The external aperture of the fissura metotica in Clupea faces downwards and backwards.

An additional difference between Salmo and Clupea is the fact that whereas in the former the glossopharyngeal and vagus nerves are separated by the basicapsular commissure, in the latter the two nerves emerge together through the tubular fissura metotica. There can, therefore, be no posterior basicapsular commissure in Clupea, and the attachment of the floor of the capsule to the lateral edge of the parachordal, in front of the glossopharyngeal, must be a basivestibular commissure.

The curious modification of the fissura metotica in Clupea is associated with two other features which distinguish it from Salmo, viz. the enormous size of the intracranial notochord in Clupea, and the invasion of the auditory capsule by a diverticulum from the air-bladder. At the 30 mm. stage in embryos of Clupea, at the level of the occipital arch the notochord has a larger diameter than the cavity of the cavum cranii. The result is that the occipital arches, mounted on the dorsolateral edges of the notochordal sheath, are relatively small while the auditory capsules which extend down to the level of the ventral edge of the notochord, are relatively large. It is probable, therefore, that a reduction in size of the occipital arch has been partly responsible for the course of the fissura metotica in Clupea.

But the posterior extension of the auditory capsule is a displacement occasioned by the formation in the anterior region of a large bulla containing a diverticulum of the air-bladder. The development of this interesting feature has been described briefly by de Beaufort (1909); Wells makes no mention of it, and the following account is to be regarded merely as a general description of the chief outline of events which would thoroughly repay a detailed study.

In Clupea harengus at the 15 mm. stage conditions are normal; the basicapsular fenestra is open, but the anterior basicapsular commissure is more massive than in Salmo, and its posterior surface is hollowed out into a cup-shaped depression concave posteriorly. Paired (and at this stage, almost solid) diverticula from the air-bladder run forwards by the side of the notochord and penetrate into the cavity of the auditory capsules through the basicapsular fenestrae. Each diverticulum runs into the cup-shaped depression and ends blindly.

At the 20 mm. stage the cup-shaped depression has deepened and expanded to form a bulla the walls of which are cartilaginous and cut off the cavity of the bulla from that of the auditory capsule except in one place where a gap allows a sac-like diverticulum of the perilymphatic space to enter the bulla and come into contact with the air-bladder diverticulum. Posteriorly, the cavity of the bulla opens on the under surface of the capsule by a tunnel which slopes backwards and downwards and represents the remnant of the basicapsular fenestra. Through this tunnel the air-bladder diverticulum runs back and continues, passing beneath the capsule, to the air-bladder.

At the 25 mm. and later stages the bulla has swollen still further, with the result that the anterior cupola of the auditory capsule, is pressed considerably farther back. It is this displacement which leads the auditory capsule as a whole to extend backwards, to the side of the occipital arch and bring about the conditions already described with regard to the fissura metotica.

At the 35 mm. stage the walls of the bulla are quite thin, and they become ossified by the prootic bone.

Another consequence of the posterior displacement of the auditory capsule is the formation of a vacuity in its wall, beneath the ampulla of the anterior semicircular canal. Through this vacuity another diverticulum of the perilymphatic space passes out on to the lateral surface of the capsule, and comes into contact with a remarkable enlargement of the supratemporal lateral line canal, which forms a sinus into which the infraorbital lateral line canal opens. This sinus and the diverticulum of the perilymphatic sac are in the bony skull lodged in another bulla formed from the pterotic. Ridewood (1904*b*) refers to a diverticulum of the air-bladder in the pterotic (his 'squamosal') bulla. According to de Beaufort (1909) this second air-bladder diverticulum arises only after the larval period is over, as a lateral outgrowth of the canal connecting the anterior air-bladder diverticulum with the air-bladder itself. This second diverticulum is at first completely enclosed in the cartilaginous wall of the capsule, bursting out laterally later. The development of this diverticulum appears to be accompanied by erosion of the cartilage, for nothing is visible at the 25 mm. stage, while at 35 mm. a short blind pit is seen in the cartilage of the posterior (basivestibular) commissure leading out of the anterior diverticulum canal near the point where the latter opens on the ventral surface of the capsule.

ii. The visceral arch skeleton

A peculiarity of the visceral arch skeleton of Clupea is the close association in which several of the cartilages arise. The quadrate and hyosymplectic cartilages chondrify together, eventually becoming distinct and separate, while there is no separate origin for the symplectic. This condition is clearly secondary.

The pterygoid process of the quadrate, on the other hand, arises, not as in Salmo by forward growth from the body of the quadrate cartilage, but as an independent chondrification immediately ventral to and articulating with the ethmoid plate by a rostropalatine attachment. Chondrification then spreads backwards until the pterygoid process and the body of the quadrate are a single cartilage. An ethmopalatine process is present, but it fails to articulate with the lamina orbitonasalis.

Compared with Salmo the pterygoid process of the quadrate is formed at a relatively earlier stage in Clupea. Perhaps correlated with this fact is the early break-down of the trabecula in Clupea, for the upper jaw then has a cartilaginous skeleton in the form of the pterygoid processes. In Salmo, on the other hand, the trabecula forms the sole skeletal element in the roof of the mouth, until a comparatively advanced stage.

IV. Gymnarchus

The little that is known concerning the development of the skull of Gymnarchus is due to Assheton (1907), and serves chiefly to whet the appetite for more information.

1. Stage 1. (9 days.) PLATE 52, FIGS. 1, 2.

The base of the skull is formed by the trabeculae fused on to the ethmoid plate anteriorly and posteriorly to the parachordals, and is markedly platybasic; no interorbital septum is formed. The internal carotid foramina are separated from the hypophysial fenestra by a narrow strip of cartilage. The orbital cartilage is attached to the nasal septum and the lamina orbitonasalis, preoptic and metoptic roots (between which the optic nerve emerges) connecting it with the trabecula of its own side. Posteriorly, the orbital cartilage is connected with the auditory capsule by a taenia marginalis. Dorsal to the ethmoid plate, in the middle line, is an isolated paraphysial cartilage. There is no pila antotica, so that the oculomotor, trigeminal and facial nerves all emerge through the same fenestra. One (unspecified) branch of the facial is described as having a special exit; this is presumably the palatine.

The auditory capsules are well formed. They are attached to the parachordal of their side by the anterior and posterior basicapsular commissures, leaving a basicapsular fenestra, through which the glossopharyngeal nerve emerges and the diverticulum of the air-bladder enters the cavity of the auditory capsule. In addition, the lateral wall of the auditory capsule is fenestrated dorsally to the lateral semicircular canal, and a vacuity is thus left through which the saccule will come into relations with the lateral line canal sinus.

The vagus nerve leaves the skull separately from the glossopharyngeal, by a jugular foramen enclosed between the auditory capsule and the occipital arch.

The pterygoquadrate is remarkable in possessing a pterygoid process which is apposed to the side of the trabecula for a long distance, from the level of the olfactory foramen to that of the trigeminal nerve. So far as can be seen, this long attachment, which is effected by a procartilaginous connexion between the pterygoquadrate and the trabecula, involves both the ethmoid and basal attachments of other forms.

Posteriorly, the pterygoquadrate is attached by procartilage to the hyosymplectic. Meckel's cartilages lie almost in the transverse plane, and meet in a ventral symphysis. The ceratohyal lies behind Meckel's cartilage; there appears to be no interhyal. In the 1st branchial arch, pharyngo-, epi-, and ceratobranchial cartilages are present. Epi- and ceratobranchials are present in the 2nd branchial arch and a ceratobranchial arch in the 3rd.

The only ossification present at this stage is the opercular bone.

2. Stage 2. (43 days.) PLATE 52, FIGS. 3, 4.

The chief advances in development shown by this stage concern the roofing of the skull and the fusion of the pterygoquadrate with the brain-case.

Anteriorly, the roofing of the cranial cavity is effected by the fusion of the paraphysial cartilage anteriorly with the nasal septum, and laterally with the orbital cartilage leaving a pair of anterolateral fontanelles. Posteriorly in the middle line the paraphysial cartilage, which may now be called the tectum cranii, extends as a taenia tecti medialis. A tectum synoticum interconnects the auditory capsules.

The fusion of the pterygoquadrate with the lateral edge of the trabecula reaches from the ethmoid region to the level of the trigeminal foramen. However, in the absence of any information concerning the courses of the nerves, it is impossible to determine the relations of the various structures or to say anything definite other than that the matter peremptorily demands reinvestigation.

The relations of the auditory capsule to the air-bladder and to the lateral line canal also demand further study. The air-bladder diverticula lose connexion with the air-bladder itself (Ballantyne, 1927). As to the lateral aperture in the wall of the capsule, there appears to be doubt as to the identity of the structure which protrudes through it; Allis (1904) regarding it as a dilated portion of the saccule while Assheton claims it to be a diverticulum from the air-bladder.

CYPRINIFORMES

V. Amiurus nebulosus

The development of the skull of the catfish was first described by McMurrich (1884*b*). Since then it has been studied with the help of wax models by Kindred (1919), while the formation of the Weberian ossicles has been followed by Matveiev (1929). Their morphology was considered by Ramsay Wright (1884).

i. The development of the chondrocranium of Amiurus

In the earliest stage described by Kindred chondrification has already advanced so far that all the cartilages are present and joined up. It is consequently impossible to say anything about their method of origin, and the following account must be restricted to a description of the morphology of the chondrocranium at the two stages studied by Kindred supplemented by observations by the present author on material kindly placed at his disposal by the late Professor Ramsay Wright.

1. Stage 1. (10 mm.) Plate 53, figs. 1, 2.

The most obvious feature of the skull is its great width: it is in fact markedly platybasic (or platytrabic). The hypophysial fenestra is large and bounded anteriorly by a trabecular plate which is almost as broad as the parachordal or basal plate posteriorly. The internal carotid arteries enter the cranial cavity in the typical way, passing medially to the trabeculae, in notches in the medial edges of those cartilages. In view of their subsequent fate, it is important to notice that each internal carotid artery before passing through the hypophysial fenestra and while still ventral to the trabecula (of its own side) gives off an artery which runs forwards and upwards *laterally* to the trabecula. After passing through the hypophysial fenestra and rising upwards medially to the trabecula, the internal carotid comes into close contact with the other artery, above the trabecula, but there is no communication here between them at this stage.

The orbital cartilages are interconnected by an epiphysial bar, and posteriorly, each orbital cartilage is continuous with the postorbital process of the auditory capsule, pierced by the otic branch of the facial nerve. Anteriorly, each orbital cartilage is connected with the trabecula and lateral edge of the trabecular plate by a very wide preoptic root, forming a plate of cartilage shielding the side of the anterior part of the brain-case. In the middle of each of these preoptic plates is a vacuity in the cartilage—the preoptic fontanelle. The oblique eye-muscles are attached to the edge of the trabecular plate opposite this fontanelle, through which, however, they do not penetrate.

Immediately in front of the preoptic fontanelle the lamina orbitonasalis rises up from the lateral edge of the trabecular plate and is fused with the external surface of the preoptic root of the orbital cartilage except for two passages: the orbitonasal foramen which conveys the profundus nerve and the orbitonasal canal which conveys a small vein out of the orbit into the region of the nasal capsule.

The nasal capsule itself has merely a floor—the lateral edge of the trabecular plate, formed in part by the cornu trabeculae, a hind wall (the lamina orbitonasalis) and a medial wall (the preoptic root of the orbital cartilage and the lateral edge of the vertical transverse plate or lamina precerebralis which forms the front wall of the cranial cavity). The foramen olfactorium (evehens) is bordered anteriorly by the lateral edge of the lamina precerebralis, posteriorly by the anterior edge of the preoptic plate of the orbital cartilage, ventrally by the trabecular or ethmoid plate, and dorsally by the sphenoseptal commissure which connects the orbital cartilage with the lamina precerebralis. It will be seen, therefore, that there is as yet no internasal septum in the ordinary sense, or, rather, the internasal septum is stretched out in the transverse plane to form the lamina precerebralis, in consequence of the great breadth of the skull, and the fact that the brain and cranial cavity extend right forward between the nasal capsules.

The orbital cartilage is pierced by foramina for the trochlear and ophthalmic nerves. There is no lateral commissure or pila antotica. The side wall of the skull in the orbital region therefore presents a large continuous fontanelle—the sphenoid fontanelle—between the hind edge of the preoptic root of the orbital cartilage and the anterior basicapsular commissure, and through which the optic, oculomotor, trigeminal, abducens, and facial nerves emerge.

The auditory capsules and parachordals present little interest save in the posterior region where modifications take place in connexion with the Weberian ossicles. The glossopharyngeal nerve emerges through the remnant of the basicapsular fenestra, the vagus through the fissura metotica between the auditory capsule and the occipital arch. The auditory capsules and occipital arches are interconnected across the middle line by a combined tectum synoticum and posterius which extends back as the superior occipital cartilage over the spinal cord and connects with the neural arches and spine of the 3rd vertebra.

On each side of the notochord the parachordals are recessed on the dorsal surface, forming a pair of deep grooves (recessus sacculi) lodging the saccules and lagenae. The grooves are roofed over by connective-tissue membranes which stretch from side to side, passing over the notochord and separating the cranial cavity from a median space, the cavum sinus imparis, which itself is separated from the recessus sacculi on each side by membranes. The cavum sinus imparis continues back, immediately above the notochord and beneath the spinal cord, to the level of the 2nd vertebral neural arch where it forks into the two lateral atria sinus imparis, which are separated by membrane from the saccus paravertebralis.

Just anteriorly to the entrances to the recessus sacculi, the saccules are interconnected across the middle line by a transverse endolymphatic canal, from which a median diverticulum, the sinus impar, projects backwards into the cavum sinus imparis.

The Weberian ossicles do not properly enter into a study of the skull, but some of them must be considered here, because they arise from vertebral elements which enter into connexion with the skull by means of the superior occipital cartilage and the inferior occipital cartilage which develops later. In fact, the neural arches of the first two vertebrae (in connexion with, or from which, the claustrum, scaphium, and intercalarium, arise) may be regarded as modified spino-occipital arches. The rib of the 3rd vertebra contributes the tripus.

Without going in detail into the composition of the Weberian ossicles it may be said that the claustrum appears to represent the neural arch of the 1st vertebra, of which the rib gives rise to the scaphium; the intercalarium is formed from the neural arch of the 2nd vertebra. The neural spines of both 1st and 2nd vertebrae appear to form the superior occipital cartilage which, as has been seen, connects the tectum posterius with the 3rd vertebra. The 1st spinal nerve passes out between the occipital arch and the scaphium; the 2nd between the scaphium and the intercalarium, and the 3rd between the intercalarium and the neural arch of the 3rd vertebra.

In the visceral arch skeleton the pterygoquadrate is curiously modified. In the first place, the body of the quadrate cartilage is apparently fused with the ventral end of the hyosymplectic. This appears to be general in Siluroidea, for the same condition is found in Ictalurus, Plate 54, fig. 3 (Ryder, 1887), and Callichthys (Ballantyne, 1930). The pterygoid process is separate from the body of the quadrate cartilage. In Ictalurus it is not only separate but segmented into two portions.

According to Edgeworth (1935, p. 86), the hyomandibular branch of the facial nerve at first runs outward *in front of* the hyosymplectic and subsequently becomes enclosed in a foramen as a result of the forward growth of that cartilage.

2. STAGE 2. (32 mm.) PLATE 53, FIGS. 3, 4.

The modifications which the chondrocranium has undergone between this and the last stage are small. The trabeculae have been interrupted. The cranial cavity does not extend quite so far forward as at previous stages, for the lamina precerebralis has now given rise to an internasal septum. The internal carotid artery now passes into the cranial cavity by running up laterally to the trabecula and inwards *dorsally* to it, as described by Allis (1908*b*). This condition has been brought about by the establishment of the connexion dorsally to the trabecula between the internal carotid and the artery which it gave off and which passed up laterally to the trabecula. The portion of the internal carotid medial to the trabecula is thus aborted. In other words, the atypical course of the carotid in Amiurus is due to the establishment of a loop round the trabecula.

A cartilage is now present dorsally to the nasal sac. This cartilage is independent of the remainder of the nasal capsule, and gives off a process which supports the nasal barbel. This cartilage has been compared by Sagemehl (1885) with the nasal cartilage of Selachii.

The cartilages supporting the barbels, three in number on each side, are well seen in a 10-day larve of Ictalurus albidus, Plate 54, fig. 3 (Ryder, 1887).

Behind the hypophysial fenestra the parachordals are interconnected above the middle line by a flat plate of cartilage at the same level as the rest of the parachordals i.e. there is no subpituitary space and the prootic bridge, if it may so be called (for it forms only a slight transverse ridge in the floor of the saccule), is merely the anterior edge of a simple parachordal or basal plate. The notochord is completely enclosed within this plate by cartilage above and beneath it.

There is thus in Amiurus no myodome nor even a myodomic space, these having being obliterated by a secondary simplification of this region. It is of interest to note in this connexion that the pituitary veins have also been lost. The pituitary is drained by a special median vein beneath the midbrain. It is of further interest to note that in another Siluroid, Silandia, a small myodome is retained (Bhimachar, 1933).

Posteriorly, the parachordals extend back on each side of the notochord to a point beneath the scaphium, and represent the fused bases of basidorsal elements. In Ictalurus these form an inferior occipital cartilage, separate from the parachordals. There is also a hypochordal cartilage ventral to the notochord, and this seems to represent basiventral elements, fused with one another in the midline.

The saccular recesses are now deeper, and they also extend forwards beyond the transverse endolymphatic canal, and lodge the anterior parts of the saccules, or processes of Comparetti.

ii. The development of the osteocranium of Amiurus

It is not proposed here to give a detailed account of the method of origin of all the bones, for many of them show similarity with those of Salmo or Gasterosteus, and the reader may be referred to Kindred's (1919) work. Some of the ossifications, however, show points of interest worthy of inclusion here.

The *basioccipital*, ossifying in the hind part of the parachordals, forms the floor of the saccular recesses, and the floor and side walls of the cavum sinus imparis. Posteriorly it is fused with the centrum of the 1st vertebra.

The *exoccipital*, ossifying in the occipital area, also spreads in the membrane forming the roof of the saccular recess, and the roof of the cavum sinus imparis, thus meeting its fellow in the midline. The foramen magnum is thus subdivided by a horizontal partition into the passage for the spinal cord above, and the cavum sinus imparis beneath. Anteriorly, the exoccipital spreads on the ventral surface of the auditory capsule and is perforated by the glossopharyngeal and vagus nerves. Posteriorly, an intramembranous extension of the exoccipital encloses the 1st spinal nerve in a foramen.

The *supraoccipital* ossifies in the tectum synoticum and has an intramembranous extension to form the dorsal spine. In addition, the supraoccipital extends on each side, and is either fused with the *parietals*, or has taken their place. On the other hand, Allis (1904, p. 424) has described as a parietal a bone which seems to be the epiotic.

The *epiotic* ossifies in part in the cartilage of the posterodorsal corner of the auditory capsule, and in part in connective tissue on the hind surface of the capsule. It is attached to one of the prongs of the post-temporal. Part of the epiotic thus comes to occupy the position of an *opisthotic*, which is otherwise absent in Amiurus.

The *pterotic* arises in a perichondral ossification in the lateral wall of the auditory capsule, and at the 32 mm. stage it is already connected with the *supratemporal*, a membrane-bone lodging the lateral line canal.

The *sphenotic* in the same way develops as a perichondral ossification which is connected with a membrane-bone, the *intertemporal*, lodging the lateral line canal.

The *lateral ethmoid*, a perichondral ossification in the lamina orbitonasalis, likewise is fused with an intramembranous lamella, the *prefrontal*.

The *ethmoid* is a median perichondral ossification in the nasal septum, fused with an overlying intramembranous ossification, the *supra-ethmoid*, which appears to be of paired origin.

The *prootic* ossifies in the anterior wall of the auditory capsule and in the prootic bridge where it meets its fellow from the opposite side.

The *pleurosphenoid* ossifies in the orbital cartilage, and also in the membranous side wall of the skull between the optic and trigeminal nerves, which are thereby separated into different foramina. The secondary looping of the carotid artery causes it to enter the cranial cavity through the optic foramen.

The *orbitosphenoid* ossifies as a perichondral lamella in the preoptic root of the orbital cartilage, and it also has a median intramembranous extension across the anterior part of the hypophysial fenestra, fusing with its fellow from the opposite side to form an unpaired ossification.

The *basisphenoid* (suprasphenoid of Kindred) ossifies in the membrane stretched across the hypophysial fenestra, dorsally to the trabeculae. It touches the fused orbitosphenoids anteriorly and the prootics posteriorly. Ventrally, it is attached to the *parasphenoid*, which covers it completely.

The bones related to the infraorbital lateral line canal include an *antorbital*, a *lachrymal*, three *suborbitals*, and two *postorbitals* of which the uppermost probably represents a *postfrontal*.

The *frontal* is of considerable interest if Kindred's description is correct, for he states that the ossifications surrounding the supraorbital lateral line canal arise independently of the membranous sheets in which the main portions of the frontals ossify, and subsequently become fused with them. It may be that the lateral line ossifications in this case represent *supraorbital* bones.

VI. Cyprinus carpio

The development of the skull of the carp has been studied only in so far as the Weberian ossicles may be said to form part of it, by Nusbaum (1908), using wax models. What little knowledge exists is therefore practically restricted to the posterior region (Plate 54, figs. 1, 2).

In a Cyprinus embryo 73 hours old the trabeculae and parachordals are present, and, attached to the hinder end of the latter, a pair of small occipital arches.

In embryos 5 days old the chondrocranium is already well formed. The occipital arches fuse with the auditory capsules, thus enclosing the vagus foramen. Behind the occipital arches there is a pair of small neural, or occipitospinal arches which will give rise to the scaphia.

On theoretical grounds, and because he accepted Sagemehl's (1891) and Fürbringer's (1897) schemes of segmentation of the occipital region, assuming a fixed position of the hind limits of the skull, Nusbaum assumed that a ventral nerve-root has completely disappeared between the occipital arch and the ventral arch of what he called the 1st vertebra, which he further assumed as indistinguishably plastered on to the hinder side of the occipital arch. Further he assumed with Sagemehl (1891) that a ventral nerve-root disappeared immediately in front of the scaphium, and that the latter is therefore formed from the 3rd vertebra. But since the so-called neural arch in front of the scaphium (Nusbaum's 2nd) is not preformed in cartilage, the 1st cartilaginous neural arch visible behind the occipital arch is what Nusbaum calls the 3rd. Its neural arch gives rise to the scaphium, its spinal process to the claustrum. Since the nerve-roots ex hypothesi have vanished, there is no reason for assuming that either Nusbaum's 1st or his 2nd vertebrae have any existence in fact. Furthermore, it is clear from his description that the rudiments of his 2nd and 3rd neural arch have a common origin, presumably representing the 1st occipitospinal neural arch. The 2nd occipitospinal neural arch (Nusbaum's 4th vertebra) contributes the intercalarium, and the rib of the 3rd (Nusbaum's 5th) the tripus.

Nusbaum's 2nd neural arch ossifies directly out of mesenchyme and forms the concha stapedis which fuses with the exoccipital.

The centrum (basidorsal and basiventral elements) of the vertebra concerned in the formation of the claustrum and scaphium is attached to the hind end of the parachordals forming the inferior occipital cartilage, while its haemal arches produce the so-called pharyngeal process. Further, the neural spines of this vertebra fuse with the tectum posterius, thus giving rise to the superior occipital cartilage.

VII. Scardinius erythrophthalmus

This Cypriniform has been studied by Matveiev (1929) solely in respect of the development of the Weberian ossicles, and this work need be considered here only in so far as the Weberian ossicles may be regarded as occipitospinal arches.

Matveiev's conclusions are that the elements of the Weberian ossicles arise from special mesenchymatous rudiments which ossify directly and only secondarily become associated with the cartilaginous elements of the first three vertebrae: only the claustrum is regarded by him as the simple product of the neural arch of the 1st vertebra.

It may be of service to express the views of different investigators regarding the Weberian ossicles in tabular form:

	Claustrum	*Scaphium*	*Intercalarium*	*Tripus*
Sagemehl (1891)	Neural arch of 1st vertebra	Neural arch of 2nd vertebra	Rib of 2nd vertebra	Rib of 3rd vertebra
Ramsay Wright (1884)	Spinal process of 1st vertebra	Neural arch of 1st vertebra	Neural arch of 2nd vertebra	Rib of 3rd vertebra
Nusbaum (1908)	Spinal process of 3rd vertebra	Neural arch of 3rd vertebra	Neural arch of 4th vertebra	Rib of 5th vertebra
Goodrich (1909)	Neural arch of 1st vertebra	Rib of 1st vertebra	Neural arch of 2nd vertebra	Rib of 3rd vertebra
Kindred (1919)	Intercalated arch	Neural arch of 1st vertebra	Neural arch of 2nd vertebra	Rib of 3rd vertebra
Matveiev (1929)	Spinal process of 1st vertebra	Neural arch of 1st vertebra special rudiment	Neural arch of 2nd vertebra special rudiment	Rib of 3rd vertebra special rudiment

ESOCIFORMES

VIII. Esox lucius

PLATE 51, FIGS. 3, 4.

From the historical point of view the skull of the pike is of interest, for it was in this form that Arendt (1822) first devoted special attention to the chondrocranium. There are some points of morphological interest in the works of Walther (1882) and Swinnerton (1902); Matveiev (1929) has figured two stages of development of the chondrocranium, and this has been further investigated by Besrukow (1928).

In the neurocranium the attachment of the pterygoid process of the pterygoquadrate cartilage with the ethmoid plate is remarkable on account of the high degree of distinctness of the rostropalatine and ethmopalatine articulations. In this respect, Esox presents an extreme case of the condition which Swinnerton has termed 'disartete'.

As far as can be made out from Walther's somewhat crude figures, chondrification of the neurocranium is very complete. The preoptic root of the orbital cartilage is present, and in the roof, a taenia tecti medialis connects the tectum cranii with the tectum synoticum, leaving a fontanelle on each side. From Matveiev's and Besrukow's figures it is clear that the chondrocranium conforms to the general type as seen in Salmo. Meckel's cartilage possesses a well-developed coronoid process.

As regards the bony skull, Esox is of interest on account of the presence of an ossification in the pre-ethmoid corner of the ethmoid plate, the *pre-ethmoid* bone of Swinnerton, or septomaxillary of other authors. It appears to be a cartilage-bone, probably homologous with the so-called septomaxillary of Amia; but as it remains to be proved that this perichondral and endochondral ossification is homologous with the dermal septomaxillary of Tetrapods, it appears to be safer to call it the pre-ethmoid.

With regard to the frontal, Walther made an observation of considerable theoretical importance, the significance of which appears to have been overlooked. He found that in Esox the frontal bones are ossified throughout their length before the supraorbital lateral line canal is developed. When the latter has been formed, a pair of bony strips rise up from the underlying frontal, arch over and fuse to form a canal. In this case, therefore, it is clear that the origin of the frontal bone as a whole is independent of the lateral line canal (cf. the condition in Amia, p. 105), and the possibility of the shifting of the canal from one bone to another appears to be given. It only requires to be assumed that the bony strips and canal will arise from whatever bone underlies the lateral line canal.

IX. Gambusia patruelis

PLATE 51, FIG. 5.

The chondrocranium of an embryo of Gambusia patruelis about 7·5 mm. long was figured, but only very briefly described by Ryder (1886). In the absence of information concerning the course of the various nerves and blood-vessels, little can be said with certainty about its morphology. However, the very slender hyosymplectic, in which no foramen is shown, prompts the question as to whether the hyomandibular branch of the facial nerve passes in front of or behind it.

ANGUILLIFORMES

X. Anguilla vulgaris

In the eel the development of the chondrocranium in its middle and later stages has been studied with the help of wax models by Norman (1926); contributions have also been made by Torlitz (1922), but little is known of the origins of the cartilages, or of the development of the bony skull.

i. Development of the chondrocranium

1. Stage 1. (5 mm. length.) PLATE 55, FIGS. 1, 2.

Most of the cartilages are already present at this stage, and have joined together. The parachordals (completely separated by the notochord) are fused with the trabeculae in front of them, and a small hypophysial fenestra is left through which the internal carotid arteries enter the skull. For the greater part of their length the trabeculae are fused to form a trabecula communis. In front, the latter is slightly expanded to form an ethmoid plate, ending in a fairly sharp rostrum without any indication of cornua trabecularum, and bearing on its dorsal surface in the middle line the nasal septum. The nasal septum is continuous on each side by means of the sphenoseptal commissure with the orbital cartilage. This is a slender curved bar which runs back and is fused with the dorsoanterior part of the auditory capsule. There is as yet no lamina orbitonasalis, and the fenestra olfactoria (bounded ventrally by the ethmoid plate, medially by the nasal septum and dorsally by the sphenoseptal commissure) has no lateral boundary. All the cranial nerves from I to VI inclusive, therefore, emerge through the large vacuity in the side wall of the skull. The lateral commissure is present, joining the lateral edge of the parachordal with the auditory capsule, forming a lateral wall to the trigeminofacialis chamber, and enclosing a foramen for the exits of the hyomandibular branch of the facial nerve together with the head vein. The palatine nerve appears at this stage to run down in front of the lateral commissure, to the side of the parachordal. Behind the lateral commissure the auditory capsule is attached to the parachordal by the anterior basicapsular commissure. There is as yet no posterior commissure, and the basicapsular fenestra and fissura metotica form a confluent slit through which the glossopharyngeal and vagus nerves emerge. The auditory capsule itself is represented only by its ventral and anterior wall. The latter is produced into a dorsally directed postorbital process which (as already mentioned) is fused with the hind part of the orbital cartilage. The occipital arches are prominent.

In the visceral arch skeleton the quadrate lacks all sign of any pterygoid process, with the result that it and Meckel's cartilage appear to lie almost in line. The quadrate is closely applied to the anteroventral edge of the hyosymplectic. The latter, pierced by the hyomandibular branch of the facial nerve, abuts broadly against the auditory capsule, and is connected by means of the interhyal with the ceratohyal. The latter in turn is in contact with a small median basihyal. There are no hypohyal elements. Ceratobranchials are present in the first two branchial arches; hypobranchials in the first, and a basibranchial lies between the latter, extending forwards to the basihyal with which it is connected by dense tissue.

2. STAGE 2. (11 mm. length.) PLATE 55, FIGS. 3, 4.

In the anterior region little change has taken place beyond a lengthening of the ethmoid plate, which produces a fairly acute rostrum. Posteriorly there have been more developments. A small cartilaginous process projects back from the trabecula communis over the anterior part of the hypophysial fenestra.

The parachordals are interconnected anteriorly by a thick prootic bridge, and posteriorly by a hyperchordal cartilaginous commissure. In the intervening region the parachordals are separated by the notochord. A posterior basicapsular commissure is present, separating the basicapsular fenestra (and the glossopharyngeal nerve) from the fissure metotica (and the vagus nerve). Further, the fissure metotica has become converted into the jugular or vagus foramen by the fusion of the occipital arch with the posterior wall of the auditory capsule which has now appeared. This posterior wall and the occipital arch are connected with their fellows of the opposite side by a combined tectum synoticum and tectum posterius. The side wall and roof of the auditory capsule are still practically absent, but the septum of the lateral semicircular canal is present as a vertical pillar.

The lateral commissure is wider than before, and the palatine nerve has become enclosed in a foramen.

Ceratobranchials are now present in the first four branchial arches, and hypobranchials in the first two.

3. STAGE 3. (31 mm. length.) PLATE 56, FIGS. 1, 2.

The lamina orbitonasalis is present connecting the orbital cartilage with the posterolateral corner of the ethmoid plate, and forming the lateral boundary to the foramen olfactorium advehens through which the olfactory nerve reaches the nasal sac. The ophthalmic branch of the trigeminal nerve does not run through the foramen olfactorium advehens, but laterally to the point of fusion between the lamina orbitonasalis and the orbital cartilage. The oblique eye-muscles are attached to the hind edge of the nasal septum.

Curious modifications have taken place in the region of the hypophysial fenestra. In the first place, the cartilaginous process which at the previous stage projected back from the trabecula communis (and to which the recti eye-muscles are attached), has now extended farther back over the hypophysial fenestra, forked, and fused with the dorsal edge of the trabeculae on each side. This T-shaped cartilage thus forms a dorsal bridge over the hypophysial fenestra and leaves a pair of antero-latero-dorsally directed foramina through which the ophthalmica magna arteries run out, dorsally to the trabeculae. Slightly behind this dorsal bridge, the ventral edges of the trabeculae are interconnected by a band of cartilage, forming a ventral bridge across the hypophysial fenestra.

The internal carotid arteries join in the middle line to form a single trunk which passes upwards and forwards through the hinder part of the hypophysial fenestra, behind the ventral bridge. It gives off a pair of cerebral arteries which run dorsally into the cranial cavity (passing behind the dorsal bridge) and itself runs forwards above the ventral bridge and beneath the dorsal bridge. It then receives the efferent pseudobranchial arteries, which have run inwards beneath the trabeculae and in front of the ventral bridge, and forks into the ophthalmica magna arteries which run outwards and forwards, dorsally to the trabeculae and in front of the dorsal bridge, separated from one another by the longitudinal piece of the T which connects the dorsal bridge with the trabecula communis.

The postorbital process has grown broader and enclosed in separate foramina the ophthalmic and otic branches of the facial nerve. The posterior aperture of the trigeminofacialis chamber becomes subdivided by a strut of cartilage separating the hyomandibular branch of the facial nerve and the head vein.

The parachordals are now interconnected dorsally to the notochord throughout their length, and form a fairly regular concave trough, lodging the brain, saccule, and other portions of the auditory vesicle. In the hindmost region the parachordals are also connected ventrally to the notochord by a hypochordal commissure, so that here the notochordal sheath is completely surrounded by cartilage. The point to which the notochord reaches is *relatively* farther back than at previous stages, but this is not due to any actual decrease in its length.

The basicapsular fenestra is obliterated by the growth of the cartilage forming the floor of the auditory capsule, and the glossopharyngeal nerve now emerges through a small foramen.

In the visceral arches the quadrate shows the small rudiment of the pterygoid process, while the posterior edge of the hyosymplectic bears a small opercular cartilage.

Ceratobranchials are present in all five branchial arches; epibranchials and pharyngobranchials in the first four. Basibranchial cartilages have appeared corresponding to the 3rd and 4th arches, while the basibranchial of the 1st arch has extended backwards and includes that of the 2nd.

There is no sclerotic cartilage in Anguilla (Stadtmüller, 1914).

The rudiment of an inferior labial cartilage lies above the hinder end of Meckel's cartilage, but does not chondrify. On the other hand, premaxillary cartilages are present, related to the hind end of the premaxillary bones and probably representing anterior upper labials.

ii. Changes consequent on metamorphosis

The chondrocranium undergoes considerable reduction. The trabeculae and trabecula communis break down and disappear between the parachordals and the ethmoid plate, except for the small ridge of cartilage overlying the hypophysial fenestra and to which the recti eye-muscles are attached. The orbital cartilage disappears, leaving the lamina orbitonasalis as a pillar ending freely above, and also undergoing reduction. The nasal septum disappears and is replaced by a great development of the supra-ethmoid bone, which separates two halves of the nasal capsule. The oblique eye-muscles are now attached to the connective tissue between the hind edge of the supra-ethmoid bone and the ethmoid plate.

The anterior part of the ethmoid plate is cleft in two by a deep V-shaped notch in which the posterior parts of the premaxillae are lodged. Eventually, the premaxillae, the prevomer, and the supra-ethmoid are fused into a single compound bone.

The larval teeth, carried by the maxilla and dentary at earlier stages are lost, and for a time the larva is edentulous, until the adult teeth are formed.

iii. Development of the osteocranium

PLATE 56.

The earliest bones to develop are the *maxilla* and *dentary* ,which are present at the 5 mm. stage, and already have larval teeth attached to them. These teeth are very long and pointed; three being carried by the maxilla and four by the dentary. In addition there are two strong teeth anteriorly, with their bases embedded in the connective tissue surrounding the rostrum.

At the 11 mm. stage the dentary bears five teeth and the maxilla four, leaving one tooth still independent anteriorly. The *supra-ethmoid* and the *parasphenoid* are present at the 31 mm. stage (Plate 56, fig. 4).

The fused *premaxilla* (lying on the dorsal surface of the ethmoid cartilage), together with the *angular* and the bones of the *opercular* series, appear in larvae 78 mm. long, while the *nasal*, *frontal*, and *prevomer* appear at the start of metamorphosis, as do the cartilage-bones of the otic series and of the occipital region.

Some information concerning the later development of the bones has been given by Torlitz (1922). In an elver (Plate 56, figs. 5 and 6) the supra-ethmoid, premaxillae, and prevomer have fused.

The *intertemporal* (and *supratemporal*?) lodging the lateral line canal, is at first free from the *pterotic*, but eventually becomes attached to it and gives rise to the long bony process which stretches forwards between the frontal and sphenotic. The *sphenotic* is free from any association with lateral line canal bones.

The *exoccipitals* meet dorsally above the foramen magnum and form its dorsal border. Similarly, the *epiotics* approximate in the middle line immediately in front of the exoccipitals, with the result that the *supraoccipital* is far removed from the foramen magnum. It is possible that the epiotics may represent fused epiotics and opisthotics.

A *basisphenoid* and paired *pleurosphenoids* are present, and it would seem that they ossify directly out of membrane.

GASTEROSTEIFORMES

XI. Gasterosteus

The development of the skull of a stickleback (Gasterosteus aculeatus) was studied by Huxley (1858) and provided him with evidence for the refutation of the vertebral theory of the skull. Since then, it has been the subject of a classical monograph by Swinnerton (1902), upon which the following account is based.

i. The development of the chondrocranium of Gasterosteus aculeatus

It will be convenient to treat the successive stages studied by Swinnerton in order.

1. STAGE 1. (5th to 7th day; 3·5 to 4·2 mm. length.) PLATE 57, FIGS. 1, 2.

Paired parachordal cartilages flank the notochord, closely adpressed to it for the hinder half of their length, and diverging away from it anteriorly. The anterior and posterior portions of the parachordals each arise from a separate centre of chondrification (anterior parachordal, posterior parachordal) with, in early stages, an intervening zone of procartilage. Quite soon this zone chondrifies, and each parachordal is then a complete bar.

The lateral edge of the anterior region of the parachordal is produced laterally to form the anterior basicapsular commissure, with which the anterior part of the auditory capsule is fused. A basicapsular fenestra is thereby outlined, but is open posteriorly. Along the dorsal surface of each parachordal a curved ridge arises, concave laterally, and marks the lateral limit of the future fossa in which the saccule will be lodged. The posterolateral corner of each parachordal bears the rudiment of the occipital arch.

The auditory capsule is represented by a curved plate of cartilage, moulded round the anterolateral surface of the otic vesicle, and bearing the postorbital process directed forwards and upwards. The dorsal end of the hyomandibula appears to be fused to the lateral surface of the auditory capsule.

The trabeculae are paired and separate from one another, and in some cases separate also from the anterior ends of the parachordals. Fusion with the latter soon takes place. The anterior end of the notochord is depressed ventrally at this stage, and so are the anterior ends of the parachordals, with the result that the trabeculae fuse with the latter at an obtuse angle, with the point turned *ventrally*.

In the visceral arch skeleton the quadrate has begun to chondrify and shows rudiments of the pterygoid process and metapterygoid process. Meckel's cartilage is fused at the symphysis with its fellow of the opposite side.

The hyomandibula and symplectic are chondrified together, as a hyosymplectic, free from the quadrate, but dorsally the hyomandibula is continuous with the auditory capsule. The hyomandibula is perforated by a foramen for the hyomandibular branch of the facial nerve. The ceratohyal is attached to the hyomandibula by a procartilaginous interhyal.

Ceratobranchial cartilages are present in each of the first four branchial arches, and the fused first three basibranchials form a median copula.

2. STAGE 2. (8th to 9th day, circ. 5 mm. length.) PLATE 57, FIGS. 3, 4, 5.

The trabeculae are fused together anteriorly, forming the ethmoid plate from the posterolateral corners of which the laminae orbitonasales arise. The anterolateral corners of the ethmoid plate are produced to form the preethmoid cornua, which represent the out-turned anterior ends of the trabeculae (cornua trabecularum). The parachordals have become broader, the occipital arches rise high, and the auditory capsules have extended farther back but posteriorly they are still free from the parachordals. The basicapsular fenestra is thus still continuous with the fissura metotica. From the anterior wall of each auditory capsule, a thin prootic process has arisen and is directed inwards and downwards, towards the point of fusion between trabecula and parachordal. Dorsally, the epiphysial cartilage has appeared as an isolated transverse bar, behind the epiphysis.

The quadrate is V-shaped, the arms being the pterygoid process, which is in procartilaginous connexion with the under surface of the pre-ethmoid cornu of the ethmoid plate (thus conforming to the type of attachment which Swinnerton has called 'acrartete'), and the metapterygoid process which lies on the antero-dorsal edge of the symplectic. Meckel's cartilage possesses prearticular and retroarticular processes. The hyosymplectic is now articulated with the wall of the auditory capsule; the symplectic portion has elongated considerably; the interhyal has now chondrified, as have the hypohyals and the unpaired basihyal, which is in procartilaginous connexion with the copula behind it.

The skeleton of the branchial arches is further developed but apparently the segments are not yet demarcated. A fused pharyngobranchial element has appeared in the 3rd and 4th branchial arches, and the 4th basibranchials have chondrified as an unpaired posterior copula.

3. STAGE 3. (11th day, 6·3 to 9 mm. length.) PLATE 57, FIGS. 6, 7, 8.

The auditory capsules are well developed, and in each the floor is continuous with the lateral edge of the parachordal by the obliteration of the basicapsular fissure. The fissura metotica has been converted into a foramen by the fusion of the dorsal end of the occipital arch with the hind wall of the capsule, and through this foramen emerge the glossopharyngeal and vagus nerves. Here, therefore, the glossopharyngeal nerve is not separated from the vagus by a basicapsular commissure, and does not traverse the basicapsular fenestra as in Salmo. The hinder connexion between the floor of the auditory capsule and the lateral edge of the parachordal, forming the anterior limit to the foramen metoticum, is therefore a basivestibular commissure.

A tectum synoticum is now present, and, in each capsule, the septa for the lateral and posterior semicircular canals have appeared.

The parachordals of the two sides are fused with one another dorsally to the tip of the notochord, thus forming a prootic bridge. The basicranial fenestra, however, is only virtual, for the space behind the prootic bridge and between the parachordals is occupied by the notochord. The tip of the notochord is relatively farther back than at earlier stages. Actually, the intra-cranial notochord increases in length during development by about threefold, but the increase in length of the chondrocranium is twentyfold.

The prootic process is fused with the lateral edge of the parachordal, thus forming a lateral commissure (lateral wall of the trigeminofacialis chamber). Behind the lateral commissure and in front of the anterior basicapsular commissure, a foramen is enclosed on each side, through which the hyomandibular branch of the facial nerve and the head vein leave the trigeminofacialis chamber. Near the base of the postorbital process is a foramen for the exit of the otic branch of the facial nerve.

The trabeculae are fused to form a trabecula communis along the greater part of their length, with the result that the hypophysial fenestra is reduced to smaller dimensions.

The orbital cartilages have now chondrified and are continuous in front with the dorsal ends of the laminae orbitonasales, and behind with the lateral ends of the epiphysial cartilage. The nasal septum rises up as a median ridge on the ethmoid plate, and dorsally the septum is in cartilaginous connexion with the lamina orbitonasalis and orbital cartilage of either side by means of the spheno-septal commissure. In this way the foramen olfactorium advehens on each side is formed.

Little change has occurred in the cartilage of the visceral arch skeleton beyond the appearance of hypobranchials in the first three branchial arches, epibranchials in the first four branchial arches and a pharyngobranchial in the second. It is to be noticed, however, that the pterygoid process of the quadrate is fused to the pre-ethmoid horn (cornu trabeculae) of the ethmoid plate.

4. STAGE 4. (11 to 25 mm. length.) PLATE 58, FIGS. 1, 2.

Ossification has now begun in the neurocranium, but for the present, attention must be confined to the few changes in the cartilage. These are: the discontinuity between the trabecula communis and the parachordals: the discontinuity between the orbital cartilages and the epiphysial cartilage and the connexion of the latter with the tectum synoticum by the cartilage of the taenia tecti medialis; and the anterior extension of the cartilage of the prootic bridge.

At late stages the labial cartilages are present in the form of a dorsal rostral cartilage associated with the premaxillae, and a ventral pair, near the tips of Meckel's cartilages.

ii. Ossification of the skull of Gasterosteus aculeatus

The ossifications of the neurocranium are all present at stage 4; some of those of the visceral arch skeleton make their appearance at stage 2.

1. The membrane-bones (bones arising without relation to, or contact with, cartilage).

The *parietals* are separated from one another by the supraoccipital.

The *frontals* are very long and extend from the parietals and sphenotics to the nasals. In front of the prootic, each frontal sends down a process which meets an ascending process of the parasphenoid and thus encloses the maxillary and mandibular branches of the trigeminal nerve in a foramen.

The *nasals* overlie the fossa nasalis and form the anterior border to the nostril.

The supra-ethmoid is of peculiar origin for which reason it will be dealt with below.

The suborbital chain of bones is much reduced and consists of an *orbital* or *lachrymal* beneath the nostril and two *suborbitals*, which fail to touch the frontal or sphenotic.

The *parasphenoid* extends from the basioccipital behind to the prevomer in front; laterally for the hinder part of its length the parasphenoid touches the prootics and forms the floor of the hinder part of the myodome. In front of the prootic, the parasphenoid sends up a (paired) ascending process which meets a downward process of the frontal, anteriorly to the exit of the trigeminal nerve.

The internal carotid arteries enter the skull by penetrating foramina through the parasphenoid, close to the suture between that bone and the prootic on each side.

The *prevomer* ossifies beneath the ventral surface of the ethmoid plate.

The *endopterygoid* and *ectopterygoid* have joined to form a (paired) V-shaped ossification, developed close to the pterygoid process of the quadrate cartilage. The ectopterygoid constituent of the bone forms one arm of the V, running from the outer surface of the quadrate bone to the autopalatine, while the endopterygoid constituent runs along the dorsal and inner surface of the quadrate cartilage.

The *maxilla* appears at stage 2 in the form of an elongated bone extending posterolaterally from the pre-ethmoid horn of the ethmoid plate, medially to the orbital.

The *premaxilla* forms the dorsal border of the mouth, bearing teeth, and sends an ascending process upwards in front of the supra-ethmoid, from which it is separated by a nodule of cartilage—the rostral cartilage, with the help of which the premaxilla is able to slide on the supra-ethmoid.

The *dentary* appears at stage 2 as a plate of bone bearing teeth and flanking Meckel's cartilage, with which, at later stages it is said to become closely related.

The remaining membrane-bones are the *preopercular*, *opercular*, *interopercular* and *subopercular* in the opercular series, and the *branchiostegal rays*.

2. Cartilage-bones (perichondral and endochondral ossifications). The *basioccipital* ossifies as perichondral lamellae on the dorsal and ventral sides of the hinder part of the basal plate. The ventral lamella grows forwards in membrane, beneath the cartilage and assists the parachordals in forming the floor of the myodome. The hinder part of the basioccipital resembles the centrum of a vertebra.

The *exoccipital* ossifies in the occipital arch and the ossification also extends forwards in the floor of the hinder part of the auditory capsule, and backwards behind the occipital arch. In this way the jugular foramen (glossopharyngeal and vagus nerves) is entirely surrounded by the exoccipital, and the first spinal nerve is enclosed in a foramen and the second in a notch by the posterior extension of the bone. Eventually this notch is also converted into a foramen.

The *supraoccipital* ossifies in the taenia tecti medialis and extends right and left in the membrane roofing the cranial cavity in front of the tectum synoticum. Posteriorly it extends, likewise in membrane, upwards and backwards to form the spine.

The *opisthotic* is absent, its place having been usurped by extension of the exoccipital.

The *prootic* appears on the internal and external surfaces of the anterior part of the auditory capsule and the lateral commissure, thus forming the lateral wall of the trigeminofacialis chamber and surrounding its posterior opening. Ultimately it extends into the prootic bridge.

The *epiotic* arises on the external surface of the dorsoposterior part of the auditory capsule, the cartilage immediately beneath it becoming resorbed.

The *pterotic* and *sphenotic* appear, the former behind the latter, along the dorsolateral edge of the auditory capsule. As in the case of the epiotic, the cartilage underlying the pterotic and sphenotic becomes resorbed.

Although Swinnerton does not specifically describe dermal additions to the pterotic and sphenotic, his figures indicate the presence of such elements, firmly fused on to the underlying cartilage-bone. These dermal elements must correspond to *supratemporal* and *intertemporal* respectively.

The *lateral ethmoid* ossifies in the lamina orbitonasalis: it does not appear to have a dermal element added to it.

The *ethmoid* arises as a perichondral lamella on the dorsal surface of the nasal septum, but its lateral edges extend in membrane away from the cartilage. It may perhaps represent combined *ethmoid* and *supra-ethmoid* ossifications.

The *quadrate* ossifies in the vicinity of the articular portion of the quadrate cartilage.

The *metapterygoid* ossifies in the tip of the metapterygoid process of the quadrate cartilage.

The *autopalatine* arises around the pterygoid process of the quadrate cartilage, near the place where the pterygoid process is cartilaginously fused with the pre-ethmoid horn (cornu trabeculae) of the ethmoid plate.

The so-called *articular*, apparently the *angular* (Haines, in press), encircles Meckel's cartilage in the region of the articulation with the quadrate.

The *retroarticular* appears to be a perichondral ossification of the processus retroarticularis of Meckel's cartilage.

The *hyomandibula* ossifies in the dorsoposterior part of the hyosymplectic cartilage, leaving dorsally two cartilaginous articular heads which fit into facets between the prootic and sphenotic and on the pterotic, respectively. The foramen for the hyomandibular nerve is carried back to the hind edge of the hyomandibula on the lateral surface.

The *symplectic* arises in the anteroventral part of the hyosymplectic cartilage, separated by an intervening cartilaginous zone from the hyomandibula. The symplectic bone extends dorsally and ventrally into membranous osseous lamellae.

The *interhyal* ossifies in the interhyal cartilage.

The *epihyal* ossifies in the posterior portion of the ceratohyal cartilage.

The *ceratohyal* ossifies in the anterior portion of the ceratohyal cartilage.

The *hypohyal* ossifies on the dorsal and ventral surfaces of the hypohyal cartilage.

The *basihyal* occupies the posterior part of the cartilage.

The *ceratobranchials* are ossified in all five branchial arches; the *epibranchials* and *pharyngobranchials* (paired) are ossified in the first four; the pharyngobranchials of the 3rd and 4th arches are fused with the tooth-bearing pharyngeum superius.

The *copula* ossifies in three places, corresponding to the basibranchials of the first three arches.

Eventually, the *hypobranchials* ossify.

iii. The posterior myodome and trigeminofacialis chamber in Gasterosteus.

In its main features the myodome of Gasterosteus is similar to that of Salmo. The differences are that in Gasterosteus the trabeculae break down relatively early; and there is no basisphenoid to separate the orbital openings of the myodome from the sphenoid fontanelle or from one another; the septum separating the dorsal from the ventral compartments of the myodome ossifies as a part of the parasphenoid, and the myodome is closed behind.

The relations of the contents of the trigeminofacialis chamber are slightly altered by the fact that the head vein in Gasterosteus is here a vena capitis medialis, passing medially to both hyomandibular and palatine branches of the facial nerve. The palatine branch, as described by Allis (1919*a*), runs down laterally to the head vein within the trigeminofacialis chamber, then forwards and inwards and makes its way into the ventral compartment of the myodome, from the anterior aperture of which it enters the orbit.

The junction between the frontal and parasphenoid bones in front of the prootic encloses V2 and V3 in a foramen, while V1, IV, III, and the head vein run out into the orbit medially to the junction.

XII. Syngnathus

The development of the skull of the pipe-fish (Syngnathus peckianus) was investigated a long time ago by McMurrich (1884*a*). More recently, another species (Syngnathus fuscus) has been studied by Kindred (1921, 1924); certain aspects have been investigated by Swinnerton (1902) with whose work on Gasterosteus the study of the more specialized member of the same group forms an interesting object of comparison.

i. The development of the chondrocranium of Syngnathus fuscus

1. Stage 1. (8 mm. length.) Plate 59, figs. 1, 2.

At this, the earliest stage studied, the various cartilages have for the most part appeared and joined up, but compared with Gasterosteus the chondrocranium presents a modified appearance. In particular there is to be noticed a marked elongation of the trabecula communis and of the symplectic and other visceral cartilages. Further, the anterior portion of the ethmoid plate is bent upwards at right angles to the general line of the trabeculae. It bears at its tip the rostral cartilage (fused upper labials) and its anterolateral corners, the pre-ethmoid cornua, articulate with the pterygoid processes of the quadrate, from which cartilage they are at this stage separated by an intervening fibrous but non-cartilaginous region.

On each side of the hinder part of the ethmoid plate, but unattached to it, are the laminae orbitonasales (ectethmoid cartilages), which extend upwards.

The trabecula communis being very long, the hypophysial fenestra is confined to the hinder part of the skull. The auditory capsules are well formed and attached to the parachordals; a tectum synoticum is present, but it is remarkably anterior in position. The postorbital process is already well developed. Posteriorly, the upper ends of the occipital arches have fused with the auditory capsules, leaving on each side a jugular foramen through which both glossopharyngeal and vagus nerves emerge.

The lateral commissure is present.

The quadrate carried at the tip of the elongated symplectic (and in cartilaginous connexion with it) is relatively far forward. It possesses a metapterygoid process, and, as already mentioned, is discontinuous with its pterygoid process. Meckel's cartilages are curiously curved, concave ventrally, and meet in a symphysis. A small prearticular (coronoid) and a retroarticular process are present.

The hyomandibula and symplectic are fused into a hyosymplectic, but form a right angle, the former being vertical and the latter horizontal.

A small interhyal connects the hyomandibula with a large ceratohyal, which in turn is connected with the median basihyal (copula communis) by a very elongated hypohyal. Ceratobranchials are present in the first four branchial arches.

The rudiments of a pair of inferior labial cartilages are to be found overlying the dorsal surface of the anterior part of Meckel's cartilages.

2. STAGE 2. (12 mm. length.) PLATE 59, FIGS. 3, 4.

The anterior end of the ethmoid plate has flattened out so that it is in direct anterior prolongation of the trabecula communis. The dorsal ends of the laminae orbitonasales are now interconnected by cartilage (the sphenoseptal commissures), so that the foramina olfactoria advehentia are now enclosed. But as there is yet no nasal septum, these two foramina are confluent. The ventral ends of the laminae orbitonasales are still free from the ethmoid plate. Posteriorly, the dorsal ends of the occipital arches are interconnected by a tectum posterius.

The rostral cartilage remains at the tip of the ethmoid plate, and the pterygoid process of the quadrate, still attached to the pre-ethmoid cornu in acrartete manner, is cartilaginously continuous with the body of the quadrate cartilage. Owing to the straightening of the ethmoid plate, the pterygoid process is now more horizontal in position. The metapterygoid process is more pronounced.

Meckel's cartilage has lost its ventral concavity and is now slightly curved upwards in front. The cartilaginous connexion between quadrate and symplectic has broken down, and the angle between the symplectic and the hyomandibula has been flattened out. The ceratohyal now runs vertically downwards from the interhyal, and not horizontally forwards as at the previous stage. The hypohyals extend away from the copula, instead of lying parallel with it as at the previous stage.

The copula extends backwards to the region of the 1st and 2nd basibranchials, and independent 3rd and 4th basibranchials have now chondrified, as have the 1st pharyngobranchial and the 5th ceratobranchial.

3. STAGE 3. (45 mm. length.)

The chief changes which have occurred at this stage are the appearance of a nasal septum, the attachment to the ethmoid plate of the laminae orbitonasales, and the appearance of short orbital cartilages projecting back from the latter. The trabecula communis has broken down. A pila lateralis has developed in front of the lateral commissure, thus enclosing V2, and V3 in a foramen, and the prootic bridge has appeared between the parachordals, in front of the notochord, and forming the roof over the short myodome. The only roof over the chondrocranium is the tectum synoticum.

ii. Comparison of the chondrocranium of Syngnathus and Gasterosteus

The specialized nature of Syngnathus is evidenced in the lack or reduction of a number of features possessed by its more generalized relative Gasterosteus, such as the epiphysial cartilage and orbital cartilages. But one of the most interesting features of Syngnathus is the precocious elongation of many elements of the chondrocranium, to such an extent indeed that some of them, the ethmoid plate and Meckel's cartilage, are at first bent and only straighten out later. At the 8 mm. stage the ratio of dorsoventral height to length of the skull is 7 to 11, while at the 12 mm. stage it is 1 to 2, and at the 45 mm. stage, 1 to 4. The independence of the laminae orbitonasales from the ethmoid plate is probably associated with this progressive elongation and stretching of bent elements.

The shape of the chondrocranium of Syngnathus at early stages is thus clearly a case of embryonic adaptation without phylogenetic significance. In its fairly close relative Siphonostoma, the bending of the ethmoid plate does not occur (Norman, 1926). On the other hand, in Hippocampus (Ryder, 1882) the flexure of the ethmoid region appears to persist.

In its more fundamental morphological respects, however, Syngnathus agrees well with Gasterosteus as regards the acrartete attachment of the pterygoid process of the quadrate, the rostral cartilage, and the passage of both glossopharyngeal and vagus nerves through the jugular foramen.

iii. Ossification of the skull of Syngnathus fuscus (based on 45 mm. stage)

1. The membrane-bones (bones arising without relation to cartilage).

The *frontals* cover most of the roof of the cranial cavity. Each bone possesses on its dorsal surface an open groove, which is the vestige of the lateral line canal of more primitive forms.

The *parietals* have not yet appeared at the stage studied and may be absent as in Fistularia.

The *nasals* are not mentioned by Kindred.

The *parasphenoid* extends from the basioccipital to the prevomer and forms the floor of the short myodome and of the cranial cavity in the orbital region. The internal carotid arteries apparently enter the cranial cavity by passing on each side of the parasphenoid.

The *prevomer* ossifies beneath the ventral surface of the ethmoid plate.

The *suborbital*, a single ossification on each side, is all that is left of the chain of suborbital bones of other fish.

The *metapterygoid*, curiously enough, ossifies as an intramembranous bone, on the dorsal side of the quadrate and symplectic. It is excluded from contact with the stump of the metapterygoid process of the quadrate cartilage by the intervening pterygoid.

The *pterygoid* probably represents a combined ectopterygoid and endopterygoid, the former being a vertical bar over the anterior face of the quadrate cartilage, the latter extending backwards dorsally to the quadrate towards the metapterygoid.

The *maxilla* lies in the transverse plane with its median end touching the rostral cartilage. It is toothless.

The *premaxilla*, likewise toothless, is articulated on the tip of the ethmoid plate by the rostral cartilage.

The *dentary*, also toothless, lies on the outer side of Meckel's cartilage.

The *preopercular* is very much elongated, and has been mistaken for a bone of the suborbital series.

The *interopercular* lies medially to the preopercular and is connected by ligaments in front with the dentary and posteriorly with the ceratohyal, and plays an important part in the movements of the mouth.

The *opercular* is large, and mostly covers over

The *subopercular*.

Two *branchiostegal rays* are present on each side.

2. The cartilage-bones. The *basioccipital* ossifies in the hinder part of the basal plate.

The *exoccipitals* ossify on the surface of the occipital arches and meet dorsally in the tectum posterius, thus excluding the supraoccipital from the border of the foramen magnum. The exoccipital encloses the jugular foramen, and a posterior intramembranous extension encloses the 1st spinal nerve.

The *supraoccipital* ossifies with its anterior part in relation to the cartilage of the taenia tecti medialis, but for the most part it is an intramembranous ossification.

The *epiotic* appears on the posterodorsal surface of the auditory capsule, hidden by the supraoccipital.

The *pterotic* arises on the dorsolateral surface of the auditory capsule.

The *sphenotic* ossifies on the postorbital process and has an intramembranous lateral extension (perhaps representing the *intertemporal*).

The *prootic* ossifies in the anterodorsal wall of the auditory capsule, in the prootic bridge, lateral commissure, and pila lateralis.

The *lateral ethmoid* covers the lamina orbitonasalis and has a lateral intramembranous extension.

The *ethmoid* ensheaths the cartilage of the ethmoid plate, beneath the *supraethmoid* (membrane-bone), with which it is partly fused.

The *quadrate* arises in the quadrate cartilage as a perichondral ossification, but it fuses with a membranous ossification around the tip of the symplectic and with a vertical sagittal plate of intramembranous bone dorsal to the symplectic. Conditions are here obviously so modified that it is difficult to form any opinion as to the nature and value of these ossifications.

The *palatine* (autopalatine) appears as a perichondral ossification round the pterygoid process of the quadrate cartilage.

The so-called *articular*, apparently the *angular*, ossifies in Meckel's cartilage in the region of the articulation with the quadrate.

The *retroarticular* ossifies at the hind end of the processus retroarticularis of Meckel's cartilage.

The *hyomandibula* ossifies in the posterior part of the hyosymplectic, but unlike Gasterosteus, it possesses only one articular facet on the auditory capsule.

The *symplectic* forms a perichondral sheath round the similarly named part of the hyosymplectic cartilage, but ossification extends into intramembranous dorsal and ventral lamellae which form the lateral skeletal walls to the greater part of the mouth cavity.

The *interhyal*, *ceratohyal*, *hypohyal* and *copula* (basihyal) ossify as perichondral lamellae (not, apparently, the epihyal).

Hypobranchials are ossified in all five branchial arches, *ceratobranchials* in the first four, *epibranchials* in the first three, and *pharyngobranchials* in the first two.

iv. Comparison of the osteocrania of Gasterosteus and Syngnathus

The bony skull of Syngnathus is much modified in connexion with the specialized method of opening and shutting the mouth for the purpose of sucking. When the mouth is shut, by raising the dentaries, the interoperculars and ceratohyals are pulled forwards and upwards, thus raising the floor of the mouth and decreasing the volume of its cavity. At the same time the hyomandibulae and operculars are drawn inwards, narrowing the cavity. By the reverse process, when the mouth is opened, its cavity is enlarged and water is sucked in. The gape of the mouth is very small, and the teeth have been lost.

From the developmental point of view, the modification shown by Syngnathus as compared with the more generalized type exhibited by Gasterosteus, involves the reduction of cartilage to such an extent that certain bones which must originally have been wholly perichondral ossifications are now partly (e.g. quadrate) or wholly (e.g. metapterygoid) intramembranous in development. Further, there has been an extension of the intramembranous outgrowths of certain bones (e.g. symplectic) which are present to a limited extent in Gasterosteus.

SCORPAENIFORMES

XIII. Sebastes

PLATE 51, FIG. 6.

The development of the chondrocranium of Sebastes marinus has been studied by Mackintosh (1923), to whose work Norman (1926) has made certain additions.

Concerning the origin of the chief cartilages, trabeculae, parachordals, &c., there is little of importance to notice. In the visceral arch skeleton Mackintosh imagined that he could make out the fused origin of the quadrate, hyomandibula and symplectic cartilages at the 4–5 mm. stage. Norman, however, has observed the mark of a suture between the quadrate and the hyosymplectic (or fused hyomandibula and symplectic) at this stage, which suture is replaced by complete fusion at the 5·5 mm. stage. Later on, the quadrate and hyosymplectic become separate again, so the fusion is a temporary larval modification.

At the 25 mm. stage full development of the chondrocranium has been reached, and in certain respects, passed, for regression has begun to set in. For instance, the trabecula communis has broken down and is discontinuous, as are the orbital cartilages between the laminae orbitonasales and the epiphysial cartilage. The latter is connected with the postorbital process of the auditory capsule by taeniae marginales, and with the tectum synoticum by a taenia tecti medialis. A rostral cartilage is present, dorsally to the nasal septum.

No information is available concerning the lateral commissure, and the exits of the glossopharyngeal and vagus nerves. However, from Allis's (1909) figures and description of the adult skull in Sebastes dactylopterus, it is clear that the lateral commissure is present, ossified as part of the prootic bone, and forming the lateral wall of the trigeminofacialis chamber. Further, the glossopharyngeal nerve issues through a foramen some distance in front of that of the vagus, so it is probable that the relations of these nerves to the chondrocranium conform to the plan shown by Salmo.

XIV. Dactylopterus

The chief interest of this form from the present point of view is the fact that according to Allis (1909) Dactylopterus is the only Teleost in which the postfrontal is regularly fused with the sphenotic. On the other hand, the fusion of the pterotic with the supratemporal is such as to suggest that these are two separate and independent ossifications which have fused, and not a dermal bone which has invaded cartilage. The reason for this is that the dermal and perichondral portions of the fused bone are not exactly superimposed; the supratemporal constituent lying farther forward than the pterotic.

XV. Trigla

Nothing is known of the development of the skull of the gurnard, beyond a small point in connexion with the trigeminofacialis chamber. In this form, the palatine branch of the facial nerve passes laterally from the pars ganglionaris into the pars jugularis of the trigeminofacialis chamber and then pierces its (bony) floor. This is in contrast to such a form as Salmo where (see p. 128) the palatine nerve never enters the pars jugularis of the trigeminofacialis chamber at all, but runs ventrally from the pars ganglionaris, pierces the prootic bridge and then the palatine foramen (between postpalatine and basitrabecular constituents of the lateral commissure). The relatively more lateral course of the nerve in Trigla is responsible for its becoming included in the pars jugularis of the trigeminofacialis chamber when the prootic ossifies.

XVI. Cyclopterus lumpus

The development of the skull of the lumpsucker has been studied by Uhlmann (1921*a* and *b*) and presents an interesting example of a Teleost in which the cartilage of the chondrocranium has undergone an abnormally large development.

i. The development of the chondrocranium of Cyclopterus lumpus

PLATE 60.

The figures and descriptions given by Uhlmann suffer from the fact that the relations of the cartilages to neighbouring nerves and blood-vessels have been neglected. It is not, therefore possible to treat of the chondrocranium of Cyclopterus with the same degree of detail and precision as is the case with many other forms.

At the first stage, 3–4 mm. long, 15 days after fertilization (Fig. 1), the trabeculae and parachordals have joined up, the auditory capsule is attached to the parachordal by the anterior basicapsular commissure, and the occipital arch is present.

At the next stage, 4–5 mm. long, 26 days after fertilization (Fig. 2), the orbital cartilages have appeared, interconnected by the epiphysial bar. The laminae orbitonasales rise up towards the orbital cartilage from the lateral edges of the trabecular plate. Posteriorly, the auditory capsules are well formed and attached posteriorly to the dorsal end of the occipital arches. The trabeculae fuse to form a broad trabecula communis.

At the 22 mm. stage (Figs. 3, 4) the chondrocranium is well formed. The orbital cartilages have joined posteriorly with the postorbital processes of the auditory capsules, and anteriorly with the lamina orbitonasalis of their side and with the nasal septum. A taenia tecti medialis projects back from the epiphysial bar. Posteriorly, the auditory capsules are interconnected by a tectum synoticum. The trabecula communis has become interrupted, while the parachordals are greatly elongated and have become interconnected by a prootic bridge, in front of the notochord. It does not appear that a lateral commissure is formed, but apparently the external recti muscles pass inwards over the anterior portions of the parachordals, which are depressed below the level of the prootic bridge. No details of a posterior myodome are given.

The cranial cavity at first extends far forwards; at the 11 mm. stage the front of the brain lies between the laminae orbitonasales. As development proceeds the cranial cavity and brain retreat, and the nasal septum becomes thick and massive, and contains pockets of resorption. The oblique eye-muscles push forwards on each side into the cartilage of the nasal septum, behind the laminae orbitonasales, thus coming to lie in a pair of anterior myodomes, which however do not (as in Salmo) meet in the midline.

Subsequent stages (38 mm., Fig. 5; 56 mm., Fig. 6; and adult, Fig. 7) show progressive development of the cartilage of the ethmoid region and of the roof, which becomes complete, except for a small slit between the huge taenia tecti medialis and the postorbital process, the tip of the latter having severed its connexion with the orbital cartilage.

The cartilages of the visceral arches also are especially well developed and massive.

ii. Development of the osteocranium of Cyclopterus lumpus

The development of the bones of the skull of the lumpsucker has been studied in considerable detail by Uhlmann (1921*b*), who has introduced a few new terms to give precision to his views and descriptions. A 'perilamella' is a bone which develops as a perichondral lamella, beneath or in the perichondrium, or in such part of the true cranial wall (dura mater) which fails to chondrify and remains membranous. Uhlmann's perilamellae therefore include elements which belong to both categories, of cartilage-bones and membrane-bones. By the term 'epilamella' is meant the extension of a perilamella externally but parallel to the perichondrium or dura mater. On the other hand, an 'apolamella' is an extension of ossification right away from the perilamella, into the surrounding connective tissue.

Owing to the fact that nearly all the bones as described by Uhlmann develop by means of what he has called perilamellae followed by epilamellae and apolamellae, it is not easy to apply the usual classification into cartilage-bones and membrane-bones.

1. Membrane-bones (i.e. bones originating outside the perichondrium). The *parietal* is present at 7 mm. as an ossification in the membrane covering the cranial cavity, and immediately overlying (or perhaps in) the perichondrium of the free edge of the roof of the auditory capsule. Uhlmann speaks of the parietal here as in part perichondral, but it seems never to lie beneath the perichondrium, nor to invade the cartilage. It is a membrane-bone closely adpressed to the chondrocranium.

The *frontal* arises at 6 mm. in the neighbourhood of the epiphysial bar, as an ossification like the parietal, in the membrane covering the roof of the cranial cavity, and close to the perichondrium of the orbital cartilage. At the 11 mm. stage, lamellar extensions are developed forming a wider roof to the orbit and surrounding the lateral line canal.

The *nasal* appears at 11 mm. as a scroll-like ossification round the lateral line canal.

The *orbital* (orbital I) arises at 6 mm. in the connective tissue laterally to the pterygoid process of the pterygoquadrate cartilage, and without any connexion with the lateral line.

The *suborbitals* (orbitals II and III) and *postorbital* arise at 6 mm. and are related to the infraorbital lateral line canal. The 2nd suborbital gives off a process ventrally and posteriorly to the preopercular.

The *preopercular*, *opercular*, *subopercular*, *interopercular*, and *branchiostegal rays* arise early, at 4–5 mm. as plate-like and ray-like ossifications in the opercular fold. The preopercular is related to the lateral line canal.

The *parasphenoid* arises at 4 mm. in the membrane spanned across the hypophysial fenestra, and close to the perichondrium of the trabecula communis and parachordals. Uhlmann speaks of it in part as perichondral. It bears no teeth.

The *prevomer* arises at 6 mm. from paired groups of osteoblasts beneath the ethmoid plate, which produce a median ossification. This lies close to the perichondrium, and at late stages gives off lamellae which actually penetrate into the cartilage. The prevomer bears no teeth.

The *premaxilla* arises at 3–4 mm. independently of the teeth to which it will become attached. The vertical anterior process of the premaxilla is separated from the ethmoid bone by the rostral cartilage, which probably represents fused upper labial cartilages.

The *maxilla* arises at 3–4 mm. It bears no teeth.

The *ectopterygoid* and *endopterygoid* arise at 11 mm. as membrane-bone underlying the pterygoid process of the pterygoquadrate cartilage.

The *dentary* is present in embryos 3–4 mm. long. Anteriorly it lies close to the cartilage as a perichondral ossification, seemingly corresponding to a *mentomeckelian bone*. Farther back, the dentary lies free from the perichondrium. The dentary bears teeth and protects the lateral line canal.

The *angular* becomes related to Meckel's cartilage and will be referred to again below.

2. Cartilage-bones (i.e. bones originating beneath the perichondrium). The *supraoccipital* arises in 11 mm. larvae as a perichondral lamella on inner and outer surfaces of the cartilage of the tectum synoticum with an intramembranous extension in the membranous roof of the skull in front of the tectum and another vertical (apolamella) one which forms the spine. The supraoccipital arises in connexion with the attachment of muscles. It does not reach the border of the foramen magnum.

The *basioccipital* appears in 7 mm. larvae as a genuine perichondral ossification in the hind part of the parachordals, on each side of the notochord. The hind part of the basioccipital is formed like the centrum of a vertebra.

The *exoccipital* ossifies at 7 mm. as perichondral lamellae in the occipital arch, and gives off apolamellae posteriorly which embrace the 1st vertebra and enclose the 1st spinal nerve in a foramen.

The *epiotic* arises at 11 mm. in the form of a perichondral lamella in the wall of the auditory capsule, with strong apolamellae to which muscles and ligaments are attached.

The so-called *opisthotic* arises at 11 mm. as a lamella in its dorsal part perichondral, but which further ventrally is lifted off the perichondrium (epilamella). In addition, apolamellae are developed and serve for the attachment of ligaments.

The *pterotic* ossifies at 6 mm. in the lateral wall of the auditory capsule as a perichondral lamella which extends as an epilamella in connexion with the attachment of the adductor hyomandibulae muscle and the formation of the posterior articular facet for the head of the hyomandibula. Later, at the 15 mm. stage, apolamellae develop around the temporal lateral line canal. This lamella possibly represents the *supratemporal* or *intertemporal*, fused on to the pterotic.

The *sphenotic* ossifies at 6 mm. as a perichondral lamella in the postorbital process, from which epilamellae extend in connexion with muscle attachments and the formation of the anterior articular facet for the hyomandibula. Only later an apolamellae is developed around the lateral line canal, possibly representing the intertemporal.

The *prootic* (Uhlmann's 'alisphenoid') ossifies at 6 mm. as a perichondral lamella behind the foramen of exit of the trigeminal and facial nerves: the inner lamella is better developed than the outer.

The *lateral ethmoid* arises at 7 mm. as two perichondral lamellae, in the lamina orbitonasalis and surrounding the olfactory foramen. Later, these become interconnected by apolamellae which extend and form a stout wall dividing the eye from the nasal sac (possibly representing the *prefrontal*).

The *ethmoid* arises at 11 mm. as perichondral lamellae on the upper and lower surfaces of the ethmoid plate and nasal septum. These lamellae eventually become interconnected by endochondral ossification, and apolamellae are developed in connexion with the attachment of ligaments.

Perichondral ossifications in the pterygoquadrate cartilage are:

The *autopalatine* arises at 7 mm. in the anterior part of the cartilage and develops apolamellae (there is no dermopalatine in Cyclopterus).

The *quadrate* arises at 7 mm. at the site of the articulation with the lower jaw, and gives off numerous apolamellae.

The *metapterygoid* arises at 11 mm. in the posterior part of the cartilage and gives off apolamellae which become attached to the hyomandibula.

Perichondral ossifications in Meckel's cartilage are:

The so-called *articular*, apparently the *angular*, arises at 5 mm. at the site of the articulation with the upper jaw, but the ossification becomes epilamellar. Apolamellae develop around the lateral line canal, and this fact was held to point to the inclusion in the so-called articular of the fused rudiment of the angular. The articular is apparently absent (Haines, in press).

The *retroarticular* arises at 6–7 mm. in the hindmost part of Meckel's cartilage and gives off apolamellae for the attachment of the mandibulohyoid ligament.

Perichondral ossifications in the hyoid arch include:

The *hyomandibula* (at 7 mm.), the *symplectic* (at 7 mm.), the *interhyal*, *epihyal*, *ceratohyal*, *hypohyal*, and *basihyal*, all at 4–5 mm. and arising as thin lamellae ensheathing their respective cartilages, which latter persist. The basihyal develops a powerful apolamella which extends ventrally in the sagittal plane, as a keel for muscle attachments. So large is this keel in the adult that the bone no longer gives the appearance of having been of perichondral origin at all. Nevertheless, cartilage is still present within its original perichondral portion. It is not yet clear whether this structure is homologous or analogous with the 'urohyal' of other Teleostei (p. 130).

In the branchial arches perichondral ossifications are found as thin sheaths surrounding the cartilages. Considerable interest attaches to the 5th ceratobranchial and 2nd epibranchial. Teeth are present in the lining of the pharynx overlying these cartilages at the 5 mm. stage, and at 6 mm. the cartilages are surrounded by perichondral ossifications, from which, later on, apolamellae are developed, to which muscles, and the teeth, become attached.

The *pharyngeum inferius* and *superius* which are formed in this way, are thus integral with the cartilage-bones of the corresponding segments of the branchial arch skeleton.

ZEORHOMBIFORMES

XVII. SOLEA VARIEGATA, PLEURONECTES PLATESSA, LIMANDA, FLESUS

The development of the chondrocranium of the sole and the plaice has been studied in its early stages by Berrill (1925), while the modifications involved in the torsion which occurs at metamorphosis, first correctly described by Traquair (1865), have been followed more recently by Williams (1902) and Mayhoff (1914).

The development of the bones in several of the flat-fish has been described by Mayhoff (1914), while the morphology of the osteocranium has been considered by Cole and Johnstone (1901).

i. Development of the chondrocranium of Solea variegata

(PLATE 61, FIGS. 1–5.)

The early development of the chondrocranium of the sole is typical of that of other Teleostei. The skull is symmetrical; the trabeculae are extensively fused, forming a long trabecula communis; already at an early stage the external rectus muscle passes inwards dorsally to the trabecula and is attached to the mesenchymatous rudiment of the prootic bridge in the form of a transverse strand stretched across the hypophysial fenestra.

The auditory capsule, already attached to the parachordal by the anterior basicapsular commissure, soon develops the posterior basicapsular commissure, so as to enclose the glossopharyngeal nerve in the basicapsular fenestra and separate it from the vagus. The occipital arches are large. The quadrate and hyosymplectic cartilages appear to be continuous at first, but soon become demarcated from one another. The pterygoid processes of the pterygo-quadrate articulate with the anterolateral corners of the ethmoid plate, while from its posterolateral corners arise the slender laminae orbitonasales. The hyomandibular branch of the facial nerve emerges freely in front of the anterior basicapsular commissure and passes outwards behind the hyo-symplectic cartilage.

At a later stage (Plate 61, fig. 4), the cartilage of the anterior wall of the auditory capsule has grown upwards and backwards and fused with the occipital arch and with the corresponding structure of the opposite side, forming a combined tectum synoticum and tectum posterius. A gap is left in the lateral wall of the capsule, and this is occupied by an independent plate of cartilage. The postorbital process is well developed, as is also the lateral commissure which is separated from the anterior basicapsular commissure by the posterior aperture of the trigeminofacialis chamber. Through this pass the hyomandibular branch of the facial nerve, the orbital artery, and the head vein, in typical Teleostean manner. The hyomandibular nerve now pierces a foramen in the hyosymplectic cartilage. The palatine nerve runs forwards over the lateral commissure and downwards laterally to the trabecula. The internal carotid arteries meet in the middle line and pass upwards through the hypophysial fenestra.

At a still later stage (Plate 61, fig. 5) the wall of the auditory capsule is complete, but the trabeculae have broken down, so that the trabecula communis is discontinuous with the parachordals. The latter are now united across the middle line by a cartilaginous prootic bridge forming a roof to the posterior myodome into which the external rectus muscles extend. The remaining rectus muscles have their attachments shifted forwards, on to the trabecula communis.

The ethmoid plate possesses a well-developed nasal septum, in front of which is the separate rostral cartilage. Anteriorly, paired processes of the ethmoid plate grow forwards and downwards and fuse with one another, forming a small foramen of doubtful significance.

In the visceral arch skeleton the pterygoid processes are still continuous with the body of the quadrate cartilage. A small opèrcular cartilage is present on the hind edge of the hyosymplectic. The interhyal retains cartilaginous continuity with the hyosymplectic. The hypohyals are perforated by foramina through which pass the efferent hyoidean arteries. There is no basihyal. In addition to the rostral cartilage, three paired cartilages appear (in Pleuronectes: submaxillary, subrostral or premaxillary, and postoral cartilages) which seem to represent the labial cartilages of the Selachii.

ii. Metamorphosis of the chondrocranium

Further development of the chondrocranium is largely affected by metamorphosis. The orbital cartilages appear late, connecting the laminae orbito-nasales with the postorbital processes on each side. However, the orbital cartilage on the future 'blind' side (in these cases, the left) immediately breaks down again, and it is important to notice that this occurs while the skull is still symmetrical, and before any torsion or migration of the eye takes place. This torsion has been particularly studied in Pseudopleuronectes americanus by Williams (1902), on whose work the following account is based.

At the stage figured on Plate 61, fig. 7, the orbital cartilage is still present on the right (the future ocular side) while it has disappeared on the left, and through the gap thus formed, the left eye will eventually move dorsally.

At the next stage (Plate 61, fig. 8) the right orbital cartilage is still complete while the gap on the left is still larger. At the same time, the left lamina orbitonasalis is connected with the dorsal edge of the nasal septum by a sphenoseptal commissure, thus enclosing a left foramen olfactorium advehens.

At the same time, changes are taking place in the jaws. The gape becomes widened by elongation of the pterygoid processes of the quadrate and of Meckel's cartilages.

At the next stage (Plate 61, fig. 9), the right foramen olfactorium advehens has also been enclosed, by the development of the sphenoseptal commissure on that side. The right orbital cartilage now begins to disappear, from behind forwards, and the anterior part of the pterygoid process becomes detached from the body of the quadrate cartilage owing to the posterior position which the latter has assumed and the great stretching to which the pterygoid process has been subjected.

A new structure has now appeared in the form of the interorbital septum. This cartilage is attached to the posterior edge of the nasal septum and to the dorsal surface of the trabecula communis, leaving a gap or fontanelle through which (except, of course, for membrane) the orbits communicate with one another, and near which point the oblique eye-muscles are attached. The head is now definitely asymmetrical and the eye of the left or future blind side is now situated in the sagittal plane, 'above' the trabecula communis.

At the last stage (Plate 61, fig. 10) rotation is seen to have affected the interorbital and nasal septa, so that these definitely incline towards the right side. At the same time, the dorsal ends of the laminae orbitonasales have been rotated forwards, so that the foramina olfactoria now no longer point straight forwards but downwards and to the left. It is therefore the posterior surface of the laminae orbitonasales that is seen in this figure. The right orbital cartilage is reduced to a stump, the hamulus ethmoideus, attached to the posterior side of the dorsal end of the right lamina orbitonasalis. This stump now separates the two eyes, for the left eye is now on the right side dorsally to the stump, while the right eye is ventral to it.

It is to be noted that the cartilage in the orbital region is very sparsely developed; there is no epiphysial bar or roof to the anterior portion of the cranial cavity.

The asymmetry which results from the migration of the eye is the result partly of absorption of cartilage (especially on the left side) and partly of unequal growth of the cartilage including interstitial growth (revealed by intracartilaginous mitoses). In this way the cartilage of the left side of the ethmoid region extends dorsally, that on the right, ventrally.

The attachments of the eye-muscles are naturally altered as a result of this migration of the eyes, but this seems to be a developmental consequence of the migration and can hardly have been its phylogenetic cause.

iii. Development of the osteocranium of Pleuronectes platessa

Ossification has already set in before the migration of the eye takes place. Among the membrane-bones there are the parasphenoid, prevomer, supra-ethmoid, frontal, parietal, lachrymal of the right side, preopercular, premaxilla, ectopterygoid, endopterygoid, angular, dentary, opercular, subopercular, interopercular, branchiostegal rays, pharyngea inferius and superius. Among the cartilage-bones are the supraoccipital, epiotic, pterotic, quadrate, retro-articular, hyomandibular, symplectic, ceratohyal and hypohyal ossifications.

The *frontals* begin symmetrically, each one overlying a postorbital process, bearing typical relations to the supraorbital lateral line canals. The right frontal extends forwards farther than the left, but both undergo resorption along their left edges. In this way that portion of the left frontal which lodges the middle portion of the lateral line canal is destroyed. The right frontal then comes to lie between the two eyes.

On the left side a strand of dense tissue, apparently corresponding to the rudiments of the string of suborbital and postorbital bones, is attached anteriorly to the left lamina orbitonasalis, and posteriorly to the left frontal, passing ventrally to the left eye. As the right frontal and the parasphenoid increase in length, this left subocular strand is stretched tighter, and the left eye appears thereby to be lifted. Eventually this subocular strand ossifies as the secondary process of the left frontal, which is attached anteriorly to the left *lateral ethmoid* which ossifies in the lamina orbitonasalis. The extension of this ossification in the subocular strand of connective tissue appears to correspond to a *prefrontal* (membrane-bone). Other cartilage-bones which now appear are the *exoccipital* (which sends back a process enclosing the 1st spinal nerve), the *sphenotic*, *prootic*, *pleurosphenoid*, *autopalatine* and *metapterygoid*.

Further differences of growth-intensity result in the right eye moving forwards relatively to the (original) left.

After completion of the eye-migration the following ossifications appear: among membrane-bones, the *orbital*, *nasal*, *postparietal*, *supratemporal*, *inter-temporal*, *opisthotic* (*intercalary*); among cartilage-bones, the *coronomeckelian*, *epihyal*, *pharyngobranchial I*, *epi-*, *cerato-*, *hypo-*, and *basibranchials*. The presence of nasals seems to be correlated with the supraorbital lateral lines (Traquair, 1865).

Some of the membrane-bones come to acquire direct contact with the cartilage of the chondrocranium. Among these are the *angular*, *prevomer*, *supra-ethmoid*, and the anterior end of the *dentary* which presumably represents the *mentomeckelian*.

In Pleuronectes limanda, P. flesus, and Arnoglossus laterna the supra-intertemporal fuses with the underlying pterotic. This takes place by ossification of strands of intervening connective tissue. In Pleuronectes platessa, the supra-intertemporal fuses with both the pterotic and the sphenotic.

GADIFORMES

XVIII. Gadus merlangus

The literature appears to be devoid of any references to studies on the development of the skull in the whiting, or indeed in any member of the cod family. In view of the features of interest which these forms present, both in the cartilaginous and in the bony skull, this lacuna is remarkable. The following account is based on investigations specially made for this book, and is to be regarded as a preliminary survey of the more important features, in no way obviating the desirability of an exhaustive study of these most interesting forms.

i. The development of the chondrocranium of Gadus merlangus

1. STAGE 1. (3 mm.) PLATE 62. FIGS. 1, 2.

In the whiting the visceral arch skeleton undergoes chondrification before the brain-case. At this stage there are visible: quadrate and Meckel's cartilages, hyosymplectic, ceratohyal, and hypohyal. The quadrate is represented only by its articular portion, and is close to yet independent of the hyosymplectic. The hyosymplectic is a slender rod, unperforated by any foramen for the hyomandibular branch of the facial nerve, for that nerve runs out laterally *in front* of the hyosymplectic cartilage—a condition unique in Teleostei so far as is known. The anterior ends of Meckel's cartilages fuse in a symphysis.

2. STAGE 2. (5 mm.) PLATE 62, FIGS. 3, 4.

The neurocranium at this stage is already well formed. The trabeculae are joined anteriorly into a trabecula communis, and posteriorly are fused with the parachordals. The lateral edge of the latter is fused with the auditory capsule except for the foramen of exit of the glossopharyngeal nerve. The vagus foramen (fissura metotica) between the posterior wall of the auditory capsule and the occipital arch forms a tunnel facing not sideways but backwards, for the hind part of the auditory capsule extends backwards to the side of and overlapping the occipital arch. Behind the occipital arch is a small nodule of cartilage on the dorsolateral surface of the notochord, and which seems to represent the rudimentary neural arch of the 1st vertebra.

The auditory capsules are interconnected by a tectum synoticum, and anteriorly each capsule bears a postorbital process. There is no lateral commissure or prootic process; all the cranial nerves as far back as the facial inclusive pass out in front of the anterior basicapsular commissure.

In the middle line dorsally to the trabecula communis in its foremost region where it expands to form the ethmoid plate, is an independent rostral cartilage.

The quadrate is represented by its articular and metapterygoid portions, which lie close in front of but separate from the hyosymplectic. The latter is very long, and thus the angle of the jaw is situated far forward. Meckel's cartilage, which has a prominent coronoid process, is in an almost vertical position, as the embryo already has the characteristic Gadoid facies with the large and upturned lower jaw. The visceral arch skeleton also includes the interhyal, ceratohyal, and hypohyal cartilages; a median copula or basibranchial, and ceratobranchials of the first two branchial arches.

The hypohyal cartilages are perforated by foramina through which pass the afferent hyoidean arteries.

3. Stage 3. (11 mm.) PLATE 62, FIGS. 5, 6.

The trabecula communis is now discontinuous with the parachordals, and the latter are interconnected in front of the notochord by a flat prootic bridge. There is no lateral commissure, and consequently no trigeminofacialis chamber. The orbital cartilages are now present in the form of slender bars attached posteriorly to the postorbital processes of the auditory capsules, and interconnected by the epiphysial bar from which a taenia tecti medialis projects backwards. Anteriorly, each orbital cartilage is attached to the dorsal end of the lamina orbitonasalis of its own side and then by means of the sphenoseptal commissure to the nasal septum. The nasal septum rises up from the dorsal surface of the ethmoid plate as a median ridge immediately beneath the rostral cartilage. Each lamina orbitonasalis arises from the postero-lateral corner of the ethmoid plate and slopes upwards and backwards to meet the orbital cartilage. No preoptic root of the orbital cartilage is chondrified, with the result that on each side the only foramen to be delimited is the foramen olfactorium advehens. (This is bounded anteriorly by the nasal septum, dorsally by the sphenoseptal commissure, posteriorly by the lamina orbitonasalis, and ventrally by the ethmoid plate.) As Goodrich (1909) showed, the cranial cavity extends right forward, dorsally to a membranous interorbital septum, and the olfactory nerves where they pass out of the membranous wall of the cranial cavity are opposite the foramen olfactorium advehens and in front of the lamina orbitonasalis. It is because the lamina orbitonasalis occupies a position relatively so far back that the olfactory nerve in the Gadoid does not traverse the orbit. Dietz' (1921) explanation is less satisfactory.

In addition to the olfactory nerve from the cranial cavity, the foramen olfactorium advehens is traversed by the orbitonasal artery passing forwards from the orbit into the nasal capsule. The ophthalmic nerve runs forwards dorsolaterally to the lamina orbitonasalis.

The quadrate now possesses an elongated and slender pterygoid process, the anterior end of which is apposed to the lateral edge of the ethmoid plate. The remainder of the visceral arch skeleton is much as in previous stages. The hyosymplectic is slender and lies behind the hyomandibular branch of the facial nerve. The ceratohyal shows a blade-like broadening of its posterior region. The skeletal elements of the remaining branchial arches are now present.

A ventral labial cartilage is present in the fold at the edge of the lower jaw, laterally to Meckel's cartilage.

A nodule of what appears to be 'secondary' cartilage in the ossifying rudiment of the maxilla, may represent the posterior upper labial cartilage of Selachians, while the anterior upper labial cartilages of the latter may be represented by the median rostral cartilage of Gadus.

ii. Development of the osteocranium of Gadus merlangus

As in the case of the chondrocranium, the following account refers merely to a few observations made on some of the more interesting ossifications, to supplement the very meagre references in the literature to the development of the bony skull of Gadoid fishes. Cole (1898) has drawn attention to the relations of the bones traversed by the lateral line canals, while Allis (1909) has considered some of the cartilage-bones. A general though antiquated account of the morphology of the bony skull is given by Brooks (1883).

The *sphenotic* and *pterotic* are both cartilage-bones, ossifying as perichondral lamellae on the inner and outer surfaces of the cartilage of the auditory capsule. Overlying them, and apparently ossifying independently, are a number of dermal bones (four), lodging the lateral line canal. These dermal bones, presumably representing the *intertemporal* and *supratemporals*, consist of a gutter enclosing the canal, and a narrow plate extending laterally from the ventral side of the gutter. This plate overlies and becomes closely attached to a similar intramembranous plate developed from the underlying cartilage-bones—the sphenotic and pterotic. The ridge overlying the socket for the articular head of the hyomandibula is thus composed of a more dorsal lamella developed in connexion with a dermal lateral line bone and a more ventral lamella extending out from a cartilage-bone. In the adult skull the sphenotic and pterotic thus appear to be fused with the overlying dermal elements, though as Cole (1898) describes, they sometimes retain their separate nature. This is particularly the case in the hinder part of the pterotic, where that bone appears quite regularly to be free from the overlying supratemporal ossicle, which remains in the form of a tubular 'canal bone'.

The *opisthotic* or *intercalary* is exceedingly large in the Gadoid, and cannot be mistaken as these forms also possess well-developed perichondral *ex-occipitals* and *epiotics*. It is therefore of interest to find that the so-called opisthotic is a definite membrane-bone, developed close to but outside the perichondrium of the posterior cartilaginous wall of the auditory capsule.

The *lateral ethmoid* ossifies as a perichondral lamella surrounding the lamina orbitonasalis, and is continuous with an intramembranous bony plate which extends sideways between the nasal sac and the orbit. This plate appears to ossify before the perichondral lamella, but it is uncertain whether the two are originally separate.

The *supra-ethmoid* arises as a dermal ossification overlying the dorsal edge of the nasal septum. At the 16 mm. stage the bone is everywhere external to the perichondrium of the cartilage. But at the 33 mm. stage it can be observed that for part of its length the bone is in direct contact with the cartilage, which it appears slightly to invade.

The *hyomandibular* bone is of interest because it is perforated by the hyomandibular branch of the facial nerve (by separate foramina for its mandibular and hyoid rami) whereas the cartilaginous hyosymplectic is not so perforated but lies entirely behind the nerve. The condition of the hyomandibular bone is arrived at by the extension of intramembranous ossification on the anterior edge of the perichondral lamella of bone which surrounds the cartilage. The morphology of the bone is thus more typical and primitive than that of the cartilage, and would be explained if that part of the cartilage which in the typical Teleost hyosymplectic lies anteriorly to the nerve foramen, ceased to chondrify.

The internal carotid arteries which passed medially to the trabeculae through the hypophysial fenestra in the typical manner to enter the cranial cavity while the trabeculae existed, now enter the cranial cavity by passing upwards and inwards over the lateral edges of the *parasphenoid*.

The posterior myodome in Gadus only develops late and is very much reduced. The eye-muscles pass inwards over the anterior ends of the parachordals now ossified by the *prootics*, but do not pass back beneath the prootic bridge at all. The myodomic space represented by the posterior portion of the original hypophysial fenestra, is in the bony skull completely occluded by the contact established between the parasphenoid and the median processes of the prootics, ossifying in the prootic bridge.

The *exoccipital* is a typical cartilage-bone, ossifying in the occipital arch, but an intramembranous posterior extension extends backwards and encloses the 1st spinal nerve in a foramen.

In addition, the 1st vertebra develops in close connexion with the posterior end of the skull, its centrum abutting against the posterior face of the basioccipital, and the anterior zygapophyses of its neural arches fitting into sockets in the posterior edge of the exoccipitals. This portion of the 1st vertebra contains the cartilaginous neural arch observed at younger stages. Faruqi (1935) refers to the latter as an 'intercalated' arch, but there appears to be no reason to suppose that it is an intercalary, i.e. interdorsal, developed from an anterior half-sclerotome.

There are no sclerotic bones.

12. DIPNOI

I. Ceratodus forsteri

The development of the chondrocranium of Ceratodus has been studied by graphic reconstruction by Sewertzoff (1902) and by means of wax models by Krawetz (1911), Greil (1913), and de Beer (1926*a*). Certain features, especially the hyoid arch, have been studied by Ridewood (1894), K. Fürbringer (1904), Edgeworth (1923*c*), Schmahlhausen (1923*a*), and Goodrich (1930).

The morphology of the chondrocranium of Ceratodus served Huxley (1876*b*) as a basis for the autostylic type in his classification of jaw suspensions. The little that is known concerning the development of the bony skull is due to Greil (1913). The morphology of the skull has been considered by K. Fürbringer (1904) and Kesteven (1931).

i. The development of the chondrocranium of Ceratodus

The earliest stage at which chondrification has taken place is that described by Sewertzoff (Plate 63, fig. 1), corresponding to an early stage 45 of Semon. Parachordal cartilages are present flanking the tip of the notochord, and continuous anteriorly with the trabeculae which are short, wide apart, and lie in the same plane as the parachordals and in the line of their prolongation. According to Greil (1913, pp. 1136, 1156) the trabeculae arise independently of the parachordals and subsequently fuse with them.

The orbital cartilages are also present in Sewertzoff's reconstruction, fused to the parachordal and trabecula of their side by a pila antotica. This bears the typical relations, lying behind the oculomotor nerve, pituitary vein, and ophthalmic artery, and in front of the profundus and trigeminal nerves.

Posteriorly, the parachordals reach back to the level of the centre of the auditory vesicles, to which point the row of myomeres extends forwards. The auditory capsules are only procartilaginous.

The quadrates are independent chondrifications apposed to the hindmost portion of the trabeculae by means of the basal process; cartilaginous fusion has not yet occurred.

Subsequent development may show considerable variation. In larvae of stage 45 or 46 (Plate 63, figs. 2–6) Greil (1913, pp. 1121, 1196) has found that the ethmoid (or trabecular) plate may chondrify independently; the trabeculae soon fuse with it, however, and thus enclose the hypophysial fenestra, through the hinder part of which the internal carotid arteries enter the cranial cavity. The orbital cartilage is attached to the trabecula by a pila metoptica, typically situated behind the optic nerve and in front of the oculomotor nerve, pituitary vein, and ophthalmic artery. The auditory capsules are chondrified and attached to the parachordals by an anterior (and soon also by a posterior) basicapsular commissure. The glossopharyngeal nerve passes out in front of the posterior basicapsular commissure and thus appears to traverse the basicapsular fenestra and the cavity of the auditory capsule. The latter has no cartilaginous medial wall. Its posterior wall occasionally chondrifies independently.

The parachordals are interconnected anteriorly above the tip of the notochord. Posteriorly they gradually extend back behind the auditory capsule in the form of dorsal and ventral longitudinal bands on each side of the notochord. The foremost myomeres lie between these bands, and come eventually to be lodged in hollow pits as the parachordal chondrification extends laterally from the dorsal and ventral bands, respectively, above and below the myomeres.

The quadrate is now attached by the otic process to the auditory capsule, which develops a small crista parotica to meet it. The visceral arch skeleton also includes Meckel's cartilages, the ceratohyals and hypohyals, ceratobranchials in the first three (or four) branchial arches, and epibranchials in the first three.

At stage 47 (6th week after hatching, Greil, p. 1237; Plate 63, figs. 6, 7; Plate 64, figs. 1–4). The trabecular horns project right and left from the trabecular plate, and end freely. They lie *dorsally* and *anteriorly* to the external nostrils. Behind the internal nostril the lamina orbitonasalis projects outwards and forwards from the trabecula on each side. The orbital cartilage is now continuous with the auditory capsule, thus enclosing the prootic foramen through which *all* the roots of the trigeminal and facial and abducens nerves emerge, though the latter may acquire an independent foramen at the base of the pila antotica. The auditory capsule is now well formed, with septa for the three semicircular canals. The quadrate is now attached to the pila antotica by the processus ascendens, and a knob on the anterior face of the basal process is held to represent the pterygoid process. The front ends of the parachordals are interconnected beneath as well as above the notochord (epichordal and hypochordal commissures).

The further development of the chondrocranium is concerned mainly with four regions, which may now be dealt with in turn, viz. the roof, the occipital region, the dorsal end of the hyoid arch, and the nasal capsule.

(i) The roof of the chondrocranium is at first very incomplete. There is a tectum synoticum which is relatively far forward. A process from each orbital cartilage towards the other forms the rudiment of the epiphysial bar. Eventually, however, the entire roof becomes cartilaginous, and persists throughout life.

(ii) Some considerable distance behind the auditory capsule there appear a number of basidorsal and basiventral elements. The foremost three pairs of basidorsals are the occipital and the two occipitospinal arches; the next posterior arch being the neural arch of the 1st vertebra. As described by Krawetz (1911, see p. 29) the occipital arch may arise in the septum between segments 8 and 9, in which case there may be two hypoglossal ventral nerve-roots in front of it, or in the septum between segments 9 and 10, in which case there may be three ventral roots in front of it. Counting the hindmost two hypoglossal ventral roots behind the occipital and in front of the two occipito-spinal arches, the total number of these roots may be four or five, and the total number of segments in the skull may be ten or eleven. The dorsal posterior extension of the parachordal fuses with the occipital arch, thus forming a posterior and ventral border to the fissura metotica through which the vagus nerve passes. The 1st occipitospinal arch is variable and may be much reduced. Corresponding to it is the foremost basiventral element, to which the ventral posterior extension of the parachordal reaches. Corresponding to the 2nd occipitospinal arch is a rib, which therefore becomes a cranial rib when the vertebral segment to which it belongs is attached to the parachordal.

The dorsal connexions across the middle line of the occipital and occipito-spinal arches of one side with those of the other give rise to the tectum posterius.

(iii) The dorsal skeletal elements of the hyoid arch appear to be subject to variability. At early stages Edgeworth (1923*c*) has shown that a tract of cells stretches forwards, inwards, and upwards from the dorsal end of the cerato-hyal. In this tract two chondrifications arise. The uppermost of these cartilages becomes attached to the floor of the auditory capsule immediately behind the prootic foramen, and constitutes the 'hyomandibula' of K. Fürbringer (1904) and Krawetz (1911), the 'otoquadrate cartilage' of Edgeworth (1923*c*) and the 'pharyngohyal' of Schmahlhausen (1923*a*), and appears to correspond to the otostapes of Tetrapoda. It is situated ventrally to the hyomandibular branch of the facial nerve, to the head vein, and to the orbital artery. Eventually it extends forwards to the hind edge of the basal process of the quadrate with which it fuses.

The lower and outer cartilage remains for a time connected with the upper by ligament; it seems to be the 'hyomandibula' of Huxley (1876*b*), van Wijhe (1882*b*), Ridewood (1894), Sewertzoff (1902), and Schmahlhausen (1923*a*) (who regards it as an epihyal), the 'symplectic' of Krawetz (1911), and the 'interhyal' of Edgeworth (1923*c*). It extends upwards towards the point of fusion between the otic process of the quadrate and the crista parotica, and is situated *posteriorly* to the hyomandibular branch of the facial nerve and laterally to the head vein. This relation to the hyomandibular nerve is important, since it shows that this 'epihyal' of Ceratodus differs from the laterohyal or the dorsal process of the columella auris of Tetrapods, for this typically lies *anteriorly* to the hyomandibular nerve. The structure in

Ceratodus may nevertheless be a laterohyal which has grown upwards in a position different from that in Tetrapoda. (See p. 413.)

However, Greil (1913), Schmahlhausen (1923*a*), and Goodrich (1930) have reported the presence of a 3rd cartilaginous nodule, lying beneath the laterohyal, termed by Schmahlhausen the 'interhyal'. This interhyal may represent the hyostapes of Tetrapoda.

(iv) For the proper appreciation of the nasal capsules, it is important to remember that in Ceratodus both external and internal nostrils are situated on the ventral side of the upper jaw.

A nasal septum is formed on the trabecular plate, and from its dorsal edge a pair of cartilages project backwards towards the orbital cartilages, with which they will ultimately fuse, forming the sphenoseptal commissure on each side. Meanwhile, the orbital cartilage develops its preoptic root enclosing the optic nerve foramen, and from its anterodorsal corner a process projects towards the nasal septum.

The trabecular horns and the lamina orbitonasalis at this stage end freely, the former lying *anterodorsally* to the external (anterior) nostril, the latter lying posteriorly to the internal nostril.

The lamina orbitonasalis apparently always ends freely in Ceratodus, but the trabecular horn eventually grows backwards, laterally to the external nostril, and joins on to the sphenoseptal commissure. Irregular chondrifications in this region give rise to a much-fenestrated roof to the nasal capsules, but there is no floor.

It will be noticed, therefore, that the nasal capsule is very incomplete. There is scarcely any partition between its cavity and that of the orbit since the lamina orbitonasalis is so feebly developed. In other words, the orbitonasal foramen is undemarcated. The foramen olfactorium evehens is very large (enclosed between the nasal septum, sphenoseptal commissure, preoptic root of the orbital cartilage, and trabecular plate).

The structure described by Huxley (1876*b*) as an upper labial cartilage, situated between the two nostrils, apparently corresponds to the subnasal cartilage of Lepidosiren and Protopterus, which in those forms fuses with the trabecular horn.

ii. The morphology of the chondrocranium of Ceratodus

In the ethmoid region the most remarkable feature is the situation of the external nostril behind and below the trabecular horn, instead of in front of and above that structure as in all other choanata. This is an important point of difference between Dipnoi and Amphibia.

In the orbital region the foramina in the orbital cartilage (optic, trochlear, oculomotor, ophthalmic artery, pituitary vein) are identical with those found in Selachii (except for the passage in Selachii of the efferent pseudobranchial artery through the ophthalmica magna artery's foramen to join the internal carotid), and the pila antotica, pila metoptica, and preoptic root of the orbital cartilage bear typical relations to the neighbouring blood-vessels and nerves. The orbitotemporal region, however, is different from that of any other (non-Dipnoan) fish owing to the autosystylic attachment of the quadrate by means of basal, otic, and ascending processes. The basal connexion is ventral to the head vein and dorsal and anterior to the palatine branch of the facial nerve. The otic connexion is lateral to the head vein and orbital artery, posterior to the trigeminal nerve (and ophthalmic and buccal branches of the facial nerves), and anterior to the hyomandibular branch of the facial nerve. The ascending connexion is lateral to a branch of the head vein, postero-lateral to the abducens nerve and the ophthalmicus profundus nerve (V_1), but antero-medial to the maxillary and mandibular branches of the trigeminal nerve (V_2 and V_3), and to the orbital artery. All these relations are typical and practically constant throughout Tetrapoda.

Because of the similarity in relations between the otic process of Ceratodus and the lateral wall of the trigeminofacialis chamber of Amia, the ascending process of Ceratodus and the pila lateralis of Amia, the basal process of Ceratodus and the floor of the trigeminofacialis chamber of Amia, Allis (1914*b*) and Schmahlhausen (1923*b*) concluded that the space contained between these processes in Ceratodus (the 'cavum epiptericum' of Gaupp (1902), or the 'antrum petrosum laterale' of Drüner (1901, 1904*a*), was homologous with the trigeminofacialis chamber of fish. It was pointed out by de Beer (1926*a*) that this conclusion was unacceptable since the processes of the quadrate of Ceratodus are visceral arch structures, whereas the walls of the trigeminofacialis chamber are part of the neurocranium. Furthermore, a remnant of the true otic process is present on the quadrate of Amia. More recently Allis (1929, 1930) has abandoned his view, and no longer regards the otic process of Ceratodus as the wall of a trigeminofacialis chamber.

The fusion between the basal process of the quadrate and the trabecula is so complete and precocious that it is not possible to determine whether a basitrabecular process is present or not.

The prootic foramen transmits the trigeminal, abducens, and facial nerves: there is no prefacial commissure. The fusions of the ascending and otic processes of the quadrate with the pila antotica and auditory capsule respectively takes place in such a way as to appear to subdivide the prootic foramen. But it is important to note that the 'foramen sphenoticum minus' (between the pila antotica and the ascending process) the 'sphenotic foramen' (or 'trigeminal foramen', between the ascending and otic processes) and the 'foramen prooticum basicraniale' (or 'facial foramen', between the otic process and the auditory capsule) are in no way subdivisions of the true prootic foramen (comprised between the pila antotica and the front wall of the auditory capsule: a perforation of the true cranial wall) but openings out of an extra-cranial space, the cavum epiptericum.

The hyoid arch provides a feature of great interest, viz. the forking of its dorsal end into a medial pharyngohyal (otoquadrate) and a lateral 'epihyal' element. It is important to notice that while the pharyngohyal corresponds in position to the articulation of the hyomandibula in Selachii (and the foot-plate of the columella auris of Tetrapods), the 'epihyal' does *not* correspond to the articulation of the hyomandibula in Teleostomi (or of the dorsal process of the columella auris of Tetrapods) because the hyomandibular branch of the facial nerve passes outwards *in front* of it.

The opercular fold is stiffened by a number of opercular cartilages, regarded by Huxley (1876*b*) as modified hyal rays. Some of the hyal rays situated on the edge of the ceratohyal are of interest since they tend to become joined together at their base; a process which is carried much further in Protopterus.

There are five ceratobranchials, and epibranchials are present in the first four arches. Three pharyngobranchials were described by van Wijhe (1882*b*), but not by Fürbringer. Owing to the fact that the epibranchials point backwards from the dorsal ends of the ceratobranchials, Allis (1915) prefers to regard them as pharyngobranchials (resembling those of Selachii, in which the branchial arches are of sigma-shape, with the pharyngobranchials pointing backwards). He assumes that the cerato- and epibranchial elements are not separate from one another.

A median basihyal element chondrifies between and in front of the hypohyals. Huxley (1876*b*) described two basibranchials, but van Wijhe (1882*b*) could only find one.

The chondrocranium of Ceratodus persists in the adult in its entirety (the exoccipital is the sole cartilage-bone) and is extraordinarily well developed. The occipital and occipitospinal arches are firmly incorporated with the auditory capsules in the hind region of the skull; the hypophysial fenestra is closed (except for the internal carotid foramen, which as in Selachii is in the midline) by the cartilage of the hypophysial plate; and the roof of the cranial cavity is completely enclosed in cartilage, from the nasal capsule to the tectum posterius which unites the dorsal ends of the occipital and occipitospinal arches. The notochord is largely persistent in the basal plate.

The sclerotic cartilage is well developed.

iii. The development of the osteocranium of Ceratodus

1. The membrane-bones. A number of bones arise beneath the denticles in the mouth, which eventually give rise to the characteristic Dipnoan tooth-plates, as shown by Semon (1901). These are the *prevomers* and *palatopterygoids* in the upper jaw, and in the lower jaw the huge *coronoids* (splenials of some authors) which lie on the medial side of Meckel's cartilages, and the very small *dentaries* which are situated above the anterior end of Meckel's cartilages, but seem to lose their teeth in the adult and apparently become attached to the (true) *splenials*.

Of the bones surrounding the brain-case some are deep beneath the surface, bear no close relations to lateral line canals, and are covered over by scales continuous with those of the trunk. Among these are the '*median anterior*' or *postrostral* (nasal of Fürbringer, frontal of Kesteven) and '*median posterior*' or *parietal* (frontoparietal of Fürbringer) which are genuine median ossifications (the latter is separated from the roof of the chondrocranium by a space in which the adductor mandibulae muscle is lodged); the '*lateral*' or *frontal* (supraorbital of Fürbringer, prefrontal of Kesteven); the so-called pterotic (squamosal of Fürbringer and Kesteven) which, not being a cartilage-bone should be called the *intertemporal*.

The lateral side of Meckel's cartilage is covered by the *supra-angular*; the opercular fold contains the *opercular* and *interopercular*, while the floor of the skull is protected by the large (toothless) *parasphenoid*.

The bones related to the lateral line canals lie along the infraorbital and the mandibular canals. These bones include the *postfrontal* and the *suborbitals* for the former, and a variable number of *mandibular canal ossicles* and the *splenial* (submandibular of Fürbringer; beneath the tip of Meckel's cartilage) for the latter. The dentaries appear to become attached to the splenials.

In addition to the above-mentioned bones, Kisselewa (1929) has figured a dorsal median and unpaired bone between the postrostral and parietal plates, and paired bones situated behind the frontals and between the intertemporals and the parietal plate: these are presumably *extrascapulars*; the nature of the median ossicle is obscure, as is the fate of these bones. There are no sclerotic bones.

2. The cartilage-bones. These will not occupy much time in description, for Ceratodus has fewer cartilage-bones in its skull than any other vertebrate in which bone occurs at all.

The *exoccipitals* are the only cartilage-bones in the brain-case. They ossify in the occipital and the 2nd occipitospinal arches; occasionally also in the 1st occipitospinal arch, as Krawetz (1911) showed.

The *ceratohyals* are the only cartilage-bones in the visceral arch skeleton.

Though not strictly part of the skull, it may be mentioned that the *cranial rib* becomes ossified.

II. Protopterus annectens and Lepidosiren paradoxa

The development of the chondrocranium of Protopterus has been studied by Winslow (1898), and by Agar (1906), who also investigated Lepidosiren and the development of the bony skull in both forms, which may be considered together owing to their great similarity. The morphology of the skull of Lepidosiren is the subject of a study by Bridge (1898).

i. Development of the chondrocranium in Protopterus and Lepidosiren

In Protopterus of (Graham Kerr's) stage 31 (Plate 66, figs. 1, 2) the trabeculae are present as short rods, wide apart, and continuous posteriorly with the anterior parachordals ('Balkenplatten') which are quite separate (and indeed some little distance) from the notochord. Rising up from the point of junction between trabecula and parachordal is the pila antotica (behind II, in front of V), and attached to the hind end of the trabecula by its basal process is the quadrate. A strand of dense tissue is continued forwards from the body of the quadrate, and this may represent the vestige of the pterygoid process. Meckel's cartilage and the ceratohyal are present.

In Lepidosiren of stage 34 (Plate 66, figs. 4, 5) the trabeculae are joined anteriorly to form an ethmoid plate and short rostrum between the nasal sacs. The auditory capsules are chondrified and firmly fused to the anterior parachordals, which are still quite free from the notochord. The quadrate is fused to the pila antotica by the processus ascendens and to the auditory capsule by the otic process, in addition, of course, to the fusion of the basal process with the hinder part of the trabecular region. All branches of the trigeminal and facial nerves emerge through the prootic incisure, but the palatine branch of the facial nerve is cut off from that incisure by the postpalatine commissure which encloses it in a small foramen of its own. The oculomotor nerve is enclosed between the pila antotica and pila metoptica of the orbital cartilage, but the optic nerve runs out freely above the trabecula. The glossopharyngeal nerve perforates the floor of the hind part of the auditory capsule.

Posteriorly, the occipital arch has appeared as an upgrowth from the occipital plate or posterior parachordal cartilage, which is quite separate from the anterior parachordal cartilage. Cells from the posterior parachordal cartilage invade the notochord sheath. The occipital arch in Lepidosiren marks the hind limit of the skull, and coincides with the septum between the 3rd and 4th myomeres. If, as in Ceratodus, two metotic somites disappear, the occipital arch in Lepidosiren arises in the septum between segments 8 and 9, and corresponds exactly to that of Ceratodus. In front of this arch in Lepidosiren are two ventral nerve-roots, and behind it there are also two, in front of the foremost neural arch. This indicates that a neural arch has disappeared, and as it corresponds exactly with the 1st occipitospinal arch of Ceratodus, it is very interesting to remember that that arch in Ceratodus may be reduced or absent. The foremost free neural arch of Lepidosiren therefore corresponds with the 2nd occipitospinal arch of Ceratodus, and the skull of Lepidosiren presumably contains two segments less than that of Ceratodus, viz. eight. A cranial rib is present corresponding to the occipital arch.

In Protopterus of stage 36 (Plate 66, fig. 3) the conditions are very similar though chondrification has advanced. Anteriorly the rostrum forms a nasal septum the dorsal edge of which is spreading outwards over the nasal sacs in the form of a roof. Further back, the lamina orbitonasalis projects from the edge of the trabecula as a slender process curving round the hind wall of the nasal sac. The prootic incisure is converted into a foramen by the development of the taenia marginalis connecting the pila antotica with the auditory capsule. The prootic foramen is subdivided by a strut of cartilage which separates the ophthalmic branch of the facial from the trigeminal nerve. This strut is, of course, not a prefacial commissure, but a secondary development.

The anterior and posterior parachordal cartilages have joined, and cells from the latter invade the notochordal sheath. Three ventral nerve-roots emerge in front of the occipital arch. Behind the occipital arch is an occipitospinal arch fused on to it and enclosing a ventral nerve-root. If the positions of the occipital arches correspond there is one segment more in the skull of Protopterus than in that of Lepidosiren and one segment less than in that of Ceratodus.

Ceratobranchials are present in the first four branchial arches. Subsequently, in Protopterus, a small cartilaginous bar arises in the hyoid arch, probably as Fürbringer (1904) showed as the result of fusion of the hyal rays.

In Lepidosiren at stage 38 (Plate 66, figs. 6, 7) the parachordals in their hinder region are continuous with one another above and below the notochord. Anteriorly, the trabecular horn grows out on each side of the rostrum and back laterally to the anterior nostril. It fuses with the subnasal cartilage which arose between the nostrils. A fenestrated roof is formed by struts of cartilage joining the trabecular horn with the already formed roofing cartilage of its side. Behind the trabecular horn is an isolated cartilage, termed by Agar the posterior upper labial cartilage. Two lower labial cartilages are present, the foremost fusing with Meckel's cartilage. Two opercular cartilages appear in the opercular fold. Projecting backwards from the nasal septum in the middle line is a rudimentary anterior tectal process. A strut of cartilage on the posterior face of the otic process of the quadrate encloses Pinkus's or Wright's organ.

The remaining changes to be made in the development of Protopterus are the formation of the tectum synoticum and the great backward extension of the anterior tectal process; neither of these developments occurs in Lepidosiren, where the tectum posterius is the only roof the chondrocranium possesses. In both forms the hypophysial fenestra remains unchondrified; the hyoid arch skeleton lacks the dorsal elements (pharyngohyal, epihyal, or 'hyomandibula') as well as hypo- and basihyals. Lepidosiren has five ceratobranchials, but as the 1st gill-slit is obliterated the 1st ceratobranchial is close behind the ceratohyal. Protopterus has five ceratobranchials, and a rod formed from the fused bases of hyal rays in front of the 1st gill-slit. No basibranchials are present.

ii. The development of the osteocranium in Protopterus and Lepidosiren

(PLATE 66.)

The development of the teeth and tooth-plates in Lepidosiren have been studied by Graham Kerr (1903) and Moy-Thomas (1934). It seems that the process is essentially similar to that which takes place in Ceratodus, and that the denticles become affixed to independently developed bony plates. These are the *prevomers*, *palatopterygoids* and *coronoids* (splenials of some authors).

The posterior part of the palatopterygoid grows back in close association with the ligamentous vestige of the pterygoid process of the quadrate.

The *parasphenoid* apparently arises from paired rudiments (Agar, 1906, p. 52).

The remaining bones are much as in Ceratodus, except that the *median parietal* is plastered down closely on the roof of the cartilaginous skull and extends far out on each side. It is covered over by the adductor mandibulae muscle.

On the other hand, the *frontals* are free posteriorly and are underlain by the lateral portions of the median parietal.

There are apparently no postfrontal, suborbitals, dentaries, or splenials (submandibulars).

The cartilage-bones are the *exoccipitals* (in Protopterus meeting dorsally beneath the tectum posterius), the *ceratohyals*, and the *cranial ribs*.

AMPHIBIA

13. URODELA

Caducibranchiata

I. AMBLYSTOMA PUNCTATUM and TIGRINUM (SIREDON)

After the early investigations of W. K. Parker (1877*a*), the skull of Amblystoma was the first object to which the Born wax-model method was applied in the study of the complete chondrocranium, at the hands of Stöhr (1879). Subsequently, other stages were similarly studied by Winslow (1898) and Gaupp (1906*a*) while Peeters (1910) investigated the development by the van

Wijhe method. The earliest appearance of the cartilage and its relations to the segmentation have been worked out by Goodrich (1911) while the columella auris has been studied by Kingsbury and Reed (1909). The development of the bony skull has not been studied systematically in Amblystoma, but the changes in the bony palate have been followed by Wintrebert (1910, 1922*b*).

i. Development of the chondrocranium of Amblystoma

The visceral arch skeleton is the first to chondrify, and at 7·25 mm. consists of quadrate, Meckel's cartilage, ceratohyal, and four ceratobranchials with a long median copula or basibranchial.

The first elements of the neurocranium to chondrify (at 7·5 mm.) are the trabeculae, each of which appears from the first to be continuous with the orbital cartilage (crista trabeculae) dorsal to it by means of the preoptic root, pila metoptica, and pila antotica, between which the foramina for the optic and oculomotor nerves are enclosed. Each trabecula is attached at its posterior end to the anterior parachordal (the 'Balkenplatte' of Stöhr), a triangular-shaped cartilage flanking the extreme anterior end of the notochord, its anterior edge being level with the notochord's tip. The internal carotid arteries enter the cranial cavity between the trabeculae. The trochlear nerve passes freely over the dorsal edge of the orbital cartilage.

The quadrate is an independent cartilage, with a processus ascendens projecting towards but not reaching the pila antotica. (Plate 67, fig. 1).

At a later stage (9 mm.), the anterior ends of the trabeculae curve outwards to form the trabecular horns which extend beneath the nasal sac on each side, behind the external and in front of the internal nostril. The processus ascendens of the quadrate is fused with the dorsal end of the pila antotica, and bears the typical relations to the trigeminal nerve-roots (behind the profundus nerve or V_1 and in front of the maxillary and mandibular branches, V_2 and V_3 and the orbital artery). The quadrate also bears an otic process which projects towards the auditory capsule. This latter has appeared as an independent cartilage on the outer surface of the auditory sac. (Figs. 2, 3).

The anterior parachordals join one another in front of the tip of the notochord and form a transverse crista sellaris which marks the hind border of the hypophysial fenestra. The abducens nerves pierce the anterior parachordals and thus emerge on the ventral surface of the skull and run forwards beneath it.

Posteriorly, the occipital arches are present, seated on the notochord without any basal plate, and exactly resembling the vertebral neural arches further back. Between the anterior parachordals, the auditory capsules, and the occipital arches there is thus an extensive unchondrified space which will ultimately be occupied by the so-called mesotic cartilage. (In another Amblystomatid form, Onychodactylus as described by Okutomi (1936), however, the occipital arch appears from the start to be connected with the basal plate.)

The basal plate is completed by a spreading of chondrification along the sides of the notochord between the anterior parachordals and the bases of the occipital arches. To the lateral edges of these (now complete) parachordals, the auditory capsules become attached, the attachment corresponding morphologically to the anterior basicapsular and the basivestibular commissures of other forms, for the hinder attachment is anterior to the glossopharyngeal nerve. (Plate 67, fig. 4).

The dorsal ends of the occipital arches fuse with the hind walls of the auditory capsules thus enclosing on each side the jugular foramen transmitting the glossopharyngeal and vagus nerves. Anteriorly, the prefacial commissure connects the anterior parachordal with the antero-medial wall of the auditory capsule and encloses the facial foramen between itself and the anterior basicapsular commissure. (Fig. 5).

The prefacial commissure may chondrify earlier than the anterior basicapsular commissure and has been mistaken for it, with results fatal to the interpretation of the relations of the facial nerve, as will be seen below.

The quadrate becomes fused to the auditory capsule by the otic process and (somewhat later) to the basitrabecular process of the basal plate by the basal process in addition to the fusion of the ascending process with the pila antotica. The larval suspension of the jaws is thus of typical autosystylic type. In the metamorphosed form, however, according to Wintrebert (1922*a*), the basal and otic connexions become discontinuous. In Amblystoma punctatum, which possesses a balancer, a small process is developed laterally from the quadrate to support its base. This disappears with the balancer at metamorphosis. Between the otic and basal attachments and the side wall of the auditory capsule is the cranioquadrate passage through which the hyomandibular branch of the facial nerve, the head vein, and the orbital artery pass. Very late, the quadrate develops a pterygoid process which projects forwards and ends freely. (Fig. 6).

The basitrabecular process is covered dorsally by the anterior part of the auditory capsule, and is fused to it. This state of affairs is due to the forward growth of the auditory capsule, of which the basitrabecular process looks like the anterior part of the floor.

The palatine branch of the facial nerve runs downwards behind the basitrabecular process and forwards beneath it in typical manner. Again, owing to the anterior growth of the auditory capsule, the facial nerve runs out of the cranial cavity forwards through a canal, of which the medial wall is the prefacial commissure and the lateral wall is the medial wall of the anterior cupola of the auditory capsule. The external aperture of this canal is divided into two by the postpalatine commissure, separating the palatine from the hyomandibular branches of the nerve. As just explained, the palatine nerve runs forwards beneath the basitrabecular process; and the hyomandibular nerve runs backwards medially to the otic process, through the cranioquadrate passage. These nerves look as if they were emerging from the cavity of the auditory capsule, but they are completely separated from it by the capsular wall. The condition thus presented by Amblystoma is primitive, and important as Gaupp (1893, p. 285) has pointed out, since the condition in other Urodela (where the capsular wall is not chondrified and the facial nerve *seems* to pass through the capsule) and in Anura can both be derived from it.

Goodrich (1911) has shown that the occipital arch lies in the septum separating the 2nd and 3rd myomeres, that is, the 3rd and 4th metotic somites. The occipital arch thus marks the posterior limit of the 6th segment. (Fig. 7).

In front of the occipital arch is a blastematous preoccipital arch connecting the parachordal with the auditory capsule across the jugular foramen, corresponding to the septum between the 1st and 2nd myomeres, and enclosing a foramen between itself and the occipital arch through which the transient ventral nerve-root of the 2nd myomere (6th segment) emerges. (Fig. 8).

The occipital condyles are added on to the hind end of the parachordals as a result of the formation of intervertebral (interdorsal) cartilages (formed according to Fortman (1918) and Mookerjee (1930) from the anterior half sclerotome of the 1st trunk segment). These subsequently split, the posterior portion becoming attached to the anterior face of the 1st vertebra (thereby enclosing the transient 1st spinal nerve, or nervus suboccipitalis, in a foramen in the vertebra); the anterior portion forming the condyles plastered on to the hind edge of the parachordals. Including the condyles, therefore, the number of segments in the Urodele chondrocranium is $6\frac{1}{4}$.

The roofs of the auditory capsules become connected by a tectum synoticum, and their medial walls become chondrified, leaving foramina for the anterior and posterior branches of the auditory nerve, and for the endolymphatic and the perilymphatic ducts. The foramen perilymphaticum faces backwards and inwards, and opens into the space between the borders of the jugular foramen. From the foramen perilymphaticum, therefore, there is a passage both into the cranial cavity and on to the outer surface of the auditory capsule.

The lateral wall of the auditory capsule is perforated by the fenestra ovalis. This is occluded by the columella auris which has a peculiar and complicated history, first elucidated by Kingsbury and Reed (1909).

At the 14–15 mm. stage, the anterior region of the fenestra ovalis becomes occluded by a cartilaginous plate of independent origin. This is the columella, and its outer surface bears a process or stylus which projects forwards and is connected by ligament with the squamosal. At early stages, according to Kingsbury and Reed (1909, p. 617), there is a train of cells connecting the columella with the ceratohyal. Subsequently the edge of the columella becomes fused with the border of the fenestra ovalis and the ligament from the stylus becomes transferred to the hind surface of the quadrate. The hyomandibular branch of the facial nerve runs downwards *in front of and beneath* this ligament.

At the time of metamorphosis, a second plate of cartilage, the operculum, becomes cut out of the wall of the auditory capsule immediately behind and below the now completely fused columella, and a muscle is stretched between the operculum and suprascapula. There is no connexion between the ceratohyal and either the columella or the operculum.

Inside the auditory capsule, the septa for the anterior and lateral semicircular canals are chondrified; so apparently also is the posterior septum at young stages, but later it disappears (Gaupp, 1906*a*, p. 694).

Turning now to the anterior region of the skull, the trabeculae become interconnected (and the hypophysial fenestra enclosed) by a trabecular plate, on which a transverse vertical plate (or planum precerebrale) arises, between the olfactory nerves. Each of the latter becomes enclosed in a foramen by the connexion of the preoptic root of the orbital cartilage with the planum precerebrale by means of the sphenoseptal commissure.

The lamina orbitonasalis (or planum antorbitale) grows out laterally from the trabecula as a process passing round behind the internal nostril, and spreads, forming the hind wall and side wall of the nasal capsule. The limit between the lamina orbitonasalis and the preoptic root of the orbital cartilage is indicated by the position of the orbitonasal foramina, through which the nasal branch of the profundus nerve (already divided into medial and lateral rami) leaves the orbit and enters the nasal capsule. In the side wall of the nasal capsule, a very slight and shallow indentation, lodging the external nasal gland, represents an exceedingly rudimentary concha nasalis.

An important feature is the fact that each nasal capsule is completely separate from the other and has its own medial wall, for there is no nasal septum, its place being taken by the cavum internasale. A connexion between the trabecular horn and the lamina orbitonasalis forms the floor of the capsule, or lamina transversalis anterior. It encloses the fenestra basalis or choanalis from in front, and a process projecting backwards from it laterally to the internal nostril forms the cartilago ectochoanalis. (Plate 68, figs. 3–5).

The lamina transversalis also forms the ventral border of the fenestra narina. The medial border of this fenestra is formed by the cartilage of the anterior cupola of the capsule, and the dorsal border is made by the cartilago obliqua which connects the anterior cupola with the dorsal portion of the side wall. From each anterior cupola there projects forwards the processus prenasalis lateralis inferior. Through the fenestra narina the external nostril and the lachrymonasal duct enter the nasal capsule.

In addition to the fenestra narina in its front wall and the fenestra basalis in its floor, the wall of the nasal capsule is perforated by a number of other foramina and vacuities, indicating incomplete chondrification. Thus there is a fenestra lateralis in its side, a fenestra dorsalis in its roof, and a foramen apicale in the anterior cupola (medially to the fenestra narina) through which the medial ramus of the nasal branch of the profundus nerve leaves the capsule. The lateral ramus of this nerve passes out of the capsule through the fenestra lateralis.

At late stages some further changes may be noted in the basal plate. These involve a fenestration giving rise to the basicranial fenestra. This is at first separated from the hypophysial fenestra by the transverse crista sellaris (remnant of the anterior edge of the basal plate), but eventually becomes confluent with it when the crista breaks down. The definitive chondrocranium thus has no cartilaginous floor.

In its anterior region, the notochord disappears; in the intermediate region it undergoes chondrification as a rod enclosed between the parachordals, while posteriorly the notochord is lifted above the basal plate by the junction of the parachordals ventrally to it in a hypochordal commissure. This hind portion of the cranial notochord becomes chondrified in connexion with the 1st vertebra, and gives rise to its large tuberculum interglenoidale ('odontoid process'). This forms a median articulation with the dorsal surface of the basal plate between the occipital condyles. The craniovertebral joint thus involves three articulations: the median just mentioned and those of the condyles. The tuberculum interglenoidale appears to contain the intravertebral cartilage corresponding to the occipital arch.

Amblystoma is, according to Stadtmüller (1929*a*), peculiar among Salamandridae in that the sclerotic cartilage which arises in the larva persists after metamorphosis. (Plate 63, fig. 9).

The fully formed hyobranchial skeleton of the larva comprises the ceratohyals, articulating at their anteroventral ends by means of the hypohyals with the median anterior copula, which appears to correspond to the fused basihyal and 1st basibranchial. Also articulating with the anterior copula are the 1st and 2nd hypobranchials, which in turn articulate with the 1st and 2nd ceratobranchials. The 3rd and 4th ceratobranchials have no hypobranchials corresponding to them, and they articulate with one another and with the 2nd ceratobranchial by their ventral ends. The dorsal ends of all the ceratobranchials are in contact (but not fused) by means of incomplete terminal commissures. Behind the anterior copula is a second one (the 2nd basibranchial), the posterior end of which (forming the 'urohyal'), is slightly forked. The dorsal end of the ceratohyal is connected with the quadrate by a ligament which may chondrify at its attachment to the quadrate and so give rise to the processus hyoideus.

The changes occurring at metamorphosis, relating to the pterygoquadrate are described above (p. 178): those relating to the hyobranchial skeleton, according to Drüner (1901, 1904*a*) and Bogoljubsky (1926) are as follows: The 2nd, 3rd, and 4th ceratobranchials disappear leaving the 1st ceratobranchial articulated with the copula ('corpus of the hyoid') by the 1st and 2nd hypobranchials. The hypohyals retain contact with the ceratohyals and with the copula, the anterior end of which is also surrounded by three other cartilages, two lateral ('radii posteriores') and one dorsal. These elements, the morphological value of which is still obscure, are the separate constituents of the single 'arcuate' cartilage found in Triton (p. 183). The hindermost part of the urohyal persists, isolated from the copula by the break-down of the intervening cartilage.

ii. The osteocranium of Amblystoma

No systematic study has been made of the development of the bony skull of Amblystoma, which can, however, hardly differ much from that of Triton (p. 183). An observation of importance has been made by Bruner (1902), which is that the septomaxilla (intranasale) is present as well as the prefrontal. Both these bones are pierced by the lachrymonasal duct. Their presence together in Amblystoma is a proof that they are not to be regarded as equivalent in those forms in which only one is present.

Further, in another member of the Amblystomatidae, Onychodactylus japonicus, the prefrontal, lachrymal, nasal, and septomaxilla, are all present, according to Chung (1931), Suzuki (1932), and Okutomi (1936), though apparently the lachrymal may become attached to the prefrontal.

According to Lapage (1928*a*) the septomaxilla is continuous at its hind end with the cartilage which forms the posteroventral border to the fenestra narina. From this Lapage concludes that the septomaxilla is a cartilage-bone: a view which cannot be accepted until the first ossification of the bone in cartilage has been demonstrated, for the mere continuity of the bone with cartilage in the adult is no evidence of its perichondral or endochondral origin.

A feature of some interest is the occurrence in certain species of Amblystoma (gracile, macrodactylum) of kinetism (see p. 426). As described by Eaton (1933) the quadrate, pterygoid, and squamosal pivot as a unit on the prootic. The amount of forward movement is limited by the ligament attaching the columella to the quadrate. A quadratojugal is present (de Villiers, 1936).

Another feature of interest in Amblystoma is the presence (described by Carrière, 1884) in the larva of a horny 'jaw' in front of the dentary. (Cf. frog).

The relations of the nerves to the Urodele chondrocranium are shown in Plate 69.

II. Triton (Molge) vulgaris, taeniatus, and cristatus

The development of the chondrocranium of the newt, as revealed by the works of Stöhr (1879), Gaupp (1906*a*), Tarapani (1909), and Peeters (1910), closely resembles that of Amblystoma which may be taken as a model. Here, attention will be confined to those features in which Triton differs from Amblystoma.

i. Development of the chondrocranium of Triton

The trabeculae and anterior parachordals are laid down (at 8·5 mm., T. cristatus) as in Amblystoma, but the anterior parachordals lie behind the tip of the notochord, which projects forwards beyond them. Eventually the tip of the notochord itself chondrifies (cf. chordal cartilage in Urodela in the region of the vertebral column).

The orbital cartilage chondrifies separately from the trabeculae and soon fuses with it. The mesotic cartilage also has an independent centre of chondrification. The auditory capsule chondrifies according to Peeters from three centres: one anterior, one lateral, and one posterior connected with the base of the occipital arch.

At the 20 mm. stage (T. taeniatus) the chondrocranium is well formed. Anteriorly, the precerebral plate of Amblystoma is represented by a pair of pilae ethmoidales with a fenestra precerebralis between them. Just as in Amblystoma, the two nasal capsules are separated from one another by a cavum internasale, and each has a complete medial wall: there is no nasal septum. The morphology of the nasal capsules in Triton is identical with that of Amblystoma; they are, however, relatively longer and narrower. (Plate 68).

The pterygoid process of the quadrate chondrifies very late, and ends freely in front without reaching the lamina orbitonasalis. The ascending otic and basal processes are fused on to the neurocranium as in Amblystoma. Stöhr (1879) observed that at metamorphosis a discontinuity appeared between the otic process and the auditory capsule. According to him, the fusion of the basal process with the basitrabecular process is permanent. In view of the reported events in Amblystoma and Salamandra, further investigation of this matter is desirable.

In the auditory capsule, one small feature in its development in Triton produces an important *apparent* effect. The structure of the anterior cupola, basitrabecular process, prefacial commissure and basicapsular commissure is as in Amblystoma, only that part of the medial wall of the capsule which is situated immediately to the side of the prefacial commissure does not chondrify: the result is that the facial nerve *appears* to traverse the cavity of the auditory capsule, from which it is, however, quite typically separated by the membranous capsular wall. There is no septum semicirculare posterius in Triton (Gaupp, 1906*a*, p. 693). The pterygoid process of the quadrate cartilage only arises very late. According to Gaupp (1906*a*, p. 704) the distal ends of Meckel's cartilage become inter-connected by a median cartilage, or basimandibular.

Following Kingsbury and Reed's (1909) observations, the appearance of the columella auris in the anterior part of the fenestra ovalis and its fusion with the edge of the fenestra, take place much as in Amblystoma, but there is no sign of a stylus or ligament connecting it with squamosal or quadrate. The operculum, on the other hand, appears to chondrify in the membrane of the hinder part of the fenestra ovalis, connected with its border, and to grow forwards. After metamorphosis the columella is completely merged in the wall of the capsule, and the operculum forms a cartilaginous disc in the mouth of the fenestra ovalis, and connected with the suprascapula by the opercular muscle. There is no connexion between ceratohyal and either columella or operculum.

But the ligament connecting the ceratohyal with the quadrate may contain an independent chondrification, as Litzelmann (1923) showed. In other Urodeles this cartilage may fuse with the hind surface of the quadrate, forming the processus hyoideus.

According to Kallius (1901), Drüner (1901, 1904*a*), Tarapani (1909), and Bogoljubsky (1926) the hyobranchial skeleton resembles that of Amblystoma. However, the first hypobranchials are fused to the copula, and the dorsal ends of the ceratobranchials are fused. At metamorphosis the posterior copula ('urohyal') and the hypohyals disappear, and the three elements of the 'arcuate' cartilage (i.e. the 'radii posteriores' and the transverse cartilage) fuse into one transverse structure, almost surrounding the anterior end of the copula. According to Drüner (1904*a*, p. 365) the constituents of the arcuate cartilage arise as outgrowths of the copula. (Plate 68, figs. 6–8).

The sclerotic cartilage which is present in the larva, disappears at metamorphosis in Triton. In neoteinic (adult) specimens of Triton alpestris, Stadtmüller (1929*a*) observed that the sclerotic cartilage persists. On the other hand, in metamorphosed yet water-living specimens of T. alpestris Reiseri, the sclerotic cartilage vanishes. Its presence is therefore not associated with aquatic life; nor is its absence a result of metamorphic agents since it persists in metamorphosed Amblystoma.

ii. Development of the osteocranium of Triton

The order of appearance of the bones of the skull in Triton vulgaris has been studied by Erdmann (1933) and found to be as tabulated on p. 184.

It is to be noted that all these bones arise before metamorphosis, and the only one to arise in later life (where and when it does) is the basioccipital discovered in adult Triton alpestris by Stadtmüller (1929*b*). It will also be observed on comparing this table with that referring to the frog (p. 209) that there is no correspondence whatever in the relative times at which homologous bones arise.

Bone	Total length of larva	Head+ trunk length
	mm.	*circa* mm.
Dentary	8·5	5
Coronoid	8·5	5
Prevomer	8·5	5
Palatine	8·5	5
Pterygoid	8·5	5
Squamosal	8·5	5
Premaxilla	9·3	..
Prearticular	9·3	..
Frontal	10	5·3
Parietal	12·7	6·9
Exoccipital	17·1	8·9
'Hyoid' copula	19·5	9
Orbitosphenoid	19·3	9
Prootic	23·9	11·6
Quadrate	31·2	14·5
Mentomeckelian	..	14·5
Nasal	30	14·8
Parasphenoid	30·5	15·7
Maxilla	..	16·9
Prefrontal	32·2	16·7
'Prenasal'	34	17·1

1. The membrane-bones. The earliest bones to arise are those which bear teeth (*dentary*, *coronoid*, *prevomer*, *palatine*, and *premaxilla*), and as in these forms the teeth precede the bones in appearance, these bones furnished the material on which Hertwig (1876) based his theory that dermal bones arise from concresced bases of teeth. Erdmann (1923) accepts Hertwig's view. However, with regard to what is known concerning the developmental and phylogenetic relations between bones and teeth in other forms, it is probable that here fusion takes place, not between the bases of the teeth themselves, but between small pediments of true bone developed under the teeth and supporting them.

The *dentary* carries teeth at its anterior end which engage with those of the premaxilla. Eventually the dentary fuses with the mentomeckelian, and is connected with its fellow of the opposite side by the basimandibular cartilage.

The *coronoid* ('splenial' of some authors) on the medial side of Meckel's cartilage bears a row of teeth which engage with those of the prevomer and palatine. After metamorphosis the coronoid vanishes completely.

The *prevomer* lies laterally to the palatine process of the premaxilla. The teeth which it bears during larval life disappear at metamorphosis, and the bone itself enlarges so as to come into contact with the premaxilla and maxilla. There is uncertainty as to the relations of the prevomer with the palatine (see below).

The *palatine* in the larva is an elongated tooth-bearing plate, attached at its hind end to the (toothless) *pterygoid* which stretches back beneath the pterygoid process to the articular region of the quadrate. (The pterygoid bone develops earlier than the cartilaginous pterygoid process.) At metamorphosis, according to previously held opinion, the attachment between palatine and pterygoid breaks down and the palatine shifts, joining the prevomer (forming the 'prevomeropalatine') and gives rise to the narrow strip of tooth-bearing bone which runs back along the lateral edge of the parasphenoid. According to Erdmann (1933), this strip is the result of fresh concrescence of bony plates underlying tooth bases in this region. Wintrebert (1910, 1922*b*), however, has shown as a result of his very careful investigations that the palatine disappears altogether at metamorphosis and is replaced by growth of the prevomer. This agrees with Aoyama's (1930) account of Cryptobranchus and Okutomi's (1936) of Onychodactylus. There is, therefore, no 'prevomeropalatine' at all. (Plate 71).

The *premaxilla* begins ossifying in its (facial) ascending portion, the (tooth-bearing) alveolar and (toothless) palatine portions arising subsequently. The two premaxillae fuse in the midline.

The *maxilla* arises very late in spite of the fact that it bears teeth. The first part to ossify is the facial portion from which ossification spreads to the alveolar and palatine portions. It may be noticed with regard to both premaxilla and maxilla that the first parts of the bones to ossify bear no relation to teeth at all.

The *squamosal* (paraquadrate of Gaupp) arises as a more or less vertical strut laterally to the quadrate; later, its dorsal end grows into the plate flanking the auditory capsule.

The *prearticular* (goniale) arises on the inner side of the hind end of Meckel's cartilage, and reaches forwards to the coronoid. The chorda tympani runs between it and Meckel's cartilage.

The *frontal* and *parietal* arise as longitudinal strips of bone above the orbital cartilage. From these centres ossification spreads downwards on to the medial wall of the orbit and medially in the roof of the cranial cavity. The parietal is perforated by the foramen for the trochlear nerve (Gaupp, 1911*c*).

The *nasal* ossifies relatively late, in front of the frontal and laterally to the long ascending process of the premaxilla.

The *parasphenoid* is surprisingly late in appearance, considering that it forms the sole skeletal floor to the skull after disappearance of the basal plate. In many cases its lateral margins appear to bear teeth, but as Wiedersheim (1877) showed these are really borne by a separate ossification (palatine or prevomer? see above) attached to the parasphenoid.

The *prefontal* ossifies on the posterodorsal surface of the lamina orbitonasalis. Between the prefrontal and the maxilla the maxillary branch of the trigeminal nerve and the lachrymonasal duct run forward. Sometimes the lachrymonasal duct pierces the lateral portion of the prefrontal.

The *prenasal* is a small ossification described by Erdmann (1933) as arising in front of the nasal and behind the premaxilla. Its homologies are uncertain, but it may represent the element which Wiedersheim (1877) has called the anterior prefrontal. On the other hand, it will be noticed that no lachrymal has been mentioned. Perhaps the bones here referred to as prefrontal, nasal, and prenasal really represent the lachrymal, prefrontal, and nasal respectively (see above, Onychodactylus, p. 181). The septomaxilla is absent in Triton though present in Amblystoma and Onychodactylus.

The *quadratojugal* ('quadratomaxilla') is described by Gaupp (1906*a*) as a small splinter of bone between the squamosal and the quadrate, failing to reach the maxilla. Its origin as a separate ossification has also been observed in Onychodactylus by Okutomi (1936).

2. The cartilage-bones. The *exoccipital* starts ossifying as a perichondral lamella in the occipital arch and spreads over the posterior half of the auditory capsule, the hind part of the basal plate and the tectum synoticum. It thus covers the region occupied in other forms by the opisthotic and epiotic, and comes into contact with the prootic.

The *prootic* starts ossifying round the facial foramen, and covers the anterior portion of the auditory capsule. It also forms a bony lateral wall to the facial canal, thus shutting it off from the cavity of the auditory capsule.

By means of the exoccipital and the prootic, the auditory capsule is almost completely ossified.

The *orbitosphenoid* starts ossifying in the anterodorsal region of the orbital cartilage and spreads downwards and backwards as far as the oculomotor foramen.

The pila antotica does not ossify, i.e. there is no pleurosphenoid.

The *quadrate* ossifies close above the articulation with Meckel's cartilage, and spreads up the otic process. The connexion of the quadrate with the neurocranium by means of the ascending, basal and otic processes remains cartilaginous. The articular portion of Meckel's cartilage also fails to ossify, i.e. the articular is absent (see Gaupp, 1912). This must be due to loss, since the Stegocephalia had an articular bone (Broom, 1913). On the other hand, the anterior end of Meckel's cartilage ossifies as the *mentomeckelian*, which eventually fuses with the dentary.

The *basioccipital*, according to Stadtmüller (1929*b*), is a small perichondral nodule of bone situated dorsally to the hind end of the parasphenoid, between the bases of the exoccipitals and attached to the dorsal surface of the hypochordal commissure uniting the hind ends of the parachordal cartilages. In other Urodeles this basioccipital may fuse with the parasphenoid.

In the hyobranchial skeleton, partial ossification takes place in the anterior *copula* ('hyoid' body), the *ceratohyal*, the *1st hypobranchial*, and *1st ceratobranchial*. The isolated part of the posterior copula, or 'urohyal', which is said to disappear in Triton and which may in other Urodela ossify into the so-called *os triquetrum* (or os thyroideum, triangulare) has been observed ossified in Triton taeniatus by Bogoljubsky (1926).

The columella auris apparently fails to ossify in Triton, though in Amblystoma there are inner and outer perichondral lamellae, and the stylus is also ossified. The operculum remains cartilaginous.

III. Salamandra maculosa

Following the early work of Parker (1882*c*), the development of both the cartilaginous and bony skull of Salamandra has been studied by Stadtmüller (1924), while some features of metamorphosis have been carefully followed by Wintrebert (1922*a*). The general course of events is so similar to what happens in Amblystoma as to obviate the necessity for detailed description. Attention will be confined to the points in which Salamandra differs from Amblystoma.

i. The development of the chondrocranium of Salamandra.

During the larval period the nasal capsules resemble in their morphology those of Amblystoma or Triton; they are separate, and each has a medial wall. In later development, however, a true nasal septum is formed, occupying the space of the cavum internasale, and projecting forwards as a median prenasal process. This is important, since the absence of such a septum is the most important difference between such Urodeles as Amblystoma or Triton and the Anura and Gymnophiona. (Plate 70, fig. 5).

The relations of the facial nerve to the auditory capsule are at first as in Triton; i.e. the lateral wall of the facial canal (medial wall of the auditory capsule) is incomplete so that the nerve appears to traverse the cavity of the capsule. Later, however, this portion of the wall chondrifies, and a condition resembling that of Amblystoma is reached, with a completely enclosed facial canal.

Of the three semicircular septa, the posterior is least well developed.

As regards the fenestra ovalis, according to Kingsbury and Reed (1909), there is a columella, attached to the anterior border of the fenestra ovalis, and bearing a stylus which is in cartilaginous continuity with the posterior surface of the quadrate. The hyomandibular branch of the facial nerve passes out *ventrally* to the stylus as in Amblystoma.

The operculum, according to Fuchs (1907*b*), arises partly from chondrification in the membrane of the hinder part of the fenestra ovalis and partly by cutting out of the wall of the capsule in this region. It eventually forms a free plate, filling the fenestra and overlapped externally at its front end by the columella. It is attached by the opercular muscle to the pectoral girdle. There is no connexion between either columella or operculum and the ceratohyal.

The quadrate has ascending otic and basal processes fused with the neurocranium as in Amblystoma and Triton.

The pterygoid process of the quadrate is represented by an isolated rod of cartilage which chondrifies late. The changes which occur in the pterygoquadrate cartilage at metamorphosis are of interest.

According to Stadtmüller (1924), there is partial or complete discontinuity between the otic process and the auditory capsule. A discontinuity appears in the basal connexion in animals 43 to 82 mm. long. Later, however (139, 145 mm. long), continuity is re-established between the basal process and the neurocranium. During the discontinuity of the basal connexion the body of the quadrate rotates slightly, into a vertical position, instead of sloping forwards and downwards.

According to Kallius (1901), Drüner (1901, 1904), and Tarapani (1909), the hyobranchial skeleton resembles that of Amblystoma. The posterior copula ('urohyal') is forked at its hind end. The 3rd hypobranchials may chondrify. Between the hypohyals and the 1st hypobranchials is an extra pair of cartilages, the radii posteriores, of obscure significance, and apparently corresponding to the lateral portions of the arcuate cartilage of Triton (Bogoljubsky, 1924).

The hypohyals are retained after metamorphosis but lose connexion with the ceratohyals; the 1st hypobranchials fuse with the 1st ceratobranchials. In the related form Spelerpes fuscus, the 1st ceratobranchials are greatly elongated in connexion with the protrusibility of the tongue (L. Smith, 1920).

The slender sclerotic cartilage of the larva disappears at the time of metamorphosis.

ii. The osteocranium of Salamandra

The bony skull of Salamandra resembles that of Triton. There is no prenasal, and the septomaxilla reported by Parker to occur in Salamandra, has not been found by either Bruner (1902), Stadtmüller (1924), or Lapage (1928*a*). The *os triquetrum* (triangulare, or thyroideum) is the only portion of the hyobranchial skeleton which undergoes ossification (in the isolated hind end of the 'urohyal') in Salamandra maculosa according to Francis (1934), but Bogoljubsky (1924, 1926) has observed ossification of the copula in Salamandra caucasica.

IV. Cryptobranchus (Megalobatrachus) japonicus, and C. (Menopoma) alleghaniensis

The early development of the chondrocranium of Cryptobranchus, has been studied with wax models by Aoyama (1930), and the morphology of the definitive chondrocranium as described by Edgeworth (1923*a*) presents certain features of importance. The development of the bony skull has been described by Parker (1882*d*) and Aoyama (1930).

i. The chondrocranium of Cryptobranchus japonicus

In the posterior region of the skull a hypoglossal foramen is present, through which a persistent ventral nerve-root emerges. This does not mean that Cryptobranchus has more segments in its skull than other Urodela. On comparison with early stages of Amblystoma (p. 179) it will be apparent that Cryptobranchus preserves the preoccipital arch, between which and the occipital arch the hypoglossal foramen is enclosed. (Plate 72, figs. 1, 2).

The formation of the occipital condyles and of the facets on the atlas vertebra to receive them, as a result of splitting of an intervertebral cartilage arising in the anterior half-sclerotome of the 1st trunk somite, has been demonstrated by Fortman (1918).

The basal plate had no basicranial fenestra, and much of the chondrocranium persists in the adult.

According to Kingsbury and Reed (1909) the fenestra ovalis of the auditory capsule is occupied by a free columella bearing a stylus which is attached by ligament to the squamosal, and also to the ceratohyal. There is *no* operculum. The hyomandibular branch of the facial nerve runs out *ventrally* to the stylus. The ligament connecting the ceratohyal with the quadrate chondrifies to form the large processus hyoideus of the quadrate.

The quadrate is further of interest in that the pterygoid process is fully chondrified and continuous with the lamina orbitonasalis. There is thus an ethmoid attachment, and the quadrate with its ascending, basal, otic, and pterygoid processes, is fused with the neurocranium at four points. Hynobius is similar in this respect, which exemplifies the autosystylic type of relation between jaw and skull in its most complete form. (Plate 72, fig. 3).

The hyobranchial skeleton has been investigated by Drüner (1904*a*, p. 594), and Edgeworth (1923*b*). In the larval state of Cryptobranchus japonicus there are hypo- and ceratohyals, hypo- and ceratobranchials of the first two branchial arches, ceratobranchials in the 3rd and 4th arches and a basibranchial connected with the hypohyals and both pairs of hypobranchials. From the basibranchial, according to Edgeworth (and the same is reported of Hynobius by Bogoljubsky (1925)), an outgrowth (the 'urohyal', or 'urobranchial') is formed projecting ventrally and backwards; its distal end remains as an isolated cartilage after the main part of the outgrowth has disappeared. (According to Aoyama (1930) the urobranchial is an independent chondrification from the start.) The adult condition is obtained as a result of the disappearance of the 3rd and 4th ceratobranchials and the ossification of the 2nd hypobranchials and ceratobranchials. There appears to be considerable variation in the fusion of certain cartilages and in the separation of small cartilaginous nodules. (Plate 72, figs. 4, 5).

In Cryptobranchus (Menopoma) alleghaniensis, as studied by Drüner (1904*a*, p. 443) the larval hyobranchial skeleton is as in C. japonicus except that the 3rd hypobranchials are also present. In the adult, however, much more of the skeleton persists, including all four ceratobranchials, and the 2nd hypobranchials. (Plate 72, fig. 6).

The sclerotic cartilage is extraordinarily thick and persists in the adult (Yano, 1926).

ii. The osteocranium of Cryptobranchus japonicus

The cartilage-bones of Cryptobranchus are the *exoccipital*, *prootic*, *orbitosphenoid* (paired, ossifying in the preoptic root of the orbital cartilage), *stapes* (in the columella auris), *quadrate*, *articular*, *mentomeckelian*, *2nd hypobranchial*, *2nd ceratobranchial*. Most of these are found in the 76 mm. stage.

Of the membrane-bones, the *parietal*, *frontal*, *parasphenoid*, *nasal*, *prefrontal*, and *angular* call for no special mention. (The only other Urodeles in which the angular is present are Onychodactylus, Ranodon, and Hynobius. (Stadtmüller, 1937.))

The *premaxilla* ossifies its prenasal portion at the 28 mm. stage; at 32 mm. the palatine process is present, but teeth do not fuse to it until the 43 mm. stage.

The *maxilla* starts ossifying in its ascending or facial portion at the 34 mm. stage: later the palatine process is formed and later still the teeth become fused with it.

The *dentary* is present at the 28 mm. stage; fusion with the teeth follows at the 35 mm. stage.

The *opercular* ('splenial' autt.) begins to ossify at 35 mm.; later, teeth become attached to its anterior end. This bone persists in Cryptobranchus.

The *palatine* and *pterygoid* ossify together at about the 34 mm. stage; the palatine portion subsequently having teeth attached to it. At the 36 mm. stage, the palatine portion is completely degenerated; on the other hand, the pterygoid is greatly enlarged: it extends posteriorly to the quadrate, medially as far as the parasphenoid, and anterolaterally nearly as far as the maxilla.

The *prevomer* ossifies at the 31 mm. stage, and teeth later become attached to it. At no time does it touch the palatine, which, as just stated, disappears.

The *squamosal* ('paraquadrate', Gaupp; 'tympanic', Wiedersheim) is ossified at the 28 mm. stage laterally to the otic process of the quadrate and the auditory capsule.

The *quadratojugal* ('quadratomaxilla', Gaupp; 'tympanic', Osawa, 1902), is ossified at the 43 mm. stage, and eventually fuses with the quadrate.

In Cryptobranchus alleghaniensis the hyobranchial skeleton includes ossifications in the *ceratohyal*, *2nd hypobranchial*, *2nd* and *3rd ceratobranchials*, and the '*urohyal*'.

V. Amphiuma means

One of the chief features of interest in the chondrocranium of Amphiuma as described by Hay (1890) and Winslow (1898) is the incomplete chondrification of the parachordals and basal plate. At an early stage this presents a pair of extensive fenestrae on either side of the notochord (though not reaching to it), only separated from the hypophysial fenestra by a slender crista sellaris. When this breaks down and the notochord and the cartilage immediately flanking it disappears, the hypophysial and basicranial fenestrae are confluent, and the floor of the skull is membranous as in Gymnophiona. (Plate 72, fig. 7).

The relations of the columella auris, according to Kingsbury and Reed (1909) are similar to Cryptobranchus. The stylus is large and is connected with the ceratohyal by a ligament, the processus hyoideus of the quadrate is large; no operculum is formed, though a commencement is seen of that absorption of cartilage which, if completed, would result in the cutting out of the operculum from the wall of the auditory capsule.

The hyobranchial skeleton comprises ceratohyal, hypohyal, copula, and four ceratobranchials. (Plate 72, figs. 8, 9).

The sclerotic cartilage according to Stadtmüller (1929*a*) is lacking in the adult.

The most remarkable feature in the bony skull of Amphiuma is the appearance of a membrane-bone on the medial surface of the ceratohyal, as described by Hay (1890) and confirmed by Gaupp (1906*a*), who called it the *parahyal*.

Perennibranchiata

VI. Necturus maculatus

The chondrocranium of Necturus has been studied in its development by Platt (1897), Winslow (1898), Wilder (1903), Terry (1904), and Edgeworth (1925) by means of wax models, and by Peeters (1910) with the van Wijhe method. Its morphology is the subject of a classical paper by Huxley (1874). The origin of the visceral cartilages has been a matter of controversy, Platt (1897) claiming their 'mesectodermal' provenance, which is denied by Buchs (1902). The problem is fully discussed below (p. 472).

i. Development of the chondrocranium of Necturus

Necturus differs from other Urodeles in that the anterior parachordals or 'Balkenplatten' flanking the tip of the notochord are not chondrified. Instead, the parachordals arise as bars of cartilage wide apart from one another and from the notochord except at their extreme hind end where they are in contact with it. Anteriorly, the parachordals are in contact with the trabeculae. The hypophysial and basicranial fenestrae are confluent, and the floor of the skull is almost membranous as in Gymnophiona. (Plate 72, fig. 10).

The orbital cartilage has an independent origin but soon fuses with the trabecula by means of its preoptic root, pila metoptica, and the pila antotica.

The cartilage uniting the trabeculae anteriorly (trabecular plate) likewise has an independent centre of chondrification, as have the medial walls of the nasal capsules. However, the independent chondrification of the tectum posterius (Platt's 'tectum interoccipitale') has not been confirmed by Winslow (1898), nor by Peeters (1910).

The quadrate has typical ascending, basal and otic processes fused with the neurocranium, but no pterygoid process is developed. The quadrate maintains throughout life the position of the quadrate in Caducibranchiata before metamorphosis: i.e. it slopes forwards and downwards to its articulation with Meckel's cartilage.

The relations of the auditory capsule to the facial nerve are clearer than in other Urodeles, for in Necturus the anterior basicapsular commissure (behind the facial nerve) is formed before the prefacial commissure. As Terry (1919) showed, the extracapsular course of the nerve is thus clear, although it traverses a canal, the external aperture of which is divided into two (for the palatine and hyomandibular branches) by contact (representing the post-palatine commissure) of the basitrabecular process with the auditory capsule between the nerve-branches.

According to Kingsbury and Reed (1909) the fenestra ovalis becomes occluded by a columella auris of independent origin, bearing a stylus connected by ligament first with the squamosal and later also with the quadrate. The external and internal mandibular rami of the hyomandibular branch of the facial nerve pass out *ventrally* to this ligament, but the jugular branch of the nerve runs *dorsally* to it. In this respect, Necturus resembles the other Perennibranchiates Typhlomolge and Proteus, but differs from the Caducibranchiate Urodeles. No operculum is developed.

The preoccipital arch is present in a blastematous condition in the posterior region of the skull, with the same relations (according to Platt, 1897) as in Amblystoma (p. 178), i.e. it lies in the septum between the 2nd and 3rd metotic somites. The occipital arch corresponds to the septum between the 3rd and 4th metotic somites, and thus marks the position of the posterior limit of the 6th segment of the whole series.

In the hyobranchial skeleton the ceratohyals are connected by means of the hypohyals with the anterior end of the 1st basibranchial. With the posterior end of this the 1st hypobranchials and the 2nd basibranchial are in contact. There are three ceratobranchials, on to the medial side of the 1st of which the 2nd hypobranchial is attached. The larval condition persists, there being no metamorphosis in these forms.

A well-developed sclerotic cartilage persists throughout life (Yano, 1926).

ii. Development of the osteocranium of Necturus

The posterior part of the auditory capsule is ossified, not by the exoccipital, but by a distinct bone which Huxley (1874) regarded as the *epiotic*, but which Wilder (1903) prefers to call the *opisthotic*.

In general the bones of Necturus resemble those of Triton (p. 184), but the *nasal*, the *maxilla*, and the *prefrontal* are absent, and it is interesting to note that in Triton these bones are practically the last to arise. The neoteny of Necturus has resulted in the ossification of these bones being retarded out of the life-history altogether.

VII. Other Urodela

In addition to the forms described above, several other genera have been investigated by W. K. Parker, including Proteus, Seironota (1877*a*), Notophthalmus, Cynops, Spelerpes, Desmognathus, Onychodactylus, Taricha, Salmonea (1882*c*) and Siren (1882*d*). While the value of his observations on the chondrocranium is limited by the methods of investigation at his disposal, these works may be consulted with the greatest profit for information concerning the adult bony skull.

Luther (1914) and Edgeworth (1925) have established the interesting fact that in Plethodon, Hynobius, Salamandrella, and Ranodon, the cartilaginous continuity between the basal process of the quadrate and the neurocranium becomes replaced by a discontinuity and the formation of a permanent joint. To this condition the term 'semistreptostylic' (Gaupp, 1902, p. 217) has been given, although strictly the difference between monimostyly and streptostyly refers solely to the bony skull. It is preferable, therefore, to refer to the above-mentioned condition (which is also found in Amblystoma, Salamandra, &c.) simply by saying that the autosystylic suspension of these Urodela becomes incomplete. In Salamandra (see above, p. 187) if Stadtmüller's account is correct, the autosystylic suspension is temporarily incomplete.

Diemictylus pyrrhogaster is the only Urodele to possess sclerotic bones Tsusaki), 1925; Yano, 1926).

14. GYMNOPHIONA

I. Ichthyophis glutinosus

The development of the chondrocranium of Ichthyophis has been studied with the help of wax models by Peter (1898), Winslow (1898), and Edgeworth (1925). The hyobranchial skeleton has been investigated by the Sarasins (1890). Peter and the Sarasins have also followed the development of the bony skull.

i. Development of the chondrocranium of Ichthyophis

The embryonic and larval stages studied by Peter, Winslow, and Edgeworth are remarkable for their similarity; a single description of the structure of the chondrocranium will therefore suffice. (Plate 73, figs. 1, 2).

The parachordals are reduced to narrow bars of cartilage, continuous in front with the trabeculae, wide apart and distant from the notochord except in their hindmost region where they form a transverse bar fused with one another ventrally to the notochord. The hypophysial and basicranial fenestrae are enormous and confluent. At early stages the notochord projects freely forward into the basicranial fenestra, but disappears before hatching.

From the parachordal there arise (in order from behind forwards) the occipital arch, basivestibular and anterior basicapsular commissures, and prefacial commissure. All these structures are attached at their dorsal or lateral ends to the auditory capsule, thus enclosing a jugular foramen, basicapsular fenestra, and facial foramen, respectively. Accompanying the glossopharyngeal and vagus through the jugular foramen is a hypoglossal ventral nerve-root.

The auditory capsule is a very flimsy structure. Its outer wall is almost completely taken up by the fenestra ovalis, its floor by the basicapsular fenestra, and its medial wall by the foramen perilymphaticum, foramen endolymphaticum, and foramen acusticum. Internally there are three slender septa for the semicircular canals. There is no tectum synoticum of any kind. The facial nerve passes out freely in front of the auditory capsule.

The fenestra ovalis contains a very small operculum continuous with a columella which is pierced by the stapedial artery and attached to the hind surface of the quadrate. The hyomandibular branch of the facial nerve passes back dorsally to the columella. The cartilage of the operculum appears to have a common origin with that of the auditory capsule.

Immediately beneath and in front of the prefacial commissure, small basitrabecular and postpalatine processes project from the lateral edge of the parachordal and fuse, enclosing a foramen for the palatine nerve. The basitrabecular process is connected with the quadrate by fibrous tissue, a relic of the basal connexion.

The side wall of the skull in the orbital region is formed only by the slender trabecular and orbital cartilages, interconnected by the orbital cartilage's preoptic root and the pila metoptica. The orbital cartilage is continuous posteriorly with the auditory capsule. In this way foramina for the optic and trigeminal nerves are enclosed and separated. (The position of the oculomotor nerve has not been recorded; presumably it emerges with the trigeminal behind the pila metoptica. Should it prove to emerge in front of this structure it would be a pila antotica in an abnormally far forward position.)

Anteriorly, the trabeculae converge and unite forming a narrow trabecular plate which bears the nasal septum dorsally and is continued anteriorly into the prenasal process. Laterally, the trabecular horns diverge from the trabecular plate and become fused with the side wall of the nasal capsule, or lamina orbitonasalis. The latter arises from the trabecula laterally to the preoptic root of the orbital cartilage between which and the lamina orbitonasalis the orbitonasal foramen is enclosed. Through this foramen the nasal branch of the profundus nerve enters the nasal capsule. The lateral ramus of this nerve soon emerges again through the foramen epiphaniale, while its medial ramus runs by the side of the septum to emerge through the foramen apicale. This is pierced through the rudimentary anterior cupola or cartilago alaris which ultimately will connect the dorsal edge of the septum with the side wall of the capsule and enclose the fenestra narina, which at this stage is open dorsally. The ventral border to the fenestra narina is formed by a strut connecting the septum with the anteroventral corner of the side wall of the capsule, and called the cartilago infranarina.

From the preoptic root of the orbital cartilage a vestigial sphenoseptal process projects towards but fails to reach the nasal septum. The olfactory foramen is therefore unenclosed dorsally. The fusion between the trabecular horn and the lamina orbitonasalis, forming the solum nasale, constitutes the anterior border of the fenestra basalis (choanalis). Between the solum nasi and the cartilago infranarina is a vacuity of no morphological significance representing an unchondrified portion of the floor of the capsule.

The quadrate possesses a long pterygoid process, at first chondrified discontinuously in three pieces, but which ultimately forms a rod and comes into contact with the lamina orbitonasalis, thus forming a (transient) ethmoid attachment. There is also a processus ascendens which eventually becomes attached by fibrous tissue to the orbital cartilage, and a small otic process which projects towards the auditory capsule (and establishing contact with it in Siphonops; Goodrich, 1930). There is no basal process in Ichthyophis, where the quadrate is connected with the basitrabecular process by a ligament, but a basal process is present in Hypogeophis and Siphonops. The suspension of the jaws in the definitive chondrocranium thus conforms to the autodiastylic type (see also Luther 1914).

Meckel's cartilage is very long and slender, with a particularly well-developed retroarticular process.

No sclerotic cartilage is developed.

In the hyobranchial skeleton a median basihyal (copula I) separates the ventral ends of the ceratohyals. Behind the basihyal are two basibranchials (I and II; copulae II and III) separating the first two ceratobranchials. The 3rd ceratobranchials are united with one another in the midline, while the 4th are very rudimentary. On the other hand, the anterior ceratobranchials are very stout and are among the earliest parts of the skull to chondrify; this is presumably associated with their function in supporting the long external gills.

The chondrocranium of Ichthyophis is characterized throughout by extreme reduction of cartilage; indeed, the procartilaginous rudiment of earlier stages is more extensive than the subsequently formed definitive chondrocranium.

At metamorphosis nearly all the cartilaginous neurocranium is broken down and mostly replaced by bone; only the hyobranchial skeleton remains cartilaginous, and more or less in the larval condition except that the basihyal and 2nd basibranchial are lost and the ceratal elements fuse with one another in the midline.

ii. Development of the osteocranium of Ichthyophis

In many cases the cartilage-bones extend beyond the limits of the cartilage in which their perichondral lamellae arise, but this is to be explained by the extreme paucity of cartilage present; the cartilage-bones occupying regions which were preformed in cartilage in ancestral forms. The bony skull is compact in adaptation to the burrowing habit and in its formation the fusion of bones is very widespread.

1. The membrane-bones. The *parasphenoid*, as shown below, enters into the composition of the os basale. It forms the only skeletal floor to the brain-case, since the chondrocranium develops no cartilage in this region.

The *squamosal* (paraquadrate of Gaupp) arises on the lateral surface of the quadrate and auditory capsule and ultimately comes into contact with the maxilla.

The *septomaxilla* ('turbinale' Sarasin, 'lachrymal' Born) ossifies by the side of the nasal sac, behind the fenestra narina and maintains an external superficial position.

The possible presence of the *quadratojugal* as a process of the quadrate has already been mentioned.

The *premaxilla* undergoes fusion with the nasal and septomaxilla.

The *maxilla* fuses with the *palatine* to form a so-called maxillopalatine. Both these bones, and the *prevomer*, bear teeth; the *pterygoid* is toothless. The teeth in the lower jaw are arranged in inner and outer rows, the latter being carried by the *dentary*, the former by the coronoid. Both the latter bones become continuous with the mentomeckelian, while the *angular* fuses with the articular as already described.

The roof of the skull is formed by the *nasal*, *prefrontal*, *frontal*, *postfrontal*, and *parietal*.

2. The cartilage-bones. The *articular* and the *mentomeckelian* are the earliest to ossify; the former appears from the start to be continuous with a membrane-bone, the angular.

The occipital, auditory, and orbitotemporal regions become ossified by means of perichondral lamellae which arise simultaneously throughout these regions and give rise to a large single ossification on each side which presumably corresponds to the *pleurosphenoid*, *prootic*, and *exoccipital*, and also to the *supraoccipital*, for the two bones are united with one another posteriorly across the midline in the place where a tectum posterius would have been if present. (This apparent absence in ontogeny of separate centres for ossifications which were separate in phylogeny, if substantiated, would be a case of 'fusion primordiale'. (See p. 505.) In Hypogeophis, however (see p. 196), many separate centres have been found.) Further, the two bones fuse ventrally with a membrane-bone, the parasphenoid, so that most of the brain-case is formed by a single compound bone, termed by Sarasin the *basale*.

The *stapes* starts ossifying in the operculum, the ossification subsequently spreading into the columella.

The *quadrate* ossifies fairly completely. An intramembranous extension of the bone between the bases of the ascending and pterygoid processes into the so-called processus jugularis is held by Peter to represent the fused rudiment of a membrane-bone, the *quadratojugal* (quadrato-maxilla of Gaupp).

The *sphenethmoid* (ethmoid autt.) is the latest bone to arise. It ossifies in the nasal septum, preoptic root of the orbital cartilage, and neighbouring regions, partly as intramembranous ossification. It appears to represent the *presphenoid* and *orbitosphenoids*.

II. Hypogeophis alternans and H. rostratus

The development of the skull of Hypogeophis has been studied by Marcus (1910, 1935) and his pupils (Marcus, Winsauer, and Hueber, 1933; Marcus, Stimmelmayr, and Porsch, 1935; Eifertinger, 1933).

i. The chondrocranium of Hypogeophis

The chondrocranium has been studied in Hypogeophis alternans and H. rostratus with the help of wax models (stage 40, 25 mm., and later stages) by Marcus, Stimmelmayr, and Porsch (1935). In general it is similar to that of Ichthyophis (see p. 192), for which reason it need not be described in detail. Mention may, however, be made of the following points. (Plate 73, figs. 3, 4).

The nasal capsule has a well-developed roof, i.e. the sphenoseptal commissure is present, joining the orbital cartilage with the nasal septum, and the olfactory foramen is thus completely enclosed. The quadrate possesses an ascending process (which ends freely dorsally), a small basal process, a pterygoid process, and a backwardly directed otic process (termed by Marcus the columellar process) which enters into contact and temporary fusion with the stylus of the columella. The quadrate is therefore not fused with the chondrocranium, so that the skull may be said to conform to the autodiastylic type. However, if the transient fusion between the quadrate and the stylus columellae be taken into account, the skull must be called amphistylic.

According to Marcus (1910) the rudiment of the stylus at early stages is in blastematous connexion with that of the ceratohyal; a primitive condition indicating that it forms a part of the true hyoid arch skeleton. The columella itself is according to Marcus (1935) a portion of the lateral wall of the auditory capsule out of which it becomes carved, but with which it remains for a time in cartilaginous continuity.

ii. The development of the osteocranium of Hypogeophis

The interest of the bony skull of Hypogeophis, as interpreted by Marcus, Stimmelmayr, and Porsch (1935) lies in the fact that they were able to find a large number of separate centres of ossification at early stages, prior to their fusion to form complex bones. (Plate 73, figs. 5–8).

At stage 40 (25 mm.) the only ossifications present (in H. alternans) are the *prevomers* and the *palatines*, both developed in relation to but quite separately from groups of teeth.

At stage 47 (H. rostratus) most of the ossifications are present. Of the cartilage-bones, the auditory capsules (*prootic*, *opisthotic*, and *epiotic*?) are ossified and fused with the *exoccipitals* (ossified in the occipital arches), the *basioccipital* which ossifies in the hypochordal commissure, and the *supra-occipital* which ossifies in membrane where the tectum posterius would be if chondrified. The *quadrate* ossification is perichondral and fused to a process of membrane-bone regarded as the *quadratojugal* (quadrato-maxillary of German authors).

In addition, there are the following membrane-bones, ossifying from independent centres: *nasals*, *frontals*, *parietals*, *prefrontals*, *squamosals* (para-quadrate of Gaupp), *premaxillae*, *maxillae*, median *supra-ethmoid*, *postparietal*, *parasphenoid*, and paired ossifications in the membranous floor of the cranial cavity, between the auditory capsules.

In the lower jaw Eifertinger (1933) made out nine centres of ossification, viz. *mentomeckelian*, *articular* (both cartilage-bones), and *dentary*, *splenial*, *coronoid*, *complementale*, *supra-angular*, *angular*, and *prearticular* (membrane-bones).

At a still later stage (68 mm., H. alternans) additional ossifications are the *orbitosphenoids* (in the preoptic roots of the orbital cartilages), the *presphenoid* (mesethmoid of some authors, in the trabeculae), the *pleurosphenoids* (in the orbital cartilages and pilae metopticae), the *stapes* (in the columella auris), being cartilage-bones; and the *septomaxilla* (not found, however, by Lapage, 1928*b*), the *lachrymal*, *periorbital* bones (i.e. *supra-*, *post-*, and *suborbitals*), and the median *interparietal* which eventually occludes the pineal foramen (between the parietals and frontals).

Soon, however, extensive fusion takes place, and the following table of the bones of the adult skull indicates their composition so far as the previous existence of separate centres is known:

'*basale*', composed of the fused bones of the auditory capsule, exoccipitals, basioccipital, supraoccipital, parasphenoid (including the paired ossifications behind it), and pleurosphenoids;
'*temporal*', composed of fused frontal and prefrontal;
'*nasopremaxilla*', composed of fused nasal, premaxilla, and septomaxilla;
'*maxillopalatine*', composed of fused maxilla, palatine, and lachrymal, sometimes including the pterygoid as well;
'*sphenethmoid*', composed of fused orbitosphenoids and presphenoid (there appears to be uncertainty whether representatives of the lateral ethmoids may not be included as well);
'*squamosal*', composed of fused squamosal and periorbital bones;
'*quadrate*', composed of fused quadrate and quadratojugal;
'*pseudodentary*', composed of fused dentary, splenial, coronoid, supra-angular, mentomeckelian;
'*pseudoangular*', composed of fused angular, prearticular, articular, and complementale.

It will be seen from the description just given that the supposition that the bones of the adult Hypogeophis are the result of fusion is susceptible of ontogenetic demonstration; i.e. in most cases here, at least, there is no question of any 'fusion primordiale' (see p. 505).

The large number of centres of ossification discernible in the young Hypogeophis leads Marcus to conclude that the affinity between Gymnophiona and Stegocephalia is closer than previously supposed, and that the complete covering to the roof of the skull in the temporal region in Gymnophiona is directly derived from the similar condition in Stegocephalia, and has not been secondarily acquired. On the other hand, it is noticeable that the postfrontal and supratemporal are lacking in Gymnophiona; which they would hardly be if these forms were primitive, since these bones are regularly present in reptiles. The complete roof in Gymnophiona is therefore probably secondarily developed in connexion with the burrowing habit.

Another feature of interest presented by the skull of Hypogeophis, is the fact, described by Marcus, Winsauer, and Hueber (1933), that the skull is kinetic in Versluys's (1910, 1912) sense, and capable of a certain amount of internal movement although the quadrate is fixed to the squamosal, and the skull must be classified as monimostylic on Stannius's (1856) scheme (see p. 425).

Two portions may be distinguished in the skull, capable of movement against one another. The first of these, the 'basal segment', consists of the basale, while the other, or 'quadrate segment', comprises the remaining bones. The two segments are capable of movement against one another at the following points: (i) the articulation between the basal process of the quadrate and the basitrabecular process of the skull (this articulation is not developed in Ichthyophis); (ii) between the anterior end of the basale (rostrum of the parasphenoid) and the prevomers; (iii) between the quadrate and the stapes; (iv) between the parietals and the basale (exoccipitals).

The importance of this case of kinetism is that it is associated with a completely roofed skull, thus showing that there is no intrinsic improbability in the view that the Stegocephalia may have been kinetic.

15. ANURA

I. Rana fusca (S. temporaria)

The development of the skull of the frog has been studied by many workers, first among whom must be mentioned Dugès (1834). The chondrocranium at one stage or another has been investigated by W. K. Parker (1871, 1876*a*, 1881), Stöhr (1881), Gaupp using wax models (1893, 1906*a*), and Peeters using the van Wijhe technique (1910), while its earliest appearance has been studied by Spemann (1898). The nasal capsule has been described by Born (1876), and the development of the hyobranchial skeleton by Gaupp (1894) and Schachunjanz (1926).

The morphology of the chondrocranium of the frog, which has played an important if muddled par in the general theory of the skull, is treated by Huxley (1858, 1874, 1875), W. K. Parker (1871 1876*a*), Parker and Bettany (1877), de Beer (1926*a*), and Pusey (in press). The morphology of the adult hyobranchial skeleton in Anura has been exhaustively studied by Trewavas (1933).

The development of the bones of the frog's skull has been the object of investigation with modern methods by Gaupp (1906*a*) and Erdmann (1933).

i. Development of the chondrocranium of Rana fusca

The following account is based largely on the work of Stöhr, Gaupp, and Spemann, supplemented by observations made by the present writer and wax models constructed by the late Dr. J. W. Jenkinson.

1. Stage 1. (5 mm. stage.) PLATE 74, FIG. 1.

The visceral arch skeleton is more advanced than the neurocranium, and at the 5 mm. stage a continuous bar of procartilage represents the rudiments of the quadrate, Meckel's, and the infrarostral cartilages. The three regions are marked off by sharp bends in the procartilaginous rudiment, in such a way that the quadrate is vertical, Meckel's cartilage extends horizontally forwards from the lower end of the quadrate, and the infrarostral points downwards from the anterior end of Meckel's cartilage. Behind these structures, and separated from them by the 1st visceral pouch, is the rudiment of the ceratohyal. In the upper jaw, far forward, the procartilaginous rudiments of the suprarostral cartilages make their appearance. It is important to note the extreme forward position of the quadrate, which actually lies in front of the eye. This is associated with the very small size and gape of the larval mouth.

2. STAGE 2. (7 mm. stage.) PLATE 74, FIG. 2.

At this stage the procartilaginous trabeculae arise, far apart, each trabecula ending anteriorly in a horn (cornu) which bends outwards and downwards and becomes continuous with the suprarostral of its side. The quadrate shows rudimentary outgrowths representing the future processus muscularis and quadratocranial commissure. It is still continuous with Meckel's cartilage, while posteriorly the ceratohyal is connected with the quadrate by ligament.

3. STAGE 3. (7·5 mm. stage.) PLATE 74, FIGS. 3, 4, 5.

The trabeculae now extend back as far as the tip of the notochord, on to the sides of which they become attached, forming the 'Balkenplatte' of Stöhr, this region representing the foremost part of the parachordal cartilages.

The quadratocranial commissure becomes attached to the trabecula fairly far forward, behind the passage between the nasal sac and the mouth (choana). According to Pusey (in press) this quadratocranial commissure represents the basal process, displaced forwards in connexion with the peculiar formation of the tadpole's larval sucking mouth, bordered as it is by special structures, the supra- and infrarostral cartilages, which support the horny 'teeth'.

Chondrification sets in at this stage, separate centres being noticeable in trabeculae and suprarostrals. (In Polypedates, Okutomi (1937) regards the suprarostrals as related to the mandibular arch).

The ceratohyal and quadrate are in cartilaginous continuity, with the result that the ventral part of the 1st visceral pouch is compressed. It is from the dorsal part that the tympanic cavity is formed, and it is important to note that at these early stages it is situated far forward, immediately behind the quadrate. The separation and formation of joints between the quadrate, Meckel's cartilage, and infrarostral begin to take place.

4. STAGE 4. (7·5 mm. stage.) (Stöhr) PLATE 76, FIG. 1.

The earliest cartilaginous stage described by Stöhr shows a slight advance over the preceding one. Anteriorly, the trabeculae are interconnected by a trabecular plate (ethmoid or internasal plate) just behind the trabecular horns. The latter form the medial and anterior borders of the choanae. The hypophysial fenestra is therefore completely enclosed, and the internal carotid arteries enter the skull by passing between them in typical manner.

The 'Balkenplatte' or anterior parachordal bears a rudimentary process pointing upwards, the pila antotica, to the lateral surface of which is now attached the ascending process of the quadrate. This process is dorsolateral to the profundus nerve, and ventromedial to the maxillary and mandibular branches of the trigeminal. The quadrate is thus firmly attached to the brain-case at two points. Its articulation with Meckel's cartilage is far forward, even lying anteriorly to the level of the quadratocranial commissure. Posteriorly, the continuity with the ceratohyal has been replaced by a joint, while on the dorsolateral edge of the quadrate the processus muscularis rises up.

In the parachordal region chondrification sets in progressively from in front backwards, with the result that on each side of the notochord a parachordal cartilage is formed, medially to the auditory sac, and continuous with the already-formed anterior parachordal.

5. STAGE 5. (14 mm. stage: head+trunk 6 mm., tail 8 mm.).

PLATE 76, FIGS. 2, 3.

The advances of chief importance at this stage concern the auditory capsules, the quadrates, and the visceral arches.

The auditory capsule chondrifies first on the outer surface of the lateral semicircular canal, from which centre it spreads to form the side wall, the roof, and the anterior and posterior cupolae which are attached to the parachordal by the anterior basicapsular and the basivestibular commissures respectively (for the latter is anterior to the glossopharyngeal nerve). There is as yet no cartilaginous medial wall nor a floor, the space of the latter representing the confluent fenestra basicapsularis and fenestra ovalis (vestibuli). The portion of this space representing the basicapsular fenestra becomes obliterated by the chondrification of a strip of cartilage which Gaupp regarded as the lateral extension of the parachordal, but which is more likely the true cartilaginous floor of the capsule. The fenestra ovalis would thus be entirely enclosed by capsular cartilage.

The parachordals flank the notochord on each side and are completely separated by it except at their extreme front end, where a transverse bar of cartilage (forming a dorsum sellae) interconnects them in front of the tip of the notochord.

On the anterior border of the quadratocranial commissure is a small process pointing forwards, and attached by ligament (the quadratoethmoidal ligament) to the trabecula just in front of the trabecular plate. This is the rudimentary pterygoid process which will play a very important part in the changes which the quadrate undergoes at metamorphosis, and will give rise to the long subocular bar connecting the quadrate with the ethmoid region of the skull.

The ceratohyals fuse with one another in the midline forming the so-called pars reuniens. This is not to be regarded as a true copula, for a basihyal or copula I will arise later on. Perhaps the pars reuniens represents the fused hypohyals, which are not otherwise accounted for. (Plate 77, fig. 9).

The cartilages of the four branchial arches arise independently (representing ceratobranchials), but like the ceratohyals they become fused to their fellows of the opposite side in the middle line, and they also become attached to a median cartilage (histologically different from the surrounding cartilages, see p. 474) known as copula II. This structure appears to represent the 1st basibranchial. To its anterior end are attached (by means of the pars reuniens) the ventral ends of the ceratohyals, while the large and expanded ventral ends the ceratobranchials (Ridewood's (1898*a*) hypobranchial plate, probably representing the fused hypobranchials) are attached to its dorsal side about half-way along its length. The posterior end of copula II projects freely ventrally to the hypobranchial plate. Close to the pars reuniens and therefore near the mid-ventral line, each ceratohyal gives off a processus anterior and a processus posterior. The ventral ends of the remaining ceratobranchials (2nd to 4th) are attached to the base of the 1st, while the dorsal ends of all the ceratobranchials of each side are interconnected by commissurae terminales.

Further, a short dorsally directed spicula is developed from the ventral end of each ceratobranchial. All the cartilages of the branchial arches and the ceratohyals, therefore, are fused together, and articulated with the quadrates by means of the joint between the latter and the ceratohyals.

6. Stage 6. (29 mm. stage: head and trunk 11 mm., tail 18 mm.).

PLATE 76, FIGS. 4, 5.

All regions of the skull have at this stage made considerable advances.

The two suprarostral cartilages hitherto paired and separate, now fuse with one another across the midline. Between the infrarostrals there appears a small unpaired cartilage, forming a mandibular copula or basimandibular.

The front wall of the brain-case arises in the form of a pair of pilae ethmoidales projecting dorsally from the intertrabecular or ethmoid plate, separated from one another by a fenestra precerebralis which faces directly forwards, and situated medially to the olfactory nerves. Laterally to the latter the preoptic root of the orbital cartilage arises on each side, and encloses the foramen olfactorium evehens between itself and the pila ethmoidalis of its side. The foramina olfactoria and the fenestra precerebralis are limited dorsally by bars of cartilage joining the dorsal ends of the preoptic roots of the orbital cartilages and of the pilae ethmoidales and corresponding to the sphenoseptal commissures of other forms.

Above the greater part of the trabeculae the orbital cartilages have not yet appeared, and the optic and trochlear nerves, emerge freely dorsally to the trabecula. The pila metoptica, however, arises from the trabecula and fuses with the pila antotica behind it, thus enclosing the foramen for the oculomotor nerve and ophthalmica magna artery. The dorsal end of the pila antotica is also connected with the roof of the auditory capsule (far back, near the tectum synoticum) by a taenia marginalis, with the result that the foramen prooticum (for the trigeminal, abducens, and facial nerves, there being no prefacial commissure) is enclosed.

The fenestra hypophyseos is reduced in size by progressive development of the cartilage of the trabeculae, and the internal carotid arteries are now enclosed each in its own foramen. Inside the skull each carotid gives off the ophthalmica magna, a small palatine artery (which leaves the skull again through the cranio-palatine foramen), and subdivides to form the anterior and posterior cerebral arteries.

The roofs of the auditory capsules are interconnected by a tectum synoticum, from which in the midline a taenia tecti medialis projects forwards. (Gaupp thought that the tectum synoticum had an independent median centre of chondrification: this has, however, not been confirmed by Peeters.) The lateral wall of the auditory capsule is perforated by the fenestra ovalis which at this stage is covered over only by membrane. The medial wall arises gradually, first by the subdivision of the original wide opening into an anterior general foramen acusticum and a posterior foramen perilymphaticum. The latter faces inwards and backwards and opens on to the fissura metotica (for the glossopharyngeal and vagus nerves), which has become enclosed by the upgrowth of the occipital arch from the hindmost part of the parachordal to the hind wall of the auditory capsule. (Plate 77, fig. 1).

Next, a subdivision occurs between the foramen acusticum proper (for both vestibular and cochlear branches of the auditory nerve) and the more dorsally situated foramen endolymphaticum. The foramen perilymphaticum becomes subdivided into two by a strut of (capsular) cartilage: the foramen perilymphaticum superius through which the ductus perilymphaticus superius passes from the cavity of the auditory capsule into that of the brain-case, and the foramen perilymphaticum inferius, through which the ductus perilymphaticus inferior leaves the auditory capsule and leads into the saccus perilymphaticus which occupies the region of the fissura metotica (corresponding to the recessus scalae tympani of higher forms). The saccus perilymphaticus communicates with the ductus perilymphaticus superior in the cranial cavity by means of the canalis perilymphaticus anastomoticus which passes through the fissura metotica. On its way, this canal is (for a time) separated from the glossopharyngeal and vagus nerves by a transient strut of cartilage which stretches like a preoccipital arch from the edge of the parachordal to the wall of the auditory capsule, and thus subdivides the fissura metotica into an anterior unfortunately so-called foramen perilymphaticum accessorium, and a posterior jugular foramen for the glossopharyngeal and vagus nerves. (Plate 77, figs. 2, 3).

Internally, the three semicircular canals are separated from the main cavity of the capsule by cartilaginous septa.

The anterior end of the notochord, in front of the foramen acusticum, has broken down and here the two parachordal cartilages are fused across the middle line into a flat basal plate which is on the same level as the trabeculae, i.e. there is no upraised edge to the dorsum sellae. It is also to be noticed that there is no perforation of the basal plate, or basicranial fenestra. The hinder part of the intracranial notochord is flanked by parachordal cartilage on each side, but is free dorsally and ventrally.

The quadrate has become more massive, the processus muscularis is larger, and posteriorly a new attachment to the neurocranium has been effected by the development of what for reasons which will appear later must be called the larval otic process. This connects the posterolateral corner of the quadrate with the anterolateral wall of the auditory capsule. The hyomandibular branch of the facial nerve and the head vein run in typical manner ventrally to the larval otic process.

It is to be noted that there are no basitrabecular or basal processes, unless the latter be represented by the quadratocranial commissure. A transient structure (better developed in other Anura) at about this stage is a cartilaginous process extending backwards from the quadratocranial commissure in the floor of the orbit, parallel with the trabecula. Gaupp has called this the pseudopterygoid process, and it may perhaps be an indication of the secondary displacement in an anterior direction which the basal process has undergone if it is represented by the quadratocranial commissure.

In the hyobranchial skeleton the chief novelty of importance at this stage is the appearance of the basihyal, or copula I, as a small independent cartilage in the middle line close in front of the pars reuniens of the ceratohyals. A small cartilaginous synapticulum (the processus branchialis) interconnects the 2nd and 3rd ceratobranchials near their ventral ends (thus restricting the aperture of the 3rd gill-slit) and serving as an attachment for the long hyobranchial muscles. All the ceratobranchials and their dorsal commissures now bear short and irregular cartilaginous rays. (Plate 77, fig. 10).

ii. Metamorphosis (stage, 29 mm.: head and trunk 13 mm., tail 16 mm.)

For the description of this stage, which approaches that of maximum chondrification, and which shows differences over preceding stages in nearly all regions, it will be convenient to adopt the regional subdivision of the subject-matter. (Plate 76, figs. 6, 7, 8).

1. Occipital region. The preoccipital arch which at the previous stage separated the canalis perilymphaticus anastomoticus from the glossopharyngeal and vagus nerves has broken down; the fissura metotica is now, therefore, uninterrupted. The hind edges of the parachordal cartilages are now articulated with the 1st vertebra by means of a proper occipito-atlantal joint, which is formed, as Mookerjee (1931) has shown, in the following manner. The occipital arch and the neural arch of the 1st vertebra arise each from the caudal sclerotomite of their respective segments. From the cranial sclerotomite of the segment immediately behind the occipital arch, paired (interdorsal) cartilages arise, similar to the intervertebral cartilages further back between the parachordals and the 1st vertebra. Each of these cartilages then becomes split into two by the invasion of a connective tissue septum, and the anterior portions become attached to the parachordals (forming the paired occipital condyles) while the posterior portions become fused on to the sides of the 1st vertebra. The 1st spinal nerve (nervus suboccipitalis) disappears in Rana, but it persists for some time in Bufo.

As at the previous stage, the hinder half of the parachordals are separated from one another by the notochord, but at the extreme hind end, the two parachordals approach one another dorsally to the notochord.

2. Auditory region. On the external surface, the lateral semicircular canal forms a prominent horizontal ridge-like bulge, or crista parotica, to which the larval otic process is attached, and ventrally to which the fenestra ovalis is situated. This fenestra is now occluded by two structures. In its hinder part is a plate of cartilage, the operculum, which arises quite independently of the wall of the auditory capsule, but which at this stage and for a short time acquires a cartilaginous continuity between its dorsal edge and the edge of the fenestra ovalis.

The anterior part of the fenestra ovalis (in front of the operculum) is open, and it is here that the second structure, the pars interna of the columella auris, or plectrum (or otostapes), arises as an independent rod of cartilage, which, however, soon acquires cartilaginous continuity with the fenestra's ventral edge. This edge is turned upwards, forming the crista preopercularis. The posterior end of the columella is connected by ligament to the operculum; its anterior end is connected likewise by ligament to the posterior surface of the quadrate. The hyomandibular branch of the facial nerve runs outwards *dorsally* to this (suspensorio-columellar) ligament. The columella at this stage has no connexion with the ceratohyal, which is still situated far forwards, close behind the quadrate. But according to Salvadori (1928) who has investigated Rana and Bufo, a transient ligament connects the ceratohyal with the pars externa plecti, which develops later (see below, p. 207) and becomes attached to the pars interna. In Rana palustris, Violette (1930) has tried to derive the plectrum from the 1st branchial arch, but this cannot be accepted, and

Gazagnaire (1922) has reaffirmed the hyoid arch derivation of the columella in Rana fusca.

On the internal surface the single foramen acusticum of the previous stage has become divided into an anterior foramen for the vestibular and a posterior foramen for the cochlear branch of the auditory nerve.

There is no doubt that the main part of the roof of the skull in this region is a tectum synoticum. The occipital arches are, however, directly continued into it, and it is therefore possible that the hindmost portion of the roof represents a tectum posterius.

3. Orbitotemporal region. The hypophysial fenestra is much reduced in size. The orbital cartilages are now well developed, connecting the preoptic root with the pila antotica, and enclosing the optic and trochlear nerves in their respective foramina. On the other hand, the foramina for the oculomotor nerve and for the internal carotid artery have become confluent, by means of the absorption of the intervening (trabecular) cartilage. This is why the internal carotid artery now appears to enter the skull through the oculomotor foramen (morphologically dorsally to the trabecula) and to give off the ophthalmica magna artery before doing so, thus spuriously resembling the condition in Polypterus or Amiurus. The two pilae antoticae may at this stage be interconnected by a transverse bar, the tectum transversum, (originating apparently from a median unpaired chondrification) to which the taenia tecti medialis may reach and become attached. There is considerable variability in the time of development of these structures.

The processus ascendens has now broken down and disappeared, as will be mentioned again in connexion with the change undergone by the quadrate.

The eyeball becomes protected on its medial side by a saucer-shaped sclerotic cartilage.

4. Ethmoid region. The changes undergone by this region are many and extensive, and relate on the one hand to the formation of the nasal capsule, and on the other to the substitution of the adult for the larval system of mouth aperture. Both these sets of changes are intimately connected with the change in mode of life of the organism at metamorphosis.

The fenestra precerebralis (between the pilae ethmoidales) has been obliterated by the formation and development forwards of a median vertical cartilaginous plate, the nasal septum, the hind edge of which separates the two olfactory foramina. Dorsally the hind end of the nasal septum is continued on each side into the flat roof of the nasal capsule or tectum nasi. This grows forwards somewhat independently of the septum, so that anteriorly on each side a sort of cupola is formed with its own medial wall which then becomes attached to the septum. This point is of importance for the comparison between Anura and Urodela. The roof of the nasal capsule is continuous with the two cartilages (on each side) which form its side wall. (Plate 77, fig. 4).

The hinder of these cartilages is the lamina orbitonasalis which projects laterally and forwards from the trabecula (immediately in front of attachment of the quadratocranial commissure) and the preoptic root of the orbital cartilage as a plate (the 'pars plana') forming the posterolateral wall of the nasal capsule, immediately behind the choana. The morphological limit between the lamina orbitonasalis and the preoptic root of the orbital cartilage is shown by the position of the orbitonasal foramen through which the nasal branch of the profundus nerve leaves the orbit and enters the nasal capsule.

The anterior of the side wall cartilages is the so-called cartilago obliqua, which curves round ventrally, forming the dorsal and lateral border of the fenestra narina (external nostril). At first the cartilago obliqua ends freely, but eventually it becomes attached to the floor of the nasal capsule thereby forming in this region a zona annularis.

The side wall of the capsule thus presents a notch between the cartilago obliqua and the lamina orbitonasalis, and through this notch the lateral ramus of the nasal branch of the profundus nerve leaves the capsule.

The anteroventral part of the lamina orbitonasalis bears two processes, one pointing forwards and the other backwards: the processus maxillaris anterior and posterior respectively. The former ends freely, but to the latter becomes attached the little pterygoid process of the quadrate, thereby forming the ethmoid attachment of the quadrate to the skull. The processus maxillaris posterior is destined to elongate to the extent of half the length of the orbital region, in connexion with the posterior displacement of the quadrate; and in preparation for this, already at this stage the quadratocranial commissure has broken down and vanished. (Plate 75, figs. 2, 3).

Likewise, the suprarostral cartilage and the tips of the trabecular horns have been resorbed and disappeared. The stumps of the trabecular horns form the floor of the hinder part of the nasal capsule; but the floor of the anterior part forming the anterior border to the fenestra basalis for the choana, and thus separating the latter from the fenestra narina (external nostril), appears to be derived from newly formed cartilage, partly contributed by the ventral edge of the nasal septum, and partly from the so-called lateral marginal cartilage, which develops on the outer edge of the trabecular horn.

The anterior wall of the nasal capsule is somewhat complicated. Its posterior surface (on each side) is projected backwards into a ridge (the crista intermedia) which runs from the septum obliquely downwards and outwards, and forks, thus marking out a cavum superius, a cavum inferius, and a cavum medium (between the arms of the fork), which lodge the three diverticula of the olfactory sac. (Plate 77, fig. 5).

The aperture of the fenestra narina (through which in addition to the external nostril pass the duct of the external nasal gland and the lachrymonasal duct) is much reduced by the projection into it from its medial border of the cartilago alaris. This also subdivides the fenestra so that the external nostril opens anterodorsally into the cavum superius, while the two gland ducts open posteroventrally into the cavum medium. On the medial side of the cartilago alaris is a small foramen for the exit of a twig of the medial branch of the nasal ramus of the profundus nerve, the remaining fibres emerging from a foramen apicale ('nasobasale', Parker) in the anterior wall of the cavum inferius.

On the anterior surface of the front wall of the nasal capsule there are on each side two cartilaginous spines, the superior and inferior prenasal cartilages. The former (Born's cartilage) is attached to the cartilago alaris; the latter (Wiedersheim's cartilage) arises from the lateral marginal cartilage. Both enter into contact with the premaxillary bone. It is clear that the nasal capsule of the frog (and of Anura generally) is highly specialized (largely owing to the fact that during larval life the trabecular horns are involved in the support of the suprarostral cartilages) and markedly different from that of Urodela. A closer approach to the condition in the latter would result if in Anura the nasal septum were to disappear. (Plate 77, figs. 6, 7).

5. The visceral arch skeleton. Some of the features of the quadrate at this stage have already been noted: the disappearance of the quadratocranial commissure and of the processus ascendens, and the attachment of the pterygoid process to the posterior maxillary process of the nasal capsule. The essence of the changes at this and subsequent stages concerns the backwards withdrawal of the jaw articulation from a position in front of the olfactory foramen to a point beneath the auditory capsule. This is brought about by a lengthening of the pterygoid process and of Meckel's cartilage, on to the anterior end of which the infrarostrals become fused as their direct anterior prolongation. The basimandibular cartilage breaks down into a ligament forming the mandibular symphysis. The suprarostrals vanish. At the same time a buckling, folding, and resorption takes place between the larval otic process and the processus muscularis.

Many changes also occur in the hyobranchial skeleton. The basihyal (copula I) disappears. The ceratohyal becomes more slender and its processus posterior disappears: its anterior process, on the other hand, becomes accentuated. The proximal (lateral) end of the ceratohyal is carried back with the quadrate to which it is no longer articulated but extends freely backwards ventrally to the tympanic cavity. All four ceratobranchials disappear, leaving only the hypobranchial plate, itself considerably reduced, continuous anteriorly with the ceratohyals, and projecting posteriorly into the paired thyroid processes. The latter, on to which the hyoglossal muscles are attached, are left as a relic of the larger hypobranchial plate of previous stages, and doubtless owe their persistence to their function as muscle attachments. (Plate 77, figs. 11, 12).

iii. The chondrocranium of the metamorphosed frog (20 mm.)

The chondrocranium of the metamorphosed frog resembles that of the last described stage except in respect of the visceral arch skeleton, including the columella auris. (Plate 75, fig. 4; Plate 77, figs. 13, 14).

The roof of the brain-case is very incomplete, owing to the persistence of the large vacuity in the orbital region, and to paired vacuities on each side of the taenia tecti medialis.

In the basal plate the notochord has been completely excluded from the hindmost region by the junction of the parachordals dorsally to it, and in front of this there has taken place the apparent chondrification of the notochord itself. In addition to the lateral articulations by means of the occipital condyles, the cranio-vertebral joint also involves a median articulation, formed by a very short anterior process on the centrum of the 1st vertebra (the tuberculum interglenoidale, chondrified on the site of the broken down notochord) which fits into the notch between the occipital condyles.

In Rana grayi, du Toit (1933*b*) has described a peculiar 'subethmoidal' cartilage of doubtful significance, lying beneath the hind part of the nasal septum. This seems to correspond to the median palatine cartilage described in Bufo by Schoonees (1930). The trabecular plate is complete and the hypophysial fenestra obliterated. The sclerotic cartilage is well developed. In the auditory capsule the crista parotica has become accentuated, and the fenestra ovalis has been reduced by growth of cartilage in its anterior part, so that the operculum now fits its rim closely.

The larval otic process has disappeared, and in its backward displacement and rotation the processus muscularis of the quadrate has come into contact and fused with the crista parotica thus forming the definitive otic process, with the same relations to nerves, &c., as its predecessor. Meanwhile a block of cartilage has arisen beneath the anterior end of the auditory capsule. As Pusey (in press) has shown, the quadrate is backed on to this and fuses with it, while it in its turn fuses with the floor of the auditory capsule *behind* the prootic foramen. The quadrate thus acquires a new attachment to the neurocranium. Gaupp (1893) called this the 'basal process', but as the palatine nerve passes down *in front* of it, de Beer (1926*a*) showed that it could not be a true basal connexion. It appears to correspond to a postpalatine commissure attached to the quadrate, forming a pseudobasal connexion, dorsally to which the hyomandibular branch of the facial nerve and the head vein run back.

Later on, a discontinuity and a joint is formed between the pseudobasal process and the auditory capsule. Thus, the suspension of the jaws in the frog (but not in Bufo according to Ramaswami) is incompletely autosystylic.

The pars interna of the columella no longer runs forwards to the quadrate but sideways to that part of the skin where the tympanum is formed and where it is fused with another cartilaginous element, the pars externa columellae or plectri (or hyostapes), which, according to Salvadori (1928), develops in a transient ligament connected with the ceratohyal. This pars externa columellae gives off a dorsal process which fuses with the crista parotica forming a 'laterohyal', (lacking in Bufo; Ecke 1935). The connexion between the columella and the edge of the fenestra ovalis breaks down into a ligament. Thus, the columella auris, apart from the operculum, itself arises from two chondrifications. The pars interna appears to correspond to the otostapes, and the pars externa to the hyostapes of other forms.

The hyomandibular branch of the facial nerve runs back between the pseudobasal and otic connexions of the quadrate, dorsally to the tympanic cavity and to the columella auris, medially to the dorsal process. Thence the ramus mandibularis internus of the nerve (chorda tympani) runs ventrally behind the columella and *behind* and below the tympanic cavity to reach the medial side of Meckel's cartilage. This course of the chorda tympani differs from that found in Amniotes where it runs *in front* of and above the tympanic cavity (though behind the spiracular cleft, as Goodrich (1915) showed). But the condition in Anura is readily understandable when it is remembered that at early stages, when the nerves grow out, the mandibular and hyoid arches (represented by the quadrate and the ceratohyal) which enclose the tympanic cavity between them are far forward, in front of the eye as Spemann (1898) showed. Under such circumstances it would be manifestly impossible for the chorda tympani to run over and in front of the tympanic cavity, with the result that when the latter is brought backwards to its definitive adult position, the course of the nerve appears to be abnormal.

A feature characteristic of and peculiar to the Anura is the cartilaginous tympanic ring, on which the tympanic membrane is stretched. According to Villy (1890) and Gaupp (1893) this is formed from cells which seem to be proliferated from the quadrate, ventrally to the anterior part of the processus muscularis, and surrounding the tympanic cavity at quite early stages (12 mm.). It is only after metamorphosis that the tympanic ring (which is carried backwards with the quadrate) chondrifies, and its dorsal edge fuses with the crista parotica. Gaupp's (1899) attempt to homologize the tympanic ring with the Selachian prespiracular cartilage is hardly acceptable, since the latter appears to represent the entire otic process of the quadrate. (The cartilage of the tympanic ring acts as an 'organizer' in the formation of the tympanic membrane. See Helff, 1928). (Plate 77, fig. 8).

The final stages in the conversion of the larval hyobranchial skeleton into the 'hyoid' of the adult involve the further thinning and lengthening of the ceratohyals, so that they run backwards as cylindrical rods beneath the tympanic cavities, upwards behind them, and become attached to the ventral surface of the auditory capsule of their side, beneath the anterior margin of the fenestra ovalis. The processus anterior of the ceratohyal becomes the antero-medial process (or anterior) of the 'hyoid' plate; the anterolateral (or alary) and posterolateral processes are formed as a result of new cartilaginous growth at the sides of the original copula II and hypobranchial plate, while the thyroid process of the latter becomes the posteromedial process of the 'hyoid' plate. (In Pelodytes punctatus according to Ridewood (1897*b*) the posterolateral process is not a new growth but the persistent stump of the 1st ceratobranchial.) It is thus evident that only little of the cartilage of the original larval hyobranchial skeleton is preserved in the adult.

iv. Development of the osteocranium of Rana fusca

The chief sources of information on this subject are Dugès (1834) and Gaupp (1906*a*), on whose work the following account is largely based, supplemented by alizarine preparations made by the present writer. The time of appearance of the various bones has been specially studied by Erdmann (1933), on whose work the table on p. 209 is based, showing the stages at which ossification starts in the different bones. (Plate 78).

1. The membrane-bones (in the order of their appearance). The *parasphenoid* (toothless) appears beneath the hypophysial fenestra and continuous ossification stretches forwards and laterally into the lateral wings.

The *frontal* arises over the orbital cartilage and spreads inwards over the large anterior vacuity in the roof of the chondrocranium. It soon fuses with the parietal.

The *parietal* appears over the vacuity in the roof on each side of the taenia tecti medialis, and soon fuses with the frontal to form the *frontoparietal*.

The *premaxilla* ossifies first in its ascending (prenasal) process, from which ossification spreads to the alveolar portion and ultimately to the palatine process. Teeth appear later than the bone and become attached to it at the end of metamorphosis.

The *septomaxilla* ('intranasal' of Gaupp, 1904) arises in the hinder part of the fenestra narina and forms part of the roof of the cavum mediale of the

Bone		*Total length of larva*	*Head and trunk length*
		mm.	mm.
Parasphenoid	Before metamorphosis	30	11·8
Frontal		33·7	12·8
Exoccipital		33·7	12·8
Parietal		32	13
Premaxilla		33·7	12·3
Septomaxilla	During metamorphosis	27·3	10
Maxilla		31·5	11·8
Prootic		30·6	12·2
Dentary		29·5	14
Angular		29·5	14
Squamosal		23·6	12·2
Quadratojugal		..	9·7
Pterygoid		..	9·7
Mentomeckelian		..	11
Nasal	After metamorphosis	..	11·3
Prevomer		..	12
Palatine		..	12
'Hyoid' (posteromedial process)		..	12
Sphenethmoid		..	26·2

nasal capsule. There appears to be no good evidence in support of Lapage's (1928*b*) contention that it is a cartilage-bone, a conclusion with which Ramaswami (1932*b*) is inclined to agree in respect of Kaloula, Microhyla, and Cacopus. However, the attachment and apparent continuity of one end of the bone with part of the cartilage of the nasal capsule does not mean that the bone was preformed in cartilage. The septomaxilla partially or completely encloses the lachrymonasal duct, for which reason Born (1876) regarded it as a lachrymal, which would otherwise be absent from the frog.

The *maxilla* arises from a single centre from which ossification spreads to the ascending (facial) and alveolar portions and to the palatine process. Teeth appear later than the bone and subsequently become attached to it.

The *dentary* (toothless) arises laterally to Meckel's cartilage and its anterior end eventually fuses with a cartilage-bone, the mentomeckelian at the symphysis.

The *angular* (angulosplenial) arises on the medial side of the posterior part of Meckel's cartilage and spreads outwards ventrally to it. The chorda tympani passes on the medial side of the angular, for which reason it cannot be regarded as representing or containing the prearticular.

The *squamosal* (paraquadrate of Gaupp) arises as a vertical strut of bone laterally to the processus muscularis of the quadrate while the latter still occupies its forward position. It gets carried backwards with the latter to its definitive position, and rotated so that its axis lies obliquely, sloping forwards and downwards. From its dorsal end a process grows backwards over the crista parotica, and, later, the zygomatic process grows forwards, thus giving rise to the T-shape of the adult bone.

The *quadratojugal* (quadrato-maxilla of Gaupp) ossifies on the outer surface of the portion of the quadrate cartilage which forms the articulation with the lower jaw, and grows forwards into contact with the maxilla. The quadratojugal is continuous with a perichondral ossification in the lateral surface of the quadrate cartilage, which probably represents a part of the quadrate bone.

The *pterygoid* (toothless) arises on the medioventral surface of the pterygoid process of the quadrate and spreads forwards into contact with the maxilla, and backwards, forking into a lateral portion which lies on the medial surface of the articular region of the quadrate cartilage, and a medial portion which underlies the pseudobasal connexion between the quadrate cartilage and the brain-case. In Hemisus, the pterygoid and squamosal fuse (de Villiers, 1931*c*).

The *nasal* overlies the roof of the nasal capsule and comes into contact laterally with the ascending portion of the maxilla. In Hemisus, the nasal fuses with the sphenethmoid.

The *prevomer* (paired) consists of four splints of bone united at their base radiating outwards and backwards beneath the floor of the nasal capsule, on the medial side of the choana. The medial splint of the prevomer ultimately bears teeth.

The *palatine* (toothless) arises as a transverse strut of bone ventrally to the lamina orbitonasalis, connecting the maxilla and the anterior end of the pterygoid with the sphenethmoid. The palatine is absent in some Anura (Microhyla, H. W. Parker, 1928; Ascaphus, de Villiers, 1934*a*; Scaphiopus, Megalophrys, Ramaswami, 1935*a*).

2. The cartilage-bones (in the order of their appearance). The *exoccipital* ossifies in the occipital arch, and the perichondral ossification invades the lateral part of the tectum synoticum and of the basal plate, and the hinder part of the auditory capsule. The jugular and perilymphatic foramina are thus enclosed by bone. The occipital condyles remain cartilaginous.

The *prootic* arises in the hinder border of the foramen prooticum and spreads (as external and internal perichondral lamellae) over the anterior part of the auditory capsule and the anterolateral region of the basal plate.

The columella auris ossifies in its middle region as the *mediostapedial* (of Parker), its lateral portion (now known as the extracolumella) and the attachment to the operculum and the operculum itself remaining cartilaginous.

The *mentomeckelian* ossifies in that anterior portion of Meckel's cartilage which originally formed in the infrarostral. It acquires contact with the anterior end of the dentary.

The posteromedial or thyroid process of the *hyoid* is the only part of the hyobranchial skeleton to undergo ossification.

The *sphenethmoid* ('os en ceinture' of Cuvier) arises as paired perichondral ossifications (on both outer and inner surfaces) of the preoptic roots of the orbital cartilages, and therefore represents the *orbitosphenoids*. Eventually they become connected with one another across the trabecular plate and across the hind edge of the roof of the nasal capsule. There appears to be no evidence for the presence in the sphenethmoid of Rana of a representative of the *mesethmoid*. But in the Indian Ranids Rhacophorus maculatus and Philautus petersi and oxyrhynchus, Ramaswami (1934) has described the sphenethmoid as involving ossification of the nasal septum. According to Schoonees (1930), in Bufo angusticeps the sphenethmoid ossifies first in the nasal septum. Yet in other species of these genera the nasal septum remains cartilaginous. Where present, such median ossification appears to represent the *presphenoid*.

According to Debierre (1895), a *basioccipital* may ossify in Bufo marinus.

It will be clear from the foregoing that most of the chondrocranium persists in the adult, especially the nasal capsule, the trabecular and basal plates, the hinder part of the auditory capsule, the tectum synoticum, pterygoquadrate cartilage, and hyobranchial skeleton.

II. Alytes obstetricans

The development of the skull of the midwife toad forms an interesting object of comparison with Rana. It has been studied by Peeters (1910), van Seters (1922) and Kruijtzer (1931), while the visceral arches have been dealt with by Ridewood (1898*a*). It may also be noted that it was in Alytes that Vogt (1842*a*) noted the vertebral nature of the occipital arch.

i. The development of the chondrocranium of Alytes obstetricans prior to metamorphosis

In the following account, based on the works of van Seters and Peeters, only those points will be stressed in which Alytes differs from Rana, since the general morphology of the chondrocranium in the two forms is the same.

1. Stage 1. (7 mm. stage.) PLATE 79, FIG. 1.

Chondrification takes place in caudorostral succession, the opposite of the conditions in Rana. The trabecular horns are fairly close together and parallel, the suprarostral cartilages diverging away strongly on each side. The main portion of the orbital cartilage ('crista trabeculae') is present at an early stage, connected with the top of the pila antotica. The floor of the auditory capsule is already chondrified and well developed, excluding the parachordal from sharing in the border of the fenestra ovalis. The auditory capsule chondrifies from two centres (anterior and posterior) instead of one as in Rana.

2. Stage 2. (53 mm. stage: tail 31 mm., hind limbs 3 mm.).

This stage closely resembles that of Rana at 29 mm. (p. 203) but with the following noteworthy difference. A prefacial commissure is present but obscured by the fact that the auditory capsule is pressed forward, and the prefacial commissure as it were duplicates its medial wall. The result is that the facial nerve on leaving the cranial cavity enters a canal running forwards and bounded medially by the prefacial commissure, laterally by the medial wall of the anterior part of the capsule, and ventrally by the anterior basicapsular commissure. The facial nerve never enters the true cavity of the auditory capsule at all; the *appearance* of the passage of the nerve through the capsule is due solely to the delay in chondrification of the capsule's medial wall (lateral wall of the facial canal). The existence of such a facial canal is of interest since a comparable state of affairs is found in Urodela (p. 178).

The pterygoid and pseudopterygoid processes are especially well developed; on the other hand, there is no larval otic process. The taeniae marginales are attached to the front instead of the back of the roof of the auditory capsule.

In the visceral arch skeleton may be noted the presence of admandibular (paramandibular) cartilages, in the form of small paired struts situated laterally to Meckel's cartilage. They may represent vestigial tentacular cartilages, such as are found in Xenopus (p. 215).

The hyobranchial skeleton at about this stage shows a well-developed basihyal or anterior copula, with distinct traces of paired origin, according to Kallius (1901), while the pars reuniens between the bases of the ceratohyals is not yet cartilaginous. The paired hyobranchial plates are separated from one another by the basibranchial or posterior copula, and the ceratobranchials are *not* in cartilaginous continuity with the hypobranchial plates.

ii. Metamorphosis of the chondrocranium of Alytes obstetricans

The changes in the chondrocranium of Alytes appear to be similar in general to those in Rana, but in some respects there are differences of considerable importance.

The most interesting of these concern the relations of the plectrum or columella auris, and the formation of the pseudobasal connexion between the quadrate and the neurocranium, as described by Kruijtzer (1931).

The pars interna plectri arises in the anterior part of the fenestra ovalis, in front of the operculum, and runs forwards towards the quadrate. Medially to the anterior free end of the plectrum, the cartilage which will give rise to the pseudobasal connexion (between the quadrate and the neurocranium) arises as an independent chondrification, which is, however, attached to the plectrum by mesenchyme. For this reason Kruijtzer regards the pseudobasal cartilage as of hyoid nature, and considers it to represent the pharyngohyal, a view which is very difficult to hold since the columella auris has a much greater claim. Further, the pseudobasal cartilage gives off a dorsal process which stretches up to the crista parotica immediately medially to the otic process of the quadrate. This dorsal process, likened by Kruijtzer to a laterohyal, corresponds, he thinks, to the isolated nodules of cartilage found in the same position in Megalophrys (Kruijtzer), Pipa (van Kampen, 1926), and to the intercalary or dorsal process of the columella auris in Lacerta. It is, however, not clear from Kruijtzer's descriptions (l.c., p. 107) whether the plectrum does not *also* possess a dorsal process connecting it with the crista parotica, and this is what would correspond to the processus dorsalis of Lacerta. At all events, it is important to note that in Alytes there is no subdivision between internal and external parts of the plectrum, and that the dorsal end of the ceratohyal, after becoming detached from the hinder surface of the quadrate, becomes attached to the plectrum.

A peculiarity of less importance in Alytes is the presence of a large independent supraorbital cartilage. This structure (known among Anura only in Alytes and Phyllomedusa) arises during metamorphosis dorsally to the bulb of the eye, as Stadtmüller (1931*c*) has shown, but is separated from the dorsal edge of the orbital cartilage by the lateral edge of the frontoparietal bone.

iii. Ossification of the skull of Alytes obstetricans

The chief peculiarity in the ossification of the skull of Alytes is the appearance of a V-shaped membrane-bone or *parahyoid* on the ventral surface of the 'hyoid' plate. The morphological value of this structure, which also occurs in Pelodytes, Ascaphus, Liopelma, Discoglossus, Bombinator, Pelobates, and

Rhinophrynus (Stadtmüller, 1936), is unknown. The parietal is sometimes perforated by the trochlear nerve, which also perforates the orbital cartilage.

III. Megalophrys montana

The chondrocranium of Megalophrys, as interpreted by Kruijtzer (1931), shows a number of features of interest. The cartilage of the pseudobasal connexion has an independent chondrification and becomes attached posteriorly to a process which projects forwards and downwards from the crista preopercularis (ventral edge of the fenestra ovalis). The cartilage of the pseudobasal connexion first forms a joint with the quadrate, and subsequently fuses with it; the autosystyly is thus complete.

A separate nodule of cartilage ('laterohyal') intervenes between the otic process of the quadrate and the crista parotica, as in Pipa. The plectrum appears to arise from internal and external centres of chondrification.

A peculiarity is the presence behind the suprarostral cartilages (which are attached by ligament to the trabecular horns) of a second pair of suprarostrals; these structures of doubtful significance, are compared by Kruijtzer with the posterior upper labial cartilages of Selachii.

The morphology of the adult skull of Megalophrys (Megophrys) sp. has been described by Ramaswami (1935*a*), who says that there are no palatine bones; but this is not corroborated by H. W. Parker.

IV. Bombinator pachypus

The development of the skull of Bombinator was studied by Goette (1875) in his classical monograph, but his descriptions are difficult to reconcile with the results obtained in other Anura by more modern methods. There is, however, little doubt that in general features Bombinator conforms to the typical Anura. Two points of interest from the work of recent investigators may be mentioned: Iwanzoff's (1894) statement that the proximal end of the ceratohyal becomes attached to the operculum has been contradicted by Kothe (1910). On the other hand, there is still uncertainty concerning the presence of a columella auris, for whereas Litzelmann (1923) claims its presence, Stadtmüller (1931*a*) has been unable to observe it. Ecke (1935) agrees that it has been reduced altogether in Bombinator, while in Pelobates the columella is represented by two cartilages, presumably the partes interna et externa plectri. The operculum is present, but no tympanic ring. Stadtmüller (1931*b*) has found that a part of the ligament connecting the ceratohyal with the quadrate may undergo chondrification as an independent para-articular cartilage.

The basihyal is especially well developed (Ridewood, 1898*b*). A parahyoid bone is present (Fuchs, 1929).

V. Ascaphus truei

The development of the skull of Ascaphus has not been studied, but its morphology as described by de Villiers (1934*a*) shows features of the greatest interest.

Ascaphus is the only Anuran known in which a true basitrabecular process is present, projecting from the hind part of the trabecular region of the skull, lying dorso-anteriorly to the palatine nerve, and ending freely. This basitrabecular process is connected posteriorly by the postpalatine commissure with the floor of the auditory capsule. The hyomandibular branch of the facial nerve runs outwards dorsally to the postpalatine commissure. In addition to these structures belonging to the neurocranium, there is a basal process projecting inwards from the quadrate, anteroventrally to the basitrabecular process and ending freely. It is possible that this basal process is the vestige of the anterior quadrate cranial commissure, carried backwards with the displacement of the quadrate at metamorphosis, and lost at that time in all other Anura.

Lastly, there is, as in Alytes, a prefacial commissure and a facial canal which, lacking its lateral wall (the medial wall of the auditory capsule) appears to lead the facial nerve through the cavity of the auditory capsule.

The otic process of the quadrate is fused with the crista parotica, but there seems in the adult to be no fusion between the basitrabecular and basal processes. The autosystyly of Ascaphus is thus incomplete.

In the wall of the auditory capsule the operculum is present, but plectrum and tympanic ring are absent.

The palatine bone is wanting; a parahyoid is present.

VI. Liopelma hochstetteri

As in the case of Ascaphus, the skull of Liopelma is known only from its adult morphology, and Wagner's (1934) descriptions are of interest since in some respects Liopelma is intermediate between Ascaphus and other Anura. For instance, the palatine nerve runs downwards in a notch cut into the anterior edge of the postpalatine commissure. It looks, therefore, as if the basitrabecular process had recently been cut through by the palatine nerve, leaving the postpalatine commissure behind that nerve as in Rana. A parahyoid bone is present (Trewavas, 1933; Stadtmüller 1936) as a membrane-bone underlying the hyoid plate.

VII. Xenopus laevis

The development of the chondrocranium of Xenopus before metamorphosis has been studied with the help of wax-models by Kotthaus (1933), while the hyobranchial skeleton has been investigated by Ridewood (1897*a*). Parker (1876*a*) described the skull of 'Dactylethra capensis (=Xenopus laevis)', but according to Kotthaus it cannot have been this species, but probably Xenopus calcaratus, which shows considerable differences.

i. Development of the chondrocranium of Xenopus

The differences in development between this Aglossan and the Phanero-glossan Anura as represented by Rana are considerable, but in the following account (based on Kotthaus's work) the conditions will as far as possible be described in terms of the frog.

1. Stage 1. (7·4 mm.; head and trunk 2·9 mm.; 5 days old.)

The parachordals and trabeculae are present and joined, the latter being interconnected by a very extensive rectangular trabecular or ethmoid plate, the anterolateral corners of which represent the trabecular horns. No separate suprarostral cartilages are present (nor are they at any stage in Xenopus).

The quadrate is connected with the side of the trabecular plate by a very wide quadratocranial commissure; a processus muscularis is present, and posteriorly the quadrate ends freely, without contact with the neurocranium. Meckel's cartilages are separated from one another in the midline by a median transverse cartilage which represents either the basimandibular, or the fused infrarostrals, probably the latter, since there is evidence of its being paired. There is as yet no contact between quadrate and ceratohyal.

The ceratohyals are very wide and triangular in shape. They are separated from one another in their hinder region by a median copula which probably represents the 1st basibranchial (perhaps containing the basihyal in its anterior region). The hind end of the copula is connected with hypobranchial plates on each side, themselves continuous with the four ceratobranchials.

2. Stage 2. (12·3 mm.; head and trunk 5·1 mm.; 8 days old.)

The auditory capsules are present (with much fenestrated walls) and the occipital arches. The pila antotica is present and attached to it laterally is the processus ascendens of the quadrate. Anteriorly, the pila antotica is continuous with the orbital cartilage which, attached to the trabecula farther forward by its preoptic root, encloses the optic and oculomotor nerves in a common foramen. Anteriorly, the trabecular plate remains flat and the olfactory nerves pass freely forward over it. The trabecular horns project freely sideways from its anterolateral corners.

The hyobranchial skeleton has become more massive, and the dorsal ends of the ceratobranchials are interconnected by terminal commissures. The dorsal ends of the ceratohyals articulate with a facet on the under surface of the quadrates.

3. Stage 3. (18·2 mm.; head and trunk 7·8 mm.; 11 days old.)

The hypophysial fenestra is now almost obliterated, leaving special foramina for the internal carotid arteries. The jugular foramen is enclosed by the contact of the occipital arch with the auditory capsule, and the prootic foramen likewise by the connexion of the pila antotica with the anterior part of the roof of the auditory capsule. There is no prefacial commissure.

On the lateral surface of the auditory capsule the crista parotica begins to appear, while from the hind end of the capsule a 'processus muscularis' (mastoid process?) grows downwards and sideways.

In the anterior region the tentacular cartilage is now well developed. It arises from the quadrate at the anterior edge of the processus muscularis (of the quadrate), runs forwards parallel to Meckel's cartilage, and touches the end of the trabecular horn, from which point it curves sideways into the tentacle.

The dorsal end of the 2nd ceratobranchial projects upwards beyond the terminal commissure into a processus cranialis which will eventually fuse with the crista parotica.

4. Stage 4. (48 mm.; head and trunk 23·3 mm.; 53 days old.)

This stage which represents the optimum development of the premetamorphic chondrocranium, presents a highly specialized appearance.

The lateral wall of the auditory capsule is now produced into an enormous crista parotica which is continuous with the processus muscularis (mastoideus), leaving a foramen for the jugular vein, and forming a roof over the branchial basket, to which its processus branchialis is attached. A tectum synoticum is present. The fenestra ovalis is occluded by an operculum which apparently arises in contact with the pars interna of the columella. The quadrate is now attached to the auditory capsule by an otic process, from the under side of which there develops the annulus tympanicus (termed by Kotthaus the processus basilaris) which almost joins the anterior corner of the crista parotica. (Plate 79, figs. 3–5).

A pila metoptica now separates the optic and oculomotor foramina. Anteriorly, the nasal capsule has appeared. So far as can be judged, this arises by the formation of a nasal septum in the midline on the dorsal surface of the trabecular plate, and a roof continuous with the dorsal edge of the septum. The olfactory nerves thus find themselves in canals, of which the lateral walls appear to be formed by anterior continuations of the orbital cartilages. The orbitonasal foramen (for the entry into the capsule of the nasal branch of the profundus nerve) is far forward; laterally to it the side wall of the capsule is formed by the lamina orbitonasalis which is continuous with the anterior edge of the quadratocranial commissure.

The branchial arches form a pair of elongated troughs or branchial baskets, markedly concave dorsally, and pierced by three gill-slits. This condition has been brought about by great widening and curving of the 1st and 4th ceratobranchials (forming the front and hind walls) and of the hypobranchial plates and terminal commissures (forming the medial and lateral walls). The processus cranialis of the branchial basket is fused with the processus branchialis of the crista parotica, so that the brain-case and the entire visceral arch skeleton are in one piece. The edge of the crista parotica fits that of the branchial basket fairly closely.

The absence of separate suprarostral cartilages, articulated with the trabecular horns (whose distal parts they seem to be) is associated with the fact that Xenopus has no horny teeth, and relies during larval life on what microscopic food it can obtain with the current of water sucked in through the mouth. The high degree of development of the branchial basket is associated with the production of this water current.

ii. Metamorphosis

The changes occurring at metamorphosis in Xenopus are known only from Parker's (1876*a*) work on Dactylethra capensis. The adult form is not unlike that of other Anura, and this is achieved by the backward displacement of the quadrate and the completion of the nasal capsule, while the annulus tympanicus becomes separated from the quadrate, and the tentacle and the crista parotica disappear. The sclerotic cartilage persists in the adult and according to Stadtmüller (1929*a*) is unique in possessing a cartilaginous process to which the inferior rectus muscle is attached.

The changes in the hyobranchial skeleton have been followed by Ridewood (1897*a*). The ceratobranchials become reduced, and the posteromedial process (thyrohyal cartilage) arises quite separately from them, on the medial side of the 4th ceratobranchial, as an outgrowth backwards on each side from the hypobranchial plate. The 'hyoid' plate is represented by large and elongated flat outgrowths from the hypobranchial plates, attached anteriorly to the ceratohyals in the midline, and posteriorly to the thyrohyal cartilages. The 'hyoid' plate is perforated in the midline by the hyoglossal foramen, in the hind border of which the hyoglossus muscle is inserted.

De Villiers (1932) has described the columella auris as consisting of a pars externa and pars interna plectri, united by an ossified pars media. Behind the pars interna plectri which blocks most of the aperture of the fenestra ovalis, is a small operculum. This latter is lacking in Pipa and Hymenocheirus. The thyrohyal cartilages are attached to the skeleton of the larynx, which shows sexual dimorphism.

In the closely related Hymenocheirus (Ridewood, 1899) the 'hyoid' plate is detached from the ceratohyals, and the latter are unique in Anura in undergoing ossification. On the other hand, in Pipa (Ridewood, 1897*a*) the ceratohyals disappear at metamorphosis.

Concerning the bony skull of Xenopus it is of interest to note that Winterhalter (1931) has found a pineal foramen between the parietals.

VIII. Other Anura

In addition to the Anura mentioned above, the development of the skull in a number of forms has been studied by Parker, including Bufo, Pipa (1876*a*), Hyla, Nototrema, Calyptocephalus, Camariolius, Pseudis, and several species of Rana. His descriptions of the chondrocrania suffer from the methods of investigation at his disposal, but these works may be consulted with great profit for information concerning the bony skull of the forms mentioned as well as of many others of which only the adult stage was studied. The same applies to the splendid series of studies on the morphology of the skull of numerous adult South African Anura by de Villiers (1929–34) and C. and G. du Toit (1930–4). Similarly some Indian forms have been described by Ramaswami (1932–5).

16. **REPTILIA.** LACERTILIA

I. Lacerta

The chondrocranium of the lizard has been the object of study of Leydig (1872), and W. K. Parker (1880*a*), albeit by primitive methods. Born's (1879) work on the nasal capsule was the first application of the wax-plate model reconstruction method, which has since been used by Gaupp (1900 and 1906*a*) for the whole skull of the 31 mm. and 47 mm. stages of Lacerta agilis. The earlier stages have been studied by de Beer (1930*a*) with the help of the van Wijhe method. The development of the columella auris has been studied by Hoffmann, (1889, 1890), Versluys (1898, 1903), Cords (1909), and Dombrowsky (1918, 1924).

The development of the bony skull is known from the works of W. K. Parker (1880*a*) and Gaupp (1906*a*).

i. Development of the chondrocranium of Lacerta agilis

The visceral skeleton is the first to chondrify, and in many embryos of 2·25 mm. (head length), Meckel's cartilage is the only one which can be discerned. In others, the parachordals, continuous posteriorly with a slender occipital arch, show faint chondrification (Plate 80, fig. 1).

At 2·5 mm. (h-l.) the canalicular portions of the auditory capsules are present, as paired independent cartilages moulded on to the lateral surfaces of the utricles. Of the three permanent roots of the hypoglossal nerve, which at the previous stage emerged freely over the parachordal, the hindmost is now enclosed in a foramen between the occipital arch and a preoccipital arch which has chondrified in front of it (Figs. 2, 3).

The trabeculae make their appearance at 3·5 mm. (h-l.). Anteriorly, they are already fused to form a trabecula communis; posteriorly, the trabeculae diverge but are separated by a considerable distance from the parachordals (Figs. 4, 5). The 2nd hypoglossal root is on one side enclosed in a foramen, on the other, in a notch.

In the visceral skeleton the quadrate is now present. Ventrally, there is a median processus lingualis, representing the basihyal, and chondrifying independently. On each side of the hind end of this are the hypohyals, likewise independent, the ceratohyals, and the 1st and 2nd ceratobranchials (Fig. 6). Meckel's cartilages show a sigmoid curvature.

At later stages (4 mm. h-l.) the trabecula communis and Meckel's cartilage begin to elongate (Plate 81, figs. 1, 2). All three hypoglossal roots are enclosed in foramina: there are therefore three preoccipital arches in front of the occipital arch, and as each of these may be regarded as the representative of a former vertebral neural arch, the hind region of the skull shows traces of segmentation into four segments.

At their extreme hind end the parachordals are in cartilaginous connexion with one another by means of a hypochordal commissure. It is from this cartilage that the median hypocentral occipital condyle will arise.

The anterior edge of the parachordals is a considerable distance behind the tip of the notochord, and is ultimately destined to form the hind border of the basicranial fenestra. The lateral borders of this fenestra are now present in the form of processes projecting forwards from the anterolateral corners of the parachordals, and extending towards (but still free from) the trabeculae.

The auditory capsule is now more completely chondrified. The roof and septa of the semicircular canals are present, as is the wall of the cochlear portion. The latter arises in continuity with the canalicular portion and is fused with the lateral edge of the parachordal near its anterior end (by the basicochlear commissure). Farther back, the auditory capsule has developed a second connexion with the parachordal (the basivestibular commissure). These attachments result in the formation of a notch between the anterior edge of the auditory capsule and the parachordal—the incisura prootica; a cleft between the hind edge of the auditory capsule and the occipital arch—the fissura metotica; and a fenestra enclosed between the basicochlear commissure, the lateral edge of the parachordal, the basivestibular commissure, and the ventral edge of the canalicular portion of the auditory capsule—the confluent fenestra ovalis (or vestibuli) and fenestra basicapsularis. The basicapsular fenestra represents the as yet unchondrified floor of the auditory capsule: the fenestra ovalis has at this stage, therefore, no cartilaginous ventral border.

In the fenestra ovalis the proximal portion (otostapes of Hoffmann, 1889, 1890) of the columella auris is present as an independent nodule of cartilage, entirely free from the auditory capsule. The columella auris is definitely a part of the hyoid arch skeleton, as Cords (1909) showed. The foramen perilymphaticum is a gap in the hind wall of the cochlear portion of the auditory capsule, opening towards the fissura metotica.

Through the fissura metotica emerge the glossopharyngeal and vagus nerves; through the incisura prootica the trigeminal nerve emerges, and it appears to be separated at a very early stage from the facial nerve by the prefacial commissure. The facial nerve thus emerges through a facial foramen, enclosed between the prefacial commissure in front and the basicochlear commissure behind.

Other novelties at this stage are the interorbital septum, an independent piece of cartilage dorsal to the trabecula communis near its anterior end, and the paired processus ascendens or columella cranii. The latter is an element of the visceral skeleton and its base is connected with the quadrate by a strand of dense mesenchymatous tissue (described by Gaupp, 1891) which may undergo chondrification, as also in Zonurus, Eremias, and Mabuia described by Broom, (1903, 1924) and in Eumeces, described by Rice (1920), thus presenting a condition resembling that found in Rhynchocephalia, Chelonia, and Crocodilia.

At the stage of 4·5 mm. (h-l.) the hind ends of the trabeculae are fused with the anterior projections of the parachordals (Plate 81, figs. 3, 4). No independent polar cartilages are to be seen, but their position is indicated by the short basitrabecular processes, which are borne by the polar cartilages in such forms as possess them. These basitrabecular processes are for a time in procartilaginous connexion with the pterygoquadrate. A hypophysial fenestra is now enclosed, but it is still confluent with the basicranial fenestra behind it. The junction between trabeculae and parachordals is effected in such a way as to form an angle pointing upwards at the place where the cartilage separating the hypophysial and basicranial fenestrae will arise.

The interorbital septum has extended and is continuous below with the trabecula communis, and above with a new structure, the planum supraseptale. The latter represents the orbital cartilages, paired and wide apart in other forms, but here forced together in the middle line. Anteriorly, the planum supraseptale is continued into a pair of processes which at this stage end freely—the sphenethmoid processes. In front of them, the roof of the nasal capsule is beginning to chondrify as the parietotectal cartilage.

In the auditory capsule the floor has now chondrified, thus obliterating the basicapsular fenestra and forming a ventral border to the fenestra ovalis. The wall of the lateral semicircular canal forms a prominence, the crista parotica, against the anterior portion of which the dorsal end of the quadrate abuts.

At the 5 mm. stage (Plate 82, figs. 1, 2) the interorbital and nasal septa are more extensive, though leaving considerable unchondrified gaps as fenestrae septi. The parietotectal cartilages are continuous with the dorsal edge of the nasal septum. The planum supraseptale has expanded to form a triangular plate underlying the cerebral hemispheres. From the hinder edge of the planum supraseptale a pair of processes extend towards but do not yet touch the rudiments of the taeniae mediales which are at this stage the isolated and sole representatives of the side wall of the skull in this region, except for the stumps of the pilae antoticae which project dorsally from the points of junction between the trabeculae and parachordals. From these same points, the rudiments of the crista sellaris extend towards one another, meeting at the tip of the notochord, and separating the hypophysial from the basicranial fenestrae.

In the visceral arch skeleton, an isolated pair of cartilages representing the posterior or dorsal portions of the 2nd ceratobranchials have appeared.

At the 5·25 mm. stage (Plate 82, figs. 3, 4) the side wall of the skull in the orbitotemporal region is complete. The taenia marginalis extends from the planum supraseptale to the auditory capsule; the taenia medialis from the planum supraseptale to the point of junction of the pila antotica with the pila accessoria, the former running down to the foremost edge of the parachordal, the latter running up to the taenia marginalis; the pila metoptica runs down from the taenia medialis until it almost touches the trabecula communis, it then makes a sharp bend forwards, joins its fellow from the opposite side to form the subiculum infundibuli, and runs into the hind edge of the interorbital septum. The pila metoptica is the postoptic root of the orbital cartilage; the contact between the planum supraseptale and the interorbital septum is to be regarded as part of the preoptic root of the orbital cartilage.

The various pilae and taeniae between them subdivide the orbitotemporal region of the side wall of the skull into four fenestrae: optica (through which the optic nerve emerges), epioptica (no nerve), metoptica (oculomotor and trochlear nerves), and prootica (trigeminal nerve). The pituitary vein enters the skull through the fenestra metoptica; the abducens nerve penetrates a foramen in the root of the pila antotica.

The basitrabecular processes are now large structures. Intercalated between each basitrabecular process and the processus ascendens of its own side is a small nodule of cartilage, the meniscus pterygoideus. This cartilage was shown by Gaupp (1891) to be connected with the base of the processus ascendens in early stages, and (Gaupp, 1902) probably to represent the basal process of the pterygoquadrate, which thus becomes detached (see p. 228).

The wall of the auditory capsule has closed in round the footplate (otostapes of Hoffmann, 1889, 1890) of the columella auris, thus restricting the aperture of the fenestra ovalis. The distal portion of the columella auris (hyostapes) has chondrified independently as Versluys (1903) showed, and subsequently becomes attached to the proximal portion (otostapes). The columella is now a rod expanded at each end, proximally covering the fenestra ovalis of the auditory capsule; distally covering the tympanic membrane with a processus inferior. Dorsal to the columella is a nodule of cartilage, in mesenchymatous connexion with it, and forming the processus dorsalis. This cartilage becomes attached to the crista parotica of the auditory capsule, and is known as the intercalary or processus paroticus.

Dorsally, the auditory capsules are interconnected by a slender tectum synoticum, which is the sole representative of the cartilaginous roof of the skull.

The chondrification of the nasal capsule has now made great progress, and by the extension of the parietotectal cartilages, much of the roof and of the anterior part of the side walls have been formed. In the regions of the hinder part of the side wall and the back wall of the nasal capsule, two additional pairs of cartilages have appeared. Of these, the paranasal cartilages are crescentic structures, the anterior horn of which is attached to the posterolateral edge of the parietotectal cartilage, enclosing at the point of junction the lateral nasal branch of the ethmoidal nerve in a small epiphanial foramen. Near this point also the sphenethmoid process is attached, thus forming the sphenethmoid commissure. In this way the fenestra olfactoria evehens is enclosed between the parietotectal cartilage anteriorly, the nasal septum medially, and the sphenethmoid commissure posterolaterally, and through which the olfactory nerve emerges from the brain-case.

The other additional pair of cartilages, the laminae orbitonasales, are at this stage isolated and lie as transverse struts on each side of (but free from) the nasal septum, beneath the sphenethmoid commissures, and medially to the hind ends of the paranasal cartilages.

The body of the paranasal cartilage forms as it were a reduplication of the side wall of the nasal capsule in this region, for the hind edge of the side wall formed from the parietotectal cartilage extends back a little way medially to the paranasal cartilage. Between these two cartilaginous plates there is therefore a space, the cavum conchale, which ends blindly posteriorly and opens anteriorly by means of an aperture, the aditus conchae, situated immediately beneath the foramen epiphaniale.

The walls of the cavum conchale form the so-called concha nasalis, which lodges the lateral nasal gland, the duct of which passes forwards and out through the aditus conchae. The cavum conchae is, of course, an extracapsular space. The isolated rudiments of the paired paraseptal cartilages have appeared, just below the ventral edge of the nasal septum.

Further progress in the formation of the nasal capsule is seen at the 5·5 mm. (h-l.) stage (Plate 82, figs. 5, 6), where the laminae orbitonasales are attached by their lateral edge to the hinder part of the paranasal cartilage of their own side, thus forming the hind wall to the nasal capsule. The side walls are still very incomplete, and indeed one large region, representing the fenestra lateralis, will never be chondrified in Lacerta (though it is in Eumeces and Lygosoma), and it is through this gap that it is possible to see the concha and part of the nasal septum.

ii. Morphology of the fully formed chondrocranium of Lacerta agilis

The chondrocranium of Lacerta at the 31 mm. (t.l.) stage has been studied very thoroughly by Gaupp (1900), whose work forms the basis of the following description (Plate 83; Plate 84, figs. 1–6).

1. The basal plate and notochord. The basal plate is formed by the parachordal cartilages which have fused beneath the notochord: the latter comes out of the tip of the odontoid process of the axis vertebra, where, according to Pusanow (1913), it undergoes endochordal chondrification, and runs forward in a groove on the dorsal surface of the basal plate, traverses the basicranial fenestra, and ends immediately behind the crista sellaris. The fact that the notochord runs dorsally to, instead of through, the basal plate at the extreme posterior edge of the latter is a fact of some importance in connexion with the nature of the occipital condyle and the occipito-atlantal joint.

It is also to be noticed that the basicranial fenestra arises as such; it is not the result of resorption in pre-existing cartilage.

The limits of the basal plate may be defined as follows. Posteriorly, the ventral border of the foramen magnum forms its hind edge. Laterally, the limit of the basal plate follows a line which may be drawn through the medial border of the three hypoglossal foramina, then along the medial border of the anterior portion of the fissura metotica, to the auditory capsule. The basal plate has been slightly compressed from side to side between the cochlear portions of the auditory capsules, and its lateral border thus skirts the inner limit of the cochlear capsule and turns laterally to form the ventral edge of the incisura prootica. It is a result of the encroachment by the cochlear capsule into space originally occupied by the basal plate, that the prefacial commissure rests on the dorsal surface of the cochlear capsule instead of on the lateral edge of the basal plate as it does in lower vertebrates.

The root of the pila antotica marks the anterolateral corner of the basal plate; the two corners being connected with one another across the middle line by the crista sellaris, which forms the anterior border of the basicranial fenestra, and corresponds to the dorsum sellae or acrochordal cartilage of other forms. The abducens nerve, which in many skulls emerges through the fenestra prootica, in Lacerta penetrates a foramen piercing the root of the pila antotica.

2. The occipital region. The occipital arches rise up from the posterolateral corners of the basal plate and are interconnected dorsally by a strip of cartilage, the tectum posterius, which is fused (though still visibly distinct from) the tectum synoticum immediately in front of it, and forms the dorsal border to the foramen magnum.

In front of the (definitive) occipital arches are three pairs of preoccipital arches, which spring from the lateral edge of the basal plate but are of very small size, and serve only to separate the three permanent hypoglossal nerve roots from one another and to enclose them in foramina by fusion with one another, dorsally to the nerves.

The fused ends of the preoccipital arches and the lateral surface of the chief occipital arch form the posteromedial border of the hinder part of the fissura metotica, which is closed dorsally by the fusion between the tectum posterius and tectum synoticum. Through this posterior part of the fissura metotica (which forms the foramen jugulare) the vagus and spinal accessory nerves and the jugular vein leave the cranial cavity.

The occipital condyle is formed from the hindmost edge of the basal plate, in the middle line. Seen from behind, the condyle takes the form of a kidney-shaped shallow prominence, the concavity dorsal (lodging the tip of the odontoid process) and the convexity facing ventrally. The lateral portions of the condyle, which articulate with lateral facets on the atlas vertebra, are slightly more prominent than the median portion which articulates with a median process on the ventral atlas ring.

The relations of the condyle to the notochord (i.e. the fact that the latter passes freely and dorsally over the former) and the study of the segmentation in early stages favour the view held by Hayek (1923–4), to the effect that the Lacertilian occipital condyle represents the hypocentral element of the vertebra next in front of the atlas (the proatlas vertebra). In the ordinary course of events, a hypocentrum fuses with the pleurocentrum immediately behind it to form a vertebra. In the case of the proatlas vertebra of Lacertilians, however, the pleurocentral element appears to fuse, not with the hypocentral element (and therefore basal plate) in front of it, but with the tip of the odontoid process of the axis vertebra behind it. The odontoid process, therefore, contains the pleurocentral elements of both the atlas and proatlas vertebrae, and as is to be expected, the notochord emerges from the centre of the tip of the odontoid process, which overlaps and lies above the occipital condyle.

The occipito-atlantal joint in Lacertilians, therefore, falls in between the hypocentral and pleurocentral elements of the proatlas vertebra: it is intravertebral and intersegmental.

3. The auditory region. The auditory capsules are large structures, as long as the basal plate, and extending the full dorsoventral height of the chondrocranium. This height is, however, not equal to that of the brain in this region, except at the extreme hind end where the auditory capsules are interconnected by a narrow tectum synoticum, beneath which the medulla oblongata connects with the spinal cord. The tectum synoticum is fused with the tectum posterius immediately behind it, and anteriorly it bears a slender process, the processus anterior tecti (called ascendens by Gaupp), which points forwards and upwards and ends blindly. To the dorsal surface of each capsule, about half-way along its length, the hind end of the taenia marginalis is fused.

The base of the auditory capsule is attached to the lateral edge of the basal plate by an extensive fusion, the basivestibular commissure, which reaches from the posteroventral corner of the fenestra prootica to the anterior end of the fissura metotica. The external surfaces present a number of prominences representing internal enlargements for the accommodation of the various parts of the membranous labyrinth. Thus, on the lateral surface of the capsule there may be distinguished the prominentiae: semicircularis anterior, lateralis, and posterior; ampullaris anterior, lateralis, and posterior; recessus utriculi, saccularis, and cochlearis. The medial surface of the capsule bears the prominentiae: ampullaris posterior; cochlearis, and utricularis; a dimple beneath the anterior semicircular canal forms the so-called fossa subarcuata. The prominentia semicircularis lateralis is accentuated to form a ridge, the crista parotica, to which the processus paroticus is fused.

The cavity of the auditory capsule is somewhat complicated, and is made up as follows. The main space is a cavum vestibulare, which is divided into anterius and posterius portions by a vertical transverse septum intervestibulare, pierced by two foramina (foramen intervestibulare, laterale, and mediale).

The cavum vestibulare anterius gives off the recessus ampullaris anterior and lateralis, which lead into the anterior orifices of the anterior and lateral semicircular canals respectively.

The cavum vestibulare posterius opens widely anteroventrally into the cavum cochleare, and posteriorly into the recessus ampullaris posterior which leads into the inferior orifice of the posterior semicircular canal. The posterior orifice of the lateral semicircular canal opens into the cavum vestibulare posterius just dorsally to the recessus ampullaris posterior. In the roof of the cavum vestibulare posterius is the foramen pro sinu superiore, which leads into the posterior orifice of the anterior semicircular canal and into the superior orifice of the posterior semicircular canal.

The semicircular canals themselves are separated off from the main cavity (cavum vestibulare anterius and posterius) of the capsule by septa, the septum for the anterior canal being horizontal and forming the roof of the cavum vestibulare anterius and posterius; the septum for the lateral canal being vertical and longitudinal, and forming the lateral wall of the cavum vestibulare posterius, while the septum for the posterior canal is vertical and transverse, and forms the hind wall of the cavum vestibulare posterius. In one small region the septum of the posterior semicircular canal is incomplete, allowing communication between the cavity of that canal and the cavity of the lateral canal near the latter's posterior orifice.

Having described the cavities of the auditory capsule, attention must now be turned to the identity of the parts of the membranous labyrinth which the various cavities accommodate. In this connexion the recessus ampullaris anterior, lateralis, and posterior, and the three canals, present no difficulty, since, as their names imply, they lodge the various ampullae and semicircular canals. The anterior portion of the utricle (connecting with the anterior and lateral ampullae) and the recessus utriculi, lie in the cavum vestibulare anterius. The posterior portion of the utricle lies in the cavum vestibulae posterius, and communicates with the sinus superior, the posterior ampulla, the posterior limb of the lateral semicircular canal, and with the anterior portion of the utricle through the foramen intervestibulare mediale. (The foramen intervestibulare laterale is occluded by membrane and transmits nothing.)

The cavum vestibulare also lodges the ductus endolymphaticus and the saccule, while the lagena (ductus cochlearis) and the cavum perilymphaticum lie in the cavum cochleare.

The apertures and foramina in the walls of the auditory capsule must now be considered. On the medial surface are to be found: the foramen acusticum anterius, which conveys the anterior branch of the auditory nerve (rami ampullae anterioris, ampullae lateralis, recessus utriculi) into the cavum vestibulare anterius; the foramen acusticum posterius, which conveys the posterior branch of the auditory nerve (rami ampullae posterioris, sacculi, neglectus, cochlearis) into the cavum vestibulare posterius; and the foramen endolymphaticum, through which the ductus endolymphaticus leaves the cavum vestibulare posterius and enters the cranial cavity.

The chondrification of the roof of the cavum cochleare is incomplete, leaving a vacuity, which is, however, not a foramen.

The facial foramen must be mentioned here, because although it has no intrinsic association with the auditory capsule, the extension of the cochlear portion of the capsule has resulted in the prefacial commissure resting directly upon it, so that the facial foramen and the foramen acusticum anterius come to be situated close to one another.

On the external surface of the capsule, the fenestra vestibuli (ovalis) is a large aperture in the lateral wall of the cavum cochleare, exactly covered by the footplate of the columella auris. The posterior wall of the cavum cochleare is perforated by the foramen perilymphaticum, which faces directly into the anterior portion of the fissura metotica. A transverse section through this region shows three free edges of cartilage: viz. the lateral border of the foramen perilymphaticum, the medial border of the foramen perilymphaticum, and the lateral edge of the basal plate. Between these edges, a space, triangular in cross-section, is enclosed—the recessus scalae tympani, which has three ways leading out of it: anterodorsally into the cavum cochleare by the foramen perilymphaticum (through which the ductus perilymphaticus enters the recessus scalae tympani where it expands to form the saccus perilymphaticus); medially into the cranial cavity by the apertura medialis recessus scalae tympani, which is occluded by a membrane of connective tissue pierced for the entry of the ductus perilymphaticus into the cranial cavity and for the exit from the latter of the glossopharyngeal nerve; laterally, towards the typanic cavity by the apertura lateralis recessus scalae tympani which is occluded by the secondary tympanic membrane, stretched across between the lateral border of the foramen perilymphaticum and the lateral edge of the basal plate, and thus separating the tympanic cavity from the recessus scalae tympani. The glossopharyngeal nerve passes out through the hinder part of the apertura lateralis recessus scalae tympani.

The recessus scalae tympani in Lacerta, therefore, is a space unenclosed by cartilage, and extracranial.

Although strictly part of the visceral arch skeleton, and arising quite independently of the auditory capsule, it will be convenient to treat the columella auris here. It is a rod-like structure, expanded proximally to form a footplate which fits over the aperture of the fenestra vestibuli, and distally into a cruciform plate (the extracolumella of Gadow, 1889) inserted on to the tympanic membrane (tympanum): its four arms being: the pars inferior (longest, and projecting forwards); pars superior (short, and projecting backwards); pars accessorius anterior (short, and projecting upwards); pars accessorius posterior (short, and projecting downwards). The latter represents the stump of the ligament, or pars interhyalis, which connected the columella with the ceratohyal at earlier stages.

Half-way along the shaft of the columella auris the processus internus arises and runs forwards to fuse with the hind surface of the quadrate. Also half-way along the shaft of the columella, (from the originally otostapedial portion) a strand of dense connective tissue (representing the processus dorsalis) runs up to the processus paroticus, which has become fused on to the crista parotica and against which the posterodorsal extremity of the quadrate rests. The processus paroticus corresponds to Parker's 'suprastapedial', and to the 'intercalary' of Versluys's (1898, 1903) terminology. But while Gaupp (1900) regarded the processus paroticus as entirely derived from the processus dorsalis, Versluys (1903, p. 127) considered that the processus paroticus is largely formed by growth of the crista parotica.

The morphological relations of the columella auris to neighbouring blood-vessels and nerves are of great importance; the following account is based on the descriptions given by Versluys (1898, 1903) and Goodrich (1915).

The hyomandibular branch of the facial nerve runs backwards from the facial foramen, and passes over the dorsal side of the shaft of the columella auris, medially to the processus dorsalis (or the strand connecting the columella with the processus paroticus). The hyomandibular nerve then gives off the chorda tympani and itself continues to run backwards and downwards, laterally to the pars interhyalis and to the ceratohyal (or cornu hyale). The chorda tympani runs forwards again over the dorsal side of the columella auris, but this time laterally to the processus dorsalis. It runs down in front of the columella auris, laterally to the processus internus, and then passes forwards to the medial surface of Meckel's cartilage.

The vena capitis lateralis runs dorsally to the columella auris, and to the hyomandibular branch of the facial nerve.

The stapedial artery arises from the internal carotid and runs dorsally, passing behind the columella auris, but in front of the hyomandibular branch of the facial nerve. The stapedial artery then turns forwards over the columella auris, medially to the processus dorsalis.

4. The orbitotemporal region. The region of the chondrocranium comprised between the nasal capsule in front and the auditory capsule behind, and through which the optic, eye-muscle, and trigeminal nerves emerge, consists for the most part of a scaffolding of slender cartilaginous rods separated by large fenestrae.

There is no roof to this part of the chondrocranium. The floor is formed by the paired trabeculae, which run forwards from the anterolateral corners of the basal plate to meet one another and form the trabecula communis, thus enclosing the triangular-shaped hypophysial fenestra which lodges the pituitary body and through which the internal carotid arteries enter the skull. From the hindmost part of the trabeculae, the paired basitrabecular (basipterygoid) processes are given off, extending laterally as triangularly shaped plates.

The trabecula communis is continuous with the interorbital septum immediately above it; a vertical sagittal plate of cartilage fenestrated by a fenestra septi. In front of and behind this fenestra (which is only an area of incomplete chondrification) the dorsal edge of the interorbital septum bears the paired orbital cartilages, here pressed together in the middle line between the relatively enormous eye-balls (which are enclosed in very delicate sclerotic cartilages), and forming the planum supraseptale, upon which the base of the cerebral hemispheres rests. The interorbital septum itself is to be regarded as the bases of the preoptic roots of the orbital cartilages, fused together in the midline.

The lateral edge of the planum supraseptale is connected with the auditory capsule of its own side by a sinuous cartilaginous rod, the taenia marginalis. Parallel with this, but at a more ventral and more medial level, the taenia medialis arises from the posterolateral corner of the planum supraseptale, and runs backwards to a point just in front of the anterior end of the auditory capsule. There, the taenia medialis connects with a vertical pillar, the pila antotica, which arises from the anterolateral corner of the parachordal at a point immediately lateral to the crista sellaris. The base of the pila antotica is pierced by the foramen for the abducens nerve. Half-way between the planum supraseptale and the pila antotica, the taenia medialis is connected with two other vertical pillars: one of these, the pila accessoria, connects with the taenia marginalis above; the other, the pila metoptica, runs ventrally and inwards, meets its fellow from the opposite side, and both then turn sharply forward to form a narrow horizontal plate—the subiculum infundibuli (upon which the base of the infundibulum rests). The subiculum infundibuli is then continued forwards as the cartilago hypochiasmatica, and fuses with the posterior edge of the interorbital septum, close to where the latter springs from the trabecula communis.

It is now possible to define the various fenestrae in the orbital region:

The optic fenestra, transmitting the optic nerve, bounded in front by the posterior edge of the interorbital septum, below by the cartilago hypochiasmatica and subiculum infundibuli, behind by the pila metoptica, above by the taenia medialis.

The metoptic fenestra, transmitting the oculomotor and trochlear nerves and the pituitary vein, bounded in front by the pila metoptica, below by the paired trabecula, behind by the pila antotica, and above by the taenia medialis.

The fenestra epioptica (transmits nothing), bounded in front by the posterolateral edge of the planum supraseptale, below by the taenia medialis, behind by the pila accessoria, and above by the taenia marginalis.

The fenestra prootica, transmitting the trigeminal nerve, bounded in front by the pila antotica and pila accessoria, below by the lateral edge of the basal plate, behind by the anterior wall of the auditory capsule and prefacial commissure, and above by the taenia marginalis.

Thus, in spite of the reduction in size of the orbital cartilages and their compression and distortion to form the planum supraseptale, they still preserve their typical connexions. The formation of the interorbital septum has only obscured the relations of the preoptic root of the orbital cartilage; this root is represented by the contact between the planum supraseptale and the interorbital septum.

A further development of the cartilage in the orbital region is seen at the 47 mm. (c.r.-l.) stage, in the formation of the supratrabecular bar. This structure connects the base of the pila metoptica (at the point where the latter turns forwards to form the subiculum infundibuli) with the base of the pila antotica. The supratrabecular bar is thus parallel with and dorsal to the paired trabecula, and subdivides the fenestra metoptica into a dorsal and a ventral portion. (Plate 86).

It will be convenient in this place to deal with the processus ascendens of the pterygoquadrate, although it strictly belongs to the visceral arch skeleton. The processus ascendens (columella cranii of Cuvier, epipterygoid of Parker, 1880*a*; antipterygoid of Gaupp, 1891) is a slender vertical pillar, situated laterally to the pila antotica, ending freely above, while its ventral end is continuous with the pterygoid process, which runs horizontally forwards and likewise ends blindly. The pterygoid process is later resorbed and has disappeared at the 47 mm. (c.r.-l.) stage. The connexion (mesenchymatous and sometimes cartilaginous) between the base of the processus ascendens and the body of the quadrate has broken down.

Immediately ventral to the base of the processus ascendens, and, as it were, intercalated between the latter and the underlying basitrabecular process, is a small nodule of cartilage—the meniscus pterygoideus. This cartilage, the rudiment of which at early stages is connected with the base of the processus ascendens, may be regarded as the representative of the basal process of the pterygoquadrate, articulating with the basitrabecular process of the neurocranium. Broom (1922) considers that it may represent an unossified metapterygoid. At later stages, after the appearance of the pterygoid bone the meniscus pterygoideus lies embedded on the medial surface of that bone and provides the articular facet in contact with the basitrabecular process.

Of great importance are the relations of the processus ascendens to the branches of the trigeminal nerve: V_1 runs forward on the medial side of the processus ascendens, while V_2 and V_3, and the orbital artery, run out behind it: the process is thus morphologically situated between V_1 and V_2.

In Lacerta the processus ascendens is lateral to the true side wall of the skull (represented by the various pilae and taeniae and the dura mater membrane which covers the fenestrae), and separated from it by a space, the cavum epiptericum, which lodges the trigeminal ganglion. It is important to note that this space, situated as it is between the processus ascendens and the true side wall of the skull, and dorsally to the basitrabecular process, is wholly extracranial in Lacerta.

5. The ethmoid region. The immediate anterior prolongation of the trabecula communis and of the interorbital septum forms the nasal septum, which divides the cavity of the nasal capsule into symmetrical right and left halves. Like the interorbital septum, the nasal septum is fenestrated by areas of incomplete chondrification.

The dorsal edge of the nasal septum passes directly into the roof of the capsule on each side, and the roof passes equally uninterruptedly into the anterior part of the side wall, thus forming the parietotectal cartilages.

The side wall of the posterior portion of the nasal capsule is situated farther laterally than that of the anterior portion, and the nasal capsule owes this greater width in this region to the fact that the concha nasalis is situated here and that the cavity of the capsule extends laterally to the concha to form the recessus extraconchalis on each side. The morphology of this region of the nasal capsule is best understood with the help of transverse sections (Pl. 85, figs. 1–8). The side wall of the anterior part of the capsule is seen to be continuous posteriorly with the medial wall of the concha nasalis. The cavity of the latter, or cavum conchale, is an extracapsular space, lodging the lateral nasal gland, and enclosed on all sides by cartilage except in front, where it opens by the aditus conchae. The cartilage forming the lateral wall of the concha is really a reduplication of the lateral wall of the capsule, and arises from the paranasal cartilage. The paranasal cartilage forms the side wall to the posterior part of the nasal capsule, but this wall is very incomplete, being perforated in Lacerta by a large fenestra lateralis, through which the side wall of the concha may be seen by looking right through the recessus extraconchalis.

Behind the fenestra lateralis, the side wall passes into the hind wall of the capsule, lamina orbitonasalis (or planum antorbitale), which is a rounded plate of cartilage, roughly in the transverse vertical plane, concave forwards, and almost touching the nasal septum. The ventrolateral corner of the lamina orbitonasalis is continued backwards into a triangular-shaped piece of cartilage, which bears two processes: one of these, the processus maxillaris anterior is blunt and points downwards; the other, the processus maxillaris posterior, is long and pointed, and appears to represent the ethmoid attachment of the pterygoquadrate to the lamina orbitonasalis such as is found in Teleostomi and Amphibia. Medially to the tip of the processus maxillaris posterior on each side is a string of small isolated nodules of cartilage, of doubtful significance. They lie on the dorsal surface of the palatine bone.

The roof of the nasal capsule ends posteriorly in a straight edge, extending transversely away from the nasal septum on each side, this edge forming the anterior boundary to the fenestra olfactoria evehens. The medial boundary to this fenestra is the nasal septum, while its lateral and posterior boundary is formed by a cartilaginous bar, the commissura sphenethmoidalis, which connects the posterolateral corner of the roof of the nasal capsule with the anterior end of the planum supraseptale on each side. The sphenethmoid commissure thus lies dorsally to the dorsal edge of the lamina orbitonasalis, and separated from it by the orbitonasal fissure. The real aperture into the nasal capsule is the fenestra olfactoria advehens, bounded in front by the hind edge of the roof of the nasal capsule, medially by the nasal septum, behind by the dorsal edge of the lamina orbitonasalis, and laterally by the dorsal edge of the side wall of the capsule, immediately in front of the root of the sphenethmoid commissure. The fenestrae olfactoriae advehens and evehens are thus in slightly different planes, the two planes being inclined to one another by an acute angle, and enclosing a wedge-shaped space opening posteriorly into the orbit through the orbitonasal fissure. This space, which may be called the cavum orbitonasale, is strictly both extracranial and extracapsular. The olfactory nerve enters the cavum orbitonasale by passing through the fenestra olfactoria evehens (therefore medially and dorsally to the sphenethmoid commissure), and then leaves it for the cavity of the nasal capsule by penetrating through the fenestra olfactoria advehens. The profundus branch of the trigeminal nerve (V_1), here known as the ethmoid nerve, enters the cavum orbitonasale from the orbit by passing through the orbitonasal fissure (therefore laterally and ventrally to the sphenethmoid commissure), passes laterally to the branches of the olfactory nerve, and also enters the cavity of the nasal capsule through the fenestra olfactoria advehens. There, the ethmoid nerve divides into two branches of which one, the ramus lateralis nasi soon emerges again on to the dorsal surface of the roof of the nasal capsule by passing through the foramen epiphaniale (situated immediately dorsally to the aditus conchae), while the other, the ramus medialis nasi, runs forwards just beside the nasal septum and emerges at the extreme front end of the nasal capsule through the foramen apicale, on each side.

The front wall of the nasal capsule is formed by a pair of dome-shaped plates of cartilage—the cupolae anteriores—marked off from one another by a groove running down the middle line, and directly continuous with the nasal septum, and with the roof of the capsule except for a small unchondrified region—the fenestra superior, on each side. The front wall is pierced by the foramen apicale, already mentioned, on each side.

The lateral edge of the front wall of the nasal capsule ends freely, and forms the anterior border to the fenestra narina. This border is indented by two processes projecting from the front wall of the capsule—the processus alaris superior and processus alaris inferior (the latter is rudimentary at the 31 mm. stage but well-formed at the 47 mm. stage). The fenestra narina is really a gap in the floor and side wall of the capsule, at its extreme front end. The hind border of the fenestra narina is therefore formed in part by the free anterior edge of the side wall. The anteroventral corner of the side wall, immediately behind the fenestra narina, is continuous with a plate of cartilage—the lamina transversalis anterior—which runs horizontally inwards and fuses with the ventral edge of the nasal septum thus forming a narrow floor, upon which Jacobson's organ rests. The anterior edge of the lamina transversalis anterior completes the hind border of the fenestra narina. From the fact that each half of the cavity of the nasal capsule is here completely ringed round by cartilage (medially by the nasal septum, dorsally by the roof, laterally by the side wall, and ventrally by the lamina transversalis anterior), this region of the capsule is known as the zona annularis. It is behind the zona annularis, in the gap between the lamina transversalis anterior and the free ventral edge of the side wall that the lachrymonasal duct passes into the olfactory sac.

The medial portions of the lamina transversalis anterior on each side, where they spring from the nasal septum, are to be regarded as the representatives of the cornua trabecularum.

The posterior edge of the lamina transversalis anterior on each side is produced backwards into two cartilaginous processes. The medial one of these extends backwards close to but freely from the ventral edge of the nasal septum, and fuses with the ventromedial corner of the planum antorbitale of its own side, forming the paraseptal cartilage. The lateral one extends back ventrally to the concha nasalis, for about half the length of the paraseptal cartilage, and ends blindly: this is the cartilage ectochoanalis.

The large gap in the floor of the nasal capsule, posterior to the lamina transversalis anterior, lateral to the paraseptal cartilage, anterior to the ventral edge of the planum antorbitale and medial to the cartilago ectochoanalis is the fenestra basalis or primary choana, through which the duct of Jacobson's organ and the internal nostril communicate with the roof of the buccal cavity. The external and internal nostrils are therefore separated from one another by the lamina transversalis anterior.

It is to be noted that the whole of the hinder part of the nasal capsule is free from the nasal septum, and that the paraseptal cartilage is complete and unbroken along the whole length of the capsule.

6. The visceral arch skeleton. Of the dorsal component of the mandibular arch, the pterygoquadrate, much has already been said in the foregoing sections. At the 31 mm. (c.r.-l.) stage, the body of the quadrate is separate from the processus ascendens and pterygoid process. Nevertheless, in spite of this discontinuity, which is a specialization conferring extra power of mobility on the quadrate in the chondrocranium, the four typical processes of a normal autodiastylic pterygoquadrate are all present. The ascending process has already been described, as has the (detached) basal process (meniscus pterygoideus) articulating with the basitrabecular process, and the pterygoid process which fails to reach the lamina orbitonasalis and form an ethmoid connexion, the remnants of which are, however, probably represented in the processus maxillaris posterior of the nasal capsule.

The otic process is part of the quadrate cartilage, and abuts against the crista parotica and processus paroticus. The body of the quadrate cartilage is crescent shaped, and curves forwards and downwards, over the columella auris, to the condylar facet by which it articulates with Meckel's cartilage. The free and curved lateral edge of the quadrate will, after further development and growth, form the rim of attachment for the tympanic membrane. The processus internus of the columella auris is attached to the hind surface of the quadrate.

Meckel's cartilage presents few features of importance, other than the presence of a processus retroarticularis behind the articulation with the quadrate, and the fact that at its front end, Meckel's cartilage is attached to its fellow of the opposite side by an extensive symphysis. The sigmoid curvature of early stages has been lost.

The dorsal constituent of the hyoid arch, the columella auris, has been described in a foregoing section (p. 225). The ventral elements of this arch consist of a median and elongated basihyal forming the processus lingualis, to the base of which are fused the paired hypohyals and ceratohyals, thus forming the cornu hyale on each side, originally connected with the columella auris by the ligamentous pars interhyalis.

The ceratobranchials of the 1st and 2nd branchial arches are fused by their anterior ends to the base of the processus lingualis. The posterior portion of the 2nd ceratobranchial is, however, discontinuous with the anterior portion.

iii. Ossification of the skull of Lacerta agilis

The origin of the bones of the lizard's skull has been followed by Gaupp (1906*a*) in the 47 mm. (c.r-l.) stage of Lacerta agilis, on whose description the following account is largely based. Additional information has been obtained from Parker's (1880*a*) work and personal observations. (Plate 84, figs. 7–9: Plate 86).

1. Membrane-bones. The *parietal* arises along the lateral surface of the hinder part of the taenia marginalis. It spreads inwards forming a roof over the hinder part of the cranial cavity and fuses with its fellow of the opposite side, leaving a pineal foramen.

The *frontal* arises along the lateral surface of the anterior part of the taenia marginalis, over the lateral edge of the planum supraseptale and sphenethmoid commissure, and reaches the hinder part of the roof of the nasal capsule. It spreads inwards and joins its fellow of the opposite side, forming a roof over the anterior part of the cranial cavity.

The *prefrontal* ('lachrymal' of Gaupp, 1910*b*) arises over the posterolateral surface of the nasal capsule and covers over the fenestra lateralis. It lies medially to the lachrymonasal duct.

The *nasal* arises over the anterior part of the roof of the nasal capsule, and covers over the fenestra superior.

The *septomaxilla* arises inside the nasal capsule as a flat horizontal plate, underlying the cavity of the nasal sac, and overlying Jacobson's organ, its medial edge resting on the nasal septum, its lateral edge on the inner surface of the side wall of the capsule.

The *supratemporal* (squamosal of Gaupp, 1895, and Watson, 1914, tabular of Broom, 1935*a*) arises just over the crista parotica of the auditory capsule.

The *squamosal* (paraquadrate of Gaupp, 1895; quadratojugal of Watson 1914) arises off the lateral surface of the auditory capsule, beneath and in front of the supratemporal, above the dorsal end of the quadrate, and behind the postorbital.

The controversy concerning the identity of the supratemporal and squamosal in Lacertilia is not easily susceptible of solution by developmental studies, because of the difficulty in finding crucial fixed points and morphological relationships to nerves, blood-vessels, &c., but Brock's (1935) investigations on Agama, Mabuia, and Gerrhosaurus strongly support Thyng's (1906) view and favour the identities here expressed. On this view,

The *quadratojugal* is absent.

The *postorbital* arises in front of the squamosal (with which it forms the superior temporal arch) and beneath the hinder part of the postfrontal. It is separated from the auditory capsule by the temporal fossa.

The *postfrontal* arises laterally to the anterior part of the parietal and the posterior part of the frontal.

The *jugal* forms the ventrolateral border to the orbit, and stretches from the posterolateral region of the nasal capsule to the postorbital.

The *lachrymal* ('adlachrymal' of Gaupp, 1910*b*), a very small bone, arises dorsally to the anterior end of the jugal, and lies laterally to the lachrymonasal duct. It is absent from some Lacertilia (Ascalabotes, Draco, Tiliqua, &c.).

The *pterygoid* arises along the ventral surface of the pterygoid process of the pterygoquadrate and extends from the quadrate to the palatine, with a lateral process directed towards the transpalatine. The pterygoid articulates with the basitrabecular process by means of the meniscus pterygoideus; on its dorsal surface rests the ventral end of the epipterygoid. It lies laterally to the palatine branch of the facial nerve.

The *palatine* separates the anterior part of the buccal cavity from the orbit. It extends from the front end of the pterygoid to a point beneath the ventral edge of the lamina orbitonasalis, and has a lateral process directed towards the maxilla. On the dorsal surface of the palatine are the isolated nodules of cartilage which lie in a line extending inwards from the processus maxillaris posterior.

The *transpalatine* (ectopterygoid) reaches from the pterygoid to the maxilla. It lies immediately ventrally to the line which the pterygoid process of the pterygoquadrate would follow were it to join the processus maxillaris posterior of the nasal capsule.

The *maxilla* is a large bone underlying the side of the nasal capsule, sending an ascending facial process up between the nasal and the prefrontal, and a palatine process inwards under the cartilago ectochoanalis.

The attachment of teeth to the maxilla occurs later.

The *premaxilla* arises beneath the front part of the floor of the nasal capsule, and possesses an alveolar portion, a palatine process directed backwards towards the prevomer of its side, and an ascending (prenasal) process. The bone soon fuses with its fellow of the opposite side to form a median bone with a median ascending process rising up in front of the anterior edge of the nasal septum. The fused premaxillae bear the (dentinal) egg-tooth in the midline (Sluiter, 1893).

The *prevomer* underlies the paraseptal cartilage and curves up to its medial side. It may fuse with its fellow of the opposite side.

The *parasphenoid* arises from one median (rostral) centre underlying the trabecula communis, and two paired lateral centres beneath and behind the basitrabecular processes. The parasphenoid covers over the ventral surface of the hypophysial fenestra (leaving a small foramen) and of the anterior part of the basicranial fenestra. The lateral wings of the parasphenoid become fused with the overlying basisphenoid to form the so-called *sphenoid*, thus enclosing the internal carotid artery and palatine nerve in a Vidian or parabasal canal on each side.

The *prearticular* arises on the medial side of the hind end of Meckel's cartilage, in front of the articular with which it eventually fuses. The chorda tympani runs between the prearticular and Meckel's cartilage.

The *angular* arises on the ventral side of the middle portion of Meckel's cartilage, in front of the prearticular.

The *supra-angular* arises on the lateral surface of the middle portion of Meckel's cartilage, above the angular.

The *splenial* arises on the medial surface of the middle portion of Meckel's cartilage, above and slightly in front of the angular.

The *coronoid* arises above the hinder part of the splenial, and curves outwards over the supra-angular.

The *dentary* arises on the lateral and ventral surfaces of the anterior part of Meckel's cartilage, and eventually has teeth attached to it.

In addition to the above-mentioned 'standard' bones, the skull of Lacerta also presents a number of dermal ossifications (osteoscutes) in the skin, which appear late and eventually become attached to the underlying bones of the skull proper. Among these dermal or integumentary ossifications are: the *supraocular*; the *supraciliar*; and the *supraorbital*, which arises laterally to the prefrontal, and runs outwards in the skin. This ossification was thought by Siebenrock (1892, 1894) to arise in cartilage; a view which Gaupp (1898) showed to be erroneous.

The *sclerotic* bones are about a dozen in number, surrounding the rim of the sclerotic cartilage.

2. Cartilage-bones. The *basioccipital* ossifies as a ring surrounding the notochord over the hinder part of the basal plate, behind the basicranial fenestra. The ossification then spreads, and two lamellae of perichondral bone are formed, one on the dorsal and the other on the ventral surface of the basal plate, eventually covering it completely. Anteriorly, the ventral lamellae extends forwards under the membrane spanning the basicranial fenestra. This part of the basioccipital therefore ossifies without cartilaginous preformation.

The *exoccipital* arises from a pair of perichondral lamellae, on the medial and lateral surfaces of the occipital arch. The exoccipital soon becomes fused with the opisthotic of its own side.

The *supraoccipital* originates from perichondral lamellae on the dorsal and ventral surfaces of the tectum synoticum and tectum posterius, and from a centre in the dorsal region of the auditory capsule. The latter centre, situated on the inner surface of the posterior semicircular canal, may represent an *epiotic*.

The *opisthotic* arises from perichondral lamellae on the outer and inner surfaces of the wall of the auditory capsule in the posteroventral region. The lamellae spread, and fusion with the exoccipital of the same side takes place across an obliterated portion of the fissura metotica to form a compound bone, the *otoccipital*.

The *prootic* arises in the anterior part of the wall of the auditory capsules, but spreads forward in the membranous side wall of the skull where no cartilaginous preformation occurs.

The *basisphenoid* ossifies in the crista sellaris and in the adjacent trabeculae and basitrabecular processes. The basisphenoid is thus dorsal to the internal carotid arteries and palatine nerves; it soon undergoes fusion with a membrane-bone, the parasphenoid, to form the so-called *sphenoid*.

All the foregoing bones may eventually become fused together to form the *os basilare commune*.

The *orbitosphenoid* ossifies in the pila metoptica.

The *presphenoid* arises in the hindmost part of the trabecula communis.

The *quadrate* ossifies in the quadrate cartilage as a perichondral sheath, and is only loosely connected with the squamosal and the pterygoid, for which reason it conforms to the streptostylic type (see p. 425).

The *epipterygoid* arises in the processus ascendens, likewise as perichondral sheaths.

The *articular* ossifies as perichondral sheaths round the hinder part of Meckel's cartilage, in the region of the articulation with the quadrate. The articular soon undergoes fusion with a membrane-bone, the prearticular, on its own side.

The *mentomeckelian* may ossify in the anterior tip of Meckel's cartilage.

The *stapes* ossifies in the proximal part of the columella auris. The term 'extracolumella' is then applied to the cartilaginous distal part of the columella auris which forms a joint with the stapes. The boundary between extra-columella and stapes is proximal to the previous boundary between the cartilaginous hyostapes and otostapes.

The *cornu branchiale primum* ossifies in the 1st ceratobranchial.

II. Ascalabotes fascicularis

The early stages of development of the skull in Ascalabotes fascicularis have been studied by Sewertzoff (1897, 1900), by the method of graphic reconstruction. In agreement with Lacerta is the appearance on each side of the hindmost region of the basal plate of four occipital arches; three preoccipital, and the definitive occipital arch, as evidence of the metameric segmentation of this region into four segments, represented by neural arches of erstwhile vertebral elements.

The pila antotica and orbital cartilage (sphenolateral of Gaupp; 'alisphenoid' of Sewertzoff) seem to appear at an earlier stage than in Lacerta, and to be pierced by the oculomotor nerve; it is, however, not quite clear from the figures given what the limits are between cartilage and dense mesenchyme.

III. Eumeces quinquelineatus

Various stages of development of the skull of the skink Eumeces have been studied and modelled in wax by Rice (1920), and provide instructive comparison with corresponding stages of Lacerta. (Stages of Eumeces studied: embryos of head-length 3·25, 3·5, 4·5, 4·75, and 6 mm.)

The general structure of the chondrocranium and of the ossifications in Eumeces is strikingly similar to that of Lacerta, and it will only be necessary here to consider the more important differences, in order to obtain a more complete and representative conception of the Lacertilian skull.

The occipital condyle has larger lateral prominences and a deeper median notch than in Lacerta, thus approximating more closely to the type found in Monotreme mammals. A feature of great interest is the transient connexion between the cartilages of the tip of the odontoid process and of the hindmost part of the basal plate, in the condylar notch. Since the condyle (of Lacertilia) is held to represent the hypocentrum, and the tip of the odontoid process the pleurocentrum, of the proatlas vertebra, the constituent elements of this vertebra are here temporarily connected in the manner typical of all vertebrae. This condition seems to imply that the intravertebral nature of the occipito-atlantal joint is a recent evolutionary novelty, subsequent to the inclusion in the hinder part of the skull of originally free, typically formed, vertebrae.

An interesting feature of the auditory region is the enclosure (in some specimens of Eumeces) of the glossopharyngeal nerve in a special foramen (the foramen glossopharyngei internum), dorsal to the fissura metotica (apertura medialis recessus scalae tympani) and cut off from it by an extension of the cartilage in this region: this cartilage being a ventral prolongation of the free edge which forms the medial border to the foramen perilymphaticum. The immediate result of this trifling variation is to give the impression that the glossopharyngeal nerve enters the cochlear portion of the auditory capsule and then emerges through the foramen perilymphaticum and the apertura lateralis of the recessus scalae tympani. By comparison with Lacerta, it is clear that what has happened in Eumeces is that the recessus scalae tympani is on the way to becoming an intracapsular space, by the extension of the cartilage on its medial side, thus restricting the apertura medialis, and enclosing the glossopharyngeal nerve in a foramen.

The side wall of the orbitotemporal region in Eumeces is not at any single given stage as complete as that of Lacerta, but in early stages it possesses some structures which Lacerta lacks (such as two additional pillars on each side, behind the pila accessoria, connecting taenia medialis and taenia marginalis).

On the other hand, in later stages of Eumeces certain structures have disappeared (notably, the pila antotica, pila accessoria, and the two additional pillars just mentioned in the previous sentence). These conditions point to the fact that the side wall of the Lacertilian chondrocranium is reduced from a previously more complete stage, and that the reduction has gone further in some Lacertilia than in others.

The ascending process of the pterygoquadrate and the pterygoid process are continuous with each other and with the quadrate at early stages, but the cartilaginous continuity is soon lost. An isolated nodule of cartilage is present at some distance in front of the tip of the processus pterygoideus, and is to be regarded as a vestige of the continuity (seen in Amphibia) between pterygoquadrate and lamina orbitonasalis.

The nasal capsule of Eumeces is, in one respect, more primitive and better developed than that of Lacerta, viz. in the completeness of the side wall in the region of the recessus extraconchalis, and absence of a fenestra lateralis, in which it resembles other reptilian orders. On the other hand, the floor and the side wall in the anterior region are less well developed than in Lacerta, and there is a large fenestra superior.

Another skink, Ablepharus pannonicus, is of interest because practically the entire chondrocranium persists in the adult (Haas, 1936), except in the auditory region where it is largely but not entirely replaced by cartilage-bones.

IV. Anguis fragilis

Zimmermann's (1913) description of the chondrocranium of an embryo of 6 mm. head-length is on the whole similar to what has been said of Lacerta and Eumeces. The planum supraseptale is composed of paired separate orbital cartilages: there is no fenestra lateralis in the nasal capsule; the pilae and taeniae of the orbitotemporal region are very incomplete.

V. Lygosoma

Interesting information concerning the later stages of ossification of the skull are given in Pearson's (1921) description of a Lygosoma embryo of head-length 4·8 mm., studied with the help of a wax-model reconstruction.

Particularly useful are the determinations of the limits to which the various bones extend. Thus, the *prootic* extends back on the outer side of the auditory capsule as far as the processus paroticus, above the fenestra vestibuli: on the inner side as far back as the foramen acusticum posterius. Anteriorly, the prootic has a process lateral to the basisphenoid, and extending as far as the pituitary fossa. The *opisthotic* extends forwards on the outer side beneath the fenestra vestibuli as far as the anterior border of that fenestra; on the inner side it forms the ventral half of the hinder part of the wall, behind the foramen acusticum posterius and beneath the foramen endolymphaticum. The position of the original suture between the opisthotic and exoccipital is indicated by the foramen jugulare. The *supraoccipital* extends downwards behind the prootic on the inner side of the auditory capsule as far as the foramen endolymphaticum; on the outer side, nearly as far as the crista parotica.

The *prefrontal* covers the orbitonasal fissure, and the ethmoid nerve therefore passes out of the orbit into the cavum orbitonasale on the medial side of this bone. The lachrymal foramen through which the lachrymonasal duct leaves the orbit is enclosed between the prefrontal on the inside and the *lachrymal* bone on the outside. The duct then runs forward dorsally to the processus maxillaris posterior of the nasal capsule and between the facial process of the maxilla and the cartilaginous side wall of the capsule, and turns in to connect with the nasal sac posteriorly to the zona annularis. The duct of the lateral nasal gland emerges from the aditus conchae and runs forwards in front of the anterior edge of the facial process of the maxilla and laterally to the cartilage in the region of the zona annularis, to enter the capsule through the fenestra narina. The duct of the lateral nasal gland is, however, separated from the external nostrillar aperture of the nasal sac by the small facial portion and processus intrafenestralis (Wegner, 1922) of the *septomaxilla*, which connects the outermost part of the lamina transversalis anterior with the processus alaris superior. The external nostril is then situated anteroventrally to this diagonal bony bar: the duct of the lateral nasal gland runs in posterodorsally to it. The *prevomers* are fused.

In connexion with the membrane-bones of the temporal region, it is interesting to note that the medial side of the *squamosal* and of the *postorbital* is quite close to the lateral edge of the *parietal*; in other words, the superior temporal fossa is not yet formed. The hind end of the squamosal projects into a fenestra in the dorsolateral edge of the *quadrate*; the hind end of the *supratemporal* is wedged between the dorsal end of the quadrate and the crista parotica.

The palatine nerve emerges from the anterior aperture of the parabasal or Vidian canal, ventrally to the *basisphenoid*, dorsally to the *parasphenoid*, and medially to the root of the basitrabecular process. The palatine nerve then runs forwards ventrally to the latter process, and dorsally to the *pterygoid* and to the *palatine*.

A curious feature of some interest in Lygosoma is the fact that the external recti eye-muscles take their origin from the dorsal surface of the trabeculae, on each side of the pituitary body. They have, therefore, pushed into the cranial cavity in a manner reminiscent of the Teleostean myodome.

VI. Hemidactylus, Platydactylus

A feature of some interest described by Versluys (1903) and confirmed by personal observations, in the skull of the Geckones, is the fact that the columella auris is pierced by a foramen, as in certain Gymnophiona and in Mammals. The foramen serves for the passage of the stapedial artery which in Geckones, therefore, runs through the columella auris, instead of dorsally to it as in Lacerta, or ventrally to it as in Sphenodon.

Another important peculiarity of the hyoid arch skeleton in Geckones is the fact that after the connexion between the extracolumella and the ceratohyal has broken down, the dorsal end of the latter grows upwards and fuses with the processus paroticus (processus dorsalis of the columella auris, fused with the crista parotica of the auditory capsule; see above p. 225), thus presenting an appearance somewhat similar to that of Mammals (p. 413) (see also Sphenodon p. 240, Crocodiles p. 264, and Birds p. 275). However, Brock (1932*b*) has given reasons for believing that this resemblance is spurious, for in early stages of Lygodactylus the pars interhyalis with the ceratohyal still attached, is looped upwards, thereby bringing both the extracolumella and the ceratohyal into connexion with the crista parotica. The connexion with the extracolumella soon degenerates into a ligament.

The chondrocranium of an embryo of Platydactylus annularis (7 mm. head-length) has been described in detail by Hafferl (1921) without, however, revealing any significant differences between this form and Lacerta. The chondrocranium of the Gecko is less well developed, especially as regards the ethmoid and orbitotemporal regions, and the interior of the auditory capsule. A feature of interest is the fact that the dorsal end of the processus ascendens (columella cranii) fuses with the anterior wall of the auditory capsule instead of remaining free as in other Lacertilia. This is further evidence in favour of the view that fusion between the processes of the pterygoquadrate and the neurocranium is a secondary and commonplace phenomenon (see p. 451).

In the bony skull the temporal region is in Geckones characterized by the presence of one only of the two bones present in this region in Lacerta, and Brock (1935) has shown reason in Platydactylus maculosus to believe that it is the uppermost medial ossification (probably the supratemporal) which persists.

17. RHYNCHOCEPHALIA

Sphenodon punctatum

The development of the skull of Sphenodon has been studied from the earliest stages to the adult by Howes and Swinnerton (1901), with the help of wax-model reconstructions. Certain later stages have been admirably modelled and described by Schauinsland (1900, 1903), while Wyeth (1924) has followed the development of the auditory capsule. On these works and on the present writer's observations on the material in the late Professor Dendy's collection the following account is based.

i. Development of the chondrocranium of Sphenodon

The earliest stage described is Stage P, where the elements are still largely procartilaginous. There is a pair of trabeculae, parallel, and wide apart (the skull is distinctly platytrabic), ending freely behind without touching the basal plate, and joined with one another anteriorly by a wide trabecular plate from which the preoptic roots of the orbital cartilages arise on each side. Farther forward still, the trabecular plate is continuous with the nasal septum, from the dorsal edge of which the roof of the nasal capsule begins to extend on each side. (Plate 87, figs. 1–3).

Flanking the notochord and separated by it are the parachordals which form a rectangular and complete basal plate (there is no basicranial fenestra at this stage) at right angles to the trabeculae, and the anterior edge of which is the acrochordal or crista sellaris. From the lateral ends of this crista sellaris the pila antotica rises up on each side and is already connected with the orbital cartilage in such a way as to enclose the oculomotor and trochlear nerves in separate foramina. The auditory capsules, in contact with the lateral edges of the basal plate, are still procartilaginous.

In the visceral arch skeleton rudiments are present of the quadrate and Meckel's cartilage, the former with procartilaginous pterygoid, ascending, and otic processes.

At the next stage (Q) the chondrocranium is well chondrified and has already assumed much of its characteristic form. The trabeculae are fused behind with the parachordals, and their extreme posterolateral portions project to the side, forming modest basitrabecular processes. In front, the trabeculae have fused into a trabecula communis, above which the preoptic roots of the orbital cartilages approximate to form a low interorbital septum, spreading out again into the broad planum supraseptale. The latter, which shows plainly its paired composition, is continuous in front with the nasal capsule by the sphenethmoid commissure, and behind with the pila antotica by means of two bars, the taenia marginalis above and the taenia medialis below on each side. Further, the taenia medialis is connected with the trabecula by the pila metoptica. There is thus a fenestra optica in front of the pila metoptica and beneath the taenia medialis (for the exit of the optic nerve); a fenestra epioptica above the taenia medialis (representing merely an unchondrified portion of the planum supraseptale). Behind the pila metoptica, where in other forms there is the single fenestra metoptica, this in Sphenodon at this stage is subdivided into four, one foramen each for the oculomotor and trochlear nerves, the pituitary vein and the ophthalmic artery. (Plate 87, figs. 4, 5).

The pila antotica forms the anterior border of the incisura prootica (for the exit of the trigeminal nerve). The abducens nerve pierces a foramen in the base of the pila antotica, while the facial nerve is separated from the incisura prootica by the prefacial commissure.

The auditory capsule is well-chondrified and possesses on its lateral surface a crista parotica against which the otic process of the quadrate rests. Posteriorly, the occipital arches arise from the posterolateral corners of the basal plate and join the hind wall of the auditory capsule, thus enclosing the fissura metotica (for the exit of the glossopharyngeal and vagus nerves). Four hypoglossal foramina are enclosed on each side for the five nerve-roots present at this stage.

The nasal capsule has a side wall in its posterior region, formed of the paranasal cartilage. This is connected with the dorsal edge of the nasal septum by the hinder part of the capsular roof (or parietotectal cartilage) and with the planum supraseptale by means of the sphenethmoid commissure. The paranasal cartilage is pierced by a large vacuity, the fenestra lateralis. The hind wall of the capsule is formed by the lamina orbitonasalis, a cartilage which is fused at its lateral end with the posterior edge of the paranasal cartilage, but remains free of the nasal septum at its medial end.

In the visceral arch skeleton the quadrate shows very well-developed pterygoid, ascending, and otic processes.

The columella auris arises as a continuous bar of procartilage whose medial end fits into the fenestra ovalis in the wall of the auditory capsule (with which it may be in temporary continuity), and whose lateral end is in direct cartilaginous continuity by means of the pars interhyalis with the ceratohyal. Later on, after chondrification (which appears to take place in one piece), the columella becomes divisible into the medial otostapes and lateral hyostapes (Gray, 1913). The tympanic membrane in which the latter (which forms the extrastapedial) is inserted is very degenerate. Extending dorsally from the stem of the columella to the quadrate is the *procartilaginous* processus dorsalis in which an independent chondrification (suprastapedial, intercalary, or processus paroticus) eventually fuses with the quadrate near the articulation of the latter with the crista parotica. From the intercalary a cartilaginous strut ('laterohyal') runs back to the extrastapedial, enclosing 'Huxley's foramen' between itself and the processus dorsalis, and apparently corresponding to the connexion found in Geckones (p. 238), Crocodiles (p. 264), Birds (p. 275), and Mammals (p. 413).

ii. The morphology of the fully formed chondrocranium of Sphenodon

The chondrocranium of Sphenodon bears a very close resemblance to that of Lacerta. On the one hand, this fact renders unnecessary an exhaustive description of Sphenodon's chondrocranium; but on the other, however, the very fact of this great similarity with Lacerta is of importance, for reasons of phylogeny. This section is therefore devoted to a review of the more important features in the morphology of the chondrocranium of Sphenodon. (Plate 88).

1. The basal plate and notochord. In forming the definitive basal plate the parachordals fuse with one another beneath the notochord. The occipital condyle, when formed, appears to be hypocentral and to represent the hypocentrum of the proatlas vertebra: it cannot be the pleurocentrum of that vertebra, for Schauinsland (1900, p. 770) has shown that this is attached to the pleurocentrum of the atlas vertebra and there forms the tip of the odontoid process, which lies dorsally to the occipital condyle. As in Lacerta, therefore, the occipito-atlantal joint in Sphenodon is intravertebral and intersegmental. But Sphenodon has the added interest that the interdorsal elements of the proatlas vertebra chondrify independently (instead of being fused with the proatlas basidorsal (=occipital) arch immediately in front) and give rise to the structures which Albrecht (1880) called the 'proatlas' (see p. 385).

In the anterior part of the basal plate, the stages figured by Howes and Swinnerton at no time show any basicranial fenestra. Of the stages studied by Schauinsland, the embryo of 4–5 mm. head-length (figured 1900, Plate 33) shows a small basicranial fenestra immediately behind the crista sellaris. This is, however, not apparent in the figures of embryos of 4·5 mm. head-length and ready to hatch, which accompany Schauinsland's later work (1903, Plates 1–3).

2. The occipital region. The occipital arches rise up behind the auditory capsules but are not fused to them, nor are they joined to one another dorsally. There is thus no tectum posterius, and the foramen magnum is bordered dorsally by the tectum synoticum.

The hypoglossal foramina are reduced to two in number on each side. The fissura metotica is extensive, and that portion of it which lies beneath the foramen perilymphaticum forms a recessus scalae tympani just as in Lacerta. The glossopharyngeal nerve (and, of course, the vagus) issues freely through the fissura metotica.

3. The auditory region. The attachment of the auditory capsule to the basal plate is a region where the cartilage is very thin, suggesting (cf. Schauinsland, 1900, p. 804) that fusion is recent. This virtual presence of a basicochlear fissure is important evidence against the view that the cochlear part of the capsule is derived from the basal plate (see p. 401). A special description of the auditory capsule has been given by Wyeth (1924), and it may be noted that the cavity of the cavum vestibulare is simple and not divided into two by a vertical partition as in Lacerta (see p. 223). Otherwise, the conditions are very similar, especially, in regard to the outer surface, with the fenestra ovalis and the crista parotica with which the processus dorsalis ('suprastapedial' or intercalary, processus paroticus) of the columella auris comes into contact (but in Sphenodon the processus dorsalis fuses with the quadrate and not with the crista parotica); the posteroventral surface with its simple round foramen perilymphaticum opening into the recessus scalae tympani; and the medial surface, pierced by anterius acusticum, posterius acusticum, and endolymphatic foramina. But the saccus endolymphaticus, after emerging from the auditory capsule into the cranial cavity is peculiar, for a diverticulum of it pierces the tectum synoticum on each side. These foramina are obliterated later. A large taenia tecti medialis extends forwards from the tectum synoticum and ends immediately behind the pineal eye.

The posterior part of the crista parotica gives rise at later stages to a marked horizontal ridge on the hind surface of the auditory capsule: this is the parotic process.

The structure of the columella auris, composed of a medial stapes and a lateral extrastapedial which is connected with the intercalary and (throughout life) with the ceratohyal, has been described above (p. 239).

The vena capitis lateralis and the hyomandibular branch of the facial nerve run back dorsally to the columella auris and medially to the processus dorsalis. The chorda tympani then runs forwards above the extracolumella and laterally to the processus dorsalis, to reach the medial side of the quadrate. The orbital artery runs forwards *anteroventrally* to the columella auris.

The persistent connexion with the ceratohyal of the columella auris is important morphological evidence in favour of the hyal nature of the latter structure. It is, however, the persistence of an embryonic feature which represents no phylogenetically primitive condition, but must be associated with the degeneration of the tympanic membrane and therefore constitutes a secondary condition, brought about by arrest of development.

4. The orbitotemporal region. From the pila antotica the taenia marginalis extends back above the roof of the auditory capsule almost as far as the root of the tectum synoticum. The pila metoptica, just before joining the trabecula communis, meets its fellow of the opposite side and forms a subiculum infundibuli, from the lateral corners of which the supratrabecular bars connect with the pila antotica of their side. The foramina for the oculomotor and trochlear nerves are now confluent above the supratrabecular bar, and the foramina for the ophthalmic artery and the pituitary vein are confluent beneath it. The relations of the fenestra optica, epioptica, and metoptica are thus exactly as in Lacerta. The same applies to the abducens foramen which pierces the base of the pila antotica, and to the processus ascendens of the pterygoquadrate, which although part of the visceral arch skeleton, may conveniently be treated here.

The processus ascendens lies behind the profundus (V_1) and in front of the maxillary (V_2) and mandibular (V_3) divisions of the trigeminal nerve, and the space which it encloses between itself and the pila antotica—the cavum epiptericum—is extracranial. The resemblance with Lacerta even extends to the little nodule of cartilage, the meniscus pterygoideus which likewise in Sphenodon is intercalated between the pterygoid and the basitrabecular process.

The interorbital septum in the fully formed chondrocranium is as high as in Lacerta. It is fused with the anterior region of the trabecula communis, but farther back there is a slit between them. The skull is now tropitrabic.

5. The ethmoid region. The nasal capsule is of simple structure. Its roof is continuous with the dorsal edge of the nasal septum, and with the planum supraseptale by means of the sphenethmoid commissure. The hind wall is formed by the lamina orbitonasalis, of which the ventrolateral corner is produced into the processus maxillaris posterior, and the side wall by the paranasal cartilage. This side wall, according to Howes and Swinnerton, has a fenestra lateralis but there is no concha. As in Lacerta, a zona annularis is formed by the lamina transversalis anterior, which is attached medially to the ventral edge of the nasal septum and laterally to the anteroventral corner of the side wall. From the posterior edge of the lamina transversalis anterior on each side there projects backwards the ectochoanal cartilage (paramaxillary process of Schauinsland), the cartilaginous roof over Jacobson's organ, and paraseptal cartilage, the latter attached posteriorly to the ventromedial corner of the lamina orbitonasalis. Jacobson's organ eventually becomes almost completely enclosed in a cartilaginous capsule of its own, opening into the roof of the mouth by a separate foramen, immediately anterior to the choana. The front wall of the nasal capsule is formed by the cupola anterior on each side.

The floor of the nasal capsule behind the lamina transversalis anterior, is occupied by the large fenestra basalis or choana. Broom (1906) has described a cartilago papillae palatinae, projecting back in the midline from the ventral edge of the foremost region of the nasal septum.

6. The visceral arch skeleton. The relations of the processus ascendens of the pterygoquadrate have been mentioned above (supra). The chief interest about this cartilage is that it is stout and continuous, the otic process and body of the quadrate are not separate from the ascending and pterygoid processes. In this respect Sphenodon differs from Lacerta and resembles other Reptiles and Amphibia. The pterygoid process (which near its anterior end is kinked outwards, and then forwards again) extends so far forwards that it almost reaches the processus maxillaris posterior. Later, however, it is reduced. The basal process is probably represented by the detached meniscus pterygoideus. The pterygoquadrate is of the typical autodiastylic type.

Meckel's cartilage has only a small retroarticular process, and anteriorly is firmly fused with its fellow of the opposite side.

The columella auris has been described above (pp. 239, 241), from which it will be seen that the conditions are very similar to those of Lacerta.

The hyobranchial skeleton consists of the ceratohyal, continuous above with the columella auris and below with a median copula, or basihyal, which is prolonged forwards into a processus lingualis. Ceratobranchials are present in the 1st and 2nd branchial arches, the 1st in contact, the 2nd fused, with the copula. (Plate 89, fig. 1).

iii. The development of the osteocranium of Sphenodon (PLATES 88, 89).

1. The membrane-bones. The *squamosal* arises on the lateral surface of the auditory capsule, but it soon comes to cover the otic process of the quadrate, to which it ultimately becomes firmly attached. The squamosal is of cruciform shape. The anterior arm makes contact with the postorbital and so forms the upper temporal bar, the lower arm makes contact with the quadratojugal. The upper arm touches the parietal, for Sphenodon has no supratemporal.

The *quadratojugal* lies laterally to the articular portion of the quadrate, with which it eventually fuses.

The *jugal* forms the ventroposterior border of the orbit; for some time the jugal fails to reach back as far as the squamosal and quadratojugal; the lower temporal bar thus arises late.

The *prevomers* arise and remain as paired bones lying to each side of the ventral edge of the nasal septum, forming the medial borders of the choanae.

The *parasphenoid* arises as a small disc in the floor of the hind part of the hypophysial fenestra. (According to Schauinsland (1900) its origin is paired, the two centres immediately fusing into one median bone.) Eventually it forms lateral wings and an anterior rostrum, and fuses with the overlying basisphenoid.

The *premaxilla* (toothed) has alveolar and ascending portions, the latter in contact but not fused with its fellow of the opposite side in the midline. (There is no proper dentinal egg-tooth; an epidermal horny papilla is present on the dorsal surface of the snout.)

The *maxilla* (toothed) has alveolar and ascending portions, but no palatine process. Anteriorly, it forms the lateral border to the choana; farther back it is in contact with the palatine and the transpalatine.

The *palatine* (toothed) is large and flat, lying in the floor of the orbit and forming the posterior border of the choana.

The *transpalatine* (ectopterygoid) lies in the hind surface of the transverse portion of the pterygoid process of the pterygoquadrate cartilage. It connects the maxilla and the pterygoid bone.

The *pterygoid* is a large bone on the medial and ventral surface of the pterygoid process of the pterygoquadrate cartilage, reaching anteriorly as far as the prevomer. The pterygoid ultimately fuses with the quadrate bone. It is separated from the basipterygoid process by the meniscus pterygoideus.

The *postorbital* forms the hind border of the orbit.

The *postfrontal* arises over the dorsolateral edge of the taenia marginalis; it is a large bone which for some time connects the frontal with the parietal.

The *prefrontal*, which forms the anterior border of the orbit lies laterally to the paranasal cartilage of the olfactory capsule.

There is no lachrymal bone.

The *frontal* overlies the lateral edge of the planum supraseptale and the sphenethmoid commissure, and gradually spreads inwards to the midline and back towards the parietal.

The *parietal* arises late, as a thin strip of bone overlying the hindmost part of the taenia marginalis. Posteriorly, the parietal sends a slender process towards the midline, along the anterior edge of the tectum synoticum. Eventually, by growth inwards and forwards, the parietals cover the hind part of the cranial cavity, leaving only the pineal foramen between them.

The *nasal* arises late, over the roof of the nasal capsule.

The *septomaxilla*, which appears to be quite definitely a membrane-bone (Osawa, 1898, called it a 'turbinal'), arises late, on the dorsal surface of the cartilaginous roof over Jacobson's organ. It forms the hind border of the fenestra narina.

The membrane-bones of the lower jaw, *dentary*, *angular*, *supra-angular*, *coronoid*, and *prearticular*, call for no special mention. There is no splenial.

The *sclerotic* bones are 16 in number.

2. The cartilage-bones. The *basioccipital* starts (according to Schauinsland) from paired perichondral lamellae on both dorsal and ventral surfaces of the basal plate in the posterior region, and ossifies the single median condyle.

The *exoccipital* ossifies in the occipital arch, and eventually fuses with the opisthotic thus forming the so-called paroccipital process of the exoccipital.

The *supraoccipital* ossifies in the tectum synoticum.

The *basisphenoid* arises (like the basioccipital) from paired dorsal and ventral perichondral lamellae in the anterior part of the basal plate and hind region of the trabeculae and spreads into the basitrabecular processes, thus forming the basipterygoid processes. Eventually the basisphenoid fuses with the parasphenoid.

The *prootic* ossifies in the anterior part of the auditory capsule, and soon forms the larger part of its bony wall.

The *opisthotic* ossifies in the hind wall of the auditory capsule, in the region of its parotic process, and eventually fuses with the exoccipital.

The *quadrate* ossifies in the body and otic process of the quadrate cartilage. It becomes attached to the squamosal and fused with the quadratojugal, thereby conforming to the monimostylic type.

The *epipterygoid* ossifies in the processus ascendens of the pterygoquadrate cartilage.

The *articular* ossifies in the posterior part of Meckel's cartilage.

The *stapes* ossifies in the proximal part of the columella auris: the distal part remains a cartilaginous extracolumella.

Perichondral ossification is also found in the *1st ceratobranchial*, and, apparently, also in the *copula*.

18. OPHIDIA

I. Tropidonotus natrix

The skull of the grass snake is of interest from the historic point of view, since it was on this form that Rathke (1839) made the first investigation into the development of a chondrocranium. It has also been studied by W. K. Parker (1879*a*), Bäckström (1931) and Tschekanowskaja (1936): some features of its morphology have been considered by Gaupp (1902) and de Beer (1926*a*).

i. The development of the chondrocranium of Tropidonotus

The earliest stages (5·3 mm. head-length) at which the blastematous rudiments of the future chondrocranium can be made out, show an almost square basal plate, formed of parachordals fused in the midline ventrally to the notochord, and continuous anteriorly with the trabeculae at an obtuse angle. The trabeculae are wide apart posteriorly, giving the skull a platytrabic appearance, and fuse anteriorly into a trabecula communis. The internal carotid arteries enter the cranial cavity through the hypophysial fenestra, but soon acquire separate foramina of their own. Posteriorly, the rudiment of an occipital arch is present, in front of which are three hypoglossal foramina on each side.

The auditory capsule is quite free of the basal plate. Rudiments of the quadrate and Meckel's cartilage are present, and of the columella auris. The blastematous rudiment of the latter arises earlier than that of the auditory capsule, with the wall of which its footplate (otostapes) appears to merge. The distal end of the columella is connected by a faint and transient ligament with the rudiment of the ceratohyal.

Chondrification when it sets in (*circ.* 6 mm.) follows the line of the procartilaginous rudiments more or less, but a basicranial fenestra remains unchondrified in the basal plate behind a crista sellaris; the auditory capsule becomes anchored on to the side of the basal plate by a basivestibular commissure; a fourth pair of hypoglossal foramina is enclosed, and the olfactory capsule is developed. These features may best be dealt with in the following section. (Plate 90).

ii. The morphology of the chondrocranium of Tropidonotus

1. The basal plate and notochord. In the fact that throughout its length the basal plate lies ventrally to the notochord, the snake resembles Lacerta and Sphenodon. Again, as in these forms, the single occipital condyle is a hypocentral element, presenting dorsally a concave surface into which the tip of the odontoid process (presumably containing the pleurocentral element of the proatlas vertebra) fits. From the centre of the odontoid process the notochord runs forwards dorsally to the basal plate. Since the occipital condyle and the occipital arches are respectively the hypocentral and basidorsal elements of the proatlas vertebra, the occipito-atlantal joint in the snake is intravertebral and intersegmental.

The anterior edge of the basal plate is the crista sellaris, on each side of which a tiny stump, perforated by a foramen for the abducens nerve, represents the vestige of the pila antotica. Behind this, on each side and level with the anterior end of the auditory capsule, there is an elongated vacuity in the basal plate close to its lateral edge. No structures pass through this enigmatical vacuity, which is known as fenestra 'x'.

2. The occipital region. The occipital arches are applied to the hind wall of the auditory capsule and enclose the fissura metotica in such a manner as to leave a small fissura occipitocapsularis dorsally, separated from the more ventral fissura metotica proper. Dorsally, however, the occipital arches fail to meet, but just join the tectum synoticum, and there is apparently no tectum posterius.

At the base of the occipital arch on each side there are four hypoglossal foramina of which the last two are peculiar in that they are situated one above the other. The segmentation of the occipital region was studied by Chiarugi (1890), but it is not yet possible to state with certainty the number of segments occupied by the snake's skull.

The ventral part of the fissura metotica is divided more or less completely into two by a strut of cartilage connecting the lateral edge of the parachordal with the floor of the posterior part of the auditory capsule. The posterior division is the jugular foramen, while the anterior division is the apertura medialis of the recessus scalae tympani. The recessus scalae tympani is the space comprised between the floor of the hindmost part of the auditory capsule and the anterior part of the fissura metotica. While the vagus nerve invariably emerges through the jugular foramen, the relations of the glossopharyngeal nerve are variable. Gaupp (1906*a*, p. 793) and Bäckström (1931) describe this nerve as perforating a small foramen glossopharyngei internum of its own (immediately above the apertura medialis recessus scalae tympani), on its way to the recessus scala tympani. The present writer has, however, observed the glossopharyngeal to pass through the jugular foramen on one side of the specimen, and through a small foramen of its own, situated between the apertura medialis recessus scalae tympani and the jugular foramen, on the other side of the same specimen.

At late stages there is developed a second strut of cartilage connecting the edge of the basal plate with the auditory capsule. There is then an additional foramen between the apertura medialis recessus scalae tympani and the definitive jugular foramen. Nothing passes through it, however.

3. The auditory region. The auditory capsule presents the same prominences and cavities on its surfaces as in Lacerta (p. 223) except that there is no crista parotica on the lateral surface (correlated with the fact that the quadrate is free from the auditory capsule), and on the medial surface the fossa subarcuata is very marked. The capsules are interconnected dorsally by a tectum synoticum, which appears to chondrify from an independent centre. The base of the auditory capsule is attached to the lateral edge of the basal plate by the extensive basivestibular commissure, which extends from the facial foramen to the anterior end of the fissura metotica. The prefacial commissure rests on the lateral edge of the basal plate (immediately above the hind part of fenestra 'x') and not on the cochlear part of the auditory capsule, for the latter is relatively small and does not compress the sides of the basal plate.

The apertures in the medial wall of the capsule are the anterior and posterior acustic and the endolymphatic foramina: in the outer wall is the fenestra ovalis, while in the ventro-posterior wall the foramen perilymphaticum faces straight back into the anterior part of the fissura metotica, the recessus scalae tympani. Across the mouth of the foramen perilymphaticum runs the foremost of the struts connecting the edge of the basal plate with the auditory capsule. The cavity of the auditory capsule therefore communicates through the foramen perilymphaticum into the recessus scalae tympani, and thence via its apertura medialis with the cranial cavity, via its apertura lateralis with the outside. The latter course is apparently sometimes taken by the glosso-pharyngeal nerve in those cases where it enters the recessus scalae tympani through its apertura medialis.

In addition to the above-mentioned standard apertures, the wall of the auditory capsule shows a number of unchondrified regions or vacuities, notably in the roof of the sinus superior and in the floor of the cavum cochleare.

The fenestra ovalis is occluded by the footplate of the columella auris, which even at late stages is still in procartilaginous connexion with the ventral edge of the fenestra. Indeed, Möller (1905) and Okajima (1915) believe that the columella is of capsular origin. This is disproved by the connexion observed at earlier stages between the columella and the ceratohyal. The connexion with the capsule is thus probably only a mechanical adaptation without derivatory significance. The columella itself seems to be represented solely by the otostapes; there is no hyostapes and no extracolumella, and the ligamentous connexion with the ceratohyal soon vanishes.

The columella auris runs outwards and backwards, while a process on its anterior edge projects forwards, chondrifies independently, and acquires contact with the hind surface of the quadrate, with which it fuses. This structure has been regarded as a 'stylohyal' by Parker (1879*a*), an extra-columella by Gadow (1901), and a processus internus by Bäckström (1931). But there is little doubt that it is the processus dorsalis or intercalary, fusing not with the crista parotica as in Lacerta, but with the quadrate as in Sphenodon and Crocodilus. This is confirmed by the relations to the columella of the facial nerve.

The hyomandibular branch of the facial nerve runs back dorsally to the columella and medially to the processus dorsalis. It thus gives off the chorda tympani which runs forwards *ventrally* to the columella to the medial side of Meckel's cartilage. In Lacerta, the chorda tympani runs forwards dorsally to that part of the columella (extracolumella) which lies laterally to the processus dorsalis, but since this part of the columella is absent in Tropidonotus, the chorda tympani is allowed to drop into its ventral position. The relation to the facial nerve would be the same in the case of a processus internus, but the detachment and fate of the intercalary in the snake shows that the structure in question can be none other than the processus dorsalis.

The vena capitis lateralis passes dorsally to the columella, but the orbital artery runs anteroventrally to it. (Cf. Sphenodon, p. 241.)

4. The orbitotemporal region. The chondrocranium in the orbito-temporal region consists of nothing but the floor, formed by the trabeculae which are peculiar in that, while wide apart posteriorly, they run together anteriorly but do not fuse. Instead, they continue forwards parallel with one another as far as the nasal septum. There is no cartilaginous interorbital septum, side wall or roof to this region of the chondrocranium. Parker's description of 'orbitosphenoid' cartilages has never been confirmed.

A problem of importance is the question as to whether a basitrabecular process is present. Such a structure is claimed by Brock (1929) to be represented by that projecting edge of the floor of the chondrocranium that lies immediately laterally to fenestra 'x'. If so, its position is unusually far back, but paralleled in Urodela (p. 178). Otherwise this region conforms to the requirements of a basitrabecular process, for it lies ventrally to the vena capitis lateralis, and anterodorsally to the palatine branch of the facial nerve. Further, in Causus (night-adder), Brock (1929) has observed a mesenchymatous connexion between the quadrate and this region of the neurocranium.

There appears to be no sclerotic cartilage surrounding the eye.

5. The ethmoid region. The nasal capsules present a certain resemblance to those of Lacerta. There is on each side a roof or parietotectal cartilage, continuous with the dorsal edge of the nasal septum and with the side wall. The latter is continuous posteriorly with a lamina orbitonasalis which fails to reach the nasal septum. Each half of the nasal capsule is marked off from the other by a deep groove in the midline on the dorsal surface of the roof. The front of the capsule is formed by a cupola anterior, the lateral edge of which forms the anterior border of the fenestra narina. The floor of the capsule is reduced to a slender and crooked lamina transversalis anterior which joins the ventral edge of the side wall about half-way along its length with the ventral edge of the cupola anterior near the nasal septum. There is thus a thin zona annularis. A processus alaris inferior projects from the lateral edge of the cupola anterior.

The lamina transversalis anterior forms a capsule for Jacobson's organ, but there is no paraseptal cartilage. A cartilago ectochoanalis is present and forms a posterior projection from the Jacobson's capsule. Immediately behind and beside the cartilago ectochoanalis, an independent rod of cartilage, called by Born (1883*b*) 'Ke', eventually becomes attached half-way along its length to the tip of the ectochoanal cartilage, and forms the 'hypochoanal cartilage' of Peyer's (1912) descriptions of Vipera. It is possible that cartilage 'Ke' represents the processus maxillaris posterior of Lacerta, and therefore a remnant of the ethmoid process of the pterygoquadrate.

The side wall of the capsule is bent inwards to form the concha nasalis, the aperture of which, the aditus conchae, points outwards and forwards. Owing to the poor development of the side wall of the capsule, the concha is visible in lateral view, projecting beneath its ventral edge. The cavity of the concha is small, and contains only connective tissue, but the lateral nasal gland is situated near its aditus. Behind the aditus conchae is a fenestra lateralis while immediately above it is the epiphanial foramen through which the lateral nasal branch of the profundus nerve emerges from the capsule.

Owing to the absence of any orbital cartilage or sphenethmoid commissure, no foramen olfactorium evehens and no orbitonasal fissure can be demarcated. Only the foramen olfactorium advehens can be defined.

6. The visceral arch skeleton. The pterygoquadrate cartilage is reduced to the body of the quadrate and the otic process. There is no pterygoid nor basal process, and apparently no ascending process of any kind, unless the latter be represented on the cartilage of the basitrabecular process, where the so-called epipterygoid bone subsequently ossifies. At first the quadrate is situated laterally to the anterior end of the auditory capsule, but during development it becomes rotated and displaced backwards until it lies laterally to the hind end of the capsule, to which it is attached by the supratemporal.

The columella auris and its relations to the quadrate have been described above (pp. 245, 247).

Meckel's cartilages present no special features beyond the fact that their anterior ends remain separate. The hyobranchial skeleton is represented solely by the ceratohyals which are very long and fused together anteriorly to form a small basihyal. Edgeworth (1935, p. 94) calls these 1st ceratobranchials, the ceratohyals having disappeared according to him.

iii. The development of the osteocranium of Tropidonotus

1. The membrane-bones. The membrane-bones develop relatively very early, and are all present at the 6·8 mm. head-length stage. (Plate 90).

The *premaxillae* are fused together in the midline, both by their alveolar portions and by their ascending processes, which form a massive knob rising up in front of the nasal septum between the anterior cupolae of the nasal capsule. There are also small palatine processes. The premaxillae bear the (dentinal) egg-tooth in the midline.

The *maxilla* is a long slender ossification which acquires contact posteriorly with the transpalatine. When the latter moves forwards, the maxilla is tilted upwards in front, and also sideways owing to the fact that it slides on the prefrontal.

The *prevomer* (paired) is large and forms the medial wall and floor of the bony capsule in which Jacobson's organ is lodged.

The *septomaxilla* is especially well developed and forms the roof and side of the capsule containing Jacobson's organ. In so doing, the septomaxilla reaches from the side wall of the cartilaginous nasal capsule to the nasal septum. Posteriorly, the septomaxilla has a facet for articulation with the frontal, thus enabling the entire nasal capsule to be raised when the mouth is opened.

The *palatine* is a long slender bone, spreading out anteriorly into a plate overlying the nasopharyngeal passage. Posteriorly it is in contact with the pterygoid.

The *pterygoid* (toothed) reaches half the length of the skull, from the quadrate to the palatine.

The *transpalatine* develops between the hind end of the maxilla and the pterygoid.

The *parasphenoid* starts ossifying as a small plate of bone ventrally to the trabeculae, and sends a lamella up between them, thus forming a low bony interorbital septum, with which the ventral edges of the frontals eventually come into contact. Posteriorly, the parasphenoid fuses with the basisphenoid in the anterior region of the hypophysial fenestra. This fusion takes place both dorsally and ventrally to the trabeculae. It will be noticed that the parasphenoid is here represented only by its median rostral portion, without any lateral wings. There are therefore no parabasal canals, and the internal carotid arteries simply perforate foramina in the basisphenoid.

The *nasal* is a small bone on the dorsal side of the roof of each half of the nasal capsule, dipping down into the median furrow between the cupolae.

The *prefrontal* forms the anterior border of the orbit and covers the hind wall of the nasal capsule. It is perforated by a foramen for the lachrymonasal duct. There is no lachrymal bone.

The *frontal* becomes a very large bone extending ventrally down to the parasphenoid and thus forms the side wall to the skull in the anterior orbital region. Anteriorly, the frontal articulates with the septomaxilla, thus enabling the whole nasal capsule to rise when the mouth is opened.

The *parietal* is likewise a very large bone forming the roof and side wall of the posterior part of the orbitotemporal region. In growing downwards to the basisphenoid, the parietal encloses an extracranial space in which lie, as Gaupp (1902) showed, the oculomotor, trochlear, abducens, and profundus nerves. This space thus represents the cavum epiptericum of other forms (at least in part). All the above-mentioned nerves emerge through a gap, the foramen orbitale magnum, between the parietal and the frontal (Gaupp, 1902, p. 184). The parietals fuse with one another in the midline.

The *supratemporal* (Thyng, 1906; squamosal of Parker 1879*a*, tabular of Brock, 1935) lies over the dorsolateral surface of the auditory capsule and medially to the dorsal end of the quadrate, which it supports. The lengthening of the supratemporal which results in pushing back the otic process of the quadrate occurs in post-embryonic life.

Parker (1879*a*, p. 406) described a 'small sub-oval scale of bone' situated medially to the supratemporal (his squamosal) and which he regarded as the supratemporal. Neither Gaupp (1895, p. 87) nor the present writer have been able to confirm its existence.

The membrane-bones of the lower jaw are the *dentary*, *splenial*, *angular*, *supra-angular*, *coronoid*, and *prearticular*. Fusion occurs between supra-angular, prearticular, and articular (cartilage-bone) to form a mixed ossification.

There are no sclerotic bones in Ophidia.

2. The cartilage-bones. The *exoccipital* ossifies at the 6·8 mm. stage round the occipital arch: dorsally the two exoccipitals meet and exclude the supraoccipital from bordering on the foramen magnum. Eventually the exoccipitals spread forwards and enclose all four hypoglossal foramina. They also fuse with the opisthotics.

The *basioccipital* ossifies as perichondral lamellae on the ventral surface of the basal plate behind the basicranial fenestra. Later on, a lamella appears on the dorsal surface of this cartilage.

The *basisphenoid* ossifies in the anterior region of the basal plate, in the crista sellaris, and in the hind part of the trabeculae. The foramina for the internal carotid arteries and for the abducens nerves thus come to perforate the basisphenoid. Anteriorly, the basisphenoid fuses with the parasphenoid. Parker's statement that the basisphenoid arises from paired centres of ossification has not been confirmed by Bäckström.

The *epipterygoid* (Brock, 1929, 'alisphenoid' of Parker, 1879*a*) arises as a perichondral ossification in the cartilage of the basitrabecular process, and extends upwards as a bone containing some hypertrophic cartilage, but not preformed in embryonic cartilage. At the 10 mm. stage the epipterygoid extends towards but does not quite touch the anterior wall of the auditory capsule, but eventually it fuses with the prootic. The epipterygoid is situated posterolaterally to the profundus and all the eye-muscle nerves, laterally to the head-vein, though medially to another branch of it, and anteromedially to the maxillary and mandibular branches of the trigeminal nerve. The ossification also extends along the cartilage forming the lateral border to fenestra 'x'.

If Brock's interpretation, here adopted, is correct, the processus ascendens must be regarded as having lost connexion with the remainder of the pterygo-quadrate, become attached to the basitrabecular process, and then suppressed so that the ossification, the epipterygoid, appears to arise intramembranously.

The *supraoccipital* arises fairly late, in the anterior and lateral parts of the tectum synoticum, extending into the roof of the auditory capsule on each side. These lateral portions of the supraoccipital were regarded by Parker as the epiotics, but his account of their independent ossification has not been confirmed.

The *prootic* covers the anterior half of the auditory capsule, as far back as the anterior border of the fenestra ovalis, and also extends into the prefacial commissure. Eventually, the epipterygoid becomes fused to it anteriorly.

The *opisthotic* ossifies in the hinder part of the auditory capsule, and eventually fuses with the exoccipital.

The *quadrate* ossifies in the quadrate cartilage, and in the intercalary or processus dorsalis of the columella auris which is fused with it. The *columella auris* ossifies completely as the stapes, and a joint is developed between it and the intercalary attached to the quadrate.

The *articular* is the only cartilage-bone in the lower jaw, and it eventually fuses with the prearticular and supra-angular.

The bony skull of Snakes is characterized by an extreme development of kinetism and streptostyly. The quadrate is almost completely free, and its movement, transmitted by the pterygoid and transpalatine to the maxilla, and by the palatine to the prevomer and septomaxilla, results in a raising of the front part of the margin of the jaw and the widening of the gape, by the lateral movement of the maxillae, and by the separation of each half of the lower jaw. The details of this mechanism in various Ophidia have been studied by Haas (1929, 1930*b*, *c*, 1931).

II. Other Ophidia

The development of the skull has also been studied in Vipera aspis by Peyer (1912) and in Leptodeira hotamboia by Brock (1929). The conditions in these forms are, however, so similar to those in Tropidonotus as to render separate descriptions unnecessary.

To the usual statement that the parietals always become fused in Ophidia Typhlops braminus provides an exception as Haas (1930*a*) and Mahendra (1936) have shown, and this condition may be correlated with the fact that this form is small. The Opoterodonta are usually described as lacking the postfrontal and supratemporal, but a small supratemporal (or tabulare?) has been found in Glauconia nigricans by Brock (1932*a*), and a postfrontal in Typhlops braminus by Mahendra (1936).

19. CHELONIA

I. Chrysemys marginata and Emys lutaria

The early development of both chondrocranium and bony skull in Chrysemys has been studied with the help of wax models by Shaner (1926), while the morphology of the chondrocranium of Emys has been described in detail by Kunkel (1912*a*). Of special regions, the nasal capsule has been dealt with by Seydel (1896), and the hyobranchial skeleton by Siebenrock (1899) and Fuchs (1907*a*) while the columella auris and related structures have been investigated by Smith (1914), and Dombrowsky (1924). The segmentation of the head of Emys has been studied by Filatoff (1908).

i. The development of the chondrocranium of Chrysemys marginata

At the earliest stage studied (6 mm., Plate 91, figs. 1, 2), the trabeculae are present, as independent rods of cartilage extending forwards parallel with one another and fairly wide apart at an angle of some 45° from the plane of the notochord beneath the forebrain. The hinder ends of the trabeculae are interconnected across the midline by a mesenchymatous but extensive acrochordal or dorsum sellae, which rises up on each side to form a pila antotica. The trabeculae are free from the notochord and parachordal cartilages at this stage; there is no sign of any independently chondrified polar cartilages.

At this stage the parachordal cartilages are absent except for the presence at the extreme hind end of the skull of three occipital vertebral elements, each with a neural arch and situated immediately behind its corresponding hypoglossal ventral nerve-root. These neural arches probably lie in the septa between the 7th and 8th, 8th and 9th, 9th and 10th segments (see pp. 25, 31).

The auditory capsules, still procartilaginous at this early stage, are free of any other skeletal structures.

The visceral arch skeleton is represented by the quadrate, Meckel's cartilage, the columella auris, and the ceratohyal and 1st ceratobranchial. The columella auris is laid down in a procartilaginous rudiment in which two chondrifications, the otostapes and the hyostapes, arise. These are continuous with the pars interhyalis which (according to Smith (1914) in Chrysemys, and to Bender (1911) in Testudo) is connected by mesenchyme with the ceratohyal at early stages. Noack (1907) claimed the derivation of the columella in Emys from the auditory capsule, but this is not confirmed by Kunkel (1912*a*) who, on the contrary, found it quite independent.

As development proceeds (6·75 mm., Plate 91, figs. 3, 4) the parachordals are developed between the auditory capsule (which become attached to them) incorporating the occipital vertebrae posteriorly, and anteriorly establishing connexion with the hind ends of the trabeculae. A basicranial fenestra, traversed by the notochord remains unchondrified immediately behind the acrochordal. The trigeminal nerve emerges through the prootic incisure, between the pila antotica and the auditory capsule. The abducens nerve pierces a foramen in the base of the pila antotica.

The orbital cartilage begins to spread forwards from the pila antotica, and the oculomotor and trochlear nerves and the ophthalmica magna artery are enclosed in foramina as a result of the formation of the pila metoptica. The preoptic roots of the orbital cartilages are present, rising up from the anterior part of the trabeculae. Anteriorly, the trabeculae meet in the midline to form a trabecular plate or trabecula communis, surmounting which is a small nasal septum.

The auditory capsule is now well chondrified, and a prefacial commissure encloses the facial foramen. The attachment of the auditory capsule to the lateral edge of the parachordal takes place in such a way as to enclose the glossopharyngeal nerve in a groove, apparently traversing the cavity of the capsule. In the lateral wall of the capsule behind the fenestra ovalis, a region has been described by Kunkel (1912*b*) as chondrifying independently (the 'stapes inferior') and subsequently becoming incorporated in the capsular wall. This structure is reminiscent of the operculum of Urodela (p. 179), but with what justification remains to be seen.

The quadrate becomes apposed by its large otic process to the lateral wall of the auditory capsule, and its pterygoid process develops in a forward direction. At this stage there is a transient connexion between the quadrate and Meckel's cartilage, and the (membranous) pars interhyalis of the columella auris is now also attached to the hind end of Meckel's cartilage.

At the next stage (8 mm.) the most important changes resulting in the definitive form of the chondrocranium have taken place. Posteriorly, a large occipital arch rises up on each side enclosing the fissura metotica from behind. From the pila antotica a taenia marginalis extends towards but fails to reach the auditory capsule. Meanwhile, the preoptic roots of the orbital cartilages have been pressed together in the midline and now form an interorbital septum, rising up from the trabecula communis, and splaying out again above into the planum supraseptale. This is, of course, formed from the orbital cartilages which are now complete, having established connexion with the pila metoptica, with the result that the optic nerves are now enclosed in their foramina. It is to be noticed that as a result of the fusion of the trabeculae in their anterior region to form a trabecula communis and the formation of an interorbital septum, the chondrocranium is abandoning the platytrabic and approximating to the tropitrabic type. (Plate 91, fig. 5).

The nasal capsules arise from two pairs of cartilages. One pair forms the roofs and side walls, the medial edges of the roofs fusing with the dorsal edge of the nasal septum. The other pair are the paraseptal cartilages. Anteriorly, these become connected with both the ventral edge of the nasal septum and the ventral edge of the side wall of their side, thus forming a lamina transversalis anterior (and a zona annularis). For a time, the paraseptal cartilages extend back freely by the sides of the nasal septum. They eventually become connected with the ventral edge of the septum near their hind ends, thus enclosing a foramen prepalatinum on each side of the septum.

The otic process of the quadrate becomes hollowed out, lapping round the tympanic cavity, and the ascending process is developed, rising up laterally to the pila antotica between V_1 and V_2. The quadrate and Meckel's cartilage are now separated by a joint. Centres of chondrification appear in the otostapes and hyostapes (Kunkel, 1912*b*), but not apparently in the pars interhyalis.

ii. The morphology of the chondrocranium of Emys lutaria

The description of the chondrocranium of an embryo of carapace-length 11 mm. has been given by Kunkel (1912*a*). (Plate 92, figs. 1–3).

1. The basal plate, notochord, and occipital region. The parachordal cartilages which form the basal plate are continuous with one another across the midline both above and beneath the notochord. In other words, the latter is completely embedded in the basal plate, and in the large single median occipital condyle which projects back in the midline from the posterior edge of the basal plate. This condyle thus represents a pleurocentrum (presumably formed from the anterior sclerotomite of the 10th segment, see p. 31), and, in fact, that of the proatlas vertebra, the occipito-atlantal joint being here intervertebral and intrasegmental. There are probably therefore $9\frac{1}{2}$ segments included in the skull. Anteriorly, the notochord extends into and through the acrochordal (dorsum sellae or crista sellaris) to its anterior surface. Of the three pairs of hypoglossal foramina present in early stages, the anterior two tend to become joined.

The lateral edges of the basal plate in the posterior region (bordering on the fissura metotica) are thick and produced into dorsal and ventral crests between which is a groove (the sulcus supracristularis) lodging the vagus ganglion. The ventral crest is particularly prominent posteriorly, where it projects to the side and gives the impression of representing a cranial rib.

Between the anterior parts of the auditory capsules, the basal plate has been slightly compressed. The anterior limit of the basal plate is marked by the acrochordal and the pila antotica on each side of it.

The occipital arches on each side are stout structures with three hypoglossal foramina at their base. Each occipital arch is applied to the posterior surface of the auditory capsule of its side, thus enclosing the fissura metotica (but not completely). It does not meet its fellow of the opposite side (i.e. there is no tectum posterius), nor does it touch the tectum synoticum.

The hindmost portion of the fissura metotica is the jugular foramen, through which the vagus nerve emerges: the anterior portion of the fissura metotica (recessus scalae tympani) is obscured by secondary developments of the auditory capsule.

2. The auditory region. The auditory capsules form the side walls of the brain-case in this region, and they do so more effectively than in Lacerta, since in the Chelonian their height is equal to that of the brain. In the capsule itself there may, for purposes of description, be distinguished a vestibular and a cochlear portion: the former lodging the semicircular canals, utricle, and saccule, the latter lodging the cochlea which in Chelonia is peculiar in that the lagena which arises from it is developed in a backward rather than in a forward direction. This has important consequences, as will be seen below.

Dorsally the auditory capsules are interconnected by the tectum synoticum from which the taenia tecti medialis projects forwards; laterally, each capsule has abutting against it the large otic process of the quadrate, behind which the lateral wall of the capsule forms a short crista parotica. The apertures in the walls of the capsule are: laterally the fenestra vestibuli (ovalis) and the external glossopharyngeal foramen; posteriorly, the foramen perilymphaticum, the relations of which are somewhat obscured by the development of the cochlear part of the capsule in this region; and medially, the internal glossopharyngeal foramen, the posterior and anterior acustic foramina, and the foramen endolymphaticum.

The cavity of the auditory capsule is similar to but simpler than that of Lacerta in that the main chamber or cavum vestibulare is not subdivided by a septum intervestibulare. Into the cavum vestibulare, then, there open the orifices of both ends of all three semicircular canals, while posteroventrally the cavum vestibulare communicates by a wide opening with the cavum cochleare.

The attachment of the auditory capsule to the basal plate is effected by means of the broad basivestibular commissures. The posterior extent of this commissure is difficult to determine, because the anterior part of the fissura metotica has been occluded by the backward extension of the cochlear capsule, lodging the lagena. As in Lacerta (p. 225) the space comprised in the anterior part of the fissura metotica is the recessus scalae tympani, communicating dorsally with the cavity of the cochlear capsule (by the foramen perilymphaticum through which the ductus perilymphaticus leaves the capsule and enters the recess); medially by the apertura medialis recessus scalae tympani (through which the glossopharyngeal nerve leaves the cranial cavity and enters the recess); and the apertura lateralis recessus scalae tympani, (through which the glossopharyngeal nerve passes out and the ductus perilymphaticus is exposed to extracapsular structures, but not in Chelonia to the tympanic cavity, for no secondary tympanic membrane is formed).

The chondrification of the floor, lateral and medial walls of the recessus scalae tympani leaves a foramen in both medial and lateral walls for the passage of the glossopharyngeal nerve, which thus appears to traverse the cavity of the capsule, instead of which the space (recessus scalae tympani) which the nerve traverses has been secondarily added to and incorporated in that of the capsule.

The so-called foramen internum glossopharyngei, therefore, marks the position of the foremost portion of the fissura metotica or apertura medialis recessus scalae tympani, just as the foramen externum glossopharyngei represents the anterior portion of the apertura lateralis. But there is a strut of cartilage immediately behind each of these foramina, with the result that the ductus perilymphaticus after passing ventrally through the foramen perilymphaticum finds itself in a canal. This canal opens towards the cranial cavity immediately in front of the jugular foramen.

The cartilage which forms the medial side of the canal for the ductus perilymphaticus (i.e. the cartilage separating the foramen internum glossopharyngei and the posterior aperture of the perilymphatic canal) is an upgrowth of the lateral edge of the basal plate (possibly representing a pre-occipital neural arch) joining the medial wall of the auditory capsule and thus stretching across and occluding this portion of the fissura metotica. But the cartilage of the lateral and ventral walls of the perilymphatic canal forms at the same time the medial and dorsal walls of a space, termed by Nick (1912, p. 87) the 'ductus hypoperilymphaticus' which at this stage is a diverticulum pointing backwards from the foramen perilymphaticum of the auditory capsule, ending blindly, and occupying the anterior part of the fissura metotica. At later stages the ductus hypoperilymphaticus opens posteriorly into the sulcus supracristularis (see above, p. 254).

The nature of the cartilaginous walls of the ductus hypoperilymphaticus is still obscure. They might be the result of the backgrowth of the floor of the cochlear capsule, but if this were so it would be difficult to account for the internal and external glossopharyngeal foramina. The probability is that the lateral wall of the ductus hypoperilymphaticus is derived from one or more cranial ribs, the proximal ends of which are attached to the ventrolateral edge of the basal plate, while the distal ends are fused with the ventrolateral wall of the posterior part of the vestibular capsule behind the exit of the glossopharyngeal nerve. That more than one cranial rib is involved is suggested by the fact that there is a gap in the lateral wall of the ductus hypoperilymphaticus in Dermochelys as described by Nick (1912). The medial wall of the ductus hypoperilymphaticus may then be derived from another preoccipital neural arch, posterior to the arch separating the foramen internum glossopharyngei and the posterior aperture of the perilymphatic canal. The cavity of the ductus hypoperilymphaticus would then be merely an enclosed portion of the recessus scalae tympani. The tip of the lagena fits into the anterior opening of the ductus hypoperilymphaticus.

The columella auris is a single structure, whose origin from separate otostapes and hyostapes rudiments is reflected only in the fact that the former gives rise to the footplate (fitting into the fenestra ovalis) and stalk, while the latter produces the insertion piece (extracolumella, affixed to the tympanic membrane). The latter bears a ventrally directed processus interhyalis, connected by a ligament with the retroarticular process of Meckel's cartilage. The absence of dorsal or internal processes facilitates the description of the relations of the columella to neighbouring structures. The hyomandibular branch of the facial nerve and the head vein run backwards dorsally to it, while the chorda tympani and the orbital artery run forwards dorsally to it.

The prefacial commissure lies close in front of the foramen acusticum anterius, with the result that the facial nerve passes, in close company with the anterior branch of the auditory nerve, through a meatus acusticus internus.

3. The orbitotemporal region. At the point where the trabeculae are fused to the anterolateral corners of the basal plate, the cartilage is produced into a short crest which projects ventrolaterally (Kunkel's 'crista basipterygoidea'). This crest is all there is of the basitrabecular process, and to it becomes attached a small and originally free cartilage, probably homologous with the meniscus pterygoideus of Sphenodon (p. 242) and of Lacerta (p. 228). The basitrabecular process does not establish contact with the (cartilaginous) pterygoquadrate, but with the pterygoid bone. The palatine branch of the facial nerve runs down behind and forwards beneath the basitrabecular process, where together with the internal carotid artery it lies in a groove (Kunkel's 'sulcus cavernosus').

The hypophysial fenestra (through which the internal carotid arteries pass freely upwards into the cranial cavity) is triangular, and the trabeculae converge rapidly to form the trabecula communis, which bears the interorbital septum. The tall dorsum sellae of previous stages has shallowed down to a considerable extent, and is flanked on each side by the pila antotica, through the base of which the abducens nerve foramen is perforated.

The pila antotica rises up in front of the auditory capsule but is not connected with it, with the result that there is no prootic foramen but a prootic incisure for the exit of the trigeminal nerve. In Emys at this stage the pila antotica is also unattached to the orbital cartilage and pila metoptica, so that the pituitary vein and the trochlear and oculomotor nerves pass out through an incisura metoptica. But in Chrysemys the foramen metopticum is enclosed in the typical manner by a taenia marginalis. The ventral ends of the pilae metoptica are bent forwards forming a subiculum infundibuli, attached to the hind edge of the interorbital septum. But at the point where they bend, the pilae metopticae are also attached each to the trabecula of their side by a strut of cartilage which encloses the ophthalmic artery in a foramen. (See p. 261).

The two preoptic roots of the orbital cartilages, pressed against one another in the midline and surmounting the trabecula communis, form the interorbital septum, above which the orbital cartilages diverge to form the large planum supraseptale. This extends over the whole region of its namesake in Lacerta, together with the space represented in the latter by the fenestra epiotica, absent in Chelonia.

The planum supraseptale is connected behind with the pila metoptica, thus enclosing the optic nerve foramen. Anteriorly, the planum is continuous with the sphenethmoid commissure on each side.

The eye-ball is surrounded by a well-developed sclerotic cartilage.

4. The ethmoid region. The nasal septum forms the direct anterior continuation of the interorbital septum. Dorsally, the nasal septum is continuous on each side with the cartilaginous roof and side wall of the nasal capsules, the parietotectal and paranasal cartilages. The posterodorsal wall of each nasal capsule is formed by the lamina orbitonasalis which comes close to but does not fuse with the interorbital septum. The fissura orbitonasalis is the gap left between the posterior edge of the parietotectal cartilage, the dorsal edge of the lamina orbitonasalis, and the sphenethmoid commissure, and through it the profundus nerve leaves the orbit and enters the cavum orbitonasale. The latter is a small space bounded dorsally by the foramen olfactorium evehens (conveying the olfactory nerve out of the cranial cavity, between the sphenethmoid commissure, the dorsal edge of the septum, and the hind edge of the parietotectal cartilage), and ventrally by the foramen olfactorium advehens (conveying the olfactory nerve and the profundus nerve into the nasal capsule, between the dorsal edge of the lamina orbitonasalis, the hind edge of the parietotectal cartilage, and the septum). The ramus lateralis of the profundus nerve leaves the capsule again through the epiphanial foramen in its roof.

Anteriorly, the nasal capsule has a floor formed of the lamina transversalis anterior which joins the anterior parts of the ventral edges of the nasal septum and of the side walls, which completes the zona annularis, and forms the ventral border of the fenestra narina on each side. These two foramina face straight forward: there is therefore no front wall to the capsule.

Backwards from the lamina transversalis anterior on each side of the ventral edge of the septum there extend the paraseptal cartilages near the midline, and the ectochoanal cartilages further to the side. Eventually, the ectochoanal cartilage becomes attached laterally to the ventral edge of the side wall and medially to the paraseptal cartilage, and the latter becomes attached to the septum (leaving the foramen prepalatinum between this attachment, the lamina transversalis anterior, the septum, and the paraseptal cartilage), thus forming a floor to this region of the capsule.

As compared with the nasal capsule of Lacerta, that of Emys is peculiar in that the fenestra basalis (for the choana) faces backwards rather than downwards; the lamina orbitonasalis is therefore dorso-posterior and not posterior; there is no processus maxillaris posterior or anterior, no concha nasalis; and the cartilaginous wall of the capsule is complete.

5. The visceral arch skeleton. The pterygoquadrate has pterygoid, ascending, and otic processes, all of them with typical relations to the neighbouring blood-vessels and nerves. There is no fusion with the neurocranium, and the condition is autodiastylic. (In Testudo graeca, however, Bender (1912) reported a fusion between the otic process and the auditory capsule. The same statement is made by Edgeworth (1935, p. 55) without mention of Bender. Fuchs (1915, p. 247) is very sceptical of this alleged condition, without parallel in Amniota, except for an equally obscure statement about the crocodile, pp. 263, 266).

The otic process is exceptionally large, and laps backwards over and under the tympanic cavity and the columella auris. On the rim of the otic process the tympanic membrane is stretched.

It will be noticed that the pterygoquadrate appears to have no basal process. However, bearing in mind the conditions in Sphenodon (p. 242) and Lacerta (p. 228), it is probable that the meniscus pterygoideus is the detached basal process of the pterygoquadrate.

Meckel's cartilage has a moderately well-developed retroarticular process, to which the ligament from the pars interhyalis of the columella auris is attached. Distally, each cartilage is attached to its fellow of the opposite side in a symphysis. (In neither Emys nor Chrysemys was a separate basimandibular cartilage found.) The columella auris has been described on p. 256.

The hyobranchial skeleton is very feebly developed. In the hyoid arch, the ventromesial ends of the ceratohyals are fused to the median copula (apparently representing the basihyal) and form the so-called hyoid cornua. Behind the basihyal are two more median elements, the 1st and 2nd basibranchials, all of which soon fuse to form the 'body' of the 'hyoid'. Ceratobranchials chondrify (from independent centres) in the 1st and 2nd branchial arches (forming the branchial cornua), and epibranchials in the 1st. The 2nd ceratobranchial is attached to the body of the 'hyoid'. Eventually, all three cornua become distinct from the body.

Anteriorly, paired processes extend forwards from the basihyal and fuse forming the processus lingualis. In addition to this, and situated ventrally to it, there is in late stages of development an entoglossal cartilage, separated from the processus lingualis by connective tissue. The entoglossal cartilage chondrifies directly out of connective tissue, and forms no part of the hyobranchial skeleton.

6. The fate of the chondrocranium. In the adult, the hinder half of the chondrocranium is replaced by cartilage-bone, the remainder is reduced to membrane and the only persisting cartilage to be found is in the nasal septum, the hyostapes, and the 'hyoid' and entoglossum.

iii. The development of the osteocranium of Chrysemys and Emys

1. The membrane-bones. Of the membrane-bones in Chrysemys the squamosal, prefrontal, maxilla, and dentary are present at the 8 mm. stage; all except the parasphenoid are present in the 19 mm. stage, while the latter is found at the 28 mm. stage. In Emys (11 mm. carapace length) all the membrane-bones are present except the parasphenoid, quadratojugal, and coronoid. (Plate 91, figs. 6, 7; Plate 92, figs. 1–3).

The *squamosal* arises immediately to the side of the quadrate, to which bone it becomes solidly attached in typical monimostylic manner. The relations of the squamosal to the quadrate at early stages well demonstrates the error of Gaupp's (1895, see p. 437) view, regarding the squamosal of Amphibia as a 'paraquadrate'.

The *quadratojugal* (which Gaupp is driven to call the 'paraquadrate') appears between the quadrate and the jugal and postfrontal. Arising late in Emys, it does not arise at all in Cistudo ornata, Chelodina longicollis, or Geoemyda spinosa.

The *jugal* borders the orbit posteroventrally immediately below the postfrontal.

The *maxilla* has ascending (facial) alveolar, and palatine proceses. It is, of course, toothless.

The *premaxilla* is small, paired, and situated ventrally to the lamina transversalis anterior of the nasal capsule. There is apparently no ascending (prenasal) process: i.e. the premaxilla does not separate the narial apertures. It may also be noted that no egg-tooth is carried by the premaxilla; instead an epidermal papilla is present on the dorsal surface of the snout.

The *postfrontal* borders the orbit posterodorsally.

The *parietal* commences ossifying dorsally to the side wall of the cranial cavity in the hinder part of the orbitotemporal region. An extension of the ventral edge of the parietal, the processus inferior, grows downwards laterally to the pila antotica, and connects with the tip of the processus ascendens of the pterygoquadrate cartilage. In this way an extracranial space, the cavum epiptericum, lodging the trigeminal ganglion, comes to be enclosed within the bony skull.

The *frontal* starts ossifying over the lateral edge of the planum supraseptale. It borders the orbit dorsally.

The *prefrontal* is very large and borders the orbit anteriorly, covering the posterolateral surface of the lamina orbitonasalis of the nasal capsule. This bone is sometimes referred to as the lachrymal but without justification, especially as in the absence of a lachrymonasal duct the typical relations of a lachrymal cannot be demonstrated.

The *pterygoid* lies along the ventral and medial sides of the pterygoid process of the pterygoquadrate cartilage, and between the latter and the base of the neurocranium, extending forwards into the floor of the orbit.

The *palatine* lies in the floor of the orbit, in front of the pterygoid and medially to the palatine process of the maxilla.

The *parasphenoid* starts ossifying as a small plate of bone under the hypophysial fenestra; it ultimately becomes fused with the basisphenoid.

The *prevomer* is at the outset a paired bone, ossifying as a splint along the ventrolateral edge of the nasal septum. Soon, the two prevomers fuse in the midline.

The *dentary* (toothless) lies along the outer side of Meckel's cartilage for most of its length.

The *supra-angular* lies behind the dentary and laterally to Meckel's cartilage in the region of its articulation with the quadrate.

The *angular* ossifies ventrally to the hinder part of Meckel's cartilage, beneath the supra-angular.

The *prearticular* ossifies on the medial side of the hinder part of Meckel's cartilage; the chorda tympani runs between the cartilage and the prearticular.

The *coronoid* (complementare) appears on the medio-dorsal side of Meckel's cartilage, in front of the prearticular.

It will be noted that the nasal bone is absent. (Among Chelonia it is apparently present in the Chelydidae, but not in Chelys.) Among the membrane-bones must be mentioned a pair of subparachordal plates (cf. Hypogeophis, p. 196) which eventually fuse with the basioccipital.

The *sclerotic* bones are about a dozen in number.

2. The cartilage-bones. The *basisphenoid* ossifies in the crista sellaris (two centres) and in the hinder part of each trabecula; it subsequently becomes fused with the underlying parasphenoid, leaving on each side a parabasal canal, for the passage of the palatine nerve and internal carotid artery.

The *basioccipital* arises from paired centres in the basal plate, extending back to the occipital condyle. Anteriorly, each of the portions of the basioccipital becomes continuous with a small subparachordal plate of membrane-bone.

The *exoccipital* ossifies in the occipital arch, is pierced by the hypoglossal foramina, and extends into the side of the occipital condyle, remaining distinct from the basioccipital.

The *supraoccipital*, ossified in the tectum synoticum, is usually said to be fused with the *epiotics*, but no traces of such alleged fusion have been found. It is possible that this ossification is a supraoccipital which has extended and occupied the place of the epiotics.

The *prootic* occupies the anterior half while the *opisthotic* occupies the posterior half of the auditory capsule.

The *quadrate* ossifies in the body of the pterygoquadrate cartilage, while a separate ossification, the *epipterygoid*, arises in the processus ascendens.

The *articular* ossifies in Meckel's cartilage near the hind end where it articulates with the quadrate.

The *stapes* becomes ossified in the proximal position of the columella auris, while its distal portion, the extracolumella, remains cartilaginous.

In the '*hyoid*' plate, four centres of ossification arise, which eventually spread over it and the processus lingualis. Ossifications also arise in the hyoid cornu and the branchial cornua.

II. Other Chelonia

Of other Chelonia in which the development of the skull has been studied, Chelone viridis has been described by W. K. Parker (1880*b*) and Gaupp (1902, 1906*a*), Chelone midas by Nick (1912), Chelone imbricata by Fuchs (1915), Dermochelys coriacea by Nick (1912), Chelydra serpentina by Nick (1912) and de Beer (1926*a*). In all these cases, investigation has been restricted to the fully formed chondrocranium. In Trionyx japanicus, the remains of the chondrocranium in the adult have been described by Ogushi (1911).

The general features of the chondrocranium are too similar to those of Chrysemys to necessitate redescription. Attention need, therefore, be paid only to certain important points of difference.

1. The orbitotemporal region. Chelone is remarkable in that the hypophysial fenestra is divided sagittally into two by a median taenia intertrabecularis, which runs straight back from the trabecula communis to the crista sellaris. The internal carotid arteries thus enter the cranial cavity through foramina on each side of the taenia intertrabecularis, between it and the trabeculae cranii. The posterior portion, projecting forwards from the crista sellaris, becomes ossified as the rostrum of the basisphenoid, for Chelone has no parasphenoid. A similar taenia intertrabecularis is found in Dermochelys at the 24 mm. (head-length) stage. At the 26 mm. stage it becomes discontinuous in the middle. (Plate 92, fig. 6).

The interest of this taenia intertrabecularis (this term of Fuchs's is preferable to 'intertrabecula' since the latter has been used for (hypothetical) elements between the trabeculae farther forward) lies in the fact that a comparable structure, the central stem, is found in Mammals.

Another feature of Dermochelys is the presence of supratrabecular bars, extending from the subiculum infundibuli (pila metoptica) to the pila antotica, between the oculomotor nerve (above) and the ophthalmic artery (beneath), just as in Lacerta (p. 227), Sphenodon (p. 241), Birds (p. 270), and certain Primates (pp. 353, 360). The bar of cartilage which in Emys encloses the ophthalmic artery in its foramen from behind is probably formed out of the anterior portion of a supratrabecular bar. In Chelone the supratrabecular bar is incomplete, and projects back as an anterior clinoid process (cf. Man, p. 360) from the pila metoptica.

The cartilaginous side wall of the skull in the orbitotemporal region in Dermochelys and Chelone is remarkably complete; there is a cartilaginous strut separating the oculomotor and trochlear nerves in the fenestra metoptica, and the taenia marginalis connects with the auditory capsule above the foramen prooticum. (In Chelone there is a taenia supramarginalis in addition to the taenia marginalis.) The chief interest of this condition lies in the fact that this great development of cartilage (which persists in the adult) is correlated with a poor development of the bone in this region. In particular, there is in Dermochelys no epipterygoid, and no processus inferior of the parietal. In Dermochelys, therefore, the cavum epiptericum is not incorporated in the skull.

Lastly, it may be mentioned that the planum supraseptale in Chelone and Dermochelys shows a vacuity corresponding to the fenestra epioptica of Lacerta, thus increasing the similarity between the chondrocrania of Chelonia and Lacertilia.

2. The fate of the chondrocranium. Dermochelys is remarkable not only for the great development of the chondrocranium, but also for its persistence in the adult.

There, large portions of the basal plate, the medial wall of the auditory capsule, the enormous tectum cranii (which forms a complete covering over all but the foremost part of the cranial cavity), the pila antotica, planum supraseptale, interorbital septum and the nasal capsule are all preserved. There is also an enigmatical plate of cartilage overlying the basal plate.

In Chelone nearly as much cartilage is retained in the adult, but the basal plate is nearly all ossified and the tectum cranii does not extend beyond the level of the crista sellaris. In Trionyx the amount of cartilage preserved is smaller, and consists of the trabeculae, interorbital septum, nasal capsule, the taenia tecti medialis, and the medial wall of the auditory capsule. In other forms (see above, p. 258) very little cartilage remains.

In the fully developed embryo of Chelone, Fuchs (1915, p. 283) has found a nodule of cartilage projecting ventrolaterally from the pterygoid bone, towards the coronoid of the lower jaw. This cartilage may be derived from the anterior end of the pterygoid process of the pterygoquadrate cartilage (Cf. Crocodilia, p. 266).

20. CROCODILIA

Crocodilus biporcatus

The development of the skull of the crocodile was studied by W. K. Parker (1883*b*); more recently Meek (1911) has investigated some stages with the help of the glass-plate method of reconstruction, while a detailed study with wax models of the fully formed chondrocranium has been made by Shiino (1914). The latter is therefore well understood whereas the development is still incompletely known. The columella auris has been studied by Versluys (1903) and Goldby (1926).

i. The morphology of the chondrocranium of Crocodilus biporcatus (13 mm. head-length). (PLATE 93).

1. The basal plate and notochord. The parachordal cartilages, fused with one another above and below the notochord, form an elongated rectangular plate, ending anteriorly in the upraised dorsum sellae or crista sellaris. There is no basicranial fenestra.

Posteriorly, the basal plate is continued into the median occipital condyle, which, surrounding the notochord as it does, represents the pleurocentrum of the proatlas vertebra. The occipito-atlantal joint here is therefore intervertebral and intrasegmental. The basidorsal elements of the proatlas vertebra are, of course, the occipital arches, but the interdorsal elements of that vertebra remain separate and form the so-called 'proatlas' (Albrecht, 1880).

On each side, the basal plate is flanked by the auditory capsules, but the fissura metotica which separates each capsule from the basal plate is very long and extends for the posterior two-thirds of the length of the basal plate. Anteriorly, each fissura metotica turns inwards, for in this region the cochlear part of the auditory capsule compresses the basal plate.

2. The occipital region. The occipital arches are massive and pierced at their base by the three hypoglossal foramina. (At younger stages there may be four foramina.) Each occipital arch fuses with the posterior wall of the auditory capsule, and dorsally is continuous with the tectum synoticum, which may thus in part represent a tectum posterius. From the lateral surface of each occipital arch, just above the hypoglossal foramina, a process is given off which extends forwards beneath the auditory capsule and the fissura metotica. This is the processus subcapsularis (corresponding to the metotic cartilage of birds, p. 272), which appears to represent a cranial rib belonging to the occipital (proatlas) vertebra. The processus subcapsularis forms a floor to the foremost part of the fissura metotica (corresponding to the recessus scalae tympani in Lacerta), and the secondary tympanic membrane is stretched (laterally to the foramen perilymphaticum) from the lateral edge of the processus subcapsularis to the lateral wall of the auditory capsule. The apertura lateralis of the recessus scalae tympani is therefore enclosed between the processus subcapsularis and the floor of the cochlear capsule. Through the hinder part of the fissura metotica, or jugular foramen, the glossopharyngeal and vagus nerves emerge.

3. The auditory region. The auditory capsules present the familiar vestibular and cochlear portions, but the vestibular portion has been rotated backwards so that it lies definitely behind the cochlear portion. This rotation has affected the position of the crista parotica on the lateral surface, for this ridge, instead of being more or less horizontal as in Lacerta, is in the vertical plane, and parallel with the occipital arch. The same is true of the 'horizontal' or lateral semicircular canal.

The otic process of the quadrate cartilage becomes attached by connective tissue to the lateral surface of the capsule. Edgeworth (1935, p. 56) states that the otic process fuses with the capsule, but supplies no figures or details. Other features of the external surface are the fenestra ovalis and the foramen perilymphaticum, for the latter faces sideways immediately above the apertura lateralis recessus scalae tympani, and not directly into the recessus as in Lacerta.

On the medial surface, the auditory capsule is perforated by the anterius and posterius acusticum and the endolymphatic foramina. Dorsally the capsules are interconnected by the tectum synoticum behind, while in front each is attached to the taenia marginalis of its side. The prefacial commissure, enclosing the facial foramen, is very massive.

The internal cavity of the auditory capsule calls for no detailed description, for it is subdivided in the usual way into a cavum vestibulare lodging the utricle and saccule with the semicircular canals separated off by septa, and a cavum cochleare lodging the ductus cochlearis. The ductus perilymphaticus does not pass through the foramen perilymphaticum (de Burlet, 1929), and thus the 'recessus scalae tympani' in the crocodile fails to contain the scala tympani of the perilymphatic space. The restricted apertura lateralis of the 'recessus scalae tympani', enclosed between the processus subcapsularis and the wall of the auditory capsule, and across which the secondary tympanic membrane is stretched, bears analogy (but *not* homology) to the mammalian fenestra rotunda, and may therefore in the crocodile be known as the fenestra pseudorotunda.

It is convenient in this place to deal with the columella auris, although part of the visceral arch skeleton. The columella auris arises from a medial otostapes and a lateral hyostapes which fuse to form a single rod (Versluys, 1903, p. 169). This latter subsequently becomes subdivisible into a medial ossified stapes and a lateral cartilaginous extracolumella, inserted in the tympanic membrane.

From the stem of the columella (from its originally otostapedial portion), a processus dorsalis ('suprastapedial' of Parker, and apparently containing an independent intercalary chondrification) rises up to the under surface of the otic process of the quadrate, with which in Crocodilus porosus it fuses. A process from the extracolumella, the pars interhyalis, ('infrastapedial' of Parker), projects downwards to a small cartilage, the epihyal, which in turn is continuous with a longer cartilaginous bar, the stylohyal ('ceratohyal' of Parker). The latter cartilage is attached to the retroarticular process of Meckel's cartilage (cf. Chelonia, p. 256). The original continuity between the dorsal (columella auris, &c.) and ventral (hyoid cornu) elements of the hyoid arch skeleton has been lost, and apparently, is not even represented in mesenchyme. But at late stages, Versluys (1903, p. 171) has shown that the epihyal is in contact not only with the pars interhyalis of the extracolumella but also with the intercalary, thus forming a 'laterohyal' between which and the processus dorsalis a foramen is enclosed, as in Sphenodon and Birds.

The relations of the columella auris to neighbouring blood-vessels and nerves are as in Chelonia. The hyomandibular branch of the facial nerve runs back dorsally to the columella medially to the processus dorsalis. The chorda tympani then runs forwards laterally to the processus dorsalis to reach the medial surface of Meckel's cartilage. The vena capitis lateralis runs back dorsally to the columella auris, and the orbital artery (arteria temporo-orbitalis, Shindo, 1914) runs forwards dorsally to it.

4. The orbitotemporal region. The trabeculae remain separate in the hinder part of the orbitotemporal region, surrounding the hypophysial fenestra; but farther forward they are fused into a trabecula communis. Ventrally from the point of attachment of each trabecula to the basal plate, a basitrabecular process is given off. But instead of projecting laterally, these processes project backwards, underlying but not touching the cochlear capsules, and form the infrapolar processes (Cf. Birds, pp. 271, 276).

Between each infrapolar process and the under surface of the basal plate, there is therefore an elongated passage opening to the side, and through this the palatine branch of the facial nerve and the internal carotid artery run medially (therefore dorsally to the infrapolar process), the former to continue forwards medially to the root of the infrapolar process (therefore morphologically ventrally to the basitrabecular process), and the latter to enter the cranial cavity through the hypophysial fenestra. Likewise through this passage, and therefore dorsally to the infrapolar process, the Eustachian tube runs laterally to the tympanic cavity.

Overlying the trabecula communis is a tall interorbital septum, from the dorsal edge of which the preoptic roots of the orbital cartilages spread out on each side, forming the planum supraseptale. This is connected with the pila metoptica and pila antotica by means of the taenia marginalis and taenia medialis, with the result that the optic, metoptic, epioptic, and prootic fenestrae are enclosed and transmit the same structures as in Lacerta (p. 227); only that the trochlear nerve has a separate foramen distinct from the fenestra metoptica (as in Sphenodon, p. 239, and Chelonia, p. 261). The abducens nerve pierces a foramen at the base of the pila antotica. The interorbital septum increases in height during development, and while at the stage studied it is a single median structure, it probably represents the bases of the preoptic roots of the orbital cartilages fused together.

Anteriorly, the planum supraseptale is continued into a pair of sphenethmoid processes which fail to reach the nasal capsule. The foramen olfactorium evehens and the fissura orbitonasalis are thus incompletely delimited.

It will be noted that in the crocodile the true side wall of the chondrocranium in the orbitotemporal region is preserved, and the cavum epiptericum (which is scarcely delimited since the ascending process of the pterygoquadrate is vestigial) is extracranial.

The eye-ball is surrounded by a well-developed sclerotic cartilage.

5. The ethmoid region. The nasal septum is the direct anterior prolongation of the interorbital septum. The dorsal edge of the nasal septum is continuous with the roof of the nasal capsule, or parietotectal cartilage on each side, which in turn passes down continuously into the side wall. The ventral edge of the latter is connected anteriorly with the very deep lamina transversalis anterior which forms a floor to the anterior half of the nasal capsule. The hind border of the fenestra narina, however, is only a little distance in front of the hind edge of the lamina transversalis anterior; the zona annularis is therefore narrow. The lachrymonasal duct runs into the cavity of the capsule immediately behind the zona annularis.

Jacobson's organ is vestigial, and the paraseptal cartilage is represented only by a small process projecting back from the lamina transversalis anterior. There is no ectochoanal cartilage.

The roof of the nasal capsule does not extend as far forwards as the floor, so that the opening of the fenestrae narinae face upwards as well as outwards. Eventually the hind part of the fenestra narina becomes separated off by a bar of cartilage as a foramen through which the lachrymonasal duct passes. In the anterior region of the floor are the paired apical foramina, serving for the exit of the medial nasal branches of the profundus nerves. Between these foramina on the ventral surface of the floor of the nasal capsule, a small prenasal process projects forwards in the midline. (Cf. Birds, p. 271.)

The hind wall of the nasal capsule is formed by the lamina orbitonasalis which joins the hind edge of the side wall of the capsule to the nasal septum with which it is fused. The fenestra basalis (for the choana) is thus completely enclosed between the ventral edges of the side wall, the nasal septum and the lamina orbitonasalis, and the hind edge of the lamina transversalis anterior.

The side wall of the capsule shows a large indentation half-way along its length. This is the concha nasalis (the 'middle turbinal' of Meek, 1911; 'upper turbinal' of Parker, 1883*b*), the mouth (aditus conchae) of which opens sideways and forwards while its cartilaginous wall bulges backwards and inwards into the cavity of the capsule. The concha nasalis contains a diverticulum of the nasal sac (Meek's 'middle turbinal sinus'), which passes out through a slit near the ventral edge of the side wall of the capsule, and in through the aditus conchae.

In front of the concha nasalis, an ingrowth of cartilage gives rise to the atrioturbinal (Meek's 'anterior turbinal'; Parker's 'inferior turbinal'). Another cartilaginous ingrowth behind the concha nasalis forms what Meek has called the 'ethmoturbinal', and Gegenbaur (1873) the 'pseudoconcha'.

It is improbable that this structure is homologous with the ethmoturbinal of Mammals although it occupies a corresponding position.

The foramen olfactorium advehens is a comparatively large slit-like aperture on each side of the nasal septum, and it owes its size to the fact that the hind edge of the roof of the nasal capsule does not extend as far back as in other forms. This is why there is no epiphanial foramen in the crocodile, for the lateral nasal branch of the profundus nerve which runs forwards over the mouth of the foramen olfactorium advehens, is able to run outwards over the aditus conchae without ever being enclosed beneath the capsular roof.

6. The visceral arch skeleton. The pterygoquadrate consists of a massive otic process and articular portion, continuous anteriorly with a slender pterygoid process which curves forwards and outwards. The pterygoid process carries a diminutive ascending process. There is no basal process, nor any contact with the basitrabecular process of the skull. The otic process is attached by connective tissue to the lateral wall of the auditory capsule (Edgeworth, 1935, p. 56, states that there is cartilaginous fusion), and to the under surface of the otic process there becomes attached (by cartilaginous fusion in Crocodilus porosus) the processus dorsalis of the columella auris.

At later stages it would seem that the anterior end of the pterygoid process becomes detached and gives rise to the 'pterygoid cartilage' which Parker described as situated on the lateral surface of the pterygoid bone, in Crocodilus palustris and Alligator mississipensis. (Cf. Chelone, p. 262.)

Meckel's cartilage has a short retroarticular process to which the stylohyal is attached. Parker (1883*b*) described the anterior ends of Meckel's cartilages as being united by an independent basimandibular cartilage, but this has not been found by Shiino. On the other hand, the coronoid cartilage which Parker described (on the medial side of the coronoid bone of the lower jaw) and which Shiino failed to find, has been found by Meek. It articulates with the 'pterygoid cartilage'. (See above.)

The columella auris and its relations to Meckel's cartilage and to the hyobranchial skeleton have been described above (p. 263).

The 'hyoid' consists of a stout wide plate or copula, apparently corresponding to the basihyal perhaps fused with the 1st basibranchial, produced anteriorly into a short processus lingualis. On each side the 'hyoid' is fused with a very short hyoid cornu (ceratohyal), articulates with a fairly long cornu branchiale primum (1st ceratobranchial), and is fused with a very short cornu branchiale secundum (2nd ceratobranchial).

ii. The development of the osteocranium of Crocodilus

1. The membrane-bones. The *quadratojugal* arises laterally to the ventral part of the body of the quadrate cartilage, and by contact with the jugal forms the inferior temporal bar.

The *squamosal* arises laterally to the posterodorsal portion of the otic process of the quadrate and to the hind part of the lateral wall of the auditory capsule.

The *jugal* forms the lower border of the orbit, between the maxilla and the quadratojugal.

The *maxilla* has large alveolar and ascending portions, and a well-developed palatine process which eventually meets its fellow of the opposite side in the midline ventrally to the prevomers and the nasopharyngeal passage, thus forming part of the false palate.

The *premaxilla* (paired) likewise has well-developed alveolar and palatine portions. There is also a facial portion which forms the lateral and posterior boundary of the external nostril, but there is no prenasal ascending process, for the nostrils are confluent on the dorsal surface of the snout. There is no egg-tooth, but instead, an epidermal papilla carried on the snout in front of the nostrils (Sluiter, 1893).

The *nasal* lies on the dorsal surface of the nasal capsule, but is practically excluded by the premaxilla from bordering the nostril.

The *lachrymal* ('adlachrymal' of Gaupp) lies laterally to the side wall of the nasal capsule and is pierced by the lachrymonasal duct, just as in mammals.

The *prefrontal* ('lachrymal' of Gaupp) forms the anteromedial border of the orbit, and overlies the hind part of the roof of the nasal capsule. The presence together in the crocodile of the lachrymal and the prefrontal, and the relations of the former to the lachrymonasal duct, disposes, as Meek (1911) and Gregory (1920) have shown, of Gaupp's (1910*b*) view that the mammalian lachrymal is the reptilian prefrontal.

The *frontal* overlies the lateral edge of the planum supraseptale and gradually extends medianwards over the cranial cavity.

The *parietal* arises late, over the hindmost part of the taenia marginalis, and gradually extends over the hinder part of the cranial cavity. In Alligator mississipensis Mook (1921) has observed a median *postparietal*, which arises separately but fuses with the parietals.

The *postorbital* arises laterally to the hind part of the taenia marginalis. By its contact with the squamosal it forms the superior temporal bar.

The *pterygoid* arises medially to the pterygoid process of the pterygo-quadrate cartilage. It eventually develops an enormous palatine process which meets its fellow in the midline ventrally to the nasopharyngeal canal, the opening of which is thus carried far back, while the apertures of the Eustachian tubes are carried back in the midline. The pterygoid bone also develops a descending process, the ventrolateral corner of which is tipped by cartilage, perhaps derived from the anterior end of the pterygoid process of the pterygo-quadrate cartilage.

The *palatine* arises ventrally to the ventral edge of the lamina orbitonasalis, and eventually forms the floor of the middle portion of the nasopharyngeal passage by uniting with its fellow in the false palate.

The *prevomer* arises (paired) on the ventrolateral edge of the nasal septum. It eventually becomes completely covered ventrally by the palatine.

The *transpalatine*, which develops late, connects the maxilla with the pterygoid, and follows the line of the anterior out-turned end of the pterygoid process of the pterygoquadrate cartilage.

The *parasphenoid* is in a condition of great interest, for it arises in three (permanently) separate portions: a median rostrum or *vomer*, appearing in the floor of the hypophysial fenestra and eventually fusing with the basisphenoid: and paired *basitemporals* appearing beneath the anterolateral corners of the basal plate, and also fusing with the basisphenoid. As the internal carotid artery and palatine nerve run ventrally to the basisphenoid but dorsally to the basitemporals, a parabasal canal is left between the fused bones on each side, for their passage. The vomer and basitemporals become covered ventrally by the pterygoids.

It will be noticed that the crocodile has no septomaxilla.

Of the dermal bones of the lower jaw, the *dentary*, *splenial*, *angular*, and *supra-angular* call for no special comment. The *coronoid* is tipped medially with cartilage, articulating with that on the pterygoid. According to Baur (1896) a small *prearticular* ('postopercular') is present but fuses with the articular.

There are no sclerotic bones, but the upper eyelid contains an ossification (de Witte, 1927).

2. The cartilage-bones. The auditory capsule apparently ossifies from three centres: the *epiotic*, *opisthotic*, and *prootic*, of which the first fuses with the *supraoccipital* (in the tectum synoticum) and the second with the *exoccipital* (in the occipital arch). The *basisphenoid* (which fuses with the vomer and basitemporals or parasphenoid) and the *basioccipital* (both ossified in the basal plate) become hollowed out and the cavities occupied by diverticula of the tympanic cavity.

The *pleurosphenoid* ('alisphenoid' autt.) is morphologically important, for as it ossifies in the pila antotica it is part of the true cranial wall, situated medially to the vena capitis lateralis and anteriorly to all the branches of the trigeminal nerve. It cannot, therefore, be homologous with the mammalian alisphenoid or with the reptilian epipterygoid (which latter is not ossified in crocodiles). A splint of bone may later separate the exits of the profundus and maxillary branches of the trigeminal nerve.

The *quadrate* ossifies in the body of the pterygoquadrate cartilage, but it remains hollow, being permeated by diverticula of the tympanic cavity. All internal movement in the upper jaw is prevented by the attachment of the quadrate to the prootic, exoccipital, pleurosphenoid, basisphenoid, quadrato-jugal, and squamosal, and by the development of the palatine process of the pterygoid. The crocodile's skull is the most markedly monimostylic and akinetic known.

The *articular* is the only ossification in Meckel's cartilage, and is extensively permeated by diverticula of the tympanic cavity. It apparently fuses with the prearticular.

The *stapes* ossifies according to Parker from two centres in the columella auris: a basistapedial centre in the footplate covering the fenestra ovalis, and a mediostapedial centre in the stalk. The extracolumella remains cartilaginous.

The *1st ceratobranchials* ossify in the 1st branchial cornua of the 'hyoid' plate.

21. AVES

I. Anas boschas

The development of the chondrocranium of the duck has been investigated with the help of the van Wijhe method by Sonies (1907) (see also van Wijhe, 1907, 1910) and by de Beer and Barrington (1934). A special study of the hyobranchial skeleton has been made by Kallius (1905*a*).

i. The development of the chondrocranium of Anas boschas

1. Stage 1. (132 hours). PLATE 94, FIG. 1.

The first element of the bird's skull to become chondrified is the acrochordal cartilage, which takes the form of a small transverse bar, at the tip of the notochord.

As maintained by van Wijhe (1922) and Jager (1926), and confirmed by de Beer (1926*a*, 1931*b*), the acrochordal arises in connexion with the mesenchyme which has been formed by the breaking down of the hollow transverse commissure between the two 1st prootic somites; it therefore represents the most anterior portion of the axial skeleton and of the basal plate.

The parachordal cartilages in the duck form a thin cylinder enclosing the notochord, about ½ mm. long, with the foremost point about 0·7 mm. behind the acrochordal. The notochord here, then, has an intrabasal course.

Behind the parachordal in the duck embryos at the stage described there can be seen three separate rings of cartilage surrounding the notochord. Comparison with later stages shows that these represent two vertebrae which will become incorporated in the skull (the 1st and 2nd occipital vertebrae; corresponding to the 8th and 9th, and the 9th and 10th segments: see p. 31) and the 1st vertebra of the neck, or atlas vertebra.

Another pair of chondrifications which appear very early are the rudiments of the cochlear portions of the auditory capsules. At this stage, they are just recognizable as faint tracts of blue-stained tissue, on each side of but distinct from the anterior end of the parachordal. These are the cartilages which Sonies called mesotic or basiotic. They clothe the medial faces of the cochlear portions of the auditory sac, and are the first part of the auditory capsules to undergo chondrification. The fact that they are independent of the parachordal is of theoretical importance, since it further invalidates Gaupp's theory that the cochlear portion of the auditory capsule is a modified derivative of the basal plate of the skull.

2. Stage 2. (138 hours). PLATE 94, FIGS. 2, 3.

The acrochordal cartilage extends at each side into a process which is the rudiment of the pila antotica. Immediately behind the acrochordal, and on each side of the notochord is a small bar of cartilage which will ultimately form part of the basal plate and will enclose the basicranial fenestra. These are the anterior parachordal cartilages.

The cochlear capsules are now in cartilaginous continuity with the front of the parachordal cylinder, although the distinction between them can still be made out. Posteriorly, the parachordal and the 1st and 2nd occipital vertebrae have become synchondrosed, but the latter can be clearly recognized. In lateral view the hypocentra of the two occipital vertebrae project ventrally below the level of the remainder of what may be called the basal plate. The atlas and axis vertebrae are well defined, and each has a well-marked hypocentrum, that of the atlas being the larger, and more or less distinct from the ring-shaped body or pleurocentrum of its vertebra, which is destined to give rise to the odontoid process.

3. Stage 3. (144 hours). PLATE 94, FIGS. 4, 5.

The trabeculae have now arisen as cartilaginous bars, paired and separate, beneath the floor of the forebrain. They are quite wide apart, as if the skull of the bird were destined to be platytrabic instead of tropitrabic. Between the trabeculae and the acrochordal, and likewise isolated, are the polar cartilages in line with the hind ends of the trabeculae.

The pila antotica is extending upwards and forwards, with a notch at its base for the abducens. Dorsal to each polar cartilage and ventro-anterior to each pila antotica, another pair of chondrifications has appeared, known as the supratrabecular, or suprapolar cartilages. They correspond to the supra-trabecular bars which in the Lacertilia, Sphenodon, Chelonia, and Primates connect the anterior with the posterior clinoid processes. The ophthalmic artery runs out beneath them, and their precocious appearance is probably related to the fact that the recti muscles are attached to them.

The anterior parachordals have established connexion with the parachordal and cochlear capsules, and the basal plate is now complete, enclosing a basi-cranial fenestra. At the same time, the basal plate is extending to the side. The hypocentra of the two occipital vertebrae are still visible, and the atlas and axis vertebrae are as before, except that their neural arches have now arisen as separate paired struts of cartilage.

The first appearance of the visceral arch skeleton is found at this stage, in the form of the rudiments of the quadrate cartilages, and of the median unpaired (basibranchial) copula.

4. Stage 4. (156 hours). PLATE 94, FIGS. 6, 7.

A band of procartilaginous tissue, staining faintly with victoria blue, is discernible stretching between the polar cartilage and the quadrate on each side. This is a vestige of the basal articulation between the mandibular arch and the neurocranium, i.e. between the basal process of the former and the basitrabecular process of the latter. (Other cases of this have been described by Filatoff (1906), but regarded as a remnant of the processus ascendens.)

The canalicular portion of the auditory capsule has now begun to chondrify as an isolated curved plate, covering the lateral and posterior surfaces of the semicircular canals.

Two pairs of occipital (strictly, occipitospinal) arches are now visible at the hind end of the basal plate. These obviously belong to the two absorbed occipital vertebrae, of which they represent the neural arches, and (see p. 31) they correspond to the septa between the 8th and 9th and the 9th and 10th segments. The hypocentrum of the 2nd (absorbed) occipital vertebra is still plainly visible.

5. Stage 5. (160 hours).

The trabeculae are joined together anteriorly, forming an ethmoid or trabecular plate which constitutes the anterior limit to the hypophysial fenestra. This fenestra is not yet completely enclosed, for the polar cartilages are still free from the acrochordal, but they are attached, each to the hind end of the trabecula of its own side, by procartilage.

From each trabecula, near its anterior end, there arises a process which projects laterally, and slightly forwards and upwards. This is the anterior portion of the orbital cartilage, with its preoptic root. It is situated medially to the nasal branch of the profundus nerve.

The canalicular part of the auditory capsule is faintly connected with the cochlear portion by a strand of tissue which marks the anterior limit of the fissura metotica. Laterally to the cochlear capsule, the rudiment of the columella auris can be seen as an isolated structure, without any connexion with the wall of the auditory capsule.

The connexion between the polar cartilage and the quadrate has broken down, but a stump is left on the polar cartilage from which the infrapolar process (see below) will develop. In the 1st branchial arch, the ceratobranchial element is just discernible as a procartilaginous rudiment.

6. STAGE 6. (7 days). PLATE 95, FIGS. 1, 2.

The hypophysial fenestra has been completed by the fusion of the polar cartilages with the acrochordal: the fusion is, however, slender, and the position of the originally isolated polar cartilages is plainly visible. From the ventro-posterior surface of each polar cartilage, the infrapolar process is directed backwards towards the under surface of the basal plate; it is an outgrowth from the stump of the basitrabecular process, and it passes ventrally and medially to the internal carotid artery, and medially to the palatine nerve.

At the hinder end of the basal plate four pairs of occipital arches are now present, corresponding to the hind septa of the 6th–9th segments. This means that in addition to the two absorbed occipital vertebrae (represented now by the occipital arches of septa 8/9 and 9/10) the hinder portion of the originally continuous and unsegmented parachordals shows evidence of metameric segmentation. Between every two occipital arches on each side, a root of the hypoglossal nerve emerges.

As regards the visceral skeleton, Meckel's cartilage has now made its appearance, but on each side it chondrifies from two centres.

7. STAGE 7. (7½ days).

The anterior portions of the trabeculae have fused in the midline to form the trabecula communis. At the same time, the somewhat indistinct region of the ethmoid plate is becoming deeper in the sagittal plane, and the nasal septum is now visible. Anteriorly, the nasal septum is prolonged into the prenasal process. All these structures have chondrified in connexion with the trabeculae, and have not arisen from independent centres.

The suprapolar cartilages have become attached to the dorsal surface of the polar cartilages of their own side, by two fusions, so as to enclose the ophthalmic artery in a foramen. Between the posterior edge of the suprapolar cartilage and the anterior edge of the pila antotica on each side, there is a notch through which the oculomotor nerve passes. The trochlear nerve has now become enclosed in a foramen in a growing plate of cartilage, the posterior portion of the orbital cartilage which chondrifies in connexion with the dorsal end of the pila antotica. Similarly, the abducens nerve has become enclosed in a foramen at the base of the pila antotica. The canalicular capsule is now more fully developed and the occipital arches have become joined distally so as to enclose three hypoglossal foramina on each side.

In the mandibular arch Meckel's cartilage is now a single element with a sigmoid curvature, while the quadrate shows both otic and pterygoid processes.

8. STAGE 8. (8 days). PLATE 95, FIGS. 3, 4.

The prenasal process is elongating, and the dorsal edge of the nasal septum extends on each side to form the parietotectal cartilages, which will give rise to the roof and to the greater part of the side walls of the nasal capsule. Behind the nasal septum, the interorbital septum is rising up, carrying the preoptic roots of the orbital cartilage with it.

Both the anterior and posterior portions of the orbital cartilages are expanding: the former spreading out on each side beneath the cerebral hemispheres, while the latter, attached by the pila antotica to the basal plate, is spreading out sideways and is concave towards the now enormous eye.

The canalicular and cochlear portions of the auditory capsule are now connected by cartilage, both anteriorly and posteriorly. Between the posterior orbital cartilage, pila antotica, anterior parachordal, and auditory capsule there is, therefore, a deep and wide notch, the incisura prootica, through which all the branches of the trigeminal nerve emerge. It is important to note that the pila antotica possesses its typical relation, lying in front of all the trigeminal roots.

The posterior or definitive occipital arch is growing upwards and forms a posterior boundary to the fissura metotica, through which the glossopharyngeal, vagus, and spino-accessory nerves emerge. Opposite the centre of the fissura metotica, and ventral to the canalicular portion of the auditory capsule, the isolated metotic cartilage has made its appearance; it seems to be developed in connexion with the cranial ribs.

At this stage of development in the duck, four cranial ribs can be recognized on each side, corresponding to the hinder septa of segments 6 to 9. They take the form of dense mesenchymatous tissue, which, at least in the posterior ones, shows some intercellular substance. In addition to the evidence of metameric segmentation which these ribs show, for they alternate with the hypoglossal nerve-roots in typical manner, these ribs are of interest because a mesenchymatous connexion can be made out between them and the metotic cartilage. In the crocodile (see p. 263) the subcapsular process shows great similarities with the metotic cartilage and is also to be regarded as a cranial rib.

In the visceral arches the otic process of the quadrate and the retroarticular process of Meckel's cartilage have developed. The copula shows a slender process directed backwards: this is really the 2nd copula which has fused on to the hind end of the 1st.

9. STAGE 9. (8½ days). PLATE 96, FIGS. 1, 2.

The developments which characterized the previous stage have become accentuated in this: the prenasal process is longer, the anterior and posterior portions of the orbital cartilage have expanded further, the parietotectal cartilages of the nasal capsule have extended, and the interorbital septum is higher. A vacuity has appeared in the posterior orbital cartilage, and with the widening of the basal plate, more hypoglossal foramina have become enclosed: of these there are now five on one side and four on the other. The walls of the auditory capsule are nearly complete, and reveal, on the outside, the foramen ovale into which the foot of the columella auris fits, and the foramen perilymphaticum which faces towards the fissura metotica; and on the inside a large opening which will eventually become narrowed down to the foramina for the auditory nerve. The facial nerve is separated from the incisura prootica by the prefacial commissure, and enclosed in the facial foramen.

But this stage also shows some important new developments. Two pairs of independent cartilages have appeared in the region of the nasal capsule; these are the paranasal cartilages and the laminae orbitonasales. The former are curved and show a marked concavity facing outwards, which will give rise to the concha nasalis. The laminae orbitonasales form the posterior wall of the nasal capsule on each side, and the lower border of the orbitonasal fissure which is, however, not completely delimited, for its lateral border will not be formed until the parietotectal cartilage becomes connected with the laminae orbitonasales. The dorsal border of the orbitonasal fissure is now present in the sphenethmoid commissure, which joins the parietotectal cartilage to the anterior orbital cartilage. The lamina orbitonasalis is situated ventrolaterally to the nasal branch of the profundus nerve.

Posteriorly, the metotic cartilage has become fused with the base of the occipital arch, from which it projects forwards and outwards beneath the auditory capsule, and presents an unmistakable resemblance to the subcapsular process of the crocodile.

In the hyoid arch the stylohyal cartilage has appeared in procartilaginous connexion with the outer end of the columella auris (by means of the pars interhyalis), but from a separate centre of chondrification. As for the columella itself, it seems to chondrify in one piece, although there appear to be separate otostapedial and hyostapedial mesenchymatous rudiments.

ii. Morphology of the fully formed chondrocranium of Anas

1. OPTIMUM STAGE. (9 days). PLATE 96, FIGS. 3, 4.

This stage is among the most important, for the chondrocranium of the bird now exhibits its maximum resemblance to that of reptiles. The orbital cartilage is complete, the anterior orbital and posterior orbital cartilages having fused on each side, so as to form a planum supraseptale and a continuous lateral wall to the cranial cavity, from the sphenethmoid commissure to the pila antotica. A large sphenoid fontanelle is thus enclosed on each side; dorsal to the suprapolar cartilage, posterior to the interorbital septum and preoptic root of the orbital cartilage, anterior to the pila antotica, and ventral to the orbital cartilage. Through this fontanelle, the optic, oculomotor, and trochlear nerves emerge, for the latter has now been released from its foramen in the posterior orbital cartilage. In reptiles the sphenoid fontanelle is commonly subdivided by a pila metoptica, pila accessoria, and taenia medialis into three fenestrae, viz. optica, epioptica, and metoptica. The duck at this stage shows a rudimentary attempt at a similar subdivision, for an isolated chondrification is present in the middle of each sphenoid fontanelle, and probably represents a portion of the taenia medialis.

The eye-ball is surrounded by a well-developed sclerotic cartilage.

Anteriorly, the nasal capsule is more complete. It is important to notice that the lateral edge of the lamina orbitonasalis juts out to the side beyond the side wall of the nasal capsule. Just in front of it the concha nasalis (superior or posterior turbinal of other authors), formed by the concave paranasal cartilage, is plainly visible.

The olfactory nerve passes out of the cranial cavity through a foramen olfactorium evehens, which is bordered by the front edge and preoptic root of the anterior orbital cartilage, the sphenethmoid commissure, the nasal septum, and part of the hind edge of the parietotectal cartilage. The olfactory nerve then finds itself for a short distance in the orbit, which it leaves by the foramen olfactorium advehens for the cavity of the nasal capsule. The latter foramen is difficult to delimit, but its boundaries may be said to be constituted medially by the nasal septum; anteriorly and laterally by the hind edge of the parietotectal cartilage; and posteriorly by the dorsal edge of the lamina orbitonasalis. The orbitonasal fissure is still more difficult to define, owing to the fact that it is so very large; its medial boundary is the interorbital septum and preoptic root of the orbital cartilage; its ventral boundary is the dorsal edge of the lamina orbitonasalis; its dorsal boundary is the sphenethmoid commissure, while its lateral boundary is formed by part of the hind edge of the parietotectal cartilage.

The metotic cartilage has become attached by its dorsal border to the side wall of the canalicular portion of the auditory capsule. The ventro-medial edge of the metotic cartilage is fused to the lateral edge of the basal plate at a point between the glossopharyngeal and vagus nerves, which are thus enclosed in separate foramina, and, with the increasing fusion between the anterior edge of the occipital arch and the posterior surface of the canalicular portion of the auditory capsule, these two foramina are all that remains of the originally extensive fissura metotica.

By its conformation and its adhesions to neighbouring cartilages as just described, the metotic cartilage forms a covering from the side and from behind to the recessus scalae tympani, into which the foramen perilymphaticum of the auditory capsule opens. The posterior edge of the metotic cartilage is fused to the occipital arch and auditory capsule; its free anterior edge forms a crest which extends forwards and upwards, almost to the otic process of the quadrate. Over this edge, which forms the ventral and posterior borders of a fenestra pseudorotunda, the tympanic membrane is stretched.

The dorsoposterior corners of the auditory capsules are connected with one another across the middle line by tectum synoticum. The dorsal ends of the occipital arches are free and take no part in its formation. The tectum synoticum seems to be of paired origin.

The hind edge of the basal plate forms the ventral border of the foramen magnum, and, immediately on each side of the notochord, a pair of prominences have developed. These are destined to form the single median occipital condyle; here, they are in the form of the paired elements of the pleurocentrum of the 2nd (absorbed) occipital vertebra; i.e. they represent the anterior sclerotomites of the 7th metotic segment, or 10th segment of the entire series. This feature is of importance in showing the invertertebral (intrasegmental) nature of the occipito-atlantal joint in birds.

In the visceral arch skeleton the chief points worthy of notice are the great elongation of Meckel's cartilage (which has lost the sigmoid curvature of early stages), the growth of which keeps pace with that of the prenasal process, the appearance of isolated coronoid cartilages, described by Kallius (1905*a*), and the separation of the stylohyal cartilage from the columella auris.

The columella itself may now be regarded as consisting of an (eventually ossifying) medial stapes and a (cartilaginous) lateral extracolumella (or 'extrastapedial') inserted in the tympanic membrane. From the shaft of the columella a diminutive processus dorsalis (or 'suprastapedial') rises up: from the extracolumella a processus interhyalis represents the original connexion with the stylohyal ('infrastapedial') and a 'laterohyal' process projects upwards and is connected with the processus dorsalis by a slender bar of cartilage, thus enclosing a ('Huxley's') foramen. These processes do not have independent centres of chondrification, but grow out of the columella auris. The hyomandibular branch of the facial nerve runs back dorsally to the columella, medially to the processus dorsalis, behind which it gives off the chorda tympani which runs forwards laterally to the laterohyal. The orbital artery passes dorsally to the columella and medially to the processus dorsalis.

2. Regression of the chondrocranium. (14 days). Plates 97, 98.

The chief feature presented by this stage is the reduction almost to disappearance of the anterior orbital and supraorbital cartilages. The connexion with the posterior orbital cartilage has vanished, and the sphenethmoid commissure has broken down. All that is left of the extensive anterior orbital cartilage of the previous stage is a planum supraseptale. Its paired nature is still obvious, and anteriorly, it runs into the sphenethmoid process on each side, representing the posterior remnant of the sphenethmoid commissure. The anterior remnant of this commissure may now be recognized on each side as the backwardly projecting processus tectalis. The boundaries of the foramen olfactorium evehens and of the orbitonasal fissure are now, therefore, indistinct.

For the greater part of its length, the planum supraseptale is more or less horizontal, underlying the telencephalon; but its hindmost portion slants downwards and backwards. This portion is marked off from the underlying interorbital septum by extensive slit-like fissures, and gives the impression of having had a paired origin separate from that of the remainder of the planum supraseptale. It is possible that this hindmost portion is derived from the pair of isolated chondrifications described on p. 273 as probably representing the taenia medialis of reptiles.

The hind edge of the roof of the nasal capsule has grown back and projects freely as the processus tectalis, the base of which is fused with the interorbital septum. The olfactory nerve runs forwards beneath this backgrowing roof and thus traverses a portion of the orbit on its way to the foramen olfactorium advehens. Thus, the gap between the backgrowing roof and the dorsal edge of the interorbital septum is not to be mistaken for the gap between the interorbital septum and the sphenethmoid commissure—really the foramen olfactorium evehens—through which the olfactory nerve passes at previous stages.

It is the disappearance of the anterior orbital cartilage which normally separates the olfactory nerve from the orbit, which is responsible for the passage of this nerve through the orbit in the duck. The posterior portion of the planum supraseptale has now assumed a vertical position, and lies just above the exits of the optic nerves. The most striking feature is the great elongation of the prenasal process, parallel with that of Meckel's cartilages. The anterior part of the nasal capsule is still very incompletely chondrified, but the posterior part is now well formed. On each side, the concha nasalis forms a deep inpushing, opening outwards, and situated immediately in front of the lamina orbitonasalis. The wall of the concha passes anteriorly into the parietotectal cartilages, which form the roof and side walls of the greater part of the capsule. Close to its ventral edge each side wall gives off an elongated scroll-like structure, curved in such a way as to form a half-cylinder, the concavity facing sideways and downwards. This is the maxilloturbinal. Posteriorly, the maxilloturbinal extends beneath the concha nasalis forming a floor to the hindmost region of the capsule. Dorsal to this floor, and ventral to the aperture of the concha and to the lateroventral edge of the lamina orbitonasalis, is the aperture through which the nasal sac communicates with the orbital sinus. The anterior cupola, fenestra narina, and atrioturbinal are not yet chondrified. Two fenestrae are to be seen in the nasal septum, and in one place beneath its ventral edge are to be seen the nodules of secondary cartilage representing the prevomer bones.

The posterior orbital cartilage on each side has established a secondary connexion with the lateral edge of the basal plate, between the profundus and maxillary branches of the trigeminal nerve. This connexion, the pila antotica spuria, thus encloses the profundus nerve in a foramen of its own, of which the true pila antotica forms the slender medial and dorsal margin. The maxillary and mandibular branches of the trigeminal nerve emerge behind the pila antotica spuria, through what may be called the incisura prootica spuria, which is almost converted into a foramen by the approximation of the orbitocapsular process of the posterior orbital cartilage to the postorbital process of the auditory capsule. The posterior orbital cartilages show a fenestra or vacuity as a result of absorption.

The lateral extension of the posterior orbital cartilage partially encloses a space in which the trigeminal ganglion is lodged. This trigeminal chamber lies outside the dura mater (here membranous) and is therefore to be regarded as intramural.

The infrapolar processes have established contact with the ventral surface of the basal plate, with the result that each of the internal carotid arteries on its way to the hypophysial fenestra has to pass through a foramen—the lateral carotid foramen. A vestige of the basitrabecular process remains as a blunt stump projecting sideways from the anterior part of the infrapolar process.

The number of hypoglossal foramina is undergoing reduction; the occipital condyle is increasing in size, and the tectum synoticum has become interrupted, so that instead of the continuous bar of cartilage which at previous stages stretched from one auditory capsule to the other, there are now two processes projecting dorsally side by side: the supraoccipital processes. The occipital arches end dorsally as blunt points medially to the base of the supraoccipital processes.

The foramina for the glossopharyngeal and vagus nerves serve to indicate the position of the original fissura metotica, and of the line of fusion between the metotic cartilage and the lateral edge of the basal plate.

In the visceral arches the chief features are the great elongation of the cerato-branchial and the appearance of epibranchial cartilages in the 1st branchial arch, and the appearance and fusion of the paired ceratohyal cartilages to form the median unpaired paraglossal cartilage of Kallius (1905*a*) (though that author considers that these structures bear no relation to the hyoid arch).

On each side of the posterior process of the copula and near its hind end there are now to be seen a pair of cartilages which embrace the aperture of the glottis. They lie at a more dorsal level than the copula and appear to be the rudiments of the cricoid cartilages; possibly they represent the cartilaginous elements of the 2nd branchial arch, hitherto not found in birds.

3. THE NASAL CAPSULE. (17 days). PLATE 97, FIG. 3.

The maxilloturbinal scrolls become still further wound, and now present a complete turn and a half. The chief advance is in the chondrification of the anterior region of the capsule, involving the formation of a cupola anterior on each side, bounding the fenestra narina from in front, and of the atrio-turbinals. The latter are plates of cartilage descending from the roof of the capsule and forming longitudinal curtains; their hinder extremity overlaps and is lateral to the anterior extremity of the maxilloturbinals. It is noticeable that the floor of the capsule is practically absent. On a level with the hind end of the atrioturbinals, a small pair of processes are to be seen jutting out to each side from the nasal septum near its ventral edge. These processes curve forwards and are joined by a strand of dense mesenchymatous tissue with the lowermost and hindmost point of the cupola anterior of their own side, thus forming a very rudimentary floor to this region of the capsule.

The side wall of the capsule is also very incomplete in the anterior and middle regions; but in the hinder region, the free ventral edge of the side wall (parietotectal cartilage) now extends down laterally to, and further ventrally than, the maxilloturbinals.

One of the most striking objects seen in a section through the nasal capsule of the duck is the relatively enormous size of the ramus medialis nasi, which is related to the development of the large flat bill with its extensive sensory surface. This nerve passes backwards beneath the rudimentary processes representing the floor, and then, as it reaches more dorsal levels at the side of the nasal septum, it is covered over by a ridge of cartilage. This ridge must be regarded as forming part of the morphological floor of the capsule, and is, indeed, connected with the cartilaginous process mentioned above by means of dense mesenchyme. It seems that the relations of the various struc-tures in this region have been distorted by the outgrowth of the prenasal process. This process arises, not from the anterior end of the nasal capsule, but from its ventral surface, some little distance behind its anterior end.

The atrioturbinals and maxilloturbinals are clothed with folds of epithelium of the nasal sac, as is the inner surface of the concha nasalis. The aperture of the lachrymonasal duct and the choana are situated ventrally to the maxillo-turbinal, and between the posterior extension of the latter (which forms the posterior floor) and the concha nasalis, an interruption in the side wall serves for the egress of the orbital sinus from the main cavity of the nasal sac. The concha nasalis is occupied only by mesenchyme.

Owing to the breaking down of the sphenethmoid commissure, the orbitonasal fissure is only virtual at this stage, and the ramus lateralis nasi does not enter the cavity of the nasal capsule at all. Instead, immediately after parting company with the ramus medialis nasi which enters the foramen olfactorium advehens, the ramus lateralis nasi runs forwards by the side of the parietotectal cartilage, immediately above the mouth or aditus of the concha nasalis.

The nasal capsule in the duck shows signs of reduction as compared with that of reptiles, and, in particular, it presents the appearance of having been squashed from behind, as if the lamina orbitonasalis, itself displaced forwards by the huge eye, had therefore ceased to form the dome-shaped cupola posterior which is characteristic of the hinder region of the nasal capsule in reptiles and in mammals.

W. K. Parker (1866*b*, 1870, 1876*b*) recognized three turbinals in the nasal capsule of birds which he named, 'nasoturbinal', 'inferior turbinal', and 'upper turbinal' ('Bagged condition of the aliethmoidal wall'). But he also observed a small 'middle turbinal', on the inner side of the root of the inferior turbinal in ostriches.

Gegenbaur (1873) also distinguished three turbinals, which he called 'anterior turbinal' or 'inferior turbinal', 'middle turbinal', and 'posterior turbinal' or 'superior turbinal' (also called 'Riechhügel'). It is clear that the expressions 'middle turbinal' and 'inferior turbinal' are being used in different senses by Parker and Gegenbaur.

Born (1879) recognized a 'primary turbinal' (middle turbinal), a 'secondary turbinal' (upper turbinal) and a 'vestibular turbinal' (nasoturbinal of Parker, anterior turbinal of Gegenbaur). T. J. Parker (1892) working on Apteryx described an 'anterior turbinal', an 'anterior accessory turbinal', a 'ventral accessory turbinal', a 'middle turbinal', and a 'posterior turbinal'. Suschkin (1899) speaks in Tinnunculus of a 'vestibular turbinal', and a 'middle turbinal', of which the 'upper turbinal' is a part. Sonies (1907) did not follow the

W. K. Parker	*Gegenbaur*	*Born*	*T. J. Parker*	*Suschkin*	*This Book*
nasoturbinal	anterior turbinal	vestibular turbinal	anterior, anterior accessory, and ventral accessory turbinal	vestibular turbinal	atrio-turbinal
inferior turbinal	middle turbinal	primary turbinal	middle turbinal	middle turbinal	maxillo-turbinal
superior turbinal	posterior turbinal	secondary turbinal	posterior turbinal	(upper turbinal)	concha nasalis

development long enough to see the chondrification of the anterior region of the nasal capsule, and speaks only of a 'concha superior' and a 'concha inferior'.

The main point which emerges from this rapid survey of previous work is the inadvisability of using comparative topographical terms such as 'upper', 'middle', 'anterior', to denote structures which, in different forms may have topographically different positions while retaining their essential morphological relations. For this reason, it seems preferable to speak of an atrioturbinal, a maxilloturbinal, and a concha nasalis, in the nasal capsule of birds.

The development of the osteocranium of the duck has not been studied, though some Anseriformes were described by W. K. Parker (1890). New work in this field would be welcome, especially as Anser ferus seems to have a postparietal bone. The existence of such a bone in other birds has been shown by Staurenghi (1900*b*).

II. Gallus domesticus

The development of the skull of the fowl was studied by W. K. Parker (1870, 1876*b*, 1891*b*) and Gadow (1891) and more recently by Tonkoff (1900) who constructed a wax model, and Sonies (1907) who employed the van Wijhe technique. The morphology of the nasal capsule was described by Born (1879).

i. The chondrocranium of Gallus domesticus

The development of the chondrocranium of the fowl is so similar to that of Anas that a complete separate description will be unnecessary, and attention may be confined to the more important points of difference. These differences amount in many cases to the chondrification of structures in Gallus in continuity with pre-existing cartilages instead of independently as in Anas. This applies to the cochlear capsule (Sonies's lamina basiotica) which arises in continuity with the parachordal, the polar cartilage which is continuous from the start with the trabecula, and the supratrabecular cartilage which is continuous with the pila antotica behind and the trabecula in front and thus resembles the supratrabecular bar of other forms (p. 227).

On the other hand, the pila antotica chondrifies independently, as do the hyostapes and otostapes rudiments of the columella auris. According to G. Smith (1904) the base of the columella auris may become temporarily continuous with the cartilaginous wall of the auditory capsule. The columella auris is further of interest because of the relations to it of the chorda tympani nerve. In Gallus this nerve is pretrematic as well as pretympanic and passes downwards clean in front of the columella.

In the fully formed chondrocranium a feature of interest is the break-down of the pila antotica, with the result that the orbital cartilage is supported by the pila antotica spuria only, which is situated between the profundus and the maxillary branches of the trigeminal nerve (see Anas, p. 276). These seemingly abnormal relations are fully explained as a result of the development.

A nodule of cartilage is intercalated between the head of the quadrate and the auditory capsule. (Gaupp, 1906*a*, p. 801). This nodule is detached from the capsule. But in view of the condition in Reptiles (p. 225) it would be interesting to know if it had any relations with the dorsal process of the columella auris.

The prenasal process in Gallus is reduced. The posterior orbital cartilage is fused to the postorbital process of the auditory capsule. The infrapolar processes ('lingulae sphenoidales' of W. K. Parker, 1870) are reduced and fail to reach the basal plate.

The tectum synoticum is attached to the auditory capsule by two roots on each side, of which the hindermost gives the impression of being the detached dorsal end of the occipital arch, now attached to the auditory capsule. The remainder of the occipital arch ends freely. (Plate 99, figs. 1–3).

ii. The development of the osteocranium of Gallus

1. The membrane-bones. Most membrane-bones arise during the middle of the 2nd week of incubation, head-length *circ.* 20 mm.

The *frontal* is large and arises over the processus tectalis (back-growing roof of nasal capsule), the planum supraseptale, and the posterior orbital cartilage. It thus overlies the position of the now broken-down connexion between anterior and posterior portions of the orbital cartilage. There is no trace of any postfrontal or prefrontal ossifications.

The *parietal* is very small and arises over the posterior part of the auditory capsule as a plate sloping upwards and outwards, i.e. it arises on the underside of the bulge of the brain, which it will eventually overgrow.

The *nasal* arises on the dorsal and lateral surfaces of the nasal capsule, sloping obliquely forwards, outwards, and downwards.

The *lachrymal* (prefrontal of Gaupp) is fairly large, and projects sideways from the lamina orbitonasalis.

The *squamosal* (supratemporal of Lakjer, 1926) arises laterally to the postorbital process of the auditory capsule and to the lateral edge of the posterior orbital cartilage. In this position it protects the trigeminal ganglion (lodged in its intramural trigeminal chamber) from the side, and thus comes to form part of the wall of the bony brain-case. Brock (1935) has shown that the squamosal of the bird probably corresponds to the outermost of the two bones in the temporal region of Lacerta.

The *premaxilla* has alveolar, palatine, and ascending processes, and fuses with its fellow of the opposite side. The ascending processes run back in contact, but not fused with one another, over the roof of the nasal capsule, between the nasals and almost as far back as the frontals. There is, as in all birds, no dentinal egg-tooth, but an epidermal papilla on the dorsal surface of the snout.

The *maxilla* is a slender bone consisting of a zygomatic process extending backwards and sideways from the side wall of the nasal capsule, and a palatine process.

The *jugal* is a slender rod joining the zygomatic process of the maxilla to the quadratojugal.

The *quadratojugal* is a slender rod joining the jugal to the quadrate, thus completing the arcade of the upper jaw.

The *parasphenoid* arises, according to Parker (1888), from three centres, which are the median anterior rostrum, and the paired lateral wings or *basitemporals*, which latter become attached to the ventral surface of the basisphenoid, thereby enclosing the internal carotid arteries and palatine nerves in parabasal canals, and also covering over the Eustachian tube and a diverticulum of the tympanic cavity. The posterolateral edge of each basitemporal thus comes to form the anterior border of a so-called 'tympanic recess', into which the internal carotid, Eustachian tube, and glossopharyngeal and vagus nerve foramina open.

The rostral portion of the parasphenoid also becomes fused to the basisphenoid.

The *prevomers* arise as paired struts underlying the nasal septum at the level of the lamina orbitonasalis. They become fused in the midline.

The *palatine* arises as an elongated bone ventrolaterally to the ventral edge of the interorbital septum. Anteriorly, the palatine is rodlike, and comes into contact with the palatine process of the premaxilla. Posteriorly, the palatine broadens into a plate, slanting upwards and inwards to articulate with the parasphenoid rostrum. The contact between the palatine and the pterygoid represents the original separation between the bones, and not a intrapterygoid joint as in other birds (p. 285).

The *pterygoid* arises almost in the transverse plane, as a bony strut running along the anterolateral surface of the pterygoid process of the pterygoquadrate cartilage towards the parasphenoid rostrum and the hind end of the palatine. The hemipterygoid which in other birds becomes detached and fused with the palatine (see p. 445) here remains as a process projecting forwards from the anterior end of the pterygoid.

According to Parker (1870, p. 781) a nodule of cartilage is intercalated between the pterygoid and the parasphenoid. This might represent the meniscus pterygoideus of other forms (p. 228).

The membrane-bones of the lower jaw are the *dentary*, *angular*, *supra-angular*, and *splenial*. Gallus has no prearticular (Voit, 1924).

The *sclerotic* bones are in the form of a dozen semi-overlapping platelets, lying on the rim of the sclerotic cartilage but separated from it by connective tissue.

2. The cartilage-bones. Most cartilage-bones arise at the beginning of the third week of incubation; head-length *circ.* 30 mm.

The *basisphenoid* arises from paired perichondral lamellae surrounding the bases of the trabeculae cranii, and becomes fused with the underlying parasphenoid (both rostrum and basitemporals).

The *exoccipitals* ossify in the occipital arches and also in the metotic cartilage, thus surrounding the glossopharyngeal and vagus nerve foramina. The posterior border of the 'tympanic recess' is thus formed by the exoccipital; the limit between exoccipital and basioccipital is indicated by the hypoglossal foramina.

The *basioccipital* arises from perichondral lamellae on the dorsal and ventral surfaces of the basal plate, and ossifies the occipital condyle.

The *supraoccipital*, according to Parker (1870, p. 772), arises from paired rudiments in the tectum synoticum, which soon fuse in the midline. Laterally, the supraoccipital spreads over the inner wall of the hind part of the anterior semicircular canal. The only form in which Parker found a single median supraoccipital rudiment is Turdus (1873*c*).

The *prootic* covers most of the anterior half of the auditory capsule, including the prefacial commissure.

The *opisthotic* ossifies in the wall of the posterior semicircular canal, and fuses with the prootic to form the periotic bone.

The *epiotic* (pterotic of Parker, 1866*b*) arises late in the wall of the dorsal part of the posterior semicircular canal, and fuses with the supraoccipital. (The 'epiotic' of Parker (1870, p. 790) is of doubtful existence, and appears to be a portion of the exoccipital.)

The *pleurosphenoid* ossifies in the posterior orbital cartilage and pila antotica spuria. The 'postfrontal' of Parker (1870, p. 790) seems to be a part of the pleurosphenoid.

The *orbitosphenoid* is apparently represented by two centres on each side, developed late in the planum supraseptale and the fibrous remains of the orbital cartilage.

The *presphenoid* (mesethmoid autt.) arises (perhaps from more than one centre) in the interorbital and the hind part of the nasal septum, and in the hinder part of the roof of the nasal capsule.

The *quadrate* ossifies in the body of the pterygoquadrate cartilage, and eventually occupies the whole of it.

The *articular* arises in the articular region and the retroarticular process of Meckel's cartilage.

The *stapes* ossifies in the footplate and medial portion of the shaft of the columella auris.

In the hyobranchial skeleton ossification takes place in the *paraglossum*, *copulae* (basibranchials 1 and 2), and in the *1st ceratobranchial* and *epibranchial*.

At an early age the various bones of the skull fuse together and the sutures are obliterated. At the same time, an extensive pneumatization takes place, for the details of which reference may be made to Columba.

III. Columba livia

The development of the pigeon's skull has not been studied in detail, but important results have been obtained by H. Strasser (1905) and Lurje (1906) in their investigations of the phenomenon of pneumatization.

The pneumatization of the skull in birds begins shortly before the time of hatching, and takes place by means of outgrowths of the tympanic cavity and of the nasal cavity. The diverticula of the tympanic cavity are numerous and complicated. From each tympanic cavity a supratubar diverticulum pushes inwards and forwards dorsally to the Eustachian tube, and meets its fellow of the opposite side. This supratubar diverticulum becomes enclosed within the bony lamellae of the parasphenoid (both basitemporal and rostral portions), and invades the cartilage of the trabecula communis and the ventral part of the interorbital septum. Another tympanic diverticulum perforates and permeates the basal plate; another again penetrates into the periotic capsule and spreads in the dermal bones of the skull roof (parietal, frontal) and side (squamosal), and in the occipital region; yet another penetrates into the body of the quadrate; and one more diverticulum makes its way into the articular, and so into Meckel's cartilage and the lower jaw. In some birds this diverticulum is specially protected by a bony cylinder, the *siphonium*.

The process of pneumatization goes hand in hand with the ossification of the bones and their fusion, thus obliterating the sutures. The air-chambers spread in the space created between the perichondral lamellae of cartilage-bones by the resorption of the cartilage, and between the inner and outer tables of the membrane-bones by the resorption and widening of the diploe.

Analogous processes occur in connexion with the diverticula of the nasal sac, which perforate the posterior wall of the nasal cavity and penetrate into the maxilla, premaxilla, nasal, and lachrymal bones, as well as expanding into the infraorbital air-sinus.

As a result of pneumatization, almost every bone of the skull is hollow. The only exceptions practically are the jugal and quadratojugal.

IV. Tinnunculus alaudarius

The development of the skull of the kestrel has been studied with the help of wax models by Suschkin (1896, 1899). In many ways the course of development follows that of the duck and chick, and attention here will be largely confined to the more important points of difference.

i. The development of the chondrocranium of Tinnunculus

At the earliest stage studied the parachordals are already chondrified, forming the basal plate surrounding the notochord completely and fused with the auditory capsules. Posteriorly, there are rudiments of three pairs of occipital arches, the hinder two being as large and distinct as the neural arches of the atlas vertebra, and obviously representing the neural arches of the two occipital vertebrae which in Anas and Gallus have been seen to become fused on to the skull. There are also mesenchymatous indications of four pairs of cranial ribs. Anteriorly, the parachordals are fused with the transverse acrochordal, from the sides of which the pilae antoticae arise. To their anterior edges the supratrabecular cartilages (the term is Suschkin's) are fused so as to enclose the oculomotor nerve in a foramen. (Plate 100, figs. 1, 2).

The trabeculae and polar cartilages are already fused and form slender bars, wide apart and as yet separate from the basal plate, with which they form an angle of some 110°. From the hindmost region of each trabecular polar bar, an elongated infrapolar process projects back freely beneath the basal plate. Between the trabeculae, anteriorly, Suschkin described an 'intertrabecula' from which the interorbital and nasal septum subsequently arises, but it is more than doubtful whether this is to be regarded as a distinct element.

In the visceral arch skeleton the pterygoquadrate, Meckel's cartilage, the median copula, and the 1st cerato- and epibranchial cartilages call for no comment. The columella auris is present, bearing a ventrally directed processus interhyalis (infrastapedial) which is connected by ligament with the independently chondrified stylohyal. The ventral end of the latter is continued as a ligament in the direction of the ceratohyals. (Plate 100, figs. 3, 5).

Later on, the stylohyal becomes fused to the processus interhyalis. There are also developed a processus dorsalis and a laterohyal which connects it with the lateral end of the columella, thus enclosing a foramen. According to Suschkin the footplate of the columella is derived from the wall of the auditory capsule, which is improbable. (Plate 101, fig. 1).

At later stages the chondrocranium presents the same general appearance as in Anas. The infrapolar processes fuse with the basal plate enclosing lateral carotid foramina, the metotic cartilage is developed, the anterior orbital cartilage has a well-developed though transient preoptic root; later, the hinder part of the anterior orbital cartilage becomes detached and joins the posterior orbital cartilage. Thus, the cartilaginous medial wall of the orbit does not appear to be all complete at one time. (Plate 100, figs. 4, 6, 7).

The nasal capsule is of interest, for it possesses a floor in the anterior region formed by the lamina transversalis anterior. There is thus a zona annularis, and the fenestra narina is completely enclosed. The anterior part of the cavity of the nasal capsule is nearly cut off from the posterior part by a vertical transverse partition formed by the upturned hind end of the lamina transversalis anterior.

The tectum covering the hinder part of the skull appears to be a tectum posterius, connecting the dorsal ends of the occipital arches. Vestiges of the epiphysial cartilage arise later in the form of a pair of rods diverging anteriorly, and situated beneath the eventual point of fusion of the frontal and parietal bones.

Instead of fusing completely to form a single median paraglossum, the ceratohyals retain their paired nature.

ii. The development of the osteocranium of Tinnunculus

The development of the bony skull of the kestrel follows very much the same lines as Gallus, especially as regards the hinder region. Anteriorly, however, in the nasal capsule the kestrel develops a number of perichondral bony centres which presumably serve to give strength and rigidity to the upper beak. These ossification centres arise comparatively late and include a centre in the nasal septum, one median and two pairs of centres in the roof of the nasal capsule dorsally to the fenestrae narinae, one pair at the ventral border of these fenestrae, one pair in the floor of the capsule, and one median and one pair of centres in the prenasal process. The ossifications in the floor of the capsule and in the prenasal process are closely associated with the palatine processes of the maxillae and premaxillae respectively.

There is also an ossification centre in each lamina orbitonasalis, giving rise to the '*ectethmoid*' ('prefrontal' of Suschkin).

In the lower jaw the *mentomeckelian* and the *prearticular* (complementare of Suschkin) are present.

As in other carinate birds, practically all the separate bones fuse together and the sutures become obliterated. The only bones to preserve their identity in the adult are the prevomer, the pterygoid, the lachrymal, the quadrate, and the stapes. There is no jugal, and the quadratojugal joins the zygomatic process of the maxilla direct.

V. Passer domesticus

No systematic study has been made of the development of the skull of the sparrow, but it presents a few features of importance, which must be mentioned here. (Plate 102, fig. 1).

The chondrocranium presents the same general features as in Tinnunculus: the infrapolar processes are well developed and fused with the basal plate, the pila antotica persists, the tectum is a tectum posterius, the fenestra narina is completely surrounded, &c. A curious condition is, however, presented by the preoptic root of the orbital cartilage. In the birds described above, this structure sooner or later disappears completely, and as a result there is no lateral wall to the foramen olfactorium evehens, and the olfactory nerve runs through the orbit. In Passer, on the other hand, the preoptic root of the orbital cartilage persists but it is plastered flat against the interorbital septum on each side, and this encloses the (very small) olfactory nerve in a long and tubular foramen olfactorium evehens. The olfactory nerve here, therefore, does not traverse the orbit.

The nasal branch of the profundus nerve passes through the foramen olfactorium advehens into the cavity of the nasal capsule, but the ramus lateralis almost immediately leaves it again, by passing through the epiphanial foramen, immediately above the mouth of the concha nasalis.

Other features worthy of notice are the well-developed basitrabecular processes, from the bases of which the infrapolar processes project back, and the presence of a sagittal bar of cartilage (taenia intertrabecularis) which divides the hypophysial fenestra into right and left halves.

In the bony skull of Passer, as a type of Passerine bird, two features call for comment. The first is that W. K. Parker's (1877*b*, p. 302) identification of a 'transpalatine' bone has been shown to be erroneous. The other is the fragmentation of the pterygoid in such a way that its foremost portion which reaches or nearly reaches the prevomer (as in Palaeognathous birds, see p. 288) becomes detached as the mesopterygoid of Parker (1875, &c.) or the hemipterygoid of Pycraft (1901) and fused on to the posteromedial end of the palatine bone. Thus, the joint between pterygoid and palatine bones is morphologically within the territory of the pterygoid. (See Plate 101, figs. 8, 9.)

VI. Struthio sp.

The development of the skull of the ostrich has been studied by Parker (1866*b*), and more recently the chondrocranium has been investigated by Brock (in press) on whose work the following account is based.

i. The development of the chondrocranium of Struthio

At the earliest stage studied (13·5 mm.) the parachordals, polar cartilages, and trabeculae are already chondrified and in continuity. The latter are fused anteriorly to form a trabecula communis. A hypophysial and a basicranial fenestra are enclosed. The angle between the planes of the parachordals and the trabeculopolar bars is acute, so that the hinder part of the basitrabecular process, forming the infrapolar process, is almost in contact with the ventral surface of the basal plate, enclosing the lateral carotid foramen.

The basitrabecular process is well developed and extends to the medial surface of the pterygoquadrate. The palatine nerve runs beneath it in typical manner. The pila antotica is present on each side of the acrochordal, and bears the typical relations to the surrounding nerves and arteries. The orbital cartilages are represented by their posterior regions, in continuity with the pila antotica, and by the rudiments of their preoptic roots, rising from the trabecula communis.

The auditory capsule is anchored to the side of the basal plate, but there is a wide space between the capsule and the occipital arch. In front of the latter are three rudimentary preoccipital arches and three ventral nerve-roots.

At the 18 mm. stage, the orbital cartilages are complete, and extend uninterruptedly from the preoptic roots to the pila antotica. The preoptic roots are pressed together in the midline forming an interorbital septum, dorsally to which the orbital cartilages spread out forming a planum supraseptale. The anterior continuation of the trabecula communis forms the nasal septum and prenasal process, but the nasal capsule is not yet present. The lateral carotid foramen and the jugular foramen are completely enclosed. (Plate 99, fig. 4).

The suprapolar (or supratrabecular) cartilage has appeared in the hinder part of the large sphenoid fontanelle. It soon fuses with the trabecula in such a way as to enclose the ophthalmic artery in a foramen. The columella auris is present, but the connexion with the hyobranchial apparatus has broken down.

The next stage (10·5 mm. head-length) the maximum development of the chondrocranium has been reached. The roof and side walls of the nasal capsules are present, connected with the orbital cartilages by sphenethmoid commissures. The orbitonasal and the evehent and advehent olfactory foramina are now enclosed. The backgrowing roof of the nasal capsules converts the foramen olfactorium evehens into a tunnel. The side wall of the nasal capsule is indented to form the concha nasalis, and the hind wall of the capsule, or lamina orbitonasalis, forms a cupola posterior and presents a convex surface to the orbit. The ramus lateralis of the profundus nerve passes outwards dorsally to the lamina orbitonasalis, over the mouth of the concha.

The orbital cartilages are now attached to the basal plate by pilae antoticae spuriae (between V_1 and V_2) in addition to the pilae antoticae, with the result that the profundus nerve is enclosed in a foramen separate from the maxillary and mandibular branches of the trigeminal.

The basitrabecular process is no longer in contact with the pterygoquadrate cartilage. A prefacial commissure is present, anteriorly to the auditory capsule, while posteriorly the metotic cartilage connects the side wall of the capsule to the edge of the basal plate behind the exits of the glossopharyngeal and vagus nerves. The capsules are interconnected dorsally by the tectum synoticum.

The columella auris now possesses a short processus dorsalis leading to an intercalary which is also connected with the extracolumella, enclosing a foramen. The hyomandibular branch of the facial nerve passes back dorsally to the columella and medially to the processus dorsalis, but the chorda tympani runs forwards immediately in front of the processus dorsalis, over which it presumably has slipped forward. The orbital artery runs forwards over the columella.

The hyobranchial skeleton is composed of a median copula, probably representing the basihyal and basibranchial, bearing anteriorly a pair of hyoid cornua, and articulating about the middle of its length with the 1st ceratobranchials, which in turn articulate with the 1st epibranchials.

At later stages (21 mm. head-length) regressive changes set in in the orbitotemporal region and the continuity between the anterior and posterior portions of the orbital cartilages is lost. The former are represented now only by the small planum supraseptale, and the disappearance of the cartilage forming the lateral boundary to the evehent olfactory tunnel results in the passage of the olfactory nerve through the orbit. The interorbital septum is now very high, and the roof of the nasal capsule grows back over it.

In the nasal capsule itself the atrioturbinal and maxilloturbinal are present, and the ramus lateralis of the profundus nerve which is now caught up in the cartilage forming the anterolateral border of the foramen olfactorium advehens, emerges from the capsule through an epiphanial foramen.

The taenia marginalis connects the posterior orbital cartilage with the auditory capsule. The metotic cartilage has spread forward and the glossopharyngeal and the vagus nerves are now enclosed in separate foramina.

ii. The development of the osteocranium of Struthio

No systematic study of the development of the bones of the skull of the ostrich has been made since Parker's (1866*b*), though the conditions in the nestling have been described by Pycraft (1900). According to him a lachrymopalatine ossicle, or 'os uncinatum', is present. Its development requires further study. Broom and Brock (1931) claim that the parasphenoid, including the median rostral portion and the lateral wings, ossifies in one piece, as in most reptiles.

It is to be noted that the sutures persist, and that the various bones tend to keep their individuality in the adult.

VII. Apteryx sp.

The development of the skull of the kiwi is fairly well known from the work of T. J. Parker (1888, 1892, 1893).

i. The development of the chondrocranium of Apteryx

At early stages of the chondrocranium the parachordals appear to preserve their paired nature, being separated by the notochord. The anterior orbital cartilages are very large, and connected with the trabecula communis by a preoptic root which, though close to its fellow of the opposite side, is not fused with it. There is no interorbital septum. The posterior orbital cartilage is attached to the basal plate by a pila antotica and a pila antotica spuria. The anterior and posterior orbital cartilages are interconnected on each side by a taenia medialis, which forms a dorsal border to a foramen pseudopticum through which the optic and the eye-muscle nerves emerge. (Plate 103, fig. 1).

The basitrabecular processes are well developed, and infrapolar processes project back from them to fuse with the under surface of the basal plate, thus enclosing the lateral carotid foramina. At the same time the infrapolar processes are interconnected across the midline, thus subdividing the hypophysial fenestra.

The occipital arches are interconnected by a tectum posterius.

The fully formed chondrocranium is remarkable for the great elongation of the nasal capsule. The nasal septum occupies the entire length of the prenasal process of other forms, and the roof and side walls of the capsule are produced forwards to its anterior end where the fenestrae narinae are situated. The nasal capsule has no floor, but vestiges of the paraseptal cartilages appear to be present ('Jacobson's cartilages' of Parker) in the form of isolated struts, on each side of the ventral edge of the nasal septum.

As development proceeds the anterior orbital cartilages become relatively smaller, but they remain connected with the pila antotica by the taenia medialis. Another important feature is the fact that the lamina orbitonasalis forms a proper cupola posterior to the nasal capsule, which reaches back as far as the level of the optic nerve. This is in marked contrast with other birds, in which the nasal capsule is pressed forwards to accommodate the huge eye. In Apteryx the eye is relatively small, which accounts for the condition of the lamina orbitonasalis, and for the absence of an interorbital septum.

ii. The development of the osteocranium of Apteryx

The development of the various bones, as studied by T. J. Parker (1892) and by Pycraft (1900), reveals several features of interest. (Plate 103, figs. 2–5).

The majority of the membrane-bones arise at Parker's Stage G. The *premaxillae* have very long ascending processes which appear to be fused in the midline from the outset of ossification.

The *prevomers* probably arise separate, and fuse with one another anteriorly, their hind ends remaining widely divergent. Of the *parasphenoid* the rostral portion ossifies first, the *basitemporal* elements arising later.

The *pterygoid* touches the prevomer anteriorly, and has an elongated suture with the *palatine*. No joint is developed between these bones. Postero-medially, the pterygoid articulates with the basipterygoid process by means of an intercalated cartilage which may represent a meniscus pterygoideus (see p. 228), or may be simply a secondarily developed articular cartilage. The latter explanation must apply to the cartilaginous facets whereby the *squamosal* and the *quadratojugal* articulate with the quadrate.

Of the remaining membrane-bones, the *maxilla*, *jugal*, *lachrymal*, *frontal*, *dentary*, *angular*, *supra-angular*, and *splenial*, call for no special mention. The *parietal* and the *prearticular* ('coronoid' of Parker) arise late.

Among the cartilage-bones, the *basioccipital*, *exoccipital*, *basisphenoid*, *pleurosphenoid*, *prootic*, *opisthotic*, *quadrate*, *articular*, *stapes*, and *ceratobranchial* call for no special mention.

The *supraoccipital* appears to be unpaired in its origin, and to take the place of the epiotics which are absent. There are no orbitosphenoids. The *presphenoid* 'mesethmoid') is the only ossification in the ethmoid region of the chondrocranium, where it spreads in the nasal septum and in the roof of the nasal capsule.

Eventually, all the bones fuse together, with the exception of the quadrate and stapes, and pneumatization takes place.

22. MAMMALIA. MONOTREMATA

I. Ornithorhynchus paradoxus

The development of the skull of the platypus has been studied with the help of wax models by Wilson (1905), Watson (1916), and by de Beer and Fell

(1936), while some special points have been investigated by Wilson and Martin (1893), Wilson (1894, 1901, 1906), Broom (1895*a*, 1895*b*, 1900), and Green (1930).

i. The development of the chondrocranium of Ornithorhynchus

1. STAGE 1. (8·5 mm.) PLATE 104, FIGS. 1–4.

The chondrocranium is represented by a number of elements of young cartilage which have already fused together. These comprise the parachordals, fused together in the midline to form the basal plate; the occipital arches, arising from the posterolateral corners of the basal plate; the trabeculae (or perhaps, more strictly, the polar cartilages), attached to the anteroventral edge of the basal plate; and the pilae antoticae ('taeniae clino-orbitales' of Gaupp), continuous with the lateral corners of the upturned anterior edge of the basal plate, or dorsum sellae. The auditory capsules are still in the procartilaginous condition, their cochlear portions in contact with the lateral edges of the basal plate.

The basal plate is elongated and rectangular, and its sides are more or less straight. This latter feature, and the fact that the trabeculae are attached to the basal plate itself and not to the cochlear part of the auditory capsule (as in higher mammals) are of importance in showing that Ornithorhynchus resembles reptiles in this respect, and that Gaupp's (1900) hypothesis of the formation of the cochlear portion of the auditory capsule in mammals out of the basal plate is erroneous (see p. 401).

The notochord lies exposed on the dorsal surface of the basal plate throughout most of its intracranial length; only at the anterior end is the notochord embedded in the cartilage of the dorsum sellae, which forms a well-marked transverse ridge overhanging the pituitary fossa from behind. It is directly continuous with the pila antotica on each side, just as it (or its homologue the crista sellaris) is in Lacerta. The pila antotica is situated behind the oculomotor and trochlear nerves and the pituitary vein, and in front of the trigeminal and abducens nerves.

The trabeculae, or rather those posterior portions of them which are represented by the polar cartilages in other forms, are attached to the ventral surface of the dorsum sellae a little distance behind the anterior edge of the latter, and they are curved slightly ventrally as compared with the plane of the basal plate. Anteriorly, the trabeculae are wide apart and enclose the remnants of the hypophysial stalk between them (in the hypophysial fenestra). A little farther back the trabeculae also enclose the internal carotid arteries in typically reptilian manner. But here, the development of the cartilage of the central stem between the trabeculae has resulted in each carotid artery being enclosed in its own foramen, not continuous any longer with the hypophysial fenestra.

The hindmost portion of each trabecula, just in front of where it is attached to the basal plate, projects to the side, forming the rudimentary processus alaris, or basitrabecular process.

The occipital arches are well chondrified, and in perfect serial continuity with the neural arches of the first three vertebrae. Their size decreases slightly from front to back, but there can be no doubt whatever that the occipital arch is the neural arch of the vertebra next in front of the atlas, i.e. of the proatlas vertebra. The pleurocentrum of the axis vertebra is attached to its corresponding neural arches, but is not yet fused with the pleurocentrum of the atlas vertebra. The latter is free of its corresponding neural arches, which are continuous with the hypocentral element of the atlas vertebra, forming the ventral part of the 'atlas ring'. The relations of the notochord to the basal plate show that the hindmost region of the latter (and perhaps most or all of it) represents a hypocentral element, for it lies wholly ventrally to the notochord, whereas the notochord is embedded in the midst of the pleurocentral elements. There is at this stage an independent procartilaginous rudiment of the pleurocentrum of the proatlas vertebra, through which the notochord runs, immediately in front of the pleurocentrum of the atlas vertebra (to which it will become attached, forming the tip of the odontoid process when the pleurocentrum of the atlas is attached to that of the axis vertebra). The proatlas pleurocentrum is free of its corresponding neural arches—the occipital arches. Since the neural arches (including the occipital arch) are derived from basidorsal elements (posterior sclerotomite or half-sclerotome) while the pleurocentra are the product of intercalary elements (anterior sclerotomite), the occipito-vertebral joint is evident already at this stage as an *inter*segmental *intra*vertebral joint. In other words, the elements of the proatlas vertebra, as well as of the atlas vertebra, retain the rachitomous, discontinuous condition.

Another point to notice at this stage is that the occipital condyles are already present as rudimentary bosses on the hinder surface of the base of the occipital arches.

2. Stage 2. (Approx. 9 mm. long.)

Anteriorly, the trabeculae are prolonged by procartilage to form the interorbital region of the central stem. The hypophysial fenestra is thereby enclosed from in front. The orbital cartilage is present on each side as a plate of cartilage continuous with the dorsal end of the pila antotica.

The processus alaris is slightly longer than previously, and a distinct condensation of procartilage at its distal end is clearly the rudiment of the ala temporalis. The relations of the palatine (or Vidian) nerve and of the orbital (or stapedial, or anterior maxillary) artery are important, the former being situated ventrally and the latter dorsally to the processus alaris. The orbital artery passes on the medial side of the ala temporalis.

The auditory capsule is a little more distinct than previously. The crista parotica is apparent, forming a lateral edge to the sulcus facialis, through which the vena capitis lateralis and the facial nerve pass.

The dorsal ends of the occipital arches are continuous anteriorly with the parietal plates; these are plates of cartilage situated dorsally to the auditory capsules with which they are almost in contact. The space between the auditory capsule and the occipital arch (through which the cranial nerves IX to XII inclusive emerge) is a well-marked fissura metotica. At no time is the base of the occipital arch perforated by a hypoglossal foramen, the hypoglossal nerve emerging through the fissura metotica.

In the visceral arch skeleton Meckel's cartilage is present, and there are procartilaginous rudiments of the stylohyal (as yet unattached to the crista parotica) and thyrohyal elements. The stylohyals of each side are interconnected ventrally in the midline by the basihyals, of obviously paired nature.

3. STAGE 3. (Approx. 9·4 mm. long.) PLATE 104, FIGS. 5, 6.

The hypophysial fenestra is obliterated as a result of the complete chondrification of the trabecular plate or central stem in the midline.

In the region of the pituitary fossa, immediately in front of the dorsum sellae, the trabecular plate is wide and thin. Farther forward it gradually becomes narrower from side to side and deep dorsoventrally, forming an interorbital septum. This is continued forwards uninterruptedly into the nasal septum. In so doing, the dorsal edge of the interorbital septum rises rather suddenly to form a crista galli, and this is continuous with the roof of the nasal capsule, which extends laterally on each side from the dorsal edge of the nasal septum. At this stage the roof is short, and the anterior half of the nasal sac is as yet uncovered by cartilage. The side wall is also very rudimentary, and is represented only by a portion, the paranasal cartilage, which is continuous with the hindmost part of the nasal capsular roof. The sphenethmoid commissure is not yet formed, though there is a process on the paranasal cartilage pointing backwards towards the orbital cartilage.

A rudimentary floor to the nasal capsule is present anteriorly in the form of paired plates projecting sideways from the ventral edge of the anterior part of the nasal septum. These plates are the laminae transversales anteriores, and their outermost parts represent the rudiments of the marginal cartilage of the snout. There is as yet no contact between the lamina transversalis and the side wall of the capsule, or, in other words, no zona annularis. The anterior edge of the laminae transversales shows a deep notch in the midline. This notch lodges the rudiments of the fused ascending or prenasal processes of the premaxillae ('os carunculae') which are already present at this early stage.

In the orbitotemporal region the orbital cartilages have begun to extend backwards, towards the postorbital processes on the auditory capsules, which they do not, however, yet reach, though a mesenchymatous connexion is present (rudiment of the orbitoparietal commissure). The ala temporalis and processus alaris are fused, but the foramen through which the orbital artery passes indicates the position of the original boundary between these structures.

On its dorso-medial surface each auditory capsule is widely open towards the cranial cavity by the primitive foramen acusticum. On its lateral surface, the future fenestra ovalis is blocked by the rudiment of the stapes. The foramen perilymphaticum, however, is not yet discernible on its postero-ventral surface. The crista parotica has become further developed, and the dorsal end of the stylohyal cartilage has become attached to it. The anterior end of the crista parotica (which projects forwards as a small tegmen tympani) is in fused contact with the incus, which in turn, is in cartilaginous continuity with both the stapes and the malleus (hind part of Meckel's cartilage).

The anterior wall of the auditory capsule is produced forwards into a short process overlying the tegmen tympani, forming the limbus precapsularis. Dorsally, the auditory capsule has fused with the parietal plate, the anterior portion of which forms the postorbital process projecting towards the orbital cartilage. The occipital arch is likewise extensively fused with the hind wall of the auditory capsule, with the result that the fissura metotica is reduced and divided into the (more ventral) jugular foramen (serving for the passage of nerves IX to XII) and the (more dorsal) fissura occipito-capsularis.

The notochord emerges from the tip of the (proatlas portion of the) odontoid process, and continues forwards on the dorsal surface of the basal plate as far as the dorsum sellae.

In the visceral arch skeleton rudiments are present of the anterior and posterior thyroid arches.

ii. The morphology of the fully formed chondrocranium of Ornithorhynchus

STAGE 4. (28 mm.) PLATE 105, FIGS. 1–4.

1. The occipital region. The occipital arches are interconnected dorsally by the tectum posterius and the foramen magnum is thus enclosed. The occipital condyles are very prominent on the ventro-posterior surface of the occipital arches, but they do not form any part of the border of the foramen magnum. One result of this is that the prominence of the condyles does *not* involve a correspondingly deep intercondylar incisure, and this feature is associated with the fact that the articular surface of the occipito-atlantal joint is continuous across the midline from one condyle to the other.

The occipito-capsular fissure is obliterated, and the jugular foramen is the only relic of the fissura metotica.

2. The auditory region. The auditory capsule comprises a canalicular portion above and a cochlear portion below, not sharply distinguished from one another. The cochlear portion lies in front of and slightly medially to the canalicular portion, but not so markedly as in higher mammals. The anchoring of the auditory capsule on to the basal plate takes place by means of the extensive basivestibular commissure, which is attached to the cochlear portion throughout its length.

The cochlear capsule compresses the basal plate slightly. The basal plate is therefore slightly concave on each side. It is important to note that the cochlear capsule does not reach as far forward as the front of the basal plate. It is for this reason that the trabeculae are attached to the basal plate and not to the cochlear capsules as in higher mammals.

On the medial side, the large primitive acustic foramen occupies a large part of the roof of both cochlear and canalicular portions. The fenestra ovalis on the lateral surface of the cochlear portion is occluded by the stapes, while the foramen perilymphaticum opens from the cavity of the cochlear portion through its posterior wall directly into the jugular foramen. A prefacial (or suprafacial) commissure is present enclosing the (primary) facial foramen. The facial nerve runs outwards through this and back through the sulcus facialis which is well marked, between the crista parotica (to which the stylohyal cartilage is attached) and the lateral wall of the capsule. The chorda tympani runs forwards laterally to the styloid process. The limbus precapsularis of the previous stage is no longer distinct, and appears to have been absorbed in the formation of the prefacial commissure. The parietal plate is connected with the orbital cartilage by means of the orbitoparietal commissure.

The stapes is perforated at early stages (Goodrich, 1915) but it becomes imperforate, and the stapedial (orbital) artery runs upwards behind it and forwards above it. The incus retains cartilaginous connexion with the stapes, with the crista parotica, and with the malleus. This continuity is quite small in extent in each case, and appears to be the result of the absence just in these places of a well-formed perichondrium. It is important to note that the incus is dorsal to the malleus, as the quadrate is to the articular in non-mammalian vertebrates, and not behind it as in higher mammals.

3. The orbitotemporal region. The orbital cartilage is now attached to the central stem by a preoptic root and by the pila antotica. Anteriorly, the orbital cartilage is attached to the roof of the nasal capsule by the sphenethmoid commissure, and posteriorly, with the parietal plate on the auditory capsule by the orbitoparietal commissure. There are therefore three apertures on each side:

(1) The orbitonasal fissure (between the preoptic root of the orbital cartilage, the sphenethmoid commissure, and the lamina orbitonasalis) which leads, not into the cranial cavity, but into a space, the supracribrous recess, which is morphologically anterior to the cranial cavity (as delimited by the dura mater). The ethmoid branch of the profundus nerve leaves the orbit through the orbitonasal fissure and enters the supracribrous recess (which it leaves again almost immediately through the epiphanial foramen).

(2) The pseudoptic foramen (between the preoptic root of the orbital cartilage and the pila antotica) transmits nerves II, III, and IV. It represents the conjoint optic and metoptic foramina of other forms, confluent owing to the absence of a pila metoptica.

(3) The foramen prooticum (between the pila antotica, orbitoparietal commissure, and prefacial commissure) through which nerves V and VI emerge.

The processus alaris and its attached ala temporalis project laterally and ventrally from the central stem beneath the base of the pila antotica and form the floor of the cavum epiptericum. This space which lodges the trigeminal ganglion, has an imperfect cartilaginous medial wall in the form of the pila antotica, but it has no lateral wall, owing to the very poor development of the ala temporalis. The ala temporalis does not, therefore, present any definite relations to the various branches of the trigeminal nerve, but lies simply beneath them. It is perforated by the orbital artery. The palatine or Vidian nerve runs from the facial foramen forwards and downwards round the side wall of the cochlear capsule, and then forwards beneath the processus alaris.

The carotid foramina are situated fairly close to the midline in the deep pituitary fossa. The dorsum sellae is very prominent. The cartilage situated immediately laterally to the carotid foramen (on each side) represents the full width of the trabecula, and is continuous posteriorly with the front of the basal plate.

In front of the pituitary fossa the central stem tapers slightly, but this is more apparent from below than from above where the preoptic roots of the orbital cartilage are fused to the central stem by wide attachments. The interorbital septum is therefore less well marked than at the previous stage, and now scarcely merits the term.

4. The ethmoid region. The nasal capsules are now well chondrified. The nasal septum is perforated near its anterior end by a fenestra. The roof shows two well-marked, dome-shaped cupolae anteriorly, and on each side is directly continuous with the side wall or paries nasi. This ends ventrally by a free edge except in front where it is fused to the lamina transversalis anterior forming a zona annularis or complete cartilaginous ring surrounding each nasal sac. The primary fenestra narina, through which the external nostril passes, is delimited by the anterior edges of the nasal septum, of the roof, of the side wall, and by the lamina transversalis, which is now continued forwards and outwards into the already extensive marginal cartilage of the snout. The marginal cartilage is thinner than the lamina transversalis anterior, which thus appears like a transverse ridge, some distance behind the front edge of the marginal cartilage. A superior alar process projects downwards from the dorsolateral border of each fenestra narina.

The lamina transversalis anterior is produced backwards on each side into the paraseptal cartilage, which runs back along the ventrolateral edge of the nasal septum for about half the length of the capsule, and ends freely. Each paraseptal cartilage is curved, being concave on each side, where it lodges Jacobson's organ. Anteriorly, it encloses it in a complete cylinder from the lateral wall of which a 'turbinal' projects inwards into the cavity (Broom, 1896*c*).

The paranasal cartilage is now connected with the sphenethmoid commissure, and also with the central stem by means of the lamina orbitonasalis, which forms a roof to the hindmost part of the cavity of the nasal capsule, which here has no hind wall and no floor of any kind. By these developments the olfactory foramina are delimited by cartilage. The foramen olfactorium evehens, which leads out of the cranial cavity into the cavity of the supracribrous recess, is delimited medially by the crista galli (of the nasal septum), anterodorsally by the hind edge of the roof of the nasal capsule, laterally and ventrally by the anterior edge of the preoptic root of the orbital cartilage. The passage out of the supracribrous recess into the cavity of the nasal capsule is the foramen olfactorium advehens (which, in Monotremes, never becomes subdivided into a cribriform plate). This has the same medial and dorsal boundaries (crista galli and hind edge of the capsular roof) as the foramen olfactorium evehens, but its lateral and ventral boundary is the medial and dorsal edge of the lamina orbitonasalis. The plane of the foramen olfactorium advehens is almost at right angles to the central stem. The supracribrous recess is thus a wedge-shaped space opening to the side by the gap between the lamina orbitonasalis and the preoptic root of the orbital cartilage, and this gap is the orbitonasal fissure. The ethmoid branch of the profundus nerve passes through this fissure into the supracribrous recess where it lies externally to the dura mater (and to the cranial cavity) and dorsally to the olfactory nerve. Instead, however, of penetrating through the foramen olfactorium advehens into the cavity of the nasal capsule (as in higher mammals), the ethmoid nerve (or rather, its lateral ramus) emerges almost immediately from the supracribrous recess through the epiphanial foramen. This foramen represents the original gap between the nasal roof and the paranasal cartilages. The ventral edge of the lamina orbitonasalis bears two processes on each side, one pointing backwards and the other forwards. These are respectively the posterior and anterior maxillary processes. The anterior maxillary process makes contact again farther forward with the ventral edge of the paries nasi, and thereby encloses a slit-like fissure through which the lachrymonasal duct passes inwards and forwards. This duct does not, however, run into the nasal sac here, but continues forwards laterally to the zona annularis, dorsally to the marginal cartilage, and enters the nasal capsule through the primary fenestra narina.

5. The visceral arch skeleton. The anterior ends of Meckel's cartilage are fused in an extensive symphysis.

Mention has already been made of the malleus, incus, and stapes (p. 293), and of the fact that the stylohyal cartilage is fused dorsally with the crista parotica. In addition to these skeletal elements of the 1st and 2nd visceral arches, this stage possesses cartilaginous structures in the 3rd, 4th, and 5th visceral (1st, 2nd, and 3rd branchial) arches. These are the thyrohyal, anterior laryngeal, and posterior laryngeal cartilages respectively, and, as is well known, they persist in the adult. Behind them the cricoid cartilages form a ring just closed dorsally. Whether they represent a tracheal ring or the 4th branchial arch it is as yet impossible to decide, though the former alternative is probably correct.

STAGE 5. (122 mm.) PLATE 106, FIGS. 1–5.

6. Final changes. The chondrocranium of Ornithorhynchus at this stage is remarkable for its extraordinarily complete and massive development. The roof is remarkably complete, and is formed by the tectum synoticum which is continuous with the tectum posterius; it extends forwards as far as the front wall of the auditory capsule.

The chondrocranium as a whole is more elongated than at the previous stage, and this is largely due to the anterior development of the nasal capsule and to the marginal cartilage. The latter not only extends farther forwards, but also farther back on each side. The ridge formed ventrally by the lamina transversalis anterior has become flattened out, and it passes smoothly forward on to the under surface of the marginal cartilage. From the hind edge of the lamina transversalis anterior on each side a cartilage extends back medially to the ventral edge of the paries nasi and laterally to the primary choana (nasopalatine canal, aperture of Jacobson's organ). These paired cartilaginous processes, called processus palatini soli nasi in Echidna by Gaupp (1908), correspond exactly to the ectochoanal cartilages of Urodela and Lacertilia. But instead of remaining separate, as in those forms, in Ornithorhynchus these cartilages fuse with one another in the midline ventrally to the paraseptal cartilages and behind the aperture of Jacobson's organ, forming a transverse commissure which Wilson (1901) termed the 'cartilaginous lamina forming the skeleton of the anterior part of the secondary palate'.

In addition to the superior alar process, there is now a processus alaris inferior, projecting laterally from the front of the nasal septum across the primary fenestra narina. When it meets the superior alar process it will fuse with it and thus enclose the definitive fenestra narina, as Wilson (1901) showed in an older embryo. (Plate 106, fig. 6).

An additional development shown by the nasal capsule is the formation of a maxilloturbinal and an ethmoturbinal on each side.

The ala temporalis remains small, but such as it is it lies between V_2 and V_3.

In the medial wall of the auditory capsule the large primary foramen acusticum has become subdivided into the definitive foramina acustica, sunk together with the medial aperture of the facial foramen in a common depression (the internal auditory meatus) and a small foramen endolymphaticum.

Eventually (250 mm. stage, Watson, 1916) a small tegmen tympani (processus perioticus superior) projects forwards from the crista parotica.

iii. The development of the osteocranium of Ornithorhynchus

1. The membrane-bones. The *premaxillae* arise very early, and are fused in the midline by a double symphysis which bears the egg-tooth. The ascending or prenasal processes are fused forming the huge 'os carunculae' which rises up in front of the anterior edge of the nasal septum. The paired nature of the os carunculae is plainly demonstrated by the presence of a median slit between its two sides. Dorsally, the os carunculae forms a large knob in which some secondary cartilage is present. From the region of the symphysis bearing the egg-tooth, each premaxilla presents two processes projecting backwards. Of these, the lateral (or 'alveolar') process runs backwards ventrally to the lateral part of the marginal cartilage. The medial or palatine process projects backwards close to the midline beneath the medial part of the marginal cartilage, and reaches almost as far back as the lamina transversalis anterior.

At later stages (stage 5, for instance; 122 mm.) the *premaxillae* are still fused with one another, but only by the foremost of the two symphyses which at the previous stage supported the anterior and posterior rims of the egg-tooth. The egg-tooth itself has now been shed. The 'os carunculae' (fused ascending processes) is still present, but shows signs of absorption. The lateral or alveolar processes of the premaxillae extend backwards and outwards beneath the large marginal cartilage and are turned inwards near their hind ends. The medial or palatine processes are separate from one another all the way behind the anterior symphysis. They run back parallel with one another beneath the marginal cartilage and lamina transversalis anterior to a point beneath the anterior end of the paraseptal cartilage where they taper out. About $\frac{1}{4}$ mm. behind this point, and therefore isolated and independent, there is on each side a small plate-like ossification. These, the rudiments of the *prevomers* ('os paradoxum' or 'dumb-bell bone' when they fuse), are enclosed in a mesenchymatous sheath which is continuous with that of the palatine process of the premaxillae. For this reason, and because he thought he could make out a thin thread of ossification connecting the palatine process of the premaxilla with the prevomer, Green (1930) concluded that the latter was in process of detachment from the former (as Symington (1896) had supposed), having previously been continuous with it.

Eventually, of course, the os carunculae and the palatine processes disappear altogether, and the premaxillae are thus widely separate from one another, and from the prevomers, which latter fuse together in the midline, giving rise to the 'dumb-bell shaped bone'.

The *septomaxilla* is a small paired bone lying isolated on the dorsal surface of the hinder part of the marginal cartilage (which separates it from the alveolar process of the premaxilla), immediately laterally to the anterior edge of the paries nasi. Eventually, the septomaxilla becomes shaped like the alveolar process of the premaxilla from which it is separated by the marginal cartilage, and with which it will fuse when the marginal cartilage disappears.

The *vomer* is an unpaired bone covering the ventral edge of the hinder part of the nasal septum. It shows no trace of paired origin, and represents the median or rostral portion of the parasphenoid.

The *maxilla* consists of an alveolar portion, an ascending portion which covers the lower part of the paries nasi, a palatine process which extends inwards beneath the ventral edge of the paries nasi and a zygomatic process which extends back as far as the similarly named process of the squamosal. The infraorbital foramen transmitting the maxillary branch of the trigeminal nerve perforates the maxilla between its ascending and alveolar portions. The lachrymonasal duct runs forwards medially to the ascending process of the maxilla: there is no lachrymal bone.

The *palatine* is represented by a plate of bone extending back from the hinder part of the ventral edge of the paries nasi almost as far as the root of the processus alaris. The palatine encloses a space, the palatine canal, which opens laterally by the sphenopalatine foramen (traversed by a branch of the maxillary division of the trigeminal nerve), ventrally by the palatine foramen, and anteriorly into the cavity of the nasal capsule. The anterolateral portion of the palatine forms the pars perpendicularis, which envelops the ventral edge of the lamina orbitonasalis.

The '*mammalian pterygoid*' (or detached lateral wing of the parasphenoid) is completely covered ventrally by the lateral edge of the palatine. It is an elongated and slightly curved plate of bone forming the side wall of the hinder part of the nasopharyngeal passage, and lying medially and ventrally to the anterior part of the ala temporalis from which it is separated by the parabasal or Vidian canal through which the Vidian or palatine nerve passes on its way forwards to the sphenopalatine ganglion. There can be no doubt that the 'mammalian pterygoid' is the detached lateral wing of the parasphenoid as Gaupp (1910*a*) contended, and this stage is very interesting as only a little extension of ossification is wanting to join this lateral wing of the parasphenoid on to the vomer (dorsally to the nasopharyngeal passage), which is generally held to represent the median dagger-like portion of the parasphenoid. A suture between the 'mammalian pterygoid' and the vomer is actually formed later, at the 250 mm. stage, as shown by Watson (1916).

The '*Echidna-pterygoid*' arises late as an ovoid nodule of bone lying postero-ventrally to the ala temporalis, and laterally to the Vidian nerve. There is little doubt that it corresponds to the pterygoid of reptiles, and it is probable that it is represented in higher mammals by the nodule of the secondary pterygoid cartilage.

The *squamosal* consists of a small squamous portion lying against the anterior part of the side wall of the auditory capsule (from which it is separated by a small space which is the future post-temporal fossa), and a well-developed zygomatic process.

The *jugal* develops late as a small bone in front of and dorsal to the anterior end of the zygomatic process of the squamosal.

The *nasal* lies over the roof of the nasal capsule immediately in front of the epiphanial foramen out of which the ethmoid branch of the profundus nerve emerges and runs forwards beneath the nasal.

The *frontal* is remarkably small and arises on the lateral surface of the sphenethmoid commissure.

The *parietal* lies along the dorsal edge of the parietal plate, before extending medially over the cranial cavity.

The *tympanic* is a small sickle-shaped ossification supporting the developing tympanic membrane, medially to the hind end of Meckel's cartilage.

The *prearticular* is a small splinter of bone on the medial side of Meckel's cartilage, behind the tympanic. The chorda tympani runs forwards between Meckel's cartilage and the prearticular.

The *dentary* ensheaths the ventrolateral surface of Meckel's cartilage throughout most of its length. It also extends over the dorsal side of the cartilage for a short distance, this portion being connected with the main part of the bone by a bridge which passes over the mandibular branch of the trigeminal nerve and thus delimits the mental foramen.

2. The cartilage-bones. The cartilage-bones present at stage 5 (122 mm.) are the *basioccipital*, *exoccipital*, *supraoccipital*, *basisphenoid*, and *alisphenoids*. The occipital bones occupy typical positions as perichondral lamellae in the basal plate, occipital arches, and tectum posterius respectively, and call for no special comment. Nor does the basisphenoid, which ossifies as perichondral lamellae on the dorsal and ventral surfaces of the trabecular region of the central stem. The ossification in the pila antotica described by Wilson (1906, p. 88) seems to represent a vestige of the *pleurosphenoid* (the 'posterior clinoid process, of van Bemmelen, 1901).

The existence and relations of the *alisphenoids*, however, are of great importance. Each alisphenoid ossifies as a perichondral lamella on the anteroventral surface of the ala temporalis, where it lies laterally to the 'mammalian pterygoid' and anterodorsally to the 'Echidna-pterygoid'. The alisphenoid is at the start quite distinct from the basisphenoid, with which, however, it fuses later on. The alisphenoid remains small and no lamina ascendens is developed; its position is occupied by the anterior process of the periotic.

The *periotic* starts ossifying late (250 mm. stage) from two centres. One of these, presumably the *opisthotic*, is situated in the posteroventral portion of the auditory capsule. The other, the *prootic*, is a perichondral lamella above the crista parotica, but it is also in direct continuity with a large intramembranous bony plate which covers over the side of the orbitotemporal region and takes the place of the lamina ascendens of the alisphenoid of other mammals. This is the processus anterior perioticus (Watson, 1916). Its relation to the lamina ascendens of the alisphenoid is still problematical.

A nodule of cartilage found by Watson (1916) laterally to the malleus may perhaps represent an endotympanic.

The *orbitosphenoid* also ossifies late, in the orbital cartilage, and gives rise to a very large bone which extends from the nasal capsule nearly as far as the auditory capsule.

The *presphenoid* ossifies in the nasal septum; as Broom (1930) showed, there is no mesethmoid.

The monotreme skull is further interesting in that no sinuses are excavated by the olfactory sac (Paulli, 1900).

II. Echidna aculeata

The development of the skull of Echidna has been the subject of a monumental monograph by Gaupp (1907*c*, 1908), while certain features of the snout have been revealed by Seydel (1899) and Wilson (1901, 1906). (Plate 107, figs. 1, 2).

The general and detailed form of the chondrocranium of Echidna is so remarkably similar to that of Ornithorhynchus that while noting the fact as evidence of the close affinity between these forms, it is unnecessary to describe it fully. However, the richness of Gaupp's material enabled him to observe a number of important facts concerning the initial chondrification of various structures, and to these attention may be turned.

The trabeculae (Gaupp, loc. cit., p. 558) are of great interest, for at their earliest appearance they are definitely independent and paired, and lie laterally to the internal carotid arteries. The posterior part of the trabecula probably represents the polar cartilage since it projects to the side to form the processus alaris. No separate ala temporalis centre chondrifies, and it is possible that it may be absent altogether. Chondrification next takes place between the trabeculae in front of the hypophysial stalk, and later still behind the stalk, thus forming the hypophysial plate perforated by the hypophysial foramen. Hypophysial plate and trabecula then become attached to the anterior edge of the basal plate.

The cochlear portion of the auditory capsule seems to chondrify in continuity with the basal plate, but the canalicular portion has an independent centre.

An independent median centre gives rise to the nasal septum which soon fuses with the trabeculae (loc. cit., p. 560), while paired independent chondrification centres have been observed for the orbital cartilages (loc. cit., p. 559), the laminae orbitonasales (loc. cit., p. 561), the paranasal cartilages (loc. cit., p. 562), the lamina transversalis anterior, and the crista marginalis (loc. cit., p. 572), the ectochoanal ('palatine') cartilage (loc. cit., p. 580), the maxilloturbinal and ethmoturbinal (loc. cit., p. 581).

The lamina orbitonasalis forming the cupola posterior of the nasal capsule extends far back and fuses on each side with the central stem (nasal septum and trabeculo-hypophysial plate), giving rise to a plate termed by Gaupp the lamina infracribrosa. The preoptic roots of the orbital cartilages are mounted on the laminae orbitonasales.

The pilae antoticae (taeniae clino-orbitales) persist in the adult; for the hinder three-quarters of their length they become fused with the side of the central stem; the anterior portion ossifies in conjunction with the orbital cartilage and gives rise to the structure (erroneously) termed by van Bemmelen (1901) the 'anterior clinoid process'.

A feature of some interest is the presence of a tectum transversum, formed of a cartilage which arises independently in the midline, and becomes temporarily connected with the anterior ends of the parietal plates. Noteworthy also is the papillary cartilage, found by Wilson (1901) beneath the ventral edge of the nasal septum. Another remarkable structure is a plate of cartilage situated laterally to the auditory capsule and which becomes embedded in the squamosal. It is clearly *not* an articular secondary squamosal cartilage, and Gaupp (1908, p. 741) believes it to be derived from the crista parotica. It may, however, represent the processus opercularis of other mammals (see p. 391).

Small points of difference between Echidna and Ornithorhynchus are the persistence in the former of the limbus precapsularis (p. 291) and the feeble development of the dorsum sellae.

Meckel's cartilage shows a slight sigmoid curvature.

Ruge's (1896) claim that the cartilage of the outer ear is derived from the hyoid arch is contradicted by Cords (1918).

In the bony skull the conditions are also very similar to those in Ornithorhynchus. The premaxillae are fused and bear a dentinal egg-tooth; they possess fused ascending processes which become detached, forming an isolated 'os carunculae'. No prevomers are present. The vomer, on the other hand, arises from paired rudiments. The dentary in its anterior region comes into relation with Meckel's cartilage, the ossification in which presumably represents a mentomeckelian element.

23. MARSUPIALIA

I. Didelphys marsupialis

The development of the skull of the opossum has been studied with the help of wax-model reconstructions at the 15 mm. stage by Levi (1909*a*) and at the 45·5 mm. stage by Töplitz (1920).

i. The development of the chondrocranium of Didelphys

The basal plate at early stages is remarkably broad, though it becomes relatively narrower as development proceeds. The occipital arches rise vertically upwards and are joined above by a tectum posterius, in front of which is a tectum synoticum, as Levi (1909*b*) first showed. There are two pairs of hypoglossal foramina. The cavity of the occipito-atlantal joint is at first continuous between the condylar facets across the midline, but subsequently becomes subdivided.

The auditory capsules are remarkable for many reasons. The canalicular and cochlear portions chondrify separately, but the latter in continuity with the basal plate, from which it *subsequently* becomes separated by the appearance of a basicochlear fissure. The canalicular capsules are very large compared with the cranial cavity, but the cochlear portions are relatively small. The medial walls of the canalicular capsules are more or less vertical and form a considerable part of the side wall of the cranial cavity in this region, the parietal plates being relatively low. The canalicular part is more or less directly above the cochlear part of the capsule. (Plate 107, fig. 8).

On the lateral surface of the capsule is a crista parotica, from which a diminutive tegmen tympani projects forwards, and to which the styloid process is fused. The facial nerve passes under the prefacial (or suprafacial) commissure and runs through the sulcus facialis to emerge behind the styloid process, but the chorda tympani which it then gives off runs forwards *medially* to the styloid process, as Gaupp (1905*c*, p. 1042) first showed.

Running forwards along the dorsal surface of the cochlear capsule is a sharp ridge, or crista cochlearis, which rises up laterally to the facial ganglion and to the palatine nerve, and thus marks the lateral boundary of the cavum supracochleare (see p. 430).

As regards foramina in the wall of the capsule, the foramen perilymphaticum shows an incipient development of a processus recessus, foreshadowing the formation of fenestra rotunda and aquaeductus cochleae. Between the foramina acustica superius and inferius, there is a small foramen acusticum medium, as in Echidna, for a branch of the auditory nerve innervating the saccule.

The trabecular region of the skull forms a central stem of cartilage perforated in the midline by the hypophysial foramen and fused behind with the basal plate. At early stages the internal carotid arteries run upwards into the cranial cavity through notches in the lateral edges of the central stem. The front wall of each of these notches is formed by the processus alaris, which projects to the side. Subsequently the carotid notches are converted into foramina by the back growth of cartilage from the root of the processus alaris to the basal plate on each side: this cartilage represents the hindmost portion of the trabeculopolar bars of Monotremes and non-mammalian vertebrates, and it also corresponds to the alicochlear commissure of Placental mammals. The conditions in Didelphys are thus of importance in establishing the homologies of the trabeculae and carotid foramina in higher mammals (see p. 379). The important point to note in Didelphys is that the carotid foramina are situated well within the trabecular plate in the fully formed chondrocranium.

The processus alaris in mammals usually bears the ala temporalis which rises up between the maxillary and mandibular branches of the trigeminal nerve. In many Marsupials the processus alaris also bears the processus ascendens which rises up between the ophthalmic and maxillary branches of the trigeminal nerve. In Didelphys, according to Fuchs (1915, pp. 147, 253), the processus ascendens only is present, and it would seem that the true ala temporalis is either absent or reduced, for Töplitz has shown that the lamina ascendens of the alisphenoid ossifies in membrane, without any cartilaginous precursor.

The orbital cartilage is attached to the central stem only by its preoptic root, i.e. there is no pila metoptica, and for this reason, as in all Marsupials, the optic foramen is confluent posteriorly with the sphenoid fissure. A characteristic feature is that the sphenethmoid and the orbitoparietal commissures are bands of cartilage of the same width as the orbital cartilage. A small ala minima projects forwards from the preoptic root of the orbital cartilage towards the lamina orbitonasalis.

The pila antotica is represented only by a transient rod of cartilage in the dura mater immediately overlying the cochlear capsule.

The nasal capsules are well chondrified, and the cupolae posteriores are firmly fused with the nasal septum. The paraseptal cartilages extend back from the lamina transversalis anterior and end freely, about half-way along the nasal septum. Their anterior portions enclose Jacobson's organ forming Jacobson's capsule, and possess an 'outer bar' enclosing Jacobson's organ from the side. (Broom, 1896*b*). A papillary cartilage is present (Broom, 1895*b*).

The foramen olfactorium advehens is subdivided into the numerous apertures of the cribriform plate. This is effected first by the formation of the crista intercribrosa, to which the 1st ethmoturbinal is attached. The interior of the nasal capsule is subdivided in typical mammalian manner (see p. 312), with a crista semicircularis projecting inwards and backwards from the side wall, an atrioturbinal, maxilloturbinal, four ethmoturbinals, and some frontoturbinals.

The floor of the capsule is represented by the laminae transversales anterior and posterior, and as each is fused with both side wall and nasal septum, there are two zonae annulares. An epiphanial foramen is present. The cupolae anteriores are remarkably distinct and there is a deep furrow between them forming a cavum internasale.

In the visceral arch skeleton a feature of interest is the position of the incus, dorsally to the malleus. Meckel's cartilages are fused anteriorly in an extensive symphysis, out of which, subsequently, a median 'basimandibular' cartilage becomes isolated.

The styloid process and the curious course of the chorda tympani have been described above (p. 301).

The stapes is elongated and columelliform, but pierced by the stapedial artery. Laterally to the stapes and presumably representing the hyostapes (see p. 412) is Paauw's cartilage, found by van Kampen (1915*b*).

ii. The development of the osteocranium of Didelphys

Without going into details concerning the separate bones, a few points of interest may be noted, first among which is the extreme precocity of appearance of the bones, especially those of the palate. This appears to be correlated with the precocity of birth and of sucking in the marsupium, entailing rapid development of the skeletal roof of the mouth.

The cavum internasale contains dense tissue which recalls the vestige of the prenasal processes of the premaxillae found in this region in Trichosurus and the 'os carunculae' of Caluromys (p. 303); in Didelphys, however, Töplitz is disinclined to accord it this significance.

Paired *interparietals* of large size are present, subsequently undergoing fusion. A nodule of secondary pterygoid cartilage is present. No trace has been found by either Töplitz or Broom (1935*b*) of the median ossification which Fuchs (1908) claimed as a representative of the parasphenoid. The vomer arises paired. Information is still wanting concerning the manner of development of the tympanic bulla formed by the alisphenoid. The *tympanic* has a horizontal limb extending forwards beside Meckel's cartilage (Goodrich, 1930, fig. 503), thereby resembling the angular of Theromorpha.

II. Caluromys philander

The development of the skull of the woolly opossum has been the subject of a short paper by Denison and Terry (1921) which has revealed certain points of great interest. (Plate 107, fig. 3).

In general form the chondrocranium resembles that of Didelphys, especially in such typical Marsupial features as the breadth of the basal plate, large size of the auditory capsules, position of the carotid foramen, absence of pila metoptica, and presence of two pairs of hypoglossal foramina.

The ala minima projecting forwards from the preoptic root of the orbital cartilage fuses with the lamina orbitonasalis and cuts off the medial portion of the orbitonasal fissure as the foramen infracribrosum on each side. The maxillary branch of the trigeminal nerve passes through a cartilaginous foramen rotundum, i.e. both processus ascendens and ala temporalis are present, attached to the processus alaris. Small basicochlear fissures are present.

In the bony skull a feature of the greatest importance is the presence of a small median bone between the cupolae anteriores of the nasal capsules, in the cavum internasale. This bone answers exactly to the isolated 'os carunculae' of Echidna, and like it must represent the prenasal processus of the premaxillae. (See also Trichosurus, p. 305.)

III. Dasyurus viverrinus

The development of the skull of the dasyure has been studied at the 8 mm. stage by Broom (1909*a*), and at the 9·3 mm. stage by Fawcett (1918*a*), while the formation of the nasal capsule has been followed in detail by Fawcett (1917, 1919), and certain other features have been described by Cords (1915). (Plate 107, figs. 4–7).

The chondrocranium is exceedingly broad, and the other typical Marsupial features are well shown (large auditory capsules with vertical medial walls, vertical occipital arches, two pairs of hypoglossal foramina, carotid foramina well within the trabecular plate, no pila metoptica). At early stages the paraseptal cartilages project back from the lamina transversalis anterior and end freely, but later they grow back and fuse with the lamina transversalis posterior thus presenting the primitive condition of complete paraseptal cartilages. An outer bar is present in Jacobson's capsule (Broom, 1896*b*). A papillary cartilage is present (Broom, 1895*b*).

The preoptic roots of the orbital cartilages are delayed in their formation, and on growing inwards from the orbital cartilage they fuse, not with the central stem, but with the lamina orbitonasalis, thereby almost obliterating the orbitonasal fissure.

The ala temporalis is large and fuses dorsally with the orbital cartilage. It lies between the maxillary and mandibular branches of the trigeminal nerve.

The chorda tympani runs forwards *medially* to the styloid process. The stapes is columelliform and imperforate.

Paauw's cartilage is present (van der Klaauw, 1923).

IV. Perameles obesula and nasuta

The development of the skull of the bandicoot has been studied with the help of model reconstructions by Esdaile (1916), while the morphology of the fully formed chondrocranium has been described by Cords (1915), and the development of the ear ossicles has been followed by Palmer (1913).

The chondrocranium of Perameles is typically Marsupial in some respects, whereas in others it no less strongly resembles the Placental mammals.

Among the Marsupial features are the equal width of the sphenethmoid and orbitoparietal commissures and the orbital cartilage, the vertical position of the occipital arches, the absence of a pila metoptica, the presence of a foramen rotundum, of two pairs of hypoglossal foramina, and the passage of the internal carotid arteries through the trabecular plate. The stapes is columelliform and imperforate; Paauw's cartilage is present.

The features in which Perameles resembles Placental mammals are the following. The tectum posterius is in the vertical transverse plane. The auditory capsules lean outwards, and the canalicular portions are definitely behind the cochlear portions, which latter are relatively large, in consequence of which the basal plate is somewhat narrow. The internal carotid arteries on emerging from their foramina find themselves in the cava epipterica, from which they enter the cranial cavity. The carotid foramina are sufficiently far apart and the cochlear capsules sufficiently large to bring it about that the hindmost portion of the trabeculae are in contact not only with the basal plate but also with the cochlear capsules, thus somewhat resembling the alicochlear commissures of Placentals. (Plates 108, 109).

The basicochlear fissure is large and confluent with the jugular foramen: i.e. the basivestibular commissure is absent.

The paraseptal cartilages are incomplete and end freely posteriorly. An outer bar is present in Jacobson's capsule, and Broom (1896*b*) has stressed the similarity between Perameles and Didelphys in respect of this region. A papillary cartilage is present (Broom, 1895*b*). The chorda tympani runs forwards laterally to the styloid process. Paauw's cartilage is present (van der Klaauw, 1923).

An interesting feature of Meckel's cartilage is the fact that at early stages it is very much bowed outwards, which condition may, Esdaile suggests, be associated with the act of sucking. The same factor probably accounts for the great precocity in ossification of the bones of the palate.

In the hyobranchial skeleton as described by Esdaile, a basihyal, hypohyals, and thyrohyals (1st branchial arches) are present, together with thyroid and cricoid cartilages. It is not clear, however, whether the thyroid cartilages are formed from the 2nd branchial arches only, or from the 2nd and 3rd. The cricoid cartilages are immediately behind the thyroid, with which they fuse, and they may therefore represent a pair of branchial arches; if homologous with the cricoids of Ornithorhynchus they must represent the 4th (6th visceral). But if the thyroid cartilages represent the 2nd branchial arches only, the 3rd must be lost. On the other hand, the cricoid cartilages form a complete ring, and are probably only the modified foremost tracheal ring.

Perameles differs from other Marsupials in lacking an interparietal bone.

V. Trichosurus vulpecula

The development of the skull of the phalanger has been studied with the help of wax models at the 10 mm. and 14 mm. stages by Broom (1909*a*), and at the 17·5 mm. stage by the present writer.

In general form the chondrocranium is of typical Marsupial form (large auditory capsules with vertical medial wall, vertical occipital arches, two pairs of hypoglossal foramina, basal plate very wide, carotid foramina well within the trabecular plate, no pila metoptica). The ala temporalis chondrifies independently of the processus alaris and is pierced by a foramen rotundum, the anterior border of which represents the processus ascendens.

The paraseptal cartilage at early stages is complete and extends from the lamina transversalis anterior to the lamina transversalis posterior; the latter cartilage is free from the nasal septum. At later stages Esdaile (1916) found that the paraseptal cartilage ends freely posteriorly.

An outer bar is present in Jacobson's cartilage (Broom, 1896*b*). A papillary cartilage is present (Broom, 1895*b*).

The stapes is perforated, and laterally to it, Paauw's cartilage is present (van der Klaauw, 1924*c*). The styloid process bears the normal relations to the chorda tympani, i.e. the nerve runs forwards *laterally* to the cartilage. At late stages a process is developed from the dorsal end of the styloid process, and passing downwards, laterally to the chorda tympani, fuses again with the styloid process (van der Klaauw, 1924*b*). The styloid process then looks as if it were perforated by the chorda tympani. But as no such process is found in Didelphys or any of the other forms in which the chorda tympani runs medially to the styloid process, it is not possible to explain the conditions in the latter as due to the substitution of the extra-chordal process for the styloid process.

The development of the thyroid cartilage has been studied by Edgeworth (1916), who showed that it is formed of the skeletal elements of both 2nd and 3rd branchial arches fused.

As regards the bony skull, a feature of interest is the precocity of appearance of the premaxillae, maxillae, palatines, pterygoids, and dentaries. As these bones are concerned with the support of the walls of the mouth it is probable that their development is associated with the precocity of suckling in these forms. The presphenoid centre ossifies independently of the orbitosphenoids (Broom, 1926*a*).

The premaxillae are remarkable in that Broom (1909*a*, p. 202) found indications of an ascending or prenasal process in the cleft between the cupolae anteriores of the nasal capsules. Such an ascending process is found in Lacertilia, and as the 'os carunculae' in Monotremes, and is associated with the carrying of the dentinal egg-tooth by the fused premaxillae. The presence of vestiges of the ascending process in Marsupials (Trichosurus, perhaps Didelphys, the definitely ossified 'os carunculae' of Caluromys) is therefore of interest since it suggests that these animals have only comparatively recently abandoned the oviparous method of development. This suggestion is further supported by the discovery in Trichosurus by Hill and de Beer (unpubl.) of a vestigial median egg-tooth papilla, immediately beneath the premaxillae.

VI. Other Marsupialia

No systematic study of the development of the skull has been made in the following forms, but a few points of interest are known concerning them.

Phascolarctus cinereus is, according to Paulli (1899), the only Marsupial in which excavation of ethmoidal cells and the formation of sinuses takes place. It is also the only Diprotodont known in which the chorda tympani runs forwards *medially* to the styloid process (van der Klaauw, 1924*c*): Paauw's cartilage is present but very small. On the other hand, in Phascolomys it is enormous.

Halmaturus possesses a complete paraseptal cartilage (Seydel, 1896).

Caenolestes has been investigated only as regards Jacobson's organ and capsule by Broom (1926*b*), who finds that it resembles Didelphys and other Polyprotodonts. An outer bar and a papillary cartilage are present. Various features of the bony skull lead to the same conclusion (Broom, 1911), the chief exception, of course, being the teeth.

24. PLACENTALIA. *RODENTIA*

I. Lepus cuniculus

The development of the chondrocranium of the rabbit has been studied in its early stages by de Beer and Woodger (1930) and in its fully formed condition by Voit (1909, 1911). It will here be described in some detail to serve as a type of Placental mammal for comparison with others.

i. The development of the chondrocranium of Lepus

The earliest parts of the chondrocranium to appear are the occipital arches, at the 15-day (12·5 mm. g.-l.; 7 mm. h.-l.) stage. Chondrification spreads ventrally, enclosing the posterior hypoglossal foramen, to the hinder part of the parachordals which are thus of paired origin. Soon afterwards paired anterior parachordal cartilages become chondrified, and at the 16-day stage all the parachordal rudiments join up to form an elongated rectangular basal plate, the anterior edge of which bears a low crista transversa. This marks the hind wall of the future pituitary fossa, but it is not the dorsum sellae. The notochord lies dorsally to the posterior portion, ventrally to the middle portion, and is embedded in the anterior portion of the basal plate. (Plate 110, figs. 1–3).

Meanwhile, the auditory capsules begin to chondrify, quite independently, in the lateral wall of the canalicular part. Anteriorly, in the midline, the interorbitonasal part of the central stem chondrifies as a median impaired rod, representing the anterior region of the trabeculae. Between this structure and the basal plate, another element soon makes its appearance, surrounding the hypophysial duct. This is the hypophysial cartilage (the 'polar plate' of Noordenbos, or 'pars trabecularis' of Fawcett) which arises independently but quickly fuses with the basal plate behind and the interorbitonasal cartilage in front. The central stem is now complete, perforated by a hypophysial foramen, and extends forwards as the nasal septum between the olfactory sacs.

Meanwhile the visceral arch skeleton is partly chondrified, and represented by the incus, and by Meckel's cartilage, the posterior portion of which, the future malleus, presents unmistakable resemblance to a retroarticular process. The basihyal is a dumb-bell lying transversely and indicating a paired origin. The stylohyal is chondrified in the hyoid arch, and the thyrohyal cartilage in the 1st branchial arch. (Plate 110, figs. 4–7).

At the 19 mm. stage a number of important changes have occurred. There are now two hypoglossal foramina enclosed at the base of each occipital arch, which gives off a paracondylar process (cranial rib). Dorsally, the occipital arches end freely, but just above them is a pair of independent supraoccipital cartilages, which soon fuse with one another forming the tectum posterius, and with the dorsal ends of the occipital arches, thus completing the enclosure of the foramen magnum. (Plate 110, figs. 8–10).

The auditory capsule is still unattached to any cartilage. Quite independent also at this stage are the parietal plates, situated above the auditory capsules and in front of the supraoccipital cartilages. The orbital cartilages are now present, and attached to the central stem by the pila metoptica on each side. The preoptic roots develop from the orbital cartilage towards the central stem which they do not yet reach, but immediately behind the point of their future fusion with it are a pair of small isolated cartilages, the alae hypochiasmaticae. These become attached to the preoptic roots from which they project backwards. Eventually their hind ends fuse with the central stem, thereby enclosing the prechiasmatic foramina.

The nasal capsule chondrifies from three sources. There is first the parietotectal cartilage which grows out from the dorsal edge of the nasal septum and forms the roof and the anterior part of the side wall on each side. The (paired) lamina orbitonasalis chondrifies independently and forms the hind wall and cupola posterior. Lastly, the paranasal cartilage, likewise paired and independent, forms the hind part of the side wall. It overlaps the parietotectal cartilage laterally and the lamina orbitonasalis, with the result that the posterior free edge of the former projects back into the cavity of the capsule as the crista semicircularis, and the free edge of the latter similarly projects forwards as the 1st ethmoturbinal. Where the paranasal cartilage comes into contact and fuses with the roof of the nasal capsule the epiphanial foramen is left as the means of exit of the ramus lateralis of the nasal branch of the profundus nerve.

The sphenethmoid commissure of the orbital cartilage becomes fused to the paranasal cartilage, thereby enclosing the orbitonasal fissure.

At the 19·5 mm. stage the cochlear part of the auditory capsule has chondrified in continuity with the canalicular part and quite independently of the basal plate from which it is separated by the basicochlear fissure. This is interrupted posteriorly (and thereby cut off from the fissura metotica) by the basivestibular commissure which provides the only contact between the auditory capsule and the rest of the neurocranium at this stage. The wall of the capsule itself presents three large openings. These are the primary foramen acusticum medially, the foramen perilymphaticum posteriorly, and the fenestra ovalis laterally, in the mouth of which the stapes is now present, having arisen quite independently and without any participation of capsular cartilage (contra Fuchs).

The ala temporalis arises as a completely isolated nodule of cartilage, before the processus alaris develops as an outgrowth from the central stem towards it. The ala temporalis soon develops a lamina ascendens which rises up between the maxillary and mandibular divisions of the trigeminal nerve, and a pterygoid process which projects forwards. Eventually but not for some time, the ala temporalis fuses with the processus alaris. (Plate 111, fig. 1).

The internal carotid arteries run upwards on each side of the central stem behind the processus alaris, in a notch which is open to the side and confluent posteriorly with the basicochlear fissure. This notch is converted into the carotid foramen by the formation of the sphenocochlear commissure behind it (cutting it off from the basicochlear fissure) and the alicochlear commissure at the side. It is to be noted, however, that here the alicochlear commissure chondrifies from the cochlear capsule forwards, towards the processus alaris, which it does not reach for some time.

The parietal plate becomes fused to the supraoccipital cartilage behind it, and then the orbital cartilage becomes connected with the auditory capsule by the orbitocapsular commissures. Lastly, the parietal plate becomes attached to this orbitocapsular commissure, thus converting it into the orbitoparietal commissure. The auditory capsule is still free from the parietal plate above and the occipital arch behind it. There is thus an extensive occipitocapsular fissure which is confluent with the fissura metotica. This becomes divided up as the anterolateral edge of the occipital arch (which is produced to form the lamina alaris) establishes contact with the hind wall of the auditory capsule in two places: between the jugular foramen and the fissura occipitocapsularis inferior, and between the latter and the fissura occipitocapsularis superior (or parietocapsularis, or foramen jugulare spurium).

At the 21 mm. stage the visceral arch skeleton is complete but its elements are still separate. In the hyoid arch the tympanohyal (or laterohyal) has appeared laterally to the stapes and dorsally to the stylohyal (with which it soon fuses). The hypohyal is present between the stylohyal and the basihyal, with which it fuses forming the anterior horn of the 'hyoid'. The thyrohyal cartilage also fuses with the basihyal and forms the posterior horn. The thyroid cartilages are present in the 2nd and 3rd branchial arches but they soon fuse together on each side, and, ventrally, with their fellows of the opposite side. The cricoid cartilages form an almost complete ring, and do not convey the impression that they are visceral arches. (Plate 111, figs. 2, 3).

ii. The morphology of the fully formed chondrocranium of Lepus (45 mm. stage) (PLATES 112, 113, FIGS. 1, 2).

1. The basal plate and notochord. The basal plate is rectangular, and extends from the shallow incisura intercondyloidea behind to the crista transversa in front. Dorsally to the crista transversa and attached to it by two struts of cartilage is the dorsum sellae or acrochordal. The unchondrified space between this and the crista transversa therefore corresponds to a basicranial fenestra. Laterally, the basal plate is separated from the auditory capsule by the basicochlear fissure.

The course of the notochord is transbasal: i.e. it runs dorsally to the hindmost part of the basal plate, then pierces it and runs ventrally to the middle part, penetrates it again and runs embedded in the thickness of its anterior part.

The basal plate forms an angle of almost 60° with the axis of the more anterior regions of the chondrocranium.

If, as there is every reason to suppose, the rabbit agrees with other mammals (see p. 290) and the pleurocentrum of the proatlas vertebra fuses with the tip of the odontoid peg, the hind limit of the basal plate, and the occipito-atlantal joint, are intravertebral and intersegmental.

2. The occipital region. The occipital arches are massive, and extend backwards and upwards from the hind end of the basal plate to the tectum posterius. There are two pairs of hypoglossal foramina. Posteriorly, the base of each occipital arch is expanded to form the occipital condyles, each of which has its own separate synovial cavity between it and the atlas vertebra (i.e. the articular facets are not continuous across the midline). Laterally each occipital arch is produced to form the lamina alaris, which extends behind and under the auditory capsule and ends ventrally in a well-developed para-condylar process. That the latter is the rib or transverse process of a vertebral element incorporated in the skull is plain not only from its form, but also because the rectus capitis lateralis muscle, stretched between it and the atlas transverse process, is clearly the foremost of the intertransversarii muscles. The lamina alaris establishes contact with the posterior wall of the auditory capsule in two places thereby separating the jugular foramen, the inferior occipitocapsular, and the parietocapsular or superior occipitocapsular fissures. On the medial surface, the groove between the lamina alaris and the auditory capsule forms the recessus supra-alaris, lodging the sigmoid sinus, which, becoming the internal jugular vein, emerges through the jugular foramen.

3. The auditory region. The auditory capsule consists of canalicular and cochlear positions, the latter being relatively large and lying anteriorly and medially to the former. The medial wall of the auditory capsule thus slopes outwards and forms part of the floor rather than the side wall of the cranial cavity. The auditory capsule is attached by the alicochlear commissure to the processus alaris and by the basivestibular commissure to the basal plate. Between these commissures is the basicochlear fissure. Dorsally, the capsule is attached to the parietal plate; laterally its wall is produced to form the crista parotica, to which the styloid process is fused. The crista parotica is produced posteriorly to form the mastoid process (which overlaps the styloid process), and anteriorly into the tegmen tympani which fuses with the lateral wall of the cochlear capsule, thus enclosing the secondary facial foramen by means of the lateral prefacial commissure. The true prefacial (or suprafacial) commissure (enclosing the primary facial foramen, leading out of the internal auditory meatus) runs from the anterior wall of the canalicular to the dorsal wall of the cochlear capsule, and represents part of the true side wall of the skull. Between it and the lateral prefacial commissure is an extracranial space, the cavum supracochleare (which subsequently becomes enclosed within the bony skull). The gap between the prefacial and lateral prefacial commissures is the hiatus Fallopii, through which the palatine branch of the facial nerve after emerging from the primary facial foramen, runs forwards and downwards, laterally to the cochlear capsule and alicochlear commissure, ventrally to the processus alaris. The remainder of the facial nerve penetrates the secondary facial foramen to the sulcus facialis and emerges behind the styloid and in front of the mastoid processes. The chorda tympani runs forwards laterally to the styloid process to reach the medial surface of the malleus.

Of the foramina in the wall of the capsule, on the medial side the foramina acustica anterius and posterius become known as superius and inferius owing to the backward rotation of the capsule as compared with lower forms. These foramina lie in a deep depression, the internal auditory meatus, which is covered over by the prefacial commissure. Posterodorsally to this is the foramen endolymphaticum. On the lateral surface of the cochlear capsule the fenestra ovalis opens beneath the crista parotica and in front of the styloid process, and lodges the footplate of the stapes. In the hind wall of the cochlear capsule is the foramen perilymphaticum, from the ventral border of which the processus recessus (intraperilymphaticus of Voit) extends back towards the floor of the canalicular capsule, this enclosing an extracranial space, the recessus scalae tympani. When the processus recessus meets the floor of the canalicular capsule, two new apertures are enclosed: laterally the fenestra rotunda, covered by the secondary tympanic membrane, and medially the aquaeductus cochlae which faces into the cranial cavity through the foremost part of the jugular foramen.

Internally, the auditory capsule contains the cavum cochleare and the cavum vestibulare in the cochlear portion, and the cavum utricoampullare in the canalicular portion, which spaces, however, communicate widely with one another. The cavum vestibulare is the part into which the foramina acustica and perilymphaticum and the fenestra ovalis open, and it lodges the saccule. The cavum cochleare lodges the ductus cochlearis and contains a spiral septum round which the ductus begins to wind. The cavum utriculo-ampullare lodges the utricle and communicates with the semicircular canals, each of which has a septum cutting it off from the main part of the cavum. The posterior septum is, however, incomplete and the posterior and lateral semicircular canals communicate with one another before opening by a joint aperture into the cavum. The cartilage in the region of the septum semi-circulare anterius is very thick, forming the massa angularis. This is, however, reduced by depressions on both outer and inner surfaces of the capsule, known as the fossae subarcuatae.

4. The orbitotemporal region. The orbital cartilages are large and connected with the central stem by preoptic and metoptic roots enclosing the optic foramen between them. The processus paropticus, developed out of the ala hypochiasmatica, forms a shelf jutting out from the central stem on each side, beneath the optic foramen. Remnants of prechiasmatic foramina are found between the central stem and the paroptic processes.

The central stem between the dorsum sellae and the metoptic roots of the orbital cartilages is broad and concave dorsally, forming the pituitary fossa or sella turcica. In front of the preoptic roots of the orbital cartilages, however, the central stem is compressed in the sagittal plane, and forms a small low interorbital septum, directly continuous with the nasal septum in front of it.

The hypophysial foramen at the bottom of the sella turcica is closed at this stage, and therefore the craniopharyngeal foramen observed by Arai (1907) in the basisphenoid of the adult must be a new and independent formation.

The orbital cartilage is connected with the parietal plate posteriorly by the orbitoparietal commissure (thus enclosing the sphenoparietal fontanelle through which nerves III, IV, V, and VI emerge), and with the nasal capsule anteriorly by the sphenethmoid commissure (thus enclosing the orbitonasal fissure through which the nasal branch of the profundus nerve enters the supracribrous recess on its way from the orbit to the nasal capsule).

In the hinder part of the orbitotemporal region the processus alaris projects from the central stem and bears the ala temporalis. The processus alaris (or basitrabecularis) is connected with the cochlear capsule by the alicochlear commissure which encloses the carotid foramen from the side and represents the hindmost part of the trabecula. The palatine nerve runs forwards laterally to the alicochlear commissure and ventrally to the processus alaris. The ala temporalis consists of a pterygoid process which projects forwards and is perforated by the orbital artery (the 'alisphenoid canal'), and a lamina ascendens which rises up between the maxillary and mandibular branches of the trigeminal nerve and marks the lateral boundary of the cavum epiptericum, an extracranial space which becomes enclosed within the bony skull. The true side wall of the cranial cavity in this region is represented at the 43 mm. stage by a number of vestigial cartilaginous rods lying in the dura mater. Of these cartilages one ('a') overlies the prefacial commissure, and may be called the supracochlear cartilage; another ('b') lies over and in front of the root of the trigeminal nerve ganglion; while a third ('c') is attached to the lateral edge of the central stem behind the pila metoptica. There is no doubt that cartilage 'b', and perhaps also cartilage 'a', represent remnants of the pila antotica which have become disconnected from the ends of the dorsum sellae (the roots of which rise up as marked stumps at the 43 mm. stage) and have become displaced backwards. Equivalent structures may be found in other forms as the so-called abducens bridge (Sus, Semnopithecus, Homo). Cartilage 'c' which is variable may represent a portion of the supratrabecular bar. All these cartilages vanish before the 45 mm. stage. (Plate 113, fig. 1).

5. The ethmoid region. The nasal capsules are large. Each consists of a roof (tectum nasi) continuous with the dorsal edge of the nasal septum and with the side wall of the capsule, and perforated anteriorly by the fenestra superior, and posteriorly by the epiphanial foramen. The hind wall of the capsule or cupola posterior, formed of the lamina orbitonasalis, projects back beside and freely from the nasal septum. The ventral edge of the cupola posterior projects forwards and forms the extensive lamina transversalis posterior, or floor to the hindmost part of the nasal capsule.

Near its anterior end the ventral edge of the side wall of the capsule gives off the lamina transversalis anterior which runs horizontally inwards towards the ventral edge of the nasal septum, with which, however, it does not fuse: there is therefore no complete zona annularis. The inner end of each lamina transversalis anterior is continuous with the paraseptal cartilage, which is complete and joined posteriorly to the lamina transversalis posterior.

The fenestra basalis is thus completely enclosed by capsular cartilage, and its elongated aperture is further restricted at late stages by the formation of an ectochoanal cartilage, situated laterally to the duct (Stenson's) of Jacobson's organ; this cartilage does not become attached to the lamina transversalis anterior, but remains free. The anterior part of the paraseptal cartilage forms Jacobson's capsule, lodging Jacobson's organ. Broom (1900) has stressed the similarity between the conditions in Lepus and those in Dasypus.

The fenestra narina is continuous ventromedially with the slit between the paraseptal cartilage and the nasal septum. There is no anterior wall to the capsules, and the two fenestrae narinae are separated in front only by the anterior edge of the nasal septum. The lateral border of the fenestra narina (the anterior edge of the side wall) bears a downwardly directed processus alaris superior which separates the lachrymonasal duct from the external nostril.

The apertures in the hinder part of the ethmoid region may be defined as follows. The orbitonasal fissure is bounded above by the sphenethmoid commissure, below by the lamina orbitonasalis, and behind by the preoptic root of the orbital cartilage; it leads from the orbit into an extracranial space, the recessus supracribrosus or cavum orbitonasale. This is the path taken by the nasal branch of the profundus nerve. The foramen olfactorium evehens is bounded laterally by the preoptic root of the orbital cartilage and spheneth-moid commissure, anteriorly by the posterior edge of the tectum nasi, and medially by the central stem (interorbital and nasal septum); it is occluded by the dura mater and leads out of the cranial cavity into the supracribrous recess. This is the path taken by the olfactory nerves.

The foramen olfactorium advehens, or fenestra cribrosa, is bounded laterally by the lamina orbitonasalis, anteriorly by the posterior edge of the tectum nasi, and medially by the nasal septum. Its aperture is subdivided by the crista intercribrosa and other cartilaginous struts into the numerous pores of the cribriform plate, which lead out of the recessus supracribrosus into the cavity of the nasal capsule. This course is followed by the olfactory nerves and the ethmoid branch of the profundus nerve of which the ramus lateralis re-emerges on the roof of the nasal capsule through the epiphanial foramen.

The recessus supracribrosus becomes enclosed within the bony skull, but its morphologically extracranial nature is important and the ethmoid branch of the profundus nerve does not traverse the cranial cavity on its course. The dorsal edge of the hindmost part of the nasal septum, between the fenestrae cribrosae, is raised to form a slight crista galli.

The cavity of the nasal capsule becomes subdivided into a number of recesses by the formation of turbinals, &c., which project into it from the side and hind walls. The atrioturbinal is the inturned anterior edge of the side wall of the capsule, while its inturned ventral edge forms the maxilloturbinal, dorsally to which is the small nasoturbinal. Behind the latter, the side wall presents a crest in the vertical plane, the posterior curved edge of which projects backwards into the cavity of the capsule. This is the crista semi-circularis, the ventral end of which is free, forming the processus uncinatus. Laterally to the crista semicircularis, therefore, is a space ending blindly in front; this is the recessus anterior. Between the processus uncinatus and the maxilloturbinal is a space lodging the lateral nasal gland, the recessus glandularis. The 1st ethmoturbinal projects forwards from the hind wall of the capsule as a vertical plate, fused below with the lamina transversalis posterior, above with the crista intercribrosa, and laterally with the side wall of the capsule by means of its horizontal so-called anterior root. Laterally to the 1st ethmoturbinal and above its anterior root is the recessus frontalis, while beneath the anterior root is the recessus maxillaris. (Plate 112, fig. 5).

From the side wall of the recessus frontalis a few frontoturbinals (conchae obtectae) are developed. The 2nd and 3rd ethmoturbinals develop medially to the 1st, between it and the nasal septum, and consist of vertical plates projecting forwards from the hind wall of the capsule.

The recessus anterior, frontalis and maxillaris, are the starting-points of the excavations which result in the formation of the various ethmoidal, frontal, and maxillary sinuses.

6. The visceral arch skeleton. The anterior portion of the pterygoquadrate cartilage is represented by the ala temporalis, described above (and see p. 421). The posterior portion is the incus, which consists of a body, a crus breve representing the otic process and articulating with the crista parotica, and a crus longum which projects downwards and backwards and articulates with the stapes. Fuchs (1906) claims that the crus longum chondrifies independently of the incus and becomes fused to it. The anterior face of the body of the incus articulates with the downturned posterior end of Meckel's cartilage which forms the malleus. The malleus lies definitely in front of the incus. The anterior ends of Meckel's cartilages are slender and run downwards and then slightly upwards before fusing in a symphysis.

The stapes is genuinely stirrup-shaped and pierced by the stapedial (or orbital artery). As already mentioned, the styloid process (or 'Reichert's cartilage') is fused to the crista parotica, but the stylohyal subsequently becomes free in the bony skull, by the non-ossification of the tympanohyal. Ventrally, the stylohyal also ends freely, and does not reach the hypohyals which form the anterior horns of the 'hyoid', or basihyal. The posterior horns are formed by the 1st branchial arch or thyrohyal cartilage, while the fused 2nd and 3rd branchial arches form the thyroid cartilage. Whether the cricoid cartilages represent the 4th branchial arch is still uncertain but improbable, for they lie within the thyroid cartilages and seem to bear a greater resemblance to tracheal rings.

iii. The development of the osteocranium of Lepus

1. The membrane-bones. All the membrane-bones are present at the 45 mm. stage.

The *dentary* is the first to ossify, and is already present at the 22 mm. stage, and possesses large nodules of secondary cartilage in the condylar and angular regions. For the greater part of its length the dentary surrounds Meckel's cartilage completely.

The *interparietal* is paired and arises immediately in front of the tectum posterius. It fuses with its fellow of the opposite side, but remains separated from the parietals and from the supraoccipital by sutures. (In the hare it is said to fuse with the supraoccipital.)

The *parietal* arises beside and above the parietal plate and the orbitoparietal commissure. Its ventral edge dips down beneath the level of the orbitoparietal commissure and helps to close over the sphenoparietal fontanelle.

The *frontal* arises beside and above the orbital cartilage and sphenethmoid commissure. Laterally it forms a ridge marking the dorsal boundary of the orbit, and ending posteriorly in the freely projecting processus supraorbitalis posterior.

The *squamosal* arises laterally to the crista parotica and the orbitoparietal commissure. Its zygomatic process is large.

The *jugal* joins the zygomatic processes of the maxilla and squamosal, lying above the former and beneath the latter.

The *lachrymal* arises in the front of the orbit, behind the facial portion of the maxilla.

The *nasal* overlies the roof of the nasal capsule and covers its fenestra superior.

The *premaxilla* is relatively large, and consists of a well-developed alveolar and facial portions, and a long palatine process. The size of the alveolar portion is due to the sockets for two large incisors, in front of which, however, are the vestiges of another pair of tooth-germs. The foremost incisor of the Duplicidentata is therefore the 2nd. The palatine process extends backwards beneath the paraseptal cartilage (but not medially to it) for almost all its length. The hinder part of this palatine or prevomerine process probably represents the *prevomer*, and it would be interesting to know whether it arises from a separate ossification centre.

The *maxilla* arises beneath the ventral edge of the paries nasi and consists of alveolar and facial portions and palatine and zygomatic processes (the latter containing a nodule of secondary cartilage).

The *vomer* arises as a median and unpaired splint beneath the ventral edge of the nasal septum, and forks in front and behind. Whether at earlier stages the vomer has completely paired rudiments (as in Mus) is unknown.

The *palatine* arises beneath the lateral edge of the central stem and forms part of the side wall and roof of the nasopharyngeal passage. The pars horizontalis in the floor of this passage (the false palate) is as yet only feebly developed.

The '*pterygoid*' arises from a horizontal plate of membrane-bone representing the lateral wing of the parasphenoid, lying ventrally to the processus alaris (separated from it by the parabasal or Vidian canal, for the Vidian or palatine nerve), and from the nodule of secondary cartilage which gives rise to the hamulus, which Fuchs (1909*c*) found to arise independently of the parasphenoid, and probably represents the reptilian pterygoid. The mammalian 'pterygoid' is thus probably a compound element (see p. 435).

The *tympanic* arises in the form of an almost closed ring, on the lateroventral surface of the hinder part of Meckel's cartilage, its relations to which substantiate van Kampen's (1905) view that the mammalian tympanic represents the angular of lower vertebrates, and is therefore originally a bone of the lower jaw.

The *prearticular* is a small splinter of bone lying medially to the hinder end of Meckel's cartilage. The chorda tympani runs between them.

2. No systematic study has been made of the development of the cartilage-bones, although the little that is known is of great interest, and a knowledge of the development of the ear-bones would be very valuable. A *mesethmoid* is developed in the nasal septum. The stylohyal becomes discontinuous from the periotic (Howes, 1896).

II. Microtus amphibius

The development of the skull of the water-rat has been studied at the 25 mm. stage by Fawcett (1917) with the help of wax models. (Plate 113, fig. 3).

The chondrocranium of Microtus shows great similarities with that of Lepus, including such features as the well-developed paracondylar and mastoid processes, two pairs of hypoglossal foramina, the foramina prechiasmatica enclosed by the alae hypochiasmaticae, the complete paraseptal cartilages, the large size of the lamina transversalis posterior and its freedom from the nasal septum. On the other hand, the lamina transversalis anterior is fused with the nasal septum, and there is a complete zona annularis. Further, in Microtus the pila metoptica is interrupted and the optic foramen is confluent with the sphenoparietal fontanelle. The fossa subarcuata communicates right through the massa angularis of the auditory capsule with its external surface, and with the cavum vestibulare. Lastly, Microtus has a plate of cartilage attached to the supraoccipital cartilage and extending forwards, laterally to the lateral wall of the auditory capsule. This is the processus opercularis, the lower edge of which serves as an attachment for muscles from the scapula.

The bony skull calls for no special mention, except for the fact that the alisphenoid, which arises as perichondral lamellae in the ala temporalis, is perfectly continuous with its large intramembranous extension. The dentary has an anterior nodule of secondary cartilage in addition to the condylar and angular nodules.

III. Mus norvegicus and musculus

No systematic study has been made of the development of the skull in either rats or mice, but from the figures given by Shindo (1915), and from a model made but not described by Jenkinson, it is clear that it conforms closely to the Rodent type. An interesting point of difference with Lepus is the fact that in Mus the notochord is entirely suprabasal (the Tourneux, 1912). Jenkinson (1911) has studied the development of the auditory ossicles in the mouse at eight stages and with the help of wax models. His results are a complete confirmation of the Reichert theory.

The stapes arise in connexion with the hyoid arch, and where it comes into contact with the mesenchymatous wall of the auditory capsule that wall is converted not into cartilage but into the membrane spanning the fenestra ovalis. The malleus and incus are formed from the mandibular arch. The crus longum of the incus grows freely back to establish contact with the stapes.

The early development of the bones of the rat has been studied by Strong (1925), who finds that they arise in the following order:

Bone	*Age*
Dentary	17 days 1 hours
Maxilla	17 ,, 1 ,,
Frontal	17 ,, 1 ,,
Palatine	17 ,, 8 ,,
Squamosal	17 ,, 8 ,,
Premaxilla	17 ,, 8 ,,
Pterygoid	17 ,, 8 ,,
Vomer	17 ,, 8 ,,
Parietal	17 ,, 8 ,,
Exoccipital	17 ,, 8 ,,
Basioccipital	17 ,, 8 ,,
Jugal	18 ,, ..
Nasal	18 ,, ..
Lachrymal	18 ,, ..
Interparietal	18 ,, 10 ,,
Tympanic	18 ,, 10 ,,
Basisphenoid	18 ,, 10 ,,
Supraoccipital	19 ,, 9 ,,
Alisphenoid	19 ,, 9 ,,
Presphenoid	19 ,, 9 ,,
Orbitosphenoid	22 ,, 19 ,,

IV. Erethizon dorsatus

The development of the skull of the porcupine has been studied by Struthers (1927). The chondrocranium shows definite resemblances to that of other Rodents, but is in some respects specialized. A short median rostrum projects forwards from the nasal septum between the cupolae anteriores. The paraseptal cartilage is incomplete. The preoptic root of the orbital cartilage is very wide, and inserts on the lamina orbitonasalis as well as on the central stem. There is no orbitoparietal commissure. The internal carotid artery enters the skull through the foremost part of the basicochlear fissure (i.e. behind the sphenocochlear commissure), which is confluent posteriorly with the jugular foramen through the absence of a basivestibular commissure. The 'carotid' foramen is present and enclosed by the alicochlear commissure, but transmits nothing. As in Microtus a processus opercularis is attached to the supraoccipital cartilage; it seems to chondrify independently.

In the bony skull a feature worthy of notice is the separate ossification of paired centres situated between the premaxillae. They may represent prevomers but if so their position is abnormally far forward. (Plate 113, fig. 4).

25. 'EDENTATA.' *XENARTHRA*

I. Tatusia novemcincta

The development of the skull of the armadillo was described by Decker (1883) and by W. K. Parker (1885), and recently it has been studied with the help of model reconstructions by Fawcett (1919, 1921, 1923).

i. The development of the chondrocranium of Tatusia

1. STAGE 1. 10 mm. stage. (PLATE 114, FIG. 1).

The hinder part of the basal plate is chondrified, ventrally to the notochord, and in continuity with the occipital arches, on the anterior edge of which are notches for the hypoglossal nerves. Anteriorly, the interorbitonasal part of the central stem is chondrified, its hinder end bifurcating, and the prongs, perhaps, representing the anterior region of paired trabeculae. To each of these prongs becomes attached the metoptic roof of the orbital cartilage. Isolated nodules of cartilage represent the alae temporales.

2. STAGE 2. 12 mm. stage. (PLATE 114, FIGS. 2, 3).

The basal plate, hypophysial cartilage, and interorbitonasal part of the central stem are all chondrified and separate, though the first two are united by procartilage. The notochord lies on the dorsal surface of the basal plate except anteriorly where it is enclosed within the cartilage. Behind the hypophysial cartilage the notochord emerges on the ventral surface, and its tip is bent downwards and backwards. The occipital arches end freely dorsally, and are produced to the side to form the paracondylar process.

The canalicular part of the auditory capsule is chondrified, the cochlear part is procartilaginous and attached to the central stem and processus alaris by procartilaginous sphenocochlear and alicochlear commissures. There is as yet no basivestibular commissure (usually the earliest to be formed, and the basicochlear fissure and the metotic fissure are confluent.

Small parietal plates are present attached to the anterodorsal corners of the canalicular capsules, and continuous with orbitoparietal commissures which almost but not quite reach the orbital cartilages.

The ala temporalis is chondrified independently as a slender lamina ascendens attached to the central stem by a procartilaginous processus alaris. The orbital cartilage is attached to the nasal capsule by the sphenethmoid commissure, and to the central stem by preoptic and metoptic roots, the latter of which bears a small paroptic process (ala hypochiasmatica).

The nasal capsule is large, simple, and pear-shaped with very expanded frontal prominences. Posteriorly, the lamina orbitonasalis fails to reach the nasal septum with the result that the foramen olfactorium advehens, the fenestra basalis (choanalis), and the orbitonasal fissure, are all in communication through the gap. A slender lamina transversalis anterior is present from which a short (anterior) paraseptal cartilage projects back.

The incus is of relatively very large size. The malleus has a small head and a very large manubrium. The stapes is still in mesenchymatous connexion with the stylohyal cartilage. The latter is quite free and at a considerable distance from the crista parotica. Basihyal and thyrohyal cartilages are present, and the thyroid cartilage is of great interest for it is bilobed dorsally, and a deep notch separates what may be the representatives of the 4th and 5th visceral arches.

3. STAGE 3. 17 mm. stage. (PLATE 114, FIGS. 4, 5).

The basal plate is fused with the hypophysial cartilage, but on the ventral surface a portion of the intervening fissure is preserved, and through this the tip of the notochord protrudes, bent downwards. There is no dorsum sellae or crista transversa of any kind. The auditory capsules are now attached to the basal plate by extensive basivestibular commissures, and the jugular foramen on each side is cut off from the basicochlear fissure, but it is continuous with the extensive occipitocapsular fissure which surrounds the canalicular capsule almost entirely.

The dorsal ends of the occipital arches are fused with the supraoccipital cartilages which are interconnected by the tectum posterius and attached to the parietal plates in front of them. There is no suprafacial commissure. The crista parotica, to which the stylohyal cartilage is attached by connective tissue, not fused, projects forwards as a small tegmen tympani.

The hypophysial cartilage and the interorbitonasal part of the central stem are still discontinuous. Anterior clinoid processes project back from the metoptic roots of the orbital cartilages and overlie the processus alares. The orbital cartilages and orbitoparietal commissures are fused.

The nasal capsules are as long as the remainder of the chondrocranium, and while relatively thin and cylindrical anteriorly, they are very bulbous posteriorly. On the dorsal surface the epiphanial foramina are sunk in grooves. Anterior and posterior laminae transversales are present, and anterior and posterior paraseptal cartilages project backwards and forwards from them respectively, just failing, however, to meet. Broom (1900) showed that in these forms the anterior paraseptal cartilage has an 'outer bar' enclosing Jacobson's organ. At later stages (60 mm.) superior and inferior alar processes are developed from the borders of the fenestra narina and meet. The ventral edge of the nasal septum in front of the laminae transversales anteriores is produced on each side to form large processus laterales ventrales.

The malleus, incus, and stapes are relatively very large. In the ligament of the stapedial muscle, Paauw's cartilage (representing the hyostapes) has been found by van Kampen (1915*a*) and van der Klaauw (1923). As mentioned above, the stylohyal is not fused with the crista parotica, and its distal end is apparently segmented off to form the ceratohyal.

The thyrohyal cartilage calls for no special mention, and the thyroid cartilage has closed the notch present at the previous stage to form a foramen.

ii. The development of the osteocranium of Tatusia

Without going into the details of ossification of all the bones, attention may here be paid to four, which show features of wide general importance.

The *septomaxilla* (nariale of Wegner) is found only in Xenarthra among Mammals. Fawcett (1919) observed it at the 60 mm. stage as a small membrane-bone the body of which lies immediately beneath the lamina transversalis anterior, and gives off a slender anterior ascending process (processus intrafenestralis of Wegner, 1922) which projects forwards and upwards into the fenestra narina, medially to the joined superior and inferior alar processes. In ventral view the hinder part of the septomaxilla is concealed by the premaxilla. (Plate 114, fig. 6).

The *premaxilla* shows the interesting feature that its palatine process ossifies independently (at the 17 mm. stage) and even precedes the ossification of the body of the premaxilla. The palatine process here without doubt represents the *prevomer* subsequently fused with the premaxilla. (See p. 434.)

The *pterygoid* is of interest, for Broom (1914) has reported the separate existence of a 'mammalian pterygoid' and an 'Echidna pterygoid'. The former represents the lateral wing of the parasphenoid, and the latter presumably (see p. 435) represents the pterygoid of reptiles and the secondary pterygoid cartilage of other mammals. In other Xenartha, e.g. young Choloepus, Lubosch (1907, p. 527) observed that the pterygoid was divided, its portions representing the lateral wing of the parasphenoid and the 'Echidna pterygoid'.

The *endotympanic* may perhaps have been observed by W. K. Parker (1885) for he figured a cartilaginous rudiment of the 'tympanic'. On the other hand, from what is now known of the stages at which the endotympanic arises it is more probable that his description of a cartilaginous rudiment was a simple mistake for the bony rudiment of the tympanic. Van der Klaauw (1922) observed the blastematous rudiment of the caudal endotympanic in an embryo of 59 mm. carapace length, where it occupies the hinder part of the floor of the tympanic cavity, and is (secondarily) connected by a ligament with the stylohyal. Later this rudiment becomes chondrified from behind forwards, and ultimately ossified along its medial border.

The cartilaginous rudiments of the rostral endotympanic was found by van der Klaauw in an embryo of 64 mm. carapace length, lying in the anterolateral part of the roof of the tympanic cavity, medially to the prearticular, with its hind end ventral to the tegmen tympani.

II. Bradypus cuculliger

The chondrocranium of a 17·5 embryo of Bradypus has been figured by de Burlet (1927). (Plate 113, figs. 5, 6).

The skull is elongated and the basal plate is very narrow. The notochord runs ventrally to it throughout its length (de Burlet, 1913*a*) and it forms almost a right angle with the axis of the nasal capsule. No dorsum sellae is formed.

The occipital arches lean backwards and upwards and are separated from the auditory capsules by a large occipitocapsular fissure. A small lamina alaris is present ending ventrally in a short paracondylar process. The tectum posterius slopes forwards and upwards.

The auditory capsules are relatively large, and the medial wall of the canalicular part forms most of the side wall of the cranial cavity, for the parietal plate is relatively low and slopes slightly inwards, thus forming part of the roof. The cochlear capsules are large and constrict the basal plate very markedly. Though attached to it, there are deep basicochlear sulci on both dorsal and ventral surfaces. The basicochlear fissure is confluent anteriorly with the carotid foramen, or rather incisure, for there is no alicochlear commissure.

The ala temporalis is perforated by a foramen rotundum. The pila metoptica has an ala hypochiasmatica attached to it; the preoptic root of the orbital cartilage is broad, and the orbital cartilage and the sphenethmoid and orbitoparietal commissures are of approximately the same width as in Marsupials. A small interorbital septum is present.

The nasal capsule is simple and in the form of a slightly tapering cylinder. A lamina transversalis anterior is present with short paraseptal cartilages. The side wall of the capsule bears two processes, resembling the 'paranasal cartilage' and 'maxillary process' of Homo.

The development of an endotympanic bone has been observed by van der Klaauw (1922).

III. Other 'Edentata'

TUBULIDENTATA

(i) Orycteropus capensis

Stages in the development of the skull of the aardvark have been described by W. K. Parker (1885), who noted its dissimilarity from other so-called Edentates. It is the only one in which an interparietal bone is present. A mesethmoid bone is also present (Broom, 1926*a*). Great advantage would certainly accrue from a study of the development of this skull by modern methods.

The cartilages surrounding Jacobson's organ have been studied by Broom (1909*b*), who showed that the paraseptal cartilage possesses an 'outer bar', enclosing Jacobson's organ. There is no cartilago ductus nasopalatini, and no ectochoanal cartilage.

PHOLIDOTA

(ii) Manis sp.

Stages in the development of the skull of the pangolin have been described by Decker (1883) and by W. K. Parker (1885). It is, however, much to be hoped that the development of this skull will be investigated by modern methods, especially since van Kampen (1905) has shown that the chorda tympani runs forwards medially to the styloid process in this form (so far as known, unique among Placental mammals). According to Edgeworth (1935), the thyroid cartilages are formed solely from the 5th visceral arches.

26. *INSECTIVORA*

I. Talpa europea

In addition to the earlier work by W. K. Parker (1885), the early development of the skull of the mole has been followed with the van Wijhe technique by Noordenbos (1905), and the fully developed chondrocranium has been studied with the help of wax models by Fischer (1901) and by Fawcett (1918*a*). The paraseptal and associated cartilages have been described by Broom (1915*b*). The segmentation of the head has been studied by Dawes (1930).

i. The development of the chondrocranium of Talpa

At the 7·5 mm. stage the hindmost part of the parachordal plate is chondrified as a transverse bar ventral to the notochord, and unpaired from the start. Soon it extends forwards and, enveloping the notochord, this middle portion of the parachordal plate lies dorsally to the notochord. The anterior border of the parachordal plate at this stage is concave, with lateral horns projecting forwards on each side of the notochord which emerges from beneath the parachordal plate and runs forwards to the hypophysis.

From the hindmost part of the parachordal plate the occipital arches grow out, and as the notch for the hypoglossal nerve-root opens absolutely laterally on each side, it is clearly enclosed between an occipital and a pre-occipital arch. (Plate 115, fig. 1).

At the 8 mm. stage the interorbitonasal part of the central stem (Noordenbos's 'trabecular plate') arises as a median and unpaired structure, very nearly in the same plane as the parachordal plate from which it is separated by a wide space. The hinder part of the stem forms a flattened horizontal plate, to the sides of which the orbital cartilages will eventually become attached. The anterior part of the stem is compressed sagittally to form the nasal septum, from the dorsal edge of which the roof of the nasal capsules grows out on each side. Eventually the front walls or cupolae anteriores of the nasal capsules are formed by extension of this cartilage. There is at first a deep groove between the two cupolae.

The sides of the nasal capsules arise from the independently chondrified paranasal cartilages, which arise at the 9 mm. stage. Also at this stage there appear Meckel's cartilages, and the canalicular parts of the auditory capsules, the latter starting with their lateral walls.

Important advances are made at the 10 mm. stage, when the hypophysial cartilages arise, in the form of two pairs of isolated nodules, the hypophysial cartilages, surrounding the hypophysial stalk. These coalesce to form Noordenbos's 'polar plate', the 'trabecular plate' of Fawcett (1916), but which may better be called the hypophysial plate. The internal carotid arteries pass upwards laterally to it. The hypophysial plate eventually becomes fused anteriorly with the hind end of the interorbitonasal part of the central stem. (Plate 115, figs. 2, 3).

Posteriorly, Noordenbos imagined that the hypophysial plate becomes attached directly to the horns which project forwards on each side from the front edge of the parachordal plate, leaving the basicranial fenestra open as a relic of the originally intervening space. On this view the basicranial fenestra is an opening *in front of* and not *in* the parachordal or basal plate. But Noordenbos's figures and descriptions show the existence of a transverse bar of cartilage behind the hindmost pair of hypophysial cartilages, and in front of the basicranial fenestra. This bar was regarded by Noordenbos as a part of the hypophysial plate, but it may equally well represent the anterior part of the parachordals (or perhaps even the acrochordal), in which case the basicranial fenestra is situated within the basal plate.

The orbital cartilages also arise at the 10 mm. stage as independent triangular-shaped chondrifications. Eventually, they become connected anteriorly with the roof of the nasal capsule by the sphenethmoid commissure, posteriorly with the parietal plate by the orbitoparietal commissure, and medially with the central stem by a splender process which, forking over the optic nerve, represents the very reduced preoptic and metoptic roots.

At the 11 mm. stage the supraoccipital cartilages appear, already interconnected by the tectum posterius, but still free from the occipital arches or parietal plates. The latter also arise at this stage as independent chondrifications. The order in which these structures become connected up is as follows:

1st the supraoccipital cartilages and tectum posterius fuse with the occipital arches;
2nd the parietal plates fuse with the auditory capsules;
3rd the parietal plates fuse with the supraoccipital cartilages;
4th the orbitoparietal commissures fuse with the parietal plates.

Another element to arise at this 11 mm. stage is the lamina orbitonasalis, which forms the hind wall of the nasal capsule on each side. (Plate 115, fig. 4).

The cochlear part of the auditory capsule also becomes chondrified at this stage, and this takes place from an independent centre according to Fawcett (1918*a*), but according to Noordenbos in continuity with the cartilage of the canalicular part; in either case without any connexion with the basal plate. It is on this material that Noordenbos based his opposition to Gaupp's view (see p. 401) that the cochlear capsules were formed out of the basal plate.

The first contact to be formed between the auditory capsule and the basal plate is by means of the basivestibular commissure, which forms the anterior limit of the fissura metotica and the posterior limit of the basicochlear fissure. Next, there arise *three* outgrowths on each side from the hypophysial region of the central stem. The foremost of these projects sideways and ends freely; it will form the ala temporalis. The next or middle outgrowth projects sideways for a distance and bends backwards to fuse with the anterior part of the cochlear capsule, forming the lateral border of the carotid foramen. The transverse portion of this process fuses with the process in front of it thus contributing to the ala temporalis, and is regarded by Noordenbos as the processus alaris, its backturned portion (Noordenbos's 'lateral sphenocochlear synchondrosis') representing the alicochlear commissure. The third or hindmost of the three outgrowths also joins with the cochlear capsule forming the medial border to the carotid foramen, thereby separating it from the basicochlear fissure, and is the (medial) sphenocochlear commissure. Lastly, an outgrowth from the lateral edge of the basal plate to the medial wall of the cochlear capsule forms the basicochlear commissure, and subdivides the basicochlear fissure into anterior and posterior portions. It is important to note that the ala temporalis has no independent chondrification centre, but arises in continuity with the central stem. (Plate 115, figs. 5, 6).

The stapes arises at the 12 mm. stage, and has no connexion whatever with the wall of the auditory capsule.

ii. The morphology of the fully formed chondrocranium of Talpa (19 mm. and 27·3 mm. stages) (PLATE 116).

1. The basal plate and notochord. The notochord lies dorsally to the hindmost part of the basal plate and lies embedded in the remainder (i.e. that portion of the notochord which at previous stages ran ventrally to the basal plate has now become enclosed within it). Fairly broad posteriorly, the basal plate is narrow anteriorly, and the anterior basicochlear fissures are large.

The posterior basicochlear fissures have become obliterated and the basal plate has a wide contact with the auditory capsule extending from the basi-vestibular to the basicochlear commissures. Posteriorly, there is a marked intercondylar incisure across which at early stages the articular facet of the occipito-atlantal joint is continuous before becoming differentiated into right and left condylar portions.

The basal plate as a whole is remarkably flat; there is no prominent dorsum sellae, and the plane of the basal plate is almost the same as that of the anterior region of the central stem, i.e. the skull is 'straight', and it is also remarkably elongated and narrow.

2. The occipital region. The occipital arches slope slightly backwards and upwards, and the tectum posterius is in the vertical transverse plane. The fusion between the occipital arch and the auditory capsule is extensive, and the inferior occipitocapsular fissure is much reduced. The lamina alaris, however, is feebly developed and there is only a very small paracondylar process. The single hypoglossal foramen on each side is situated a considerable distance behind the jugular foramen.

3. The auditory region. The canalicular parts of the auditory capsules are situated well behind the relatively small cochlear parts, and contribute only half the height of the chondrocranial wall in this region, the remainder being formed by the parietal plate. Between the parietal plate and the capsule is a large superior occipitocapsular fissure (or foramen jugulare spurium), the opening of which is almost covered by the dorsal edge of the processus opercularis. This is a sagittal plate of cartilage lying laterally to the lateral wall of the capsule and fused ventrally to the crista parotica. The latter has the styloid process fused to it, and is produced forwards into the well-developed tegmen tympani, from which the lateral prefacial commissure connects with the cochlear capsule. A true or medial prefacial commissure is also present. There is no mastoid process.

As regards the interior of the capsule, the septum of the posterior semi-circular canal is incomplete.

4. The orbitotemporal region. The processus alaris, or root of the ala temporalis is very broad. The ala temporalis consists only of a lamina ascendens; there is no pterygoid process. The preoptic and metoptic roots of the orbital cartilage are very slender and join one another before becoming attached to the central stem. A small ala minima projects forwards from the anterior edge of the orbital cartilage towards the lamina orbitonasalis. There is no ala hypochiasmatica. According to Fischer the secondary pterygoid cartilage becomes attached to the under surface of the root of the orbital cartilage, but this has not been confirmed by Noordenbos. The spheneth-moid commissure connecting the orbital cartilage with the roof of the nasal capsule is broad, but the orbitoparietal commissure, between the orbital cartilage and the parietal plate is slender.

5. The ethmoid region. The nasal capsules occupy half the total length of the chondrocranium. They are markedly elongated and pear-shaped with large frontal prominences. The lamina transversalis anterior forms a broad connexion or floor between the ventral edges of the nasal septum and of the side walls of the capsules. From the lamina transversalis anterior on each side the paraseptal cartilages extend back and end freely. Anteroventrally, each paraseptal cartilage is produced into a small cartilago ductus nasopalatini which projects forwards and downwards medially to the duct of Jacobson's organ and ventrally to the premaxilla.

The lamina orbitonasalis forms the cupola posterior or hind wall to the nasal capsule on each side, and is fused with the nasal septum. The ventral edge of the cupola posterior is bent forwards to form a floor or lamina transversalis posterior, to the hinder part of the nasal capsule. The hind border of the fenestra narina is produced into a well-developed superior alar process. The front wall of each nasal capsule is formed by the cupola anterior, but the previously deep groove between the two cupolae has now been filled up.

In the roof of the nasal capsule there is on each side a small fenestra dorsalis, situated far forward, and an epiphanial foramen far back.

6. The visceral arch skeleton. The auditory ossicles call for no special mention. The anterior ends of Meckel's cartilages fuse to form a median element which Parker termed the 'basimandibular'. The thyroid cartilages are very slender.

II. Sorex vulgaris and (Crocidura) araneus

The chondrocranium of the shrew has been studied by W. K. Parker (1885), Levi (1909*a* and *b*), and with the help of wax models by de Beer (1929). Certain points of morphology were treated by Ärnbäck-Christie-Linde (1907).

The chondrocranium of Sorex is less elongated than that of Talpa, but presents close resemblances to it. Thus, the basal plate and the anterior parts of the central stem are almost in line; there is no marked dorsum sellae; a processus opercularis is present, fused to the crista parotica which projects forwards as a large tegmen tympani; both medial and lateral prefacial commissures are present; the orbital cartilage has a small ala minima; the nasal capsule is pyriform; anterior and posterior laminae transversales are present, and the paraseptal cartilage is incomplete.

The chief points of difference are that in Sorex a small posterior paraseptal cartilage is present; the preoptic and metoptic roots of the orbital cartilage are attached not to the central stem but to the cupola posterior of the nasal capsule which extends very far back; the secondary pterygoid cartilage is attached to the ala temporalis; the basicochlear and the inferior occipitocapsular fissures are obliterated. The anterior part of the lamina transversalis anterior lies at a more ventral level than its hinder part.

According to Levi the ala temporalis has an independent chondrification.

III. Erinaceus europaeus

The chondrocranium of the hedgehog has been studied by W. K. Parker (1885), and, with the help of wax models, by Fawcett (1918*a*) and by Michelsson (1922). (Plate 117).

The chondrocranium of Erinaceus closely resembles that of Talpa and

Sorex, but is relatively broader and shorter. As in them, the skull base is straight and there is no dorsum sellae, the nasal capsules have very expanded frontal prominences, and an ala minima is present. The preoptic and metoptic roots of the orbital cartilage are attached to the central stem and a small ala hypochiasmatica is attached to the metoptic root. The orbitoparietal commissure is stout; the basicochlear fissure is obliterated, and there is no processus opercularis.

The ala temporalis has a bifid lamina ascendens and a processus pterygoideus in addition to the secondary pterygoid cartilage.

Two regions of the chondrocranium are especially interesting. Posteriorly, the occipital arch is fused with the auditory capsule above and below, enclosing a long inferior occipitocapsular fissure. But the dorsal end of the occipital arch is not fused to the supraoccipital cartilage and tectum posterius, so that the superior occipitocapsular fissure is open posteriorly. Anteriorly, the posterolateral corners of the lamina transversalis anterior are produced sideways on each side into Fawcett's transverse processes, which lie ventrally to the lachrymonasal ducts. As in Sorex, the anterior part of the lamina transversalis anterior is at a more ventral level than the hinder part. The thyroid cartilages are very wide.

IV. Tupaja javanica

The chondrocranium of the tree-shrew at the 20 mm. stage has been studied with the help of a wax model by Henckel (1928*b*). (Plate 130, figs. 1, 2).

In spite of the flatness of its floor, in general form the chondrocranium of Tupaja differs from those of Talpa, Sorex, or Erinaceus not only in the fact that the brain-case is relatively much greater than the nasal capsules, but also in several morphological points. The occipital arches slope backwards at an angle of some 45° from the plane of the basal plate; the superior and inferior occipitocapsular fissures are almost obliterated. The preoptic and metoptic roots of the orbital cartilage are inserted separately on the central stem, the preoptic root being very broad. An ala minima is present. The central stem beneath these roots forms a shallow interorbital septum. The roof of the nasal capsule projects back in the midline above the nasal septum and between the olfactory foramina as a 'protuberantia ethmoidalis posterior', spina mesethmoidalis, or crista galli not unlike that of the bat Miniopterus (see p. 349). The nasal capsules are not pear-shaped, but lozenge-shaped with very narrow anterior and posterior laminae transversales. The paraseptal cartilages are incomplete behind, and they possess an 'outer bar' enclosing Jacobson's organ (Ärnbäck-Christie-Linde, 1914; Broom, 1915*a*). There is no cartilago ductus nasopalatini, but an isolated ectochoanal cartilage is present. A large papillary cartilage is present.

No study has been made of the development of the bony skull of Tupaja, but Clark (1925) has advanced a number of reasons for believing (with Broom) that Tupaja and the Menotyphla have little affinity with the Lipotyphla or 'true' Insectivora. In possessing an endotympanic, Tupaja (and Macroscelides) shows similarity with Cheiroptera and Dermoptera (van der Klaauw, 1922).

V. Other 'Insectivora'

W. K. Parker's (1885) work contains descriptions and figures of late stages of development of the skulls of Centetes, Microgale, Ericulus, and Rhynchocyon. Broom has studied the paraseptal and associated cartilages in Macroscelides (1902*b*), and in Gymnura, Centetes, and Chrysochloris (1915*a* and *b*). In Macroscelides and in Chrysochloris an 'outer bar' is present enclosing Jacobson's organ. Macroscelides has an endotympanic bone (van Kampen, 1905).

Wortman's (1920) description of the skull of Rhynchocyon claims the existence of a number of reptilian bones (prefrontal, postfrontal, &c.) the existence of which, and percentage occurrence, require further study and confirmation.

The skull of a new-born Chrysochloris hottentota has been studied by Broom (1916). At this stage the chondrocranium has passed its optimum stage of development, but the bony skull shows many features of interest, such as the presence of paired tabulars, supra-angulars, extensions of the basisphenoid ventrally to the ala temporalis to form basipterygoid processes with which the pterygoids articulate, and an enigmatical dermal ossification ventral to the cupola posterior of the nasal capsule. Chrysochloris has no mesethmoid bone, an ossification which is present in Lipotyphla and Menotyphla.

27. *CARNIVORA*

I. Felis domestica

The development of the chondrocranium of the cat has been studied by Decker (1883) and by Wincza (1896) among the earlier workers, and more recently by Kernan (1915) and Terry (1917), using the van Wijhe technique and wax models. A model has also been reconstructed by Fawcett (1918*a*), and by the present writer. The cartilages of the nasal capsule have been studied by Zuckerkandl (1909) and by Steinberg (1912). The development of the bony skull has been followed by Drews (1933).

i. The development of the chondrocranium of Felis

At the 10 mm. stage the hinder parts of the parachordals are present as paired and separate cartilages, united anteriorly by a transverse commissure passing ventrally to the notochord. Subsequently, another such commissure is developed between the hindmost parts of the parachordals, and later still according to Kernan (1915) two more intermediate commissures may be developed, making four in all. It is possible, as Terry suggests, that these structures may represent the metameric segmentation of the hypochordal (hypocentral) elements of four occipital vertebrae. (Plate 118, fig. 1).

The occipital arches develop in continuity with the hindmost part of the parachordals, and one hypoglossal foramen becomes enclosed on each side. Its anterior border is clearly a preoccipital arch, but not the one immediately preceding the occipital arch as three nerve-roots pass through the foramen. Kernan (1915) has observed three mesenchymatous preoccipital arches at the

9 mm. stage. Beneath and in front of the tip of the notochord is the crescent-shaped hypophysial cartilage, embracing the hypophysial duct from behind.

The only other chondrification at the 10 mm. stage is the rudiment of the side wall of the canalicular part of the auditory capsules.

At the 12 mm. stage the parachordals are still imperfectly connected posteriorly, but anteriorly they extend forwards along the notochord, fused above and below it, towards the independent procartilaginous acrochordal or dorsum sellae which arise at its tip. The supraoccipital cartilages ('lamina parietales' of Terry) are present as isolated structures above and slightly in front of the occipital arches. A medially directed process from each will give rise to the tectum posterius. The interorbitonasal part of the central stem is now chondrified as a slender bar in the midline, as yet unconnected with the hypophysial cartilage. The ala temporalis is present as an isolated cartilage on each side, and in the visceral arch skeleton, the incus and Meckel's cartilage are present. (Plate 118, fig. 2).

At the 15 mm. stage the central stem is continuous from front to back. It is very thin in the region of the nasal septum and of the anterior part of the basal plate, but it is expanded laterally in the interorbital and hypophysial regions. Anterodorsally to the auditory capsules the parietal plates ('orbito-parietal commissures' of Terry) are present as independent chondrifications. Meanwhile, the supraoccipital cartilages have fused with the occipital arches, each on its own side. The orbital and paranasal cartilages are present as independent chondrifications, and the parietotectal cartilages of the nasal capsule grow out on each side from the dorsal edge of the nasal septum. (Plate 118, fig. 3).

At the 17 mm. stage numerous advances have been made. The basal plate is at last continuous from side to side throughout its length. The supra-occipital cartilages are interconnected by the tectum posterius, and fused with the parietal plate of their own side. The cochlear part of the auditory capsule chondrifies in continuity with the canalicular part, and quite independently of the basal plate. The auditory capsule becomes connected with the basal plate by basivestibular and sphenocochlear commissures, enclosing a large basicochlear fissure. A processus alaris is also present, connected with the cochlear capsule by the alicochlear commissure, but the ala temporalis is still independent. The orbital cartilage is attached to the central stem by its preoptic and metoptic roots, with the nasal capsule by the sphenethmoid commissure, and with the parietal plate by the orbitoparietal commissure. The connexion between the parietal plate and the auditory capsule is still procartilaginous, but the dorsal end of the prefacial commissure is in perfect cartilaginous continuity with the orbitoparietal commissure. For some time the connexion between the dorsal end of the suprafacial commissure and the canalicular capsule is imperfect. (Plate 118, fig. 4).

The ventral end of the prefacial commissure is continuous with a tract of cartilage which is fused on to the roof of the cochlear capsule and reaches forwards almost to the alicochlear commissure. This tract must represent the lateral part of the basal plate almost obliterated by the great expansion of the cochlear capsule, and is evidence that the cochlear capsule is *not* formed out of the basal plate. (Cf. the lamina supracochlearis in Cetacea, p. 341;

and p. 379.) The lamina orbitonasalis is present in the hind wall of the nasal capsule, but quite independent. The paranasal cartilage overlaps the lamina orbitonasalis behind and that part of the side wall of the capsule that is formed by the parietotectal cartilage in front. The lateral edge of the lamina orbitonasalis which thus projects freely into the cavity of the capsule medially to the paranasal cartilage forms the 1st ethmoturbinal; similarly, the freely projecting posterior edge of the parietotectal cartilage forms the crista semicircularis. The epiphanial foramen represents the original space between the paranasal cartilage and the parietotectal cartilage.

ii. The morphology of the fully formed chondrocranium of Felis

(23·1 mm. stage, Terry; 32 mm. stage, Fawcett) (PLATE 119).

1. The basal plate and notochord. The basal plate is wide posteriorly, but exceedingly narrow anteriorly, where the cochlear capsules approach one another so closely as almost to touch. Another result of this very great development of the cochlear capsules is that the basicochlear fissure, which is large posteriorly, is obliterated in front.

The basal plate ends anteriorly in a conical dorsum sellae, overlying a shallow sella turcica. With the anterior region of the central stem the basal plate forms an angle of some 50°, but with a gradual curvature. The notochord enters the hind part of the basal plate and at first lies immediately beneath the perichondrium, but for the remainder of its course it is completely enclosed within the basal plate.

2. The occipital region. The occipital arches arise perpendicularly from the hind part of the basal plate, and the condyles into which the base of each arch is produced posteriorly have faceted cavities which are paired and separate from the start. The form of the condyles shows that they are arcual and not central elements.

Each occipital arch is broadly connected with the posterior wall of the auditory capsule, and is produced into a large paracondylar process. According to Kernan (1915) this structure represents the transverse process and rib, not of the occipital, but of a preoccipital arch.

The single hypoglossal foramen transmits three nerve-roots. Its anterolateral margin is raised into a high crest, the tuberculum jugulare, which also forms the posterior border of the jugular foramen.

The tectum posterius is narrow, and the foramen magnum has a large incisura occipitalis posterior. The lateral borders of this incisure are formed by the supraoccipital cartilages, while the borders of the foramen magnum proper are formed by the occipital arches.

3. The auditory region. The auditory capsules are characterized by the great development and medial position of the cochlear portions, and by their rounded contours. Externally the crista parotica has the styloid process fused to it, but there is neither mastoid process nor tegmen tympani. The suprafacial commissure is very massive, and lies in a continuation of the orbitoparietal commissure. Dorsally, a superior occipitocapsular fissure is enclosed between the capsule, the parietal plate, and the supraoccipital cartilage. It is sometimes subdivided into two. The processus recessus develops from the medial edge of the foramen perilymphaticum and projects posterolaterally, ultimately separating a fenestra rotunda from an aquæductus cochleae. In the interior of the capsule, the cavum vestibulare is constricted into anterior and posterior portions by a crista intravestibularis (cf. Lacerta, p. 223) which projects inwards from the lateral wall.

The parietal plates are at first low, but they increase in height with development, and at later stages they curve inwards over the cranial cavity.

4. The orbitotemporal region. The processus alaris and alicochlear commissure curve around a large carotid foramen on each side. The ala temporalis has a lamina ascendens which is perforated by a foramen rotundum (for the ramus maxillaris trigemini) and a pterygoid process; it is in contact but not fused with the processus alaris.

The metoptic roots of the orbital cartilages rise from the central stem on each side of the sella turcica and run diagonally forwards and outwards; the optic foramina are very large and the preoptic roots of the orbital cartilages have exceptionally broad attachments to the central stem.

At the 32 mm. stage, an ala minima from each preoptic root fuses with the lamina orbitonasalis in front of it, enclosing a small foramen infracribrosum ('prechiasmaticum' of Fawcett, see p. 389).

From the dorsal surface of the base of each metoptic root at the 32 mm. stage an anterior clinoid process arises as a blunt knob projecting upwards, while a small paroptic process (ala hypochiasmatica) projects outwards from its anteroventral surface, and serves as a point of origin for the superior, internal, and external recti muscles. The inferior rectus and retractor bulbi arise from the central stem ventrally to the metoptic root; the superior oblique arises from the preoptic root, and the inferior oblique from the lamina orbitonasalis.

A vestige of the true side wall of the neurocranium in the orbitotemporal region is found at late stages (70 mm.) in the form of a cartilaginous rod situated in the dura mater dorsally to the roots of the trigeminal nerve and representing a part of the pila antotica (see p. 390).

The central stem is very flat between the roots of the orbital cartilages, and there is no semblance of an interorbital septum.

The sphenethmoid and orbitoparietal commissures are both stout.

5. The ethmoid region. The nasal capsules are relatively short and occupy about one-third of the total length of the chondrocranium. Their outlines are smooth and rounded, the maximum width being about half-way along their length. The cupola posterior, formed by the lamina orbitonasalis, is fused medially with the central stem. Two epiphanial foramina are present on each side.

Anterior and posterior laminae transversales are present. The paraseptal cartilages are at first isolated, but at late stages (70 mm.) they come into contact but do not fuse with the lamina transversalis anterior, while posteriorly they reach back and fuse with the lamina transversalis posterior. Complete paraseptal cartilages are thus present. A cartilago ductus nasopalatini projects forwards from each paraseptal cartilage ventrally to the lamina transversalis anterior and medially to the duct of Jacobson's organ. Laterally to the latter, a small ectochoanal cartilage projects back from the lamina transversalis anterior.

A processus maxillaris anterior projects forwards from the ventral edge of the side wall of the capsule, ventrally to the lachrymonasal duct. The latter runs forwards ventrolaterally to the lamina transversalis anterior, and enters the capsule through the posteroventral corner of the fenestra narina.

The arrangement of the interior of the nasal capsule, of the crista semicircularis, and of the atrio-, naso-, maxillo-, and ethmo-turbinals, and the various recesses, is similar to that in Lepus (see p. 312).

6. The visceral arch skeleton. An interesting feature of the visceral arch skeleton is the presence of a small cartilage first described by Spence (1890) whose name has been given to it by van der Klaauw (1923). Spence's cartilage is an isolated horizontal rod of cartilage lying in the sagittal plane immediately in front of the styloid process, and dorsally to the tympanic membrane, pointing towards the malleus. The chorda tympani after running forwards laterally to the styloid process, runs laterally and then ventrally to Spence's cartilage on its way to the medial side of the malleus. Spence's cartilage may represent the processus internus of the columella auris.

iii. The development of the osteocranium of Felis

The order of appearance of the bones of the skull can be made out as follows, from Drew's (1933) researches:

46 mm., 28 days p.c.:
parietal, frontal, vomer (paired), *premaxilla, maxilla, pterygoid* (dermal bone and ossification in secondary pterygoid cartilage), *palatine, jugal, lachrymal, dentary.*

63 mm., 31 days p.c.:
squamosal, nasal, tympanic.

65 mm., 32 days p.c.:
interparietal (paired); *basioccipital, exoccipital.*

71 mm., 35 days p.c.:
supraoccipital, alisphenoid.

84 mm., 38 days p.c.:
orbitosphenoid.

101 mm., 47 days p.c.:
basisphenoid (paired?); *presphenoid* (paired).

115 mm., 49 days p.c.:
cochlear centre of periotic; malleus, incus, tentorium cerebelli (according to Bayer (1897) developed in connexion with the parietal).

130 mm., 53 days p.c.:
stapes, ethmoturbinals.

154 mm., 2 days p.p.:
basihyal.

169 mm., 5 days p.p.:
mastoid centre of periotic; stylohyal.

178 mm., 10 days p.p.:
epihyal.

215 mm., 20 days p.p.:
thyrohyal.

216 mm., 25 days p.p.:
ceratohyal.

310 mm., 81 days p.p.:
mesethmoid.

The *endotympanic* as described by Wincza ('metatympanic') (1896) appears to correspond to the caudal endotympanic element described by van der Klaauw (1922) in Canis. It develops as an ossification in a cartilage which arises in the floor of the tympanic cavity shortly before birth. It was in the Felidae that the endotympanic was first discovered, by Flower (1869) as a 'tympanic' bone preformed in cartilage.

II. Canis familiaris

The chondrocranium of the dog has been studied with the help of model reconstructions by Olmstead (1911). The development of the bones has been followed by Drews (1930).

i. The morphology of the fully formed chondrocranium of Canis (27 mm. stage) (PLATE 120).

The basal plate is wide posteriorly and fairly narrow anteriorly where it is compressed between the cochlear capsules. Basicochlear fissures are present. According to Tourneux and Tourneux (1912), the notochord lies on the dorsal surface of the hinder part of the basal plate and on the ventral surface of the anterior part, rising up again to end in an acrochordal or dorsum sellae which is independent at early stages. The basal plate makes an angle of over 90° with the plane of the anterior region of the central stem.

The occipital arches slant markedly backwards; each has a well-developed lamina alaris, ending ventrolaterally in a large paracondylar process. Between the lamina alaris and the auditory capsule is an extensive inferior occipitocapsular fissure which is confluent with the jugular foramen; both being relics of the fissure metotica. A small tectum posterius is present. The parietal plates, between which and the auditory capsules there are extensive parietocapsular (superior occipitocapsular) fissures, end freely in front, for there are no orbitoparietal commissures.

The styloid process is fused with the crista parotica, but there is no tegmen tympani.

The ala temporalis is in contact but not fused with the processus alaris, and is pierced by the foramen rotundum for the maxillary branch of the trigeminal nerve, and by the 'alisphenoid canal' for the orbital artery. The orbital cartilages are reduced and end freely posteriorly, but they have well-developed preoptic and metoptic roots, inserted in the central stem wide apart, the latter bearing a small paroptic process (ala hypochiasmatica).

The nasal capsules are long and tapered in front and behind. The fenestrae narinae are large and there are no cupolae anteriores. The lamina transversalis anterior is wide, resulting in a broad zona annularis, but there is only a very small lamina transversalis posterior. The anterior paraseptal cartilages are fused anteriorly, not with the lamina transversalis anterior, but with the ventral edge of the nasal septum behind the lamina. Behind these and separate from them are small posterior paraseptal cartilages which project forwards from the lamina transversalis posterior.

Spence's cartilage has been found by van Kampen (1915*a*).

ii. The development of the osteocranium of Canis

The order of appearance of the bones of the skull can be made out as follows, from Drews' results:

In 5th week p.c.:
premaxilla.

70 mm., 40 days p.c.:
parietal, *frontal*, *nasal* (from two centres on each side), *maxilla*, *palatine*, *pterygoid*, *jugal*, *lachrymal*, *squamosal*, *tympanic*, *dentary*, *basioccipital*, *exoccipital*, *supraoccipital*, *alisphenoid*, *orbitosphenoid*, *presphenoid* (paired).

76 mm., 42 days p.c.:
vomer (paired).

105 mm., 45 days p.c.:
interparietal (median, and immediately fused with supraoccipital) *preinterparietal* (paired), *basisphenoid*.

122 mm., 54 days p.c.:
malleus, *incus*, *stapes*.

189 mm., 62 days p.c.:
tentorium cerebelli (developed according to Bayer in connexion with the interparietal), *pre-preinterparietal* Wormians, *ossiculum Kerckringii* (fused with supraoccipital), *cochlear* centre of *periotic*, *ethmoturbinals*.

298 mm., 10 days p.p.:
mastoid centre of *periotic*.

First year:
mesethmoid.

The blastematous rudiment of the cartilage in which the *caudal endotympanic* subsequently ossifies was observed by van der Klaauw (1922) in an embryo of 43 mm. head-length.

III. Poecilophoca weddelli

The chondrocranium of the Seal at the 27 mm. stage has been studied with the help of wax models by Fawcett (1918*b*). A brief description without figures of the skull of the embryo of Phoca groenlandica of 17 mm. head-length was given by Decker (1883). (Plate 121, figs. 1, 2).

In general form the chondrocranium is remarkably wide and short, the cranial cavity being roughly spherical.

The notochord lies on the dorsal surface of the hind end of the anterior region of the basal plate, and dips beneath the perichondrium in the intermediate region. Wide posteriorly, the basal plate is narrow anteriorly, between the cochlear capsules, from which, however, it is separated by wide basicochlear fissures. A conical dorsum sellae is present.

The occipital arches are stout and short and end freely dorsally, where they are covered over by the supraoccipital cartilages, themselves interconnected by the tectum posterius. Only at the hindmost point of the contact between the occipital arches and the supraoccipital cartilage is there fusion between these elements. Each occipital arch has a lamina alaris and a fairly well developed paracondylar process.

The auditory capsules are well rounded and have relatively small cochlear portions which lie much farther in towards the midline than the canalicular portions, i.e. the axes of the capsules diverge very widely behind.

The auditory capsule is remarkably free from neighbouring structures, to which it is attached only by the basivestibular, sphenocochlear, alicochlear, and parietocapsular commissures. The jugular foramen is confluent with an occipitocapsular fissure which thus extends all round the canalicular part of the capsule as far as the parietocapsular commissure. There is no tegmen tympani, and, at this stage, at least, no suprafacial commissure.

The parietal plate is high, and where it joins the supraoccipital cartilage behind it, a plate of cartilage projects downwards, laterally to the auditory capsule, forming the processus opercularis. The orbitoparietal commissure is stout, and where it joins the orbital cartilage a process rises up on each side and projects towards a pair of fused cartilages in the roof of the cranial cavity. These structures, the 'tectum anterius' of Fawcett, are relics of the tectum transversum.

The ala temporalis is small and imperforate; it is in contact but not fused with the processus alaris. The orbital cartilage has well-developed preoptic and metoptic roots, the latter bearing a backwardly projecting anterior clinoid process; the former bearing an ala minima which fuses with the lamina orbitonasalis and cuts off the medial portion of the orbitonasal fissure as a foramen infracribrosum ('prechiasmaticus' of Fawcett). On each side of the sella turcica a middle clinoid process rises up from the central stem.

The nasal capsule is broad and short. The lamina transversalis anterior projects sideways from the ventral edge of the nasal septum, but fails to reach the side wall; there is thus no zona annularis. The paraseptal cartilages end freely behind; anteriorly, each projects into a small cartilago ductus nasopalatini.

The epiphanial foramen is large and slit-like.

The cavity of the nasal capsule at this stage is simple, there being only a shallow crista semicircularis and two ethmoturbinals.

IV. Other Carnivora

The chondrocranium of a thirty-day embryo of Mustela domestica (the ferret) has been modelled and figured, but not described, by Fawcett (1918*b*). It shows numerous close similarities with that of Felis. Thus, the hypoglossal foramen is bordered anteriorly by a crest-like tuberculum jugulare; the auditory capsule has rounded surfaces, with the cochlear part pushed well inwards; the suprafacial commissure lies in the curved continuation of the orbitoparietal commissure; a conical dorsum sellae is present; the ala temporalis is perforated by a foramen rotundum; an ala minima projects forwards from the preoptic root of the orbital cartilage and, fusing with the lamina orbitonasalis, encloses a foramen infracribrosum. The turbinals in the nasal capsule are remarkably well developed. (Plate 121, fig. 3).

The relations of the styloid process and its attachments to the crista parotica have been described for Putorius putorius by van Kampen (1907), who found them to be normal (contrary to previous descriptions). The same is true of Mustela (Brauer, 1906).

A brief description without figures of the chondrocranium of a bear embryo of 15 mm. head-length was given by Decker (1883). In the polar bear, Wincza (1896) showed that the ala temporalis remains separated from the processus alaris by a partition of connective tissue.

28. *UNGULATA ARTIODACTYLA*

I. Sus scrofa

In addition to Spöndli's (1846), W. K. Parker's (1874), and Decker's (1883) early work, the development of the chondrocranium of the pig has been studied with the van Wijhe technique by Noordenbos (1905), and with the help of wax models, in its early stages by Lebedkin (1918), and in its fully formed state by Mead (1909). The roof has been studied by Augier (1936), the nasal capsule by Sturm (1937), and the visceral arches by Kallius (1910). The development of the bony skull has been followed in detail by Augier (1934*a*), while certain points have been studied by Forster (1902), Engelmann (1910), Andres (1924), Vogler (1926), and Neukomm (1933).

i. The development of the chondrocranium of Sus

The notochord is completely enclosed within the parachordal plate which, according to Noordenbos (1905), is unpaired from the start. The tectum posterius arises as a transverse bar, the ends of which subsequently fuse with a supraoccipital cartilage ('lamina suprapilaris') on each side, which in turn fuses with the dorsal end of the occipital arch. The parietal plates chondrify in continuity with the orbitoparietal commissures, and subsequently fuse with the supraoccipital cartilages, but for a long time without establishing contact with the auditory capsule. In this manner, the sphenoparietal fontanelle is confluent with the superior occipitocapsular fissure. The auditory capsule begins to chondrify in its lateral wall through which at the start are many small perforations.

A transverse septum of connective tissue cutting through the central stem indicates the position of the original discontinuity between the interorbitonasal and the hypophysial cartilages. The hind end of the interorbitonasal cartilage at its earliest appearance is forked, according to Noordenbos. The ala temporalis has an independent chondrification centre.

ii. The morphology of the fully formed chondrocranium of Sus

(30 mm. stage; head-length 12 mm.) (PLATE 122, FIGS. 1, 2).

1. The basal plate and notochord. The notochord conforms to the type described by the Tourneux (1912) as intrabasal. The basal plate is fairly wide, and ends anteriorly in a well-marked dorsum sellae from the sides of which cartilaginous clinocochlear commissures (abducens bridges) connect with the cochlear capsules, dorsally to the abducens nerves. These structures correspond in part to the supracochlear cartilages found in other forms (see p. 390) and represent portions of the original side wall of the neurocranium, possibly of the pila antotica.

The sides of the basal plate are at early stages in contact with the cochlear capsules, but at later stages extensive basicochlear fissures are found separating them.

The plane of the basal plate forms an angle of some 50° with the anterior regions of the central stem.

2. The occipital region. The occipital arch is fused to the auditory capsule above and below, thus enclosing an extensive inferior occipitocapsular fissure. The supraoccipital cartilage is also connected with the auditory capsule by a commissure which thus cuts off the hindmost part of the superior occipitocapsular fissure, forming the 'mastoid' foramen.

The lamina alaris of the occipital arch is well developed, and there is a relatively enormous paracondylar process. One pair of hypoglossal foramina are present, but these may be subdivided by connective tissue septa (representing unchondrified preoccipital arches) and transmit more than one nerve-root.

The tectum posterius is broad and lies in the vertical transverse plane. Immediately in front of it the supraoccipital cartilages are interconnected by a slender tectum intermedium of which the median portion persists as an isolated ('supratectal') cartilage after the break-down of its lateral connexions.

3. The auditory region. The cochlear part of the auditory capsule is relatively large; the canalicular part is perforated right through its thickness from the fossa subarcuata to the outside by a number of canals occupied by veins. The crista parotica has the styloid process attached to it at first by connective tissue and later (38 mm.) fused, and projects forwards into a small tegmen tympani. The abducens bridge and the basicochlear fissure have been mentioned above. There is no mastoid process.

The parietal plate is very large, and separated from the capsule by an extensive parietocapsular fissure (superior occipitocapsular or foramen jugulare spurium). The parietal plate grows progressively during development until at the 100 mm. (head-length 38 mm.) stage it meets its fellow of the opposite side above the hind brain but does not fuse with it (Augier, 1936). For practical purposes, therefore, there is a tectum synoticum. Eventually the parietal plate is destroyed and absorbed.

4. The orbitotemporal region. The pituitary fossa is very deep. The ala temporalis is slender and consists of a lamina ascendens and a pterygoid process; there is no alicochlear commissure, and the internal carotid artery lies in an incisure, not a foramen. The orbital cartilage has stout preoptic and metoptic roots, inserted wide apart on the central stem which in this region assumes the form of an interorbital septum. A small ala hypochiasmatica is attached to the metoptic root. The orbitoparietal commissure is broad; the sphenethmoid commissure slender.

5. The ethmoid region. The nasal capsule occupies about one-third of the total length of the chondrocranium. Anterior and posterior laminae transversales are present; from the former, very short paraseptal cartilages project back. On the side wall of the capsule, about one-third of its length from the hind end, is a processus maxillaris posterior.

6. The visceral arch skeleton. The styloid process, or Reichert's cartilage, is really free for it fails to reach the body of the hyoid distally, and proximally it is only attached to the crista parotica by connective tissue. Mead has established some interesting relations of magnitude of the auditory ossicles. When the head is only 12 mm. long, the incus has reached nearly one-half and the malleus one-third of the length of these structures in the adult. The auditory ossicles are full sized in the newborn.

Paauw's cartilage (representing the hyostapes, see p. 412) has been found laterally to the stapes in the stapedial muscle tendon by van der Klaauw (1923). Spence's cartilage, perhaps representing the processus internus of the columella auris, has been described by Bondy (1907). The thyroid cartilage represents the skeleton of the 4th visceral arch only, according to Kallius (1910).

iii. The development of the osteocranium of Sus

(in the order of their appearance: it will be noted that the last membrane-bone to arise appears at 44 mm., and the earliest cartilage-bone at 40 mm. greatest length.) All measurements are greatest length in direct line unless otherwise stated. (Plate 122, figs. 3, 4, 5).

1. The membrane-bones. The *dentary* is the earliest ossification to appear in the whole body. It arises at 31 days (48 mm. dorsal contour length) as a frail splinter; at 33 days another splinter arises in the angular process.

The *frontal* arises at 24 mm. from a single centre on each side according to Augier (contra Neukomm).

The *premaxilla* arises at 36 days (50 mm. dorsal contour length). According to Biondi (1886) the palatine process has an independent centre of ossification, presumably representing a *prevomer*.

The *palatine* also arises at 36 days.

The *maxilla* arises at 38 days (55 mm. dorsal contour length), and apparently from four centres on each side; in the facial portion, in the anterior and middle parts of the alveolar portion, and in the palatine process. At about 49 days a 5th centre appears in the hinder part of the palatine process and gives rise to the tuber maxillae.

The *vomer* also arises at 38 days, from paired centres.

The *jugal* likewise arises at 38 days.

The *lachrymal* arises at 40 days (59 mm. dorsal contour length).

The *parietal* and the *squamosal* arise at 35 mm., from single centres on each side according to Augier (contra Neucomm).

The *prearticular* arises at 38 mm. as a narrow strip apposed to the hinder part of Meckel's cartilage; it becomes lamelliform and curves round the medial surface of Meckel's cartilage and develops a large processus internus (which may, if it has an independent ossification centre, represent the supra-angular). The prearticular fuses with the malleus at 88 mm.

The *tympanic* arises at 38 mm. and forms a ring which at 130 mm. extends inwards to form the bulla, and at 170 mm. outwards to form the external auditory meatus. At 190 mm. it fuses with the epitympanicomastoid process of the squamosal.

The *pterygoid* arises at 40 mm. in the secondary pterygoid cartilage and extends as an intramembranous plate above it.

The *nasal* arises at 44 mm. (60 dorsal contour) from two centres on each side.

The *prenasal* (os rostri) does not ossify until after birth.

2. The cartilage-bones. The *exoccipital* arises from a main centre at 40 mm. in the occipital arch, and from accessory centres at 76 mm. in the pre-occipital arch ('suprahypoglossal centre'), at 100 mm. in the paracondylar process, and at 140 mm. at the tip of the paracondylar process. The separate ossification in the preoccipital arch may be regarded as a manifestation of the metameric segmentation of the occipital region, reflected in the ossification of the neural arch of an absorbed vertebra.

The *basioccipital* arises from a median main endochondral centre in the hinder part of the basal plate at 43 mm. A perichondral lamella on the ventral surface of the basal plate appears at 45 mm. Additional exceptional endo-chondral centres may be found in the basiotic centre at 70 mm. (in front of the main centre) and the 'interdorso-occipital' centre at 152 mm. (between the basioccipital and basisphenoid).

The *basisphenoid* ('basipostsphenoid') arises from a main median endo-chondral centre at 47 mm. in the hypophysial cartilage and floor of the sella turcica, and invades the dorsum sellae. At 95 mm. accessory alar endochondral centres ossify in the processus alares, and soon fuse with the central basi-sphenoid centre and the alisphenoids to form the 'postsphenoid' bone.

The *orbitosphenoid* ('alipresphenoid') arises in the orbital cartilage laterally to the optic foramen at 55 mm., and fuses with the presphenoid centre of its own side at 80 mm. An occasionally independent perichondral 'internal orbital' centre may arise at 100 mm. on the lateral surface of the preoptic root of the orbital cartilage, and spread over the lamina orbitonasalis and cupola posterior of the nasal capsule. It remains to be determined whether this structure is comparable with the ossiculum Bertini of Man (see p. 372).

The *alisphenoid* ('alipostsphenoid') arises at 60 mm. as an ossification partly endochondral in the lamina ascendens of the ala temporalis, and partly perichondral and intramembranous. It fuses with the basisphenoid at 95 mm. There is no evidence of any 'intertemporal' membrane-bone fused with the alisphenoid.

The *supraoccipital* arises as a main median endochondral centre in the tectum posterius at 65 mm. This develops on its external surface only a perichondral lamella which extends forwards as an intramembranous ossification. There is no interparietal at all. An endochondral centre may ossify in the supratectal cartilage at 73 mm., and become fused to the under surface of the intramem-branous lamella; it disappears at 160 mm. (Augier, 1937).

The *presphenoid* ('basipresphenoid') arises at 67 mm. as a paired endochondral centre in the central stem at the root of the pila metoptica. At 80 mm. it fuses with the orbitosphenoid of its own side, and at 85 mm. with its fellow of the opposite side. The 'presphenoid bone' is then complete.

The *malleus* begins to ossify at 85 mm. by a centre in the neck and another at 100 mm. in the head of the cartilaginous malleus.

The *thyrohyal* ossifies at 95 mm. in the cartilage of similar name.

The *cochleo-canalicular* centre of the *periotic* appears at 120 mm. round the fenestra rotunda and extends towards the fenestra ovalis, in the posteroventral region of the canalicular capsule, the posterior semicircular canal, and the cochlear capsule. Accessory centres may be found in the lining of the internal auditory meatus, at the apex of the cochlea, and there may be an intermediate centre for the posterior semicircular canal.

The *anterior canalicular* centre of the *periotic* appears at 125 mm. in the dorsal border of the fenestra ovalis, and extends in the lateral wall of the canalicular capsule, the crista parotica and tegmen tympani, and in the anterior and lateral semicircular canals.

The *incus* also ossifies at 125 mm.

The *posterior canalicular* centre of the *periotic* appears at 130 mm. in the medial wall of the canalicular capsule, and extends around the rim of the internal auditory meatus.

It may be noted that while the cartilaginous cochlear capsule appears later than the canalicular, it becomes ossified sooner.

The *stapes* ossifies at 130 mm. from one centre in the footplate, and one in each of the arms of the stirrup.

The *stylohyal* usually ossifies in the middle portion of the styloid process at 185 mm., but it may appear as early as the 85 mm. stage, or as late as the time of birth. It remains separated by cartilage from

The *tympanohyal*, which ossifies in the uppermost part of the styloid process at 187 mm.

The *maxilloturbinal*, *ethmoturbinals*, and *nasoturbinal* begin to ossify at 190 mm. The cribriform plate and crista galli are not ossified before birth. There is no mesethmoid.

The *basihyal* is sometimes ossified at birth, the *ceratohyal* and *epihyal* not until later.

II. Bos taurus

The chondrocranium of the ox was studied by Decker (1883), and two stages in its development have been modelled and figured, but not described by Fawcett (1918*a*). The nasal capsule has been studied by Sturm (1937).

The chondrocranium of Bos

At the 19 mm. stage the basal plate, hypophysial cartilage, and interorbitonasal part of the central stem are chondrified and fused, but wide fissures remain to indicate the limits of their former extents. These three parts of the skull floor are almost in line, i.e. the floor is nearly flat. (Plate 123, fig. 3).

The basal plate is very wide, and completely fused with the cochlear capsules; there are no basicochlear fissures. The notochord is enclosed within the basal plate throughout its length (Tourneux, 1912). The occipital arches lean slightly forwards, and are surmounted by the supraoccipital cartilages, as yet unconnected by any tectum posterius. The lateral wall of the cochlear capsule is still incomplete. The suprafacial commissure develops from in front backwards. The parietal plate is fused to the anterior wall of the canalicular capsule, and attached to the orbital cartilage by the orbitoparietal commissure.

The alicochlear commissures are stout, and the alae temporales are quite independent.

The orbital cartilages are large and attached to the central stem, which is very wide in this region, by stout preoptic and metoptic roots, to the latter of which alae hypochiasmaticae are attached.

The sphenethmoid commissure is incomplete and grows back from the roof of the nasal capsule. The latter has a fairly well developed roof and side wall, but the lamina orbitonasalis is still quite free on each side.

At the 40 mm. stage, where the chondrocranium is fully formed, the most notable advances on the previous stage are the following: (Plate 123, fig. 4).

Very large paracondylar processes are developed from the occipital arches. A tectum posterius is present and each supraoccipital cartilage is fused with the parietal plate of its own side, but without touching the auditory capsule, and the superior and inferior occipitocapsular fissures are large and confluent.

The ala temporalis has a simple lamina ascendens and a pterygoid process. The orbitonasal fissure is remarkably elongated and horizontal. The nasal capsule has an elongated fenestra dorsalis in its roof; a lamina transversalis anterior is present and large anterior paraseptal cartilages.

Pauuw's cartilage is present (van der Klaauw, 1923).

No systematic study has been made of the development of the osteocranium. Wilhelm (1921) has shown the existence of *interparietals* and *preinterparietals*. The *supraoccipital* usually arises from paired centres (Augier, 1931*b*).

PERISSODACTYLA

III. Equus caballus

The chondrocranium of the horse has been studied with the van Wijhe technique by Noordenbos (1905), and by Limberger (1925), Arnold (1928), and Muggia (1931*b*); the two last with the help of wax models. The cartilages of the nasal capsule were described by Spurgat (1896). (Plate 123, figs. 1, 2).

The chondrocranium of Equus (36 mm. and 40 mm. stages)

The basal plate is fairly wide and rectangular, and completely fused with the cochlear capsules; there are no basicochlear fissures. Anteriorly, the basal plate has a small dorsum sellae on its dorsal surface. The notochord runs dorsally to the posterior part of the basal plate and enclosed within the anterior part (Tourneux, 1912). The basal plate and the hypophysial cartilage are in the same plane, but the nasal capsule is bent sharply ventrally.

The occipital arches rise fairly straight up from the basal plate and bear marked paracondylar processes. Dorsally, they are fused with the supra-occipital cartilages which are themselves joined by the tectum posterius and fused with the auditory capsules and the parietal plates. The superior and inferior occipitocapsular fissures are separate and small. The tectum posterius lies in the vertical transverse plane, and immediately above it is a small median isolated supratectal cartilage.

The crista parotica of the auditory capsule has the styloid process fused to it, and it projects anteriorly to form a tegmen tympani, and posteriorly into a mastoid process.

The suprafacial commissure is massive. The parietal plates are high, and bend in towards one another over the cranial cavity.

The ala temporalis, which according to Noordenbos chondrifies independently, is a simple knob, pierced by the orbital artery (passing through the 'alisphenoid canal'), and fused to a slender processus alaris, from the base of which a stout alicochlear commissure runs to the cochlear capsule.

The central stem in the interorbital region is very wide, and the preoptic and metoptic roots of the orbital cartilage are inserted on it fairly far apart. The orbitoparietal commissure is stout; the sphenethmoid slender.

The nasal capsule is elongated and tubular, and bent downwards about half-way along. The epiphanial foramen is large and slit-like. The ventral edge of the lamina orbitonasalis is bent forwards to form a small lamina transversalis posterior. A slender lamina transversalis anterior is present at the 40 mm. stage, from which long (but incomplete) paraseptal cartilages extend backwards. In the adult, Spurgat described a cartilago ductus nasopalatini projecting forwards ventrally to the premaxilla, a paraseptal cartilage projecting back medially, and a 'cartilago lateralis ventralis', which appears to be a detached part of the ventral edge of the side wall of the capsule (not an ectochoanal cartilage). It may be noted that these terms which are in general use for the nasal capsule are Spurgat's.

The development of the auditory ossicles has been studied by Coyle (1909), but his interpretation of the stapes as a derivative of the mandibular arch is unacceptable.

IV. Other 'Ungulata'

Of the development of the skull in the elephant, all that is known is Eales's (1932) observation that the mandible at early stages resembles the early stages of less specialized forms, and that the prealveolar portion of the mandible increases in length during ontogeny. There is no mesethmoid bone (Broom, 1935*c*).

In the Hyracoidea the relations of the styloid process, although described by Howes (1896) and van Kampen (1905), are still obscure, as is the question of the endotympanic bone perhaps observed by W. K. Parker (1874, p. 320). A mesethmoid bone is present in Hyrax (Broom, 1926*a*).

29. *CETACEA. MYSTACOCETI*

I. Megaptera nodosa

The development of the chondrocranium of the hump-back whale has been studied with the help of wax models by Honigmann (1915, 1917), while the fate of the chondrocranium and the development of the bones have been followed by Ridewood (1922).

i. The chondrocranium of Megaptera (92 mm. d.c.l. stage) (PLATE 124).

The basal plate is broad and flat, and its anterolateral corners form laminae supracochleares overlying and fused with the cochlear capsules. Basicochlear fissures are present at earlier stages; there is no marked dorsum sellae but only a low crista transversa. The notochord enters the basal plate posteriorly on its ventral surface, emerges on to the dorsal surface of the middle portion of the basal plate, dips beneath the dorsal surface for a short distance, and continues on the dorsal surface of the anterior portion of the basal plate. The basal plate and trabecular cartilage are more or less in line.

The occipital arches lean backwards from the basal plate and have laminae alares ending ventrally in very large paracondylar processes. The supraoccipital cartilages are large and convex, and fused with the parietal plates; the tectum posterius is small. No hypoglossal foramina are present, and the roots of that nerve emerge through the jugular foramen. The occipital condyles are at first paired and separate, but they ultimately fuse in the midline to form a single hemispherical knob.

The auditory capsules are rotated so that the cochlear portions are practically medial to the canalicular, and are depressed beneath the basal plate (laminae supracochleares). A further consequence of this rotation is the fact that the 'lateral' surface of the canalicular capsule faces forwards, and the fenestra ovalis faces downwards. The canalicular capsule is relatively small. The auditory capsules are almost free from neighbouring cartilages, for the superior and inferior occipitocapsular fissures are confluent with the jugular foramen. As is well known, in the adult the tympano-periotic bone becomes free (see Boenninghaus, 1904). A crista parotica is present with the styloid process fused to it and continued forwards as a tegmen tympani; both true (medial) and lateral suprafacial commissures are present, the former being continuous dorsally with the parietal plate, the latter stretching across between the tegmen tympani and the cochlear capsule.

The parietal plates are very high and connected with the orbital cartilages by orbitoparietal commissures.

The ala temporalis consists only of a slender lamina ascendens. The 'alicochlear' commissure extends widely back from the processus alaris, enclosing the carotid foramen, and fuses not with the cochlear capsule but with the lamina supracochlearis. These relations are important in showing that the lamina supracochlearis is a part of the parachordal or basal plate which is suppressed by the cochlear capsule in most mammals, and that the alicochlear commissure represents the hinder region of the trabeculae. The orbital cartilages are fairly large and horizontal in position, and connected with the wide central stem by preoptic and metoptic roots, the latter bearing a 'preoptic process' (Honigmann) or ala hypochiasmatica. Orbitoparietal and sphenethmoid commissures are present.

The nasal capsule is very reduced, especially anteriorly where the roof is thin and discontinuous with the dorsal edge of the nasal septum, and the side wall is represented only by a bar of cartilage which projects forwards and inwards and gives off the cartilago ductus nasopalatini forwards and the small para-septal cartilage backwards. This bar of cartilage, which in part therefore represents the lamina transversalis anterior, is in contact but not fused with the nasal septum. The hinder part of the side wall of the capsule is very bulbous, and from its ventral edge a processus maxillaris anterior (processus paranasalis) is given off. The nasal septum is continued forwards as a rostrum, which is nearly as long as the remainder of the skull. Posteriorly, a small spina mesethmoidalis rises up from the nasal septum between the olfactory foramina. Internally, the capsule contains only those turbinals typical of the hinder region. At later stages, what little there is of the roof of the capsule in the anterior region, is rotated into a vertical position, parallel with its fellow of the opposite side. This change is associated with the fact that the external nostrils come to face vertically upwards.

ii. The development of the osteocranium of Megaptera

(Plate 124, figs. 4, 5).

At the 152 mm. stage (6 inches) the general appearance of the bony skull is quite comparable to that of other mammals, except for the exaggerated snout. All the membrane-bones are present at this stage except the lachrymal; of the cartilage-bones, the supraoccipital, basioccipital, and basisphenoid are present.

The *maxilla* is large and elongated, and has from 35 to 37 teeth. The palatine processes do not meet. The facial portion of the maxilla has a wide contact with the frontal, over which it grows at later stages.

The *premaxilla*, toothless, lies medially to the maxilla, parallel with its fellow of the opposite side, dorsolaterally to the rostrum. Posteriorly, the premaxillae embrace the external nostrils between them.

The *nasal* is small, and lies medially to the hind end of the premaxilla and in front of the frontal.

The *frontal* is large, and becomes elongated transversely. It possesses a postorbital process which eventually almost reaches the zygomatic process of the squamosal, and a palatine process which descends in the front wall of the orbit, covering the cupola posterior of the nasal capsule. Eventually the frontal grows forwards beneath the nasal bone, and is itself overgrown by the maxilla from in front and the parietal from behind.

The *parietal* is large, and eventually overlaps the frontal. There is no interparietal.

The *jugal* is slender, and connects the maxilla with the squamosal.

The *lachrymal* is elongated and rod-like, and imperforate. There is no lachrymonasal duct.

The *squamosal* is small at first, with a large zygomatic process and a pterygoid process which projects towards the pterygoid (enclosing the foramen pseudo-vale (mandibular branch of trigeminal) between itself and that bone). Eventually the squamosal attains a large size, with a postglenoid process which becomes displaced farther and farther backwards.

The *vomer* is elongated and thin, and lies on the ventral surface of the nasal septum and rostrum.

The *palatine* is elongated and flat, and does not fuse with its fellow of the opposite side even in the adult.

The *pterygoid* has a very well developed hamular process. It comes into relation with the squamosal and the basisphenoid.

The *tympanic* is at first a simple flat semicircle, lying in the roof of the tympanic cavity, but it soon becomes modified in a remarkable way by the development of processes (sigmoid, posterior pedicle), and the tympanic membrane becomes drawn out into the shape of a cone. The processus gracilis of the malleus becomes fixed to the tympanic, which in turn fuses with the periotic. (See Plate 125, fig. 4).

The *prearticular* early fixes itself to the processus gracilis (or Folii) of the malleus.

The *dentary* bears about 36 teeth and has a small coronoid process.

The *supraoccipital*, ossifying in the tectum posterius, eventually forms an elevated dome which hides the parietals in a posterior view of the skull. There is no interparietal.

The *basioccipital* becomes fused in the adult with the *basisphenoid*. Both start as endochondral centres eventually reaching the surfaces. A small spur is developed from the basisphenoid ventrally to the processus alaris on each side and comes into contact with the pterygoid, thus forming a basipterygoid process.

The *exoccipital* ossifies in the occipital arch and the paracondylar process.

The *presphenoid* ossifies separately from the *orbitosphenoids*.

The *alisphenoid* is a purely cartilage-bone, without any intramembranous extensions.

The *periotic* arises as a diffuse ossification in the walls of the auditory capsule, and not from any particular centres. Eventually the periotic and tympanic become free from the remaining bones of the skull, for which reason they may be found isolated as so-called 'cetoliths'. There appears to be no mesethmoid.

Of the cartilage-bones of the visceral arch skeleton, the *malleus*, *incus*, *stapes*, and *thyrohyal* call for no special mention; the stylohyals eventually fuse with the basihyal. Part of Meckel's cartilage within the dentary becomes ossified, as a sort of *mediomeckelian*.

II. Balaenoptera rostrata

The development of the chondrocranium of the lesser rorqual has been studied with the help of wax models at the 105 mm. stage by de Burlet (1914*a*). Observations on the development of the bony skull in different species of Balaenoptera have been made by Ridewood (1922).

i. The chondrocranium of Balaenoptera (Plate 125).

The chondrocranium of Balaenoptera is in general lines similar to that of Megaptera (p. 341). What remains of the notochord at this stage (the hindmost portion which enters the basal plate in its posteroventral edge and runs obliquely forwards to its dorsal surface) suggests that its relations are the same as in Megaptera. There are no prominent paracondylar processes. The hypoglossal foramina are variable, and there may be one on each side, or on one side, or none, in which case the hypoglossal nerve emerges through the jugular foramen. The occipital condyles are still paired at this stage. The auditory capsule shows the characteristic Cetacean transverse position, but its contacts with neighbouring structures (occipital arch, parietal plate, basal plate) are at this stage more extensive than in Megaptera.

The relations of the (relatively very large) cochlear capsules to the basal plate are important. Laminae supracochleares are well developed, overlying the capsules, and fused with them by commissures which correspond in position with the alicochlear and sphenocochlear commissures in other forms. But the carotid foramina are well within the trabecular plate and it is easy to see how, upon reduction of the basal plate and occupation of its position by the cochlear capsules, as in other mammals, the carotid foramen becomes bordered posteriorly by the cochlear capsule, and virtually confluent with the basicochlear fissure. A medial (but no lateral) suprafacial commissure is present.

The orbital cartilage is large and its preoptic and metoptic roots are very wide. The metoptic root bears an ala hypochiasmatica and an anterior clinoid process. The orbitoparietal and sphenethmoid commissures are slender. The central stem in this region is extraordinarily wide, but it projects ventrally as a sagittal crest which presents the appearance of an interorbital septum, and is continuous anteriorly with the nasal septum and the enormous rostrum.

A high spina mesethmoidalis is present. The nasal capsule is much reduced, especially anteriorly. A lamina transversalis anterior runs inwards from the side wall to a paraseptal cartilage, which is prolonged anteriorly as a cartilago ductus nasopalatini. A lamina transversalis posterior is present.

Laterally to the side wall of the nasal capsule is a cartilage (termed by de Burlet the 'processus paranasalis') which appears to represent the parethmoidal cartilage of other forms. Behind and below it is a complicated structure in which perhaps the processus maxillares anterior and posterior may be recognized.

ii. The development of the osteocranium of Balaenoptera

In general, the development of the bones of the skull of Balaenoptera is the same as that of Megaptera, but attention may be called to two points of special interest.

In Balaenoptera borealis (the pollack whale) Ridewood (1922) found a large median *interparietal* separating the parietals. In B. musculus (the blue whale), however, there was no interparietal at all, and the parietals met in the midline in the specimens examined by him. On the other hand, a small interparietal was described in this species by Smets (1885). An interparietal is present in B. rostrata. This marked difference in related, and perhaps in the same, species, involving the presence or absence of such a large bone, might well receive further study.

The second point refers to the ossiculum accessorium mallei, possibly representing the *supra-angular*, which Ridewood discovered in B. musculus, situated dorsally to the prearticular. (Plate 125, Fig. 4).

ODONTOCETI

III. Globiocephalus melas

The chondrocranium of the pilot whale at the 133 mm. stage has been studied by Schreiber (1915) with the help of a wax model. (Plate 127, figs. 3, 4).

The chondrocranium is remarkably broad and short, its floor is very wide and flat, and it terminates anteriorly in a short rostrum directed forwards and downwards. There is no dorsum sellae, but a fissure marks the limit between the basal plate and hypophysial cartilage.

The basal plate is separated from the cochlear capsules by extensive basicochlear fissures, and anteriorly, extends over the capsules in the form of laminae supracochleares. The occipital arches rise up more or less vertically, and bear fairly large paracondylar processes. The tectum posterius is in the vertical plane. One pair of hypoglossal foramina is present.

The auditory capsules are orientated transversely, the cochlear capsules, which are very large, approaching one another closely, ventrally to the basal plate. Both medial (true) and lateral prefacial (suprafacial) commissures are present. The crista parotica, to which the styloid process is fused, is prolonged forwards into a large tegmen tympani. The auditory capsules are fairly firmly anchored to the occipital arches and parietal plates by fusions which break down when the capsules become free, as in the adult. The parietal plates are high and fenestrated and connected with the orbital cartilage by a wide orbito-parietal commissure.

The ala temporalis is slender and simple, and the alicochlear commissure, connecting the processus alaris with the lamina supracochlearis (basal plate) and enclosing the carotid foramen from the side, is very broad.

The orbital cartilage is fairly large and connected with the exceptionally wide central stem by preoptic and metoptic roots, the latter bearing a paroptic process (ala hypochiasmatica). There is no sphenethmoid commissure.

The nasal capsule is very incomplete, and composed of a hind wall (lamina orbitonasalis) from which a slender cartilage extending forwards represents all that is left of the side wall. The roof is formed of a small bar connecting the lamina orbitonasalis with the nasal septum, and of a pair of isolated cartilages situated a little farther forward. A lamina transversalis posterior is present, from which a paraseptal cartilage projects freely forwards, ending in a cartilago ductus nasopalatini. The crista galli is very high, without actually forming a spina mesethmoidalis. There is no lamina transversalis anterior.

Meckel's cartilages are curved and twisted in a remarkable manner.

IV. Phocaena communis

The chondrocranium of the porpoise has been studied by de Burlet (1913*b*, 1916) at the 48, 58, and 92 mm. stages, with the help of wax models.

In general form the chondrocranium of Phocaena closely resembles that of Globiocephalus (supra) though the rostrum is longer and the alicochlear commissure is broken down and represented by a small nodule. Among additional items of interest may be mentioned the following. The notochord is entirely enclosed within the basal plate except at its extreme anterior end where it emerges on to the dorsal surface. The pleurocentrum of the proatlas (occipital) vertebra may be observed to become attached to and form the anterior end of the odontoid peg of the axis vertebra. One pair of hypoglossal foramina is present. The auditory capsules show in very marked degree the orientation in the transverse plane and the relatively small size of the canalicular portion. The tegmen tympani is very prominent; both medial and lateral prefacial commissures are present. The metoptic root of the orbital cartilage, present at early stages, breaks down.

Remnants of the true side wall of the skull, possibly representing the pila antotica are found in a stump rising up from the floor on each side, between the cranial cavity and the cavum epiptericum, and in a down-growing process from the orbital cartilages. Anterior, as well as posterior, paraseptal cartilages are present. (Plates 126, 127, figs. 1, 2).

V. Lagenorhynchus albirostris

The chondrocranium of this dolphin has been studied by de Burlet (1914*b*). In general form it resembles the chondrocrania of Globiocephalus and Phocaena closely. The notochord enters the posterior end of the basal plate on its ventral surface, traverses it, and emerges on to its dorsal surface over which it runs forwards for a short distance before diving into the cartilage, enclosed in which it runs before finally emerging again on to the dorsal surface at its extreme anterior end.

The occipital condyles are conjoined in the midline. An interesting feature of this skull is the fact that the chondrocranium already shows asymmetry. Thus, on the left side, two hypoglossal foramina are present (one only on the right), the parietal plate and lamina supracochlearis are larger than on the right, and a sphenethmoid commissure is present (absent on the right).

In the olfactory region a short spina mesethmoidalis is present. (Plate 127, fig. 5).

30. *SIRENIA*

I. Halicore dugong

The chondrocranium of the dugong has been studied with the help of wax model reconstructions at the 150 mm. (back-length) by Matthes (1921*a*, *b*). The development of the osteocranium has been described by Freund (1908).

i. The chondrocranium of Halicore (Plate 128).

The basal plate is wide and flat, and fused on each side with the cochlear capsules by what appear to be basicochlear commissures; in the absence of basivestibular and sphenocochlear fissures the hinder part of the basicochlear fissure is confluent with the jugular foramen, and the anterior part with the carotid fissure (there is no alicochlear commissure). The anterolateral corners of the basal plate overlap the cochlear capsules dorsally, and form small laminae supracochleares. There is only a very low crista transversa behind a shallow pituitary fossa.

The basal plate forms almost a right angle with the anterior regions of the central stem.

The occipital arches slope somewhat forwards from the basal plate. Each is produced laterally into a wide lamina alaris, which ends ventrally in a blunt paracondylar process. Dorsally, the occipital arches are fused with the supraoccipital cartilages which are interconnected by the wide tectum posterius. Two hypoglossal foramina are present on each side.

The auditory capsules are arranged almost in the transverse plane, i.e. the cochlear capsules lie medially to the canalicular. The crista parotica, to which the styloid process is fused, is produced anteriorly into an enormous tegmen tympani. A stout medial suprafacial commissure is present (there is no lateral suprafacial commissure). In addition to the normal superior and inferior acustic foramina, there is a foramen accusticum intermedium for the ramus saccularis intermedius, and a foramen singulare for the ramus ampullaris posterior. The parietal plates are completely fused with the canalicular capsules, there is no superior occipitocapsular fissure. Between the parietal plates a transverse bar of cartilage in front of the tectum posterius represents the tectum synoticum, discontinuous on both sides.

The ala temporalis is composed of lamina ascendens and pterygoid process, but at this stage ossification of the alisphenoid is already far advanced. The processus alaris is depressed slightly ventrally.

Remnants of the original true side wall of the skull in this region are found on the right-hand side in the supracochlear cartilage which is fused on to the roof of the cochlear capsule, and in an isolated horizontal bar medial to the trigeminal ganglion, between the supracochlear cartilage and the anterior clinoid process of the orbital cartilage, which seems to represent the pila antotica (or perhaps part of the taenia interclinoidea, or supratrabecular bar).

The orbital cartilage is slender, but its preoptic and metoptic roots are stout; the former is inserted partly on the central stem and partly on the cupola posterior of the nasal capsule which is here fused with the central stem. From the metoptic root an anterior clinoid process projects backwards. The central stem in this region is very broad, but ventrally it is produced into a ridge which may be regarded as a diminutive interorbital septum. The orbitoparietal and sphenethmoid commissures are stout.

The nasal capsules are much reduced, especially anteriorly where the roof is absent, and the side wall is represented by a thin bar fused with the paraseptal cartilage which itself is fused with the nasal septum. A zona annularis is therefore present, but the lamina transversalis anterior is unrecognizable. Each paraseptal cartilage is produced forwards into a cartilago ductus nasopalatini, while the nasal septum ends anteriorly in paired processus incisivi. A small processus maxillaris anterior (processus paranasalis) and a paraethmoidal cartilage are present. Posteriorly, the nasal capsule has a well-developed cupola posterior, with a lamina transversalis posterior. Two ethmoturbinals and a crista semicircularis are present, but none of the anterior turbinals are formed. The dorsal edge of the nasal septum immediately behind the hind edge of the roof of the nasal capsules is produced dorsally into two median processes; the very large spina mesethmoidalis anterior, and the smaller spina posterior. They appear to be developments of the crista galli.

Meckel's cartilage is composed of separate proximal and distal portions.

ii. The development of the osteocranium of Halicore

From the few late stages in the development of the bony skull described by Freund (1908), it may be seen that the characteristic Sirenian appearance becomes accentuated during development, and probably throughout life in the case of the large anterior down-bent portions of the premaxilla and of the dentary. The nasal bone is absent. An interparietal is present and fused with the supraoccipital and with the parietals. There is no mesethmoid.

The periotic and tympanic are fairly closely associated with one another, and loosely with the surrounding bones of the skull, somewhat as in Cetacea. But the Sirenian skull differs markedly from the Cetacean in that its bone is compact and dense instead of spongy.

II. Manatus latirostris

No complete study has been made of the development of the skull of the manatee, but Hirschfelder (1936) has described the chondrocranium and Matthes (1912) the cartilages of the nasal capsule in a 68·5 mm. embryo.

As in the case of Halicore, the nasal capsule of Manatus is much reduced, the roof and side wall in the anterior region being absent altogether. The nasal septum is prolonged anteriorly as a processus incisivus, and posteriorly a spina mesethmoidalis rises up. A feature of some interest is the fact that the paraseptal cartilages are complete and continuous with the lamina transversalis posterior. Anteriorly, the paraseptal cartilages are fused with the nasal septum, for there is no lamina transversalis anterior. Anteriorly to this each paraseptal cartilage gives off the cartilago ductus nasopalatini. The paranasal process (processus maxillaris anterior) is exceptionally large.

The later stages of development of the bony skull have been studied by Dilg (1909). Of interest is the fact that the nasal bone is not only vestigial but variable. There is no mesethmoid bone (Broom, 1935*c*).

31. *CHEIROPTERA*

I. Miniopterus schreibersi

The only chondrocranium known among the bats is that of Miniopterus at the 17 mm. stage, modelled and studied by Fawcett (1919). The nasal capsules of various bats have been described by Grosser (1902) and by Zuckerkandl (1893), while the endotympanic bones have been studied by van Kampen (1915*b*) and van der Klaauw (1922). There is ample scope for a systematic description of the early stages of development of the skull in Cheiroptera, which might provide important evidence in regard to mammalian phylogeny.

i. The chondrocranium of Miniopterus (Plate 129, figs. 1, 2).

The basal plate is very slender between the cochlear capsules, which are separated from it by extensive basicochlear fissures, confluent with the carotid foramina (i.e. there is at this stage no basicochlear commissure). Anteriorly, the basal plate ends in a fairly well developed crista transversa, and it forms a shallow angle with the anterior regions of the central stem. The occipital arches have fairly well developed paracondylar processes, and are separated from the auditory capsule by the extensive inferior occipitocapsular fissure. The superior occipitocapsular (or parietocapsular) fissure is also large, and the auditory capsule is only in contact with the neighbouring structures by means of four slender commissures: the basivestibular, alicochlear, parieto-capsular, and supraoccipitocapsular.

The most striking feature of the auditory capsule is the fact that the fossa subarcuata (beneath the anterior semicircular canal) is widely perforated right through to the outside. This fossa lodges the paraflocculus of the cerebellum.

The suprafacial commissure projects dorsally as a sharp crest ending in a free point. The tegmen tympani is slender and is bent inwards at the tip, as if to join the cochlear capsule and form a lateral suprafacial commissure. The crista parotica is prominent, and the sulcus facialis (and fossa for the stapedial muscle) is remarkably deep.

In addition to the usual superior and inferior foramina for the branches of the auditory nerve, there is a foramen singulare for the ramus ampullaris posterior.

The ala temporalis is small. The orbital cartilage is also slender, and connected with the central stem by a broad preoptic root, but the metoptic root is interrupted. The sphenethmoid and orbitoparietal commissures are thin.

The nasal capsule is relatively small, and pear-shaped with widely diverging cupolae anteriores. The hind edge of the roof projects upwards over the nasal septum to form a prominent crista galli or spina mesethmoidalis. The cribriform plate is remarkably upright. The ventral regions of the nasal capsule are extremely complicated.

From the lateral border of the fenestra narina the inferior alar process projects prominently upwards, and it is also joined to the anteromedial wall of the fenestra, ventrally to the external nostril, by the alicupolar commissure. The lachrymonasal duct enters the nasal cavity through an incisure postero-ventrally to the alicupolar commissure. From the foremost region of the ventral edge of the nasal septum the processus lateralis ventralis extends to each side.

The lamina transversalis anterior projects inwards from the ventral edge of the side wall but fails to reach the nasal septum. The lamina transversalis gives off an anterior process which runs forwards to (but does not fuse with) the processus lateralis ventralis. Posteriorly, the lamina transversalis gives off the anterior paraseptal cartilage and the ectochoanal cartilage. The anterior paraseptal cartilage is continued forwards as a long cartilago ductus nasopalatini; behind, it is discontinuous from the slender posterior paraseptal cartilage. Between the anterior paraseptal cartilage is an isolated papillary cartilage. The ventral edge of the side wall of the capsule shows prominent anterior and posterior maxillary processes.

The malleus is exceptionally large, the incus small. The stapes is perforated by a stapedial artery which is larger than the internal carotid, and the stapedial muscle is enormous.

ii. The development of the osteocranium of Miniopterus

Without describing in detail all the ossifications found by Fawcett, attention may be confined to certain important points.

There appear to be no separate nasals, at least at the stage studied.

The *premaxilla* is at this stage continuous with the palatine process which according to Broom (1895*c*) becomes detached later, and presumably represents the *prevomer*.

The interparietals are large and paired.

The *tympanic* forms a half ring round the anterior part of the tympanic cavity.

The *endotympanic* at this stage is a cartilaginous plate in the floor of the tympanic cavity and in contact with the tympanic bone in front of it. Fawcett suggests that the formation of an endotympanic bone is associated with the necessity for protection of the tympanic cavity from the digastric muscle, and that it is comparable with the cartilage surrounding the Eustachian tube as to the cause of its origin. In Roussettus amplexicaudatus, van der Klaauw (1922) has described two endotympanics on each side. The cartilage of the caudal endotympanic element acquires continuity with the styloid process (as van Kampen (1915*b*) showed for Pteropus edulis) while the cranial endotympanic becomes fused with the Eustachian tubal cartilage.

DERMOPTERA

II. Galeopithecus sp.

The chondrocranium of Galeopithecus temmincki at the 28 mm. stage has been studied by Henckel (1929) with the help of a wax model. (Plate 129, figs. 3, 4).

The basal plate is fairly broad, as is the trabecular region of the central stem, between the axes of which there is an angle of about 60°. There is no marked dorsum sellae. The occipital arches slope backwards, but the plane of the tectum posterius slopes forwards and upwards. There are no paracondylar processes. The auditory capsules form most of the side wall of the cranial cavity in the posterior region, and the parietal plates are low immediately above the capsules, but rise to a considerable height anteriorly to them.

The preoptic root of the orbital cartilage is very broad and inserted on the lamina orbitonasalis; the pila metoptica is interrupted, leaving a large stump projecting from the central stem, and the optic foramen is consequently confluent with the sphenoid fissure. There is no interorbital septum whatever.

The paraseptal cartilages are long but incomplete; a lamina transversalis anterior and a zona annularis are present, but there is no lamina transversalis posterior. The ala temporalis is simple. An alicochlear commissure is present, but the carotid foramen is confluent with the anterior part of the basicochlear fissure.

Spence's cartilage (p. 413) has been found by van der Klaauw (1923).

Various stages in the formation of the bony skull have been described by W. K. Parker (1885). Points of special interest concern the pterygoids and 'mesopterygoids', which appear to be separate and to correspond to the 'Echidna-pterygoid' and the lateral wing of the parasphenoid respectively (see p. 435).

A mesethmoid is developed, as is an interparietal.

Rostral and caudal endotympanic elements are present (van der Klaauw, 1922).

A curious feature of the membrane-bones of Galeopithecus is, as Gaupp (1907*a*) showed, the very extensive preformation in secondary cartilage which they exhibit. This applies to the dentary throughout almost all its length, the maxilla, palatine, squamosal, pterygoid, vomer, jugal, and frontal.

32. *PRIMATES. LEMUROIDEA*

I. Nycticebus tardigradus

The chondrocranium of the slow loris at the 30 mm. stage has been modelled and figured by Henckel (1928*a*). (Plate 131, fig. 1).

The basal plate is broad and fused with the cochlear capsules, there being no basicochlear fissures. Anteriorly, the basal plate ends in a marked dorsum sellae from the lateral ends of which posterior clinoid processes project upwards. The parietal plates end freely anteriorly, there being no orbito-parietal commissures. The ala temporalis is large but not perforated. The orbital cartilages are small and form a transversely placed crescent, the centre of the crescent being a small planum supraseptale formed above the joined preoptic roots. The metoptic roots are slender and inserted on the sides of the central stem which is here broad, but farther forward, beneath the pre-optic roots, forms a low interorbital septum. The sphenethmoid commissures are remarkable in that they diverge slightly as they run forwards to the roof of the nasal capsule, enclosing both foramina olfactoria advehentia in a common rectangular depression. The nasal capsule is long and angular. At the stage studied there is no complete lamina transversalis anterior, but according to Frets (1914) such a structure is present in Propithecus and in Lemur, together with a cartilago ductus nasopalatini.

TARSIOIDEA

II. Tarsius spectrum

The chondrocranium of Tarsius at the 24 mm. stage, first modelled and figured by Fischer (1905), is described by Henckel (1927). (Plate 130, figs. 3, 4).

The basal plate is fairly wide posteriorly but very narrow and rod-like anteriorly, between the cochlear capsules, from which it is separated by extensive basicochlear fissures. A well-marked dorsum sellae is present. The basal plate makes a small angle with the anterior regions of the central stem.

The occipital arches lean slightly backwards. The medial walls of the auditory capsules contribute a considerable portion of the side wall of the cranial cavity. The parietal plates slope upwards and inwards over the cranial cavity, and are connected with the orbital cartilages by rod-like orbitoparietal commissures. On the right side a small isolated nodule seems to represent the supracochlear cartilage (see p. 390).

The ala temporalis is simple; the alicochlear commissure is incomplete and the carotid artery runs through an incisure that is almost converted into a foramen.

The orbital cartilages are small and lie almost in a transverse plane, widely separated from the nasal capsule: there is no sphenethmoid commissure. The metoptic roots of the orbital cartilages are inserted on the central stem close in front of the processus alares; the preoptic roots join one another to form a small planum supraseptale which is attached to the dorsal surface of the central stem. The optic foramina face forwards and outwards. The ventral part of the central stem in this region forms a large interorbital septum, which lies in the direct anterior prolongation of the rod-like anterior portion of the basal plate, and continues forwards into the nasal septum.

The nasal capsule is very compressed and small and the foramina olfactoria advehentia are combined and sunk beneath a joint rim. The lamina transversalis anterior is complete, and there is a zona annularis; according to Frets (1914) the nasal capsule of Tarsius resembles that of Catarrhinae more than that of Lemuroidea, but this is not borne out by Wen's (1930) work. It seems from Frets' descriptions as though the paraseptal cartilage were complete.

PLATYRRHINI

III. Chrysothrix sciurea

The chondrocranium of a 24 mm. embryo of Chrysothrix has been modelled and described by Henkel (1928). The nasal capsule has been studied by Frets (1913), and by Wen (1930). (Plate 131, fig. 2).

The basal plate is concave dorsally, wide behind and narrow in front where it ends in a wide and prominent dorsum sellae. The occipital arches lean backwards, but the tectum posterius slopes upwards and forwards. The auditory capsules form a considerable portion of the side walls of the cranial cavity. The jugular foramen is elongated and slit-like, and the basicochlear and superior and inferior occipitocapsular fissures are extensive.

The parietal plate is low and ends freely anteriorly, there being no orbitoparietal commissure. The ala temporalis is slender, but perforated by a foramen rotundum; there is no alicochlear commissure. The orbital cartilages are large and connected with the nasal capsule by a very broad sphenethmoid commissure whose attachment extends for the whole length of the foramen olfactorium advehens. The preoptic roots join one another in the midline to form a small planum supraseptale which is attached to the central stem beneath it. The central stem in this region forms a small interorbital septum. (No interorbital septum is found in Mycetes or Cebus, though it is well developed in Hapale and in Ateles.)

The nasal capsule is small, and situated ventrally as well as anteriorly to the large cranial cavity.

According to Frets, a complete lamina transversalis anterior and zona annularis is formed. The paraseptal cartilage seems to be incomplete in Chrysothrix, but it is complete in Mycetes (Fawcett, unpub.).

A cartilago ductus nasopalatini is present, and an ectochoanal cartilage, which in other Platyrrhini is independent.

A peculiarity of Chrysothrix is the presence of a transverse cartilaginous bar, the 'commissure supracribrosa' which overlies the hinder part of the foramina olfactoria advehentia, and disappears at later stages.

CATARRHINI

IV. Macacus cynomolgus

The chondrocranium of Macacus at the 25 mm. stage has been studied by Fischer (1903) with the help of a wax model. The most remarkable feature is its external compression from front to back so that it is literally broader than it is long. (Plate 131, figs. 3, 4).

The basal plate is broad and concave dorsally. Anteriorly it ends in a well-marked crista transversa, dorsally to which, but quite free from it, is a dorsum sellae. The ends of the dorsum sellae are continuous by means of the supra-trabecular bars with the pilae metopticae.

The occipital arches lean backwards, and the tectum posterius is in the transverse vertical plane. A well-developed lamina alaris is present, and the jugular foramen is really a slit-like fissura metotica. Superior and inferior occipitocapsular foramina are present. The auditory capsule forms a considerable part of the side wall of the cranial cavity. A parietal plate is present but no orbitoparietal commissure.

The processus alaris carries, but is not fused with, a large ala temporalis, which consists of a lamina ascendens perforated by a foramen rotundum for the maxillary branch of the trigeminal nerve. That part of the cartilage in front of this foramen is thus the processus ascendens of the pterygoquadrate. There is no alicochlear commissure.

The orbital cartilage is connected with the central stem by wide metoptic and preoptic roots, the latter forming a flat planum supraseptale on the top of the fairly high interorbital septum. Whether an ala hypochiasmatica is or was present cannot be determined from the evidence available. Anteriorly, the orbital cartilage is continuous with the roof of the nasal capsule by means of the sphenethmoid commissure.

The nasal capsule is very short and in the form of a triangular pyramid the base of which is completely open, for there is no floor; the lamina transversalis anterior is incomplete, failing to reach the side wall, and there is no zona annularis. The paraseptal cartilage is well developed although there is no Jacobson's organ (Frets, 1914). An ectochoanal cartilage is present.

V. Semnopithecus maurus

The chondrocranium of Semnopithecus (the sacred langur monkey) at the 53 mm. stage has been studied by Fischer (1903) with the help of a wax model, and it bears such close resemblance to that of Man, that close observation is necessary to distinguish them. (Plate 131, figs. 5, 6).

The basal plate is rectangular and concave dorsally, and the course of the notochord is transbasal (see p. 382). Anteriorly, the basal plate ends in a crista transversa overlying which is a dorsum sellae. The lateral edges of the dorsum sellae are continuous anteriorly with the supratrabecular bar (taenia interclinoidea), which is complete, and posteriorly with the cartilage of the abducens bridge, which almost fuses with the cochlear capsule, and probably represents a remnant of the pila antotica.

The basal plate makes an angle of some 65° with the anterior part of the central stem.

The occipital arches lean backwards and the tectum posterius is in the vertical transverse plane. The foramen magnum extends into a marked superior occipital incisure. One pair of hypoglossal foramina is present in S. maurus, but in S. pruinus, Fischer found two foramina on one side and three on the other. Superior and inferior occipitocapsular fissures are present, but the lamina alaris is not well developed and there is no paracondylar process.

The auditory capsule is strongly rotated backwards, but its canalicular portion still forms part of the side wall of the cranial cavity. The crista parotica and tegmen tympani are very well developed, but the mastoid process is not marked. The septum between the lateral and posterior semicircular canals is incomplete.

Semnopithecus was the first form in which the processus recessus, separating the fenestra rotunda from the aquaeductus cochleae, was recognized (though not named) by Fischer (1903).

A small parietal plate surmounts the auditory capsule, but there is no orbitoparietal commissure.

The processus alaris is fused to the ala temporalis which has a lamina ascendens but no pterygoid process, no cartilaginous foramen rotundum, and there is no alicochlear commissure. The orbital cartilage is large, and its attachment to the central stem by the metoptic root is continuous with the anterior end of the supratrabecular bar. The preoptic root of the orbital cartilage is continuous with the dorsal edge of the central stem (which in this region forms a high interorbital septum), and also with the anterior end of the ala hypochiasmatica (which was regarded by Fischer as the preoptic root of the orbital cartilage) which is fused with the central stem on each side so as to enclose a prechiasmatic foramen, through which the dorsal edge of the interorbital septum may be seen. Immediately in front of this the preoptic roots form a flat planum supraseptale. Anteriorly, the orbital cartilage is fused with the roof of the nasal capsule by the sphenethmoid commissure, and also with the lamina orbitonasalis by the ala minima. Thus, an orbitonasal fissure and a foramen infracribrosum are enclosed, and a lamina infracribrosa is formed, continuous posteriorly with the planum supraseptale.

The nasal capsule is short and triangular in transverse section, the apex being on top. There is no hind wall to the capsule, and the floor is represented by a very slender lamina transversalis anterior, which connects the side wall with the paraseptal cartilage. There is no Jacobson's organ (Frets, 1912).

There appears to be no interparietal bone.

VI. Homo sapiens

The development of the human chondrocranium has been studied by a number of investigators, the earlier of whom, however, such as Dursy (1869) did not use model reconstruction. Those who have done so have usually been concerned only with one stage. It is necessary, therefore, in order to get a consecutive impression of the processes of development of the human chondrocranium to refer to several works, which for the sake of convenience are drawn up here in tabular form:

Stage: Length in mm.; greatest length	*Age*	*Author*
13		Levi, 1900, 1909*a*
13·6		Fawcett, 1918*a*
14		Fawcett, 1918*a*
14	37–8 days	Levi, 1900, 1909*a*
17	42–5 days	Levi, 1900, 1909*a*
17		Hagen, 1900
17	50 days	von Noorden, 1887
18·5	7½ weeks	von Noorden, 1887
19		Fawcett, 1910*a*
20		Bardeen, 1910
20		Kernan, 1916
21	8 weeks	Lewis, 1920
23	8½ weeks	von Noorden, 1887
28	58–62 days	Levi, 1900
30		Jacoby, 1895
30		Fawcett, 1910*b*
43		Macklin, 1921
40 (crown-rump)	63 days	Macklin, 1914
80 (greatest length)		Hertwig, 1898

A selection of these stages will be used to describe the development of the human chondrocranium.

i. The development of the chondrocranium of Homo sapiens

At the *13·6 mm. stage* (Fawcett, 1918*a*) paired parachordals are present, flanking the notochord and connected with one another above the latter only at their extreme anterior end. Posteriorly each parachordal is continuous with an occipital arch, on the anterior surface of the base of which is an incisure for the hypoglossal nerve. The auditory capsules are quite free from the parachordals, as is the rudiment of the pars trabecularis. The latter is somewhat variable in the degree of its chondrification, for at the 14 mm. stage (Fawcett, 1918*a*) it has been found to consist of separate paired nodules of cartilage (hypophysial cartilages) flanking the hypophysial duct. Otherwise, this 14 mm. stage is similar to the 13·6, except that the hypoglossal incisure has become a foramen, on each side, and the parachordals have fused with one another to form the basal plate. This takes place partly above and partly beneath the notochord, so that the latter has a transbasal course: it lies dorsally to the hindmost part of the basal plate, then pierces it and runs forwards ventrally to its middle portion, pierces it again and becomes embedded in its anterior part (Linck, 1911). (Plate 132, fig. 1).

At the *19 mm. stage* (Fawcett, 1910*a*) the basal plate is bounded anteriorly by the crista transversa, dorsally to which the dorsum sellae or acrochordal, is an independent cartilaginous bar. Anteriorly, the crista transversa of the basal plate is fused with the pars trabecularis or central stem, which is still perforated by the hypophysial foramen.

From the central stem on each side there projects the processus alaris, in front of the carotid artery and dorsally to the palatine (or greater superficial petrosal) nerve. The processus alaris appears to have had an independent centre of chondrification, as certainly has the ala temporalis, laterally to it. The orbital cartilage on each side also has an independent chondrification centre.

At the *20 mm. stage* (Kernan, 1916) the chondrocranium is already fairly well developed. Each auditory capsule is anchored to the basal plate by an extensive basivestibular commissure, and also posteriorly, to the occipital arch by its lamina alaris, with the result that the fissura metotica is represented by the jugular foramen below and by the fissura occipitocapsularis above. Surmounting the auditory capsule and fused with it (except for the region of the fissura occipitocapsularis superior, or parietocapsularis) is the parietal plate; this is continuous posteriorly with the supraoccipital cartilage, which in turn is fused ventrally with the dorsal end of the occipital arch, and medially with its fellow of the opposite side to form the tectum posterius. The occipital condyles begin to show as bulges at the base of the hinder surface of each occipital arch, on each side of a deep intercondylar incisure. The latter lodges the odontoid process, of which the tip, formed by the pleurocentrum of the proatlas vertebra, occasionally chondrifies separately (Gladstone and Powell, 1914; Augier, 1928). The lateral surface of each occipital arch is produced to form the paracondylar process (cranial rib), immediately in front of which the lateral wall of the auditory capsule bears the small mastoid process (which seems to contain an independently chondrified nodule). Continuing forwards on the lateral wall of the auditory capsule is the crista parotica, to which the tympanohyal and stylohyal are attached, and in front of this, the tegmen tympani. (Plate 132, figs. 2–4).

The apertures of the auditory capsule now comprise the fenestra ovalis, the foramen perilymphaticum, the foramen endolymphaticum, and the foramen acusticum. The latter is overlain by the suprafacial (prefacial) commissure, which thus forms the dorsal border to the internal auditory meatus, through which the facial nerve passes to reach the sulcus facialis (in the groove between the crista parotica and the lateral wall of the capsule) in which it runs backwards and outwards behind the tympanohyal.

The dorsum sellae is fused with the crista transversa at the front of the basal plate. The hypophysial foramen is closed. The processus alaris on each side bears a backwardly directed processus alicochlearis which almost but not quite reaches the cochlear part of the auditory capsule (passing laterally to the internal carotid artery). The orbital cartilage is still unattached to the central stem, and ventromedially to it the ala hypochiasmatica has appeared as an independent (paired) chondrification.

The central stem is continued anteriorly into the nasal septum, the dorsal edge of which is produced upwards to form the crista galli. On each side of the septum and quite separate from it is the paranasal cartilage, forming the lateral wall of the nasal capsule.

Meckel's cartilage is well developed, with a massive retroarticular process forming the manubrium of the malleus. Equally massive is the incus, in contact with the mallear part of Meckel's cartilage, and by means of its crus breve with the crista parotica and of its crus longum with the stapes.

The development of the hyoid arch skeleton has been studied by Broman (1899). The blastematous rudiment is forked at its upper end, with the facial nerve passing between the prongs. The medial prong is also in connexion with the crus longum of the incus. The stapes chondrifies independently in the medial prong, and loses connexion with the remainder of the arch. The lateral prong chondrifies as the tympanohyal (laterohyal, Broman, or intercalary of Dreyfuss, 1893) which fuses with the crista parotica above, and with the more distal part of the arch (stylohyal, or Reichert's cartilage) beneath, thus forming the styloid process. The facial nerve runs medially to this and gives off the chorda tympani which runs forwards laterally to the styloid process to reach the medial surface of the malleus.

At the *21 mm. stage* (Lewis, 1920) the orbital cartilages are in contact with the central stem by means of their metoptic roots (pilae metopticae), with which the alae hypochiasmaticae have fused, each on its own side.

The roof of the nasal capsule is procartilaginous, connecting the dorsal edge of the nasal septum with the dorsal edge of each paranasal cartilage. Ventrally, the paraseptal cartilages are present as independent chondrifications.

The ala temporalis is attached to the processus alaris by mesenchyme. In regard to the tectum posterius, Lewis's embryo is less well developed than Kernan's, for the supraoccipital cartilages (squamae) are not yet interconnected. A feature of some interest is the fact that here the dorsal end of the occipital arch is still free, which makes the similarity between the occipital and posterior neural arches very apparent. (Plate 132, fig. 5; Plate 133, figs. 1–4).

Lewis has calculated the relative increases in size of certain dimensions, between the 21 mm. stage and the adult. An interesting feature of these ratios is that they seem to fall into two groups, one of which is roughly twice the magnitude of the other.

	21 mm.	*Adult*	*Ratio*
Length of basal plate	2·0	26	1 : 13
Width of central stem (including processus alaris)	2·1	30	1 : 14
Width between optic foramina	2·0	28	1 : 14
Width between hypoglossal foramina	2·0	29	1 : 14
Length from foramen magnum to anterior end of nasal septum	4·2	93	1 : 22
Width of basal plate	1·0	25	1 : 25
Width between parietal plates	4·0	110	1 : 28
Length of nasal septum	1·8	54	1 : 30

At the *30 mm. stage* (Fawcett, 1910*b*, 1923) the chief advances in chondrification are as follows. The orbital cartilage is attached to the roof of the nasal capsule (now chondrified) of its side by the sphenethmoid commissure. The alicochlear commissure is fused with the cochlear capsule thus enclosing the carotid foramen.

The chief interest in this stage, however, refers to the roof of the chondrocranium in the auditory and occipital region. The tectum posterius (between the supraoccipital cartilages) is very broad, and bears small processes in the midline projecting forwards (ascending) and backwards (descending). In front of this is a slender tectum intermedium connecting the hindmost portions of the parietal plates. Farther forward again, the tectum synoticum is represented by a median and two paired cartilages, arranged transversely. These cartilages are subject to variation; there may be a single median cartilage, or only the paired cartilages (Muggia, 1932*a*), or they may be continuous with the parietal plates on one or both sides or they may be absent altogether (Augier, 1935). The descriptions given by Bolk (1904) refer to later stages where calcification and ossification have eroded the tectum posterius. The tectum posterius is also subject to variations (Muggia, 1932*b*).

ii. The morphology of the fully formed chondrocranium of Homo sapiens (based on 43 mm. greatest length, 40 mm. crown-rump, and 80 mm. crown-rump) (PLATE 133, FIGS. 4, 5; PLATE 134).

1. The basal plate and notochord. The basal plate is oblong and almost rectangular in shape. Its anterior edge is marked by the very well-developed dorsum sellae, while posteriorly its edge forms the intercondylar incisure of the foramen magnum.

The notochord lies on the dorsal surface of the hindmost part of the basal plate, then runs anteroventrally through it, and into contact with the pharyngeal bursa and mucosa of the roof of the mouth, and lastly turns anterodorsally into the basal plate again, in the thickness of the anterior part of which it ends, a little distance behind the root of the dorsum sellae. The basal plate makes an angle of about 65° with the plane of the anterior part of the central stem.

As an abnormality Augier (1928) has observed the presence of a hypocentrum belonging to the proatlas vertebra, and attached to the posteroventral region of the basal plate, in the midline, beneath the intercondylar incisure. This structure is a vestige of the original median hypocentral condyle (as found in Lacertilia, Ophidia, and Sphenodon: see p. 386).

2. The occipital region. The occipital arches lean back and merge dorsally into the supraoccipital cartilages and tectum posterius, the plane of which slopes upwards and backwards. The dorsoposterior margin of the foramen magnum is indented by the superior occipital incisure, which becomes partly obliterated later by the development of a processus descendens (Lacoste, 1929; Augier, 1931*b*) of the tectum posterius. A small ascending (anterior) process is also developed.

Laterally, each occipital arch spreads out into the lamina alaris, which forms a crest of which the lowest point projects as the paracondylar (or jugular) process. The lamina alaris is fused with the hind wall of the auditory capsule, leaving the fissura metotica represented only by the jugular foramen (ventrally) and the occipitocapsular fissure (dorsally). On the internal surface of the skull, the fusion between the occipital arch and the auditory capsule is marked by the recessus supra-alaris, lodging the sigmoid venous sinus.

The base of each occipital arch is perforated by a (sometimes wholly or partly subdivided) hypoglossal foramen. The occipital region is often slightly asymmetrical, and on one side there may be a paracondylar foramen, immediately laterally to the hypoglossal foramen.

3. The auditory region. The auditory capsule shows a marked subdivision into vestibular (including canalicular) and cochlear portions, and as it lies not beside but under the brain, the canalicular portion is markedly lateral to the cochlear portion in position. Furthermore, the backward rotation of the capsule is so marked that the upper wall of the cochlear part is level with the top of the canalicular part.

Dorsally, the auditory capsule is surmounted by the parietal plate (the fissura occipitocapsularis superior, or parietocapsularis is enclosed between them, sometimes termed 'foramen jugulare spurium', or 'mastoid foramen'), which is fused with the supraoccipital cartilage (and tectum posterius) posteriorly, but there is no orbitoparietal commissure (remnants of which may perhaps be represented by certain nodules described by Augier, 1934*b*). The tectum synoticum is usually at most represented by a median (or paired, according to Muggia, 1932*a*) isolated nodule (which marks the position of the preinterparietal bones). Occasionally, at late stages (161 mm.), Augier (1935) has found the tectum synoticum complete, uniting the parietal plates.

The external surface of the capsule bears the mastoid process, the crista parotica (to which the styloid process is attached), and the tegmen tympani. The latter may have attached to it a small suprategminal cartilage (Augier, 1930), and eventually gives rise to the lateral prefacial commissure by fusing with the cochlear capsule (Fischer, 1901). The suprafacial (or true prefacial) commissure joins the anterior surface of the canalicular to the dorsal surface of the cochlear capsule, and forms the dorsal border to the primary facial foramen, and indeed to the whole internal auditory meatus. Between the true and the lateral prefacial commissures is a gap, the hiatus facialis or Fallopii, through which the palatine (greater superficial petrosal) nerve passes. The facial nerve then runs back in the sulcus facialis.

The medial surface of the capsule bears a well-marked fossa subarcuata which, however, becomes shallowed out and lodges nothing but connective tissue, for the paraflocculus of the cerebellum is vestigial.

The foramina in the medial wall of the capsule are the endolymphatic foramen, and the superior and inferior acustic foramina which open out of the internal auditory meatus: the former conveying fibres innervating the utricle, the anterior and lateral ampullae and part of the saccule: the latter innervating the remainder of the saccule, the ductus cochlearis, and the posterior ampulla. These last fibres eventually become enclosed in a separate foramen singulare.

On the lateral surface the fenestra ovalis (or vestibuli) opens into the sulcus facialis; on the ventroposterior surface of the cochlear capsule, the foramen perilymphaticum opens into the jugular foramen. The aperture of the foramen perilymphaticum is in process of becoming arched over by the anterior and posterior processus recessus (or 'intraperilymphatici'), projecting respectively backwards and forwards from its anterior and posterior margins. When these processes meet there will be formed laterally the fenestra rotunda and medially (leading through the jugular foramen into the cranial cavity) the aquaeductus cochleae.

The cavity of the auditory capsule is roughly constricted into cavum cochleare and cavum vestibulare. The former bears in its floor the spiral septum which curves forwards and outwards. The latter contains the cavities of the semicircular canals and their ampullae, which are separated from the main cavity by septa. Streeter (1917, 1918) has shown that the volume of these canals increases by erosion of their walls on the inner surface, in the course of which phenomenon the cartilage cells dedifferentiate into mesenchyme. This process may be the means whereby the utricle and semicircular canals rotate within the capsule, for Lebedkin (1924) has pointed out that at the 118 mm. stage the lateral semicircular canal is in a plane parallel with the basal plate, whereas in the fully formed skull these planes diverge by 40°.

4. The orbitotemporal region. The ala temporalis, perforated by the foramen rotundum (ramus maxillaris trigemini) is fused to the processus alaris; the alicochlear commissure projects towards, but now no longer reaches, the cochlear capsule. The carotid foramen is thus emarginated. The palatine (greater superficial petrosal) nerve runs from the facial foramen antero-ventrally over the lateral wall of the cochlear capsule and ventrally to the processus alaris, laterally to the alicochlear process (which represents the trabecula cranii, see p. 379).

The orbital cartilage is large and forms a practically horizontal plate, connected with the nasal capsule by the wide sphenethmoid commissure, and with the central stem by both metoptic and preoptic roots, enclosing the optic foramen between them. From the anterior surface of each preoptic root (at the 80 mm. stage) a cartilaginous process (the ala minima of Luschka, 1857) projects towards but does not quite reach the lamina orbitonasalis.

From the base of each metoptic root, the small ala hypochiasmatica projects forwards beneath the optic foramen and fuses with the preoptic root of the orbital cartilage, thereby enclosing a couple of small foramina prechiasmatica between itself and the central stem.

The central stem in this region is excavated on its dorsal surface to form the deep sella turcica, lodging the pituitary body and bordered posteriorly by the dorsum sellae, anteriorly by a transverse ridge or tuberculum sellae.

Each metoptic root of the orbital cartilage bears a backwardly directed anterior clinoid process pointing towards the forwardly directed posterior clinoid process at the side of the dorsum sellae. These processes are occasionally found interconnected (on each side) by a cartilaginous taenia interclinoidalis or supratrabecular bar, identical with the similarly named structure in reptiles (p. 227) and lower Primates (p. 353).

The sphenoparietal fontanelle is not enclosed dorsally, for there is no orbitoparietal commissure. It is to be noted that in the chondrocranium the orbital cartilage (in which the 'small wing of the sphenoid' will ossify) is much larger than the ala temporalis (which gives rise to the 'great wing of the sphenoid' of the bony skull).

The cavum epiptericum and cavum supracochleare become added to the cranial cavity, the lateral limit of which now becomes the ala temporalis. But interesting relics of the true side wall of the skull in this region, representing the pila antotica, are found in the posterior clinoid processes and in an isolated supracochlear cartilage, situated in the dura mater beneath the trigeminal or Gasserian ganglion.

Ventrally to the roots of the orbital cartilage, the central stem becomes flattened into a plate in the sagittal plane, and thus constitutes a small interorbital septum.

5. The ethmoid region. The later development of the nasal capsule has been studied by Zuckerkandl (1893) Kallius (1905*b*), Peter (1906, 1912), Schaeffer (1910*a*, *b*, and *c*, 1916) Wen (1930), and Schultz (1935).

The nasal septum, which is the direct anterior prolongation of the interorbital septum, rises up between the olfactory foramina to form the crista galli, and then drops again into the roof of the nasal capsule. In addition to this roof the nasal capsule has a side wall and a hind wall or cupola formed by the lamina orbitonasalis (which is free from the nasal septum), but beyond a diminutive lamina transversalis posterior there is no floor, and the rudimentary processus transversalis anterior which juts out from the nasal septum (and from which the diminutive paraseptal cartilage projects backwards) fails to reach the side wall (Macklin's 'processus paraseptalis'). The result is that there is no zona annularis, and the fenestra narina is not separated from the fenestra basalis (choanalis).

The foramen epiphaniale (for the exit of the lateral nasal branch of the profundus nerve) is in the roof of the capsule, just in front of the root of the sphenethmoid commissure.

Internally, the lateral wall of the capsule presents three longitudinal ingrowing shelves, one above the other: the inferior, middle, and superior turbinals. The inferior appears (Peter, 1906, 1912) to be the maxilloturbinal, the middle the 1st ethmoturbinal, and the superior, the 2nd ethmoturbinal. Between the maxilloturbinal and the 1st ethmoturbinal the space is known as the middle meatus, that between the 1st and 2nd ethmoturbinals is the superior meatus. Subsequently, two or three more vestigial ethmoturbinals are developed. An important ridge in the side wall of the capsule is the crista semicircularis (in Man usually called the uncinate process), forming the anterior border of the aperture leading from the main cavity of the nasal capsule to the recessus anterior (the hiatus semilunaris). Behind the crista semicircularis a number of frontoturbinals develop, and the two foremost of these fuse to form the ethmoidal bulla, which narrows the aperture of the hiatus semilunaris. In front of the crista semicircularis the nasoturbinal (agger nasi, mistaken by Killian (1896) for the 1st ethmoturbinal) forms a shallow shelf overlying the maxilloturbinal. The anterior ethmoidal sinuses penetrate into the frontal bone, nasoturbinal, and other parts of the side wall of the nasal capsule. They open into the hiatus semilunaris. The middle ethmoidal sinus inflates the ethmoidal bulla and opens at its summit. The two posterior ethmoidal sinuses open behind the hiatus semilunaris, in the superior meatus, while the maxillary sinus (antrum of Highmore) opens in the middle meatus. The sphenoidal sinus develops as an expansion of the posterodorsal portion of the nasal sac, which eventually perforates the lamina orbitonasalis.

The aperture of the foramen olfactorium advehens becomes subdivided by cartilaginous struts extending from the nasal septum to the side wall of the capsule. In this manner the cribriform plate is formed, in the floor of the supracribrous recess; the foramen olfactorium evehens (roof of the supracribrous recess) remains simple and bounded by the dura mater. Laterally, the supracribrous recess opens into the orbit by the orbitonasal fissure, through which the nasal branches of the profundus nerve pass to enter the nasal cavity.

On the outer surface of the side wall of the capsule is a small 'paranasal cartilage', and, above it, a small process (Macklin's 'processus maxillaris posterior', but certainly not the homologue of the structures of this name in other forms). Between these structures the lachrymonasal duct runs downwards and forwards, and passes under the ventral edge of the side wall into the nasal sac, behind the paraseptal process. An isolated parethmoidal cartilage is found beside the posterodorsal region of the side wall of the capsule.

6. The visceral arch skeleton. The distal ends of Meckel's cartilages are bent inwards and upwards, at right angles to the main shaft of the cartilages. The proximal portions of Meckel's cartilage, or malleus, and the incus and stapes, call for no special mention except to point out that the articulations between them are already formed at the 27 mm. stage (Hesser, 1925) when the eardrum is still thick and no vibrations can pass. The primary differentiation of these joints, therefore, is independent of function.

The tympanohyal (laterohyal) and stylohyal cartilages, fused to the crista parotica, forming the styloid process, end freely at the distal end, which is, however, connected by ligaments with the hypohyal, which forms the lesser cornu of the 'hyoid'. The number of segments in the hyoid arch is, however, variable (Bruni, 1909). The 'hyoid' is formed of the basihyal, in the form of a transverse bar of cartilage, to which on each side the greater cornua (thyrohyals, 3rd visceral or 1st branchial arch) are attached.

Behind the 'hyoid' are the thyroid cartilages, fused together in the midline ventrally to the larynx, and probably representing the 4th and 5th visceral arches, as Kallius (1897) and Grosser (1912) showed, the limit between them being indicated by the thyroid foramen (Dubois, 1886), when present. On the other hand, Soulié and Bardier (1907) and Frazer (1910) are disinclined to admit any contribution from the 5th arch.

The cricoid cartilage forms a complete ring round the trachea and bears the arytenoid cartilages. It is still questionable whether the cricoid is a modified and enlarged tracheal ring, or a (6th) visceral arch.

iii. The development of the osteocranium of Homo sapiens

To give a complete list of the investigators who have studied the development of the human bony skull in one or more of its aspects would require a bibliography as large as that referring to the development of the bony skull in all other Vertebrates. Fortunately, however, there is now no need for the provision here of an exhaustive human osteocranial bibliography, for not only has Augier (1931*a*) provided one, but, more important, he has reinvestigated the development of the bones of the human skull using the most modern methods (Spalteholz and Schultze) on a magnificent range of material (over 250 specimens), which has enabled him to distinguish between 'normal' development and exceptional variations: the latter have sometimes been regarded as 'normal' by workers who had not investigated a sufficient number of specimens of each stage. The following account is based largely on his work (to which the reader is referred for further details), with a few alterations of terminology to bring the human skull into line with other Vertebrates.

According to Augier the order of appearance of the centres of ossification of the human skull is as follows:

Centre	*Crown-rump; length in mm.*	*Age*
Dentary	15	
Maxilla	15	
Frontal	25	
Palatine	25	
Squamosal (zygomatic process)	25	End of 2nd month
Jugal	25	
Premaxilla	26	
Supraoccipital	30	
Prearticular	30	
Pterygoid (hamular centre)	30	
Vomer (paired centres)	30	
Interparietal	32	
Tympanic (angular) (anterior centre)	32	
Nasal	34	
*Squamosal (squamous portion)	36	
Lachrymal	36	
Parietal (two centres)	37	
Exoccipital	37	
Alisphenoid (ala temporalis)	37	
Basioccipital	51	
Tympanic (inferior centre)	54	
*Pterygoid (parasphenoid centre)	54	
Orbitosphenoid	60	
Basisphenoid (median centres)	65	
*Tympanic (posterior centre)	65	
*Preinterparietal	70	
*Paraseptal	70	End of 3rd month
Basisphenoid (processus alaris)	90	
Presphenoid (lateral centres)	90	
Opisthotic (cochleo-canalicular centre)	110	
Incus	110	
Malleus (inferior centre)	117	
Prootic (anterior canalicular centre)	117	
Presphenoid (median centres)	117	
Maxilloturbinal	117	End of 4th month
Prootic (posterior canalicular centre)	130	
*Pleurosphenoid (supracochlear centre)	130	
Ossiculum Bertini	130	
Malleus (superior centre)	130	
Inferior ethmoturbinal	130	
Opisthotic (cochlear centres)	130	
Prootic (intermediate centre, posterior s.s.c.)	134	
Stapes	139	
Prootic (intermediate centre, anterior s.s.c.)	150	
Superior ethmoturbinal	158	
Epiotic (mastoid centre)	161	
Epiotic (intermediate centres, lateral s.s.c.)	166	
Presphenoid (median unpaired centre)	168	End of 5th month
Thyrohyal		,, 6th ,,
Basihyal		,, 9th ,,
Laterohyal (tympanohyal)		,, ,, ,,
Mental ossicles		,, ,, ,,
Mesethmoid		9th month after birth
Stylohyal		3rd year ,, ,,
*Ceratohyal		,, ,, ,,
Hypohyal		9th ,, ,, ,,

(* Denotes ossification centres of inconstant occurrence.)

A characteristic feature of the human skull is the extensive fusion which takes place between the various ossification centres, giving rise to large composite bones, the limits (sutures) between the constituents of which may become completely obliterated.

The following table, adapted from Augier, gives the stages at which the various fusions occur:

Fusion between:	*Crown-rump; length in mm.*	*Age*
Maxilla and Premaxilla	27	
Supraoccipital and Interparietal	34	
Paired rudiments of Vomer	60	
Tympanic centres	65	
Basisphenoid (median and alar centres)	110	
Prearticular and Malleus	134	
Orbitosphenoid and Presphenoid (paired centres)	135	
Pterygoid and Alisphenoid	158	
Pre- and Orbitosphenoid and Basisphenoid	165	
Presphenoid paired and Presphenoid median unpaired centre	180	
Periotic centres	200	
Squamosal and Tympanic	..	End of 9th month
Squamoso-tympanic and Periotic	..	Birth
Laterohyal and Periotic	..	,,
Alisphenoid and Basisphenoid	..	,,
Basisphenoid and Pterygoid	..	,,
Ethmoturbinals (lateral plate)	..	3rd month after birth
Frontals	..	9th ,, ,,
Exoccipital and Supraoccipito-interparietal	..	2nd year ,,
Basioccipital and Exoccipital	..	4th ,, ,,
Presphenoid and ossicula Bertini	..	12th ,, ,,
Basisphenoid and Basioccipital	..	18th month after birth
Stylohyal and Laterohyal	..	Adult

To facilitate the comparison of the bones of the human skull with those of other Vertebrates, the various ossifications will now be described: first the membrane-bones and then the cartilage-bones, and this will be followed by a Table showing the composition of the compound bones of the fully formed bony skull.

1. The membrane-bones (in the order of their appearance). (For times of origin and fusion, see above.) (PLATES 134, 135).

The *dentary* starts ossifying laterally to Meckel's cartilage, about one-third of its length from the symphysis. From there, ossification extends backwards and forwards and on the medial side of Meckel's cartilage. (Fawcett, 1910*c*).

The condyle and the coronoid process of the dentary are preformed in nodules of secondary cartilage: thereby performing mechanical functions (attachment of muscles: articulation) in anticipation of the dentary. In Man there is no nodule of secondary cartilage in the angular region.

The extreme distal end of Meckel's cartilage becomes ossified, but it is not clear whether this is an extension of the dentary, or an independent endochondral ossification corresponding to a *mentomeckelian* (Fawcett, 1905*c*). At all events, it is intimately continuous with the dentary. In addition there is a variable number of symphysial cartilages which become absorbed, and, unconnected with them, of *mental* membranous ossifications in the ventro-anterior region of the symphysis between the dentaries.

The two dentaries become completely fused at the symphysis.

The *maxilla* starts ossifying in its alveolar portion, and extends up beside the wall of nasal capsule (ascending process), backwards in the floor of the orbit, and inwards in the false palate (palatine process).

The *premaxilla* at its earliest appearance so far studied (26 mm.) is an independent centre (Augier, 1932*a*) but it almost immediately fuses with the maxilla (Golling, 1915). Thus, after endless controversy as to whether Man possesses a premaxilla (or 'intermaxillary bone'), it is now possible to assert that the premaxilla has a separate centre of ossification. But the precocity of its fusion with the maxilla is of interest, for the cessation of growth thereby caused results in the orthognathism typical of human skulls.

The premaxilla starts in its alveolar portion and spreads outwards and upwards (forming the side of the external nostril), and also inwards forming the palatine process. From the medial end of the palatine process a prevomerine (subvomerine: Augier) process projects back medially to the anterior palatine foramen and to the paraseptal cartilage. According to Fawcett (1911) and to Jarmer (1922) this process originates as an independent ossification, which would correspond to a *prevomer*.

In the subsequent development of the maxilla, an uprising ridge is formed (internal maxillary wall) ventrally to the maxilloturbinal, and medially to the side wall of the (cartilaginous) nasal capsule, and to the ascending process of the maxilla. In between this ascending process and the internal maxillary wall a diverticulum of the olfactory sac is enclosed which will give rise to the maxillary sinus. After the disappearance of the cartilaginous wall of the nasal capsule, the epithelium of the maxillary sinus is closely adpressed to the surrounding bone, the growth of which it follows: the process being accompanied by erosion of the internal surface (thus giving rise to the antrum of Highmore) and by appositional growth to the external surface of the bone.

The *frontal* arises in the region of the orbital arcade and spreads rapidly over the forehead and the inner wall of the orbit. The (metopic) suture between the right and left frontals soon disappears completely. Neither Augier nor Inman and Saunders (1937) find any support whatever for Ashley-Montagu's (1931) contention (perpetuating an old view of Serres, 1819) that a postfrontal centre is constantly present. Over the orbit the frontal becomes hollowed out, accommodating a diverticulum of the olfactory sac forming the frontal sinus. The cavity of the frontal sinus becomes enlarged after puberty by internal erosion, while external apposition here gives rise to the brow-ridges. In this respect the female retains the juvenile condition.

The nasal spine of the frontal bone appears to have a separate origin, as an endochondral ossification in the dorsal part of the nasal septum.

The *palatine* arises immediately behind the side wall of the nasal capsule as a strut from which the perpendicular portion rises up forming the side wall of the nasopharyngeal passage, while the horizontal or palatine process in the false palate forms its floor, immediately behind the palatine process of the maxilla (Fawcett, 1906). Part of the space between the dorsal edge of the perpendicular portion of the palatine and the presphenoid and ectethmoid above it, persists as the sphenopalatine foramen.

The *squamosal* ossifies first in its zygomatic portion, and the ossification then extends into the squamous portion. But very often this squamous portion has an independent centre of ossification which immediately fuses with the other. This appears to be an incipient case of fragmentation of a bone which in other Vertebrates arises from a single centre.

The squamosal lies laterally to the incus and malleus and to the auditory capsule, with the ossification (periotic, petromastoid) in which it eventually fuses, and contributes to the formation of the mastoid process. The squamosal also fuses with the tympanic. Squamosal, periotic, and tympanic thus combine to form the 'temporal' bone (Sutton, 1883; Gaupp, 1915).

A nodule of secondary cartilage is found in the region of the glenoid fossa.

The *jugal* arises as a small centre of ossification in the floor of the orbit, from which a temporal process develops towards the squamosal, an infraorbital process towards and above the maxilla, and a frontal process which forms the postorbital bar (lateral border of the orbit) towards the frontal. A lamina orbitalis is thus developed, which, growing inwards and backwards makes contact with the alisphenoid and thereby shuts off the orbit from the temporal fossa. While these processes are developing, resorption takes place in the original bony plate (Toldt, 1902), and should this resorption proceed beyond a certain point, division of the bone may result (Toldt, 1903).

The *prearticular* (goniale) arises on the medial side of the posterior region of Meckel's cartilage and becomes fused with the malleus, forming its processus Folii.

The '*pterygoid*' (internal pterygoid plate) ossifies in a nodule of secondary cartilage (the pterygoid or hamular cartilage) in the hind part of the side wall of the nasopharyngeal passage. The tendon of the tensor veli palatini muscle is bent round the hamular cartilage, which fact explains the existence of secondary cartilage in this region of the pterygoid membrane-bone. From the hamular region the ossification extends dorsally into a thin horizontal plate which may have an independent centre of ossification (Fawcett, 1905*a*, *b*, 1910*a*) corresponding to the lateral wings of the parasphenoid, while the hamular ossification appears to represent the reptilian pterygoid (see p. 435).

The pterygoid becomes fused with the external pterygoid plate of the alisphenoid and with the basisphenoid. The fusion with the latter encloses the Vidian canal, for the palatine branch of the facial nerve runs ventrally to the basisphenoid but dorsally to the pterygoid (or parasphenoid).

The *vomer* arises from paired centres of ossification along the ventral edge of the nasal septum, behind the paraseptal cartilages. At the 50–60 mm. stage the centres fuse to form the median bone which is regarded as the homologue of the rostrum of the parasphenoid (see p. 433). The vomer becomes higher in the sagittal plane, and ultimately replaces the lower portion of the nasal septum. Occasionally the hindmost parts of the paraseptal cartilages may undergo endochondral ossification (Fawcett, 1911) and the resulting bones fuse with the vomer.

The *interparietal* and *preinterparietal* are variable in their occurrence, independence as paired or median bones, and in the fusions which they undergo with neighbouring bones. The interparietals arise from paired centres immediately in front of the tectum synoticum, with the endochondral ossification of which (the supraoccipital) they fuse, as also with one another. In front of the interparietals and also medially to them, there may arise paired preinterparietal ossification centres (immediately over the cartilaginous nodule representing the tectum synoticum), which likewise fuse with one another and with the interparietal.

The interparietals may (rarely) preserve their identity as paired (or three, or four) bones, not fused with the supraoccipital, or as a median bone when fused with one another. Similarly, the preinterparietals may sometimes preserve their identity as paired or median ossifications, in the latter case the preinterparietal being sometimes known as the Inca bone. (For the morphological significance of the interparietal and preinterparietal bones, see p. 444). Meanwhile it must be noted that when the interparietal appears as a separate element, its posterior border does not coincide with the original suture between the interparietal and supraoccipital ossifications, but lies farther forward and within the territory of the true interparietal. The hindmost, portion of that bone remains fused to the front edge of the supraoccipital forming the 'lamella triangularis', and the incisura lateralis may cut right through in front of it as the sutura mendosa, separating it from the anterior part of the interparietal, which when separate in Man is therefore only a fragment (Debierre, 1895).

The *tympanic* presents a remarkable case of fragmentation of a bony element, for while it corresponds without doubt to the tympanic of other Mammals and to the angular of reptiles, which are simple ossifications, in Man the tympanic has two and sometimes three ossification centres (anterior, inferior, and posterior) which soon fuse to form the characteristic tympanic ring, which at the time of birth fuses with the squamosal. The (Glaserian) fissure between these bones serves to let the chorda tympani out of the tympanic cavity. The lateral extension of the ring to form the tubular wall of the external auditory meatus takes place later. The plane of the tympanic ring undergoes many changes of position during development (see Forster, 1925).

The *nasal* arises as a single (paired) centre in the connective tissue immediately overlying the cartilaginous roof of the nasal capsule, and eventually forms (with its fellow of the opposite side) the dorsal border of the external nostrillar aperture.

The *lachrymal* arises as a single (paired) centre laterally to the hind part of the side wall of the cartilaginous nasal capsule.

The *parietal* of each side arises from *two* centres which normally fuse very soon to form one. Occasionally, however (Ranke, 1899*b*), the fusion does not take place, and there are then two bones together corresponding to the parietal of lower forms, which arises as a single centre. This appears, therefore, to be a case of 'division primordiale' (see p. 507). During birth, the parietals are able to overlap one another, the frontals and the interparietal, thus altering the shape of the head.

Wormian bones are small ossifications, irregular in occurrence, shape, number, position, and time of appearance, for which reason no phylogenetic significance can be ascribed to them. Wormians may be dermal or endochondral ossifications (for the latter, see p. 373). They appear to be fortuitous embryonic ossifications concerned with the enclosure in a bony case of the enormous human brain. (A parallel occurrence is found in the Polish breed of poultry where the hole between the frontals through which the cerebral hernia protrudes is closed over by means of Wormian bones; see p. 486.) When they arise in a space between three or more 'real' bones, the Wormians are known as fontanellar, or sutural when they arise between two closely adjacent bones.

Augier has drawn attention to the fact that if Wormians were adaptively developed as bungs to stop up large fontanelles, they should occur most frequently in the fontanelles which remain open longest. But this is not the case. In the majority of cases Wormian bones appear to be due to the detachment of marginal 'islands' of ossification from the edges of the normal bones, possibly owing to the stretching of the connective tissue membranes (dura mater) surrounding the rapidly expanding brain. This is particularly evident in cases of hydrocephaly when the expansion of the brain and the consequent stretching of the membranes are much exaggerated. But in otherwise normal skulls, it is difficult to see why there should be such abnormal stretching as to detach Wormian bones.

In other cases, Wormian bones appear to be real intercalary ossifications, not derived by detachment from any neighbouring bone (e.g. the lachrymo-maxillary ossicle; or perhaps the 'epipteric' bone of Sutton, 1884*b*). In such cases, the question arises as to whether such ossifications are homologous with elements in other Vertebrates (in which case they should take their name, and cease to deserve the term Wormian), or whether they represent new elements altogether. The latter is the more probable explanation, though each case must be tested on its own merits. At all events it must be noted that the existence in some lower Vertebrates of a separate bone in a position corresponding to that of a human Wormian bone is no justification for the assumption of a homology between them. This can only be established if a bone in the position in question is constantly found throughout a reasonable phylogenetic series.

In 0·03 per cent. of cases (in 6,000) human skulls present according to Augier a small membrane-bone between the lachrymal and the frontal. Such a bone is present according to Cornevin (1883) in a greater percentage of cases in Ungulates (10 per cent. in Bos, 5 per cent. in Ovis), and Wortman (1925) has described it in certain Carnivora and Insectivora. It may therefore represent a *prefrontal*, which in Mammals would then be normally crowded out of existence by the frontal. Before this can be regarded as established, however, it will be necessary to determine its occurrence, and its frequency of occurrence, in other members of the Primate stock. The same is true of a bone found with extreme rarity in the human skull between the frontal and the jugal, and which might represent a *postfrontal*. At all events, it may be taken as certain that the prefrontal and postfrontal of reptiles are *not* absorbed into or fused with the frontal as some authors have tried to contend (e.g. Serres, 1819; Rambaud and Renault, 1864; Maggi, 1897, 1903; see also above, p. 365).

The formation of *sutures* between adjacent bones has been studied by Sitsen

(1933). In the case of dermal bones this phenomenon involves the thickening of the edges of the bones where they come into contact (sometimes violently as during birth) and the formation of facets of contact which become joined by ossification of the intervening connective tissue. An attempt has been made by Aichel (1915) and Troitsky (1932) to show that the position of the sutures is predetermined in the dura mater by the existence of non-osteogenic strips separating the fields of ossification of the future bones. This question is discussed further on p. 483.

2. The cartilage-bones (in the order of the appearance in the various regions of the chondrocranium. For times of origin and fusion see pp. 363, 364).

(a) In the occipital region. The *supraoccipital* ossifies in the tectum synoticum, from paired centres according to Mall (1906), though Augier was able to find only one. The dorsal (posterior) occipital incisure in the tectal cartilage eventually becomes ossified, partly by approximation of its cartilaginous and ossifying sides, and partly by the formation of an intramembranous extension of the supraoccipital bone in the form of a backwardly directed process (process of Kerkringius), sometimes with the intervention of a separate Kerkringial ossicle which fuses with the body of the bone. This bone may arise intramembranously or in the cartilage of the processus descendens (posterior) of the tectum posterius (Augier, 1931*b*).

Anteriorly, the supraoccipital fuses with the (membrane-bone) interparietal and sometimes, through it, with the preinterparietal (p. 367). On each side, the supraoccipital eventually fuses with the exoccipitals, but only after birth and after the absorption of the intervening cartilage, Budin's (1876) 'obstetric hinge', which permits a certain amount of flexibility of the occipital region during the act of birth, and alteration of the length of the maximum mento-supraoccipital diameter of the head.

The *exoccipital* ossifies in the occipital arch, around the hypoglossal (condylar) foramen, and in the dorsal and major part of the condyle. It eventually fuses, dorsally with the supraoccipital and ventrally with the basioccipital.

The *basioccipital* arises normally from a single median centre in the posterior part of the basal plate, ventrally to the notochord. Occasionally, however, there may be another median centre in front of the basioccipital but soon fusing with it. This anterior or basiotic centre arises in that part of the basal plate which is dorsal to the notochord (Augier, 1926, 1928). The basioccipital extends into the ventral part of each condyle, when it meets and fuses completely with the exoccipital. Very late, the basioccipital fuses with the basisphenoid.

The *alisphenoid* (great wing of the sphenoid, alipostsphenoid of Augier) ossifies from a centre in the lateral part of the ala temporalis and its processus pterygoideus, surrounding the foramen rotundum in the cartilage (Sutton, 1885; Fawcett, 1905*a* and *b*, 1910*a*). At the same time there appears to be another centre of ossification developed in close relation to the first, which it engulfs, and then spreads away from the cartilage of the ala temporalis as an intramembranous ossification giving rise to the external pterygoid plates and the orbital or squamous portion of the alisphenoid, gradually surrounding the mandibular branch of the trigeminal nerve and enclosing it in the foramen ovale. The nature of this intramembranous ossification is still obscure. Ranke (1899*c*) considered it to represent the intertemporal bone; Nicola (1903) a Wormian ossification: in the latter case it may represent a portion of the alisphenoid temporarily detached from the parent bone because of the stretching and extension which it has to undergo to cover the brain in this region.

The alisphenoid eventually fuses by means of its external pterygoid plate with the pterygoid bone (or internal pterygoid plate), and also with the basisphenoid (in the processus alaris).

(b) In the orbitotemporal region. The *orbitosphenoid* (small wing of the sphenoid, alipresphenoid of Augier) ossifies in the metoptic root of the orbital cartilage, and spreads forwards into the preoptic root (thus surrounding the optic foramen from above) and backwards into the anterior clinoid process. The orbitosphenoid fuses with the presphenoid, the lateral portion of which forms the floor of the optic foramen or canal. Part of the space between the orbitosphenoid and the alisphenoid persists as the sphenoidal fissure or foramen lacerum anterius of English authors (for the exit of oculomotor, trochlear, ophthalmic, and abducens nerves).

The *basisphenoid* (basipostsphenoid of Augier) arises from paired centres in the cartilage of the central stem (in the pituitary fossa) which become fused together and with another pair of ossification centres which arise subsequently in the processus alaris (also referred to by the unfortunate name of 'sphenotic'). The latter, or alar, centres extend backwards in the alicochlear commissures thus giving rise to the lingulae sphenoidales. These subdivide the space between the basisphenoid and the periotic on each side (the foramen lacerum medium of English authors (anterius of Continental writers)) into a more median carotid foramen and a more lateral sphenopetrous fissure for the palatine (greater superficial petrosal) nerve. The dorsum sellae ultimately becomes ossified by extension of ossification from the basisphenoid.

The basisphenoid becomes fused with the presphenoid in front, with the alisphenoid on each side, with the pterygoid beneath, and, ultimately, with the basioccipital behind.

The *presphenoid* arises from no less than five centres. Of these, one pair (lateral paired centres) arises in the ala hypochiasmatica, and another (median paired centres) in the cartilage of the central stem immediately medially to the ala hypochiasmatica. These centres soon fuse on each side, and with the orbitosphenoid of their side. Lastly, the median unpaired presphenoid centre ossifies in the hind part of the nasal septum, and fuses with the paired presphenoid centres on each side. The latter then fuse with the basisphenoid behind, thus enclosing an unossified space in the middle, which becomes smaller and smaller.

Apposed to the anterior surface of the presphenoid are the ossicula Bertini, which eventually become fused with it. The sphenoidal sinuses perforate the ectethmoids, and then invade the presphenoid which becomes hollowed out as a result of internal erosion, and thus comes to form a shell enclosing the paired sphenoidal sinuses (see Augier, 1927).

The *pleurosphenoid* is a vestigial ossification occasionally found in the supracochlear (or infra-Gasserian) cartilage, which itself is a vestige of the pila antotica. The pleurosphenoid may either become fused with the periotic, or remain as an isolated nodule in the dura mater, representing the original partition between the cavum epiptericum (primitively extracranial, lodging the trigeminal ganglion) and the cranial cavity.

(c) In the auditory region. The *opisthotic* appears to be represented by the centres which Augier has described as cochleo-canalicular and cochlear. They arise in the posteroventral wall of the auditory capsule, forming the hind wall of the posterior semicircular canal and of the hinder part of the lateral semicircular canal; the ossification which they form extends into the wall of the cochlear part of the capsule, and forms the ventral border of the fenestra ovalis and the dorsal border of the foramen perilymphaticum.

The opisthotic fuses with the other otic ossifications to form the periotic or petrous portion of the petromastoid.

The *prootic* appears to correspond to the centres which Augier has described as anterior and posterior canalicular, situated in the walls of the anterior and posterior limbs of the anterior semicircular canal, together with an intermediate centre (at the base of this canal) and (occasionally) a centre in the prefacial commissure, and an intermediate centre on the posterior semicircular canal. The ossification resulting from the fusion of these centres forms the anterior, dorsal, and lateral walls of the canalicular portion of the auditory capsule, extending ventrally as far as the fenestra ovalis on the outside and the foramina acustica (internal auditory meatus) on the inside; anterolaterally it extends into the tegmen tympani, where, according to Broman (1899), there may be an independent ossification centre.

Eventually the prootic fuses with the remaining ossifications of the auditory capsule to form the periotic.

The *epiotic* appears to correspond to the ossification centres which Augier has described as the mastoid centre and the intermediate centre of the lateral semicircular canal. Together with the squamosal with which it fuses, the epiotic contributes to the formation of the mastoid process. This mastoid region of the periotic is penetrated by diverticula of the tympanic cavity; internal erosion of the bone sets in, thus giving rise to the antrum and mastoid cells.

It is interesting to note that, as reported by Bast (1930), the ossification of the auditory capsule does not begin until the labyrinth has reached its maximum life, i.e. there is no growth.

The *periotic* is the term given to the fused ossifications in the walls of the auditory capsule, and which also fuses with the squamosal and the tympanic. The fusion with the tympanic (which is already fused with the squamosal) involves the formation of the roof and floor of the bony tympanic cavity, by the outgrowth of bony plates from the periotic to the tympanic ring. In this process, the internal carotid artery becomes enclosed in a long canal which it has to traverse before it reaches its real entry into the cranial cavity at the carotid foramen. Similarly the facial nerve, after emerging through the facial foramen and finding itself in the facial groove or hiatus of Fallopius becomes enclosed in the inner wall of the tympanic cavity in a canal which opens at the stylomastoid foramen. All that is then left of the hiatus of Fallopius is the small opening from the anterior part of the canal through which the petrosal (palatine of facial, pharyngeal of glossopharyngeal) nerves emerge, dorsally to the lateral prefacial commissure which is formed by the fusion between the tegmen tympani and the lateral wall of the cochlear capsule.

(d) In the ethmoid region. The *paraseptal* ossifications (Fawcett, 1911; Augier, 1931*c*) are occasionally seen to arise in the hindmost part of the paraseptal cartilages and to become fused with the vomer.

The *maxilloturbinal* ossifies in the cartilage of similar name, and becomes fused with the maxilla.

The *ossiculum Bertini* ossifies in the cartilage of the lamina orbitonasalis, forming the hind wall of the nasal capsule on each side. It becomes fused on to the anterior face of the presphenoid, and pierced by the sphenoidal sinus. The ossiculum Bertini is sometimes referred to as the 'sphenoidal turbinal'.

The *ethmoturbinals* ossify in the inferior and superior cartilages of similar name, and the ossifications, spreading laterally, meet and fuse with other centres (in the ethmoidal bulla and crista semicircularis) in the side wall of the nasal capsule to form the lateral plate (papyraceous plate; os planum), which in turn forms part of the medial wall of the cavity of the orbit. Irregular ingrowths from the hinder region of the lateral plate, together with certain independent ossification centres, contribute to form the cribriform plate. The nasal septum is ossified in part from extensions of the ossifications of the cribriform plate, and in part from an independent *mesethmoid* ossification. Occasionally the region of the crista galli may ossify from a separate centre.

The ethmoturbinals, lateral plate, cribriform plate, and mesethmoid fuse together to form the so-called ethmoid, which is extensively perforated and permeated by diverticula from the olfactory sac. The lateral plate is riddled with such cavities, some of them extending farther into neighbouring bones and forming the frontal, sphenoidal, and maxillary sinuses. (See Seydel, 1891; van Gilse, 1926.)

The lateral plate in the orbit is in contact anteriorly with the lachrymal, posteriorly with the orbitosphenoid. Dorsally, its contact with the frontal is interrupted by the ethmoido-frontal foramina, which convey the nasal branch of the ophthalmic nerve from the orbit to the cavity of the nasal capsule, and mark the position of the orbitonasal fissure.

(e) In the visceral arch skeleton. The *incus*, ossifying in the quadrate cartilage, calls for no special mention.

The *malleus* ossifies from two centres, one above the other, in the posterior region of Meckel's cartilage. These centres fuse together, and, soon with the prearticular which thus gives rise to the processus Folii, or processus gracilis of the malleus.

The *stapes* ossifies from two centres, one in each arm of the cartilaginous stapes. These centres then fuse both medially and laterally to the foramen through which the stapedial artery passes.

The *laterohyal* (tympanohyal) ossifies in the most dorsal portion of Reichert's cartilage, and almost immediately fuses with the periotic. The more ventral portion of Reichert's cartilage, or *stylohyal* ossifies later, and only very late becomes fused with the laterohyal (and therefore with the periotic). Occasionally, a *ceratohyal* also becomes ossified and fused with the stylohyal. There may also be an *accessory* ossification which fuses either with the ceratohyal, or with the next ventral segment, the *hypohyal* (apohyal) which gives rise to the lesser cornu of the hyoid.

The *basihyal* or body of the hyoid, appears to ossify from a single median centre.

The *thyrohyal* ossifies in the cartilage of similar name and gives rise to the greater cornu of the hyoid.

(f) Wormian cartilage-bones. An ossification centre is sometimes found in the sphenethmoid commissure, giving rise to the so-called sphenoethmoido-frontal (or alipresphenofrontal of Augier), which may fuse with the orbitosphenoid. It may be present only on one side, and is usually larger on the inner than on the outer surface.

Other examples of endochondral Wormians are to be found in the ossicles which are (rarely) developed between the supraoccipital and the exoccipitals in the dorsal part of the occipital arches.

3. The compound bones of the fully formed human skull.

The *occipital* is composed of the basioccipital, exoccipitals, and supraoccipital, cartilage-bones, and of the interparietals and sometimes also the pre-interparietals, membrane-bones.

The *sphenoid* is composed of the basiphenoid, alisphenoids, presphenoid, orbitosphenoids, ossicula Bertini, cartilage-bones, and of the pterygoids, membrane-bones.

In adult Man the sphenoid and the occipital fuse to form the *spheno-occipital.*

Each (paired) *temporal* is composed of the periotic bones (prootic, opisthotic, epiotic), cartilage-bones, and of the squamosal, and tympanic, membrane-bones. In addition, some cartilage-bones of the visceral arch skeleton (laterohyal, stylohyal) become fused with the temporal.

The *ethmoid*, is composed of the ethmoturbinals, cribriform plate, and mesethmoid, cartilage-bones.

III. COMPARATIVE SECTION

33. GENERAL MORPHOLOGICAL CONSIDERATION OF CERTAIN CARTILAGES

I. The Problem of the Trabeculae Cranii and Polar Cartilages

THE trabeculae cranii, first recognized by Rathke (1839) in the grass-snake, were regarded by Huxley (1874, 1875*b*) as belonging to the visceral arch skeleton, pressed into the service of the brain-case. Parker (1879*b*) inclined more or less to the same view, which was further supported by Platt (1897), Howes and Swinnerton (1901), and by Allis (1923*b*, 1925, and 1931) who identified the trabeculae and polar cartilages as the pharyngal elements of the premandibular and mandibular arches respectively. The polar cartilages have been recognized as distinct elements in Squalus, Scyllium, Scymnus, Acipenser, Lepidosteus, Anas, and Echidna.

The evidence in favour of this view may be considered under three heads: morphological, developmental, and experimental.

(i) Allis has pointed out that a space—the subpituitary space, traversed by the pituitary vein (p. 57)—is enclosed ventrally to the dura mater in the floor of the cranial cavity, and dorsally to the trabeculae themselves. This indicates that the trabeculae are not themselves a part of the original cranial wall which must be taken as represented by the dura mater. The subpituitary space is therefore strictly extracranial.

(ii) In development, it has been shown by van Wijhe (1922) in Selachians and by Jager (1926) in birds that the cross commissure between the two premandibular somites (the most anterior of the entire series) marks the position of the future acrochordal cartilage, or extreme front edge of the basal plate (parachordal plate). Consequently the trabeculae, which lie in front of the parachordals, cannot be regarded as of axial origin, i.e. derived from the sclerotomes of the mesodermal somites. This accords with Sewertzoff's (1916) and Matveiev's (1925) statements that in Acipenser the sclerotomes of the foremost somites contribute to the formation of the parachordals. These results are supported by those of Sewertzoff and of Platt (1897) on Amphibia and of Filatoff on Reptilia.

In an attempt to trace the origin of the trabeculae themselves, de Beer (1931*b*) concluded (from a study of Scyllium, Salmo, Rana, and Amblystoma) that the trabecular rudiment condenses out of mesenchyme in situ in the maxillary process, in close association with the rudiment of the pterygoquadrate. There is no evidence of derivation of this mesenchyme from the somites: if there were it would necessitate the view that the pterygoquadrate also is of axial origin.

(iii) In the experiments performed by Stone and others involving removal of the neural crest, it has been consistently found that the trabeculae (as far back as the optic or oculomotor foramen) are subject to the same variation as the visceral arches. These experiments are described in detail on p. 472, and without prejudice to the separate problem as to whether the neural crest cells give rise to visceral cartilage, these results are mentioned here solely because of the identical behaviour of the trabeculae and visceral arches under the conditions of the experiments.

If then the trabeculae represent visceral arch cartilages, they must be the skeletal supports of a whilom premandibular arch, separating the originally anterior mouth from a pair of mandibular visceral clefts, and corresponding in the scheme of segmentation to the premandibular somite and the profundus nerve. There is evidence of the existence of such an arch and cleft in the Ostracoderms, as shown by Stensiö (1927).

Attention may now be turned to the Cyclostomes, for in these forms Sewertzoff (1916) has contended that the floor of the skull is entirely of axial origin and formed of parachordal cartilages (anterior and posterior). Koltzoff (1902) found that these cartilages arose from the sclerotomes of the first three segments, and those results have been confirmed by Filatoff (in Sewertzoff, 1916).

The question therefore arises: have the Cyclostomes any element corresponding to the trabeculae? Sewertzoff (1917) and Balabai (1935) attempted to identify the posterodorsal cartilage as the trabeculae fused together, but this view neglects the objections that the position of the posterodorsal cartilage is the result of the very specialized development of the upper lip (see de Beer, 1923), and is reversed end for end as compared with the relations of the true trabeculae; i.e. if the posterodorsal cartilage is the trabeculae, then they have become attached to the parachordals by their *morphologically anterior* end.

The trabeculae have been looked for in Petromyzon by de Beer (1931*b*) in the premandibular arch of the Ammocoete. The conditions are not as simple as could be desired, since it is now known that the mucocartilaginous premandibular arch of the Ammocoete does not become, but is rather replaced by, the definitive premandibular arch of the adult Petromyzon. The latter, then, would represent the trabeculae, attached by their dorsal end to the front of the parachordals, and flanking the buccal cavity on each side. (Plate 136).

The development of the Gnathostome type involved a great development of the brain, which bulges forwards in front of its morphologically anterior end (recessus neuroporicus); and it is presumably to assist in enclosing the correspondingly enlarged cranial cavity that the trabeculae come to form the floor of the prechordal region of the brain-case. The mouth is pressed backwards on to the ventral surface of the head, and its 'angles' occupy the position of the former mandibular visceral clefts, between the premandibular and mandibular arches. In many fish it is interesting to note that the trabeculae form the sole cartilaginous skeleton of the upper jaw, before the pterygoid processes of the quadrates develop.

There is thus much to be said in favour of Allis's view that the trabeculae represent the premandibular arch, though it is an excess of refinement to try and recognize a pharyngal as apart from epal or other elements. Quite possibly, the participation of the premandibular arch in the formation of the floor of the skull preceded the segmentation of the visceral arches into their pharyngal, epal, and other elements.

But as to Allis's contention that the polar cartilage is a pharyngomandibular there is need for caution. It is, for instance, claimed by Sewertzoff and Disler

(1924) that the pharyngomandibular is represented by a cartilaginous nodule situated between the pterygoquadrate and the skull, and which may either fuse on to the pterygoquadrate forming part of the basal process, or remain isolated in the ligament stretched between the basal process and the skull. Matveiev (1925) believes that the polar cartilage is of axial origin like the parachordal (the anterior part of which it may possibly represent in some forms), and indeed, the case for the polar cartilage being of visceral nature has not yet been established.

Reference may be made to a peculiarity of the trabeculae in Elasmobranchs, which may, perhaps, owe its explanation to the visceral, or at least non-axial, origin of the trabeculae. In Elasmobranchs the internal carotid arteries enter the skull in the typical manner between the trabeculae, but they receive the efferent mandibular arteries (efferent pseudobranchial arteries) *after* entering the skull, instead of before as in all other vertebrates. The efferent mandibular arteries in Elasmobranchs (after giving off the ophthalmic arteries) enter the skull through special foramina situated *dorsally* to the trabeculae, whereas in all other vertebrates they run inwards ventrally to the trabeculae (and the ophthalmic arteries emerge through foramina dorsally to the trabeculae). It is, of course, possible that this result is due to a change in the course of the blood-vessel, but in the absence of any evidence this view can hardly be regarded as morphologically satisfactory.

Allis (1923*b*) has advanced an explanation of this state of affairs, based on the possibility that the attachment of the hind end of the trabecula to the polar cartilage may have occurred beneath (Elasmobranchs) or above (remaining vertebrates) the efferent mandibular artery. He has shown that the relations of the efferent branchial arteries to the branchial arches supply evidence for this view, for in Selachii with their sigma-shaped branchial arches the artery runs anterodorsally to the pharyngobranchial, whereas in Teleostomes with their V-shaped arches the artery runs behind the pharyngobranchial.

The trabeculae of different vertebrates may vary considerably in their distance apart. When wide apart the skull is said to be platybasic; when close together and extensively fused in the midline the skull is said to be tropibasic, and the trabeculae form a trabecula communis. It was pointed out by van Wijhe (1922) that as this variation is concerned not with the basal plate but with the trabeculae it would be more appropriate to refer to the two conditions as 'platytrabic' and 'tropitrabic' respectively.

The platytrabic or tropitrabic form of a skull is correlated with the relative sizes of the eyeballs and of the brain at the level of the eyes. It has been found possible to obtain the tropitrabic condition by experimental treatment of embryos which would normally produce platytrabic skulls with agents conducive to cyclopia (see p. 477). It is probable that the correlation between the condition of the trabeculae and the width of the brain is due to the fact that both these structures are dependent on the 'organizer' in the form of the roof of the gut.

The originally primitive conditions of the skull must have been platytrabic, as probably represented actually by the Selachii. Since the condition of the trabeculae is controlled by an agency which is easily susceptible to modification in its action, it is not surprising that the tropitrabic condition should be found in a number of different groups where it was probably independently evolved, e.g. Teleostei and Reptilia. For the same reason, the fact that a skull is platytrabic is not *ipso facto* evidence that this condition is primitive, for it may have been secondarily reacquired. This is certainly the case in the Cypriniformes and Gymnarchus among the Teleostei, and probably in the Anura and Urodela, for there is reason to believe that the primitive Stegocephalia were tropitrabic. There can be no hard and fast distinction between the platytrabic and tropitrabic conditions, for even in the former type the trabeculae are joined anteriorly to form a trabecular plate.

Tropitrabic skulls are commonly associated with the formation of an interorbital septum (see p. 388), and wherever there is a median nasal septum it appears to be a direct anterior prolongation of the trabecular plate or trabecular communis.

Three vertebrate groups call for special mention in connexion with their trabeculae: the Ophidia, the Chelonia, and the Mammalia.

In the Ophidia the trabeculae are very long and close together, and the curious feature is that they do not fuse to form a trabecula communis, but remain separate, except at their extreme front end.

Among the Chelonia, Chelone and Dermochelys possess a taenia intertrabecularis which divides the hypophysial fenestra sagittally into two, between the trabeculae. A similar structure is present among birds in Passer (pp. 261, 285). This is in position comparable with the mammalian hypophysial plate, as will immediately be seen.

The Placental mammals are characterized by the possession of a median central stem, *laterally* to the hinder part of which the internal carotid arteries enter the cranial cavity: in all other forms (for exceptions see p. 494) these arteries run *medially* to the trabeculae. The mammalian condition is explained by the Monotremes, and a series can be traced passing through the following stages (see Plate 137):

(i) *Non-mammals.* The trabeculae and polar cartilages form bars which join the parachordal plate *laterally* to the internal carotid arteries; the hypophysial fenestra between these bars eventually becomes chondrified, forming the hypophysial plate. In Chelone and Dermochelys this takes the form of a taenia intertrabecularis.

(ii) *Monotremes.* The hind ends of the trabeculae and the polar cartilages arise *laterally* to the internal carotid arteries, and become interconnected in the midline by the formation of the hypophysial plate. The junction with the parachordal plate takes place both *laterally and medially* to the internal carotid arteries.

(iii) *Marsupials.* Trabeculae are not independent, but arise in continuity with the wide hypophysial plate, which is perforated by the paired carotid foramina. Posteriorly the hypophysial plate and trabeculae fuse with the anterior edge of the parachordal plate. The hindmost part of the trabecula may be delayed in its appearance (e.g. Didelphys, where for a time the carotid foramen has no lateral boundary), and develop later as a backgrowth exactly resembling the alicochlear commissure of Placental mammals except that it fuses with the basal plate.

(iv) *Placental Mammals.* The hypophysial plate develops first, *medially* to the internal carotid arteries, as the central stem, much resembling the Chelonian taenia intertrabecularis; and it may arise from paired centres of chondrification (improperly called polar cartilages), to which the term hypophysial cartilages should be applied. The true polar cartilages and hind ends of the trabeculae are delayed in their appearance (sometimes altogether), and are represented by the alicochlear commissures, which are situated *laterally* to the internal carotid arteries, and medially to the palatine (greater superficial petrosal) nerves. In most Placentals the alicochlear commissure connects the base of the processus alaris with the front of the cochlear capsule, which latter has usurped the position of the anterolateral corner of the parachordal plate and suppressed it. In certain whales, however (Balaenoptera, Megaptera, Lagenorhynchus), a lamina supracochlearis is present, representing the original anterolateral part of the parachordal, and to which the alicochlear commissure is attached (see p. 341).

The objection raised by Voit (1909) to the homology of the alicochlear commissure with the trabecula, on the grounds that in Lepus the internal carotid artery enters the cavum epiptericum and lies (topographically) laterally to the abducens nerve, and that therefore the mammalian carotid and its foramen are not homologous with their namesakes in lower vertebrates, is not valid. Gaupp (1910*a*) has drawn attention to the fact that in other mammals the carotid does not traverse the cavum epiptericum, and yet the artery and its foramen are clearly the same. In Lepus the cavum epiptericum bulges medially, which accounts for the course of the artery through it. The same is presumably the case in Perameles.

The difference between the Placental mammal and other vertebrates in respect of the trabeculae is therefore not fundamental, but the result of heterochrony in development as regards the lateral trabeculae and polar cartilages and the median hypophysial plate, and of a migration towards the side on the part of the internal carotid artery. This migration is further reflected in the fact that in primitive mammals (e.g. Monotremes, Marsupials, Erinaceus, Centetes, Vespertilio, Orycteropus, Herpestes; van Kampen, 1905) the internal carotid foramina perforate the basisphenoid, while in other mammals (including the Marsupial Acrobates; Wincza, 1896) the arteries pass laterally to the basisphenoid, between it and the periotic, through the foramen lacerum medium (anterius of continental authors).

The anterior parts of the trabeculae in Placental (and other) mammals are, of course, fused and merged in the interorbitonasal part of the central stem. At early stages, however, e.g. in Tatusia, this structure may show a bipartition of which the prongs may represent the paired trabeculae (p. 317).

II. The Parachordals

The parachordal cartilages, first so called by Huxley (1874, p. 198), form the basal plate of the chondrocranium. They are traceable throughout the vertebrate series from the Cyclostomes (where they seem to be the sole constituents of the skull-floor) to Man as paired cartilages flanking the notochord and (in the highest forms) obliterating it in the process of their mutual fusion.

While there can be no question of the homology of the parachordals throughout the vertebrates, they may show considerable differences in their method of chondrification and in their posterior extension.

i. Details of chondrification

The differences in method of chondrification refer to the position of the parachordals with regard to the notochord. They may lie closely apposed to it throughout their length (Urodela, Anura, Mammalia); or their anterior ends may diverge away from the notochord leaving a basicranial fenestra at early stages (Selachii, Teleostomi, Reptilia, Aves); or they may fail altogether to touch the notochord at their first appearance (Protopterus).

Another difference relates to the possibility of chondrification of the parachordal cartilage from more than one centre on each side. This is found to be the case in Petromyzon (p. 42), Amia (p. 98), Gasterosteus (p. 147), Protopterus (p. 175), Urodela (p. 177), and mammals (p. 306). It is, of course, obvious that in those cases where the addition of vertebral elements can be observed to take place during ontogeny, such vertebral elements have separate centres of chondrification. This is found, e.g. in Acipenser, Ceratodus, and birds. In the fish in which such incorporation of vertebral elements takes place, it is usual to find extensions of the parachordals growing back as dorsolateral and ventrolateral bands and establishing contact with the basidorsals and basiventrals respectively (e.g. Lepidosteus, pp. 108, 111. See also pp. 50, 59, and 170).

In nearly all cases the parachordals are of paired origin, the subsequent fusion to form a median basal plate being secondary. In many birds, however, at their first chondrification the parachordals form a cylinder ('the investing mass') surrounding the notochord.

A question which is as yet unsolved concerns the possibility that in some forms where the polar cartilages do not appear as separate chondrifications they may arise already fused on to the anterior end of the parachordals. This is apparently true in Scyllium and Acipenser and may be true in Teleostei (p. 115), where the relations of the anterior end of the parachordals to the trigeminofacialis chamber suggest that they may correspond to polar cartilages.

ii. Segmentation in the parachordals

The differences in extension of the parachordals are related to the process of cephalization, since additional vertebral elements, when they are added on to the back of the skull, lengthen the parachordals. The ontogenetic details of this process have been preserved in many primitive fish (Acipenser, Amia, Lepidosteus, Ceratodus) and in birds, but are obscured in other forms. Reference must be made to pp. 28–31 for the different numbers of vertebral bodies tacked on to the hind end of the parachordals. Here attention must be paid to the question as to how far forward along the parachordals there is evidence of segmentation, i.e. how much of the parachordals may be regarded as representing segmental vertebral elements free in previous phylogeny.

In the duck the five pairs of hypoglossal foramina indicate the existence of occipital and preoccipital arches corresponding to segments 5 to 9 inclusive. The originally vertebral nature of these arches is further supported by the existence of cranial ribs corresponding to segments 6 to 9 inclusive (de Beer and Barrington, 1934). In Felis four pairs of occipital and preoccipital arches and four hypocentral elements are indicated (Kernan, 1915), but the hypoglossal nerve-roots pass through one foramen (see p. 384). In Scyllium the three pairs of hypoglossal nerve foramina indicate the presence of occipital and preoccipital arches in segments 5 to 7 inclusive (Goodrich, 1918; de Beer, 1931*a*). In Urodeles the preoccipital and occipital arches correspond respectively to segments 5 and 6 (Goodrich, 1911). The occipital arch of Anura seems to correspond to segment 5 (van Seters, 1922).

It is to be noted that there is no evidence of segmentation (involving neural or occipital arches, hypoglossal or interarcual foramina, ribs) in the parachordals anteriorly to segment 5; or, in other words, while that portion of the parachordals which lies in or behind segment 5 may be regarded as representing originally (phylogenetically) free vertebrae, there is no evidence of the existence (in ontogeny or phylogeny) of free vertebrae in front of segment 5. This conclusion is in full agreement with the state of affairs in Cyclostomes, where the parachordals occupy the first four or five segments of the head; there is no occipital arch, and the vertebrae are only incipient structures. Two further sets of facts concur in this conclusion; the almost universal existence of a basicranial fenestra, resulting in the separation of the anterior region of the parachordals from the notochord: a feature incompatible with any originally vertebral nature of this region; and the restriction of the zone of invasion of the cranial notochord in Selachii to the posterior region of the parachordal.

iii. The relations of the parachordals to the notochord

Apart from the question (referred to above) of the position of the parachordals relatively to the notochord at their earliest chondrification, there is considerable variation in the fate and position of the notochord relative to the completed basal plate. The notochord persists as a more or less well-developed structure in the adult skull of Cyclostomes, Selachii, Holocephali, Acipenseroidei, and Dipnoi (in all of which except Cyclostomes the notochordal sheath is invaded by cells from the parachordals, though this has not yet been proved for Acipenseroidei).

In the remainder, however, the notochord is either reduced and restricted to the hinder region of the basal plate (e.g. Polypterus), or partly chondrified by notochordal cartilage (e.g. Anura, Urodela) or completely destroyed (e.g. mammals). In some forms (Amia, Teleostei) the notochord persistently separates the parachordals from one another, at least in this hind region; in other forms (Selachii, Polypterus, Dipnoi), the parachordals fuse with one another both above and below the notochord: in yet others, the notochord or what remains of it lies dorsally to the basal plate, the parachordals having fused beneath it (Holocephali, Lepidosteus, Urodela, Gymnophiona, Lacertilia); while in others again the notochord lies beneath the basal plate, having been excluded from it by the fusion of the parachordals above the notochord (Anura). Within the mammals alone, there is great variation. As shown by Tourneux and Tourneux (1907, 1912), and by Bolk (1922), the notochord (or its vestige) may run all the way from the odontoid peg to the hypophysis dorsally to the basal plate ('retrobasal' or 'suprabasal' type, e.g. Mus); or dorsally to the hindmost part of the basal plate and embedded in the middle of it for the remainder of its course (e.g. Bos); or dorsally to the hindmost part of the basal plate, ventrally to the middle portion, and embedded in the anterior portion ('transbasal' type, e.g. Homo, Lepus). Talpa at early stages resembles this third type, but later conforms to the second when the sub-basal portion of the notochord becomes enveloped in the cartilage of the basal plate. In Bradypus (de Burlet, 1913*a*) the notochord lies wholly ventrally to the basal plate ('antebasal' or 'infrabasal' type). In Sus the notochord lies wholly within the basal plate ('basal' or 'intrabasal' type) but may vary (Augier, 1923, 1924).

Similarly in Phocaena the notochord is intrabasal. In Capra the notochord runs through the hindmost part of the basal plate, dorsally to the middle part, and is embedded in the anterior part. In Lagenorhynchus the course is somewhat similar. The attempt has been made by Levi (1909*b*) to use the position of the notochord as a criterion for determining whether the parachordals represent vertebral pleurocentral or hypocentral elements, or both. However, there can be little doubt that these variations reflect nothing but the different time-relations of the process whereby the notochord at very early stages becomes separated off from the notochord plate of the foregut, and are of little value for the estimation of phylogeny and affinity. They are of some interest, however, in connexion with the formation of tumours in this region (see Williams, 1908).

iv. The acrochordal and the basicranial fenestra

An unchondrified region of the basal plate, or basicranial fenestra, between the anterior ends of the parachordals is present in all groups of vertebrates except Dipnoi, Anura, and Crocodilia. Some Urodela develop a complete basal plate in which a basicranial fenestra subsequently becomes carved out by absorption (p. 180).

In all forms (except Cyclostomes where the conditions are obscure, and Gymnophiona where the basal plate is extremely reduced) the basicranial fenestra is delimited anteriorly by a transverse bar of cartilage, the acrochordal, dorsum sellae, prootic bridge, or crista sellaris (the 'median trabecula' of Rathke, 1839) which overlooks the pituitary fossa from behind, forming the hind bolster to the sella turcica, extends more or less far into the plica encephali ventralis (between the floors of the mid- and forebrains), is more or less continuous with the base of the pila antotica on each side, and constitutes the anterior edge of the basal plate. In Selachii (van Wijhe, 1922) and in birds (Jager, 1926) the acrochordal arises in connexion with the transverse commissure between the premandibular somites, and thus represents the foremost part of the metamerically segmented axial skeleton.

Allis (1928) has recently attempted to deny the homology of the acrochordal, crista sellaris, &c., in the different vertebrate groups, and he claims that there are really two distinct transverse cartilaginous bars at the front of the basal plate.

The foremost of these he calls the commissura acrochordalis and supposes it to be preparachordal in the premandibular segment; the hinder bar he terms the commissura transversa, supposed to represent the front of the parachordals proper, and to lie in the mandibular segment. As a consequence of this distinction, Allis is further led to distinguish two different morphological vacuities in place of what was previously recognized as the basicranial fenestra. These Allis calls (i) the fenestra prootica medialis between his commissura acrochordalis and transversa, and (ii) the fenestra mesotica medialis behind his commissura transversa. The former is supposed to exist in Squalus, Acipenser, Polypterus, Amia, and Lepidosteus; while the latter is supposed to occur in Salmo, Amiurus, Syngnathus, and Lacerta.

In rejecting this view de Beer (1931*a*) has pointed out that in order to establish it, it would be necessary to demonstrate the existence of both transverse bars or both fenestrae in any one form, which has never been done. It is also inadmissible that the fenestra behind the prootic bridge in Amia and Lepidosteus should be different from that in Salmo and other Teleostei.

In some mammals at early stages of development a space is found and described as lying between the parachordal plate and the hypophysial cartilage, and identified with the basicranial fenestra by Noordenbos (1905), Terry (1917), and Fawcett (1917). But if this space lies *in front* of the anterior edge of the parachordal, as identified by the crista transversa or the dorsum sellae instead of *behind* it, it would seem that it could only represent a hinder part of the hypophysial fenestra, separated from its anterior part by the precocious and specialized development of the hypophysial cartilages. The true basicranial fenestra in mammals is to be found in the space between the dorsum sellae and the crista transversa in all those cases where the dorsum sellae chondrifies independently (e.g. Lepus, Felis, Macacus, Semnopithecus, and Homo). In those forms in which no separate dorsum sellae is formed and the hypophysial plate passes insensibly back into the basal plate (e.g. Marsupials, Xenarthra, Insectivora, Halicore), the basicranial fenestra must be sought elsewhere. In this connexion it should be remembered that in those forms in which the notochord pierces the basal plate to get from the ventral to the dorsal surface, the basal plate may present a perforation which seemingly also represents a basicranial fenestra, which, after all, in lower vertebrates is a space traversed by the notochord. It is just possible that this is the fenestra claimed by Noordenbos as the basicranial and that it really lies *within* and not in front of the basal plate, for behind the hypophysial cartilages a transverse cartilage may be found which soon fuses with them but probably represents the foremost part of the parachordal plate, or an acrochordal. In this case Noordenbos and Fawcett would be correct in their identification of the basicranial fenestra, but wrong in the definition of its relations (p. 321).

III. The Occipital Arches

In Cyclostomes the skull ends posteriorly with the auditory capsules, but in all Gnathostomes the parachordals project behind the auditory capsules and bear a varying number of neural arches, of which the hindmost is usually termed the occipital arch and those preceding it, the preoccipital arches. The latter vary in number according to the number of segments included in the skull (pp. 27–31). Thus, the preoccipital arches are: none in Anura, one in Urodela, two in Scyllium, three in Lacerta, &c.

In addition to the preoccipital and occipital arches which arise in continuity with the parachordals (although they may precede them in time, as in Amia, Dipnoi, and Urodela), the next posterior neural arches in varying number may become attached to the parachordals during development, as in Acipenser, Amia, Lepidosteus, Ceratodus, and in birds, and for these late additions van Wijhe (1922) proposed the term occipitospinal arches. There is, of course, no morphological difference between the preoccipital, occipital, occipitospinal, and free vertebral neural arches, and it is usual to refer to the single product of fusion (on each side) of the preoccipital, occipital, and occipitospinal arches as the definitive occipital arch. Where this is attached to the hind wall of the auditory capsule, it is sometimes narrowed down to a thin strip known as the lamina alaris.

In vertebrates below birds and Placental mammals, the occipital arches occupy a more or less upright position, the plane of the foramen magnum being vertical and transverse. In birds and Placental mammals, on the other hand, the occipital arches lean backwards and the plane of the foramen magnum is rotated backwards and downwards, until in Man it lies horizontally in respect of the head.

Dorsally, the occipital arches may end freely, as in Chelonia or Anas, or more usually they are interconnected by a tectum posterius which may be merged with a tectum synoticum.

The occipital arches being of the same nature as the neural arches of the vertebrae, it is not surprising to find that they may bear transverse processes or ribs, which then become known as cranial ribs. These remain distinct in Acipenser and Ceratodus, but in birds where rudiments of four pairs of cranial ribs are found (p. 272) they appear to give rise to the metotic cartilage. This is clearly homologous with the subcapsular process of Crocodilus and the paracondylar process of mammals (e.g. Lepus), which structures are equally clearly transverse processes or reduced ribs of one or more occipital arches. According to Kernan (1915) the paracondylar process represents the rib of the preoccipital (but not of the occipital) vertebra.

The number of hypoglossal foramina varies greatly between different forms and at different stages of development. Information concerning these in lower vertebrates is given by Fürbringer (1897). Among living Amphibia Cryptobranchus is unique in possessing one pair of hypoglossal foramina; in reptiles and birds the number varies between four and two; in Monotremes and in Megaptera there are none; in Marsupials usually two pairs; in Placental mammals usually one pair (two pairs in Lepus, and occasionally in Halicore and Lagenorhynchus). It should be noted, however, that the presence of only a single pair of hypoglossal foramina *in the cartilage* is no evidence that only a single pair of preoccipital arches is present, for some of these arches may fail to chondrify, and are found as connective tissue septa subdividing an apparently single foramen (e.g. in Sus, Lebedkin, 1918).

IV. The Cranio-vertebral Joint

The problem of the cranio-vertebral joint does not present itself in the Cyclostomes, where there is, properly speaking, no vertebral column. In the Gnathostomes the most primitive forms are characterized by the total or partial persistence of the notochord in many cases, and by the unmistakable resemblance of the hindmost part of the skull to a vertebra, with the result that the flexibility between the skull and the vertebral column is no greater than that between adjacent vertebrae. Further, owing to the absorption of several vertebrae in the hind part of the skull, the transition from skull to vertebral column is gentle. This is the case in many Selachii, Acipenseroidei, and Dipnoi, in which the notochord persists, and in other Selachii and Teleostomi in which the notochord is intersegmentally obliterated by the formation of the vertebral centra. The neural arches of the vertebrae which are absorbed in the hind part of the skull by incorporation *during ontogeny* are called occipitospinal arches, following van Wijhe (1922).

The first examples of a definite and special articulation between the skull and the vertebral column is to be found in Batoid Selachii and in Holocephali. In both these cases, and apparently independently in each, a special joint has been rendered necessary by the rigidity of the anterior portion of the vertebral column owing to the anchylosis and fusion of vertebrae. The joint involves articulation between the centrum of the 1st free vertebra and the corresponding region of the basal plate, and between the bases of the neural arches of the 1st vertebra and the occipital arches. This joint is intervertebral, and therefore intrasegmental.

In the Amphibia, both Urodela and Anura (and perhaps also the Gymnophiona), the articulation between skull and vertebral column involves the formation of ball and socket joints out of the interdorsal cartilages developed in the anterior sclerotomite of the segment whose posterior sclerotomite produces the basidorsals of the 1st free or atlas vertebra. The interdorsal cartilages are split in such a way that the cranial portions become attached to the hind end of the skull forming its occipital condyles, and the caudal portions are attached to the anterior face of the atlas vertebra. In addition, in Urodela, there is an articulation between the median portion of the basal plate and the chondrified portion of the notochord which becomes attached to the atlas vertebra and forms the tuberculum interglenoidale ('odontoid process'). This tuberculum interglenoidale seems to contain the intervertebral (notochordal) cartilage belonging to the vertebra next in front of the atlas, i.e. of the proatlas vertebra whose neural arches are the occipital arches.

The cranio-vertebral joint in Urodela is therefore complicated, and is intervertebral (intrasegmental) as regards the median articulation and also the lateral articulations, but in different segments; the former lies in the segment anterior to the occipital arch, the latter lie in the segment behind it.

In Anura the tuberculum interglenoidale is exceedingly short and probably does not contain any element corresponding to the segment of the occipital arch.

Before considering the condition of the cranio-vertebral joint in Amniota, as interpreted by Hayek (1923, 1924), it is necessary to turn to the question of the so-called proatlas, especially as the nature of such an element, first so called by Albrecht (1880), has been frequently misunderstood and often confused. The proatlas vertebra is simply the vertebra serially next in front of the 1st *complete free* vertebra, or atlas. In the majority of cases, the proatlas neural arch is the occipital arch, as in Selachii, Amphibia, or Amniota; or it may be the hindmost occipitospinal arch as in Acipenser, Lepidosteus, Ceratodus, or birds. Sometimes, however, the centrum of the proatlas vertebra is included in the hindmost part of the basal plate of the skull in the ordinary way, but the neural arch (*either* basidorsal *or* interdorsal) corresponding to it may be free. Thus, there is a free proatlas basidorsal arch in Polypterus and Salmo. On the other hand, it may be an interdorsal element which is free from the body of its (proatlas) vertebra, and in this way there may be a free proatlas interdorsal arch as in Squalus. In the Amniota it seems that the rule is for the interdorsal arches to become fused on to the basidorsal arches immediately in front of them. But they may fail to do so, and such failure on the part of the interdorsal arch which normally should fuse on to the hind surface of the occipital (proatlas basidorsal) arch gives rise in these forms also to a free proatlas interdorsal arch, as in Sphenodon, Crocodilus, or Erinaceus. According to Barge's (1918) studies on Ovis, the proatlas interdorsal arch in mammals normally fuses with the atlas basidorsal arch immediately behind it.

Amniota are characterized by the greater development of the pleurocentral elements, formed from the anterior (intercalary) half-sclerotomes, while the hypocentral elements formed from the posterior (basal) half-sclerotomes are reduced. Now, the pleurocentrum of the proatlas vertebra in Chelonia, Crocodilia, and birds, is tacked on to the hind end of the skull, pierced by the notochord, and forms the median single condyle characteristic of those forms. The cranio-vertebral joint here, therefore, is intervertebral and intrasegmental (See Plate 138).

On the other hand, in Sphenodon (Schauinsland, 1900), Lacertilia (probably Ophidia) and mammals (rat, Weiss, 1901; Echidna, Gaupp, 1907*b*; Bradypus, de Burlet, 1913*a*; Phocaena, de Burlet, 1913*b*; Ornithorhynchus, de Beer and Fell, 1936; Man, Dixon, 1906; Gladstone and E. Powell, 1914; Weigner, 1911), the pleurocentrum of the proatlas vertebra is fused on to the front end of the pleurocentrum of the atlas vertebra, which itself is fused on to the front of the pleurocentrum of the 2nd or axis vertebra to form the odontoid peg. The single median condyle of Sphenodon, Lacertilia, and Ophidia lies ventrally to the notochord and represents the hypocentral element of the proatlas vertebra, fused on to the back of the skull. This condyle is therefore not homologous with that of Chelonia, Crocodilia, or Aves.

In mammals the hypocentrum of the proatlas vertebra appears to be reduced (although vestiges of it may persist, perhaps even in Man: Augier, 1928; see also Brachet, 1909, and Bolk, 1921, on 'condylus tertius') and new paired articular facets are formed on the hind surface of the bases of the occipital arches. These paired occipital condyles of the mammals are therefore not homologous with those of Amphibia, and may be called arcual condyles, as distinct from central condyles.

The cranio-vertebral joint in Sphenodon, Lacertilia, Ophidia, and mammals is intravertebral and intersegmental. The detailed study of the chondrocranium thus produces the unexpected evidence that the paired condition of the cartilaginous condyles in Amphibia and in mammals on the one hand, and the single condition of the condyle in living reptiles and birds on the other, in each case conceals fundamental differences of structure and origin.

Gaupp (1907*b*, 1911*a*) has drawn attention to the fact that alongside the problem of the number and position of the condyles, there is the question of the single or multiple nature of the synovial cavities separating the articular facets of the skull and of the atlas vertebra. Thus, in the Amphibia, the synovial cavities are separated from one another, whereas in reptiles and birds there is a single median cavity. In primitive mammals (Echidna, Halmaturus, Petrogale, Erinaceus, Canis, Felis, Dasypus, Pteropus, Lemur, Stenops, Bos) the articular facet is continuous across the middle line from one condyle to the other. On the other hand, in other forms (Lepus, Sus, Equus, Bradypus, Homo) the articular facets of the two condyles are discontinuous in the middle line.

It is clear from the distribution of these conditions that the continuity or non-continuity across the midline of the articulation of the condylar facets does not reflect affinity or phylogeny. On the other hand, the developmental nature of the condyles themselves must be of phylogenetic significance. Thus, the duplicity of the mammalian condyle cannot be derived from a splitting of any pleurocentral condyle such as is present in Chelonia, crocodiles, or birds, for in mammals this proatlas pleurocentrum is already accounted for in the odontoid process of the axis vertebra. The mammalian condition could only be derived from one in which there was no pleurocentral condyle, but a hypocentral condyle (as in Lacerta) which has been gradually replaced by the development of paired bosses at the bases of the occipital arches.

This investigation leads to one further conclusion of far-reaching importance and significance, namely, that *the evolution of the cartilaginous condyles has been independent of that of the bones* (exoccipitals and basioccipital) *which may ossify in them*. Thus, the basioccipital may ossify in a hypocentral condyle (Lacerta) or in a pleurocentral condyle (crocodile). Further, a pleurocentral condyle may be ossified either solely by the basioccipital (crocodile, birds) or jointly by the basioccipital and exoccipitals (Chelonia). And here it is important to note that the pleurocentral nature of the cartilage in which these bones ossify in Chelonia prevents the so-called triple (bony) condyle found in these forms from being regarded (as by Osborn, 1900; Mead, 1906) as any stage on the way to the evolution of the mammalian condition, by reduction of the median basioccipital and enlargement of the lateral exoccipital elements.

Lastly, it may be noticed that in the Cetacea the dicondylar condition may be present in early stages, and replaced later by a monocondylar condition by the development of a large median dome-shaped mass of cartilage in which the paired condyles are merged.

V. The Side Wall of the Neurocranium

The side wall of the neurocranium, reduced to its simplest terms, consists of an *orbital cartilage* which is attached to the floor of the brain-case by three pillars:

(i) anteriorly, the *preoptic root* (the term 'pila prooptica' is liable to confusion with 'pila prootica'), situated anteriorly to the optic foramen (optic nerve and arteria centralis retinae); attached to the trabecula (sometimes to the lamina orbitonasalis, as in Echidna, Dasyurus, Sorex);

(ii) the *pila metoptica*, behind the optic nerve and artery, in front of the metoptic foramen (oculomotor nerve, the ophthalmic artery and the pituitary vein) attached to the trabecula (absent in Marsupials, for which reason the optic foramen and sphenoid fissure are confluent in these);

(iii) the *pila antotica* (the term 'pila prootica' is discarded as liable to convey the false impression that the structure has anything to do with the prootic bone), behind the oculomotor nerve, ophthalmic artery, and pituitary vein, in front of the prootic foramen (trigeminal and abducens nerves, though the latter may pierce its base); attached to the parachordals in the anterolateral corners of the basal plate on each side of the dorsum sellae or acrochordal. (The trochlear nerve usually has its own foramen through the orbital cartilage, or runs dorsally to it (Urodela), or accompanies the oculomotor (reptiles).)

The orbital cartilage is connected anteriorly with the lamina orbitonasalis or the nasal capsule by the sphenethmoid commissure (Gaupp, 1900), enclosing the orbitonasal foramen or fissure; and posteriorly with the auditory capsule by the taenia marginalis (or orbitoparietal commissure of Decker, 1883) enclosing the foramen prooticum. The latter is separated from the facial foramen by the prefacial commissure, connecting the basal plate with the auditory capsule.

In the higher groups, where the orbital cartilage tends to become fenestrated, the pila metoptica is often joined to the pila antotica by the supratrabecular bar (taenia interclinoidalis, e.g. Lacerta, Sphenodon, Dermochelys, birds, Semnopithecus, Homo), which is dorsal to the ophthalmic artery and pituitary vein, and ventral to the oculomotor nerve. Vestiges of the supratrabecular bar are found in the anterior and posterior clinoid processes.

Apart from the Cyclostomes where the conditions may conform to this plan, but are still obscure (because the preoptic and metoptic roots are attached to the trabeculae of Gnathostomes but, apparently, to the anterior parachordals of Cyclostomes), orbital cartilages with these relations are very constant throughout the vertebrates (e.g. Selachii, Dipnoi, Amphibia, Reptilia, Monotremata). The only important modifications met with are the formation of an interorbital septum, which is associated with great size of the eye, and loss of the pila antotica, which may be associated with the formation of the posterior myodome, or with the expansion of the cranial cavity.

The formation of an *interorbital septum* has occurred independently on numerous occasions in vertebrate evolution. In Holocephali (p. 75) the septum is formed dorsally to the brain by the approximation in the midline of the dorsal portions of the orbital cartilages. In many Teleostei the septum involves the pressing together of the preoptic roots of the orbital cartilages, either anteriorly to the brain (e.g. in Salmo where the olfactory nerves traverse part of the orbit to reach the nasal sacs) or ventrally to the brain (e.g. in Gadus where the cranial cavity reaches as far forwards as the nasal capsules, and the olfactory nerves do not have to traverse the orbit). The absence of any interorbital septum in Gymnarchus and in Cypriniformes must be secondary and due to loss.

Living Amphibia have no interorbital septum, but here also its absence would seem to be due to loss since it was present in Stegocephalia. From these ancestors presumably it has been inherited and enlarged by reptiles and birds. In them the interorbital septum has reached such large dimensions that the anterior portions of the orbital cartilages are reduced to the planum supraseptale, surmounting the septum.

The interorbital septum itself is a median structure, and the question arises whether it is to be regarded as the product of fusion of the ventral portions of the preoptic roots of the orbital cartilages, or as some new structural element. As Gaupp (1898) has stated, the former is the more probable explanation, although it is not always possible to demonstrate the formation of the ventral part of the septum from the paired orbital cartilages during development.

In many birds the eye reaches such dimensions that in the fully formed chondrocranium the preoptic roots of the orbital cartilages are either destroyed (e.g. in Anas, where the olfactory nerve has to traverse the orbit to reach the nasal capsule), or plastered against the side of the septum (e.g. in Passer, where the olfactory nerves are enclosed in long evehent olfactory tunnels). The absence of an interorbital septum in Ophidia among reptiles and in Apteryx among birds must be secondary and due to loss.

The presence of a definite though low interorbital septum in some mammals raises interesting problems, for in these forms it seems to be correlated not with the size of the eye but with reduction in size of the nasal capsule. Since the interorbital septum in mammals is directly continuous with the nasal septum in front of it, a portion of nasal septum would become interorbital if the hind wall of the nasal capsule were to be moved forward, but such an interorbital septum would not be homologous with that of the reptilian ancestor, since it would be composed, not of the orbital cartilages, but of median trabecular cartilage.

The interorbital septum is absent in: Ornithorhynchus, Echidna, Dasyurus, Tatusia, Erinaceus, Talpa, Sorex, Tupaja, Miniopterus, Galeopithecus, Felis, Poecilophoca, Equus, Mycetes, and Cebus. It is feebly developed in: Didelphys, Bradypus, Canis, Bos, Sus, Phocaena, Balaenoptera, and Lagenorhynchus. It is well developed in: Microtus, Lepus, Lemur, Nycticebus, Tarsius, Ateles, Nasalis, Hapale, Chrysothrix, Macacus, Semnopithecus, and Homo.

From this distribution it is clear that in mammals the interorbital septum is commonest in Primates, where the nasal capsule is least well developed. It is probable, therefore, that the mammalian interorbital septum has been independently evolved.

In many mammals paired cartilages of independent origin are found, the *alae hypochiasmaticae*, which project to the side beneath the root of the optic nerve as it emerges from the optic foramen. The ala hypochiasmatica appears to correspond to part of the cartilago hypochiasmatica of Lacerta (p. 227). The ala hypochiasmatica may become attached to the preoptic root of the orbital cartilage (e.g. Lepus), or to the pila metoptica (e.g. Tatusia), or to both (e.g. Homo), and thus give rise to the so-called paroptic process, on each side of the skull. Between the ala hypochiasmatica and the central stem it is common for one or more small foramina to be enclosed. These were termed 'prechiasmatic' foramina by Macklin (1914) who regarded them as isolated portions of the optic foramen. Fawcett (1918*b*, p. 425) has, however, applied the term 'prechiasmatic foramen' to the medial portion of the orbitonasal fissure, cut off from the remainder by a small cartilaginous strut, the ala minima of Luschka (1857), which joins the preoptic root of the orbital cartilage with the lamina orbitonasalis in Caluromys, Felis, Mustela, Poecilophoca, Manatus, and Semnopithecus. A diminutive ala minima also occurs in Didelphys, Erinaceus, Talpa, Tupaja, and Man, and it is clear, therefore, that Fawcett's 'prechiasmatic foramen' is not homologous with Macklin's, and should be called 'foramen infracribrosum'. Semnopithecus has both.

In birds the trigeminal ganglion becomes enclosed by a lateral extension of the orbital cartilage, which acquires a new lateral supporting pillar (the pila antotica spuria, between the profundus and maxillary branches of the trigeminal nerve) while the pila antotica may or may not be preserved (p. 276).

The formation of the posterior myodome in Teleostome fish involves the disappearance of the pila antotica and pila metoptica. In those forms which have lost the myodome, the cartilaginous side wall may be reformed, but it shows atypical features (Polypterus, p. 81; Acipenser, p. 89; Lepidosteus, p. 112). In Amia the orbital cartilage becomes supported posteriorly by a pila lateralis.

In Ophidia the orbital cartilage and its pillars have been lost entirely, their function being previously taken on by the frontal and parietal bones.

In mammals the orbital cartilage is well developed, and most forms possess preoptic and metoptic roots. But the pila antotica (taenia clino-orbitalis) is preserved only in the Monotremes. In Marsupials and Placental mammals, the inclusion of the cavum epiptericum (see p. 430) in the cranial cavity by the development of a new side wall formed from the ala temporalis (= processus ascendens of the pterygoquadrate cartilage) rendered obsolete that portion of the original wall (chondrification in the dura mater) represented by the pila antotica.

It is, therefore, of great interest to find that isolated cartilaginous fragments of the pila antotica have been found in many higher mammals: e.g. Lepus (Voit, 1909); Man ('supracochlear cartilage', Macklin, 1914; Augier, 1934*b*); Felis (Terry, 1917); Didelphys (Töplitz, 1920); Halicore (Matthes, 1921*a*); Tarsius (Henckel, 1927). In some mammals (Sus, Mead, 1909, Augier, 1934*a*; occasionally Man, Voit, 1919, Augier, 1931*a*) the supracochlear cartilage is in cartilaginous connexion with the dorsum sellae, forming a commissura clinocochlearis, beneath which the abducens nerve passes. It may be regarded as a displaced vestige of the pila antotica.

It was noticed by Sutton (1888) that in mammals the eye-muscle nerves perforate the dura mater in places which are not directly opposite their (eventually formed) foramina of exit from the skull. This remarkably acute observation can now be explained: the nerves perforate the dura mater as they did in the ancestral mammalian forms when the original side wall of the chondrocranium (pila antotica) was regularly chondrified.

The *prefacial commissure* is also a remarkably constant structure, since it can be traced through most groups from Selachii (or even Cyclostomes to Man. In mammals it is usually known as the suprafacial commissure, from its topographical position in these vertebrates, where, furthermore, the great development of the cochlear portion of the auditory capsule results in the commissure being seated on the latter instead of on the edge of the parachordal. (See, however, Felis, p. 327).

In Teleostomes a duplication of the side wall is formed, laterally to the prefacial commissure, by the *lateral commissure*. This structure, which lies laterally to the head vein and orbital artery (the prefacial commissure, as a part of the primitive side wall, of course lies medially to these vessels), is formed by the junction of the prootic process developed from the auditory capsule, and the basitrabecular and postpalatine processes, developed from the edge of the trabeculae and basal plate. In Teleostei, where the lateral commissure forms the side wall to the trigeminofacialis chamber, the prefacial commissure is lost.

A duplication of the side wall in the posterior part of the orbitotemporal region is found in mammals. The *tegmen tympani* (processus perioticus of Gradenigo, 1887) which projects forwards from the crista parotica, laterally to the true side wall of the skull (here represented by the suprafacial commissure), seems to have been evolved within the mammalian class. It is feebly developed in Monotremes and in Didelphys (Töplitz, 1920).

In some mammals (e.g. Talpa, Homo, Fischer, 1901; Lepus, Voit, 1909; Phocaena, de Burlet, 1913*b*; Sorex, de Beer, 1929) the anteroventral part of the tegmen tympani fuses with the lateral wall of the cochlear capsule, thus forming a lateral prefacial commissure in the posterolateral wall of the cavum supracochleare (see p. 430). The main branch of the facial nerve passes back through this lateral facial foramen into the sulcus facialis, while the palatine (greater superficial petrosal) branch runs anteriorly to it, in the gap (hiatus Faloppii or facialis) between it and the true prefacial commissure.

Yet another structure which duplicates the side wall in some mammals is the *processus opercularis*, situated above the tegmen tympani, and so far found only in Talpa (Fischer, 1901), Microtus (Fawcett, 1917), Poecilophoca (Fawcett, 1918*b*), Erethizon (Struthers, 1927), and Sorex (de Beer, 1929). Remnants of it may perhaps have been found in Man (Augier, 1930). It is somewhat variable, being attached to the supraoccipital cartilage in some cases, and seems to function as an insertion for the trapezius muscle, while at the same time protecting the lateral jugular vein on its emergence from the fissura occipitocapsularis (foramen jugulare spurium). Perhaps the plate of cartilage in the squamosal of Echidna may represent the processus opercularis.

The *parietal plate* in mammals, on the other hand (lamina parietalis, Spöndli, 1846), is a part of the true cranial wall, and is developed out of the hindmost part of the taenia marginalis, in correlation with the increased size of the brain and the inadequacy of the auditory capsule to provide a sufficient side wall. It is, therefore, of interest to note that Chelone has a taenia supramarginalis in a comparable position.

There remains to be discussed the *basitrabecular* (basipterygoid) *process*, for although it does not enclose the side of the cranial cavity, it is most conveniently treated here. The basitrabecular process is a lateral projection of that hindmost portion of the trabecular bar which is formed of the polar cartilage when that element chondrifies independently. It always lies ventrally to the head vein and anterodorsally to the palatine nerve, and typically makes contact with the basal process of the pterygoquadrate, with which it may articulate movably, or undergo cartilaginous fusion. In Selachii it is represented by a portion of the subocular shelf; in Teleostomes it contributes (together with the postpalatine process or commissure, an extension of the basal plate behind the palatine nerve) to the formation of the lateral commissure, but in Lepidosteus it nevertheless preserves its primitive contact with the basal process of the pterygoquadrate. The existence of a basitrabecular process in Dipnoi and in Urodela is difficult to establish or refute owing to the precocity with which the pterygoquadrate undergoes autosystylic fusion with the neurocranium (and which might involve either a basal or a basitrabecular process, or both). In Anura the basitrabecular process has been lost (except in Ascaphus) for the basal process of the pterygoquadrate fuses with the neurocranium *behind* the palatine nerve, thus forming a pseudobasal connexion (see p. 207).

In many reptiles (e.g. Lacertilia) and birds (e.g. Hirundo, Struthio, Apteryx, Passer) the basitrabecular process is well developed, and ossified by the basisphenoid, thus giving rise to the basipterygoid process. A movable articulation between the latter and the pterygoid bone is characteristic of the type of internal mobility known as kinetism (p. 426). In many of these cases a nodule of cartilage, the meniscus pterygoideus, is intercalated between the pterygoid bone and the basipterygoid process (p. 419).

In crocodiles and birds a posteriorly directed process from the root of each basitrabecular process gives rise to the infrapolar processes which underlie the anterior part of the basal plate, and, in birds, fuse with it thus enclosing the lateral carotid foramina (pp. 264, 276).

In mammals the basitrabecular process has been recognized in the processus alaris (Hannover, 1880), which bears identical relations to neighbouring nerves and blood-vessels (Voit, 1909; Fuchs, 1915). Gaupp (1902) contended that the processus alaris *plus* the ala temporalis of mammals represented the basitrabecular process (for which reason he regarded the processus ascendens of the pterygoquadrate cartilage as absent from mammals), and he was probably influenced by the fact that in the only mammalian chondrocrania at that time described (Ovis, Wincza, 1896; Talpa, Fischer, 1901), and Echidna at which he was then working (published Gaupp 1905*a*, 1908), the ala temporalis chondrifies in continuity with the processus alaris.

In every other mammalian chondrocranium studied (except Sorex, de Beer, 1929) the ala temporalis has a separate centre of chondrification, and is to be regarded as the homologue of the processus ascendens of the pterygoquadrate cartilage (p. 420). Its development in continuity with the processus alaris in the above-mentioned forms must therefore be a developmental modification devoid of morphological significance.

VI. The Roof of the Chondrocranium

The roof is a usually neglected though none the less interesting part of the chondrocranium, which shows extreme variation in the degree to which it is developed in even closely related forms. As a rule, the roof is composed of the following constituent cartilages, all of which, however, are by no means always present:

(i) a tectum synoticum, connecting the auditory capsules;
(ii) a tectum posterius, connecting the occipital arches;
(iii) an epiphysial cartilage, connecting the orbital cartilages half-way along their length;
(iv) a paraphysial cartilage connecting the orbital cartilages anteriorly;
(v) a tectum transversum, connecting the orbital cartilages posteriorly, where they or the taeniae marginales join the auditory capsules or parietal plates;
(vi) a sagittal taenia tecti medialis, which may connect two or more of the above-mentioned transverse cartilages, or which may be reduced to a process projecting backwards or forwards from one or other of them.

The distinction between tectum synoticum and tectum posterius is not always absolute, for, as Gaupp (1906*b*) remarks, the chondrocranial roof in this region often chondrifies independently and may fuse with the occipital arches or the auditory capsules, or both.

The following account refers to the maximum development of the roof and whether it persists in the adult.

The Cyclostomes have no cartilaginous roof at all in Myxine, and only a tectum synoticum in Petromyzon. In Selachii, on the other hand, the chondrocranium has a roof which is complete (and, of course, persists in the adult) except for the foramina in the posterior region through which the endolymphatic ducts leave the auditory capsule and (without traversing the cranial cavity) reach the parietal fossa and the exterior. The prefrontal or precerebral fontanelle in the anterior region of the skull is, according to Allis (1923*a*), and probably rightly, to be regarded as an unchondrified part of the front wall, and not of the roof of the skull. In some Selachii (e.g. Squalus) there may be a transient epiphysial (pineal) foramen.

The Holocephali also have a complete and persistent cartilaginous roof, except for the endolymphatic foramina. There is also a transient epiphysial foramen. The roof is lifted upwards in the formation of the peculiar interorbital septum of these forms, and the ethmoidal canal also becomes roofed over.

In Polypterus the chondrocranium is roofed only by the slender epiphysial cartilage and tectum synoticum, whereas in Acipenser the cartilaginous roof is massive, persistent, and quite complete. Amia also has a persistent complete cartilaginous roof (a transient epiphysial foramen), while that of Lepidosteus is persistent and complete except for a pair of fontanelles on each side of the taenia tecti medialis.

The variations of the chondrocranial roof in the Teleostei have been reviewed by Norman (1926, p. 435), and they range from a condition in which the roof is persistent, massive, and complete, as in Cyclopterus, to one in which it is represented only by the tectum synoticum, as in Syngnathus.

The chondrocranial roof is persistent and complete in Ceratodus, but very incomplete in Protopterus and Lepidosiren.

In Urodela the roof is reduced to the tectum synoticum while in Gymnophiona it is reduced altogether. In Anura, on the other hand, it is better developed, consisting of the tectum transversum, tectum synoticum, and taenia tecti medialis.

In reptiles generally, as in birds, the roof is reduced to the tectum synoticum, sometimes (Lacerta) associated with a tectum posterius. But certain Chelonia have a cartilaginous roof which is either almost complete (Dermochelys) or half complete (Chelone), and persists throughout life. Remnants of the epiphysial cartilage are found in Tinnunculus among birds.

In mammals the development of the parietal plates and of the supraoccipital cartilages renders difficult the distinction between tectum posterius and tectum synoticum (see Levi, 1909*b*). In general, however, it seems that a tectum posterius is present (joining the supraoccipital cartilages), to which there may be added, as in Man, a tectum intermedium and a tectum synoticum. In Echidna, Poecilophoca, and in Halicore, there are remnants of a tectum transversum. The chondrocranium of Ornithorhynchus is very remarkable, for its roof is exceedingly massive, and almost complete.

A very little consideration of the above-mentioned facts suffices to show that the degree of development of the chondrocranial roof has no taxonomic significance; it is exceedingly poorly developed in Lepidosiren and Chrysemys, but very well developed in their respective close relatives Ceratodus and Dermochelys. On the other hand, it looks as if the degree of development of the chondrocranial roof were correlated in some as yet unknown way with the mode of life; for it is remarkable that all the forms in which the roof is particularly well developed, or shows a better development than in related forms (Selachii, Holocephali, Acipenser, Amia, Lepidosteus, Cyclopterus, Ceratodus, Dermochelys, Chelone, Ornithorhynchus, Poecilophoca, Halicore), are associated with an aquatic mode of life. It would be interesting to know whether the same phenomenon occurs in penguins. At the same time, as Fuchs (1920) has pointed out, there are very many aquatic forms without excessive chondrification. The question is therefore by no means solved.

At all events, the apparent ease with which a complete chondrocranial roof can be formed secondarily (for such must be its interpretation in the higher vertebrates) raises the question whether in Selachii, the complete roof is to be regarded as phylogenetically primitive. This question has been treated by Swinnerton (1902, p. 562) and by Norman (1926, p. 439), but it is doubtful whether it can be solved by embryology, for a case can be made out either way. Perhaps palaeontology may supply the necessary evidence.

VII. The Nasal Capsule

The protection of the olfactory sac in a cartilaginous capsule is a characteristic feature of all skulls, but whereas in most forms the nasal capsule and the brain-case are fused with one another, there are good reasons for believing that they were originally separate.

In Petromyzon conditions are somewhat obscured by the secondary fusion of the olfactory capsules in the midline, but the most important feature is the fact that the capsule is attached to the brain-case only by connective tissue. In Myxine the state of affairs is even more difficult to interpret, and there the much-fenestrated capsule is connected with the brain-case by cartilaginous struts (Cole, 1905, 1909).

In all Gnathostomes the nasal capsule is fused with the brain-case, and it is formed typically of a lamina orbitonasalis (also called planum antorbitale) contributing the side and hind walls, a front wall, or cupola anterior, and a medial wall which in Selachii and Urodela is lateral to and distinct from the median rostrum (anterior prolongation of the trabecular plate). In other forms, however, the true medial wall of the capsule disappears and is replaced by the rostrum, which then forms the nasal septum. A stage in this substitution may be seen in the anterior part of the nasal capsule of Rana (p. 204).

Further vestiges of the medial wall are to be found in the paraseptal cartilages. These structures, first so called by Spurgat (1896), represent the ventral portion of the medial walls of the capsules of Urodela, and are to be found in Reptilia and Mammalia extending beside the ventral edge of the nasal septum (and free from it, except in Chelonia). Primitively the paraseptal cartilage extends all the way from the lamina transversalis anterior (see below) to the hind wall (cupola posterior) of the capsule, as in Lacerta, Sphenodon, and Chrysemys, and, among mammals, in Halmaturus (Seydel, 1896), Trichosurus (Broom, 1909*a*), Lepus (Voit, 1909), Felis, Erinaceus and Cavia (Zuckerkandl, 1909), Manatus (Matthes, 1912), Dasyurus and Microtus (Fawcett, 1917), and Mycetes (Fawcett, unpubl.). This structure may be referred to as the cartilago paraseptalis communis (Zuckerkandl). In other forms the paraseptal cartilages are discontinuous and reduced. There may then be anterior and posterior paraseptal cartilages. The anterior paraseptal cartilage is concerned in providing a capsule in which Jacobson's organ is lodged (see below, p. 398). Among birds, remnants of the paraseptal cartilages are apparently preserved in Apteryx (p. 288).

The degree of development of the front wall or cupola anterior varies with the position of the fenestra narina. Thus, it is well formed in Selachii where the nasal aperture is ventral; fairly well developed in Amphibia and most Reptilia, Aves, and Mammalia, where the fenestra narina is anterolateral; and absent altogether in Chelonia, Crocodilia, and higher Primates, where the nostrils open directly forwards.

In the Selachii, the incurrent and excurrent nasal apertures are surrounded by special nasal cartilages, in the form of a '6' or figure of '8'. Such cartilages are probably represented among the numerous rostrolabial cartilages of Holocephali. Among Teleostei, nasal cartilages are present in Cypriniformes (Sagemehl, 1885).

The roof joins the medial wall or nasal septum with the lamina orbitonasalis.

The lamina orbitonasalis is attached to the preoptic root of the orbital cartilage, but in such a way as to allow several structures (nasal branch of profundus nerve, orbitonasal artery, veins) to pass between them, through an orbitonasal foramen, fissure, or canal. These relations show that the lamina orbitonasalis is not part of the true cranial wall, here represented by the preoptic root of the orbital cartilage, and in many forms (e.g. reptiles, birds) it does not even touch the true cranial wall, or interorbital septum.

Primitively the nasal capsule has no floor, for the nasal sac is there in communication with the outside. There is, however, a rudiment of it in the form of the trabecular horns, which project laterally from the rostrum and, in Selachii, end freely. In Tetrapoda, the trabecular horn becomes joined to the ventral edge of the side wall of the capsule on each side, *posteriorly* and *ventrally* to the external nostril, thus forming the lamina transversalis anterior, or solum nasi, which separates the fenestra narina (for the external nostril) in front from the fenestra basalis or choanalis behind. In this region, therefore, the capsule forms a complete ring or zona annularis, round the nasal sac on each side.

The lamina transversalis anterior is not formed in some Aves (e.g. Anas, p. 277), nor in the higher Primates (e.g. Macacus (Frets, 1914), Chimpanzee (Virchow, 1914), and Man).

In the Dipnoi, on the other hand, the incurrent and excurrent nostrils are both on the ventral surface of the snout, and the trabecular horn passes *anteriorly* and *dorsally* to the incurrent (or external) nostril.

In Polypterus and Acipenser the nostrilar apertures are in the side wall of the capsule, which has well-developed cartilaginous walls, floor, and roof. In the Holostei and Teleostei, however, the nostril apertures are on the dorsal side of the snout, and the nasal capsule has neither roof nor side wall, and is nothing more than a fossa with a hind wall (lamina orbitonasalis), medial wall (nasal septum), and floor (trabecular or ethmoid plate).

The aperture of the fenestra basalis is narrowed in many Tetrapoda by the cartilago ectochoanalis which extends back from the lamina transversalis anterior laterally to the choana. The cartilago ectochoanalis is typically developed in Urodela, Lacerta, Sphenodon, Chelonia, while in Ophidia it may be independent as the hypochoanal cartilage.

In mammals the false palate carries the opening of the secondary choana far back and the position of the primary choana is marked by Stenson's duct, the duct of Jacobson's organ, laterally to which the cartilago ectochoanalis is found in many forms (Echidna, Seydel, 1899; Gaupp, 1908; Ornithorhynchus, Broom, 1900; Dasyurus and Trichosurus, Broom, 1896*b*; Sorex, Ärnbäck Christie-Linde, 1907; Tupaja, Broom, 1915*a*; Macroscelides, Broom, 1902*b*; Vesperugo, Grosser, 1902; Miniopterus, Felis, and Lepus, Broom, 1900; Cheirogaleus and Galago, Ärnbäck Christie-Linde, 1914; Cebus, Ateles, Mycetes, and Chrysothrix, Frets, 1913). In some forms the cartilago ectochoanalis may be independent; in Monotremes it fuses with its fellow of the other side ventrally to the nasal septum and paraseptal cartilages. As Matthes (1922) has shown, great confusion prevails in the nomenclature of the cartilago ectochoanalis in mammals, where it is also known as the palatine cartilage, processus posterior lateralis, outer nasal floor cartilage, &c. An attempt was made by Jacobson (1928) to prove in Talpa that the palatine cartilage is really the nodule of secondary cartilage in the palatine process of the maxilla. But in Sorex, de Beer (1929) showed that both cartilages were present and distinct, and therefore cannot be homologous.

A further source of confusion arises from the presence in some mammals (Ovis, Lemur, &c., Herzfeld, 1889; Equus, Spurgat, 1896; Bos, Mihalkovics, 1899) of a 'cartilago ductus nasopalatini' which projects downwards and forwards from the lamina transversalis anterior, and lies immediately in front of Stenson's duct. Unfortunately, the name of this cartilage was applied by Spurgat (1896) in Lepus (where it does not exist) to the ectochoanal or palatine cartilage.

Mention must also be made of the median and unpaired cartilago papillae palatinae, situated beneath the nasal septum and found in Echidna (Wilson, 1901), Marsupials (Broom, 1896*b*), Cheiroptera (Grosser, 1902; Zuckerkandl, 1909), Tupaja, (Broom, 1915*a*), Macroscelides (Broom, 1902*b*), and certain rodents (Broman, 1920).

Among reptiles a papillary cartilage has been described in Sphenodon by Broom (1906) as a backgrowth from the anterior part of the nasal septum, beneath the prevomers. Broom states that a similar cartilage occurs in crocodiles. Perhaps the median palatine cartilage of Bufo (p. 206) may be of the same nature as the papillary cartilage.

Lastly, while dealing with the floor of the capsule it must be mentioned that in some mammals the ventral edge of the hind wall or cupola posterior (lamina orbitonasalis) is bent forwards, forming a lamina transversalis posterior. In Trichosurus, Halmaturus, Canis, Lepus, and Talpa, this is free from the nasal septum, but it is fused with it in many other mammals.

The side wall of the nasal capsule becomes complicated in connexion with the formation of the concha nasalis: an inpushing with its mouth or aditus conchae facing outwards, which increases the internal surface area and which lodges either the lateral nasal gland (e.g. Lacerta) or a diverticulum of the nasal sac (e.g. Crocodilus) or nothing but connective tissue. The concha nasalis is present in Urodela, reptiles (except Sphenodon and Chelonia) and birds, while in mammals it seems to be represented by the crista semicircularis, for the recessus anterior in mammals (the recess of the nasal cavity lateral to the crista semicircularis) seems to correspond exactly to the recessus extraconchalis in reptiles.

In reptiles, birds, and mammals the middle region of the side wall of the capsule chondrifies from a separate centre, the paranasal cartilage, and the epiphanial foramen marks the line of fusion between the paranasal cartilage and the roof or parietotectal cartilage. The latter develops in continuity with the dorsal edge of the nasal septum, and also grows down to form the anterior part of the side wall. It is, however, overlapped by the anterior edge of the paranasal cartilage, with the result that the hind edge of the anterior side wall cartilage projects into the cavity of the capsule, and then gives rise to the concha nasalis or crista semicircularis, as the case may be. (See Plate 139.)

Similarly, in mammals, the hind edge of the paranasal cartilage overlaps the anterolateral edge of the lamina orbitonasalis, which thus projects into the cavity of the capsule as the 1st ethmoturbinal. Other ethmoturbinals develop on the medial side of the 1st, and all these, together with the frontoturbinals which arise in the recessus frontalis laterally to the 1st ethmoturbinal and the nasoturbinal (anterior to the crista semicircularis), are found only in mammals.

It is a matter of great difficulty to equate the other turbinals in the different groups. The maxilloturbinal, an inrolling of the ventral edge of the side wall in front of the concha nasalis or crista semicircularis, seems to be represented in crocodiles (turbinal 'e' of Gegenbaur, 1873), birds (see p. 278 for synonyms), and mammals. The same is probably true of the atrioturbinal in crocodiles (anterior turbinal of Meek, 1911), birds (anterior, or vestibular turbinal of authors), and mammals. The posterior turbinal of crocodiles (Meek, 1911; the 'pseudo-concha' of Gegenbaur, 1873) is probably an independent development, not homologous with the mammalian ethmoturbinals.

The foramen through which the olfactory nerve enters the nasal capsule (foramen olfactorium advehens) is not the same as that through which the olfactory nerve leaves the cranial cavity (foramen olfactorium evehens). The key to the interpretation of these relations, and of the facts that the nasal branch of the profundus nerve enters the nasal capsule without passing through the cranial cavity, and that in some forms the olfactory nerve traverses the orbit, is to be found in the recognition of an extracranial space, the cavum orbitonasale, the morphology of which is specially considered on page 432.

It is a curious fact that the relations of the lachrymonasal duct to the nasal capsule are variable. Whereas in Reptilia the duct opens into the nasal sac *behind* the zona annularis, in Amphibia and mammals it passes *in front* of the cartilage and therefore runs in through the fenestra narina, or a small foramen cut off from it, partially or completely, by the superior and inferior alar processes. These processes (termed by some authors the accessory processes, or processus laterales septi anteriores), surrounding the apertura nasalis externa, are remarkably constant in Tetrapoda. They are absent in Cetacea, Sirenia, and reduced in Man.

In a long series of papers Broom (1895*c*, 1896*a*, *b*, and *c*, 1897, 1900, 1902*b*, 1909*b*, 1915*a* and *b*, 1926*b*) has examined that part of the paraseptal cartilage which encloses Jacobson's organ and forms Jacobson's capsule in different mammals. Starting from the condition in Monotremes where, anteriorly, Jacobson's capsule forms a complete cylinder with a 'turbinal' projecting into its lumen, Broom distinguishes a condition which is a simplification of this, with the turbinal reduced to an 'outer bar', and found in Marsupials, Rodents, Xenarthra, Tubulidentata, Tupaja, Macroscelides, Chrysochloris (the 'Archaeorhinata'), from a condition in which there is no outer bar and Jacobson's capsule is produced forwards beneath the lamina transversalis anterior to form a cartilago ductus nasopalatini. This group, the 'Caenorhinata', includes Artiodactyla, Perissodactyla, Carnivora, Insectivora, Cheiroptera, Hyracoidea, and Primates.

VIII. The Auditory Capsule

The auditory capsule is one of the most important constituents of the chondrocranium, and may indeed have been instrumental in its evolution if van Wijhe's (1906) hypothesis is correct, that not only the cartilaginous auditory capsule but also the parachordal cartilages arose in connexion with the necessity for the provision of a rigid anchorage for the ears as organs of balance. Originally, then, the auditory vesicle may have been surrounded by a connective tissue capsule, unconnected with the dura mater, and with the vena capitis medialis running between them. With the evolution of the semicircular canals, the auditory capsule became cartilaginous, and the basicapsular commissures united it rigidly with the basal plate.

In the development of living vertebrates the auditory capsule almost always arises from a centre (or centres) of chondrification independent of the parachordals, and subsequently becomes attached to them. In this connexion a number of morphological points call for consideration: the identity of the commissures of attachment, the morphology of the space enclosed between them, the relations of the capsule to the facial and glossopharyngeal nerves, and the relations of the capsule to the basal plate. Attention may then be turned to the structure of the fully formed capsule itself, and to the part which it plays in providing a portion of the side wall of the cranial cavity.

(i) The commissures by means of which the auditory capsule is attached to the parachordal of its side were termed basicapsular by Gaupp (1900, p. 446; 1906*a*, p. 663), but his use of the term has been attended by some confusion since he referred to both the prefacial commissure of Lacerta and the lateral commissure of Salmo as 'prefacial basicapsular commissure'. Actually, there is no need to consider the prefacial commissure in this connexion at all, and attention may be confined to the anterior and posterior basicapsular commissures. The anterior is always situated behind the facial nerve, the posterior basicapsular commissure in Salmo taken as a type lies behind the glossopharyngeal and in front of the vagus nerve. Between the anterior and posterior basicapsular commissure in Salmo, therefore, there is a space composed of the basicapsular fenestra (see below) and the foremost portion of the fissura metotica (lodging the glossopharyngeal nerve). The posterior basicapsular commissure, therefore, joins the parachordal in such a way as to cut off a portion of the fissura metotica and incorporate it with the basicapsular fenestra. Polypterus, Acipenser, Amia, Lepidosteus, Gymnarchus, Amiurus, Anguilla, Solea, Gadus, and Ceratodus, agree with Salmo in this respect.

In other Teleosts, however (Gasterosteus, Syngnathus, Clupea), and in Amphibia, while the anterior basicapsular commissure is as in Salmo, the hinder attachment joins the parachordal *in front of* the glossopharyngeal nerve, and thus has the relations of the basivestibular commissure of Amniota. In the latter, the relations of the anterior basicapsular commissure are often obscured because of the development of the cochlear part of the auditory capsule, which is commonly attached to the parachordal by an elongated suture extending from the fissura metotica behind to the front of the basal plate. But in many mammals the basicapsular fenestra is represented by the basicochlear fissure, and the anterior basicapsular commissure becomes the sphenocochlear commissure. In some forms (e.g. Talpa) the basicochlear fissure becomes further subdivided into two by the formation of a basicochlear commissure (p. 322).

In Petromyzon the auditory capsule is at first attached to the parachordal by a single commissure which is probably the anterior basicapsular; later it broadens out and there is no basicapsular fenestra. In Selachii the anterior and posterior basicapsular commissures are present, but the latter are obscured by the development of the lamina hypotica (see p. 60).

(ii) The basicapsular fenestra (or basicochlear fissure) is a space included between the ventromedial edge of the auditory capsule and the lateral edge of the parachordal cartilage, bounded anteriorly by the anterior basicapsular (or sphenocochlear) commissure, and posteriorly by the basivestibular commissure, or it may be confluent with the foremost part of the fissura metotica which transmits the glossopharyngeal nerve and is limited behind by the posterior basicapsular commissure.

The basicapsular fenestra was regarded by W. K. Parker (1871), Stöhr (1881), and Gaupp (1906*a*, pp. 583, 663, 721, 725) as the equivalent of the fenestra ovalis (or vestibuli) of the auditory capsule, becoming obliterated in fish but persisting in Tetrapoda. As it stands, however, this view is misleading, for the medial border of the basicapsular fenestra is formed of parachordal (i.e. axial) cartilage, whereas the medial (or ventral) border of the fenestra ovalis is composed of capsular cartilage. In early stages of development it is common for the floor of the auditory capsule to remain unchondrified, and under these circumstances the fenestra ovalis has no medial border, and may be confluent with the basicapsular fenestra. Evidence of the truth of this view is provided by Alytes (p. 211), where the floor of the capsule develops early and excludes the parachordal from participating in the limitation of the fenestra ovalis.

The position of the fenestra ovalis in Tetrapoda is further defined by the fact that the footplate of the columella auris (otostapes) fits into it. In the fish, of course, no fenestra ovalis is formed, but a perforation of the capsular wall immediately beneath the articular facet of the hyomandibula has been found in Heptanchus by van Wijhe (1924).

(iii) The typical relations of the facial nerve to the auditory capsule are that the nerve passes out in front of the capsule and never traverses any part of its cavity, although in Cyclostomes the front wall of the capsule seems to represent the prefacial commissure, and the facial nerve does seem to traverse a small portion of the cavity of the capsule. If the auditory nerve is to be regarded as a derivative of the facial, a close association of the roots of these nerves is not surprising. In mammals the facial and auditory nerves may for a short distance run together through a special region of the cranial cavity forming the internal auditory meatus. In Amphibia, however, a peculiar modification is found, in that the facial nerve on leaving the cranial cavity, appears to traverse the cavity of the auditory capsule on its way out. The explanation is that in these forms the auditory capsule is pressed forwards, and its medial wall overlaps and duplicates the prefacial commissure. In Amblystoma the facial nerve running between these cartilages, emerges as from a canal (p. 178). In Triton (p. 182) and Alytes (p. 211), however, the medial wall of the auditory capsule in this region remains permanently or temporarily membranous, so that the cavity of the capsule *appears* to be bounded medially by the prefacial commissure and to be traversed by the facial nerve. Actually, the nerve is excluded from the capsular cavity by the membranous capsular wall, which can usually be distinguished.

(iv) The typical relations of the glossopharyngeal nerve to the auditory capsule are that the nerve passes out behind the capsule, and never traverses any part of its cavity. But in Selachii, Teleostomi, and Dipnoi, and in some reptiles, modifications are found which obscure these relations.

In Selachii the hinder part of the lateral edge of the parachordals projects on each side forming a lamina hypotica (p. 63) beneath the auditory capsule. The glossopharyngeal nerve runs out dorsally to the lamina hypotica and ventrally to the floor of the auditory capsule, which, however, in this region, remains membranous. The capsular cavity thus *appears* to be bounded here by the cartilaginous lamina hypotica and to be traversed by the nerve. Actually, the nerve is excluded from the capsular cavity by the membranous capsular wall, which can usually be distinguished.

The glossopharyngeal nerve has a typical course in the Holocephali, although in these forms the capsule has no cartilaginous medial wall. On the other hand, the absence of a cartilaginous medial wall appears to be responsible for the passage of the glossopharyngeal nerve through the cavity of the capsule in Teleostomi and Dipnoi.

In reptiles the glossopharyngeal nerve never traverses the cavity of the capsule, but it runs through that anterior part of the fissura metotica which forms the recessus scalae tympani, and as the walls of that recess tend to become enclosed in cartilage in some forms (e.g. Eumeces, p. 235; Chelonia, p. 255), the nerve appears to traverse the capsule.

(v) The relations of the auditory capsule to the basal plate present no difficulty in the lower vertebrates, where the two structures are visibly distinct in origin. In Amniotes, however, the development of the ductus cochlearis and the formation of a special cochlear portion of the auditory capsule, which projects into the territory of the basal plate, led Gaupp (1900, p. 583) to conclude that the cochlear part of the auditory capsule is actually formed out of the basal plate; i.e. by part of the parachordal cartilages. Gaupp based his argument on the facts that in mammals the trabeculae (alicochlear commissures) make contact with the cochlear capsule and not with the parachordal, the prefacial commissure becomes a suprafacial commissure mounted on the top of the cochlear capsule, and that the mesenchyme surrounding the ductus cochlearis is continuous with that out of which the basal plate subsequently chondrifies.

In opposition to this view Noordenbos (1905) pointed out that the cochlear part of the auditory capsule chondrifies in continuity with the canalicular part and independently from the basal plate from which it is usually separated, at least for a time, by the basicochlear fissure. It may further be noticed that the reason why the trabecula (alicochlear commissure) in mammals establishes contact with the cochlear capsule instead of with the basal plate is simply because the former has replaced the latter; this is no evidence for the view that the former has been constructed out of the latter. In reptiles, birds, and Monotremes, the cochlear capsules are not so large, and there the trabeculae are attached to the parachordal in the typical manner. It is, therefore, of the greatest interest to find that in the Cetacea studied by de Burlet (1916) and Honigmann (1917) the anterolateral portions of the basal plate persist, overlying the cochlear capsules as the laminae supracochleares, and it is with these that the alicochlear commissures make contact. Comparable conditions are found in Halicore (p. 346) and in Felis (p. 327).

There can, therefore, be no doubt that the cochlear capsules are extensions of the true auditory capsules which displace the anterolateral parts of the basal plate, but are in no way formed out of it.

(vi) The term foramen perilymphaticum is applied to the posterior opening in the wall of the auditory capsule of Tétrapoda, through which the ductus perilymphaticus leaves the cavity of the capsule and enters the recessus scalae tympani. The foramen perilymphaticum is entirely surrounded by capsular cartilage. Occasionally it has been referred to by the term 'fenestra cochleae', but this is best discarded for it has also been applied to the fenestra rotunda of mammals, with which the foramen perilymphaticum is not homologous.

The formation of the fenestra rotunda and of the aquaeductus cochleae in mammals is the result of the backgrowth of the cartilaginous floor of the cochlear capsule from the anterior border of the foramen perilymphaticum. This structure, the processus recessus (de Beer, 1929; 'processus intraperilymphaticus' of Voit, 1909), forms a floor to the recessus scalae tympani (see p. 431) and joins the ventral surface of the canalicular part of the auditory capsule, which may (e.g. Homo, p. 359) develop a process to meet it. The processus recessus does not lie in the plane of the mouth of the foramen perilymphaticum at all, and cannot therefore be said to subdivide it. On the contrary, the fenestra rotunda and the aquaeductus cochleae are formed out of the apertura lateralis and medialis respectively of the recessus scalae tympani.

The relations of these structures is best understood in the light of a transverse section passing through the foramen perilymphaticum and the anterior part of the fissura metotica. In this region three free edges of cartilage may be seen: (i) the ventral edge of the lateral wall of the auditory capsule; (ii) the ventral edge of the medial wall of the auditory capsule; and (iii) the lateral edge of the basal plate. The aperture between (i) and (ii) is the foramen perilymphaticum; that between (ii) and (iii) is the apertura medialis of the recessus scalae tympani, which becomes the aquaeductus cochleae in mammals; and that between (i) and (iii) is the apertura lateralis of the recessus scalae tympani, over which the secondary tympanic membrane is stretched, and part of which becomes the fenestra rotunda. The processus recessus (iv) lies laterally to (iii), and therefore the fenestra rotunda is bordered dorsally by (i) and ventrally by (iv).

In Monotremes the processus recessus is not formed, and consequently there is no fenestra rotunda.

In crocodiles and birds the recessus scalae tympani develops a cartilaginous floor, but this is formed by the subcapsular process of metotic cartilage, presumably representing modified cranial ribs, and in no way homologous with the mammalian processus recessus. Consequently, the aperture between the ventral edge of the lateral wall of the capsule (= lateral border of the foramen perilymphaticum) and the lateral edge of the subcapsular process or metotic cartilage, and across which the secondary tympanic membrane is stretched, is only analogous to the fenestra rotunda, and may be termed the fenestra pseudorotunda (see pp. 263, 274).

It may therefore be concluded that Gaupp (1900, p. 515) was wrong in regarding the fenestra rotunda and aquaeductus perilymphaticus as the product of subdivision of the foramen perilymphaticum, for they are not its equivalent. It is because Voit's (1909, p. 445) 'processus intraperilymphaticus' (first observed but not named by Fischer, 1903) does *not* lie in the plane of the foramen perilymphaticum, but in the floor of the recessus scalae tympani that de Beer (1929) substituted the term processus recessus.

(vii) The structure of the auditory capsule is closely related to the form of the auditory vesicle, and the various fossae and prominences on its surface reflect the semicircular canals, ampullae, utricle, saccule, and ductus cochlearis beneath it (except in Cyclostomes where the capsule is ovoid in shape and bears less relation to the form of the contained vesicle). At the same time the medial surface of the capsule may become adapted to the form of the cranial cavity, and the fossa subarcuata may accommodate the paraflocculus of the cerebellum. Sometimes (e.g. Cheiroptera, Microtus) the fossa subarcuata may penetrate right through to the outside surface of the capsule. But while the capsule surrounds the semicircular canals and ampullae very closely, space is left between the utricle, saccule, and ductus cochlearis and the wall of the capsule, and this is later occupied by the perilymphatic system. The cavities of the capsule may become enlarged by erosion and absorption of the cartilage of the internal surface (see Homo, p. 359; Streeter, 1917, 1918), but it appears that ossification does not set in until the definitive size has been reached (Bast, 1930).

The cavity of the capsule becomes partitioned off into various chambers by septa. These do not exist in Cyclostomes, but from the fish upwards there are three septa for the semicircular canals. A curious feature, found in the chondrocranium of Urodela, Lacertilia, Chelonia, birds, and many mammals (e.g. Didelphys, Lepus, Sorex, Talpa, Macacus, Semnopithecus) is the incompleteness of the posterior septum, whereby the posterior and lateral semicircular canals communicate with one another. According to Denker (1899) this condition persists in some adult mammals (e.g. Ornithorhynchus, Ursus, Equus, &c.). The main chamber of the cavity or cavum vestibulare becomes divided in Lacertilia into anterior and posterior portions by a transverse partition: the septum intervestibulare. This also occurs in Felis. In mammals a spiral septum is developed from the wall of the cochlear part of the capsule, and separates the spirally wound coils of the cochlea.

The size of the auditory capsule is closely related to the volume of its cavity, and the ratio between it and the volume of the cranial cavity has been estimated by Voit (1909, p. 523) as follows in the fully formed chondrocranium:

Rana	1 to 4
Lacerta	1 ,, 10
Lepus	1 ,, 20
Echidna	1 ,, 30
Homo	1 ,, 70

At the same time it must be remembered that absolute size plays an important part in the ratio, for as Watson (1928) has pointed out, a lion is fifty times as heavy as a cat, yet the volume of its internal ear is only three times as great. Nevertheless, it is clear that at increasing levels of vertebrate evolution the auditory capsule becomes *relatively* smaller. Thus, while in fish and Amphibia, the capsule is high enough to form a complete side wall to the cranial cavity; in reptiles the height of the capsule is raised by a backward extension of the taenia marginalis or taenia supramarginalis, and in mammals the parietal plate is developed in this region. According to Gaupp (1906*a*, p. 664) the non-chondrification of the medial wall of the capsule in Teleostomi is associated with a deficiency of space for the cranial cavity. However, the medial wall also fails to chondrify in Holocephali and Dipnoi, but probably not for the same reason. It might be asked whether non-chondrification of the medial wall may not be a primitive feature, but this is hardly probable in view of the conditions in Selachii.

Another result of the relative increase in size of the cranial cavity in higher forms is seen in the position of the auditory capsule. In Lacertilia the long axes of the capsules *diverge* anteriorly, i.e. the cochlear parts are wider apart than the canalicular parts. In mammals the axes of the capsules *converge* anteriorly, and the greater width between the canalicular parts reflects a widening of the cranial cavity. Next, while in fish, Amphibia, and lower mammals, the medial wall of the capsule is more or less vertical and the tectum synoticum is horizontal, in higher mammals the medial wall is splayed outwards and comes to form part of the floor rather than the side wall of the cranial cavity. At the same time, the whole capsule is rotated backwards, so that the canalicular part, which in reptiles (and in Monotremes) lies more or less above the cochlear part, is situated definitely behind it in mammals, and the tectum synoticum lies in the vertical transverse plane. In birds the medial wall of the capsule retains its sagittal position, but the backward rotation is very marked, and the tectum synoticum may slope upwards and backwards, and the relative positions of the foramina are altered. In mammals and birds, therefore, the foramina acustica anterius and posterius become known as superius and inferius respectively.

The foramen acusticum is primitively single as in Cyclostomes and Selachii. In Rana a subdivision takes place as follows:

foramen acusticum anterius		*foramen acusticum posterius*
nervus utriculo-ampullaris =	ramus ampullaris anterior	ramus ampullaris posterior
	,, ,, lateralis	,, saccularis
	,, recessus utriculi	,, lagenae

In Lacerta the conditions are similar though the ramus lagenae becomes known as the ramus cochlearis. In Salamandra this ramus passes through a separate foramen acusticum medium.

In mammals as a result of the above-mentioned rotation, the typical conditions are as follows (e.g. Perameles, Lepus, Talpa):

foramen acusticum superius	ramus superior =	nervus utriculo-ampullaris
foramen acusticum inferius	ramus inferior =	ramus ampullaris posterior
		,, saccularis
		,, cochlearis

Voit (1907) has drawn attention to the fact that in some mammals the sacculus may be innervated by the ramus superior as well as the ramus inferior. Other complications may arise from the fact that the ramus saccularis and the ramus ampullaris posterior may have independent foramina. These conditions have been diagrammatically summarized as follows by Matthes (1921*b*):

Lepus:

foramen acusticum superius . . .	nervus utriculo-ampullaris
	ramulus saccularis superior
foramen acusticum inferius . . .	ramulus saccularis inferior
	ramus ampullaris posterior
	,, cochlearis

Echidna:

foramen acusticum superius . . .	nervus utriculo-ampullaris
foramen acusticum medium . . .	ramus saccularis
foramen acusticum inferius . . .	ramus ampullaris posterior ,, cochlearis

Didelphys:

foramen acusticum superius . . .	nervus utriculo-ampullaris ramulus saccularis superior
foramen acusticum medium . . .	ramus saccularis inferior anterior
foramen acusticum inferius . . .	ramulus saccularis inferior posterior ramus ampullaris posterior ,, cochlearis

Halicore:

foramen acusticum superius . . .	nervus utriculo-ampullaris ramulus saccularis superior
foramen acusticum intermedium. . .	ramulus saccularis intermedius
foramen acusticum inferius . . .	ramulus saccularis inferior ramus cochlearis
foramen singulare	ramus ampullaris posterior

Homo:

foramen acusticum superius . . .	nervus utriculo-ampullaris
foramen acusticum inferius . . .	ramus saccularis ,, cochlearis
foramen singulare	ramus ampullaris posterior

It is to be noted that the rotation of the auditory capsule as a whole may be independent of the rotation of the planes of the semicircular canals, which may, as apparently in Man (p. 360), take place within the capsule. As Lebedkin (1924) has pointed out, the lateral semicircular canal lies horizontally at normal postures of the head. Turkewitsch (1936) has shown that in birds the axis of the ductus cochlearis and the planes of the semicircular canals may vary independently.

The rotations of the capsule induced experimentally are described on p. 479.

On the lateral surface of the auditory capsule, a remarkably constant feature is the crista parotica forming a ledge, horizontal in Pisces, Amphibia, and Reptilia, sloping backwards and downwards in Aves and Mammalia, accentuating the prominence of the lateral semicircular canal, and against which abut the head of the hyomandibula, hyosymplectic, or laterohyal, and the otic process of the pterygoquadrate, or crus breve, of the incus. In mammals the crista parotica is produced forwards to form the tegmen tympani (or processus perioticus superior of Gradenigo (1887)), which, by establishing connexion with the cochlear capsule may give rise to the lateral suprafacial (or prefacial) commissure; posteriorly it may be produced downwards to form the mastoid process. The intermediate portion of the crista parotica in mammals, or crista facialis, forms the lateral wall of the sulcus facialis, and typically has the stylohyal fused to it, forming the styloid process.

IX. The Sclerotic Cartilage

The sclerotic cartilage develops in the connective tissue surrounding the eyeball. Since the eyeball itself is morphologically part of the brain, the sclerotic cartilage is to be regarded as comparable to the chondrocranium on Gaupp's (1906*b*) view. Indeed, it has a closer morphological connexion with the neurocranium than the olfactory or auditory capsules, for the sense-organs which they surround are not parts of the brain.

The function of the sclerotic cartilage is doubtless to maintain the eye in constant shape, and it is probably also associated with the provision of firm points of insertion for the extrinsic eye-muscles. (See Xenopus, p. 216.)

The sclerotic cartilage is free and movable, though in Selachii it is in contact with a cartilaginous outgrowth from the neurocranium: the eyestalk. The sclerotic cartilage may thus be held to represent the original condition of the olfactory and auditory sense capsules before fusion with the neurocranium occurred. It is interesting to note that the primitive vein of the head passes between the sense capsule and the primitive side wall of the skull, represented by the dura mater. Thus, the vena capitis medialis runs medially to the auditory capsule for as long as it exists; the orbital sinus is medial to the sclerotic, and the orbitonasal vein passes medially to the lamina orbitonasalis.

The distribution of the sclerotic cartilage has been studied by Stadtmüller (1914). A sclerotic cartilage is present in all groups of vertebrates except: Cyclostomes (where the eyes are in any case degenerate), Siluroid Teleosts, Anguilla, Gymnophiona, Marsupial and Placental mammals. Typically the sclerotic cartilage is shaped like a kettle-drum, enveloping the entire proximal portion of the eyeball. It is, however, often reduced, and may then take the form of a shallow cylinder round the greatest circumference of the eyeball, or of a shallow plate surrounding the exit of the optic nerve (see also Chatin, 1910).

The conditions in Amphibia are peculiar, for whereas the sclerotic cartilage arises fairly late (towards the end of metamorphosis) in all Anura, in the Urodela it arises early and in many forms disappears at metamorphosis. The conditions controlling the formation and disappearance of the sclerotic cartilage are discussed elsewhere (p. 479), as are the sclerotic bones which often assist the cartilage in surrounding the eyeball (p. 443).

X. The Visceral Arch Skeleton

The skeleton of the branchial arches

The development and morphology of the hyobranchial skeleton has been reviewed with great thoroughness by Gaupp (1905*c*), since which time little in the way of fresh evidence has come to light in regard to the main problems which they present.

In particular, the question of the median copulae or basibranchials is unsettled. Gegenbaur (1872) originally regarded them as elements situated between the bases of the branchial arches, but later (1898) he came to regard them as derived from the bases of the branchial arches in front of them. This is the more probable view, and it has practically been confirmed embryologically, for while the basibranchials regularly *chondrify* independently of the branchial arches, Gibian's (1913) researches in Selachii go to show that the mesenchyme from which this cartilage is differentiated may have been derived from the base of the arches. Certain it is from Stone's (1922, 1926) experiments on Amblystoma that the 1st basibranchial is not formed when the branchial arches are absent; but on the other hand, the 2nd basibranchial *is* formed even in their absence (see p. 474).

In Anura it is found that absence of the branchial arches entails absence of the pars reuniens, but not of the basihyal nor of the basibranchial.

During the course of evolution, the function of the hyobranchial apparatus changes from supporting gill-arches between open gill-slits to surrounding the larynx and the front of the trachea, and supporting the tongue. These changes are accompanied by changes in the number of skeletal arches present, and in the segmentation of each arch.

As regards the number of arches present, the tendency is to reduction. Not only is the number greatest in Cyclostomes, but in young stages of primitive Gnathostomes like Heptanchus, Daniel (1916) has shown that as many as nine branchial (i.e. eleven visceral) arches may be present. In Chlamydoselache rudiments of a 7th branchial (9th visceral) arch may be present (Goodey, 1910). In Ophidia and in birds only the first pair of arches is left. On the other hand, in the mammals three pairs of branchial arches are preserved. While there is no doubt concerning the contribution to the thyroid cartilage of both 2nd and 3rd branchial arches in Monotremes and Marsupials (Dubois, 1886; Gegenbaur, 1892; Symington, 1899; Edgeworth, 1916), the conditions in Placentals are not so clear. The thyroid cartilage appears to be formed of the 2nd and 3rd branchial arches in Man, Lepus, Tatusia (Edgeworth, 1935); of the 2nd only in Sus (Edgeworth, 1916); of the 3rd only in Manis (Edgeworth, 1935). The 1st branchial arch in mammals is the thyrohyal. (The claim of the cricoid cartilages to represent the 4th branchial arches cannot be regarded as established.) On the number of branchial arches, of course, depends the extent of the visceral arch skull or splanchnocranium. The various numbers of segments concerned in its composition are given on page 26. The manner in which the reduction in number of arches is brought about in phylogeny has led Versluys (1922) to some interesting conclusions discussed on page 448. In Anura and Urodela the number of branchial arches present in the larva is greatly reduced at metamorphosis.

Similarly, in regard to the number of pieces into which each arch is segmented, the tendency in the Tetrapoda is towards reduction. In fish of all groups the segmentation of the branchial arches into the four familiar elements appears to be general. It need only be noticed that in Selachii and Holocephali the arches are Σ-shaped, both pharyngo- and hypobranchials pointing backwards; while in Teleostomi the arches are >-shaped, pharyngo- and hypobranchials pointing forwards. The Dipnoi resemble the Selachii as regards the hypobranchials, but the condition of the pharyngobranchials is not clear.

In the higher vertebrates it is common to find only two elements in the branchial arches, usually regarded as epi- and ceratobranchials; or these arches may be unsegmented.

Two points need consideration in this connexion. First, it is only with doubtful legitimacy that the segments of the arches of higher vertebrates are regarded as homologous with those of fish, for it may well be that they have been independently segmented. For instance, as Gaupp (1905*c*) and Bruni (1909) have pointed out, it is improbable in the extreme that the four pieces of the hyoid arch skeleton in mammals were really derived from and represent the four segments of the hyoid arch of ancestral fish, for the hyoid arches in Amphibia and reptiles are quite differently segmented. In fact, this case may be taken as proof of the contrary. And this leads to the second point, namely, whether the primitive condition of the branchial (or visceral) arches was segmented as in fish or unsegmented as in Cyclostomes.

In the fish, it is clear (Braus, 1905; van Wijhe, 1904*a*) that the various elements chondrify independently in a continuous mesenchymatous rudiment. As Gaupp (1905*c*, p. 850) points out, this evidence from embryology cannot be straightway applied to phylogeny. It may be that the ancestral form had continuous mesenchymatous or procartilaginous arches, from which condition the Cyclostomes evolved the continuous cartilaginous arch, and Gnathostomes the *ab initio* segmented cartilages. Or the Gnathostomes may have been derived from forms with continuous cartilaginous arches, and have evolved their segmentation into different pieces in adaptation to muscular requirements. (For examples of such fragmented chondrification see p. 509.) But on either view the arches of fish might be expected to-day to show independent chondrification of the various segments, and therefore no reliable information regarding phylogeny is supplied by embryology.

The branchial arches in fish generally are quite independent of the neurocranium. This condition is clearly primitive, and a consequence of the fact that the visceral arch skeleton or splanchnocranium is phylogenetically older than the neurocranium: a visceral arch skeleton is already present in Balanoglossus, Amphioxus, and Tunicates, where there is no brain-case at all. It is for this reason that all contacts between the brain-case and the mandibular or hyoid arches, which are merely specialized branchial arches, must be regarded as secondary.

In some forms the branchial arches show how this contact with the braincase may be brought about. Thus, in the first three branchial arches in Polypterus (p. 84), the first two in Acipenser (p. 92), and in Lepidosteus (p. 113), the pharyngobranchial and epibranchial have extended upwards, presenting a forked appearance, and the two prongs of the fork (the 'infrapharyngobranchial' and 'suprapharyngobranchial' respectively, of van Wijhe, 1882*b*) may make contact with the auditory capsule. It is also of great interest to note that in this upgrowth, varying relations to the dorsal nerve corresponding to the arch may be obtained. Thus, in Polypterus the nerve runs behind both prongs; in Acipenser it runs in front of them, while in Lepidosteus it runs between them, in front of the supra- and behind the infrapharyngobranchial. The importance of this case for the interpretation of the hyoid arch is mentioned elsewhere (pp. 411, 493).

While the primitive function of the hyobranchial skeleton is to surround the pharynx, with the evolution of a tongue it often acquires the function of supporting it. In cases where the tongue is protrusible this may entail modification of the skeletal elements of the 1st branchial arch, which become greatly elongated. This is the case in the Urodele Spelerpes and among birds in the woodpecker (e.g. Gecinus, W. K. Parker, 1875). In the latter the cartilages are so long that they coil up behind the skull and forwards over it as far as the nasal bones.

XI. THE LABIAL CARTILAGES

In Selachii small cartilaginous bars are occasionally found laterally and in front of the upper and lower jaws. Usually they consist of two dorsal pairs and one ventral pair. In Holocephali the nasal apertures and the mouth are surrounded by a number of cartilaginous structures some of which in all probability represent the Selachian labials, while others perhaps correspond to the nasal cartilages of Selachians (which W. K. Parker also called 'labial' cartilages).

In Teleostomi three pairs of cartilages are also occasionally found, and for want of evidence to the contrary, regarded as homologous with the labial cartilages of Selachii. Of these, the 'premaxillary cartilages' (anterior upper labials), found in Salmo (Gaupp, 1906*a*), Gasterosteus (Swinnerton, 1902), Anguilla (Norman, 1926), and Cyclopterus, arise late and are closely associated with the premaxillary bones. In Pleuronectes and Solea (Berrill, 1925) these cartilages fuse to form a median rostral cartilage, such as is also found in Syngnathus (Kindred, 1921).

The 'maxillary cartilages' (posterior upper labials) develop in the same way in relation to the maxillary bones (Sagemehl, 1885, 1891).

The 'labial cartilages' (lower labials) are formed in the angle of the mouth, e.g. Polypterus, Gadus, &c. In Amia and Lepidosteus this region is occupied by the coronoid process of Meckel's cartilage which according to van Wijhe (1882*b*) may represent the labial cartilage.

In the Siluroidea one or more of these cartilages may be greatly elongated and support the barbel.

In Dipnoi there is a 'subnasal' cartilage (upper labial?) and two pairs of lower labials, one of which fuses with Meckel's cartilages.

In Anuran larvae tentacular cartilages are found in Xenopus, and the so-called admandibular cartilage of Alytes may be a vestige of such a structure.

It may be noted that 'coronoid' cartilages are present in crocodiles and in birds, but they can hardly be regarded as relics of the labial cartilages of fish.

Sewertzoff (1916) has attempted to homologize the labial cartilages with the anterolateral and posterolateral cartilages of Cyclostomes, and further to regard them as vestiges of two pairs of preoral visceral arches. While the former of these propositions may perhaps be correct, the latter is in opposition to all the evidence of the segmentation of the head (see p. 20), and in particular to the evidence that the premandibular somite is the original foremost somite of the body. To this the trabeculae cranii may correspond (see p. 376) as premandibular visceral arches, but anterior to these, there cannot have been any visceral arches.

There remains the possibility, put forward by Pollard (1894, 1895), that these various cartilages may represent, together with the tentacles of Myxine, the oral cirrhi such as are found in Amphioxus, or in Palaeospondylus (Bulman, 1931).

XII. The Hyoid Arch Skeleton

The conditions of the hyoid arch skeleton in Cyclostomes, where there is a substitution of an unjointed cartilaginous rod for a mucocartilaginous predecessor are too peculiar to permit of profitable comparison with higher vertebrates. Until further fossil evidence is available the Cyclostomes must be left aside.

In most Selachii the skeleton of the hyoid arch is seen to be composed of two elements: the hyomandibula and the ceratohyal; and since the branchial arches each present four elements, the question arises whether the hyoid arch skeleton ever was divided into four elements, and if so, which of the elements of the branchial arches correspond to the hyomandibula and ceratohyal.

In answer to this question, it may be noted that in some forms the hyoid arch skeleton contains more elements than the typical two. Thus, hypohyal cartilages have been recorded in Laemargus (White, 1890, 1895) Heptanchus (Braus, 1906; Sewertzoff, 1927) and perhaps in Squalus (Gibian, 1913), among Selachii; fairly constantly in Teleostomi, Urodela, Lacertilia (de Beer, 1930*a*), Lepus (de Beer and Woodger, 1930). In other forms where the hypohyals are apparently absent (e.g. Scyllium, p. 62) it seems that they are fused to the basihyal.

Pharyngohyal cartilages have been described in Holocephali (Schauinsland, 1903; de Beer and Moy-Thomas, 1935), Stegostoma, Mustelus, and Galeus (Luther, 1909).

It is therefore possible that originally the hyoid arch skeleton did not differ from that of the remaining visceral arches behind it, in possessing pharyngo-, epi-, cerato-, and hypohyal elements. The Holocephali may have retained this condition, only having lost the hypohyal. The apparently primitive nature of the hyoid arch skeleton in Holocephali is associated with the fact that it is non-suspensorial, for the pterygoquadrate is of the extreme autostylic type known as holostylic (unless the 'otic process' should after all turn out to be the hyomandibula: see p. 76).

The close affinity between Holocephali and Selachii requires that if the hyoid arch skeleton of the former is primitive and non-suspensorial, then the latter must originally also have had non-suspensorial hyoid arches; and it would follow that the suspensorial function of the hyomandibula, or hyostyly, must have evolved independently in Selachii and in Teleostomi: a conclusion which is necessary for other reasons, as will be seen immediately.

The hyomandibula of Selachii is, following Gegenbaur (1872) generally held to be the epihyal, corresponding serially to the epibranchials, though Allis (1915, 1918) believes the hyomandibula of Batoidea to be a pharyngohyal on grounds which do not appear to be sufficient.

The hyoid arch skeleton in Selachii has the following peculiarities:

(i) The dorsal element or hyomandibula always chondrifies in one piece.

(ii) The dorsal end of the hyomandibula articulates with the wall of the auditory capsule *medially* and *ventrally* to the head vein.

(iii) The hyomandibular branch of the facial nerve passes out in front of and dorsally to the hyomandibula.

(iv) The contact between the ventral end of the hyomandibula and the mandibular arch in Selachii takes place *dorsally* to the ramus mandibularis internus facialis, i.e. this nerve runs ventrally to the hyomandibular ligament on its way to the medial surface of Meckel's cartilage.

(v) The ceratohyal remains directly attached to the hyomandibula, but in Batoids the ceratohyal undergoes reduction, and is replaced more or less completely by a pseudohyoid rod formed from the fusion of the hyal rays, as Krivetski (1917) and de Beer (1932) showed (p. 70).

The hyoid arch skeleton of Teleostomes is peculiar in the following respects:

(i) The dorsal element sometimes chondrifies in two portions, a dorsal hyomandibula and a ventral symplectic, which may remain separate (Acipenser, p. 90), or fuse to form a single hyosymplectic (Salmo, p. 118), or it may chondrify as a single hyosymplectic (Polypterus, p. 79; Amia, p. 98; Gasterosteus, p. 147; Anguilla, p. 143). Even in the latter cases there are subsequently separate and independent hyomandibular and symplectic ossifications.

(ii) The dorsal end of the hyomandibula articulates with the wall of the auditory capsule *laterally* and *dorsally* to the head vein.

(iii) The contact between the ventral end of the hyosymplectic or symplectic and the mandibular arch takes place *ventrally* to the ramus mandibularis internus facialis, i.e. this nerve runs dorsally to the quadratosymplectic ligament to the medial side of the pterygoquadrate cartilage, and thence to the medial side of Meckel's cartilage.

(iv) The hyomandibular branch of the facial nerve passes out *either* behind the hyomandibula (Acipenser, p. 93), *or* through it (Amia, p. 98; Lepidosteus, p. 109; Salmo, p. 123, and most Teleostei), *or* in front of it (Gadus, p. 165), *or* forks round it, the ramus mandibularis in front and the ramus hyoideus behind (Polypterus, p. 84).

(v) The ceratohyal is attached to the hyomandibula or hyosymplectic, not directly, but by the intermediary of an independent cartilage, the interhyal (stylohyal).

It will be noted that in each of these five respects the Selachii differ from the Teleostomi. In attempting to find an explanation for this fact it may be observed that all the Teleostomi possess a jugular canal formed by the lateral commissure, and Stensiö (1925, p. 145) suggested that the dorsal end of the hyomandibula might move from the Selachian position (medially) to the Teleostome position (laterally) with regard to the head vein, by passing over the lateral commissure as over a bridge. (See also below, p. 414).

But this scheme will not account for the varying relations of the facial nerve. Apart from whether the hyoid arch skeleton of Holocephali is primitive or not, the suspensorial functions must have been independently acquired in Selachii and Teleostomi, and it may be imagined that the dorsal element of the hyoid arch grew vertically upwards to the auditory capsule. In so doing it might grow behind, in front of, or between the branches of the nerve, or it might grow up on both sides of the nerve and enclose it in a foramen.

In favour of this view, proposed by van Wijhe (1922), it may be noted that this is in fact the way in which the hyomandibula may be seen to develop (Squalus, p. 50; Acipenser, p. 90; Salmo, p. 117). Comparable developments on the part of the 1st branchial arch in growing up to the auditory capsule may be seen in Polypterus, Acipenser, and Lepidosteus (p. 408). The part played by the lateral commissure would then have been to protect the head vein in the jugular canal from being constricted between the head of the hyomandibula and the auditory capsule.

All the other differences between the hyoid arch skeletons of Selachii and Teleostomi may be regarded as the result of divergent evolution prior to the independent acquisition of hyostyly in both groups. And while the actual skeletal element, the hyomandibula, or epihyal, may be homologous in both groups, the independent history of the suspensorial function may be expressed by the use of the terms hyostylic and methyostylic for Selachii and Teleostomi respectively.

In Dipnoi and in the Tetrapoda the dorsal end of the hyoid arch skeleton is typically forked (Broman, 1899; Versluys, 1903; Kallius, 1910; Schmalhausen, 1923*a*). The medial prong is directed towards the auditory capsule, where it either fuses with the capsular wall (Ceratodus), or occludes the opening of the fenestra ovalis (Tetrapoda), forming the columella auris (or stapes). The columella auris lies ventrally to the head vein, and ventrally to the facial nerve in all except certain Urodela (e.g. Amblystoma) where it lies dorsally to the nerve, and in other Urodela (e.g. Necturus) where it lies ventrally to the hyoidean (jugular) branch but dorsally to the mandibular branches of the nerve.

The relations to the columella auris of the orbital artery are variable: the artery running forwards under the cartilage in Sphenodon and Ophidia: though the cartilage in certain Gymnophiona and most Mammalia, and over the cartilage in the remaining types. While the hyoid arch skeleton was originally behind the spiracular visceral cleft, with the development out of the latter of the tympanic cavity, the relations of the columella auris or stapes to the visceral pouch may vary. The position of the stapes in Equus is thus no justification for denying, as Coyle (1909) does, the derivation of the stapes from the hyoid arch.

In most forms the columella auris chondrifies from two centres: a medial otostapes (or pars interna plectri)[1] and a lateral hyostapes (pars externa plectri), which either remain separate, as in Ceratodus, or join, as in most other forms, to give rise to a single rod with a footplate fitting into the fenestra ovalis and an insertion plate, or 'extra-stapedial' fitting on to the tympanic membrane. The term 'extracolumella' is applied to the unossified distal portion of the columella auris, which may not be coextensive with the hyostapes. The hyostapes appears to be absent in Urodela, Gymnophiona, Ophidia, and Mammalia, except in the latter for Paauw's cartilage, found in Cheiroptera (Bondy, 1907), Didelphys, Tatusia (van Kampen, 1915*b*), Dasyurus, Perameles, Dasypus, Sus, Bos, Ovis (van der Klaauw, 1923), Trichosurus, Phascolarctus (van der Klaauw, 1924*c*). It is possible that the flat base of the footplate fitting on to the membrane of the fenestra ovalis may have a separate origin from the otostapes, as claimed for Man by Schmid (1928), but there is no good evidence that this base has any connexion with the auditory capsule.

[1] It may be observed that Hoffmann (1889, 1890) who introduced these terms was under the erroneous impression that the otostapes was derived from the auditory capsule. The terms may, however, be usefully retained to denote the positional relations and not the derivation of the rudiments.

Any connexion that there may be between the columella auris and the ceratohyal or stylohyal (present in a blastematous condition in many forms: persistent in cartilage in Sphenodon, Crocodilus, Aves) takes place via the hyostapes and the pars interhyalis which projects ventrally from it (or 'infrastapedial'). The existence of this connexion and the fact that the otostapes chondrifies independently of the wall of the auditory capsule, are the evidence that the columella auris is part of the hyoid arch skeleton. On the other hand, the operculum in Amphibia appears to be a part of the capsule wall. Any connexion that there may be between the columella auris and the body of the quadrate (cartilaginous in Ceratodus and Lacertilia, ligamentous in Anura and Urodela) takes place by means of a processus internus given off from the lateral part of the otostapes. (See Plates 140, 141.)

In Gymnophiona and in mammals the connexion between the quadrate (or incus) and the otostapes appears to be effected not by a process of the latter but by a process of the former (the processus columellaris of the quadrate of Hypogeophis; the crus longum incudis of mammals). In mammals it is possible that the cartilage described by Spence (1890) in Felis and Ovis, by Bondy (1907) in Sus, Bos and Viverra, by van Kampen (1915*b*) in Canis and Paradoxurus, and by van der Klaauw (1923) in Galeopithecus, Tatusia, and Ateles, may be homologous with the processus internus of the columella auris, in which case, of course, the crus longum of the incus cannot be. However, Fuchs (1905, 1906) has contended that the crus longum incudis chondrifies independently, and becomes attached to the stapes before it fuses with the incus. Dombrowsky (1925) sees the processus internus in Platner's ligament in birds.

The lateral prong of the hyoid arch skeleton (or 'suprastapedial') may be formed by the processus dorsalis of the otostapes (e.g. Lacerta, Ophidia), or by a (laterohyal) process of the hyostapes (e.g. Rana), or by both together, thereby enclosing a foramen (e.g. Sphenodon, Crocodilus, Aves).

The lateral prong of the hyoid arch skeleton as just related, may therefore include the 'intercalary' of Dreyfuss (1893), the laterohyal of Broman (1899), and the processus dorsalis and the 'intercalary' of Versluys (1898, 1903), and is directed dorsally towards the crista parotica of the auditory capsule (or to the otic process of the quadrate in that position) where it may either fuse with the crista parotica (Rana, p. 207; Lacerta, p. 225: most Mammalia, p. 292) or with the quadrate (Ceratodus, p. 170; Sphenodon, p. 240; Ophidia, p. 247; Crocodilus, p. 264), or it may end freely as in Aves (p. 275) and Lepus (definitive condition; p. 313). It is absent in Urodela, Gymnophiona, and Chelonia. As explained above, the lateral prong may be connected via the processus dorsalis with the otostapes, or via the laterohyal with the hyostapes, or with both, or with neither as in mammals where the laterohyal is continuous with the stylohyal. It may be noted that Dreyfuss's 'intercalary', as found in mammals, becomes attached to the stylohyal, and that Versluys's 'intercalary', as found in reptiles, is related to the otostapes. They may nevertheless be homologous, differing in their attachment to the remainder of the hyoid arch skeleton.

The lateral prong, processus dorsalis or laterohyal, is always lateral to the head vein, and to the orbital artery when the latter runs over the columella auris; the laterohyal is typically lateral and anterior to the hyomandibular branch of the facial nerve, except in Ceratodus (p. 170), and certain mammals (e.g. Hyrax) where in the ossified state it is postero-medial to the nerve (Howes, 1896; van Kampen, 1905). The exceptional relations in these cases appear to be due to the different positions in which the laterohyal may grow up, for in Lepus it becomes disconnected from the crista parotica and the facial nerve runs directly over its free dorsal end, whence it might slip off either way.

The chorda tympani is the homologue of the ramus mandibularis internus facialis, and like it is post-trematic, yet pretympanic as Goodrich (1915) showed. It is typically given off from the main trunk of the facial nerve immediately behind the laterohyal, or processus dorsalis, and runs forwards laterally to the latter and over the columella auris to the medial side of Meckel's cartilage. Exceptions to this course are provided by Rana (p. 207) where the chorda tympani is post-tympanic and runs forwards ventrally to the columella; Struthio (p. 286) where it is pretympanic but appears to have slipped forwards over the small laterohyal; Gallus (p. 279) where it is pretrematic altogether, and certain mammals: Manis (van Kampen, 1905); Didelphys (Gaupp, 1905*c*, p. 1042; 1910*c*; Fuchs, 1906; van der Klaauw, 1924*b*), Dasyurus (Cords, 1915); Phascolarctus (van der Klaauw, 1924*c*) where it runs forwards medially to the laterohyal. Attempts to explain this discrepancy in mammals have been made by van Kampen (1905) on the supposition that the chorda tympani might have slipped over the top of the tympanohyal before the latter became attached to the crista parotica, and by Gaupp (1913) on the assumption that the tympanohyal may have grown upwards in an abnormal position. There remains the possibility that the chorda tympani itself may have grown forwards in an abnormal position. The possibility that another cartilage has been substituted for the tympanohyal is unlikely, in spite of the conditions in Trichosurus. (See p. 305).

Schmahlhausen (1923*a*) has drawn attention to the similarity in relations between the Selachian hyomandibula and the medial (columellar) prong, and between the Teleostome hyomandibula and the lateral (laterohyal) prong of the Dipnoan and Tetrapod hyoid arch skeleton. He has further pointed to the forked dorsal extremity of the 1st branchial arch in some Teleostomes (Polypterus, Acipenser, Lepidosteus) as an example of how the forking of the hyoid arch may have occurred.

It may thus be imagined that the forked dorsal end of the hyoid arch might be obtained by the growth of the pharyngohyal towards the auditory capsule ventrally to the head vein, and the epihyal (overlapping the pharyngohyal posteriorly) dorsally to the head vein. This is apparently the condition in Ceratodus and Tetrapoda, where the otostapes, hyostapes, or columella auris may be regarded as the pharyngohyal, and the laterohyal as the epihyal. This is also the condition in Teleostomi where it only needs to be assumed that the pharyngohyal has been lost. Perhaps fossil fish may be found in which the forked condition of the hyoid arch is preserved.

There remains, however, one problem concerning the hyomandibula of Selachii. From its relation to the head vein and facial nerve, it might be imagined that the Selachian hyomandibula corresponds to the columella auris, and thus to the pharyngohyal. Against this view, however, there are two important considerations. In the first place, the serially perfect correspondence between the Selachian hyomandibula and the epibranchials behind it, bearing hyal rays corresponding to the branchial rays. Next, there is the fact that in Elasmobranchs the visceral arch skeleton is Σ-shaped, and not >-shaped as in Teleostomes. The hyoid arch of Holocephali is ⋝-shaped, and from this condition an elongation of the epihyal rather than of the pharyngohyal would lead to the Selachian hyomandibula.

It must be assumed, therefore, that in the Elasmobranchs the pharyngohyal has been reduced and the epihyal preserved; in the original Osteichthyes both pharyngohyal and epihyal were present, the hyoid arch becoming forked by the upgrowth of the free dorsal end of the epihyal, this condition persisting in Dipnoi and Tetrapoda, but the pharyngohyal has been lost in Teleostomi.

That the epihyal of Selachii should bear relations to neighbouring structures similar to those of the pharyngohyal and different from those of the epihyal of other forms, would be perfectly possible in the course of independent acquisition of the suspensorial function and upgrowth of the hyoid arch skeleton to the auditory capsule on the part of Selachii and Teleostomi.

A problem of some difficulty is raised by the relation of the basihyal, or anterior copula, to the hyoid arch. In Scyllium the fact that the basihyal is pierced by a foramen (for the stalk of the thyroid gland) suggests that it is of paired origin, and its apparent continuity with the hypohyals (separate in Heptanchus) suggests further that the basihyal is the product of fusion in the midline of the ventral ends of the hyoid arch skeletons. Similarly, in mammals the basihyal betrays a paired origin.

On the other hand, the development of the basihyal in Salmo (p. 119), Amblystoma (p. 181), Rana (p. 202), Lacerta (p. 218), and Anas (p. 270), reveals it as an essentially median chondrification. This does not, however, invalidate the view that this median chondrification takes place in tissue which has been contributed from the bottom of each hyoid arch in the mesenchymatous state. On the other hand, it must be noted (see p. 474) that in Anura the basihyal appears to arise from mesendoderm.

The phylogeny of the basihyal is, however, unlikely to be elucidated from embryology but from palaeontology. In reptiles generally the basihyal is prolonged forwards to form the pointed processus lingualis (processus entoglossus). In Chelonia, ventrally to the processus lingualis there is a separate plate of cartilage, the entoglossum, which does not appear to form part of the hyobranchial skeleton at all. In birds there is developed in front of the basihyal a large cartilaginous process, often known as the entoglossum, which arises from paired rudiments. Kallius (1905*a*), who gave to this structure the name paraglossum, denies its derivation from the hyobranchial skeleton. On the other hand, Gegenbaur (1898) and de Beer and Barrington (1934) consider that the paraglossum of birds represents the ceratohyals and/or the hypohyals. Gaupp (1905*c*) draws attention to the existence round the basihyal of structures unconnected with the hyobranchial skeleton, particularly the radii posteriores and arcuate cartilage of Urodela (see p. 181), and considers that the paraglossum may correspond to these.

Huxley (1876*a*) and Parker (1883*a*) considered the lingual cartilage of Cyclostomes to be a basihyal, but this (see p. 47) must be considered very doubtful.

The so-called 'urohyal' of Teleostei does not appear to form part of the hyoid arch skeleton at all, but to be a tendon ossification developed between the sternohyoid muscles (p. 130). It appears to correspond to the Y-shaped 'parahyoid' of evidently paired origin found in Polypterus (see p. 87), and Fuchs's term parahyoid for this structure is to be preferred to all those which suggest that it forms part of either the hyoid arch or the pectoral girdle skeleton (i.e. 'episternal' E. Geoffroy St. Hilaire; 'urohyal', Owen, Günther; 'interclavicle', Cole and Johnstone; 'sternum', Allis; 'jugular', Cunningham).

The parahyoid of Polypterus and Teleostei may be homologous with the similarly named membrane-bone present in many Anura, e.g. (Discoglossus, Alytes, Bombinator, Ascaphus, Liopelma, Pelodytes, Pelobates, Rhinophrynus. (See Stadtmüller, 1936). This parahyoid seems to have nothing in common with the 'parahyals' (dermal bones on the medial side of the ceratohyals) of Amphiuma (see p. 190).

Ruge's (1898) and Broman's (1899) claim that the cartilage of the pinna of the external ear is derived from the hyoid arch, has not been substantiated by Baum and Dobers (1905) or Cords (1918).

XIII. The Mandibular Arch

i. Historical: The processes of the quadrate in Amphibia

The relations of the processes of the quadrate in Amphibia have played an important part in the classification of the types of jaw-suspension by Huxley (1876*b*). An historic investigation into their study in the light of present knowledge reveals, however, a remarkable tangle of confusion and mistaken identity, as Pusey (in press) has pointed out. In view of the importance and interest of this problem, the following may not be out of place, in order to facilitate consultation of the works of the earlier authors, and the recognition of the structures to which they were (or thought they were) referring.

Huxley (1858) stated (p. 588) that in Amphibia, 'The hyoid and mandibular arches are thus suspended to the skull by a common "*peduncle*" which, to avoid all theoretical suggestions, I will simply term the "suspensorium".'

Referring to the frog tadpole Huxley continues (p. 575): 'Behind the eye and just in front of the auditory capsule (in the posterior part of the first visceral arch therefore) a cartilaginous process lies, which is connected proximally with the root of the trabecula, close to the basal plate [erroneous, the ascending process is connected with the end of the pila antotica] while at its distal end it sends a prolongation upwards to unite with the posterior end of the ethmovomerine cartilage' [quadratocranial commissure].

'The posterior crus of the palato-suspensorial or suborbital arch, is not yet united with that portion of the cranial wall which encloses the auditory capsule' [larval otic process] (p. 576).

Parker (in 1871) assumed that the upper end of the hyoid arch was fused on to the posterior side of the mandibular arch (as the ceratohyal is, for a time, attached to the quadrate), and he therefore referred to the larval otic process as the 'supra-hyomandibular' and regarded it as the true dorsal end of the hyoid arch. The ascending process he referred to as either the 'metapterygoid connexion', the 'mandibular root', or the 'top of the mandibular pier', and he thought (erroneously) that its lateral portion gave rise to the definitive otic process. He described the trigeminal nerve as running 'over and in front of the mandibular root', which can only apply to V_2 and V_3.

Parker saw the pseudobasal process and termed it the 'infrahyomandibular', describing the relations to it of the hyomandibular (but not of the palatine) branch of the facial nerve.

Huxley (in 1874) defined the ascending process in Menobranchus and correctly described its relations to the trigeminal nerve. He also defined the otic process giving its relations to the facial nerve, and the basal process, calling it the 'pedicle of the suspensorium'. Unfortunately, in the frog, he homologized the ascending process with the 'pedicle of the suspensorium of Necturus', i.e. with the basal process, and thought (erroneously) that its base became the pseudobasal process in the adult frog. Parker had thought that the stump of the ascending process gave rise to the definitive otic process. Huxley correctly described the definitive otic process and (in 1875*a*) he corrected Parker's view of the 'suprahyomandibular', showing that it is the larval otic process.

Parker (1876*a*) then accepted Huxley's views with regard to the pseudobasal and larval otic processes, and (in 1877*a*) he correctly described the otic, ascending, and basal processes in Amblystoma. But as he regarded the 'pedicle' (basal process) of Amblystoma as equivalent to the 'pedicle' (ascending process) of the frog, he found himself forced to say that the frog had no ascending process.

Stöhr (1881, p. 97) unravelled the tangle by showing that no trigeminal root runs ventrally to the 'pedicle' (basal process) of the Urodele, while a trigeminal root (the profundus) does run ventrally to the 'pedicle' of the frog, which must therefore be an ascending process.

There remains the question as to whether the frog possesses basal and basitrabecular processes. Gaupp (1893, p. 451) maintained that a true basal process was present. He neglected, however, to consider the relations of the palatine nerve to the structure in question, and since this nerve runs downwards in front of this cartilaginous process, de Beer (1926*a*) showed that it was not a true basal connexion but an attachment to the neurocranium involving a postpalatine commissure, and for which the term pseudobasal was more appropriate.

The (postlarval) connexion between the pterygoid process of the quadrate of the frog and the lamina orbitonasalis corresponds to the ethmoid attachment in Urodeles. There thus remains only one attachment in the frog unaccounted for, and that is the quadratocranial commissure. Pusey (in press) has advanced reasons to show that it may (at least in part) correspond to a basal process precociously formed and displaced forwards in association with the specialized and peculiar mechanical requirements of the frog's larval jaw apparatus. This basal process is lost at metamorphosis except apparently in Ascaphus.

The pseudobasal connexion has been regarded by Kruijtzer (1931) as a portion of the hyoid arch skeleton, and in fact as the pharyngohyal, corresponding to the otoquadrate cartilage of Ceratodus. He supports this view by the facts that the pseudobasal cartilage arises wholly separate from the quadrate but (in Alytes) connected to the plectrum and to a dorsal process on the medial side of the quadrate, which fuses with the crista parotica and on this view corresponds to the laterohyal (processus dorsalis of Lacertilia). The plectrum or columella auris on this view becomes the epihyal (hyomandibula of fish). This involves *three* dorsal ends to the hyoid arch: a contingency which is not impossible, but which will require further evidence to become established. Meanwhile, it is interesting to note that Kruijtzer's view of the pseudobasal connexion as a pharyngohyal gets back to Parker's notion of the 'infrahyomandibula'.

Lastly, the basitrabecular process with typical relations to the palatine nerve is present only in Ascaphus among Anura, and has been lost in the others.

ii. Descriptive

While there is no doubt of the derivation of the mandibular arch skeleton from a condition in which all the visceral arches had similar skeletal structures, the question arises with the mandibular no less than with the hyoid arch whether it was originally segmented into four pieces like the branchial arches of fish.

The pterygoquadrate and Meckel's cartilages, universally present in all Gnathostomes, are generally regarded as the epimandibular and cerato-mandibular elements respectively. A paired *hypomandibular* has been described in Hexanchus by Sewertzoff (1927), and in the fossil Acanthodes by Jaekel (1927). The two centres of chondrification in Meckel's cartilage found in Squalus, in birds, and in Halicore, may perhaps be explained, as van Wijhe pointed out, by regarding the distal element as the hypomandibular. Similarly, the infrarostrals of Anura may perhaps be interpreted in a similar manner.

The problem of the hypomandibulars is obscured by the fact that in a variety of forms (Hexanchus, Odontaspis, Scymnus) cartilages have been described between the distal ends of Meckel's cartilages and the basihyal. These may be hypomandibulars, but they are often variable in number, and K. Fürbringer (1903) is inclined to regard them as remnants of mandibular rays. Such rays were well developed in Acanthodes (Jaekel, 1927).

Further confusion is due to the fact that a median cartilage between the distal ends of Meckel's cartilages, and presumably therefore a *basimandibular*, is found in Holocephali (Schauinsland, 1903; K. Fürbringer, 1903; Dean, 1906), Hexanchus and Laemargus (White, 1890, 1895), Chlamydoselachus (K. Fürbringer, 1903), Triton (Gaupp, 1906*a*, p. 704), Rana (Gaupp, 1893, p. 451), Alytes (van Seters, 1922), and Hyla (Howes, 1891).

The median cartilage described by W. K. Parker in Crocodilus and in Chelonia in this position has not been found by later investigators. On the other hand, such a cartilage has been found in the parrot Melopsittacus by Kallius (1906), in Perameles by Cords (1915), and in Didelphys by Töplitz (1920) where the median cartilage is derived from the symphysis between the Meckel's cartilages, and is not an originally independent chondrification. Other examples of median mandibular cartilages are given by Baumüller (1879).

Turning now to the question of the existence of a *pharyngomandibular*, such an element has been claimed for Laemargus by Sewertzoff and Disler (1924). Further, Sewertzoff found that in Mustelus, Squalus, and Scyllium, the pharyngomandibular arises as a separate cartilage which soon fuses with the medial surface of the basal process of the pterygoquadrate. A separate pharyngomandibular is described by Sewertzoff (1923) in Scaphirhynchus and by Bugajew (1929) in Acipenser. It is further possible that the meniscus pterygoideus cartilage found in Lacerta (p. 228), Sphenodon (p. 242), and Chelonia (p. 256) may represent the pharyngomandibular. Perhaps the cartilage intercalated between the pterygoid and the basipterygoid process in Apteryx may represent the same element (p. 288).

It should be added that on Allis's (1923*b*) view, the pharyngomandibulars are the polar cartilages (see p. 376), but the difficulties in accepting this homology are too great.

It is probable, therefore, that, like the hyoid arch, the mandibular arch was originally segmented into four elements, like the branchial arches behind them.

In some forms (e.g. Amia, Lepidosteus) Meckel's cartilage possesses a large coronoid process. As this is the position of the posterior ventral labial cartilage of Polypterus, van Wijhe (1882*b*) considered that it might be homologous with it, especially as, according to him, the coronoid process of Amia may perhaps have an independent origin.

A common feature of Meckel's cartilage in many forms is the retroarticular process. In mammals the posterior portion of Meckel's cartilage becomes detached as the *malleus*, and the retroarticular process seems in them to have become the manubrium, inserted in the tympanic membrane.

The pterygoquadrate cartilage may have as many as four processes, which will now be treated in turn.

The *basal process* was regarded by Huxley (1876*b*) as the morphological top end of the mandibular arch (contra Gegenbaur (1872) for whom the otic process was the top end), and this view is further supported by the discovery of cartilages interpreted as pharyngomandibulars associated with it (see above).

The basal process is directed towards the neurocranium (and in particular the subocular shelf or that particular portion of it which forms the basitrabecular process) with which it may make contact without permanent fusion (Selachii, Lepidosteus, some Urodela, Gymnophiona, Ascaphus, Lacerta, Rhynchocephalia, Struthio), or with permanent fusion (Cyclostomes, Holocephali, Dipnoi, some Urodela), or which it may fail to reach (Amia, Salmo). The basal process of Selachii was long known as the orbital process.

The contact between the basal and basitrabecular processes forms the palatobasal articulation or fusion, which always lies ventrally to the head vein, and dorsally and anteriorly to the palatine branch of the facial nerve. Only in the Anura is the pterygoquadrate connected with the neurocranium behind the palatine nerve, forming the pseudobasal connexion. The question of the existence of a true palatobasal connexion in Anura is discussed elsewhere (p. 417).

The *otic process*, or postorbital articulation of the pterygoquadrate, is directed towards the auditory capsule, with which it may make contact without permanent fusion (some Selachii, Gymnophiona, Lacertilia, Rhynchocephalia, Chelonia, Crocodilia, Aves, Mammalia), or with permanent fusion (Holocephali, Dipnoi, Urodela, Anura), or which it may fail to reach (Heterodontus, Amia). In some Selachii the otic process becomes detached as the prespiracular cartilage. Suschkin (1927) denied the homology of the reptilian otic process with that of lower vertebrates, but on seemingly insufficient grounds.

The otic process always lies laterally to the head vein, the orbital artery, and the efferent pseudobranchial artery, and anterolaterally to the hyomandibular branch of the facial nerve. Between the articular head of the otic process and the wall of the auditory capsule there is in many forms a cartilaginous nodule. This, the 'intercalary' or processus paroticus, seems to be derived from the columella auris in Lacerta (p. 225), Sphenodon (p. 241), and Crocodilus (p. 264). On the other hand, a cartilage in a similar position in Gallus (p. 279) seems to be derived from the auditory capsule.

In some forms (Lacerta, perhaps Ophidia, Mammalia) the otic process becomes discontinuous with the more anterior regions of the pterygoquadrate. In mammals it gives rise to the crus breve of the *incus*.

Between the side wall of the skull and the basal and otic processes of the pterygoquadrate is a passage, termed by Goodrich (1930) the cranioquadrate passage, which is a very useful morphological landmark.

The *ascending process*, or processus ascendens palatoquadrati, rises upwards from the pterygoquadrate cartilage anteriorly to the otic process, from which it may be discontinuous, as in Lacertilia and Mammalia. It lies laterally to the head vein, posterolaterally to the profundus branch of the trigeminal nerve, and anteromedially to the maxillary and mandibular branches, and to the orbital artery. It is found in all forms including and above the Dipnoi, and may end freely dorsally (Gymnophiona, Lacerta, Sphenodon, Chelonia, Crocodilus, Didelphys) or may fuse with the pila antotica or orbital cartilage (Dipnoi, Anura, Urodela), or with the auditory capsule (Gecko).

The similarity in relations between the processus ascendens and the pila lateralis (or 'pedicle of the alisphenoid') of Amia led Allis (1897) to conclude that they were homologous, but as de Beer (1926*a*) showed, this cannot be, for the pila lateralis is a neurocranial and not a visceral arch structure (p. 102).

The conditions in most mammals require a certain amount of explanation, for there it is often found that the *ala temporalis* rises up between the maxillary and mandibular divisions of the trigeminal nerve. For this reason Gaupp (1900, p. 541; 1902, p. 191) believed that the ala temporalis was not homologous with the processus ascendens, which he regarded as absent in mammals and replaced by a new upgrowth from the basitrabecular (or alar) process. On the other hand, Broom (1907, 1909*a*, 1914) showed that the transition from the epipterygoid of reptiles (ossified in the processus ascendens) to the alisphenoid of mammals (ossified in the ala temporalis) could be traced through an unbroken series. This view, that the ala temporalis has evolved out of the processus ascendens, is supported by the following considerations:

(i) In practically all mammals that have been adequately studied, the ala temporalis chondrifies from an independent centre: this is as it should be if the ala temporalis represents a part of the pterygoquadrate, and proves that it is not merely an upgrowth of the basitrabecular (alar) process.

(ii) In Didelphys Fuchs (1915, pp. 147, 253) found that the ala temporalis lies between the profundus and maxillary nerves, exactly as in Lacerta.

(iii) In many mammals (e.g. Trichosurus, Mus, Mustelus) the cartilage is to be found not only between the profundus and maxillary, but also between the maxillary and mandibular branches of the trigeminal nerve. The maxillary branch is thus enclosed in a (cartilaginous) foramen rotundum. The cartilage in front of this foramen is the processus ascendens, that behind it is the ala temporalis of those forms in which the maxillary branch emerges freely in front of it.

Fuchs (1912) believed that the relations of the nerves to the cartilages could be explained if the maxillary nerve slipped forwards over the processus ascendens. Although less probable, this view may be possible.

The processus ascendens must thus be imagined as having lapped back round the maxillary nerve, and given rise to the typical mammalian ala temporalis. It appears to have been lost in Echidna; in Dasyurus its dorsal end fuses with the orbital cartilage; in all other mammals, so far as known, it ends freely.

The anterior end of the pterygoquadrate cartilage may extend forwards as the *pterygoid process*. This may fuse with its fellow of the opposite side, as in Selachii and Acipenser. The pterygoid process may come into contact with the neurocranium in the ethmoid region, as in Chlamydoselachus (Allis, 1923*a*), Scymnus (Bugajew, 1930), and Teleostomes. In the latter it may develop an anterior rostropalatine and a posterior ethmopalatine articulation (Swinnerton, 1902). The pterygoid process may actually fuse with the ethmoid region of the neurocranium, as in Holocephali, the Urodela, Cryptobranchus, and Hynobius (Edgeworth, 1923*a*), and in Anura.

In many Urodela, in Gymnophiona, Lacertilia, Sphenodon, Chelonia, Crocodilia, and birds, the pterygoid process ends freely, but it is not impossible that the processus maxillaris posterior may represent its extreme anterior end, fused to the lamina orbitonasalis.

The pterygoid process has been reduced to nothing in Dipnoi and Ophidia; and in mammals it is, when present at all, a small process projecting forwards from the base of the ala temporalis (thereby further supporting the claim of the latter to represent the ascending process of the pterygoquadrate cartilage.

A curious feature of the form of Meckel's cartilage in young stages of development has been noted by Lebedinsky (1917) in reptiles, birds, and to a slight extent in Echidna. This is a sigmoid curvature in the horizontal plane, the cartilage running forwards and inwards and then outwards before turning inwards again to the symphysis. Later, this curvature becomes straightened out and its significance is unknown.

XIV. THE NEUROCRANIUM AND THE VISCERAL ARCH SKELETON

i. The relations of the jaws to the brain-case in the chondrocranium

As is well known, Huxley (1876*b*) subdivided the types of suspension of the jaws into three categories: viz. amphistylic, hyostylic, and autostylic. As it is a matter of great importance to ascertain exactly what he meant, some extracts from his paper will be quoted here.

(1) 'In the amphistylic skull the palato-quadrate cartilage is quite distinct from the rest of the skull; but it is wholly, or almost wholly, suspended by its own ligaments, the hyomandibular being small and contributing little to its support' (e.g. Notidanus = Heptanchus).

(2) 'The palato-quadrate cartilage is no longer continuous with the chondrocranium (though the bony elements of that arch (i.e. mandibular) may unite suturally with those of the skull, as in the Plectognathi), but is, at most, united with it by ligament. Moreover, the dorsal element of the hyoidean arch, or the hyomandibular, usually attains a large size and becomes the chief apparatus of suspension of the hinder end of the palato-quadrate cartilage with the skull. Skulls formed upon this type, which is exemplified in perfection in Ganoidei, Teleostei, and ordinary Plagiostomes, may therefore be termed *hyostylic*.'

(3) 'The part of the palato-quadrate cartilage which is united with the skull, between the exits of the fifth and second nerves, answers to the "pedicle of the suspensorium" of the amphibian, while its backward and upward continuation on to the periotic cartilage corresponds with the otic process. As in the Amphibia and in the higher Vertebrate, the mandibular arch is thus attached directly to the skull by that part of its own substance which constitutes the suspensorium. It may thus be said to be *autostylic*.'

With the extension of knowledge it is only natural that this classification should require amplification and modification; but there are also a few ambiguities to which attention must be turned.

Taking first the amphistylic type, it is necessary to analyse it further, for the pterygoquadrate cartilage may have no less than four processes, viz. ethmoid, basal, ascending, and otic, by means of which suspension may be effected, and it becomes necessary to distinguish between them. Heptanchus is amphistylic and its jaw is attached by the basal and otic processes (and hyomandibula); in Heterodontus (Cestracion) the attachment is only by the basal process (and, of course, the hyomandibula).

As regards the autostylic type, Huxley's definition does not make it quite clear whether its essential feature is the *non-participation of the hyomandibula* in the jaw-suspension, or the actual cartilaginous *fusion* of one or more of the processes of the pterygoquadrate with the neurocranium; for no living fish presents the one of these conditions without the other. While the latter alternative has come to be widely held as the meaning of the term autostylic, the former connotation, being wider, appears to be the more proper use. Therefore, all Gnathostomes in which the hyomandibula plays no suspensorial function may be termed autostylic whether or not there be cartilaginous fusion between the jaw and the brain-case. To distinguish between these two conditions, however, de Beer and Moy-Thomas (1934) have introduced the modifications *autodiastylic* for the non-fused type possessed by ancestral fish, Gymnophiona and Amniota[1] and *autosystylic* for the fused type represented by Cyclostomes, Holocephali, Dipnoi, Urodela, and Anura. Gregory's (1904) term holostylic is useful to denote the peculiar autosystylic condition

[1] In mammals a portion of the pterygoquadrate—the processus ascendens—becomes detached from the rest and fused on to the basitrabecular process (processus alaris) to form the ala temporalis. In Dasyurus the dorsal end of the ala temporalis fuses with the taenia marginalis. This is clearly a secondary fusion strengthening the side wall of the skull in compensation for the loss of the pila antotica. In Geckonid Lacertilia the dorsal end of the processus ascendens may fuse with the front wall of the auditory capsule.

found in Holocephali. Similarly, the term parautostylic is useful to denote the very specialized type of fusion found in Cyclostomes. It is probable that the autosystylic type found in *modern* Dipnoi and Amphibia was independently acquired, for the Gymnophiona retain the autodiastylic type.

That the autodiastylic type is the more primitive in phylogeny there can be no doubt, since the jaws are merely modified branchial arches, and these are free from the axial skeleton in all forms save the obviously specialized Cyclostomes. There is, indeed, reason to believe that the visceral arch skeleton is phylogenetically older than that of the brain-case, since visceral clefts were in existence in Hemichordata, Urochordata, and Acrania before a brain and paired sense-organs had been evolved. As K. Fürbringer (1904, p. 585) remarked, a belief in the primitive freedom of the jaw is 'a morphological necessity'. This opinion is shared by Gaupp (1906*a*, p. 589), Fuchs (1909*b*, 1910*b*, 1915), Versluys (1912, p. 702), Luther (1914), and Lakjer (1927). The contrary view, held by Edgeworth (1925, 1935) is based on the supposition that embryonic conditions necessarily represent phylogenetic states; a supposition discredited in modern morphology (see p. 449).

The hyostylic type has been up till recently imagined to include both the Selachii and the Teleostomi. It has, however, been shown that in addition to such differences as the presence of symplectic and interhyal elements in Teleostomi, there are important morphological differences in the relations of the hyomandibula to the facial nerve and head vein in these two groups, for the articulation of the hyomandibula with the auditory capsule is medioventral to the head vein and posteroventral to the facial nerve in Selachii, whereas in Teleostomi it is dorsolateral to the head vein and (typically) anterolateral to the facial nerve (or at least to some of its branches; see p. 410). These and other considerations (p. 411) have rendered necessary the view that the suspensorial function of the hyomandibula has been evolved independently in Selachii and in Teleostomes. It is further probable that the autostylic type is phylogenetically older than the amphistylic, for primitive forms already have otic and basal attachments, and that the hyostylic is derived from the amphistylic type by reduction of the pterygoquadrate's processes.

The distinction between the suspensorial hyoid arch skeleton in Selachii and Teleostomi may be expressed using Gregory's (1904) terms, and confining *hyostylic* to Selachii and *methyostylic* to Teleostomi. This latter type was presumably derived from the autodiastylic type, remnants of the attachment of the pterygoquadrate by its own processes being found in the ethmoid articulation of Polypterus and Holostei, the basal articulation of Lepidosteus, and the diminutive otic process of Amia.

Another question which Huxley's classification raises is the fact that it is essentially a functional classification, since it is primarily based, not on whether a particular structure, say, the otic process, exists, but on whether such a structure actually serves in the suspension of the jaws. Now, in some forms, the otic process exists but it is too small to effect a functional suspension, as in Scymnus or Amia, or it may be detached from the pterygoquadrate cartilage to form the spiracular cartilage, as in Scyllium. Clearly, Scymnus, Scyllium, and Amia cannot be placed in the same group as Heptanchus on Huxley's classification, and yet the possession of an otic process (albeit diminutive) in Amia and Scyllium deserves a recognition of the affinity expressed betweeen these forms and Heptanchus by the possession of a homologous structure. It seems, therefore, that while Huxley's scheme may most certainly be retained in order to indicate the type of suspension of the jaws, another classification based purely on morphological considerations is necessary to express the phylogeny of the suspension.

An attempt was made to revise the classification of the types of jaw-suspension by Gregory (1904), according to the following system:

Type	*Palato-quadrate*	*Hyomandibula*	*Term*
Ancestral	little or no attachment	not suspensorial	Palaeostylic
Chlamydoselachus	loose articulation (basal)	suspensorial	Hyostylic
Scyllium	no articulation	suspensorial	Hyostylic
Heterodontus	close articulation (basal)	suspensorial	Hyostylic
Heptanchus	close articulation (basal and otic)	suspensorial	Amphyostylic
Raja	quite free	suspensorial	Euhyostylic
Acipenser	quite free	suspensorial	Methyostylic
Polypterus	articulation (ethmoid)	suspensorial	Methyostylic
Salmo	articulation (ethmoid)	suspensorial	Methyostylic
Chimaera	close fusion	not suspensorial	Holostylic
Ceratodus	fusion (basal, ascending, otic)	not suspensorial	Autostylic
Amphibia	fusion (ethmoid, basal, ascending, otic)	not suspensorial	Autostylic

Only slight alterations are necessary to adapt this scheme to the present state of knowledge. After the palaeostylic stage (equivalent to Lakjer's (1927) astylic condition), with no attachment between jaws and brain-case, there must have been an autodiastylic stage represented by ancestral Elasmobranchs and ancestral Osteichthyes, and by Amniota. Amphistylic and hyostylic may be retained instead of Gregory's 'amphyostylic' and 'euhyostylic', and the condition of the Dipnoi and Amphibia becomes autosystylic. The specialized and peculiar form of attachment of the visceral arches in Cyclostomes may be termed parautostylic. The primitive nature of the basal connexion is well expressed in Suschkin's (1910) term protostylic, for autodiastylic forms in which the basal was the only connexion between pterygoquadrate and brain-case; such forms are, however, still hypothetical.

The final analysis of the relations of the jaws to the brain-case in the chondrocranium is as follows (See plate 142.):

Palaeostylic: primitive ancestral condition; no relation between brain-case (if present) and visceral arch skeleton:
e.g. Amphioxus.

Parautostylic: foremost visceral arches fused to brain-case:
e.g. Cyclostomes.

Autostylic: the pterygoquadrate is articulated *or* fused with the brain-case by its own processes without participation of the hyoid arch:
e.g. Ancestral Elasmobranchs, Holocephali, Ancestral Osteichthyes, Dipnoi, Amphibia, Tetrapoda.

Autodiastylic: Autostylic forms in which the pterygoquadrate is apposed but *not* fused to the brain-case:
e.g. Ancestral Elasmobranchs and Osteichthyes, Tetrapoda.

Autosystylic: Autostylic forms in which the pterygoquadrate is fused to the brain-case by its own processes:
e.g. Holocephali, Dipnoi, Amphibia. [The otic process in Testudo is said by Bender (1912) to be fused with the auditory capsule; the same is said of Crocodilus by Edgeworth (1935).]

Holostylic: peculiar autosystylic type found in Holocephali.

Amphistylic: condition combining autodiastylic relations of the pterygoquadrate with suspensory function of the hyomandibula:
e.g. Heptanchus.

Hyostylic: pterygoquadrate suspended not by its own processes but by a hyomandibula articulated with the auditory capsule medioventrally to the head vein:
e.g. most Selachii.

Methyostylic: pterygoquadrate suspended not by its own processes (except ethmoid) but by a hyosymplectic or hyomandibula articulated with the auditory capsule dorsolaterally to the head vein:
e.g. Teleostomi.

The use of Stannius's (1856) terms 'monimostylic' and 'streptostylic' *in connexion with the chondrocranium* is illogical and to be deprecated, as Fuchs (1909*b*, 1910*b*, 1915) has shown, since they refer to the relations of the quadrate bone to neighbouring bones *in the bony skull*. Lakjer's (1927) application of Versluys's terms 'kinetic' and 'akinetic' indiscriminately to chondrocrania and osteocrania gives them a functional rather than a morphological meaning. The same applies to Lakjer's terms: pleurostylic (pterygoquadrate cartilage or pterygoid and quadrate bones articulated or fused with the brain-case by two or more processes); ethmostylic (pterygoquadrate cartilage or palatine and pterygoid bones attached to the ethmoid region of the skull).

ii. The relations of the jaws to the brain-case in the bony skull

As long ago as 1856 Stannius drew attention to the fact that in some vertebrates the quadrate bone is movable relatively to neighbouring bones, particularly the squamosal. This condition, which he called *streptostylic*, involves a mobility relative to the brain-case of the whole upper jaw, the motion of the quadrate being imparted to the more anterior bones of the jaw margin by means of the pterygoid, which can slide forwards on the basipterygoid (ossified basitrabecular) process. The quadratojugal is often absent in such forms, as one of the conditions of the mobility of the quadrate bone. This condition is met in Lacertilia, Ophidia, and Aves, among living forms. Contrasted with this is the condition termed *monimostylic* in which the quadrate bone is firmly and immovably attached to its neighbours as in Amphibia, Rynchocephalia, Chelonia, and Crocodilia. These forms have a well-developed bony temporal roof, and the upper jaw is fixed to the brain-case (Stannius, 1856, p. 45).

It was shown by Versluys (1910, 1912) that a proper understanding and classification of the relations between the upper jaw and the brain-case in the bony skull must be conceived on lines broader than those merely involving the possibility of the quadrate bone being movable relatively to its neighbours. Versluys showed that in a skull where the quadrate is firmly fixed to the squamosal, the upper jaw may nevertheless move *as a whole* (involving quadrate, squamosal, quadratojugal if present, jugal, maxilla, premaxilla, pterygoid, transpalatine, palatine, prevomer, nasal, frontal, and parietal—all together constituting the 'maxillary segment') relatively to the brain-case (involving the prootic, opisthotic, basisphenoid, parasphenoid, basi-, ex-, and supraoccipitals —all together constituting the 'occipital segment'). The points where the paroccipital processes of the exoccipitals touch the quadrate bones act as pivots, and sliding contacts are established between the pterygoids and the basipterygoid processes and between the under surface of the parietals and the supraoccipital. The two 'segments' are moved on one another by the protractor and levator pterygoid muscles.

A skull possessed of this type of mobility is called *kinetic*, and is found in Gymnophiona, Lacertilia, Ophidia, many Dinosaurs, and Aves. The immovable condition of the upper jaw is then known as *akinetic*.

It will be noticed that not all monimostylic skulls are akinetic, nor are all kinetic skulls streptostylic, since the 'maxillary segment' may move relatively to the 'occipital segment' (criterion of kinetism) although the quadrate may be rigidly fused with the squamosal (criterion of monimostyly), as in Hypogeophis (see p. 197).

It might be thought that all akinetic skulls must be monimostylic or that all streptostylic skulls must be kinetic, but the conditions in mammals renders such a conclusion untenable. For there, the quadrate bone, which has become the incus, is clearly free from the squamosal, movable, and streptostylic; but the upper jaw is firmly fused with the brain-case, and akinetic.

The various possibilities may be shown in the form of a table:

	Monimostylic	*Streptostylic*
Akinetic	Most Amphibia Chelonia Crocodilia	Mammalia
Kinetic	Gymnophiona (Hypogeophis)	Lacertilia Aves

It will be clear that the relations of the upper jaw to the brain-case in the bony skull are independent of the relations of the pterygoquadrate cartilage in the chondrocranium. Thus, the animals in which the monimostylic and akinetic conditions are most extremely developed are the Chelonia and Crocodilia, in which the pterygoquadrate cartilage is not fused with the neurocranium (autodiastylic). Conversely, in such a Urodele as Amblystoma gracile (see p. 182) where the pterygoquadrate cartilage is fused to the neurocranium (autosystylic), a kinetic bony skull is eventually formed. It is, of course, true that animals whose chondrocrania are so specialized as to be permanently autosystylic (like the living Dipnoi, many Urodela, and Anura) must have bony skulls conforming to the akinetic and monimostylic types.

The light which a study of the development of the skull in all vertebrates throws on these problems, falls in two directions. On the one hand, it is possible in some cases to detect the recent loss of kinetism in an akinetic skull; on the other, an indication may be obtained as to which condition is the more primitive. A case in point appears to be provided by Sphenodon. In this animal not only is a basitrabecular process present, but also there is a meniscus pterygoideus cartilage and a detached processus dorsalis of the columella auris. These cartilages arise just where the friction and pivoting of the 'maxillary segment' on the 'occipital segment' takes place in kinetic skulls. Furthermore, the embryos of Sphenodon possess rudiments of the protractor and levator pterygoidei muscles. The contact between the parietal and the supraoccipital is very slender.

These facts can only be satisfactorily explained on the assumption that Sphenodon has inherited these embryonic features from ancestors in which they persisted in the adult stage, i.e. were kinetic.

Without committing the error of supposing that embryonic conditions necessarily represent primitive phylogenetic conditions, it is possible to show that the very wide distribution among vertebrates of the basipterygoid process as an articular facet against which the pterygoid bone can move, points towards the kinetic condition as being the primitive one for the bony skull. It is obvious that in such skulls the chondrocranium cannot have been autosystylic in the adult, whatever the embryonic condition may have been like.

It is further to be noticed that in the fish the various bones are not rigidly fixed to one another; like the scales of the body (or the bones of the operculum) they must have been able to slide, and Watson (1925*a*) has shown that in Osteolepis the frontals were able to hinge on the parietals, so that a considerable amount of internal mobility was provided for.

It is probable that the kinetic condition characterized primitive Stegocephalia and Reptilia. In most living Amphibia the akinetic condition has been imposed on the bony skull by the development and persistence of the autosystylic chondrocranium. In Crocodilia, Chelonia, and Rhynchocephalia the bony skull has become akinetic because the quadrate bone has become extremely monimostylic and fused with the prootic. On the other hand, in Lacertilia, Ophidia, and Aves the kinetism has been increased by the quadrate bone becoming streptostylic. This increase in mobility is associated either with a reduction of the epipterygoid or a discontinuity between it and the quadrate as a result of the absorption of the intervening region of the pterygoquadrate cartilage. This latter condition, associated with the development of a new hinge between the upper and lower jaws, has led to the Mammalia, in which a streptostylic quadrate (incus) is associated with an immovable akinetic bony upper jaw.

XV. The Morphology of Certain Extracranial Spaces

To Bland Sutton (1888) belongs the credit of recognizing 'that morphologically the true cranium is limited by the dura mater, and that any structure quitting the brain becomes extracranial at the spot where it perforates the dura mater. If this be so, then the Gasserian ganglion and its branches, including the nasal nerve, the structures in the cavernous sinus, the geniculate ganglion of the seventh with its petrosal nerves, the tympanic branch of the ninth and tenth nerves, the internal carotid artery, from the anterior clinoid process downwards, and the meningeal vessels, must all be for the morphologist extra-cranial.'

It is no exaggeration to say that this view has provided the key to a number of problems in cranial morphology, and fifty years of research have done nothing but substantiate its correctness, and extend its application.

i. The posterior myodome and the trigeminofacialis chamber

The morphological interpretation of these regions of the chondrocranium is largely due to the work of Allis (1897, 1903*a*, 1909, 1914*a*, 1919*a*). Starting from the conditions in Selachii (see p. 56) where actually neither posterior myodome nor trigeminofacialis chamber as such are present, Allis (1914*a*) has recognized the subpituitary space, and the acustico-trigeminofacialis recess, together with the jugular canal, as their partial respective homologues.

The most primitive type of posterior myodome is that of Amia (see p. 103). Here, the external recti muscles push their point of origin inwards dorsally to the trabeculae, which they are able to do since the side wall of the chondrocranium in this region is membranous. Probably, as Gegenbaur (1872) suggested, the path followed by the external recti muscles corresponds to the foramen traversed by the pituitary vein in the Selachian skull. At all events, having passed over the trabeculae, the muscles are still separated from the true cranial cavity by the dura mater, and they lie in the subpituitary space. This space no longer lodges the pituitary body, which is as it were lifted up by the presence of the muscles. (Plate 143, figs. 2, 4.)

Posteriorly, the subpituitary space leads out on to the ventral surface of the skull through the hinder part of the hypophysial fenestra, in which region it is covered dorsally by the prootic bridge (or front of the basal plate). In this manner is formed the so-called dorsal (posterior) myodome, lodging the external recti muscles.

The trigeminofacialis chamber of Amia (see p. 103) is bounded laterally by the lateral commissure and the pila lateralis (both subsequently ossified) and medially by the (persistently) membranous dura mater, except for a certain anterior region where the trigeminofacialis chamber is confluent with the dorsal myodome. The chamber is not internally subdivided, and the blood-vessels and nerves pass through it together, as they would in Squalus if the acustico-trigeminofacialis recess and the jugular canal were thrown into one.

In the Teleostei, e.g. Salmo (see p. 122), the posterior myodome and the trigeminofacialis chamber are separate, and each is subdivided into two. The dorsal myodome, lodging the external recti muscles, and traversed by the pituitary vein, corresponds to the entire myodome of Amia. But there is in these forms also a ventral myodomic compartment, lodging the internal recti which have likewise pushed inwards dorsally to the trabeculae and lie in a space traversed in part by the internal carotid arteries and the palatine branches of the facial nerves, and therefore corresponding to the parabasal canals (see p. 104) of Amia. (Plate 143, fig. 3.)

The trigeminofacialis chamber of Salmo (see p. 128) is bounded laterally by the lateral commissure, subsequently ossified (there is no pila lateralis) and medially by the membranous dura mater. In the chondrocranium the trigeminofacialis chamber of Salmo resembles that of Amia, (except that it does not communicate with the dorsal myodome). But when ossification sets in, the head vein and orbital artery occupy the space of the resorbed cartilage of the lateral commissure and become enclosed between the inner and outer lamellae of the prootic bone, in the space termed the pars jugularis of the trigeminofacialis chamber, which is thus separated from the pars ganglionaris (lodging the trigeminal and facial ganglia) by the inner lamella of the prootic. In Salmo the palatine branch of the facial nerve does not enter the pars jugularis, which it does, however, in Trigla.

In some Teleostei (e.g. Siluroids), the posterior myodome is absent: and this can only be due to loss, since it would otherwise be necessary to assume that it had been independently evolved in all those Teleosts which possess it. This fact is of interest in connexion with the conditions in Polypterus, Acipenser, and Lepidosteus, in none of which is there a functional posterior myodome, but all of which possess well-formed subpituitary spaces, trigeminofacialis recesses, and jugular canals (see pp. 83, 93, 112). On palaeontological grounds there are reasons for believing that each of these forms must have lost the myodome, since the reputed ancestral forms (the Palaeoniscoids) certainly possessed one (Watson, 1925*b*). It is therefore of interest to find that the morphology of the chondrocranium in this region shows signs of secondary modification such as would be expected if a functional myodome had been present and subsequently lost (see p. 390).

In particular, the side wall of the skull in the orbitotemporal region, is the place of interest in this connexion, since that is where the external recti muscles would have penetrated the cartilage. In Polypterus the pila antotica is absent, and the internal carotid artery takes an atypical course, entering the skull laterally and dorsally to the trabeculae. In Acipenser also, at early stages, the internal carotid artery enters the skull by passing inwards over the trabeculae. In Lepidosteus the pila antotica is atypical in that it lies much too far forwards.

All these abnormalities of morphological relations would receive an explanation if these forms had possessed a functional myodome (involving a breakdown of the pila antotica and discontinuity of the trabeculae, which would enable the internal carotid arteries to slip upwards), and had subsequently lost it and reformed the cartilaginous side wall of the skull.

As thus defined, the posterior myodome and the trigeminofacialis chamber are partly intramural and partly extracranial spaces; they never form part of the true cranial cavity as delimited by the dura mater, although they may be contained within the cartilage or bone of the dried skull.

The similarity in relations of the trigeminofacialis chamber of Amia with another extracranial space, the antrum petrosum laterale or cavum epiptericum of Ceratodus (see p. 172), led Allis (1914*b*) (for a time) and Schmahlhausen (1923*b*) to regard them as homologous. It has been pointed out by de Beer (1926*a*) that the bounding structures of the trigeminofacialis chamber are part of the neurocranium, whereas those of the cavum epiptericum of

Ceratodus are part of the visceral arch skeleton, and that therefore the two spaces are analogous and not homologous; a view with which Allis (1928, 1929) has since concurred.

In birds the trigeminal ganglion becomes lodged in a chamber which is bounded laterally by the posterior orbital cartilage and the squamosal bone, and medially by the membranous dura mater, or part of it. It seems that in this region the side wall of the skull has split away from the dura mater leaving an epidural or intramural trigeminal chamber. The facial nerve is not concerned since it is lodged in the long and tubular facial foramen.

ii. The cavum epiptericum and cavum supracochleare

The cavum epiptericum (Gaupp, 1902; antrum petrosum laterale of Drüner, 1904*b*) is an extracranial space situated laterally to the side wall of the orbitotemporal region of the skull, dorsally to the basitrabecular process, and medially to the processus ascendens of the pterygoquadrate cartilage. The cavum epitericum thus lodges the ganglia of the trigeminal and facial nerves, and is traversed by their branches and by the abducens nerve, the head vein, and the orbital (or stapedial) artery. It includes the sinus cavernosus in mammals (Shindo, 1915). It is therefore analogous, but not homologous, with the trigeminofacialis chamber (see p. 429).

A cavum epiptericum is recognizable in Dipnoi, Amphibia, Lacertilia and Rhynchocephalia as a permanently extracranial space, bounded medially by the pila antotica. The space is also found in Chelonia, but here it becomes included in the bony skull by the downgrowth of the parietal to meet the epipterygoid (processus ascendens). In Ophidia a space corresponding to the cavum epiptericum is likewise included by the parietal, but in the chondrocranium there is uncertainty about the presence of a processus ascendens. In Crocodilia this process is so small that the cavum epiptericum in these forms cannot be said to be demarcated.

In Monotremes the cavum epiptericum is well defined, its medial wall being indicated by the pila antotica. But the space becomes included within the bony skull by the development of the anterior process of the periotic (Watson, 1916) in its outer wall (the membrana spheno-obturatoria). Similarly in Marsupials and Placental mammals the cavum epiptericum becomes included within the bony skull, but here it is by the development of the lamina ascendens of the alisphenoid, and the incorporation is all the more complete since the pila antotica is no longer present, and the boundary between the cavum and the cranial cavity is indicated only by the dura mater (Dürer's membrane), in which in some forms (p. 390) remnants of the original side wall may be found in the form of isolated cartilaginous nodules.

The cavum supracochleare (Voit, 1909, p. 529) is a space found in mammals, immediately behind and communicating with the cavum epiptericum, laterally to the prefacial (suprafacial) commissure and dorsally to the cochlear capsule. Laterally, the cavum supracochlear is bounded by the tegmen tympani, crista cochlearis, and lateral prefacial commissure. The cavum supracochleare lodges the ganglion of the facial nerve which enters the cavum through the primary facial foramen (under the prefacial commissure), leaves it for the sulcus facialis through the secondary facial foramen (under the lateral prefacial commissure), and gives off the palatine nerve in the cavum itself, i.e. in the gap between the prefacial commissure which thus constitutes the hiatus facialis. (The stylomastoid, or tertiary facial foramen, leads the facial nerve out of the sulcus facialis.)

iii. The recessus scalae tympani

In all Tetrapod vertebrates a diverticulum of the perilymphatic spaces of the ear emerges through the foramen perilymphaticum in the wall of the auditory capsule, and finds itself in an extracranial and extracapsular space, the recessus scalae tympani, situated in the anterior region of the fissura metotica. The recessus scalae tympani communicates with the cranial cavity through its apertura medialis which coincides with the anterior portion of the fissura metotica, and over its apertura lateralis is stretched the secondary tympanic membrane, which separates the recessus scalae tympani from the tympanic cavity. The apertura medialis is usually marked off from the posterior part of the fissura metotica, or jugular foramen.

The recessus scalae tympani remains extracranial in Urodela, Gymnophiona, Anura, Lacertilia, Rhynchocephalia, Ophidia, and Monotremata. In the other groups the recessus tends to become enclosed by cartilage, but in different ways (see p. 402).

In Chelonia both the apertura lateralis and apertura medialis are restricted by the development of cartilaginous walls, with the result that the glossopharyngeal nerve is trapped in the recess, into which it enters and from which it emerges by special foramina. The cartilaginous walls of the recess behind these foramina appear to be formed by a preoccipital neural arch on the inside, and by one or more cranial ribs on the outside (see p. 255).

In Crocodilia a well-developed cranial rib gives rise to the processus subcapsularis which forms a floor to the recessus scalae tympani. The apertura lateralis is now comprised between the wall of the auditory capsule and the processus subcapsularis, which between them border a fenestra pseudorotunda, for it is not homologous with the fenestra rotunda of mammals. The same is true in Aves, where the metotic cartilage (cranial rib) is fused with the edge of the basal plate medially and with the auditory capsule laterally. Versluys (1898, p. 357) recognized the importance of the distinction in the fowl between the foramen perilymphaticum and the fenestra pseudorotunda (his rotunda). (See pp. 263, 274.)

In mammals the recessus scalae tympani becomes enclosed within cartilaginous walls but by a totally different method. The ventral wall of the foramen perilymphaticum becomes extended backwards in the form of a process, the processus recessus (processus intraperilymphaticus of Fischer, 1903, and of Voit, 1909, p. 445), which fuses with the floor of the auditory capsule farther back, while in Man there is also developed a process which projects forwards to meet it. The processus recessus not only forms a floor to the recessus scalae tympani, but also it restricts the medial and lateral apertures of the recessus scalae tympani, to give rise to the aquaeductus cochleae and fenestra rotunda respectively. These foramina are therefore not subdivisions of the foramen perilymphaticum, which can still be made out, leading from the cavum vestibulare of the auditory capsule into the now enclosed space of the recessus scalae tympani (see p. 402). The secondary tympanic membrane stretched over the fenestra rotunda is therefore homologous with a part of the synonymous membrane stretched over the apertura lateralis of the recessus scalae tympani in Lacerta (Rice, 1920; de Beer, 1929; contra Gaupp, 1900, p. 514).

The mammalian skull thus circumscribes a fenestra rotunda without the intervention of a cranial rib.

iv. The cavum orbitonasale and the supracribrous recess

A rigorous analysis of the morphology of the structures which bound the passage of the olfactory nerve out of the cranial cavity into that of the nasal capsule, shows that in many forms there is an extracranial space, the cavum orbitonasale, of considerable importance for the interpretation of certain phenomena, such as the passage of the olfactory nerve through the orbit in Teleostei and Aves, or the apparent re-entry into the skull of the nasal branch of the profundus nerve in mammals.

The foramen in the true wall of the cranial cavity leading the olfactory nerve out of that cavity is the foramen olfactorium evehens. It is bounded posteriorly by the preoptic root of the orbital cartilage, ventrally by the trabecular or ethmoid plate, anteriorly by the nasal septum or pila ethmoidalis, and dorsally by the sphenoseptal commissure. In some forms (Selachii, Holocephali, Polypterus, Acipenser, Gadus, Dipnoi, Amphibia) the foramen olfactorium evehens opens directly into the nasal cavity or nasal fossa. This is not the case in other forms.

The hind wall of the nasal cavity or fossa is formed by the lamina orbitonasalis, between which and the preoptic root of the orbital cartilage is a passage, the orbitonasal foramen or fissure, through which the profundus nerve typically passes out of the orbit and into the nasal cavity or fossa. The orbitonasal fissure is bounded dorsally by the sphenethmoid commissure, and ventrally by the ethmoid plate or by the ventral portion of the lamina orbitonasalis.

In some forms a relative shifting of position has taken place between the lamina orbitonasalis and the preoptic root of the orbital cartilage, the former moving forwards or (or and) the latter backwards. The result is that the foramen olfactorium evehens then no longer opens directly into the nasal cavity, but into a space, the cavum orbitonasale, which may be regarded as either an acquisition to the orbit or a loss from the nasal cavity. The cavum orbitonasale is bounded laterally by the lamina orbitonasalis, anteromedially by the anterior border of the foramen olfactorium evehens (nasal septum), and posteromedially by the preoptic root of the orbital cartilage. The cavum orbitonasale is thus an extracranial space, which opens into the nasal cavity by a wide-mouthed passage, the foramen olfactorium advehens, bounded laterally by the lamina orbitonasalis, ventrally by the ethmoid plate or by the ventral portion of the lamina orbitonasalis, medially by the nasal septum, and dorsally by the roof of the nasal capsule. (See Plate 143, fig. 1.)

The olfactory nerve passes through the cavum orbitonasale (thus appearing to traverse the orbit) in many Teleostomes, on its way from the foramen olfactorium evehens to the foramen olfactorium advehens. At the same time the cavum orbitonasale is traversed by the nasal branch of the profundus nerve and by the orbitonasal artery on their way from the orbit to the foramen olfactorium advehens through which they too enter the nasal cavity or fossa (p. 395).

The passage of the olfactory nerve through the orbit in certain birds (e.g. Anas, p. 276) is due to the disappearance of the preoptic root of the orbital cartilage and consequent emargination of the foramen olfactorium evehens.

In reptiles and in mammals the cavum orbitonasale becomes a wedge-shaped space, with the apex in contact with the nasal septum, and situated in front of the cranial cavity and behind the nasal cavity. This position has been achieved owing to the widening of the gap between the preoptic root of the orbital cartilage and the lamina orbitonasalis, the lifting up of the preorbital root of the orbital cartilage in connexion with the formation of an interorbital septum, and the inward extension of the lamina orbitonasalis to the nasal septum, almost to the midline.

In the chondrocranium of reptiles and mammals the relations of the cavum orbitonasale remain typical; in mammals the foramen olfactorium advehens becomes subdivided by the cribriform plate, for which reason the cavum orbitonasale may in them be known as the supracribrous recess. In the mammalian bony skull this supracribrous recess becomes surrounded by bone and incorporated, so to speak. The orbitonasal fissure is reduced to small dimensions. But the morphologically extracranial nature of the recess is still demonstrated by the fact that the profundus nerve, after leaving the orbit through the orbitonasal fissure, lies ventrally and externally to the dura mater, which is stretched over the opening of the foramen olfactorium evehens and marks the boundary between the cranial cavity and that of the cavum orbitonasale or supracribrous recess. The profundus nerve, therefore, does not re-enter the cranial cavity.

34. GENERAL MORPHOLOGICAL CONSIDERATIONS OF CERTAIN BONES

This section does not pretend to provide a general consideration of the morphological value of all the bones of the skull, but only of such as owe their interpretation to developmental (as apart from adult anatomical) studies.

i. The mammalian vomer

Ever since Sutton (1884*a*) and Smets (1885) put forward the view that the mammalian vomer represents the median portion of the reptilian parasphenoid (immediately underlying the trabecula communis and interorbital septum), evidence has accumulated in its favour. A corollary of this hypothesis is that on the one hand the mammalian vomer is not the same as that paired element in non-mammals which is now called prevomer (see below, p. 434), and on the other that the lateral portions of the parasphenoid in non-mammals are detached from the median portion; what they produce will be discussed below (p. 435). Meanwhile, it is of the greatest interest to note that in Lacerta (Gaupp, 1906*a*, p. 774) occasionally the parasphenoid may ossify from three centres, one median and two paired, which thus foreshadows the mammalian condition. It should, however, be remembered that in some mammals the vomer arises from paired centres of ossification. This, however, does not necessarily invalidate the conclusions arrived at, although it requires explanation.

The condition in birds (Parker, 1870) is parallel to what is found in mammals, for there also the parasphenoid appears to be split into three: a median vomer, and paired lateral basitemporals which fuse with the basisphenoid. In the crocodiles (Parker, 1883*b*) likewise, the lateral portions of the parasphenoid are detached as basitemporals which fuse with the basisphenoid, but the median rostral portion, the vomer, is reduced.

Fuchs (1908) claimed the existence in Didelphys of a median 'parasphenoid', separate from and behind the vomer. Broom (1935*b*) has, however, been unable to confirm this in spite of detailed investigation; nor has Töplitz (1920).

ii. The prevomers

The prevomers are paired, originally tooth-bearing, bones situated in the anterior region of the palate, close behind the premaxillae. In some forms (Teleostei, Chelonia, Lacertilia, birds, Ornithorhynchus) the prevomers in the adult appear to be fused into a single median bone, but the development reveals traces of paired origin (Schleip, 1904; Kunkel, 1912*a*; Gaupp, 1906*a*; Broom and Brock, 1931; de Beer and Fell, 1936; respectively). The characteristic feature of the prevomers in the Amniota is that they form little plates underlying the paraseptal cartilages (in which Jacobson's organs are lodged), with short processes extending backwards on each side of the ventral edge of the nasal septum. In this position, separate ossifications have been found by Parker (1885) in Tatusia (confirmed by Fawcett, 1921), Dasypus, Choloepus, Manis, Orycteropus, and Erinaceus, which then become fused on to a process directed backwards from the premaxilla of their own side. This was direct support for Sutton's (1884*a*) contention that the 'palatine process' of the premaxilla in mammals is in reality a separate morphological unit, representing the element which Broom (1895*a*, 1902*a*) called the 'prevomer'. Broom showed that in Miniopterus a distinct ossification is present in the position of the dumb-bell bone of Ornithorhynchus, the paired nature of which latter was recognized by Wilson and Martin (1893) while the independent ossification of its paired rudiments has been observed by de Beer and Fell (1936). In other mammals separate ossification of the 'palatine process' of the premaxilla has been observed in the rabbit (Burne, quoted from Howes by Broom, 1895*a*), sheep (Schwink, 1888), pig (Biondi, 1886), and Man (Fawcett, 1911; Jarmer, 1922).

The earliest stages so far studied in Miniopterus (Fawcett, 1919) show the 'palatal process' already continuous with the body of the premaxilla. It is therefore unknown whether they have separate centres of ossification in this form, and it is clear that their separate existence in the adult is the result of subsequent detachment.

There is, therefore, good reason to believe (in spite of Gaupp's (1906*b*) view) that the prevomers of non-mammalian vertebrates are totally different structures from the vomer in mammals, and are represented in the latter by the dumb-bell bone of Ornithorhynchus, the separate bone in Miniopterus, and (as Broom, 1895*c*, and Wilson, 1901, suggested) by the palatine (or paraseptal (Fawcett, 1921), or prevomerine) process of the premaxilla in those mammals in which the prevomers lose their separate individuality.

iii. The mammalian pterygoid

The question of the homology of the mammalian pterygoid dates from Gaupp's (1908, 1910*a*) discovery of the existence in Echidna of separate dorsal and ventral pterygoid bones on each side. He observed that the more ventral of these (the 'Echidna-pterygoid') resembled the pterygoid of reptiles, while the more dorsal (the 'mammalian pterygoid') he regarded as corresponding to the pterygoid of ditrematous mammals. The result was, of course, on this view that the so-called pterygoids in reptiles and in ditrematous mammals were not homologous. At the same time it became necessary to assume that the reptilian pterygoid has vanished in ditrematous mammals, and that the pterygoid in the latter, by the argument of exclusion, can represent nothing else than the detached lateral portion of the parasphenoid (represented in crocodiles and birds by the basitemporals, and occurring separate occasionally in Lacerta).

Gaupp was under the impression that the pterygoid in ditrematous mammals was a single element, for so it was described in its development in Erinaceus by Wincza (1896) and in Talpa by Fischer (1901).

But there is now considerable evidence that the pterygoid of ditrematous mammals is a compound structure, ossifying from two centres. Of these, the more dorsal is a plate of membrane-bone, close to the ventral surface of the processus alaris, the Vidian nerve passing in between (i.e. dorsolaterally to the bone) through what in fact is nothing but a parabasal canal (see p. 233). This plate therefore corresponds to the detached lateral portion of the parasphenoid and to the dorsal pterygoid of Echidna. The more ventral ossification arises in a nodule of *secondary* cartilage and forms the so-called hamulus of the pterygoid. On de Beer's (1929) suggestion, this secondary pterygoid cartilage represents the reptilian pterygoid, and the 'Echidna-pterygoid' or ventral element of Echidna. It lies laterally to the Vidian nerve.

The existence of a nodule of secondary cartilage in the pterygoid has been found in every mammal examined except the Monotremes (Gaupp, 1905*a*; de Beer and Fell, 1936), where the element which it is held to represent is present as a bony rudiment from the start; in all the others, its precocious manifestation as secondary cartilage is presumably associated with the mechanical functions which it performs in connexion with the formation of the false palate and nasal passage, the hind wall of which it supports. In particular, the pterygoid secondary cartilage is related to the tensor palati muscle. (Other examples of the precocious representation of *membrane-bones* by *secondary* cartilage are given on page 502.)

Separate ossifications in the dorsal plate and in the pterygoid cartilage have been observed in Man (Fawcett, 1905*b*, 1910*a*, 1910*b*; Augier, 1931*a*), the rabbit (Fuchs, 1909*c*), the cat (Terry, 1917; Drews, 1933); while the existence of elements distinct from the 'mammalian pterygoid', and presumably representing the 'Echidna-pterygoid', was shown in Petrogale, Tatusia, and Tamandua by Broom (1914). There is uncertainty about the true nature of the elements called 'basitemporals' or 'mesopterygoids' by Parker in the pig, the fox, and the hedgehog (1874, pp. 316, 324), in Cavia (1883*b*, p. 289), and Galeopithecus (1885, p. 253), but they are evidence of the presence of two separate pterygoid elements in those forms. The composite nature of the pterygoid was suspected in Choloepus by Lubosch (1907, p. 527) who then inclined towards the view here adopted that the pterygoid of ditrematous mammals is the product of fusion between the reptilian pterygoid and the detached lateral portion of the parasphenoid. The persistent separation between the 'Echidna-pterygoid' and the 'mammalian-pterygoid' as found in Monotremes would then be a primitive, reptilian, feature.

While this view has the merits of simplicity and of fitting the facts so far as known, it should be mentioned that other authorities have proposed different solutions, which are set out in the following table. It should, however, be noted that Gaupp's term 'mammalian pterygoid' refers solely to the dorsal element in Monotremes, and that, therefore, it is necessary to use a distinctive term for the pterygoid of ditrematous mammals. The simplest way to do this seems to be simply to call it the mammalian pterygoid, reserving inverted commas for Gaupp's use of it.

Authority	*Reptile*	*Monotreme*	*Ditrematous mammal*
Gaupp, 1908, 1910*a*	lateral wing of parasphenoid reptilian pterygoid	= 'mammalian pterygoid' = 'Echidna-pterygoid'	= mammalian pterygoid = absent (or ? endotympanic
Fuchs, 1910*a*	 reptilian pterygoid	'mammalian pterygoid' = 'Echidna-pterygoid'	= pars perpendicularis of palatine = mammalian pterygoid
Watson, 1916	reptilian pterygoid epipterygoid	= 'mammalian pterygoid' = 'Echidna-pterygoid'	= mammalian pterygoid = alisphenoid
Broom, 1914	reptilian pterygoid transpalatine (ectopterygoid)	= 'mammalian pterygoid' = 'Echidna-pterygoid'	= mammalian pterygoid = ectopterygoid of Tatusia, &c.
Parker, W. K., 1874, 1885	lateral wing of parasphenoid reptilian pterygoid		= basitemporal of Cavia = mammalian pterygoid
van Kampen, 1922	basitemporal reptilian pterygoid	= 'mammalian pterygoid' = 'Echidna-pterygoid'	= processus tympanici = mammalian pterygoid
Lubosch, 1907		'mammalian pterygoid' 'Echidna-pterygoid'	= } mammalian = } pterygoid
de Beer, 1929	lateral wing of parasphenoid reptilian pterygoid	= 'mammalian pterygoid' = 'Echidna-pterygoid'	= } mammalian = } pterygoid

The view held by van Kampen is based on the possibility (for which there is, however, little evidence) that the basitemporal may have been a separate element all the way from the fish stage. The remaining views are largely based on possibilities and argument by exclusion. A strong objection to both Broom's and Watson's schemes is that they deny the homology of the 'mammalian pterygoid' of Echidna with the lateral wing of the parasphenoid, for which the evidence is really very strong. Furthermore, in connexion with Watson's (and Kesteven's, 1918) view that the Echidna-pterygoid is the alisphenoid, de Beer and Fell (1936) have shown that Ornithorhynchus has an alisphenoid in addition to the 'Echidna-pterygoid' (see p. 298).

iv. The squamosal and the 'paraquadrate'

Gaupp (1895) defined the squamosal as recognized by him in birds and mammals as a membrane-bone related to the lateral surface of the auditory capsule. On the other hand, in Amphibia he failed to find a bone answering this description, for the squamosal of current literature in these forms is related to the lateral surface of the quadrate. Hence Gaupp called this bone in Amphibia the paraquadrate and concluded that it was not homologous with the squamosal: a view in which he is followed by de Villiers (1936). This led to a confusing and quite unnecessary shifting of homologies of bones in different forms, for which there is no good evidence at all. By giving to the supratemporal and squamosal respectively of Stegocephali and Lacertilia the names squamosal and paraquadrate, Gaupp was able to show, as he thought, the co-existence of squamosal and paraquadrate in one and the same form and thus establish their separate identity. Since the supratemporal and squamosal can be traced by their lateral line canals from Osteolepis to Stegocephalia, and from the latter to Anapsid reptiles, this is inadmissible. Applying the argument to Chelonia and Crocodilia, Gaupp found his squamosal and 'paraquadrate' in the squamosal and quadratojugal of current literature. In Sphenodon, however, he regarded the squamosal and quadratojugal as 'paraquadrate' and quadratojugal respectively. In other words, Gaupp denied the presence of a squamosal in Sphenodon, and his 'paraquadrate' which is the squamosal of Amphibia becomes the quadratojugal of Chelonia and Crocodilia, in spite of the fact that the Amphibia also possess a quadratojugal. Faced with the problem of what has become of the 'paraquadrate' in birds and mammals, Gaupp was driven to the assumption that it had disappeared in the former and was represented by the tympanic in the latter: an untenable view, since palaeontological and embryological studies have established the derivation of the mammalian tympanic from the angular (van Kampen, 1905; Broom, 1912; Watson, 1921*b*; Goodrich, 1930). It should be added that later Gaupp (1906*b*) admitted the existence of the squamosal in Sphenodon.

But there is no need whatever for the complicated 'general post' which Gaupp's view would necessitate, for the premisses from which it starts neglect the facts that the living Amphibia with their large quadrate cartilages are highly specialized, and that with the reduction of the quadrate to the incus which becomes protected by the tegmen tympani, the squamosal which covered the quadrate in lower vertebrates, becomes a covering bone of the auditory capsule in the higher forms, as pointed out by Thyng (1906) and supported by Brock (1935). It is also to be noted that in mammals the squamous part of the squamosal develops later than the zygomatic process; i.e. at early stages the squamosal is *not* closely related to the wall of the auditory capsule at all.

There is thus no reason to doubt the homology of the squamosal of current literature in all Tetrapoda, and there is no need for any term 'paraquadrate'.

v. The squamosal and supratemporal of Lacertilia

A problem is presented by the Lacertilia in which the two bones of the temporal region here referred to (following Parker and Bettany, 1877) as supratemporal (upper and innermost) and squamosal (outer and lowermost) have been regarded as representing respectively the tabular and squamosal by Broom (1924, 1935*a*), the squamosal and quadratojugal by Watson (1914), or the squamosal and 'paraquadrate' by Gaupp (1895). This last view has been shown in the previous section to be untenable since it involves the alleged presence as a separate unit of a 'paraquadrate' for the existence of which there is no evidence. The possibility must, however, be considered that the bones represent the squamosal and the quadratojugal.

The matter must ultimately be decided one way or the other by means of fossil series. But in so far as it is possible to arrive at conclusions from embryonic stages, Brock (1935) has shown from a consideration of the relations to the auditory capsule, the quadrate, and the external adductor muscles, that the outermost bone in Lacertilia is the same as the single bone in this position in birds and in mammals and which it is classical to call the squamosal. The inner bone of Lacertilia is more closely related to the hinder region of the wall of the auditory capsule, and is therefore either a supratemporal or a tabular. Some Lacertilia (e.g. Pachydactylus) have only one bone in this position and this corresponds to the inner one of Lacerta. For this reason Brock considers that the single bone in Ophidia is likewise a supratemporal (or tabular): the squamosal having been lost.

vi. The lachrymal

Consideration of the lachrymal in the developing mammalian skull led Gaupp (1910*b*) to the conclusion that the lachrymal is a membrane-bone related to the lateral wall of the cartilaginous nasal capsule. Therefore, he proposed to homologize it with the prefrontal of reptiles, which occupies this position. A difficulty immediately arises, of course, since in addition to the prefrontal a lachrymal is present in Lacerta and the crocodile; this Gaupp is driven to call an 'adlachrymal', because it lies closer to the maxilla than to the nasal capsule. In many Lacertilia it is absent, as also in Anura, Urodela (except Hynobius and Ranidens), and in Sphenodon, Ophidia, Chelonia, and birds. When the lachrymal is absent, the lachrymonasal duct is often enclosed in the prefrontal. Gaupp's argument rests, however, on an abuse of refinement, far in excess of the capacity of the material to supply critical evidence of such detailed topographical relations. The fact remains that the lachrymal whenever present has constant relations to the lachrymonasal duct, which it encloses from the side. The lachrymal of Lacerta and Crocodilus is really identical in relations with that of mammals. There is, therefore, no need to question the homology of the elements generally accepted as lachrymals, and traceable by means of the lateral line canal through the Stegocephalia back to Osteolepis, as Gregory (1920) has shown. Investigations would, however, be welcome on the nasal, adnasal, and prefrontal and lachrymal of Urodeles, which are still obscure. (See Noble, 1921.)

vii. The alisphenoid, pleurosphenoid, and pterosphenoid

It was for a long time assumed that the hindmost bone of the orbital region of the skull, behind the orbitosphenoid and in front of the auditory capsule was homologous in all vertebrates from fish to mammals, and it was called the alisphenoid. This view is still held by Kesteven (1918, 1926). But since the recognition by Gaupp (1902) and Allis (1919*b* and *d*) of the fact that the cavum epiptericum in mammals (enclosed by the alisphenoid) is extracranial in reptiles, and the identification by Broom (1907, 1909*a*), Fuchs (1912), and Gregory and Noble (1924) of the reptilian epipterygoid with the mammalian alisphenoid, it is clear that the bone previously called 'alisphenoid' in non-mammals has no right to that title; for in these forms its position and relations to neighbouring nerves and blood-vessels (see p. 268) show that it is an ossification of the true cranial wall (pila antotica) whereas the mammalian (and true) alisphenoid is an ossification in a visceral arch structure, the ala temporalis which corresponds to the processus ascendens of lower forms (see p. 420). The ossification of the alisphenoid often extends out of the cartilage of the ala temporalis into membrane: in some forms (e.g. Didelphys) it is entirely intramembranous. The relations of this to the anterior process of the periotic of Monotremes is still obscure. Ranke (1899*c*) has tried to see in the lamina ascendens of the alisphenoid an independent membrane-bone, the intertemporal, but this view is hard to accept. At all events it is clear that the hindmost bone of the orbital region, ossified in the pila antotica, in non-mammals must therefore be distinguished as a pleurosphenoid (Goodrich, 1930). This may perhaps be the same bone that is found in bony fish, which is well developed in Amia, and forms the hinder part of the 'sphenoid' in Polypterus. The bony fish do not, however, possess a pila antotica, and the bone in question in these forms has occasionally been known as the pterosphenoid, pending the demonstration of its homology with the pleurosphenoid.

The pleurosphenoid has been lost in mammals (though a remnant may perhaps be found in Ornithorhynchus and in Man, pp. 298, 370) following on the reduction and loss of the cartilaginous true wall of the skull in this region (pila antotica). At the same time the gap then formed became filled at a more lateral plane (thus including the cavum epiptericum in the skull) by the epipterygoid. This in the mammals is expanded in a posterior direction, and the mandibular branch of the trigeminal nerve is enclosed in the foramen rotundum. Thus, while the epipterygoid of reptiles lies between V_1 and V_2, the alisphenoid of mammals primitively separates all three branches of the trigeminal from one another. In other forms, the cartilage and bone intervening between V_1 and V_2 may be reduced, with the result that the ala temporalis (and alisphenoid) separate V_2 from V_3.

In birds the formation of the pila antotica spuria leads to the ossification of the pleurosphenoid between V_1 and V_2, but it is nevertheless a part of the true cranial wall, and has nothing to do with the alisphenoid.

viii. The endotympanic and the bony floor of the tympanic cavity

The endotympanic (entotympanic, Mivart, 1881; metatympanic, Wincza, 1896) is an ossification (or perhaps two, see below) which forms part of the skeletal protection to the floor of the tympanic cavity in some Marsupials, Insectivora, Cheiroptera, Dermoptera, Edentata, Hyracoidea, and Carnivora. It differs in origin from the tympanic in that it ossifies in cartilage (secondary cartilage?), as first described for Felis by Flower (1869), for which reason a study of its development is essential for an appreciation of its morphology.

According to van Kampen (1905, 1915*a*) the endotympanic has evolved within the class Mammalia, and it is therefore impossible to find homologues for it in other vertebrates. His reasons are that the endotympanic can only arise in forms with a well-developed floor to the tympanic cavity; and that such a floor is absent in non-mammals and very imperfect in primitive mammals. It is possible that the cartilage found by Watson (1916, p. 330) laterally to the malleus in Ornithorhynchus may represent an endotympanic.

The development of the endotympanic has been studied systematically in many forms by van der Klaauw (1922) who came to the conclusion that two elements, one rostral and the other caudal, which subsequently fuse, are included under the term endotympanic. The former is present well developed while the latter is small in Procavia. The rostral element is smaller and the caudal larger in Roussettus (bat) and Galeopithecus. In Pteropus and in Tatusia the rostral endotympanic remains cartilaginous, and the endotympanic bone in these forms, as in Canis and Felis, is formed from the caudal element. The rostral endotympanic ossification arises in association with the secondary cartilage surrounding the Eustachian tube in Procavia, but separated from it in other forms. It is, therefore, difficult to decide what the nature of the relation between these structures is. The view suggested by van Kampen (1915*a*) that the caudal endotympanic is related to the laterohyal is not supported by van der Klaauw (1922).

It is probable, therefore, that the endotympanics are essentially mammalian structures, ossifying in secondary cartilage, and Fawcett (1919) has suggested that the endotympanic is called into being as a result of the pressure exerted on the tympanic cavity by the digastric muscle.

The floor of the tympanic cavity presents an excellent example of a region which may be ossified by different bones. In addition to the endotympanic where it occurs, the floor may become ossified by the alisphenoid (Marsupials), basisphenoid (Insectivores), tympanic (Carnivores, Ungulates, Rodents, Cetacea), or periotic (Primates). The bony floor has been lost in Sirenia.

ix. The mammalian auditory ossicles

It is not intended here to repeat the evidence in favour of Reichert's theory of the derivation of the mammalian ear ossicles, since the literature has been reviewed by Gaupp (1899, 1913) Broom, (1912), and van der Klaauw (1924*a*), and a complete account based on personal observations has been given by

Goodrich (1915, 1930). Suffice it to say that the present work is in entire agreement with the homology of the articular, quadrate, and columella auris, of reptiles, with the malleus, incus, and stapes, respectively, of mammals.

Attention may, however, here be paid to the development of certain diminutive ossifications occurring in mammals.

The *processus anterior* (Folii, gracilis) of the malleus may with reasonable certainty be regarded as in part formed of the homologue of the *prearticular* (*postopercular* of Gaupp, 1905*b*, or *goniale* of Gaupp, 1908) of reptiles, and birds (Voit, 1924). As Gaupp (1911*b*) has shown, it is a *membrane-bone* situated on the medial side of Meckel's cartilage, and the chorda tympani nerve runs forwards between the two, which is a characteristic relation. Later, the prearticular fuses with the articular in birds (Cords, 1904; Voit, 1924), or with its homologue the malleus in mammals. It is to be noted that some authors regard the terms 'processus anterior (gracilis or Folii)' and 'prearticular' as equivalent. However, it should be noted that the processus anterior is primarily a part of the malleus representing all that is left of its connexion with Meckel's cartilage and ossifying as cartilage-bone, to which the prearticular membrane-bone subsequently becomes fused. In some forms, however (Ornithorhynchus, Echidna, Erinaceus, Pteropus, Halicore, many Myomorpha, and Cetacea), the prearticular fuses with the tympanic.

The *ossiculum accessorium mallei* has been described by van Kampen (1905) as a small membrane-bone situated on the dorsal surface of Meckel's cartilage near its proximal end, in Ovis, Sus, Tragulus, and Choloepus. Sometimes it fuses with the prearticular, of which Voit (1923) believes that it is merely a part. It has been suggested by Watson (1916), and by Broom (1916) and Ridewood (1922) who found the ossiculum in Chrysochloris and Balaenoptera respectively, that it may represent the supra-angular, which is probable. Van Kampen thought it might represent the coronoid.

The *os quartum* sometimes found in the tendon of the stapedial muscle, appears to be an ossification of Paauw's cartilage, probably representing the extracolumella (p. 412).

x. The tympanic

The homologue of the mammalian tympanic was sought in the paraquadrate by Gaupp (1895) and in the quadratojugal by Gegenbaur (1898). Van Kampen (1905) at first homologized it with the supra-angular, but when the processus Folii was recognized as the prearticular and not the angular, he suggested that the tympanic could be homologized with the angular of non-mammals, a view which is generally accepted. From the developmental point of view, a feature of interest and importance concerning the tympanic is the rotation which it undergoes. The plane of the tympanic membrane in Amphibia and Reptilia is vertical, but in primitive mammals the tympanic membrane (and therefore the tympanic ring) is nearly horizontal (Monotremes, Insectivora). According to van Kampen this change in position is due in phylogeny to the broadening of the base of the skull and brought about by a rotation of the membrane outwards about a sagittal axis passing through its ventral border. In addition to this, there is in higher mammals a second change observable in ontogeny, resulting in a partial reacquisition of the vertical position by means of a rotation about a sagittal axis passing through the lateral (originally dorsal) border of the tympanic membrane. This rotation is ascribed by van Kampen to the enlargement of the cochlear part of the auditory capsule and to the development of a floor to the tympanic cavity, which then becomes protected by a bulla formed by the tympanic, endotympanic, or alisphenoid.

The position of the tympanic membrane in various mammals has been summarized by Denker (1899).

xi. The septomaxilla

The septomaxilla (W. K. Parker, 1871), intranasal (Gaupp, 1904), or nariale (Wegner, 1922), has been recognized in many reptiles and some mammals (Monotremes, Edentates) as a membrane-bone lying behind the external narial aperture and extending inwards as a plate overlying Jacobson's organ. This bone is also found in Gymnophiona, Urodela, and Anura, though in the two latter groups, the problem is complicated by Lapage's (1928*a* and *b*) contention on seemingly insufficient evidence that the septomaxilla ossifies as a cartilage-bone. The matter is of importance, for an undoubted cartilage-bone (the pre-ethmoid) is found in a more or less comparable position (the posterior surface of the trabecular horn) in Amia and Esox. While the derivation of a membrane-bone from an ossification which in previous phylogeny was a cartilage-bone is not impossible, the homology of the pre-ethmoid of Amia with the septomaxilla of reptiles cannot be regarded as proved. It must be remembered that the early Osteolepid fishes possessed a large number of small rostral membrane-bones, any one of which might have become the septomaxilla. Further researches are therefore necessary, not only in fossil fish and Stegocephalia, but also into the embryology of the septomaxilla in living Amphibia.

xii. The mesethmoid

Contrary to previous notions, it is now clear from Broom's (1926*a*, 1927, 1930, 1935*c*) investigations that the mesethmoid was evolved within the Mammalia, and that only some and not all of the Orders of the Class possess this bone.

No mesethmoid is developed in: Monotremata, Marsupialia, Xenarthra, Artiodactyla, Perissodactyla, Proboscidea, Sirenia, Chrysochloridea, Cetacea (?) (Broom's 'Palaeotherida').

A mesethmoid is present in: Pholidota, Tubulidentata, Hyracoidea, Rodentia, Insectivora, Cheiroptera, Dermaptera, Carnivora, Primates (the 'Neotherida').

The ossification in the nasal septum in those forms which lack a mesethmoid is therefore the presphenoid in Tetrapods. The bone which develops in or invades the nasal septum in certain Teleostei cannot apparently be homologous with the mammalian mesethmoid, and may therefore perhaps be distinguished as the ethmoid. It will not be difficult to distinguish this from the similarly named compound bone found in man. In some birds, especially those in which the beak is subjected to severe stress as in woodpeckers and hawks, the nasal septum becomes the seat of secondary ethmoid ossifications.

xiii. The lateral ethmoid and other ossifications of the lamina orbitonasalis

The lateral ethmoid is a cartilage-bone developed in the lamina orbitonasalis of fish, and to be distinguished from the prefrontal (see p. 498), a membrane-bone which overlies the lamina orbitonasalis and is especially well developed in reptiles. The lamina orbitonasalis does not appear to ossify in amphibia or in reptiles, but in some birds there arises in it an important bone which forms the front wall of the orbit. Since it would seem that this bone has been independently evolved within the class Aves and has not been inherited directly from the fish it should not be called the lateral ethmoid. Perhaps the term ectethmoid may be restricted to mean the ossification in birds, without increasing the nomenclature.

In many birds there is a bone, the lachrymo-palatine ossicle, or uncinate bone, extending from the ectethmoid to the jugal or to the palatine. It appears to be a portion of the ectethmoid segmented off but requires further study. It is of fairly wide distribution among birds, and is reported in: Tubinares (Diomedea), Steganopodes (Fregata), Musophagi (Corythaix), Cuculidae (Scythrops), Limicolae (Larus), Columbae (Goura), Grues (Cariama).

Another, and seemingly independent, ossification of the lamina orbitonasalis occurs in Man, in the form of the ossiculum Bertini (see p. 372).

xiv. The sclerotic bones

Sclerotic bones are ossifications developed either on the sclerotic cartilage (e.g. Teleostei) or in the connective tissues around and in front of it (e.g. Acipenser, Reptilia, Aves). In Acipenser the two sclerotic bones are dorsal and ventral to the pupil; in those Teleostei which have them, the two bones are anterior and posterior. Sclerotic bones are absent in Polypterus, Amia, Lepidosteus, and Dipnoi. The only Amphibian in which sclerotic bones have been reported is Diemictylus pyrrhogaster, where they are few in number, irregular and inconstant; Tsusaki (1925) considered them to represent ossified parts of the sclerotic cartilage (which disappears in the adult), but Yano (1926) regards them as of intramembranous origin.

Among reptiles and birds the sclerotic plates vary in number between ten and seventeen, and are disposed round the pupil, overlapping and movable on one another (see Slonaker, 1918). No sclerotic bones are present in Ophidia, Crocodilia, or mammals.

The attempt has been made by Dabelow (1927) to homologize the sclerotic bones with the circumorbital bones present in certain fish. This view was shown to be untenable by Edinger (1928, 1929) who demonstrated the existence of both sclerotic bones and circumorbital plates in Dapedius pholidotus.

xv. The supraoccipital and interparietal

The supraoccipital is a cartilage-bone ossifying in the tectum synoticum or posterius, and normally forming the dorsal border to the foramen magnum. It is present in Teleostei and in Amniota. It appears to ossify from a median unpaired centre in Salmo, Gasterosteus, Lacerta, Turdus (W. K. Parker, 1873*c*), Ovis (Lacoste, 1927), Felis (Drews, 1933), Sus (Augier, 1934*a*), and apparently in Man (Augier, 1931*b*, who was unable to confirm the existence of paired centres reported by Mall, 1906). On the other hand, the supraoccipital arises from paired centres of ossification in Gallus (W. K. Parker, 1870), Bos (Augier, 1931*b*), Canis (Drews, 1933).

Among reptiles and birds the supraoccipital seems either to fuse with or to occupy the place of the epiotics in the walls of the auditory capsules.

Anteriorly to the supraoccipital and behind the parietals, intramembranous ossification centres may arise, as the so-called postparietal or interparietal in Alligator (Mook, 1921), Anser, Columba, Meleagris, Strix, and Athene among birds (Staurenghi, 1900*a* and *b*), and in many mammals. Attention was drawn to these centres by Meckel (1809). There appears, however, to be no interparietal in Monotremes, Perameles, Dasypus, or Tatusia, nor in Sus where a small cartilage-bone ossifies in the remnant of the tectum synoticum, in front of the supraoccipital, and fuses with it (Forster, 1902; Engelmann, 1910; Claus, 1911; Augier, 1934*a*). The interparietal centres fuse with the supraoccipital (e.g. Man, Canis, Felis, Bos, &c.) or with the parietals (e.g. Sirenia) or remain separate (e.g. Lepus).

The problem of the morphological identification of these centres is rendered more difficult by the claims that the interparietal may exceptionally arise from two pairs of centres in Man (Maggi, 1896; Ranke, 1899*a*, 1913; Aichel, 1913) which if they remain separate give rise to the condition known as os Incae quadripartitum (Hepburn, 1907; W. R. Smith, 1908). A further complication is introduced by the existence of paired preinterparietal centres found occasionally in Man (Ficalbi, 1890; Bianchi, 1893), Bos, (Wilhelm, 1924), Canis (Drews, 1933), &c. Stieda (1892) endeavoured to explain the phenomena on the view that the median pair of interparietal centres were the preinterparietal centres of other authors, migrated backwards. On the other hand, Maggi (1896) and Ranke (1899*a*) asserted the possibility of co-existence of the preinterparietal and quadripartite interparietal centres. Ranke therefore regarded the preinterparietal centres as Wormian ossifications devoid of morphological significance, while the two pairs of interparietal centres were equated by von Huene (1912) and Broili (1917) with the paired postparietals and tabulars (medial and lateral extrascapulars) of lower forms.

In a general way, the homology of the interparietal of mammals with the postparietal of reptiles is highly probable. The existence in mammals of tabulars has been shown in Microgale (W. K. Parker, 1885), in Phoca, and in Chrysochloris (Broom, 1916; Cooper, 1928). But it is not safe to conclude that the lateral elements of the abnormally fragmented interparietal in Man are homologous with tabulars, especially as (see p. 367) an independent interparietal in man does not correspond to the entire interparietal of other forms.

xvi. The palate of Aves

It is well known that T. H. Huxley (1867) devised a classification of birds according to the different sizes and contacts of the various bones of the palate. The subdivisions were as follows:

1. *Dromoeognathous*. Prevomers only partly fused, very large, extending far back; pterygoids extending forwards between and in fused contact with prevomers and palatines; palatines wide apart, anteriorly in contact with palatine processes of maxillae which extend inwards to the prevomers.
2. *Schizognathous*. Prevomers completely fused, fairly large; palatines

anteriorly fused with premaxillae, posteriorly articulate by joint with pterygoids, and both in contact with parasphenoid rostrum (pterygoids do not touch prevomers); palatine processes of maxillae separate.

3. *Desmognathous*. Prevomers fused and very small; palatine processes of maxillae join in front of prevomers, palatines join in midline behind prevomers, and articulated with pterygoids and parasphenoid rostrum.
4. *Aegithognathous*. As Schizognathous but prevomers truncated anteriorly; palatines extend forwards ventrally to palatine processes of maxillae.

To these classes W. K. Parker (1875) added:

5. *Saurognathous*, to cover the (erroneously supposed) paired and separate prevomers in Picidae.

It was soon recognized that while the desmo-, schizo-, sauro-, and aegithognathous types merge into one another, the only distinct type is the dromoeognathous. Pycraft (1900) emphasized this, calling the dromeognathous type *palaeognathous* (exemplified in Ratites and Tinamus) in contradistinction to all other types referred to as *neognathous*. The latter, and in particular the schizognathous and aegithognathous appeared to be modifications of the palaeognathous type, while the desmognathous could be derived from the schizognathous. Krassowsky (1936) has shown that the so-called saurognathous type has nothing to distinguish it from the schizognathous. (Parker's 'medio-palatine' is the fused prevomers, and his 'paired vomers' are secondary ossifications giving rigidity to the woodpecker's palate.)

The essential feature of the distinction between the palaeognathous and neognathous types is that *in the adult* in the former the pterygoid makes contact with the prevomers, and has a long suture with the palatine, while in the latter the pterygoid makes contact with the parasphenoid rostrum and articulates by a joint with the palatine which excludes the pterygoid from the prevomers.

But Pycraft (1901, 1907) also showed that the distinction between the palaeognathous and neognathous types was not as absolute as had been supposed, because many neognathous birds pass through a 'palaeognathous' stage in their development. The pterygoid in many nestlings actually, or nearly, touches the prevomers, but its anterior end subsequently becomes broken off and fuses with the palatine as the so-called 'mesopterygoid' of Parker (1875, 1876*b*, 1877*b*, 1879*c*) and the hemipterygoid of Pycraft. This is why in the adult the pterygoid appears to fail to touch the prevomers, and this state of affairs prevails in the following forms (see Plate 101, figs. 8, 9):

Genus	*Authority*	*Order*	*Adult type*
Podiceps	Pycraft, 1901	Colymbiformes	schizognathous
Pygoscelis	,, ,,	Sphenisciformes	schizognathous
Tetrapteryx	,, ,,	Gruiformes	schizognathous
Larus	Parker, 1876*b*	Limicolae	schizognathous
Steatornis	Pycraft, 1901	Caprimulgi	desmognathous
Falco	Parker, 1876*b*	Accipitres	desmognathous
Scythrops	,, ,,	Cuculi	desmognathous
Opisthocomus	Parker, 1891*a*	Opisthocomi	schizognathous
Corvus	Parker, 1872	Passeres	aegithognathous
Menura	Parker, 1877*b*	Passeres	aegithognathous
Oriolus	Parker, 1879*c*	Passeres	aegithognathous
Parus	Parker, 1873*b*	Passeres	aegithognathous

On the other hand, in the following forms the hemipterygoid never becomes separated off from the pterygoid: i.e. the 'palaeognathous' type is retained:

Genus	*Authority*	*Order*	*Adult type*
Megalaema	Pycraft 1901	Picidae	aegithognathous
Bucco	,, ,,	,,	desmognathous
Gecinus	Parker 1875	,,	schizognathous

Lastly, in the following the anterior end of the pterygoid is reduced, i.e. there is no hemipterygoid at all:

Genus	*Authority*	*Order*	*Adult type*
Gallus	Pycraft 1901	Galliformes	schizognathous
Anas	,, ,,	Anseriformes	desmognathous

It is clear from these facts that all the neognathous types can be formally 'derived' from the palaeognathous, and it has been almost universally assumed that because the neognathous birds pass through a palaeognathous-like stage in their ontogeny, the palaeognathous type must have been phylogenetically primitive. Such a dogmatic application of the theory of recapitulation is now discredited (see p. 449) since it is known that embryonic features may secondarily be retained in the adult and represent no phylogenetically primitive condition at all.

In this case, without prejudice to such primitive features as palaeognathous birds possess (e.g. large parasphenoid rostrum, basipterygoid processes), it is quite possible that the persistent contact between the pterygoid and the prevomers in these forms is due to the secondary retention of an immature condition (as in Picidae), or in other words, that the palaeognathous type is not phylogenetically primitive at all within the class Aves. In support of this view are the facts that the palaeognathous birds show retarded development and the retention of immature characters of other birds in the persistence of sutures between bones and in the high degree of separation of the prevomers. Further, the presence of well-developed basipterygoid processes suggests a kinetic skull (see p. 426) in which the pterygoid could slide; in this case the neognathous hinge between palatine and pterygoid would be more primitive phylogenetically than the broad squamous suture between these bones in the palaeognathous type.

Further, if the dromoeognathous type were really phylogenetically primitive, its occurrence in adult birds would be expected to be more widespread than it is.

Should these considerations be well founded, the study of the development of the palate in birds would support the evidence from other sources (structure of the wing, feathers) to the effect that Ratites and Tinamus have degenerated from a flying type of birds.

IV. GENERAL SECTION

35. EMBRYOLOGY AND EVOLUTION OF THE SKULL

THE vertebrate skull provides particularly favourable material on which to study the problem of the relations of ontogeny to phylogeny, for not only is the development of the chondrocranium now fairly well known, but the main lines of evolution within the vertebrates have been established by morphological research, and also, from the nature of the tissues involved, the assistance of palaeontology may be invoked.

In this chapter, therefore, it is proposed to compare the embryology with what is known of the phylogeny of the skull, and in particular, the chondrocranium. But first it is necessary to come to a decision regarding an important point, namely, what (if not all) stages of development of the chondrocranium are suited for such a comparison.

i. The effects of heterochrony

On comparing the developments of different chondrocrania, it is soon apparent that while the very initial stages may present certain fundamental resemblances (e.g. parachordals, trabeculae, auditory capsules, &c., see p. 453), subsequent intermediate stages in the construction of the chondrocrania commonly show greater differences than are presented by the fully formed chondrocrania. These differences are largely due to different relative times of chondrification, and different sequences of development.

So, to take but a few examples, the brain-case begins to chondrify before the visceral arch skeleton in Selachii, Salmo, Lacerta, Anas, and Lepus; the reverse is the case in Gadus, Amblystoma, and Rana. Whichever be the phylogenetic priority of these constituents of the skull, it is obvious that comparisons between chondrocrania at these stages might be misleading in the extreme.

Other examples may illustrate the point further. Within the form Scyllium canicula differences in method of chondrification may be found; embryos developed at Naples show a tendency to delayed chondrification in the antero-dorsal region of the skull, as compared with embryos developed at Plymouth (de Beer, 1931*a*). In Squalus acanthias van Wijhe (1922) found that cartilage first appears in embryos 22 mm. long from Heligoland, but 32 mm. long from Boston (Mass.).

Very varying relations may be found between the notochord and the parachordals in mammals (see p. 381), but the distribution of these variations throughout the Class shows that they have no phylogenetic significance.

In Anguilla the chondrocranium when well developed has a nasal septum, but the laminae orbitonasales only appear very late; in Gadus, on the other hand, the laminae orbitonasales are formed early, and the nasal septum very late. In Rana the chondrocranium develops in craniocaudal succession; in Alytes the successional order of chondrification of parts is caudocranial. In Lepus the alicochlear commissure chondrifies in a postero-anterior direction; in Man the direction of development is the reverse. In Lepus the occipital arches chondrify before the basal plate; in Talpa the order of chondrification is the reverse.

The orbitoparietal commissure in mammals may arise in a variety of different ways. In Talpa the parietal plate fuses first with the auditory capsule, next with the supraoccipital cartilage, and lastly with the orbital cartilage; in Felis the parietal plate fuses first with the supraoccipital cartilage, next with the orbital cartilage, and lastly with the auditory capsule; in Lepus the parietal plate fuses with the supraoccipital cartilage, but with the orbital cartilage only after the latter has established connexion with the auditory capsule; in Sus the parietal plate arises in continuity with the orbital cartilage and then fuses with the supraoccipital cartilage, but not with the auditory capsule for a considerable time.

As regards the developing bony skull, a consideration of the difference between the sequence of ossification of the various bones in Triton (p. 184) and Rana (p. 209) will show how difficult it is to compare early stages.

If, then, comparisons between incompletely developed chondrocrania are useless, *it is the fully formed chondrocranium which must* (and does, admirably) *serve as material for comparison.* That is the reason why throughout the systematic part of this work a distinction has been drawn between the *development* and the *morphology* of the chondrocranium.

The unreliability of the developmental stages of the chondrocranium to serve as materials for comparison is due to the effects of heterochrony in the development of the various parts, itself the manifestation of different rates of histogenetic activity in different regions. It may be observed that these cases form exceptions to von Baer's principle of decreasing resemblance between different forms at increasing stages of morphogenesis, for fully formed chondrocrania often resemble one another more than when incompletely developed.

In some cases, Evolution seems to have made direct use of the effects of heterochrony. Thus, in non-mammals the trabeculae cranii appear before the intertrabecular cartilage; but in mammals the central stem (taenia intertrabecularis) develops long before the alicochlear commissures (trabeculae cranii). Monotremes are intermediate in this respect.

The phenomenon of greater difference between intermediate than later stages of morphogenesis may be referred to as embryonic variation or caenogenesis (see de Beer, 1930*b*), and is of importance; for embryonic variation may be the starting-point of variations also affecting the fully formed structure. Meristic variations illustrate this point well. The difference between Selachii with seven, six, or five branchial arches is due not to excalation but to transposition (Goodrich, 1913). Versluys (1922) has, moreover, given reasons to suppose that the change is due to the persistence in the adult of an embryonic variation which resulted in the laying down *ab initio* of either x, or $x-1$ or, $x+1$ branchial arches.

ii. The skull and phylogeny

In the previous section it has been shown that the very first stages and the fully formed chondrocranium provide the most satisfactory material for comparison, and it now remains to perform such comparison.

In all bony vertebrates the chondrocranium is an embryonic structure, while it persists as the sole adult skull only in Elasmobranchs and Cyclostomes. If, then, there is any truth in the theory of recapitulation, it should be possible to find some higher vertebrate whose chondrocranium (necessarily embryonic) resembles the adult Selachian skull more than it resembles the embryonic Selachian skull. It may be suggested that Salmo, Acipenser, Ceratodus, or even Chelone, provide examples, but it is soon apparent that such resemblances as these forms show are of no significance; first because there is evidence (see p. 394) that in Acipenser, Ceratodus, and Chelone the completeness of the chondrocranium is secondary, and secondly because the similarity between Salmo and a Selachian in having a complete roof to the chondrocranium is far outweighed by differences in other directions. Even allowing for the fact that the Selachii are not the direct ancestors of higher vertebrates, it is quite clear that no embryo of a higher vertebrate possesses a chondrocranium with a complete roof, enclosing the brain. Nor do the fully formed chondrocrania of lower bony vertebrates present any particular resemblance to the incompletely formed chondrocrania of higher vertebrates. On the contrary, the fully formed chondrocrania of higher vertebrates present more resemblance to the incompletely developed chondrocrania of younger stages of development of lower vertebrates.

If the embryonic chondrocranium of higher forms were merely the inherited relic of some ancestral adult stage pressed back into earlier and earlier stages of development, it should be less well developed in higher than in lower vertebrates. But this is not the case. It is true that the human chondrocranium presents a somewhat pathetic spectacle, but that is only because of the *relatively* huge size of the brain. On the contrary, in mammals the chondrocranium is in many respects (especially as regards the nasal capsule and the parietal plate) better developed than in the lowest forms. In other words, the chondrocranium is not a relic, but a *functional structure* progressing in Evolution along its own lines. It is present in the embryonic stages of all Craniates as a repeated inheritance, modified in all sorts of ways, following the changes in the structures which it is its function to support and protect. In Marsupials, which are born in a very immature condition, the chondrocranium is precociously developed (Töplitz, 1920); in Man, with his prolonged gestation period and nursed childhood, the membrane-bones have time to develop and take the place of the dorsal portions of the chondrocranium. At the same time the bones must not be too precociously developed, or the skull would be too rigid at birth. There is no evidence that the chondrocranium started in phylogeny as an adult structure: it may have originated as an embryonic structure which persists throughout life in some forms and is lost in the adult in others.

The only possible conclusion is that the chondrocrania have evolved on their own, without any 'recapitulation', or hypothetical pressing back of adult stages of ancestral forms into embryonic stages of development. And this divergent evolution affects the very earliest stages of development. Thus, the parachordals are no longer paired in the duck; paired trabeculae are no longer found in placental mammals.

The fact that the chondrocranium of higher vertebrates cannot be regarded as 'representing' the adult Selachian skull was known to Reichert nearly a century ago (1838), although he was concerned, not with phylogenetic evolution, but with the *scale of beings*. He recognized clearly that in each group the skull has developed on lines of its own: 'The Holocephala, Plagiostomata, and Cyclostomata appear to us to be lower developmental stages individually differentiated, so that the other fully differentiated vertebrates cannot easily be referred directly to their type' (translated by E. S. Russell).

That the form of the chondrocranium may be devoid of phylogenetic significance is in many cases obvious, e.g. in Man, where it barely covers the floor of the brain, or in the premetamorphic stages of Anura, where it is adapted to the larval methods of feeding.

There are, moreover, a number of cases in which embryonic features in the form of the chondrocranium and its developmental processes can be proved to represent secondary modifications and no ancestral adult features at all.

(*a*) First, there may be considered those cases in which a cartilaginous structure 'grows out of' another structure, or arises in continuity with it, although in phylogeny such structures were completely independent. In a few mammals (Ovis, Sorex) the ala temporalis develops as a continuous outgrowth of the processus alaris. Yet the ala temporalis is without doubt the representative of the processus ascendens of the pterygoquadrate cartilage, and therefore part of the visceral arch skeleton, which in phylogeny was completely distinct in origin from the processus alaris or basitrabecular process (part of the neurocranium). Similarly, the ethmoid process of the pterygoquadrate cartilage in Lacerta appears to be represented by the processus maxillaris posterior, which chondrifies in continuity with the nasal capsule.

The fact is, as Spemann (1898), Gaupp (1900), and Lubosch (1909), have pointed out, that developmental processes, formation of procartilage, independent or continuous chondrification, ligamentous connexions, &c., are not infallible guides to the morphology, and therefore to the phylogenetic history of structures.

(*b*) Polypterus has no (posterior) myodome, and the subpituitary space (which a myodome would occupy) is filled with a pituitary body that retains its open hypophysial connexion with the roof of the mouth, the parasphenoid being perforated to allow the passage of the hypophysial stalk. This open hypophysial connexion is a remarkable *embryonic* feature, but since it is now certain (Watson, 1925*b*) that the Palaeoniscoids, from which Polypterus was derived, possessed a myodome, it is clear that the hypophysis in those forms cannot have had a persistent connexion with the roof of the mouth, and therefore that the persistence of this embryonic feature in Polypterus far from being primitive, is *secondary*, and of no phylogenetic significance.

(*c*) Sphenodon is remarkable in that the entire hyoid arch skeleton, from the columella auris to the basihyal, is cartilaginous and continuous, throughout life. No other reptiles or even Amphibia retain a complete cartilaginous hyoid arch skeleton in the adult, whatever their embryonic condition may be. It is therefore clear that the continuity of the skeletal elements of the hyoid arch in Sphenodon (that they should be cartilaginous or mesenchymatous is of no moment) is an embryonic feature. But there again, the close affinity between Sphenodon and Lacertilia shows that this embryonic feature is not primitive but a secondary retention. Indeed, it is to a certain extent a degeneration, for the tympanum in this form is degenerate, and it is clear that the continuity of the columella auris with the distal elements of the hyoid arch would be incompatible with the function of the columella as a vibration-transmitter.

(*d*) In many Teleostomes (Polypterus, Clupea, Anguilla) there is a temporary cartilaginous fusion between the pterygoquadrate and the hyosymplectic. A comparable fusion occurs between the incus and stapes (and malleus) in Ornithorhynchus.

The relations of the spiracular visceral cleft, the blood-vessels and nerves, prove beyond doubt that the pterygoquadrate and the hyosymplectic belong to separate (mandibular and hyoid) visceral arches. In the primitive Gnathostomes (with autostylic jaw-suspension, see p. 422) these arches never came into contact at all, as is proved by the fact that the spiracular slit extended the full height of the arch (e.g. in Acanthodes). The embryonic contacts and fusions in modern Teleostomes, therefore, cannot have phylogenetic significance. It is for this reason (together with the fact that the visceral arch skeleton must have preceded the neurocranium in phylogeny) that no importance can be assigned to cases (see p. 423) in which there may be a transient fusion between the processes of the pterygoquadrate and the brain-case, or between the columella auris and the auditory capsule.

(*e*) In Acipenser and in Ceratodus the chondrocranium is preserved in the adult almost *in toto*; there are only four (paired) cartilage-bones in the neurocranium of the former, and one in the latter. Yet this persistence of an embryonic condition and resemblance to Selachii is clearly secondary, for the fossil record has established that both these forms belong to groups whose members were fully ossified. Furthermore, there is evidence that in Acipenser the development of the cartilage has been secondarily intensified, for it may envelop membrane-bones (p. 94).

(*f*) In Apteryx there is no median cartilaginous interorbital septum, and the preoptic roots of the orbital cartilages arise directly from the trabecula communis, as in the early stages of development of the duck, for instance. If this were a primitive condition, Apteryx would have to be regarded as more primitive not only than any bird, but even than any living reptile. It is clear that the chondrocranium of Apteryx shows a secondary retention of a feature which is only transient in the development of other birds, and associated with a reduction in the size of the eye consequent upon loss of the power of flight.

There is thus considerable evidence that embryonic conditions in the skull may be secondarily retained in the adult. This may occur in Man as an abnormality (e.g. the persistence in cartilage unossified of the styloid process; Corsy, 1920), but in many respects it characterizes the normal morphology of his skull. In other words, the *adult* descendant resembles the *embryonic* stages of its ancestors, which is the reverse of 'recapitulation'. It is, further, probable that this process of foetalization or paedomorphosis (see de Beer, 1930*b*) played an important part in the evolution of the human skull, for the adult stages of the non-human Primates appear to have been too specialized. For instance, Leche (1912) showed that whereas the embryonic skull of Mycetes resembles that of higher Primates, the adult resembles the skull of lower mammals. Bolk (1926) has drawn attention to the fact that whereas the embryonic cranial flexure becomes flattened out in most mammals, in Man it is retained in the adult and this is the reason why the foramen magnum in Man is situated far forward on the ventral surface. Further, many of the cranial features of *adult* Man show resemblances to *embryonic* monkeys and apes, e.g. relative brain-size, retarded closure of sutures between bones, flatness of the face (orthognathy) and absence of brow-ridges. On the same lines Buxton and de Beer (1932) have pointed out that the similarities between the skulls of adult Homo sapiens and juvenile Homo neanderthalensis, as regards chin, dentition, frontal ridges, position of the foramen magnum and bone-sutures, are such as to suggest that modern Man may have been descended from the Neanderthal race by retention of embryonic characters and consequent discarding of the specialized adult cranial features which characterize Neanderthal.

It has also been shown (p. 445) that the relation of the palaeognathous to the neognathous type of palate in birds is probably to be explained on the view that the former is not primitive but secondarily derived from the latter by paedomorphosis.

Conclusions of great interest can be reached as a result of a comparison of the different times which related forms require to reach morphologically corresponding stages of development. This has been done for the skulls of Gallus domesticus and Tinnunculus alaudarius by Suschkin (1899) by correlating the ages and morphogenetic values of W. K. Parker's (1870) stages of the former with his own stages of the latter.

Tinnunculus hatches after 28 days' incubation at its stage *7* which corresponds as regards the skull to a stage of Gallus midway between *3* and *4* and reached by it in 12 days. At this period, therefore, Gallus develops relatively much faster than Tinnunculus. Gallus reaches stage *5* two days after hatching (or 11 days after the stage corresponding to that at which Tinnunculus hatches, or 23 days in all) and then corresponds to stage *11/12* of Tinnunculus, reached 11 days after hatching (39 days in all). During these 11 days, therefore, Gallus and Tinnunculus are undergoing morphogenesis at similar rates. Thereafter, however, Gallus's rate becomes relatively much slower, and Gallus stage *9* reached 8 months after hatching (261 days in all) corresponds to Tinnunculus stage *16* reached 24 days after hatching (52 days in all). The skull of Tinnunculus is completely developed in 10 months, that of Gallus not before 2 or 3 years.

This can be expressed thus in tabular form:

		Gallus stage		*Tinnunculus stage*		
		From start of incubation				
12 days		12 days	*3*	hatches 28 days	*7*	28 days
11 ,,	hatches and	21 ,,	*4*			11 ,,
	leaves nest	23 ,,	*5*	39 ,,	*11*	
238 ,,		81 ,,	*7*	41 ,,	*12*	13 ,,
		111 ,,	*8*	44 ,,	*13*	
		261 ,,	*9*	52 ,,	*16*	
				leaves nest 60 ,,	*18*	

It is clear that either Gallus has been greatly accelerated in its earlier and retarded in its later stages of development, or that the development of Tinnunculus has been slightly retarded during the earlier and accelerated during the later phases. The former appears the more probable explanation.

The interest of this case lies in the fact that Gallus is nidifugous and has a precocious 'chick' stage capable of independent life, while Tinnunculus is nidicolous with a helpless nestling stage. Further work in the way of comparisons between the rates of morphogenesis of other nidifugous and nidicolous birds would probably help to solve the problem as to which type is the more primitive within the Class Aves. The indications are that the accelerated development of the 'chick' phase is a secondary interpolation in the life-history of birds, especially such as have largely abandoned the arborial for a terrestrial mode of life. The question is of importance in view of the possibility that the Ratites have retained in the adult several features of the chick stage.

There remains to be considered the undoubted fact that the *initial* stages of development in many different forms present a similarity to the extent of possessing corresponding centres of chondrification—trabeculae, parachordals, sometimes polar cartilages, auditory and nasal capsules, and the several elements of the visceral arch skeleton.

First and foremost, it must be understood that these resemblances are manifestations of genetic affinity, which do not of themselves convey any information as to whether the inherited structure in question was originally embryonic or adult in the ancestor. The cartilages of the visceral arches must have been adult structures, but equally certainly, they must in each ontogeny have been preceded by developmental stages, and it is the latter which have been inherited by and are manifested in embryos of modern vertebrates. It is on these lines that Eales's (1931) discovery of the existence in foetal elephants of a prealveolar region of the dentary may be explained. It remains for palaeontology to say whether there ever were vertebrates in which the brain-case was composed in the adult of nothing but trabeculae and parachordals. The nearest approach so far known is provided by Palaeospondylus (Bulman, 1931) and the Cyclostomes, but here again the adult structures have embryonic precursor stages, and there would still be no justification in claiming that the trabeculae and parachordals of a modern vertebrate embryo represent the *adult* skull of the ancestral forms. That both should be cartilaginous is still no evidence of 'recapitulation' since there are (see p. 454) considerations of developmental mechanics which show that embryonic skulls cannot very well be composed of any other material.

In conclusion, therefore, it may be said that the study of the development of the skull lends no support to the view that embryonic stages of descendants recapitulate ancestral adult stages, but that on the contrary, embryonic structures tend for developmental-mechanical reasons to be repeated in successive ontogeneses at roughly corresponding stages of development, and may or may not persist in the adult.

iii. The phylogenetic origin of the chondrocranium

There remain to be considered numerous problems relating to the early evolutionary history of the chondrocranium. First among these is the question as to whether the chondrocranium is phylogenetically older than the osteocranium, or vice versa. On the basis of the old recapitulation theory, such a question would not even arise, for the chondrocranium, preceding the osteocranium in ontogeny, must have done so in phylogeny! But the matter is not so simple. First of all it seems clear that the visceral arch skeleton preceded the neurocranium, for gill-slits but no skulls are found in Hemi-, Uro-, and Cephalochordata. It may be that the original skeletal material was bony, in the form of scales, and that thereafter the neurocranium was formed in cartilage as an embryonic adaptation for protecting the brain during its enlargement, or, as van Wijhe (1906) supposes, for providing a rigid support for the organs of balance. Lastly, (phylogenetically speaking) parts of the neurocranium itself might become replaced by bone. On the other hand, in some groups, the chondrocranium might be retained and bone lost.

As for the visceral arch skeleton, it is difficult to resist the conclusion that its original material was cartilaginous, which would have enabled it to function and grow simultaneously. (See Haines, 1934). It, too, could then become bony in the adult.

It must straightway be admitted that a final decision on this very important question lies outside the scope of embryology, but in that of Palaeontology. It is perhaps significant that Ostracoderms and Acanthodians possessed bone while living Cyclostomes and Selachians do not, for this would appear to indicate that these latter groups have lost their bone. This is possible, but against the view that bone was evolved before cartilage it must be remembered that cartilage is present in Amphioxus (in the buccal cirrhi) but, of course, no bone. It is therefore possible that the brain was originally protected by bone (in the form of scales) while the visceral arches were supported by cartilage.

While waiting for the palaeontologist to establish the phylogeny and elucidate the structure of early Craniates, the embryologist must content himself with a consideration of the relative fitness of cartilage and bone for developmental and adult stages of the skull. As regards the neurocranium the requirements are a hollow box which can expand. The degree of expansion has been calculated in respect of the dogfish and of Man (p. 470), at opposite extremes of the scale, and in square measure of surface to be covered is found to be about $\times 72$ in the former case and $\times 500$ in the latter. The essential difference between the two materials, cartilage and bone, from the present point of view is the fact that the former is capable of interstitial growth, and the latter not. It follows that cartilage can make a completely continuous box practically from the start, and expand it by intussusceptive growth. Whether this box is able to enclose the brain completely depends on the degree of development of the brain. On the other hand, if bone were the material used in the embryo, and the bones were all in contact with one another to form a complete box (i.e., the sutures obliterated), expansion would be impossible (except by means of internal erosion and external apposition, which could, however, not take place on such a scale) since growth of bones can apparently only take place at their surfaces and edges. Thus, while cartilage is admirably suited for embryonic stages, bone is not, and if the embryo is to have a neurocranium it must be of cartilage, in which cartilage-bones may afterwards arise (but not too early, see p. 491), separated by 'epiphyses' (see p. 39). The chondrocranium and osteocranium, therefore, may have evolved as it were side by side.

An example of the independent evolution of chondro- and osteocranium is provided by the occipital condyles (p. 386). The chondrocranium of crocodiles, Chelonia, and birds presents a median pleurocentral condyle, which may become ossified either by the basioccipital alone, or by that bone and the exoccipitals. In mammals, however, the exoccipitals (and the basioccipital) ossify in basidorsal condyles, while in Lacertilia and Ophidia the basioccipital ossifies in a hypocentral condyle.

Further, homologous cartilages are not always ossified by the same bones. Thus, the exoccipitals may join above the foramen magnum (i.e. in the tectum posterius normally ossified by the supraoccipital) as in Amia, Tropidonotus, and some Chelonia; or below the foramen magnum (i.e. in the basal plate normally ossified by the basioccipital) as in Salmo, some Ophidia, and Crocodilia; or both above and below as in Amphibia and Chamaeleo.

The ordinal sequence of chondrification of cartilages by no means necessarily runs parallel with the order of ossification of these structures. Thus, the cochlear part of the auditory capsule chondrifies later but ossifies earlier than the canalicular part; in Man the supraoccipital is the earliest cartilage-bone, but the tectum posterius in which it ossifies is a tardy element of the chondrocranium.

The independence of the chondrocranium and bony skull does not, however, prevent the possibility of co-operation between them, in the function of enclosing the brain-case and sense-capsules. Thus, it is remarkable that in Lacerta the nasal bone fits exactly over the fenestra dorsalis in the roof of the cartilaginous nasal capsule; in the same way in Rana the parietal exactly covers over the posterior vacuity in the roof, behind the tectum transversum and laterally to the taenia tecti medialis. In Chelonia the development of the cartilaginous side wall to the orbitotemporal region (pila antotica) in the adult is inversely proportional to that of the parietal and especially to its descending portion. In mammals the development of the cartilaginous floor to the nasal capsule is inversely proportional to that of the palatine processes of the premaxillae and maxillae, and the interruption of the paraseptal cartilage appears to be correlated with the development and forward position of the vomer. Further, in a general way, it may be noticed throughout the vertebrates that the dermal bones which form the roof of the brain-case start ossifying as strips immediately or closely overlying some feature of the chondrocranium such as the nasal capsule, the edge of the orbital cartilage, planum supraseptale, taenia marginalis, parietal plate, or tectum synoticum. On Kokott's (1933) view the chondrocranium plays an important part in the formation of the bony skull by providing certain fixed points which, by resisting the expansion of the dura mater, set up local fields of stress and strain to which the spicules of the bones which develop there become orientated (see p. 485).

The conditions in the visceral arch skeleton resemble those of the appendicular skeleton rather than the neurocranium, since they involve rod-like structures stiffening mobile regions instead of the enclosure of a box. Here again, cartilage would provide the best embryonic material, whether or not it is replaced in the adult by bone.

In general, the chondrocranium tends to be absorbed and to disappear in the adult in vertebrates possessing bone. Here and there, however, throughout the vertebrate scale, cases are found in which the chondrocranium persists in the adult (Acipenser, Ceratodus, Chelone) for reasons which are unknown. It may, however, be noted that these forms (and Ornithorhynchus, in which the chondrocranium is better developed than in any other mammal) are all either exclusively or largely aquatic in mode of life. (See p. 394).

The conclusions as to the actual course of phylogeny, and the affinities of the several groups of vertebrates, in the light of the study of the development of the skull, follow in the next chapters.

36. THE PHYLOGENY OF THE CHONDROCRANIUM, AND CHARACTERISTICS OF THE VARIOUS CLASSES

i. Pisces

The development of the chondrocranium in the Selachii, both of the Squaloid and Batoid types, shows not only that this group is a very compact one but that it differs quite considerably from the Teleostomes, and notably in the following points:

(i) The trabeculae make a large angle with the parachordals.
(ii) The nasal capsules have no floor.
(iii) The hyomandibula articulates with the auditory capsule ventromedially to the head vein.
(iv) The auditory capsules have medial walls.
(v) There is no interhyal or symplectic.
(vi) The branchial arches are of the sigma-type.
(vii) The efferent pseudobranchial artery passes dorsally to the trabecula (see p. 66).

On the other hand, the evidence of the chondrocranium goes to show that the Selachii and Holocephali are closely related, in spite of the specializations of the latter.

A striking similarity between these two groups of fish is provided by the histological picture presented by sections. The staining reactions of the tissues, the size of the cells, the nature of the cartilage, and several other features impress the observer with the closeness of the affinity between these two groups. Other points of similarity are to be found in the arrangement of the semicircular canals of the auditory capsule, the persistence of the foramina for the endolymphatic ducts, the sigma-shape of the branchial arches, the fact that the trabecular plate arises more or less at right angles to the parachordal plate and is attached to the under surface of the latter, the invasion of the notochordal sheath by cartilage cells, and the passage of the efferent pseudobranchial artery dorsally to the trabecula.

The most remarkable feature of the Holocephalian is the apparent presence of a pharyngohyal element in the hyoid arch, which, if confirmed, would mean that the ancestors of these fish never were hyostylic or amphistylic (see p. 410). Another primitive feature is the complete cartilaginous floor to the auditory capsule: in Holocephali the glossopharyngeal nerve does not traverse the cavity of the auditory capsule and there is no lamina hypotica such as exists in Selachians (de Beer, 1931*a*). The otic and basal connexions of the pterygoquadrate must have been inherited from the common ancestor with the Selachians. The Holocephali have departed from what must have been the primitive condition by converting these connexions into permanent cartilaginous fusions, thus becoming what Gregory (1904) has called holostylic. Further specializations are the formation of the interorbital septum and ethmoidal canal, and the break-down of the medial wall of the auditory capsule. None of these specializations is to be found in Selachians. Nevertheless, in spite of the shiftings and distortions which the Holocephalian structures must have undergone, they conform to the general plan of morphological relations as found in other vertebrates.

It seems that the suspensorial function of the hyoid arch has been independently acquired in Selachii and in Teleostomi, which accounts for the differences in the morphological relations of the hyomandibula in the two groups (see p. 410). If it should be confirmed that the hyoid arch skeleton of Holocephali is primitive, which would mean that the ancestor of Selachii and Holocephali had a non-suspensorial hyoid arch, the independent acquisition of the suspensorial function would be proved.

The chondrocrania of the Teleostomi have sufficient points in common to show that they, too, form a natural group. These include the following:

(i) The trabeculae and parachordals are in line.
(ii) The nasal capsules have cartilaginous floors.
(iii) The hyomandibula (or hyosymplectic) articulates with the auditory capsule laterally to the head vein.
(iv) The auditory capsules have no medial walls.
(v) An interhyal and symplectic are present.
(vi) The branchial arches are V-shaped.
(vii) The efferent pseudobranchial artery passes ventrally to the trabecula (see p. 122).

In addition to these points there is, of course, the fact that the skulls of Teleostomi undergo ossification. An attempt has been made by Sewertzoff (1928) to show that the Acipenseroidea were derived from Selachii independently of the remaining Teleostomi, largely on the grounds of the following points, in which Sewertzoff believes that Acipenser resembles the Selachii more than the Teleostomes:

(i) The trabeculae fuse with the underside of the front edge of the parachordals.
(ii) The parachordals touch the notochord all the way along.
(iii) The hypophysial fenestra becomes obliterated from behind forwards.
(iv) The pterygoquadrate has a median symphysis.

None of these points, however, is of great morphological importance, and none invalidates the list of characters given above as common to all Teleostomi.

Moreover, there are a number of features in which the chondrocranium of Acipenser resembles that of Polypterus in some detail, and this is of particular interest since the fossil record indicates that both are derived from some Palaeoniscoid ancestor. These features are:

(i) The enclosure in the cartilaginous rostrum of the transverse ethmoidal lateral line canal.
(ii) The similarity in form of the nasal capsules.
(iii) The atypical relation of the internal carotid artery to the trabecula (at early stages in Acipenser, persistently in Polypterus).
(iv) The similarity in relations of the jugular canal.
(v) The forking of the dorsal ends of the anterior branchial arches into infra- and suprapharyngobranchials (feature shared with Lepidosteus).

Acipenser also shares with Amia and Lepidosteus the detailed relations of the dorsal spiracular recess (or Wright's organ) to the base of the postorbital process. This perforation at the base of the postorbital process was also present in Palaeoniscoids (Watson, 1925*b*).

The chondrocranium of Acipenser may be said, therefore, to confirm the opinion that Acipenser is a proper member of the group Teleostomi, and that such resemblances as it has with Selachii, are superficial and secondary.

The existence of a posterior myodome, or of spaces homologous with it (see p. 429), is further evidence of general affinity between Teleostomi, and of close affinity between Holostei.

The Dipnoi present a problem of great interest. On the one hand, they differ considerably from all other living fish; on the other, they present remarkable similarities with Amphibia, especially Urodela. It is therefore a matter of importance to analyse the features of the chondrocranium of Dipnoi.

In the following respects the chondrocranium of Dipnoi resembles that of other Pisces and differs from that of Amphibia:

(i) The nasal capsules (as in Selachii) have no floor.
(ii) The glossopharyngeal nerve (as in Teleostomi) passes through the auditory capsule.
(iii) The auditory capsule (as in Teleostomi and Holocephali) has no medial wall.
(iv) The abducens nerve leaves the skull posteriorly to the pila antotica.
(v) The notochord is invaded (as in Selachii) by cells from the posterior parachordal cartilages.
(vi) The chondrocranial roof is complete (in Ceratodus).

The features in which the Dipnoi resemble the Urodela are enumerated on p. 459 and show that while there is a general affinity between them, there is no justification for the view that the Urodela were derived independently of the Anura from the Dipnoi. Furthermore, in the fact that the trabecular horns in Dipnoi lie anterodorsally to the external nostrils, they differ radically from Urodela, but this condition could easily be derived from that found in Selachii.

ii. Amphibia

It is important to consider the evidence of the development of the skull on the question of the relative affinities of the Amphibian groups, for recently,

Säve-Söderbergh (1934) has attempted to show that the Urodela have been separately derived from the Dipnoi, and that consequently they are less closely related to the Anura than has hitherto been believed. It may be said straight away that the study of the development of the skull in Dipnoi and in the Amphibian groups gives no support to such a view, and stresses on the contrary the correctness of the classical opinion that Urodela, Gymnophiona, and Anura form a natural group, and that what similarities they show with Dipnoi are due to convergence.

In order to bring out these facts clearly, it will be simplest first to consider the Urodela and Anura without the Gymnophiona. The points of similarity between them and also shared by the Dipnoi may then be enumerated, followed by the points of similarity not possessed by the Dipnoi. Here reference should be made to the points shared between Dipnoi and other fish (p. 458) and lacking in Amphibia. To this may then be added an enumeration and explanation of the differences between Urodela and Anura. Lastly, it can be shown that Urodela and Gymnophiona are closely related, and that the latter possess features which also occur in the Amniotes and which cannot be regarded as having been independently developed, which would have to have been the case if Urodela (and therefore also Gymnophiona) had been independently derived from the Dipnoi.

1. Characters common to Urodela, Anura, and Dipnoi:

(i) histology of the cartilage, in particular, large size of cells;
(ii) platybasic (platytrabic) skull, absence of basicranial fenestra (applicable only to Anura and Dipnoi);
(iii) autosystylic suspension of the jaws involving ascending, basal, and otic processes;
(iv) the associated chondrification of anterior parachordals and trabeculae as 'Balkenplatten', flanking the tip of the notochord.

The large size of the cells affects all the tissues, and is the result of a smaller number of cleavage divisions of the egg. On general grounds there is reason to believe that this feature is a sign of degeneration, which would therefore be no evidence of special affinity between the living representatives of the groups in question.

The platybasy of the skull is correlated with small eye-size, large eyes involving the formation of an interorbital septum, and a tropibasic skull. Further, it appears that this is a very variable phenomenon. In the single group Teleostei, for instance, the Cypriniformes are platybasic, the Clupeiformes tropibasic. Platybasy is thus no evidence of special affinity between Urodela, Anura, and Dipnoi, especially as there is reason to believe that primitive Amphibia possessed a (moderate) interorbital septum, as Watson (1925*a*) has shown. In the matter of the basicranial fenestra, it will be noted that the Urodela actually differ from the Dipnoi.

The autosystylic suspension (i.e. processes of quadrate fused with neurocranium) is probably an adaptation to prolonged larval life with a cartilaginous skull, and, as such, a specialized feature, and independently developed. Comparable fusion of processes occurs in Holocephali, and temporarily in many fish as Edgeworth (1925) has shown. The fact that ascending, basal, and otic processes are involved is evidence of *general* affinity between the groups in question, and of the existence of a common ancestor possessing these structures (e.g. Osteolepidoti). This is presumably the explanation of the mode of chondrification of parachordals and trabeculae. There is, however, no evidence of special affinity between any Amphibian group and the Dipnoi. This will be all the more obvious after consideration of the following table.

2. Characters common to Urodela and Anura, but *not* to Dipnoi:
 (i) The shortness of the occipital region of the skull.
 (ii) The chondrification of the medial wall of the auditory capsule.
 (iii) The formation of fenestra ovalis and foramen perilymphaticum.
 (iv) The existence of a columella and an operculum in the fenestra ovalis.
 (v) The formation of the occipito-atlantal joint as a result of splitting of an intervertebral cartilage, the anterior portion forming the paired condyles.
 (vi) The position of the trabecular horns, *ventrally* to the external nostrils.
 (vii) The larval hyobranchial skeleton, with fusion of the dorsal ends of the ceratobranchials.
 (viii) The number and position of the bones, especially the paired bones.

This list will already be sufficient to show that in respect of the skull, Urodela and Anura have more characters in common with one another than with other groups. Similarly the list on p. 458 shows that Dipnoi have more characters in common with fish than with Amphibia. There may be added certain curious features which all members of Anura and Urodela do not possess, but which nevertheless occur in some forms in both groups, such as:

 (ix) The prefacial commissure and facial canal (Ascaphus, Alytes).
 (x) The membrane-bones associated with the hyoid (Amphiuma, Alytes).

These similarities must, however, be weighed against the differences.

3. Differences between Urodela and Anura:
 (i) The suprarostral and infrarostral cartilages of Anura.
 (ii) The complete transformation of the quadrate at metamorphosis in Anura.
 (iii) The annular tympanic ring of Anura.
 (iv) The usual absence of a nasal septum and presence of separate medial walls to the nasal capsules in Urodela.
 (v) The form of the hyobranchial skeleton after metamorphosis.
 (vi) The course of the hyomandibular branch of the facial nerve over the columella auris in Anura and under the stylus columellae in Caducibranchiate Urodela.
 (vii) The existence of a basal connexion between the quadrate and the neurocranium in Urodela, and a pseudobasal connexion in Anura.
 (viii) The fusion of the parachordals in the hindmost region dorsally to the notochord in Anura and ventrally to it in Urodela.
 (ix) The absence of a basicranial fenestra in Anura.

These differences reflect the divergent evolution of Urodela and Anura, but within the group Amphibia. The first two are associated with the peculiar feeding habits and buccal mechanism of the Phaneroglossan Anura; (iii) is a unique Anuran development; (iv) is not an absolute difference since Salamandra possesses a nasal septum and Rana shows that the nasal capsules have rudimentary medial walls in the anterior region. The only important morphological differences are (vi) and (vii); but the fact that in Perennibranchiate Urodela the jugular branch of the facial nerve runs dorsally to the stylus would seem to show that since the loss of contact between the lateral end of the stylus and the skin (associated with the absence of a tympanic membrane) in Urodela, the facial nerve has slipped off the end of the stylus, partially in Perennibranchiates, completely in Caducibranchiates. As to the pseudobasal connexion of Anura, the Anuran Ascaphus is similar to Urodela and bridges the gap in this respect.

The differences between the skulls of Urodela and Anura are therefore of minor morphological importance.

The special characters common to Urodela and Gymnophiona as regards the chondrocranium are perhaps not very numerous, but this is largely because there is so little of the chondrocranium in the Gymnophiona. However, the large unchondrified floor of the skull in these is paralleled in other forms only among the Urodela, where, as in Necturus, this region may never chondrify, or where, as in Amphiuma (or even Amblystoma) the hypophysial and basicranial fenestrae become confluent as a result of break-down of cartilage. The hypochordal commissure at the hindmost end of the basal plate is also common to both groups.

The features wherein the Gymnophiona show similarities with the Amniota relate to their autodiastyly, and to the columella auris. This at early stages is said to be in blastematous continuity with the ceratohyal, and in some forms (e.g. Ichthyophis) is perforated by the stapedial artery. The latter feature is paralleled in Geckones and in mammals, while the former is characteristic of all Amniotes. It would be strange to find such features in Gymnophiona if they (together with the Urodela) had had a derivation directly from the Dipnoi, independently of the Anura and Amniota.

Lastly, with regard to affinities within the groups, the presence of a hypoglossal foramen and a complete pterygoid process in Cryptobranchus suggests that it is among the most primitive Urodeles; while Ascaphus's possession of basitrabecular and basal processes and a prefacial commissure points to it as being the most primitive Anuran. Xenopus, on the other hand, appears to be very specialized, and the study of the development of the skull therefore supports the modern tendency to abandon the view that Anura can be classified according to their sternum or tongue.

iii. Reptilia

The chondrocranium of living reptiles presents a number of features of similarity which argue in favour of a fairly close general affinity between their respective groups. These features include:

(i) The structure of the orbitotemporal region, involving the interorbital septum, planum supraseptale, pila antotica, pila metoptica, taenia marginalis, and taenia medialis.

(Except Ophidia.)

(ii) The presence of a processus dorsalis or intercalary of the columella auris which becomes attached either to the crista parotica or to the otic process of the quadrate.
(Except Chelonia.)

(iii) The presence of a cartilaginous meniscus pterygoideus between the basitrabecular process and the pterygoquadrate.
(Except Crocodilia and Ophidia.)

(iv) The presence of a concha nasalis.
(Except Sphenodon and Chelonia.)

(v) The presence of an ectochoanal cartilage.
(Except Crocodilia.)

(vi) The presence of a supratrabecular bar.
(Except Ophidia and Crocodilia.)

(vii) The presence of an autodiastylic pterygoquadrate, with otic, pterygoid, and ascending processes.

Within the Reptilia, the existing Orders fall naturally into two groups: on the one hand, the Lacertilia, Rhynchocephalia and probably Ophidia; on the other, the Chelonia and Crocodilia.

Characters common to Lacertilia, Rhynchocephalia and Ophidia are:

(i) The supraparachordal course of the notochord.
(ii) The fusion of the pleurocentrum of the proatlas vertebra with the tip of the odontoid process.
(iii) The formation of a median hypocentral occipital condyle.
(iv) The fusion of the premaxillae and of their ascending processes in connexion with the carrying of the dentinal egg-tooth.
(Except Rhynchocephalia).

Characters common to Chelonia and Crocodilia are:

(i) The intraparachordal course of the notochord.
(ii) The fusion of the pleurocentrum of the proatlas vertebra with the basal plate, where it forms:
(iii) The median pleurocentral occipital condyle.
(iv) The connexion of the columella auris via the pars interhyalis with Meckel's cartilage.
(v) The absence of a true egg-tooth and its substitution by a dorsal epidermal papilla.

In particular, the study of the development of the skull in Reptilia emphasizes the following points:

The affinities between Lacertilia and Rhynchocephalia are very close indeed; this is evidence of the derivation of Lacertilia from Diapsida.

The skull of Ophidia is extremely modified, but its more conservative portions, the nasal and auditory capsules and the basal plate suggest the Lacertilia as their nearest relatives.

The similarities between the chondrocrania of Crocodilia and Chelonia are greater than those between Lacertilia and Chelonia. Since Crocodilia are undoubted Diapsida and Lacertilia are almost certainly derived from Diapsida, it follows that Chelonia must also have been derived from Diapsida and that their complete temporal roofing of bones is secondary, and not directly derived from Anapsida.

iv. Aves

That the affinities between birds and reptiles in general and crocodiles in particular is close is common knowledge. It is, however, of interest to notice how very strongly the evidence obtained from the study of the development of the skull supports this conclusion. The following is a list of features characteristic of and peculiar to crocodiles and birds:

(i) The presence of a median prenasal process.
(ii) The formation of infrapolar processes.
(iii) The formation of subcapsular processes or metotic cartilages.
(iv) The fragmentation of the parasphenoid into a median rostrum and paired basitemporals.
(v) The pneumaticity of the bones of the skull.

In addition, birds share the following features with Crocodilia and Chelonia:

(i) The intraparachordal course of the notochord.
(ii) The fusion of the pleurocentrum of the proatlas vertebra with the basal plate, where it forms
(iii) The median pleurocentral occipital condyle.

With Crocodilia, Chelonia, and Rhynchocephalia, birds share:

(i) The absence of a dentinal egg-tooth, and the presence of a dorsal epidermal papilla.
(ii) The connexion between the processus dorsalis of the columella auris and the extracolumella by means of the laterohyal, thus enclosing 'Huxley's' foramen (except Chelonia).

With reptiles generally, birds agree in

(i) the persistence of continuity between the columella auris and the stylohyal, via the pars interhyalis (except Lacertilia and Ophidia);
(ii) the formation of the supratrabecular bar (except Ophidia);
(iii) the disposition of the orbital cartilage (planum supraseptale) and pila antotica.

As regards the relative affinities of Palaeognathous and Neognathous birds, it is pointed out elsewhere (p. 445) that the evidence from the development of the palate cannot be taken as proving the primitive nature of the former, since the fact that they remain in a state through which the latter pass may equally be accounted for by a secondary retention of embryonic conditions. The same explanation must be given to the persistence of the sutures between the bones in many Palaeognathous birds, since the fusion of bones (which is universal in Neognathous forms and in some Palaeognathous, e.g. Apteryx) is closely associated with pneumaticity and rigidity, which features in turn are intimately related to the power of flight which the Palaeognathous birds have lost. It may further be noted that there is reason to believe that the precociously developed juvenile 'chick' stage of nidifugous Neognathous forms which adult Palaeognathous birds resemble in many respects, is a secondary intercalation in the life cycle of birds which have largely abandoned the arborial mode of life.

The chondrocranium of Palaeognathous types so far as it is known (in Struthio and Apteryx) is very similar to that of Neognathous forms, and it is not easy to pick out features indicative of a primitive or specialized character. Two points, however, call for special mention. One is the exaggerated precocity in the development of the infrapolar processes in Struthio, which is clearly secondary. The other is the formation of a cupola posterior, presenting a convex surface to the orbit, at the hind end of the nasal capsules. In this respect Struthio and Apteryx appear to resemble reptiles more than do the Neognathous birds in which there is no cupola posterior, for the lamina orbitonasalis presents a concave surface to the orbit for the accommodation of the large eye.

There is, however, evidence that the cupola posterior of the nasal capsule in Struthio and Apteryx is the result of a secondary reduction in the size of the eye correlated presumably with loss of power of flight. The break-down of the preoptic root of the orbital cartilage and the passage of the olfactory nerve through the orbit can only mean that Struthio and Apteryx are descended from forms in which the eyes were so large as to necessitate the destruction of this cartilage as in all Neognathous forms.

The absence of an interorbital septum in Apteryx, if primitive, would make Apteryx not only more primitive than any other birds, but more primitive than any living reptile. This feature must therefore be a secondary retention of an embryonic condition, likewise associated with a reduction in eye-size.

It is to be concluded, therefore, that the so-called Palaeognathous birds are not more but less primitive than the Neognathous forms.

v. Mammalia

The study of the mammalian chondrocranium enables two generalizations to be made. The first, already known to Decker (1883), is that the chondrocrania of all mammals are remarkably uniform and conform to an easily recognizable type. The second point, discovered and stressed by Gaupp (1900, 1908), is that the mammalian type is most similar to and could easily be derived from a reptilian type, not very dissimilar from that presented by the Lacertilia.

The *essential features of the developing mammalian skull* are the following:

(i) Paired basidorsal occipital condyles are present.
(ii) The pleurocentrum of the proatlas vertebra fuses with the odontoid peg and not with the skull.
(iii) The laterohyal and stylohyal fuse with the crista parotica of the auditory capsule, forming the styloid process.
(iv) The anterior portion of the pterygoquadrate (processus ascendens, ala temporalis) is discontinuous from the posterior portion (incus) and fused with the processus alaris (basitrabecular process).
(v) The posterior portions of the pterygoquadrate and Meckel's cartilages form the incus and malleus.
(vi) Ethmoturbinals are developed in the nasal capsule.

vii.. The cartilaginous skeleton of three branchial arches persists: the thyrohyal, and the two thyroid arches.

The lowest group of mammals, the Monotremes, are characterized by the possession of a number of features which are distinctly reptilian, and which no other mammals possess. These *primitive Monotreme features* are:

(i) The presence of distinct paired trabeculae.
(ii) The presence of the pila antotica (taenia clino-orbitalis).
(iii) The foramen olfactorium advehens is undivided, i.e. there is no cribriform plate.
(iv) The foramen perilymphaticum is undivided, i.e. no fenestra rotunda or aquaeductus cochleae.
(v) The (reptilian) pterygoid ('Echidna-pterygoid') and the lateral wings of the parasphenoid remain separate.
(vi) The presence of the septomaxilla (also occurs in Xenarthra).
(vii) The persistence of separate cartilages in the 2nd and 3rd branchial arches.
(viii) The presence of the sclerotic cartilage.
(ix) The suprabasal course of the notochord (also occurs in Mus).

In addition, *Monotremes share with Marsupials the following reptilian features*, absent in Placentals:

(i) The basal plate is broad.
(ii) The cochlear capsules are small.
(iii) The canalicular capsules lie directly above the cochlear, and the medial walls of the auditory capsules thus form the main part of the lateral walls of the cranial cavity in the posterior region.
(iv) The parietal plates are low.
(v) The occipital arches are vertical.
(vi) The tegmen tympani is very small and there is no lateral prefacial commissure.
(vii) The incus lies dorsally to the malleus.
(viii) The stapes is columelliform.
(ix) The orbital cartilage and the sphenethmoid and orbitoparietal commissures form a wide band.
(x) The carotid foramina pierce the trabecular plate: i.e. no alicochlear commissures.
(xi) The internal carotid arteries enter the cranial cavity directly.
(xii) No mesethmoid bone is developed (feature shared by certain Placentals).
(xiii) No ethmoidal cells or sinuses are excavated (except in Phascolarctus).
(xiv) The premaxillae bear a dentinal egg-tooth and possess prenasal processes which fuse to form an 'os carunculae'. (This feature is shown only in vestigial fashion in Marsupials: Caluromys, and Trichosurus.)

It is clear, therefore, that the Monotremes deserve their reputation as the most primitive living mammals, that the existing Monotremes are very closely allied, and that they are fairly closely related with the Marsupials. At the same time, there are a number of *specialized Monotreme features* such as:

(i) The great development of the crista marginalis.

(ii) The fusion of the ectochoanal cartilages ventrally to the nasal septum.
(iii) The loss of the pila metoptica.
(iv) The loss of the hypoglossal foramina.
(v) The loss of the interparietal bone.

The *characteristic features of the developing Marsupial skull* apart from those shared with Monotremes (see above) and with Placentals (see below) are neither numerous, nor clear cut. Such as they are, they may be enumerated as follows:

(i) The presence of two pairs of hypoglossal foramina (feature shared with some Rodents and Sirenia).
(ii) The presence of a foramen rotundum between the ala temporalis and processus ascendens (feature not present in Didelphys, Dasyurus; shared with some Carnivora and Primates).
(iii) The loss of the pila metoptica.
(iv) The diagonal position of the plane of the foramen olfactorium.
(v) The small size of the frontals.
(vi) The large size of the lachrymals.
(vii) The great length of the jugals.
(viii) The poor development of the palatine processes of the maxillae.
(ix) The extension of the alisphenoid to form a bulla surrounding the tympanic cavity.
(x) The inflection of the posterior angle of the dentary.
(xi) The outer bar to Jacobson's capsule (shared by Dasypus, Tupaja, Macroscelides, Chrysochloris, Orycteropus).
(xii) The papillary cartilage (shared with Tupaja, Macroscelides, Miniopterus, and some Rodentia).

The *features shared by Marsupials and Placentals* are the following:

(i) The subdivision of the foramen olfactorium advehens to form the cribriform plate.
(ii) The substitution of the fenestra rotunda and aquaeductus cochleae for the foramen perilymphaticum.
(iii) The presence of a crista semicircularis in the nasal capsule.
(iv) The loss of the pila antotica.
(v) The loss of the septomaxilla (except in Xenarthra).
(vi) The fusion of the reptilian pterygoid (secondary pterygoid cartilage) with the lateral wing of the parasphenoid.
(vii) The fusion of the cartilaginous elements of the 2nd and 3rd branchial arches to form the thyroid cartilages.

It will be seen, therefore, that the Marsupials occupy a true intermediate position between Monotremes and Placentals. Of particular importance is the fact that vestiges of an 'os carunculae' and even of an egg-tooth are found in Marsupials, indicating that they can only recently have abandoned the oviparous method of reproduction. Within themselves, as regards the skull, the Marsupials form a remarkably compact group, and no criterion has been found to distinguish the chondrocranium of Polyprotodontia from Diprotodontia, or of Syndactyla from Diadactyla. Thus, the peculiar course of the chorda tympani, medially to the styloid process, found in Didelphys and Dasyurus (Polyprotodont and Diadactyl) also occurs in Phascolarctus (Diprotodont and Syndactyl), but not in Trichosurus (Diprotodont and Syndactyl) nor in Perameles (Polyprotodont and Syndactyl). The complete paraseptal cartilage is found in Dasyurus, Trichosurus, and Halmaturus, but not in Didelphys nor in Perameles.

The only conclusion which emerges from a consideration of the development of the Marsupial skull is that Perameles differs most from the general type, and in respect of the first nine items of Placental characteristics (see below) it resembles Placentals more than Marsupials. It is interesting, therefore, that Perameles should be the one anomalous Marsupial in respect of the combination of Polyprotodont and Syndactyl features, and of the allantoic placenta.

What is known of Caenolestes suggests that it is allied to the Polyprotodonts.

The features characteristic of Placental mammals are as follows:

(i) The basal plate is elongated and constricted by the cochlear capsules.
(ii) The cochlear capsules are relatively very large.
(iii) The cochlear capsule is in front of the canalicular capsule.
(iv) The medial wall of the auditory capsule forms little of the side wall and more of the floor of the cranial cavity in this region.
(v) The tectum posterius forms a transverse bar in the vertical plane.
(vi) One pair of hypoglossal foramina is present.
(vii) The carotid foramina are far apart, and the hindmost portions of the trabeculae form the alicochlear commissures.
(viii) The internal carotid arteries traverse the cavum epiptericum before entering the cranial cavity.
(ix) The cribriform plate is in the horizontal plane.
(x) The tegmen tympani is well developed and a lateral prefacial commissure is present.
(xi) The pila metoptica is present.
(xii) The sphenethmoid and orbitoparietal commissures are slender.
(xiii) The ala hypochiasmatica is usually present.
(xiv) The occipital arches lean backwards and the lamina alaris is well developed.
(xv) The parietal plates are high.
(xvi) The stapes is 'stapediform'.
(xvii) There is rarely a foramen rotundum in the chondrocranium: i.e. no processus ascendens (except in Carnivora and Primates).
(xviii) The incus lies behind the malleus.

In each of the Orders of Placental mammals, the chondrocranium usually conforms fairly well to a type. Thus, in the *Rodentia*, Lepus closely resembles Microtus, and the type is characterized by some well-marked features, e.g. curved skull base, distinct dorsum sellae, large nasal capsule, complete paraseptal cartilages, mastoid and paracondylar processes. In the *Carnivora*, Felis, Mustela, and Poecilophoca form a fairly compact group in which the auditory capsules have rounded contours, the dorsum sellae is conical, there is no tegmen tympani, the paracondylar process is large, a foramen infra-cribrosum is present, the ala temporalis is free and (as a rule) perforated by the foramen rotundum. Canis, on the other hand, has a number of specializations.

In the *Primates*, Homo is astonishingly similar to Semnopithecus, and as regards the nasal capsule, Catarrhini are closer to Man, especially Negroid, than to Platyrrhini. Platyrrhini and Catarrhini agree in the general form of the cranial box with its acutely bent floor, interorbital septum, ala temporalis perforated by the foramen rotundum, small nasal capsule partly covered over by the large cranial cavity. Tarsius is considerably specialized in connexion with the enormous development of its eyes, but it appears to bear closer resemblances to the Catarrhini than to the Lemuroidea.

In the case of some Orders it is uncertain whether the group is a natural one. Thus, in the Cetacea, it has been contended that the Mystacoceti have little to do with the Odontoceti. As regards the chondrocranium, however, they agree in the flatness of the skull floor, the reduction of the nasal capsule, the formation of a rostrum and a spina mesethmoidalis, the position of the cochlear capsules, medial to the canalicular capsules and ventral to the basal (lamina supracochlearis), and the eventual exclusion of the auditory capsules from the cranial cavity. These resemblances seem to outweigh the differences between the two groups which show that while related to one another within the Order Cetacea, each has evolved on its own. In this connexion, a feature of some interest is the fact that while Palaeontology proclaims the Mystacoceti to be a more recent group than the Odontoceti, the distinctive features appear in ontogeny relatively earlier in the former than in the latter.

The Order *Insectivora*, on the other hand, appears to be a composite group. The Lipotyphla: Erinaceus, Talpa, and Sorex, are very uniform and make up a type from which Tupaja, and what is known of Macroscelides and of Chrysochloris, differ in many respects. The *Menotyphla* and *Chrysochloridea* may therefore represent separate Orders.

The Order *Edentata* also shows evidence of not being a natural group. The chondrocrania of Tubulidentata and of Pholidota have not been studied by modern methods, but what little is known of them concerning Jacobson's cartilage, the passage of the chorda tympani medially to the styloid process in Manis, and the presence of an interparietal in Orycteropus, gives every encouragement to the view that they will be found to differ considerably from the chondrocranium of Xenarthra, thus necessitating a subdivision of the 'Edentata'.

The problem of the relative affinities of the various Orders of the Placental Mammals is one on which the study of the development of the skull can unfortunately throw but little light, and when it comes to attempt a grouping of the various Orders *inter se*, grave difficulties are encountered. Broom has attempted to classify the mammals along two different lines. The structure of Jacobson's organ and of the paraseptal cartilage led him to distinguish a group of *Archaeorhinata* represented by Marsupials, Xenarthra, Tubulidentata Rodentia, Tupaja, Macroscelides, and Chrysochloris, and characterized by a simple type of capsule with an 'outer bar' and without a cartilago ductus nasopalatini; from a group of *Caenorhinata*, represented by Insectivora (s.s.), Artiodactyla, Perissodactyla, Cheiroptera, Carnivora, Hyracoidea, and Primates, and characterized by possession of a cartilago ductus nasopalatini and absence of an 'outer bar'.

Broom has also subdivided Placental mammals according as to whether a mesethmoid bone is absent or present, respectively, into *Palaeotherida* including (in addition to Monotremata and Marsupialia), Xenarthra, Artiodactyla, Perissodactyla, Chrysochloridea, Proboscidea, Sirenia, and Cetacea (?), and *Neotherida*, including Tubulidentata, Pholidota, Hyracoidea, Rodentia, Insectivora, Cheiroptera, Dermoptera, Carnivora, and Primates.

It will be observed that these schemes do not tally; nor so far has it been possible to find any chondrocranial feature which is exclusive to any Order. Thus, the processus opercularis is found in Talpa, Microtus, and Poecilophoca; the foramen prechiasmaticum is found in Lepus and in Homo, the foramen infracribrosum in Caluromys, Poecilophoca, and Semnopithecus. The ala temporalis is perforated by a foramen rotundum in Marsupials, Carnivora, and Primates.

All that can be done in the present state of knowledge is to indicate certain general resemblances and differences between some Orders.

The little that is known concerning Dermoptera and Cheiroptera suggests that these groups resemble one another more than they resemble any others.

The resemblances between Cetacea and Sirenia, on the other hand, appear to be due to convergence and adaptation to a similar mode of life. They include the following features: width and shortness of the cranial cavity; formation of a rostrum; reduction of the nasal capsule; formation of a spina mesethmoidalis, transverse position and ultimate freedom of the auditory capsule; large size of the auditory ossicles.

The Rodents and the Primates have a number of features in common (interorbital septum, alae hypochiasmaticae with prechiasmatic foramina, separate dorsum sellae, mastoid and paracondylar processes) which indicate not only their general affinity, but also their primitive nature among the Placentals.

The Rodents and Primates illustrate in some respects better than any other mammals the fact that the mammalian chondrocranium, for all its characteristic features, is definitely a derivative of that of the reptiles, with which as typified by the Lacertilia, it agrees in its fundamental architecture. It is quite impossible to derive the mammalian chondrocranium from that of the Amphibia as at present known.

In conclusion, therefore, it may be said that the study of the chondrocranium emphasises the primitive nature of the Primates, a fact which is particularly evident in the case of Man.

37. THE GROWTH OF THE SKULL

The development of the skull is accompanied by changes in magnitude which provide interesting information concerning the power of growth of the tissues involved (cartilage and bone), and also concerning the phenomenon of heterogonic growth.

First of all, it is desirable to gain an idea of the order of size-increase which takes place in the development of the skull, and for this purpose two examples will be taken from widely different levels of the vertebrate phylum.

i. The growth of the chondrocranium in Scyllium canicula

The dogfish forms a convenient object for the study of the growth of the chondrocranium. After the 30 mm. stage the brain-case is a complete box, with floor, sides, and roof, of cartilage. The extent to which this neurocranium expands may be gauged by comparing a few simple measurements of the 30 mm. stage and the adult; in particular, the length of the chondrocranium, its width between the medial walls of the auditory capsules, and its height from the basal plate to the tectum synoticum.

	Chondrocranium		
Total length of fish	*Length*	*Width*	*Height*
34 mm.	4 mm.	1·5 mm.	1·6 mm.
500 mm. (adult)	45 mm.	10·0 mm.	10·0 mm.

Thus, the increase in linear dimensions is between six and tenfold, which is of the same order of magnitude as the linear increase of the body as a whole. It must be remembered that this increase in the size of the chondrocranium is the result of interstitial growth of the cartilage, for once the floor, sides, and roof of the brain-case have joined up, there is no addition or intercalation of new cartilages. A better idea of the extent of this growth is obtained by calculating the area represented by the floor, roof, and sides of the brain-case. At the 34 mm. stage, this area is about 25 sq. mm., while in the adult it amounts to 1,800 sq. mm. The increase in area is thus 72-fold.

ii. The growth of the human skull

The human skull provides suitable material for the study of the problem of growth in skulls and of the relative appropriateness of cartilage and bone to perform the function of protecting the brain. The human skull fits the brain closely, and the dimensions of the brain at successive stages of development are given by Streeter (1920) and by Dunn (1921). From Dunn's figures, the surface area of the brain or dura mater may be calculated, on the assumption that the brain is a sphere.

Crown-heel length, cm.	*Mean volume of brain, c.c.*	*Surface s.c.*
0–5	0·3	1·26
5–10	1·2	5·45
10–15	5·4	14·90
15–20	11·9	25·80
20–25	30·8	47·50
25–30	45·8	61·80
30–35	76·5	87·10
35–40	132·1	125·00
40–45	191·2	160·50
45–50	299·1	216·50
Adult	1,500·0	630·00

The increase in the area which the skull has to cover is thus 500-fold, and the chondrocranium does not attempt to keep pace with it, for its roof is restricted to the extreme posterior region, over the medulla oblongata, where the expansion is least.

iii. The allometric (heterogonic) growth of the skull

Since attention has been focused on the phenomenon of allometric (heterogonic) growth by J. S. Huxley (1932), some examples have come to light in the study of the skull.

The elongation of the snout of Lepidosteus is peculiar and differs from that of other long-snouted fish (Belone, Hemirhamphus) in that the nasal pits do not stay close to the orbit, but get carried along to the tip. The olfactory tracts are therefore very long. The growth of the snout has been studied mathematically by Needham (1936). By plotting the length of the snout y (from the tip to the angle of the jaw-hinge) against rest-of-body length x (from the angle of the jaws to the tip of the tail), it is seen that the elongation of the snout is due to a constant differential growth-rate during early stages, the growth-rate coefficient k being 1·9 ($y = bx^{1\cdot 9}$). This elongation principally affects the cartilage of the ethmoid plate (nasal septum) and Meckel's cartilage. But at a total length of about 60 mm. the snout growth-rate drops to about the same level as that of the rest of the body ($k = 0\cdot 95$). The decline in growth-rate of the snout appears to coincide with the onset of ossification, and bone, growing only at the surface, presumably grows more slowly than cartilage which can expand by intussusception.

Two heterogonic growth periods are also found in the elongation of the lower jaw of Belone, the value of k being 3·5 up to a total body-length of 0·6 cm. and 0·7 thereafter. On the other hand, the upper jaw grows with a coefficient of 2·4 up to a body-length of 0·9 cm., and then drops to 0·74. Thus, the lower jaw far outstrips the upper in early stages, and both grow much faster than the body. But subsequently the jaws grow more slowly than the body, and the lower jaw more slowly than the upper which thus becomes able to catch up with it.

In Hemirhamphus the lower jaw grows much as in Belone, the coefficients being 1·2 up to a total body-length of 0·8 cm. and 0·6 thereafter. But here, the upper jaw has no positive heterogonic growth phase, and even in the adult, remains quite short; i.e. it retains permanently a condition through which Belone passes in its development.

In the Chondrostean Polyodon spathula ('spoonbill'), the cartilaginous rostrum undergoes marked elongation, and Thompson (1934) has found that the growth-rate coefficient seems to fall gradually from an initial high value. This apparently inconstant differential growth-rate may be due, as suggested by J. S. Huxley, to the effect of a localized growth centre in the rostrum which as growth goes on, represents a progressively shorter proportion of the total length of the rostrum.

A study of the growth of the skull in the Sheep-dog shows that the facial region increases heterogonically relatively to the brain-pan, the growth-rate coefficient being 1·4 (Huxley, 1927). Similarly, in the baboon Papio porcarius, the length of the face increases much more rapidly than that of the skull as a whole; the coefficient being 4·25 (Zuckerman, 1926). Young has shown that this method of study is also applicable to the study of different types of a Family; thus if the facial and total lengths of skulls of different species of Edentates are compared, it is found that the larger species have relatively longer snouts in accordance with the effects of a constant differential growth coefficient.

In the human skull, a large number of different measurements at different ages have been made and compared by Hellman (1926). From these it is clear that the different dimensions do not all grow at the same rates, and it is probable that a particular distribution of an allometric growth-pattern is responsible, not only for the bony skull but also for the chondrocranium (see p. 357).

38. CAUSAL RELATIONSHIPS IN THE DEVELOPMENT OF THE SKULL

It is only natural that the interest now universally aroused in problems of causal relationship in development should be extended to the skull. For various reasons, however, this type of work, involving as it does experimentation by making transplantations and deficiencies, has hitherto largely neglected the skull; and of the countless questions which suggest themselves from a consideration of the descriptive study of the origin, fates and form-changes of the various cranial cartilages and bones, only a very limited number have so far been experimentally investigated, let alone answered. Some facts already known concerning the skeleton other than the skull (such as Nassonov's (1934) discovery that fibroblasts may be induced to form cartilage by an aqueous extract of cartilage; or Anikin's (1929) 'field formulae' for the distribution and shape of nuclei in cartilage) await their application to craniogenetic problems, many of which they can hardly fail to clarify.

It will be convenient to deal with the various aspects of this work in separate sections, concerning first cartilage and then bone.

I. Experimental Evidence Concerning the Development of the Cranial Cartilages

1. The origin of the visceral cartilages.

The problem of the origin of the cartilage of the visceral arch skeleton is a special part of the wider problem of the origin of mesenchyme-mesectoderm (in addition, of course, to ganglion-cells and sheath-cells) from the neural crest.

Mesenchyme formed as a derivative of the mesoderm of the segmented vertebral plate (sclerotomes, dermotomes) and of the unsegmented lateral plate would then have to be distinguished as mesendoderm.

The possibility of the origin of mesectoderm from neural crest cells was first envisaged by Kastschenko (1888) in the course of his work on the development of the head in Selachians. Of the large proliferations of the neural crest which migrate downwards and fill the regions of the upper jaw and visceral arches, not all the cells seemed to give rise to neural elements (ganglion-cells of the cranial dorsal-root ganglia). Some appeared to give rise to mesenchyme.

This opinion was shared by Platt (1894, 1896, 1897) working on Necturus

Dohrn (1902) on Torpedo, Koltzoff (1902) on Petromyzon, Brauer (1904) on Hypogeophis, Greil (1913, p. 1121) on Ceratodus, Landacre (1921) on Plethodon, Stone (1929) on Rana, Raven (1931) and Starck (1937) on the axolotl, and, in higher vertebrates, by Veit (1919) and Bartelmez (1923) on Man, and by Holmdahl (1928) on various birds and mammals. (Platt's original conclusions are partially invalidated from the fact that she regarded the ectoderm of the side of the head as contributing to the mesectoderm, *in addition* to the neural crest. This is certainly erroneous as Buchs (1902) showed).

On the other hand, the origin of mesectoderm from neural crest has been contested by Rabl (1894), Corning (1899), and Buchs (1902), and others, while Adelmann (1925) could find no evidence for it in the rat. It should be noted, however, that Amniote material presents considerably greater difficulties for study than Amphibian, owing to the very small size of the cells and the absence of pigment and yolk-granules in ectoderm and mesoderm respectively. On the other hand, limiting membranes in Amphibia are often hard to make out.

The nature of the evidence in these investigations consists in the recognition of persistent histological differences in the cells derived from neural crest and from mesodermal sources. As expressed in Landacre's (1921) description of his most careful and painstaking study of a very close series of stages of development of the Urodele Plethodon, the cells derived from the neural crest are smaller, deeply staining, pigmented, and contain very fine yolk-granules, which soon disappear, whereas the cells derived from the mesoderm are larger, lightly staining, unpigmented, and contain large yolk-granules which are retained for a long time. Precisely similar conclusions have been reached by Stone (1929) on Rana palustris and by Raven (1931) on the axolotl.

A further problem is presented, once the possibility of the formation of mesenchyme (mesectoderm) from the neural crest is granted, namely, the question of the origin of the cartilages in those regions where mesectoderm is found. By means of the study of the histological features of the cells, it has been claimed by Platt, Landacre, Stone, and Raven (working on Necturus, Plethodon, Rana, and Amblystoma respectively) that the cartilages of the trabeculae and of the visceral arches (excepting only the 2nd copula, or basibranchial) are formed from cells (mesectoderm) derived from the neural crest.

The first to apply the experimental method to this problem was Stone (1922, 1926), who proceeded on embryos of Amblystoma punctatum, to extirpate the cells of the neural crest in the head region. The extirpation was effected at the late neurula or early tail-bud stages, when the neural crest cells have freed themselves from the dorsal surface and are growing down into the visceral arches. Their colour and size enables these cells to be scraped away without damage to other tissue-elements. Embryos thus treated are found to show large reductions in, or to lack, the trabecula (as far back as the optic foramen), the pterygoquadrate, Meckel's cartilage, ceratohyal, 1st basibranchial and branchial arches on the operated side. The 2nd basibranchial is not affected.

Stone (1929) repeated these experiments on the Anuran Rana palustris, with precisely similar results. Removal of the neural crest in the head region prevents the formation of trabeculae (as far back as the optic foramen) and suprarostral cartilages, pterygoquadrate with ascending and otic processes, Meckel's and infrarostral cartilages, ceratohyals, 1st basibranchial, and branchial arches. The basihyal and the basibranchial immediately behind the *pars reuniens* of the ceratohyals are not affected.

Raven (1931) working on the axolotl (Amblystoma tigrinum) rendered the conditions of the experiment still more rigorous by operating, not at the late but at the early neurula stage, before the closure of the neural folds. Removal of the neural fold of one side (therefore including the future neural crest) in the head region (the gap being closed by transplantation of indifferent belly-epidermis) resulted in reduction or absence of the visceral cartilages on the operated side.

To avoid the possibility of regeneration of neural crest, or of participation of neural crest cells from the unoperated side (as a result of the irregular closing of the wound and formation of the neural tube), Raven next proceeded to excise a piece of neural fold in the head-region as before, but he filled the gap, not with belly-epidermis, but with a piece of *trunk* neural-fold, which co-operated in the formation of a normal tube. Here again, the cartilages of the trabeculae and of the visceral arch skeleton were absent on the operated side. This shows, not only that the neural crest of the head is intimately connected with the formation of these cartilages, but also that the neural crest of the trunk lacks this capacity.

The remarkable exogastrulation experiments of Holtfreter (1933) on the axolotl have a bearing on the present problem. The result of exogastrulation of the axolotl embryo is the complete or almost complete separation of the ectodermal and mesendodermal germ-layers. The latter develop by self-differentiation into the various mesodermal and endodermal structures; but no neural crests are ever formed, and it is therefore interesting to find a great diminution in the quantity of mesenchyme present (that which is present is, of course, of mesendodermal origin), coupled with absence of visceral cartilage.

Altogether, it may be said that the experiments of Holtfreter support, and that those of Raven fully confirm, those of Stone and the conclusions which Platt and Landacre had drawn from the study of preserved and sectioned material, namely, that the cartilage of the trabeculae and of the visceral arches is derived from the cells of the neural crest. (According to Raven, the trabeculae are of mesectodermal origin as far back as the oculomotor foramen.)

It may be objected that this contention is not proved by these experiments. It might be supposed (allowing that the experimental technique was as definite and rigorously specific as claimed, and that results obtained under experimental conditions may legitimately be extended to normal development) that the neural crest cells were essential for visceral cartilage formation, not by themselves turning into cartilage cells, but in some way by 'organizing' and inducing the adjacent (mesendoderm) mesenchyme cells to do so. It might be claimed that no proof would be acceptable other than the direct demonstration of the derivation of the cartilage cells of the visceral cartilages from the neural crest in an unoperated embryo, by some such method as *intra vitam* staining of the neural crest, and later finding the stain in the cartilage cells.

It may be remarked that the histological criteria (size, pigmentation, and yolk-content) which distinguish neural crest from mesodermal derivatives are themselves a sort of natural experiment of *intra vitam* staining. However, attempts have been made by application of *intra vitam* stains to the neural crest at the early neurula stage, to trace the fates of the neural crest cells. By this means, the fact of the migration of large masses of neural crest cells into the visceral arches has been established beyond doubt by Vogt (1929) in Bombinator, Stone (1932*a*) in Amblystoma, and Ichikawa (1933) in Hynobius. Unfortunately, the appearance of cartilage takes a considerable period, by which time the stain has largely faded, so that the results cannot be regarded as positive.

A possibility of overcoming this difficulty is given in the grafting at very early stages of portions of embryos of different species differing in the character and pigmentation of their cells. By this method, so widely and fruitfully used by Spemann and his colleagues in other lines of research, it should be possible to identify in the visceral cartilage cells certain specific characters which, if the grafting is suitably planned, could not have been derived from the embryo's mesoderm, and could only have come from its neural crest. These experiments have quite recently been performed by Holtfreter (1935) by means of interchange of tissue between the Anuran Limnodynastes and the Urodele Triton. The cartilage of the trabeculae and of the pterygoquadrate of the Urodele host are clearly derived from the epidermal Anuran graft, which gave rise to the neural tube and the neural crest.

It has further been found by Harrison (unpublished experiments kindly communicated personally to the present writer) that by making interchanges of neural crest between Amblystoma punctatum and tigrinum, that the cartilaginous visceral arches which ultimately develop conform to the type of the species of the donor of the neural crest material, and not of the host.

Meanwhile, the case for the origin of the visceral cartilages from neural crest cells is made yet stronger by other experiments in which portions of mesectoderm (neural crest) and mesendoderm tissue are taken from the visceral arches and are grafted elsewhere, to test their powers of cartilage formation.

Stone (1926) removed some mesectoderm from embryos of Amblystoma punctatum at the tail-bud stage and grafted it into the side of the trunk of other embryos, where it differentiated into irregular struts and plates of cartilage. These experiments were repeated by Stone (1929) on Rana palustris, with the same results. Further, Raven (1933*a*) working on the axolotl found that neural crest material taken at the early neurula stage and grafted into other embryos is capable of differentiating into cartilage. Not only that, but pieces of neural crest tissue removed from embryos of the Anuran Rana fusca, and implanted in embryos of the Urodeles Triton taeniatus and the axolotl, differentiate into cartilage, and this cartilage is of Anuran type (Raven, 1933*b*).

On the other hand, when the mesenchyme of mesodermal origin—mesendoderm— from the visceral arches is removed and grafted, as was done by Stone (1929, 1932*b*) working on Amblystoma punctatum, no cartilage develops from it, except when the graft was taken from the midventral line of the embryo. In that case a small piece of cartilage develops, and this is of great interest, for if left in situ this grafted tissue would have produced the 2nd basibranchial, which as described above, is not affected by removal of the neural crest.

It will be seen, therefore, that from the existing state of knowledge it is difficult to resist the conclusion that the cartilage of the trabeculae and visceral arches is derived from the neural crest, strange as it may seem, and great as are the difficulties presented by an attempt to frame a general theory of the origin of cartilaginous material in terms of the germ-layers.

Meanwhile, two points may be considered from the morphological interest which they present. The first is the fact that, whether the source of origin be the neural crest or not, the trabeculae are subject to the same conditions of development as the cartilages of the visceral arches. This is of interest because of the evidence described elsewhere (p. 375) which tends to show that the trabeculae themselves are visceral in nature, and probably represent the cartilaginous skeleton of the premandibular visceral arch.

The second point refers to the curious fact apparent from Landacre's (1921) study of normal development, and both from deficiency experiments and grafting experiments, on both Urodele and Anuran material, that the 2nd basibranchial cartilage differs from the remainder of the visceral arches in being formed from mesendoderm and not from mesectoderm, and that the Anuran basihyal is likewise formed from mesendoderm.

As on the ontogenetic reasons why this should be so, Landacre (1921) has pointed out that the 2nd basibranchial arises in a position to which the downward migration of the neural crest material does not extend; it is therefore formed from the available material in situ which happens to be of mesodermal origin.

But this difference between the methods of formation of the 1st and 2nd basibranchials may reflect a morphological difference. The paired origin of the neural crest material that migrates down the visceral arches and meets in the middle line between the ventral ends of the hyoid and first two branchial arches, lends support to the view that the anterior copula or 1st basibranchial is derived from the fusion of originally separate basibranchial elements. The posterior copula or 2nd basibranchial (between the ventral ends of the 3rd–5th branchial arches) has no such paired origin, but arises as an unpaired median structure, and the same appears to be true of the Anuran basihyal. These structures may always have been unpaired in phylogeny (see p. 406).

2. The formation of the neurocranium.

In spite of the absence of direct experimental evidence on the subject, it is clear from other experiments that the formation of the cartilages of the neurocranium is dependent in some way on the presence of the neural structures: i.e. the brain and the sense-organs such as nasal sac, ear, and eye. These structures become surrounded by a layer of connective tissue—the meningeal membranes, sclerotic, &c., in connexion with which chondrification sets in.

The evidence in favour of this consists in the fact that when by experimental or 'natural' causes the anterior end of an organism is bifid, forming a *duplicitas anterior* with two brains, each becomes surrounded by a neurocranium. This is apparent in the artificial production of duplicitas anterior in Triton by Spemann (1902), and in those cases in Teleosts described by Swett (1921).

In all these cases the extra neurocranium would never have arisen had it not been for the extra brain; and since the formation of the latter is known to be due to the underlying organizer (gut-roof), the neurocranium itself must be dependent on the organizer, or on the brain or on both.

Lastly, in the tissue-culture experiments, Törö (1935) has shown that an implant of (chick) brain-tissue induces chondrification in cultures of fibroblasts derived from the embryonic sclerotic, or lung-buds.

The conformation of the neurocranium may be influenced by a number of factors. First of all it may be noted that those conditions which lead to the production of *cyclopia* (i.e. fusion of the two eyes into one in the middle line), or *otocephaly* (i.e. approximation of the auditory vesicles) produce abnormalities in the neurocranium, and may affect the trabeculae in such a way as to cause them to fuse together in the middle line anteriorly or posteriorly (or both), and form a sort of artificial trabecula communis. For instance, in Triton the trabeculae are normally separate and wide apart. But embryos treated at certain phases of the gastrula stage with depressants such as chloretone or phenol, show more or less marked otocephaly, and/or cyclopia, accompanied by fusion of the trabeculae in the middle line, as discovered by Lehmann (1933). Otocephaly involves absence of the cranial notochord and narrowing of the parachordals (Lehmann, 1936). Similarly, Adelmann (1934) has shown that treatment of Amblystoma embryos with LiCl and $MgCl_2$ produces cyclopic embryos in which the trabeculae instead of being separate and wide apart are fused in the middle line. Meckel's cartilages may also be represented by a median structure.

Cyclopia and otocephaly seem to be due to a defective action of the organizer in the head region, and therefore the disposition of the trabeculae, approximating to what morphologists call the platybasic (e.g. Amphibian) or tropibasic (e.g. Salmo) types, is linked with the activity of the organizer.

The actual shape of the neurocranium can be affected by removal of the eyes, as discovered by Steinitz (1906) working on Rana fusca. If the eyes are removed from early larvae before chondrification has started, the chondrocranium when it develops is abnormal in that the orbit is too short, the brain-case is wider, and the pterygoquadrate lies nearer the middle line. It is interesting to note that in spite of the absence of the optic nerve, the optic foramen is nevertheless present, although smaller than normal.

That the shape and size of the orbit of the skull are controlled by the eye is further shown by the experiments of Stone (1930) in which eyes from Amblystoma tigrinum are grafted into the place of the eyes of Amblystoma punctatum. The larger size to which the grafted eyes grow leads to the enlargement of the orbital region of the skull.

The shape of the brain-case itself is to a large extent dependent on the volume of its contents. As Nicholas (1930) has shown, this may be enlarged by grafting rapid-growing foreign tissue such as pronephros rudiment, in place of the midbrain. The rudiments of the cranial cartilages simply give way before the rapidly expanding pronephros. On the other hand, when the internal pressure subsides, mesenchyme invades the spaces released and forms new skeletal elements there.

An additional result of great interest was of constant occurrence in these experiments: viz. the formation of a continuous median dorsal plate of cartilage extending back from the occipital region of the skull most of the way down the vertebral column, attached to the dorsal ends of the neural arches. (Cf. the superior occipital cartilage in Cypriniform Teleosts p. 137.)

This strange state of affairs appears not to be a direct result of the graft of foreign tissue into the midbrain, but to be associated with, or consequent upon, the limitation of movement manifested in the later stages by embryos which have been exposed to this treatment. If this is so, it is possible that some analogous factor may have been at work in the formation of the skull-roof, or in that hinder part of the parachordals which is regarded as composed of originally separate vertebral elements.

3. The formation of the nasal capsule.

The evidence concerning the causal relations in the development of the nasal capsule is meagre, and partly indirect. That the capsule is in the first instance dependent on the presence of the nasal sac seems to follow from certain observations of Ekman (1934) on an abnormal embryo of Salmo, which presented a partial duplication of the anterior structures of the head (*duplicitas anterior*) involving the eyes and nasal sac. From the present point of view the importance of this case resides in the fact that the extra nasal sac which this embryo possesses is surrounded by an extra cartilaginous nasal capsule.

However this may be, it is clear from experiments by Burr (1916) that the presence of the nasal sac is necessary for the normal morphological differentiation of the cartilages of the nasal capsule. If the nasal sac is extirpated from Amblystoma larvae 5–6 mm. long, the nasal cartilages, instead of forming the typical structure of a nasal capsule, are collapsed in a heap. Thus, the cartilages have differentiated histologically, but not morphologically.

4. The formation of the auditory capsule.

The conditions controlling the formation of the auditory capsule have come to constitute one of the classical fields of research in experimental embryology.

In Amphibia it has been shown conclusively that the formation of the auditory capsule is dependent on the presence of the auditory vesicle or otic sac. Removal of the otic sac from an embryo of Bufo results in the absence of any auditory capsule (Filatoff, 1916). Analogous experiments on Rana esculenta enabled Luther (1925) to show that while the capsule itself, the operculum, and the pars interna plectri are dependent on the presence of the otic sac, the pars media and pars externa plectri (of the columella auris) are independent of the otic sac and develop in its absence. However, the dependence of the operculum and pars interna plectri on the presence of the auditory capsule is no evidence for the derivation of those structures from the capsule.

It is clear, therefore, that the cartilage of the auditory vesicle is in Amphibia a product of the 'organizing action' of the otic sac. Even a small portion of the otic sac is sufficient to induce the formation of cartilage (Eisinger and Sternberg, 1924).

The converse experiment to extirpation is, of course, that of transplantation of the otic sac to other regions. This has been done by Luther (1925) in Rana esculenta, with the result that a cartilaginous capsule was induced surrounding the sac in the new position, provided that that position was in the head and not in the trunk. Judging from Stone's (1926) experiments, this would seem to be because the auditory capsule develops from mesodermal mesenchyme in the region just above the hyoid arch, and the otic sac appears to be powerless to cause induction of a cartilaginous capsule from the mesenchyme surrounding the trunk myomeres.

That the action of the otic sac is a true induction of surrounding tissues is proved by the experiments of Lewis (1906) in which the otic sac of Rana palustris was grafted into an embryo of Amblystoma, and induced the formation of a cartilaginous capsule out of Amblystoma tissue.

In other vertebrates, it seems that the condition in birds resembles that in Amphibia, for in the chick, Reagan (1915, 1917) found that extirpation of the otic sac resulted in the absence of the cartilaginous capsule. When displaced into a new position, the otic sac induced the formation of a capsule there. At the same time it was found that when the capsule is absent, the footplate of the columella auris is improperly developed, although it is not formed from the capsule.

In Acipenser, on the other hand, conditions are somewhat different, for Filatoff (1930) found that if the otic sac is removed, an amorphous mass of cartilage develops nevertheless, but bears no resemblance to the form of an auditory capsule. In other words, the histological development of cartilage is independent of the otic sac, but its morphological differentiation is dependent on it. In this respect, therefore, the auditory capsule of Acipenser resembles the nasal capsule of Amblystoma.

The fact that the lateral semicircular canal tends to lie in the horizontal plane at normal posture of the head (Lebedkin, 1924) raises problems of great interest. Experiments of grafting ear vesicles in abnormal positions in Amphibian embryos (Spemann, 1910; Streeter, 1914; Ogawa, 1921, 1926; Tokura, 1924) have shown that the vesicle may 'right' itself by rotation. The cause of such rotation is still unknown, and while direct action of gravity could account for it in Anamnia, it cannot be invoked in Amniotes owing to the haphazard orientation of the embryo, fledgling, or suckling.

5. The formation of the sclerotic cartilage.

Experimental investigation into the method of formation of the sclerotic cartilage is practically restricted to the Amphibia.

That the sclerotic cartilage, which is of mesodermal origin, is dependent on the presence of the eye-cup follows from experiments of Mangold (1931) on Triton alpestris in which the optic vesicle was removed at the tail-bud stage, without damaging the mesodermal tissues. No sclerotic developed. These results are confirmed by the experiments of Törö (1935) on the power of the retina to induce the chondrification of fibroblasts *in vitro*.

The sclerotic cartilage appears to persist throughout life in Anura, and in Perennibranchiate Urodela (see Yano, 1926, 1927). In Caducibranchiate Urodela it disappears about the time of metamorphosis, except in some forms such as Hynobius where it persists in reduced form, and in Cryptobranchus and Amblystoma where it persists unreduced. Stadtmüller's (1929*a*) investigations (see p. 406) go to show that this discrepancy in the fate of the sclerotic cartilage can hardly be ascribed to ecological factors (life in water, life on land), nor to the general metamorphic agents, since the sclerotic persists in metamorphosed Amblystoma and Cryptobranchus. The problem has been investigated experimentally by Yano (1928) who removed the sensory lining of the eye-cup in embryos of Bufo and Diemictylus. In the former the sclerotic cartilage arises towards the end of metamorphosis and normally remains small. Removal of the contents of the eye-cup, however, leads to a great increase in size of the sclerotic cartilage. In Diemictylus the sclerotic cartilage arises during larval life and disappears at the time of metamorphosis, but here again removal of the contents of the eye-cup results in an increase in the size of the cartilage, which then persists after metamorphosis. But in some cases the contents of the eye regenerate, and then the sclerotic cartilage is not so large. Yatabe (1931) obtained similar results with Hynobius.

The sensory layers of the eye-cup thus seem to exert an inhibiting action on the development of the sclerotic cartilage. What the significance of this can be from the developmental point of view remains completely obscure. Meanwhile, it is not improbable that the sclerotic cartilage is in some relation to the intrinsic and/or extrinsic eye-muscles, but experimental evidence on this point is completely lacking.

6. The balancer and the pterygoquadrate.

The balancer is a sensory process found on each side of the head in some Urodela, and consists of a long ectodermal cylinder containing a mesenchymatous core. At later stages of development the balancer drops off, but its base becomes embedded in the side of the head, where a cartilaginous process from the pterygoquadrate cartilage is developed to meet it. This is the normal course of affairs in Amblystoma punctatum which possesses a balancer, but Amblystoma tigrinum does not, and therefore normally lacks the process of the pterygoquadrate. But if the rudiment of a balancer from Amblystoma punctatum is grafted on to Amblystoma tigrinum, the process on the pterygoquadrate is developed (Harrison, 1925). The development of this process on the pterygoquadrate cartilage is therefore dependent on the presence of the balancer.

7. Endocrine effects on chondrogenesis.

As described in the section dealing with the development of the skull in the Anura (p. 206), the transition from tadpole to frog is accompanied by far-reaching changes in the chondrocranium: disappearance of suprarostrals and infrarostrals, reorganization of the suspension of the pterygoquadrate involving crumpling and reduction of the posterior part and great elongation of the anterior portion, as of Meckel's cartilage. Similarly, in Urodela, the transformation of the hyobranchial skeleton from the larval to the adult condition involves considerable absorption of cartilage by phagocytosis and fresh development (L. Smith, 1920). It follows from the control which the thyroid and pituitary hormones are known to exert on metamorphosis, that these changes in the cartilaginous jaw skeleton are in some way dependent on the endocrine mechanism.

On the other hand, Hofmann's (1923) experiments on Salamandra larvae treated with calcium chloride show that metamorphosis may take place in connexion with the processes of ecdysis, tongue-development, pulmonary breathing, and the acquisition of the terrestrial habit, while the larval structure of the visceral arch skeleton is retained.

II. Experimental Evidence Concerning the Development of the Cranial Bones

1. Phosphatase activity and ossification.

It has been shown by Robison (1923) that developing rudiments of bone contain an enzyme known as phosphatase, the function of which is believed to be the raising of the concentration of inorganic phosphate in the tissue-fluids by means of the hydrolysis of phosphoric esters in the blood. The result is a deposition of calcium phosphate in the tissue possessing phosphatase. (Kay and Robison, 1924; Martland and Robison, 1926.)

Phosphatase is secreted in the region of the osteoblasts, and is present in the rudiments of membrane-bones such as the dentary, and, after a certain time, in those portions of the cartilaginous skeleton in which perichondral and endochondral ossification occur. Phosphatase has thus been shown by Fell and Robison (1930) to exist in the pterygoquadrate cartilage of the chick, but not to exist in the greater part of Meckel's cartilage (which does not ossify except for the articular at its proximal end).

A point of further interest is the fact, likewise discovered by Fell and Robison (1930), that the phosphatase is synthesized in the tissue which contains it. Thus, grafts of cartilage from the pterygoquadrate and from the distal end of Meckel's cartilage taken from a 6-day chick and cultured *in vitro* contain no phosphatase at the outset of the experiment. The enzyme appears later, however, in the pterygoquadrate grafts, but never in those of Meckel's cartilage. Ossification (*in vitro*) takes place in the former case, but not in the latter. The onset of ossification in cartilage is preceded by the conversion of the cartilage from the small-celled, thick-capsuled type, to the so-called hypertrophic type, with large cells and very thin intercellular capsules. Synthesis of phosphatase appears to coincide with the appearance in the cartilage of the hypertrophic cells.

It is clear, therefore, that phosphatase plays an important part in the origin of a bony rudiment, although the action of the enzyme is probably indirect, and the mechanism of ossification is more complicated than a simple precipitation of inorganic phosphate salts (Fell and Robison, 1934). However, presence or absence of phosphatase seems to be the limiting factor governing respectively the appearance or non-appearance of ossification in a cartilage. This has an important bearing on the speculations concerning the possibility that during phylogeny an ossification which was originally a membrane-bone, might 'sink in' and become a cartilage-bone. The convenient phylogenetic expression of 'sinking in' has, of course, to be translated into ontogenetic terms, and in the light of the foregoing experiments, it must be held to imply the acquisition of the power to synthesize phosphatase in a specified portion of the chondrocranium. The problem therefore becomes one of the localization of a biochemically specific type of metabolism.

Related to the experiments just described are those in which Fell (1933) showed that culture *in vitro* of a portion of endosteum of a cartilage-bone (the femur) containing no cartilage whatever, might subsequently differentiate into cartilage of the hypertrophic type (never into cartilage of the primary embryonic type). Such cartilage in the cultures is always associated with bone, merging into it by means of a belt of intermediate tissue. Further, under the conditions of the experiment, this hypertrophic cartilage becomes directly transformed into bone.

Whether this metaplasia occurs normally *in vivo* as well as *in vitro* is a separate question, as is that of the relation of hypertrophic cartilage to the so-called secondary cartilage of certain membrane-bones (see p. 502). At all events, these experiments illustrate the close association between hypertrophic cartilage and ossification. Indeed, more recent experiments have led Fell and Landauer (1935) to conclude that osteoblastic activity in cartilage does not occur without, and is in some way dependent on, the presence of hypertrophic cartilage. The ossification of a cartilage-bone can be prevented by delaying the growth of the cartilage and inhibiting the formation of hypertrophic cartilage.

2. The regeneration of bones.

The question of the limitation of growth of bones is one of great interest and importance. Cases where a single bone in one form appears to correspond to two separate bones in another may be explained by the fusion of the two bones in the former case, or by the subdivision of one bone in the latter case, or by the disappearance of one bone and the extension of the other. So far as it is possible to get evidence on this matter, it is clear that when in one animal a bone develops from more than one centre of ossification and these fuse, and in a related animal the ossification centres remain separate, there is at any rate some support for the view that the single bone owes its origin to fusion of bones which were separate in previous phylogeny. This applies to the nasal of Amia and the nasal, adnasal, and terminale of Polypterus; the single median parietal of Sinamia and the paired parietals of Amia; and perhaps the extrascapular of Amia and the extrascapulars of Sinamia, Polypterus, and Lepidosteus. (See p. 505).

On the other hand, if in phylogeny a bone has disappeared and become replaced by the extension and expansion of another, it would imply that the limits of growth of a bone are not fixed and predetermined *ab initio*, but are capable of modification. This is precisely what happens in the results of experiments conducted by Tatarko (1934) in the carp, after reduction or removal of one or other of the opercular bones.

In the first experiments, the opercular fold was cut in such a way as to remove parts of the opercular, subopercular, and first two branchiostegal rays. Regeneration takes place in such a way that the opercular fold is again stiffened by bone, but not necessarily by the exact restitution of the same bones. For instance, the subopercular remained very small and its place was taken by extensions in some cases of the opercular, in others by the 1st branchiostegal ray, or in others again by the 2nd ray.

It is clear, therefore, that bony elements can extend beyond their normal limits. The same conclusion emerges from experiments in which the subopercular is simply excised whole, with or without damage to the neighbouring bones. In such cases, the place of the missing subopercular, for instance, is taken partly by extension beyond their normal limits of the opercular (which was damaged) and of the interopercular (which was untouched). If the opercular is removed, a small new opercular is regenerated from the periosteum of the joint between the old opercular and the hyomandibula, and in addition there is considerable expansion of the preopercular and subopercular.

It appears in these cases that extension beyond the normal limit only occurs when the normal neighbour to a bone is absent, or is not regenerating. A further point of interest is the fact that the normal function of the opercular fold (expansion and contraction of the branchial chamber) involving movement of the free edges of the bones on one another, must be maintained if regeneration, or even maintenance is to take place. Thus, if the interopercular is removed, the action of the mandibulo-hyo-interopercular ligament is paralysed, and, far from regenerating, the opercular degenerates. Similarly, after removal of the first two branchiostegal rays (which normally impart movement to the opercular apparatus from the ceratohyal and hyoid arch) the subopercular degenerates. This degeneration is, of course, a separate problem.

It is to be noted that the opercular bones differ from those of the skull-roof in that they overlap and do not form sutures with one another. Experiments on bones of the latter type have been made by Troitsky (1932), who removed the parietal from young mammals. If the dura mater was left in situ, a new parietal bone was regenerated and grew to normal size, the sutures being formed in normal positions. But if the osteogenic layer of the dura mater was scraped off, no bone was regenerated and the neighbouring bones extended their area slightly over the gap. Troitsky has taken his results to mean that the position of the sutures is predetermined by non-ossifying strips of the dura mater, or in other words, that the cessation of growth is not due to contact with a neighbouring bone. But the fact that under certain conditions, these bones can expand beyond their normal limits provides an experimental basis for the view that in evolution, some bony elements may have disappeared and become replaced by expansion of others. This conclusion is supported by observations of the variability of ossification of adjacent bones in the human skull.

3. Hyper- and hypodevelopment of adjacent bones in normal-sized skulls.

Certain abnormalities of ossification of the human skull in which adjacent bones may show hyper- and hypodevelopment can be taken as natural experiments of variation in the extent of bones. They have been noted and summarized by Augier (1931*a*).

For instance, the lachrymal may be entirely lacking, in which case its place may be taken by an extension of the maxilla, or of the frontal, or of the nasal. Conversely, the lachrymal may extend out of the orbit on to the face, in place of a small portion of the maxilla. (This normally occurs in Marsupials.) Or the nasals may be reduced or absent, in which case they are replaced by extensions of the maxillae. The parietals may encroach on reduced squamosals, thus presenting a simian appearance. A sort of balance is found between the frontal, orbitosphenoid, alisphenoid, and parietal, which as it were, contest the available space between them, and sometimes arrive at a result which resembles the normal apportioning of the territories of these bones found in other mammals. It is clear, therefore, that the difference in bone-contacts in the orbitotemporal region between different Primates (e.g. frontosquamosal contact in Chimpanzee, separating parietal and alisphenoid; alisphenoparietal contact in Man, separating frontal from squamosal) are of no fundamental importance. Indeed, as Petersen (1930) has pointed out, the position of the sutures is without effect on the form of the skull.

The reasons for these variations are obscure, because of the ignorance that prevails regarding the attendant conditions. The variations might be explained by the precocious or tardy appearance of one bone or another, which would, therefore, be able to cover more or less area before establishing contact with the neighbouring bones. On the other hand, it is possible that during the increase in volume of the brain, the expansion of the membrane in which the ossifications arise (i.e. of the osteogenic zones of the dura mater) may not be uniform in all areas, and in view of the conditions in microcephaly and hydrocephaly (*infra*) this may be expected to influence the relative expansion of the different bones. The results of Troitsky's (1932) experiments on bone-regeneration seem to favour the latter view, for the regenerated bone although developed later than its neighbours, often occupies the same area as its extirpated predecessor (p. 483).

4. The dependence of the bones of the brain-case on the brain.

By means of tissue-culture experiments Törö (1935) has shown that implants of living brain or retina tissue, or of brain-juice or desiccated retina, have the power of inducing ossification in cultures of fibroblasts derived from the chick embryonic sclerotic or heart. It is clear, therefore, that the neural tube possesses general bone-inducing powers.

For further information on this subject, recourse may be had to indirect evidence. Certain abnormalities occur occasionally in Man, involving on the one hand reduced size or even absence of the brain, and on the other distension of the meningeal membrane because of excessive internal pressure of fluid. These conditions are known respectively as microcephaly, anencephaly, and hydrocephaly, and they provide as it were natural experiments illustrating the dependence of the bones of the brain-case on the size of the brain. They have been admirably summarized by Augier (1931*a*).

In microcephaly the brain is present but very small, and may occupy only one-sixth of the volume of a normal brain. The bones of the brain-case thus have only a reduced surface area to cover, and are consequently smaller; the sutures between them usually persist, and this fact is important since it invalidates the view once held to the effect that the small size of the brain in microcephalic specimens is due to the precocious closing of the sutures, preventing the possible expansion of the brain. The shape of the skull departs from the normal and comes to resemble that of the Neanderthal race, owing to the fact that while the brain-case is reduced, the bones of the face and jaws are fairly normal in size.

It is therefore clear from these microcephalic skulls that the dermal bones of the brain-case (frontal, parietal, squamosal, supraoccipital) are dependent for their morphological differentiation and growth on the brain, whereas the bones of the skull-base, face, and jaws (nasal, maxilla, &c.) are independent of the brain. (For the dependence of certain tooth-bearing bones on the degree of development of the teeth, see p. 490.)

These conclusions are still further supported and confirmed by the cases of anencephaly in which the brain is quite absent. Developmentally these cases are due to the failure of the neural folds to close in the anterior region (*spina bifida*). In such skulls there is no cranial vault at all, and the bones which would normally form it are present but extremely small and abnormal. The frontal has its pars orbitalis and pars nasalis, but no pars frontalis at all. The parietal is a small plate of bone lying beneath the tip of the alisphenoid; the supraoccipital is represented by the two paired supraoccipital centres which fail to join. On the other hand, the bones of the skull-base (basioccipital, basisphenoid, periotic, &c.) are fairly normal, as are those of the face.

Anencephalic skulls therefore show, as Weinnold (1922) contends, that while the growth and morphological differentiation of the bones of the brain-case are dependent on the brain, their origin is an independent self-differentiation, and the result of 'inheritance'.

The dura mater of the vault of the human skull has been shown by Kokott (1933) to possess a system of fibre-tracts which are related to the five prominent points of the chondrocranium (crista galli, orbital cartilages, parietal plates), and it is certainly remarkable that the five dermal bones of the skull roof (frontals, parietals, interparietal) lie between these tracts. The mechanical conditions involved may therefore play a part in the localization of the centres of origin of these bones, but cannot be responsible for their origin.

The nature of the action exerted by the brain on its surrounding bones is the mechanical action of stretching or area-increase. The classical researches of Roux (1895) into the functional differentiation of bones have shown that the origin and orientation of the spicules of developing bone are a direct result of the stresses and strains to which it is exposed (see Petersen, 1930). The expansion of the area of the meningeal membranes may similarly be regarded as setting up a zone of strain (see Bluntschli, 1925; Popa, 1936) leading to radial development of the bones from their centres of origin. As Benninghoff (1925), Kokott (1933), and Ahrens (1936) have shown, the orientation of the spicules alters when sutures close and growth ceases. It must, of course, be remembered that the expansion of a membrane-bone when once formed takes place by growth only round its edge, and not interstitially, although the contrary has been claimed by Thoma (1923) and Popa (1936). The idea that bone grows interstitially has been effectively disproved by Troitsky (1932), who inserted pieces of silver wire into the bones of developing dogs, and observed by X-rays that the distances between the pieces of wire remained constant. These problems are very usefully discussed by Murray (1936).

Attention may now be turned to an abnormal condition which is the reverse of microcephaly or anencephaly, namely hydrocephaly. In hydrocephaly the size of the brain (strictly, of the meningeal membranes surrounding it) may be so much enlarged as to occupy more than twice the normal volume. It also tends to assume a more or less spherical form. The effect of this dilatation on the bones of the brain-case depends largely on the time at which the abnormality starts. If the hydrocephaly or macrocephaly manifests itself early, while the various bones are separated by wide interossicular spaces, these spaces become stretched still wider, and the ossifying spicules forming the edge of the developing bones may become torn away from the bones and give rise to independent centres of ossification or Wormian bones. These effects are reduced in intensity the later the macrocephaly starts. Nevertheless, the bones do eventually form a complete bony brain-case, which they are only able to do by growing to cover much larger areas than normally, but at the same time they are much thinner than normal. It is presumably a continuation of the mechanical process of stretching consequent upon area-increase, leading in normal development to the ordinary growth of the bones, which in hydrocephalic skulls leads to their hypertrophy.

The dependence of the skull on the brain is also apparent in the phenomenon observed by Darwin (1875, vol. 1, p. 121) that the skull is *relatively* larger in wild than in domesticated varieties of a species, and Klatt (1912) has shown that there is reason to believe that this effect is due to a larger sized brain, itself conditioned by the presence of better-developed sense-organs in wild forms.

Altogether it may be concluded that mechanical conditions prevailing in the dura mater play an important part in determining the form of the bones of the skull-roof, but these mechanical factors cannot be held responsible for the first origin of the bones, as Thoma (1911, 1913) contended. The anencephalic skull proves this, and the first appearance of the centres of ossification must be regarded as a phenomenon of self-differentiation.

5. Cerebral hernia in fowls.

In his *Animals and Plants under Domestication* Darwin described the curious condition of the head and skull in the Polish breed of poultry where the frontals are prevented from meeting properly by a bulging upwards of the brain, and the gap thus formed in the skull is closed over by a number of separate ossifications or Wormian bones. This phenomenon, interesting as it is, has acquired additional significance from the fact, discovered by R. A. Fisher (1935), that the Mendelian factor controlling the peculiar configuration of feathers on the head, forming a 'crest', is also responsible for the production of cerebral hernia. In these cases, in young chickens of wild stock into which the hernia factor has been introduced, the frontal bones are pressed apart by the extension of the brain, leaving a gap.

It is not yet clear exactly how this distortion of the frontals is brought about. From information very kindly communicated by Dr. Fisher, it would seem that in some cases, the appearances suggest that the two frontal bones had formerly been adjacent, but had been pushed apart and curved backward by the extension of the brain. On the other hand, other specimens look as if the ossified frontals had never been in contact.

The embryonic development may therefore have taken one of two courses. Either the extension of the brain takes place by abrupt rupture after the frontals have begun to ossify; or, the protrusion of the brain occurs prior to ossification, the membranous rudiments of the frontal bones then conforming to the shape of the brain.

In either case, the phenomenon would be of interest, as illustrating the dependence of the form of the frontal bone on the brain—more strictly, on the dura mater surrounding the brain. But an additional feature to be noted is that these abnormalities seem to show that the *area* of the frontal bones appears to be limited. So far as is known, these cases of hernia in wild stock retain the open gap between the frontals, without showing any trace of the secondary ossification of Wormian bones. And in the Polish breed, where the hernia condition is established, it is by means of these secondary ossifications that the gap is filled, and not by any extension of the frontals.

It is of interest to compare these results with those obtained from experiments of regeneration in the bones of the operculum of Teleostei (p. 482), and with the observations of abnormal ossification (p. 483) and of hydrocephaly (p. 485) in Man. There, it has been shown that bones can and do extend to cover areas greater than normal, or areas normally occupied by other bones. In poultry possessing the hernia factor, on the other hand, the frontals seem unable to do this, and this condition is probably due to a general retardation of the process of ossification in these bones, resulting in a weakness which makes hernia itself possible.

These phenomena lead naturally on to a consideration of the general aspect of the relations of brain and skull, which has recently been treated by Ariens Kappers (1932).

6. General relations between skull and brain.

It will have become clear from the preceding sections of this chapter that the skull, both cartilaginous and bony, stands in a very close relation to the brain, or to the membranes surrounding it. In a very general way it may be said that the morphological differentiation of the skull is dependent on the brain. It is, for instance, a result of the greater size of the cerebral hemispheres in higher mammals that the tentorium cerebelli becomes fused to the occipital instead of to the parietal as in lower mammals (Dabelow, 1931). In other words, the hard parts are dependent on the soft, in this case.

There are, however, a number of facts which show that the relations between brain and skull may be a mutual one; i.e. that certain spatial conditions imposed by the skull, acting as mechanical factors may react on the brain. Some of these have been considered by Kappers (1932).

In Teleostomes the condition in the *early* embryo are such that the head is plastered down on to the yolk and the brain is very much compressed within its membranes. The effect of this external pressure on the rudiment of the forebrain appears to be that instead of its dorsal portion forming a mantle extending outwards, it grows inwards forming a thickened mass. The cerebellum grows inwards instead of expanding on the dorsal surface of the brain, and the cavity of the ventricle of the hind brain is much reduced.

When chondrification sets in, the pressure within the cranial cavity is according to Gaupp (1906*a*, p. 664) responsible for the fact that in these fish the medial wall of the auditory capsule remains membranous, the lateral parts of the hind brain thus becoming crammed up against the auditory vesicles. The chondrocranium of these fish has a more or less well-developed cartilaginous roof, which prevents extension of the brain in a dorsal direction.

In *later* development the skull grows considerably, and allows the brain to lie in a pericerebral space which may be very extensive indeed (e.g. Lophius piscatorius), but the compact structure of the brain, imposed upon it by the mechanical conditions of early stages, persists.

Another group of vertebrates to which Kappers has drawn attention is the birds. Here, however, the mechanical factors compressing the brain are effective not in the early, but in the later stages of development. Chief among these factors is the large size of the eyes, which in turn is responsible for the fact that the orbital wall of the skull faces forwards instead of laterally. The elements of the primordial skull (chondrocranium and cartilage-bones ossifying therein) are no longer able to enclose the brain-cavity, and the squamosal, a dermal bone, is called in to close a large gap between the pleurosphenoid and the periotic. Another feature to be noticed is the early closure of the sutures, for this event practically precludes further expansion of the cranial cavity. The result of these conditions is that whereas the telencephalon of the chick at the 5th day of incubation has wide ventricles covered by a mantle, as development proceeds the mantle, unable to extend outwards, grows inwards and thickens to form the characteristic avian epistriatum. The ventricles are nearly obliterated.

So closely is a bird's brain pressed against the internal surface of its skull that an endocranial cast shows nearly all the details of the brain-surface morphology, including the fine cerebellar fissures.

In mammals the brain at early stages does not appear to be subjected to space-restriction, and the sutures between the bones close relatively late. Occasional examples can be found of the effect of the skull on the brain, such as the formation of the cruciate sulcus in Carnivora as a result of the pressure of the frontal bones. There are also the cases of artificial skull-deformation in savage human races which result in alteration of the general form of the brain.

In general, however, it seems that in the mammalian embryo the surface of the brain is free from its skeletal envelope, and it is only later that the endocranial surface of the skull becomes moulded. Even here it is noticeable in Man that the parietals carry only very shallow and imperfect markings, which is presumably due to the fact that in man's erect position, the weight of the brain causes it to press its frontal, occipital, and temporal lobes against the skull-wall, but the parietal lobe is separated from the skull-roof by a space occupied by fluid.

In old age the volume of the brain diminishes, and there may then be observed appositional growth on the internal surface of the bony brain-case, which leads to a thickening of the wall, unless this process is accompanied by erosion on the external surface, leading to a general diminution in skull size (Augier, 1932*b*).

7. The dependence of the bones on muscle attachments.

A simple comparison between the adult skulls of a Gorilla and of a Man suffices to illustrate the importance of muscle attachments in determining the form of bones, for the modifications of the frontals, parietals, and squamosals in the former to form the huge median crest and its absence in the latter are dependent on the degree of development of the temporal muscles. Experiments by Neubauer (1925) involving section of cranial muscles in the rat, and the study of the development of the bones in the fish Cyclopterus (p. 160) and in the abnormal 'grey-lethal' mouse (p. 490) likewise shows the importance of muscle-attachments in the differentiation of certain bones. A close study of the development of muscle-attachments in Man has led Augier (1931*a*) to some conclusions of interest. Instead of a process, excrescence or crest being developed by a bone *at* the point of insertion of a muscle, it seems that these structures arise *around* the point of insertion which thus comes to lie in a depression. The bone becomes as it were moulded by the attachments.

8. The problem of the relation between the lateral line canals and dermal bones.

This section is included here, not to convey information for there is little to give, but to present a problem which is open to attack on experimental lines.

In the development of certain of the dermal bones in fish, it is found that the rudiments may arise in definite relations to the sense-organs of the lateral line. The nasals in flat-fishes (Traquair, 1865), scales of the trunk in Salmo (Neave, 1936), the nasal, frontal, intertemporal, postparietal, &c., of Amia, and the rostrals of Lepidosteus are cases in point. It is further to be noted that in bones lodging lateral line canals, the centre of ossification (as revealed by the lines of growth) coincides in position with the lateral line. This is probably still true, even when, as Nielsen (1936) has shown for Coelacanthids, the supraorbital lateral line canal runs *between* the frontal and supraorbital bones. The question arises whether this topographical relation, which has been stressed for Amia by Pehrson (1922) and for Amiurus by Moodie (1922), is also a causal one, as is the case, for instance, with the eye-cup and the lens. The loss of the squamosal in Actinopterygian fish may be associated with the loss of the cheek lateral line canal (Westoll, 1937). The matter is of morphological importance, for if the formation of the bone was dependent on the lateral line sense-organ, then the identity of the bone through which a given lateral line canal passes must always be the same. Klaatsch's (1895) contention that the osteoblasts are derived from the epidermis beside the lateral line sense-organs has been conclusively disproved by Schleip (1904).

It is, however, unlikely that the bone is dependent *for its origin* on the sense-organ, for if it were there would have to be another explanation for the origin of the same bone in higher vertebrates where there is no lateral line canal, and for the other dermal bones in fish through which no lateral line canal passes. Further, it will be remembered that in Esox (see p. 142) the frontal bone is said to ossify prior to the appearance of the supraorbital lateral line canal.

The matter could, however, be settled by the extirpation of the lateral line placode in a young embryo.

Without anticipating the result of such an experiment (which it is hoped that these lines may induce an experimenter to perform), it may be pointed out that the lateral line sense-organs may serve not to *cause* the bone to arise (as the eye causes the lens), but to determine the *localization* of its rudiment (as the ectoderm of the pluteus larva controls the localization of the micromeres which secrete the larval skeleton. See Huxley and de Beer, 1934). Presumably this effect could be exerted on any bone and may be due to some trophic stimulus associated with the presence of the lateral line sense-organ. The state of affairs might be comparable to the innervation of the Urodele limb. Normally, this is supplied by certain definite spinal nerves, but if the limb is displaced, the innervation is performed by other nerves. Similarly, the phylogenetic shifting of fins up or down the body (see Goodrich, 1913) has not taken the fin-forming elements with it, but has been brought about by the contribution of other segmental elements to the formation of the fin.

Stone (1922) has shown that reversal of a lateral line placode leads to its growth in an abnormal direction. If this could be done in a fish, it would be of the greatest interest to observe the effects of an ectopic lateral line canal on the bones.

Palaeontological researches have demonstrated the greater constancy in a type of the bones through which the lateral line canals pass, and the greater susceptibility to variation of the bones through which they do not. As it stands, this is evidence for Natural Selection, eliminating the forms whose variation led to insufficient protection of the lateral line sense-organs; it is not conclusive evidence of the morphogenetic function of lateral line sense-organs in 'causing' the ossification of the bones which surround them.

A somewhat analogous problem arises in Holocephali, where modified denticles arise as half-gutters closely related to the lateral line grooves and giving the impression of having been 'caused' by them.

9. The 'grey-lethal' mouse.

The phenotypic effects on the skull of the mouse produced by the recessive Mendelian gene known as 'grey-lethal' have been studied by Grüneberg (1935). Not only does this gene suppress all yellow pigment in the coat, but it affects growth and has profound effects on the differentiation and calcification of the bones and the teeth, with consequent direct and indirect effects on the morphology of the skull.

The cartilage-bones are characterized by very feeble development of the perichondral lamellae accompanied by excessive development and persistence of the endochondral spongiosa. This condition, seemingly associated with deficient activity of the osteoclasts, is suggestive of the histological character of the bones in the Cetacea. Some bones in the 'grey-lethal', in particular the hyoid, the vomer, the turbinals, and the vertical part of the nasals, are excessively developed and heavily ossified. On the other hand, the deficient calcification of the dentine and lack of growth of the upper incisors have deprived the premaxillae of the stimulus necessary for their growth, with the result that the snout is abnormally short.

This direct correlation between tooth-development and bone-growth is of great interest, but no less important are the secondary effects of the fact that in the 'grey-lethal' mouse the teeth are insufficiently developed to allow the animal to feed as a Rodent at all. This has resulted in a reduction of certain masticatory muscles and consequent abnormal development and morphological differentiation of certain bones, such as the zygomatic process, the pterygoid, the condylar process of the dentary, and the infraorbital foramen.

10. Achondroplasia (Chondrodystrophia) in Birds and Mammals.

The effects on the skull of the pathological condition known as achondroplasia or chondrodystrophia, found in fowls, cattle, and Man, are marked and characteristic. In the hen, as studied by Landauer (1927), the skull becomes highly domed, and, the upper jaw overlapping the lower, presents a parrot-like appearance. The base of the skull is very short indeed, and the cartilage-bones of the skull-base and occipital region are precociously developed and fused together leaving no trace of sutures. At quite early stages no cartilage whatever is left inside the perichondral lamellae of the cartilage-bones, and it is this loss of cartilage and consequent loss of the power of growth between the cartilage-bones, which brings about the shortness of the skull-base and the reduction in diameter of the foramen magnum. This in turn leads to the bulging of the skull-roof, to accommodate the brain. There is also precocious modification of the cartilage (hypertrophy, calcification) in those regions of the chondrocranium where cartilage-bones subsequently arise (e.g. in the nasal septum). The membrane-bones are also precocious in their appearance, and the parietal, squamosal, and angular are all well ossified at the 11-mm. stage (when normally they have not yet appeared).

In the skull of achondroplasic cattle Crew (1923) has also noted the extreme shortness of the skull-base, the precocious fusion of all the occipital and sphenoid bones, and the high doming of the roof, with bulging forehead and parietal eminences. Further, the membrane-bones are abnormally well developed: the frontals and parietals are thicker and the vomer is massive and fused with the palatine processes of the premaxillae. On the other hand, the nasals and maxillae are very short, and as the latter and the palatines develop no palatine processes, there is no false palate, which feature gives the skull a reptilian appearance.

39. GENERAL MORPHOLOGICAL CONSIDERATIONS

I. The Problem of Morphological Relations

One of the chief conclusions which emerge from a comparative study of the development of the skull in all vertebrates is the constancy of morphological relations. The pila antotica always rises up in front of the trigeminal nerve; the palatine nerve always passes down behind and forward beneath the basitrabecular process; the vagus nerve always passes out behind the auditory capsule and in front of the occipital arch (except, of course, in Cyclostomes where there is no occipital arch). So constant are these and other such relations that many cartilages are identified and defined or non-identified on the basis of their morphological relations.

The logical status of these conclusions must not, however, be lost from view. The morphological relations of a given cartilage are established by observation of as many forms as possible, and the identity of a cartilage is then determined according to its relations. It is only because these relations are so obviously constant in so many different forms that the use of the criteria of relations to establish the identity of a cartilage in any single form is justified. Thus, the pila antotica has been recognized throughout the vertebrate series from Selachians to mammals, as has the basitrabecular process, the orbital cartilage, the lamina orbitonasalis, and many other cartilages. Similarly it has been possible to trace several bones from the most primitive Osteichthyes up to Man.

On the basis of their morphological relations the homologies of the auditory ossicles in mammals have been established beyond doubt; perhaps one of the most sensational achievements of research in vertebrate morphology.

In many cases, the non-concordance of the morphological relations of a given structure has drawn attention to the suspicion of its non-homology with another structure, and this suspicion has then subsequently been confirmed. This is the case with the skeleton of the hyoid arch in sharks and skates; the ventral element of the arch in the latter being situated laterally instead of medially to the afferent pseudobranchial artery, has been recognized not to be a ceratohyal at all, but to be a special structure formed from fused hyal rays and termed a pseudohyoid. (See p. 70).

The fact that in the fully formed chondrocranium of the fowl a cartilage is situated between the profundus and maxillary branches of the trigeminal nerve, suggested that this cartilage could not be a pila antotica. This suspicion has been confirmed by a detailed study of the development in the duck, where the pila antotica has been seen to become replaced by a secondary cartilaginous strut in the fully formed chondrocranium. Similarly, the pleurosphenoid (so-called 'alisphenoid'), ossifying in the pila antotica of reptiles, has been recognized as non-homologous with the alisphenoid of mammals, which ossifies in the ala temporalis. (See p. 439).

Sometimes, however, the non-concordance of morphological relations present problems of special difficulty, for the answer in such cases does not appear to be as simple as mere non-homology of the structure in question with other structures. For instance, in the various fish, the hyomandibular branch of the facial nerve may pass out in front of the hyomandibula (or hyosymplectic) as in Selachii or in Gadus, or behind it as in Acipenser, or through it as in Salmo, or partly in front and partly behind as in Polypterus. In the fossil Sinamia which bears such strong resemblances to Amia, Stensiö (1935) has shown that the hyomandibula is not pierced; the facial nerve can therefore not have pierced it, as it does in Amia. In mammals the facial nerve typically runs out behind the styloid process: in Hyrax it runs in front of it. The chorda tympani nerve runs down in front of the tympanum in all Tetrapoda except the Anura where it runs behind it. The same nerve is post-trematic in all vertebrates except Gallus when it is pretrematic; it runs behind the processus dorsalis of the columella auris except in Struthio where it runs in front of it. The same is true of certain Marsupials.

The efferent pseudobranchial artery in Selachians joins the internal carotid after passing dorsally to the trabecula instead of ventrally to it as in other fish. The orbital artery may pass forwards beneath the columella auris as in Sphenodon and Ophidia, or through it as in Hemidactylus, or above it as in Lacerta. The internal carotid arteries enter the cranial cavity by passing between the trabeculae except in Polypterus, Acipenser, Amiurus, and Rana.

If the morphological relations which are found to hold in the vast majority of forms are to have any validity or meaning, some explanation must be found for these exceptional cases.

The hyomandibula of fish and its relations to the hyomandibular branch of the facial nerve have formed the subject of a special discussion (p. 411), the result of which has been to conclude, both from the facts of development and the analogy of the branchial arches in certain fish, that variation in relations has been brought about by a change in the direction and position of upgrowth of the dorsal end of the hyomandibular cartilage, the cartilage growing upwards in a different position in regard to the nerves. Thus, it may grow up behind all branches of the nerve (Selachii, Gadus), between its branches (Polypterus), in front of it (Acipenser) or both in front of and behind it, enclosing it in a foramen (most Teleosts). On the other hand, the course of the chorda tympani in Didelphys, Dasyurus, Phascolarctus, and Manis (medially to the styloid process instead of laterally to it as in all other mammals, see p. 414) appears to be an effect of the growth of the nerve in an atypical position—the same is probably true in Gallus (p. 279).

These cases introduce the important fact that morphological relations are the result of the more or less independent origin of two or more structures in a three-dimensional geometric field. Each of these structures, be they nerves, blood-vessels, or cartilages, develop under certain morphogenetic influences which determine their position. Very little is known as to the nature of these influences, though in the case of nerves and blood-vessels, it is beginning to look as if the course which they take is imposed upon them by physical factors, for it must be remembered that these structures do not develop in situ, but thread their way during development from one point to another. Rudiments of cartilage and bone, on the other hand, owe the localization of their position to previous migration and concentration of mesenchyme cells, under the influence of what are now coming to be known as morphogenetic fields. (See Huxley and de Beer, 1934.)

Apparently, the primary architecture of the vertebrate head, governed by such major features as the positions of the brain, nose, eye, ear, mouth, and visceral clefts, is such that the various structures, when they do arise, find themselves in a particular geometrical relationship which is remarkably constant. It is, however, capable of modification, as a result of slight relative shifts in position of parts of the primary architecture, and the hyomandibula and hyomandibular nerve appear to be a case in point. Another example seems to be provided by the chorda tympani and the tympanum of Anura (p. 207) where the precocious and extreme forward position of the spiracular visceral pouch makes it practically impossible for the nerve to take its 'typical' course.

For reasons already stated (p. 423) the *exact* homology of the hyomandibula (or hyosymplectic) in different fish is questionable owing to the probability that the suspensional function of the hyoid arch skeleton has been independently acquired in Selachii and in Teleostomi. This difficulty does not, however, arise with the columella auris of reptiles, which must be homologous in them all. Yet here, the orbital artery may vary in its relations to the columella. If the conditions in Sphenodon and Lacerta (in which the artery runs respectively under or over the cartilage) were the only ones known, one might be led to believe that the variation was due to the establishment of the arterial channel through different capillaries in the two cases. But the condition in Hemidactylus (and also in Gymnophiona and in mammals) shows that this cannot be the case, for here the artery actually perforates the cartilage. The differentiation of the artery precedes that of the cartilage, and the explanation must be that the relative positions of the artery and of the mesenchymatous rudiment of the cartilage have shifted so that the former pierces the latter.

In a similar way it can be shown that the maxillary branch of the trigeminal nerve, which passes behind the ascending process in reptiles, but in front of the ala temporalis in mammals, has penetrated through the cartilage, since in some forms it is found in a foramen.

With regard to the variation between the internal carotid and the trabeculae, each case must be studied on its own. In Amiurus there is definite evidence that in young stages the internal carotids bear typical relations to the trabeculae, passing up between them, and that in later stages new vessels develop, looping round each trabecula. The vessel which passes into the skull laterally and dorsally to the trabecula in Amiurus is thus not the internal carotid. In Rana, similarly, at early stages the carotid enters the skull in the typical manner between the trabeculae. But later, each carotid foramen becomes confluent with the oculomotor foramen of its side, by resorption of the cartilage, with the result that the carotid appears to pass in the 'wrong' place.

In Polypterus, unfortunately, the relations of the internal carotid to the trabeculae at early stages is unknown. In Acipenser, however, it seems that at the earliest stage the carotid is lateral to the trabecula (or rather, to that portion of the trabeculae which chondrifies first), or in other words, that a variation in position has taken place. During development the carotid becomes enclosed in the lateral extension of the trabecula. In possible explanation of this it may be noticed that from the very earliest stage in Acipenser the carotids are very wide apart, as indeed they are in the adult where they enter the skull through foramina laterally to the edge of the parasphenoid, which itself is also very wide.

In this case there seems no good reason to doubt the homology of either carotid artery or trabeculae with the similarly named structures in other forms.

In other cases, it appears that the abnormality of relations is to be explained by simple displacement, where this has been geometrically possible. Thus, in Struthio the chorda tympani must have slipped forwards over the processus dorsalis of the columella auris; the facial nerve probably slipped forward over the unattached dorsal end of the laterohyal before its attachment to the crista parotica in Hyrax; the facial nerve in Urodeles may have slipped off the end of the stylus of the columella auris and thus come to pass ventrally instead of dorsally to that cartilage.

It is to be concluded from this discussion, therefore, that while morphological relations between different structures are remarkably constant, this constancy must not be taken as implying a hard-and-fast rule. It would be interesting to know what course the carotids take in those abnormal cyclopic Amphibia (see p. 477) in which the trabeculae are close together instead of wide apart, as normally. Morphological relations are *usually* of great value in determining the identification of a structure, but not always. It is, however, reasonable to require that some explanation be found to account for the exceptional cases.

Serious difficulties emerge from a consideration of the relations between certain dermal bones and the lateral line canals of the head. In *general* it may be said that the bones of the head through which the various canals of the lateral line system pass are remarkably constant. The supraorbital canal usually threads its way through the nasal, frontal, and postfrontal; the infraorbital through the lachrymal, jugal, postorbital, intertemporal, and supratemporal; and the transverse occipital canal through the extrascapulars (postparietals).

Cases occur in which the lateral lines pass through what appear to be the supraorbitals, or the parietal, or even the maxilla. A major difficulty is encountered here from the fact that the argument may be used either way. On the one hand, it may be held that the lateral line canal has changed its position and lies in a 'different' bone. On the other hand, it might be said that the lateral line canal is the primary structure round which the bone forms, and that the bone is always the same although its position may be different. On this view the maxilla of Polypterus is really a suborbital; the parietal of Polypterus is really an intertemporal; the parietal of Cyprinus is really a postparietal, &c.

There is in addition the possibility that bones may have fused, so that the parietal of Polypterus may be a parietal-plus-intertemporal, &c.

The evidence, so far, is insufficient to decide finally between these possibilities. (See also p. 508.)

II. The Possibility of Membrane-bones becoming Cartilage-bones and vice versa

The reader of those sections of this book dealing with the development of the bones in the different groups of fish will have been struck by the difficulty of strictly applying the criterion of cartilage-bones as ossifications developed solely beneath the perichondrium, or of membrane-bones as ossifications developed solely outside the perichondrium.

Examples are numerous. The prevomer (membrane-bone) of Salmo invades the cartilage of the ethmoid plate: the epiotic (cartilage-bone) develops intramembranous extensions or apolamellae outside and projecting right away from the perichondrium: the supraoccipital (cartilage-bone) in many fish develops partly as a perichondral ossification in the tectum synoticum and partly as an intramembranous plate in the membranous roof of the skull; the membranous medial wall of the trigeminofacialis chamber in Lepidosteus becomes ossified by the prootic. Further study and further consideration of already known facts would provide countless further instances of this fact: namely, that the distinction between cartilage-bones and membrane-bones as developed within or without the perichondrium cannot be applied with absolute strictness. As Gaupp (1906*a*, p. 620) has pointed out, some cartilage-bones are 'mimimorphic' and exactly copy the form of the pre-existing cartilage; others are 'automorphic', and their definitive form bears very little relation to that of the cartilage. (See also Gaupp, 1901, p. 926.)

It would, however, be a mistake to conclude from these results that the distinction between cartilage-bones and membrane-bones is unreal. On the contrary, the *general* uniformity throughout the vertebrates with which the greater part of some bones regularly develop within the perichondrium while others regularly develop outside it, and the universal regularity found in the Tetrapoda, must reflect a fundamental distinction, to which the cases confined to the fish and referred to above are exceptions.

The matter is best approached by a consideration of ossification as exhibited on the one hand by such a structure as a vertebra, on the other by such structures as the superficial bony scales.

There is every reason to believe that throughout phylogeny, the ossification of a vertebra has always been a deep-seated cartilage-bone. Similarly, the bony scale has certainly never been anything other than a superficial membrane-bone. Now, turning to the skull, it is clear that the basioccipital, exoccipital, epiotic, prootic, pleurosphenoid, orbitosphenoid, quadrate, metapterygoid and articular, hyomandibula, &c., have always been cartilage-bones. On the other hand, those bones which bear teeth (premaxilla, maxilla, prevomer, parasphenoid, dermopalatine, ecto- and endopterygoid, dentary, and splenials), or through which lateral line canals pass (nasal, frontal, inter- and supratemporal, postparietal, sub- and postorbitals, angular, and preopercular) and some other superficial bones (parietal, opercular, sub- and interopercular, branchiostegal rays), have always been membrane-bones.

As specialization sets in, the membrane-bones may sink deeper and come into such close contact with the chondrocranium as either to lie directly on the the perichondrium (e.g. parietal, frontal) or actually to invade it in part (e.g. supraethmoid). On the other hand, muscle attachments may form, requiring firm anchorage provided by extraperichondral intramembranous extensions (apolamellae) of certain cartilage-bones. The sclerotic bones (see p. 443) appear to provide a good case of ossifications which may vary in their method of histogenesis.

Another fact which must be borne in mind is that occasionally the cartilaginous phase in the development of a particular structure may be suppressed. This takes place frequently in the case of the vertebrae of Teleostei, where basidorsal and basiventral cartilages may fail to chondrify, and the bony vertebra ossifies directly out of mesenchyme. No one would think of denying the homology of such a vertebra ossified in mesenchyme with a vertebra ossified in cartilage. Another example is the lamina ascendens of the ala temporalis; in Perameles (Cords, 1915) this ossifies at least in part in the cartilage of the ala temporalis as in most mammals; in Didelphys, however, it ossifies wholly in membrane without any cartilaginous precursor at all.

The truth is that the expression 'cartilage-bone' is apt to convey the erroneous impression that the bone is invariably in some way dependent on or consequential upon the cartilage in which it ossifies. On the contrary, there is reason to believe that cartilaginous rudiments and bony rudiments are fundamentally independent manifestations of differentiation which may, however (in the case of the cartilage-bones) coincide in time and space, and may further become causally related since it appears (in some cases at least) that the presence of hypertrophic cartilage is a prerequisite for the ossification of a cartilage-bone (Fell and Landauer, 1935. See p. 482).

The scaffolding provided by the perichondrium of the previously formed cartilage may, further, lead the subsequently formed cartilage-bone to be in some geometric measure 'dependent' on the cartilage. This phenomenon is manifested in those cases of abnormality of the bony human skull which are attributed to corresponding abnormalities of the chondrocranium. A large number of these cases have been noted and classified by Augier (1931*a*); only a few need therefore be considered here. The basioccipital may preserve a remnant of the notochordal canal in those cases in which the canal has not been obliterated by the cartilage of the basal plate; the exoccipital ossification may extend into such cartilaginous rudiments of the proatlas arch which may be attached to the occipital arches; the paired rudiments of the supraoccipital may remain separate in those cases in which the two halves of the tectum posterius likewise remain separate; the supracochlear ossification may remain separate when its cartilaginous precursor (a remnant of the pila antotica) preserved its independence; the supratrabecular bar may ossify should it (abnormally) have become chondrified; an orbitospheno-frontal ossification may arise in the sphenethmoid commissure. (See also Augier, 1937).

In spite of these cases, however, it is clear in others that should the cartilaginous rudiment be suppressed or retarded, or the bony rudiment accelerated in its appearance, the 'cartilage-bone' may nevertheless appear, without cartilaginous preformation. This seems to be the explanation of the intramembranous ossification of the basisphenoid in most Teleostei, and it is especially interesting to note that in Exocoetus, the basisphenoid contains a cartilaginous core, a remnant of the previous phylogenetically extensive cartilaginous side wall of the skull, pressed into the middle line in the formation of the interorbital septum. This is also the explanation of the ossification in membrane of the metapterygoid in Syngnathus and of part of the 'sphenoid' in Polypterus. The case of the metapterygoid of Polypterus is more difficult, however, owing to the fact that the cartilage there has not been reduced; and it is possible that this bone is not a true metapterygoid, but an additional dermal ossification.

There are, however, certain bones which still present a problem.

The supraoccipital in Teleostei presents a difficulty, because the more primitive Teleostomes lack it and possess instead the membrane-bone known as dermosupraoccipital. It is possible that the supraoccipital may have originated within the Teleostei as a cartilage-bone, although recent palaeontological research suggests that a supraoccipital cartilage-bone may have been present in primitive bony fish. At all events, the reason why it develops partly in membrane is probably the reduction in size of the cartilaginous tectum synoticum.

With other bones again, the question of identity arises, doubt concerning which naturally invalidates any conclusion which may be arrived at. This is the case, for instance, with the *opisthotic*. This bone in the Tetrapoda is a cartilage-bone. But the bone which ossifies in a corresponding position in Teleostomes appears to ossify in whole or in part in membrane, for which reason it has been called the *intercalary*. It still remains to be proved that these bones are homologous before any conclusion can be arrived at concerning the possibility that the intracartilaginous opisthotic of Tetrapoda was a membrane-bone in the fish. Further research is also required to discover whether the so-called opisthotic or intercalary ossifies as a cartilage-bone in any bony fish.

There remain a number of bones which are of interest because their relations have been used as evidence for the conversion during phylogeny of dermal bones into cartilage-bones. These elements may conveniently be considered in pairs, the first member of the pair in each case being a membrane-bone, and the second a cartilage-bone: supra-ethmoid and ethmoid, prefrontal and lateral ethmoid, postfrontal and sphenotic, intertemporal—supratemporal and pterotic.

The distribution, so far as it is known, of these bones is assembled in the accompanying table (pp. 500, 501).

Nomenclatural confusion has not missed its chance of interfering in this problem, and attempts have been made by various authors to ensure clarity by referring to bones with the prefixes 'dermo', and 'auto' or 'chondro', to indicate the dermal or chondral methods of ossification respectively. This applies particularly to the sphenotic and pterotic, which appear in the literature as 'dermosphenotic', 'autosphenotic', 'dermopterotic' and 'autopterotic'. The dermopterotic has also, wrongly, been called the squamosal.

Actually, the term dermosphenotic has come to be used as a synonym for postfrontal, for before the distinction between cartilage-bones and membrane-bones was clearly realised, the 'sphenotic' of fish was regarded as homologous with the postfrontal of reptiles: the postfrontal of fish then became known as the dermosphenotic. This is unfortunate, because in only two fish, Polypterus and Dactylopterus, is the postfrontal known to be fused with the sphenotic: there is no reason to think that the postfrontal ever was the 'dermal representative' of the sphenotic: the postfrontal is the uppermost bone of the postorbital series, and it is the intertemporal which comes in question as the dermal representative of the sphenotic. Postfrontal, intertemporal, and supratemporal are all present and separate in Osteolepis (Goodrich, 1930, p. 286).

An example of the confusion which has arisen on this point is the apparent conflict between the statements of Ridewood (1904*a*) that the sphenotic and postfrontal are nearly always fused, and of Allis (1909) that the sphenotic and postfrontal are almost invariably separate. If intertemporal is read for Ridewood's postfrontal, the discrepancy disappears.

Here, in this book, the terms *sphenotic* and *pterotic* are used to denote cartilage-bones, while the terms *intertemporal* and *supratemporal* are used to denote the dermal ossifications sometimes referred to as dermosphenotic and dermopterotic.

On considering the various pairs of bones given in the accompanying table, it will be noticed that the dermal element is in general present and the chondral element absent in the more primitive fishes, while this condition is somewhat reversed in the more specialized Teleostei.

It is clear that in the phylogeny of the bony fish there has been a gradual retreat of dermal bones to a deeper level beneath the surface, and it does certainly look as if the supraethmoid had gradually invaded the cartilage and given rise to the ethmoid: as if the supratemporal had similarly sunk in and given rise to the pterotic, at the same time leaving on the surface a lateral line canal tubular ossicle. On the other hand, the sphenotic and the lateral ethmoid are already present in Polypterus. It is possible that the so-called prefrontal of Acipenser may be a fortuitous dermal ossicle. The lateral ethmoid, therefore, can hardly have arisen from the prefrontal.

A difficulty in the way of regarding the sphenotic or pterotic as a sunk membrane-bone is the fact that these cartilage-bones arise from perichondral lamellae on the internal as well as the external surface of the cartilage. Mechanical sinking in of a membrane-bone could not give rise to the internal lamella, and therefore if these cartilage-bones have been derived from membrane-bones, it must have been by 'infective' spread of ossification. (See p. 481).

Altogether, the evidence is insufficient to prove either that these cartilage-bones have arisen in phylogeny by the sinking in of membrane-bones, or alternatively that they have always been cartilage-bones, though in the case of the ethmoid, the former view seems the more probable. Nor is any comfort to be found in a study of the processes of origin of the bones. In Salmo the supratemporal and pterotic arise separately and fuse later. In Exocoetus they arise separately and remain separate. In Amiurus so far as known they appear to arise fused from the start. In Cyclopterus the supratemporal (or what must be taken to represent it, since it surrounds the lateral line) arises as an intramembranous outgrowth from the pterotic. But what can be concluded from that?

The problem is still further complicated by the possibility that in the process of sinking beneath the surface, the membrane-bone may have become divided into two portions: a superficial gutter-like bone lodging the lateral line canal, and beneath it, the plate-like portion of the membrane-bone: the latter portion being distinct from, though possibly fusing with, the still deeper perichondral ossification. Allis (1899), for instance, says of the supratemporal-pterotic group (his 'squamosal') in Teleostei that it consists of 'a canal component and a deeper-lying component which may be either a so-called membrane-bone, or such a bone fused with a so-called primary ossification'.

There would thus be three elements: one cartilage-bone and two membrane-bones, and it is apparently to distinguish the latter from one another that Allis (1935) uses the prefixes dermo- and membrano-: the former for the lateral line bone.

In many fish the pterotic, although arising as a perichondral ossification, develops intramembranous extensions or apolamellae which serve for the attachment of muscles, or the articulation of the head of the hyomandibula. In those cases where the lateral line component is left as a small tubular ossicle near the surface, one might be tempted to regard the apolamellae of the pterotic as representing the plate-like portion of the supratemporal.

To clear up this point, it would be necessary to know that a dermal ossification, such as the supratemporal, *can* split into two portions, and in order to avoid the danger of mistaking the deeper plate-like portion of the supratemporal for any part of the underlying pterotic, it would be necessary to investigate a bone through which the lateral line canal passes and which is not underlain by a cartilage-bone. A bone of the extrascapular series would be suitable material for such a study.

Dabelow (1927) has attempted to show that the perichondral sclerotic bones were derived phylogenetically from the circumorbital membrane-bones of primitive fish. Edinger (1929) has shown the untenability of this view (see p. 443), while at the same time demonstrating the fact that the sclerotic bones may vary in their histogenesis, and arise as cartilage-bones in some forms and membrane-bones in others. (See also Chatin, 1910.)

The fact is that the best evidence on which these questions can be solved is

	Supra-ethmoid	*Ethmoid*	*Prefrontal*	*Later ethmo*
Polypterus (Lehn, 1918; Allis, 1922)	present	absent	absent	prese
Acipenser (W. K. Parker, 1882*a*; Sewertzoff, 1926*a*)	present (probably represented by rostrals	absent	present	prese
Amia (Allis, 1897; Pehrson, 1922)	present	absent	absent	prese
Lepidosteus (W. K. Parker, 1882*b*; Veit, 1907)	present	absent	absent	prese
Salmo (Schleip, 1904; Böker, 1913)	present	absent	absent	prese
Amiurus (Kindred, 1919)	fused *ab initio*? (supra-ethmoid and ethmoid)		fused *ab initio*? (prefrontal and lateral ethmoid)	
Gasterosteus (Swinnerton, 1902)	develops as lateral extension of	present	absent	prese
Cyclopterus (Uhlmann, 1921*b*)	? develops as lateral extension of	present	develops as extension of lateral ethmoid	prese
Pleuronectes platessa (Cole and Johnstone, 1901; Mayhoff, 1914)	present	develops as invasion of cartilage by supra-ethmoid	develops as extension of lateral ethmoid	prese
Syngnathus (Kindred, 1924)	present — fused (with ethmoid)	present — fused (with supra-ethmoid)	develops as extension of lateral ethmoid	prese
Anguilla (Norman, 1926)	present, develops on site of removed and disappeared cartilage	?	absent	?
Exocoetus (Lasdin, 1913)				
Scomber (Swinnerton, 1902; Allis, 1903*a*)	absent	present	absent	prese

<table>
<tr><th>Postfrontal</th><th>Sphenotic</th><th>Intertemporal</th><th>Pterotic</th><th>Supratemporal</th></tr>
<tr><td colspan="2">fused ab initio</td><td>present</td><td>absent</td><td>present</td></tr>
<tr><td>present</td><td>absent</td><td>present</td><td>absent</td><td>present</td></tr>
<tr><td>present</td><td>present</td><td>present</td><td>absent</td><td>present (probably fused with intertemporal)</td></tr>
<tr><td>present</td><td>present</td><td>present</td><td>absent</td><td>present (probably fused with intertemporal)</td></tr>
<tr><td>present (fused with postorbital)</td><td>present</td><td colspan="2">ultimately free</td><td>present (probably fused with intertemporal)</td></tr>
<tr><td>present</td><td colspan="2">fused ab initio?</td><td colspan="2">fused ab initio?</td></tr>
<tr><td>absent</td><td>present</td><td>represented by apolamella of sphenotic?</td><td>present</td><td>represented by apolamella of pterotic?</td></tr>
<tr><td>absent</td><td>present</td><td>represented by apolamella of sphenotic?</td><td>present</td><td>represented by apolamella of pterotic?</td></tr>
<tr><td>absent</td><td colspan="2">present present
ultimately fuse</td><td colspan="2">present present
ultimately fuse</td></tr>
<tr><td>absent</td><td>present</td><td>represented by apolamella of sphenotic</td><td>present</td><td>?</td></tr>
<tr><td>absent</td><td>present</td><td colspan="2">present present
ultimately fuse</td><td>present (probably fused with intertemporal)</td></tr>
<tr><td></td><td>present</td><td>present</td><td>present</td><td>present</td></tr>
<tr><td></td><td>present</td><td>absent (unless fused with supratemporal)</td><td colspan="2">present present
fused</td></tr>
</table>

that provided by fossils, for which reasons it is earnestly to be hoped that palaeontologists will continue to devote their attention to such problems.

Special mention under this section should perhaps be made of the tendency presented by several undoubted membrane-bones in mammals of ossifying in regions previously occupied by *secondary* cartilage. Secondary cartilage differs histologically from chondrocranial 'primary' cartilage in that its cells are very large while the cell-capsules, or intercellular matrix, are very thin. Perhaps the best example of this phenomenon is that presented by Galeopithecus, reported by Gaupp (1907*a*). In this form secondary cartilage is present in the dentary, not only in its angular, condylar, and coronoid portions, but also extending for a considerable distance forwards; it is present in the body and in the palatine process of the maxilla, in the palatine, pterygoid, and squamosal. Further, a tissue resembling secondary cartilage is found in the jugal, the vomer, and the frontal. Strasser (1905) found secondary cartilage in the pterygoid bone of the pigeon.

Fuchs (1909*a*) attempted to make a case for the view that secondary cartilage was really comparable to primary cartilage, and therefore represented parts of the original chondrocranium. But the known phylogenetic history of the bones in question shows without shadow of doubt that they are membrane-bones, never having had anything to do with the chondrocranium. The only tenable view, therefore, is that which has been generally held since Kölliker (1849), and supported by Gaupp (1906*a*, p. 613), Palmer (1913), de Beer (1929), &c., to the effect that secondary cartilage is morphologically absolutely independent of the primary cartilage of the chondrocranium, and appears to be an adaptation for resistance to precocious strains and stresses on the part of the rudiments of membrane-bones (Schaffer, 1930; Weidenreich, 1930).

A problem of some interest arises from the fact that the secondary cartilage referred to above presents remarkable resemblances to the hypertrophic type of cartilage which develops in those parts of the 'primary' cartilage of the embryonic skeleton in which ossification subsequently takes place. Fell's (1933) experiments (described on p. 482) in which tissue culture of a cartilage-bone (endosteum of femur) was found to be capable of differentiating into hypertrophic cartilage, lends support to the view that this type of cartilage is a special histological manifestation of those cells which normally produce bone.

If secondary cartilage is commonly associated with the formation of membrane-bone, and hypertrophic cartilage (in the chondrocranium) with cartilage-bone, the question requiring solution is the method of origin of the hypertrophic type of cartilage in the primary cartilage of the chondrocranium. More especially is this necessary since it has been claimed that the cells of hypertrophic cartilage *can* become transformed into osteoblasts *in vivo* (Levi, 1930) or *in vitro* (Fell, 1933), and that no ossification in cartilage takes place in the absence of hypertrophic cartilage. (Fell and Landauer, 1935).

III. The Identity of Morphological Units

i. Discussion

Almost throughout this work, as in most others, various terms such as frontal, parietal, &c., have been in constant use, as if their connotation were definite and unquestionable. The success with which general comparisons between skulls of different adult forms can be made, and the recognition of bony units which appear to correspond, tend to create the point of view which regards the various 'bones' as standard and definite entities, and thus to conceal the awkward fact that there is considerable difficulty in arriving at a suitable definition of 'a bone'.

The attempt might be made to define a bone according to embryological criteria as the product of a single embryonic centre of ossification, but this leads into serious difficulties. In the first place, many single bones may arise from more than one centre of ossification. In the case of tooth-bearing bones, as Moy-Thomas (1934) has shown, there may be a separate little plate of bone beneath each tooth, the plates soon fusing into a single structure. Obviously the individual centres of ossification in this case cannot have the morphological value of originally separate bones.

The nasal in Amia is said to arise from three centres of ossification, and as in Polypterus these centres arise in a corresponding position, remain separate, and give rise to three bones (nasal, adnasal, and terminale), it is reasonable to regard the three centres in Amia as representing three originally separate bones. But so far as is known, the nasal in other fish arises from a single centre.

The frontal in Amia arises from four centres, but in other forms it seems to arise from a single centre. Whatever the significance of the four centres in Amia may be (four originally separate bones?), they may be represented by a single centre in other forms.

An additional complication is introduced by the fact according to Pehrson, that each of the four centres of the frontal of Amia is double: the lateral portions he regards as representing the supraorbital bones and the medial portions the frontal. He therefore regards the frontal of Amia as fused frontal plus supraorbitals.

The parietal in Man arises on *each* side from two centres of ossification which (nearly always) fuse very soon. The phylogenetic history of the parietal in the mammals does not permit of the view that it is a compound bone, and therefore the duplicity of its centres of origin cannot be the reflection of the previous existence of separate bones. The parietal in Man must, therefore, be regarded as affording evidence of the subdivision of an originally single bone, further examples of which are provided by the squamosal, tympanic, malleus, stapes, and certain Wormian ossifications in Man, or the suborbital and opercular ossicles in Lepidosteus.

It is also to be remembered that cartilage-bones often show more than one centre of ossification. Thus, the mammalian femur has a diaphysis and two epiphyses, but cannot for that reason be regarded as three bones. Most of the bones of the auditory capsule in fish (e.g. prootic) start from separate internal and external perichondral lamellae, but are nevertheless clearly single bones.

It is, therefore, clear that a separate centre of ossification cannot be taken as a safe criterion of a single bone, for allowance has to be made for the fact that in some cases a single centre may represent more than one centre in another form, or more than one centre may represent a single centre in other forms. In other words, *it is not possible to arrive at a satisfactory definition of what constitutes a bone along embryological lines, in terms of separate centres of ossification.*

It has then to be admitted that bones may have been subjected to fusion or fragmentation, and the difficulty in any given case is to decide which of these possibilities has occurred. Is the nasal of Amia to be regarded as the product of fusion of the nasal, adnasal, and terminale of Polypterus, or are the latter to be regarded as the products of fragmentation of an originally single bone such as the nasal of Amia? To these two possible explanations must be added two more: viz. that the single bone of Amia corresponds to *one* of those of Polypterus, the other two having disappeared; or, that they do not correspond at all.

At the same time it must be remembered that the formation of bone is in ontogeny in the nature of a response to stimuli of stress and strain (see p. 485), and the appearance, non-appearance, fusion, or division of bones, must be considered from this aspect. For the homology of bones must be based on the homology of the causes of their formation.

To illustrate this problem, it may be convenient to formulate it as follows:

If in two related forms, A and B, similar positions are occupied by two (or more) bones in A and one bone in B, this may be due to the possibility that:

(1) the bones in A and B are not homologous, all comparison between them being fruitless; or
(2) the two (or more) bones in A, representing the primitive condition, are fused into a single bone in B; or
(3) the single bone in B, representing the primitive condition, has subdivided into the two (or more) bones in A; or
(4) of the two bones in A representing the primitive condition, one has disappeared and the other has expanded and is the single bone in B.

Examples of these possibilities may now be considered.

(1) The squamosal of Osteolepis occupies more or less the same position as the preopercular and subopercular of Polypterus, but all attempts to derive the former from the latter or vice versa are invalidated from the start, since the fossil history of the Actinopterygian fish (including Polypterus) shows that in them the squamosal has disappeared altogether.

The converse to the disappearance of a bone is the appearance of a new one, for which reason no comparisons are possible involving such fortuitous ossifications as certain of the Wormian bones in man. Perhaps the spiracular ossicles of Polypterus, the endotympanic of mammals, the siphonium of birds, the 'supraorbitals' which W. K. Parker (1866*a*) found in the Tinamu, or the 'tympanic' bone which he described in Pavo (Parker, 1862), and the ethmoid ossifications in the nasal capsule of Tinnunculus, may be of this nature.

(2) The fusion of two (or more) centres to form one bone in place of two (or more) bones which in previous phylogeny remained separate, occurs in most demonstrative manner in Man. Each of the frontals has its own centre, but the two bones soon fuse into one so completely that all traces of the suture that once separated them may disappear. Comparable fusions occur in the formation of the human 'sphenoid', which combines the basisphenoid, presphenoid, orbitosphenoids, alisphenoids, and pterygoids, of other mammals.

Similarly the supraoccipitals fuse with the postparietals and with the exoccipitals and basioccipitals. The premaxillae fuse with the maxillae. In placental mammals, it is probable that the premaxillae and prevomer are fused, and that the so-called pterygoid is the product of fusion between the reptilian pterygoid and the lateral wings of the parasphenoid. In Hypogeophis the 'basale' is composed of the fused parasphenoid, pleurosphenoids, occipitals, and auditory capsules. Similarly the nasal fuses with the premaxilla, the maxilla with the palatine, the frontal with the prefrontal, and most of the lower jaw-bones fuse together.

In Amia the nasal, adnasal, and terminale appear to be fused into the single (paired) so-called nasal (for it is very difficult to believe the opposite, viz. that the three bones are the results of splitting of an originally single nasal). Similarly the extrascapular or postparietal of Amia seems to represent three separate extrascapulars in other fish, and the two postorbitals of most Amia may fuse into one in some specimens. On the other hand, Amia usually preserves separate parietals, although they may be exceptionally fused into one median bone (Bridge, 1877), as is apparently normally the case, in the closely related fossil Sinamia (Stensiö, 1935). In the frog the fusion between frontal and parietal is well known, and the sphenethmoid seems to represent fused orbitosphenoids.

In all these cases there is definite ontogenetic evidence of fusion, since separate centres of ossification have been observed, which ultimately become confluent. In other cases, when this evidence is lacking, the matter is much more difficult. In some cases, bones which were undoubtedly paired in previous phylogeny, may arise from a single median rudiment. This seems to be the case with the parietals in Ceratodus and the prevomers in Teleostei. In Ichthyophis, the prootic and exoccipital appear to ossify in one piece (which then fuses with the parasphenoid). If it is true, following the argument set out on page 503, that the four centres of ossification of the frontal of Amia are evidence of the derivation of this frontal from four originally separate head-scales, then the origin of the frontal from a single centre in other forms must mean that a bone which is the product of fusion of elements which were separate in previous phylogeny, can arise from a single centre. Attention was directed to this possibility as long ago as 1834 by Dugès, who termed it 'fusion primordiale', as distinct from 'fusion secondaire' in cases where there is embryological evidence of fusion. But even if this were true, it would be a dangerous precedent to go on, for there would be no check to speculative assumptions of fusion in order to explain all difficulties.

Attention may now be turned to some of the doubtful cases.

The parietals may be fused with the supraoccipital in Amiurus; or, the parietals have vanished and the supraoccipital has expanded in their place.

The so-called maxilla of Polypterus is another case in point. Not only does it occupy the position of the suborbitals in other fish, but it is pierced by the infraorbital lateral line canal, which is normal for suborbitals but exceptional for a maxilla. It should also be mentioned that this bone does not lie in the maxillary fold as the maxilla does in other fish. Three explanations appear possible. One is that the maxilla and suborbitals have fused; the second is that there is no maxilla at all, but teeth have become attached to the suborbitals;

and the third, that the suborbitals have gone and that the maxilla has taken their place. These possibilities will be discussed again below (p. 508).

The same difficulty applies to the so-called parietal of Polypterus which occupies the position of the parietal and intertemporal-supratemporal of other fish, and is traversed by the lateral line canal (normal for an intertemporal-supratemporal, but exceptional for a parietal); or to the so-called dentary of Teleostei, which bears teeth and lodges the lateral-line canal, or to the so-called parietal of Cyprinus through which the occipital lateral-line cross commissure passes (normal for an extrascapula but exceptional for a parietal. See Allis, 1904, p. 439).

In the absence of embryological evidence of separate centres of ossification, it becomes a matter of extreme difficulty to decide whether to regard these bones of Polypterus, Amiurus, and Cyprinus as results of fusion or of substitution (alternative 4 of the present discussion).

In general, it may be said that embryonic fusion of separate centres of ossification is detectable in the adult, by means of the lines of growth surrounding each rudiment, even when the suture between them has vanished. But if the fusion takes place very early (e.g. frontal of Amia), even this criterion may not be applicable to ordinary observation, for growth takes place round the fused centres as if they were one. It is possible that microscopic study of X-ray photographs, in amplification of Görke's (1904) work, might in these cases reveal the multiplicity of original ossification centres.

On the basis of comparisons between the adult anatomies of fossil and living forms, some investigators have been led to believe in fusions of a most extensive character. Thus, Stensiö and Säve-Söderbergh speak of a premaxillo-rostro-interrostro-nasal in Osteolepis. Allis (1935) refers to a parieto-dermosphenopterotico-extrascapular in Saurichthys and a supraorbito-supra-postorbito-membranosphenotic in Cheirolepis. Nielsen (1936) describes a fronto-parieto-intertemporal in Whiteia.

Watson (1921*a*) has drawn attention to the danger of assuming fusion of bones simply because the area of one bone is coextensive with the area of several bones in another form. He admits that fusion does sometimes occur, and shows that it has taken place between the parietal and supratemporal in Macropoma. But in reptiles and mammals generally, there can be no doubt that the reduction in number of bones in the brain-case is brought about not by fusion but by loss of some bones and extension of others. This matter will be discussed below (p. 507).

(3) In attempting to find cases in which a bone which was single in previous phylogeny has split and come to possess more than one centre of ossification, the difficulty is to be quite certain that it really was undivided in its previous history. This, however, seems to be the case with the parasphenoid, a single ossification in fish and Amphibia, but which occasionally in Lacertilia, always in crocodiles and birds and almost certainly in mammals, splits into three pieces; the median portion becomes the vomer, and the two lateral wings, which in crocodiles and birds form the basitemporals, in Monotremes form the so-called mammalian pterygoids, and in Placentals fuse with the reptilian pterygoids to form the definitive pterygoids.

Another case is that already mentioned of the parietal in Man, which possesses two centres of ossification (on each side). It is true that these normally fuse to form one bone, but occasionally they remain separate. At all events, it is clear in this case that the two parietal centres of ossification cannot represent any phylogenetic condition, for the parietal is a single ossification on each side from the fish upwards.

Some of the Wormian bones in Man, especially in cases of hydrocephaly (p. 486), are clearly the result of fragmentation of an originally single centre, and abnormal splitting of most of the bones has at one time or another been observed (see Augier, 1931*a*). Thus, the splitting of the lachrymal has been observed in Elephas and Lagenorhynchus by Muller (1934). The presence in Rhynchocyon and Tupaja of 'prefrontal', 'postfrontal', and other bones, claimed by Wortman (1920), is far more probably explicable as a result of fragmentation. In all such cases it is desirable to know the percentage frequency of the occurrence before a decision can be taken as to its significance.

The fragmentation of the opercular ossicles in Lepidosteus may be a true case of splitting of an originally single bone, and the same may be true of the row of rostral or suborbital bones in this fish, the rudiments of these bones having been as it were torn into eight by the great elongation of the snout. The fact that the infraorbital lateral line canal runs through them has led to the abandonment of the view that they are maxillae, but even if they were maxillae they would still be evidence of fragmentation. The variable postrostrals of Osteolepids (Westoll, 1936) may owe their variability to their being the result of fragmentation.

Still more problematical are the cases in which it has been supposed that part of a dermal bone such as the intertemporal has sunk in and invaded the cartilage and given rise to a cartilage-bone such as the sphenotic. If this be the true history of the sphenotic (which is more than doubtful), then the separate occurrence of a sphenotic and intertemporal would have to be regarded as a subdivision of an originally single bone. (See p. 499).

There is, further, the possibility that certain of the dermal bones through which lateral lines pass may become subdivided into a tubular ossicle lodging the canal and remaining near the surface, and the lamellar plate of the bone which sinks beneath the surface.

It may be noted, that the fragmentation of an originally single bone ('division primordiale' of Dugès, 1834) constitutes an exception to Williston's (1914) 'law' that 'new bones have not appeared in the skulls of reptiles, birds or mammals' (p. 24).

Lastly, mention must be made of the possibility that a single centre of ossification may by resorption become divided at a later stage of ontogeny. This may, perhaps, happen to the jugal in Man (Toldt, 1903).

(4) The substitution by extension of area of one bone for another is really only a special case of the variation which can take place in the sizes of two adjacent bones (see p. 483). It has taken place in all those cases where a reduction in the number of bones has occurred and this reduction has not been due to fusion. Examples may be found in the Actinopterygian fish, where the place of the lost squamosal has been taken by the preopercular and opercular. In the higher reptiles and mammals the original complete covering of dermal bones of the Stegocephalian has been gradually reduced by loss

of certain bones, such as the prefrontal, postfrontal, postorbital, and supratemporal. This case is all the more demonstrative as series showing progressive stages in the reduction of these bones have been discovered among fossil forms. The extension of the mammalian dentary after loss of the other lower jaw-bones is another example. In Acipitrine birds the jugal is lost, and the zygomatic arch is completed without it by the maxilla and quadratojugal.

In Man, substitution after loss sometimes occurs as an abnormality. For instance, the lachrymal may be entirely absent, in which case there is either a gap in the bony skull, or the gap may be filled by extensions of the maxilla, or of the frontal, or of both. The same may happen in connexion with complete absence of the nasal.

There may be considerable differences between closely related species. Thus, Ridewood (1922) showed that in Balaenoptera borealis the parietals are separated by a large interparietal; but in B. musculus the latter is absent, and the parietals extend to the midline and meet one another.

It will be evident from what has already been said that if the so-called 'maxilla' of Polypterus is not the product of fusion between the maxilla and the suborbital (which remains to be determined by investigation of the first origin of the ossification centres), then it must be either a maxilla which has extended into the suborbital's territory and become pierced by the lateral line, or it is a suborbital which has invaded the maxilla's territory (not quite, it would seem, since it does not lie in the maxillary fold) and has come to bear teeth. Since there would be two unusual features about such a maxilla but only one about such a suborbital, one may regard the so-called 'maxilla' of Polypterus as a suborbital, especially as the suborbitals or rostrals of Lepidosteus and of the Palaeoniscid Glaucolepis (Nielsen, 1936) also bear teeth.

Similarly, in Polypterus, if the so-called parietal is not fused with the intertemporal-supratemporal, then it must be either a parietal which has invaded the territory of the intertemporal-supratemporal and come to be traversed by a lateral line, or it is an intertemporal-supratemporal which has invaded the territory of the parietal and met its fellow from the opposite side in the middle line. Preference for one or other of these possibilities is likely to be determined by the view taken as to whether the lateral line canals can change their relations from one bone to another. If it be granted that they can, then there would be nothing inherently improbable in regarding Polypterus's so-called parietal as a true parietal; but as a matter of fact, the lateral line canals in Polypterus are normal in position with regard to the whole fish, and it is therefore more probable that the bone is an intertemporal-supratemporal.

Altogether, it may be noted that one of the main difficulties in arriving at conclusions in these matters is insufficient knowledge. It seems that although a bone, in the sense of a single structure which can be traced from one form to another, is not always the product of a single centre of ossification, it often is, and to this extent the embryological criterion may be admitted. But it rests with adult morphology to decide whether it is the centres of ossification or the product of their fusion which constitute 'the bone'. Thus, in the case of the three centres which give rise to the nasal of Amia, the anatomy of Polypterus with its separate nasal, adnasal, and terminale, which together occupy the same region and bear the same morphological relations as the nasal of Amia, indicates that each of the three centres of Amia represents a bone. On the other hand, in the case of the two ossification centres of the human parietal, a comparison between the parietals of Man and other mammals, indicates that the parietal centres in Man do not each of them represent a bone in previous phylogeny, although they may possibly become established in future evolution.

It follows, therefore, that the connotation of the expression a 'bone' is arbitrary to the extent that underlying the entire argument is the concept of an organism—ideally, a primitive bony vertebrate—in which the bony tissue was disposed in certain discrete pieces. These are 'the bones', and the task of morphology is to recognize the corresponding structures in other forms; correspondence, of course, means homology or genetic affinity based on representation of the bone in question by one bone in a common ancestor.

It is possible that the earliest bony vertebrates did not possess separate bones, but a continuous undivided structure, which only in later evolution became split up into separate bones. If this was so, then the further possibility must be granted that this original splitting may have taken place more than once, in different lines of descent, and in different ways. In such cases the 'bones' in one stock may be totally incomparable with those in another, and it then becomes the task of morphology to establish the distinction between these lines of descent not only by the uniformity of pattern within them but by the dissimilarity between them. For instance, the pattern of the bony plates in Ostracoderms or in Arthrodira seems to be quite incomparable with that found in Teleostomes.

The same sort of incomparableness would be found if in the original vertebrates separate bones arose *de novo* independently in different stocks. A further implication of this view is that if the dermal bones of the head were originally similar to the bony scales, then independent evolution of bones in different patterns could only result from independent evolution of scales. As in so many other respects, progress here is dependent on palaeontological discoveries, for while embryology may assist in supporting or rejecting any particular view, it cannot from its own data provide evidence for a reconstruction of the morphology of the original vertebrate ancestors.

So far, in this discussion, attention has been confined to the rudiments of bones, but similar problems apply to cartilaginous rudiments. The difference between these tissues is, however, that whereas bony rudiments usually give rise to separate bones, cartilaginous rudiments usually fuse up to such an extent that the cartilaginous neurocranium (and often the upper jaw as well) is one undivided piece of cartilage.

With cartilage, the problem is to know what significance to attach to the origin of a structure from two centres of chondrification ('chondrites', T. J. Parker, 1892) in one case and from one centre in another. Cases in point are the separate hyomandibula and symplectic cartilages in early stages of Salmo, and the single hyosymplectic rudiment in other bony fish; the separate origins of the auditory capsule and parachordal in Squalus and their continuous chondrification in Scyllium; the two centres of chondrification of Meckel's cartilage in Squalus, Rana, Anas, and Halicore, as against the single centre in other forms; the separate trabecula and polar cartilage in Squalus, Lepidosteus or Anas, and the single trabecular-polar bar in other forms; the separate chondrification of the pterygoid process of the pterygoquadrate cartilage in some Teleosts and Urodeles as compared with continuous chondrification in others; the separate chondrification centres of the otostapes and hyostapes in the columella auris of Lacertilia, as compared with their (initial) continuity in Sphenodon; the separate chondrification of the processus ascendens and pterygoquadrate cartilage in Lacertilia and Mammalia, compared with their continuity in Amphibia, Sphenodon, Chelonia, and Crocodilia.

In all these cases, the centres of chondrification are connected by a continuous blastema of mesenchyme or procartilage, so that the separation between the trabecula and polar cartilage, for instance, is not of the same nature as the distinctness between the cartilaginous rudiments of, say the incus and the stapes.

At the same time, it must be recognized that the various cartilages of the branchial arches (pharyngo-, epi-, cerato-, and hypobranchials) commonly chondrify in a continuous blastema of procartilage, and that these cartilages represent definite units, traceable through phylogeny. It therefore becomes a matter of great difficulty to decide whether any phylogenetic or morphological significance should be attached to the distinctness (where it occurs) between trabecula and polar cartilage, hyomandibula and symplectic, otostapes and hyostapes. These questions are related to the wider problem of the phylogenetic significance of the chondrocranium as a whole. It must also be remembered as Gaupp (1900, p. 533) remarked, that the *chondrification* of a structure is not necessarily its *origin*, nor need it represent its phylogenetic history.

Before arriving at a solution it would be necessary to know definitely whether there is in all cases of separate centres of chondrification a correlation between these centres and such structures as muscle attachments or ligaments. The existence of these centres would then have to be regarded as developmental ontogenetic adaptations, and may have no phylogenetic significance at all. This indeed appears to be true of the cases of appearance of secondary cartilage in the rudiments of membrane-bones. Information on this subject is, however, very meagre. Thus, in regard to the polar cartilage, in Squalus it appears that the external rectus muscle is not attached to it but to the parachordal; in Lepidosteus on the other hand, the external rectus muscle is attached to the polar cartilage. If this is so, then it looks as though the polar cartilage has some phylogenetic significance, and the positive fact of its independent chondrification in forms as widely separated as Squalus, Lepidosteus, Anas, and certain mammals (?) must be taken as carrying more weight than its failure to chondrify independently in other forms.

This leads on to the opposite problem, concerning the significance to be attached to cases in which cartilaginous rudiments of what appear to be quite distinct structures chondrify in continuity. This applies to the (original or temporary) fusion between the quadrate and hyosymplectic cartilages in Polypterus and some Teleosts. Since there is no doubt whatever (from the evidence of blood-vessels, nerves, and visceral clefts) that the quadrate and hyosymplectic are phylogenetically fundamentally distinct structures, belonging to different visceral arches, their combined chondrification can only be a developmental accident devoid of phylogenetic significance.

Similar phenomena have been dealt with (p. 451) in connexion with the temporary fusion of the pterygoquadrate cartilage with the brain-case, of the columella auris with the auditory capsule (pp. 279, 412), of successive segmental vertebral elements (pp. 51, 59), and between the incus and stapes (p. 291).

It is likely, therefore, that separate chondrification is of greater importance than initial or temporary synchondrosis.

At all events, it may be noted that the limits between separate centres of chondrification need not coincide with the limits between centres of ossification. Thus, the supraoccipital in Man is a single ossification centre arising in a cartilage, the tectum posterius, which is derived from paired centres of chondrification. On the other hand, in Teleosts, the separate ceratohyal and epihyal ossifications both arise in the single ceratohyal chondrification. The basisphenoid ossifies in the territories of the parachordals, polar cartilages, and trabeculae, all of which may have been separate cartilages.

Similarly, in the columella auris, there are two chondrification centres, the otostapes and the hyostapes; but they do not correspond with the limit between the stapes and the extracolumella—the boundary between which lies within the territory of the otostapes.

ii. Conclusions: terminology of bones

It should be realized that consideration of the matters discussed in the foregoing section of this chapter is of far more than mere academic interest, for the system of nomenclature of bones must be based on the closest possible approximation to their homologies. A sound system of nomenclature is therefore an essential prerequisite not only for the co-ordination of existing knowledge, but also for all future research. It may, perhaps, be useful to attempt a formulation of certain simple rules of nomenclature, in the light of the present state of knowledge of the development of the skull.

1. In no case should fusion of bones be inferred and hyphenation of their names practised merely from co-extensiveness of territory without evidence of such fusion provided by:

(*a*) observation of separate centres of ossification in development; or
(*b*) observation of distinct growth-centres as indicated by X-ray microphotography, if practicable; or
(*c*) observation of distinct centres of radiating growth-lines.

2. Passage of a lateral line canal through or in close morphological relations with a bone should be regarded as strong but not infallible evidence of the identity of the bone. The passage of a lateral line canal through an 'unusual' bone should not be accepted without evidence of:

(*a*) the abnormal position of the lateral line canal; or
(*b*) the abnormal form of the head and position of the bones; or
(*c*) the abnormally large size of the lateral line canal.

3. The criterion of homology should not be strained beyond the limits of its useful application set by possibilities of variation in segmental correspondence and transposition. This difficulty does not arise in the Tetrapods and the Crossopterygian fish which gave rise to them. Here the homologies of bones can be established with reasonable certainty by means of their morphological

relations to neighbouring structures: a refinement of the 'principle of connexions' of Etienne Geoffroy St. Hilaire. But a comparison of different fish soon reveals bones arranged in series, either longitudinally (e.g. nasals and frontals series; splenial-angular series) or transversely (e.g. rostral series, extrascapular series). This is only what is to be expected if dermal bones were originally similar to the scales of the trunk, equal in size and regularly arranged. Owing to the possibility of transposition and variation in segmental number, demonstrated for several structures by Goodrich (1913) care must be taken in the equating of elements of corresponding series in different fish, for they may at best be only partially homologous. The importance of this caveat will be obvious if it be asked, for instance, whither the hindmost scale of the trunk lodging the lateral line canal is homologous in all fish. An additional possibility to which Westoll (1936) has drawn attention is that the groupings and fusions of the elements of a series may vary.

The solution of these difficulties appears to be in the concentration of attention on the homology of the series rather than of their members.

4. In general, cartilage-bones should be homologized with cartilage-bones and membrane-bones with membrane-bones, without prejudice to the possibility that cartilage-bones may develop in the absence of their cartilage-precursors and may develop intramembranous extensions, and that membrane-bones may invade cartilage.

5. The principle of priority must not be permitted to lead to absurdities, such as the substitution of 'incus' and 'tympanic' for the 'quadrate' and 'angular' of fish.

40. AGENDA

As made clear in the Preface, the object of this book is no less to serve as a starting-point for further work than to supply information concerning what is already known. By formulating some problems, as below, it is hoped that such research may be facilitated. Page numbers in brackets refer to places in the body of this book where the problem in question is raised.

Information of any kind regarding the development and structure of the chondrocranium is required in Myxinoids, Polyodon, Varanus, Chamoeleo, Tinamu, Caenolestes, Hyrax, Elephas, Manis, Orycteropus, Macroscelides, Chrysochloris, Cheiroptera, and Anthropoid Apes.

i. Special problems relating to the morphology of the chondrocranium

1. Does the tongue skeleton in Cyclostomes represent any part of the mandibular arch? (p. 47.)

2. Are the basidorsals and basiventrals of the anterior region of the vertebral column of Selachii continuous strips of cartilage before becoming segmented? (pp. 50, 59.) Are they continuous with the parachordals?

3. How is the occipito-vertebral joint formed in Torpedo? (p. 71.) Is an additional vertebral element absorbed by the skull during development?

4. What is the position of the spiracle in Holocephali, relative to the surrounding cartilages? (p. 76.)

5. What is the course of the internal carotid arteries during the earliest stages of the development of the skull in Polypterus? (p. 83.)

6. What are the details of the relations of the pterygoquadrate to the brain-case in Gymnarchus? (p. 135.)
7. What are the relations of the hyomandibular branch of the facial nerve to the hyosymplectic cartilage in Gambusia, or in any Teleost in which the cartilage is not perforated by the nerve? (p. 142.)
8. Do the 'radii posteriores' of Urodela represent morphologically any part of the hyobranchial skeleton? (p. 181.)
9. What are the details of the formation of the true basal articulation in Ascaphus? (p. 213.) Is this condition found in any other Anuran?
10. Is the 'stapes inferior' of the auditory capsule of Chelonia comparable with the operculum of Urodela? (p. 253.)
11. Is the intercalary cartilage between the quadrate and the crista parotica in birds derived from the columella auris or from the crista parotica? (p. 279.)
12. Do the paraglossal cartilages of birds represent morphologically any part of the hyobranchial skeleton? (p. 277.)
13. Can the cricoid cartilages in mammals be considered to represent visceral arches? (p. 313.)
14. Is there a rotation of the membranous labyrinth within the auditory capsule, resulting in the ultimate horizontal position of the lateral semicircular canal? (p. 360.) If so, how and when does it take place?
15. Is the so-called basicranial fenestra in mammals morphologically within or in front of the basal plate? (p. 383.)

ii. Special problems relating to the morphology of the bony skull

1. Does a perichondral or endochondral opisthotic bone exist in any Teleostome fish? (p. 497.)
2. Do the nasal, frontal, intertemporal, or extrascapular bones of any Teleostome fish develop as in Amia from separate (segmental?) ossification centres subsequently fused? (p. 503.) Are such centres constant?
3. Is the so-called basisphenoid of Teleostome fish homologous with that of Tetrapoda? (p. 497.)
4. Can any fusion of originally separate bones be detected during development of the so-called 'maxilla' or 'parietal' of Polypterus, 'supraoccipital' of Amiurus, 'parietal' of Cyprinus? (p. 505.)
5. Can the separation of the lateral line tubular element from the plate of a membrane-bone be demonstrated during development in any Teleostome fish in a place where there is no immediately underlying cartilage-bone (e.g. in the extrascapular series)? (p. 499.)
6. Can the fundamental segmentation of the head be made out in the distribution of lateral line sense-organs and associated ossification centres of dermal bones? (p. 33.)
7. What is the nature of the so-called parahyoid or 'urohyal' of Teleostei? (p. 416.) Is it morphologically part of the hyobranchial skeleton or of the pectoral girdle, or sui generis?
8. Can the configuration of bones (e.g. the 'prenasal') in the anterior region of the snout of Urodela be brought into line with that of other forms? (p. 185.)
9. What is the nature of the so-called epipterygoid of Ophidia? (p. 250.)

10. Does the interparietal occur in any birds besides those mentioned on p. 444?

11. Is it the rule that the palate of 'neognathous' birds passes through a 'palaeognathous' stage during development? (p. 445.)

12. Has the 'os uncinatum' of birds any morphological significance? (p. 287.)

13. Is the relative acceleration of the earlier stages of morphogenesis of the skull of nidifugous birds as compared with nidicolous birds a constant phenomenon? (p. 452.) Is the relative retardation of the later stages likewise constant?

14. Does the rudiment of an endotympanic bone occur in Ornithorhynchus? (p. 440.)

15. Do rudiments of prenasal processes of premaxillae occur in Marsupials besides Trichosurus and Caluromys? (p. 466.)

iii. Special problems relating to experimental morphogenesis

1. Can the normal morphological relations of structures be altered experimentally? What is the course of the internal carotid arteries relatively to the trabeculae in frog embryos rendered tropitrabic by induced cyclopia? (pp. 477, 494.)

2. Can secondary cartilage cells be demonstrated *in vitro* to turn into osteoblasts and secrete phosphatase? (p. 482.)

3. Is secondary cartilage of the same nature histologically and physiologically as hypertrophied primary cartilage? (pp. 38, 502.)

4. What determines the persistence or loss of the sclerotic cartilage in Amphibia? (p. 479.)

5. Can extracts of auditory vesicles induce the formation of cartilage *in vitro*?, *in vivo*? (pp. 472, 477, 478.)

6. Can heteroplastic grafting experiments demonstrate the derivation of visceral arch cartilage from neural crest cells in other groups besides Amphibia? (p. 475.)

7. Can the initial existence of separate centres of ossification in fused and composite bones be detected by X-ray microphotography? (p. 506.)

8. Why in Amia may the postorbitals be either two or one in number, and the parietals paired, or median, and fused? (p. 505.) In the latter alternatives does fusion occur during development, or is only one ossification centre formed?

9. Can a bone with a normally fixed territory bounded by sutures extend its area, with or without alteration to its thickness, at all, to a limited extent, or indefinitely, as a result of extirpation of a neighbouring bone? (pp. 482, 487.)

10. Can a constant topographical relation be established between a centre of ossification and a lateral line sense-organ or nerve in any form besides Amia? (p. 489.)

11. Can a bone through which a lateral line canal normally passes (e.g. frontal) develop if the placode of the lateral line is previously extirpated? (pp. 489, 508.)

12. Can a bone which normally lodges no lateral-line canal (e.g. parietal) enter into relations with a canal, (*a*) as a result of grafting the placode of the

canal in an abnormal situation; or (*b*) inserting the parietal in the place of a bone (e.g. frontal) through which a canal normally passes? (pp. 490, 508.)

13. Can a cartilage-bone arise in the absence of its precursor cartilage? (pp. 482, 496.) Has this occurred with the metapterygoid of Syngnathus, the basi-sphenoid of Teleostei?

14. What is the basis and extent of variation in number of bones in, e.g. rostrals of Acipenser (p. 96); opercular ossicles of Lepidosteus (p. 115); spiracular ossicles of Polypterus (p. 86); interparietal of Balaenoptera (p. 508)?

15. Can the time-relations of the appearance of bones be modified? What is the sequence of ossification in (*a*) thyroidectomized frog tadpoles; (*b*) precociously metamorphosed frog or (*c*) axolotl? (pp. 184, 209).

BIBLIOGRAPHY AND INDEX OF AUTHORS

Author	*Reference*	*Subject*	*Page where quoted*
Addens, J. L.	1928. *Proc. Acad. Sci. Amst.* **31**, p. 733.	Petromyzon, eye-muscle nerves.	21.
Adelmann, H. B.	1925. *J. comp. Neurol.* **39**, p. 19.	rat, neural folds.	473.
——	1934. *J. exp. Zool.* **67**, p. 217.	cyclopia, trabeculae.	477.
Agar, W. E.	1906. *Trans. roy. Soc. Edinb.* **45**, p. 49	Protopterus, Lepidosiren, dev. of skull.	29, 174, 176.
Agassiz, L.	1844. *Recherches sur les poissons fossiles.* Neuchâtel.		4.
Ahrens, H. J.	1936. *Morph. Jahrb.* **77**, p. 357.	bone structure.	485.
Aichel, O.	1913. *Anat. Anz.* **43**, p. 463.	man, interparietal.	444.
——	1915. *Arch. Anthrop. Braunschw.* **13**, p. 130.	man, dev. of sutures.	369.
Albrecht, P.	1880. *Zool. Anz.* **3**, pp. 450, 472.	proatlas.	240, 262, 385.
Allis, E. P. jun.	1889. *J. Morph.* **2**, p. 463.	Amia, lateral line.	33.
——	1897. *Ibid.* **12**, p. 487.	Amia, morph. of skull.	97, 103, 104, 420, 428, 500.
——	1898. *Ibid.* **14**, p. 425.	Amia, snout bones.	86, 97, 107.
——	1899. *Anat. Anz.* **16**, p. 49.	Amia, post. skull bones.	97, 104, 499.
——	1900. *Ibid.* **18**, p. 257.	Polypterus, maxilla.	86.
——	1903*a*. *J. Morph.* **18**, p. 45.	Scomber, morph. of skull.	428, 500.
——	1903*b*. *Anat. Anz.* **23**, pp. 359, 321.	Bdellostoma, morph. of skull.	41.
——	1904. *Int. Mschr. Anat. Physiol.* **21**, p. 401.	lateral line canals.	95, 135, 139, 506.
——	1908*a*. *Anat. Anz.* **33**, p. 217.	Polypterus, carotids.	83.
——	1908*b*. *Ibid.* **33**, p. 256.	Amiurus, carotids.	138.
——	1909. *Zoologica, Stuttgart*, **57**.	mail-cheeked fishes, morph. of skull.	108, 156, 157, 166, 428, 498.
——	1913. *Anat. Anz.* **44**, p. 322.	Selachii, ethmoid region.	54, 55.
——	1914*a*. *Ibid.* **46**, p. 225.	trigemino-facialis chamber.	56, 57, 103, 428.
——	1914*b*. *Ibid.* **46**, p. 625.	Ceratodus, chondrocr.	172, 429.
——	1915. *J. Morph.* **26**, p. 563.	hyomandibula.	71, 72, 173, 410.
——	1917. *Proc. zool. Soc. Lond.* p. 105.	Chimaera, morph. of skull.	72.
——	1918. *Anat. Rec.* **15**, p. 257.	hyomandibula.	85, 410.
——	1919*a*. *J. Morph.* **32**, p. 207.	myodome & trigemino-facialis chamber.	57, 72, 103, 152, 428.
——	1919*b*. *Anat. Rec.* **17**, p. 73.	pterotic, supratemporal.	439.
——	1919*c*. *Amer. J. Anat.* **25**, p. 349.	Polypterus, maxilla.	86.
——	1919*d*. *J. Anat. Lond.* **53**, p. 209.	alisphenoid.	439.
——	1920. *Proc. zool. Soc. Lond.* p. 245.	Lepidosteus, chondrocr.	108.
——	1922. *J. Anat. Lond.* **56**, p. 189.	Polypterus, morph. of skull.	78, 86, 87, 500.
——	1923*a*. *Acta zool. Stockh.* **4**, p. 123.	Chlamydoselachus, morph. of chondrocr.	55, 64, 393, 421.
——	1923*b*. *J. Anat. Lond.* **58**, p. 37.	trabeculae & polar cart.	77, 375, 377, 419.
——	1924. *Ibid.* **58**, p. 256.	Cyclostomes, morph. of skull.	41.
——	1925. *Ibid.* **59**, p. 333.	trabeculae & polar cart.	375.
——	1926*a*. *Ibid.* **60**, p. 164.	Cyclostomes, morph. of skull.	41.
——	1926*b*. *Ibid.* **60**, p. 335.	Holocephali, morph. of skull.	72.

Author	Reference	Subject	Page where quoted
Allis, E. P. jun.	1928. *J. Anat. Lond.* **63**, p. 95.	pituitary fossa, myodome.	58, 382, 430.
——	1929. *Ibid.* **63**, p. 282.	Ceratodus, otic process.	172, 430.
——	1930. *Acta zool. Stockh.* **11**, p. 1.	morph. of chondrocr.	172.
——	1931. *J. Anat. Lond.* **65**, p. 247.	trabeculae & polar cart.	375.
——	1934. *Ibid.* **68**, p. 361.	lateral line canals.	33.
——	1935. *Ibid.* **69**, p. 233.	fishes, membrane-bones.	499, 506.
Andres, J.	1924. *Morph. Jahrb.* **53**, p. 259.	Sus, dev. of bony skull.	334.
Anikin, A. W.	1929. *Arch. EntwMech.* **114**, p. 549.	cartilage, fields.	472.
Aoyama, F.	1930. *Z. Anat. EntwGesch.* **93**, p. 107.	Cryptobranchus, dev. of skull.	185, 188, 18
Arai, H.	1907. *Anat. Hefte*, **33**, p. 411.	craniopharyngeal passage.	311.
Arendt, E.	1822. *De capitis ossei Esocis lucii structura singularis.* Regiomonti.		3, 142.
Ärnbäck Christie-Linde, A.	1907. *Morph. Jahrb.* **36**, p. 463.	Sorex, morph. of skull.	324, 396.
——	1914. *Ibid.* **48**, p. 343.	mammals, palatine cart.	325, 396.
Arnold, E.	1928. *Morph. Jahrb.* **60**, p. 47.	Equus, chondrocr.	339.
Ashley-Montagu, M.	1931. *J. Anat. Lond.* **65**, p. 446.	man, 'post-frontal'.	365.
Assheton, R.	1907. *Budgett Mem. Vol.* Cambridge, p. 293.	Gymnarchus, dev. of skull.	134.
Augier, M.	1923. *C. R. Ass. Anat.* **18**, p. 57.	Sus, notochord.	382.
——	1924. *Ibid.* **19**, p. 47.	Sus, notochord.	382.
——	1926. *Ibid.* **21**, p. 25.	man, basioccipital.	369.
——	1927. *Ibid.* **22**, p. 9.	morph. of sinuses.	370.
——	1928. *Ibid.* **23**, p. 10.	man, basioccipital.	356, 358, 36 386.
——	1930. *Ibid.* **25**, p. 18.	man, suprategminal cart.	359, 391.
——	1931*a*. *Squelette céphalique*, in Poirier & Charpey: *Traité d'anatomie humaine*. Paris.		362, 390, 43 483, 484, 48 497, 507.
——	1931*b*. *Arch. Anat. Strasbourg*, **13**, p. 33.	man, supraoccipital.	339, 358, 36 443, 444.
——	1931*c*. *C. R. Ass. Anat.* **26**, p. 3.	man, paraseptal bone.	372.
——	1932*a*. *Ibid.* **27**, p. 18.	man, premaxilla.	365.
——	1932*b*. *Anthrop. Paris*, **43**, p. 315.	brain & skull, senile.	488.
——	1934*a*. *Arch. Anat. Strasbourg*, **19**, p. 89.	Sus, dev. of bony skull.	334, 390, 44 444.
——	1934*b*. *C. R. Ass. Anat.* **29**, p. 24.	man, vestigial carts.	359, 390.
——	1935. *Ibid.* **30**, p. 16.	man, tectum synoticum.	358, 359.
——	1936. *Arch. Anat. Strasbourg*, **21**, p. 15.	Sus, man, roof of chondrocr.	334, 335.
——	1937. *Ibid.* **23**, p. 249.	Sus, tectum posterius.	337, 497.
Aumonier, F. J.	In press.	Lepidosteus, dev. of membrane-bones.	108, 113.
Bäckström, K.	1931. *Acta zool. Stockh.* **12**, p. 83.	Tropidonotus, dev. of chondrocr.	245, 246, 24
von Baer, K. E.	1826. *Meckels Arch. Anat. Physiol. Lpz.* p. 327.	cartilage-bone & membrane-bone.	3.
Balabai, P. P.	1935. *Bull. Acad. Sci. Ukr. Trav. Inst. Zool.* **3**, p. 167.	Petromyzon, dev. of chondrocr.	41, 48, 4 376.
Balfour, F. M.	1876. (Collected papers 1878. *A monograph on the development of Elasmobranch fishes.* London.)		8, 12, 16.
Ballantyne, F. M.	1927. *Trans. roy. Soc. Edinb.* **55**, p. 371.	air-bladder & skull.	135.
——	1930. *Ibid.* **56**, p. 437.	Callichthys, chondrocr.	138.
Bardeen, C.	1910. in Keibel & Mall, *Manual of human embryology*. London.	man, 20 mm. chondrocr.	355.

Author	Reference	Subject	Page where quoted
Barge, J. A. S.	1918. *Anat. Hefte*, **55**, p. 415.	Ovis, occip. joint.	386.
Bartlemez, G. W.	1923. *J. comp. Neurol.* **35**, p. 231.	neural crest & mesect.	473.
Bast, T. H.	1930. *Contr. Embryol. Carn. Instn.* **21**, p. 53.	man, dev. of periotic.	371, 403.
Baum & Dobers,	1905. *Anat. Hefte*, **28**, p. 587.	Sus, Ovis, ext. ear cart.	416.
Baumüller, B.	1879. *Z. wiss. Zool.* **32**, p. 466.	mandib. symph. cart.	418.
Baur, G.	1896. *Anat. Anz.* **11**, p. 410.	Reptilia, mandible.	268.
Bayer, F.	1897. *Jena. Z. Naturw.* **31**, p. 100.	tentorium cerebelli.	330.
de Beaufort, L. F.	1909. *Morph. Jahrb.* **39**, p. 536.	Clupea, air-bladder divert.	133, 134.
Beccari, N.	1922. *Arch. ital. Anat. Embriol.* **9**, p. 1.	Salmo, head-segmentation.	23, 29, 116.
de Beer, G. R.	1922. *Quart. J. micr. Sci.* **66**, p. 457.	Squalus, head-segmentation.	17, 27.
——	1923. *Ibid.* **67**, p. 257.	Petromyzon, hypophysis.	376.
——	1924*a*. *Ibid.* **68**, p. 17.	Amia, head-segmentation.	23, 28.
——	1924*b*. *Ibid.* **68**, p. 39.	Heterodontus, dev. of chondrocr.	49, 65.
——	1924*c*. *Ibid.* **68**, p. 287.	morph. of chondrocr.	21, 77, 97.
——	1924*d*. *J. Anat. Lond.* **59**, p. 97.	Petromyzon, nasal caps.	44.
——	1925. *Quart. J. micr. Sci.* **96**, p. 671.	Acipenser, dev. of chondrocr.	88.
——	1926*a*. *Ibid.* **70**, p. 263.	morph. of chondrocr.	78, 97, 108, 115, 132, 168, 172, 198, 207, 245, 260, 269, 417, 420, 429.
——	1926*b*. *Ibid.* **70**, p. 669.	Torpedo, chondrocr.	69.
——	1926*c*. *The comparative anatomy, histology, and development of the pituitary body*. Edinburgh.		20, 84.
——	1927. *Quart. J. micr. Sci.* **71**, p. 259.	Salmo, dev. of chondrocr.	115.
——	1929. *Philos. Trans.* B. **217**, p. 411.	Sorex, dev. of chondrocr.	324, 391, 392, 396, 402, 432, 435, 436, 502.
——	1930*a*. *Quart. J. micr. Sci.* **73**, p. 707.	Lacerta, dev. of chondrocr.	217, 410.
	1930*b*. *Embryology and Evolution*. Oxford.		448, 451.
——	1931*a*. *Quart. J. micr. Sci.* **74**, p. 591.	Scyllium, dev. of chondrocr.	49, 58, 74, 381, 383, 447, 457.
——	1931*b*. *Ibid.* **74**, p. 701.	trabeculae.	13, 41, 49, 59, 269, 375, 376.
——	1932. *Ibid.* **75**, p. 307.	Batoids, hyoid arch.	69, 70, 411.
—— & Barrington, E. J. W.	1934. *Philos. Trans.* B. **223**, p. 411.	Anas, dev. of chondrocr.	13, 22, 31, 268, 381, 415.
—— & Fell, W. A.	1936. *Trans. zool. Soc. Lond.* **23**, p. 1.	Ornithorhynchus, dev. of skull.	289, 386, 434, 435, 437.
—— & Moy-Thomas, J. A.	1935. *Philos. Trans.* B. **224**, p. 287.	Holocephali, chondrocr.	72, 78, 410, 422.
—— & Woodger, J. H.	1930. *Ibid.* B. **218**, p. 373.	Lepus, dev. of chondrocr.	306, 410.
van Bemmelen, J. F.	1889. *Anat. Anz.* **4**, p. 244.	Lacerta, head-segmentation.	25.
——	1901. *Denkschr. med-naturw. Ges. Jena*, **6**, p. 727.	Monotreme, morph. of skull.	298, 299.
Bender, O.	1911. *Anat. Anz.* **40**, p. 161.	Testudo, colum. aur.	252.
——	1912. *Abh. bayr. Akad. Wiss. Math.-Phys. Kl.* **25**, (10) p. 1.	Testudo, visc. arch skel.	258.

Author	Reference	Subject	Page where quoted
Benninghoff, A.	1925. *Verh. anat. Ges. Jena,* **34**, p. 189.	bone-structure.	485.
Bergmann, C.	1846. *Einige Beobachtungen und Reflexionen über die Skelettsysteme der Wirbeltiere.* Göttingen.		4.
Berrill, N. J.	1925. *Quart. J. micr. Sci.* **69**, p. 217.	Solea, dev. of chondrocr.	161, 409.
Besrukow, E. A.	1928. *Rev. zool. russe,* **8**, p. 89.	Esox, dev. of chondrocr.	142.
Bhimachar, B. S.	1933. *Half-yrly J. Mysore Univ.* **7**, (2) p. 1.	Siluroids, bony skull.	139.
Bianchi, S.	1893. *Monit. zool. ital.* **4**, p. 11.	preinterparietal.	444.
Biondi, D.	1886. *Arch. Anat. Physiol. Lpz. Physiol. Abt.* p. 550.	Sus, prevomer.	336, 434.
Bluntschli, H.	1925. *Arch. EntwMech.* **106**, p. 303.	dura mater, sutures.	485.
Boeke, J.	1904. *Petrus Camper,* **2**, p. 439.	Muraena, head-segmentation.	19.
Boenninghaus, G.	1904. *Zool. Jahrb. Abt. Anat.* **19**, p. 189.	Cetacea, ear.	341.
Bogoljubsky, S. N.	1924. *Rev. zool. russe,* **4**, p. 89.	Salamandra, hyobranch. skel.	187, 188.
——	1925. *Ibid.* **5**, (4) p. 3.	Hynobius, hyobranch. skel.	189.
——	1926. *Ibid.* **6**, (3) p. 39.	Amblystoma, hyobranch. skel.	181, 183, 186, 188.
Bojanus, L. H.	1819. *Anatome testudinis Europaeae,* Vilnae.		7.
Böker, H.	1913. *Anat. Hefte,* **49**, p. 359.	Salmo, dev. of bony skull.	115, 124, 500.
Bolk, L.	1904. *Petrus Camper,* **2**, p. 315.	man, roof of chondrocr.	358.
——	1921. *Anat. Anz.* **54**, p. 335.	man, occip. condyles.	386.
——	1922. *Bijdr. Dierk.* **22**, p. 1.	basal plate & notoch.	381.
——	1926. *Das Problem der Menschwerdung.* Jena.		452.
Bondy, G.	1907. *Anat. Hefte,* **35**, p. 293.	audit. ossic.	336, 412, 413.
Born, G.	1876. *Morph. Jahrb.* **2**, p. 577.	wax model method, & nas. caps. of Amphibia.	14, 198, 209.
——	1879. *Ibid.* **5**, pp. 62, 401.	Sauropsida, nas. caps.	217, 278, 279.
——	1883*a*. *Arch. mikr. Anat.* **22**, p. 584.	wax model method.	14.
——	1883*b*. *Morph. Jahrb.* **8**, p. 188.	Tropidonotus, nas. caps.	248.
——	1888. *Z. wiss. Mikr.* **5**, p. 433.	wax model method.	14.
Brachet, A.	1909. *Bull. Soc. Anthrop. Brux.* **27**, p. lvii.	man, occip. condyles.	386.
Brauer, A.	1904. *Zool. Jahrb. Suppl.* **7**, p. 381.	neural crest & mesect.	473.
Brauer, M.	1906. *Zool. Anz.* **29**, p. 674.	Mustela, stylohyal.	334.
Braus, H.	1899. *Morph. Jahrb.* **27**, p. 415.	Selachii, head-segmentation.	19, 71.
——	1905. *Denkschr. med-naturw. Ges. Jena,* **11**, p. 377.	Selachii, visceral arches.	52, 408.
——	1906. *Anat. Anz.* **29**, p. 545.	Heptanchus, hyobranch. skel.	410.
Bridge, T. W.	1877. *J. Anat. Lond.* **11**, p. 605.	Amia, morph. of bony skull.	97, 505.
——	1898. *Trans. zool. Soc. Lond.* **14**, p. 325.	Lepidosiren, morph. of skull.	174.
Brock, G. T.	1929. *Quart. J. micr. Sci.* **73**, p. 289.	Leptodeira, dev. of skull.	47, 248, 250, 251.
——	1932*a*. *Anat. Anz.* **73**, p. 199.	Glauconia, morph. of skull.	251.
——	1932*b*. *S. Afr. J. Sci.* **29**, p. 508.	Gecko, dev. of chondrocr.	238.
——	1935. *Anat. Anz.* **80**, p. 241.	Lacerta, temporal bones.	232, 238, 250, 280, 438.
——	In press.	Struthio, dev. of chondrocr.	285.

Author	*Reference*	*Subject*	*Page where quoted*
Brohmer, P.	1909. *Jena. Z. Naturw.* **44**, p. 647.	Spinax, head-segmentation.	18.
Broili, F.	1917. *Anat. Anz.* **49**, p. 561.	median skull-bones.	444.
Broman, I.	1899. *Anat. Hefte*, **11**, p. 507.	man, dev. of audit. ossic.	357, 371, 412, 413, 416.
——	1920. *Ibid.* **58**, p. 137.	Jacobson's organ.	397.
Brooks, H. J.	1883. *Proc. roy. Dublin Soc.* **4**, p. 166.	Gadus, morph. of skull.	167.
Broom, R.	1895*a*. *Proc. linn. Soc. N.S.W.* **10**, p. 477.	prevomers.	289, 434.
——	1895*b*. *Ibid.* **10**, p. 555.	prenasal & papillary cart.	289, 302, 303, 304, 305.
——	1895*c*. *Ibid.* **10**, p. 571.	Miniopterus, nas. caps.	350, 398, 435.
——	1896*a*. *Ibid.* **21**, p. 9.	Equus, nas. caps.	398.
——	1896*b*. *Ibid.* **21**, p. 591.	Marsupials, nas. caps.	302, 303, 304, 305, 396, 397, 398.
——	1896*c*. *J. Anat. Lond.* **30**, p. 70.	Monotremes, nas. caps.	294, 398.
——	1897. *Ibid.* **32**, p. 709.	Hyrax, nas. caps.	398.
——	1900. *Trans. roy. Soc. Edinb.* **39**, p. 231.	Ornithorhynchus, nas. caps.	289, 312, 318, 396.
——	1902 *a*. *Proc. linn. Soc. N.S.W.* **27**, p. 545.	prevomers.	434.
——	1902*b*. *Proc. zool. Soc. Lond.* (i) p. 224.	Macroscelides, nas. caps.	326, 396, 397, 398.
——	1903. *J. Anat. Lond.* **37**, p. 107.	Lacerta,pterygoquad.	219.
——	1906. *J. linn. Soc. Lond. Zool.* **29**, p. 414.	Sphenodon, nas. caps.	242, 397.
——	1907. *S. Afr. Assoc. Adv. Sci.* p. 114.	alisphenoid.	420, 439.
——	1909*a*. *Proc. linn. Soc. N.S.W.* **34**, p. 24.	Trichosurus, Dasyurus, dev. of chondrocr.	303, 305, 395, 420, 439.
——	1909*b*. *Proc. zool. Soc. Lond.* p. 680.	Orycteropus, nas. caps.	320, 398.
——	1911. *Proc. linn. Soc. N.S.W.* **36**, p. 315.	Caenolestes, morph. of skull.	306.
——	1912. *Proc. zool. Soc. Lond.* p. 419.	audit. ossic.	437, 440.
——	1913. *Anat. Anz.* **45**, p. 73.	lower jaw, Amphibia.	186.
——	1914. *Philos. Trans.* B. **206**, p. 1.	origin of mammals.	319, 420, 435, 436.
——	1915*a*. *Proc. zool. Soc. Lond.* p. 157.	Tupaja, Gymnura, nas. caps.	325, 326, 396, 397, 398.
——	1915*b*. *Ibid.* p. 347.	Talpa, Centetes, Chrysochloris, nas. caps.	320, 326, 398.
——	1916. *Ibid.* p. 449.	Chrysochloris, morph. of skull.	326, 441, 444.
——	1922. *Ibid.* p. 455.	metapterygoid in reptiles.	228.
——	1924. *Bull. Amer. Mus. Nat. Hist.* **51**, p. 39.	Lacerta, chondrocr.	219, 438.
——	1926*a*. *Proc. zool. Soc. Lond.* p. 257.	mesethmoid.	305, 320, 340, 442.
——	1926*b*. *Ibid.* p. 419.	Caenolestes, nas. caps.	306, 398.
——	1927. *Ibid.* p. 233.	mesethemoid.	442.
——	1930. *Philos. Trans.* B. **218**, p. 345.	mesethmoid, prevomer.	299, 442.
——	1935*a*. *Ann. Transv. Mus.* **18**, p. 13.	Lacerta, temporal bones.	438.
——	1935*b*. *Ibid.* **18**, p. 23.	prevomer.	302, 434.
——	1935*c*. *Ibid.* **18**, p. 33.	mesethmoid, presphenoid.	340, 348, 442.
—— & Brock, G. T.	1931. *Proc. zool. Soc. Lond.* p. 737.	prevomer, Aves.	287, 434.
Bruner, H. L.	1902. *Morph. Jahrb.* **29**, p. 317.	Amphibia, septomaxilla.	181, 188.
Bruni, A. C.	1909. *Arch. ital. Biol.* **51**, p. 11.	man, hyoid arch.	362, 408.
Buchs, G.	1902. *Morph. Jahrb.* **29**, p. 582.	Necturus, chondrocr.	190, 473.

Author	Reference	Subject	Page where quoted
Budgett, J. S.	1902. *Trans. zool. Soc. Lond.* **16**, p. 315.	Polypterus, chondrocr.	78.
Budin, P.	1876. *Bull. Soc. Anthrop. Paris*, **11**, p. 553.	man, obstetric hinge.	369.
Bugajew, J.	1929. *Anat. Anz.* **67**, p. 98.	Acipenser, phar-mandib.	92, 419.
——	1930. *Ibid.* **68**, p. 385.	Scymnus, mandib. arch.	421.
Bujor, P.	1881. *Rev. biol. Nord France*, **4**.	Petromyzon, dev. of skull.	41, 45.
Bulman, O. M.	1931. *Ann. Mag. nat. Hist.* **8**, p. 179.	Palaeospondylus.	409, 453.
de Burlet, H. M.	1913*a*. *Morph. Jahrb.* **45**, p. 393.	Bradypus, proatlas centrum.	319, 382, 386.
——	1913*b*. *Ibid.* **45**, p. 523; **47**, p. 645.	Phocaena, chondrocr.	14, 345, 386, 391.
——	1914*a*. *Ibid.* **49**, p. 119.	Balaenoptera, chondrocr.	343.
——	1914*b*. *Ibid.* **49**, p. 393.	Lagenorhynchus, chondrocr.	346.
——	1916. *Ibid.* **50**, p. 1.	Cetacea, chondrocr.	345, 401.
——	1927. in M. Weber's *Die Säugetiere*. Jena.	Bradypus, chondrocr.	319.
——	1929. *Acta oto-laryng. Stockh.* **13**, p. 153.	perilymph. duct.	263.
Burr, H. S.	1916. *J. exp. Zool.* **20**, p. 27.	Amblystoma, nas. caps. exper.	478.
Butcher, E. O.	1929. *Amer. J. Anat.* **44**, p. 381.	rat, head-segmentation.	25.
Buxton, L. H. D. & de Beer, G. R.	1932. *Nature, Lond.* **129**, p. 940.	Neanderthal, paedomorph.	452.
Carus, K. G.	1828. *Von den Urteilen der Knochen und Schalen Gerüstes*. Leipzig.		7.
Carrière, J.	1884. *Arch. mikr. Anat.* **24**, p. 19.	Amblystoma, horny jaws.	182.
Chabanaud, P.	1936. *Ann. Inst. océanogr. Monaco*, **16**, p. 223.	Pleuronectidae, morph. of skull.	107, 129.
Chatin, J.	1910. *C. R. Acad. Sci. Paris*, **151**, p. 185.	sclerotic cart. & bone.	406, 499.
Chiarugi, G.	1890. *Arch. ital. Biol.* **13**, pp. 309, 423.	Lacerta, head-segmentation.	24, 25, 31, 246.
Chung, I.	1931. *Keijo J. Med.* **2**, pp. 254, 368.	Urodela, morph. of skull.	181.
Clark, W. E. Le G.	1925. *Proc. zool. Soc. Lond.* p. 559.	Tupaja, morph. of skull.	325.
Claus, T.	1911. *Anat. Anz.* **39**, pp. 293, 364.	Ovis, preinterparietal.	444.
Cole, F. J.	1898. *Trans. linn. Soc. Lond.* **7**, p. 115.	Gadus, dermal bones, nerves.	166, 167.
——	1905. *Trans. roy. Soc. Edinb.* **41**, p. 749.	Myxine, morph. of skull.	38, 41, 394.
——	1909. *Ibid.* **46**, p. 669.	Myxine, morph. of skull.	41, 394.
—— & Johnstone, J.	1901. *Liverpool Marine Biological Committee Memoirs*. VIII.	Pleuronectes, morph. of skull.	161, 500.
Cooper, C. F.	1928. *Philos. Trans.* B. **216**, p. 265.	Chrysochloris, audit. ossic.	444.
Cords, E.	1904. *Anat. Hefte*, **26**, p. 49.	Aves, morph. of nerves.	441.
——	1909. *Ibid.* **38**, p. 219.	Lacerta, dev. of colum. aur.	217, 219.
——	1915. *Ibid.* **52**, p. 1.	Perameles, dev. of skull.	303, 304, 414, 418, 496.
——	1918. *Ibid.* **56**, p. 243.	Echidna, ext. ear cart.	300, 416.
Cornevin, C.	1883. *Rev. Anthrop. Paris*, **6**, p. 661.	wormian bones.	368.
Corning, H. K.	1899. *Morph. Jahrb.* **27**, p. 173.	origin of mesenchyme.	473.
——	1900. *Ibid.* **28**, p. 28.	Lacerta, head-segmentation.	25.
Corsy, F.	1920. *C. R. Soc. Biol. Paris* **83**, p. 228.	man, hyoid arch, cart.	451.
Coyle, R. F.	1909. *Proc. roy. Soc. Edinb.* **29**, p. 582.	Equus, dev. of audit. ossic.	340, 412.

Author	Reference	Subject	Page where quoted
Crew, F. A. E.	1923. *Proc. roy. Soc. Lond.* B. **95**, p. 228.	achondroplasia.	491.
Cuvier, F.	1836. *Leçons d'Anatomie Comparée.* 3 edit. (posth.) Bruxelles.		7.
Dabelow, A.	1927. *Z. Morphol. Anthrop.* **26**, p. 305.	sclerotic bones.	443, 499.
——	1931. *Morph. Jahrb.* **67**, p. 84.	brain & skull.	487.
Damas, H.	1935. *Arch. Biol. Paris,* **46**, p. 171.	Petromyzon, metamorph. of chondrocr.	39, 41, 45.
Daniel, J. F.	1915. *J. Morph.* **27**, p. 447.	Heterodontus, morph. of skull.	66.
——	1916. *Univ. Californ. Publ. Zool.* **16**, p. 349.	Heptanchus, visc. arches.	407.
Darwin, C.	1875. *The variation of animals and plants under domestication.* 2nd ed. London.		486.
Dawes, B.	1930. *Philos. Trans.* B. **218**, p. 115.	Mus, head-segmentation.	25, 31, 320.
Dean, B.	1906. *Publ. Carneg. Instn.* **32**.	Holocephali, chondrocr.	72, 418.
Debierre, C.	1895. *J. Anat. Physiol. Paris,* **31**, p. 385.	man, Bufo. occip. bones.	12, 211, 367.
Decker, F.	1883. *Z. wiss. Zool.* **38**, p. 190.	mammals, chondrocrania.	316, 320, 326, 332, 334, 338, 388, 464.
van Deinse, A. B.	1916. *Anat. Anz.* **49**, p. 417.	Squalus, visc. skel.	55.
Denison, W. & Terry, R. J.	1921. *Washington Univ. Studies,* **8**, *Sci. Ser.* p. 161.	Caluromys, chondrocr.	303.
Denker, A.	1899. *Ergebn. Anat. EntwGesch.* **9**, p. 297.	morph. of audit. caps.	403, 442.
Dietz, P. A.	1921. *Mitt. zool. Sta. Neapel,* **22**, p. 433.	Gadus, morph. of skull.	166.
Dilg, C.	1909. *Morph. Jahrb.* **39**, p. 83.	Manatus, dev. of bony skull.	348.
Dixon, A. F.	1906. *Trans. roy. Acad. Med. Ireland,* **24**, p. 465.	man, proatlas.	386.
Dohrn, A.	1884*a*. *Mitt. zool. Sta. Neapel,* **5**, p. 102.	Selachii, visc. arches.	49, 52.
——	1884*b*. *Ibid.* **5**, p. 152.	Petromyzon, visc. arches.	41.
——	1886. *Ibid.* **6**, p. 1.	Selachii, visc. arches.	69, 71.
——	1901. *Ibid.* **15**, p. 1.	Torpedo, head-segmentation.	10, 19.
——	1902. *Ibid.* **15**, p. 555.	neural crest, mesect.	41, 473.
Dombrowsky, B.	1918. *Rev. zool. russe,* **2**, p. 204.	Lacerta, colum. aur.	217.
——	1924. *Ibid.* **4**, (i) p. 77.	Reptilia, colum. aur.	217, 252.
——	1925. *Ibid.* **5**, (i) p. 31.	Aves, colum. aur.	413.
Drews, M.	1933. *Morph. Jahrb.* **73**, p. 185.	Felis, Canis, dev. of bony skull.	326, 330, 331, 435, 443, 444.
Dreyfuss, R.	1893. *Morph. Arb.* **2**, p. 607.	man, audit. ossic.	357, 413.
Drüner, L.	1901. *Zool. Jahrb. Abt. Anat.* **15**, p. 435.	Urodela, hyobranch. skel.	172, 181, 183, 187.
——	1904*a*. *Ibid.* **19**, p. 361.	Urodela, hyobranch. skel.	172, 181, 183, 187, 189.
——	1904*b*. *Anat. Anz.* **24**, p. 257.	man, middle ear.	430.
Dubois, E.	1886. *Ibid.* **1**, pp. 179, 225.	mammals, laryngeal cart.	362, 407.
Dugès, A.	1834. *Mém. Acad. roy. Sci. Inst. France, Math. Phys.* **6**.	Rana, dev. of bony skull, cart.- & memb.-bones.	3, 198, 208, 505, 507.
Dunn, H. L.	1921. *J. comp. Neurol.* **33**, p. 405.	man, growth of brain.	470.
Dursy, E.	1869. *Zur Entwicklungsgeschichte des Kopfes.* Tübingen.		354.
Eales, N.	1931. *Proc. zool. Soc. Lond.* (i) p. 115.	Elephas, dev. of mandible.	340, 453.
Eaton, T. H.	1933. *Univ. Californ. Publ. Zool.* **37**, p. 521.	Amblystoma, streptostyly.	182.

Author	Reference	Subject	Page where quoted
von Ebner, V.	1911. *Verh. anat. Ges. Jena*, **25**, p. 3.	cart.- & memb.- bones.	6.
Ecke, H.	1935. *Z. Morph. Ökol. Tiere*, **29**, p. 79.	Bufo, audit. caps.	207, 213.
Edgeworth, F. H.	1899. *J. Anat. Lond.* **34**, p. 113.	Lepus, head-segmentation.	19.
——	1903. *Ibid.* **37**, p. 73.	Scyllium, head-segmentation.	19.
——	1907. *Quart. J. micr. Sci.* **51**, p. 511.	Gallus, head-segmentation.	19.
——	1911. *Ibid.* **56**, p. 167.	mammals, head-segmentation.	19.
——	1916. *Ibid.* **61**, p. 383.	mammals, hyobranch. skel.	305, 407.
——	1923*a*. *J. Anat. Lond.* **57**, p. 238.	Cryptobranchus, chondrocr.	188, 421.
——	1923*b*. *Ibid.* **57**, p. 97.	Cryptobranchus, hyobranch. skel.	189.
——	1923*c*. *Quart. J. micr. Sci.* **67**, p. 325.	Ceratodus, hyoid arch.	168, 170.
——	1925. *J. Anat. Lond.* **59**, p. 225.	jaw-suspensions.	61, 65, 190, 192, 423, 459.
——	1926. *Trans. roy. Soc. Edinb.* **54**, p. 719.	Lepidosiren, Protopterus, muscles.	23.
——	1931. *J. Anat. Lond.* **66**, p. 104.	Batoids, hyoid arch.	69.
——	1935. *The cranial muscles of vertebrates.* Cambridge.		91, 138, 249, 258, 263, 266, 320, 407, 423, 425.
Edinger, T.	1928. *Anat. Anz.* **66**, p. 172.	sclerotic bones.	443.
——	1929. *Zool. Jahrb. Abt. Anat.* **51**, p. 164.	sclerotic bones.	6, 96, 443, 499.
Eifertinger, L.	1933. *Z. Anat. EntwGesch.* **101**, p. 534.	Hypogeophis, dev. of mandible.	195, 196.
Eisinger, K. & Sternberg, H.	1924. *Arch. mikr. Anat. EntwMech.* **100**, p. 542.	ear & audit. caps.	478.
Ekman, G.	1934. *Ann. Soc. Zool. Bot. fenn.* **14**, p. 156.	Salmo, nose & nas. caps.	478.
Elliot, A. I. M.	1907. *Quart. J. micr. Sci.* **51**, p. 647.	Rana, head-segmentation.	24, 30.
Engelmann, O.	1910. *Anat. Anz.* **35**, p. 485.	Sus, supraoccip.	334, 444.
Erdmann, K.	1933. *Z. Anat. EntwGesch.* **101**, p. 566.	Triton, Rana, dev. of bony skull.	183, 184, 185, 198, 208.
Esdaile, P.	1916. *Philos. Trans.* B. **217**, p. 439.	Perameles, dev. of chondrocr.	304.
Faruqi, A. J.	1935. *Proc. zool. Soc. Lond.* p. 313.	Gadus, dev. of vertebrae.	168.
Fawcett, E.	1905*a*. *Anat. Anz.* **26**, p. 280.	man, dev. of pterygoid.	366, 369.
——	1905*b*. *J. Anat. Lond. Proc. anat. Soc.* p. vi.	man, dev. of pterygoid.	366, 369, 435.
——	1905*c*. *J. Anat. Lond.* **39**, p. 439.	man, dev. of mandible.	365.
——	1906. *Ibid.* **40**, p. 400.	man, dev. of palatine.	366.
——	1910*a*. *Ibid.* **44**, p. 207.	man, dev. of sphenoid.	355, 357, 366, 369, 435.
——	1910*b*. *Ibid.* **44**, p. 303.	man, chondrocr. 30 mm.	355, 357, 435.
——	1910*c*. 16 *Int. Congr. Med.* **1**, p. 170.	man, dev. of mandib.	364.
——	1911. *J. Anat. Lond.* **45**, p. 378.	man, dev. of maxilla, vomer.	365, 372, 434.
——	1916. *J. Anat. Lond. Proc. Anat. Soc.* p. 14.	mammals, chondrocr.	321.
——	1917. *J. Anat. Lond.* **51**, p. 309.	Microtus, chondrocr.	303, 315, 383, 391, 395.
——	1918*a*. *Ibid.* **52**, p. 211.	Erinaceus, chondrocr.	303, 320, 322, 324, 326, 338 355.

Author	*Reference*	*Subject*	*Page where quoted*
Fawcett, E.	1918*b*. *J. Anat. Lond.* **52**, p. 412.	Poicilophoca, chondrocr.	332, 333, 389, 391.
——	1919. *Ibid.* **53**, p. 315.	Miniopterus, chondrocr.	303, 316, 318, 348, 434, 440.
——	1921. *Ibid.* **55**, p. 187.	Tatusia, chondrocr.	316, 434, 435.
——	1923. *Ibid.* **57**, p. 245.	man, roof of chondrocr.	316, 357.
Fell, H. B.	1933. *Proc. roy. Soc. Lond.* B. **112**, p. 417.	cartilage & bone.	2, 38, 482, 502.
—— & Landauer, W.	1935. *Ibid.* B. **118**, p. 133.	cartilage & bone.	482, 496, 502.
—— & Robison, R.	1929. *Biochem. J.* **23**, p. 767.	phosphatase.	7.
—— & ——	1930. *Ibid.* **24**, p. 1905.	phosphatase.	7, 481.
—— & ——	1934. *Ibid.* **28**, p. 2243.	phosphatase.	481.
Ficalbi, E.	1890. *Monit. zool. ital.* **1**, pp. 119, 144.	accessory ossifications.	444.
Filatoff, D.	1906. *Anat. Anz.* **29**, p. 623.	Aves, basal articulation.	270.
——	1908. *Morph. Jahrb.* **37**, p. 289.	Emys, head-segmentation.	25, 31, 252.
——	1916. *Rev. zool. russe,* **1**, p. 48.	ear & audit. caps.	478.
——	1930. *Arch. EntwMech.* **122**, p. 546.	Acipenser, ear & audit. caps.	479.
Fischer, E.	1901. *Anat. Hefte,* **17**, p. 467.	Talpa, chondrocr.	320, 359, 391, 392, 435.
——	1903. *Z. Morphol. Anthrop.* **5**, p. 383.	Semnopithecus, Macacus, chondrocr.	353, 354, 402.
——	1905. *Proc. Acad. Sci. Amst.* **8**, p. 397.	Tarsius, chondrocr.	351.
Fisher, R.	1935. *Philos. Trans.* B. **225**, p. 195.	Gallus, cerebral hernia.	486.
Flower, W. H.	1869. *Proc. zool. Soc. Lond.* p. 4.	Felis, endotympanic.	331, 440.
Forster, A.	1902. *Z. Morphol. Anthrop.* **4**, p. 99.	Sus, supraoccip.	334, 440.
——	1925. *Arch. Anat. Strasbourg,* **4**, p. 295.	inclination of tympanic.	367.
Fortman, J. P. de G.	1918. *Tijdschr. ned. dierk. Ver.* **16**, p. 121.	Urodela, head-segmentation.	30, 179, 188.
Francis, E. T. B.	1934. *The anatomy of the Salamander.* Oxford.		188.
Fraser, E. A.	1915. *Proc. zool. Soc. Lond.* p. 299.	Marsupials, prootic somites.	25.
Frazer, J. E.	1910. *J. Anat. Lond.* **44**, p. 156.	man, thyroid cart.	362.
Freund, L.	1908. *Denkschr. med-naturw. Ges. Jena,* **7**, p. 557.	Halicore, dev. of bony skull.	346.
Frets, G. P.	1912. *Morph. Jahrb.* **44**, p. 409.	Primates, nas. caps.	354.
——	1913. *Ibid.* **45**, p. 557.	Platyrrhini, nas. caps.	352, 396.
——	1914. *Ibid.* **48**, p. 239.	Lemurs, Catarrhini, nas. caps.	351, 352, 353, 396.
Froriep, A.	1882. *Arch. Anat. Physiol. Lpz. Anat. Abt.* p. 279.	morph. of skull.	9, 13.
——	1902. *Verh. anat. Ges. Jena,* **16**, p. 34.	head-segmentation.	19.
——	1917. *Arch. Anat. Physiol. Lpz. Anat. Abt.* p. 61.	Salamandra, head-segmentation.	22.
Fuchs, H.	1905. *Ibid.* Suppl. p. li.	auditory ossicles.	413.
——	1906. *Ibid.* p. 1.	auditory ossicles.	313, 413, 414.
——	1907*a*. *Anat. Anz.* **31**, p. 33.	Emys, hyobranch. skel.	252.
——	1907*b*. *Verh. anat. Ges. Jena,* **21**, p. 8.	Urodela, operculum.	187.
——	1908. *Anat. Anz.* **32**, p. 584.	Didelphys, vomer, parasphen.	302, 434.
——	1909*a*. *Arch. Anat. Physiol. Lpz. Suppl.* p. 1.	secondary cart.	3, 502.

Author	Reference	Subject	Page where quoted
Fuchs, H.	1909*b*. *Anat. Anz.* **35**, p. 113.	streptostyly.	423, 425.
——	1909*c*. *Verh. anat. Ges. Jena*, **23**, p. 85.	Lepus, dev. of pterygoid.	314, 435.
——	1910*a*. *Anat. Anz.* **36**, p. 33.	pterygoid, morph.	436.
——	1910*b*. *Ibid.* **37**, p. 250.	streptostyly.	423, 425.
——	1912. *Verh. anat. Ges. Jena*, **26**, p. 81.	alisphenoid & epipterygoid.	421, 439.
——	1915. *Voeltzkow Reise Ostafrik. Stuttg.* **5**.	Chelone, chondrocr.	258, 260, 262, 301, 392, 421, 423, 425.
——	1920. *Anat. Anz.* **52**, pp. 353, 449; **53**, p. 353.	Chelonia, dev. of bony skull.	394.
——	1929. *Morph. Jahrb.* **63**, p. 408.	Anura, parahyoid.	87, 213.
Fürbringer, K.	1903. *Ibid.* **31**,, pp. 360, 620.	Selachii, visc. arch. skel.	71, 418.
——	1904. *Denkschr. med-naturw. Ges. Jena*, **4**, p. 423.	Holocephali, Dipnoi, morph. of skull.	72, 168, 170, 175, 423.
Fürbringer, M.	1897. *Festschr. Gegenbaur*, **3**, p. 349.	head-segmentation.	10, 140, 384.
——	1903. *Morph. Jahrb.* **31**, p. 623.	Batoids, extra-sept. cart.	69.
Gadow, H.	1889. *Philos. Trans.* **179**, p. 451.	colum. auris.	225.
——	1891. *Bronn's Thierreich*, VI, (4) pp. 920, 983.	Aves, dev. of skull.	279.
——	1901. *Anat. Anz.* **19**, p. 396.	auditory ossicles.	247.
Gaskell, W. H.	1908. *The origin of vertebrates*. London & New York.		41.
Gast, R.	1909. *Mitt. zool. Sta. Neapel*, **19**, p. 269.	head-segmentation.	19.
Gaupp, E.	1891. *Anat. Anz.* **6**, p. 107.	Lacerta, ascend. proc.	219, 220, 227.
——	1893. *Morph. Arb.* **2**, p. 275.	Rana, dev. of chondrocr.	11, 15, 178, 198, 207, 208, 417, 418.
——	1894. *Ibid.* **3**, p. 399.	Rana, dev. of hyobranch. skel.	198.
——	1895. *Ibid.* **4**, p. 77.	morph. of temporal bones.	232, 250, 259, 437, 438, 441.
——	1897. *Ergebn. Anat. EntwGesch.* **7**, p. 793.	head-segmentation.	1, 11.
——	1898. *Verh. anat. Ges. Jena*, **12**, p. 157.	Lacerta, chondrocr.	233, 389.
——	1899. *Ergebn. Anat. EntwGesch.* **8**, p. 990.	auditory ossicles.	208, 440.
——	1900. *Anat. Hefte*, **14**, p. 433.	Lacerta, chondrocr.	15, 217, 221, 225, 289, 388, 399, 401, 402, 420, 432, 450, 464, 510.
——	1901. *Ergebn. Anat. EntwGesch.* **10**, p. 847.	morph. of skull.	1, 495.
——	1902. *Anat. Hefte*, **19**, p. 155.	ala temporalis.	172, 192, 220, 245, 250, 260, 392, 420, 430, 439.
——	1903. *Verh. anat. Ges. Jena*, **17**, p. 113.	cart.- & memb.-bone.	115.
——	1904. Edition of Ecker-Wiedersheim: *Anatomie des Frosches*. Braunschweig.		208, 442.
——	1905*a*. *Anat. Anz.* **27**, p. 273.	Echidna, chondrocr.	392, 435.
——	1905*b*. *Verh. anat. Ges. Jena*, **19**, p. 125.	morph. of lower jaw.	441.
——	1905*c*. *Ergebn. Anat. EntwGesch.* **14**, p. 808.	hyobranch. skel.	26, 301, 406, 408, 414, 415

Author	*Reference*	*Subject*	*Page where quoted*
Gaupp, E.	1906*a*. *Die Entwicklung des Kopfskelettes*, in Hertwig's *Handbuch der vergleichenden und experimentellen Entwicklungslehre*. Jena. **3**, (2) p. 573.		1, 115, 124, 176, 179, 182, 183, 185, 190, 198, 208, 217, 231, 246, 260, 279, 399, 403, 409, 418, 423, 433, 434, 487, 495, 502.
——	1906*b*. *Verh. anat. Ges. Jena*, **20**, p. 21.	morph. of chondrocr.	393, 406, 434, 437.
——	1907*a*. *Ibid*. **21**, p. 251.	secondary cartilage.	38, 351, 502.
——	1907*b*. *Denkschr. med-naturw. Ges. Jena*, **6**, p. 483.	Echidna, occip. joint.	386.
——	1907*c*. *Verh. anat. Ges. Jena*, **21**, p. 129.	Echidna, chondrocr.	299.
——	1908. *Denkschr. med-naturw. Ges. Jena*, **6**, p. 539.	Echidna, dev. of chondrocr.	15, 295, 299, 300, 392, 396, 435, 436, 441, 464.
——	1910*a*. *Anat. Hefte*, **42**, p. 313.	morph. of pterygoids.	297, 379, 435, 436.
——	1910*b*. *Anat. Anz*. **36**, p. 529.	morph. of lachrymal.	231, 232, 267, 438.
——	1910*c*. *Ibid*. **37**, p. 352.	hyoid arch skel.	414,
——	1911*a*. 8 *int. Zool. Congr. Graz*, p. 215.	mammals, chondrocr.	386.
——	1911*b*. *Anat. Anz*. **39**, pp. 97, 433, 609.	morph. of lower jaw bones.	441.
——	1911*c*. *Ibid*. **38**, p. 401.	skull foramina.	185.
——	1912. *Ibid*. **40**, p. 561.	Amphibia, lower jaw.	186.
——	1913. *Arch. Anat. Physiol. Lpz. Anat. Abt. Suppl*. **5**, p. 1.	auditory ossicles.	414, 440.
——	1915. *Arch. Anat. Physiol. Lpz. Anat. Abt*. p. 62.	morph. of temporal bones.	366.
Gazagnaire, C.	1932. *C. R. Soc. Biol. Paris*, **110**, p. 1076.	Rana, dev. of colum. aur.	204.
Gegenbaur, C.	1864. *Jena. Z. Naturw*. **1**, p. 343.	cartilage & bone.	2, 5.
——	1867. *Ibid*. **3**, pp. 54, 206.	cartilage & bone.	2, 5.
——	1870. *Grundzüge der vergleichenden Anatomie*. Leipzig.		5.
——	1872. *Untersuchungen zur vergleichenden Anatomie der Wirbeltiere*. III. *Das Kopfskelett der Selachier*. Leipzig.		4, 8, 11, 55, 63, 64, 406, 410, 419, 428.
——	1873. *Jena. Z. Naturw*. **7**, p. 1.	turbinals, Sauropsida.	265, 278, 397.
——	1878. *Morph. Jahrb*. **4**, *Suppl*. p. 1.	Alepocephalus, memb.-bones.	5.
——	1887. *Festschr. Kölliker*, p. 1.	morph. of skull.	9.
——	1888. *Morph. Jahrb*. **13**, p. 1.	head-segmentation.	9.
——	1892. *Die Epiglottis*. Leipzig.		407.
——	1898. *Vergleichende Anatomie der Wirbeltiere*. Leipzig.		406, 415, 441.
van Gelderen, C.	1924. *Anat. Anz*. **58**, p. 472.	meningeal membranes.	35.
——	1925. *Ibid*. **60**, p. 48.	meningeal membranes.	35.
Geoffroy St.-Hilaire, E.	1818. *Philosophie anatomique*. Paris.		7.
Gibian, A.	1913. *Morph. Jahrb*. **45**, p. 57.	Selachii, hyobranch. skel.	49, 53, 406, 410.
van Gilse, P. H. G.	1926. *J. Anat. Lond*. **61**, p. 153.	sphenoidal sinuses.	372.
Gladstone, R. J. & E.-Powell, E.	1914. *J. Anat. Lond*. **59**, p. 190.	man, proatlas centrum.	356, 386.

Author	Reference	Subject	Page where quoted
Goethe, J. W.	1820. *Zur Naturwissenschaft überhaupt, besonders zur Morphologie*. I.		7.
Goette, A.	1875. *Die Entwickelungsgeschichte der Unke*. Leipzig.		24, 213.
——	1901. *Z. wiss. Zool.* **69**, p. 533.	visceral arches.	41.
Goldby, F.	1925. *J. Anat. Lond.* **59**, p. 301.	Crocodile, colum. aur.	262.
Göldi, E. A.	1882. *Jena. Z. Naturw.* **17**, p. 401.	morph. of bones.	5.
Golling, J.	1915. *Z. Morphol. Anthrop.* **17**, p. 1.	man, premaxilla.	365.
Goodey, T.	1910. *Proc. zool. Soc. Lond.* (1) p. 540.	Chlamydoselachus, morph. of skull.	407.
Goodrich, E. S.	1907. *Proc. zool. Soc. Lond.* p. 751.	scales & bones.	6.
——	1909. *Treatise on Zoology*. Edited by E. Ray Lankester. *Cyclostomes and Fishes*. London.		6, 40, 47, 76, 124, 141, 166.
——	1911. *Proc. zool. Soc. Lond.* (for 1910) p. 101.	Amblystoma, head-segmentation.	11, 21, 30, 177, 178, 381.
——	1913. *Quart. J. micr. Sci.* **59**, p. 227.	metamery & homology.	11, 31, 448, 490, 512.
——	1915. *Ibid.* **61**, p. 13.	dev. of audit. ossic.	207, 226, 293, 414, 441.
——	1917. *Ibid.* **62**, p. 539.	hypophysis & premandib. somites.	20, 22.
——	1918. *Ibid.* **63**, p. 1.	Scyllium, head-segmentation.	11, 17, 18, 19, 27, 58, 59, 381.
——	1930. *Studies on the structure and development of vertebrates*. London.		12, 77, 168, 171, 194, 302, 420, 437, 439, 441, 498.
Görke, O.	1904. *Arch. Anthrop. Braunschw.* **1**, p. 91.	bone-structure.	506.
Gradenigo, G.	1887. *Mitt. embryol. Inst. Univ. Wien*, **9**, p. 85.	mammals, audit. caps.	391, 405.
Gray, A. A.	1913. *J. Anat. Lond.* **47**, p. 391.	auditory ossicles.	240.
Gray, P.	1929. *Museums J.* **28**, p. 341.	alizarine stain.	15.
Green, H. L.	1930. *J. Anat. Lond.* **64**, p. 512.	Ornithorhynchus, prevomer.	289, 196.
Gregory, E. H.	1904. *Denkschr. med-naturw. Ges. Jena*, **4**, (i) p. 641.	Ceratodus, head-segmentation.	23.
Gregory, W. K.	1904. *Biol. Bull. Wood's Hole*, **7**, p. 55.	jaw-suspensions.	76, 422, 423, 424, 457.
——	1920. *Bull. Amer. Mus. nat. Hist.* **42**, p. 95.	morph. of lachrymal.	267, 439.
—— & Noble, G. K.	1924. *J. Morph.* **39**, p. 435.	alisphenoid, pleurosphenoid.	439.
Greil, A.	1913. *Denkschr. med-naturw. Ges. Jena*, **4**, (i) p. 661.	Ceratodus, dev. of chondrocr.	24, 168, 169, 171, 473.
Grosser, O.	1902. *Morph. Jahrb.* **29**, p. 1.	Cheiroptera, nas. caps.	348, 396, 397.
——	1912. *Development of the pharynx, etc.* in Keibel & Mall, *Manual of Human Embryology*. London.		362.
Grüneberg, H.	1935. *Proc. roy. Soc.* B. **118**, p. 321.	Mus, 'grey-lethal'.	490.
Haas, G.	1929. *Anat. Anz.* **68**, p. 358.	Ophidia, streptostyly.	251.
——	1930*a*. *Zool. Jahrb. Abt. Anat.* **52**, p. 1.	Typhlops, morph. of skull.	251.
——	1930*b*. *Ibid.* **52**, p. 95.	Ophidia, streptostyly.	251.
——	1930*c*. *Ibid.* **52**, p. 347.	Ophidia, streptostyly.	251.
——	1931. *Ibid.* **54**, p. 333.	Ophidia, streptostyly.	251.
——	1936. *Acta zool. Stockh.* **16**, p. 409.	Ablepharus, chondrocr.	236.

Author	*Reference*	*Subject*	*Page where quoted*
Hafferl, A.	1921. *Z. Anat. EntwGesch.* **62**, p. 433.	Platydactylus, chondrocr.	238.
Hagen, W.	1900. *Arch. Anat. Physiol. Lpz. Anat. Abt.* p. 1.	man, chondrocr. 17 mm.	355.
Hague, F. S.	1924. *J. Morph.* **39**, p. 267.	Amia, dev. of chondrocr.	97.
Haines, R. W.	1934. *J. Quart. micr. Sci.* **77**, p. 77.	growth of branchial bones.	454.
——	In press.	fishes, lower jaw bones.	128.
Hammarberg, F.	1937. *Acta zool. Stockh.* **18**, p. 209.	Lepidosteus, dev. of skull.	108, 113.
Hannover, A.	1880. *Primordialbrusken og dens Forbening i det menneskelige Kranium.* Kjøbenhavn.		392.
Harrison, R. G.	1895. *Arch. mikr. Anat.* **46**, p. 500.	Salmo, segmentation.	23, 115.
——	1925. *J. exp. Zool.* **41**, p. 349.	pterygoquad. & balancer.	480.
Harvey, L. & Burr, H. S.	1924. *Proc. Soc. exp. Biol. N.Y.* **22**, p. 52.	endomeninx & neural crest.	36.
Hatschek, B.	1892. *Verh. anat. Ges. Jena*, **6**, p. 136.	Petromyzon, chondrocr.	41, 42.
Hay, O. P.	1890. *J. Morph.* **4**, p. 11.	Amphiuma, chondrocr.	190.
Hayek, H.	1923. *SitzBer. Akad. Wiss. Wien.* **130**, p. 25.	proatlas.	223, 385.
——	1924. *Morph. Jahrb.* **53**, p. 137.	proatlas.	223, 385.
Helff, O. M.	1928. *Physiol. Zoöl.* **1**, p. 463.	Anura, tymp. ring.	208.
Hellman, M.	1926. *Proc.* **1***st int. Congr. Orthodont. New York*, p. 475.	man, skull, differential growth.	472.
Henckel, K. O.	1927. *Z. Morphol. Anthrop.* **26**, p. 365.	Tarsius, chondrocr.	351, 390.
——	1928*a*. *Morph. Jahrb.* **59**, p. 105.	Chrysothrix, Nycticebus, chondrocr.	351.
——	1928*b*. *Z. Anat. EntwGesch.* **86**, p. 204.	Tupaja, chondrocr.	325.
——	1929. *Morph. Jahrb.* **62**, p. 179.	Galeopithecus, chondrocr.	350.
Hepburn, D.	1907. *J. Anat. Lond.* **42**, p. 88.	man, interparietal.	444.
Hertwig, O.	1876. *Morph. Jahrb.* **2**, p. 328.	teeth & memb.-bones.	6, 184.
——	1879. *Ibid.* **5**, p. 1.	teeth & memb.-bones.	6.
——	1898. *Lehrbuch der Entwicklungsgeschichte.* Jena.	man, chondrocr. 80 mm.	355.
Herzfeld, P.	1889. *Zool. Jahrb. Abt. Anat.* **3**, p. 551.	mammals, nas. caps.	396.
Hesser, C.	1925. *Morph. Jahrb.* **55**, p. 489.	man, dev. of joints.	362.
Hirschfelder, H.	1936. *Z. Anat. EntwGesch.* **106**, p. 497.	Manatus, chondrocr.	348.
Hoffmann, C. K.	1889. *Zool. Anz.* **12**, p. 338.	Lacerta, head-segmentation.	25, 217, 219, 220, 412.
——	1890. *Verh. Akad. Wet. Amst.* **28**, p. 1.	Lacerta, dev. of colum. aur.	217, 219, 220, 412.
——	1897. *Morph. Jahrb.* **24**, p. 209.	Selachii, head-segmentation.	19, 27.
——	1898. *Ibid.* **25**, p. 250.	Selachii, head-segmentation.	27.
Hofmann, K.	1923. *Anat. Anz.* **56**, p. 432.	Salamandra, branch. skel. metamorph.	480.
Holmdahl, D. E.	1928. *Z. mikr-anat. Forsch.* **14**, p. 99; **15**, p. 191.	neural crest & mesect.	473.
Holtfreter, J.	1933. *Arch. EntwMech.* **129**, p. 669.	Amblystoma exogastr.	474.
——	1935. *Ibid.* **133**, p. 427.	neural crest & visc. cart.	475.
Honigmann, H.	1915. *Anat. Anz.* **48**, p. 113.	Megaptera, dev. of chondrocr.	340.
——	1917. *Zoologica, Stuttgart,* **69**.	Megaptera, dev. of chondrocr.	340, 401.

Author	Reference	Subject	Page where quoted
Howes, G. B.	1891. *Proc. zool. Soc. Lond.* p. 148.	Hyla, basimandib.	418.
——	1896. *J. Anat. Lond.* **30**, p. 513.	Lepus, Hyrax, styloid proc.	315, 340, 414.
—— & Swinnerton, H. H.	1901. *Trans. zool. Soc. Lond.* **16**, p. 1.	Sphenodon, dev. of chondrocr.	238, 375.
Hubrecht, A. A. W.	1877. *Niederl. Arch. Zool.* **3**, p. 255.	Holocephali, morph. of skull.	72, 78.
von Huene, F.	1912. *Anat. Anz.* **42**, p. 522.	interparietal.	444.
Hunter, R. M.	1935. *J. Morph.* **57**, p. 501.	Lepus, head-segmentation.	25.
Huxley, J. S.	1927. *Biol. Zbl.* **47**, p. 151.	Canis skull, allometric growth.	471.
——	1932. *Problems of relative growth.* London.		471.
—— & de Beer, G. R.	1934. *Elements of experimental embryology.* Cambridge.		490, 493.
Huxley, T. H.	1858. *Proc. roy. Soc. Lond.* **9**, p. 381.	segmental theory of skull.	8, 146, 198, 416.
——	1864. *Lectures on the elements of comparative anatomy.* London.		4, 5.
——	1867. *Proc. zool. Soc. Lond.* p. 415.	Aves, morph. of palate.	444.
——	1874. *Ibid.* p. 186.	Necturus, morph. of skull.	190, 191, 198, 375, 379, 417.
——	1875*a*. *Nature, Lond.* **11**, p. 68.	Amphibia, col. aur.	198, 417.
——	1875*b*. *Proc. roy. Soc. Lond.* **23**, p. 127.	Amphioxus, morph. of head.	375.
——	1876*a*. *J. Anat. Physiol. Lond.* **10**, p. 412.	Cyclostomes, morph. of head.	41, 47, 415.
——	1876*b*. *Proc. zool. Soc. Lond.* p. 24.	jaw-suspensions.	76, 168, 170, 171, 172, 173, 416, 419, 421.
Ichikawa, M.	1933. *Proc. imp. Acad. Japan,* **9**, p. 117.	neural crest & mesect.	475.
Inman, V. & Saunders, J.	1937. *J. Anat. Lond.* **71**, p. 383.	man, frontal.	365.
Iwanzoff, N.	1894. *Anat. Anz.* **9**, p. 578.	colum. aur.	213.
Jacobson, L.	1842. *Forh. scand. Natforsk. Möte.* Stockh.	primordial cranium.	4.
Jacobson, W.	1928. *Z. Anat. EntwGesch.* **88**, p. 405.	Talpa, nas. caps.	396.
Jacoby, M.	1895. *Arch. mikr. Anat.* **44**, p. 61.	man, chondrocr. 30 mm.	355.
Jaekel, O.	1927. *Ergebn. Anat. EntwGesch.* **27**, p. 815.	morph. of skull.	418.
Jager, J.	1926. *Morph. Jahrb.* **56**, p. 1.	Aves, acrochordal.	269, 375, 382.
Jarmer, K.	1922. *Z. Anat. EntwGesch.* **64**, p. 56.	man, prevomer.	365, 434.
Jenkinson, J. W.	1911. *J. Anat. Lond.* **45**, p. 305.	Mus, dev. of audit. ossic.	315.
Johnson, C. E.	1913. *Amer. J. Anat.* **14**, p. 119.	Chelydra, head-segmentation.	25.
Kaensche, C.	1890. *Schneider's Zool. Beitr. Breslau,* **2**, p. 219.	Petromyzon, dev. of skull.	41, 45.
Kallius, E.	1897. *Anat. Hefte,* **9**, p. 301.	man, dev. of hyobranch. skel.	362.
——	1901. *Ibid.* **16**, p. 531.	Amphibia, dev. of hyobranch. skel.	183, 212.
——	1905*a*. *Ibid.* **28**, p. 305.	Anas, dev. of hyobranch. skel.	268, 275, 277, 415.
——	1905*b*. In Bardeleben's *Handbuch der Anatomie des Menschens.* Jena, **5**, (**1**, ii), p. 194.	man, dev. of nas. caps.	361.
——	1906. *Anat. Hefte,* **31**, p. 603.	Melopsittacus, dev. of hyobranch. skel.	418.

Author	Reference	Subject	Page where quoted
Kallius, E.	1910. *Anat. Hefte*, **41**, p. 173.	Sus, dev. of hyobranch. skel.	334, 336, 412.
van Kampen, P. N.	1905. *Morph. Jahrb.* **34**, p. 321.	mammals, tympanic region.	314, 320, 326, 340, 379, 414, 437, 440, 441.
——	1907. *Zool. Anz.* **31**, p. 695.	Putorius, stylohyal.	334.
——	1915*a*. *Tijdschr. ned. dierk. Ver.* **14**, p. xxiv.	endotympanic.	318, 332, 440.
——	1915*b*. *Ned. Tijdschr. Geneesk.* **59**, p. 2444.	colum. aur. & stapes.	302, 348, 350, 412, 413.
——	1922. *Bijdr. Dierk.* **22**, p. 53.	basitemporal.	436.
——	1926. *Tijdschr. ned. dierk. Ver.* **20**, p. 59.	Pipa, hyoid arch.	212.
Kappers, C. U. Ariëns,	1926. *Arch. Neurol. Psychiat. Chicago*, **15**, p. 281.	meningeal membranes.	35.
——	1929. *The evolution of the nervous system.* Haarlem.	meningeal membranes.	35.
——	1932. *Philos. Trans.* B. **221**, p. 391.	skull & brain.	487.
Kastschenko, N.	1886. *Arch. Anat. Physiol. Lpz. Anat. Abt.* p. 388.	graphic reconstruct. method.	15.
——	1887. *Anat. Anz.* **2**, p. 426.	graphic reconstruct. method.	15.
——	1888. *Ibid.* **3**, p. 445.	neural crest & mesect.	19, 472.
Kay, H. D. & Robison, R.	1924. *Biochem. J.* **18**, p. 755.	phosphatase.	481.
Kerckring, T.	1717. *Theodori Kerckringii opera omnia anatomica.* Leiden.	cartilage & bone.	1.
Kernan, J. D.	1915. *Anat. Rec.* **10**, p. 213.	Felis, occip. region.	31, 326, 328, 381, 384.
——	1916. *J. Morph.* **27**, p. 605.	man, chondrocr. 20 mm.	355, 356.
Kerr, J. Graham,	1902. *Quart. J. micr. Sci.* **45**, p. 1.	glass plate reconstr. method.	15.
——	1903. *Ibid.* **46**, p. 417.	Lepidosiren, dev. of tooth-plates.	176.
——	1907. *Budgett Mem. Vol.* Cambridge, p. 195.	Polypterus, dev. of chondrocr.	78.
Kesteven, H. L.	1918. *J. Anat. Lond.* **52**, p. 449.	alisphenoid.	437, 439.
——	1926. *Ibid.* **61**, p. 112.	alisphenoid.	439.
——	1931. *Rec. Austral. Mus.* **18**, p. 236.	Ceratodus, morph. of skull.	168.
Killian, G.	1891. *Verh. anat. Ges. Jena*, **5**, p. 85.	head-segmentation.	19.
——	1896. *Arch. Laryng. Rhin. Berl.* **3**, p. 17; **4**, p. 1.	man, nas. caps.	361.
Kindred, J. E.	1919. *Illinois biol. Monogr.* **5**, p. 1.	Amiurus, dev. of skull.	136, 139, 141, 500.
——	1921. *J. Morph.* **35**, p. 425.	Syngnathus, dev. of chondrocr.	152, 409.
——	1924. *Amer. J. Anat.* **33**, p. 421.	Syngnathus, dev. of bony skull.	152, 500.
Kingsbury, B. F. & Reed, K. D.	1909. *J. Morph.* **20**, p. 549.	Urodela, colum. aur.	177, 179, 183, 187, 188, 190, 191.
Kisselewa, Z. N.	1929. *Trav. Inst. Zool. I Univ. Moscou*, **3**, p. 1.	Ceratodus, morph. of skull.	174.
Klaatsch, H.	1895. *Verh. anat. Ges. Jena*, **9**, p. 122.	lateral line & bones.	489.
van der Klaauw, C. J.	1922. *Tijdschr. ned. dierk. Ver.* **18**, p. 135.	endotympanic.	319, 320, 325, 331, 332, 348, 350, 440.
——	1923. *Z. Anat. EntwGesch.* **69**, p. 32.	Paauw's & Spence's cart.	303, 304, 318, 330, 336, 339, 350, 412, 413.

Author	Reference	Subject	Page where quoted
van der Klaauw, C. J.	1924*a*. *Ergebn. Anat. EntwGesch.* **25**, p. 565.	audit. ossicles.	440.
——	1924*b*. *Ned. Tijdschr. Geneesk.* **68**, (ii).	Marsupials, chorda tymp.	305, 414.
——	1924*c*. *Anat. Anz.* **57**, p. 240.	Paauw's cart.	305, 406, 4 414.
Klatt, B.	1912. *SitzBer. Ges. nat-forsch. Fr. Berl.* p. 153.	domestication & skull-size.	486.
Kokott, W.	1933. *Morph. Jahrb.* **72**, p. 341.	bone-structure.	455, 485.
von Kölliker, A.	1849. *Allgemeine Betrachtungen über die Entstehung des knöchernen Schädels der Wirbeltiere.* Leipzig.	cartilage and bone.	4, 502.
Koltzoff, N. K.	1902. *Bull. Soc. imp. Nat. Moscou,* **15**, p. 259.	Petromyzon, head-segmentation.	20, 27, 4 376, 473.
Kothe, K.	1910. *Arch. Naturgesch.* **76**, (i) p. 29.	Anura, hyoid arch.	213.
Kotthaus, A.	1933. *Z. wiss. Zool.* **144**, p. 510.	Xenopus, dev. of chondrocr.	214.
Koumans, F. P.	1936. *Temminckia,* **1**, p. 272.	scales, increase in number.	33.
Krassowsky, S. K.	1936. *Anat. Anz.* **82**, p. 112.	Picidae, palate.	445.
Krawetz, L.	1911. *Bull. Soc. Imp. Nat. Moscou,* **24**, p. 332.	Ceratodus, dev. of chondrocr.	24, 29, 16 170, 173.
Krivetski, A.	1917. *Rev. zool. russe,* **2**, p. 16.	Batoids, hyoid arch.	69, 70, 411.
Kruijtzer, E. M.	1931. *De ontwikkeling van het chondrocranium van Megalophrys.* Proefschr. Leiden.		211, 212, 21 417.
Kunkel, B. W.	1912*a*. *J. Morph.* **23**, p. 693.	Emys, dev. of skull.	252, 253, 43
——	1912*b*. *Anat. Rec.* **6**, p. 267.	Emys, dev. of audit. caps.	253.
von Kupffer, C.	1894. *Studien zur vergleichenden Entwickelungsgeschichte des Kopfes der Kranioten.* II. München & Leipzig.	Petromyzon, dev. of skull.	41.
——	1895. *Verh. anat. Ges. Jena,* **9**, p. 105.	Petromyzon, dev. of skull.	41.
Kurz, E.	1924. *Z. Anat. EntwGesch.* **75**, p. 36.	Acipenser, morph. of skull.	88.
Lacoste, A.	1927. *C. R. Soc. Biol. Paris,* **97**, p. 1403.	Ovis, supraoccip.	443.
——	1929. *C. R. Ass. Anat.* **24**, p. 339.	man, tectum posterius.	258.
Lakjer, T.	1926. *Studien über die Trigeminus-versorgte Muskulatur der Sauropsiden.* Kopenhagen.	reptiles, streptostyly.	280.
——	1927. *Zool. Jahrb. Abt. Anat.* **49**, p. 57.	jaw-attachments.	423, 424, 42
Landacre, F. L.	1921. *J. comp. Neurol.* **33**, p. 1.	neural crest & mesect.	473, 476.
Landauer, W.	1927. *Arch. EntwMech.* **110**, p. 195.	Gallus, chondrodystrophia.	491.
de Lange, D.	1936. *J. Anat. Lond.* **70**, p. 515.	head-segmentation.	19.
Lapage, E. O.	1928*a*. *J. Morph.* **45**, p. 441.	Urodela, septomaxilla.	181, 188, 44
——	1928*b*. *Ibid.* **46**, p. 399.	Anura, septomaxilla.	209, 442.
Lasdin, W.	1913. *Trav. Soc. imp. Nat. Petersbourg,* **14**, pp. 12, 75, 110.	Exocoetus, dev. of skull.	130, 500.
Lebedinsky, N. G.	1917. *Anat. Anz.* **49**, p. 33; **50**, p. 313.	Meckel's cart.	421.
Lebedkin, S.	1918. *Ibid.* **50**, p. 539.	Sus, dev. of chondrocr.	334, 384.
——	1924. *Ibid.* **58**, p. 449.	axis of audit. vesic.	360, 405, 47
Leche, W.	1912. *Zool. Jahrb. Suppl.* **15**, (2), p. 1.	Mycetes, postembr. changes.	451.
Lehmann, F. E.	1933. *Rev. suisse Zool.* **40**, p. 251.	cyclopia & trabeculae.	477.
——	1936. *Ibid.* **43**, p. 535.	cyclopia & otocephaly.	477

Author	Reference	Subject	Page where quoted
ehn, C.	1918. *Z. angew. Anat.* **2**, p. 349.	Polypterus, morph. of skull.	78, 87, 500.
evi, G.	1900. *Arch. mikr. Anat.* **55**, p. 341.	man, chondrocr.	355.
—	1909*a*. *Monit. zool. ital.* **20**, p. 159.	man, chondrocr.	300, 324, 355.
—	1909*b*. *Arch. ital. Anat. Embriol.* **7**, p. 615.	morph. of occip. region.	300, 324, 382 394.
evi, G. M.	1930. *Ibid.* **27**, p. 519.	secondary cartilage.	3, 38, 502.
ewis, W. H.	1906. *Anat. Rec.* **1**, p. 141.	ear & audit. caps.	479.
—	1915. *Ibid.* **9**, p. 719.	reconstruct. method.	14.
—	1920. *Contr. Embryol. Carn. Instn.* **39**, p. 299.	man, chondrocr. 21 mm.	355, 357
exer, E.	1924. *Neue deutsche Chirurg.* **26** (ii), p. 1.	bone-growth.	39.
eydig, F.	1854. *Z. wiss. Zool.* **5**, p. 40.	Polypterus, mem.-bones.	5.
—	1872. *Die in Deutschland lebenden Arten der Säurier*. Tübingen.	Lacerta, dev. of skull.	217.
mberger, R.	1925. *Morph. Jahrb.* **55**, p. 240.	Equus, chondrocr.	339.
nck, A.	1911. *Anat. Hefte*, **42**, p. 605.	man, notochord.	355.
tzelmann, E.	1923. *Z. Anat. EntwGesch.* **67**, p. 457.	Amphibia, visc. arch skel.	183, 213.
ıbosch, W.	1907. *Denkschr. med-naturw. Ges. Jena*, **7**, p. 519.	Marsupials, Edentates, morph. of skull.	319, 436.
—	1909. *Biol. Zbl.* **29**, p. 738.	phylogeny of cartilage.	450.
ındborg, H.	1899. *Morph. Jahrb.* **27**, p. 242.	Salmo, trabeculae.	117.
ındvall, H.	1905. *Anat. Anz.* **27**, p. 520.	cartilage & bone, stain.	15.
ırje, M.	1906. *Anat. Hefte*, **31**, p. 1.	Columba, pneumaticity.	282.
ıschka, H.	1857. *Z. wiss. Zool.* **8**, p. 123.	man, ala minima.	360, 390.
ıther, A.	1909. *Acta Soc. Sci. fenn.* **37**, (6) p. 1.	Holocephali, morph. of skull.	71, 72, 410.
—	1913. *Ibid.* **41**, (8) p. 1.	Acipenser, morph. of skull.	88.
—	1914. *Ibid.* **44**, (7) p. 1.	Amphibia, jaw-suspension.	192, 194, 423.
—	1925. *Comm. biol. Soc. Sci. fenn.* **2**, p. 1.	ear & audit. caps.	478.
acBride, E. W.	1932. *Biol. Rev.* **7**, p. 108.	Amniota, head segments.	32.
ackintosh, N. A.	1923. *Proc. zool. Soc. Lond.* p. 501.	Sebastes, dev. of chondrocr.	156.
acklin, C. C.	1914. *Amer. J. Anat.* **16**, pp. 317, 387.	man, chondrocr. 40 mm.	355, 389, 390
—	1921. *Contr. Embryol. Carn. Instn.* **48**, p. 57.	man, chondrocr. 43 mm.	355.
cMurrich, J. P.	1884*a*. *Quart. J. micr. Sci.* **23**, p. 623.	Syngnathus, dev. of skull.	152.
—	1884*b*. *Proc. Canad. Inst.* **2**, p. 270.	Amiurus, morph. of skull.	136.
aggi, L.	1896. *Arch. ital. Biol.* **26**, p. 301.	man, interparietal.	444.
—	1897. *Ibid.* **28**, p. 329.	mammals, prefrontal.	368.
—	1903. *Ibid.* **39**, p. 477.	mammals, postfrontal	368.
ahendra, B. C.	1936. *Proc. Indian Acad. Sci.* **3**, p. 128.	Typhlops, morph. of skull.	251.
air, R.	1926. *Z. mikr-anat. Forsch.* **5**, p. 625.	bone growth.	39.
akushok, M. E.	1925. *Rev. zool. russe*, **5**, p. 91.	Salmo, occip. region.	115.
—	1928. *Ibid.* **8**, p. 3.	Salmo, occip. arches.	115, 116.
all, F. P.	1906. *Amer. J. Anat.* **5**, p. 433.	man, dev. of bony skull.	369, 443.
angold, O.	1931. *Ergebn. Biol.* **7**, p. 193.	eye & sclerotic cart.	479.
arcus, H.	1910. *Festschr. Hertwig*, **2**, p. 373.	Gymnophiona, head-segmentation.	24, 30, 195, 196.
—	1935. *Anat. Anz.* **80**, p. 142.	Hypogeophis, dev. of audit. caps.	195, 196.
—, Stimmelmayr, E. & Porsch, G.	1935. *Morph. Jahrb.* **76**, p. 373.	Hypogeophis, dev. of skull.	12, 195, 196.

Author	Reference	Subject	Page where quoted
Marcus, H., Winsauer, O. & Hueber, A.	1933. *Z. Anat. EntwGesch.* **100**, p. 149.	Hypogeophis, kinetism.	195, 197.
Marshall, A. M.	1881. *Quart. J. micr. Sci.* **21**, p. 72.	Selachii, head-segmentation.	16.
Martland, M. & Robison, R.	1926. *Biochem. J.* **20**, p. 847.	phosphatase	481.
Matthes, E.	1912. *Jena. Z. Naturw.* **48**, p. 489.	Manatus, nas. caps.	348, 395.
——	1921*a*. *Z. Anat. EntwGesch.* **60**, p. 1.	Halicore, chondrocr.	346, 390.
——	1921*b*. *Anat. Anz.* **54**, p. 209.	Halicore, Meckel's cart.	346, 404.
——	1921*c*. *Ergebn. Anat. EntwGesch.* **23**, p. 669.	mammals, chondrocr.	1.
——	1922. *Ibid.* **24**, p. 117.	mammals, chondrocr.	1, 396.
Matveiev, B.	1915. *J. Sect. Zool. Soc. imp. Amis Sci. nat. Anthrop. Ethnogr. Moscou*, **2**, p. 203.	Selachii, head-segmentation.	19.
——	1922. *J. russe Zool.* **3**, p. 34.	Amblystoma, head-segmentation.	22.
——	1925. *Bull. Soc. Nat. Moscou*, **34**, p. 416.	Selachii, head-segmentation.	13, 19, 28, 375, 377.
——	1929. *Zool. Jahrb. Abt. Anat.* **51**, p. 463.	Scardinius, Weberian ossic.	136, 141, 1
Mayhew, R. L.	1924. *J. Morph.* **38**, p. 315.	Lepidosteus, morph. of skull.	108, 113.
Mayhoff, H.	1914. *Zool. Anz.* **43**, p. 389.	Pleuronectes, metamorph. of skull.	161, 500.
Mead, C. S.	1906. *Amer. Nat.* **40**, p. 475.	mammals, occip. condyles.	387.
——	1909. *Amer. J. Anat.* **9**, p. 167.	Sus, chondrocr.	334, 390.
Meckel, J. F.	1809. *Beiträge zur vergleichenden Anatomie*. Leipzig. **1** (2), p. 34.	man, interparietal.	444.
——	1820. *Handbuch der menschlichen Anatomie*. Halle & Berlin. **4**.		3, 7.
Meek, A.	1911. *J. Anat. Lond.* **45**, p. 357.	Crocodile, chondrocr.	262, 265, 2 397.
Michelsson, G.	1922. *Z. Anat. EntwGesch.* **65**, p. 509.	Erinaceus, chondrocr.	324.
von Mihalkovics, V.	1899. *Anat. Hefte*, **11**, p. 1.	nas. caps.	396.
Mivart, St. G.	1881. *The Cat*. London.	endotympanic.	440.
Möller, W.	1905. *Arch. mikr. Anat.* **65**, p. 439.	Ophidia, colum. aur.	247.
Moodie, R. L.	1922. *J. comp. Neurol.* **34**, p. 319.	lateral line & bone.	489.
Mook, C. C.	1921. *Bull. Amer. Mus. nat. Hist.* **44**, p. 101.	Alligator, postparietal.	267, 444.
Mookerjee, H. K.	1930. *Philos. Trans.* B. **218**, p. 415.	Urodela, dev. of vertebrae.	30, 179.
——	1931. *Ibid.* **219**, p. 165.	Anura, dev. of vertebrae.	30, 203.
Mori, O.	1924. *Z. Anat. EntwGesch.* **73**, p. 389.	Squalus, dev. of chondrocr.	49.
Moy-Thomas, J. A.	1933. *Quart. J. micr. Sci.* **76**, p. 209.	Polypterus, dev. of chondrocr.	78.
——	1934. *Ibid.* **76**, p. 481.	teeth & memb.-bones.	6, 176, 503
——	1936. *Geol. Mag.* **73**, p. 488.	Helodus, morph. of skull.	76.
Muggia, G.	1931*a*. *Monit. zool. ital.* **42**, p. 275.	man, head-segments.	25.
——	1931*b*. *Z. Anat. EntwGesch.* **95**, p. 297.	Equus, chondrocr.	339.
——	1932*a*. *Ibid.* **99**, p. 384.	man, roof of chondrocr.	358, 359.
——	1932*b*. *Monit. zool. ital.* **42**, p. 345.	man, tectum posterius.	358.
Mulder, J. D.	1928. *Anat. Anz.* **65**, p. 413.	Acipenser, dev. of rostrum.	94.
Müller, H.	1858. *Z. wiss. Zool.* **9**, p. 147.	cart.- & memb.-bones.	4.

Author	Reference	Subject	Page where quoted
üller, J.	1843. *Arch. Anat. Physiol. wiss. Med. Lpz.* p. ccxxxviii.	primordial cranium.	4.
uller, J.	1934. *The orbitotemporal region of the skull in mammals.* Diss. Leiden.		507.
urray, P. D. F.	1936. *Bones.* Cambridge.		485.
ıssonov, N. W.	1934. *C. R. Acad. Sci. U.R.S.S.* **3**, p. 205.	induction of cartilage.	472.
eal, H. V.	1897. *Anat. Anz.* **13**, p. 441.	Squalus, head-segmentation.	19.
—	1898. *Bull. Mus. comp. Zoöl. Harv.* **31**, p. 145.	Selachii, head-segmentation.	19.
—	1918. *J. Morph.* **30**, p. 433.	Selachii, head-segmentation.	19.
eave, F.	1936. *Nature, Lond.* **137**, p. 1034.	lateral line & scales.	489.
eedham, A. E.	1936. *Proc. zool. Soc. Lond.* p. 773.	Lepidosteus, jaw-growth.	471.
esbitt, R.	1736. *Human osteogeny explained in two lectures.* London.	cartilage & bone.	1.
estler, K.	1890. *Arch. Naturgesch.* **56**, (i) p. 81.	Petromyzon, dev. of skull.	41.
eubauer, G.	1925. *Z. Morphol. Anthrop.* **23**, p. 411.	muscle & bone.	489.
eukomm, A.	1933. *Arch. Anat. Strasbourg.* **17**, p. 49.	Sus, dev. of bony skull.	334.
eumayer, L.	1898. *SitzBer. Ges. Morph. Physiol. München*, **13**, p. 69.	Petromyzon, morph. of skull.	41.
—	1910. *Verh. anat. Ges. Jena*, **24**, p. 94.	Bdellostoma, visc. arches.	41.
—	1932. *Acta zool. Stockh.* **13**, p. 305.	Acipenser, early dev.	23, 89.
icholas, J. S.	1930. *J. exp. Zool.* **55**, p. 1.	chondrocr. & brain-grafts.	477.
ick, L.	1912. *Zool. Jahrb. Abt. Anat.* **33**, p. 1.	Dermochelys, morph. of skull.	255, 256, 260.
icola, B.	1903. *Arch. ital. Biol.* **40**, p. 313.	alisphenoid.	370.
ielson, E.	1936. *Med. Grönland*, **112**, (3) p. 1.	Coelacanthini, lat. line.	489, 506, 508.
oack, H.	1907. *Arch. mikr. Anat.* **69**, p. 457.	Emys, dev. of colum. aur.	252.
oble, G. K.	1921. *Bull. Amer. Mus. nat. Hist.* **44**, p. 1.	Urodela, lachrymal.	439.
n Noorden, W.	1887. *Arch. Anat. Physiol. Lpz. Anat. Abt.* p. 241.	man, chondrocr. 17, 18.5, 33 mm.	355.
oordenbos, W.	1905. *Petrus Camper*, **3**, p. 367.	Talpa, dev. of chondrocr.	320, 334, 339, 383, 401.
orman, J. R.	1926. *Philos. Trans.* B. **214**, p. 369.	Anguilla, dev. of chondrocr.	132, 143, 154, 156, 393, 394, 409, 500.
orris, H. W.	1929. *J. Morph.* **48**, p. 543.	Selachii, parietal fossa.	64.
usbaum, J.	1908. *Anat. Anz.* **32**, p. 513.	Cyprinus, Weberian ossic.	140, 141.
gawa, C.	1921. *J. exp. Zool.* **34**, p. 17.	ear, rotation.	479.
—	1926. *Fol. anat. japon.* **4**, p. 413.	ear, rotation.	479.
gushi, K.	1911. *Morph. Jahrb.* **43**, p. 1.	Trionyx, morph. of skull.	260.
kajima, K.	1915. *Anat. Hefte*, **53**, p. 325.	Trigonocephalus, colum. aur.	247.
ken, L.	1807. *Über die Bedeutung der Schädelknochen.* Jena.	morph. of skull.	7.
kutomi, K.	1936. *Zool. Jahrb. Abt. Anat.* **61**, p. 1.	Onychodactylus, chondrocr.	177, 181, 185, 186.
—	1937. *Z. Anat. EntwGesch.* **107**, p. 28.	Polypedates, chrondocr.	199.
lmstead, M. P.	1911. *Anat. Hefte*, **43**, p. 335.	Canis, chondrocr.	331.
sawa, G.	1898. *Arch. mikr. Anat.* **51**, p. 481.	Sphenodon, morph. of skull.	244.

Author	*Reference*	*Subject*	*Page where quoted*
Osawa, G.	1902. *Mitt. med. Fak. Univ. Tokio,* **5**, p. 221.	Cryptobranchus, morph. of skull.	190.
Osborn, H. F.	1900. *Amer. Nat.* **34**, p. 943.	occip. condyles.	387
Owen, R.	1846. *Lectures on comparative anatomy and physiology of vertebrate animals.* London.		4, 7.
Palmer, R. W.	1913. *Anat. Anz.* **43**, p. 510.	Perameles, audit. ossic.	304, 502.
Parker, H. W.	1928. *Ann. Mag. nat. Hist.* **2**, p. 473.	Microhyla, morph. of skull.	210.
Parker, T. J.	1888. *Proc. roy. Soc.* **43**, p. 391.	Apteryx, dev. of skull.	287.
——	1892. *Philos. Trans.* B. **182**, p. 25.	Apteryx, dev. of skull.	278, 287, 28[illegible] 509.
——	1893. *Ibid.* **183**, p. 731.	Apteryx, dev. of skull.	287.
Parker, W. K.	1862. *Trans. zool. Soc. Lond.* **4**, p. 269.	Balaeniceps, Pavo, morph. of skull.	504.
——	1866*a*. *Ibid.* **5**, p. 149.	Tinamu, Columba, morph. of skull.	504.
——	1866*b*. *Philos. Trans.* **156**, p. 113.	Struthio, dev. of skull.	278, 282, 28[illegible] 287.
——	1870. *Ibid.* **159**, p. 758.	Gallus, dev. of skull.	5, 278, 27[illegible] 280, 281, 28[illegible] 434, 444, 45[illegible]
——	1871. *Ibid.* **161**, p. 137.	Anura, dev. of skull.	198, 399, 41[illegible] 442.
——	1872. *Month. micr. J.* **8**, p. 217.	Corvus, dev. of skull.	445.
——	1873*a*. *Philos. Trans.* **163**, p. 95.	Salmo, dev. of skull.	115, 124.
——	1873*b*. *Month. micr. J.* **9**, pp. 6, 45.	Parus, Accipiter, dev. of skull.	445.
——	1873*c*. *Ibid.* **9**, p. 102.	Turdus, dev. of skull.	281, 443.
——	1874. *Philos. Trans.* **164**, p. 289.	Sus, dev. of skull.	334, 340, 43[illegible]
——	1875. *Trans. linn. Soc. Lond.* **1**, p. 1.	Aves, dev. of skull.	285, 408, 44[illegible] 446.
——	1876*a*. *Philos. Trans.* **166**, p. 601.	Anura, dev. of skull.	198, 214, 21[illegible] 417.
——	1876*b*. *Trans. linn. Soc. Lond.* **1**, p. 99.	Aves, dev. of skull.	278, 279, 44[illegible]
——	1877*a*. *Philos. Trans.* **167**, p. 529.	Urodela, dev. of skull.	176, 192, 41[illegible]
——	1877*b*. *Trans. zool. Soc. Lond.* **9**, p. 289.	Aves, dev. of skull.	285, 445.
——	1879*a*. *Philos. Trans.* **169**, p. 385.	Tropidonotus, dev. of skull.	245, 247, 25[illegible]
——	1879*b*. *Trans. zool. Soc. Lond.* **10**, p. 189.	Scyllium, Torpedo, dev. of skull.	49, 58, 63, 6[illegible] 375.
——	1879*c*. *Ibid.* **10**, p. 251.	Aves, dev. of skull.	445.
——	1880*a*. *Philos. Trans.* **170**, p. 595.	Lacerta, dev. of skull.	217, 227, 23[illegible]
——	1880*b*. *Challenger Reports*, **1**, V.	Chelone, dev. of skull.	260.
——	1881. *Philos. Trans.* **172**, p. 1.	Anura, dev. of skull.	198.
——	1882*a*. *Ibid.* **173**, p. 139.	Acipenser, dev. of skull.	88, 94, 9[illegible] 500.
——	1882*b*. *Ibid.* **173**, p. 443.	Lepidosteus, dev. of skull.	108, 500.
——	1882*c*. *Trans. linn. Soc. Lond.* **2**, p. 165.	Urodela, dev. of skull.	186, 192.
——	1882*d*. *Trans. zool. Soc. Lond.* **11**, p. 171.	Urodela, dev. of skull.	188, 192.
——	1883*a*. *Philos. Trans.* **174**, pp. 373, 411.	Cyclostomes, dev. of skull.	41, 47, 415.
——	1883*b*. *Trans. zool. Soc. Lond.* **11**, p. 263.	Crocodilus, dev. of skull.	262, 265, 26[illegible] 434, 436.
——	1885. *Philos. Trans.* **176**, p. 1.	mammals, dev. of skull.	316, 319, 32[illegible] 324, 326, 35[illegible] 434, 436, 44[illegible]
——	1888. *Proc. roy. Soc. Lond.* **43**, p. 397.	general, dev. of skull.	280.

Author	*Reference*	*Subject*	*Page where quoted*
Parker, W. K.	1890. *Roy. Irish Acad. Cunningham mem.* **6**.	Aves, dev. of skull.	279.
——	1891*a*. *Trans. zool. Soc. Lond.* **13**, p. 43.	Opisthocomus, dev. of skull.	445.
——	1891*b*. *Trans. linn. Soc. Lond.* **5**, p. 213.	Aves, dev. of skull.	279.
—— & Bettany, G.	1877. *The morphology of the skull.* London.		198, 438.
Paulli, S.	1900. *Morph. Jahrb.* **28**, p. 147.	mammals, pneumaticity.	299, 306.
Pearson, H. S.	1921. *J. Anat. Lond.* **56**, p. 20.	Lygosoma, dev. of skull.	236.
Peeters, J. L. E.	1910. *Over de ontwikkeling van het chondrocranium van eenige Urodelen en Anuren.* Proefschr. Leiden.		176, 182, 190, 191, 198, 211.
Pehrson, T.	1922. *Acta zool. Stockh.* **3**, p. 1.	Amia, dev. of skull.	6, 34, 78, 85, 97, 104, 489, 500.
Peter, K.	1898. *Morph. Jahrb.* **25**, p. 555.	Ichthyophis, chondrocr.	192.
——	1906. *Entwicklung des Geruchsorgans*, in Hertwig's *Handbuch der vergleichenden und experimentellen Entwicklungslehre.* Jena. **2**, (2) p. 1.	dev. of nas. caps.	361.
——	1912. *Arch. mikr. Anat.* **79**, p. 427; **80**, p. 478.	mammals, dev. of nas. caps.	361.
——	1922. *Rekonstruktionsmethoden.* Greifswald.		14.
Petersen, H.	1930. In Möllendorf's *Handbuch der mikroskopischen Anatomie des Menschens.* Berlin. II. (2) p. 521.	bone & function.	484, 485.
Peyer, B.	1912. *Morph. Jahrb.* **44**, p. 563.	Vipera, dev. of skull.	248.
Platt, J. B.	1891*a*. *J. Morph.* **5**, p. 79.	head-segmentation.	19.
——	1891*b*. *Anat. Anz.* **6**, p. 251.	head-segmentation.	19.
——	1894. *Arch. mikr. Anat.* **43**, p. 911.	neural crest & mesect.	472.
——	1896. *Quart. J. micr. Sci.* **38**, p. 485.	neural crest & mesect.	472.
——	1897. *Morph. Jahrb.* **25**, p. 377.	Necturus, dev. of chondrocr.	190, 191, 375, 472.
Pollard, H. B.	1892. *Zool. Jahrb. Abt. Anat.* **5**, p. 387.	Polypterus, chondrocr.	78, 87.
——	1894. *Anat. Anz.* **9**, p. 349.	oral cirrhi.	41, 409.
——	1895. *Zool. Jahrb. Abt. Anat.* **8**, p. 379.	oral cirrhi.	41, 409.
Popa, G. I.	1936. *Morph. Jahrb.* **78**, p. 85.	dura mater.	485.
Pusanow, J.	1913. *Anat. Anz.* **44**, p. 262.	Lacerta, chordal cart.	222.
Pusey, H. K.	In press.	Rana, metamorph. of chondrocr.	15, 198, 207, 416, 417.
Pycraft, W. P.	1900. *Trans. zool. Soc. Lond.* **15**, p. 149.	Aves, morph. of palate.	287, 288, 445.
——	1901. *J. linn. Soc. Lond. Zool.* **28**, p. 343.	Aves, dev. of palate.	285, 445, 446.
——	1907. *Nat. antarct. Exp. nat. Hist. Zool.* **2** (3).	penguins, palate.	445.
Rabl, C.	1892. *Verh. anat. Ges. Jena*, **6**, p. 104.	head-segmentation.	19.
——	1894. *Ibid.* **8**, p. 163.	origin of mesenchyme.	473.
——	1897. *Theorie des Mesoderms.* Leipzig.		19.
Ramaswami, L. S.	1932*a*. *Half-yrly J. Mysore Univ.* **6**, p. 32.	Engystomatidae, hyobranch. skel.	217.
——	1932*b*. *Ibid.* **6**, p. 45.	Engystomatidae, morph. of skull.	209.
——	1932*c*. *Ibid.* **6**, p. 176.	Elypsoglossus, morph. of skull.	217.

Author	*Reference*	*Subject*	*Page where quoted*
Ramaswami, L. S.	1934. *Proc. Indian Acad. Sci.* **1**, p. 80.	Rhacophorus, morph. of skull.	210.
——	1935*a*. *Anat. Anz.* **81**, p. 65.	Pelobatidae, morph. of skull.	210, 213.
——	1935*b*. *Proc. Indian Acad. Sci.* **2**, p. 1.	Ranidae, morph. of skull.	217.
——	1935*c*. *Current Sci.* **3**, p. 306.	Pelobates, hyobranch. skel.	217.
Rambaud, A. & Renault, C.	1864. *Origine et développement des os.* Paris.		368.
Ranke, J.	1899*a*. *Abh. bayr. Akad. Wiss. II. Cl.* **20**, p. 275.	man, postparietal & tabular.	444.
——	1899*b*. *SitzBer. bayr. Akad. Wiss. Math. Phys. Cl.* **29**, p. 415.	man, divided parietal.	367.
——	1899*c*. *Ibid.* **28**, p. 227.	alisphenoid.	370, 439.
——	1913. *Ibid.* p. 223.	interparietal.	444.
Rathke, H.	1839. *Entwicklungsgeschichte der Natter.* Königsberg.	Tropidonotus, dev. of skull.	4, 244, 3[illegible] 382.
Raven, C. P.	1931. *Arch. EntwMech.* **125**, p. 210.	neural crest & mesect.	473, 474.
——	1933*a*. *Ibid.* **129**, p. 179.	neural crest & mesect.	475.
——	1933*b*. *Ibid.* **130**, p. 517.	neural crest & mesect.	475.
Reagan, F. P.	1915. *Anat. Rec.* **9**, p. 114.	Aves, colum. aur.	479.
——	1917. *J. exp. Zool.* **23**, p. 85.	ear & audit. caps.	479.
Regan, C. T.	1923. *Proc. zool. Soc. Lond.* p. 445.	Lepidosteus, morph. of skull.	108, 113.
Reichert, C.	1838. *Vergleichende Entwicklungsgeschichte des Kopfes der Amphibien.* Königsberg.	morph. of visc. arches.	3, 450.
Remak, R.	1850. *Untersuchungen über die Entwickelung der Wirbeltiere.* Berlin.	segmentation.	8.
Retterer, E.	1917. *C.R.Soc.Biol.Paris,* **80**, p. 291.	cartilage & bone.	38.
Rice, E. L.	1920. *J. Morph.* **34**, p. 119.	Eumeces, dev. of chondrocr.	219, 235, 4[illegible]
Ridewood, W. G.	1894. *Proc. zool. Soc. Lond.* p. 632.	Ceratodus, hyoid arch.	168, 170.
——	1897*a*. *J. linn. Soc. Lond. Zool.* **26**, p. 53.	Xenopus, Pipa, hyobranch. skel.	214, 217.
——	1897*b*. *Proc. zool. Soc. Lond.* p. 577.	Pelodytes, hyobranch. skel.	208.
——	1898*a*. *Ibid.* p. 4.	Alytes, hyobranch. skel.	200, 211.
——	1898*b*. *J. linn. Soc. Lond. Zool.* **26**, p. 474.	Anura, hyobranch. skel.	213.
——	1899. *Ibid.* **27**, p. 454.	Hymenocheirus, hyobranch. skel.	217.
——	1904*a*. *Proc. zool. Soc. Lond.* p. 35.	Teleost, morph. of skull.	107, 498.
——	1904*b*. *Ibid.* p. 448.	Clupeiform, morph. of skull.	134.
——	1922. *Philos. Trans.* B. **211**, p. 209.	Balaenoptera, dev. of bony skull.	341, 343, 34[illegible] 441, 508.
Robison, R.	1923. *Biochem. J.* **17**, p. 286.	phosphatase.	481.
Rosenberg, E.	1886. *SitzBer. nat-forsch. Ges. Dorpat,* **7**, p. 31.	Carcharias, occip. vertebr.	9.
Roux, W.	1895. *Gesammelte Abhandlungen über Entwicklungsmechanik.* Leipzig.	bone, functional differentiation.	485.
Ruge, G.	1896. *Morph. Jahrb.* **25**, p. 202.	Monotreme, ear cart.	300, 416.
Rutherford, N. C.	1909. *Brit. med. J.* **2541**, p. 691.	Salmo, dev. of chondrocr.	115.
Ryder, J. A.	1882. *Bull. U.S. Fish. Commission for 1881*, p. 191.	Hippocampus, chondrocr.	154.
——	1886. *Proc. U.S. nat. Mus.* **8**, p. 128.	Gambusia, chondrocr.	142.
——	1887. *Rep. U.S. Comm. Fish.* **13**, p. 489.	Ictalurus, chondrocr.	138.

Author	*Reference*	*Subject*	*Page where quoted*
Sagemehl, M.	1884. *Morph. Jahrb.* **9**, p. 177.	Amia, morph. of skull.	9, 97.
——	1885. *Ibid.* **10**, p. 1.	morph. of skull.	138, 395, 409.
——	1891. *Ibid.* **17**, p. 489.	morph. of skull.	9, 130, 140, 141, 409.
Salensky, V.	1881. *Arch. Biol. Paris.* **2**, p. 233.	Acipenser, dev. of skull.	88.
Salvadori, G.	1928. *Ric. Morf. Biol. anim.* **1**, p. 379.	Anura, colum. aur.	203, 207.
Sarasin, P. & F.	1890. *Ergebn. naturwiss. Forsch. Ceylon.* Wiesbaden, **2**, p. 1.	Ichthyophis, dev. of skull.	192.
Saunderson, E. C.	1935. *Proc. Nov. Scot. Inst. Sci.* **19**, p. 121.	Salmo, dev. of chondrocr.	115.
Schachunjanz, R.	1926. *Rev. zool. russe*, **6**, p. 109.	Anura, hyobranch. skel.	198.
Schæffer, J. P.	1910*a*. *Amer. J. Anat.* **10**, p. 313.	man, nasal caps.	361.
——	1910*b*. *J. Morph.* **21**, p. 613.	man, nasal caps.	361.
——	1910*c*. *Anat. Rec.* **4**, p. 167.	man, ethmoid sinuses.	361.
——	1916. *Amer. J. Anat.* **20**, p. 125.	man, nasal caps.	361.
Schaffer, J.	1896. *Z. wiss. Zool.* **61**, p. 606.	Petromyzon, mucocart.	41.
——	1930. In Möllendorf's *Handbuch der mikroskopischen Anatomie des Menschens.* Berlin. **2**, (ii) pp. 146, 210.	cartilage.	3, 37, 38, 39, 502.
Schalk, A.	1913. *Arch. mikr. Anat.* **83**, p. 43.	Petromyzon, dev. of skull.	41.
Schauinsland, H.	1900. *Ibid.* **56**, p. 747.	Sphenodon. dev. of chondrocr.	238, 240, 241, 243, 386.
——	1903. *Zoologica, Stuttgart*, **16**.	Callorhynchus, Sphenodon, dev. of chondrocr.	72, 78, 238, 240, 410, 418.
Schleip, W.	1904. *Anat. Hefte*, **23**, p. 331.	Salmo, dev. of bony skull.	115, 124, 434, 489, 500.
Schmäh, R.	1934. *Morph. Jahrb.* **74**, p. 364.	Polypterus, dev. of lower jaw.	78, 86, 87.
Schmahlhausen, J.	1923*a*. *Anat. Anz.* **56**, p. 534.	hyomandibula.	93, 168, 170, 171, 412, 414.
——	1923*b*. *Rev. zool. russe*, **3**, p. 239.	trigeminofacial chamber.	172, 429.
Schmid, H.	1928. *Anat. Anz.* **66**, p. 109.	man, dev. of stapes.	412.
Schmidt-Monnard, C.	1883. *Z. wiss. Zool.* **39**, p. 97.	cartilage & bone.	2.
Schneider, A.	1879. *Beiträge zur vergleichenden Anatomie und Entwicklungsgeschichte der Wirbeltiere.* Berlin.		38, 41.
Schoonees, D. A.	1930. *S. Afr. J. Sci.* **27**, p. 456.	Bufo, morph. of skull.	207, 210.
Schreiber, K.	1915. *Zool. Jahrb. Abt. Anat.* **39**, p. 201.	Globiocephalus, chondrocr.	345.
Schreiner, K. E.	1902. *Z. wiss. Zool.* **72**, p. 467.	Amia, Lepidosteus, head-segmentation.	23, 28, 97, 99, 108.
Schultz, A. H.	1935. *Amer. J. phys. Anthrop.* **20**, p. 205	Primates, nas. caps.	361.
Schwink, J.	1888. *Über den Zwischenkiefer und seine Nachbarorgane bei Säugetieren*, München.	prevomer, Ovis.	434.
Semon, R.	1901. *Denkschr. med-naturw. Ges. Jena*, **4**, p. 113.	Ceratodus, tooth-dev.	173.
Serres, E.	1819. *Les Lois de l'ostéogénèse. C.R Acad. Sci. Paris.*	mammals, prefrontal, postfrontal.	365, 368.
van Seters, W. H.	1922. *Arch. Biol. Paris*, **32**, p. 373.	Alytes, dev. of chondrocr.	24, 30, 211, 381, 418.
Sewertzoff, A. N.	1895. *Bull. Soc. imp. Nat. Moscou*, **2**, p. 186.	Acipenser, Pelobates, head-segmentation.	23, 24, 28.
——	1897. *Anat. Anz.* **13**, p. 409.	Squalus, Ascalabotes, head-segmentation.	25, 49, 234.
——	1898*a*. *Bull. Soc. imp. Nat. Moscou*, **13**, p. 1.	Torpedo, head-segmentation.	19.

Author	*Reference*	*Subject*	*Page where quoted*
Sewertzoff, A. N.	1898*b*. *Anat. Anz.* **14**, p. 278.	Torpedo, head-segmentation.	19.
——	1899. *Festschr. Kupffer*, p. 281.	Squalus, Pristiurus, dev. of chondrocr.	27, 28, 49, 65
——	1900. *Anat. Anz.* **18**, p. 33.	Ascalabotes, chondrocr.	234.
——	1902. *Ibid.* **21**, p. 593.	Ceratodus, chondrocr.	23, 29, 168, 170.
——	1911. *Ibid.* **38**, p. 487.	Acipenser, nerves.	88.
——	1914. *Ibid.* **45**, p. 280.	Petromyzon, dev. of visc. arches.	41.
——	1916. *Arch. russ. Anat. Hist. Embryol.* **1**, p. 1.	Petromyzon, dev. of chondrocr.	13, 27, 28, 41, 89, 375, 376, 409.
——	1917. *Ibid.* **1**, p. 425.	morph. of skull.	41, 48, 376.
——	1923. *Anat. Anz.* **56**, p. 389.	Selachii, visc. arches.	52, 419.
——	1925*a*. *Bull. Soc. Nat. Moscou*, **34**, p. 87.	Amia, teeth & memb.-bones.	6, 97, 106.
——	1925*b*. *Anat. Anz.* **59**, p. 271.	Polypterus, maxilla.	85.
——	1925*c*. *Ibid.* **60**, p. 427.	dermal bones & scales.	33.
——	1926*a*. *Quart. J. micr. Sci.* **70**, p. 451.	Acipenser, dev. of bony skull.	88, 94, 500.
——	1926*b*. *Palaeont. Z.* **8**, p. 75.	Pisces, chondrocr.	78, 80.
——	1927. *Pubbl. Staz. zool. Napoli*, **8**, p. 475.	Selachii, dev. of visc. arches.	49, 52, 410, 418.
——	1928. *Acta zool. Stockh*, **9**, p. 193.	Acipenser, dev. of chondrocr.	28, 88, 457.
—— & Disler, J.	1924. *Anat. Anz.* **58**, p. 345.	Selachii, pharyngomandib.	52, 62, 377, 419.
Seyd, E. L.	In press.	Petromyzon, metamorph. of chondrocr.	41, 45.
Seydel, O.	1891. *Morph. Jahrb.* **17**, p. 44.	man, nas. caps.	372.
——	1896. *Festschr. Gegenbaur*, **2**, p. 385.	Chelonia, nas. caps.	252, 306, 395
——	1899. *Denkschr. med-naturw. Ges. Jena*, **6** (I, iii), p. 445.	Echidna, nas. caps.	299, 396.
Shaner, R. F.	1926. *Anat. Rec.* **32**, p. 343.	Chrysemys, dev. of skull.	251.
Sharpey, W. S.	1848. In Quain's *Anatomy*. London.	cartilage & bone.	2.
Shiino, K.	1914. *Anat. Hefte*, **50**, p. 253.	Crocodilus, chondrocr.	262.
Shindo, T.	1914. *Ibid.* **51**, p. 267.	Reptilia, cranial arteries.	264.
——	1915. *Ibid.* **52**, p. 319.	mammal, cranial veins.	315, 430.
Siebenrock, F.	1892. *Ann. nat-hist. Mus. Wien*, **7**, p. 163.	Lacerta, morph. of skull.	233.
——	1894. *SitzBer. Akad. Wiss. Wien*, **103** (i), p. 205.	Lacerta, morph. of skull.	233.
——	1899. *Ann. nat. HofMus. Wien*, **13**, p. 423.	Chelonia, hyobranch. skel.	252.
Sitsen, A. E.	1933. *Z. Anat. EntwGesch*, **101**, p. 121.	man, dev. of sutures.	369.
Slonaker, J. R	1918. *J. Morph.* **31**, p. 351.	sclerotic bones, Passer.	443.
Sluiter, C. P.	1893. *Morph. Jahrb.* **20**, p. 75.	Lacertilia, Ophidia, egg-tooth.	233, 267.
Smets, G.	1885. *Ann. Soc. sci. Brux.* **9** (2), p. 187.	vomer & parasphenoid.	344, 433.
Smith, G.	1904. *Quart. J. micr. Sci.* **48**, p. 11.	Gallus, dev. of colum. aur.	279.
Smith, L.	1920. *J. Morph.* **33**, p. 527.	Spelerpes, hyobranch. skel.	187, 480.
Smith, L. W.	1914. *Anat. Anz.* **46**, p. 547.	Chrysemys, dev. of colum. aur.	252.
Smith, W. R.	1908. *Proc. roy. Soc. Edinb.* **28**, p. 586.	man, interparietal.	444.

Author	*Reference*	*Subject*	*Page where quoted*
Söderbergh, G. Säve,	1934. *Ark. Zool.* **26** A. (17.)	Urodela, affinities.	459.
Sonies, F.	1907. *Petrus Camper*, **4**, p. 395.	Anas, Gallus, dev. of chondrocr.	268, 278, 279.
Soulié, A. & Bardier, E.	1907. *J. Anat. Physiol. Paris.* **43**, p. 137.	man, thyroid cart.	362.
Spemann, H.	1898. *Zool. Jahrb. Abt. Anat.* **11**, p. 389.	Rana, dev. of chondrocr.	198, 207, 450.
——	1902. *Arch. EntwMech.* **15**, p. 448.	Triton, dupl. ant.	476.
——	1910. *Ibid.* **30** (2), p. 437.	ear, rotation.	479.
Spence, T. B.	1890. *Proc. Amer. Soc. Micr.* **14**, p. 146.	mammals, middle ear.	320, 413.
von Spix, J. B.	1815. *Cephalogenesis.* Monachii.	morph. of skull.	7.
Spöndli, H.	1846. *Über den Primordialschädel der Säugetiere und des Menschen.* Zürich.	cartilage & bone.	4, 334, 391.
Spurgat, F.	1896. *Morph. Arb.* **5**, p. 555.	nas. caps., parasept. cart.	339, 395, 396.
Stadtmüller, F.	1914. *Anat. Hefte*, **51**, p. 427.	sclerotic cart.	145, 406.
——	1924. *Z. Anat. EntwGesch.* **75**, p. 149.	Salamandra, dev. of skull.	186, 187, 188.
——	1929*a*. *Morph. Jahrb.* **61**, p. 221.	eye & sclerotic cart.	181, 183, 190, 216, 479.
——	1929*b*. *Z. Anat. EntwGesch.* **90**, p. 144.	Triton, basioccip.	12, 183, 186.
——	1931*a*. *Morph. Jahrb.* **66**, p. 196.	Bombinator, operculum.	213.
——	1931*b*. *Z. Anat. EntwGesch.* **94**, p. 792.	Bombinator, para-artic. cart.	213.
——	1931*c*. *Anat. Anz.* **72**, p. 261.	Alytes, supraorb. cart.	212.
——	1936. *Morph. Jahrb.* **78**, p. 1.	Liopelma, parahyoid.	213, 214, 416.
——	1937. *Ibid.* **79**, p. 436.	Urodeles, angular.	189.
Stannius, H.	1846. *Lehrbuch der vergleichenden Anatomie der Wirbeltiere.* Berlin.	morph. of skull.	4.
——	1856. *Handbuch der Anatomie der Wirbeltiere.* Berlin.	streptostyly & monimostyly.	197, 425, 426.
Starck, D.	1937. *Morph. Jahrb.* **79**, p. 358.	mesectoderm.	473.
Staurenghi, C.	1900*a*. *Arch. ital. Biol.* **34**, p. 460.	supraoccip.	444.
——	1900*b*. *Ibid.* **34**, p. 466.	interparietal, Aves.	279, 444.
van der Steen, J. C.	1930. *De ontwikkeling van de occipitovertebrale spleet bij Microhyla.* Proefschr. Leiden.	Microhyla, occip. joint.	24, 30.
Steinberg, H.	1912. *Anat. Anz.* **42**, p. 466.	Felis, nas. caps.	326.
Steinitz, E.	1906. *Arch. EntwMech.* **20**, p. 537.	Rana, eye & chondrocr.	477.
Stensiö, E. A.	1925. *Field Mus. nat. Hist. Chicago, Geol. Ser.* **4**, p. 85.	Macropetalichthys, morph. of hyomandibula.	411.
——	1927. *Skr. Svalbard Nordishavet, Oslo*, **12**.	Cephalaspis.	41, 376.
——	1932. *Med. Grönland.* **83**.	triassic fishes.	88, 107.
——	1935. *Palaeont. sin.* **3**, p. 1.	Sinamia.	492, 505.
Stephan, P.	1900. *Bull. biol.* **33**, p. 281.	cartilage & bone.	2.
Stieda, L.	1892. *Anat. Hefte*, **2**, p. 59.	man, supraoccip.	444.
Stöhr, P.	1879. *Z. wiss. Zool.* **33**, p. 477.	Urodela, dev. of chondrocr.	9, 14, 176, 182.
——	1881. *Ibid.* **36**, p. 68.	Anura, dev. of chondrocr.	9, 198, 399, 417.
——	1882. *Festschr. 300-J-Feier Jul.-Max. Univ. Würzburg.*	Salmo, dev. of chondrocr.	9, 115.
Stone, L. S.	1922. *J. exp. Zool.* **35**, p. 421.	neural crest & mesect.	407, 473, 490
——	1926. *Ibid.* **44**, p. 95.	neural crest & mesect.	407, 473, 475 479.
——	1929. *Arch. EntwMech.* **118**, p. 40.	neural crest & mesect.	473, 475.

Author	Reference	Subject	Page where quoted
Stone, L. S.	1930. *J. exp. Zool.* **55**, p. 193.	eye & orbit.	477.
——	1932*a*. *Anat. Rec.* **51**, p. 267.	neural crest & mesect.	475.
——	1932*b*. *J. exp. Zool.* **62**, p. 109.	visc. cart.	475.
Strasser, F.	1905. *Verh. anat. Ges. Jena*, **19**, p. 139.	secondary cart.	3, 38, 502.
Strasser, H.	1905. *Ibid.* **19**, p. 139.	Columba, pneumaticity.	282.
Streeter, G. L.	1914. *J. exp. Zool.* **16**, p. 149.	ear, rotation.	479.
——	1917. *Amer. J. Anat.* **22**, p. 1.	man, dev. of audit. caps.	37, 359, 403.
——	1918. *Contr. Embryol. Carn. Instn.* **7**, p. 5.	man, dev. of audit. caps.	37, 359, 403.
——	1920. *Publ. Carneg. Instn.* **274**, p. 143.	man, head-growth.	470.
Strong, R. M.	1925. *Amer. J. Anat.* **36**, p. 313.	Mus, dev. of bony skull.	316.
Struthers, P. H.	1927. *J. Morph.* **44**, p. 127.	Erethizon, chondrocr.	316, 391.
Studnička, F. K.	1912. *Anat. Anz.* **42**, p. 529.	Petromyzon, audit. caps.	41.
Sturm, H.	1937. *Z. wiss. Zool.* **149**, p. 161.	Sus, Bos, nas. caps.	334, 338.
Suschkin, P. P.	1896. *Anat. Anz.* **11**, p. 767.	Tinnunculus, dev. of skull.	283.
——	1899. *Nouv. Mem. Soc. Nat. Mos.* **16**, p. 1.	Tinnunculus, dev. of skull.	278, 283, 452.
——	1910. *Biol. Z. Moskau*, **1**, p. 241.	jaw-attachments.	424.
——	1927. *Palaeont. Z.* **8**, p. 263.	jaw-suspensions.	420.
Sutton, J. Bland,	1883. *J. Anat. Physiol. Lond.* **17**, p. 498.	man, temporal bone.	366.
——	1884*a*. *Proc. zool. Soc. Lond.* p. 577.	vomer & parasphenoid.	433, 434.
——	1884*b*. *J. Anat. Physiol. Lond.* **18**, p. 219.	man, epipteric bone.	368.
——	1885. *Proc. zool. Soc. Lond.* p. 577.	man, sphenoid.	369.
——	1888. *J. Anat. Physiol. Lond.* **22**, p. 28.	morph. of skull & extracran. spaces.	390, 427.
Suzuki, T.	1932. *Kaibogaku Zassi*, **5** (6).	Onychodactylus, dev. of skull.	181.
Swinnerton, H. H.	1902. *Quart. J. micr. Sci.* **45**, p. 503.	Gasterosteus, dev. of skull.	123, 142, 146, 152, 394, 409, 421, 500.
Swett, F. H.	1921. *Anat. Rec.* **22**, p. 183.	Salmo, dupl. ant.	476.
Symington, J.	1896. *J. Anat. Physiol. Lond.* **30**, p. 420.	Ornithorhynchus, pre-vomer.	296.
——	1899. *J. Anat. Lond.* **33**, p. 31.	Marsupials, laryng. cart.	407.
Tarapani, H.	1909. *Jena. Z. Naturw.* **45**, p. 57.	Urodela, hyobranch. skel.	182, 183, 187.
Tatarko, K.	1934. *Zool. Jahrb. Abt. allg. Zool.* **53**, p. 461.	bone regeneration.	482.
——	1936. *Acad. Sci. Ukr. Trav. Inst. Zool. Biol.* **3**, p. 47.	Acipenser, opercular.	96.
Tchernavin, V.	1937. *Proc. linn. Soc. Lond.* **149**, p. 11.	Salmo, breeding changes in skull.	115.
Terry, R. J.	1904. *Amer. J. Anat.* **3**, p. xi.	Necturus, chondrocr.	190.
——	1917. *J. Morph.* **29**, p. 281.	Felis, dev. of chondrocr.	326, 383, 390, 435.
——	1919. *Anat. Rec.* **17**, p. 235.	facial nerve & audit. caps.	191.
Thoma, R.	1911. *Virchow's Arch. path. Anat. Physiol.* **206**, p. 201.	skull, mechanical factors.	486.
——	1913. *Ibid.* **212**, p. 1.	skull, growth.	486.
——	1923. *Zbl. Path.* **33**, p. 4.	bone growth.	39, 485.
Thompson, D. H.	1934. *Illinois nat. Hist. Surv. Biol. notes*, **2**.	Polyodon, growth of rostrum.	471.
Thyng, F. W.	1906. *Proc. Boston Soc. nat. Hist.* **32**, p. 387.	squamosal.	232, 250, 437.
Tichomiroff, A.	1885. *Zool. Anz.* **8**, p. 533.	Salmo, chondrocr.	115, 119.
du Toit, C. A.	1930. *S. Afr. J. Sci.* **27**, p. 426.	Heleophryne, morph. of skull.	217.

Author	*Reference*	*Subject*	*Page where quoted*
du Toit, C. A.	1933*b*. *Proc. zool. Soc. Lond.* p. 715.	Rana, morph. of skull.	206.
——	1934. *Ibid.* p. 119.	Crinia, morph. of skull.	217.
du Toit, G. P.	1933*a*. *S. Afr. J. Sci.* **30**, p. 394.	Phrynobatrachus, morph. of skull.	217.
Tokura, R.	1924. *Fol. anat. jap.* **2**, p. 197.	ear, rotation.	479.
Toldt, K.	1902. *SitzBer. Akad. Wiss. Wien, Math. Nat.* (3), **111**, p. 241.	man, dev. of jugal.	366.
——	1903. *Ibid.* (3), **112**, p. 485.	man, jugal, split.	366, 507.
Tonkoff, W.	1900. *Anat. Anz.* **18**, p. 296.	Gallus, chondrocr.	279.
Töplitz, C.	1920. *Zoologica, Stuttgart*, **70**.	Didelphys, dev. of skull.	300, 390, 391, 418, 434, 449.
Torlitz, H.	1922. *Z. Fischerei*, **21**, p. 1.	Anguilla, dev. of bony skull.	143, 146.
Törö, E.	1935. *Anat. Anz.* **80**, p. 285.	induction of cart. & bone.	477, 479, 484.
Tourneux, F. & J. P.	1907. *C. R. Ass. Anat.* **9**, p. 180.	notochord & basal plate.	381.
—— & ——	1912. *J. Anat. Physiol. Paris*, **48**, p. 57.	notochord & basal plate.	315, 331, 335, 339, 381.
Traquair, R. H.	1865. *Trans. linn. Soc. Lond.* **25**, p. 263.	Pleuronectes, asymmetry.	161, 164, 489.
——	1871. *J. Anat. Physiol. Lond.* **5**, p. 166.	Polypterus, morph. of skull.	78.
Tretjakoff, D.	1926. *Z. wiss. Zool.* **128**, p. 267.	Petromyzon, morph. of skull.	41.
——	1929. *Ibid.* **133**, p. 470.	Petromyzon, dev. of chondrocr.	41, 45.
Trewavas, E.	1933. *Philos. Trans.* B. **222**, p. 401.	Anura, hyoid.	198, 214.
Troitsky, W.	1932. *Z. Morphol. Anthrop.* **30**, p. 504.	bone regeneration & growth.	369, 483, 484, 485.
Tschekanowskaja, O. V.	1936. *Arch. russ. Anat. Hist. Embryol.* **15**, (3) pp. 3, 133.	Tropidonotus, dev. of skull.	245.
Tsusaki, T.	1925. *Fol. anat. jap.* **3**, p. 345.	Diemyctylus, sclerotic bones.	192, 443.
Turkewitsch, B. G.	1936. *Zool. Jahrb. Abt. Anat.* **61**, p. 121.	Aves, ear & posture.	405.
Uhlmann, E.	1921*a*. *Jena. Z. Naturw.* **57**, p. 275.	Cyclopterus, dev. of chondrocr.	157.
——	1922*b*. *Ibid.* **57**, p. 315.	Cyclopterus, dev. of bony skull.	157, 158, 500.
Veit, O.	1907. *Anat. Hefte*, **33**, p. 155.	Lepidosteus, morph. of skull.	108, 113, 500.
——	1911. *Ibid.* **44**, p. 93.	Lepidosteus, dev. of chondrocr.	29, 108.
——	1919. *Ibid.* **56**, p. 305.	neural crest & mesect.	473.
Versluys, J.	1898. *Zool. Jahrb. Abt. Anat.* **12**, p. 161.	Sphenodon, Lacerta, morph. of audit. region.	217, 225, 226, 413, 431.
——	1903. *Ibid.* **19**, p. 107.	Lacerta, dev. of colum. aur.	217, 220, 225, 226, 237, 262, 263, 264, 412, 413.
——	1910. *Zool. Jahrb.* **30**, p. 175.	streptostyly.	197, 426.
——	1912. *Zool. Jahrb. Suppl.* **15**, (2) p. 545.	streptostyly.	423, 426.
——	1922. *Bijdr. Dierk.* **22**, p. 95.	Selachii, visc. arches.	407, 448.
de Villiers, C.	1929. *S. Afr. J. Sci.* **26**, p. 481.	Arthroleptella, morph. of skull.	217.
——	1930. *Quart. J. micr. Sci.* **73**, p. 667.	Phrynomerus, morph. of skull.	217.
——	1931*a*. *Anat. Anz.* **72**, p. 164.	Breviceps, morph. of skull.	217.

Author	Reference	Subject	Page where quoted
de Villiers, C.	1931*b*. *Quart. J. micr. Sci.* **74**, p. 275.	Cacosternum, morph. of skull.	217.
——	1931*c*. *Anat. Anz.* **71**, p. 305.	Hemimisus, morph. of skull.	217.
——	1931*d*. *Anat. Anz.* **71**, p. 331.	Anhydrophyne, morph. of skull.	217.
——	1932. *Ibid.* **74**, p. 33.	Aglossa, colum. aur.	217.
——	1933. *Ibid.* **75**, p. 257.	Probreviceps, morph. of skull.	217.
——	1934*a*. *Bull. Mus. comp. Zoöl. Harv.* **77**, p. 3.	Ascaphus, morph. of skull.	210, 213.
——	1934*b*. *Anat. Anz.* **78**, p. 295.	Rhombophryne, morph. of skull.	217.
——	1936. *Ibid.* **81**, p. 225.	paraquadrate.	182, 437.
Villy, F.	1890. *Quart. J. micr. Sci.* **30**, p. 523.	Rana, tymp. ring.	208.
Violette, H. N.	1930. *Anat. Rec.* **45**, p. 280.	Anura, colum. aur.	203.
Virchow, H.	1914. *Z. Ethnol.* **46**, p. 673.	chimp. nas. caps.	396.
Vogler, A.	1926. *Morph. Jahrb.* **55**, p. 568.	Sus, dev. of bony skull.	334.
Vogt, C.	1842*a*. *Untersuchungen über die Entwicklungsgeschichte der Geburtshelferkröte.* Solothurn.	Alytes, dev. of skull.	8, 211.
——	1842*b*. *Embryologie des Salmones.* Neuchâtel.		8.
Vogt, W.	1929. *Arch. EntwMech.* **120**, p. 384.	intra vitam staining.	475.
Voit, M.	1907. *Anat. Anz.* **31**, p. 635.	mammals, audit. nerve.	404.
——	1909. *Anat. Hefte,* **38**, p. 425.	Lepus, chondrocr.	306, 379, 390, 391, 392, 395, 402, 403, 430, 431.
——	1911. *Anat. Anz.* **38**, p. 341.	Lepus, chondrocr.	306.
——	1919. *Ibid.* **52**, p. 36.	man, abducens foramen.	390.
——	1923. *Verh. anat. Ges. Jena,* **32**, p. 68.	Meckel's cart.	441.
——	1924. *Z. Morphol. Anthrop.* **24**, p. 75.	Aves, prearticular.	281, 441.
Vrolik, A. J.	1871. *Niederl. Arch. Zool.* **1**, p. 219.	cartilage & bone.	5, 6, 130.
Wagner, D. E.	1934. *Anat. Anz.* **79**, pp. 20, 65.	Liopelma, morph. of skull.	214.
Walther, J.	1882. *Jena. Z. Naturw.* **16**, p. 59.	Esox, dev. of bone.	5, 142.
Watson, D. M. S.	1914. *Ann. Mag. nat. Hist.* **14**, p. 84.	temporal bones.	232, 438.
——	1916. *Philos. Trans.* B. **207**, p. 311.	Ornithorhynchus, dev. of skull.	288, 296, 297, 298, 430, 436, 440, 441.
——	1921*a*. *Ann. Mag. nat. Hist.* **8**, p. 320.	Coelacanthini, bone-fusion.	506.
——	1921*b*. *Proc. zool. Soc. Lond.* p. 35.	Cynognathus, angular, tympanic.	437.
——	1925*a*. *Philos. Trans.* B. **214**, p. 189.	Stegocephalia, morph. of skull.	427.
——	1925*b*. *Proc. zool. Soc. Lond.* p. 815.	Palaeoniscid skull.	83, 429, 450, 458.
——	1928. *Palaeontology and the evolution of man.* Oxford.		403.
Wegner, R. N.	1922. *Morph. Jahrb.* **51**, p. 413.	septomaxilla (narial).	237, 318, 442.
Weidenreich, F.	1923. *Z. Anat. EntwGesch.* **69**, pp. 382, 558.	cart.- & memb.-bones.	6.
——	1930. In Möllendorf's *Handbuch der mikroskopischen Anatomie des Menschen.* Berlin. **2** (ii), p. 391.	bone & ossification.	2, 38, 502.
Weigner, K.	1911. *Anat. Hefte,* **45**, p. 81.	man, proatlas.	386.

Author	*Reference*	*Subject*	*Page where quoted*
Weinnold, H.	1922. *Beitr. path. Anat. allg. Path.* **70**, pp. 311, 345.	anencephaly.	485.
Weiss, A.	1901. *Z. wiss. Zool.* **69**, p. 492.	rat, head-segmentation.	386.
Wells, F. R.	1923. *Proc. zool. Soc. Lond.* p. 1213.	Clupea, chondrocr.	132.
Wells, G. A.	1917. *J. Morph.* **28**, p. 417.	Squalus, morph. of skull.	55.
Wen, I. C.	1930. *Contr. Embryol. Carn. Instn.* **22**, p. 109.	Primates, nas. caps.	352, 361.
Westoll, T. S.	1936. *Geol. Mag.* **73**, p. 157.	Osteolepis, skull.	507, 512.
——	1937. *J. Anat. Lond.* **71**, p. 362.	squamosal, fish.	489.
White, P. J.	1890. *Anat. Anz.* **5**, p. 259.	Laemargus, visc. arches.	410, 418.
——	1895. *Ibid.* **9**, p. 57.	Laemargus, hypohyal.	410, 418.
von Wiedersheim, R.	1877. *Morph. Jahrb.* **3**, p. 352.	Urodela, morph. of bony skull.	185.
van Wijhe, J. W.	1882*a*. *Über die Mesodermsegmente und die Entwicklung der Nerven des Selachierkopfes.* Amsterdam.		8, 12, 16, 18, 71.
——	1882*b*. *Niederl. Arch. Zool.* **5**, p. 207.	Ganoids, Dipnoi, morph. of skull.	7, 85, 87, 88, 97, 100, 111, 113, 170, 173, 408, 409, 419.
——	1889. *Anat. Anz.* **4**, p. 558.	segmental theory of skull.	13.
——	1902. *Verh. Akad. Wet. Amst.* **5**, p. 47.	methylene blue method.	15.
——	1904*a*. *C. R. 6 Congr. int. Zool. Berne*, p. 319.	Selachii, dev. of chondrocr.	49, 52, 58, 59, 408.
——	1904*b*. *Petrus Camper*, **1**, p. 109.	Amphioxus.	41.
——	1906. *Ibid.* **4**, p. 61.	metamery, origin of skull.	20, 398, 454.
——	1907. *C. R. Ass. Anat.* **9**, p. 117.	Aves, dev. of chondrocr.	268.
——	1910. *Proc. 7 int. Congr. Zool. Boston, Mass.* p. 533.	Aves, chondrocr.	268.
——	1922. *Bijdr. Dierk.* **22**, p. 271.	Squalus, dev. of chondrocr.	15, 19, 49, 269, 375, 377, 382, 384, 385, 411, 447.
——	1924. *Verh. Akad. Wet. Amst.* **26**, p. 727.	Heptanchus, audit. caps.	400.
Wilder, H. H.	1903. *Mem. Boston. Soc. nat. Hist.* **5**, p. 387.	Necturus, morph. of skull.	190, 191.
Wilhelm, J.	1924. *Anat. Anz.* **59**, p. 1.	Bos, postparietal.	339, 444.
Willcox, M. A.	1899. *Zool. Bull. Wood's Hole*, **2**, p. 151.	Salmo, head-segmentation.	23, 29, 115, 116.
Williams, L. W.	1908. *Amer. J. Anat.* **8**, p. 251.	notochord & basal plate.	382.
Williams, S. R.	1902. *Bull. Mus. comp. Zoöl. Harv.* **40**, p. 1.	Pleuronectes, metamorph.	161, 163.
Williamson, W. C.	1850. *Philos. Trans.* **140**, p. 435.	scales.	6.
——	1851. *Ibid.* **141**, p. 643.	scales.	6.
Williston, S. R.	1914. *Water reptiles of the past and present.* Chicago.	reduction of bones.	507.
Wilson, J. T.	1894. *Proc. linn. Soc. N.S.W.* **9**, p. 129.	Ornithorhynchus, dumbbell bone.	289.
——	1901. *Ibid.* **26**, p. 717.	Monotremes, nas. caps.	289, 295, 299, 300, 397, 435.
——	1905. *J. Anat. Lond. Proc. anat. Soc.* p. v.	Ornithorhynchus, chondrocr.	288.
——	1906. *J. Anat. Lond.* **40**, p. 85.	Monotremes, pila antotica.	289, 298, 299.
—— & Martin, C. J.	1893. *Macleay mem. Vol. linn. Soc. N.S.W.* p. 179.	Ornithorhynchus, dumbbell bone.	289, 434.
Wincza, H.	1896. *Anz. Akad. Wiss. Krakau*, p. 326.	mammals, morph. of skull.	326, 331, 334, 379, 392, 435, 440.

Author	*Reference*	*Subject*	*Page where quoted*
Winslow, G. M.	1898. *Tufts Coll. Stud.* **5**, p. 147.	Ichthyopsida, chondrocr.	78, 115, 174, 176, 190, 191, 192.
Winterhalter, N. P.	1931. *Acta zool. Stockh.* **12**, p. 1.	Xenopus, pineal foramen.	217.
Wintrebert, P.	1910. *C. R. Soc. Biol. Paris,* **68**, p. 178.	Urodela, palatine.	177, 185.
——	1922*a*. *Bull. biol.* **66**, p. 277.	Urodela, quadrate, metamorph.	178, 186.
——	1922*b*. *C. R. Acad. Sci. Paris,* **175**, p. 239.	Salamandra, palatine.	177, 185.
de Witte, G. F.	1927. *Rev. zool. afr.* **14**, p. 65.	Crocodilus, eyelid bone.	268.
Wortman, J. L.	1920. *Proc. U.S. nat. Mus.* **57**, p. 1.	mammals, prefrontal, etc.	326, 368, 507
Wright, R. Ramsay	1884. *Zool. Anz.* **7**, p. 248.	Amiurus, Weberian ossic.	136, 141.
Wyeth, F. J.	1924. *Philos. Trans.* B. **212**, p. 259.	Sphenodon, dev. of audit. caps.	238, 241.
Yano, F.	1926. *Fol. anat. jap.* **4**, p. 57.	Urodela, sclerotic cart.	189, 191, 192, 443, 479.
——	1927. *Ibid.* **5**, p. 169.	Anura, sclerotic cart.	479.
Yano, F.	1928. *Fol. anat. jap.* **6**, p. 103.	sclerotic cart., exper.	480.
Yatabe, T.	1931. *Keijo J. Med.* **2**, p. 1.	Hynobius, sclerotic cart.	480.
Young, J. Z.	1933. *Quart. J. micr. Sci.* **75**, p. 571.	Scyllium, nerves.	62, 74.
Zanichelli, W.	1909. *Boll. Soc. zool. Roma,* **10**, p. 239.	Salmo, dev. of chondrocr.	115.
Zawisch-Ossenitz, C.	1927. *Z. mikr-anat. Forsch.* **10**, p. 473.	ossification in cart.	38.
Ziba, S. I.	1911. *Z. Morphol. Anthrop.* **13**, p. 157.	ossification & cart.	38.
Ziegler, H. E.	1908. *Jena. Z. Naturw.* **43**, p. 653.	Torpedo, head-segmentation.	18, 19.
Zimmermann, S.	1913. *Anat. Anz.* **44**, p. 594.	Anguis, chondrocr.	236.
Zuckerkandl, E.	1893. *Ergebn. Anat. EntwGesch.* **2**, p. 273.	nasal caps.	348, 361.
——	1909. *SitzBer. Akad. Wiss. Wien.* **117**, (iii), p. 493.	nasal caps., paraseptal cart.	326, 395, 397
Zuckerman, S.	1926. *Proc. zool. Soc. Lond.* p. 843.	baboon, allometric growth.	472.

ADDITIONAL BIBLIOGRAPHY

This list contains for the most part works published since 1937; a few, published before that date, became known to the author too late for inclusion in the text.

Author	*Reference*	*Subject*
Aumonier, F. J.	1941. *Quart. J. micr. Sci.* **82**, p. 1.	Lepidosteus, dev. of membrane-bones.
Bahl, K. N.	1937. *Rec. Indian Mus.* **39**, p. 133.	Varanus, morph. of skull.
Brash, J. C.	1934. *Edinb. med. J.* **41**, p. 305.	osteocranium, growth.
Brock, G. T.	1937. *Proc. zool. Soc. Lond.* B. (2), p. 225.	Struthio, dev. of chondrocr.
Devillers, C.	1944. *Ann. Sci. nat. Zool.* **6**, p. 25.	Salmo, lateral line & dev. of membrane-bones.
——	1947. *Ann. Paléont.* **33**, p. 1.	ditto.
——	1950. *Ann. biol.* **26**, p. 145.	evolution of skull.
Findlay, G. H.	1944. *Proc. zool. Soc. Lond.* **114**, p. 91.	Elephantulus, Ericulus, Eremitalpa, dev. of auditory ossicles.
Haas, G.	1936. *Acta zool. Stockh.* **17**, p. 55.	Chalcides, morph. of skull.
——	1937. *J. Morph.* **61**, p. 433.	Chamaeleo, nasal caps.
Handbuch der vergleichenden Anatomie, 4. *Kranium und Viszeralskelett*, 1936 (edited by Bolk. L., Göppert, E., Kallius, E., Lubosch, W.)		general.
Hauschild, R.	1937. *Z. Morph. Anthrop.* **32**, p. 215.	racial differences in man, chondrocr.
Hubendick, B.	1942. *Ark. Zool. Stockh.* 34A, No. 7.	Leuciscus, dev. of chondrocr.
Kim, B.	1933. *Z. Morph. Anthrop.* **32**, p. 486.	racial differences in pig chondrocrania.
van der Klaauw, C. J.	1929. *Proc. zool. Soc. Lond.* p. 491.	Macroscelides, dev. of tympanic region.
——	1930. *J. Mammalol.* **11**, p. 55.	Hyrax, Galeopterus, Icticyon, Tapirus, tympanic bulla.
Limson, M.	1932. *Contr. Embr. Carn. Instn.* **23**, p. 205.	white & negro, osteocr.
Matthes, E.	1938. *Arq. Mus. Bocage, Lisboa,* **9**, p. 1.	vertebral theory of skull.
McClain, J. A.	1939. *J. Morph.* **64**, p. 211.	Didelphys, dev. of ear ossicles.
Millot, J. & Anthony, J.	1958. *L'Anatomie de* Latimeria chalumnae, Paris.	Coelacanth, morph. of skull.
Moy Thomas, J. A.	1941. *Nature, Lond.* **147**, p. 681.	Salmo, exp., dev. of frontal bone.
Neumayer, L.	1938. *Arch. ital. Anat. Embriol.* **40**, Suppl., p. 1.	Bdellostoma, dev. of skull.
Pehrson, T.	1940. *Acta zool. Stockh.* **21**, p. 1.	Amia, dev. of membrane-bones.
——	1944*a*. *Ibid.* **24**, p. 27.	Acipenser, Polyodon, dev. of membrane-bones.
——	1944*b*. *Ibid.* **24**, p. 135.	Esox, dev. of membrane-bones.
——	1945. *Ibid.* **26**, p. 157.	Chelonia, dev. of skull.
——	1958. *Ibid.* **39**, p. 241.	Polypterus, lat. line & dev. of membrane-bones.
Pusey, H. K.	1938. *Quart. J. micr. Sci.* **80**, p. 479.	Rana, metamorph. of chondrocr.
——	1939. *J. roy. micr. Soc.* **59**, p. 252.	reconstruction method.
——	1943. *Quart. J. micr. Sci.* **84**, p. 105.	Ascaphus, morph. of chondrocr.

Author	*Reference*	*Subject*
Roth, O.	1938. *Z. Morph. Anthrop.* **33**, p. 409.	pituitary & skull-growth.
Salvadori, G.	1928. *Pubbl. Clin. otorinolaringol. Univ. Napoli*, **13**, p. 1.	Anura, columella auris.
Sewertzoff, A. N.	1893. *Bull. Soc. imp. Nat. Moscou*, **6**, p. 99.	Pelobates, head segmentation.
Smith, J. L. B.	1939. *Trans. roy. Soc. South Africa*, **28**, p. 1.	Latimeria, morph. of skull.
Tchernavin, V.	1938*a*. *Trans. zool. Soc. Lond.* **24**, p. 103.	Salmo, breeding changes in skull.
——	1938*b*. *Proc. zool. Soc. Lond.* B. **108**, p. 347.	Salmo, visceral arches.
Todd, T. W. & Schweikher, F. P.	1933. *Amer. J. Anat.* **52**, p. 81.	Hyaena, growth of skull.

SUBJECT INDEX

(NOTE: The reader is reminded of the very detailed Table of Contents and of the numerous cross-references from the Comparative and General Sections, which have rendered exhaustive indexing of the Systematic Section unnecessary.)

INDEX OF GENERA

REFERRED TO IN THE SYSTEMATIC SECTION

HEAD SEGMENTATION

HEAD SEGMENTATION

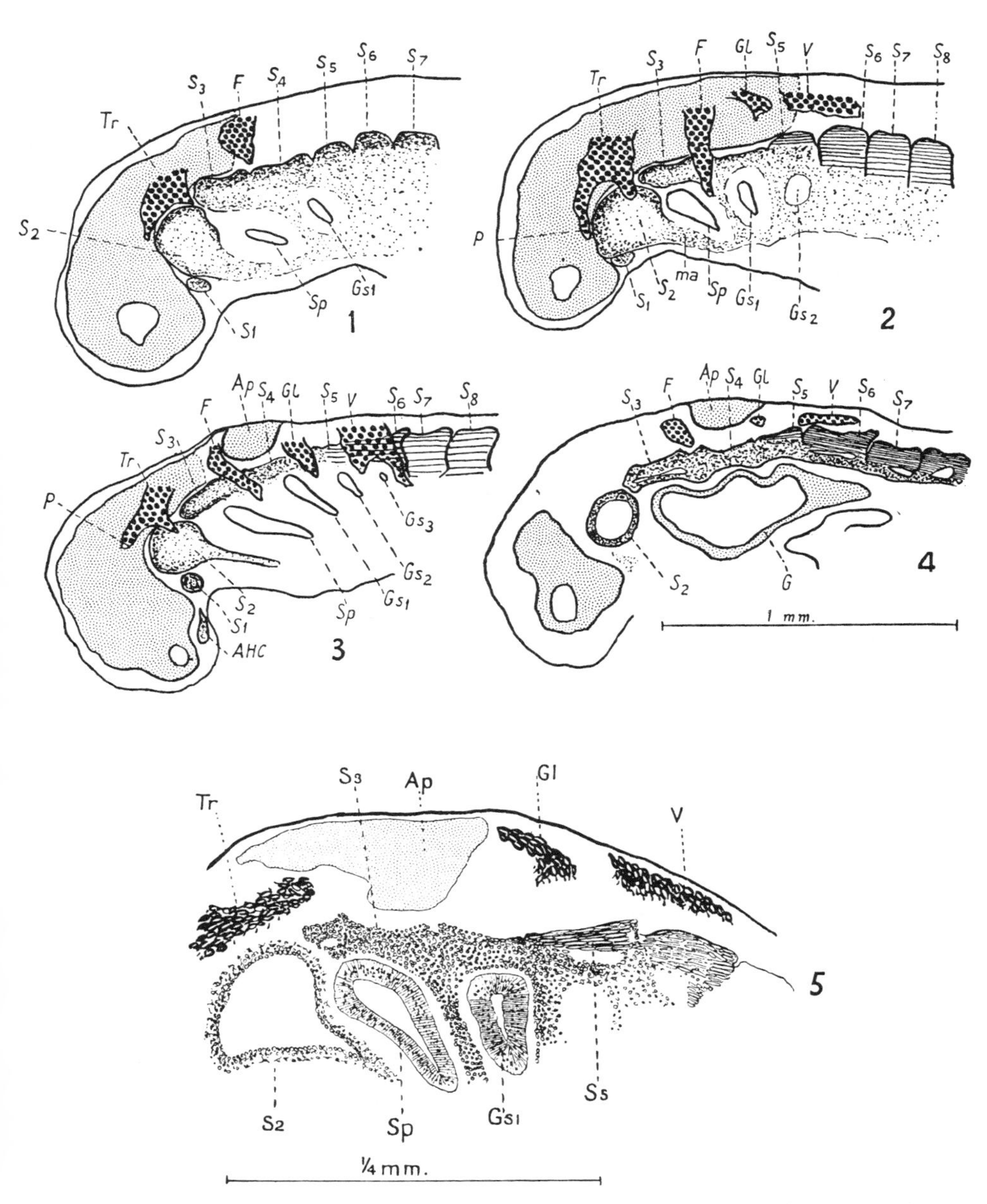

1, Longitudinal section Squalus acanthias 4·5 mm.; **2**, 5 mm.; **3, 4,** 6 mm ; **5**, enlarged view of auditory region of 6 mm. stage (from de Beer).

AHC, anterior head cavity; Ap, auditory placode; b, brain; E, eye; F, facial nerve; g, gut; Gs 1–4, gill-slit 1–4; Gl, glossopharyngeal nerve; H, heart; ma, mandibular arch; P, profundus nerve; S 1–10, somite 1–10; Sp. spiracle; SpGn, spinal ganglion; Tr, trigeminal nerve; V, vagus nerve; v7–10, ventral nerve-root of segment 7–10.

PLATE 2

HEAD SEGMENTATION

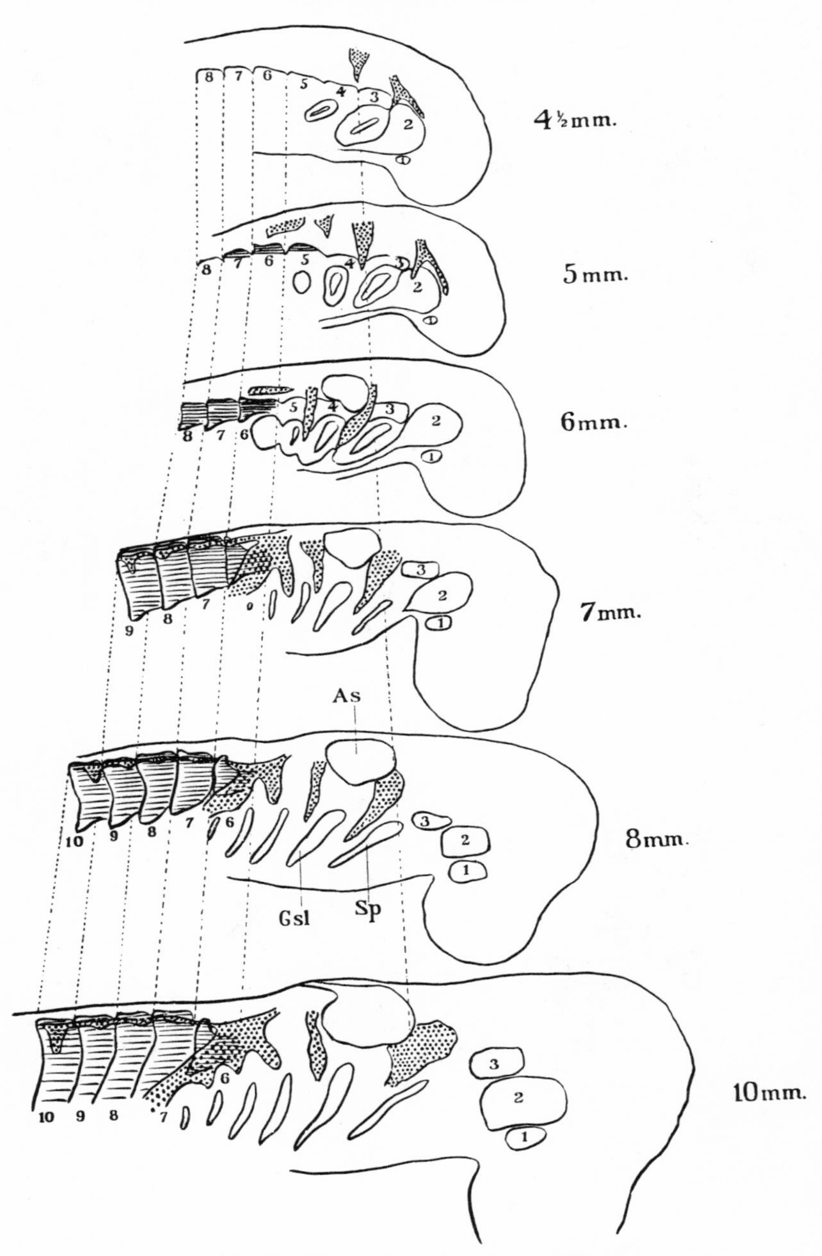

Squalus acanthias; somites are numbered (from de Beer).

Lettering explained on Plate 1

HEAD SEGMENTATION

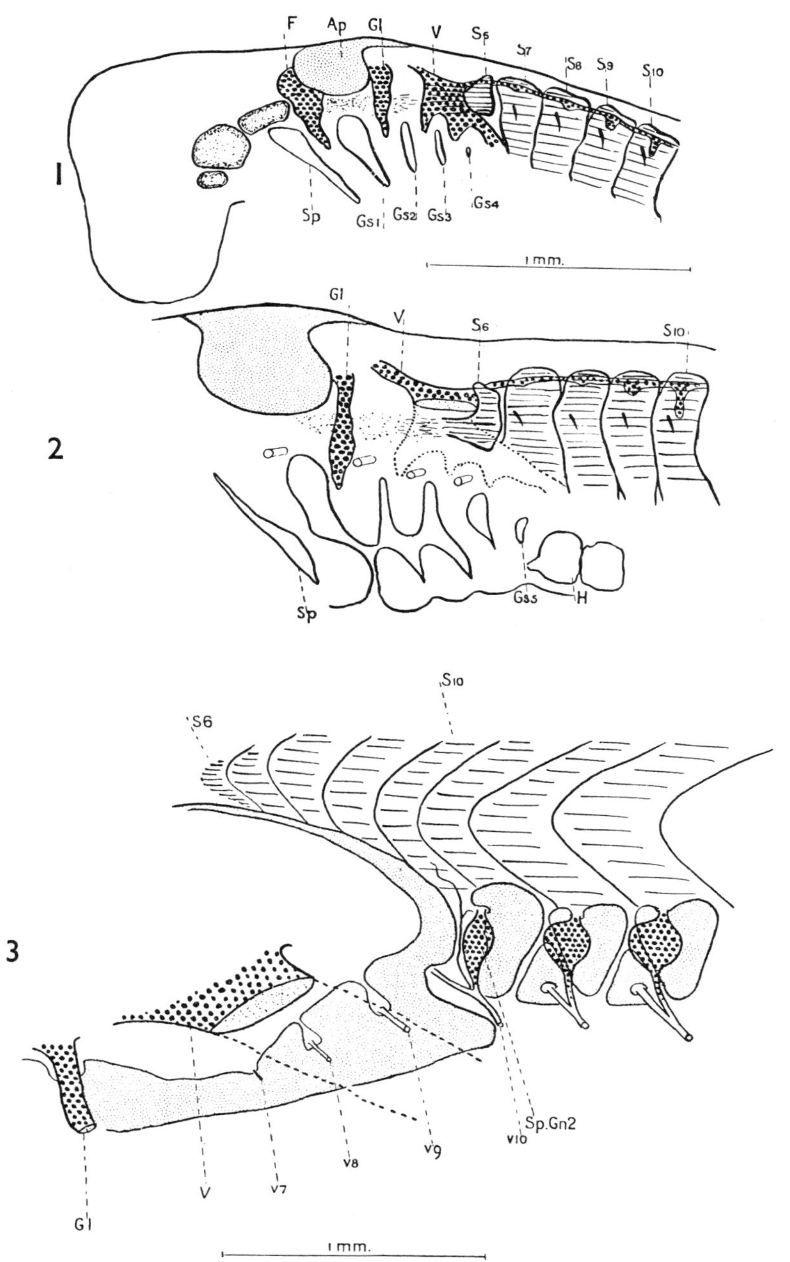

Squalus acanthias; **1**, 8 mm.; **2**, 10 mm.; **3**, 50 mm. (from de Beer).

Lettering explained on Plate 1

HEAD SEGMENTATION

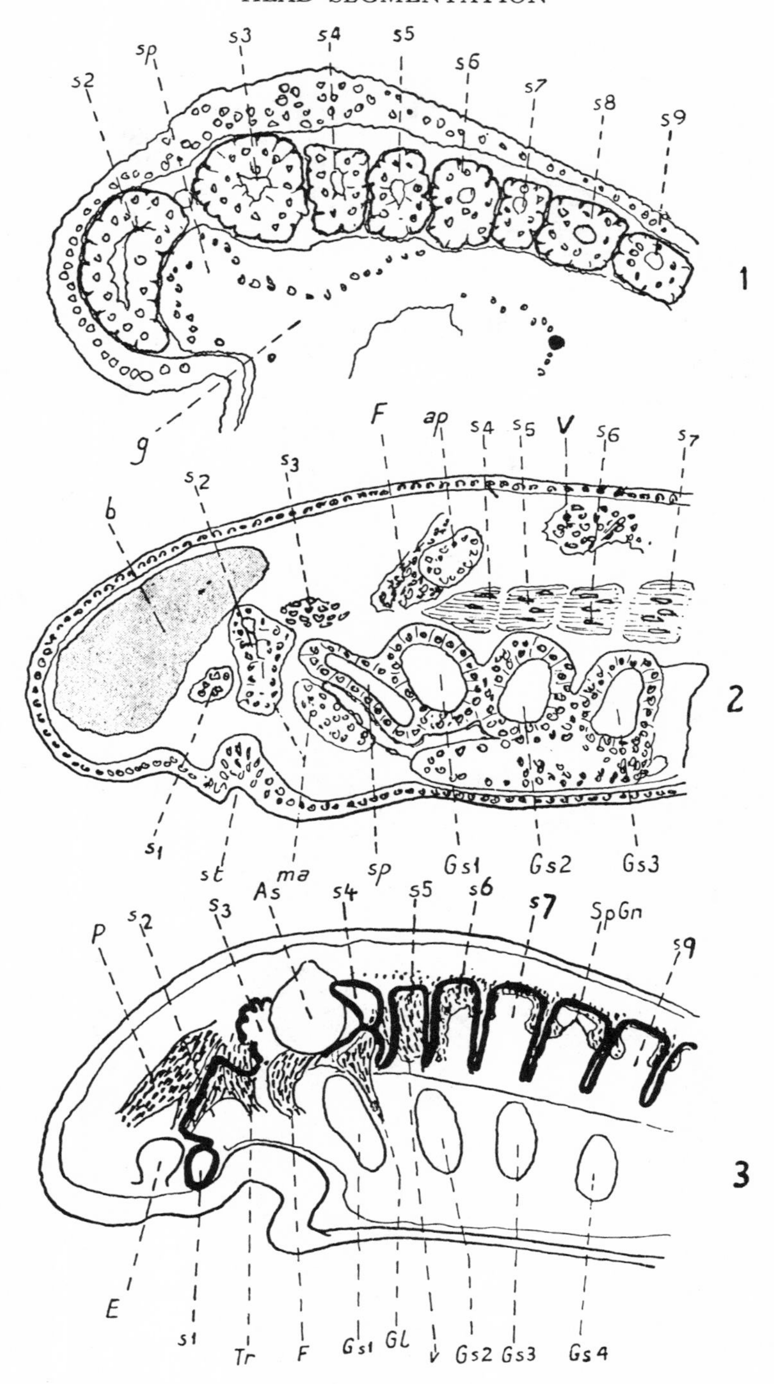

Petromyzon planeri; **1**, 8 days; **2**, 11 days; **3**, at hatching (after Koltzoff)

Lettering explained on Plate 1

HEAD SEGMENTATION

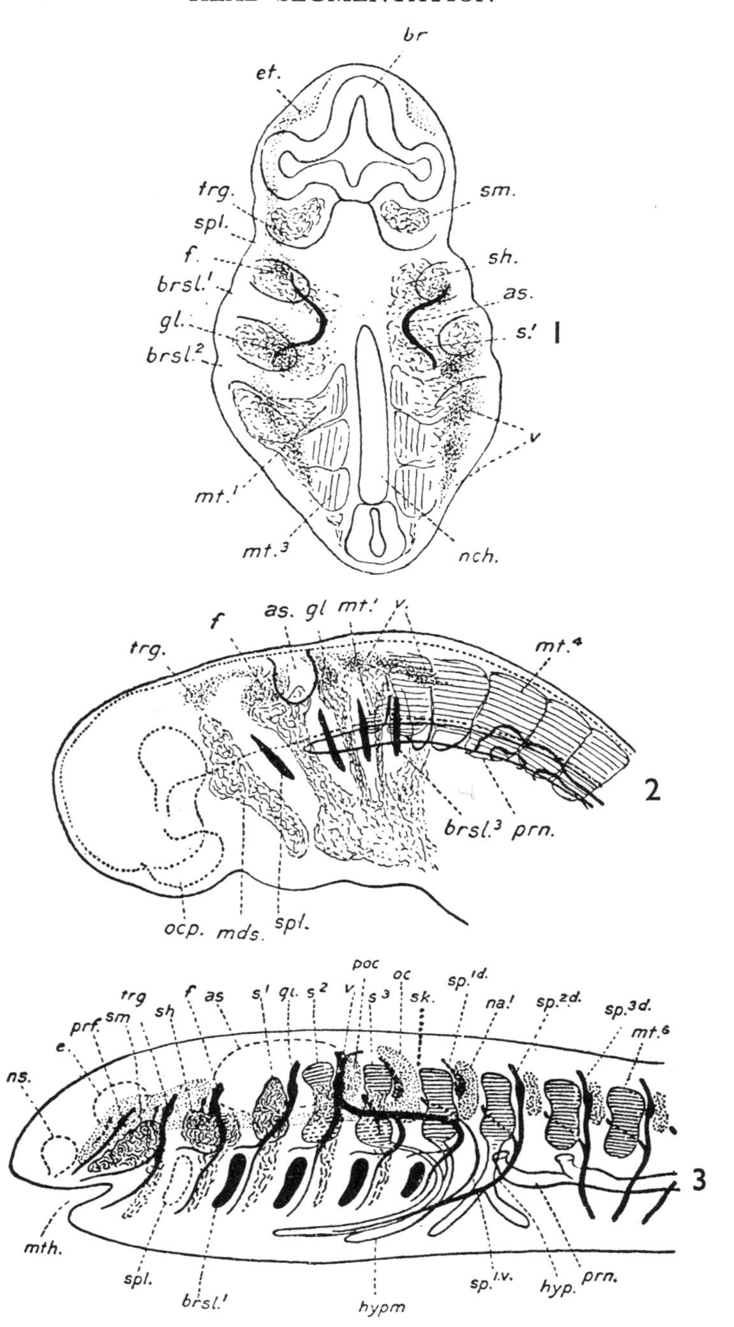

Amblystoma tigrinum; 1, 3 mm.; 2, 5 mm.; 3, scheme of segmentation of the head in Urodela (from Goodrich, by courtesy of the Zoological Society).

as, auditory sac; brsl, gill-slit 1–4; e, eye; et, epidermal thickening; f, facial nerve; gl, glossopharyngeal nerve; hyp, hypoglossal nerve; hypm, hypoglossal muscle; mds, mandibular arch; mt, myomere 1–6; mth, mouth; na 1, neural arch of 1st vertebra; nch, notochord; ns, nasal sac; oc, occipital arch; ocp, optic cup; poc, preoccipital arch; prf, profundus nerve; prn, pronephros; s 1–2, metotic somite 1–2; sh, hyoid somite; sk, limit of skull; sm, mandibular somite; sp d, v, dorsal & ventral roots of spinal nerve; spl, spiracle; trg, trigeminal nerve; v, vagus nerve

HEAD SEGMENTATION

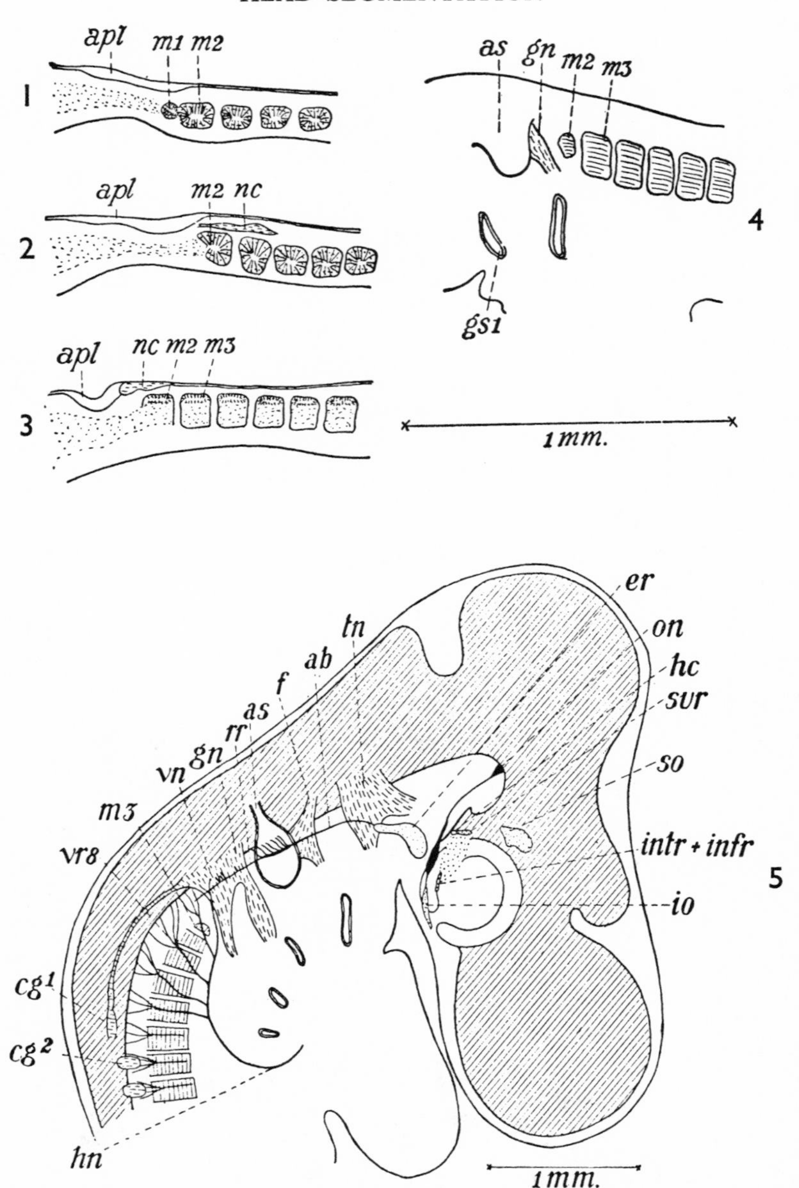

Anas boschas; **1**, 7-somite stage; **2**, 14-somite; **3**, 18-somite; **4**, 23-somite; **5**, 108-hour stage (from de Beer and Barrington).

ab, abducens nerve; apl, auditory placode; as, auditory sac; cg 1–2, ganglion of cervical nerve 1–2; er, external rectus; f, facial nerve; gn, glossopharyngeal nerve; gs 1–2, gill-slit 1–2; hc, premandibular somite; hn, hypoglossal nerve; infr, inferior rectus; intr, internal rectus; io, inferior oblique; m 1–3, metotic somite 1–3; nc, neural crest; on, oculomotor nerve; rr, ramus recurrens of abducens; so, superior oblique; sur, superior rectus; tn, trigeminal nerve; vn, vagus nerve; vr 6–9, ventral nerve-root of segment 6–9.

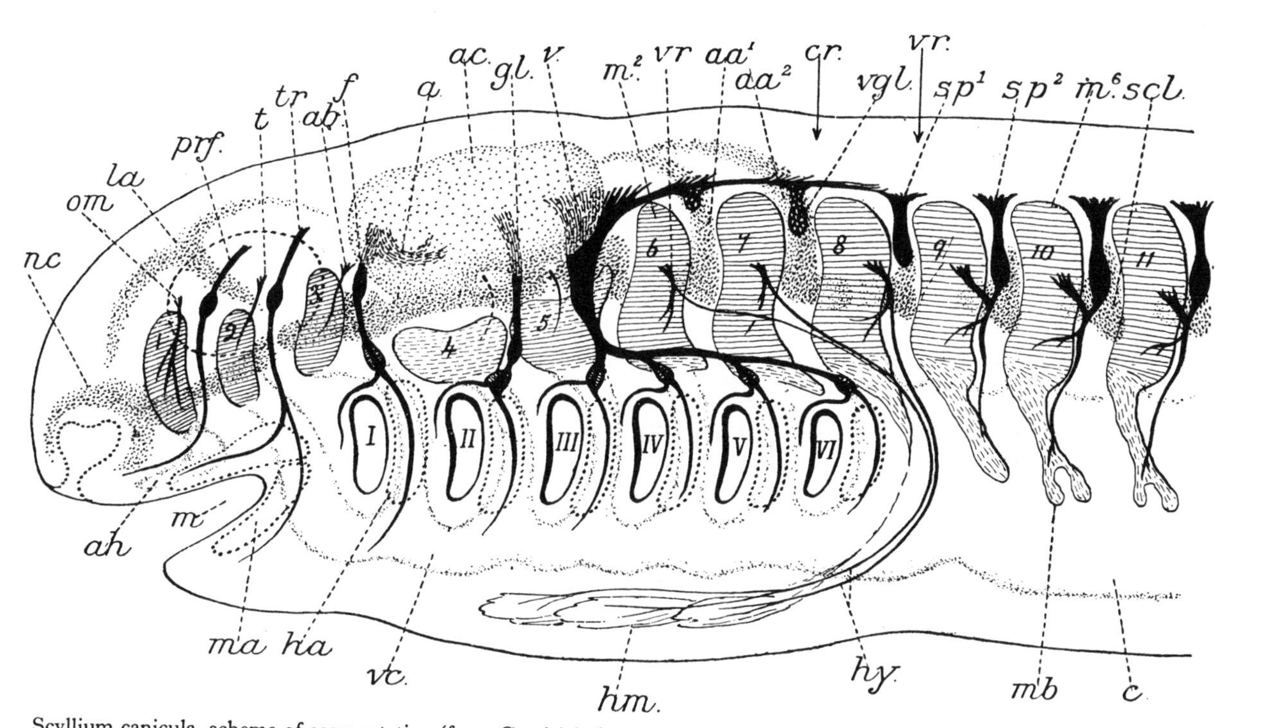

Scyllium canicula, scheme of segmentation (from Goodrich, by courtesy of the Quarterly Journal of Microscopical Science).

cr, limit of neurocranium; vr, limit of visceral arch skeleton; a, auditory nerve; aa 1, preoccipital arch; aa 2, occipital arch; ab, abducens nerve; ac, auditory capsule; ah, anterior head cavity; c, coelom; f, facial nerve; gl, glossopharyngeal nerve; ha, hyoid arch; hm, hypoglossal muscles; hy, hypoglossal nerve; la, pila antotica; m, mouth; m 2–6, myomere 2–6; ma, mandibular arch; mb, muscle-bud; nc, nasal capsule; om, oculomotor nerve; prf, profundus nerve; scl, sclerotome of segment 10; sp 1–2, ganglion of spinal nerve 1–2; t, trochlear nerve; tr, trigeminal nerve; v, vagus nerve; vgl, vestigial ganglion of segment 7; vc, ventral coelom; vr, ventral root of segment 6.

PLATE 8

HEAD SEGMENTATION

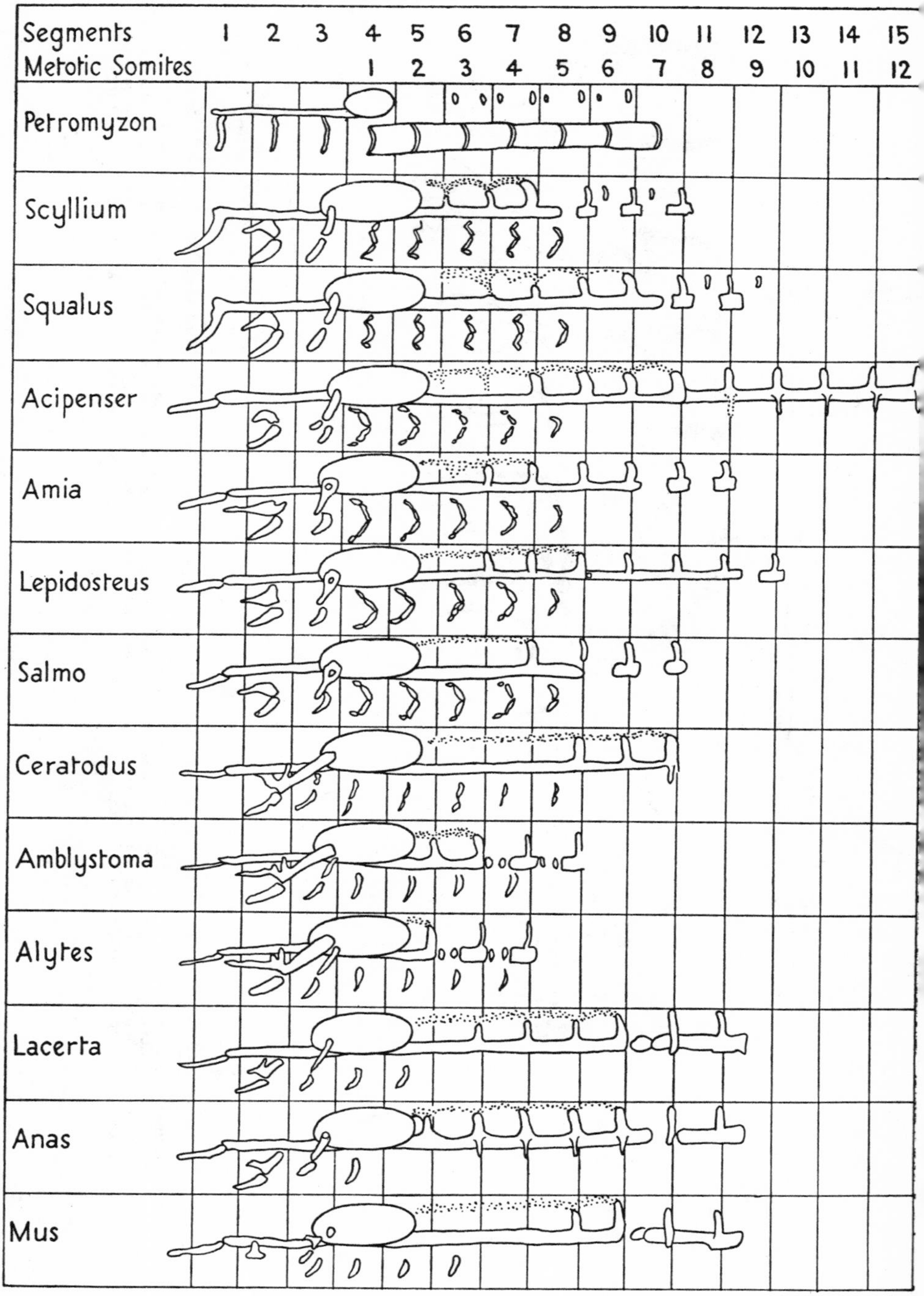

Segments	1	2	3	4	5	6	7	8	9	10	11	12	13	14	15
Metotic Somites				1	2	3	4	5	6	7	8	9	10	11	12
Petromyzon															
Scyllium															
Squalus															
Acipenser															
Amia															
Lepidosteus															
Salmo															
Ceratodus															
Amblystoma															
Alytes															
Lacerta															
Anas															
Mus															

CYCLOSTOMATA, CHONDRICHTHYES

Explanation of Lettering

a, articular facet for hyomandibula
aba, afferent branchial artery
ac, auditory capsule
aca, auditory capsule, anterior cartilage
acp, auditory capsule, posterior cartilage
acr, arteria centralis retinae
adc, anterodorsal cartilage
aec, anterior opening of ethmoidal canal
aha, afferent hyoidean artery
alc, anterolateral cartilage
an, abducens nerve
anc, annular cartilage
anteba, anterior efferent branchial artery
aoc, antorbital cartilage
aom, arteria ophthalmica magna
ap, anterior parachordal
apa, afferent pseudobranchial artery
as, auditory sac
asc, anterior semicircular canal
aul, anterior upper labial
aun, auditory nerve

b, basal plate
ba 1–5, branchial arch 1 to 5
baa, basilar artery
bcf, basicapsular fenestra
bd 1, basidorsal 1
bh, basihyal
bp, basal process
br 1, branchial rays of 1st arch
bv 1, basiventral 1

c, occipital condyle
cac, cavity of auditory capsule
cb 1–5, ceratobranchial 1 to 5
cc, cross commissural vessel
ccp, core of central papilla (mucocartilage)
ce 1, centrum of 1st cervical vertebra
cf, foramen for internal carotid
ch, ceratohyal
cnc, cavity of nasal capsule
cop, copula
cqp, cranioquadrate passage
csp, cartilage connecting basal process to lamina orbitonasalis
ct, cornu trabeculae

da, dorsal aorta
dcs, ductus canalis semicircularis posterioris
del, ductus endolymphaticus
dm, dura mater
dp, dorsal process of parachordal
ds, dorsum sellae

e, eye
eb 1–5, epibranchial 1 to 5
eba 1–4, efferent branchial artery 1–4
ec, ethmoidal canal
ee, elastica externa
ef, epiphanial foramen
eh, epihyal cartilage
eha, efferent hyoidean artery
ei, elastica interna
ep, ethmoid process
epa, efferent pseudobranchial artery
epf, eps, epiphysial fontanelle
epi, epiphysis
es, eye-stalk
et 1, 2, 3, epitrematic process or bar of branchial arch 1, 2, 3
exd 1, 1st dorsal extrabranchial
exv 1, 1st ventral extrabranchial

f, foramen magnum
fa, foramen antoticum
fac, foramen acusticum
fb, basicranial fenestra
fec, floor of ethmoidal canal
fel, foramen for ductus endolymphaticus
fep, foramen for efferent pseudobranchial artery
ff, foramen for facial nerve
fg, foramen for glossopharyngeal nerve
fh, hypophysial fenestra
fic, foramen for pituitary vein (interorbital canal)
fll, foramen enclosing lateral line canal
fm, fissura metotica
fn, facial nerve
fnc, front wall of nasal capsule
fo, foramen for optic nerve
foa, foramen for orbital artery
foc, foramen for oculomotor nerve
fol, olfactory foramen
fon, orbitonasal foramen
fos, preorbital canal (for superficial ophthalmic nerve)
fp, foramen for profundus nerve
fpf, foramen for palatine branch of facial
fpr, foramen prooticum
fsf, foramen for superficial ophthalmic of facial

fso, superfical ophthalmic branch of facial
fst, foramen for superficial ophthalmic of trigeminal
ft, foramen for trochlear nerve
ftr, foramen for trigeminal nerve
fv, basal communicating canal (of Gegenbaur)
fvr, fused cervical vertebrae

g, gap between lamina hypotica and true floor of auditory capsule
gc, canal for passage of glossopharyngeal nerve
gn, glossopharyngeal nerve
gp 1, gill-pouch 1
gs 1–5, gill-slit 1 to 5

h, hyomandibula
ha, hyoid arch
hac, hyoid arch (cartilage)
hb, hindbrain
hbc, hypobranchial rod
hbr 1–3, hypobranchial 1 to 3
hc, hypochordal posterior cartilaginous extension
hf, hypoglossal foramen or foramina
hh, hypohyal
hmf, hyomandibular branch of facial
hn, hypoglossal nerve-root or roots
hr, hyal rays
hrd, dorsal group of hyal rays
hrv, ventral group of hyal rays
ht 1, 2, 3, hypotrematic process or bar of branchial arch 1, 2, 3
hy, hypophysis
hyf, hyoideus branch of facial

i 1, interdorsal 1
ic, internal carotid artery
inp, internasal plate
ip, incisura prootica
is, invaded sheath of notochord

lbo, lamina basiotica
lc, labial cartilage
lda, lateral dorsal aorta
lho, lamina hypotica
llv, lateralis branch of vagus
lnc, lateral wall of nasal capsule
lng, lingual cartilage
lon, lamina orbitonasalis
lr, ligamentous vestige of ventral portion of true hyoid arch skeleton.
lrp, lateral rostral process

ma, mandibular artery
mac, mandibular arch (cartilage)
max & buc, maxillary branch of trigeminal & buccal branch of facial
Mc, Meckel's cartilage
Mca, Meckel's cartilage anterior centre
Mcp, Meckel's cartilage posterior centre
md, mandibular branch of trigeminal
mdin, mandibular branch of trigeminal, inferior division
mds, mandibular branch of trigeminal, superior division
mexf, ramus mandibularis externus facialis
mh, mucohyoid arch
minf, ramus mandibularis internus facialis
mm, mucomandibular arch
mnc, medial wall of nasal capsule
mp, mandibular process
mpm, mucopremandibular arch
mrp, median rostral process
mtb, mucocartilaginous transverse bar
mv, mandibular vessel

n, notochord
na 1, neural arch 1
nc, nasal cartilage
nci, inner process of nasal cartilage
ncm, middle process of nasal cartilage
nco, outer process of nasal cartilage
nep, notch for efferent pseudobranchial artery
no, notch representing cranioquadrate passage
non, notch for optic nerve
nop, notch for profundus nerve
nos, notch for superficial ophthalmic nerve
npv, notch for pituitary vein
ns, nasal septum
nst, nostril

oa, occipital arch
oc, orbital cartilage
oca, opercular cartilage
oec, posterior opening of ethmoidal canal
on, oculomotor nerve
onc, orbitonasal canal
op, otic process of pterygoquadrate
opn, optic nerve
ora, orbital artery
os, orbital sinus

p, parachordal
pa, pila antotica
pb 1–5, pharyngobranchial 1–5
pc, polar cartilage
pcc, precarotid commissure
pcf, precerebral fontanelle
pch, chordal commissure
pcv, posterior canal vacuity
pdc, posterodorsal cartilage
pev, plica encephali ventralis
pf, palatine branch of facial
pfc, prefacial commissure
ph, pharyngohyal
phd, dorsal constituent of pseudohyoid
phv, ventral constituent of pseudohyoid
pl, lamina orbitonasalis
plc, posterolateral cartilage
pma, premandibular arch (cartilage)
pn, profundus nerve
pnp, prenasal plate (mucocartilage)
poa 1, preoccipital arch 1
poc, preoptic root of orbital cartilage
pof, preoptic foramen
por, preorbital ridge
pp, posterior parachordal
ppc, postpituitary commissure
pq, pterygoquadrate
pr 1, pharyngobranchial 1
prgl, pretrematic branch of glossopharyngeal
prr, preoptic root of orbital cartilage
prv, pretrematic branch of vagus
ps, pseudohyoid
pul, posterior upper labial
pv, pituitary vein
ptangl, anterior post-trematic branch of glossopharyngeal
pteba, posterior efferent branchial artery
ptptgl, posterior post-trematic branch of glossopharyngeal

r, rostrum
rcc, roof of cranial cavity
rec, roof of ethmoidal canal
rem, external rectus muscle
rhf, ramus hyoideus facialis
rif, ramus mandibularis internus facialis
rnc, roof of nasal capsule
rop, ramus ophthalmicus profundus
ros, ramus ophthalmicus superficialis
rpf, ramus palatinus facialis
rsm, superior rectus muscle
ru, recessus utricularis

s, saccule
sb, subocular bar
sbo, subocular shelf
sc, spiracular cartilage
scr, subchordal rod
sg 1, spinal ganglion 1
sof, supraorbital foramina
spo, supraorbital cartilage
spp, spiracular pouch
sps, subpituitary space
sso, spiracular sense-organ
st, styliform cartilage
sv 2, spinal ventral nerve-root 2
soc, supraorbital cartilage

t, trabecula cranii
tep, tectum posterius
tf, foramen for thyroid gland-stalk
tm, taenia marginalis
tn, trigeminal nerve
tp, trabecular plate
trn, trochlear nerve
ts, tectum synoticum
tso, superficial ophthalmic branch of trigeminal

u, utricle
uc, unpaired cartilage representing hypobranchials 4 & 5

va, ventral aorta
vap, vein accompanying profundus nerve
vc, ventral crest (mucocartilage)
vcl, vena capitis lateralis
vf, vagus foramen
vl, ventral labial
vn, vagus nerve
vr 1–3, ventral-roots 1 to 3
vs, ventral shield (mucocartilage)
vw, vacuity between lamina orbitonasalis and preoptic root of orbital cartilage

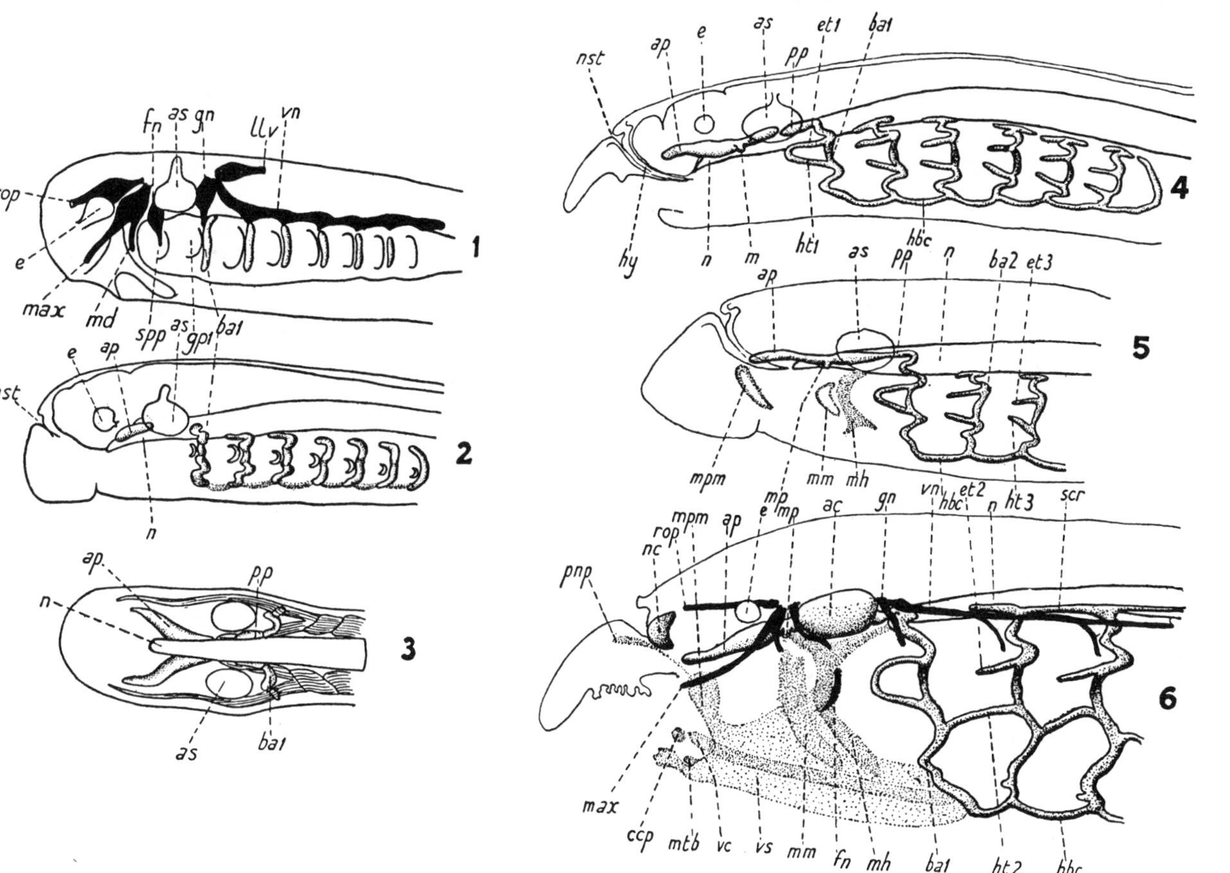

1, early stage; **2**, 24 days; **3**, circ. 25 days; **4**, 26 days; **5**, 45 days **6**, Ammocoete. All lateral views except **3**, which is dorsal (after Sewertzoff).

Explanation of Lettering opposite

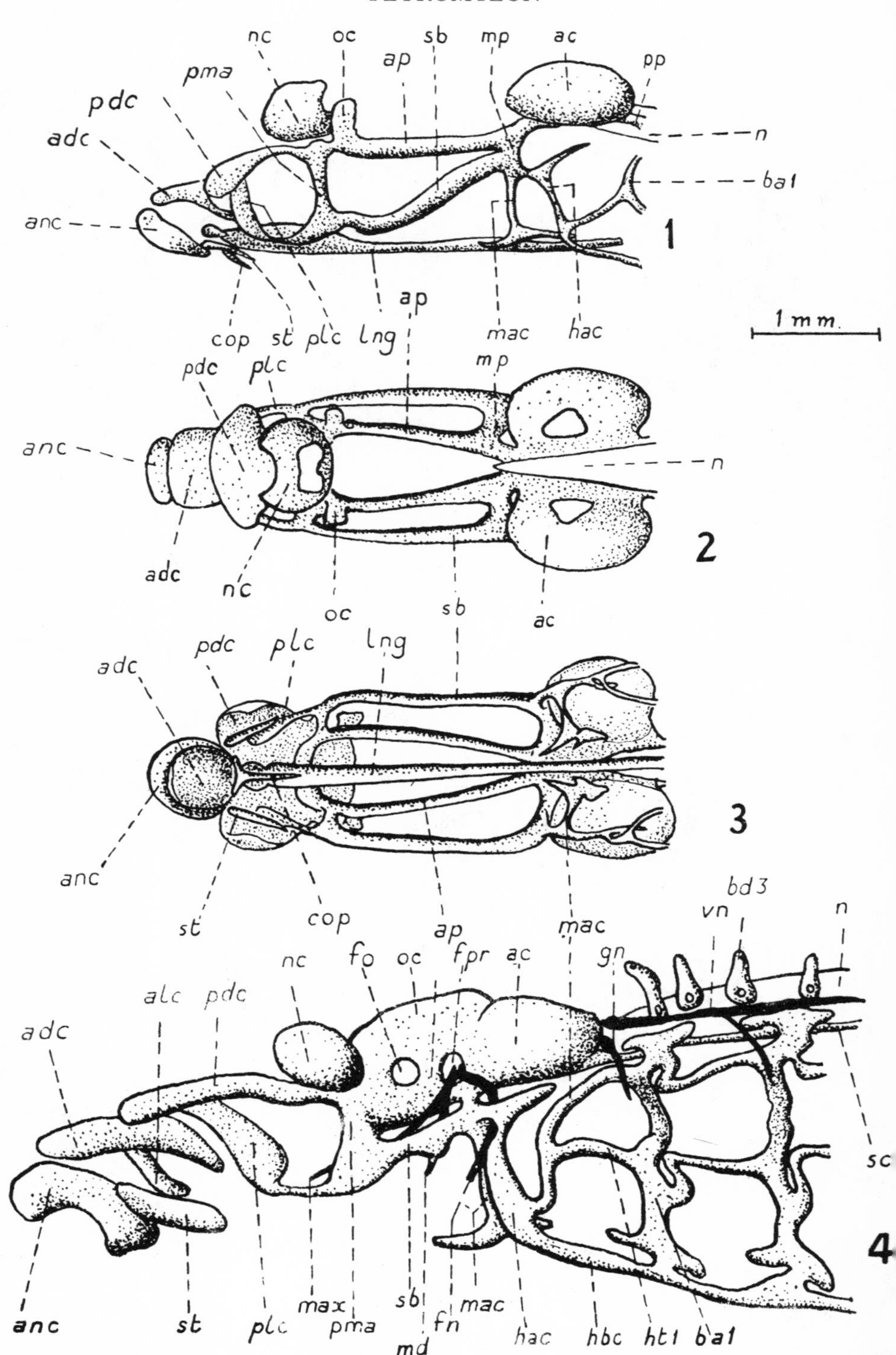

1, lateral, **2**, dorsal, and **3**, ventral view, during metamorphosis (after Damas); **4**, adult (after Sewertzoff).

Explanation of Lettering precedes Plate 9

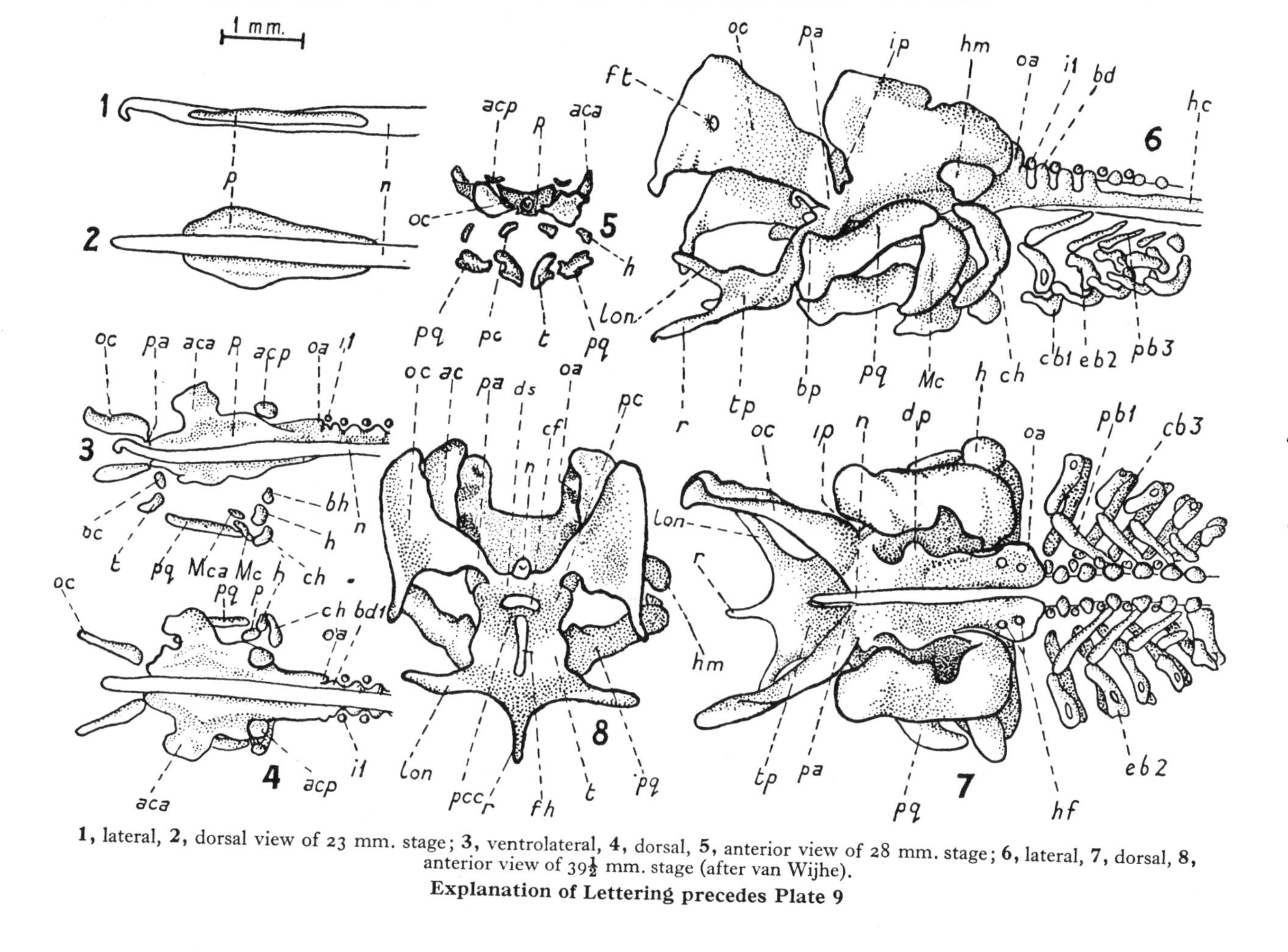

1, lateral, 2, dorsal view of 23 mm. stage; 3, ventrolateral, 4, dorsal, 5, anterior view of 28 mm. stage; 6, lateral, 7, dorsal, 8, anterior view of 39½ mm. stage (after van Wijhe).

Explanation of Lettering precedes Plate 9

SQUALUS

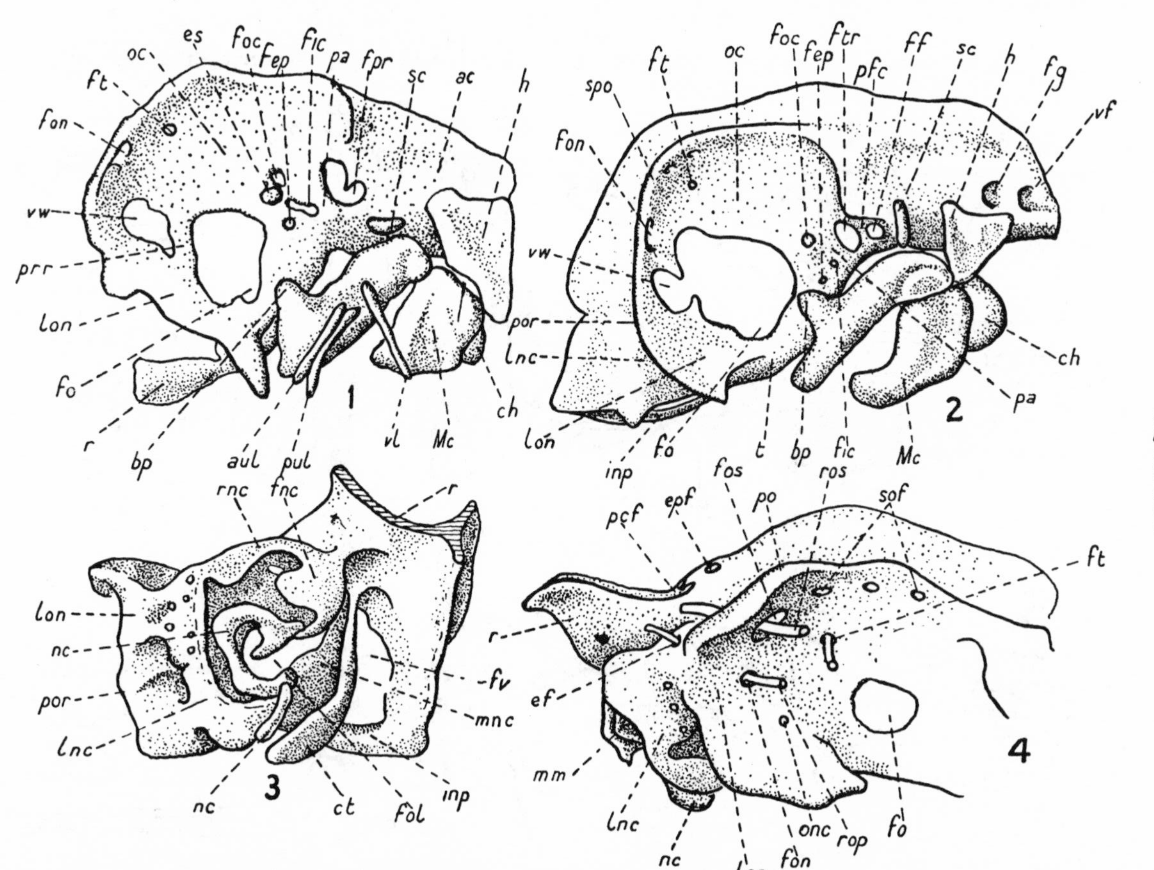

1, 44 mm. stage (after Mori); **2**, 50 mm. stage (after Sewertzoff); **3**, nasal region of 90 mm. stage (after Mori); **4**, adult.

Explanation of Lettering precedes Plate 9

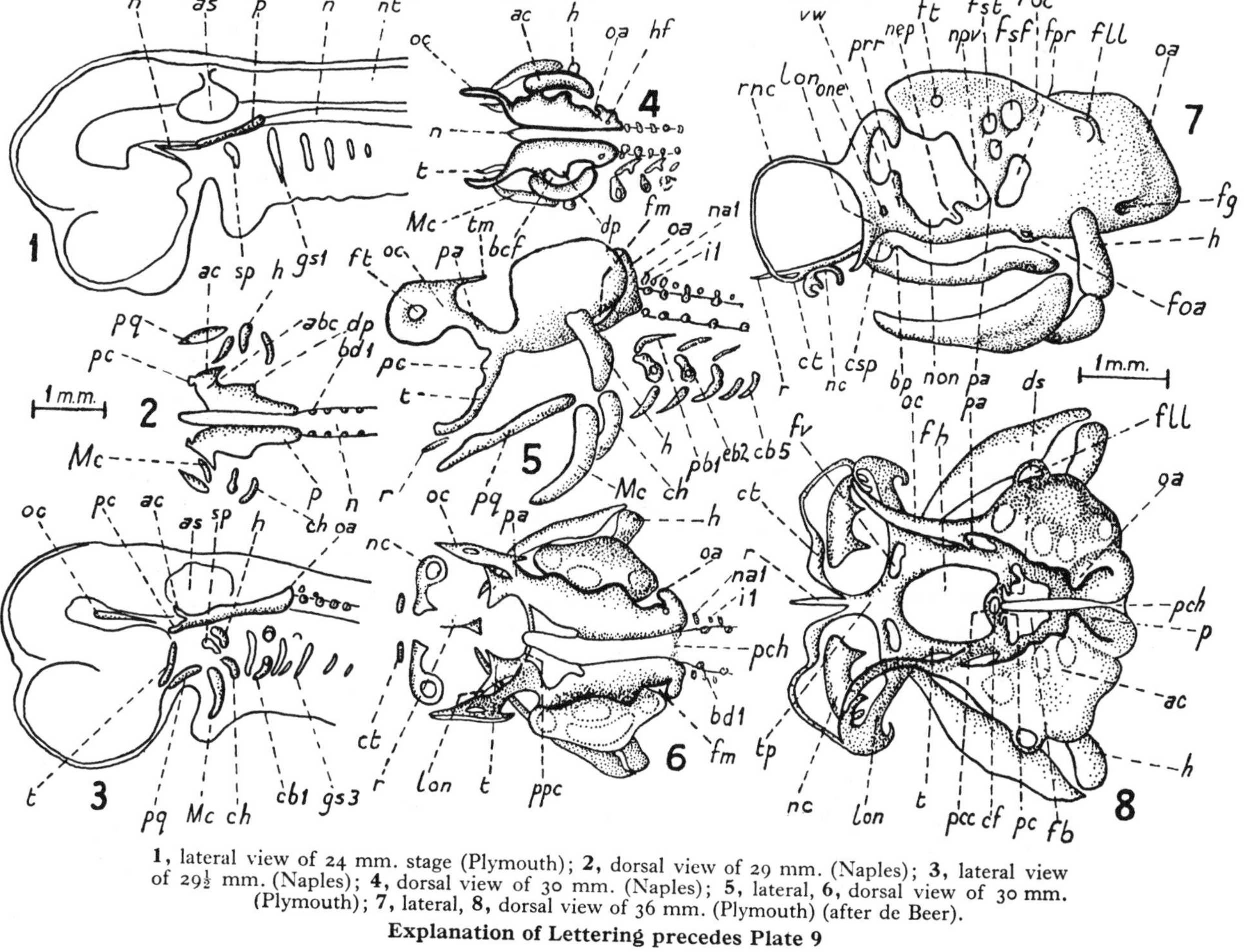

1, lateral view of 24 mm. stage (Plymouth); **2**, dorsal view of 29 mm. (Naples); **3**, lateral view of 29½ mm. (Naples); **4**, dorsal view of 30 mm. (Naples); **5**, lateral, **6**, dorsal view of 30 mm. (Plymouth); **7**, lateral, **8**, dorsal view of 36 mm. (Plymouth) (after de Beer).

Explanation of Lettering precedes Plate 9

SCYLLIUM

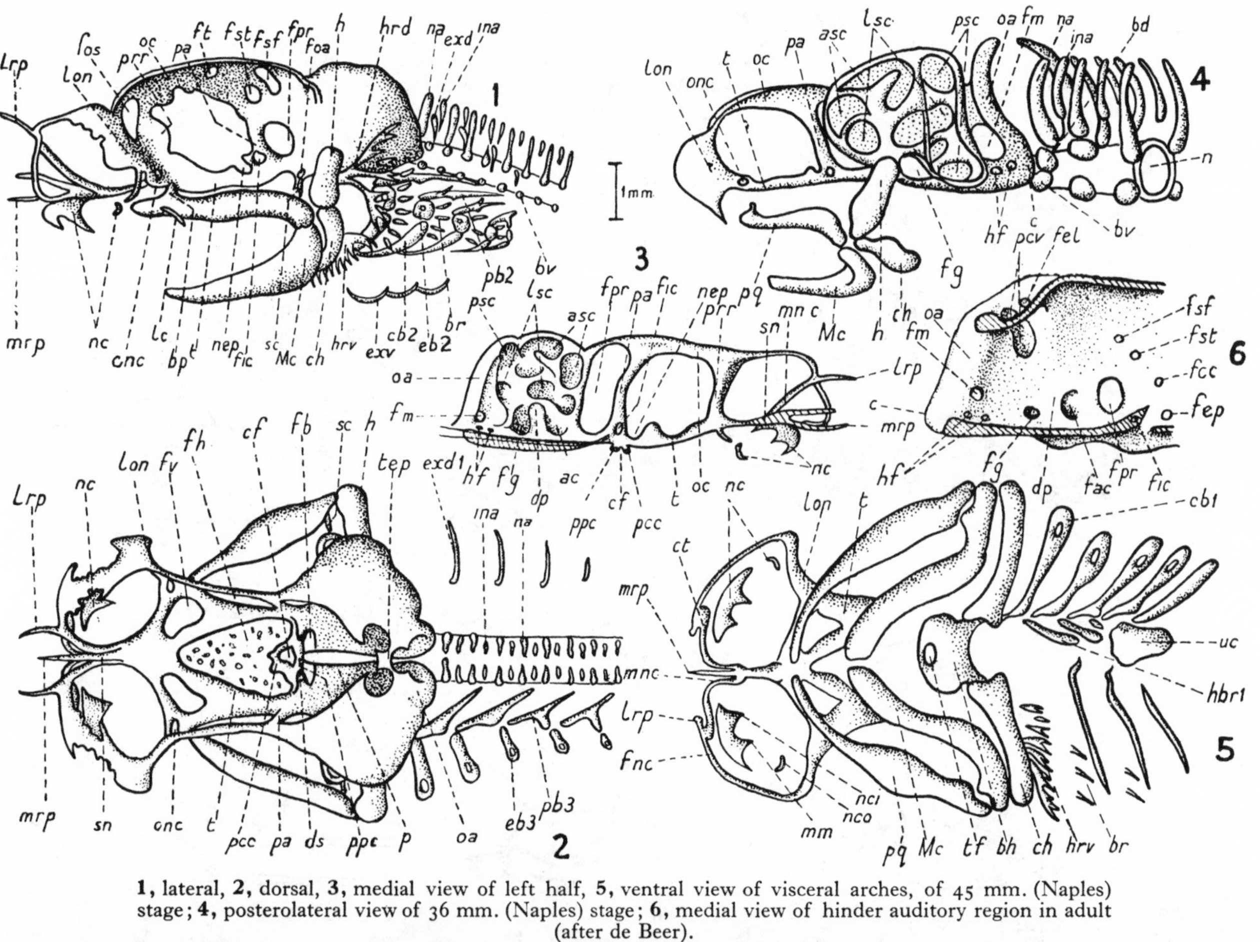

1, lateral, **2**, dorsal, **3**, medial view of left half, **5**, ventral view of visceral arches, of 45 mm. (Naples) stage; **4**, posterolateral view of 36 mm. (Naples) stage; **6**, medial view of hinder auditory region in adult (after de Beer).

SCYLLIUM

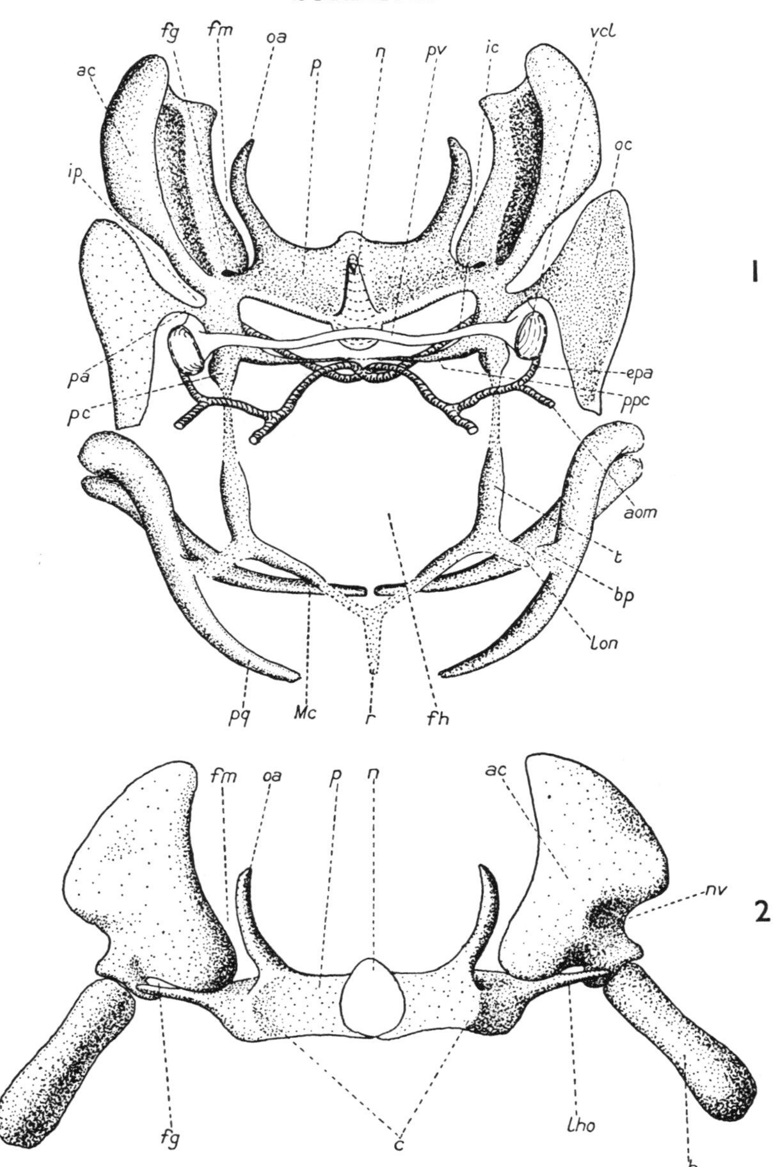

1, anterior, and **2**, posterior view of 30 mm. (Plymouth) stage (from de Beer).

Explanation of Lettering precedes Plate 9

PLATE 16

SCYLLIUM

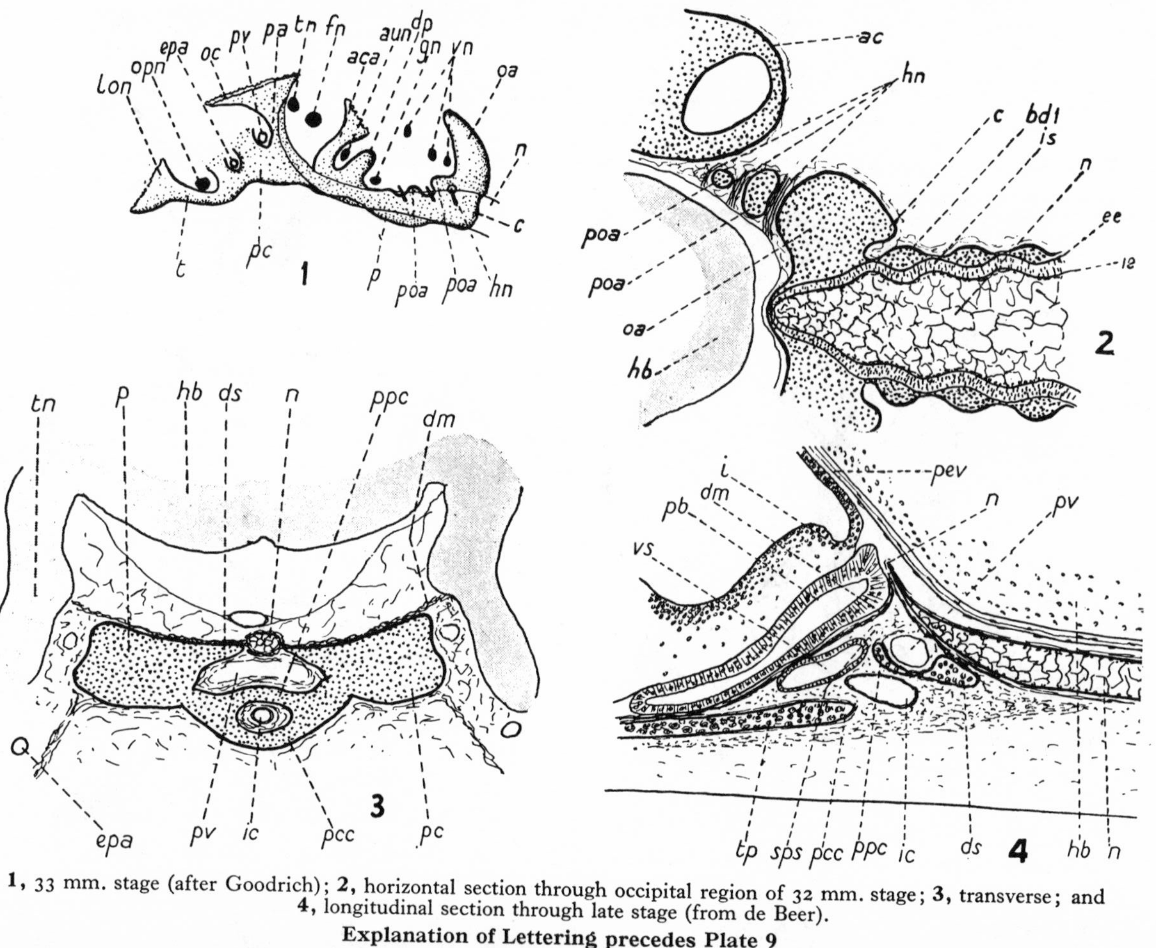

1, 33 mm. stage (after Goodrich); **2**, horizontal section through occipital region of 32 mm. stage; **3**, transverse; and **4**, longitudinal section through late stage (from de Beer).

Explanation of Lettering precedes Plate 9

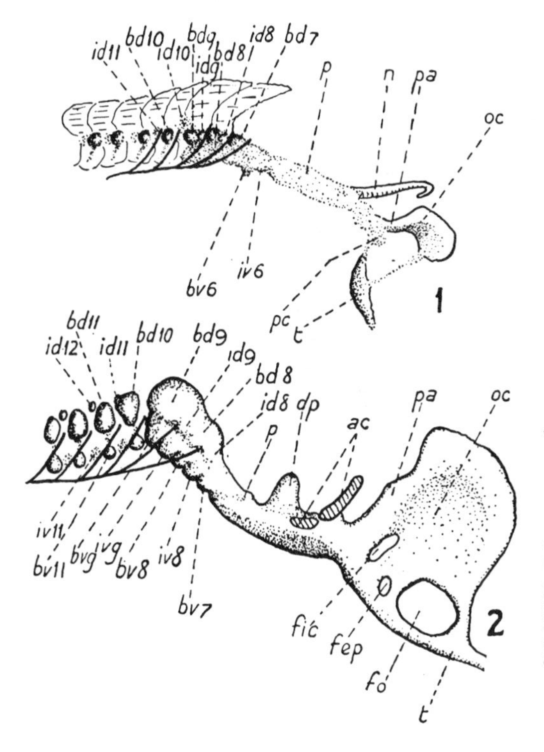

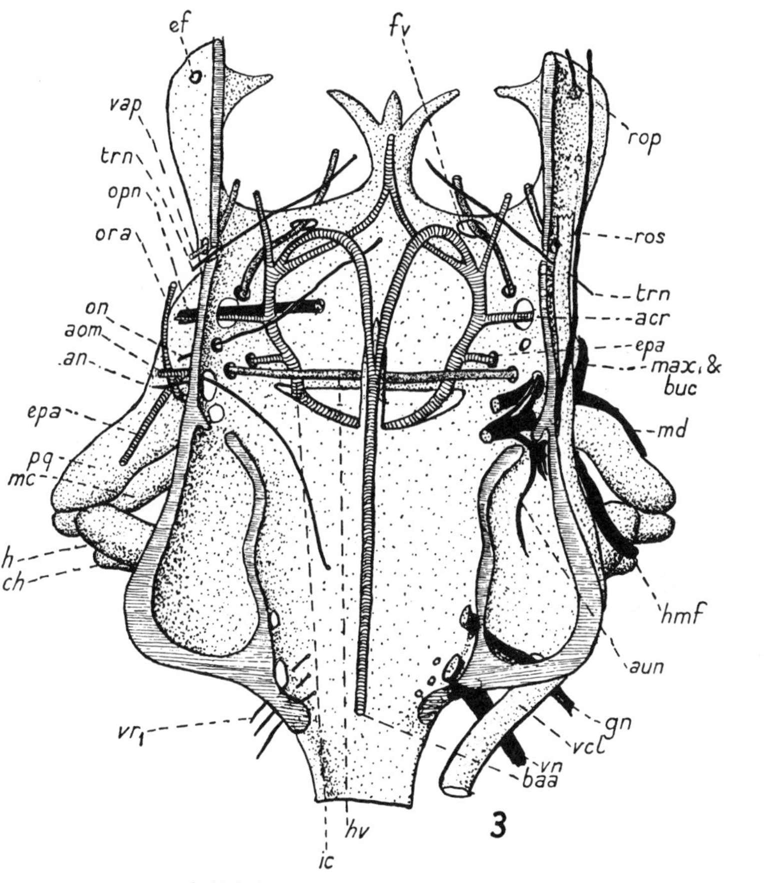

1, Pristiurus, lateral view of 23 mm. stage; **2,** Pristiurus, lateral view of 25 mm. stage (after Matveiev); **3,** Heterodontus, dorsal view of 70 mm. stage to show relations of blood-vessels and nerves (after de Beer).

Explanation of Lettering precedes Plate 9

PLATE 18

HETERODONTUS

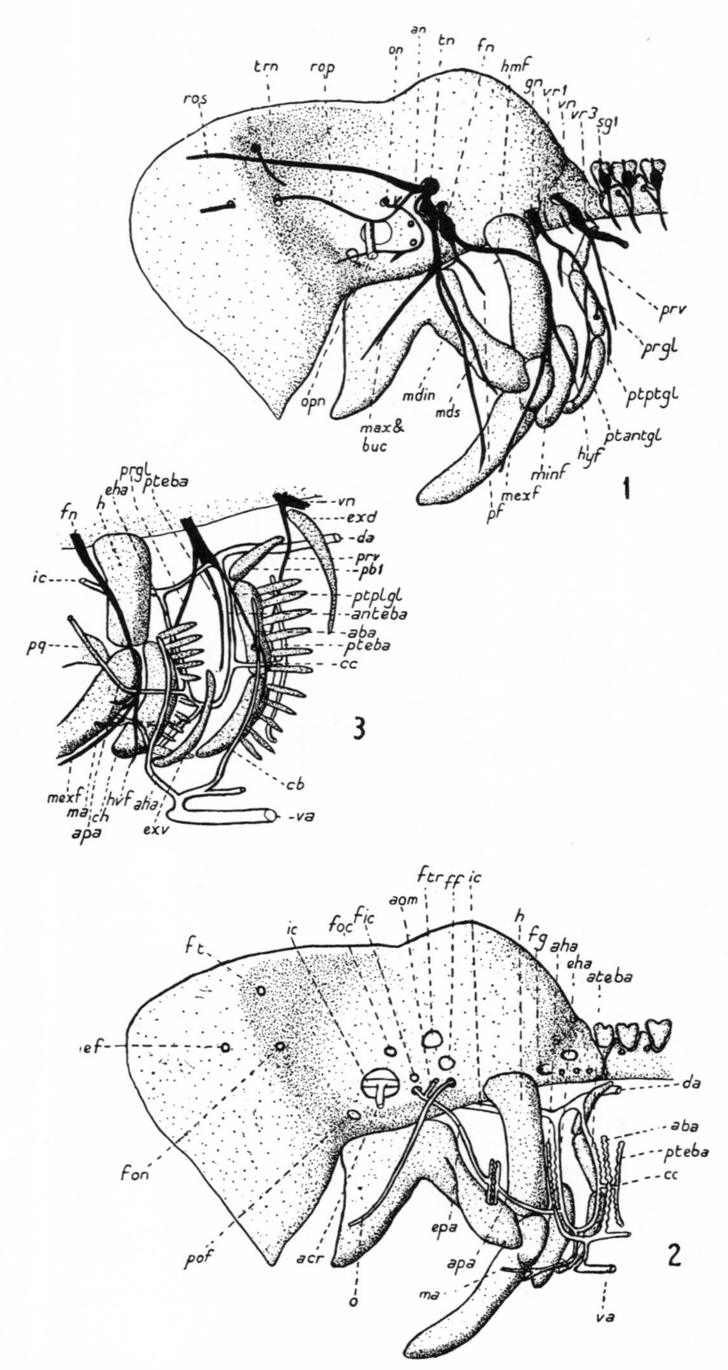

1, relations of chondrocranium to nerves; **2**, relations of chondrocranium to blood-vessels; **3**, relations of hyoid and branchial arches to nerves and blood-vessels (from de Beer).

Explanation of Lettering precedes Plate 9

TORPEDO

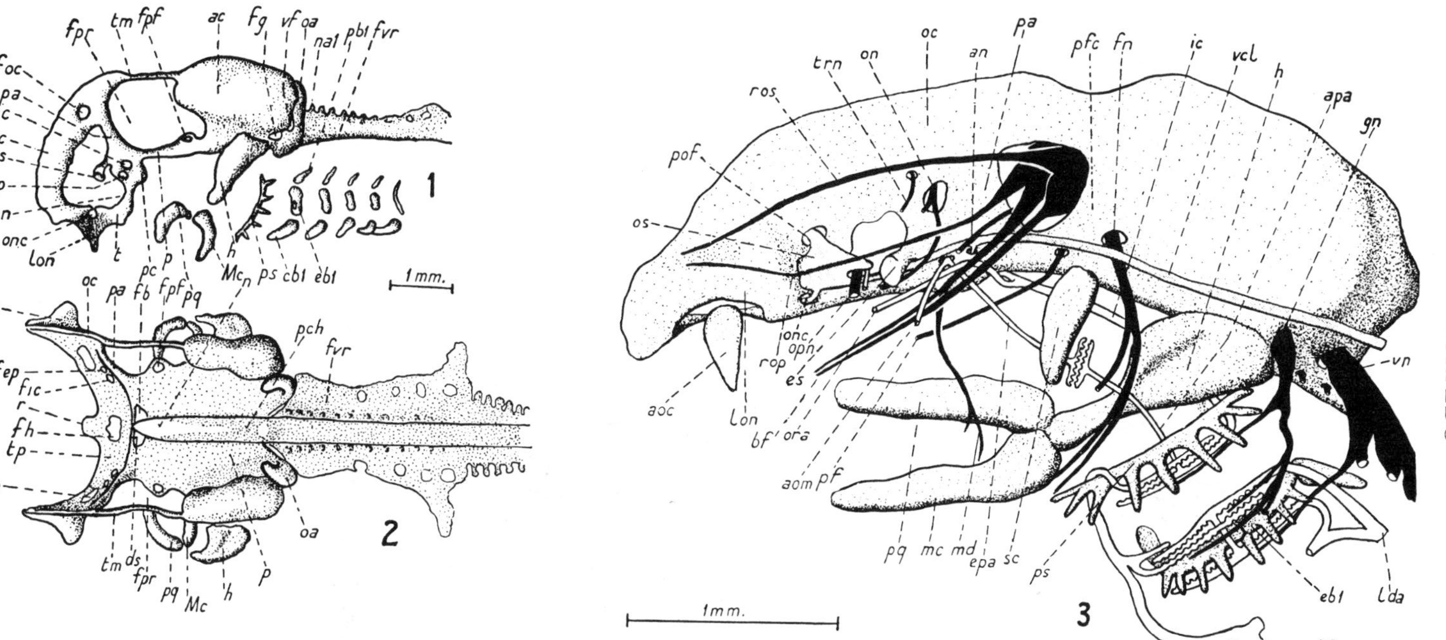

1, lateral, **2,** dorsal view of 24 mm. Torpedo ocellata; **3,** relations of blood-vessels and nerves to skull, 24 mm. Torpedo marmorata (from de Beer).

Explanation of Lettering precedes Plate 9

TORPEDO

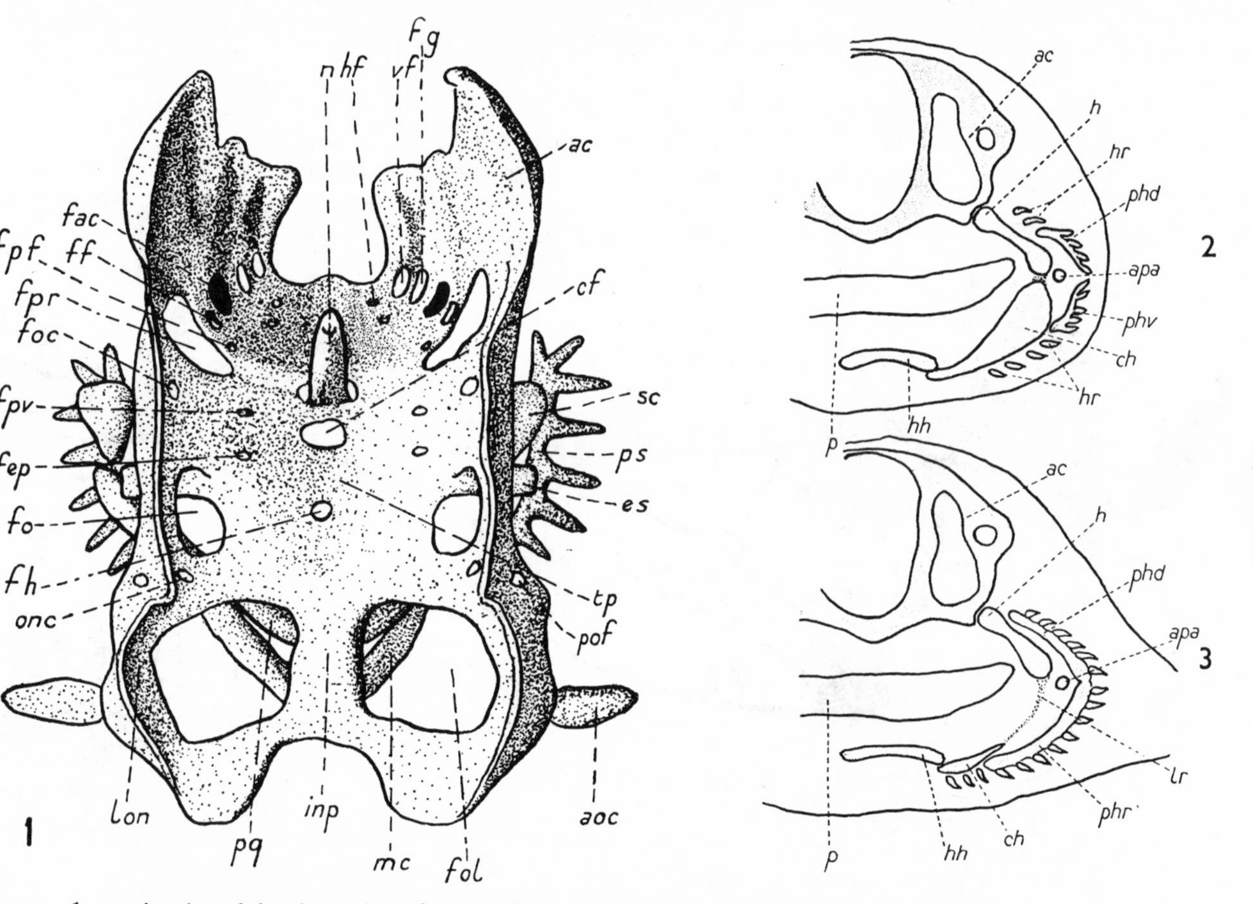

1, anterior view of chondrocranium of 24 mm. Torpedo marmorata; **2**, **3**, diagrams illustrating the formation of the pseudohyoid of Rays (3) from the fused hyal rays of Sharks (2) (from de Beer).

Explanation of Lettering precedes Plate 9

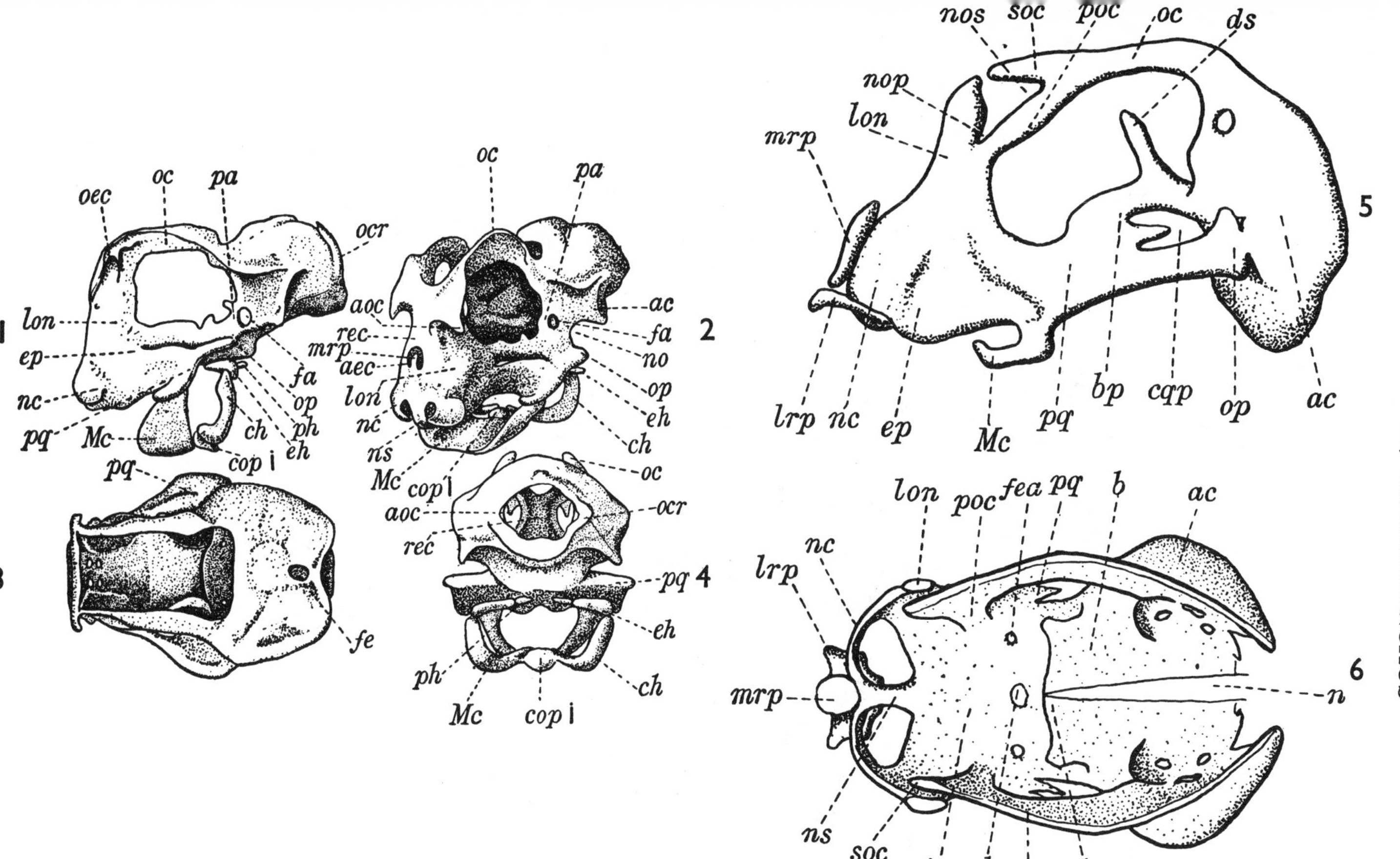

1, lateral, **2**, anterolateral, **3**, dorsal, and **4**, posterior view of chondrocranium of 6-months embryo of Chimaera collei (after Bashford Dean); **5**, lateral, **6**, dorsal view of chondrocranium of 60 mm. embryo of Callorhynchus antarcticus (after Schauinsland).

Explanation of Lettering precedes Plate 9

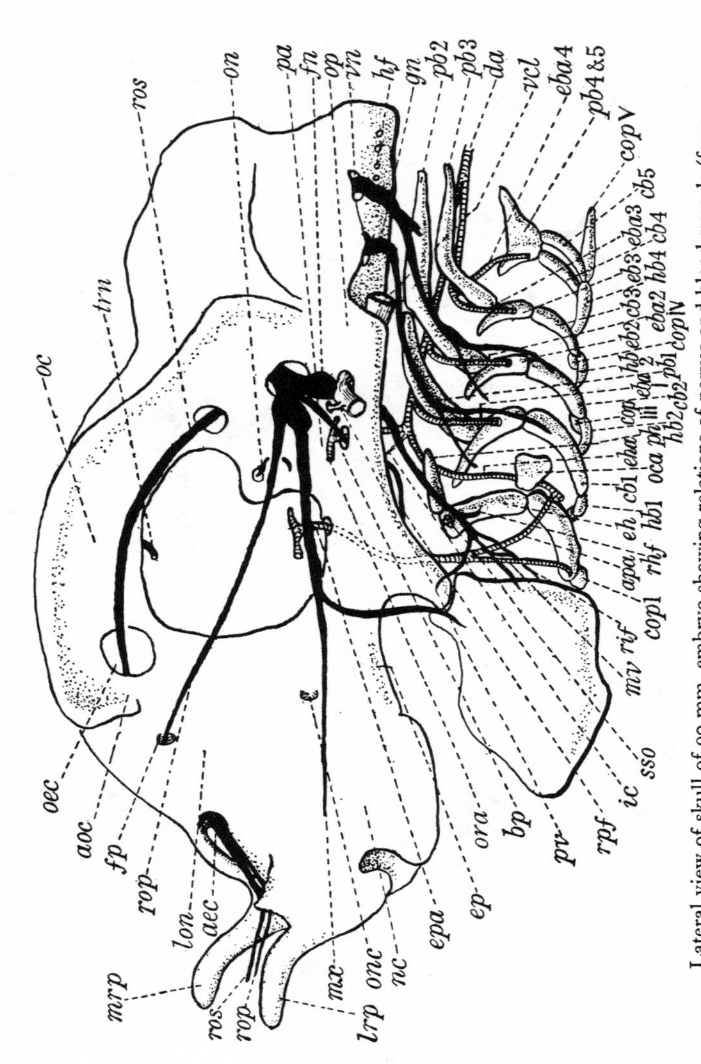

Lateral view of skull of 90 mm. embryo showing relations of nerves and blood-vessels (from de Beer and Moy-Thomas).

Explanation of Lettering precedes Plate 9

CALLORHYNCHUS

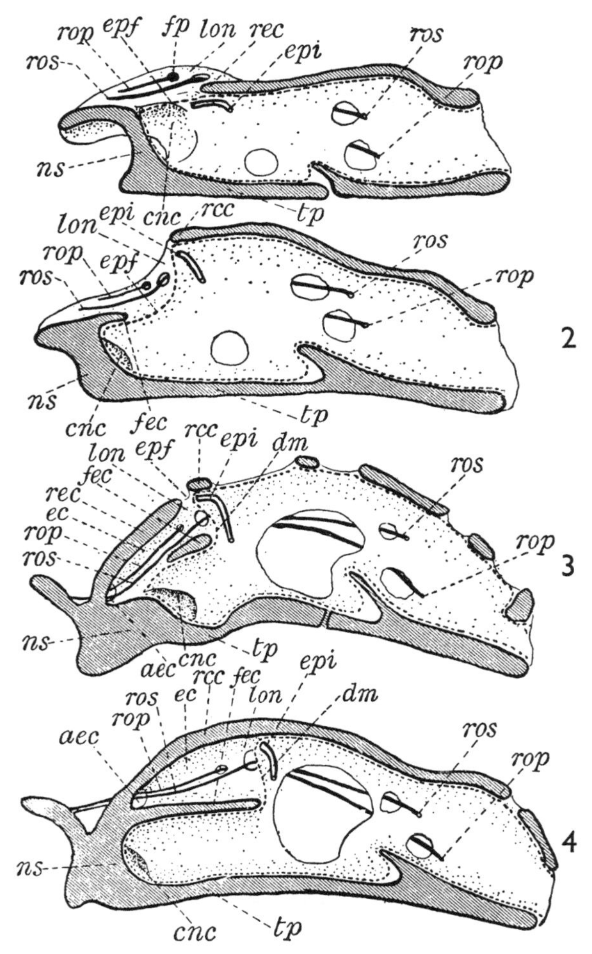

Diagrams showing transition from Selachian condition (**1**) through hypothetical stage (**2**) to early Callorhynchus (**3**) and adult Holocephalian condition (**4**) (from de Beer and Moy-Thomas).

Explanation of Lettering precedes Plate 9

OSTEICHTHYES

Explanation of Lettering

abc, anterior basicapsular commissure
abn, abducens nerve
ac, auditory capsule
aca, anterior cerebral artery
acl, independent cartilage forming lateral wall of auditory capsule
acr, arteria centralis retinae
amy, anterior myodome
an, abducens nerve
anb, adnasal bone
ang, angular bone
ant, antorbital bone
aoc, aortic canal
aof, foramen olfactorium advehens
apa, autopalatine bone
apc, anterior parachordal cartilage
ar, internal rectus muscle
arc, anterior rudiment of auditory capsule
art, articular bone
as, auditory sac
asc, anterior semicircular canal
asp, ascending process
atfc, anterior opening of trigeminofacialis chamber

ba 1, 1st branchial arch
bab, bulla containing air-bladder diverticulum
bao, basioccipital bone
bb, basibranchial cartilage
bc, barbel cartilages
bcf, basicapsular fenestra
bd, basidorsal cartilage
bf, basicranial fenestra
bfa, buccal branch of facial nerve
bh, basihyal cartilage
bp, basal process of pterygoquadrate
bsc, cartilaginous process in region of basisphenoid
bsp, basisphenoid bone
bsr, branchiostegal rays
btp, basitrabecular process
bv, basiventral cartilage

c, claustrum
cb 1–5, ceratobranchial cartilage 1 to 5
cbb, ceratobranchial bone
cg, ciliary ganglion
ch, ceratohyal cartilage
chb, ceratohyal bone
cmr, cartilages in roof of mouth
cn, ciliary nerve
com, coronomeckelian
con, cavum orbitonasale
cop 1, 2, copula 1, 2
cor, coronoid bone
cp, coronoid process of Meckel's cartilage
cr, cranial rib
crp, crista parotica
csp, posterior process of auditory capsule
ct, cornu trabeculae

da, dorsal aorta
den, dentary bone
dpa, dermopalatine bone
dpe, dorsal posterior extension of parachordal cartilage
ds, depression lodging saccule
dso, dermal ossification of supraoccipital

e, eye
eb 1–5, epibranchial cartilage, 1 to 5
eba 1, efferent branchial artery, 1st
ebb, epibranchial bone
ec, epiphysial cartilage
ecc, epichordal commissure
ecp, ectopterygoid bone
ect, lamina orbitonasalis
egc, cartilage supporting external gill
eh, epihyal cartilage
eha, efferent hyoidean artery
ehb, epihyal bone
ema, efferent mandibular artery
enp, endopterygoid bone
ent, entoglossum
eof, foramen olfactorium evehens
ep, ethmoid plate
epa, ethmopalatine articulation
epb, epiphysial bar
epo, epiotic bone
epp, ethmopalatine process
eps, efferent pseudobranchial artery
er, external rectus muscle
exo, exoccipital bone

f, facial nerve
fab, foramen for abducens nerve
fe, foramen epiphaniale
fep, foramen for efferent pseudobranchial artery
fes, epiphysial foramen
ff, foramen for facial nerve
fh, foramen in hyomandibula for facial nerve
fhv, foramen for head-vein
fhy, foramen (foramina) for hypoglossal nerve
fm, foramen magnum
fme, fissura metotica
fna, nasal fossa
fnr, fenestra narina
fo, foramen for ophthalmic nerve
foa, foramen olfactorium advehens
foe, foramen olfactorium evehens
fot, foramen for otic branch of facial nerve
fpo, combined foramen for profundus and oculomotor nerves
fpr fenestra precerebralis
fpv, foramen for pituitary vein
fro, frontal bone
fsn, foramen for 1st spinal nerve
ft, foramen for trochlear nerve
ftr, foramen for trigeminal nerve
fv, foramen for vein

g, glossopharyngeal nerve
gf, foramen for glossopharyngeal nerve
gfs, foramen for supratemporal branch of glossopharyngeal nerve

hb 1–4, hypobranchial cartilage 1 to 4
hc, hypochordal commissure
hf, hypophysial fenestra
hfa, hyomandibular branch of facial nerve
hh, hypohyal cartilage
hm, hyomandibular or hyosymplectic cartilage
hmf, hyomandibular branch of facial nerve
hn, hypoglossal nerve
hr, rudiment of hyosymplectic cartilage
hs, hyosymplectic cartilage
hv, head-vein
hyf, facet for head of hyomandibula
hyo, hyomandibular bone

ic, internal carotid artery
icf, foramen for internal carotid artery
icp, incisura prootica
id, interdorsal cartilage
ig, rudimentary occipital dorsal root ganglion
infr, inferior rectus muscle
inr, internal rectus muscle
ins, nasal septum
int, intercalary (opisthotic) bone
io, iob, inferior oblique muscle
iop, interopercular bone
ios, interorbital septum
ip, notch for palatine nerve
ipb, infrapharyngobranchial cartilage
ir, inferior rectus muscle
it, intercalarium
itt, intertemporal bone

jf, jugular (vagus) foramen
ji, jugular incisure
jug, jugal bone

la, lachrymal bone
lao, lateral dorsal aorta
lbc, lca, labial cartilage
let, lateral ethmoid bone
llf, foramina for branches of superficial ophthalmic nerve
lon, lamina orbitonasalis
lsc, lateral semicircular canal

max, maxillary bone
max & buc, maxillary branch of trigeminal & buccal branch of facial

Mc, Meckel's cartilage
Mcr, rudiment of Meckel's cartilage
md, mediomeckelian bone
mes, mesethmoid bone
mm, mentomeckelian bone
mpt, metapterygoid bone
mw, median wall of nasal capsule
mxt, maxillary branch of trigeminal nerve
my, posterior myodome

n, notochord
na, neural arch
na 3, neural arch of 3rd vertebra
nas, nasal bone
nav, neural arch of 1st vertebra
nh, notch in young hyomandibula for facial nerve
nhy, notch in old hyomandibula for facial nerve
nic, notch for internal carotid artery
nl, notch for lateral line canal
nol, notch for olfactory nerve
ns 1, 2, 3, spinal nerve 1, 2, 3

oa, occipital arch
oat, ophthalmic artery
ob, occipital bone
oc, oculomotor nerve
ocf, foramen for oculomotor nerve
ocp, copular bone
of, foramen for olfactory nerve
ol, olfactory sac
om, arteria ophthalmica magna
omf, foramen for ophthalmica magna artery
omy, opening of posterior myodome
on, olfactory nerve
ona, orbitonasal artery
onc, orbitonasal canal (for vein)
onf, orbitonasal foramen (for nerve)
oo, opercular ossicle
op, otic process of pterygoquadrate cartilage
opc, opercular cartilage
opf, foramen for optic nerve
opl, opercular bone
opi, pharyngeum inferius bone
opn, optic nerve
ops, pharyngeum superius bone
ora, orbital artery
orb, orbital bone
orc, orbital cartilage
orf, foramen for orbital artery
ors, orbitosphenoid bone
os, nasal sac
osa 1, 2, occipitospinal arch 1, 2
osc, superior occipital cartilage
osi, inferior occipital cartilage

pa, pila antotica (prootica)
paa, proatlas neural arch
pab, paraphysial bar
pal, palatine bone
par, parietal bone
pas, secondary pila antotica
pat, prearticular bone
pb 1, 3, pharyngobranchial cartilage 1 to 3
pbc, posterior basicapsular commissure
pc, parachordal cartilage
pca, posterior cerebral artery
pet, pre-ethmoid bone
pf, palatine branch of facial nerve
pft, prefrontal bone
pg, pterygoid bone
pgb, bony pectoral girdle
pgc, cartilaginous pectoral girdle
ph, pharyngohyal cartilage
pl, pila lateralis
plf, foramen for palatine branch of facial nerve
plm, pila metoptica
pls, pleurosphenoid bone
pm 1, 2, permanent myomere 1, 2
pmx, premaxillary bone
po, postorbital process of auditory capsule
poc, preoptic root of orbital cartilage
pof, preoptic fontanelle
pol, polar cartilage
pop, preopercular bone
por, postorbital bone
pp, palato-pterygoid bone
ppa, postparietal or extrascapular bone
ppp, postpalatine process or commissure
pq, pterygoquadrate cartilage
prb, prootic bridge
prc, posterior rudiment of auditory capsule
prf, prootic foramen
pro, prootic bone
prp, prootic process of auditory capsule
prs, postrostral bone
ps, pseudobranch
psc, posterior semicircular canal
psf, postfrontal (dermosphenotic) bone
psp, parasphenoid bone
pst, post-temporal bone
pt, profundus nerve
ptfc, posterior opening of trigeminofacialis chamber
ptp, pterygoid process of quadrate cartilage
ptr, pterotic bone
pv, pituitary vein
pvf, foramen for pituitary vein
pvo, prevomer

qb, quadrate bone
qr, rudiment of quadrate cartilage
qu, quadrate cartilage
quc, portion of quadrate bone not preformed in cartilage

r, rostrum
ra, retroarticular bone
rac, roof of auditory capsule
rc, rostral cartilage
rds, rudiment of dorsum sellae
rhf, hyoidean branch of facial nerve
rmf, mandibular branch of facial nerve
rmt, mandibular branch of trigeminal nerve
rof, superficial ophthalmic branch of facial nerve
rop, profundus nerve
ros, superficial ophthalmic nerve
rot, superficial ophthalmic branch of trigeminal nerve
rp, retroarticular process of Meckel's cartilage
rpa, rostropalatine articulation
rpc, rudiment of parachordal cartilage
rt, rudiment of trabecula cranii

s, scaphium
sa, anterior coronoid bone
san, supra-angular bone
sap, posterior coronoid bone
sb, sphenoid bone
sca, supraorbital cartilage
scl, supracleithrum bone
sec, sphenethmoid commissure
set, supraethmoid bone
sf, sphenoid fissure
sg 1, ganglion of first mixed nerve
sh, interhyal (stylohyal) cartilage
sm, supramaxillary bone
sn, nasal septum
so, sob, superior oblique muscle
soc, supraoccipital bone
sof, supraorbital fontanelle
sop, subopercular bone
sor, suborbital bone
sp, septum of posterior semicircular canal
spb, suprapharyngobranchial cartilage
spc, spiracular canal
spf, sphenoid fontanelle
sph, sphenotic bone
spo, supraorbital bone
sps, subpituitary (myodomic) space
sr, superior rectus muscle
ss, spiracular ossicles
ssc, sphenoseptal commissure
sut, supratemporal bone
sy, symplectic cartilage or process
syb, symplectic bone

t, tectum cranii
tb, terminal bone
tc, trabecula cranii
tcom, trabecula communis
tcr, trabecula cranii
tm, orbital cartilage or taenia marginalis
tm 1, 1st transient myomere
tma, taenia tecti medialis anterior
tn, trigeminal nerve
tp, tectum synoticum
tpl, trabecular plate
tpp, tectum posterius
tr, trabecula cranii
trn, trochlear nerve
trp, tripus
ttm, taenia tecti medialis

v, vagus nerve
var, rudiments of ceratobranchials
vcl, head-vein
vdf, foramen for dorsal branch of vagus nerve
vf, foramen for vagus nerve
vn, vagus nerve
vpe, ventral posterior extension of parachordal cartilage
vr 7, &c., ventral nerve-root of 7th segment, &c.
vwc, vacuity in wall of auditory capsule

POLYPTERUS

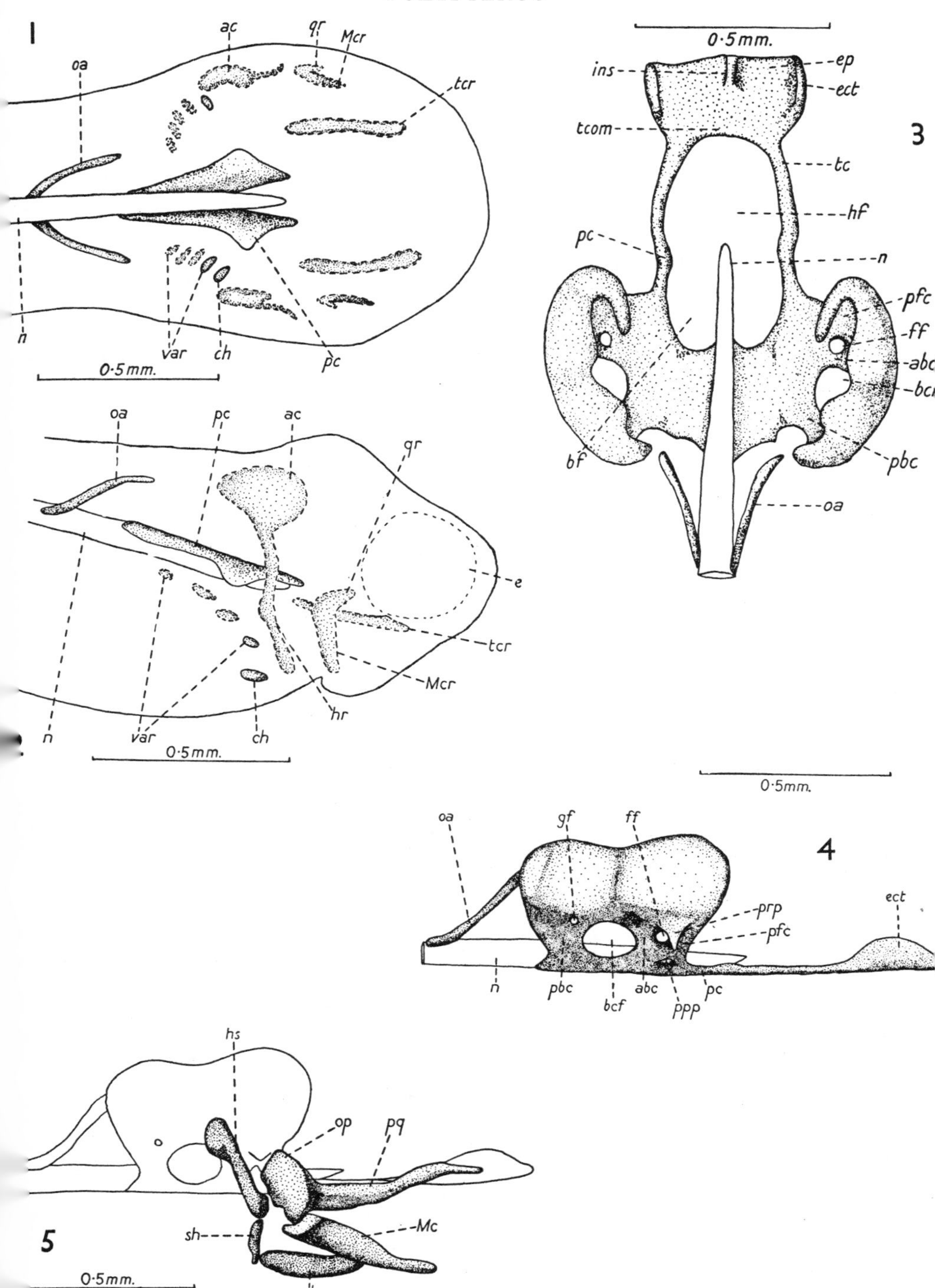

1, dorsal, **2**, lateral view of 6·75 mm. stage; **3**, dorsal, **4**, lateral view of neurocranium; and **5**, lateral view of visceral arches of 8 mm. stage (from Moy-Thomas).

Explanation of Lettering opposite

POLYPTERUS

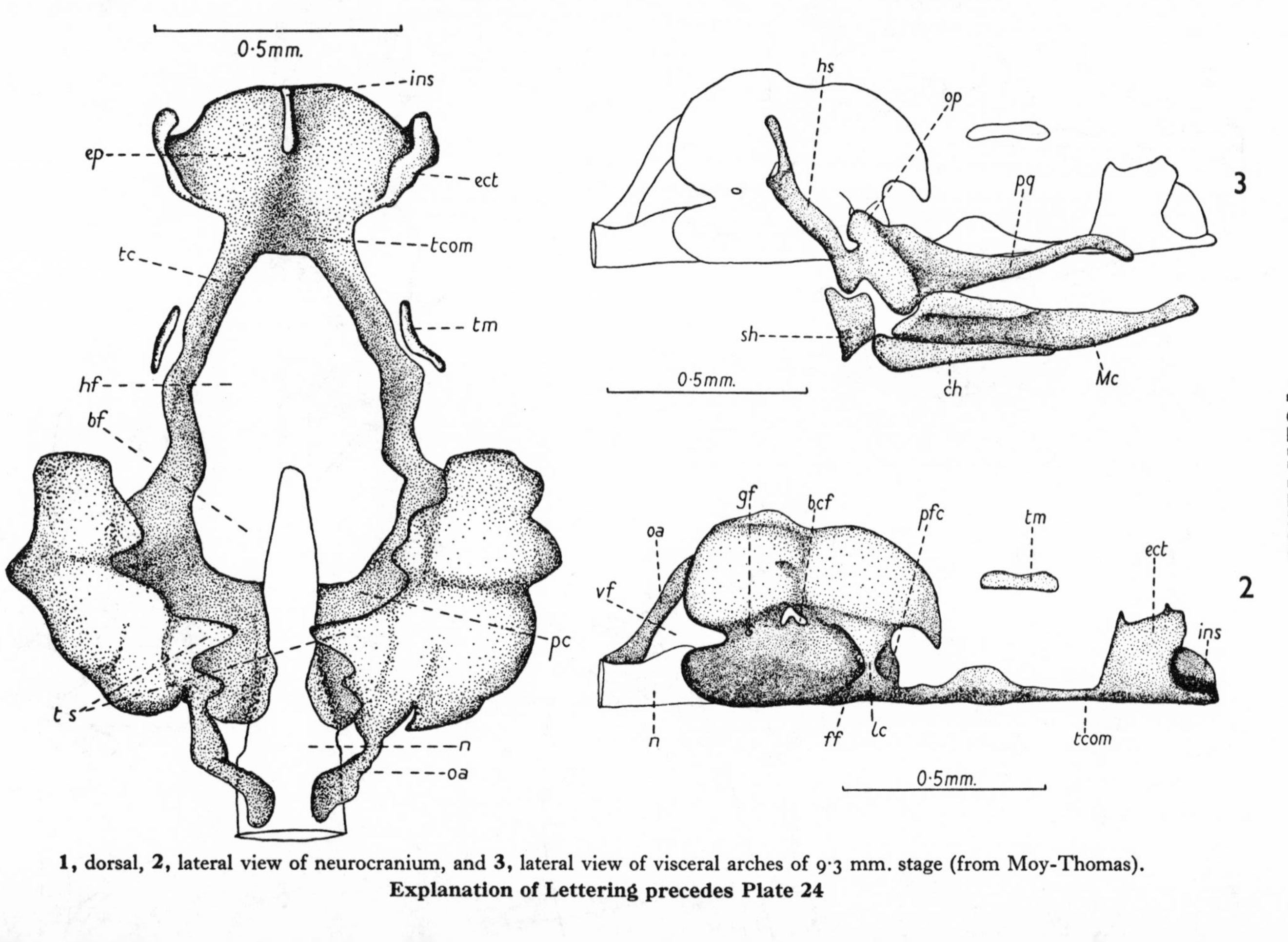

1, dorsal, 2, lateral view of neurocranium, and 3, lateral view of visceral arches of 9·3 mm. stage (from Moy-Thomas).
Explanation of Lettering precedes Plate 24

POLYPTERUS

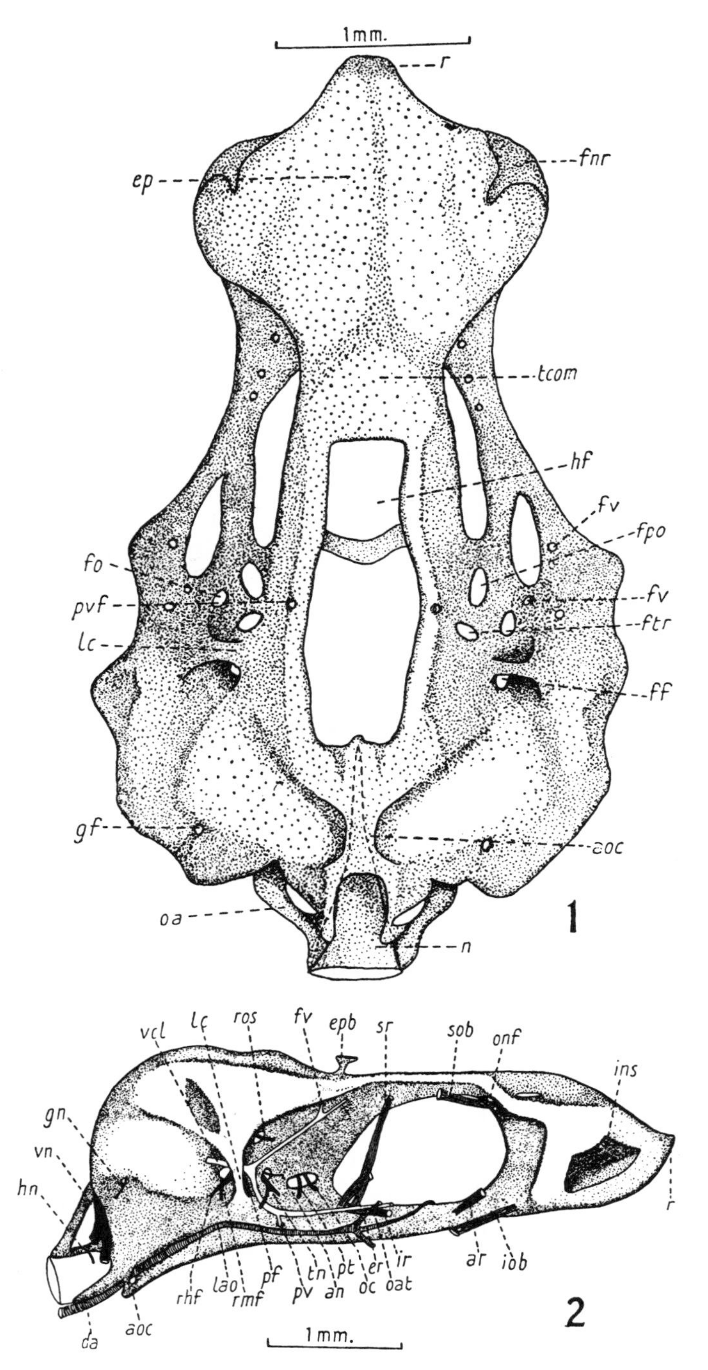

1, ventral, **2**, lateral view of neurocranium of 30 mm. stage (from Moy-Thomas).

Explanation of Lettering precedes Plate 24

POLYPTERUS

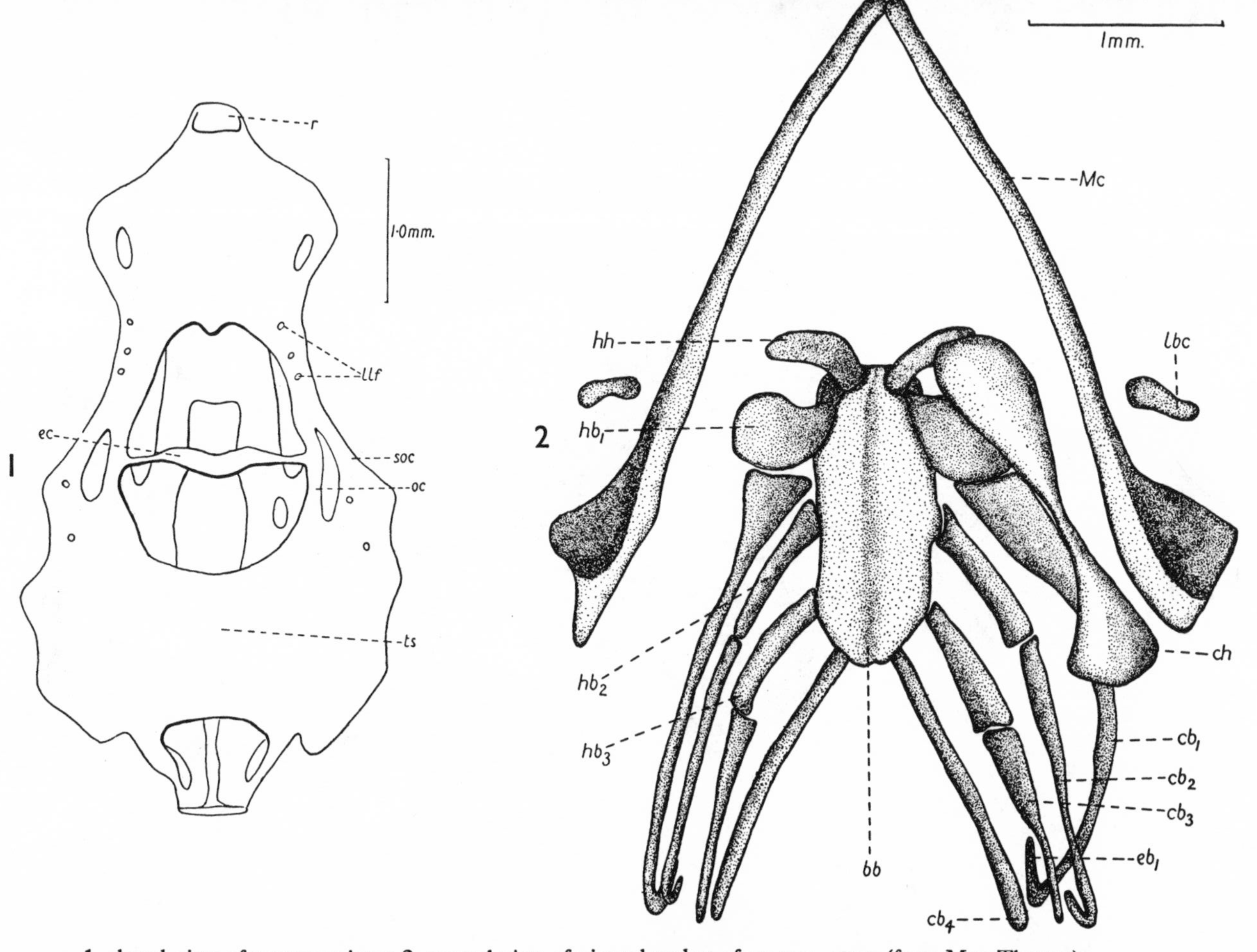

1, dorsal view of neurocranium; **2**, ventral view of visceral arches of 30 mm. stage (from Moy-Thomas).

Explanation of Lettering precedes Plate 24

POLYPTERUS

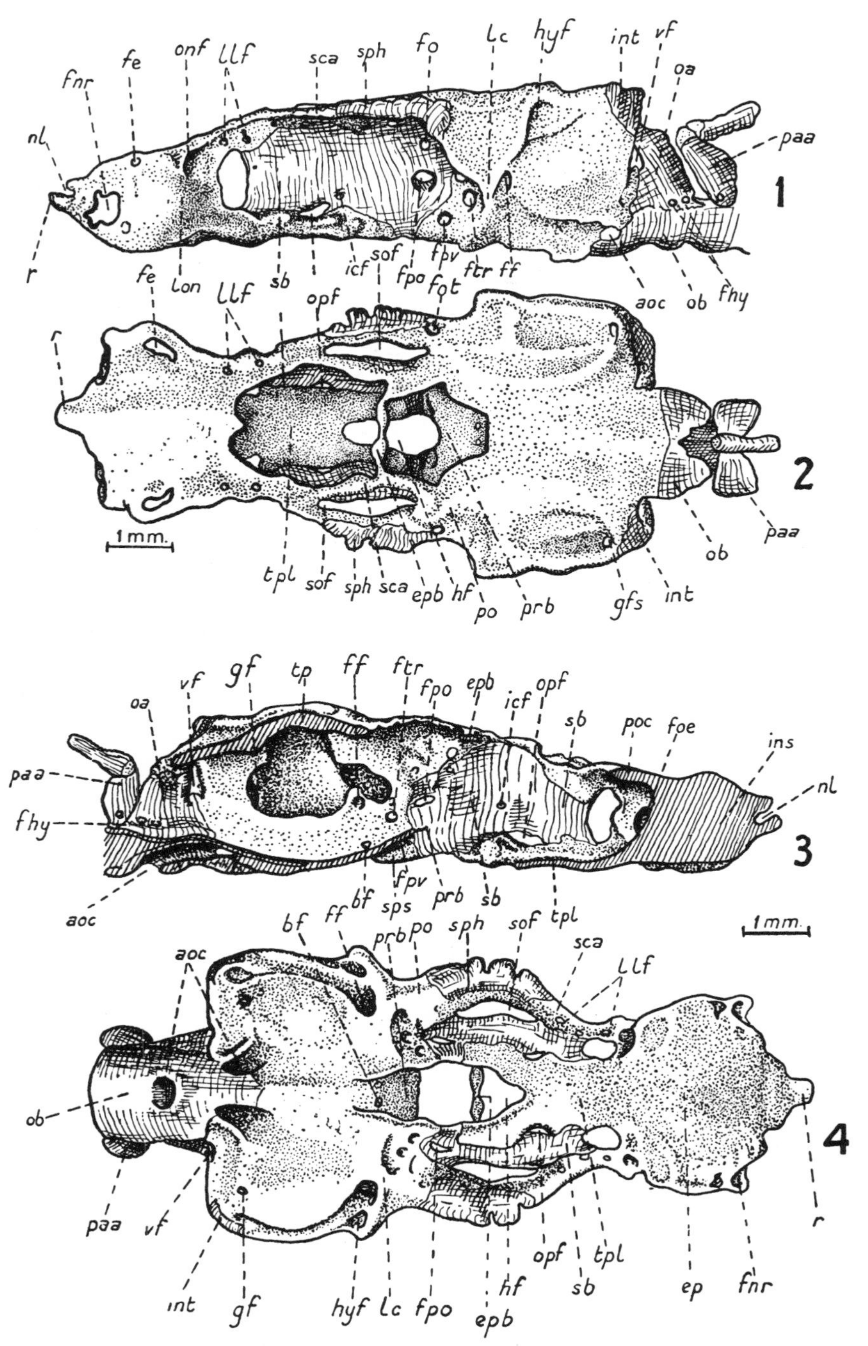

1, lateral, **2**, dorsal, **3**, medial, and **4**, ventral view of neurocranium of 76 mm. stage (after Lehn).

Explanation of Lettering precedes Plate 24

PLATE 29

POLYPTERUS

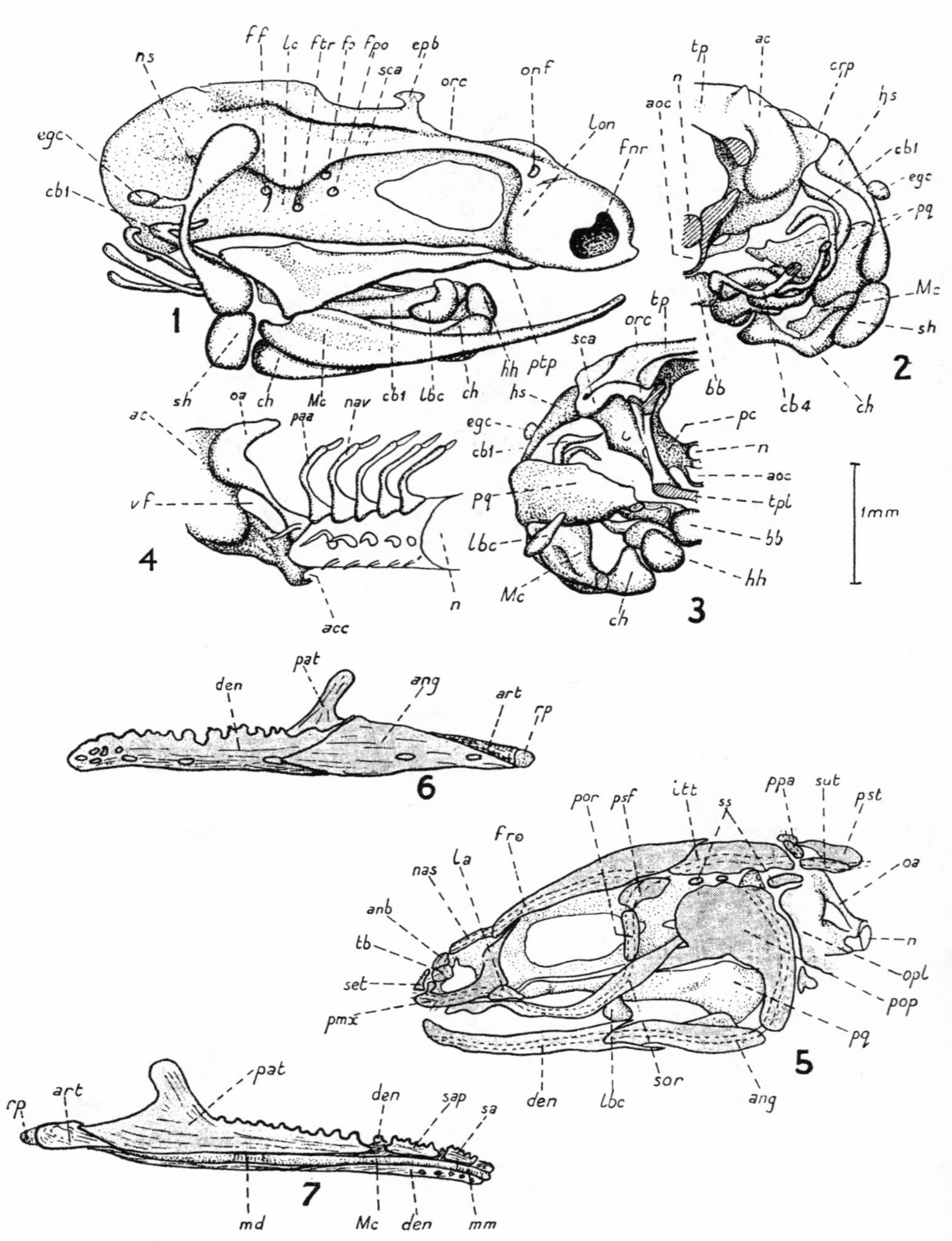

1, lateral, **2**, posterior, **3**, anterior view of skull; **4**, occipital region of 30 mm. stage (after Budgett); dermal bones of 35 mm. stage (after Pehrson); **6**, lateral, and **7**, medial view of lower jaw (after Schmäl

Explanation of Lettering precedes Plate 24

ACIPENSER

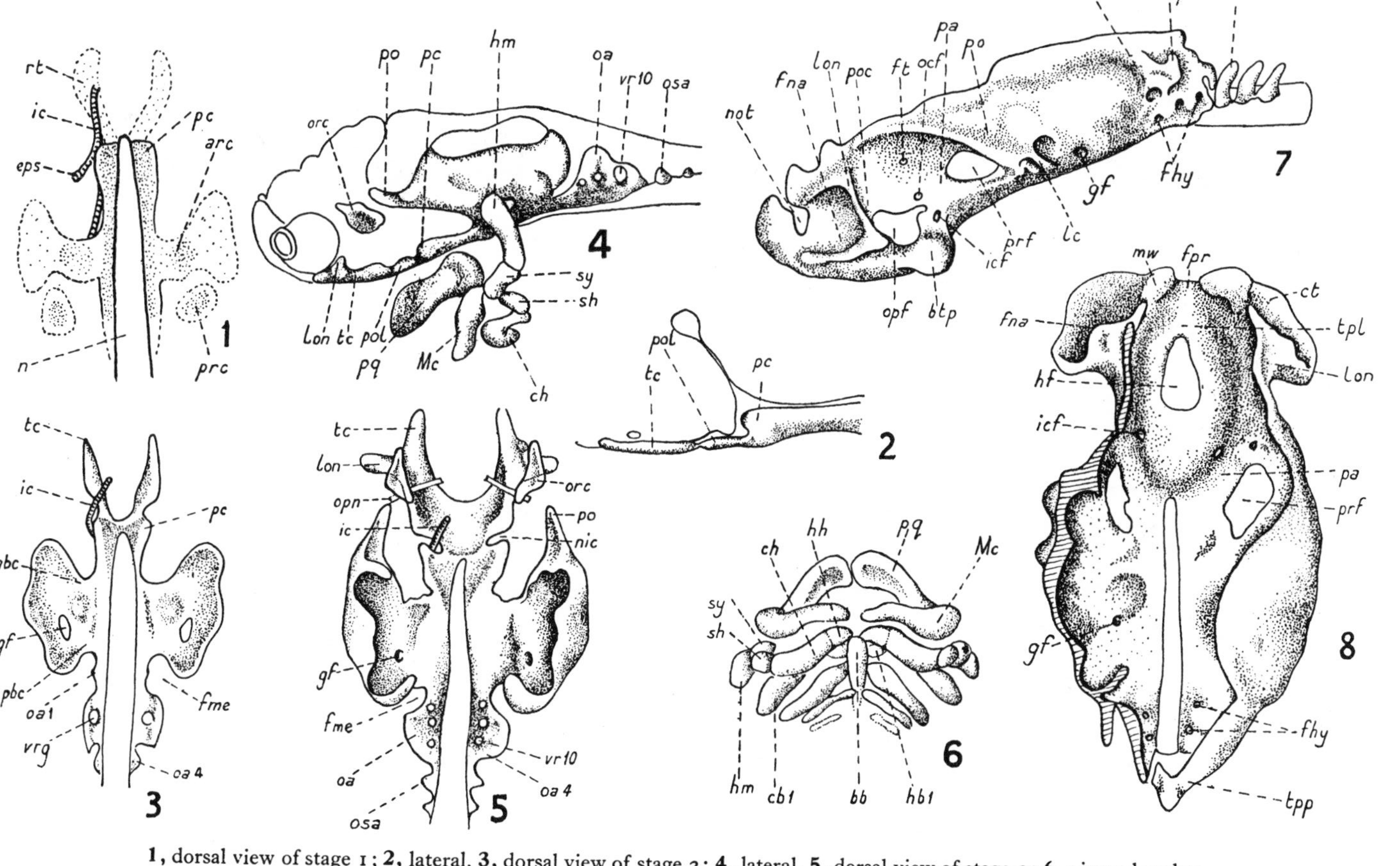

1, dorsal view of stage 1; **2**, lateral, **3**, dorsal view of stage 2; **4**, lateral, **5**, dorsal view of stage 3; **6**, visceral arches, **7**, lateral, **8**, dorsal view of stage 4 (after Sewertzoff).

Explanation of Lettering precedes Plate 24

ACIPENSER

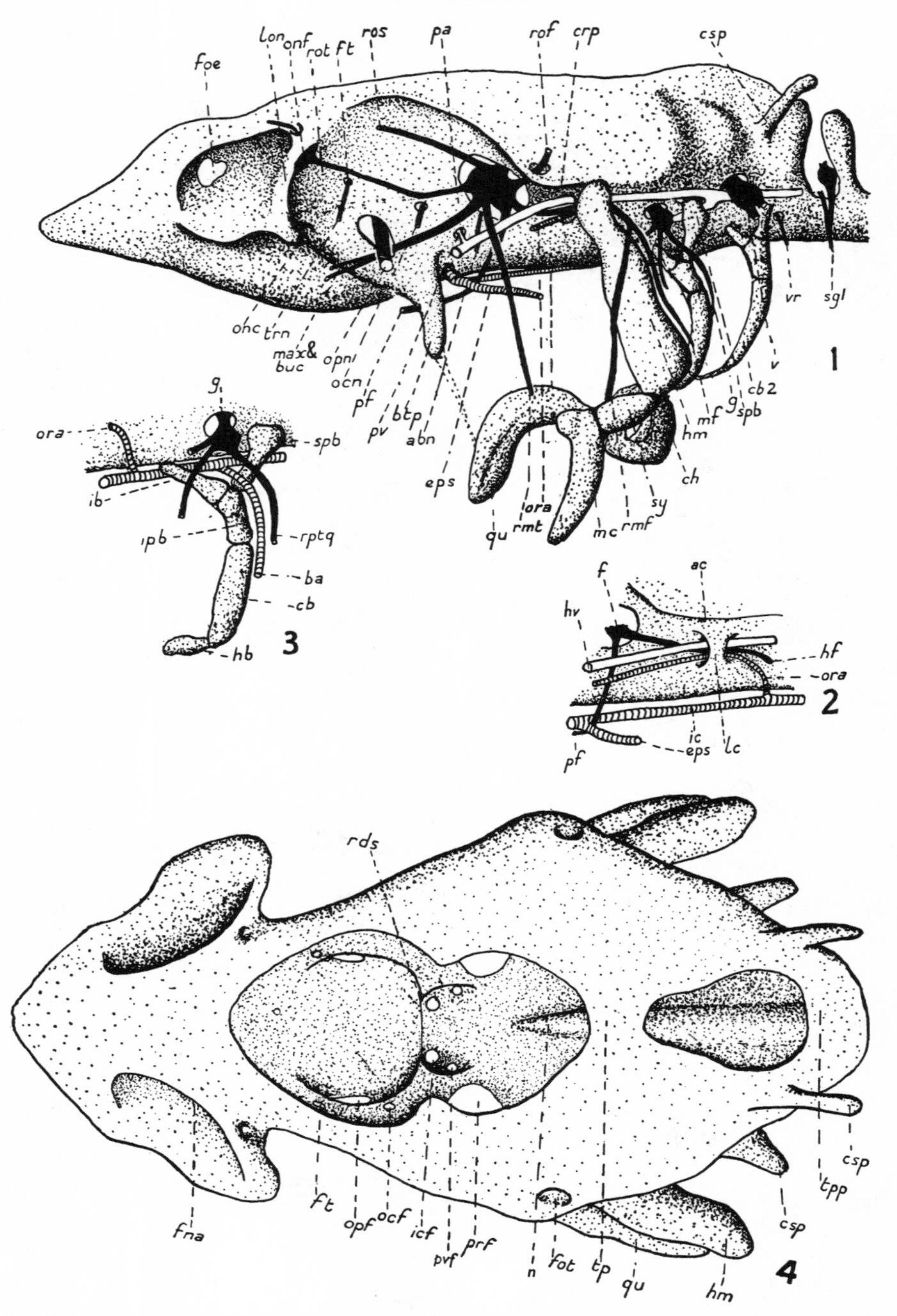

1, relations of nerves to chondrocranium (Acipenser stellatus 10-day stage); **2**, relations of jugular canal; **3**, relations of 1st branchial arch; **4**, dorsal view of chondrocranium (from de Beer).

Explanation of Lettering precedes Plate 24

ACIPENSER

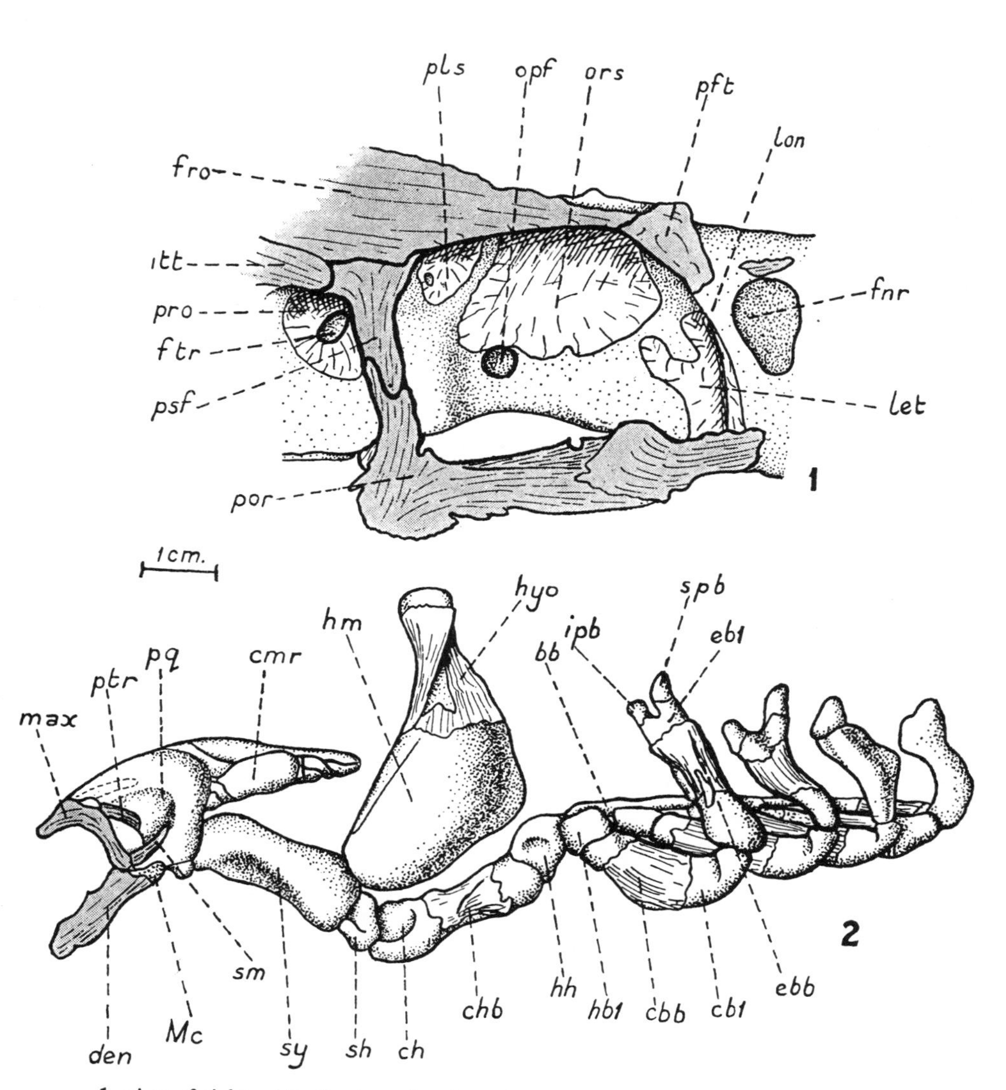

1, view of right orbit; **2,** visceral arches of young Acipenser sturio (after Parker).

Explanation of Lettering precedes Plate 24

AMIA

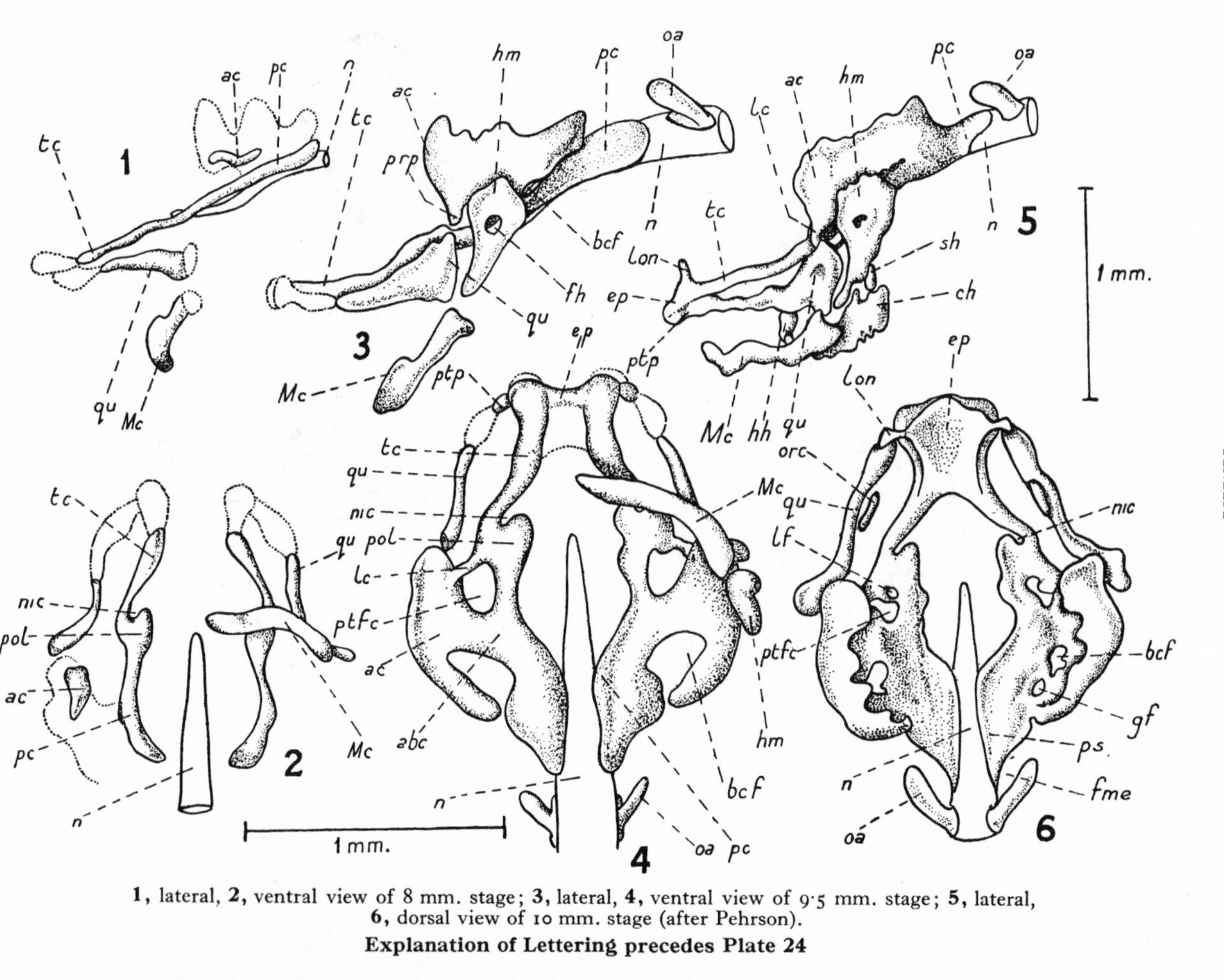

1, lateral, **2**, ventral view of 8 mm. stage; **3**, lateral, **4**, ventral view of 9·5 mm. stage; **5**, lateral, **6**, dorsal view of 10 mm. stage (after Pehrson).

Explanation of Lettering precedes Plate 24

AMIA

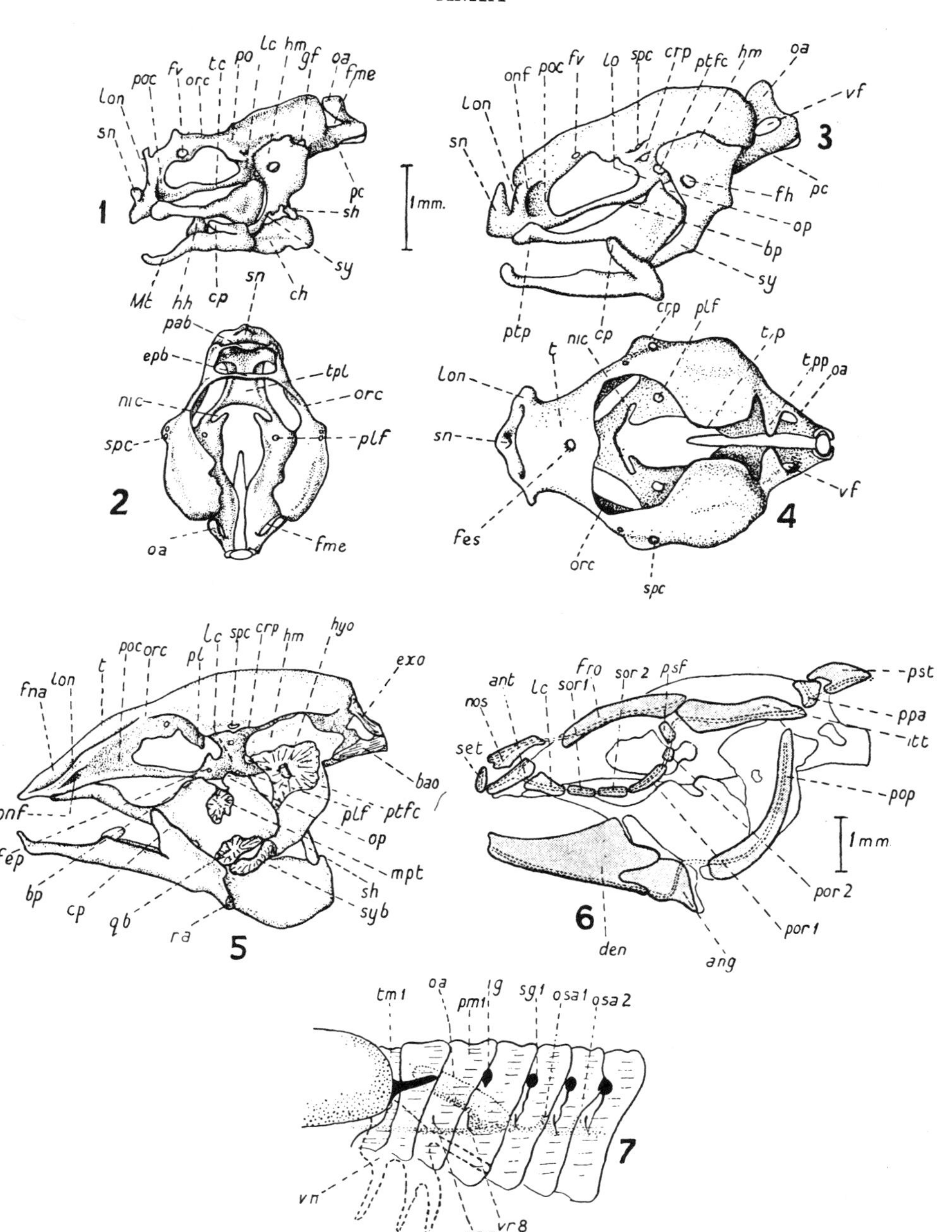

1, lateral, **2,** dorsal view of 14 mm. stage; **3,** lateral, **4,** dorsal view of 19·5 mm. stage; **5,** lateral view of 31·5 mm. stage showing cartilage-bones; **6,** membrane-bones of 31·5 mm. stage (after Pehrson); **7,** segmentation of occipital region (from de Beer).

Explanation of Lettering precedes Plate 24

AMIA

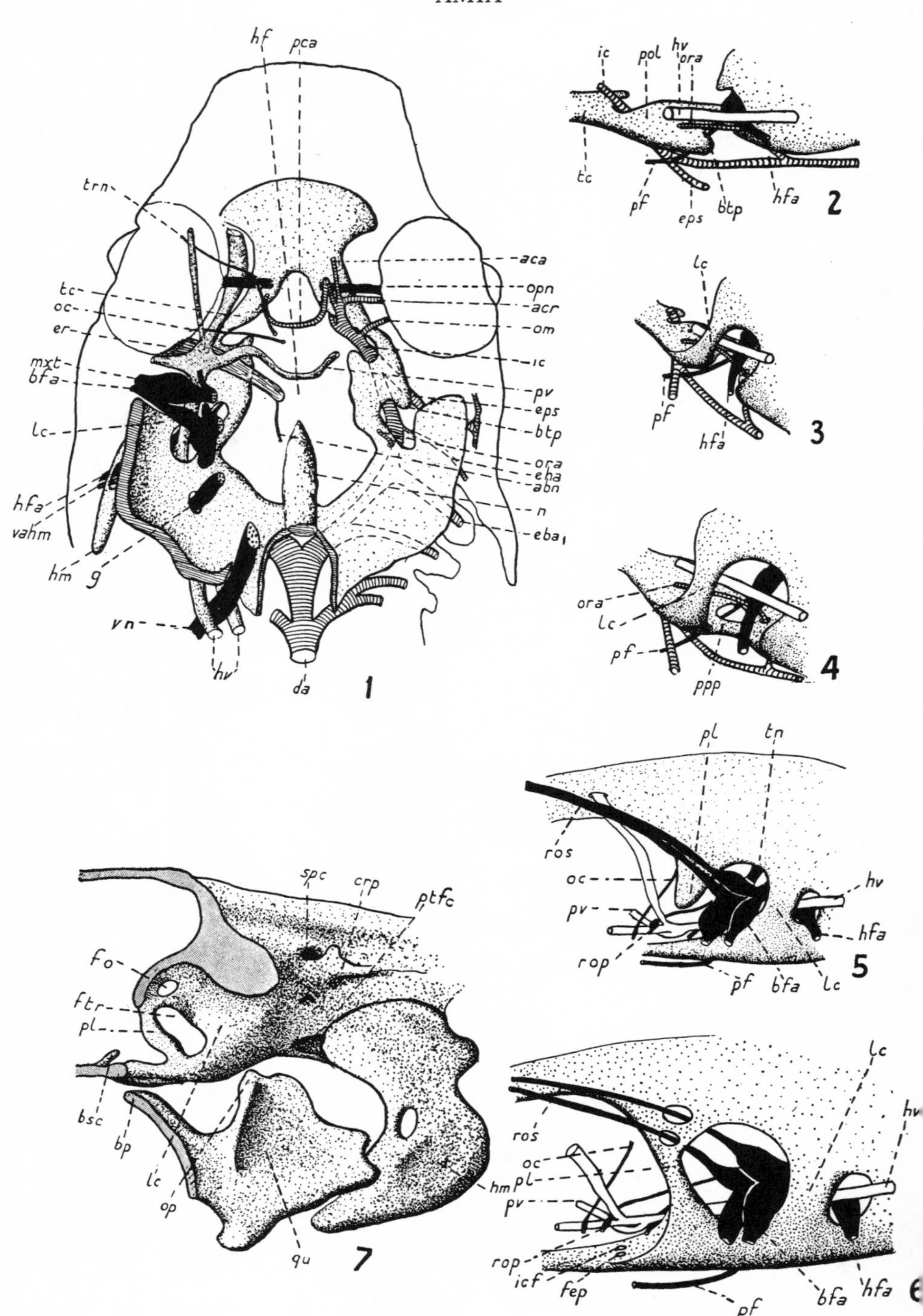

1, relations of blood-vessels and nerves to chondrocranium of 9·5 mm. stage; **2**, relations of trigeminofacialis chamber at 9·5 mm. stage; **3**, at 11 mm.; **4**, at 14 mm.; **5**, at 20 mm.; and **6**, at 50 mm. stage; **7**, trigeminofacialis chamber, pterygoquadrate, and hyomandibula at 41 mm. stage (from de Beer).

Explanation of Lettering precedes Plate 24

AMIA

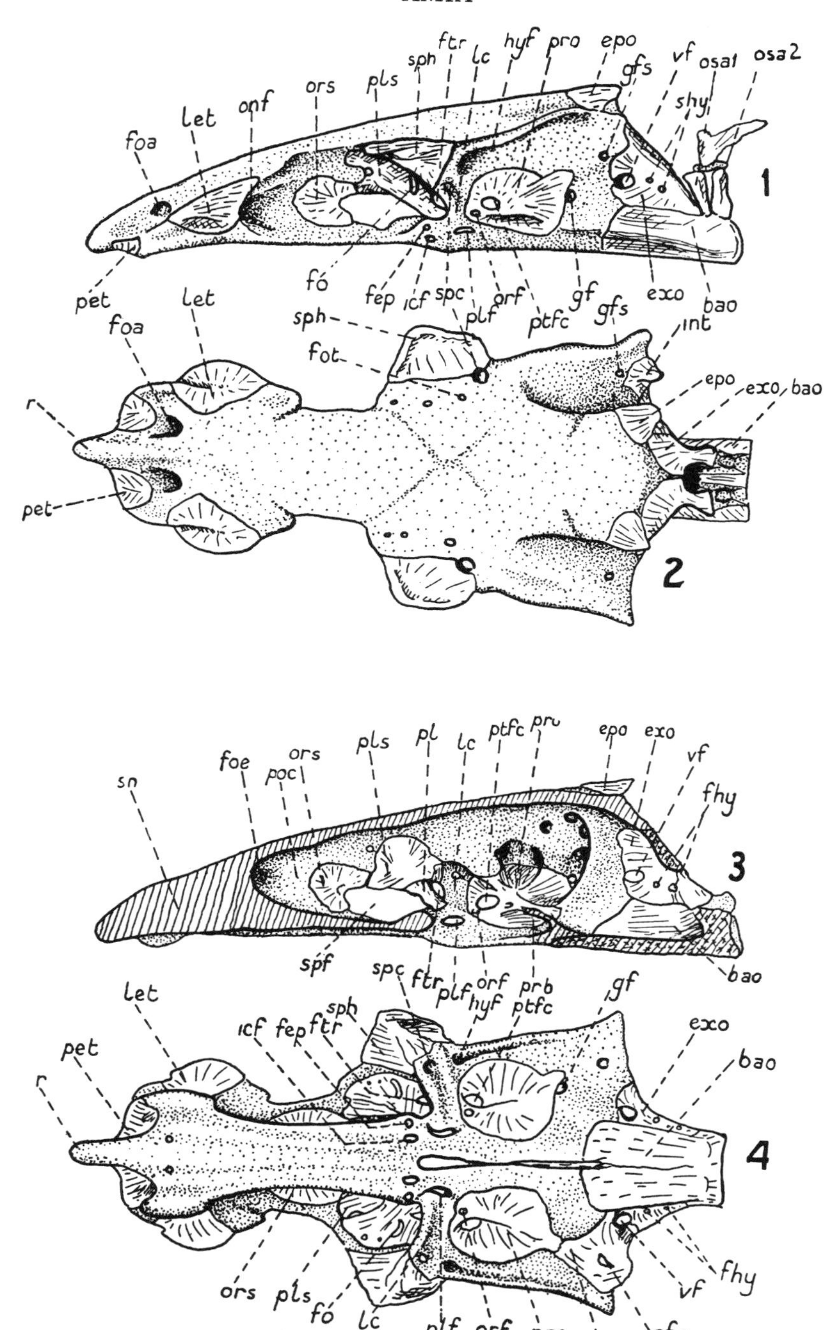

1, lateral, **2,** dorsal, **3,** medial, and **4,** ventral view of the neurocranium of adult (after Allis).
Explanation of Lettering precedes Plate 24

PLATE 37

AMIA

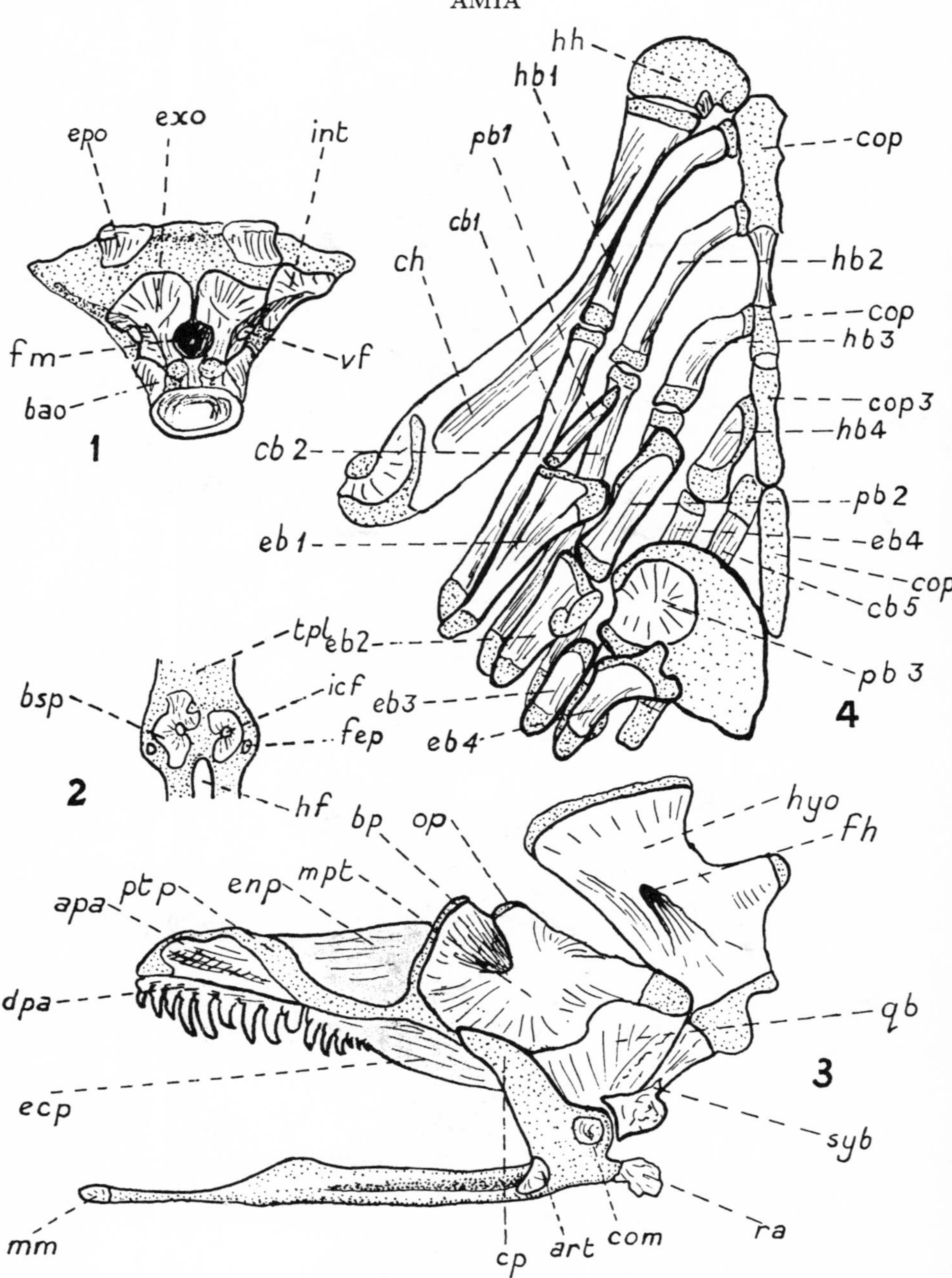

1, posterior view of neurocranium; **2**, dorsal view of basisphenoid; **3**, medial view of mandibular and hyoid arches of adult; **4**, dorsal view of visceral arches (after Allis).

Explanation of Lettering precedes Plate 24

LEPIDOSTEUS

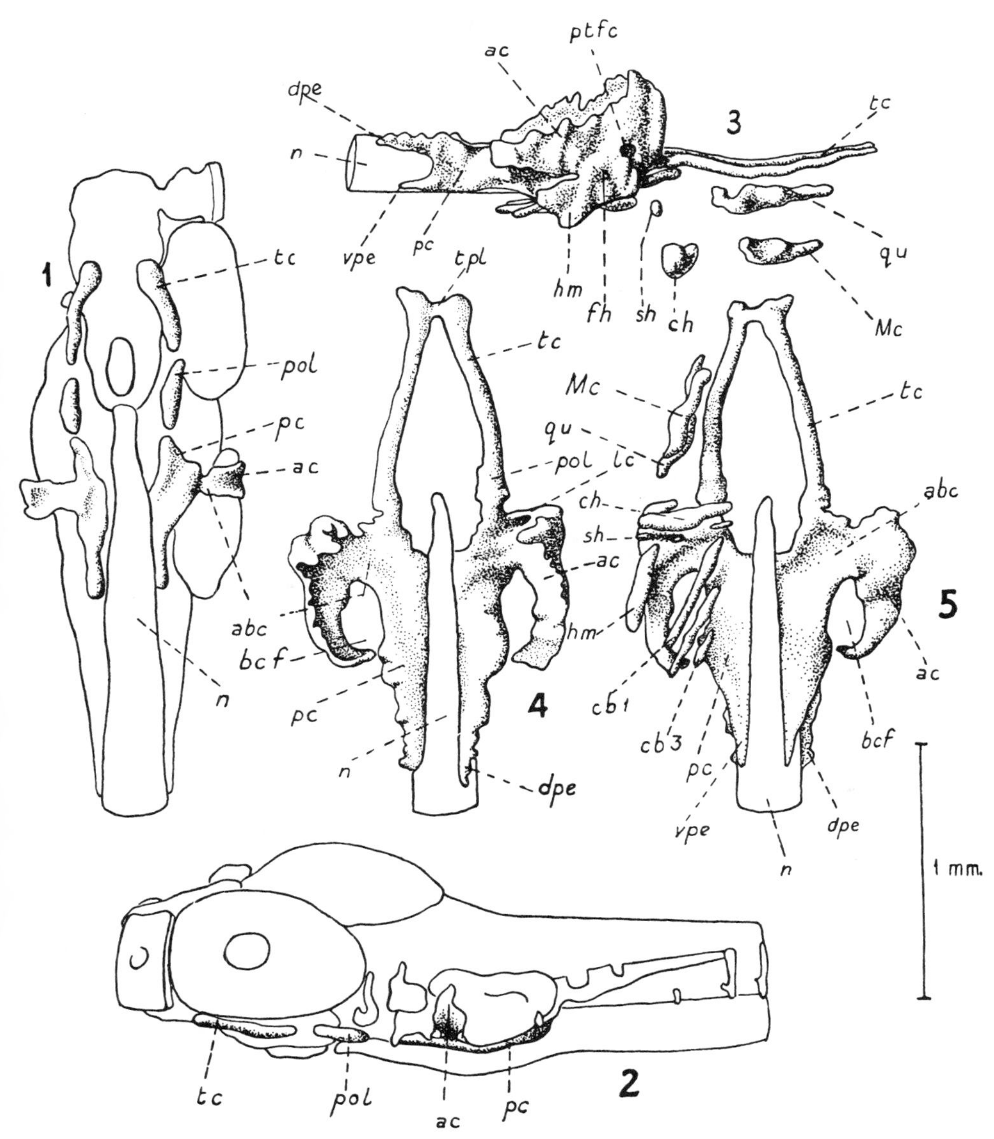

1, ventral, **2**, lateral view of 10 mm. stage; **3**, lateral, **4**, dorsal, and **5**, ventral view of 11 mm. stage (after Veit).

Explanation of Lettering precedes Plate 24

LEPIDOSTEUS

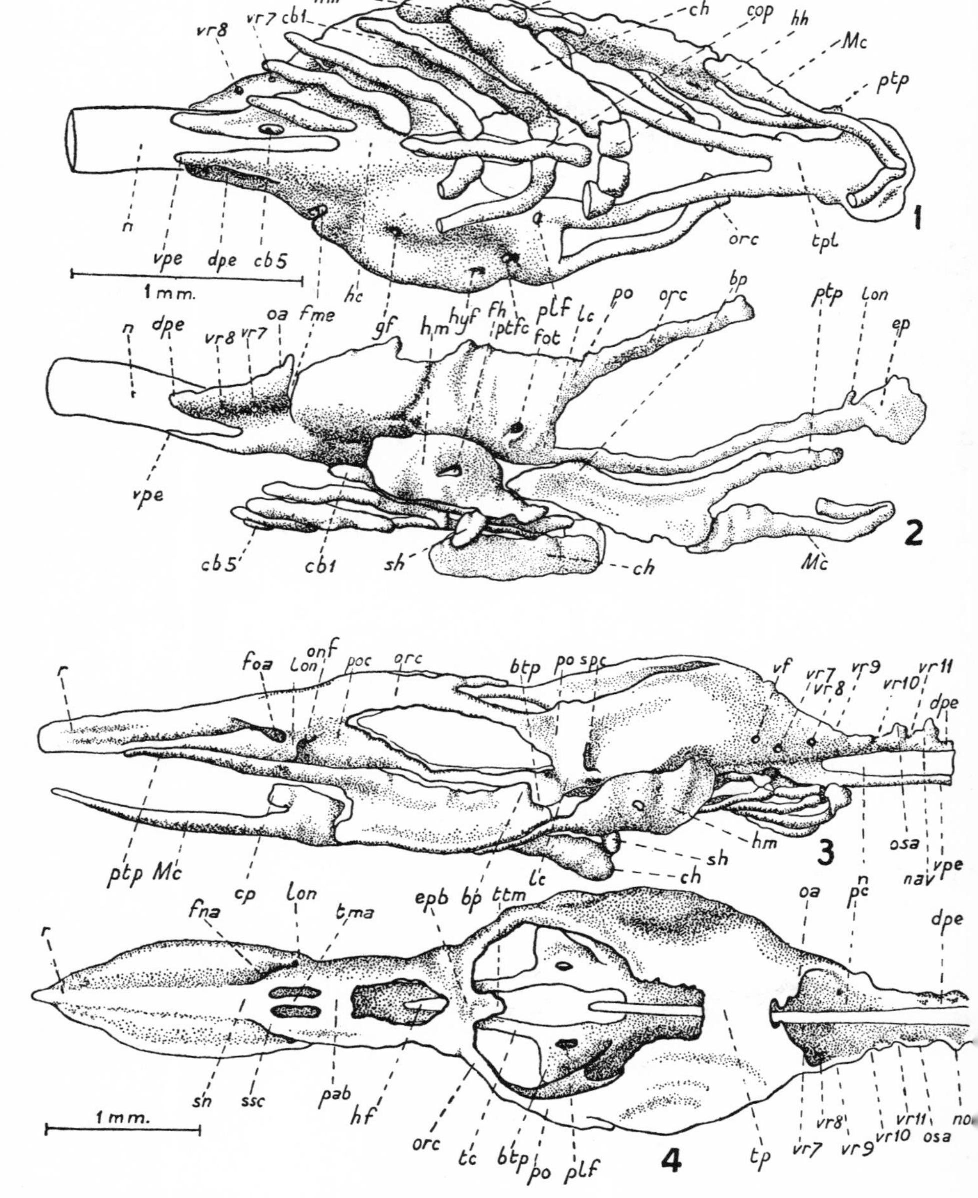

1, ventral, **2**, lateral view of 14 mm. stage; **3**, lateral, **4**, dorsal view of 20 mm. stage (after Veit).

Explanation of Lettering precedes Plate 24

LEPIDOSTEUS

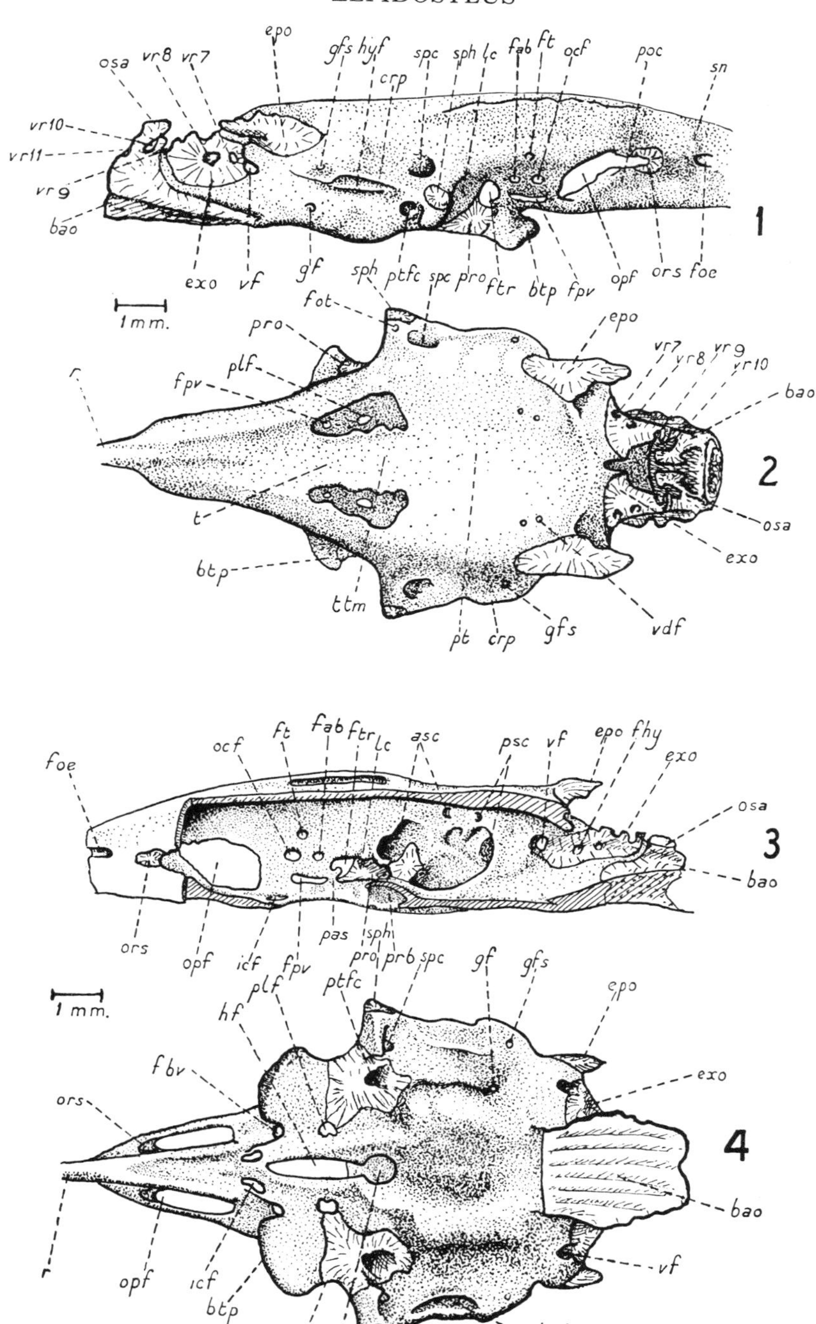

1, lateral, **2**, dorsal, **3**, medial, and **4**, ventral view of neurocranium of 15 cm. stage (after Veit).

Explanation of Lettering precedes Plate 24

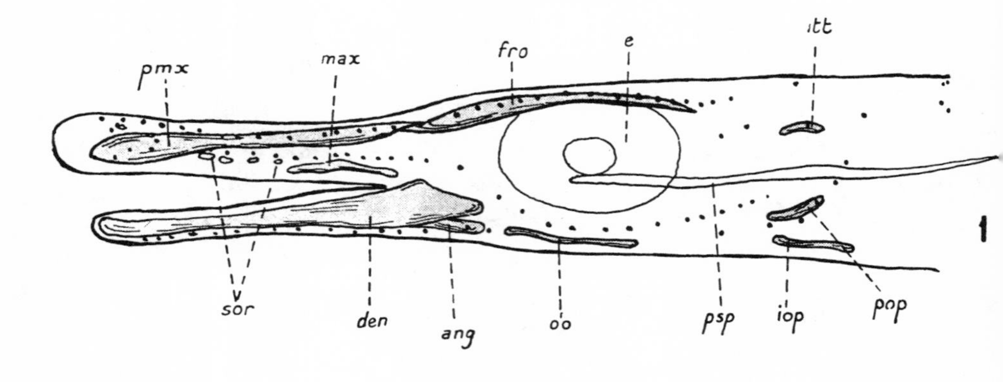

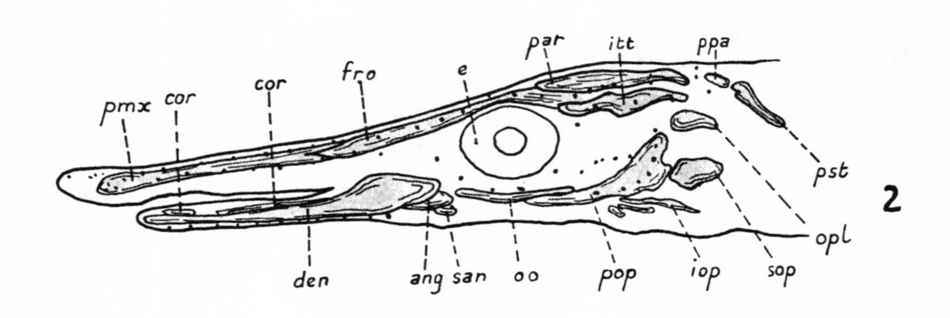

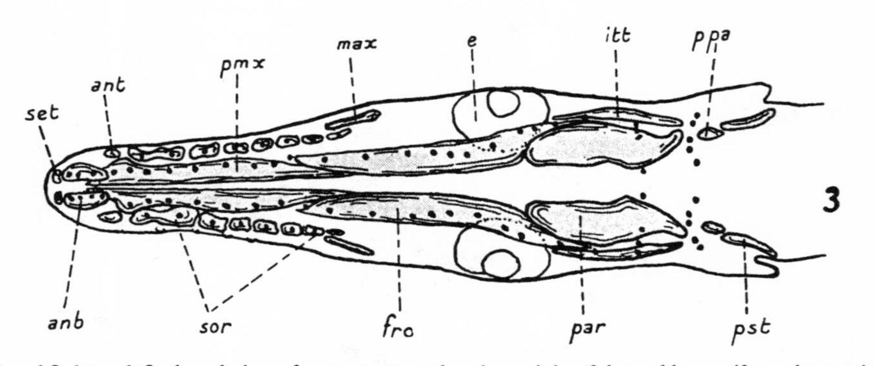

1 and **2**, lateral, **3**, dorsal view of 25 mm. stage showing origin of dermal bones (from Aumonier).

Explanation of Lettering precedes Plate 24

SALMO

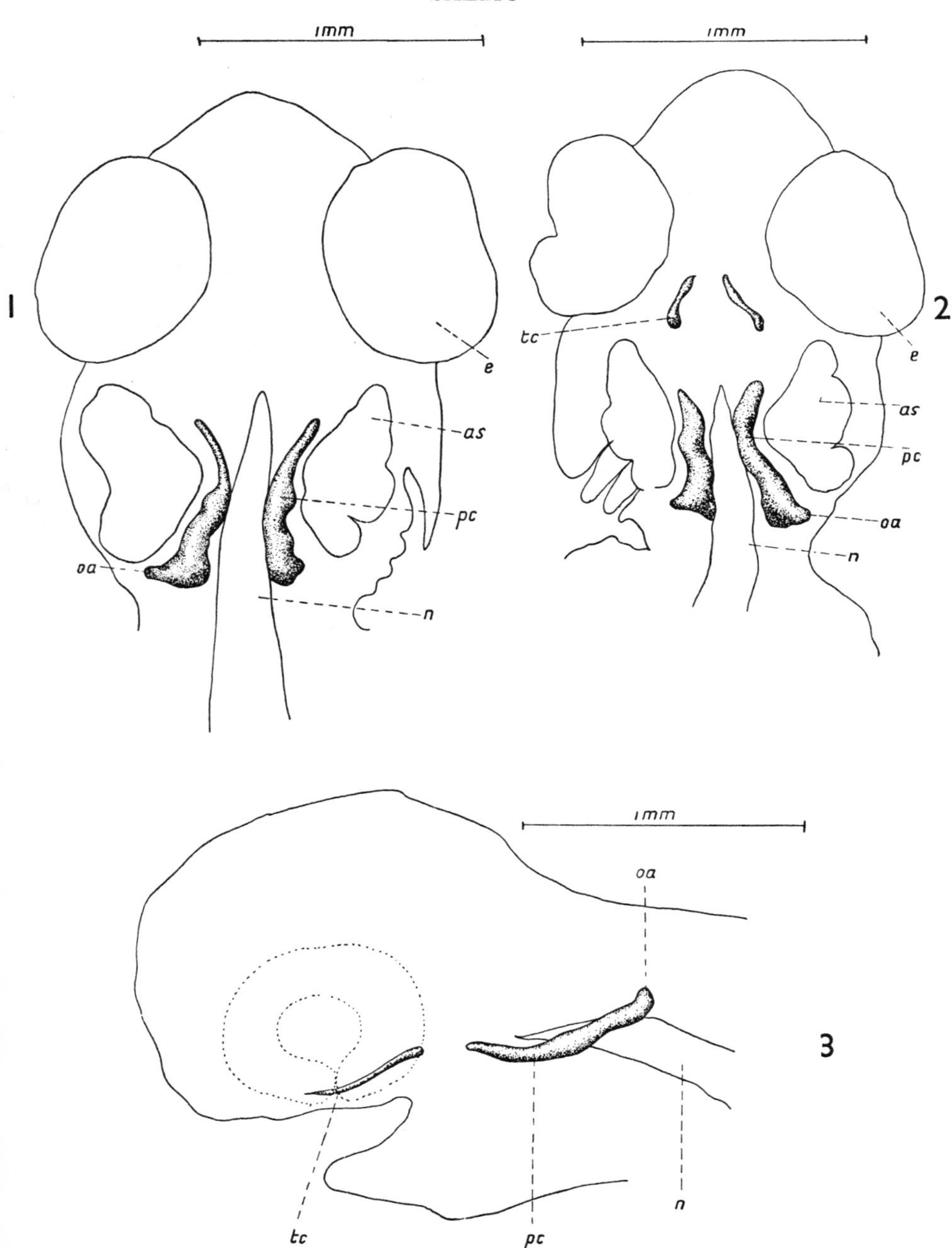

1, dorsal view of 9·3 mm. stage; **2,** dorsal, and **3,** lateral view of 9·6 mm. stage (from de Beer).
Explanation of Lettering precedes Plate 24

PLATE 43

SALMO

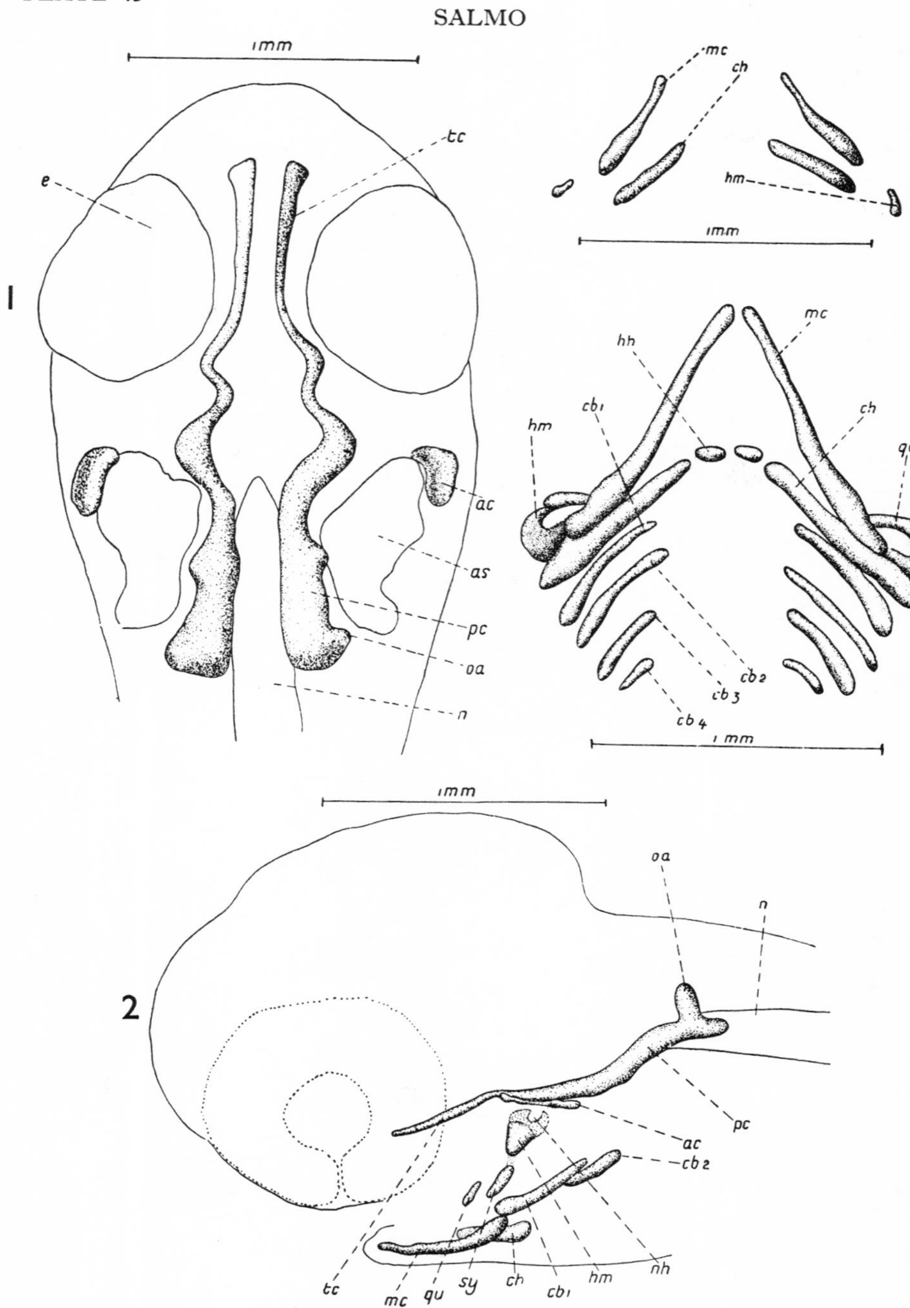

1, dorsal, and **2,** lateral view of 10·5 mm. stage; **3,** visceral arches of 9·3 mm. stage; **4,** visceral arches of 10·8 mm. stage (from de Beer).

Explanation of Lettering precedes Plate 24

SALMO

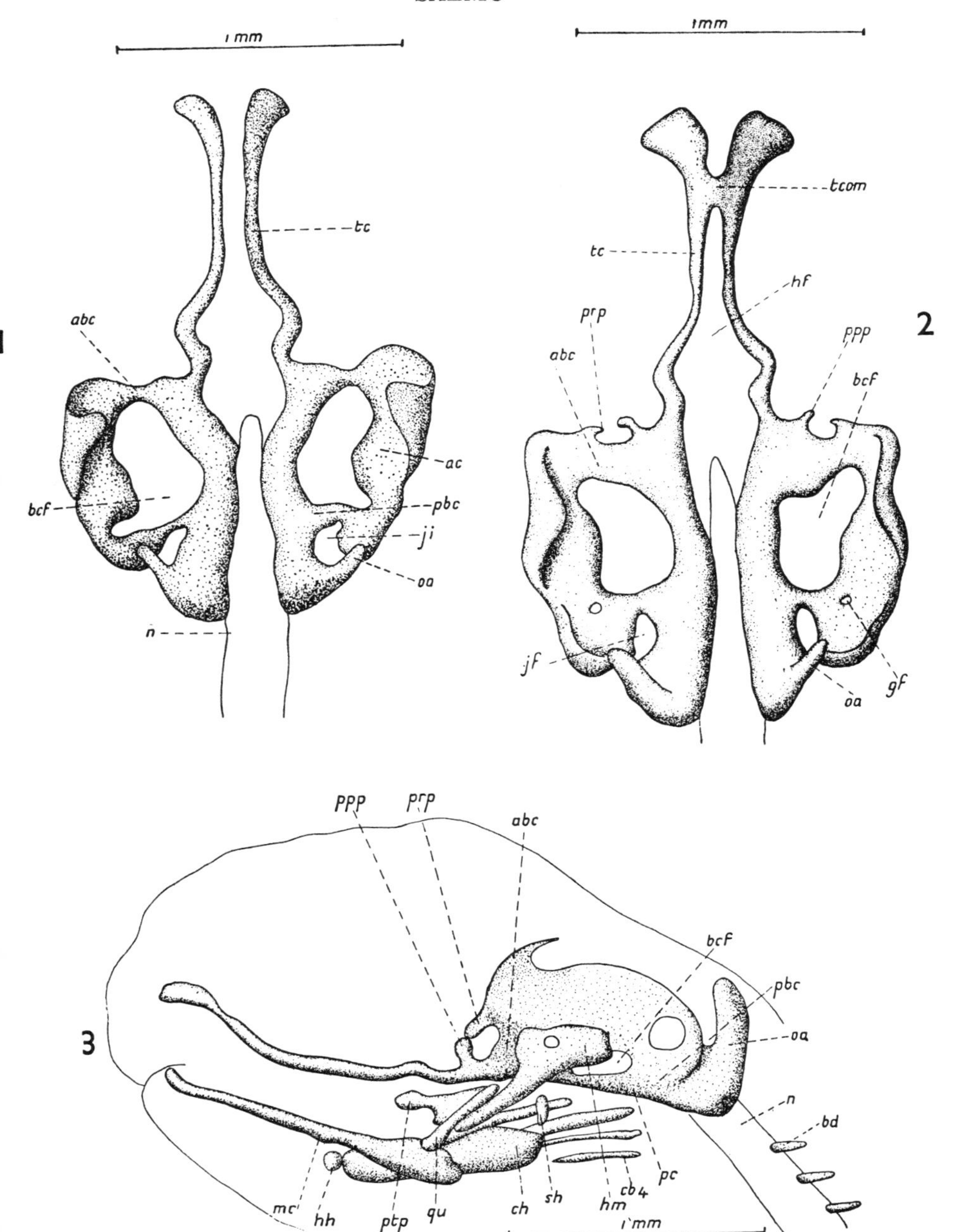

1, dorsal view of 11·4 mm. stage; **2**, dorsal, and **3**, lateral view of 12·3 mm. stage (from de Beer).

Explanation of Lettering precedes Plate 24

SALMO

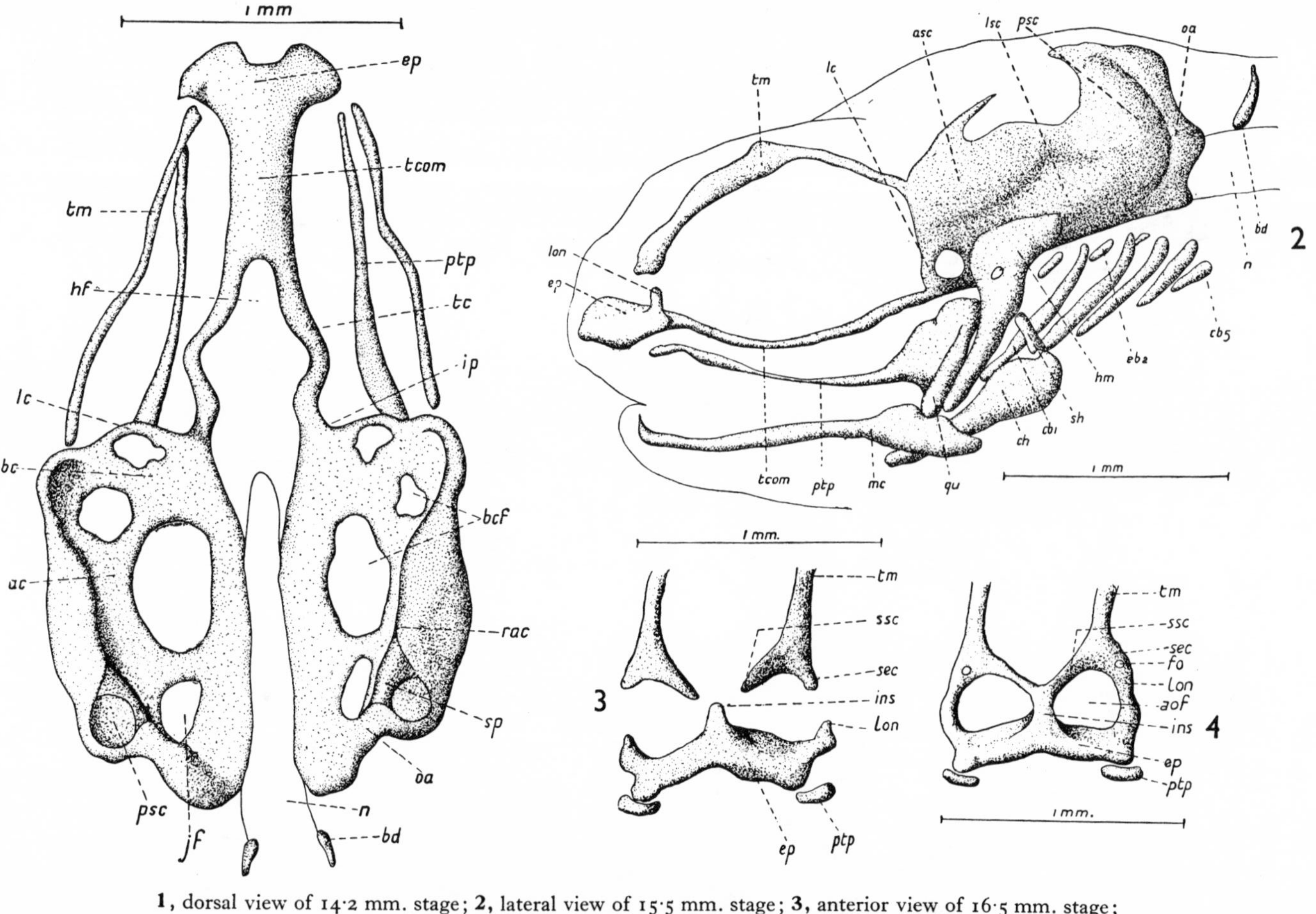

1, dorsal view of 14·2 mm. stage; **2**, lateral view of 15·5 mm. stage; **3**, anterior view of 16·5 mm. stage; **4**, anterior view of 17·5 mm. stage (from de Beer).

Explanation of Lettering precedes Plate 24

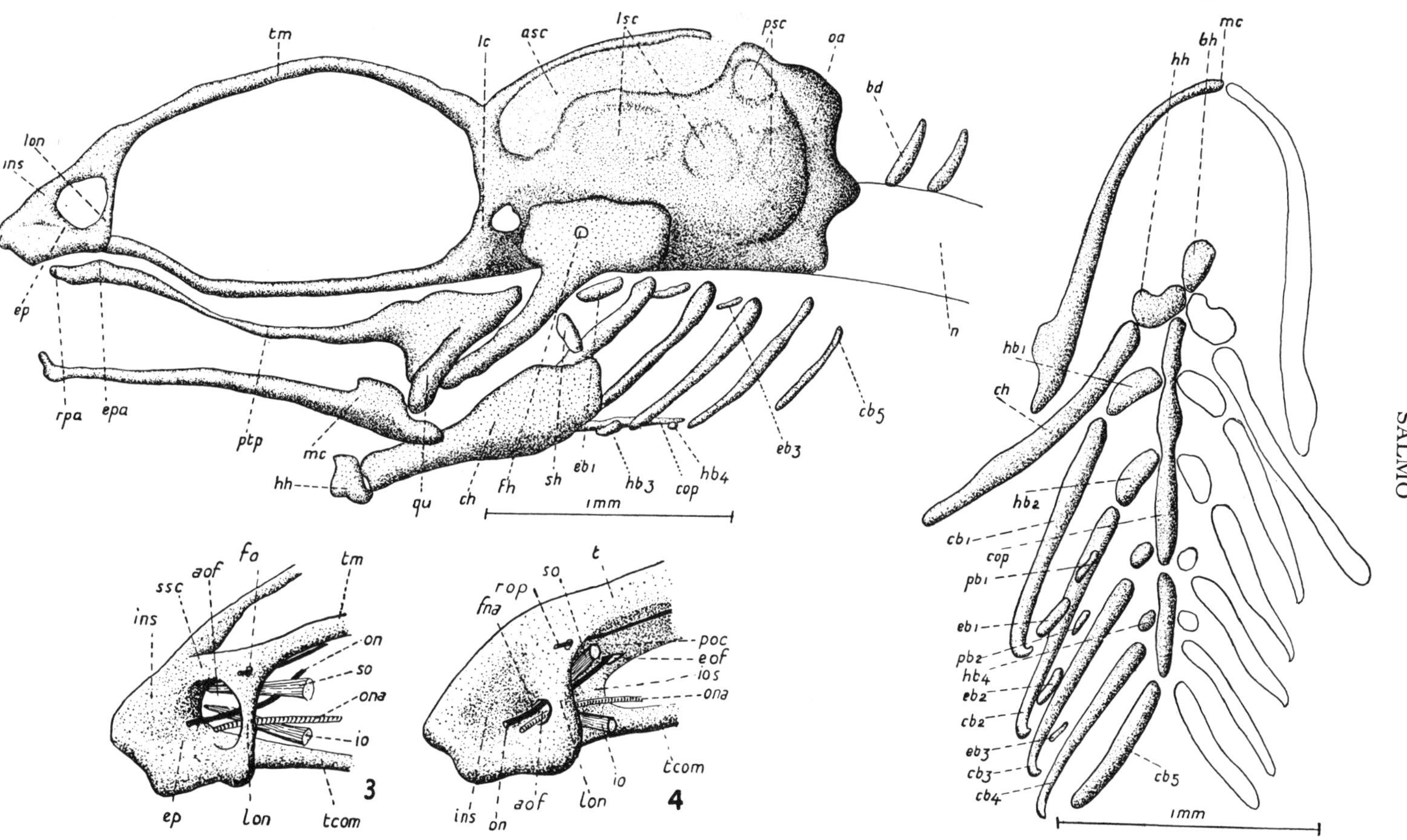

1, lateral view of 16·5 mm. stage; **2,** dorsal view of visceral arches of 17·5 mm. stage; **3,** lateral view of nasal region of 14 mm. stage; **4,** lateral view of nasal region of 30 mm. stage (from de Beer).

Explanation of Lettering precedes Plate 24

PLATE 47

SALMO

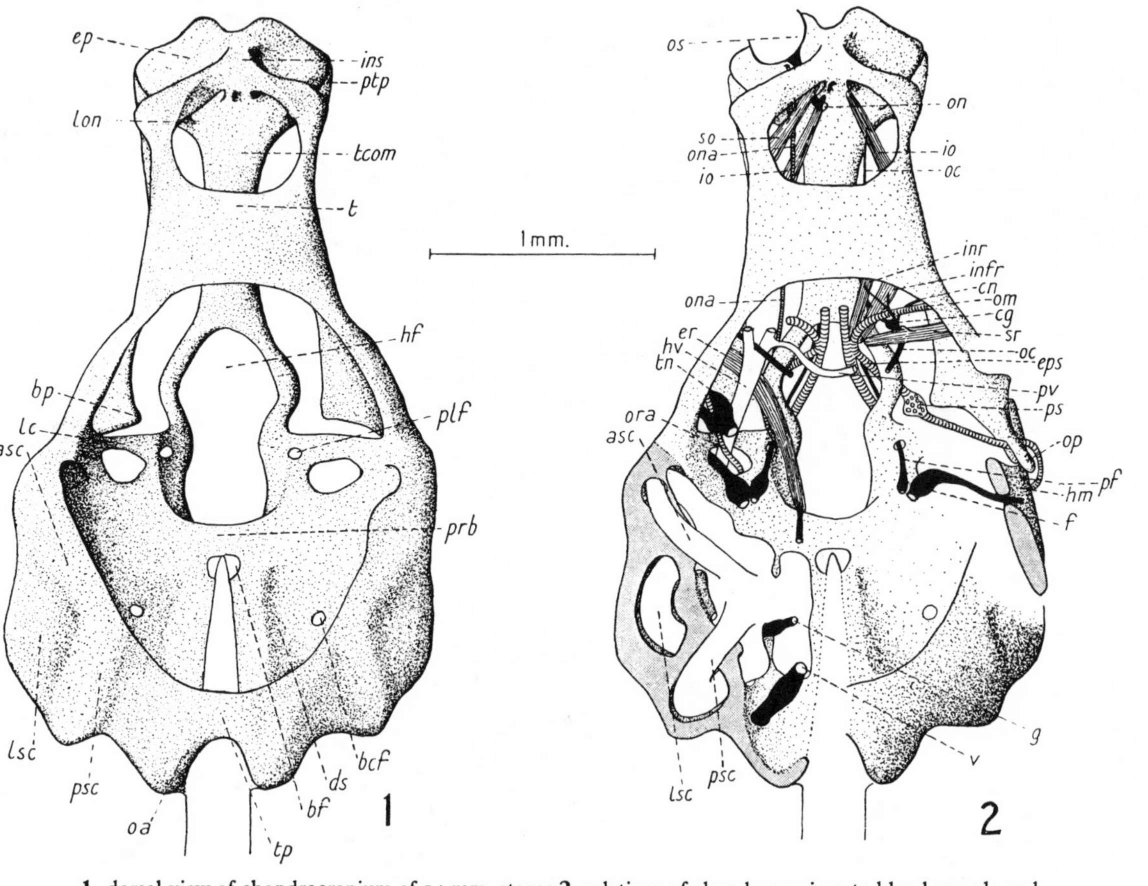

1, dorsal view of chondrocranium of 14 mm. stage; **2**, relations of chondrocranium to blood-vessels and nerves (from de Beer).

Explanation of Lettering precedes Plate 24

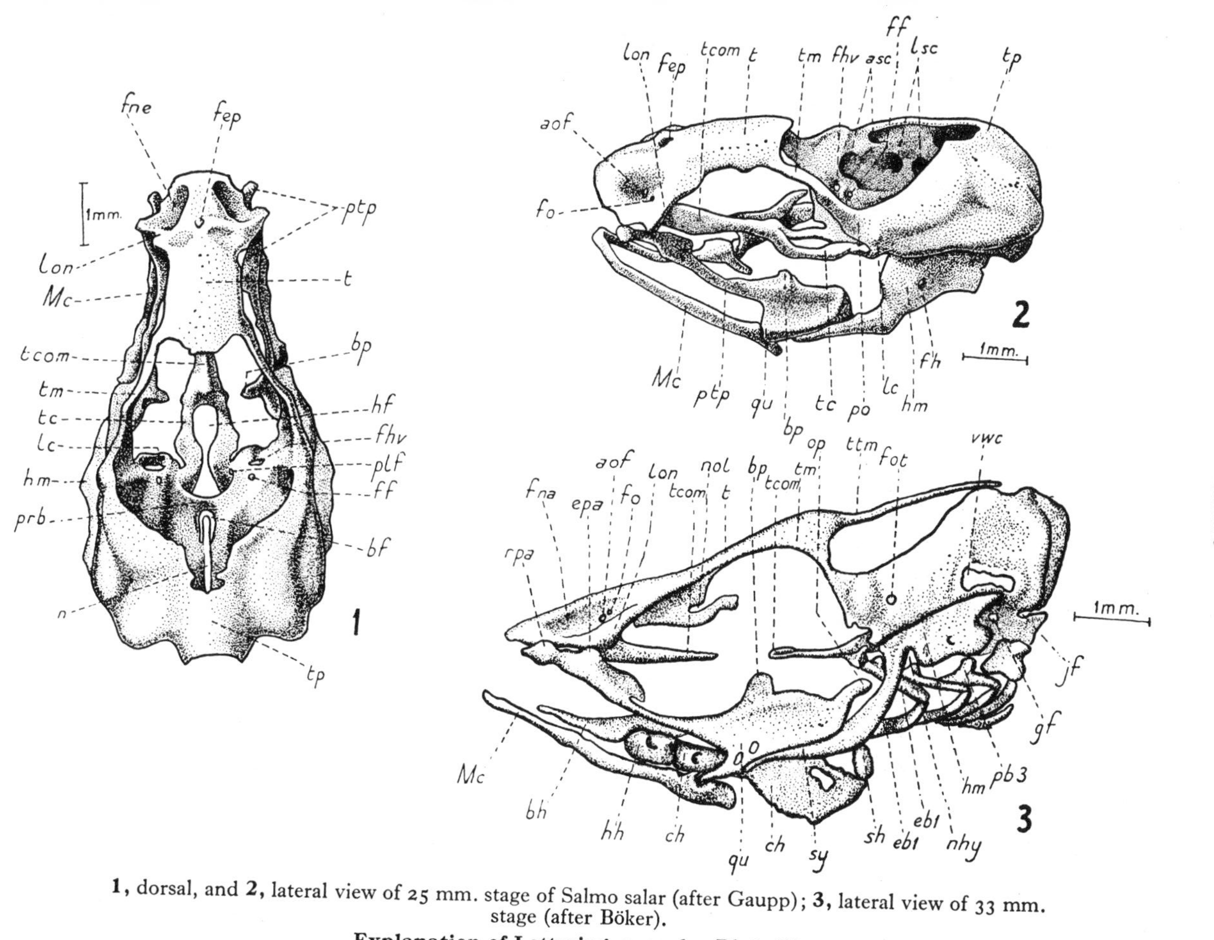

1, dorsal, and **2**, lateral view of 25 mm. stage of Salmo salar (after Gaupp); **3**, lateral view of 33 mm. stage (after Böker).

Explanation of Lettering precedes Plate 24

SALMO

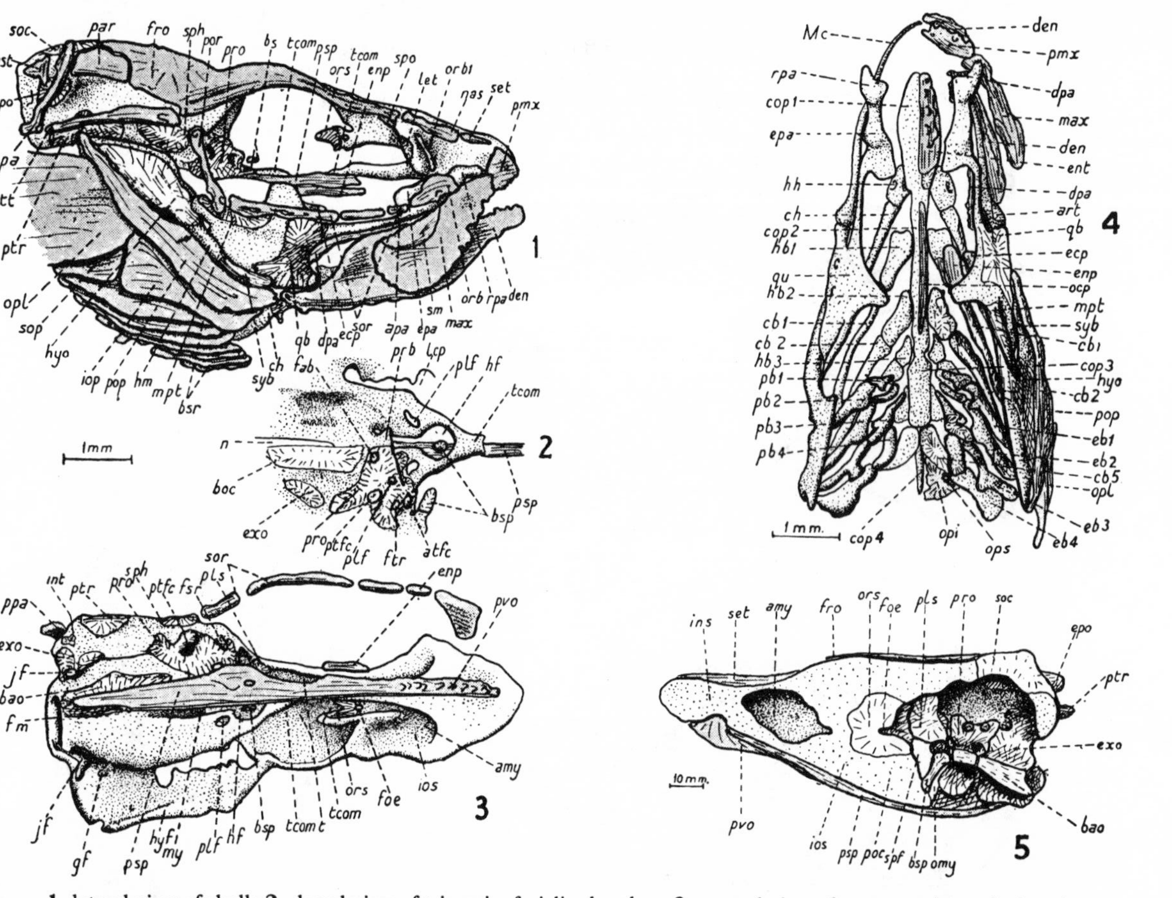

1, lateral view of skull; **2**, dorsal view of trigeminofacialis chamber; **3**, ventral view of neurocranium; **4**, dorsal view of visceral arches of 33 mm. stage (after Böker); **5**, medial view of section of adult skull (after W. K. Parker).

Explanation of Lettering precedes Plate 24

EXOCOETUS

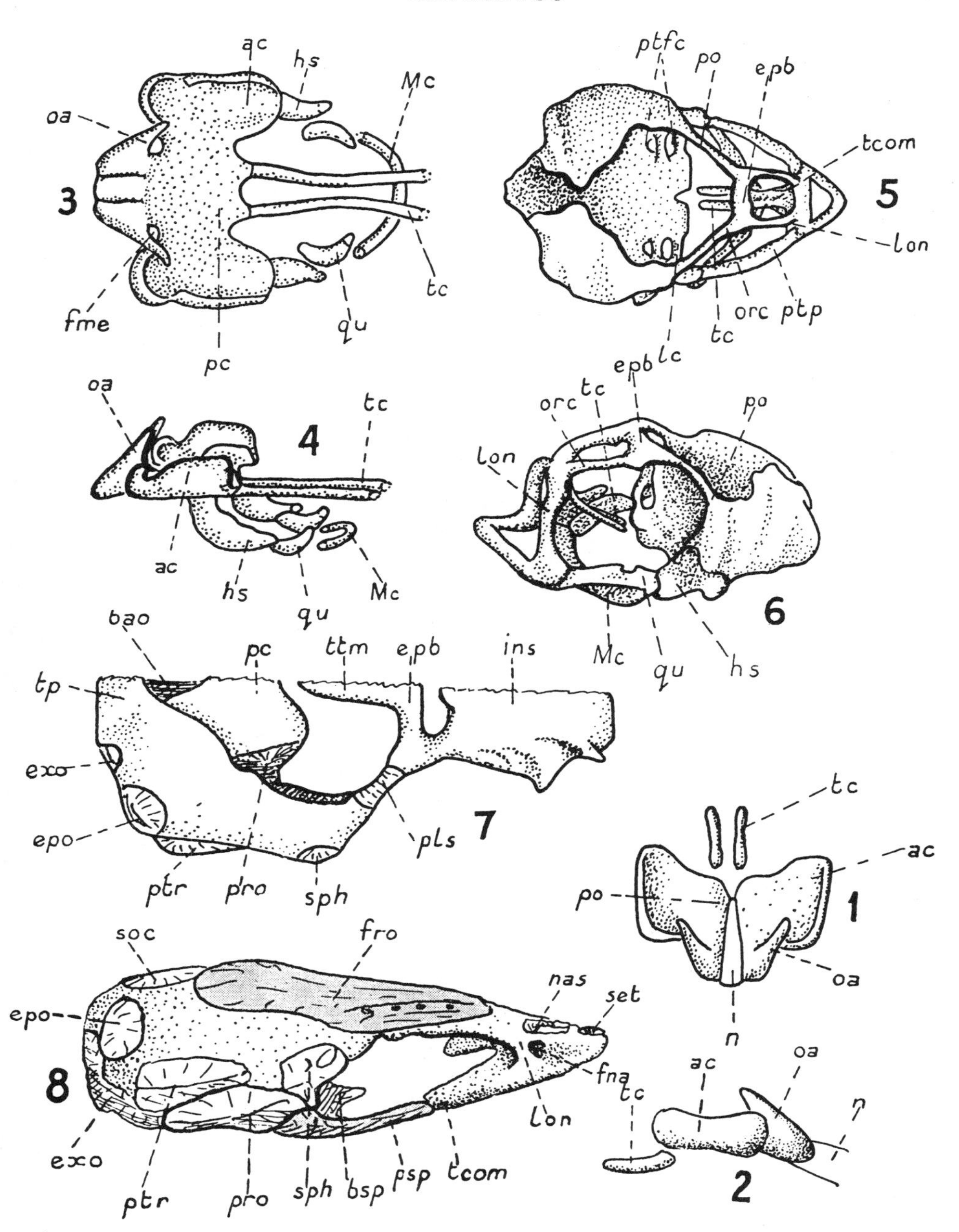

1, dorsal, and **2**, lateral view of 2·25 mm. stage; **3**, dorsal, and **4**, lateral view of 2·4 mm. stage; **5**, dorsal, and **6**, lateral view of 5 mm. stage; **7**, dorsal, and **8**, lateral view of 23 mm. stage (after Lasdin).

Explanation of Lettering precedes Plate 24

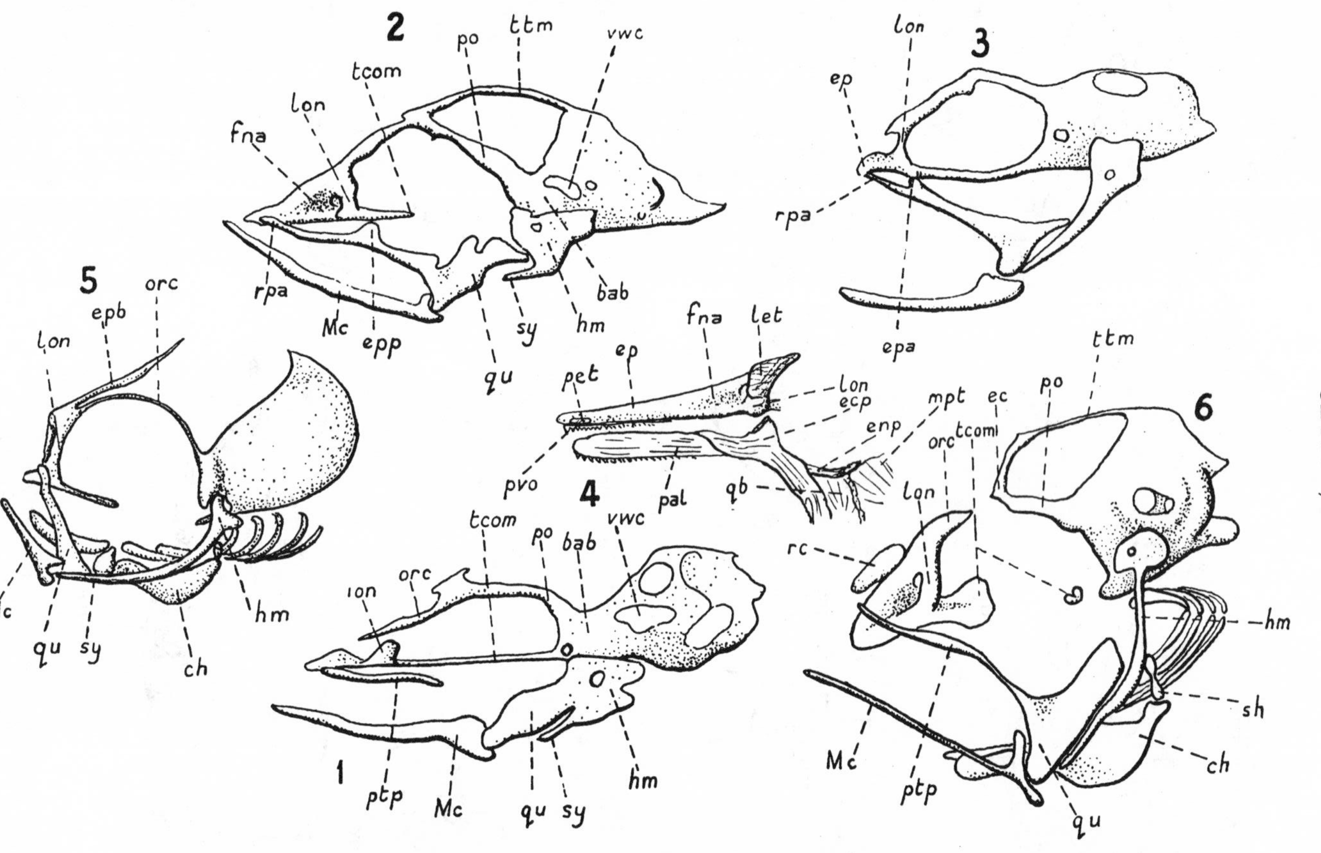

1, lateral view of 20 mm. stage of Clupea; **2,** lateral view of 50 mm. stage of Clupea (after Wells); **3,** lateral view of 10 mm. stage of Esox (after Walther); **4,** lateral view of ethmoid region of adult Esox (after Swinnerton); **5,** lateral view of 7·5 mm. stage of Gambusia (after Ryder); **6,** lateral view of 25 mm. stage of Sebastes (after Mackintosh).

Explanation of Lettering precedes Plate 24

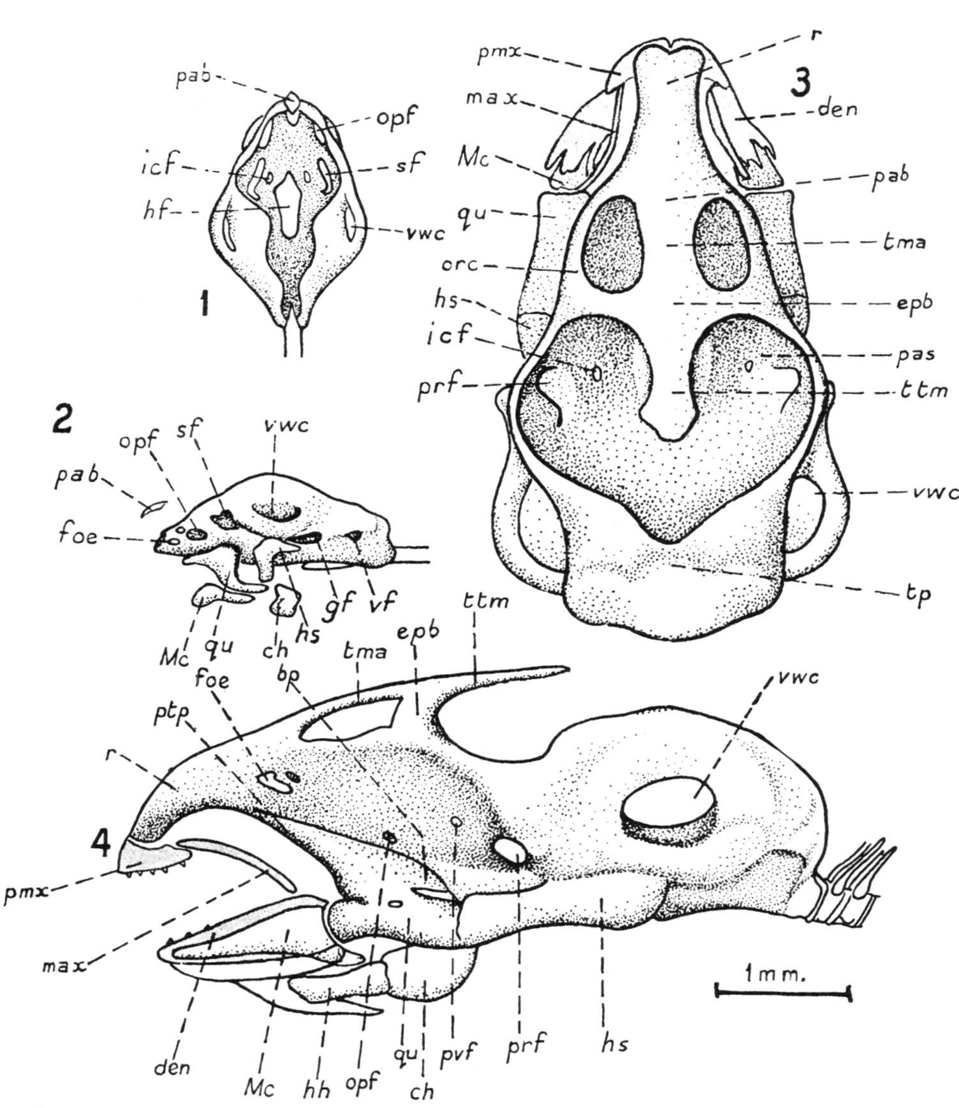

1, dorsal, and **2,** lateral view of 9-day stage; **3,** dorsal, and **4,** lateral view of 43-day stage (after Assheton).

Explanation of Lettering precedes Plate 24

AMIURUS

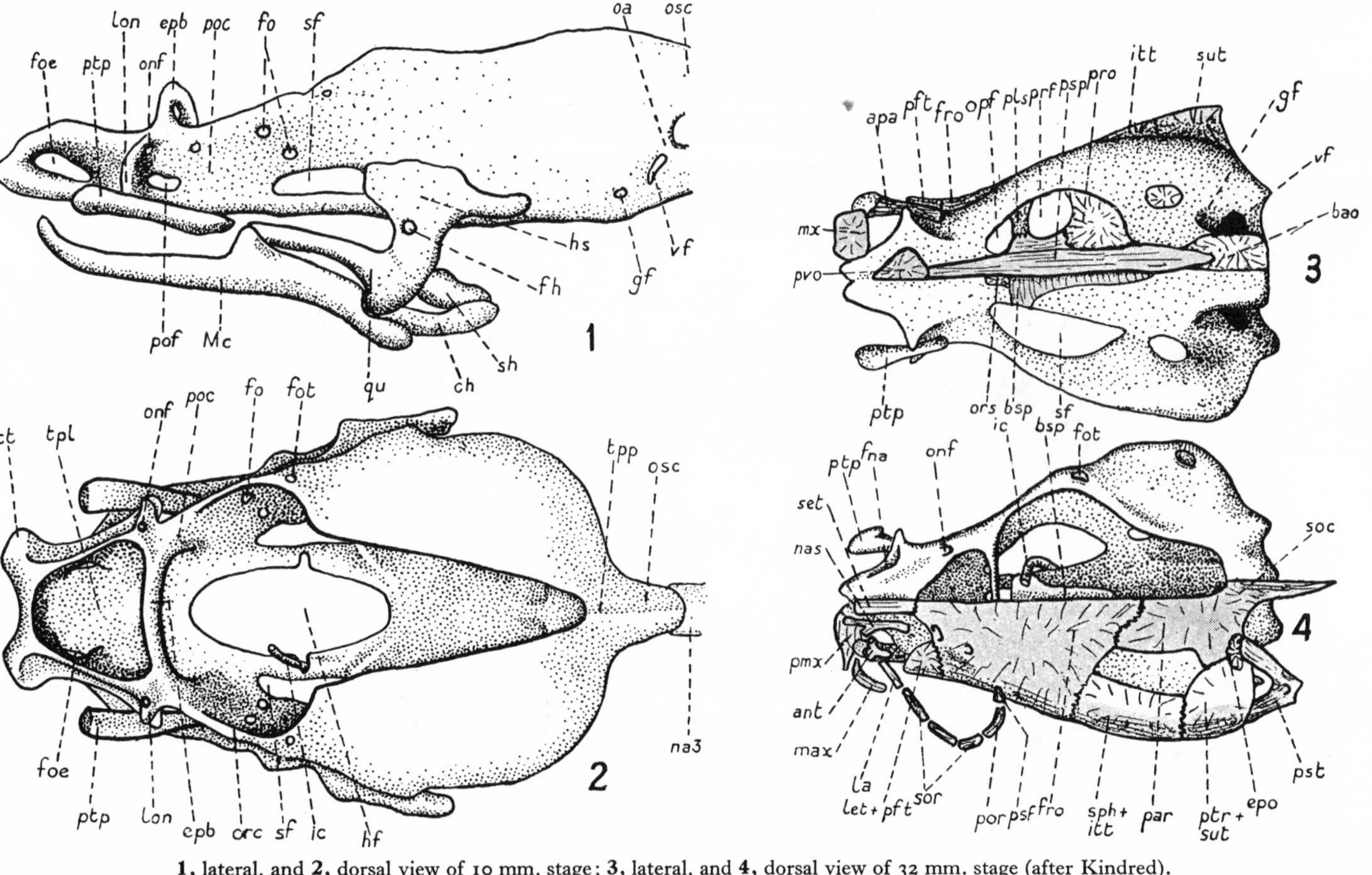

1, lateral, and **2,** dorsal view of 10 mm. stage; **3,** lateral, and **4,** dorsal view of 32 mm. stage (after Kindred).

Explanation of Lettering precedes Plate 24

CYPRINUS, ICTALURUS

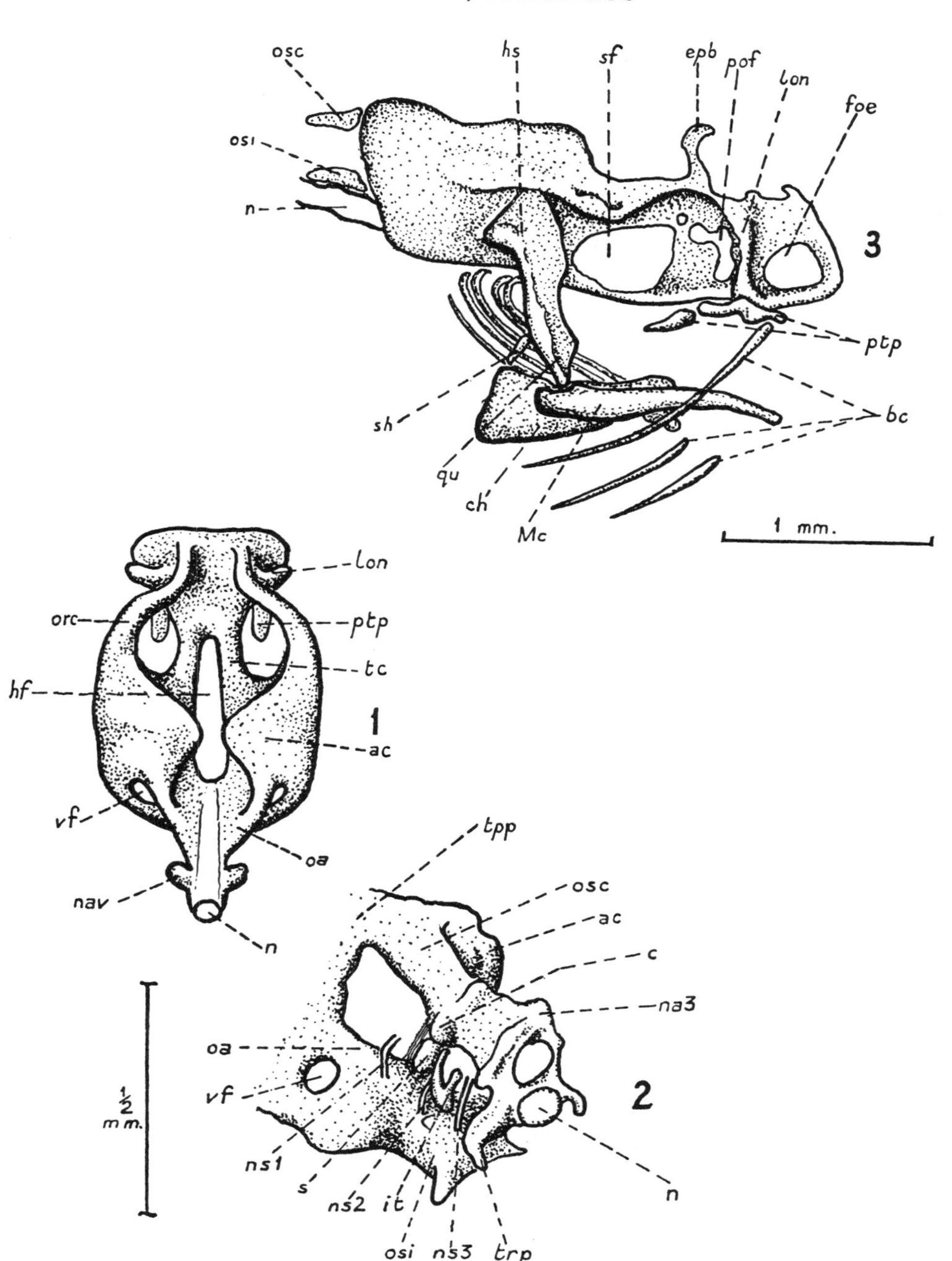

1, dorsal view of 5-day stage of Cyprinus; **2**, occipital region of 17-day stage of Cyprinus (after Nusbaum); **3**, lateral view of 10-day stage of Ictalurus (after Ryder).

Explanation of Lettering precedes Plate 24

PLATE 55

ANGUILLA

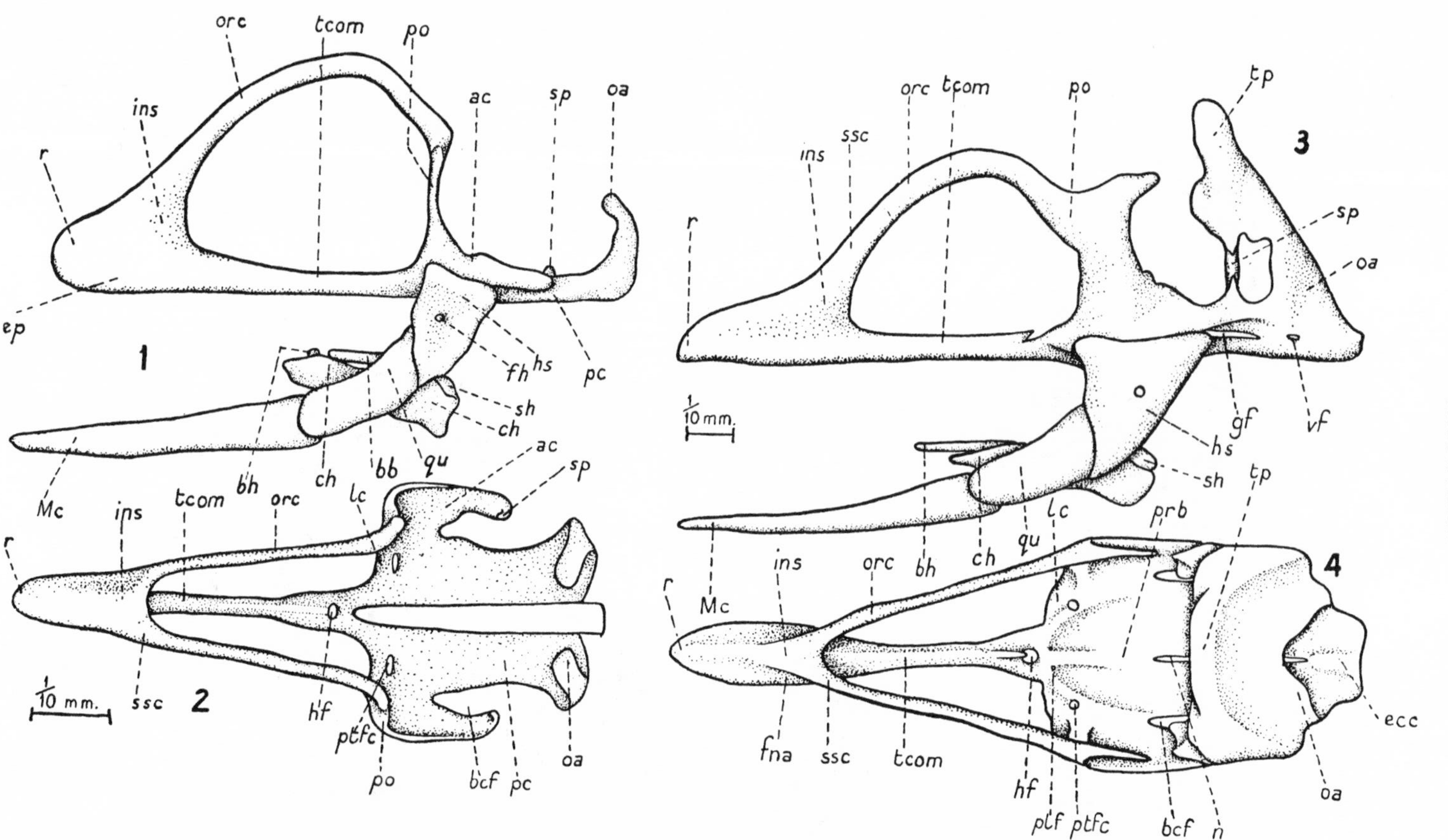

1, lateral, and **2**, dorsal view of 5 mm. stage; **3**, lateral, and **4**, dorsal view of 11 mm. stage (after Norman).

Explanation of Lettering precedes Plate 24

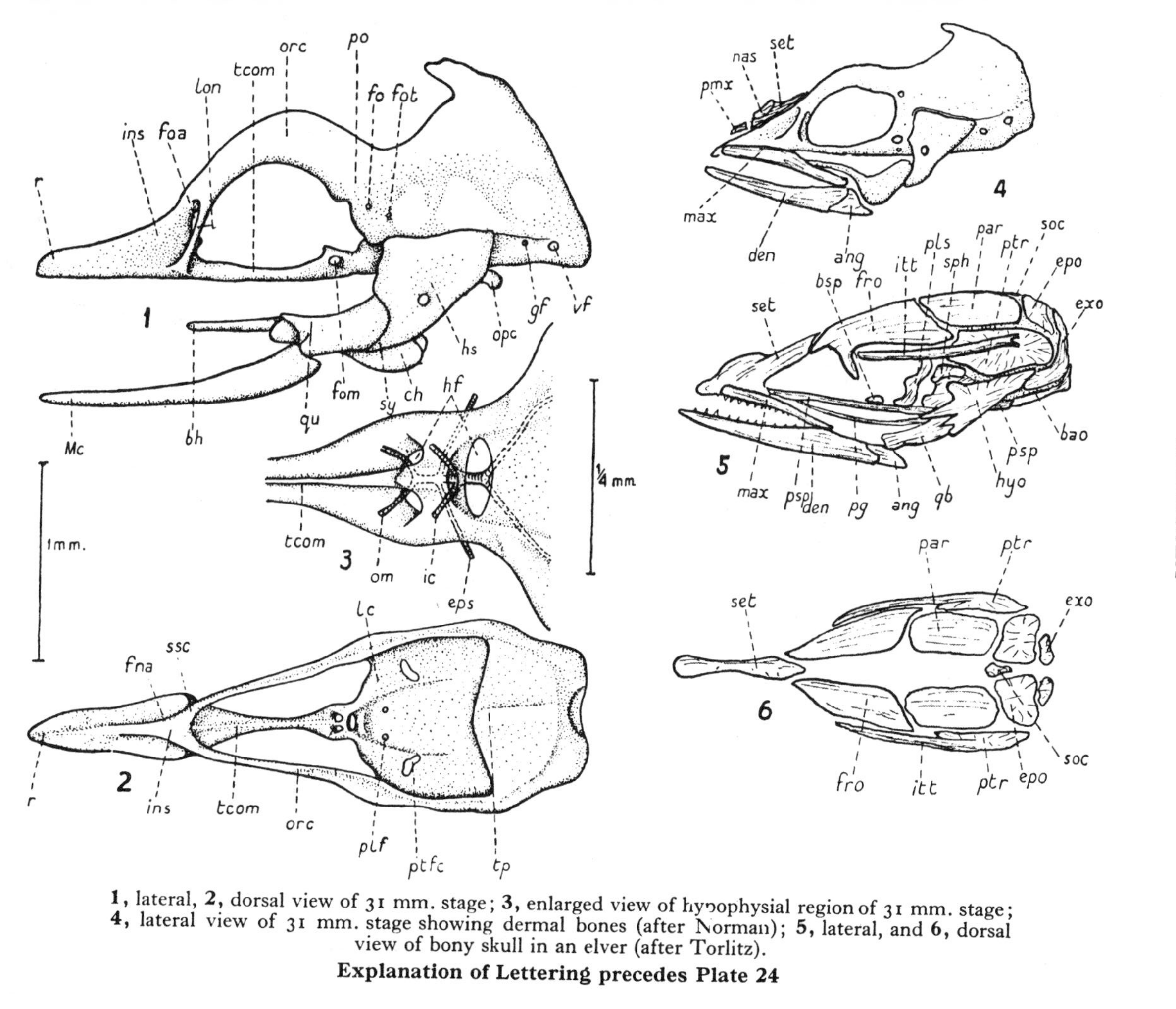

1, lateral, **2**, dorsal view of 31 mm. stage; **3**, enlarged view of hypophysial region of 31 mm. stage; **4**, lateral view of 31 mm. stage showing dermal bones (after Norman); **5**, lateral, and **6**, dorsal view of bony skull in an elver (after Torlitz).

Explanation of Lettering precedes Plate 24

GASTEROSTEUS

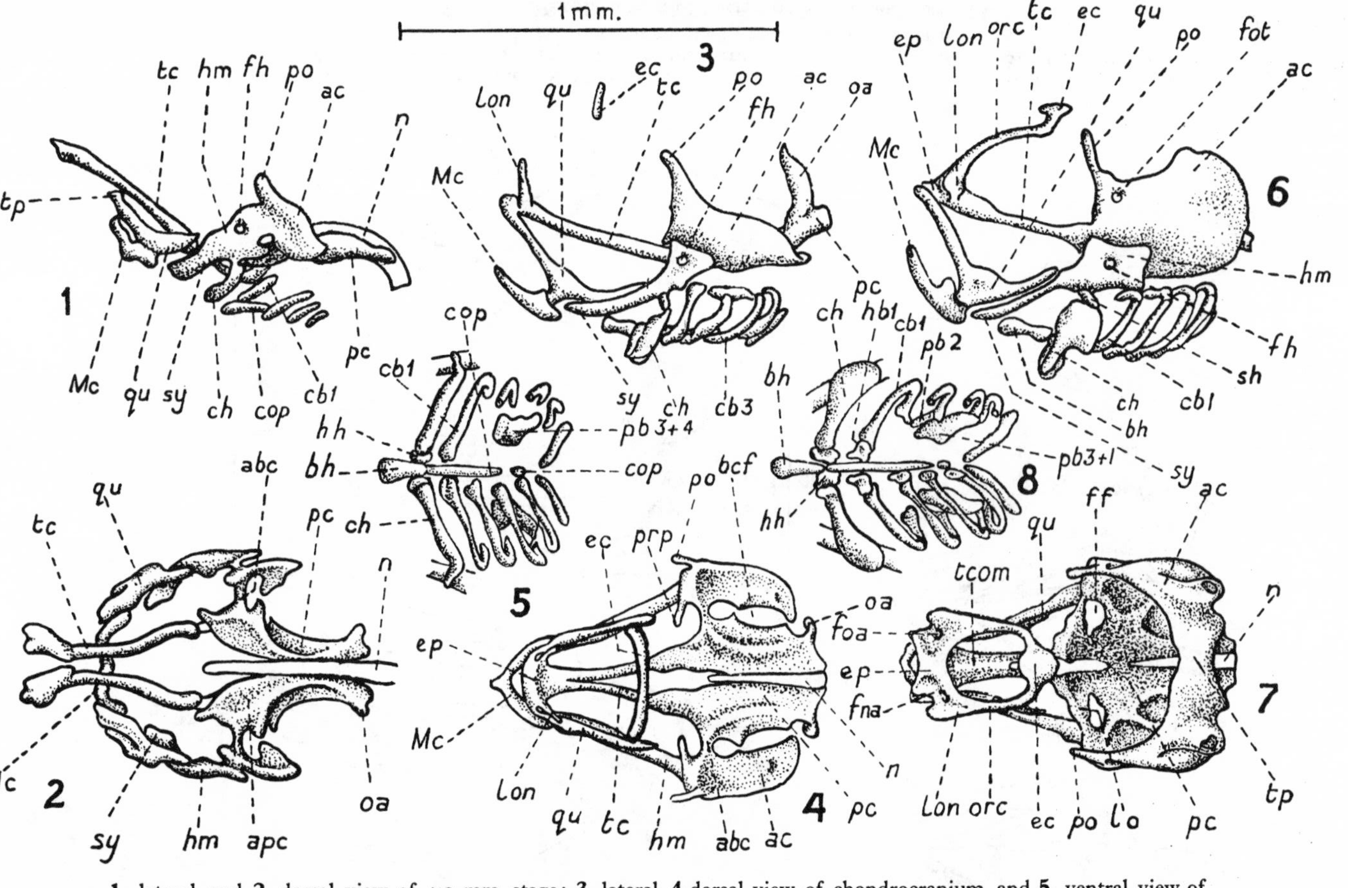

1, lateral, and **2**, dorsal view of 4·2 mm. stage; **3**, lateral, **4** dorsal view of chondrocranium, and **5**, ventral view of visceral arches of 5 mm. stage; **6**, lateral, **7**, dorsal view of chondrocranium, and **8**, ventral view of visceral arches of 9 mm. stage (after Swinnerton).

Explanation of Lettering precedes Plate 24

GASTEROSTEUS

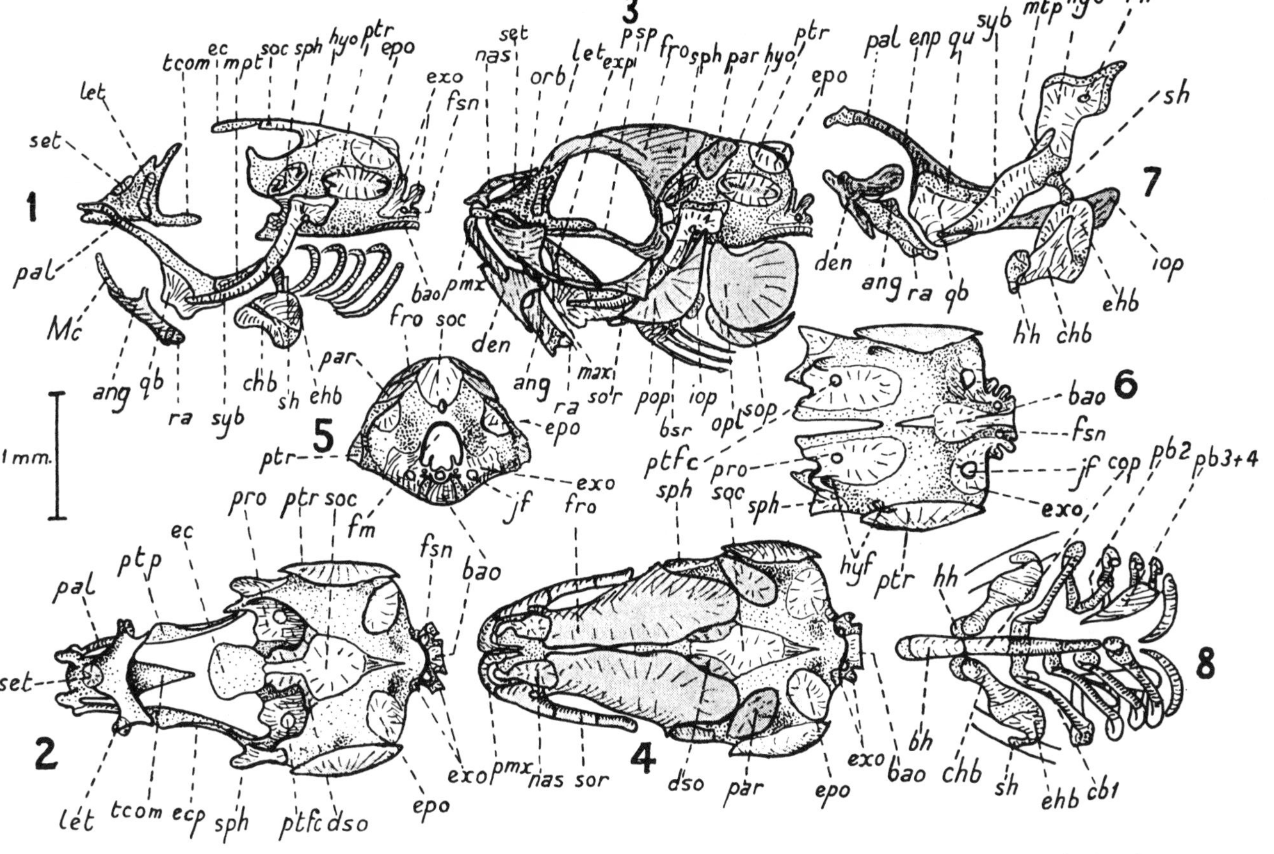

1, lateral, and **2**, dorsal view of chondrocranium and cartilage-bones of 25 mm. stage; **3**, lateral, **4**, dorsal, and **5**, posterior view of skull at 25 mm. stage; **6**, ventral view of basal plate; **7**, medial view of jaws; **8**, ventral view of visceral arches of 25 mm. stage (after Swinnerton).

Explanation of Lettering precedes Plate 24

SYNGNATHUS

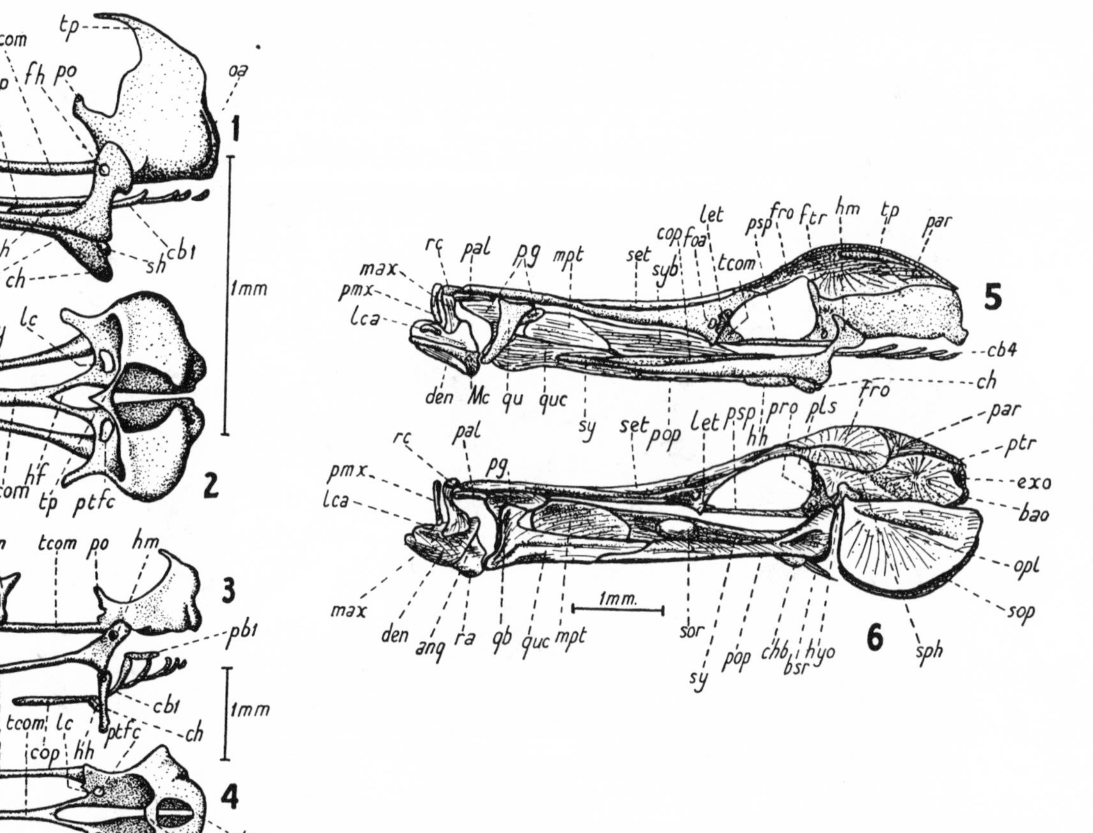

1, lateral, and 2, dorsal view of 8 mm. stage; 3, lateral, and 4, dorsal view of 12 mm. stage; 5, lateral view of chondrocranium with dermal bones on the right (distant) side of 45 mm. stage; 6, lateral view of skull of 45 mm. stage (after Kindred).

Explanation of Lettering precedes Plate 24

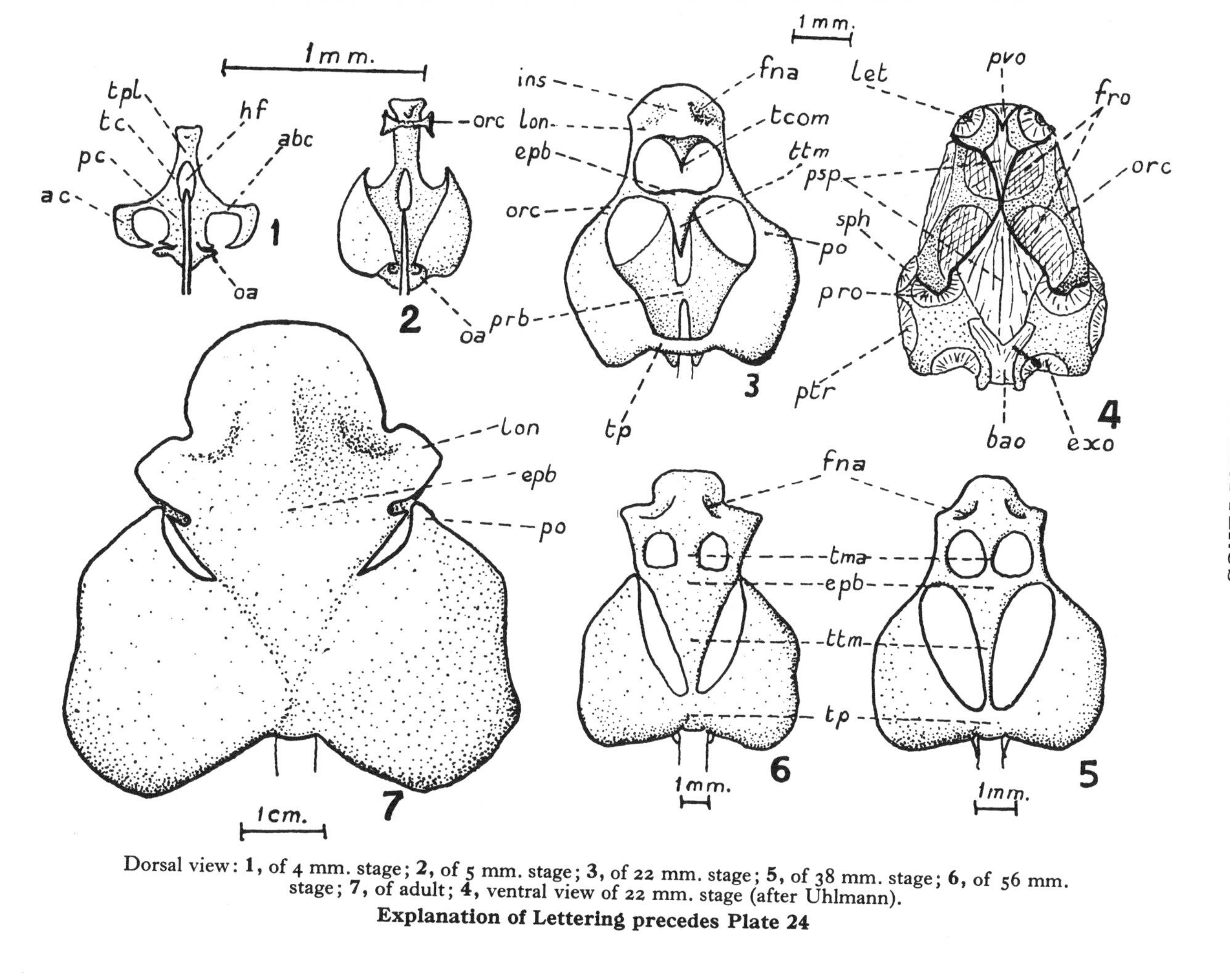

Dorsal view: **1**, of 4 mm. stage; **2**, of 5 mm. stage; **3**, of 22 mm. stage; **5**, of 38 mm. stage; **6**, of 56 mm. stage; **7**, of adult; **4**, ventral view of 22 mm. stage (after Uhlmann).

Explanation of Lettering precedes Plate 24

SOLEA, PSEUDOPLEURONECTES

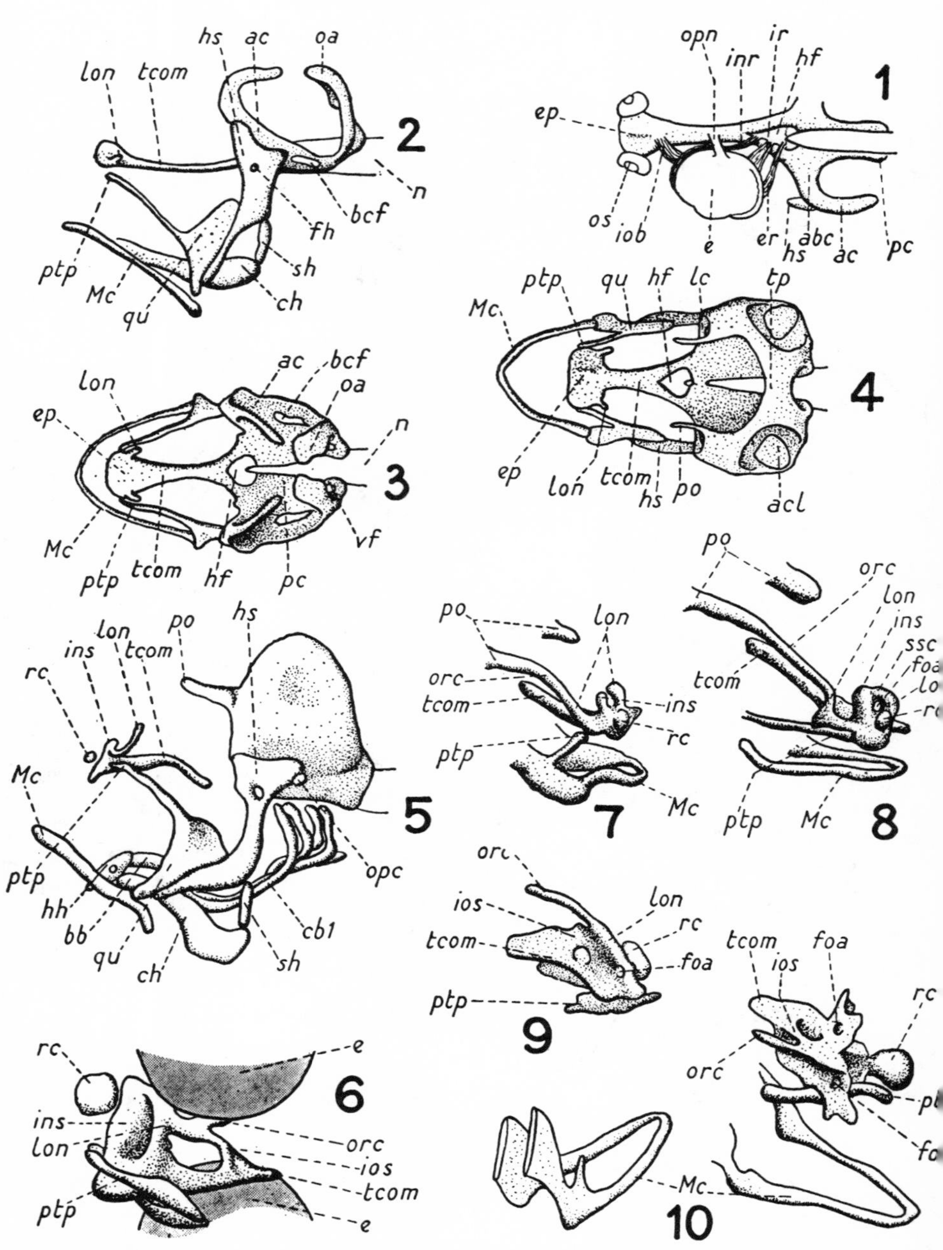

1, dorsal view of stage 2; **2**, lateral, and **3**, dorsal view of stage 4; **4**, dorsal view of stage 5; **5**, lateral view of stage 8; **6**, lateral view of anterior region after metamorphosis, of Solea variegata (after Berrill) **7–10** stages in metamorphosis of Pseudopleuronectes americanus (after Williams).

Explanation of Lettering precedes Plate 24

GADUS

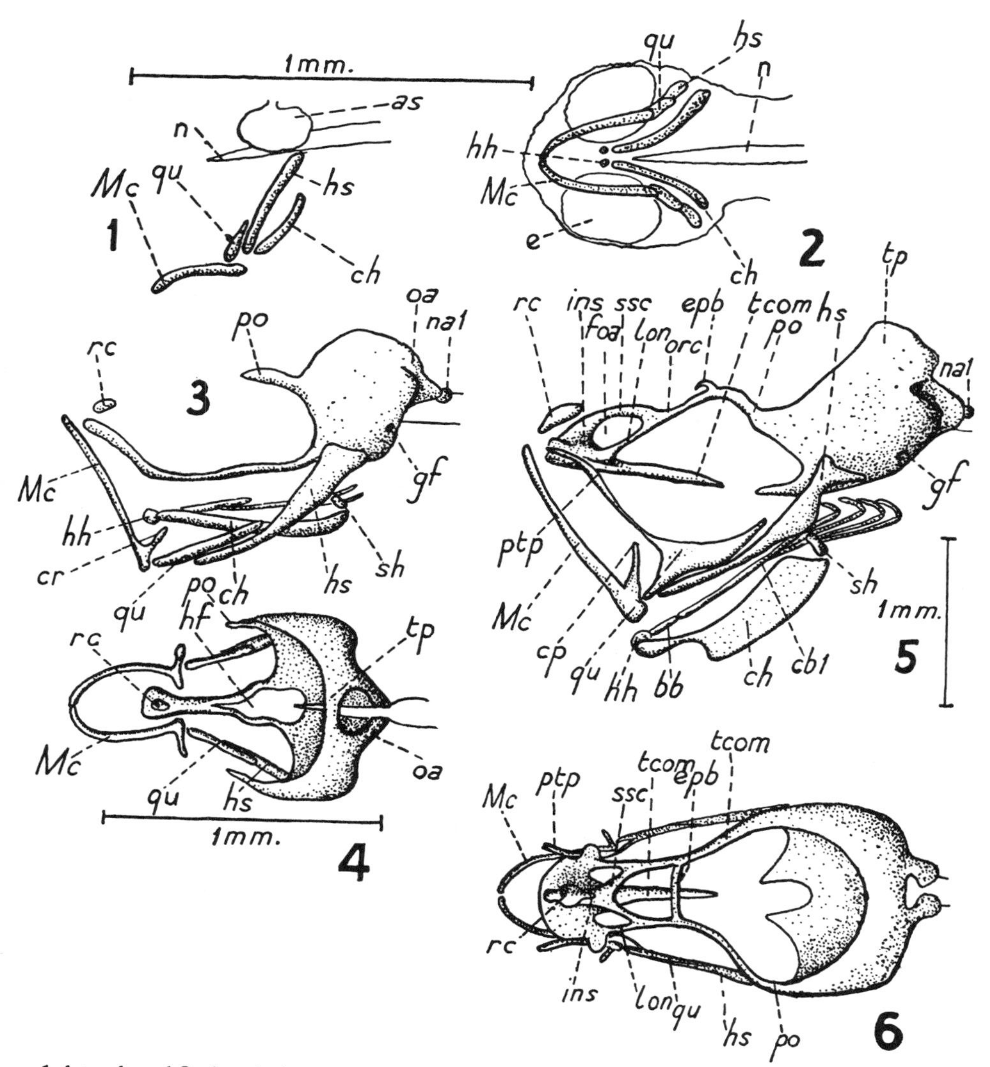

1, lateral, and **2**, dorsal view of 3 mm. stage; **3**, lateral, and **4**, dorsal view of 5 mm. stage; **5**, lateral, and **6**, dorsal view of 11 mm. stage.

Explanation of Lettering precedes Plate 24

CERATODUS

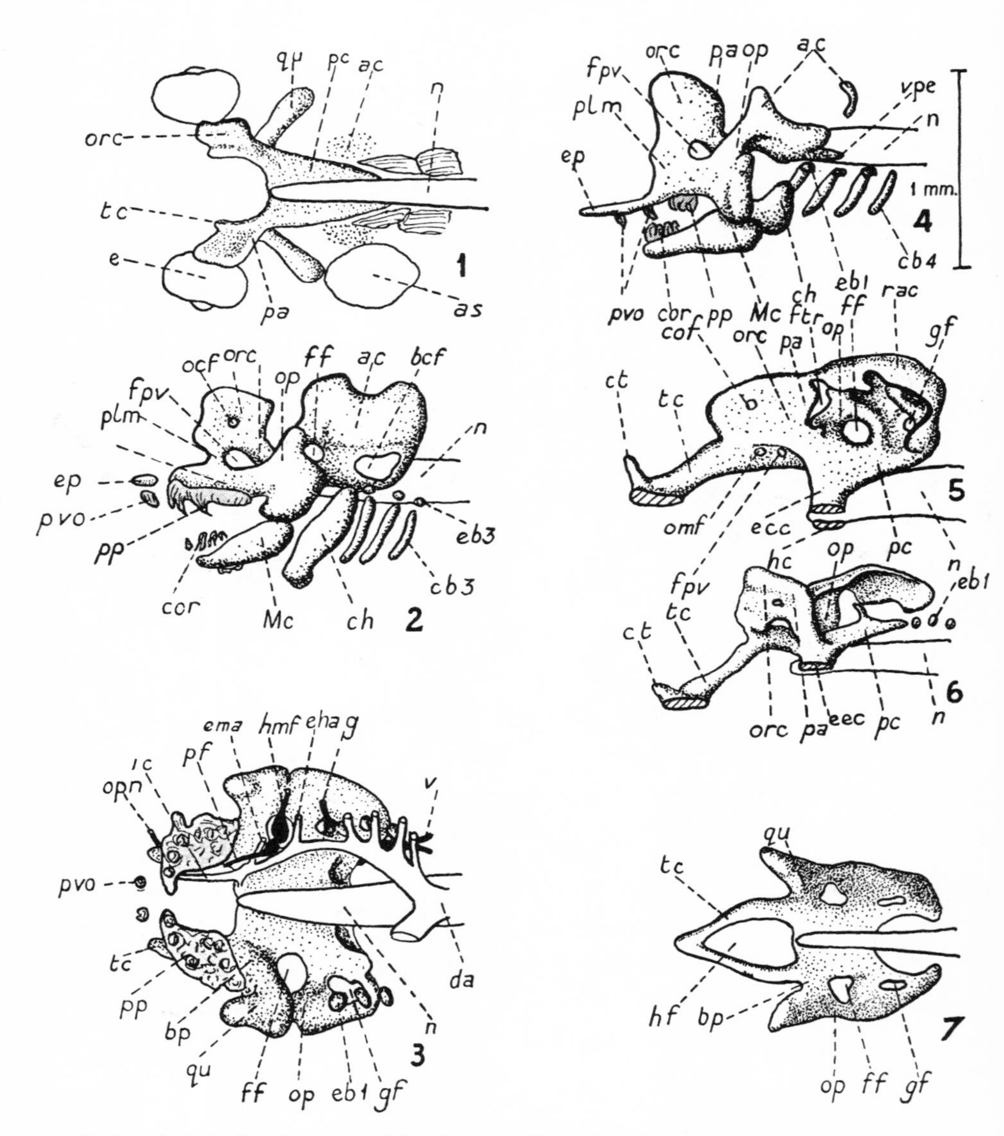

1, dorsal view of earliest stage (after Sewertzoff); **2,** lateral and **3,** ventral view of stage 45; **4,** lateral, and **5,** medial view of stage 46; **6,** medial view of stage 47 (after Greil); **7,** ventral view of stage 47 (after Krawetz).

Explanation of Lettering precedes Plate 24

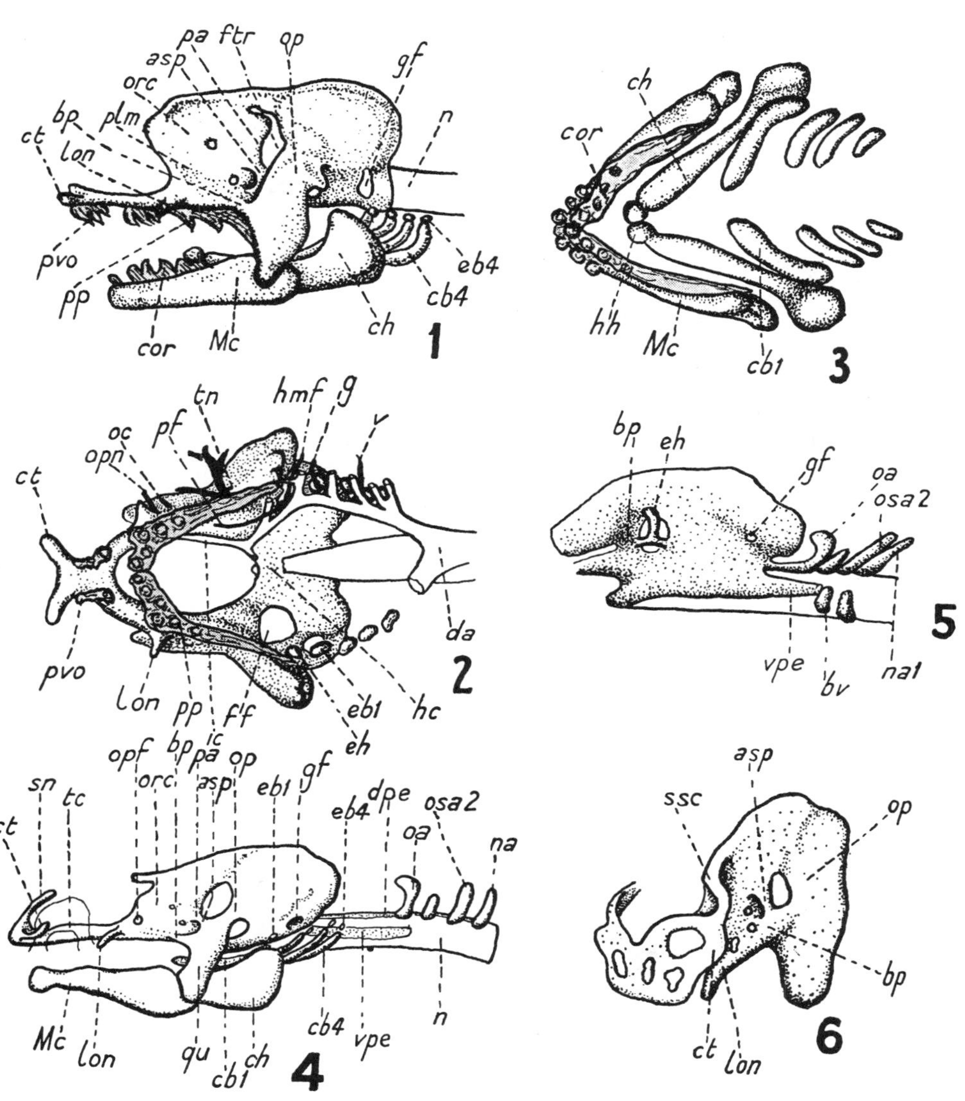

1, lateral, and **2,** ventral view of stage 47; **3,** visceral arches of stage 47 (after Greil); **4,** lateral view of late stage (after Sewertzoff); **5,** ventrolateral view of auditory region, and **6,** anterior view of late stage (after Krawetz).

Explanation of Lettering precedes Plate 24

CERATODUS

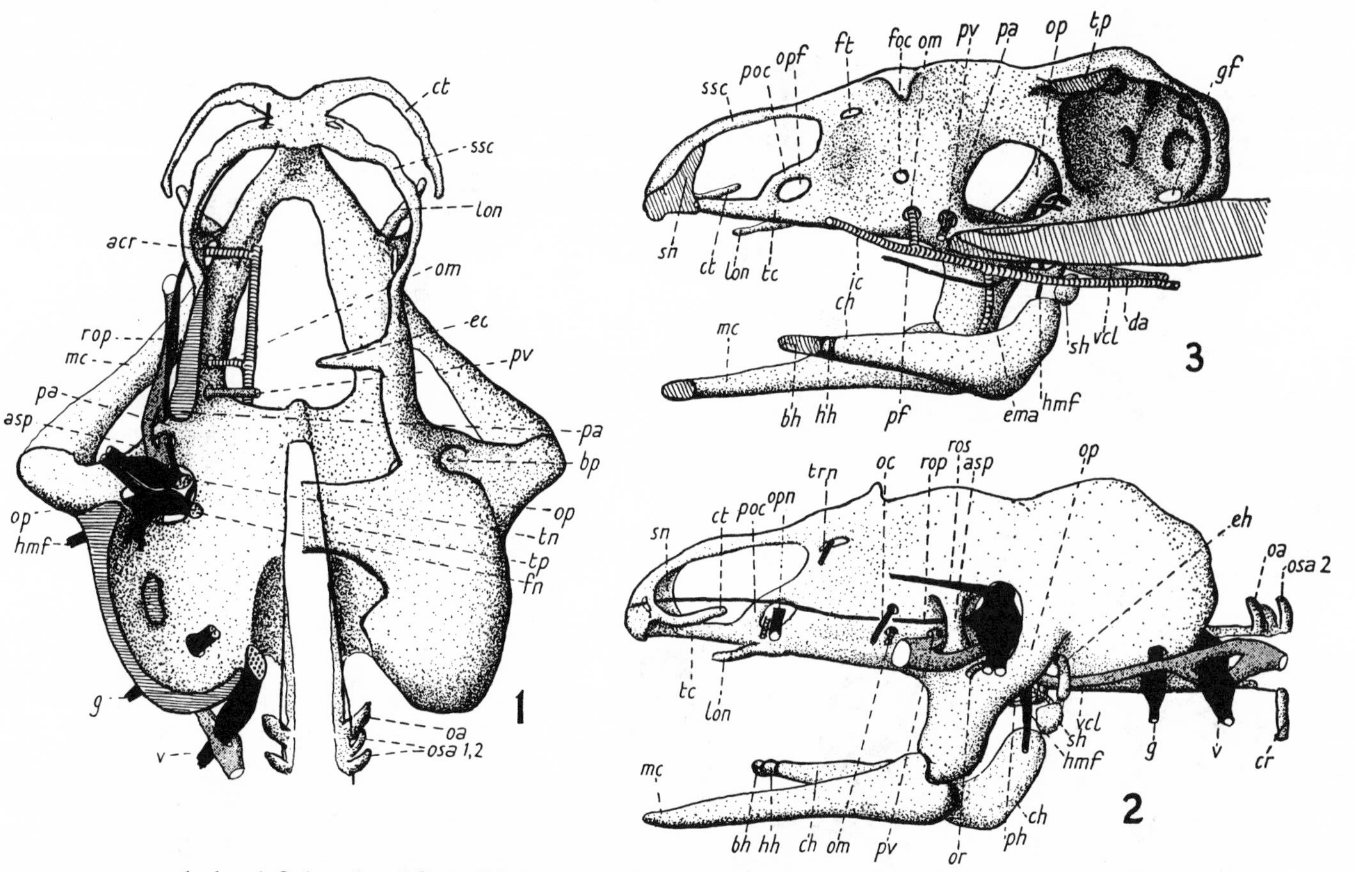

1, dorsal, **2**, lateral, and **3**, medial view of 20 mm. stage showing relations to blood-vessels and nerves.

Explanation of Lettering precedes Plate 24

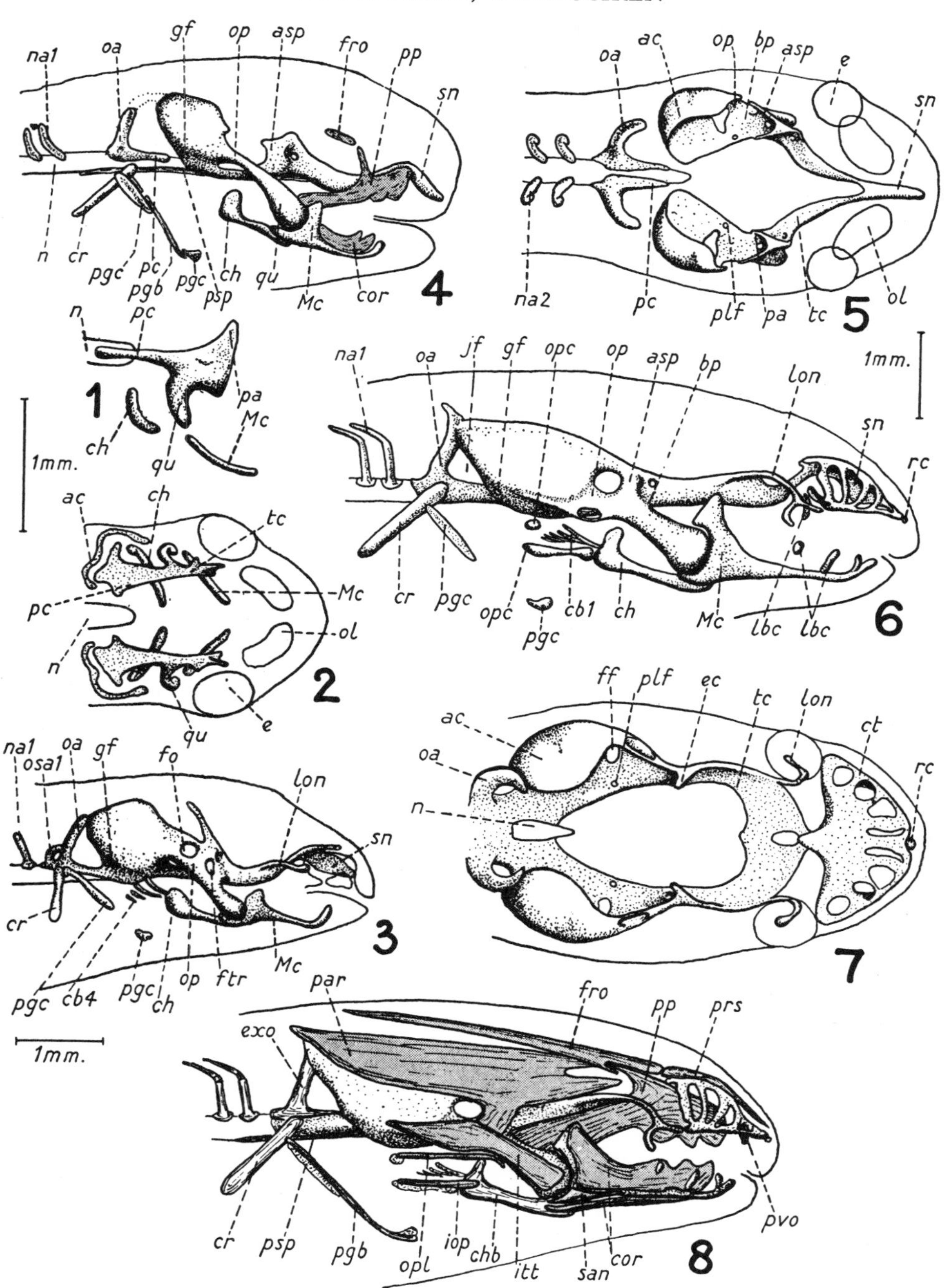

1, lateral, and **2,** dorsal view of Protopterus stage 31; **3,** lateral view of Protopterus stage 36; **4,** lateral, and **5,** dorsal view of Lepidosiren stage 34; **6,** lateral, **7,** dorsal view, and **8,** dermal bones of Lepidosiren stage 38 (after Agar).

Explanation of Lettering precedes Plate 24

AMPHIBIA

Explanation of Lettering

abc, anterior basicapsular commissure
ac, auditory capsule
alp, anterolateral process
amp, anteromedial process
an, neural arch of atlas vertebra
ang, angular
ap, anterior parachordal
arc, arcuate cartilage

as, auditory sac
asc, anterior semicircular canal
aty, tympanic ring

Bca, Born's cartilage
bm, basimandibular
boc, basioccipital
bp, basal plate
bpr, basal process (of quadrate)
bt, basitrabecular process
bvc, basivestibular commissure

ca, columella auris
cal, cartilago alaris nasi
cb 1–4, ceratobranchial 1–4
cec, ectochoanal cartilage
cf, carotid foramen
ch, ceratohyal cartilage
ci, crista intermedia nasi
cin, cavum internasale
cob, cartilago obliqua nasi
cop 1, 2, copula 1, 2
cor, coronoid
cp, crista parotica
cpi, cartilago prenasalis inferior
cps, cartilago prenasalis superior
cq, commissura quadratocranialis anterior
ct, cornu trabeculae

den, dentary

e, eyeball
exo, exoccipital

f, facial nerve
fa, foramen apicale
faa, foramen acusticum anterius
fac, foramen acusticum
fap, foramen acusticum posterius
fb, basicranial fenestra
fc, foramen craniopalatinum
fch, fenestra choanalis
fen, foramen endolymphaticum
ff, facial foramen
fj, jugular foramen
fl, fenestra lateralis nasi
fm, fissura metotica
fn, fenestra narina
fnb, foramen nasobasale
fo, optic foramen
focn, oculomotor foramen
fol, foramen olfactorium
forn, orbitonasal foramen
fov, foramen ovale
fp, foramen prooticum
fpa, foramen perilymphaticum accessorium
fpe, foramen perilymphaticum
fpi, foramen perilymphaticum inferius
fps, foramen perilymphaticum superius
fro, frontal
frp, frontoparietal
fsu, fenestra superior (or dorsalis) nasi
ftrn, trochlear foramen

gn, glossopharyngeal nerve

hb 1, 2, hypobranchial 1, 2
hbp, hypobranchial plate
hf, hypophysial fenestra
hh, hypohyal

ir, infrarostral

lon, lamina orbitonasalis
lsc, lateral semicircular canal

m 1, 1st myomere
max, maxilla

Mc, Meckel's cartilage
mm, mentomeckelian

n, notochord
nas, nasal

oa, occipital arch
oc, orbital cartilage
op, operculum
os, olfactory sac
osp, orbitosphenoid

p, parachordal
pa, processus ascendens quadrati
pal, palatine bone
paln, palatine nerve
pao, pila antotica
paq, pars articularis quadrati
par, parietal
pb, processus branchialis
pf, palatine foramen
pfc, prefacial commissure
pft, parietal fontanelle
ph, parahyoid
phl, parahyal
plp, posterolateral process
pma, processus maxillaris anterior
pmp, processus maxillaris posterior
pmq, processus muscularis quadrati
pmx, premaxilla
pn, paries nasi
pnp, prenasal process
po, preoccipital arch
poa, adult otic process
pob, preoptic root of orbital cartilage
pol, larval otic process
ppa, postparietal
ppn, pars plana of lamina orbitonasalis
pqe, processus quadratoethmoidalis
pr, processus retroarticularis
pra, prearticular
prf, prefrontal
pro, prootic
pru, pars reuniens
psc, posterior semicircular canal
psb, pseudobasal process
psp, parasphenoid
ptg, pterygoid
ptp, pterygoid process of quadrate
pvo, prevomer

q, quadrate
qj, quadratojugal

sc, sclerotic cartilage
se, supraethmoid
smx, septomaxilla
sn, septum nasi
so, supraoccipital
son, solum nasi
sos, subocular shelf
squ, squamosal
sr, suprarostral
ssc, sphenoseptal commissure
st, stapes

t, trabecula cranii
tc, tentacular cartilage
th, thyroid process
tm, taenia marginalis
tp, trabecular plate
ts, tectum synoticum
tt, tectum transversum
ttm, taenia tecti medialis
ty, tympanic cavity

v, vagus nerve

Wc, Wiedersheim's cartilage

AMBLYSTOMA

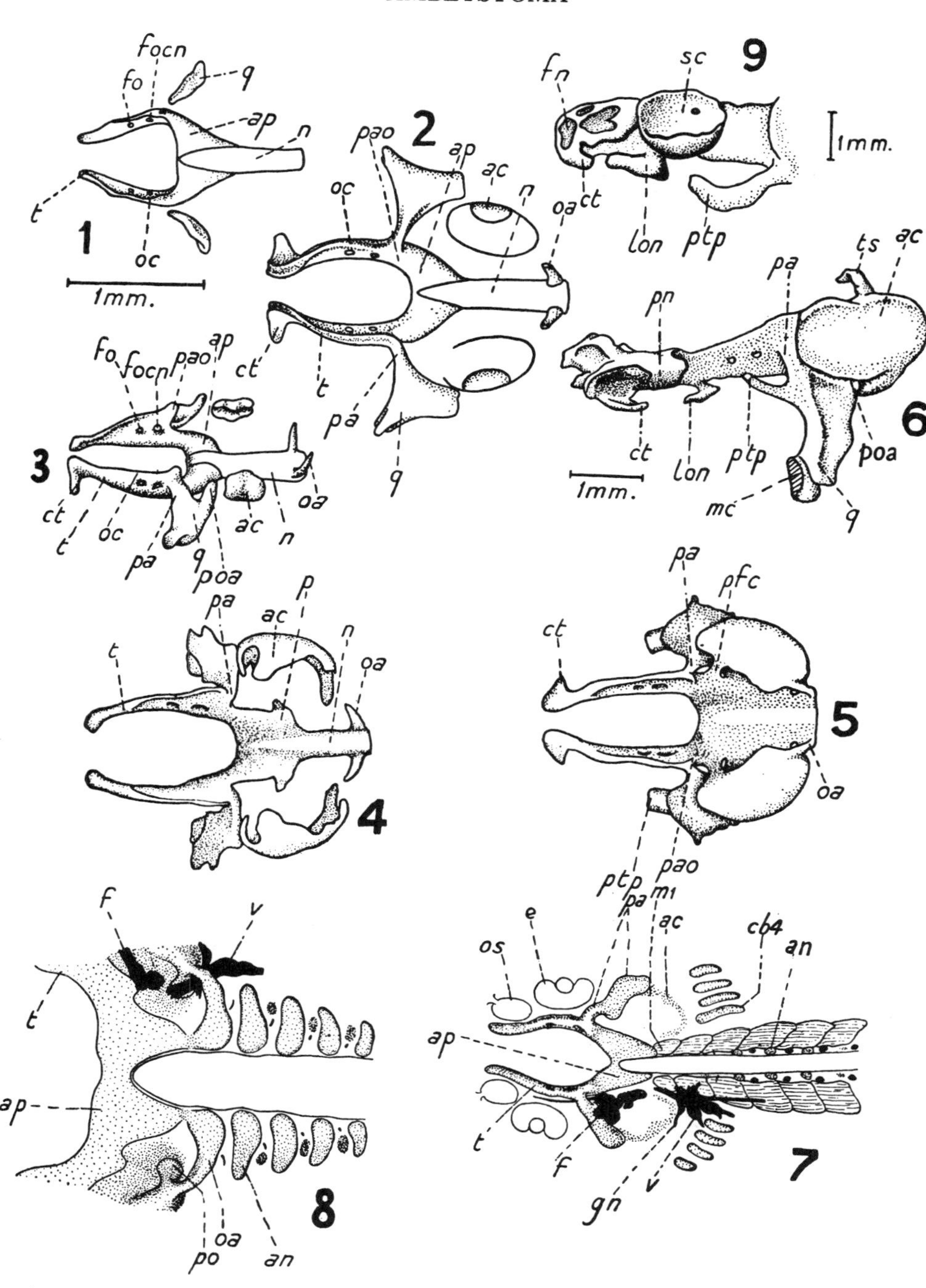

1, dorsal view of 7·5 mm. stage; **2**, dorsal, and **3**, lateral view of 9 mm. stage (after (Stöhr); **4**, dorsal view of 11 mm. stage; **5**, dorsal view of 12 mm. stage (after Winslow); **6**, lateral view of 23 mm. stage (after Gaupp); **7**, dorsal view of early stage showing myomeres; **8**, dorsal view of occipital region of later stage (after Goodrich); **9**, lateral view of orbit showing sclerotic (after Stadtmüller).

Explanation of Lettering opposite

TRITON

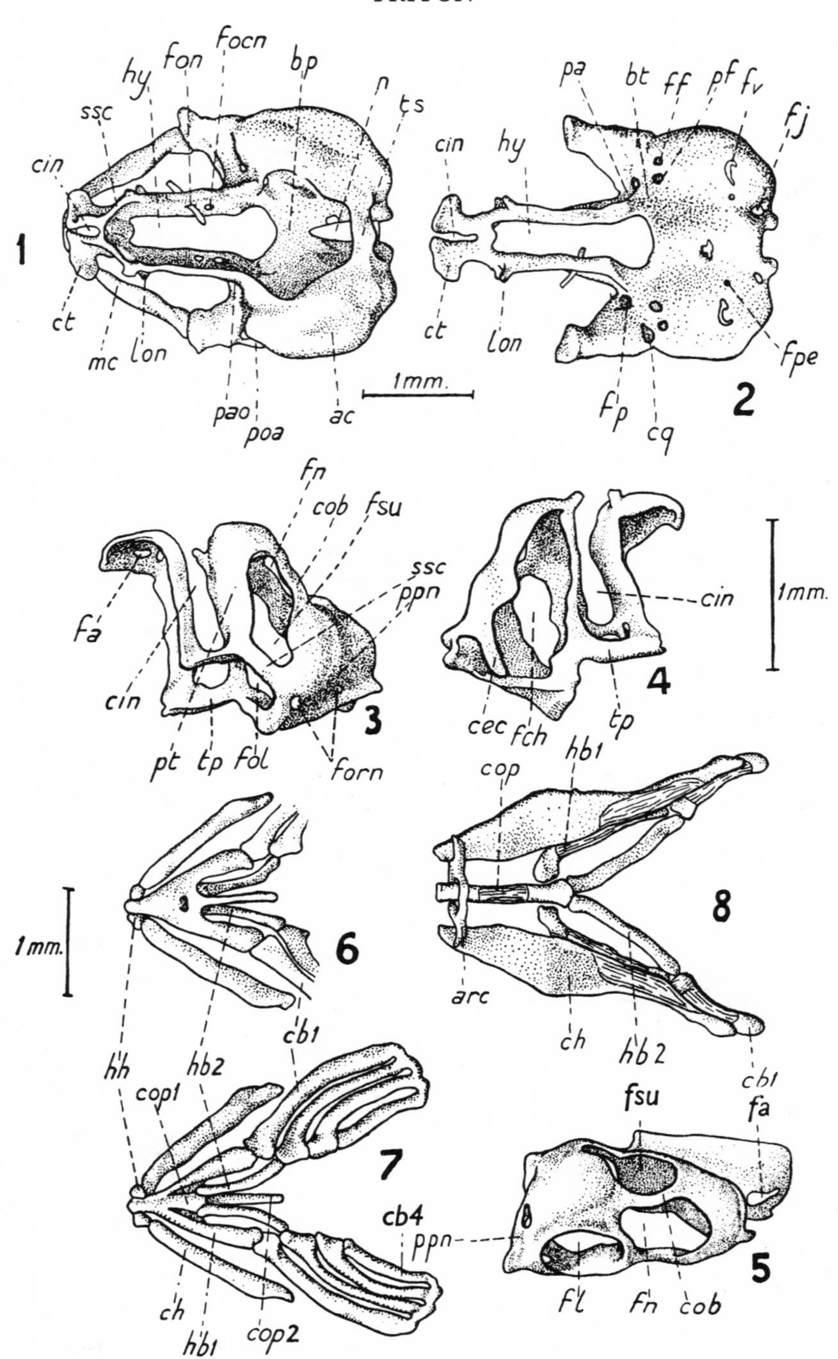

1, dorsal, and **2**, ventral view of 20 mm. stage; **3**, dorsal, **4**, ventral, and **5**, lateral view of nasal capsule of (metamorphosed) 34 mm. stage; **6**, ventral, and **7**, dorsal view of visceral arches of 20 mm. stage; **8**, dorsal view of visceral arches of (metamorphosed) 34 mm. stage (after Gaupp).

Explanation of Lettering precedes Plate 67

SALAMANDRA

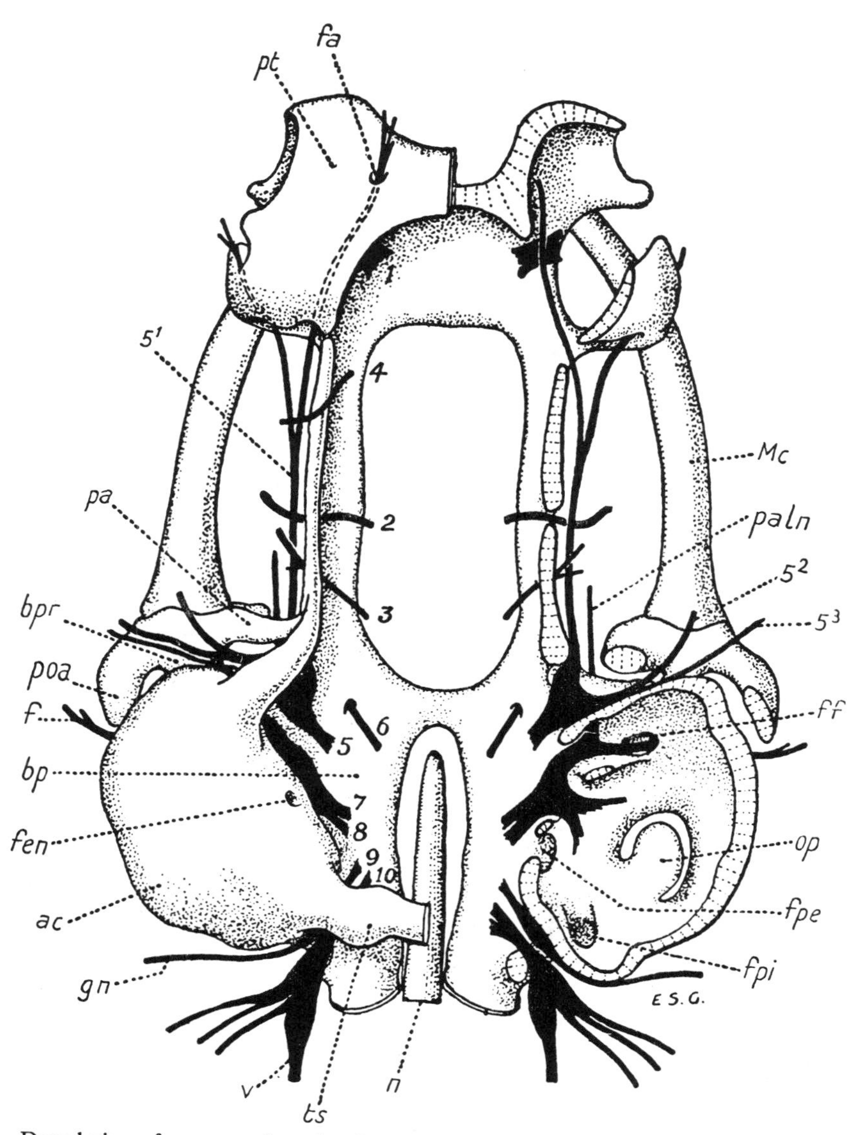

Dorsal view of neurocranium showing relations to nerves (from Goodrich, '*Studies on the Structure and Development of Vertebrates*'; by courtesy of Messrs MacMillan & Co. Ltd.).

Explanation of Lettering precedes Plate 67

SALAMANDRA

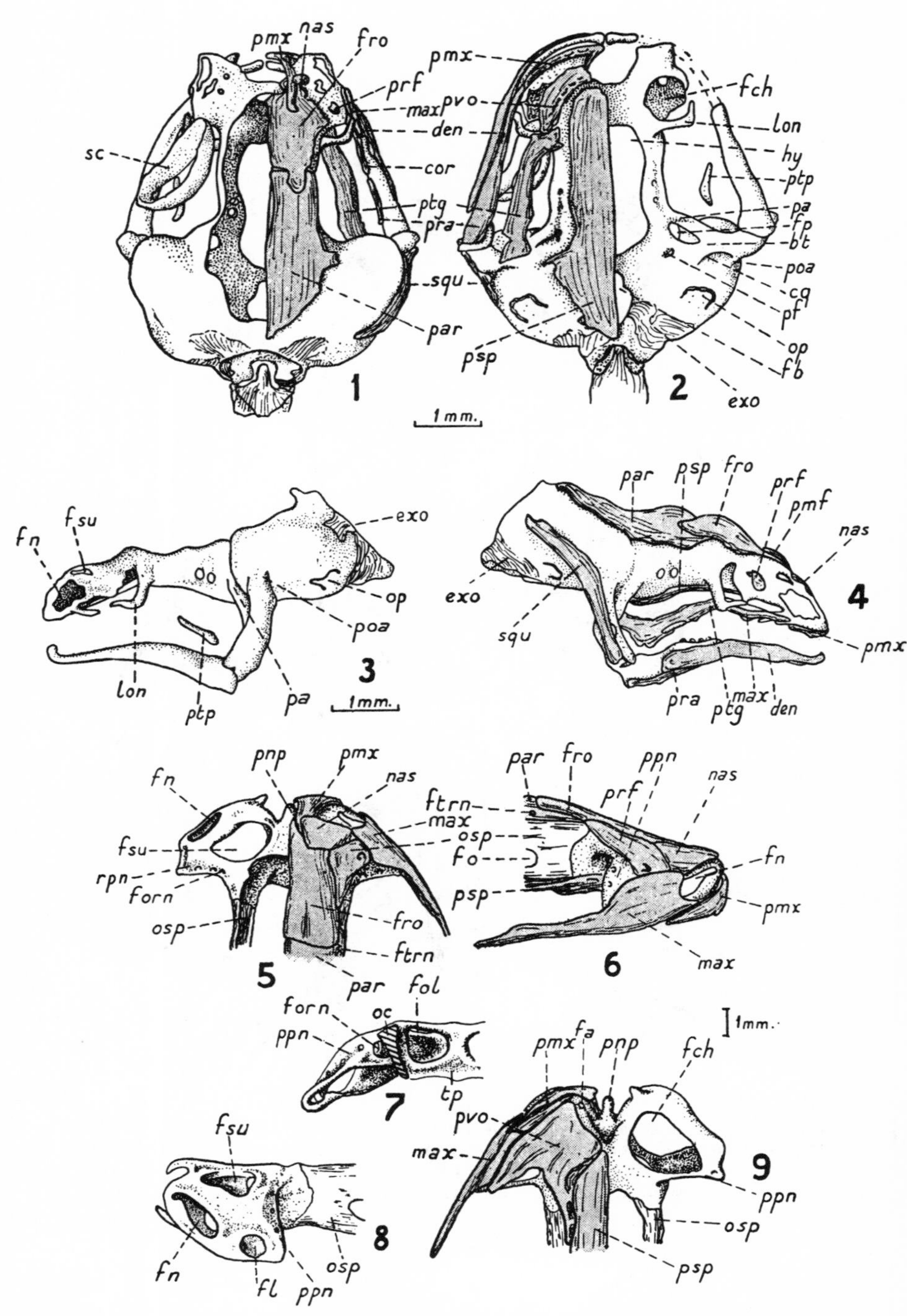

1, dorsal, **2,** ventral, **3** and **4,** lateral view of 35 mm. stage; **5,** dorsal, **6,** lateral, **7,** posterior, **8,** lateral, and **9,** ventral view of anterior region of neurocranium of 82 mm. stage (after Stadtmüller).

Explanation of Lettering precedes Plate 67

SALAMANDRA

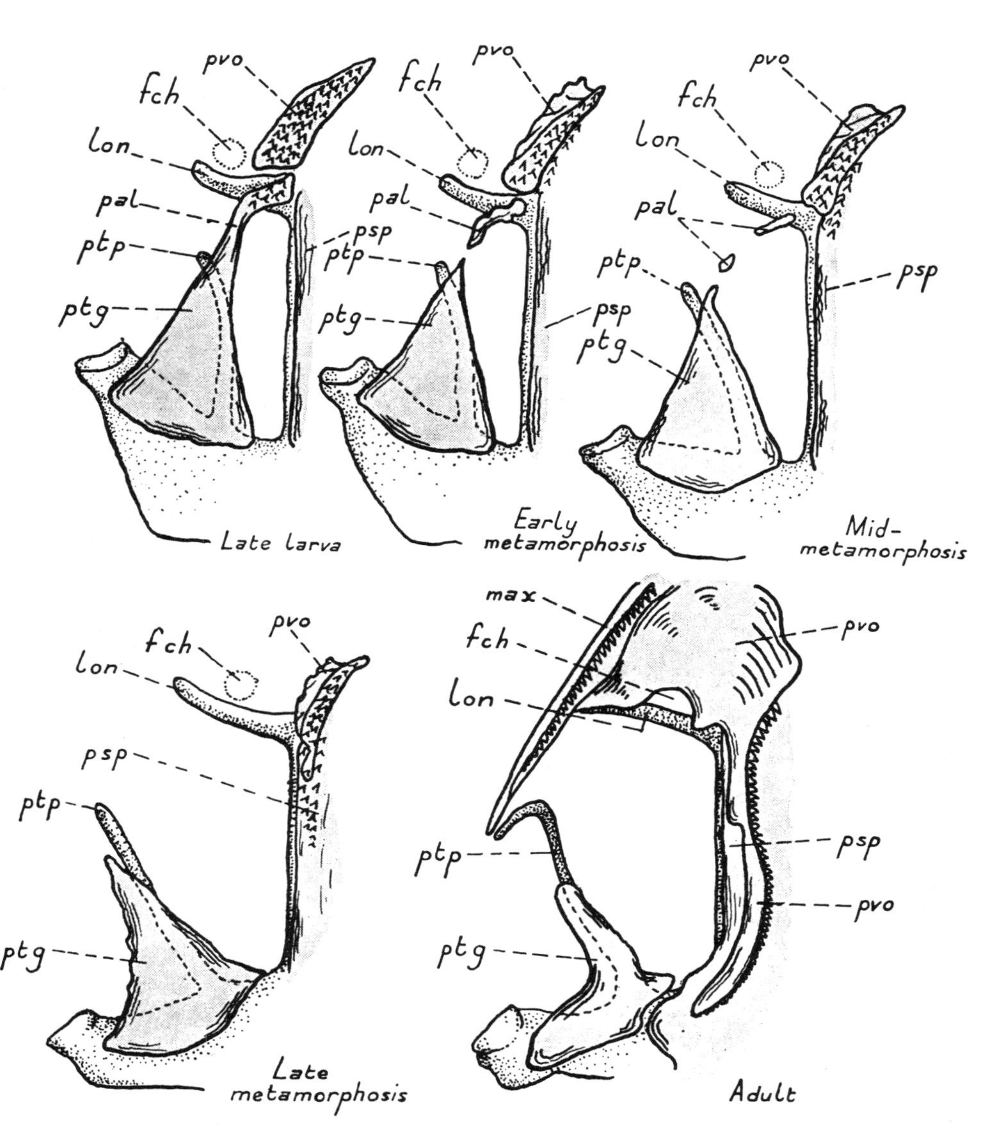

Stages in development of bony palate (after Wintrebert).

Explanation of Lettering precedes Plate 67

CRYPTOBRANCHUS, ETC.

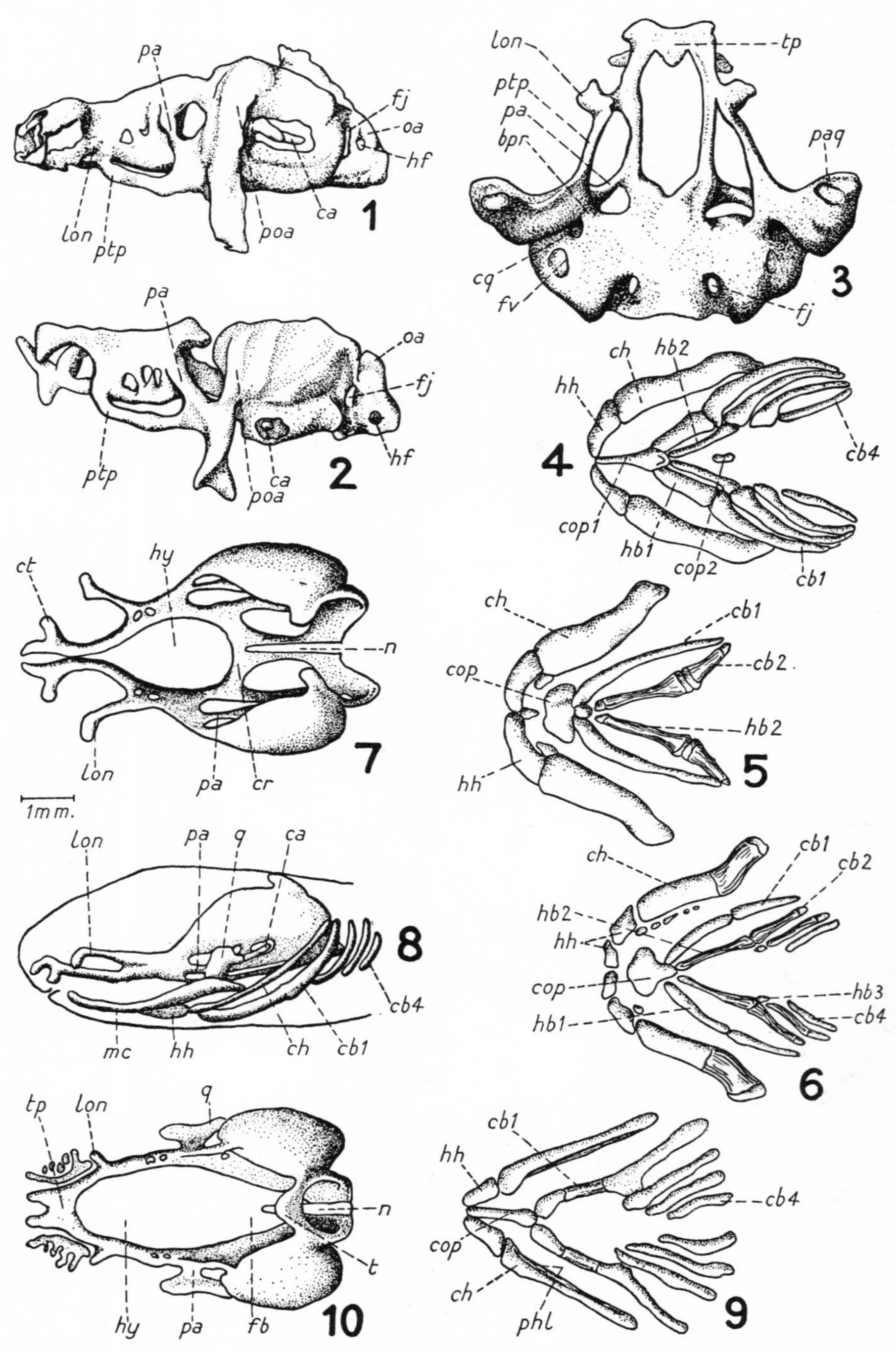

1, lateral view of 37 mm. Cryptobranchus japonicus; **2,** lateral view of 26 mm. C. (Menopoma) alleghaniensis; **3,** ventral view of 20 mm. Hynobius; **4,** visceral arches of 37 mm. C. japonicus; **5,** visceral arches of adult C. japonicus; **6,** visceral arches of adult C. (Menopoma) alleghaniensis (after Edgeworth); **7,** dorsal, **8,** lateral view, and **9,** visceral arches of 45 mm. Amphiuma (after Hay); **10,** dorsal view of 45 mm Necturus (after Winslow).

Explanation of Lettering precedes Plate 67

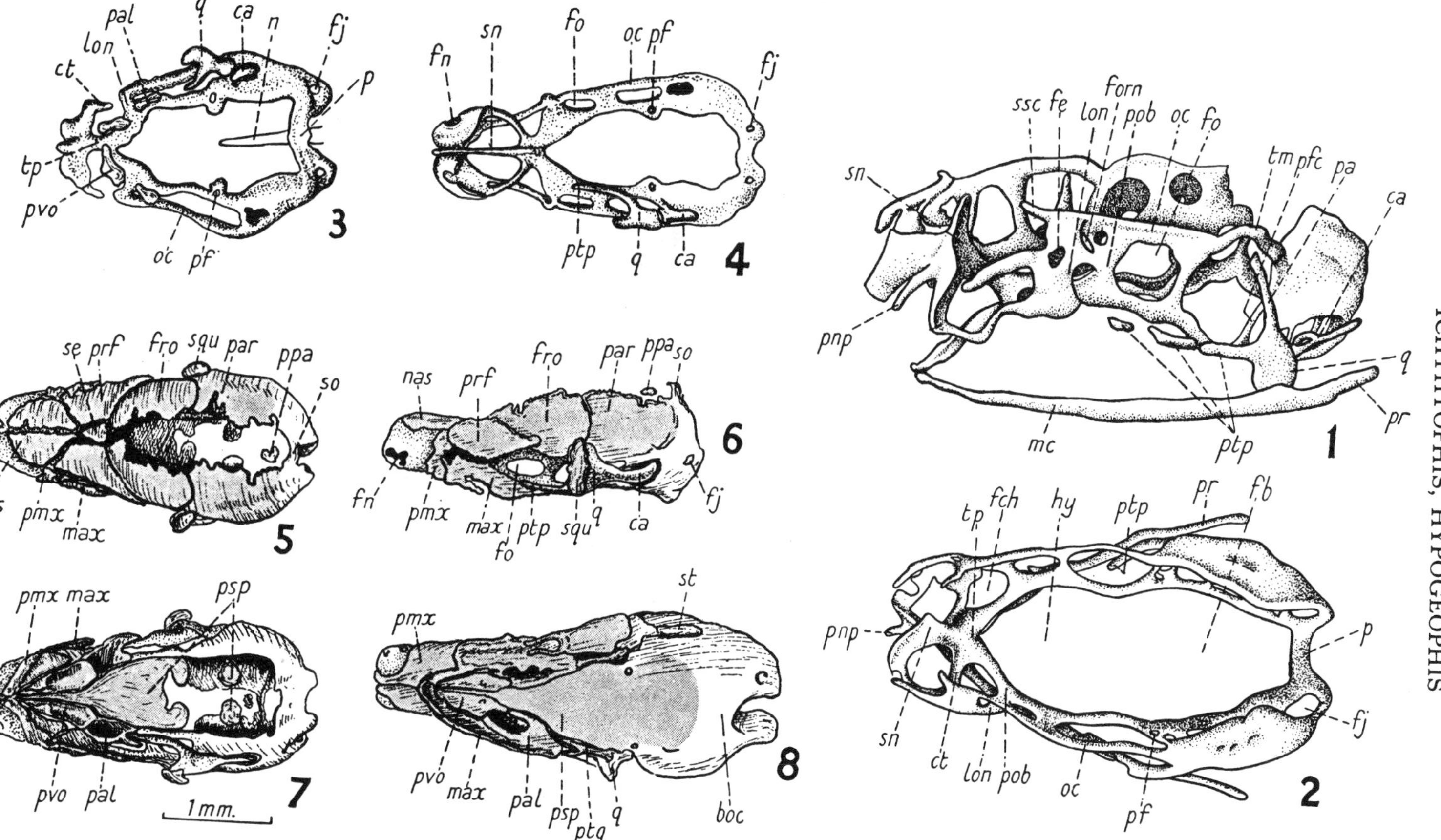

1, lateral, and **2**, dorsal view of neurocranium of Ichthyophis (after Peter); **3**, ventral view of 25 mm. Hypogeophis; **4**, maximum development of chondrocranium in Hypogeophis; **5**, dorsal, **6**, lateral, and **7**, ventral view of Hypogeophis stage 47; **8**, ventral view of 68 mm. Hypogeophis (after Marcus, Stimmelmayr and Porsch).

Explanation of Lettering precedes Plate 67

RANA

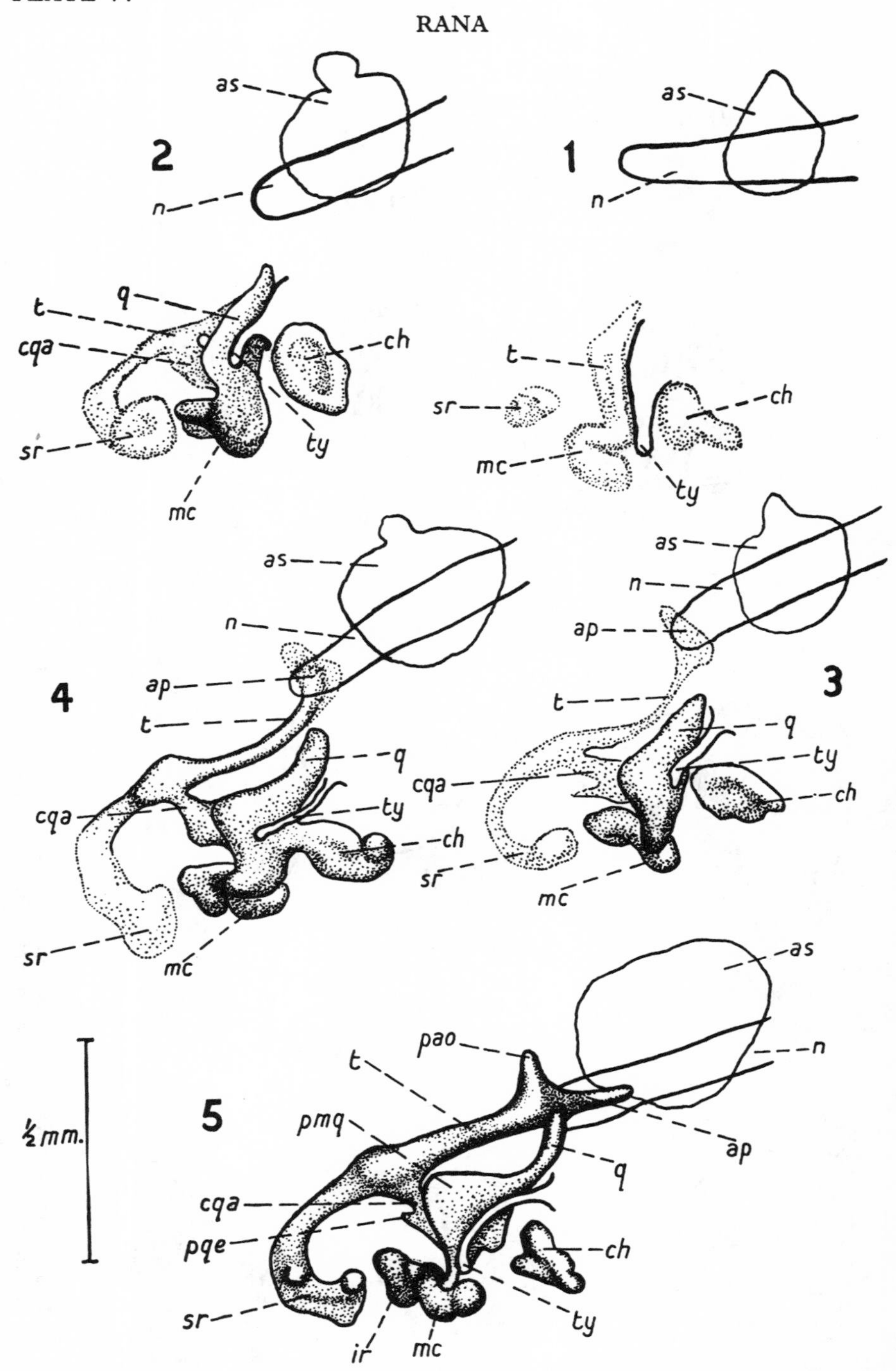

Lateral view of: **1**, 5 mm., **2**, 7 mm., **3**, 7·5 mm., **4**, 8 mm., and **5**, 10 mm. stages (after Spemann).

Explanation of Lettering precedes Plate 67

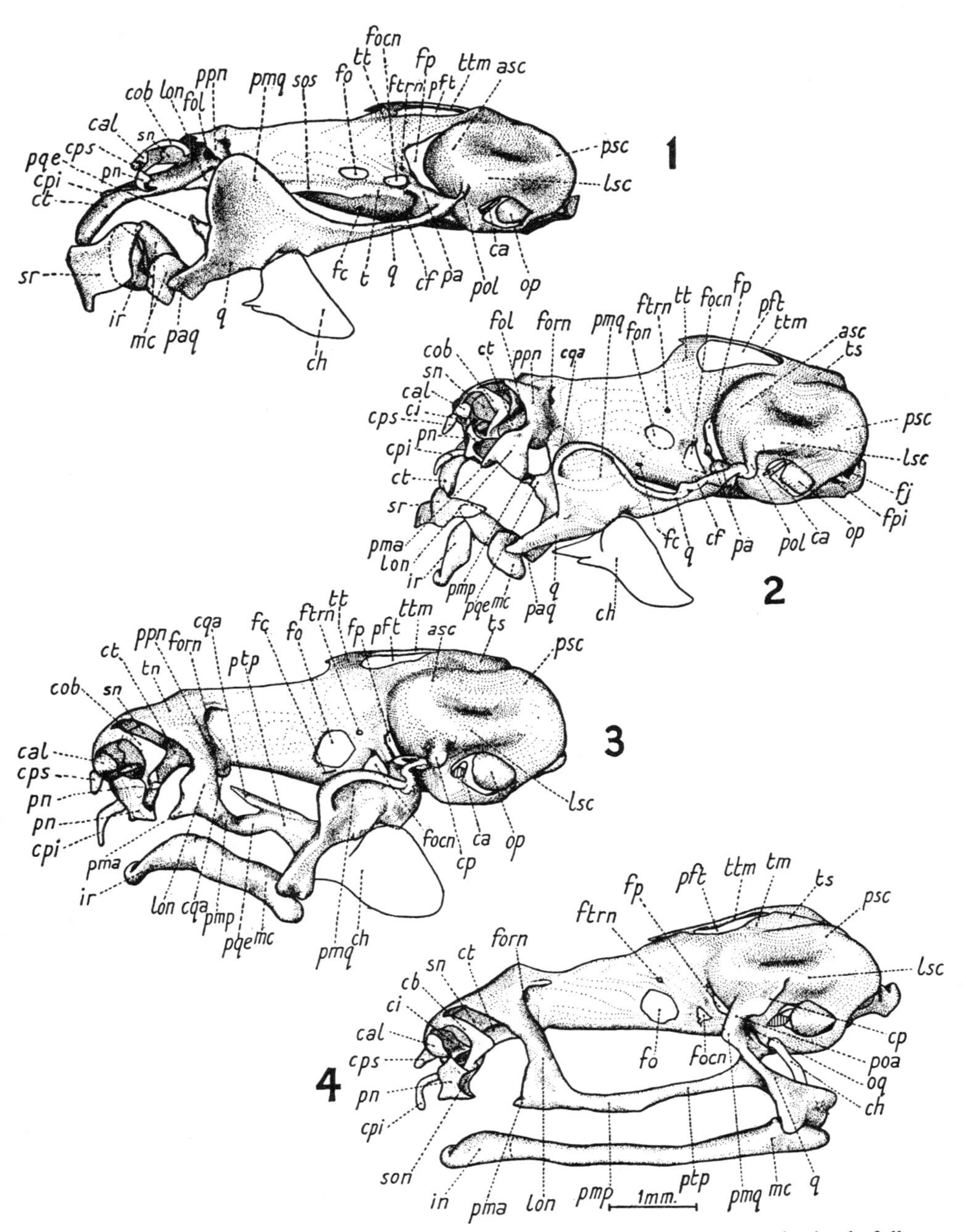

Lateral view of chondrocrania to illustrate changes consequent on metamorphosis; **1**, fully developed larval condition; **2** and **3**, intermediate stages; **4**, adult (from Pusey).

Explanation of Lettering precedes Plate 67

RANA

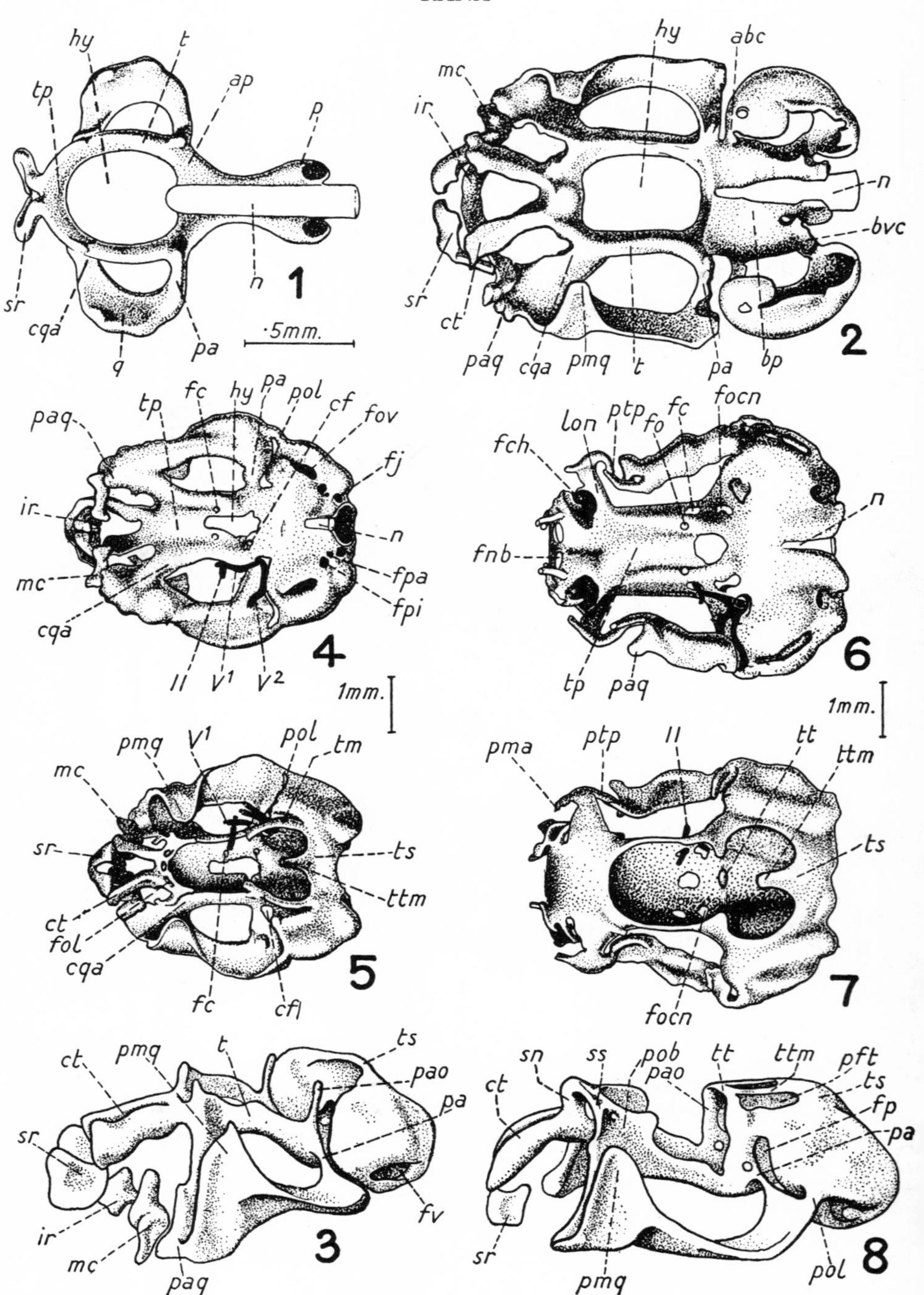

1, dorsal view of 7·5 mm. stage (after Stöhr); **2**, dorsal, and **3**, lateral view of 14 mm. stage; **4**, ventral, and **5**, dorsal view of 29 mm. stage; **6**, ventral, **7**, dorsal, and **8**, lateral view of fully developed larva (**2**, **4**, **5**, **6**, and **7** after Gaupp; **3** and **8** from Jenkinson's models).

Explanation of Lettering precedes Plate 67

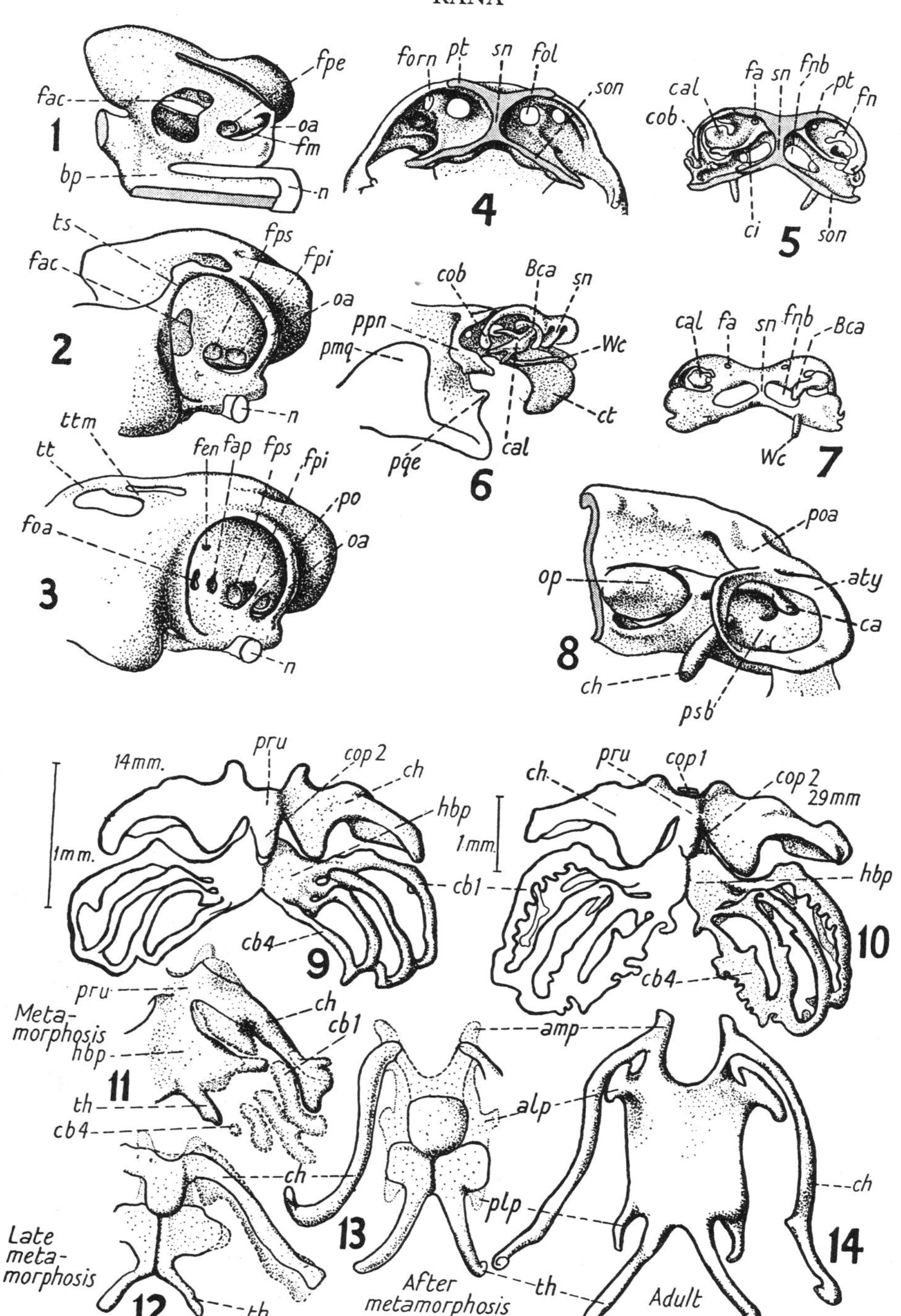

1, 2, and **3,** stages in the formation of the medial wall of the auditory capsule and perilymphatic foramina; **4,** anterior view of hinder half of nasal capsule; **5,** posterior view of front half of nasal capsule; **6,** lateral, and **7,** anterior view of nasal capsule of fully developed larva; **8,** lateral view of right tympanic ring, columella auris and operculum; **9–14,** stages in the development of the hyobranchial skeleton (all after Gaupp except **2, 3,** and **8** which are from Jenkinson's models).

Explanation of Lettering precedes Plate 67

RANA

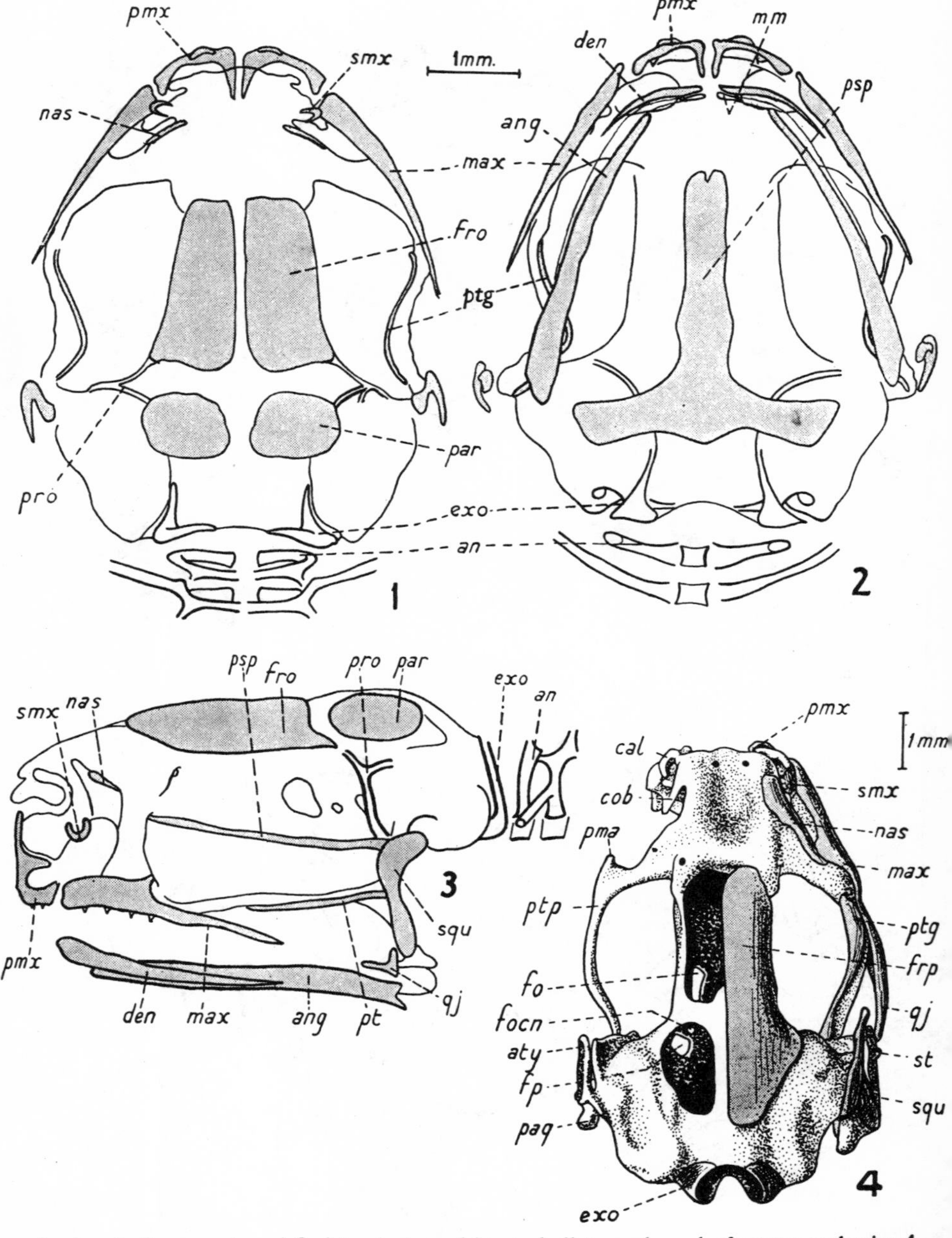

1, dorsal, **2**, ventral, and **3**, lateral view of bony skull towards end of metamorphosis; **4**, dorsal view of young metamorphosed frog (after Gaupp).

Explanation of Lettering precedes Plate 67

ALYTES, XENOPUS

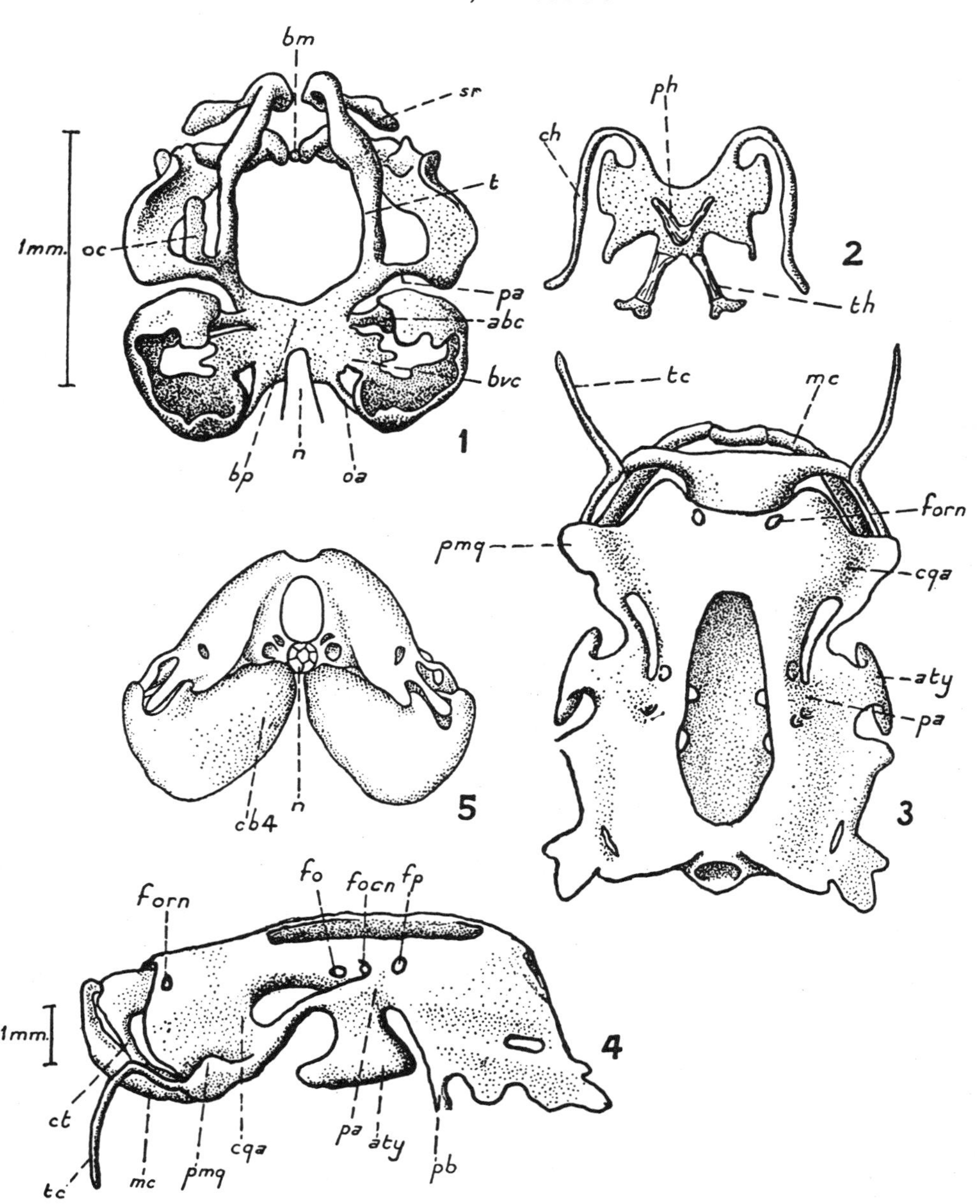

1, dorsal view of 7 mm. stage of Alytes (after van Seters); **2,** ventral view of hyoid plate of Alytes (after Ridewood); **3,** dorsal, **4,** lateral, and **5,** posterior view of 48 mm. stage of Xenopus (after Kotthaus).

Explanation of Lettering precedes Plate 67

REPTILIA

Explanation of Lettering

a, aditus conchae
ac, auditory capsule
ang, angular
art, articular
asc, anterior semicircular canal

bc, basicapsular commissure
bd, basidorsal cartilage
bf, basicapsular fenestra
boc, basioccipital
bpt, basipterygoid process
bsp, basisphenoid
bt, basitrabecular process

ca, columella auris
cb 1, cornu branchiale 1 (ceratobranchial 1)
cb 2, cornu branchiale 2 (ceratobranchial 2, distal portion)
cb 2a, cornu branchiale 2 (ceratobranchial 2, proximal portion)
cec, cartilago ectochoanalis
ch, cornu hyale (ceratohyal)
co, concha nasalis
cor, coronoid
cp, crista parotica
cr, crista sellaris
cs, sphenethmoid commissure
cv, cavum conchale

den, dentary
dgl, duct of lateral nasal gland
dl, dental lamina

e, eye
en, external nostril
ept, epipterygoid
exc, extracolumella
exo, exoccipital

fa, foramen apicale
faa, foramen acusticum anterius
fab, foramen for abducens
fap, foramen acusticum posterius
fb, basicranial fenestra
fbr, forebrain
fc, carotid foramen
fe, foramen epiphaniale
fen, foramen endolymphaticum
fep, fenestra epioptica
ff, foramen faciale
fg, foramen for glossopharyngeal
fil, foramen intervestibulare laterale
fim, foramen intervestibulare mediale
fj, foramen jugulare
fl, fenestra lateralis
fm, fissura metotica
fmo, fenestra metoptica
fn, fenestra narina
fo, fenestra optica
foa, foramen for ophthalmic artery
foc, foramen for oculomotor
fol, fenestra olfactoria
fp, fenestra prootica
fpe, foramen perilymphaticum
fpp, foramen prepalatinum
fro, frontal,
fs, fenestra septalis
fsu, fenestra superior
ftr, foramen for trochlear
fv, foramen vestibulare (ovale)
fx, foramen 'x'

g, gland
gl, lateral nasal gland

hc, hypochordal commissure
hf, hypoglossal foramina
hh, hypohyal
Huf, 'Huxley's' foramen
hy, hypophysial fenestra

in, internal nostril
ip, incisura prootica
ipp, infrapolar process

Jo, Jacobson's organ
jug, jugal

lac, lachrymal
lh, laterohyal
lsc, lateral semicircular canal
lta, lamina transversalis anterior

max, maxilla
Mc, Meckel's cartilage
mp, meniscus pterygoideus

n, notochord
nas, nasal

oa, occipital arch
oc, cavity of olfactory sac
on, olfactory nerve
op, otic process
opo, opisthotic
orc, orbital cartilage

p, parachordal cartilage
pa, processus ascendens of pterygoquadrate
paa, prominentia ampullaris anterioris
pac, pila accessoria
pai, processus alaris inferior
pal, palatine
pan, lamina orbitonasalis
pao, pila antotica (prootica)
par, parietal
pas, processus alaris superior
pat, processus anterior of tectum synoticum
pc, paraseptal cartilage
pco, prominentia cochlearis
pd, processus dorsalis
pe, processus lingualis
pfc, prefacial commissure
pi, pars interhyalis
pla, prominentia ampullaris lateralis
pls, pleurosphenoid
pm, pila metoptica
pma, processus maxillaris anterior
pmp, processus maxillaris posterior
pmx, premaxilla
pn, paranasal cartilage
po, preoptic root of orbital cartilage
poa, preoccipital arch
por, postorbital
pp, parotic process
pr, processus retroarticularis
pra, prearticular
prf, prefrontal
prn, prenasal process
pro, prootic
ps, planum supraseptale
psa, prominentia saccularis
psc, posterior semicircular canal
psp, parasphenoid
pt, parietotectal cartilage of nasal capsule
ptf, postfrontal
ptg, pterygoid
ptp, pterygoid process
pvo, prevomer

q, quadrate cartilage
qj, quadratojugal
qu, quadrate bone

r, raphe between medial and lateral nasal processes
re, recessus extraconchalis
ret, ramus ethmoidalis profundi
rl, ramus lateralis ethmoidalis
rm, ramus medialis ethmoidalis

san, supra-angular
scp, subcapsular process
sh, stylohyal
si, interorbital septum
sin, subiculum infundibuli
siv, septum intervestibulare
smx, septomaxilla
sn, nasal septum
soc, supraoccipital
spl, splenial
squ, squamosal
ss, sinus superior
stb, supratrabecular bar
stp, supratemporal

t, trabecula cranii
tc, trabecula communis
ti, taenia intertrabecularis
tl, true side wall of nasal capsule = inner wall of cavum conchale
tm, taenia marginalis
tme, taenia medialis
tp, trabecular plate
trp, transpalatine
ts, tectum synoticum
tsm, taenia supramarginalis

vr, vacuity in roof of cochlear capsule

za, zona annularis

LACERTA

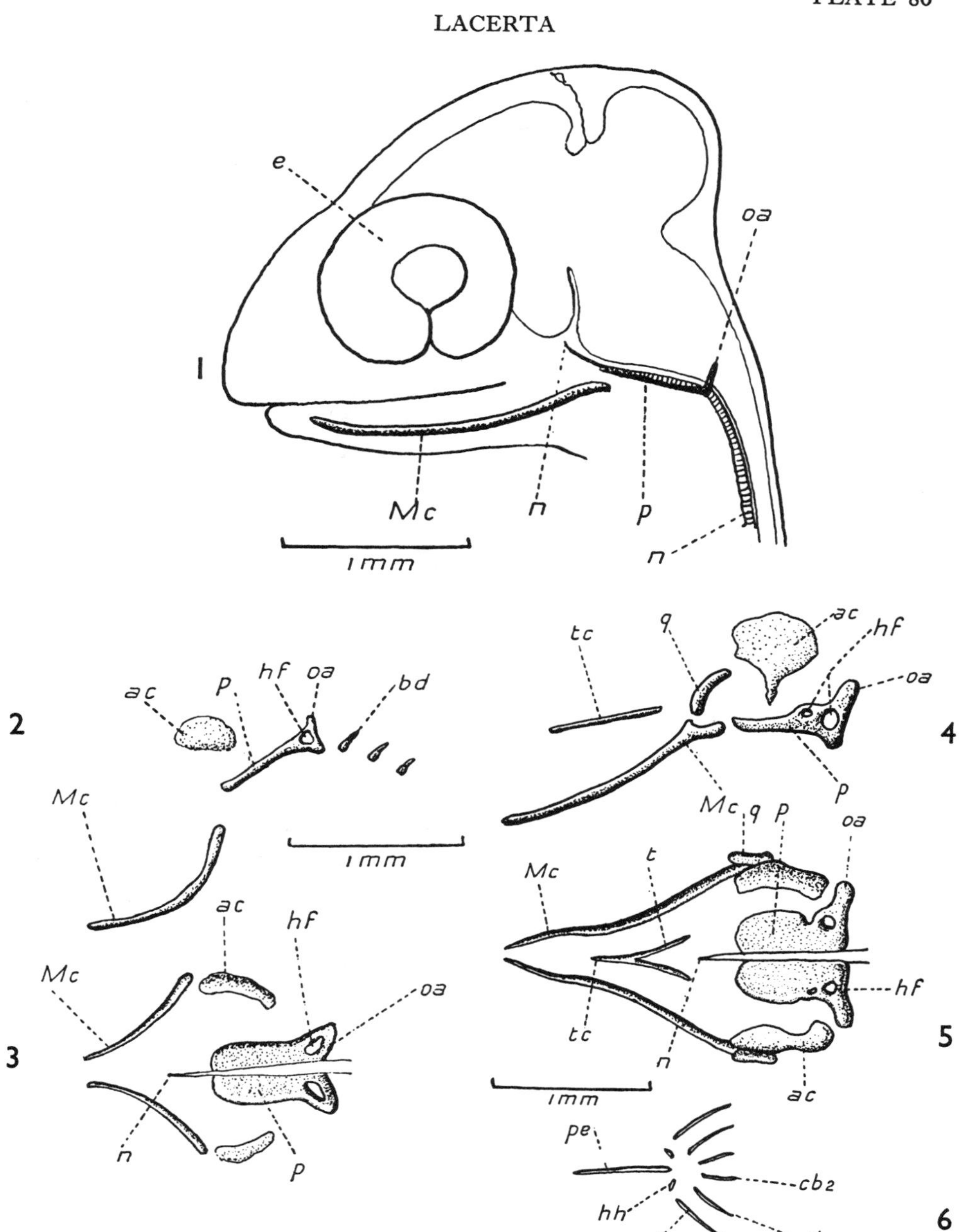

1, lateral view of 2·25 mm. (h.-l.) stage; **2,** lateral, and **3,** dorsal view of 2·5 mm. stage; **4,** lateral, and **5,** dorsal view of 3·5 mm. stage; **6,** visceral arches of 3·5 mm. stage (from de Beer).

Explanation of Lettering opposite

LACERTA

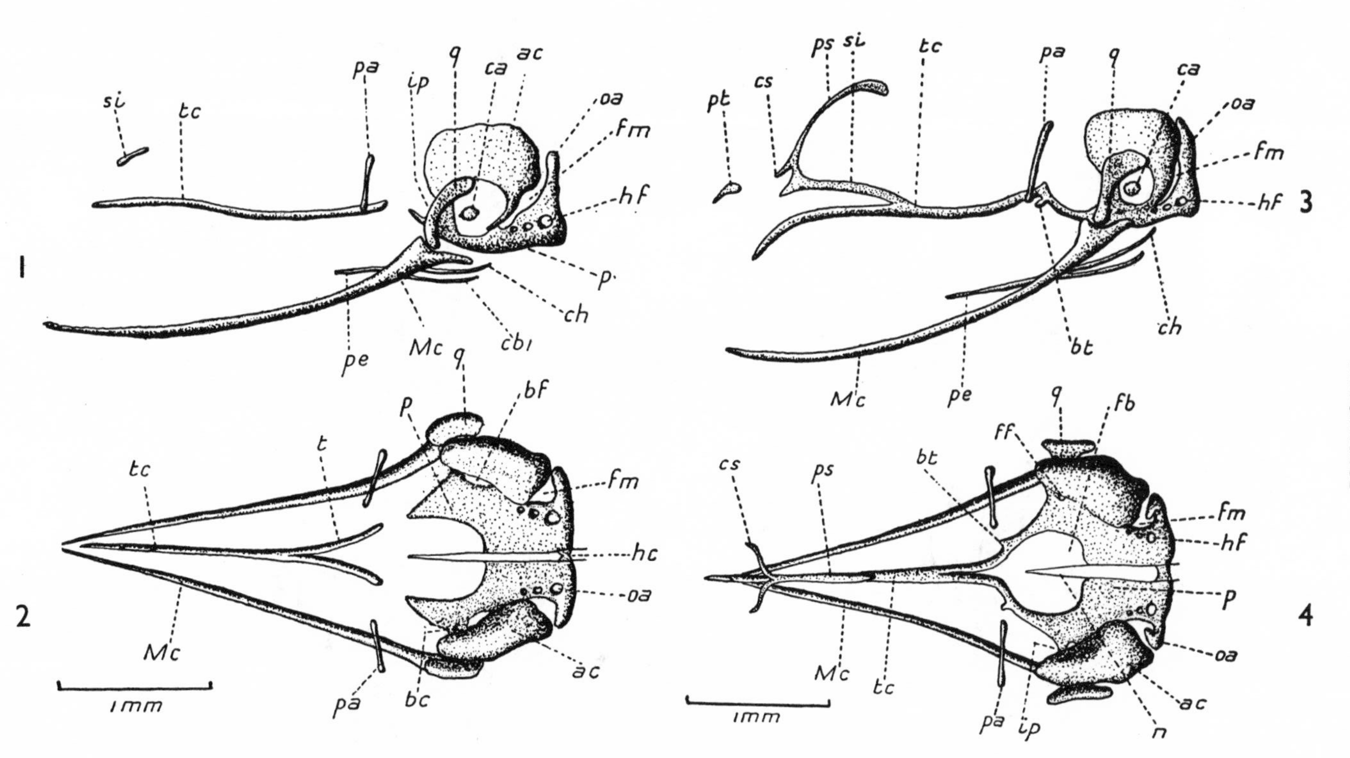

1, lateral, and **2**, dorsal view of 4 mm. stage; **3**, lateral, and **4**, dorsal view of 4·5 mm. stage (from de Beer).

Explanation of Lettering precedes Plate 80

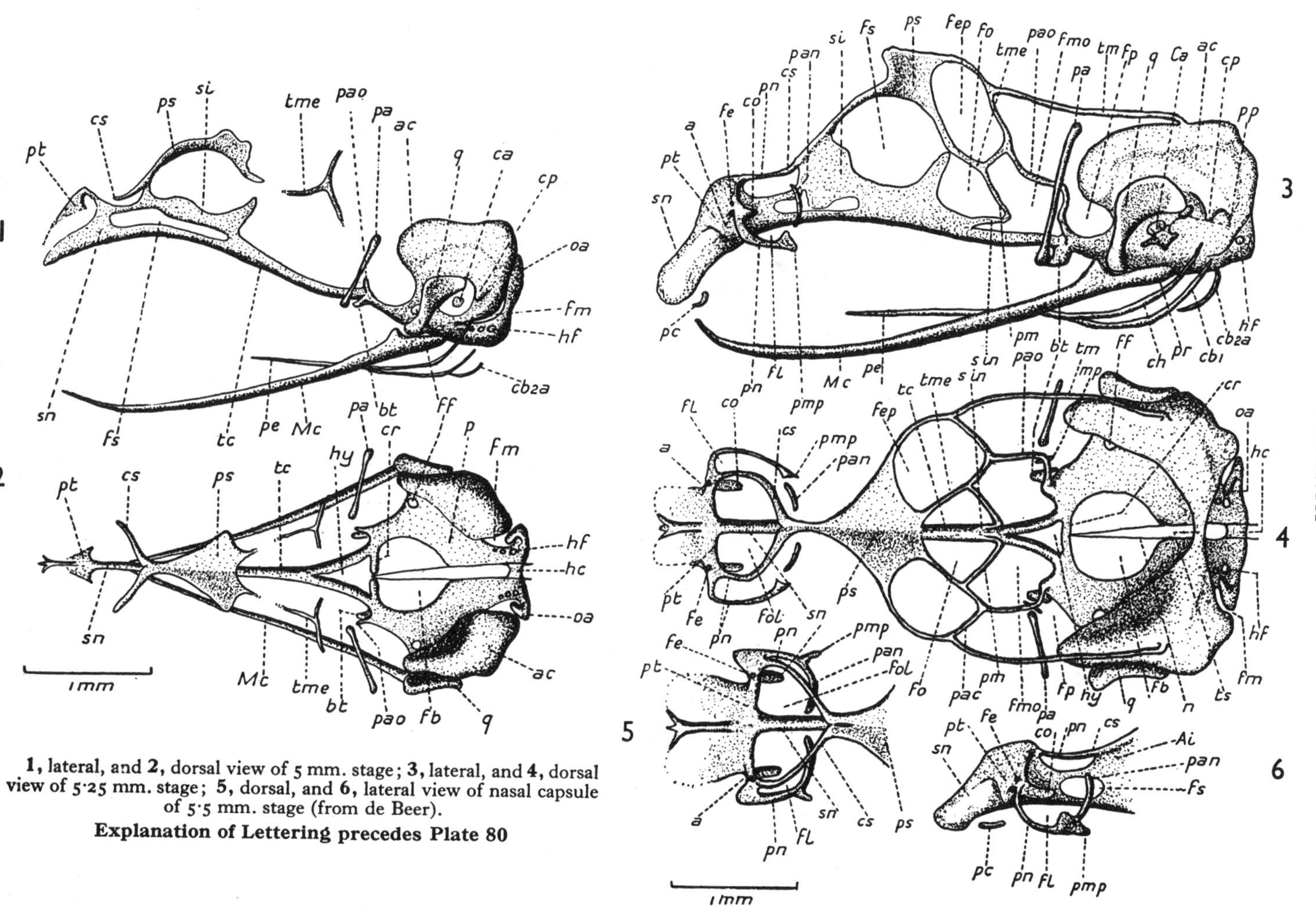

1, lateral, and **2,** dorsal view of 5 mm. stage; **3,** lateral, and **4,** dorsal view of 5·25 mm. stage; **5,** dorsal, and **6,** lateral view of nasal capsule of 5·5 mm. stage (from de Beer).

Explanation of Lettering precedes Plate 80

PLATE 83

LACERTA

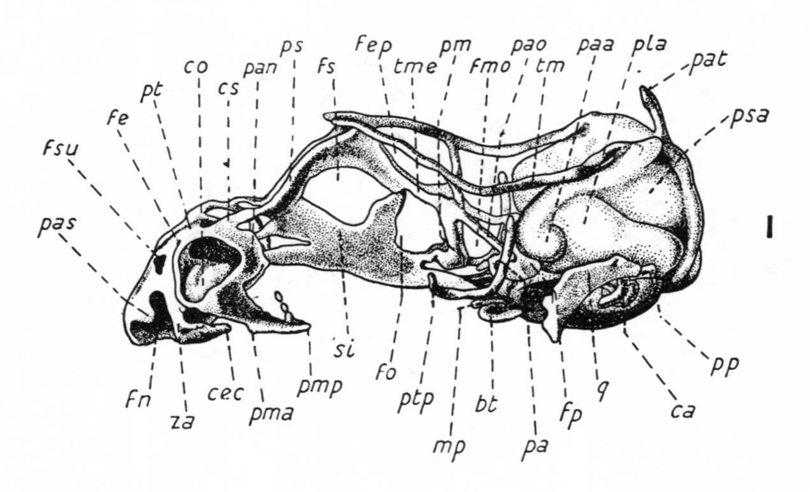

1mm

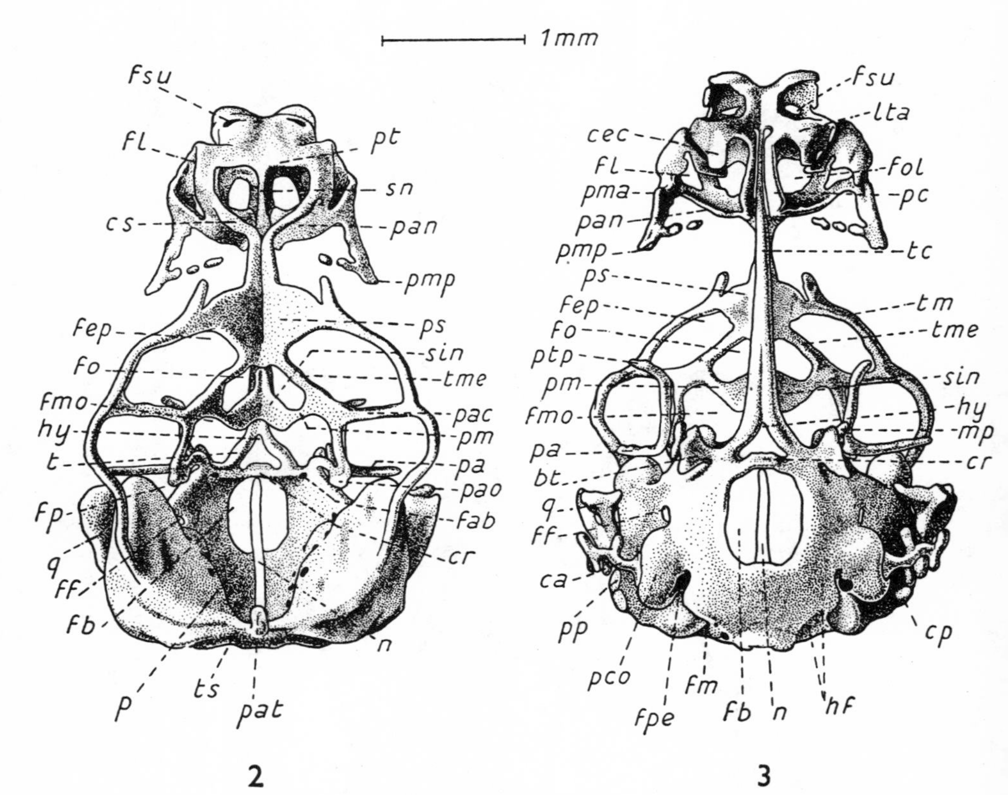

2 3

1, lateral, **2**, dorsal, and **3**, ventral view of 31 mm. (t.l.) stage (after Gaupp).

Explanation of Lettering precedes Plate 80

LACERTA

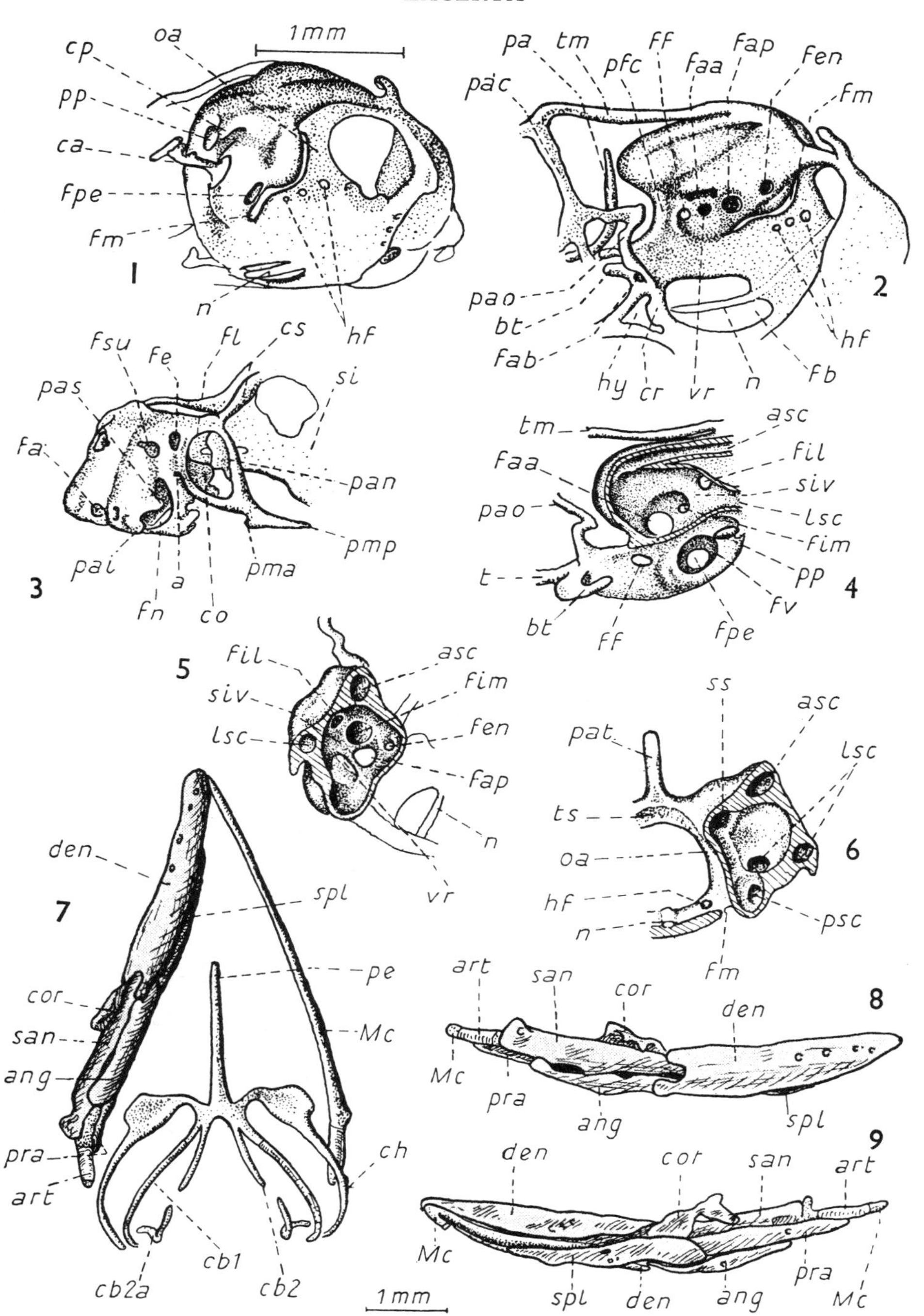

1, occipital region; **2,** medial view of auditory region; **3,** lateral view of nasal capsule, **4,** lateral view of sagittal section through auditory capsule, **5,** posterior view of anterior half of auditory capsule, **6,** anterior view of posterior half of auditory capsule, all of 31 mm. stage; **7,** ventral view of lower jaw and hyobranchial skeleton, **8,** lateral, and **9,** medial view of lower jaw, of 47 mm. stage (after Gaupp).

Explanation of Lettering precedes Plate 80

PLATE 85

LACERTA

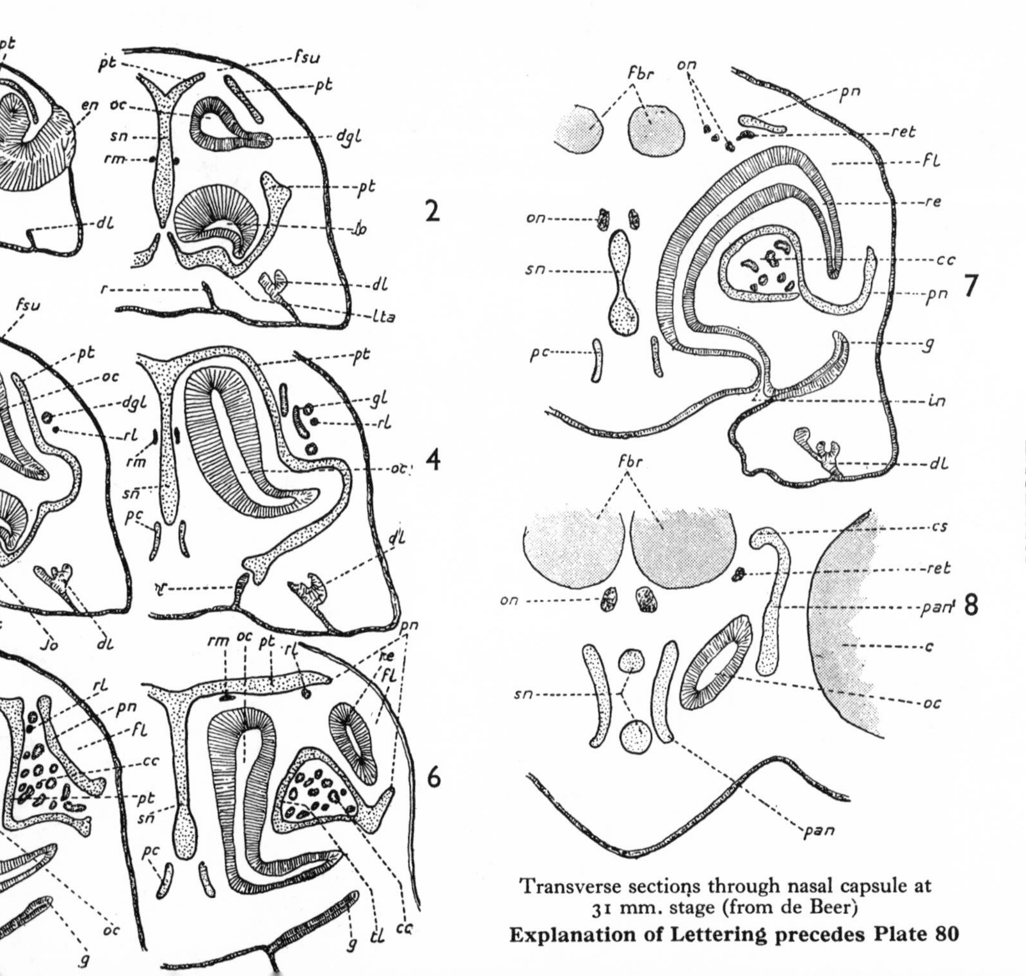

Transverse sections through nasal capsule at 31 mm. stage (from de Beer)

Explanation of Lettering precedes Plate 80

LACERTA

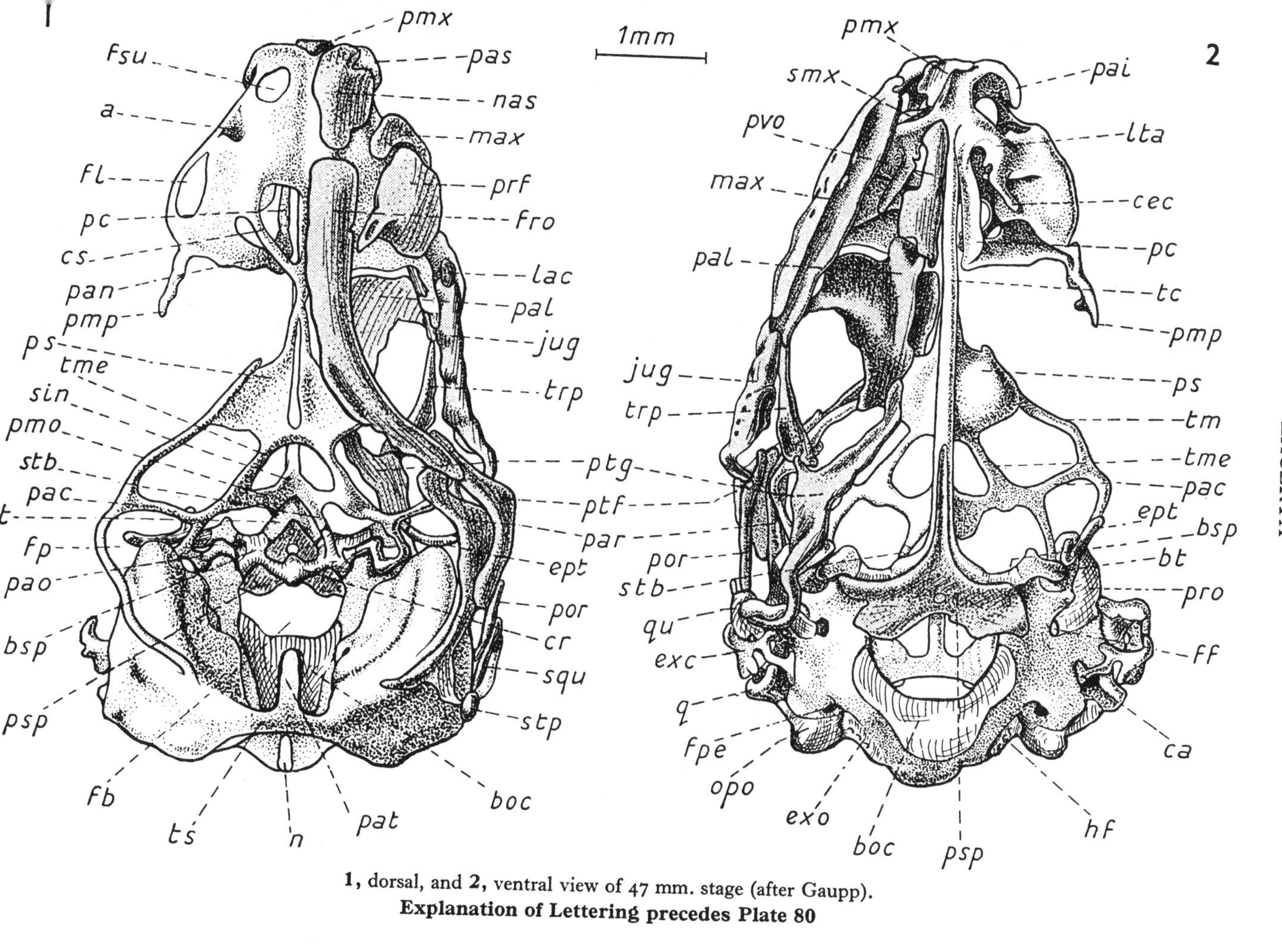

1, dorsal, and **2**, ventral view of 47 mm. stage (after Gaupp).

Explanation of Lettering precedes Plate 80

SPHENODON

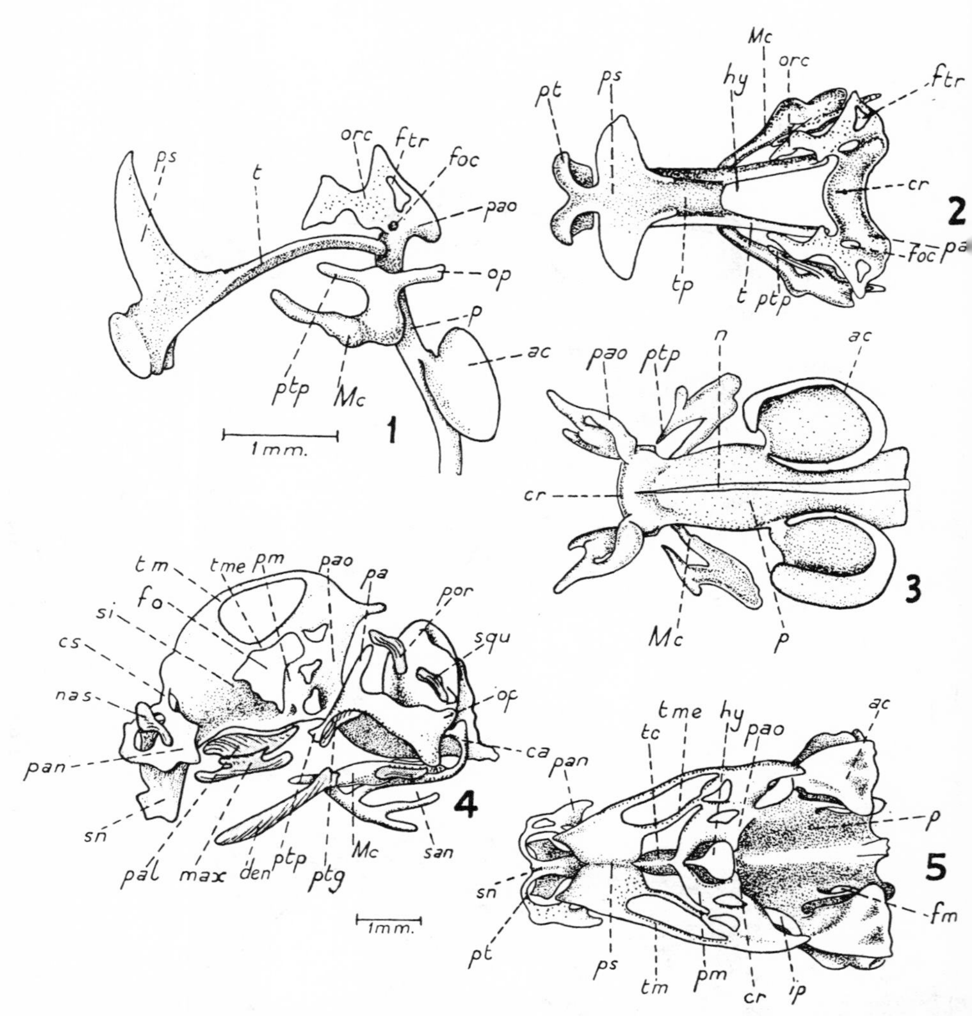

1, lateral, **2,** anterodorsal, and **3,** posterodorsal view of stage P; **4,** lateral, and **5,** dorsal view of stage Q (after Howes and Swinnerton).

Explanation of Lettering precedes Plate 80

SPHENODON

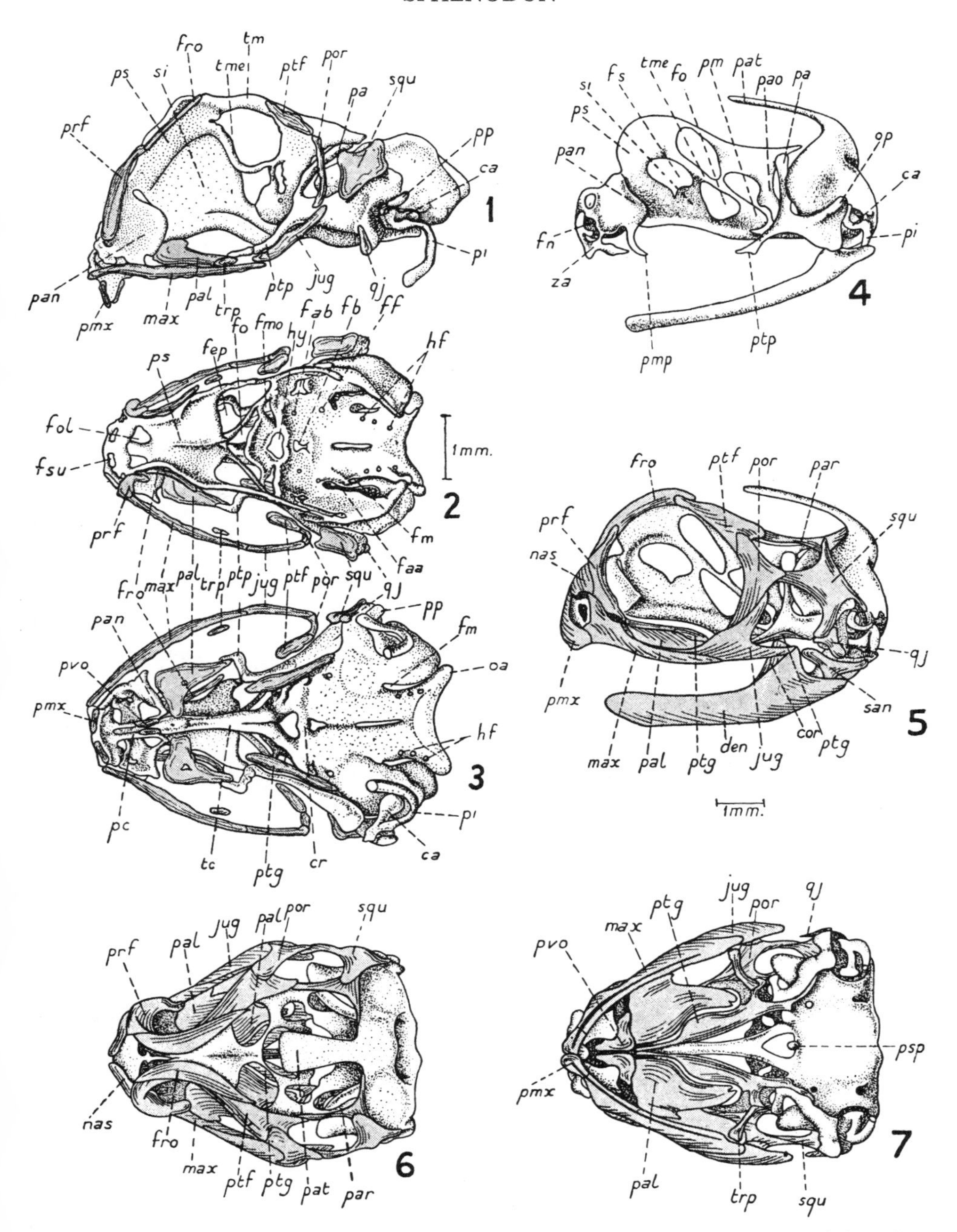

1, lateral, **2**, dorsal, and **3**, ventral view of 4·5 mm. (h.-l.) stage (after Schauinsland); **4**, lateral view of chondrocranium, **5**, lateral, **6**, dorsal, and **7**, ventral view of chondrocranium and dermal bones at stage R (after Howes and Swinnerton).

Explanation of Lettering precedes Plate 80

SPHENODON

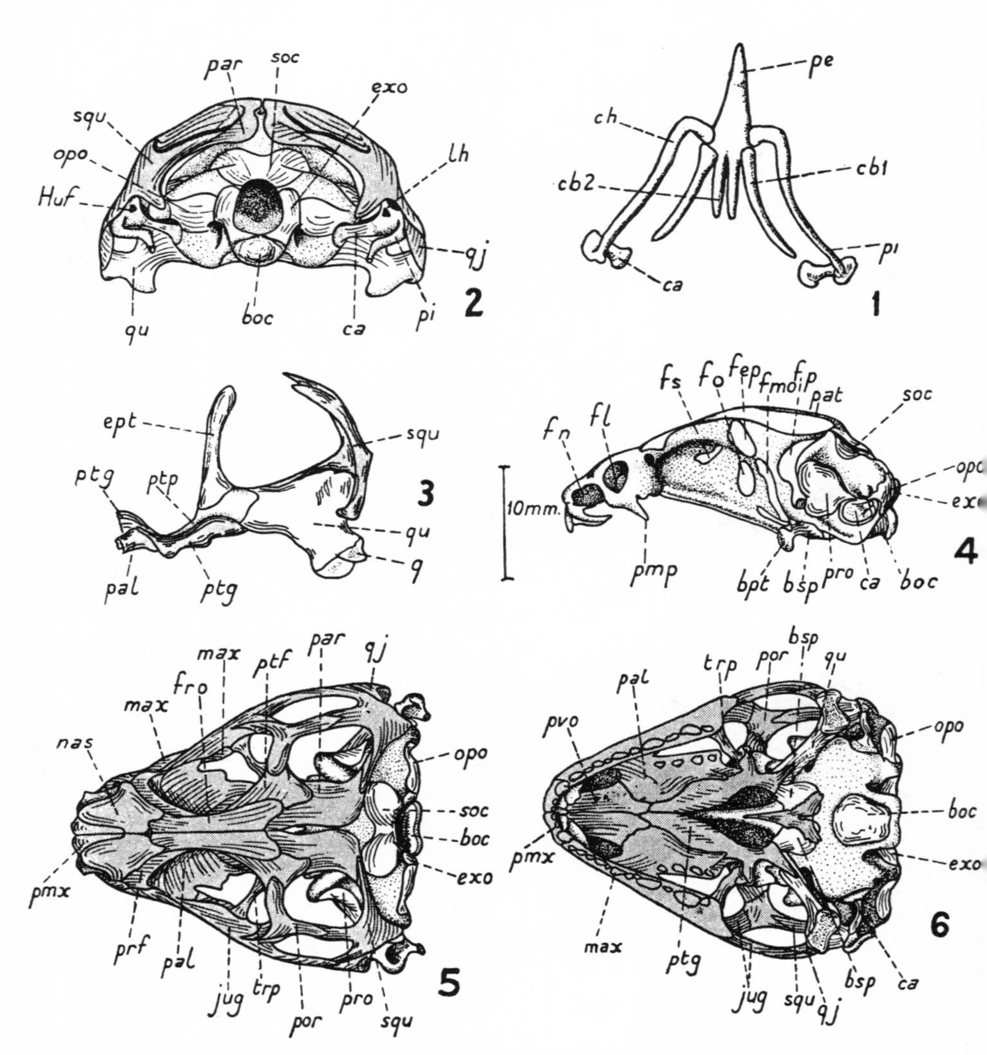

1, ventral view of hyobranchial skeleton (after Schauinsland); **2**, posterior view of stage R; **3**, pterygoquadrate of stage R; **4**, lateral, **5**, dorsal, and **6**, ventral view of stage S (after Howes and Swinnerton).

Explanation of Lettering precedes Plate 80

TROPIDONOTUS

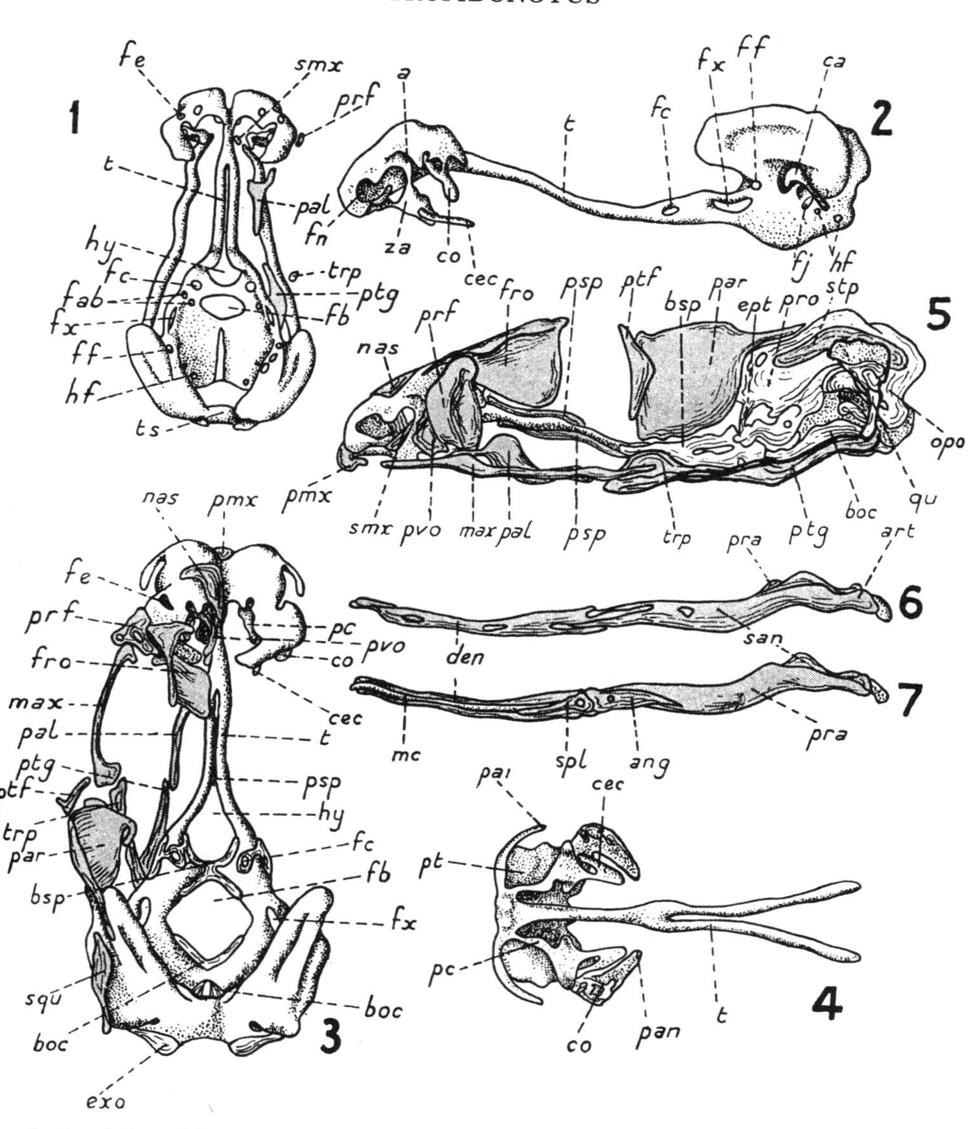

1, dorsal view of 6 mm. stage; **2,** lateral, and **3,** dorsal view of 6·8 mm. stage; **4,** ventral view of nasal capsule; **5,** lateral view of skull; **6,** lateral, and **7,** medial view of lower jaw of 13 mm. stage (after Bäckström).

Explanation of Lettering precedes Plate 80

CHRYSEMYS

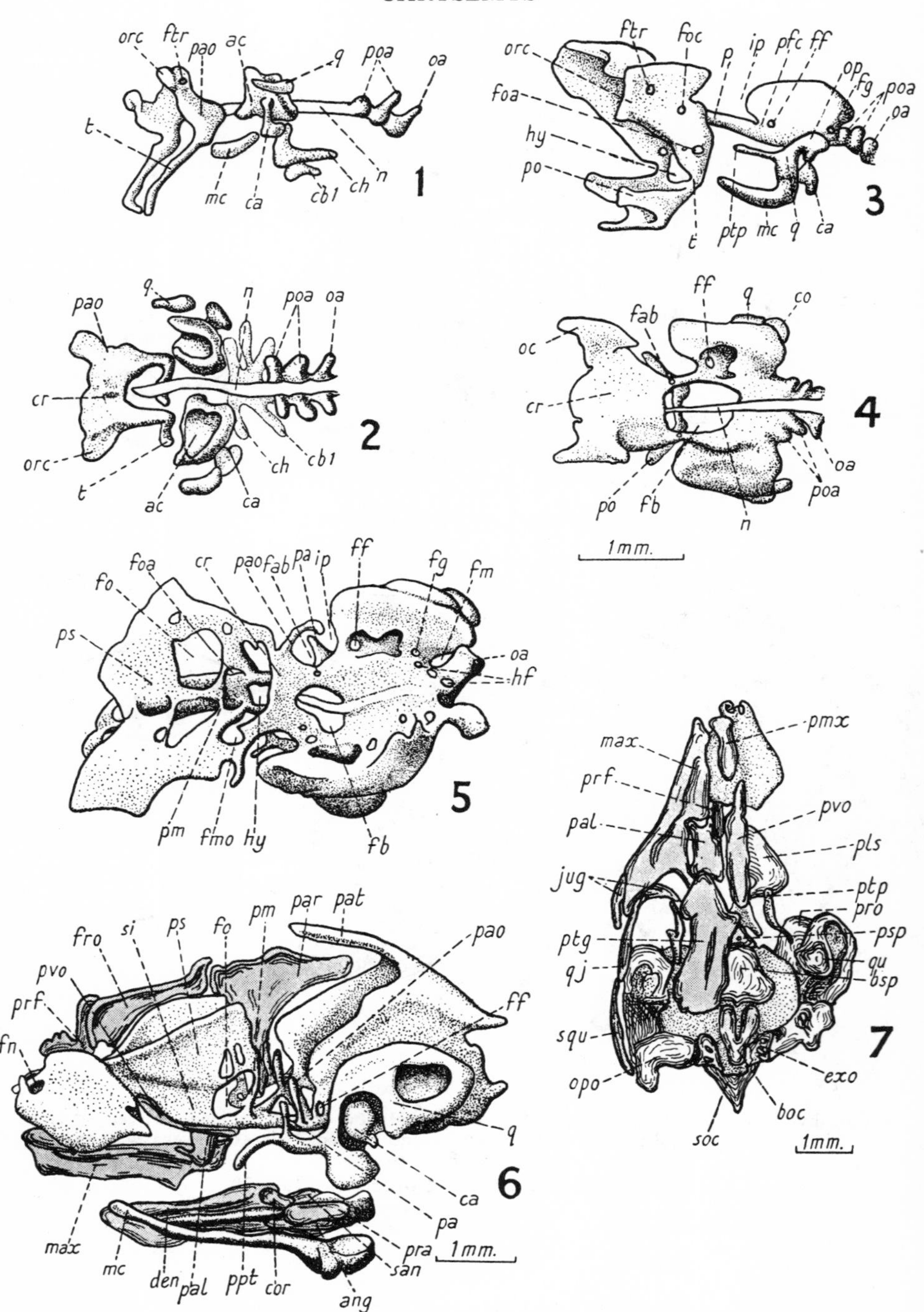

1, lateral, and **2,** dorsal view of 6 mm. stage; **3,** lateral and **4,** dorsal view of 6·75 mm. stage; **5,** dorsal view of 8 mm. stage; **6,** lateral view of 19 mm. stage; **7,** ventral view of 28 mm. stage (after Shaner).

Explanation of Lettering precedes Plate 80

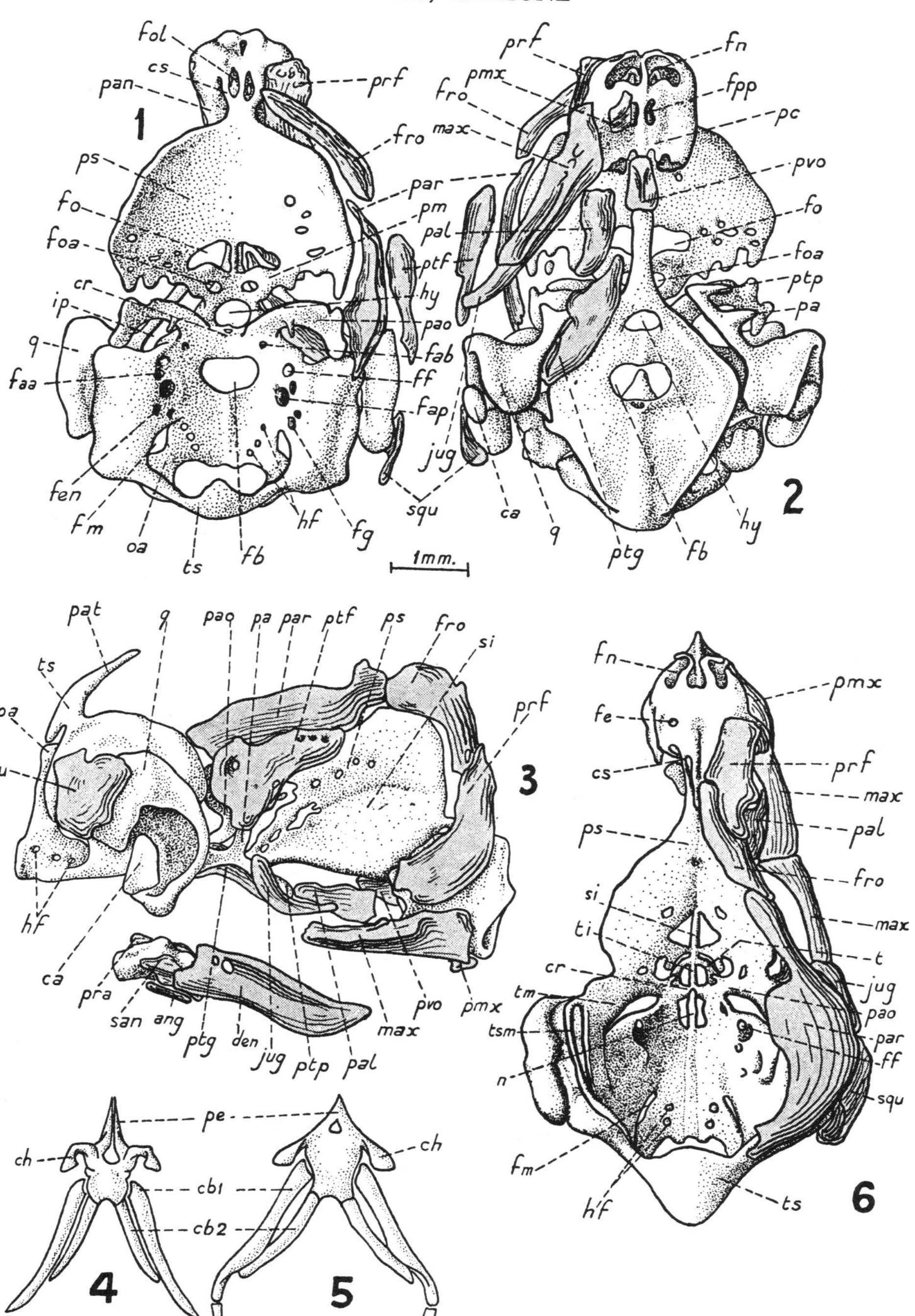

1, dorsal, **2**, ventral, and **3**, lateral view of Emys 11 mm. carapace-length (after Kunkel); **4** and **5**, stages in development of hyobranchial skeleton of Emys (after Fuchs); **6**, dorsal view of Chelone 26 mm. stage (after Fuchs).

Explanation of Lettering precedes Plate 80

CROCODILUS

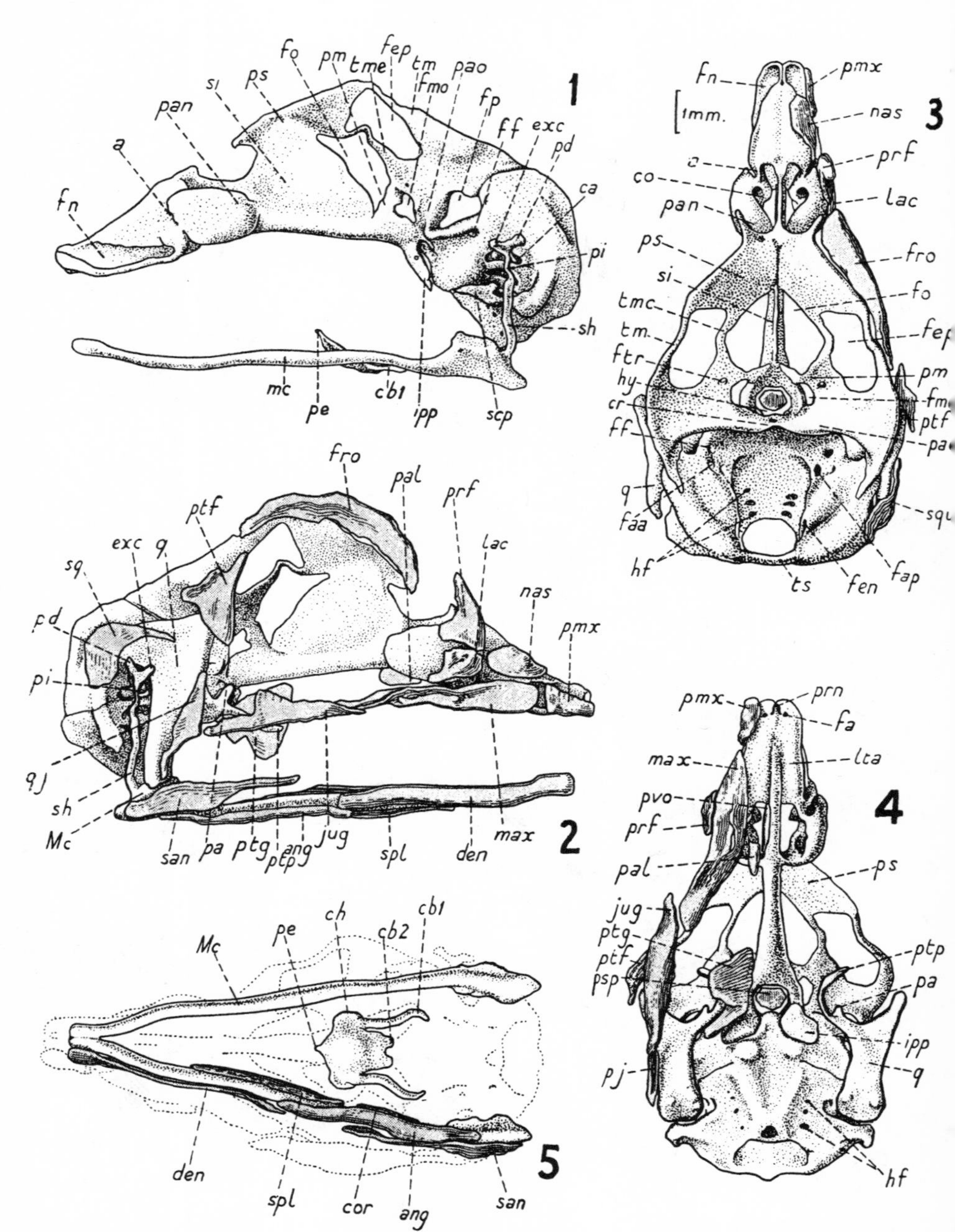

1 and **2**, lateral **3**, dorsal, **4**, ventral view of skull, and **5**, ventral view of lower jaw and hyobranchial skeleton of 13 mm. (h.-l.) stage (after Shiino).

Explanation of Lettering precedes Plate 80

AVES

Explanation of Lettering

a, acrochordal cartilage
adc, aditus conchae
ang, angular
ap, anterior parachordal cartilage
ar, articular
as, auditory sac
asor, aperture communicating with orbital sinus
at, pleurocentrum (body) of atlas vertebra: odontoid process
atn, neural arch of atlas vertebra
att, atrioturbinal
ax, pleurocentrum of axis vertebra
axn, neural arch of axis vertebra

bcf, basicranial fenestra
bo, basioccipital
br, backgrowing roof of nasal capsule
bs, basisphenoid

bt, basitemporal
btp, basitrabecular process

c, occipital condyle
ca, columella auris
cac, canalicular portion of auditory capsule
cb 1, ceratobranchial of 1st branchial arch
cc, cochlear portion of auditory capsule
cn, concha nasalis
cop, copula
cu, cupola anterior of nasal capsule

den, dentary

e, eye
ec, entoglossal cartilage (ceratohyals fused)
ee, ectethmoid
ep, ethmoid plate
exc, extracolumella
exo, exoccipital

fa, foramen for abducens nerve
fac, foramen for auditory nerve
fen, foramen endolymphaticum
ff, foramen for facial nerve
fg, foramen for glossopharyngeal nerve
fH, 'Huxley's' foramen
fm, fissura metotica
fn, fenestra narina
fo, foramen ovale of auditory capsule
foa, fenestra olfactoria advehens
foc, foramen for oculomotor nerve
foe, fenestra olfactoria evehens
fop, foramen for ophthalmic artery
fp, foramen prooticum (for complete trigeminal nerve)
fpl, foramen perilymphaticum of auditory capsule
fpr, foramen for profundus branch of trigeminal nerve
fps, foramen prooticum spurium (for maxillary and mandibular branches of trigeminal nerve)
fro, frontal
fs, fossa subarcuata
fsna, fenestra septi nasalis anterior
fsnp, fenestra septi nasalis posterior
ft, foramen for trochlear nerve
fv, foramen for vagus nerve

ha, hypocentrum of atlas vertebra
hax, hypocentrum of axis vertebra
hf, foramina for hypoglossal nerve-roots
hfe, hypophysial fenestra
ho 1, hypocentrum of 1st occipital vertebra
ho 2, hypocentrum of 2nd occipital vertebra (proatlas vertebra)
hp, hemipterygoid

ios, interorbital septum
ip, incisura prootica
ipj, intrapterygoid joint
ipp, infrapolar process
is, interorbital septum

jug, jugal

lcf, lateral carotid foramen
lch, lachrymal
lh, laterohyal

max, maxilla
Mc, Meckel's cartilage
mcc, metotic cartilage
mm, mentomeckelian
mt, maxilloturbinal

n, notochord
nab, notch for abducens nerve

nas, nasal
no, notch for oculomotor nerve
nop, notch for optic nerve
np, notch for profundus branch of trigeminal nerve
ntr, notch for trochlear nerve

oa, definitive occipital arch
oa 6–9, occipital arch of the 6th–9th segment
oc, orbital cartilage
oca, preoptic root of orbital cartilage
occ, orbitocapsular commissure
ocp, posterior portion of orbital cartilage
ocpr, orbitocapsular process
of, optic foramen
onf, orbitonasal fissure
oo, opisthotic
op, otic process of quadrate
ors, orbitosphenoid
ov 1, 1st occipital vertebra
ov 2, 2nd occipital vertebra (proatlas vertebra)

p, parachordal cartilage
pa, pila antotica
pal, palatine
pan, lamina orbitonasalis
par, parietal
pas, pila antotica spuria
pc, polar cartilage
pd, process dorsalis
pf, prefacial commissure
pls, pleurosphenoid
pmx, premaxilla
pnc, paranasal cartilage
pnp, prenasal process
pop, postorbital process
pot, prootic
pp, pterygoid process of quadrate
pr, prearticular
pra, processus retroarticularis of Meckel's cartilage
pro, prootic process
prs, presphenoid
ps, planum supraseptale
psp, posterior portion of planum supraseptale
psph, parasphenoid
pt, processus tectalis
ptc, parietotectal cartilage of nasal capsule
ptg, pterygoid
pv, prevomer
pvc, prevomer cartilage

q, quadrate
qj, quadratojugal

san, supra-angular
scp, subcapsular process
sec, sphenethmoid commissure
sep, sphenethmoid process
sh, stylohyal cartilage
sn, nasal septum
soc, supraoccipital
sop, supraoccipital process
sp, posterior portion of floor of nasal capsule
spl, splenial
squ, squamosal
st, supratrabecular cartilage

t, trabecula cranii
tc, trabecula communis
tm, taenia marginalis
tme, rudiment of taenia medialis
ts, tectum synoticum

vo, vacuity in orbital cartilage
vs, vacuities in interorbital septum
vt, vacuity in backgrowing roof of nasal capsule

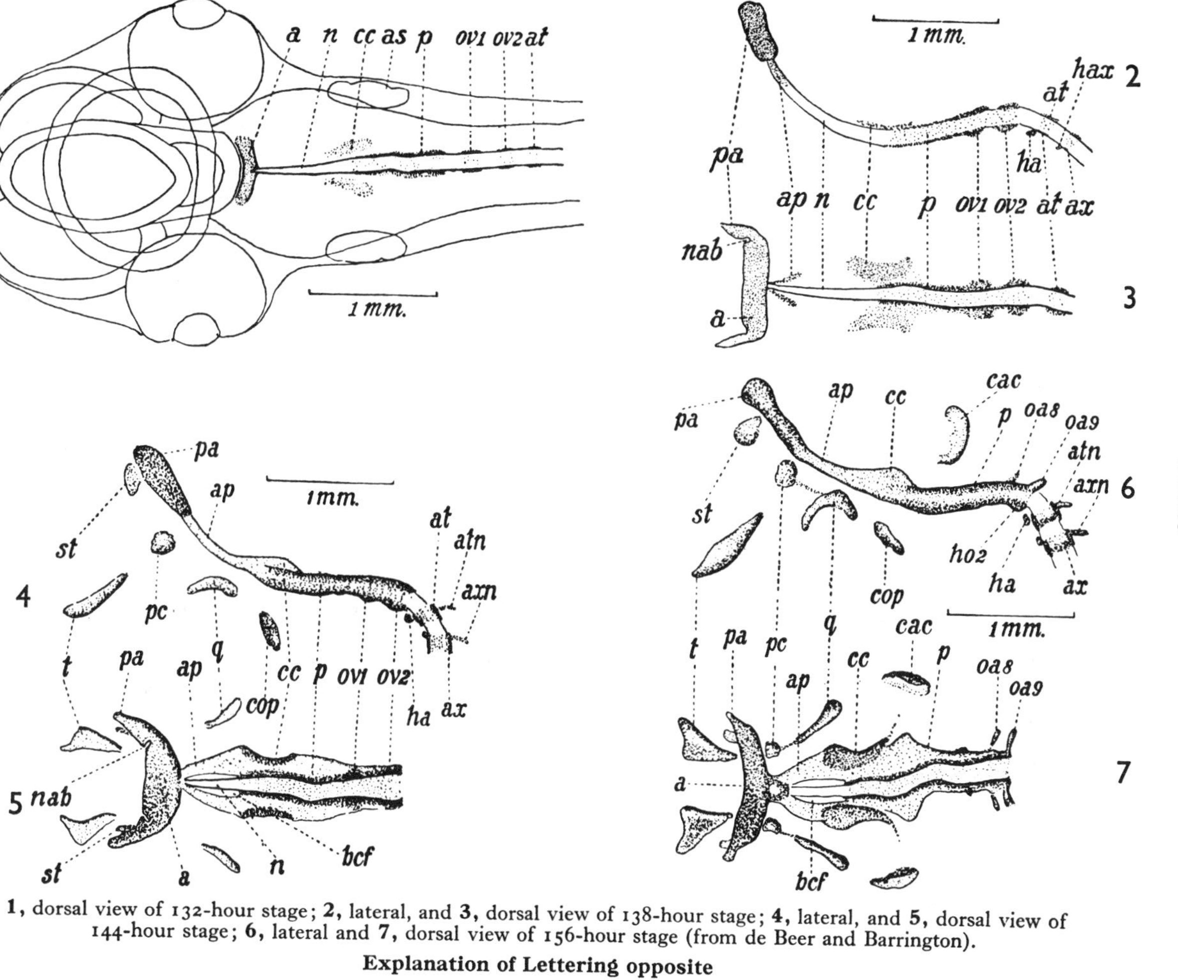

1, dorsal view of 132-hour stage; **2**, lateral, and **3**, dorsal view of 138-hour stage; **4**, lateral, and **5**, dorsal view of 144-hour stage; **6**, lateral and **7**, dorsal view of 156-hour stage (from de Beer and Barrington).

Explanation of Lettering opposite

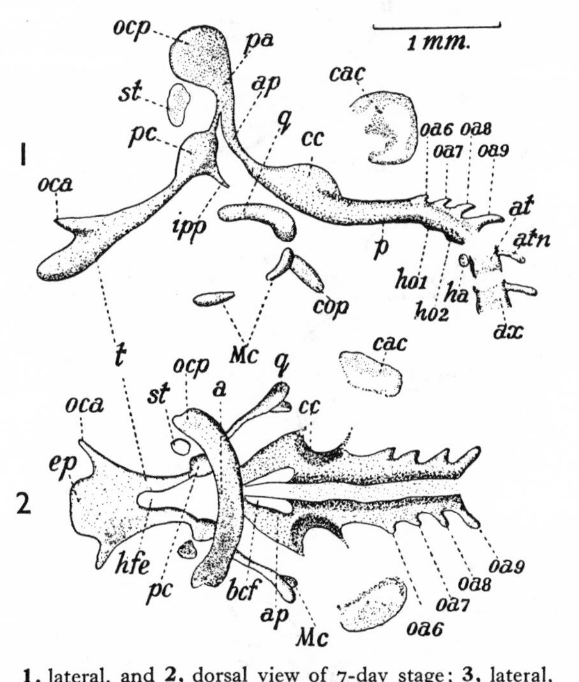

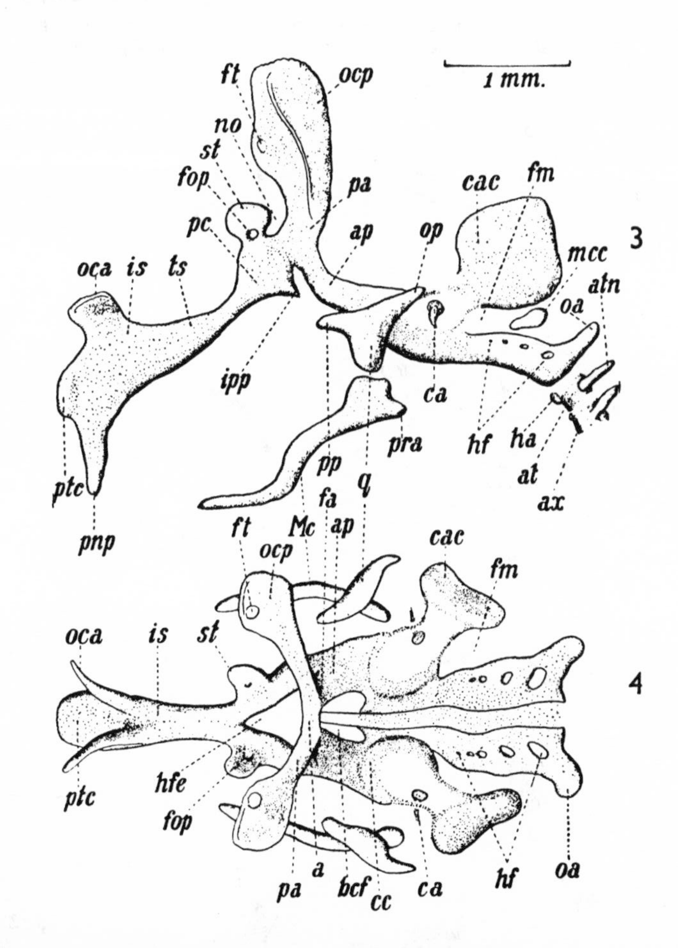

1, lateral, and **2**, dorsal view of 7-day stage; **3**, lateral, and **4**, dorsal view of 8-day stage (from de Beer and Barrington).

Explanation of Lettering precedes Plate 94

ANAS

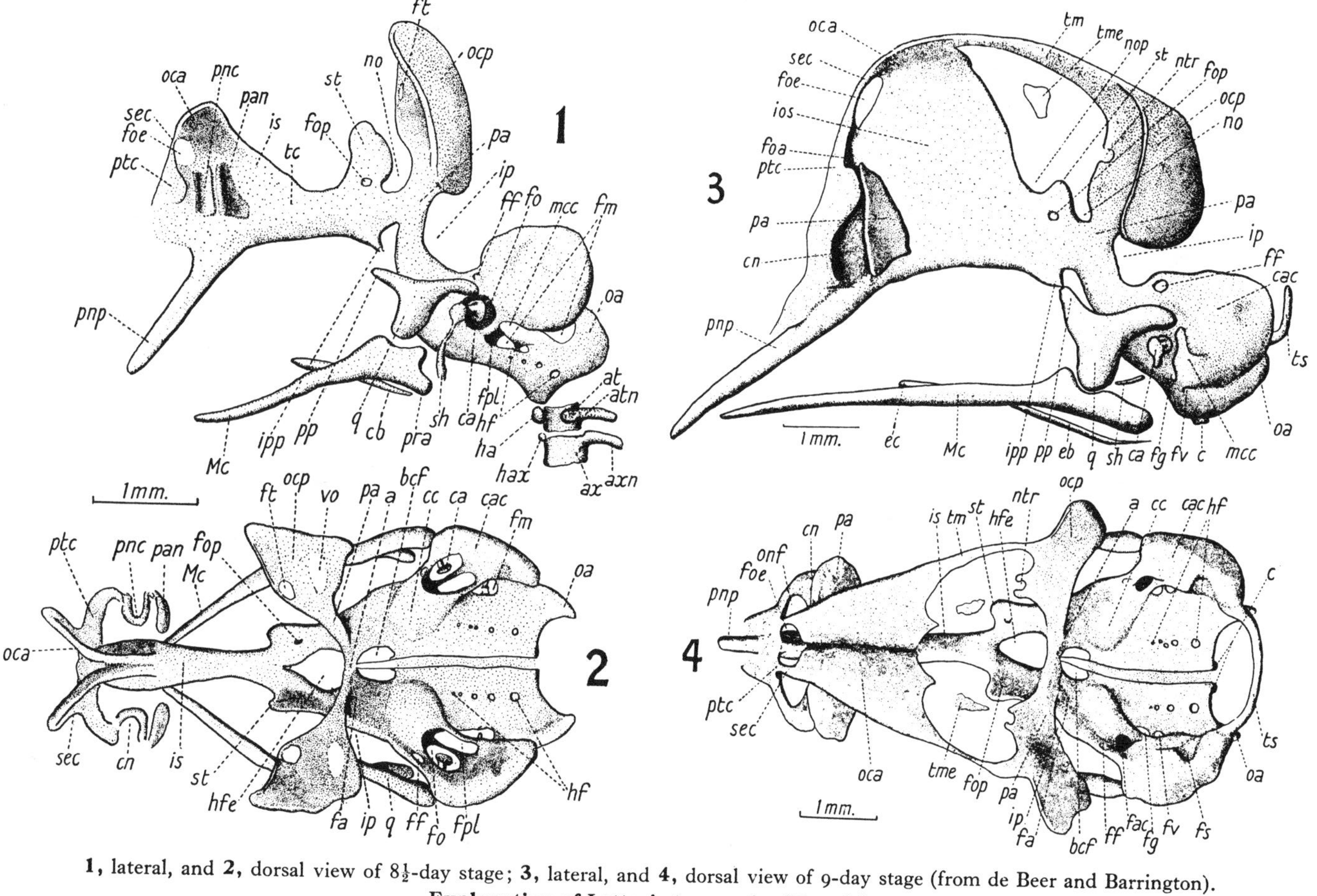

1, lateral, and **2**, dorsal view of $8\frac{1}{2}$-day stage; **3**, lateral, and **4**, dorsal view of 9-day stage (from de Beer and Barrington).

Explanation of Lettering precedes Plate 94

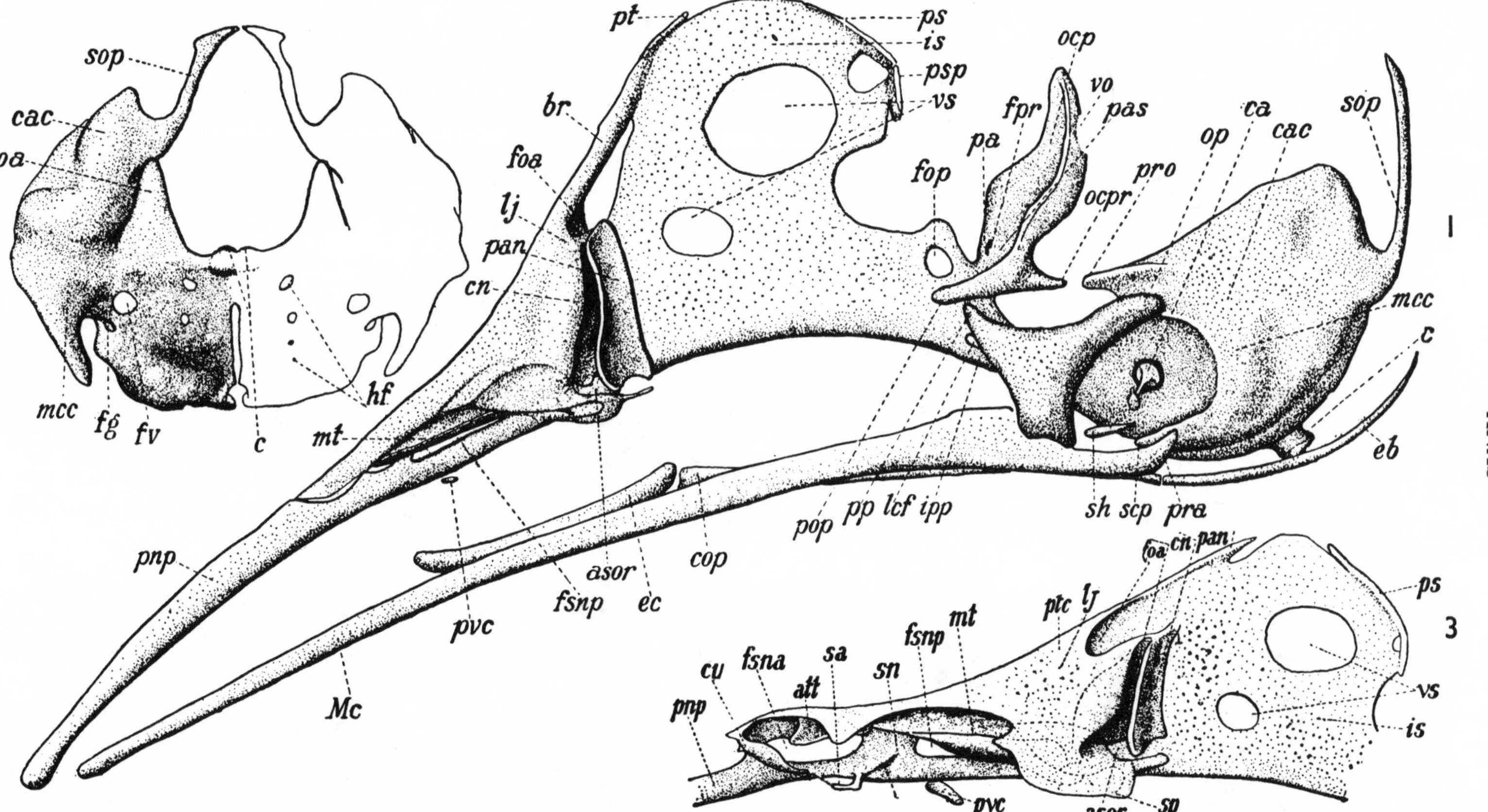

1, lateral, and 2, posterior view of 14-day stage; 3, lateral view of nasal capsule of 17-day stage (from de Beer and Barrington).

Explanation of Lettering precedes Plate 94

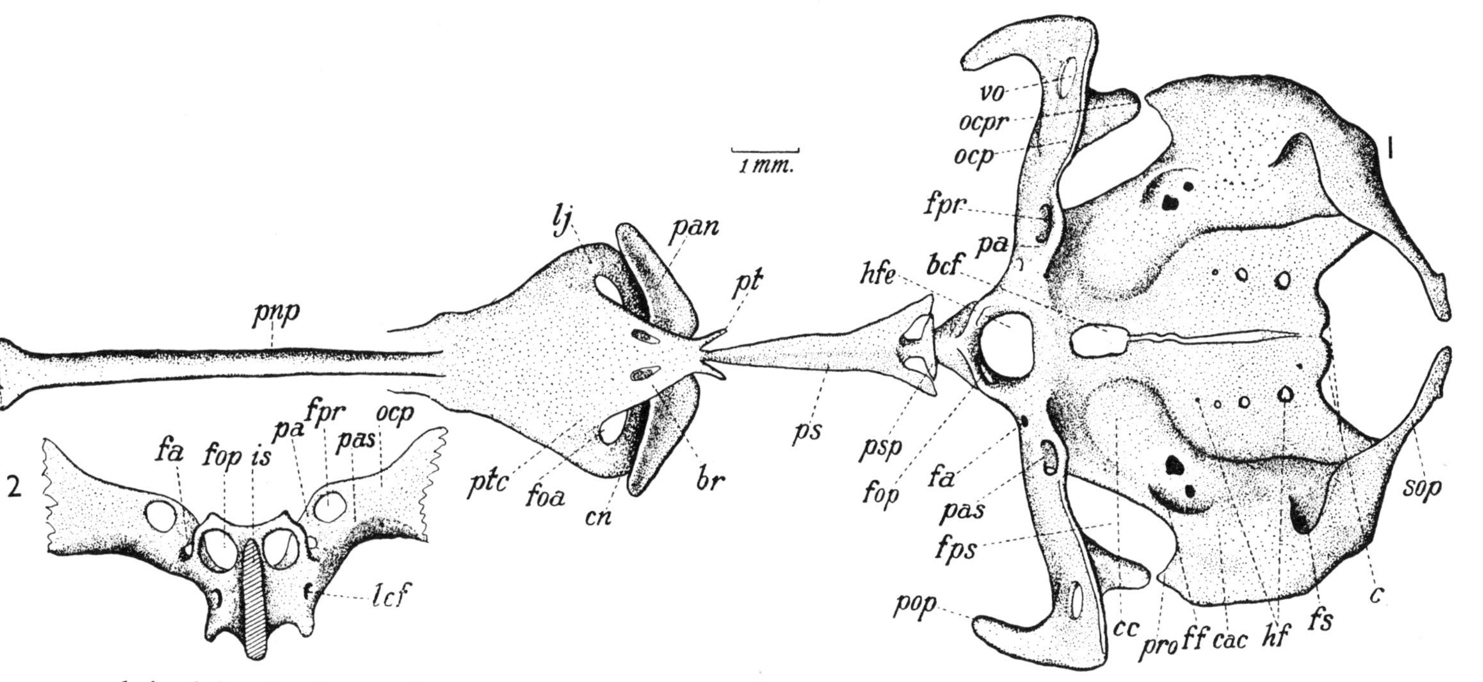

1, dorsal view of 14-day stage; **2**, anterior view of orbitotemporal region (interorbital septum cut) (from de Beer and Barrington).

Explanation of Lettering precedes Plate 94

GALLUS, STRUTHIO

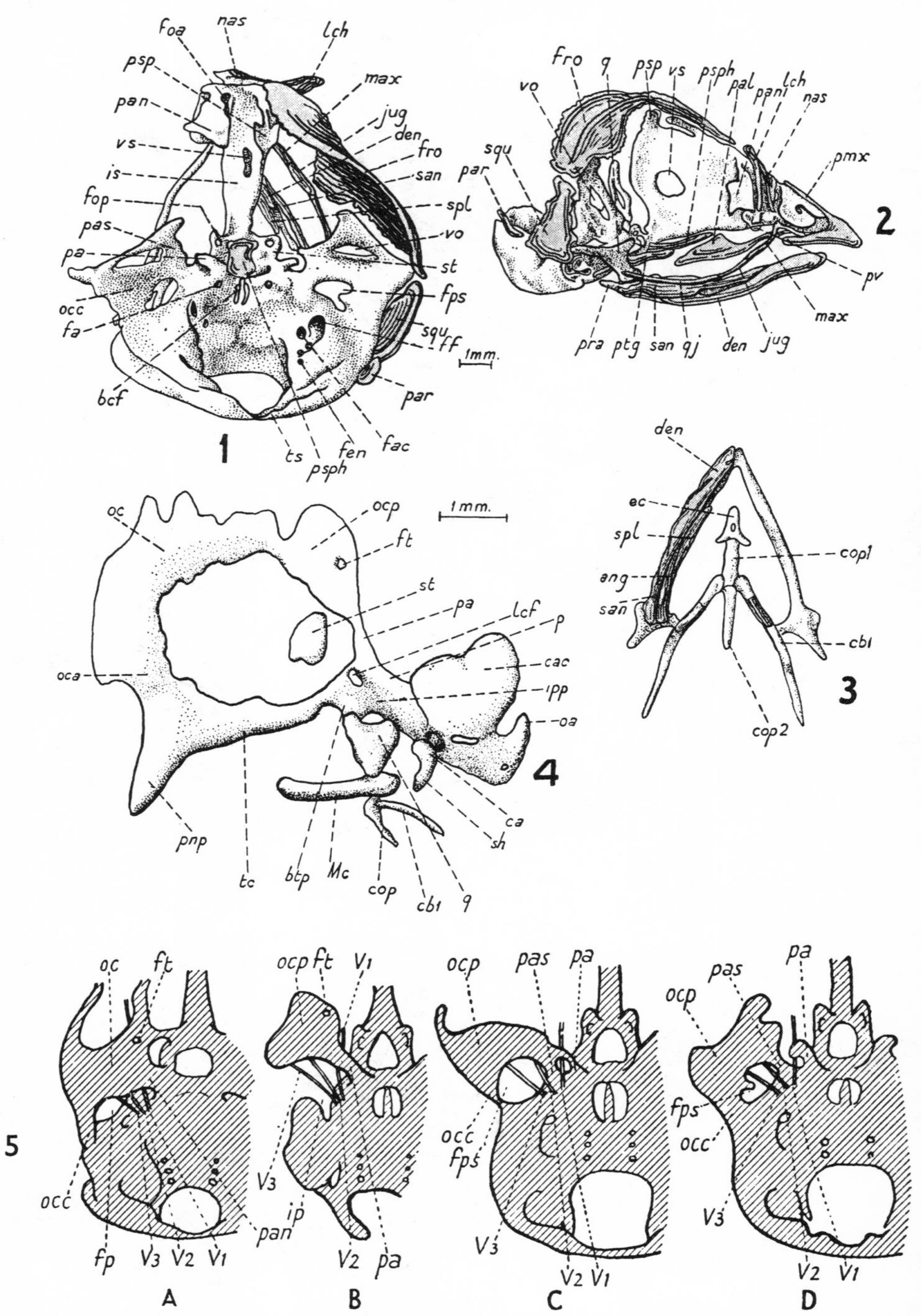

1, dorsal, **2**, lateral view of skull; and **3**, ventral view of lower jaw and hyobranchial skeleton of Gallus 65 mm. stage (after Tonkoff and Gaupp); **4**, lateral view of Struthio 18 mm. stage (from Brock); **5**, diagram of relations of pila antotica and pila antotica spuria in A, Crocodilus, B, Anas early stage, C, Anas late stage, and D, Gallus (from de Beer and Barrington).

Explanation of Lettering precedes Plate 94

TINNUNCULUS

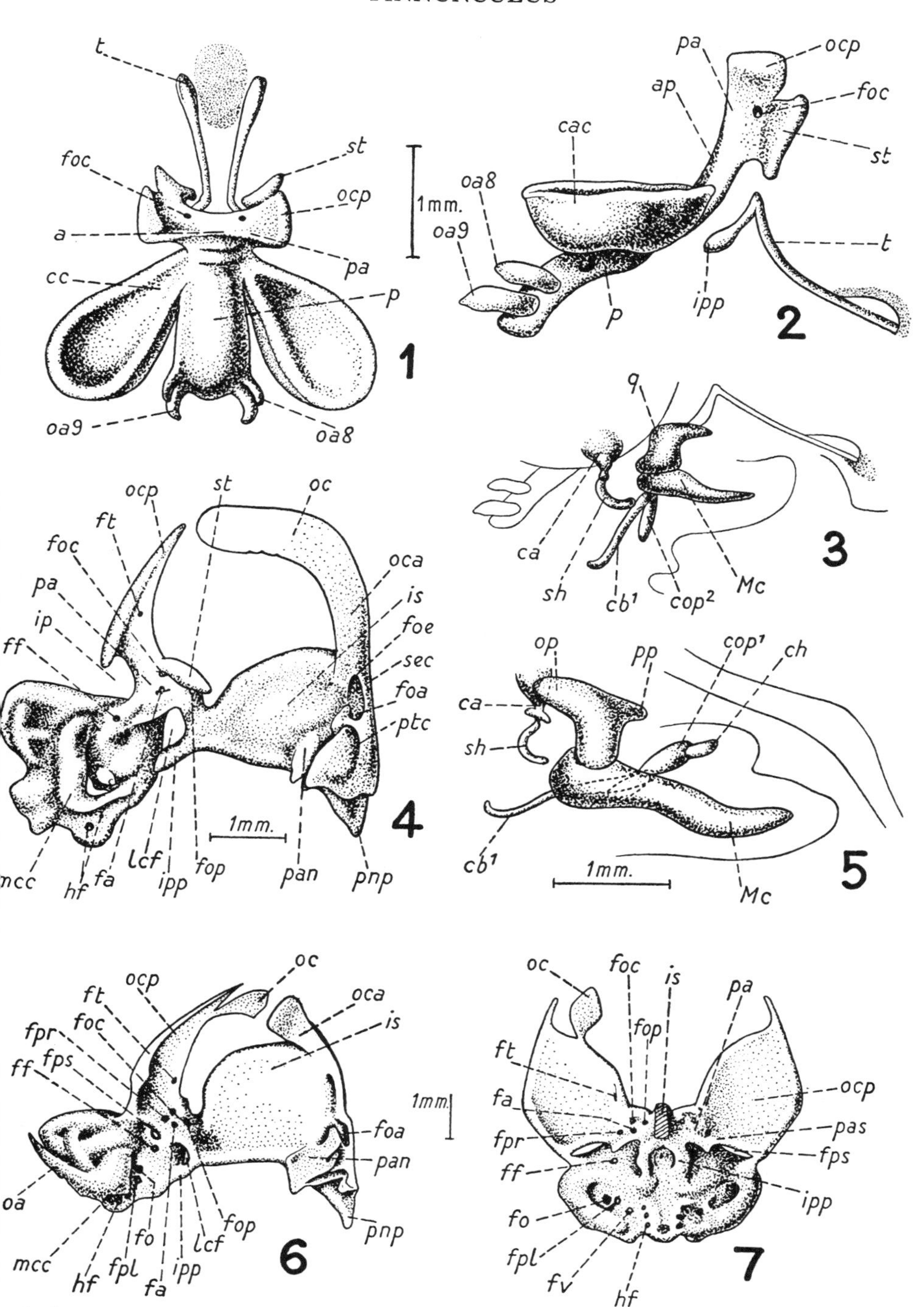

1, dorsal, and **2**, lateral view of neurocranium at stage 1; **3**, visceral arches of stage 1; **4**, lateral view of stage 2; **5**, visceral arches of stage 2; **6**, lateral, and **7**, ventral view of stage 3 (in **7** the interorbital septum has been cut and the anterior part of the neurocranium removed (after Suschkin).

Explanation of Lettering precedes Plate 94

TINNUNCULUS

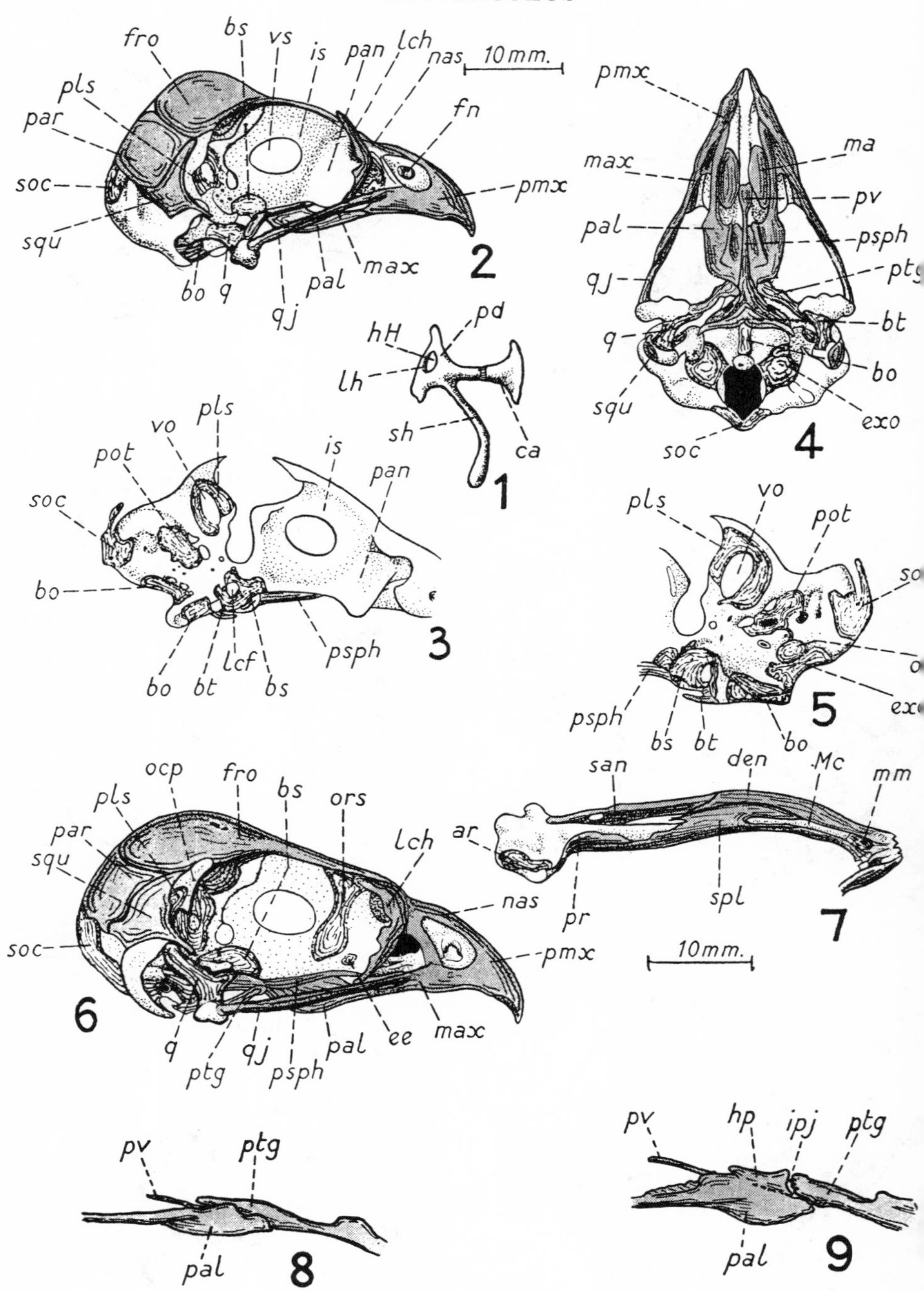

1, columella auris; **2,** lateral view of skull, **3,** lateral view of chondrocranium and cartilage-bones, **4,** ventral view of skull, **5,** medial view of auditory region, of stage 8; **6,** lateral view of skull, and **7,** medial view of lower jaw, of stage 13 (after Suschkin); **8,** lateral view of pterygoid and palatine of nestling Steatornis; **9,** ditto of adult Steatornis (after Pycraft)

Explanation of Lettering precedes Plate 94

LARUS, HIRUNDO, PASSER

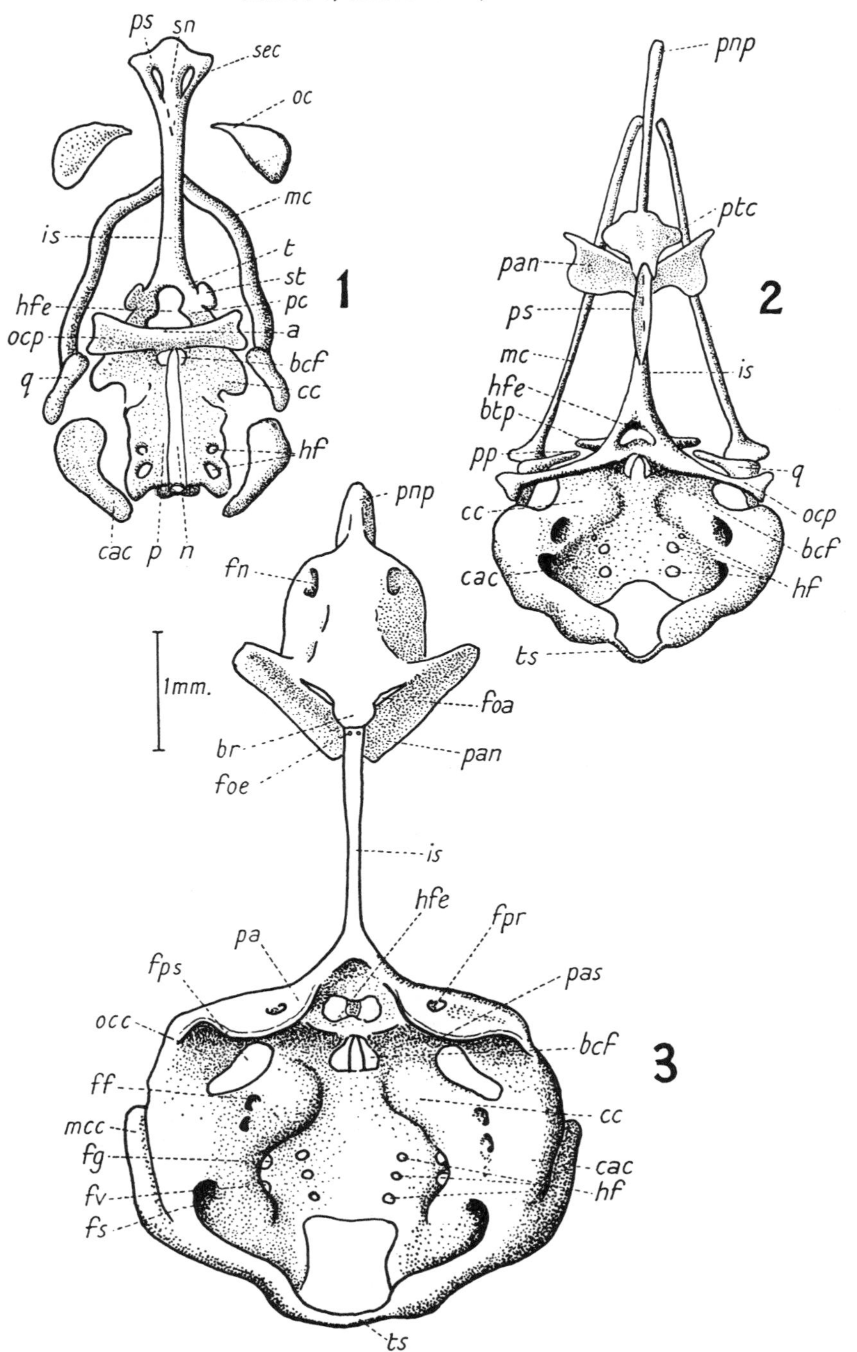

1, dorsal view of early chondrocranium of Larus sp.; **2,** dorsal view of chondrocranium of Hirundo sp.; **3,** dorsal view of chondrocranium of full-time embryo of Passer domesticus (after de Beer and Barrington).

Explanation of Lettering precedes Plate 94

APTERYX

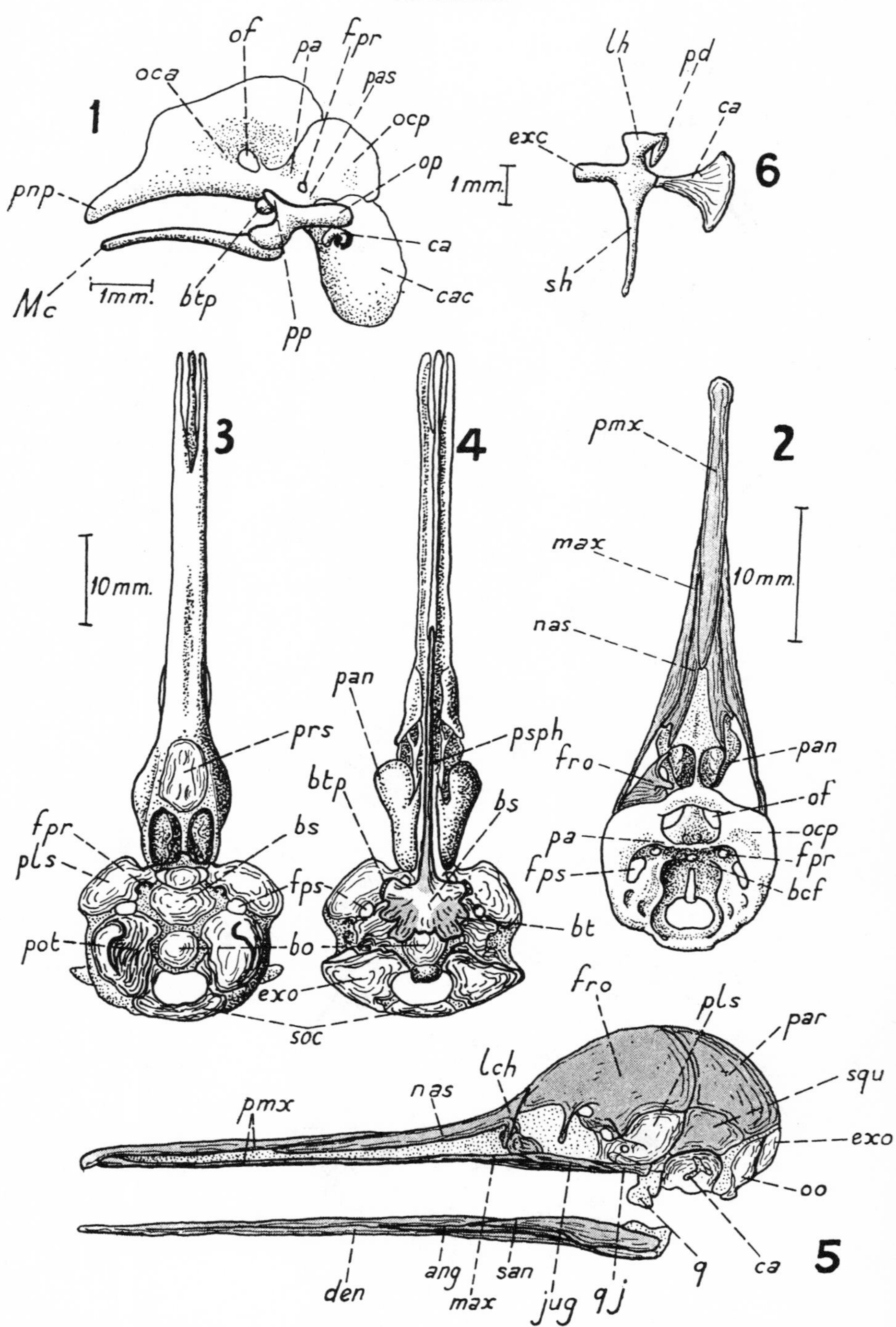

1, lateral view of chondrocranium of 43 mm. stage; **2**, dorsal view of skull at stage E[1] (frontals and parietals removed); **3**, dorsal, **4**, ventral, and **5**, lateral view of skull of stage K; **6**, columella auris (after T. J. Parker).

Explanation of Lettering precedes Plate 94

MAMMALIA

Explanation of Lettering

a, auditory capsule
aa, articular facet of atlas vertebra
ac, anterior canalicular centre
acc, alicochlear commissure
acp, anterior clinoid process
ah, ala hypochiasmatica
ali, alisphenoid bone
am, ala minima
arc, arytenoid cartilage
asc, anterior semicircular canal
at, ala temporalis
ath, hypocentrum of atlas vertebra
atn, neural arch of atlas vertebra
atp, pleurocentrum of atlas vertebra
atr, rib of atlas vertebra
att, atrioturbinal
axn, neural arch of axis vertebra
axp, pleurocentrum of axis vertebra
axr, rib of axis vertebra

bcc, basicochlear commissure
bcf, basicochlear fissure
bf, basicranial fenestra
bh, basihyal cartilage
bo, basioccipital bone
bs, basisphenoid bone
bvc, basivestibular commissure

caa, cartilage 'a'
cab, cartilage 'b'
cac, cartilage 'c'
cc, canalicular part of auditory capsule
ccc, cochleo-canalicular centre
cdn, cartilago ductus nasopalatini
ch, ceratohyal
clc, clinocochlear commissure
cm, marginal cartilage of snout (crista marginalis)
coc, cochlear part of auditory capsule
cop, orbitoparietal commissure
crc, cricoid cartilage
crg, crista galli
crp, crista parotica
cs, crista semicircularis
cse, sphenethmoid commissure

den, dentary bone
ds, dorsum sellae

e, eyeball
ecc, ectochoanal cartilage
ent, endotympanic
Epg, 'Echidna-pterygoid'
et, ethmoturbinal
exo, exoccipital bone

fac, foramen acusticum
fc, carotid foramen
fcr, fenestra cribrosa
fe, foramen epiphaniale
fen, foramen endolymphaticum
ff, foramen for facial nerve
fh, hypophysial fenestra

fic, foramen infracribrosum
fim, fissura metotica
fio, infraorbital foramen
fj, foramen jugulare
fln, foramen for lachrymonasal duct
fm, foramen magnum
fn, fenestra narina
fo, optic foramen
foa, foramen for orbital artery ('alisphenoid canal')
foc, fissura occipitocapsularis
foi, fissura occipitocapsularis inferior
fol, foramen olfactorium advehens
fop, foramen for lateral nasal branch of ophthalmic nerve
fos, fissura occipitocapsularis superior
fov, fenestra ovalis
fp, foramen perilymphaticum
fpa, palatine foramen
fpc, foramen prechiasmaticum
fpo, foramen pseudopticum
fpr, foramen prooticum
fr, foramen rotundum
fro, frontal bone
fs, fenestra superior nasi
fsa, fossa subarcuata
ft, frontoturbinal

h, hypophysis
hc, hypophysial cartilage
hf, hypoglossal foramen
hFa, hiatus Faloppii

i, incus
ic, internal carotid
icc, inferior cochlear centre
ios, interorbital septum
ip, interparietal
is, incisive suture

jug, jugal bone

la, lamina ascendens
lac, lachrymal
lh, laterohyal
lic, lamina infracribrosa
lnd, lachrymonasal duct
lon, lamina orbitonasalis
lp, limbus precapsularis
lpc, lateral prefacial commissure
lsc, lateral semicircular canal
lsu, lamina supracochlearis
lta, lamina transversalis anterior
ltp, lamina transversalis posterior

m, malleus
max, maxillary bone
Mc, Meckel's cartilage
mp, mastoid process
mpf, maxillopalatine foramen
mr, metoptic root of orbital cartilage
mt, maxilloturbinal
mxa, alveolar process of maxilla
mxp, palatine process of maxilla
mxz, zygomatic process of maxilla

n, notochord
nas, nasal bone
nh, notch for hypoglossal
ns, nasal septum
nt, nasoturbinal

oa, occipital arch
oam, ossiculum accessorium mallei
oc, occipital condyle
onf, orbitonasal fissure
ora, orbital artery
orc, orbital cartilage
ors, orbitosphenoid
osc, 'os carunculae' (premaxilla)

p, parachordal or basal plate
pab, palatine bone
pai, processus alaris inferior (of nasal capsule)
pal, processus alaris (basitrabecularis)
pan, pila antotica
par, parietal bone
pas, processus alaris superior (of nasal capsule)
pc, posterior canalicular centre
pcp, paracondylar process
pcs, palatine commissure (of ectochoanal cartilages)
pec, parethmoidal cartilage
pfc, prefacial commissure
pi, pituitary
pip, preinterparietal
pma, processus maxillaris anterior
pmp, processus maxillaris posterior
pmx, premaxillary bone
pn, paranasal cartilage
pns, paries nasi
pop, processus opercularis
por, postorbital process
ppc, papillary cartilage
ppl, parietal plate
pr, preoptic root of orbital cartilage
pra, prearticular bone
pri, processus incisivus
prp, pleurocentrum of proatlas vertebra
prr, processus recessus ('intraperilymphaticus')
prs, presphenoid
ps, paraseptal cartilage
psp, posterior paraseptal cartilage
pss, planum supraseptale
pt, parietotectal cartilage
ptc, pterygoid cartilage
ptg, pterygoid ('mammalian') bone
ptp, pterygoid process
pv, prevomer bone
pvp, palatine or prevomerine process of premaxilla

r, rostrum

s, stapes
sc, secondary cartilage
scc, sphenocochlear commissure
sh, stylohyal cartilage
sme, spina mesethmoidalis
smx, septomaxilla
sob, supraoccipital bone
soc, supraoccipital cartilage
spm, symphysis of premaxillae (bearing egg-tooth)
sq, squamosal bone
sqz, zygomatic process of squamosal
stb, supratrabecular bar
su, supracochlear cartilage
suc, superior cochlear centre

t, central stem (trabecular plate)
ta, tectum anterius
tca, anterior wing of thyroid cartilage
tcp, posterior wing of thyroid cartilage
th, thyrohyal cartilage
ti, tectum intermedium
tn, tectum nasi
tp, tectum posterius
trp, transverse process
tsy, tectum synoticum
tt, trabecula cranii
ttr, tectum transversum
tty, tegmen tympani
ty, tympanic bone

v I, ophthalmic branch of trigeminal nerve
Vn, Vidian nerve (palatine)
vb, vomer bone

za, zona annularis of nasal capsule

ORNITHORHYNCHUS

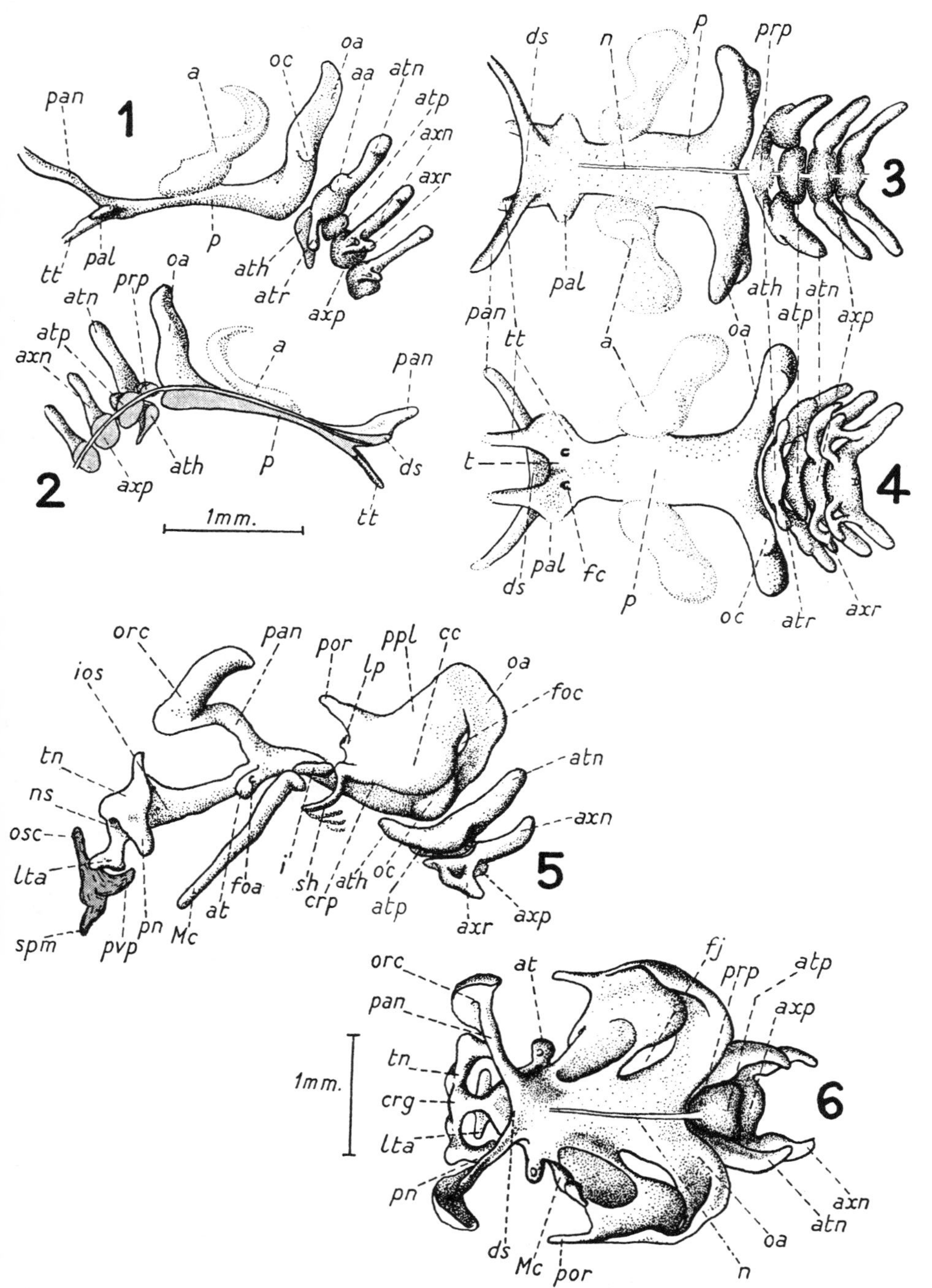

1, lateral, **2**, medial view of left half, **3**, dorsal, and **4**, ventral view of 8·5 mm. stage; **5**, lateral, and **6**, dorsal view of 9·4 mm. stage (from de Beer and Fell).

Explanation of Lettering opposite

ORNITHORHYNCHUS

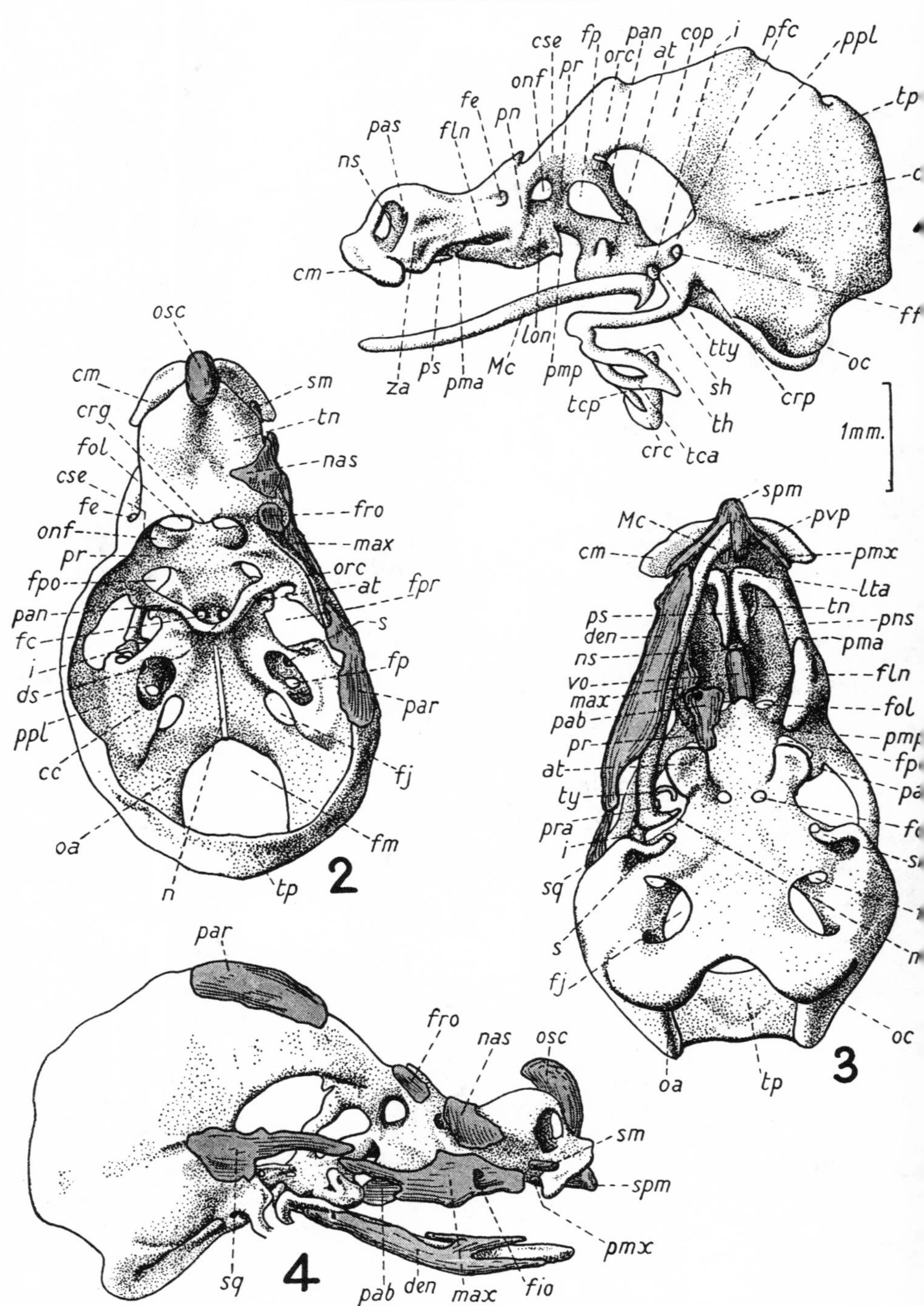

1, lateral view of chondrocranium; 2, dorsal, 3, ventral, and 4, lateral view of skull of 28 mm stage (from de Beer and Fell).

Explanation of Lettering precedes Plate 104

ORNITHORHYNCHUS

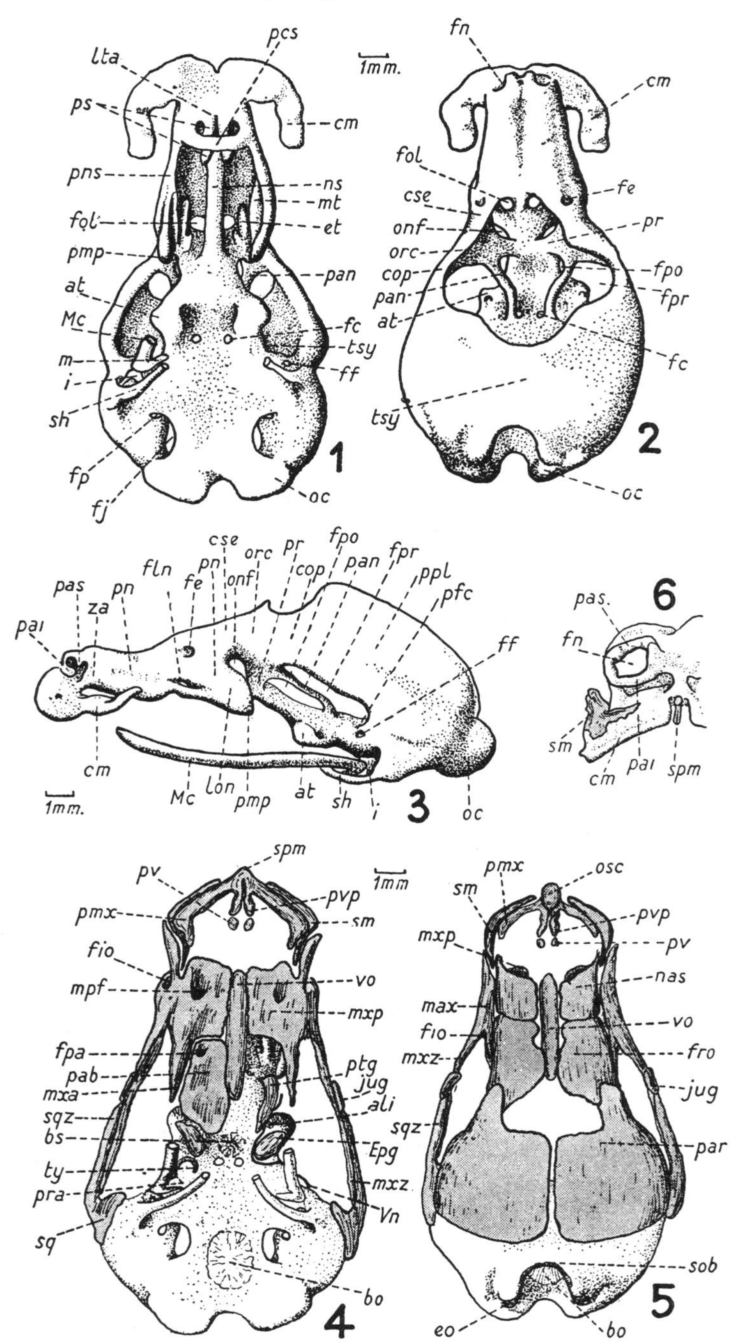

1, ventral, **2**, dorsal, and **3**, lateral view of chondrocranium of 122 mm. stage; **4**, ventral, and **5**, dorsal view of skull of 122 mm. stage (from de Beer and Fell); **6**, anterior view of nasal capsule of mammary foetus (after Wilson).

Explanation of Lettering precedes Plate 104

ECHIDNA, DASYURUS, ETC

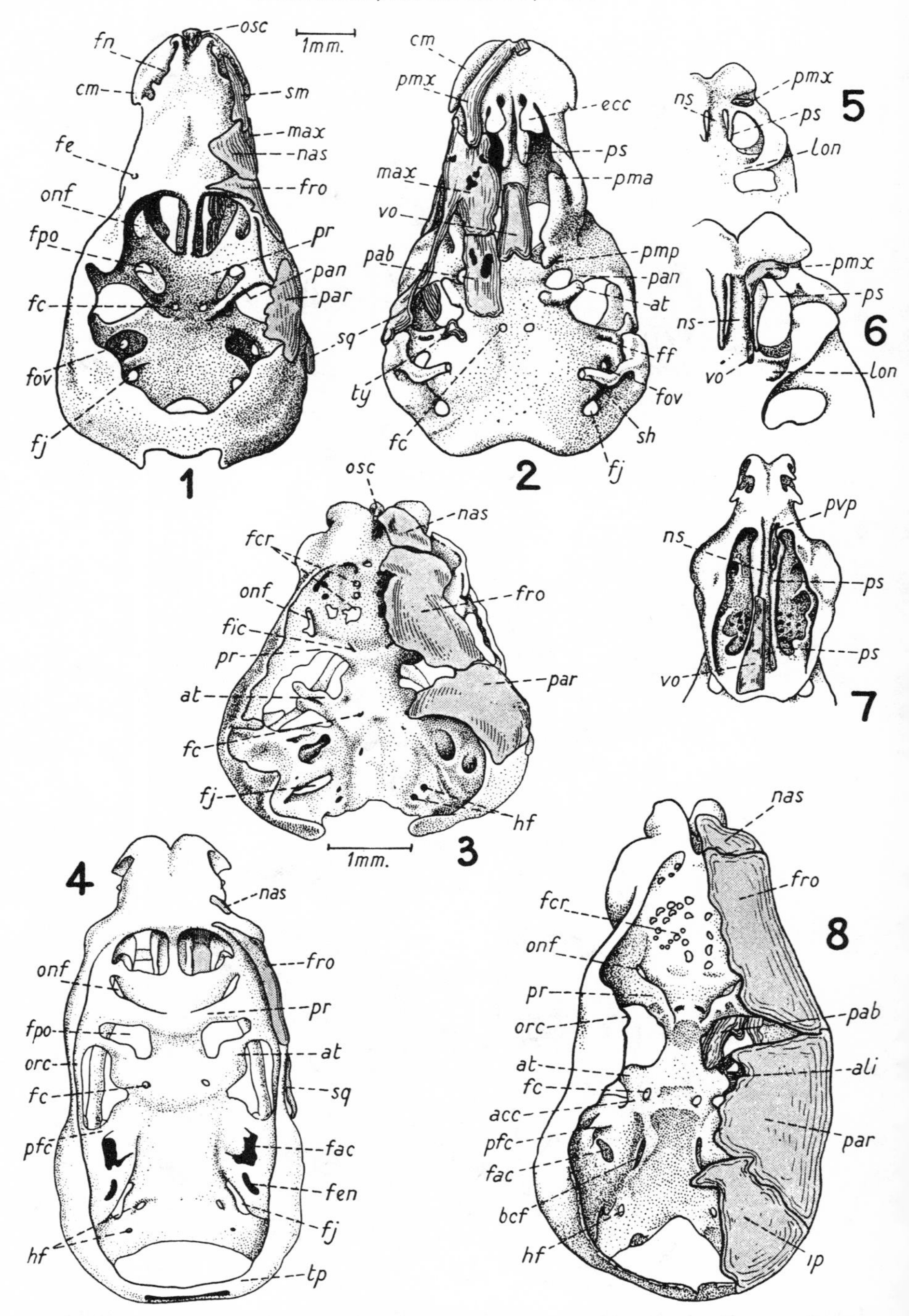

1, dorsal, and **2**, ventral view of Echidna stage 48A (after Gaupp); **3**, dorsal view of 17 mm. Caluromys (after Denison and Terry); **4**, dorsal view of 9·3 mm. Dasyurus; **5, 6, 7**, ventral views of nasal capsule of Dasyurus of 7 mm., 9·5 mm., and 25 mm. stages respectively (after Fawcett); **8**, dorsal view of 45·5 mm. Didelphys (after Töplitz).

Explanation of Lettering precedes Plate 104

PERAMELES

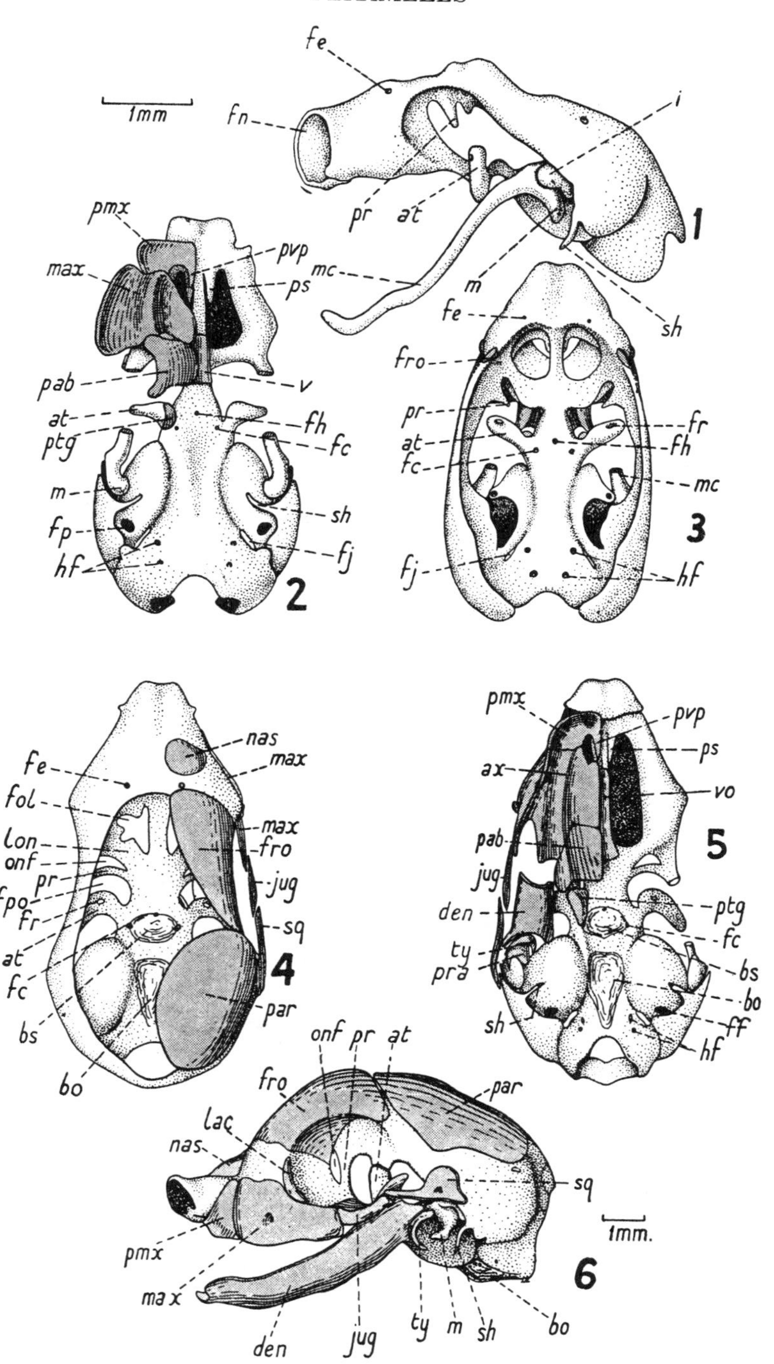

1, lateral, **2**, ventral, and **3**, dorsal view of 15·5 mm. stage; **4**, dorsal, **5**, ventral, and **6**, lateral view of 23 mm. stage (after Esdaile).

Explanation of Lettering precedes Plate 104

PERAMELES

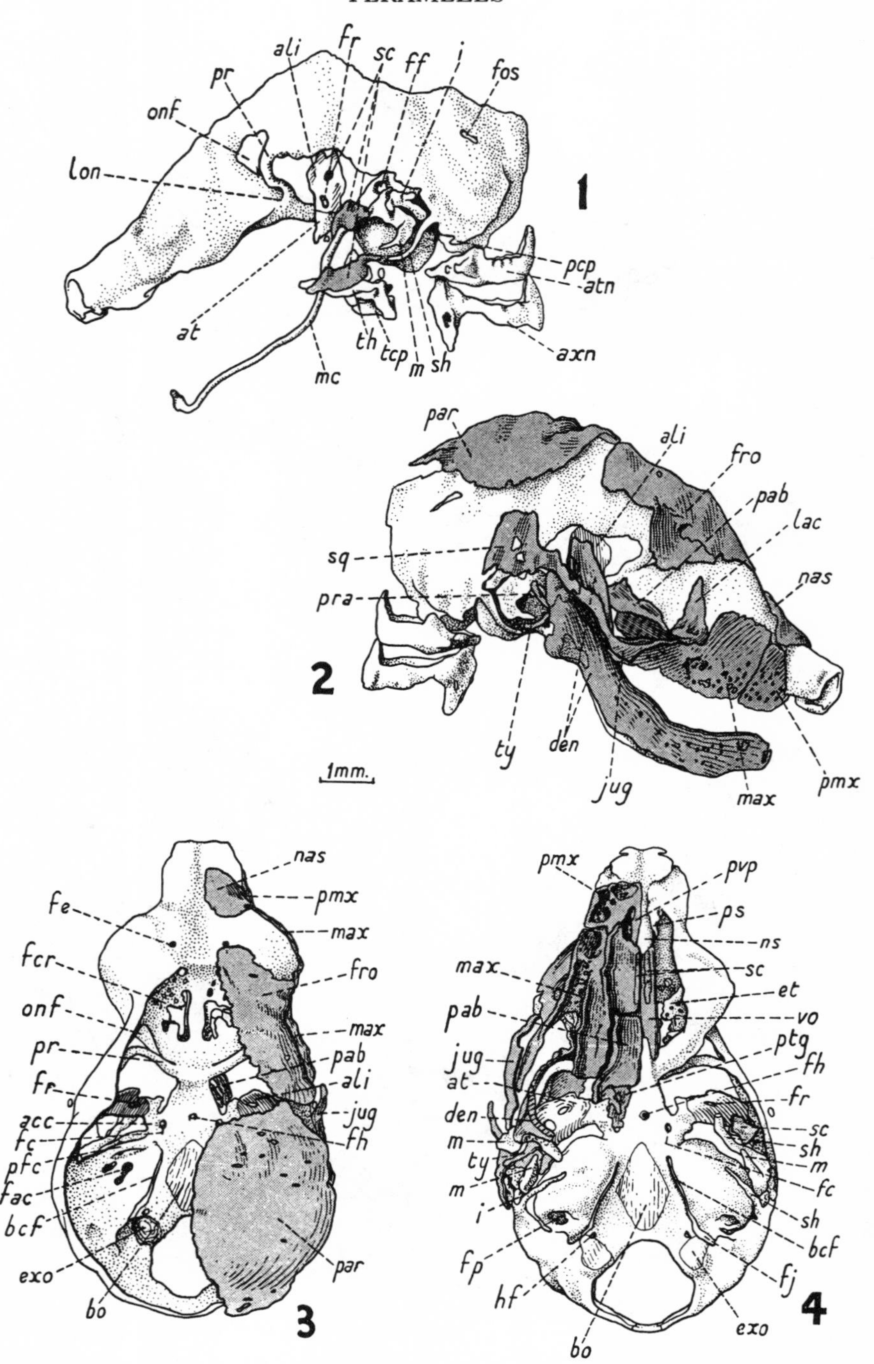

1, lateral view of chondrocranium; **2,** lateral, **3,** dorsal, and **4,** ventral view of skull of 42 mm. stage (after Cords).

Explanation of Lettering precedes Plate 104

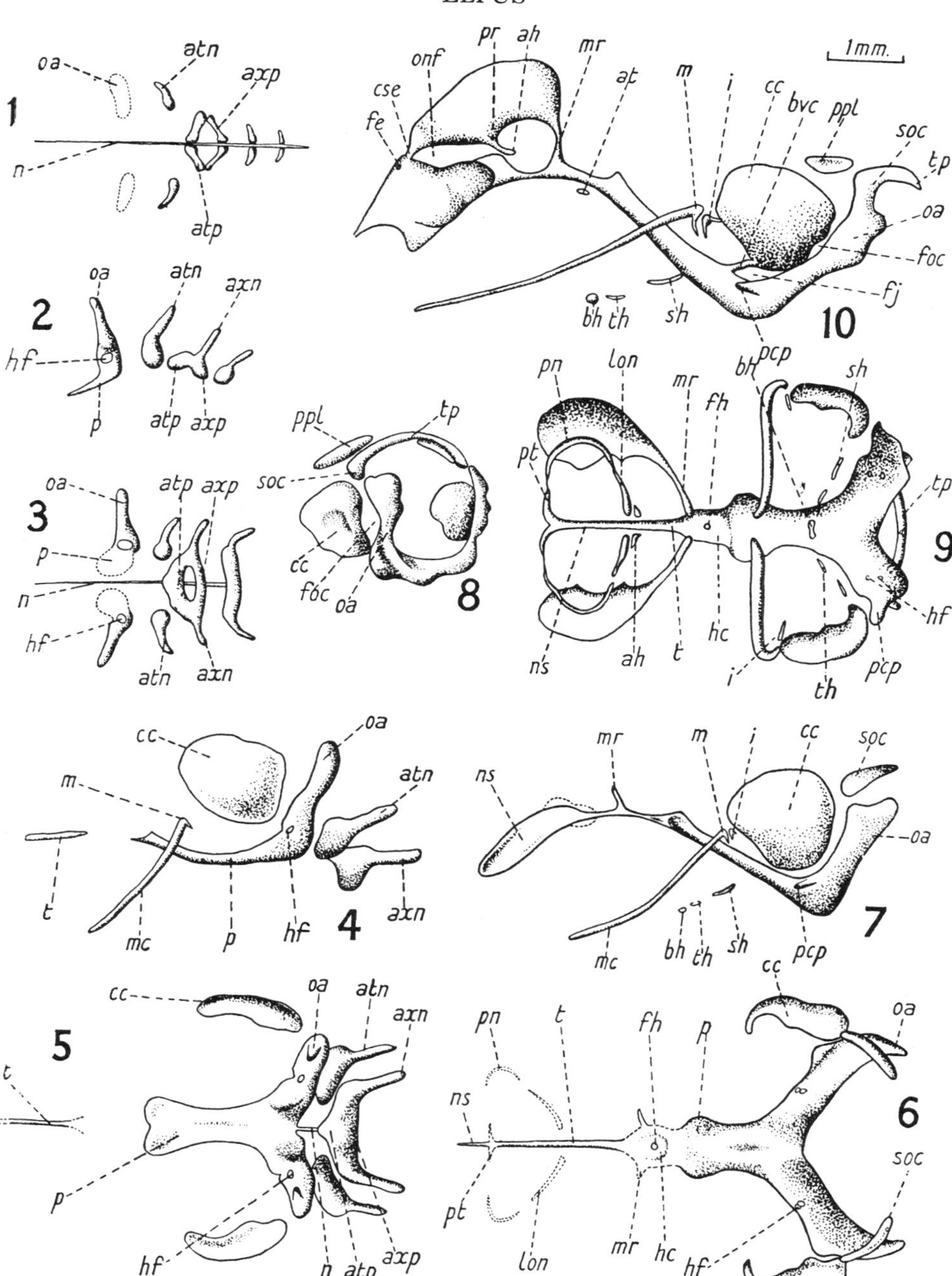

1, dorsal view of 12·5 mm. stage; **2,** lateral, and **3,** dorsal view of 13 mm. stage; **4,** lateral, and **5,** dorsal view of 16 mm. stage; **6,** dorsal, and **7,** lateral view of 18·5 mm. stage; **8,** postero-lateral, and **9,** ventral view of 19 mm. stage; **10,** lateral view of 19·5 mm. stage (after de Beer and Woodger).

Explanation of Lettering precedes Plate 104

LEPUS

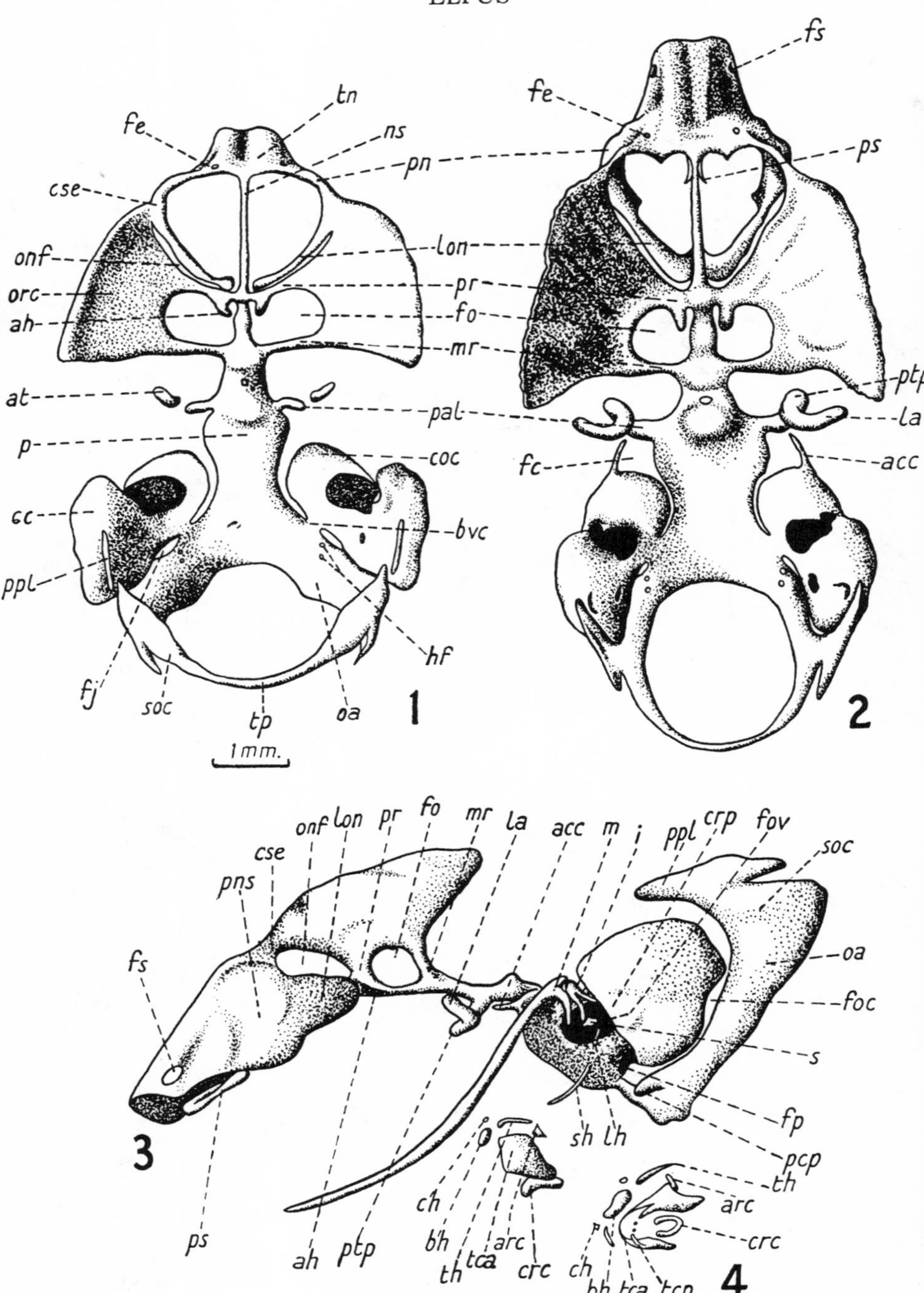

1, dorsal view of 19·5 mm. stage; **2,** dorsal, and **3,** lateral view of 21 mm. stage; **4,** ventro-lateral view of visceral arches of 21 mm. stage (after de Beer and Woodger).

Explanation of Lettering precedes Plate 104

LEPUS

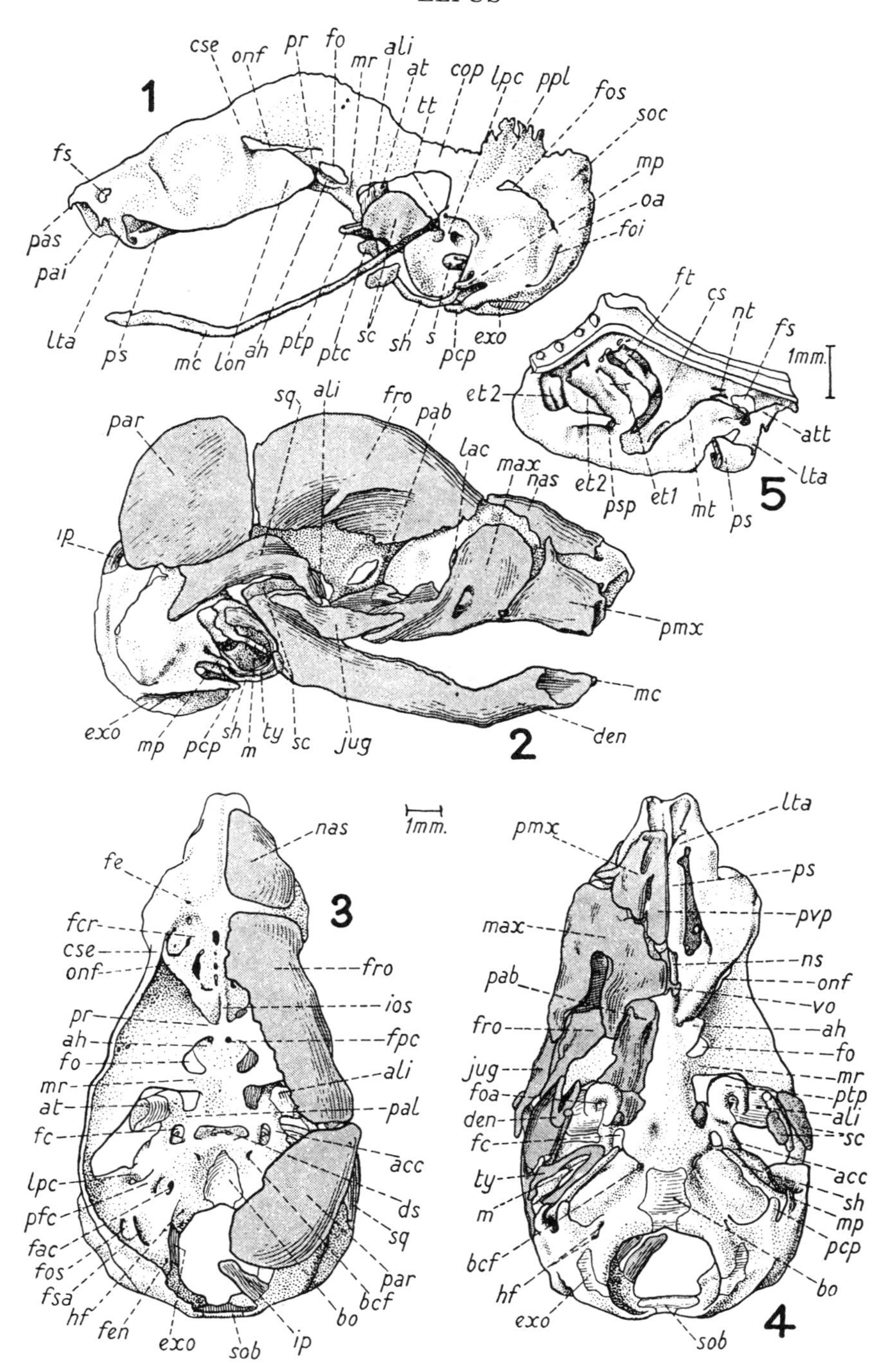

1, lateral view of chondrocranium; **2,** lateral, **3,** dorsal, and **4,** ventral view of skull of 45 mm. stage; **5,** medial view of interior of left nasal capsule (after Voit).

Explanation of Lettering precedes Plate 104

LEPUS, MICROTUS, ETC.

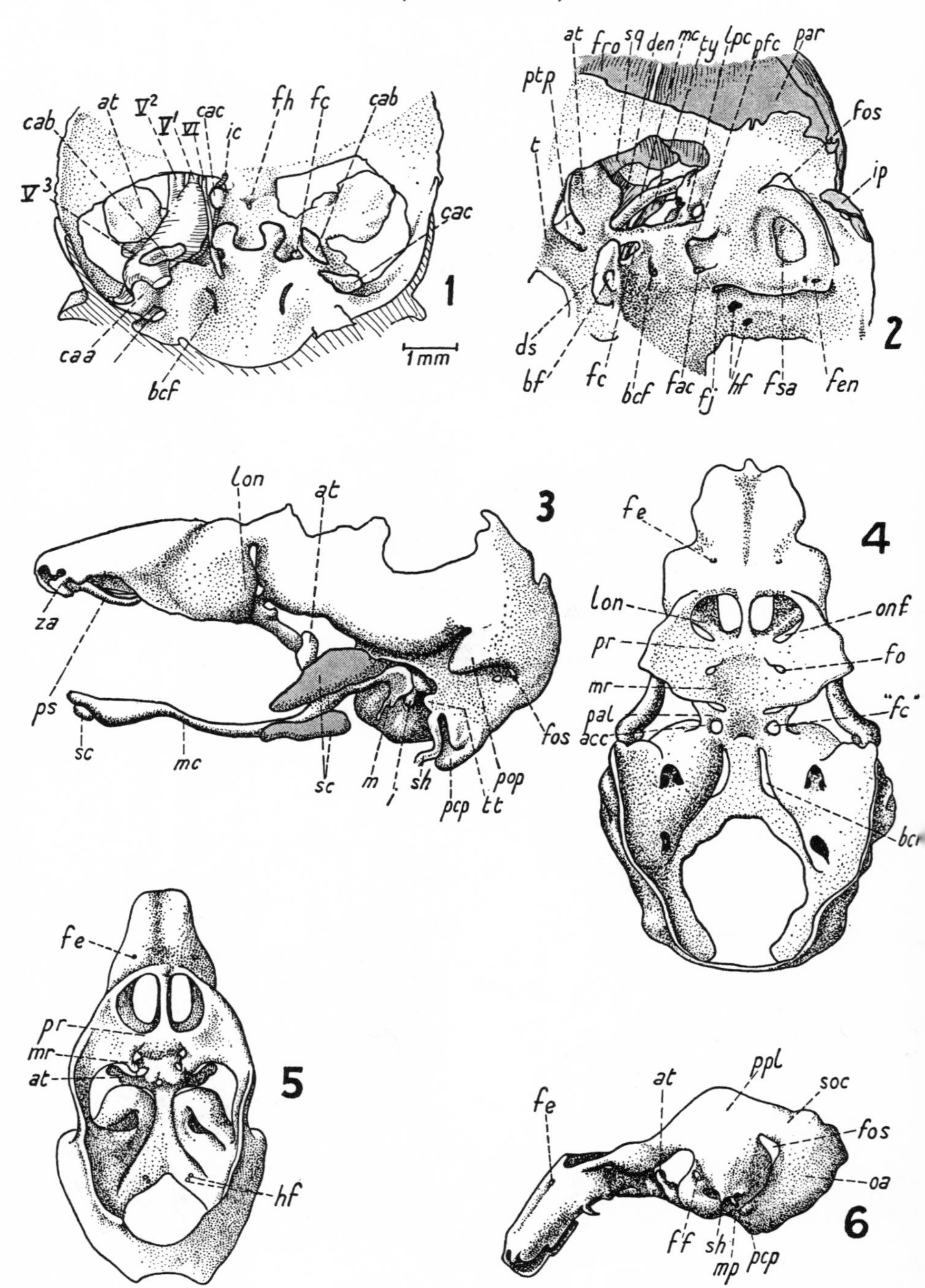

1, dorsal view of orbitotemporal region of chondrocranium of 43 mm. stage of Lepus; **2,** medial view of auditory region of 45 mm. stage of Lepus (after Voit); **3,** lateral view of 25 mm. stage of Microtus (after Fawcett); **4,** dorsal view of 28 mm. stage of Erethizon (after Struthers); **5,** dorsal, and **6,** lateral view of 17·5 mm. stage of Bradypus (after de Burlet).

Explanation of Lettering precedes Plate 104

TATUSIA

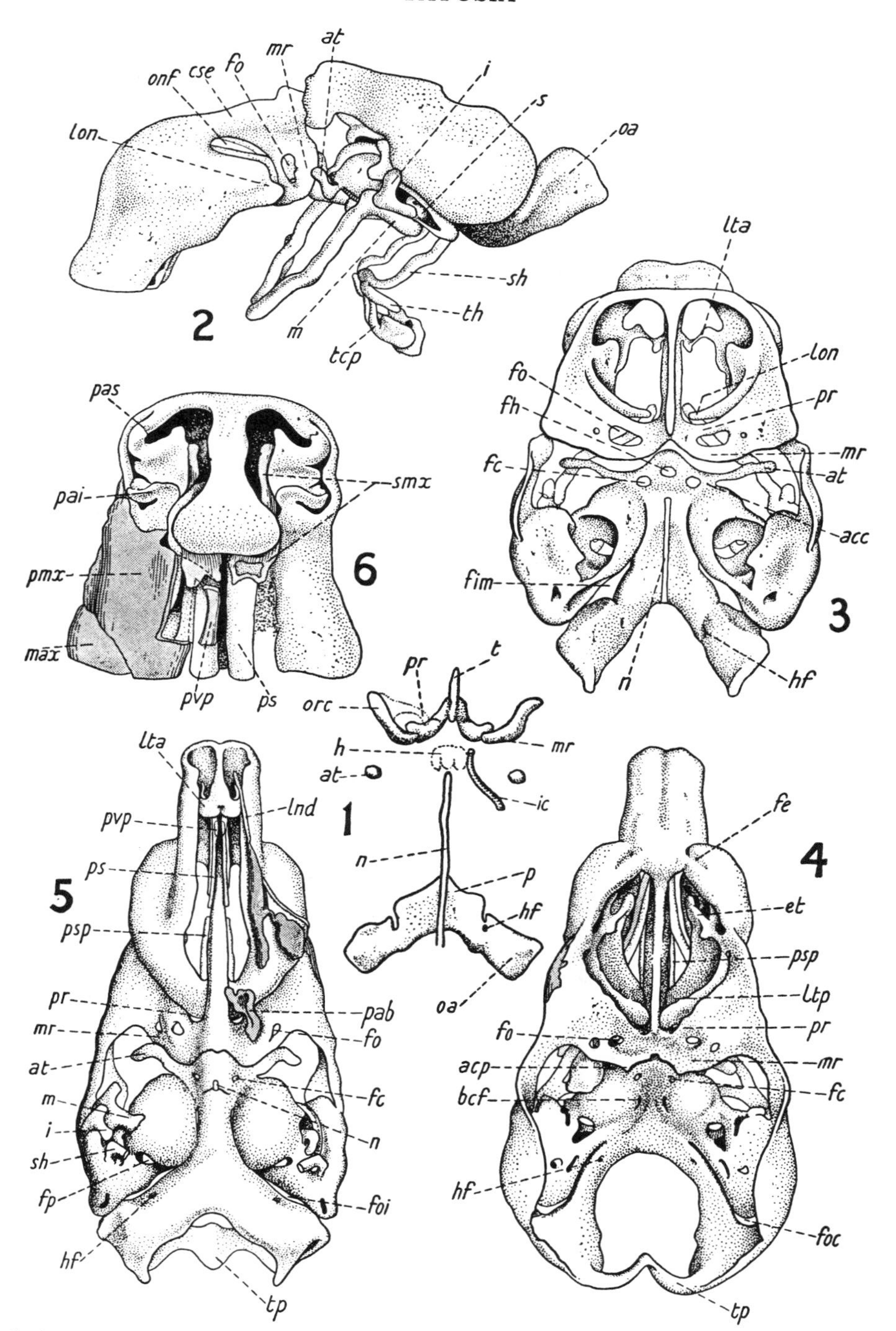

1, dorsal view of 10 mm. stage; **2**, lateral, and **3**, dorsal view of 12 mm. stage; **4**, dorsal, and **5**, ventral view of 17 mm. stage; **6**, ventral view of snout of 60 mm. stage (after Fawcett).

Explanation of Lettering precedes Plate 104

TALPA

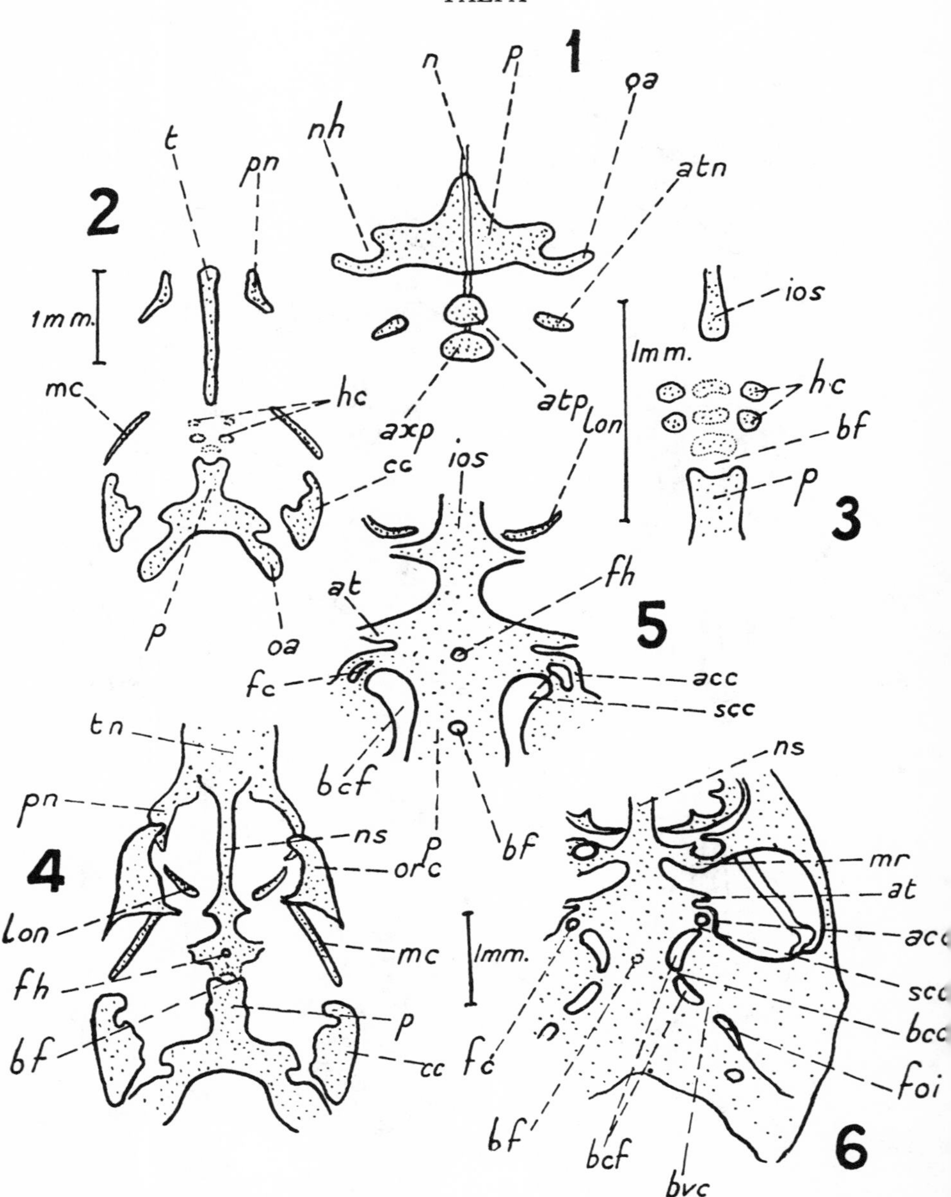

1, dorsal view of 10 mm. stage; **2**, dorsal view of 11 mm. stage, and **3**, enlargement of its hypophysial region; **4**, dorsal view of 12 mm. stage; **5**, hypophysial and surrounding regions of more advanced 12 mm. stage; **6**, dorsal view of 14 mm. stage (after Noordenbos).

Explanation of Lettering precedes Plate 104

TALPA

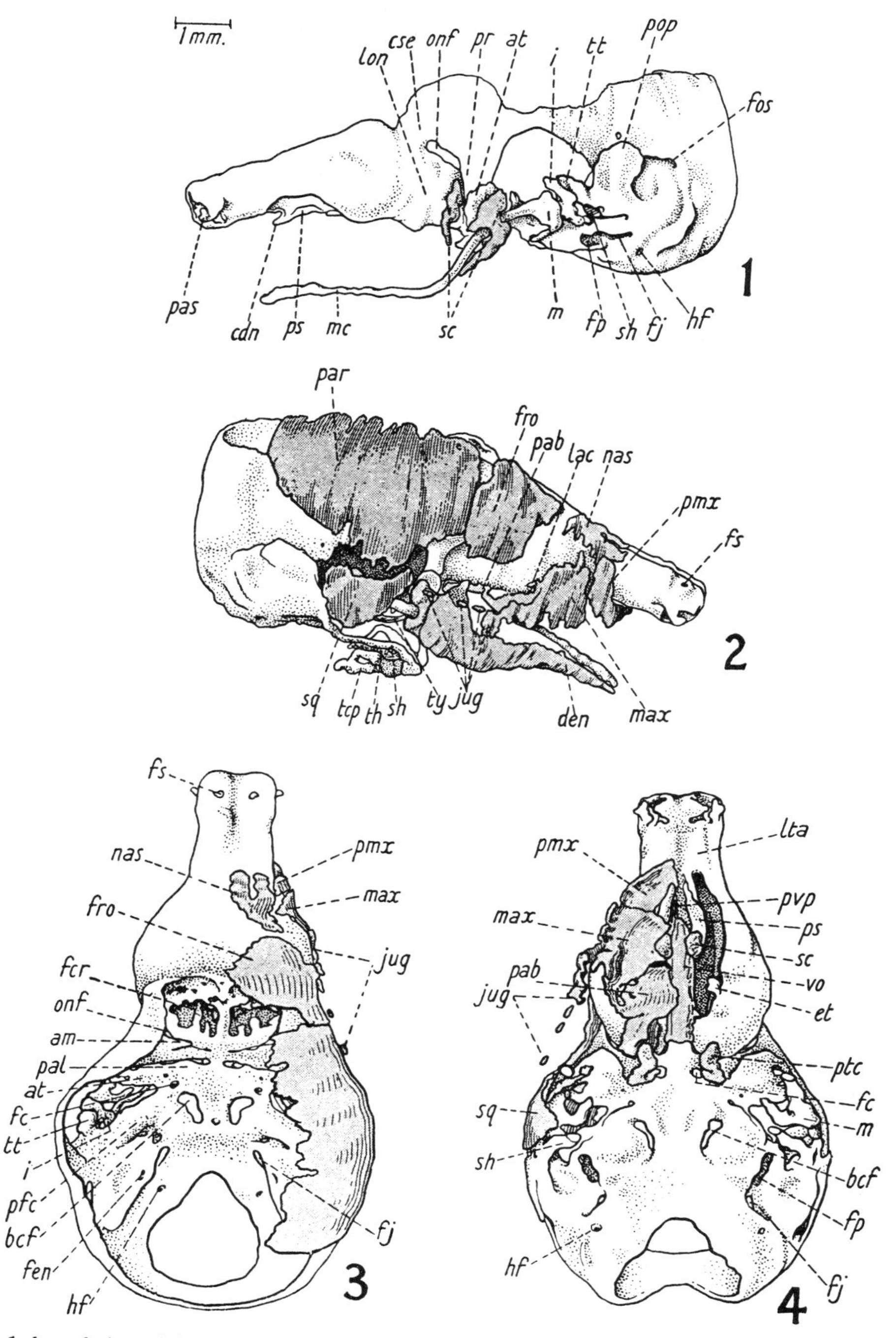

1, lateral view of chondrocranium, **2,** lateral, **3,** dorsal, and **4,** ventral view of skull of 19 mm. stage (after Fischer).

Explanation of Lettering precedes Plate 104

ERINACEUS

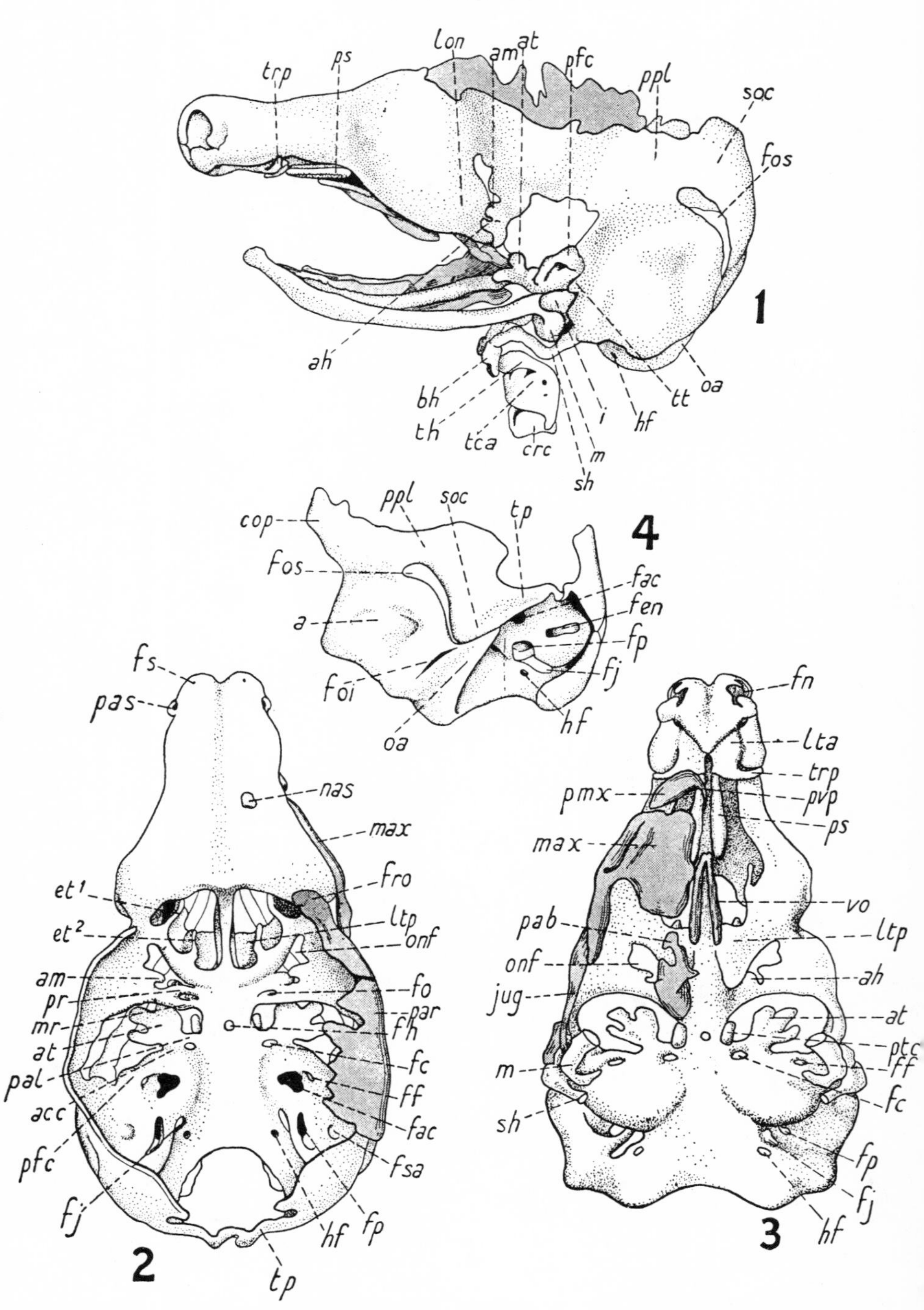

1, lateral, **2**, dorsal, **3**, ventral, and **4**, posterolateral view of 19 mm. stage (after Fawcett).
Explanation of Lettering precedes Plate 104

FELIS

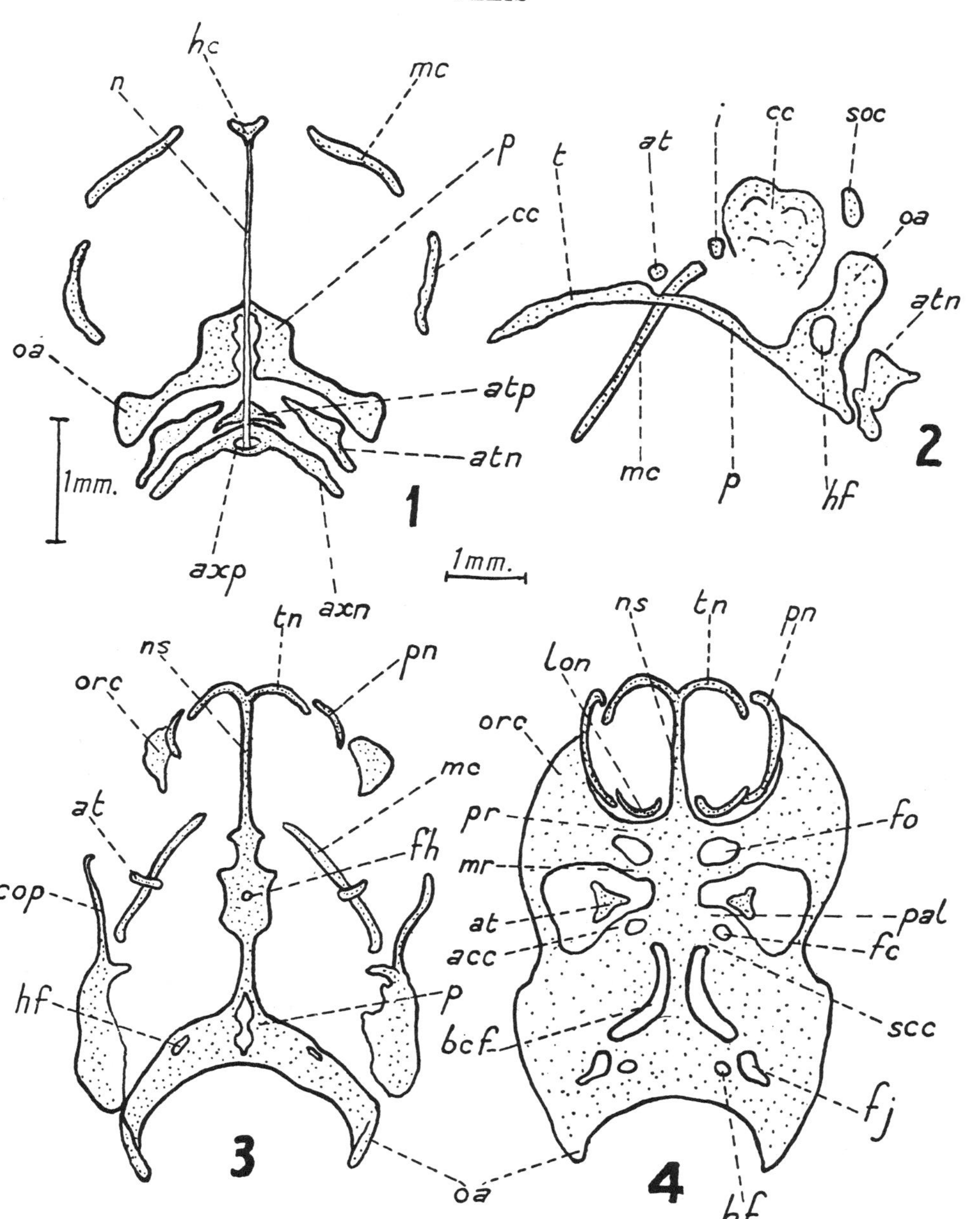

1, dorsal view of 10·5 mm. stage; **2**, lateral view of 12 mm. stage; **3**, dorsal view of 15 mm. stage; **4**, ventral view of 20 mm. stage (after Terry)

Explanation of Lettering precedes Plate 104

FELIS

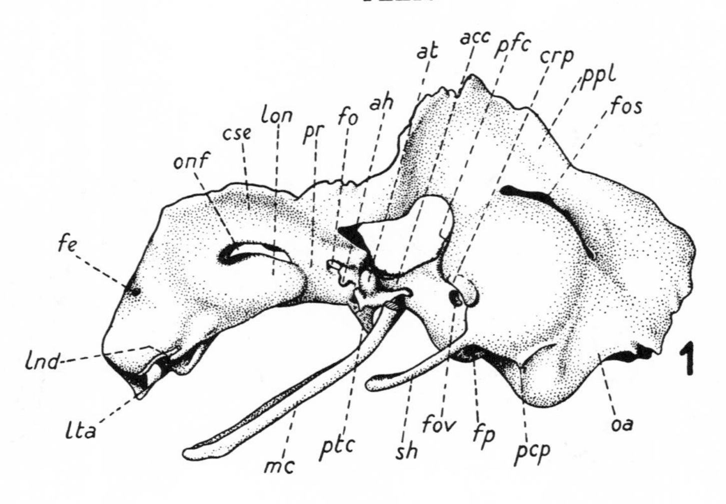

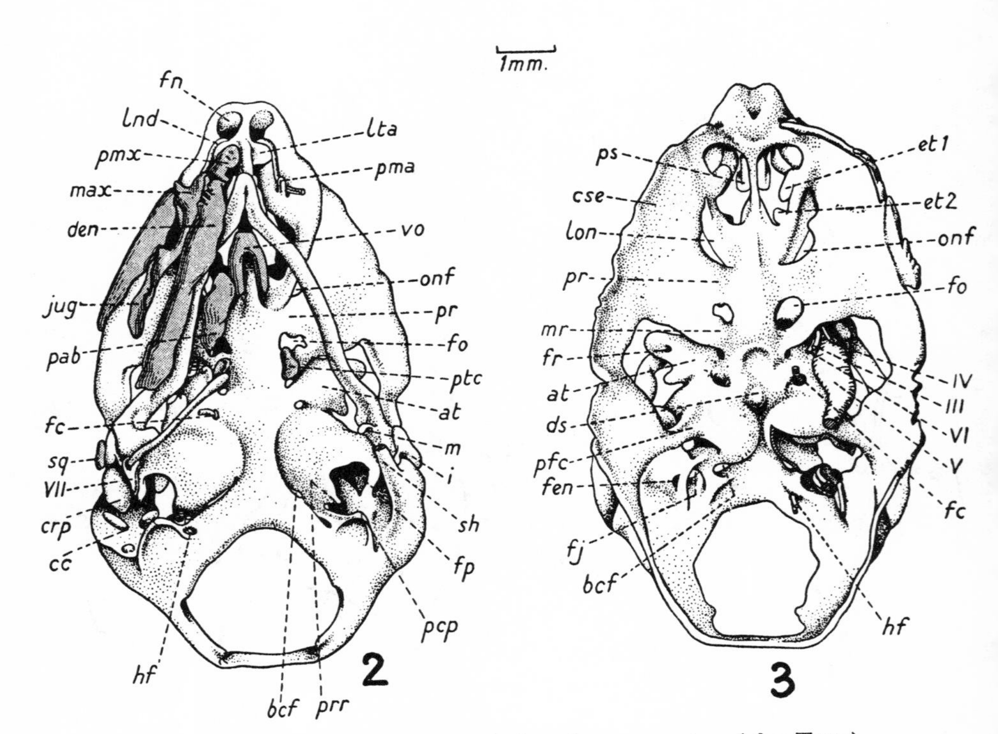

1, lateral, **2,** ventral, and **3,** dorsal view of 23·1 mm. stage (after Terry).

Explanation of Lettering precedes Plate 104

CANIS

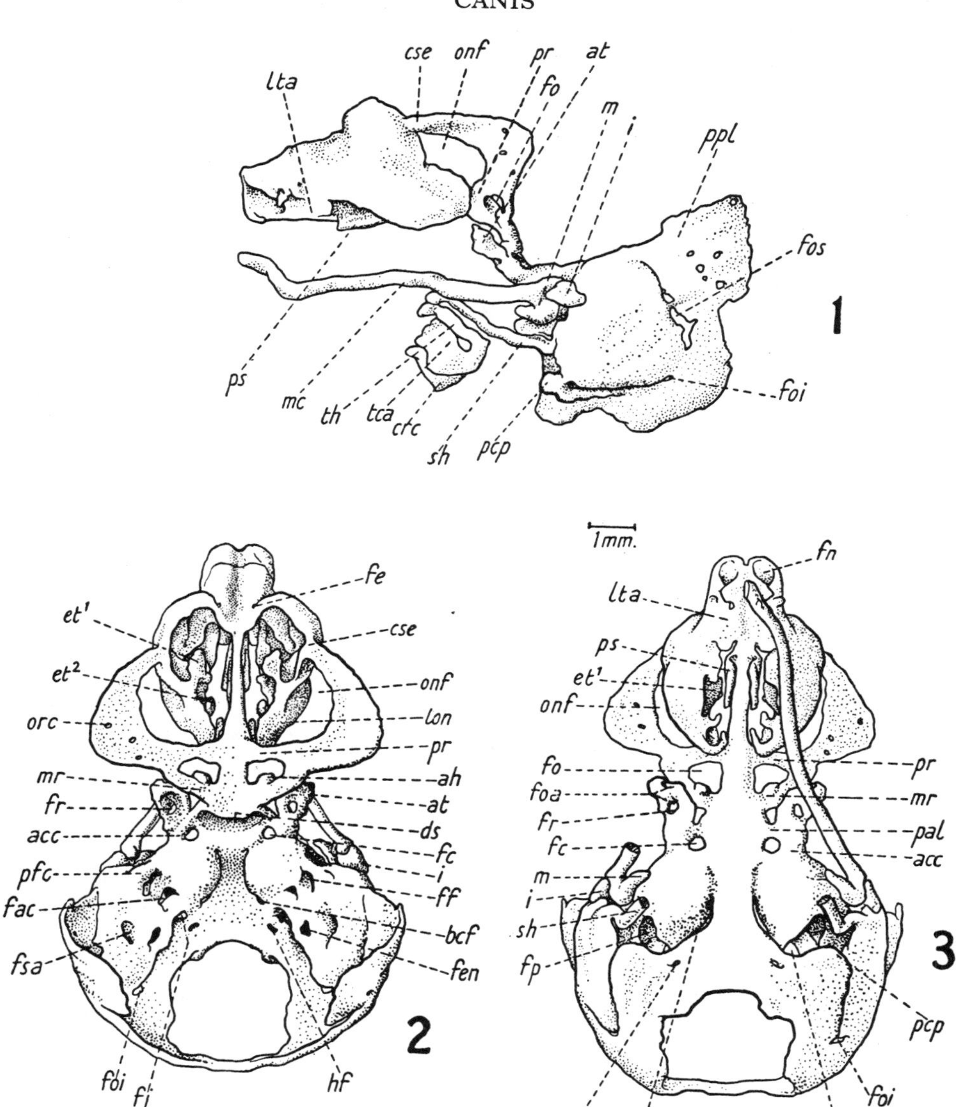

1, lateral, **2**, dorsal, and **3**, ventral view of chondrocranium of 27 mm. stage (after Olmstead).

Explanation of Lettering precedes Plate 104

POECILOPHOCA, MUSTELA

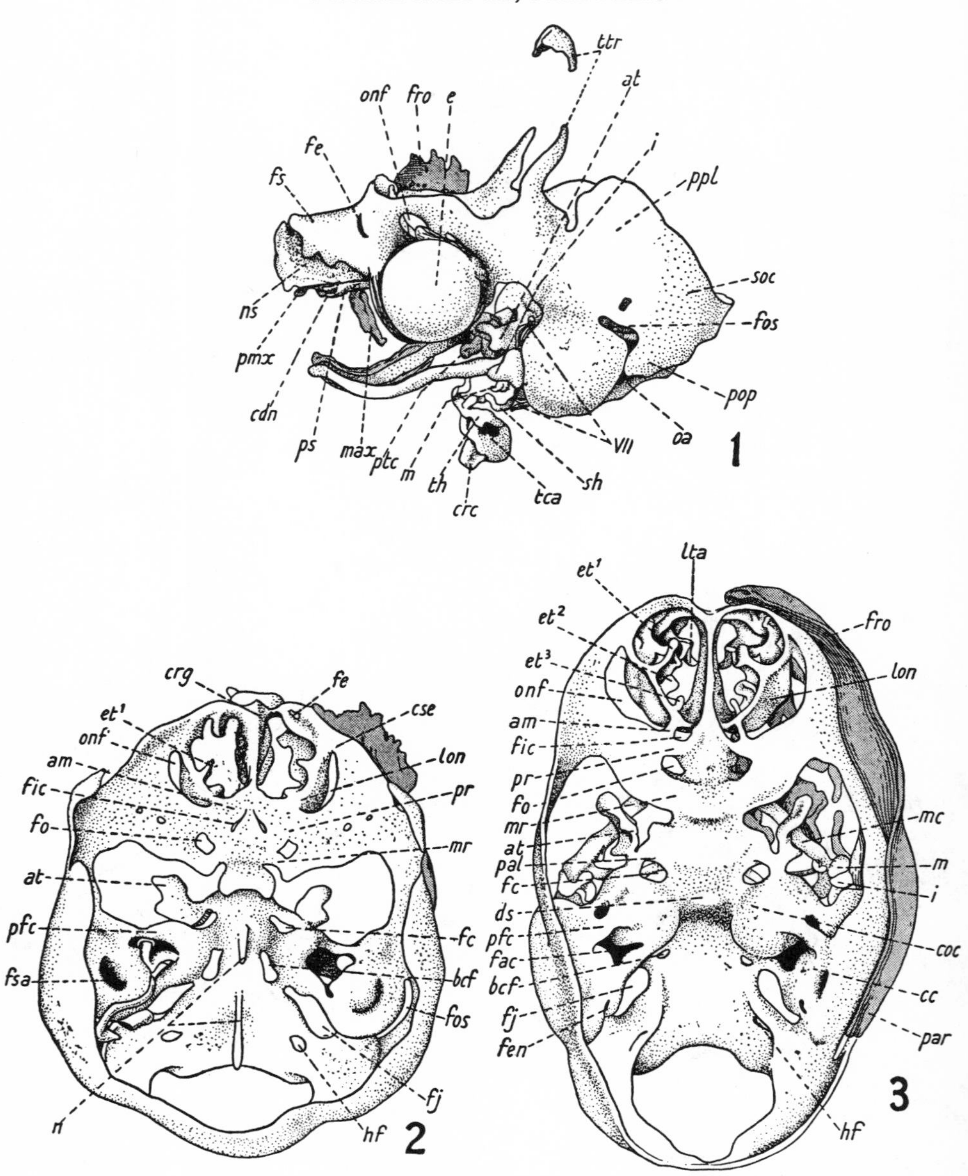

1, lateral, and **2,** dorsal view of 27 mm. stage of Poecilophoca; **3,** dorsal view of 30-day stage of Mustela (after Fawcett).

Explanation of Lettering precedes Plate 104

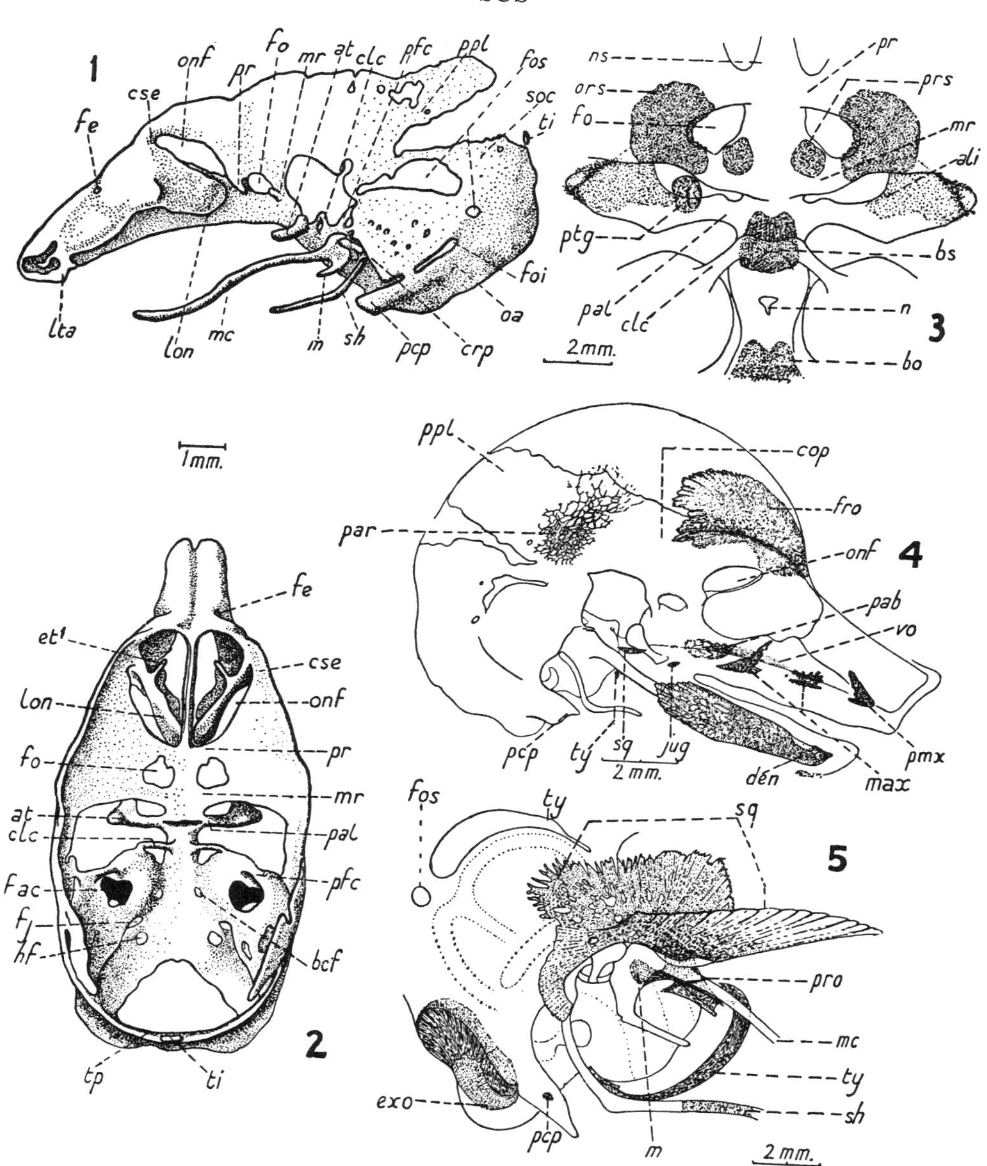

1, lateral, and **2**, dorsal view of 30 mm. stage (after Mead); **3**, dorsal view of bones of skull-base of 85 mm. stage; **4**, lateral view of skull of 38 mm. stage; **5**, lateral view of auditory region of 110 mm. stage (after Augier).

Explanation of Lettering precedes Plate 104

EQUUS, BOS

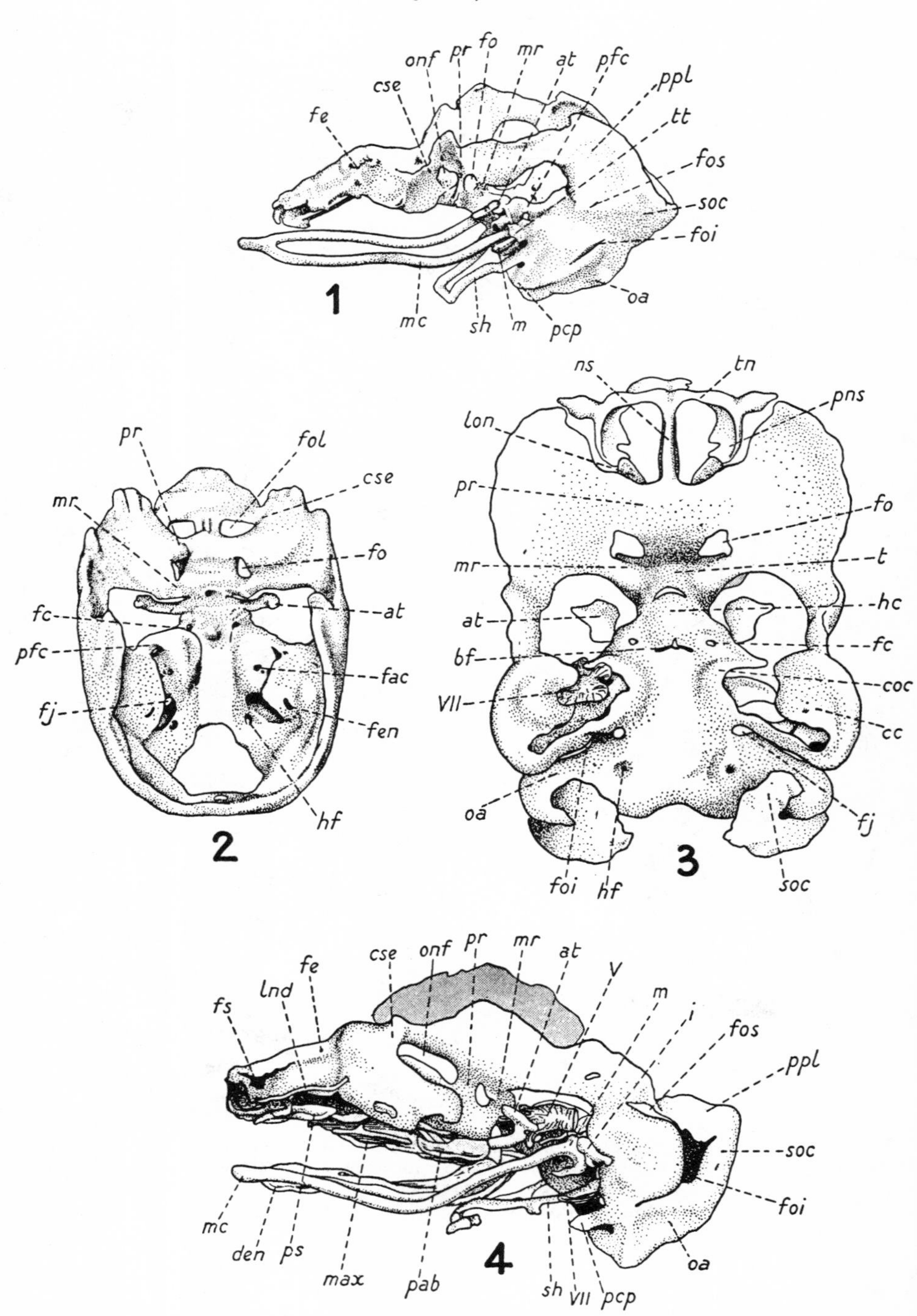

1, lateral, and **2,** dorsal view of 40 mm. stage of Equus (after Muggia); **3,** dorsal view of 19 mm. stage of Bos; **4,** lateral view of 40 mm. stage of Bos (after Fawcett).

Explanation of Lettering precedes Plate 104

MEGAPTERA

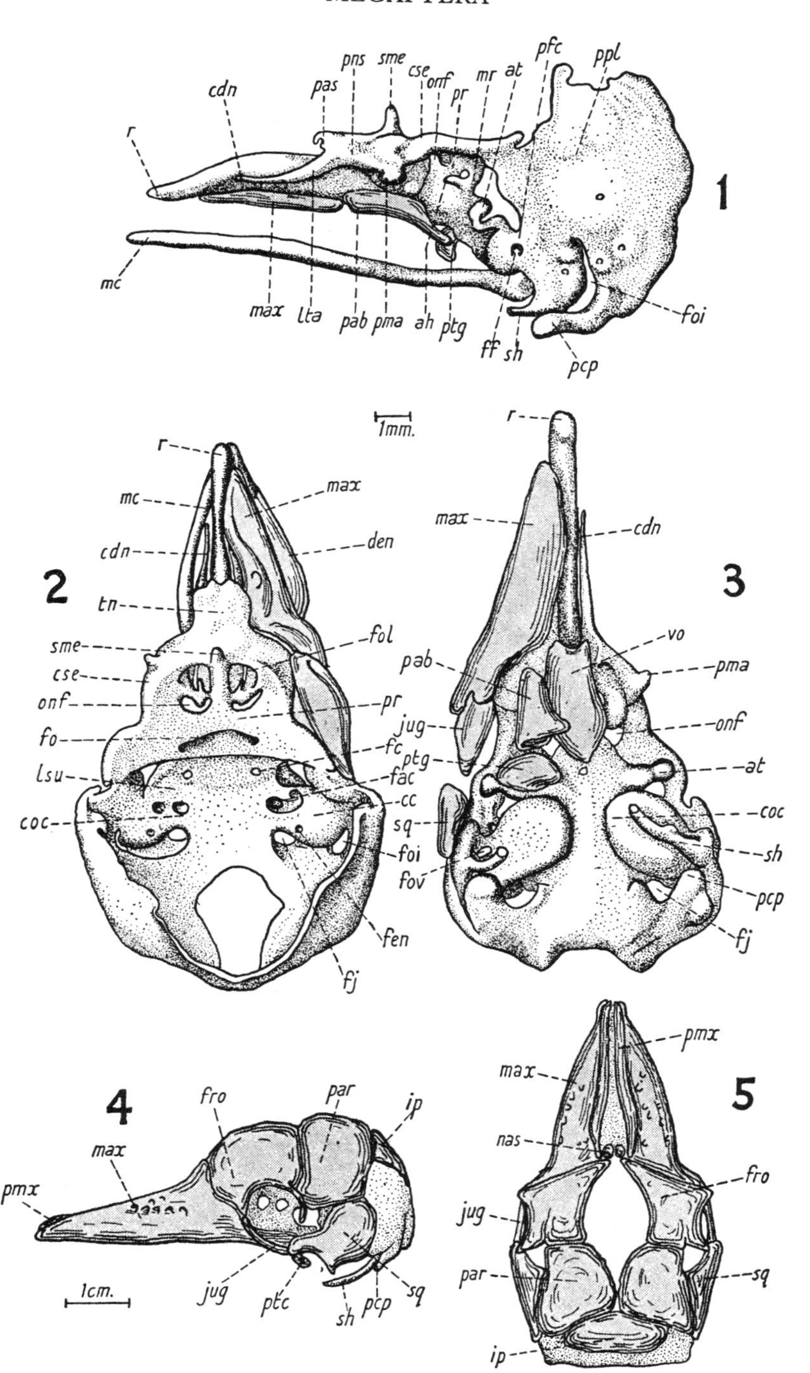

1, lateral, **2**, dorsal, and **3**, ventral view of 92 mm. stage (after Honigmann); **4**, lateral, and **5**, dorsal view of skull of 6-inch stage (after Ridewood).

Explanation of Lettering precedes Plate 104

BALAENOPTERA

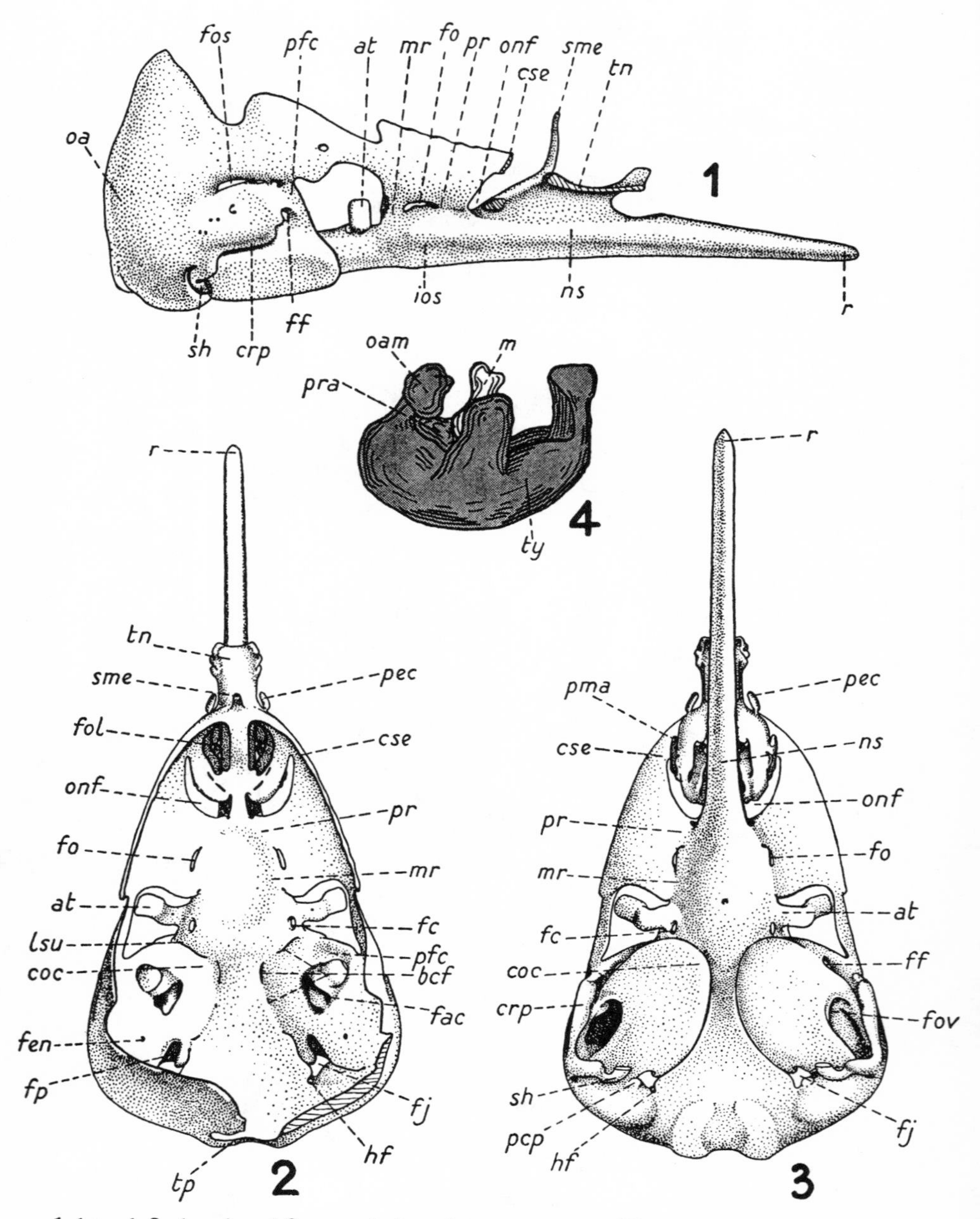

1, lateral, **2,** dorsal, and **3,** ventral view of 105 mm. stage of B. rostrata (after de Burlet); **4,** lateral view of tympanic of 6 ft. 4 in. stage of B. musculus (after Ridewood).

Explanation of Lettering precedes Plate 104

PHOCAENA

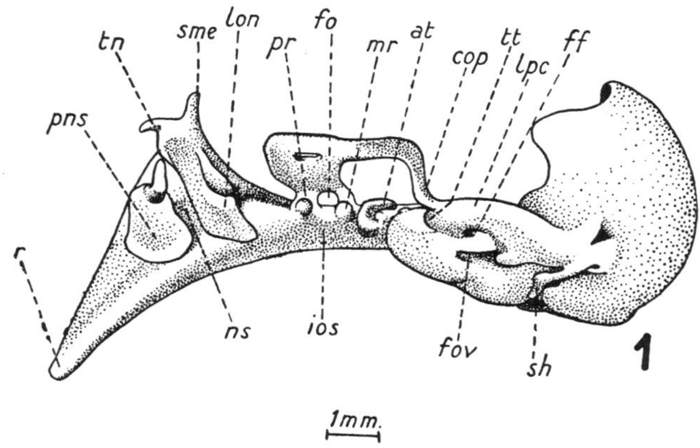

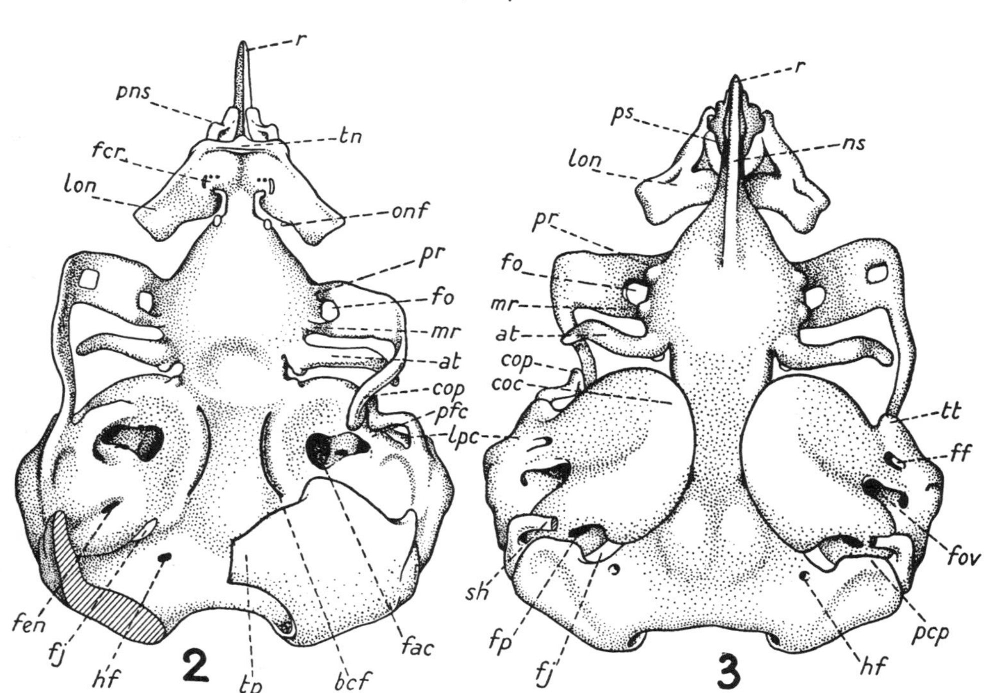

1, lateral, **2**, dorsal, and **3**, ventral view of 48 mm. stage (after de Burlet).

Explanation of Lettering precedes Plate 104

PHOCAENA, GLOBIOCEPHALUS, ETC.

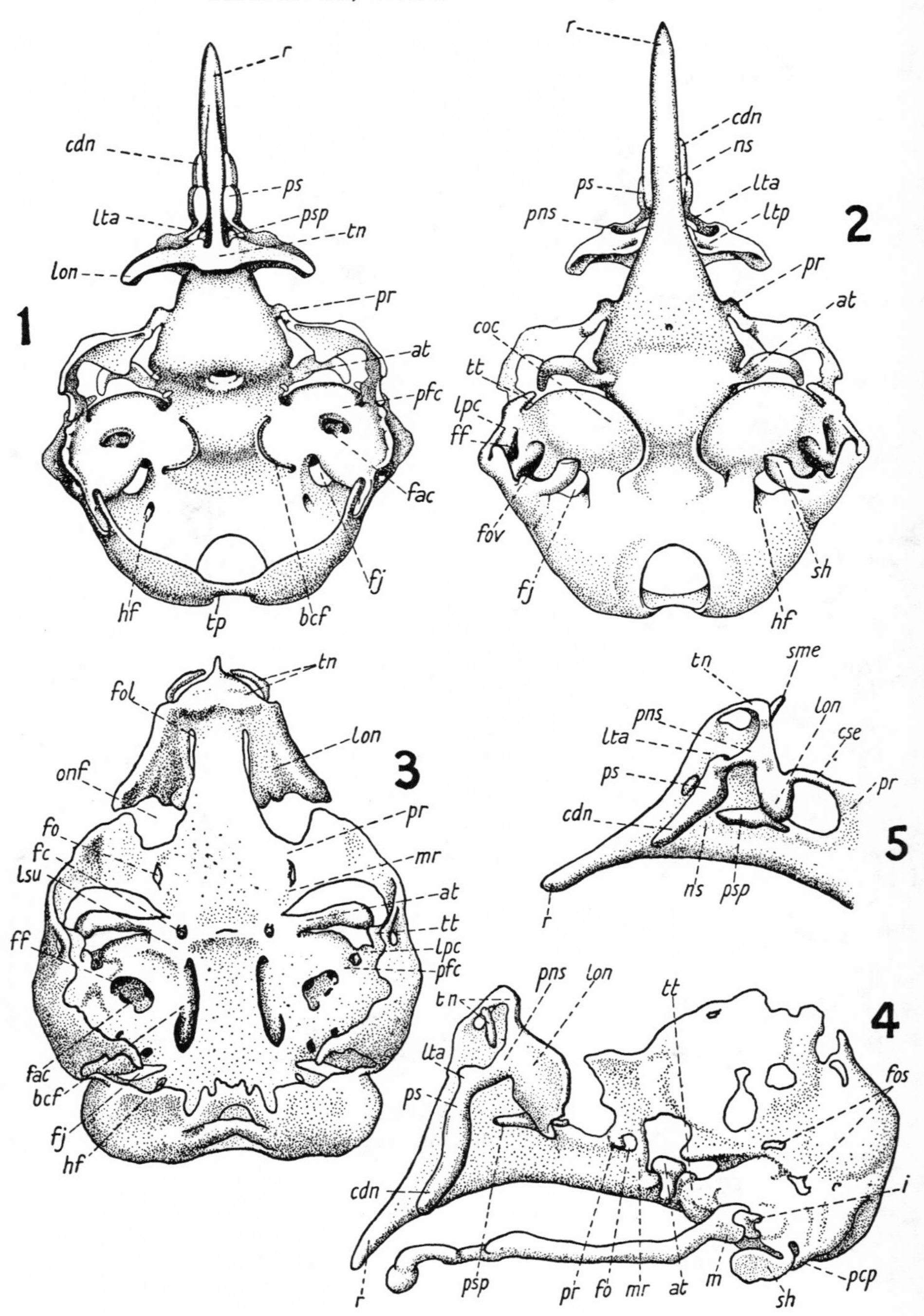

1, dorsal, and **2,** ventral view of 92 mm. stage of Phocaena (after de Burlet); **3,** dorsal, and **4,** lateral view of 133 mm. stage of Globiocephalus (after Schreiber); **5,** lateral view of nasal region of Lagenorhynchus (after de Burlet).

Explanation of Lettering precedes Plate 104

HALICORE

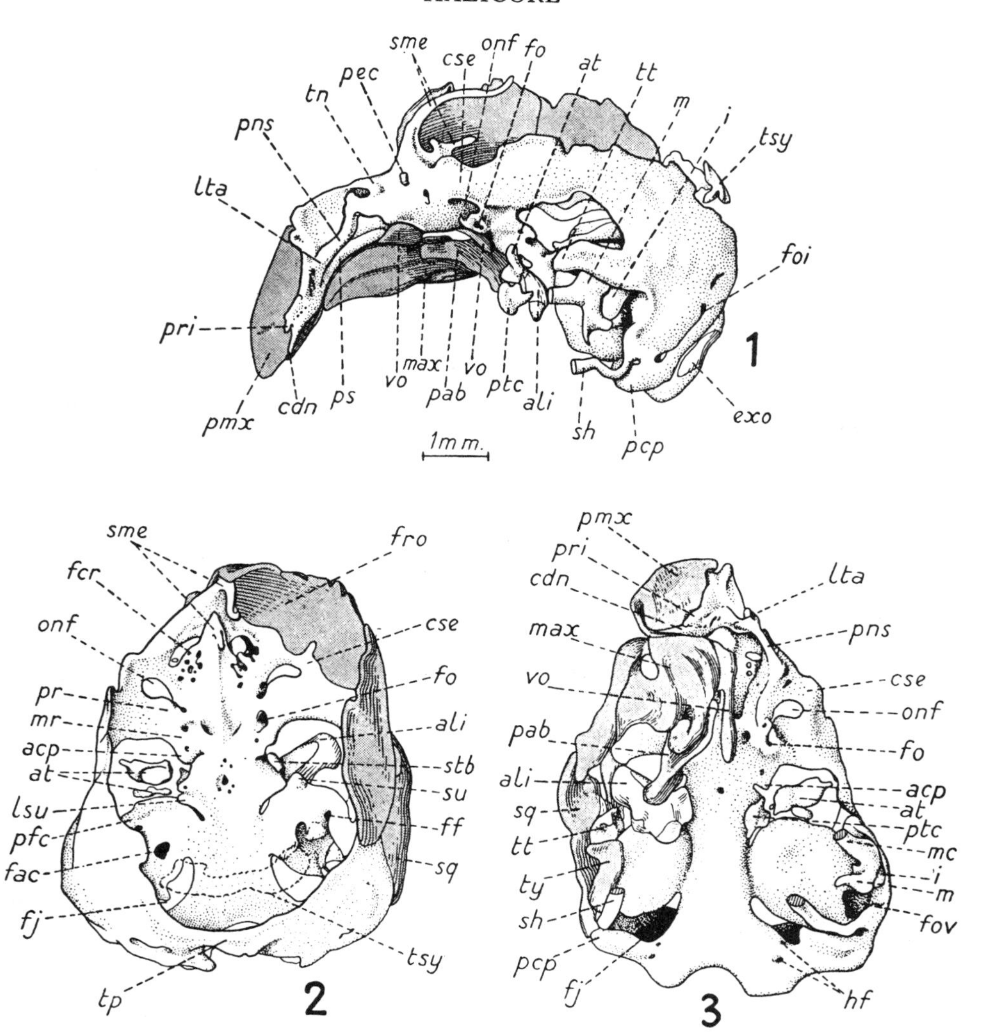

1, lateral, **2,** dorsal, and **3,** ventral view of 150 mm. stage (after Matthes).

Explanation of Lettering precedes Plate 104

MINIOPTERUS, GALEOPITHECUS

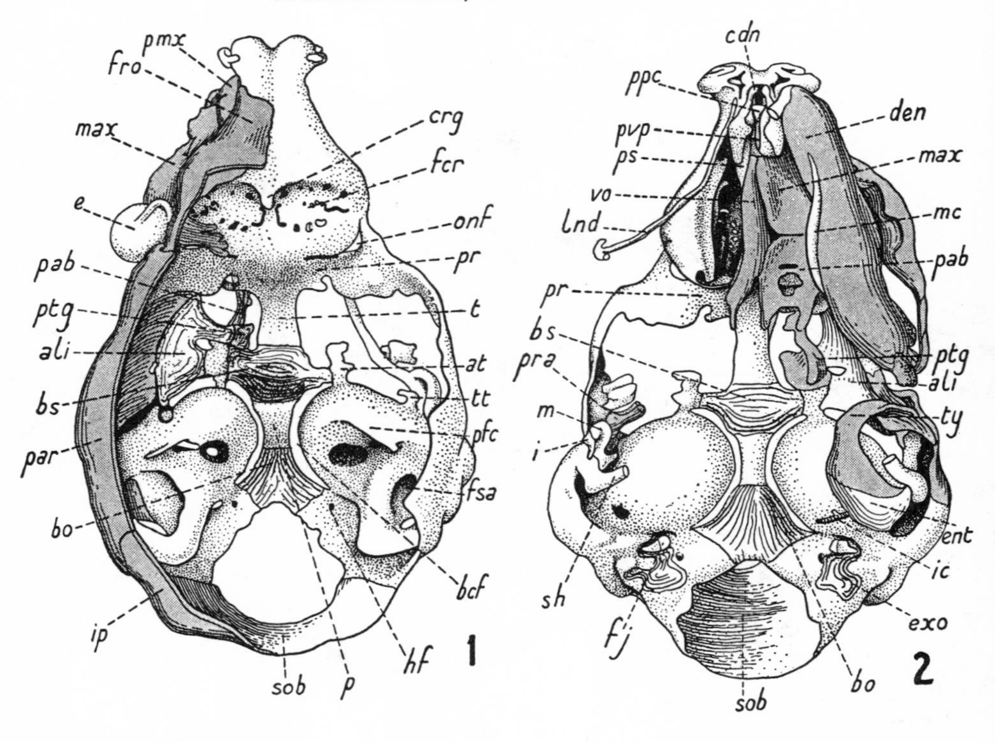

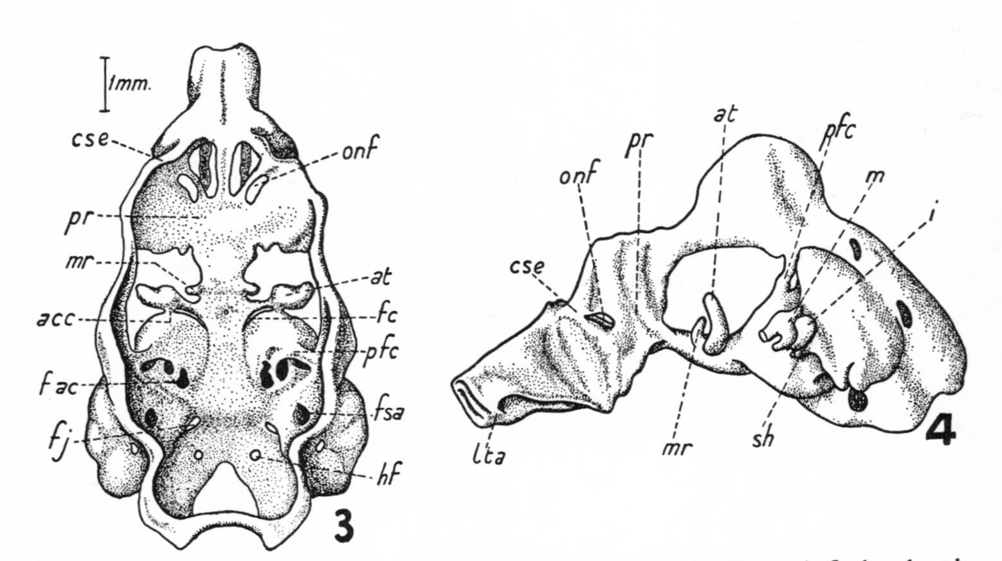

1, dorsal, and **2**, ventral view of 17 mm. stage of Miniopterus (after Fawcett); **3**, dorsal, and **4**, lateral view of 28 mm. stage of Galeopithecus (after Henckel).

Explanation of Lettering precedes Plate 104

TUPAJA, TARSIUS

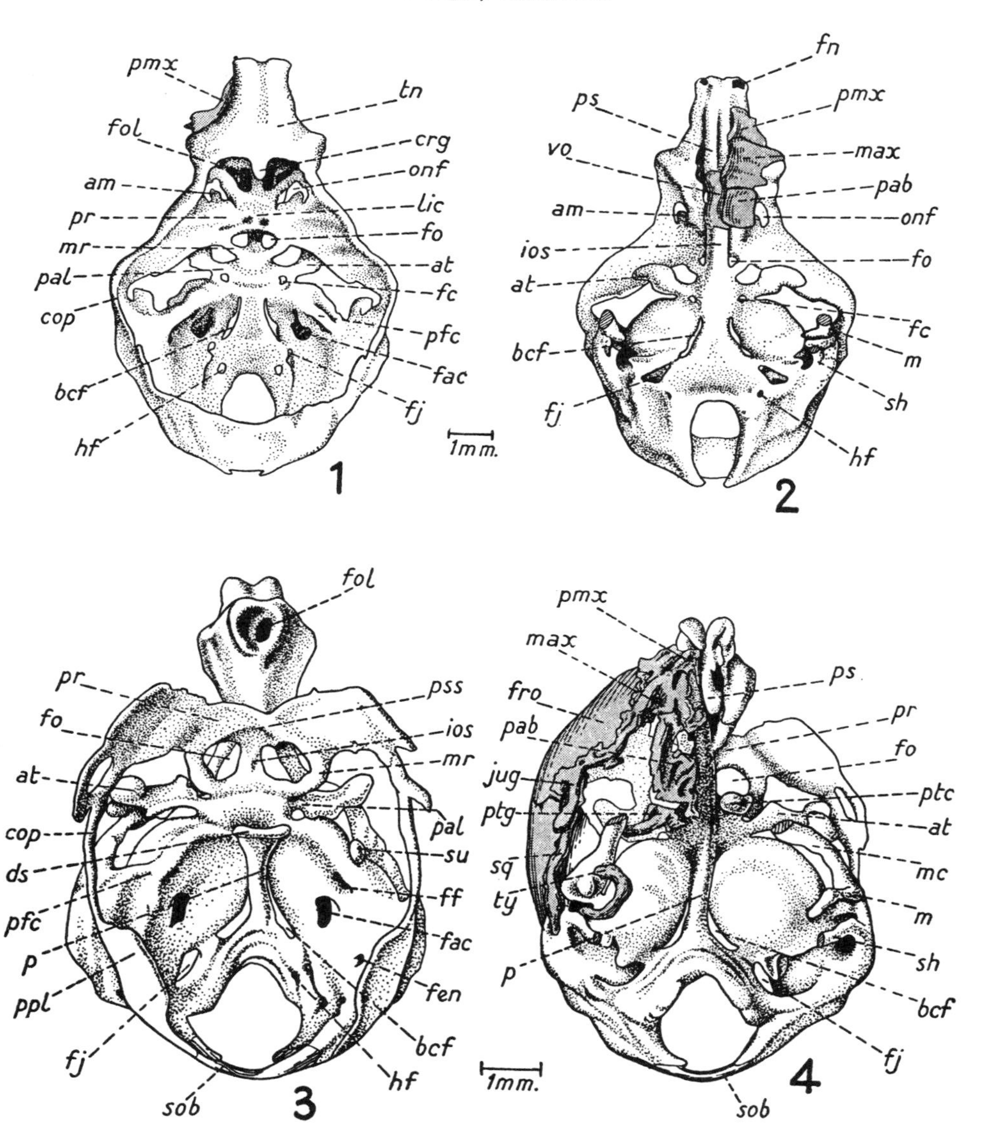

1, dorsal, and **2**, ventral view of 20 mm. stage of Tupaja; **3**, dorsal, and **4**, ventral view of 24 mm. stage of Tarsius (after Henckel).

Explanation of Lettering precedes Plate 104

NYCTICEBUS, SEMNOPITHECUS, ETC.

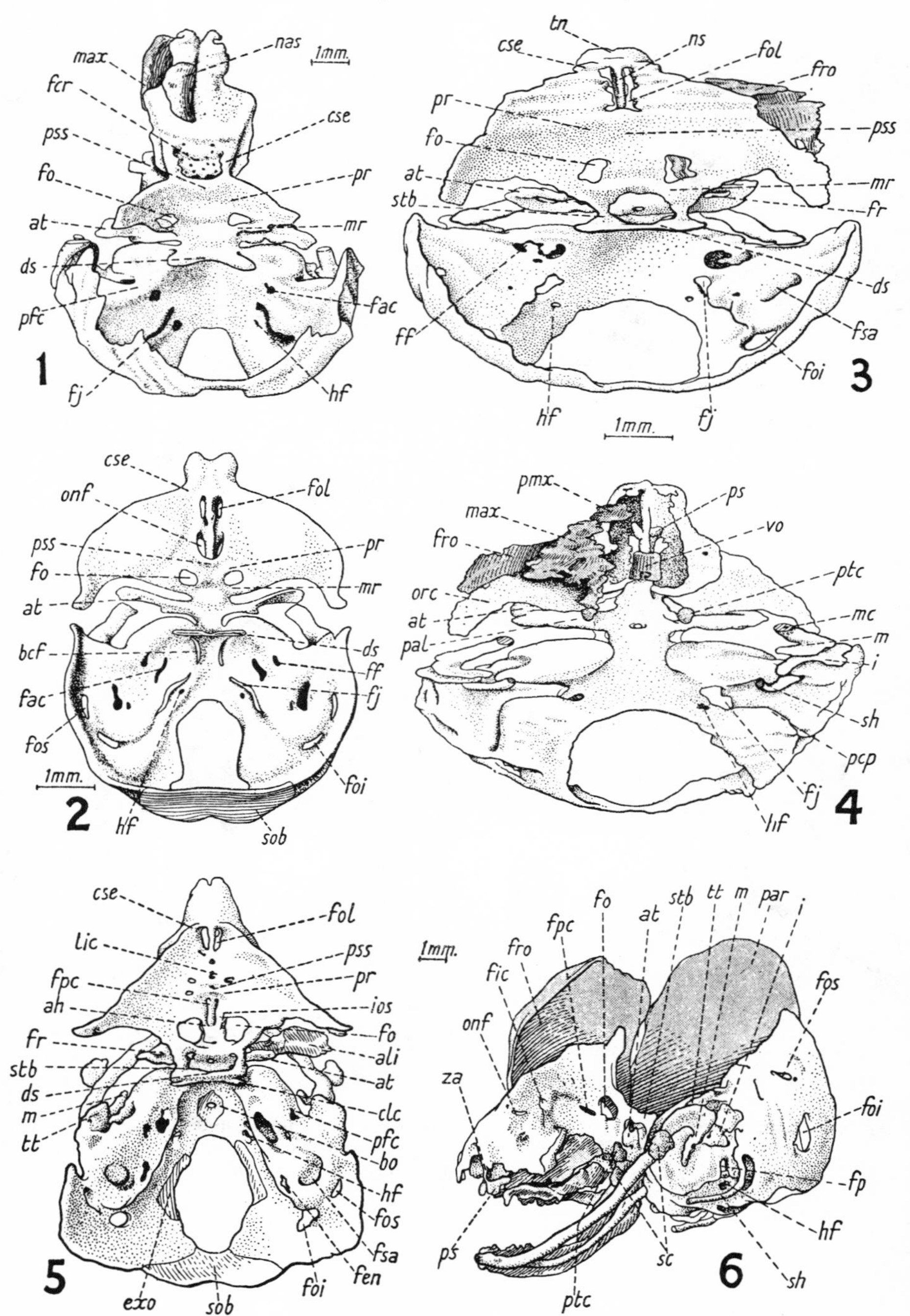

1, dorsal view of 30 mm. stage of Nycticebus; **2,** dorsal view of 24 mm. stage of Chrysothrix (after Henckel); **3,** dorsal, and **4,** ventral view of 25 mm. stage of Macacus; **5,** dorsal, and **6,** lateral view of 53 mm. stage of Semnopithecus (after Fischer).

Explanation of Lettering precedes Plate 104

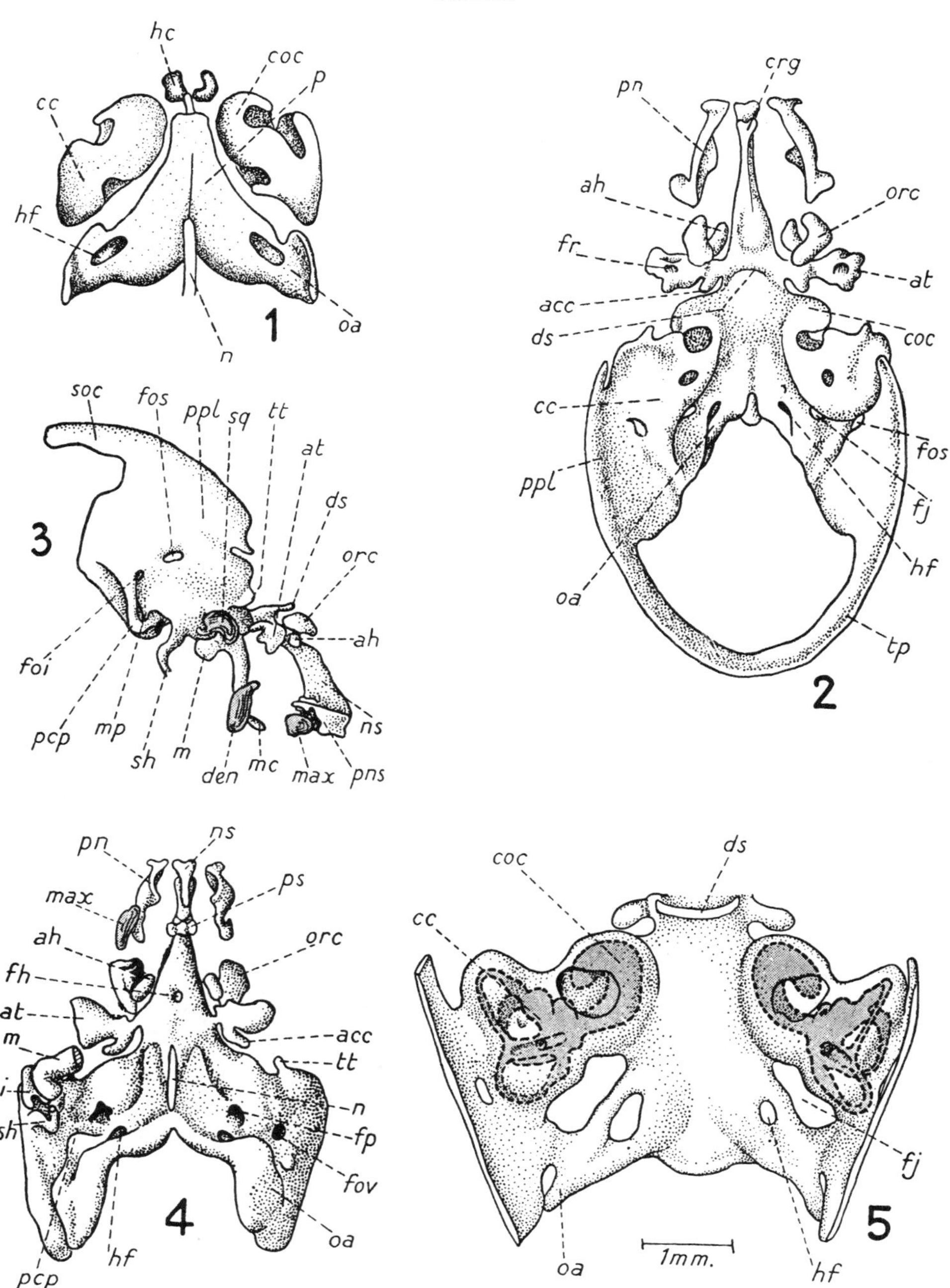

1, dorsal view of 14 mm. stage (after Fawcett); **2**, dorsal, **3**, lateral, and **4**, ventral view of 20 mm. stage (after Kernan); **5**, auditory capsules of 21 mm. stage (after Lewis).

Explanation of Lettering precedes Plate 104

HOMO

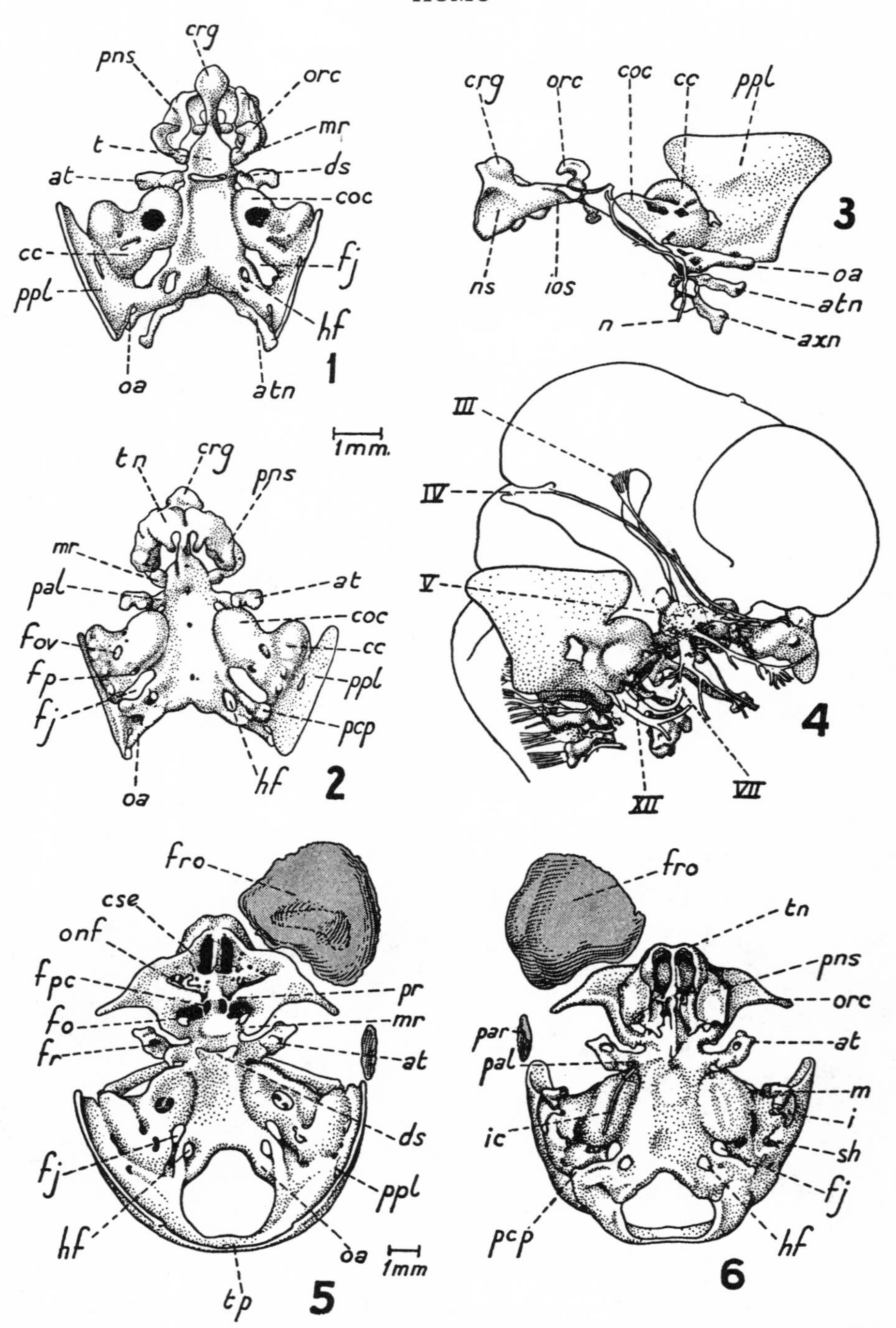

1, dorsal, **2**, ventral, **3** & **4**, lateral views of 21 mm. stage (after Lewis); **5**, dorsal, and **6**, ventral view of 43 mm. stage (after Macklin).

Explanation of Lettering precedes Plate 104

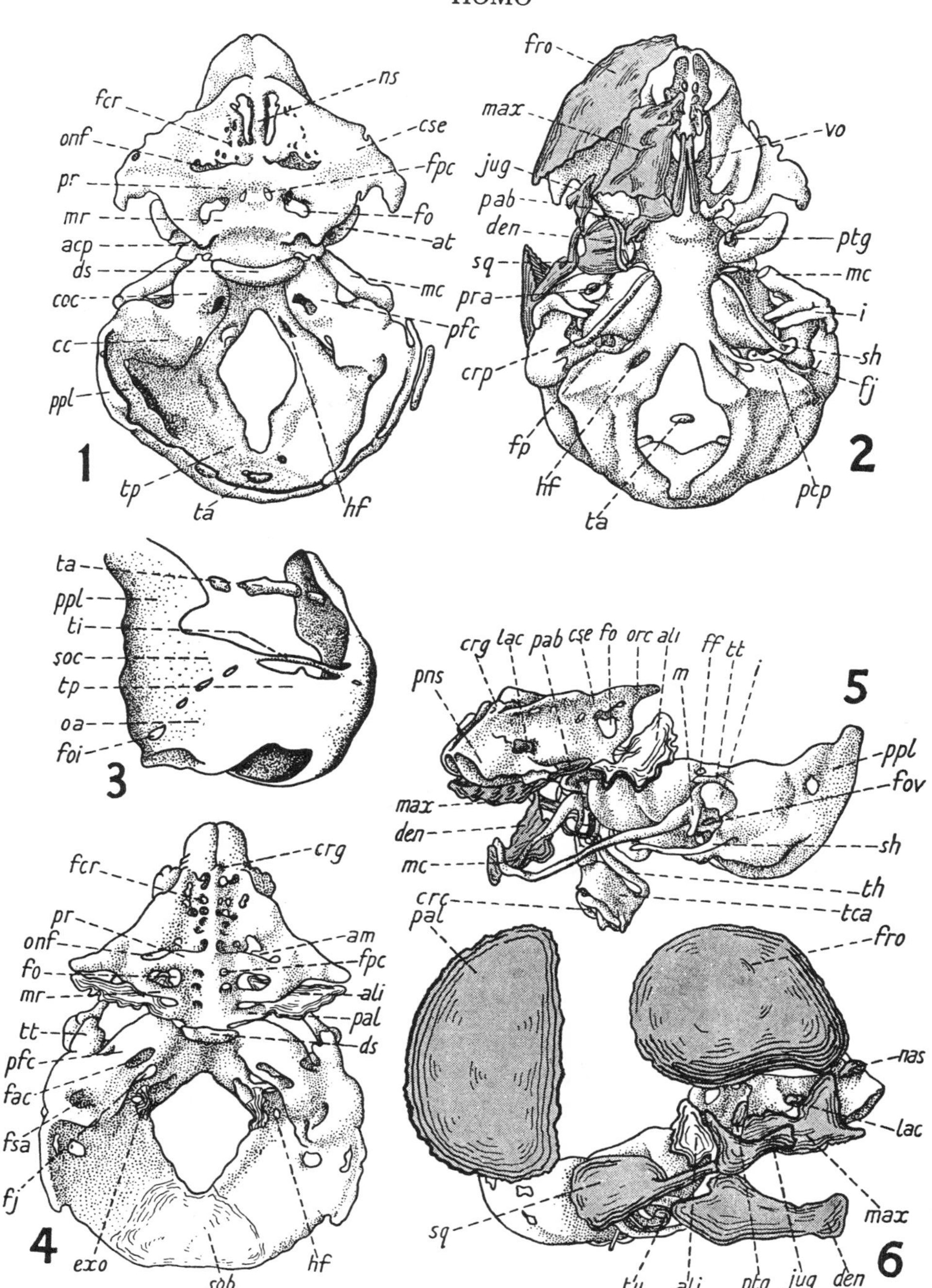

1, dorsal, and **2**, ventral view of 40 mm. (c.-r.) stage (after Macklin); **3**, posterolateral view of 30 mm. stage (after Fawcett); **4**, dorsal, and **5** and **6**, lateral views of 80 mm. stage (after Hertwig).

Explanation of Lettering precedes Plate 104

HOMO

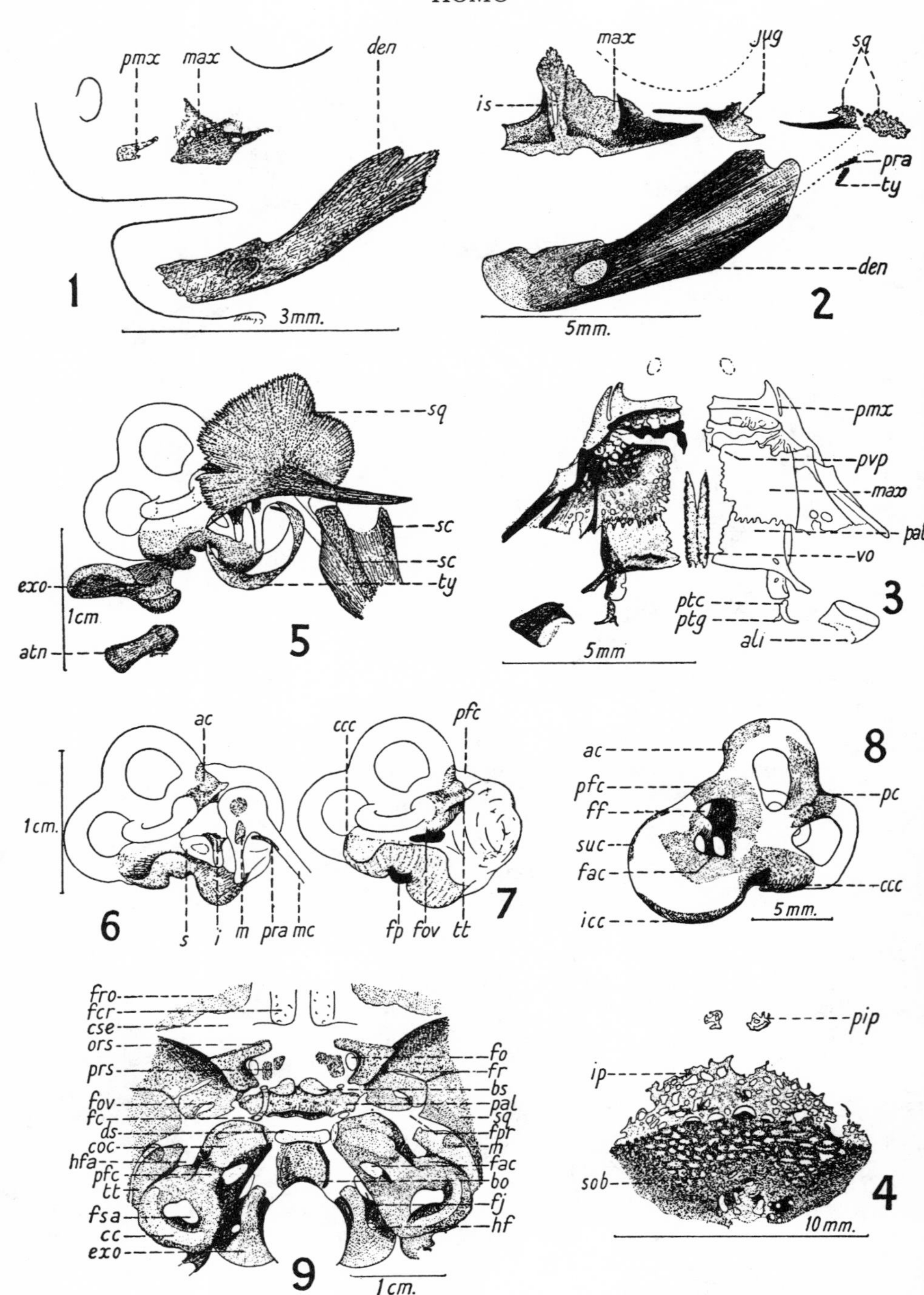

1, jaws of 26 mm. stage; **2**, jaws of 37 mm. stage; **3**, palate of 54 mm. stage; **4**, supraoccipital region of 75 mm. stage; **5**, lateral view of auditory and tympanic region of 138 mm. stage; **6**, ditto with membrane-bones removed, **7**, ditto with visceral arches removed; **8**, medial view of auditory capsule of 138 mm. stage; **9**, dorsal view of 170 mm. stage (after Augier from Poirier and Charpey's *Traité d'Anatomie Humaine*, by permission of MM. Masson).

Explanation of Lettering precedes Plate 104

TRABECULAE

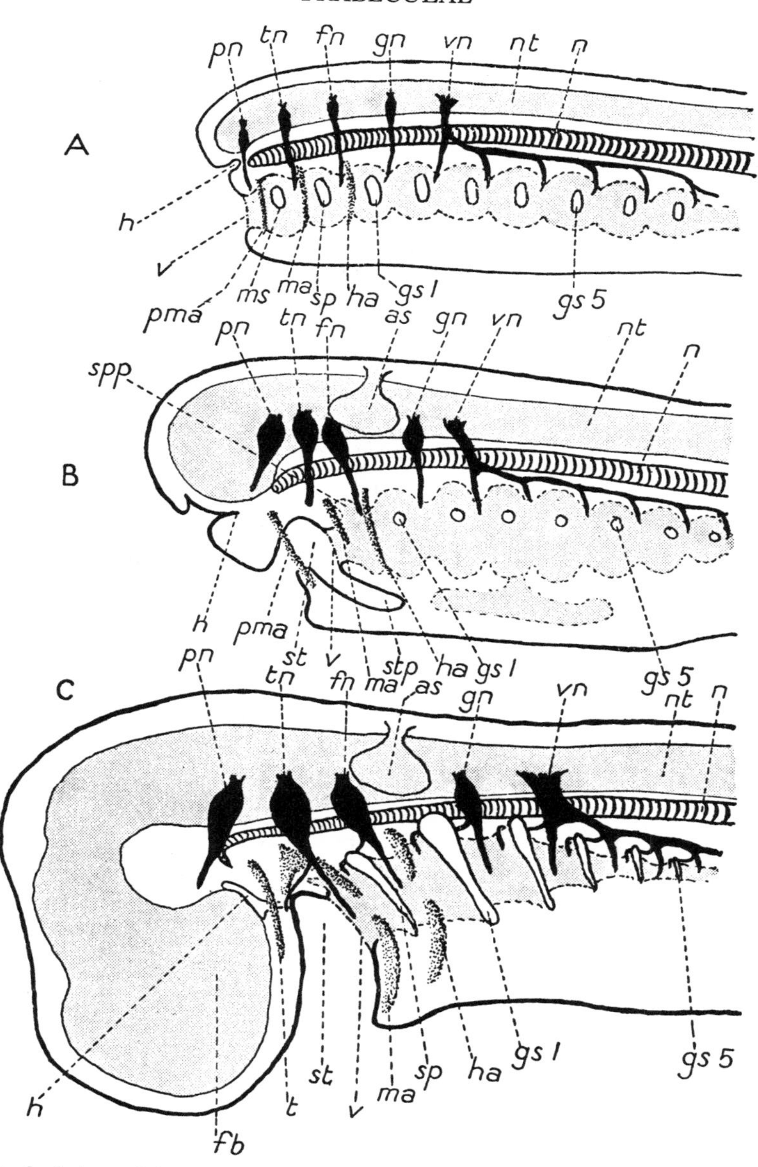

Diagram of relations of the mouth, visceral arches and clefts, and dorsal cranial nerve-roots in *A* ancestral Chordate, *B* Ammocoete larva of Petromyzon, *C* Scyllium embryo (from de Beer).

as, auditory sac; fb. forebrain; fn, facial nerve; gn, glossopharyngeal nerve; gs 1–5, gill-slit 1–5; h, hypophysis; ha, hyoid arch; ma, mandibular arch; ms, mandibular visceral slit; n, notochord; nt, neural tube; pma, pre-mandibular arch; pn, profundus nerve; sp, spiracular slit; spp, spiracular pouch; st, stomodaeum; stp, stomodaeal pouches; tn, trigeminal nerve; v, velum; vn, vagus nerve.

TRABECULAE

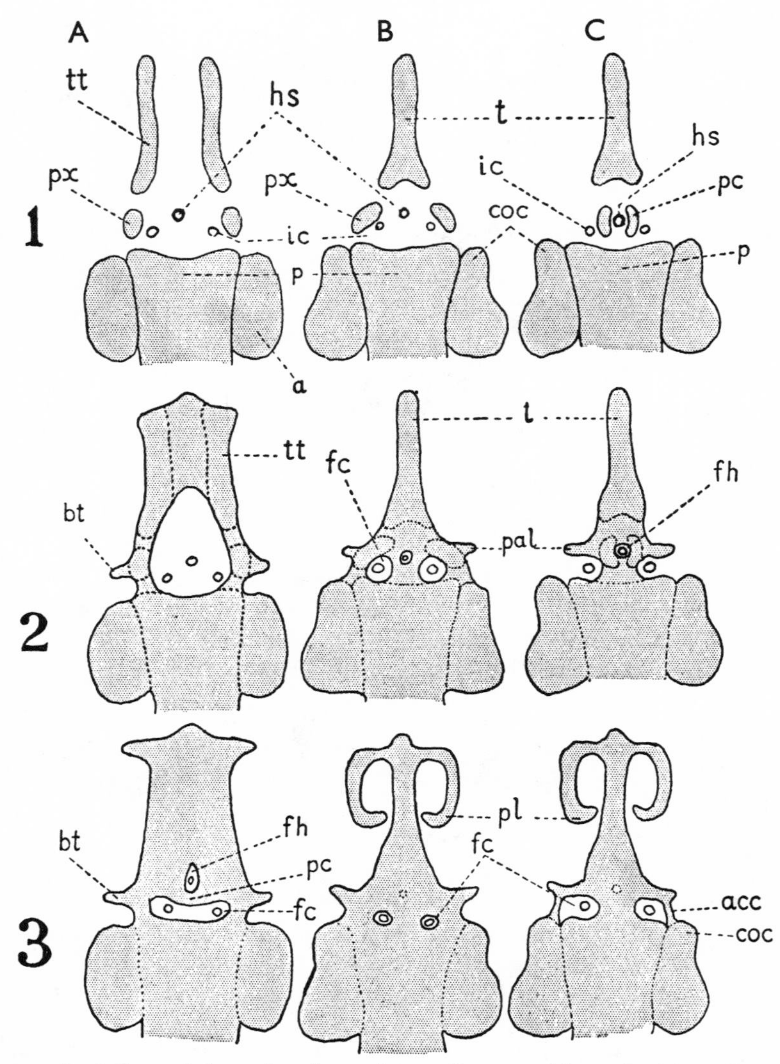

Diagram of relations of the trabeculae, polar cartilages, and hypophysial cartilages, in *A* Pisces, *B* Monotreme, *C* Placental mammal, at 3 successive stages (**1**, **2**, **3**) (from de Beer and Woodger).

a, auditory capsule; acc, alicochlear commissure; bt, basitrabecular process; coc, cochlear capsule; fc, carotid foramen; fh, hyophysial foramen; hs, hypophysial stalk; ic, internal carotid; p, parachordal; pal, processus alaris; pc, hypophysial cartilage; pl, lamina orbitonasalis; px, polar cartilage; t, central stem; tt, trabecula cranii.

OCCIPITO-ATLANTAL JOINT

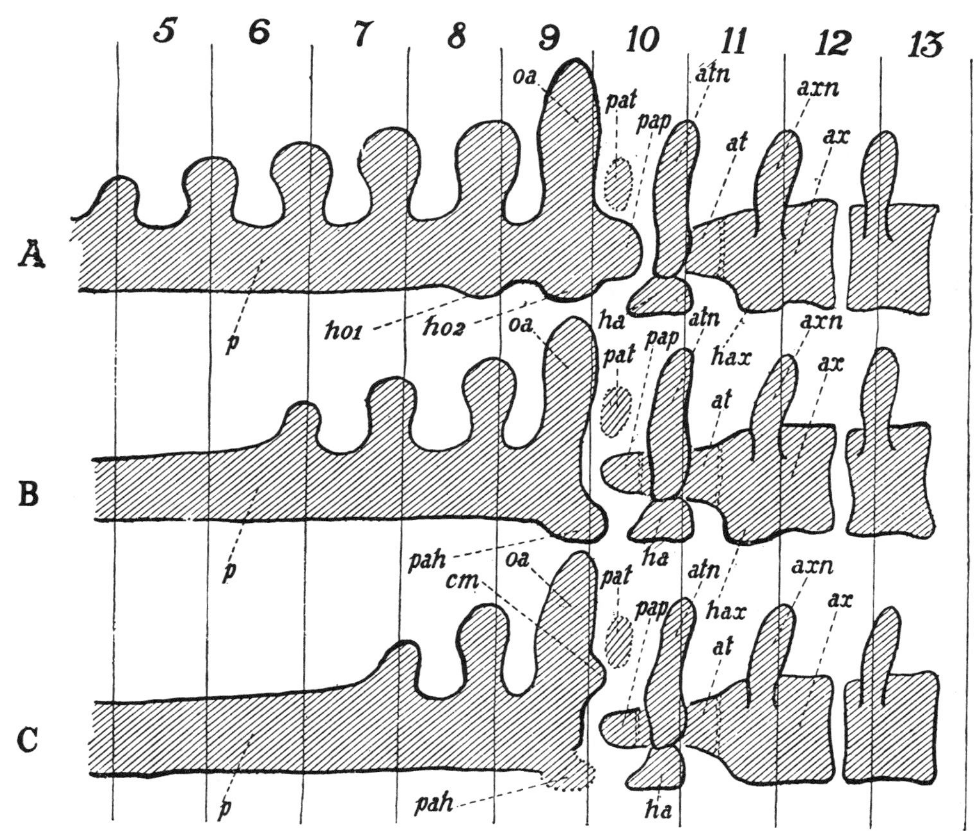

Diagram of relations of the skull, occipito-atlantal joint, and anterior cervical vertebrae in *A* Chelonia, Crocodilia, and Aves; *B* Squamata and Rhynchocephalia; *C* Mammalia (from de Beer and Barrington).

at, atlas pleurocentrum; atn, atlas neural arch; ax, axis pleurocentrum; axn, axis neural arch; cm, paired arcual condyle; ha, atlas hypocentrum; hax, axis hypocentrum; ho 1–2, hypocentrum of 1st–2nd occipital vertebra; oa, occipital arch; p, parachordals; pah, proatlas hypocentrum; pap, proatlas pleurocentrum; pat, proatlas interdorsal arch.

NASAL CAPSULE

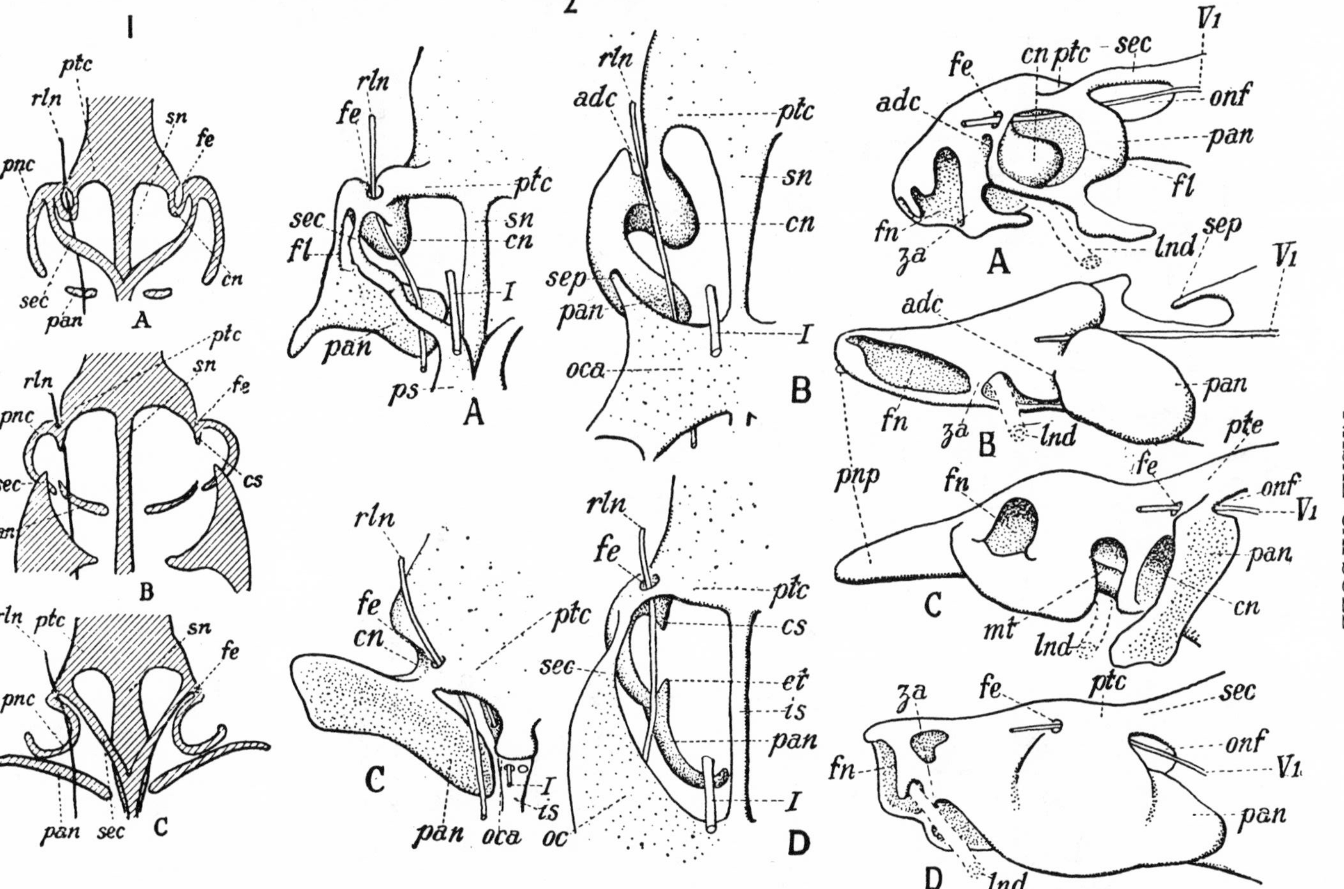

1, dorsal views of early stage in *A* Lacerta, *B* Lepus, *C* Anas; **2,** dorsal views of nasal capsule in *A* Lacerta, *B* Crocodilus, *C* Passer, *D* Lepus; **3,** lateral views of nasal capsule in *A* Lacerta, *B* Crocodilus, *C* Passer, *D* Lepus (from de Beer and Barrington).

Explanation of Lettering precedes Plate 94

In addition: adc, aditus conchae; cs, crista semicircularis; et, ethmoturbinal; fe, epiphanial foramen; fn, fenestra narina; lnd, lachrymonasal duct; rln, ramus

HYOID ARCH

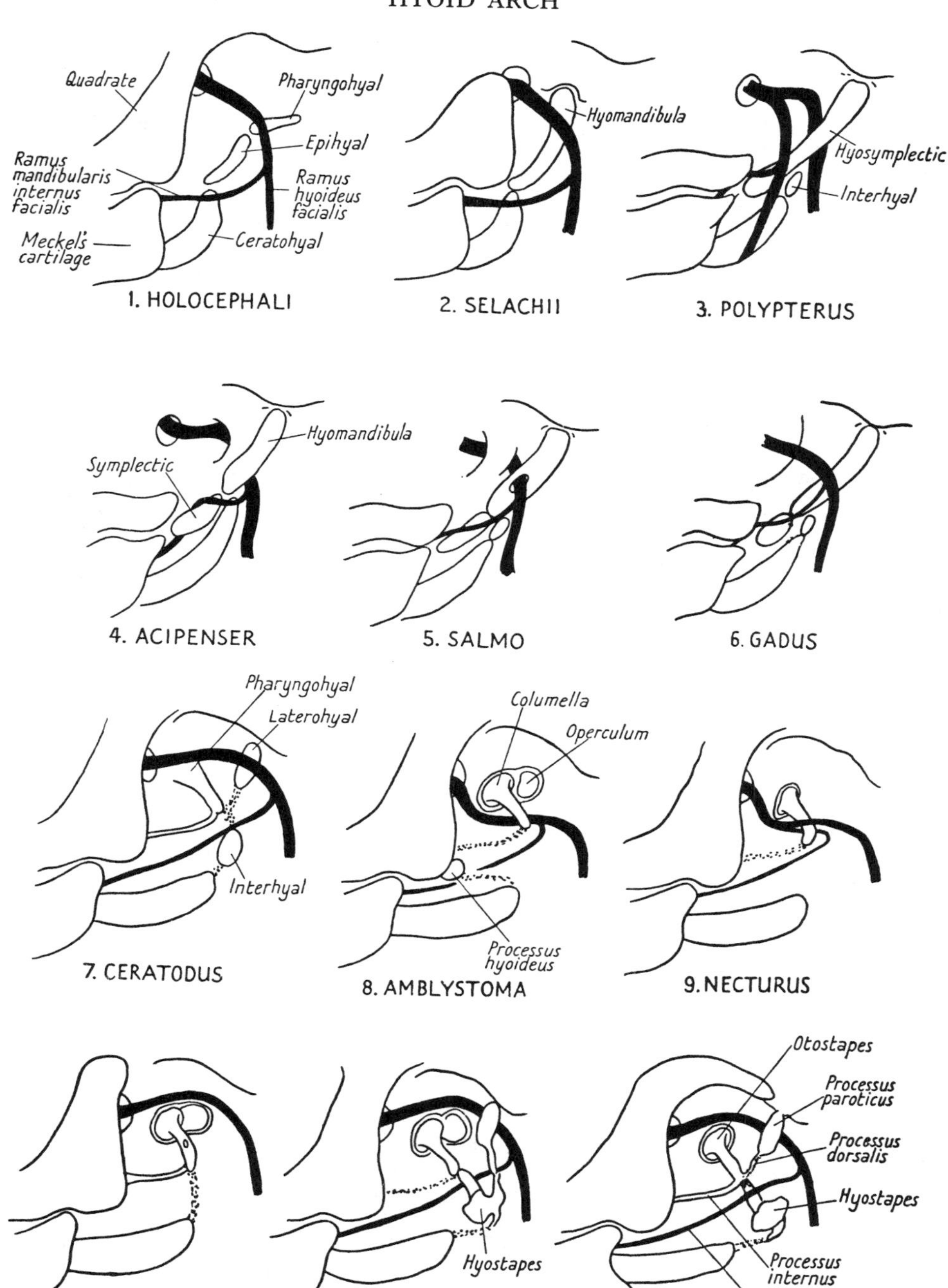
Quadrate
Pharyngohyal
Epihyal
Ramus mandibularis internus facialis
Ramus hyoideus facialis
Meckel's cartilage
Ceratohyal
1. HOLOCEPHALI
Hyomandibula
2. SELACHII
Hyosymplectic
Interhyal
3. POLYPTERUS
Hyomandibula
Symplectic
4. ACIPENSER
5. SALMO
6. GADUS
Pharyngohyal
Laterohyal
Interhyal
7. CERATODUS
Columella
Operculum
Processus hyoideus
8. AMBLYSTOMA
9. NECTURUS
10. HYPOGEOPHIS
Hyostapes
11. RANA
Otostapes
Processus paroticus
Processus dorsalis
Hyostapes
Processus internus
Chorda tympani
12. LACERTA

HYOID ARCH

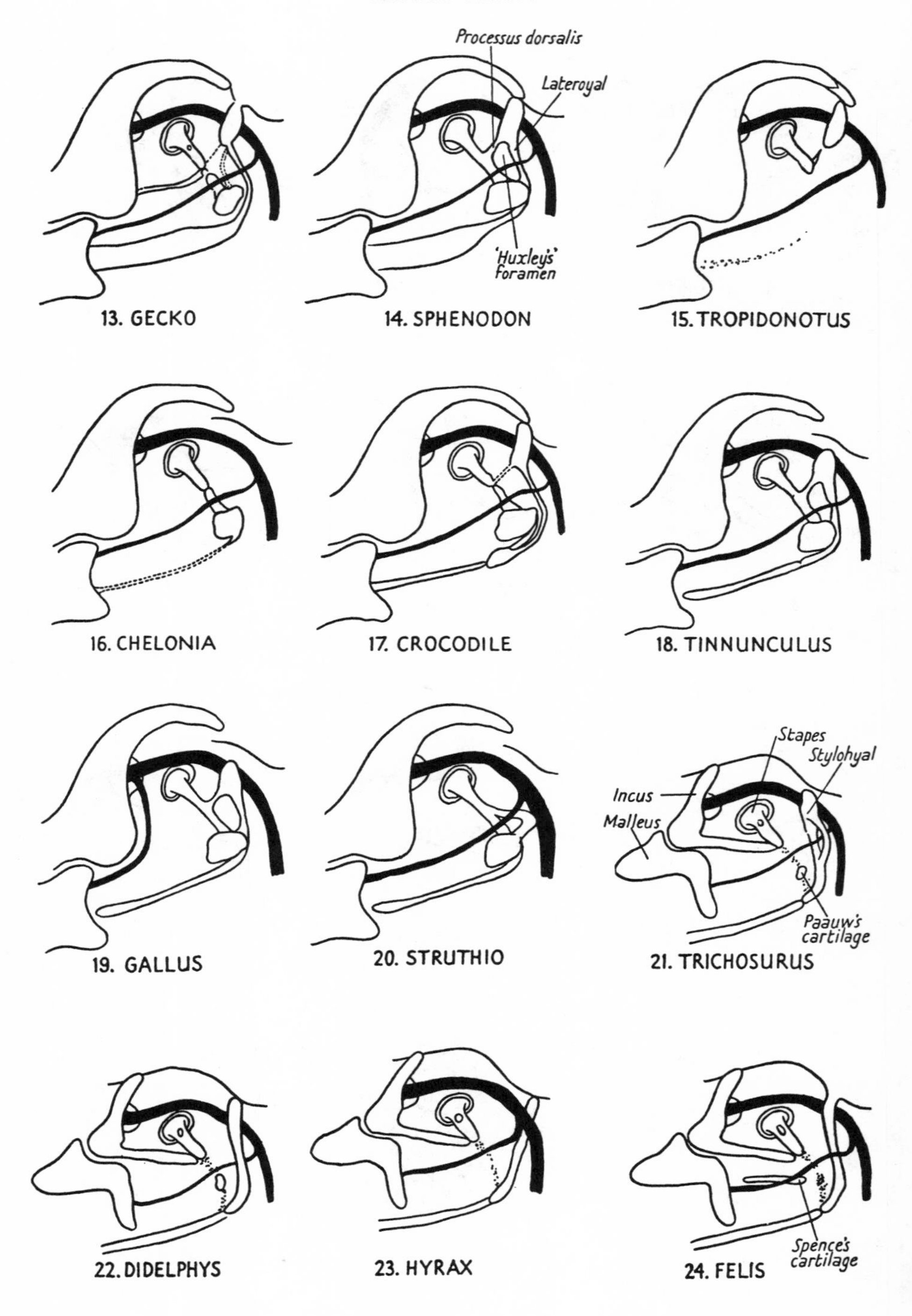

13. GECKO
14. SPHENODON
15. TROPIDONOTUS
16. CHELONIA
17. CROCODILE
18. TINNUNCULUS
19. GALLUS
20. STRUTHIO
21. TRICHOSURUS
22. DIDELPHYS
23. HYRAX
24. FELIS

JAW-SUSPENSIONS

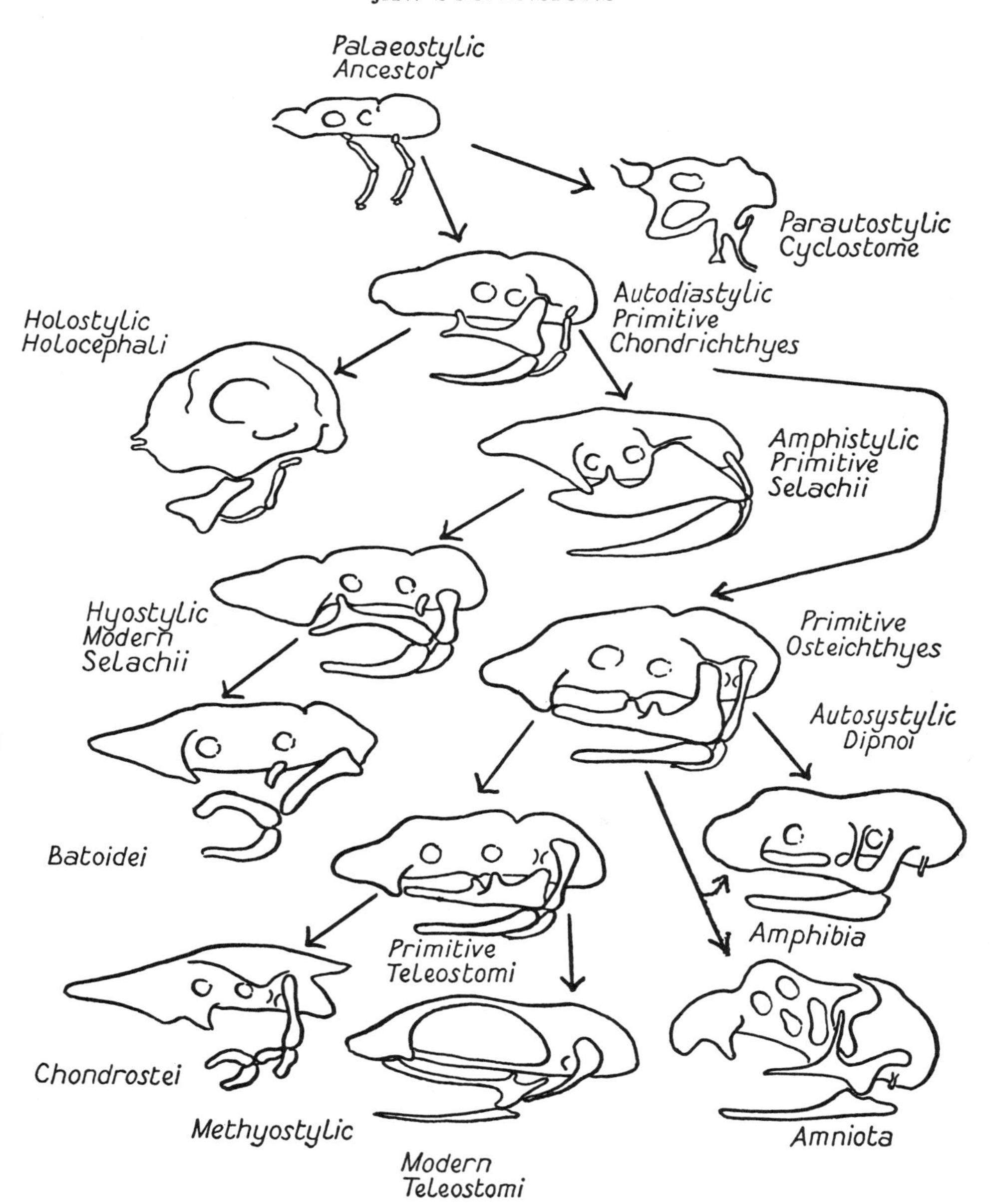

EXTRACRANIAL SPACES

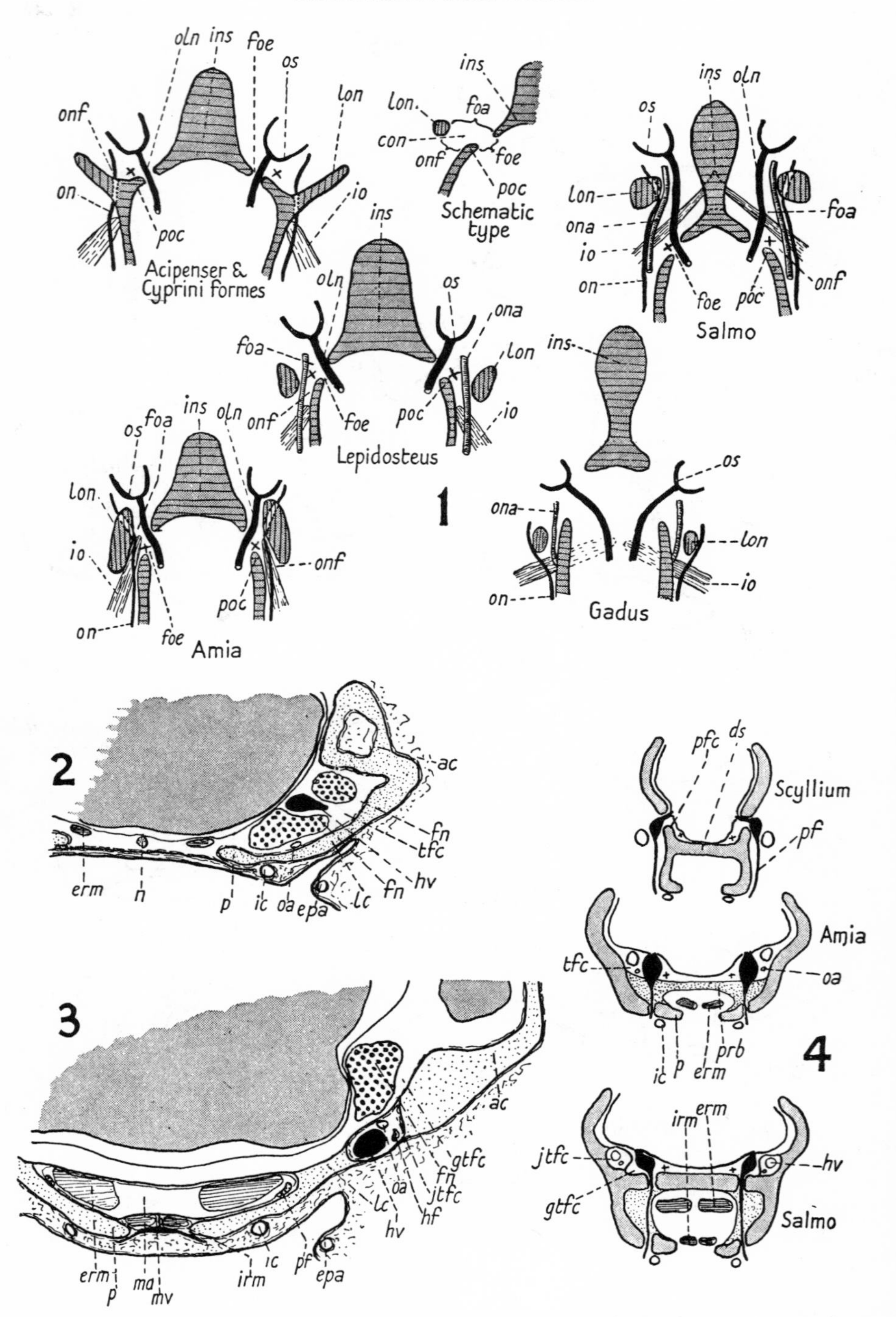

1, relations of cavum orbitonasale; **2,** transverse section through myodome and trigemino-facialis chamber of Amia, **3,** ditto of Salmo; **4,** diagram of relations of subpituitary space and myodome.

ac, auditory capsule; con, cavum orbitonasale; ds, dorsum sellae; epa, efferent pseudobranchial artery; erm, external rectus; fn, facial nerve; foa, foramen olfactorium advehens; foe, foramen olfactorium evehens; gtfc, pars ganglionaris of tfc; hf, hyomandibular nerve; hv, head-vein; ic, internal carotid; ins, internal nasal septum; io, inferior oblique; irm, internal rectus; jtfc, pars jugularis of tfc; lc, lateral commissure; lon, lamina orbitonasalis; md, dorsal myodome; mv, ventral myodome; n, notochord; oa, orbital artery; oln, olfactory nerve; on, ophthalmic nerve; ona, orbitonasal artery; onf, orbitonasal fissure; os, olfactory sac; p, parachordal; pf, palatine nerve; pfc, prefacial commissure; poc, preoptic root of orbital cartilage; prb, prootic bridge; tfc, trigeminofacialis chamber.